Analogue and Numerical Modelling of Crustal-Scale Processes

Geological Society books refereeing procedures

The Society makes every effort to ensure that the scientific and production quality of its books matches that of its journals. Since 1997, all book proposals have been refereed by specialist reviewers as well as by the Society's Books Editorial Committee. If the referees identify weaknesses in the proposal, these must be addressed before the proposal is accepted.

Once the book is accepted, the Society Book Editors ensure that the volume editors follow strict guidelines on refereeing and quality control. We insist that individual papers can only be accepted after satisfactory review by two independent referees. The questions on the review forms are similar to those for *Journal of the Geological Society*. The referees' forms and comments must be available to the Society's Book Editors on request.

Although many of the books result from meetings, the editors are expected to commission papers that were not presented at the meeting to ensure that the book provides a balanced coverage of the subject. Being accepted for presentation at the meeting does not guarantee inclusion in the book.

More information about submitting a proposal and producing a book for the society can be found on its web site: www.geolsoc.org.uk.

It is recommended that reference to all or part of this book should be made in one of the following ways:

Buiter, S. J. H. & Schreurs, G. (eds) 2006. *Analogue and Numerical Modelling of Crustal-Scale Processes*. Geological Society, London, Special Publications, **253**.

Rey, P. F. & Houseman, G. 2006. Lithospheric scale gravitational flow: the impact of body forces on orogenic processes from Archaean to Phanerozoic. *In*: Buiter, S. J. H. & Schreurs, G. (eds) 2006. *Analogue and Numerical Modelling of Crustal-Scale Processes*. Geological Society, London, Special Publications, **253**, 153–167.

GEOLOGICAL SOCIETY SPECIAL PUBLICATION NO. 253

Analogue and Numerical Modelling of Crustal-Scale Processes

EDITED BY

S. J. H. BUITER

Geological Survey of Norway, Trondheim, Norway

and

G. SCHREURS

University of Bern, Switzerland

2006
Published by
The Geological Society
London

THE GEOLOGICAL SOCIETY

The Geological Society of London (GSL) was founded in 1807. It is the oldest national geological society in the world and the largest in Europe. It was incorporated under Royal Charter in 1825 and is Registered Charity 210161.

The Society is the UK national learned and professional society for geology with a worldwide Fellowship (FGS) of 9000. The Society has the power to confer Chartered status on suitably qualified Fellows, and about 2000 of the Fellowship carry the title (CGeol). Chartered Geologists may also obtain the equivalent European title, European Geologist (EurGeol). One fifth of the Society's fellowship resides outside the UK. To find out more about the Society, log on to www.geolsoc.org.uk.

The Geological Society Publishing House (Bath, UK) produces the Society's international journals and books, and acts as European distributor for selected publications of the American Association of Petroleum Geologists (AAPG), the American Geological Institute (AGI), the Indonesian Petroleum Association (IPA), the Geological Society of America (GSA), the Society for Sedimentary Geology (SEPM) and the Geologists' Association (GA). Joint marketing agreements ensure that GSL Fellows may purchase these societies' publications at a discount. The Society's online bookshop (accessible from www.geolsoc.org.uk) offers secure book purchasing with your credit or debit card.

To find out about joining the Society and benefiting from substantial discounts on publications of GSL and other societies worldwide, consult www.geolsoc.org.uk, or contact the Fellowship Department at: The Geological Society, Burlington House, Piccadilly, London W1J 0BG: Tel. +44 (0)20 7434 9944; Fax +44 (0)20 7439 8975; E-mail: enquiries@geolsoc.org.uk.

For information about the Society's meetings, consult *Events* on www.geolsoc.org.uk. To find out more about the Society's Corporate Affiliates Scheme, write to enquiries@geolsoc.org.uk.

Published by The Geological Society from:
The Geological Society Publishing House
Unit 7, Brassmill Enterprise Centre
Brassmill Lane
Bath BA1 3JN, UK

(*Orders*: Tel. +44 (0)1225 445046
Fax +44 (0)1225 442836)
Online bookshop: www.geolsoc.org.uk/bookshop

British Library Cataloguing in Publication Data

A catalogue record for this book is available from the British Library.

ISBN-10 1-86239-191-2
ISBN-13 978-1-86239-191-8

Typeset by Techset Composition, Salisbury, UK
Printed by MPG Books Ltd, Bodmin, UK

Distributors

USA
AAPG Bookstore
PO Box 979
Tulsa
OK 74101-0979
USA
Orders: Tel. +1 918 584-2555
Fax +1 918 560-2652
E-mail bookstore@aapg.org

India
Affiliated East-West Press Private Ltd
Marketing Division
G-1/16 Ansari Road, Darya Ganj
New Delhi 110 002
India
Orders: Tel. +91 11 2327-9113/2326-4180
Fax +91 11 2326-0538
E-mail affiliat@vsnl.com

Japan
Kanda Book Trading Company
Cityhouse Tama 204
Tsurumaki 1-3-10
Tama-shi, Tokyo 206-0034
Japan
Orders: Tel. +81 (0)423 57-7650
Fax +81 (0)423 57-7651
E-mail geokanda@ma.kcom.ne.jp

Contents

Models of sedimentary basins

Models of surface processes and deformation

Models of faults and fluid flow

Preface

As Earth's crust is the most accessible part of our planet, a wealth of information on its structure and composition is available from geological observations and geophysical measurements. Its shape and composition today are, however, the result of processes occurring at different scales in time and space, rendering study of the crust a complex and challenging undertaking. Field observations and seismic data, for example, confront us with the present day structures and must, for an in-depth understanding of their origin, be reconciled with the processes of deformation that created them. Reconstructing these processes in the past is commonly hindered by lack of continuous outcrop, limited depth resolution and little to no constraints on time. Analogue and numerical models may help improve our understanding of crustal-scale processes through their ability to simulate the birth and evolution of deformational structures at different scales. In this context, the link between theory and observations makes modelling a fundamental tool for the study of processes that alter the Earth's crust.

This volume presents new analogue and numerical model studies of natural processes contributing to the state of the Earth's crust. It shows the insights offered by models over a broad range of tectonic problems, in all cases carefully documenting the model approach used. By 'model' we mean an approximation of a physical process that will help further our understanding of the process itself, the measurable or observable structures that result from it, and the parameters that control it. Models allow us to set the boundary conditions, study the effects of parameter variations and perform sensitivity analyses. By their nature, they are, however, always a simplification of reality. As the words imply, an analogue model represents a scaled analogue of a natural structure, scaling not only the dimensions, but also the forces and material (for example, sand or clay) properties. A numerical model is a mathematical description of the natural situation, whereby aspects of processes are chosen that are of interest and can also feasibly be captured in equations. Both modelling techniques allow us to generate hypotheses regarding crustal-scale processes that can be tested against the results of field observations and geophysical studies.

The volume is divided into five sections:

- *Analogue and numerical sandbox models* tests the reproducibility of models between different laboratories and the suitability of different numerical solution methods for reproducing sandbox results.
- *Models of orogenic processes* describes processes influencing styles of orogenic deformation at convergent plate margins.
- *Models of sedimentary basins* discusses different aspects of extension, inversion and hydrocarbon migration within extensionally-formed sedimentary basins.
- *Models of surface processes and deformation* focuses on links between crustal-scale deformation and erosion processes at the Earth's surface.
- *Models of faults and fluid flow* discusses faults, important for short-term seismicity and their long-term effects on crustal strength, and fluid flow, which is crucial for oil reservoir studies.

The origin of this volume lies in GeoMod2004, an international meeting on analogue and numerical modelling of geological processes, held in Switzerland in June 2004. We would like to thank our co-organizers, Giovanni Bertotti and Paolo Ruffo, and all the participants of GeoMod2004, who provided the stimulus for editing this volume. Sixteen of the 23 papers in this book were presented or initiated at this conference. We thank the Geological Society Publishing House and especially Angharad Hills, Staff Editor, Helen Floyd-Walker, Production Editor, and Bob Holdsworth, Society Books Editor, for all their help in the publication of this book.

Susanne Buiter
& Guido Schreurs

Acknowledgements

The editors thank the following colleagues and friends who kindly helped with the reviewing the papers submitted for this volume:

Jürgen Adam
Atilla Aydin
Andrey Babeyko
Maria E. Benardinelli
Antonio Benedicto
Paul Bons
Roger Buck
Douglas Burbank
Nestor Cardozo
Giacomo Corti
Dickson Cunningham
Olivier Dauteuil
Roberta de Franco
Alexander Densmore
Tim Dooley
Elizabeth Eide
Susan Ellis
Claudio Faccenna
Kevin Furlong
Denis Gapais
Lykke Gemmer
Muriel Gerbault
Rob Govers
Lyal Harris
Kes Heffer
Kaj Johnson
Hemin Koyi
Erik Lundin
Yvonne Martin
Othmar Müntener
Paul Nadeau
Thorsten Nagel
Dick Nieuwland
Marleen Nyst
Francis Odonne
Nick Oliver
Christophe Pascal
David Pedreira
Fred Pollitz
Russ Pysklywec
Antonio Ravaglia
François Roure
Mike Sandiford
Wouter Schellart
Fritz Schlunegger
Michael Seyferth
Guy Simpson
Rudy Slingerland
Blanka Sperner
Colin Stark
Randell Stephenson
Drew Stolar
Fabrizio Storti
Kurt Stüwe
Phaedra Upton
Ioannis Vardoulakis
Bruno Vendeville
John Walsh and
Martha Withjack.

The cost of color figures in this book were in part sponsored by Statoil ASA. Their support is kindly acknowledged.

STATOIL

Analogue benchmarks of shortening and extension experiments

GUIDO SCHREURS[1], SUSANNE J. H. BUITER[2], DAVID BOUTELIER[3], GIACOMO CORTI[4], ELISABETTA COSTA[5], ALEXANDER R. CRUDEN[3], JEAN-MARC DANIEL[6], SILVAN HOTH[7], HEMIN A. KOYI[8], NINA KUKOWSKI[7], JO LOHRMANN[7], ANTONIO RAVAGLIA[9,*], ROY W. SCHLISCHE[10], MARTHA OLIVER WITHJACK[10], YASUHIRO YAMADA[11], CRISTIAN CAVOZZI[5], CHIARA DELVENTISETTE[12], JENNIFER A. ELDER BRADY[10], ARNE HOFFMANN-ROTHE[7], JEAN-MARIE MENGUS[6], DOMENICO MONTANARI[12] & FARAMARZ NILFOROUSHAN[8]

[1]*Institute of Geological Sciences, University of Bern, Baltzerstrasse 1-3, CH-3012 Bern, Switzerland (e-mail: schreurs@geo.unibe.ch)*

[2]*Centre for Geodynamics, Geological Survey of Norway, 7491 Trondheim, Norway*

[3]*Department of Geology, University of Toronto, 22 Russell St., Toronto, Ontario M5S 3B1, Canada*

[4]*CNR-Istituto di Geoscienze e Georisorse, Sezione di Firenze, via G. La Pira 4, I-50121 Firenze, Italy*

[5]*Dipartimento di Scienze della Terra, Università di Parma, Parco Area delle Scienze 157/A, I-43100 Parma, Italy*

[6]*Institut Français du Pétrole, 1 et 4 avenue de Bois Préau, F-92500 Rueil Malmaison, France*

[7]*GeoForschungsZentrum Potsdam, Telegrafenberg, D-14473 Potsdam, Germany*

[8]*Hans Ramberg Tectonic Laboratory, Department of Earth Sciences, Uppsala University, Villavägen 16, S-75326 Uppsala, Sweden*

[9]*Dipartimento di Scienze della Terra, Università di Pavia, via Ferrata 1, I-27100 Pavia, Italy*

[10]*Department of Geological Sciences, Rutgers University, Piscataway, NJ, 08854, USA*

[11]*Department of Civil and Earth Resources Engineering, Kyoto University, Kyoto 606-5801 Japan*

[12]*Dipartimento di Scienze della Terra, Università di Firenze, via G. LaPira 4, 1, I-50121 Firenze, Italy*

**Current address: Midland Valley Exploration Ltd., 14 Park Circus, Glasgow, G3 6AX, UK*

Abstract: We report a direct comparison of scaled analogue experiments to test the reproducibility of model results among ten different experimental modelling laboratories. We present results for two experiments: a brittle thrust wedge experiment and a brittle-viscous extension experiment. The experimental set-up, the model construction technique, the viscous material and the base and wall properties were prescribed. However, each laboratory used its own frictional analogue material and experimental apparatus. Comparison of results for the shortening experiment highlights large differences in model evolution that may have resulted from (1) differences in boundary conditions (indenter or basal-pull models), (2) differences in model widths, (3) location of observation (for example, sidewall versus centre of model), (4) material properties, (5) base and sidewall frictional properties, and (6) differences in set-up technique of individual experimenters. Six laboratories carried out the shortening experiment with a mobile wall. The overall evolution of their models is broadly similar, with the development of a thrust wedge characterized by forward thrust propagation and by back thrusting. However, significant variations are observed in spacing between thrusts, their dip angles, number of forward thrusts and back thrusts, and surface slopes. The structural evolution of the brittle-viscous extension experiments is similar to a high degree. Faulting initiates in the brittle layers above the viscous layer in

From: BUITER, S. J. H. & SCHREURS, G. (eds) 2006. *Analogue and Numerical Modelling of Crustal-Scale Processes.* Geological Society, London, Special Publications, **253**, 1–27.
0305-8719/06/$15.00

close vicinity to the basal velocity discontinuity. Measurements of fault dip angles and fault spacing vary among laboratories. Comparison of experimental results indicates an encouraging overall agreement in model evolution, but also highlights important variations in the geometry and evolution of the resulting structures that may be induced by differences in modelling materials, model dimensions, experimental set-ups and observation location.

Geoscientists have used scaled experimental models (from here on also referred to as analogue models) for more than a century to gain insight into the kinematic and dynamic evolution of geological structures (e.g., Cadell 1888; Ramberg 1981; Koyi 1997). The ability to observe structures while they form is both attractive and insightful, and the ease with which model experiments can be performed has certainly contributed to their widespread application. Although many experiments have investigated similar geological processes, a direct comparison of experiments using the same prescribed set-up has until now been lacking. Testing the reproducibility of model results among different laboratories is important as it allows for documentation of the degree to which variations in model apparatus, modelling materials and experimental techniques can affect the outcome of experiments. Assessing the reliability of analogue models also has implications for comparisons between structures formed in experimental models, results of numerical simulations and natural field examples.

We present the results of two analogue 'benchmarks': a shortening experiment and an extension experiment. Ten modelling laboratories participated in these experiments. The experimental set-up, the construction technique, the viscous material (employed in the extension experiment) and the material covering walls and base were prescribed. Since each laboratory used its own frictional analogue material (sand, microbeads, wet clay) and experimental apparatus, our study is not a benchmark in its strictest sense. Our aims are to (1) indicate the degree of reproducibility of model results between different laboratories, (2) highlight the influence of material properties, model boundary conditions and experimental set-up, and (3) use our results to suggest further benchmark experiments.

The two experimental set-ups were designed to simulate upper-crustal processes, and can be placed among a wide range of laboratory experiments studying contractional and extensional deformation. Numerous studies document experiments of thrust wedges under normal gravity conditions (e.g., Davis *et al.* 1983; Malavieille 1984; Mulugeta & Koyi 1987, 1992; Mulugeta 1988; Colletta *et al.* 1991; Lallemand *et al.* 1994; Koyi 1995; Storti & McClay 1995; Mugnier *et al.* 1997; Gutscher *et al.* 1996, 1998; Storti *et al.* 2000; Schreurs *et al.* 2001; Turrini *et al.* 2001; Kukowski *et al.* 2002; Costa *et al.* 2004; Hoth *et al.* 2005). Deformation in these experiments occurs either by displacing a back wall or by pulling a basal sheet below a fixed wall. Such indenter-type or basal-pull-type models are not entirely equivalent, as we will discuss below. Parameters investigated in published thrust wedge experiments include basal friction, the presence of décollement layers, dip of the initial basal surface, surface slope, and the effects of erosion. In general, such models display a series of forward-propagating, in-sequence thrusts, whereby changes in, for example, basal friction or erosion may cause renewed hinterland activity. Many authors have investigated brittle-viscous extension in a normal gravity field (e.g., Tron & Brun 1991; Vendeville & Jackson 1992*a*, *b*; Brun *et al.* 1994; Keep & McClay 1997; Brun 1999; Gartrell 2001; Bahroudi *et al.* 2003). In these models, a viscous layer typically overlies a velocity discontinuity at the base of the model and is itself overlain by brittle materials. In all experiments the viscous basal layer exerts a strong control on fault nucleation and location.

In a companion paper, Buiter *et al.* (2006) present numerical equivalents of both analogue experiments and compare the results qualitatively and quantitatively. This study shows that numerical models employing different techniques (finite-element, finite-difference and distinct element methods) are able to reproduce results of analogue models successfully. The combination of analogue and numerical modelling methods may help establish the robust features of models of crustal-scale processes (e.g., Ellis *et al.* 2004; Pysklywec & Cruden 2004; Panien *et al.* 2006).

Methods

Material properties

We focus on experiments performed in a natural gravity field. In such experiments dry sand or wet clay are commonly used to simulate brittle rocks in the upper crust, and viscous materials to simulate salt or lower crustal rocks. In the shortening experiment, the models consisted of granular materials (sand and glass microbeads) only, whereas for the extension experiment both brittle and viscous materials were used. All laboratories used the same polydimethylsiloxane (PDMS) as viscous material. This is a linear viscous material (silicone polymer) with a viscosity of 5×10^4 Pas at room temperature and at strain rates below 3×10^{-3} s^{-1} (Weijermars 1986; see also Ten Grotenhuis *et al.* 2002; Cruden *et al.* 2006). In contrast, each laboratory used its own brittle materials, which generally differed in density, frictional properties, grain sizes and grain shape (Table 1). Several laboratories measured the frictional properties of their granular materials using either a Hubbert-type shear box (Hubbert 1951), a ring-shear tester (Schulze 1994) or a Casagrande shear box (Casagrande 1932). Errors in ring-shear measurements are usually small (less than 1%). Measurements in Hubbert-type shear boxes are less precise, because of the larger uncertainty associated with the determination of the exact timing of the onset of shearing. As the Casagrande shear box measurements in this study employed higher normal loads than occur in the experiments, the measured values for cohesion are not considered representative for the experimental conditions.

Although experimental modellers have often assumed that their brittle materials deform according to the Coulomb failure criterion with constant frictional properties, several studies demonstrate that materials such as quartz sand, corundum sand and glass microbeads are characterized by elastic/frictional plastic behaviour with strain-hardening preceding failure (at peak strength) and subsequent strain-softening until a stable value is reached (Lohrmann *et al.* 2003; Panien 2004). Such mechanical behaviour is very similar to the one exhibited by experimentally deformed rocks (e.g., Jaeger & Cook 1979; Marone 1998; Barnhoorn *et al.* 2004). Values for the angle of internal friction (at peak strength) range from 33° to 45° for quartz sand and from 21° to 29° for microbeads (Table 1). These values are generally comparable to those determined experimentally for upper crustal rocks (Byerlee 1978). Strain-softening may be expressed as the percentage difference between peak friction and dynamic stable friction. For the materials used in this study, it varied from 4% to 20% for quartz sand and from 0% to 14% for microbeads (Table 1). In each case, strain softening in microbeads was lower than in sand. Cohesion values of granular materials display a wide range varying between −25 and 150 Pa. The negative values and the large variation in cohesion values may be related to uncertainties in the linear extrapolation of the failure envelope to very low normal stresses. At low normal stresses, the failure envelope is no longer a straight line, but has a convex-leftward shape with negligible cohesion (Schellart 2000). One laboratory (Piscataway) used wet clay instead of dry sand for the extension experiment. Their clay has an elasto-plastic rheology with an angle of peak friction of 27°.

To reduce the influence of the variable properties of sidewalls and base of the modelling apparatus, they were covered with a self-adhesive, transparent PVC foil (brand Alkor, article number 120010). The influence of this foil was investigated by measuring the friction between sand and the foil in the GeoForschungs Zentrum ring-shear tester. For fine-grained sand (20–200 μm) the angle of peak friction was $19.8° \pm 0.1°$ and the angle of stable friction was $16.5° \pm 0.3°$. For coarser-grained sand (20–650 μm) the angles of peak and stable friction were $19.7° \pm 0.1°$ and $16.7° \pm 0.3°$, respectively. This implies that the use of different sands in the experiments probably had only little influence on values of boundary friction, as long as Alkor foil was used on all sidewalls and the base.

Differences in frictional properties of the model materials may be attributed to the applied measuring technique (e.g., Casagrande shear box, Hubbert-type shear box or ring-shear tester), the filling technique and/or intrinsic material properties such as grain size distribution, grain shape and composition (Table 1). Variations in material properties may be important for differences in the evolution of experimental models. Lohrmann *et al.* (2003) found that the kinematics of thrust wedges are largely a function of their material properties. Adam *et al.* (2005) point out that the strength increase of the granular materials during the strain-hardening phase towards peak strength influences the localization and early development of faults. A detailed investigation of the exact causes of the differences in material properties is beyond the scope of our study, but it should be kept in mind that our modelling results are influenced by the use of different granular materials.

Scaling

Our models aim at representing upper-crustal conditions and can be scaled up to natural dimensions by observing geometric, kinematic and dynamic similarity relationships (Hubbert 1937; Ramberg 1981). We assume that 1 cm in the models corresponds to 1 km in nature and that (dimensionless) angles of internal friction can

Table 1. *Material properties and grain characteristics*

Laboratory	Density (kg m^{-3})	C at ϕ_{peak} (Pa)	ϕ_{peak} (°)	ϕ_{stable} (°)	Strain soft. (%)	Grain size (μm)	Grain shape	Composition
Bern								
Quartz sand*	1560	21 ± 18	35.5	31.2	15	80–200	angular	99% SiO_2
Corundum sand*	1890	39 ± 10	37	32.2	16	88–125	angular	95% Al_2O_3
Microbeads*	1480	25 ± 4	22.3	20.6	8	70–110	well-rounded	72.5% SiO_2 13.8% Na_2O 8% CaO
Florence								
Quartz sand	1550	66	39	n.d.	n.d.	<250	rounded	n.d.
Kyoto								
Quartz sand	1390	n.d.	n.d.	34	9	m = 203	angular	93% SiO_2 4% Al_2O_3
Microbeads	1420	n.d.	n.d.	25	0	m = 193	spherical	Na_2O–CaO–SiO_2
Parma								
Quartz sand†	1500	†	34.1	n.d.	n.d.	60–250	Sub-rounded	99.8% SiO_2
Microbeads	1540	n.d.	n.d.	n.d.	n.d.	m = 180	rounded	n.d.
Pavia								
Quartz sand†	1500	†	33.2	32.1	4.1	60–250	angular to sub-angular	99.8% SiO_2
Microbeads†	1550–1600	†	26.4	26.2	1	150–250	well-rounded	n.d.
Piscataway								
Clay‡	1600	~40	27	n.d.	n.d.	<100	platy	Kaolinite + 40–50% H_2O
GFZ Potsdam								
Quartz sand*	1740	94 ± 7	35.4 ± 0.3	28.2 ± 0.05	20.2	20–630 (m = 350)	well-rounded	95% SiO_2
Microbeads*	1590	−6 ± 19	29.1 ± 0.14	22.5 ± 0.04	13.6	300-400	well-rounded	n.d.

IFP Rueil-Malmaison								
Quartz sand	1500	n.d.	45	39	19	m = 116	angular	>98% SiO_2
Corundum sand	1880	n.d.	39	34	16.7	m = 108	angular	>98% Al_2O_3
Microbeads	1400–1600	n.d.	n.d.	n.d.	n.d.	m = 121	well-rounded	72.5% SiO_2 13.8% Na_2O 8% CaO
Toronto								
Quartz sand*	1625	3	35.7	29.2	22.2	150–200	angular	>98% SiO_2
Microbeads*	1575	41	25.5	23.4	9.2	~100	rounded	n.d.
Uppsala								
Quartz sand§	1350	~150	33	30	11	60–250 m = 125	angular	97% SiO_2
Microbeads§	1440	~50	21	20	5	53–105	rounded	73 % SiO_2 12.5% Na_2O 8.5% CaO 4.5% MgO

C = cohesion; m = median of grain size; ϕ_{peak} = angle of peak friction; ϕ_{stable} = angle of stable-dynamic friction; strain-softening is measured as the difference between coefficient of peak friction and coefficient of stable-dynamic friction divided by the coefficient of peak friction.

n.d. = not determined.

*Angles of peak friction, stable-dynamic friction and inferred cohesion determined with a ring-shear tester (Schulze 1994) at GFZ Potsdam.

†Angles of peak friction and stable-dynamic friction determined with a Casagrande shear box (Casagrande 1932) at high normal loads; these normal loads are much higher than those in the experiments. Values of cohesion obtained by linear regression of high normal stress Casagrande tests are questionable (Pavia: −686 Pa for microbeads, and 21378 Pa for sand; Parma: 12000 Pa for sand).

‡Angle of peak internal friction and inferred cohesion determined with a ring-shear tester (Sims 1993).

§Angles of peak friction, stable-dynamic friction and inferred cohesion determined with a Hubbert-type shear box (Hubbert 1951) at low normal loads comparable to those produced in analogue experiments presented here.

be taken to be equal at both scales. Then,

$$\frac{C_p}{\rho_p} = 10^5 \frac{C_m}{\rho_m} \quad (1)$$

where subscript p denotes natural conditions, subscript m analogue model conditions, C is cohesion and ρ is density. Using representative densities ρ_m *c.* 1600 kgm^{-3} and ρ_p *c.* 2800 kgm^{-3}, a cohesion of 10 Pa in the model corresponds to around 2 MPa in nature. The scaling up of viscous properties depends on the assumed velocity v (v_m was prescribed in our experiments):

$$\frac{\eta_p v_p}{\rho_p} = 10^{10} \frac{\eta_m v_m}{\rho_m} \quad (2)$$

for viscosity η.

Shortening experiment

Model set-up

In the shortening experiment a thrust wedge was built in alternating layers of different strength. Six laboratories (Bern, Parma, Pavia, IFP Rueil-Malmaison, Toronto, Uppsala) deformed their model by moving the backwall (right-hand wall in our figures) of the apparatus towards the interior of the model (leftward). For ease of description these models are referred to as mobile back wall models (Fig. 1a). Two laboratories (Kyoto and GFZ Potsdam) produced shortening by pulling an underlying sheet horizontally beneath the vertical backwall. These models are referred to as fixed back wall models (Fig. 1b). The applied shortening rate varied between the models. This was not expected to influence the results as only rate-independent brittle model materials were used.

The model dimensions (length and width) varied between the laboratories (Table 2). The mechanical stratigraphy was the same in all models. A 5 mm thick layer of sand at the base of the apparatus was overlain by a 5 mm thick layer of weaker microbeads, which in turn was overlain by additional layers of sand (Fig. 1a, b). To visualize deformation, the sand layers above the microbeads were either coloured differently or alternated with corundum sand, which has similar mechanical properties to quartz sand (laboratories of Bern and IFP Rueil-Malmaison, Table 1). The total thickness of the layers was 3.5 cm. The granular materials were sieved from a height of 20 cm with a filling rate of approximately 250 grams/minute. The model was not pre-compacted before the onset of shortening. A wedge of sand, 10 cm long and with a 10° surface slope was sieved adjacent to the right-hand vertical wall to facilitate the build up of the thrust wedge. The base of the modelling apparatus remained horizontal.

As the Alkor foil created visualization problems across glass sidewalls, the laboratories of Kyoto, Parma and Pavia did not apply foil on the sidewalls, while GFZ Potsdam applied the foil only on the sidewall which was not used for visualization. Two laboratories (Kyoto and Parma) tested the sensitivity of the experimental results to the presence of Alkor foil on the sidewalls. For most laboratories no significant material loss through a gap at the base of the right-hand vertical wall was reported. In the two fixed back wall experiments (Kyoto and GFZ Potsdam) there was a small exit slot (<1 mm) above the conveyor belt. In the Toronto experiment, there was initially no slot below the right-hand mobile wall, but at the end of the experiment a small slot (<1 mm) appeared.

All laboratories monitored the cross-sectional evolution, either by photography through transparent sidewalls (GFZ Potsdam, Kyoto, Parma, Pavia, Toronto and initial stage of Uppsala) or by X-ray CT imaging through the centre of the model (Bern and IFP Rueil Malmaison) (Table 2). The laboratory from Uppsala monitored mainly the surface evolution of the model and cut a cross-section through the central part of the model at the end of the experiment.

Comparison of mobile back wall model results

Figure 2 shows the cross-sectional structural evolution of the eight thrust wedge models after 0, 2, 6, 10 and 14 cm of displacement. A comparison of surface structures of three of the models (Fig. 3) shows a difference in drag of the structures along the sidewalls (walls parallel to the shortening direction). Thrusts are convex to the hinterland for the fixed back wall models and convex to the foreland for the mobile back wall models. Because the two shortening set-ups led to differences in structural evolution and given that the majority of the experiments were mobile back wall models, we only compared quantitative parameters related to the cross-sectional geometry for the six mobile back wall models.

To compare model results quantitatively, we measured the following parameters for up to 14 cm of displacement (Table 3, Figs 4 and 5): the amount of displacement at which forward thrusts form (Fig. 4a), their basal and upper dip

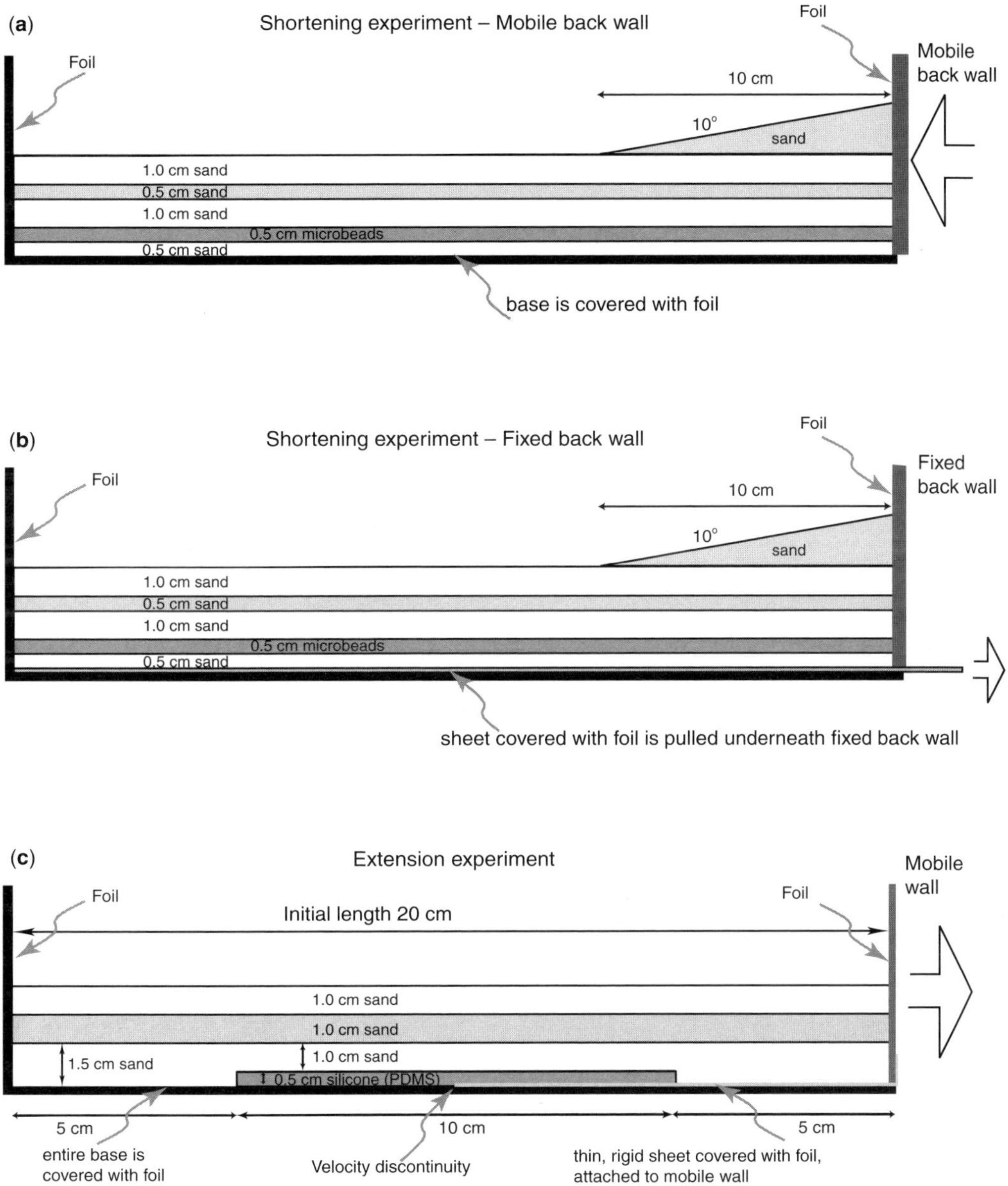

Fig. 1. (**a**) Setup of the shortening experiment with a mobile back wall. The mobile back wall moves leftward (towards the foreland). (**b**) Setup of the shortening experiment with a fixed back wall. Shortening is achieved by pulling a basal sheet below the fixed wall. (**c**) Setup of the extension experiment. A thin viscous slab of PDMS, 10 cm wide and 5 mm high, lies in the central part of the model. Half of the PDMS overlies the thin rigid sheet that is attached to the mobile wall. The mobile wall moves rightward.

angle (Figs 4b and c, respectively; see Fig. 4 for definition), the dip angle of back thrusts (Fig. 4d), the spacing between forward thrusts (measured as the distance to the previous thrust at the moment of initiation of a new thrust) (Fig. 4e), the surface slope of the wedge (Fig. 5), the amount of fault-controlled shortening, the displacement at which the microbeads layer was activated and the length of the detachment in the microbeads layer (Table 3). The amount of shortening accommodated by faults is determined from the offset of

Table 2. *Model dimensions and details of model set-up*

Laboratory	Model dimensions* (cm) (shortening experiment)	Model dimensions* (cm) (extension experiment)	Fixed or mobile back wall (shortening experiment)	Alkor foil sidewalls parallel to displ. direction	Small exit slot (shortening experiment)	Thickness rigid basal sheet (extension experiment)	Monitoring method	Other changes with respect to prescribed set-up
Bern	35.8 × 80	20 × 80	Mobile	Yes	No	0.5 mm	CT scan	
Florence[†]	n.a.	20 × 90	n.a.	Yes	n.a.	?	Surface, final sections	
		20 × 25		Yes				
Kyoto	94 × 50	n.a.	Fixed	No	0.5 mm	n.a.	Surface, side	
				Yes				
Parma[‡]	60 × 33	n.a.	Mobile	Yes	No	n.a.	Surface, side, final sections	
	60 × 33			No	No	n.a.		
Pavia	70 × 30	n.a.	Mobile	No	No	n.a.	Surface, side	
Piscataway	n.a.	20 × 61	n.a.	No	n.a.	0.2 mm	Surface, final sections	Extension to 3 cm
GFZ Potsdam	340 × 20	n.a.	Fixed	No	1 mm	0.3 mm	Surface, side	
IFP Rueil-Malmaison	37 × 78	20 × 78	Mobile	Yes	No	0.5 mm	CT scan	
Toronto	41.9 × 45	29.2 × 45	Mobile	Yes	Late stage, <1 mm[§]	1.6 mm	Surface, side (short), final sections	Initial width ext. expt. 29.2 cm
Uppsala	41 × 30	n.a.	Mobile	Yes	No	n.a.	Surface, final sections	

*Dimensions of model (length × width): length is measured parallel and width perpendicular to shortening or extension direction.
[†]Two extensional experiments were performed with different model widths.
[‡]Two shortening experiments were performed with and without Alkor foil on the sidewalls (see Fig. 4 for results with Alkor foil).
[§]A small exit slot (<1 mm) formed at the end of the experiment as the piston-wall was slightly uplifted.
n.a. = not applicable.

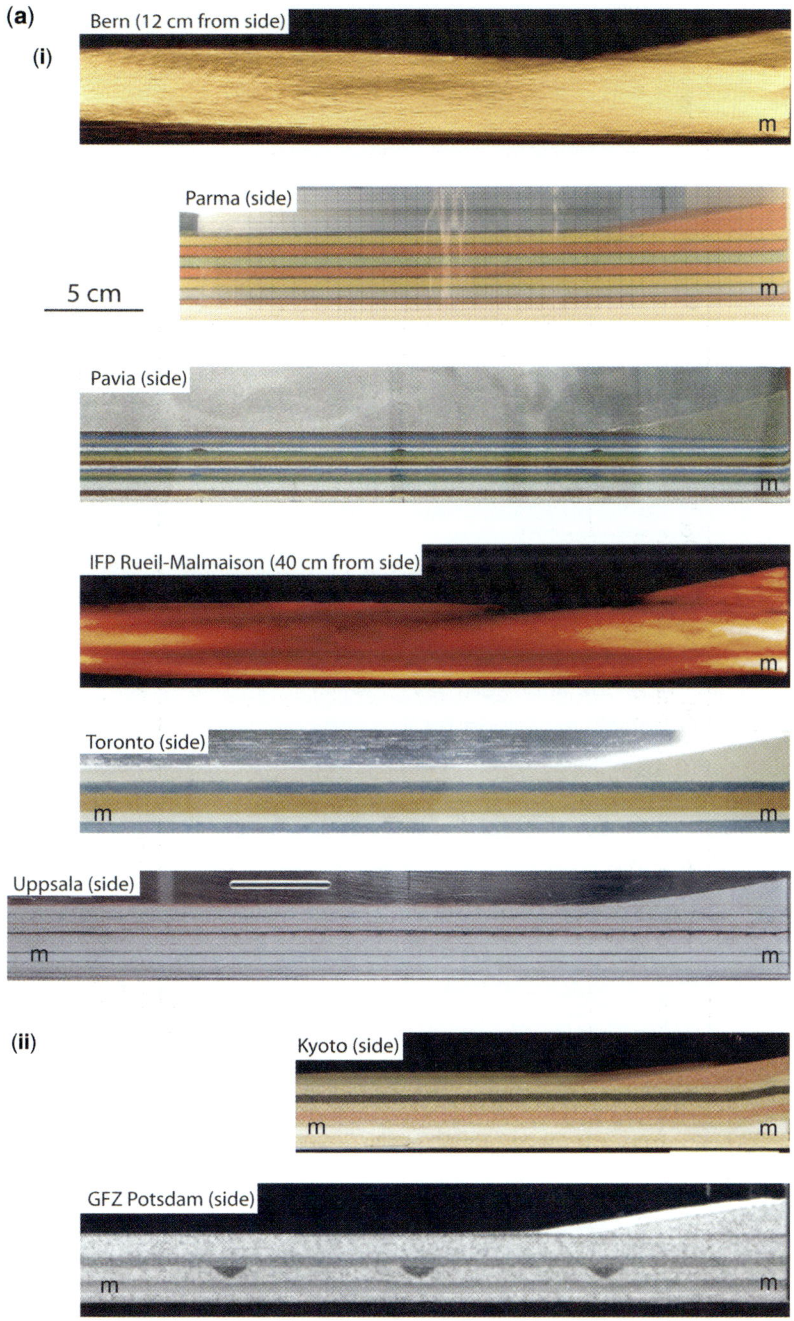

Fig. 2. Results for the shortening experiments for mobile back wall (i) and fixed back wall (ii) models. The microbeads layer is indicated by an 'm'. Model length is not shown completely for all laboratories (see also Table 2). Note slightly oblique views for the Toronto model. The microbeads layer is not well visible in CT images of the early stages of the Bern and IFP Rueil-Malmaison experiments. (**a**) Initial geometries. The locations of the cross-sections are indicated with the laboratory names.

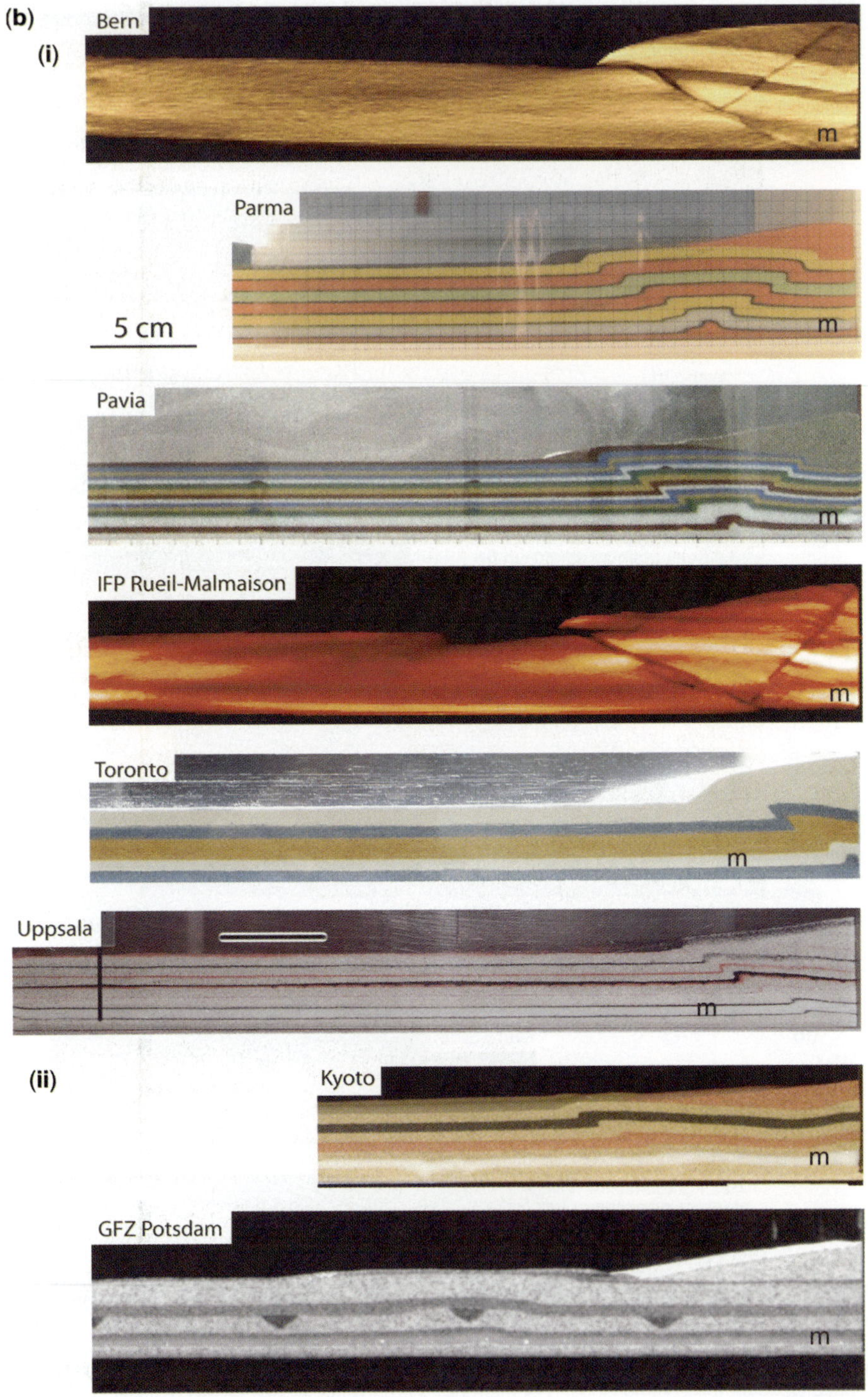

Fig. 2. (**b**) Geometries after 2 cm displacement.

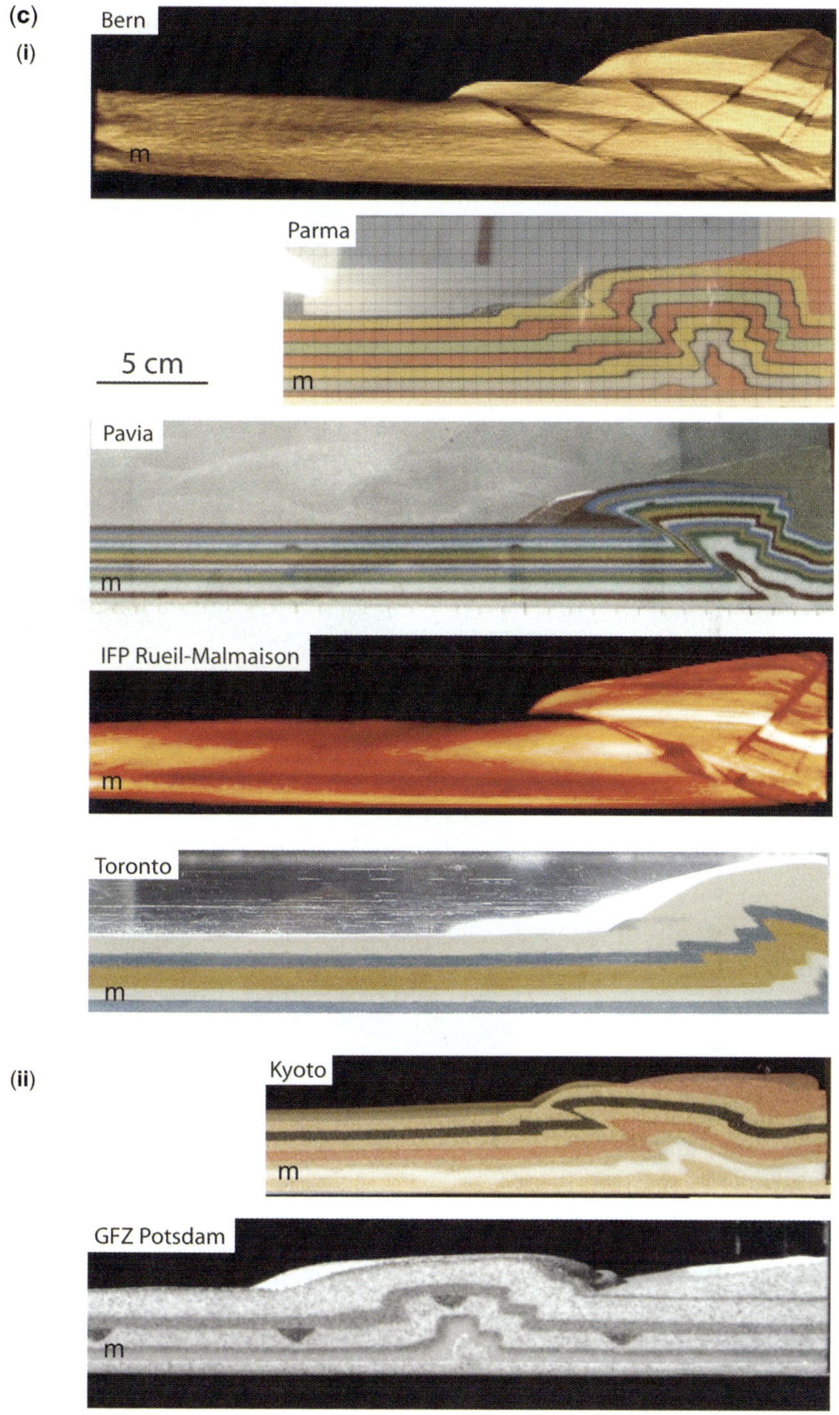

Fig. 2. (**c**) Geometries after 6 cm displacement.

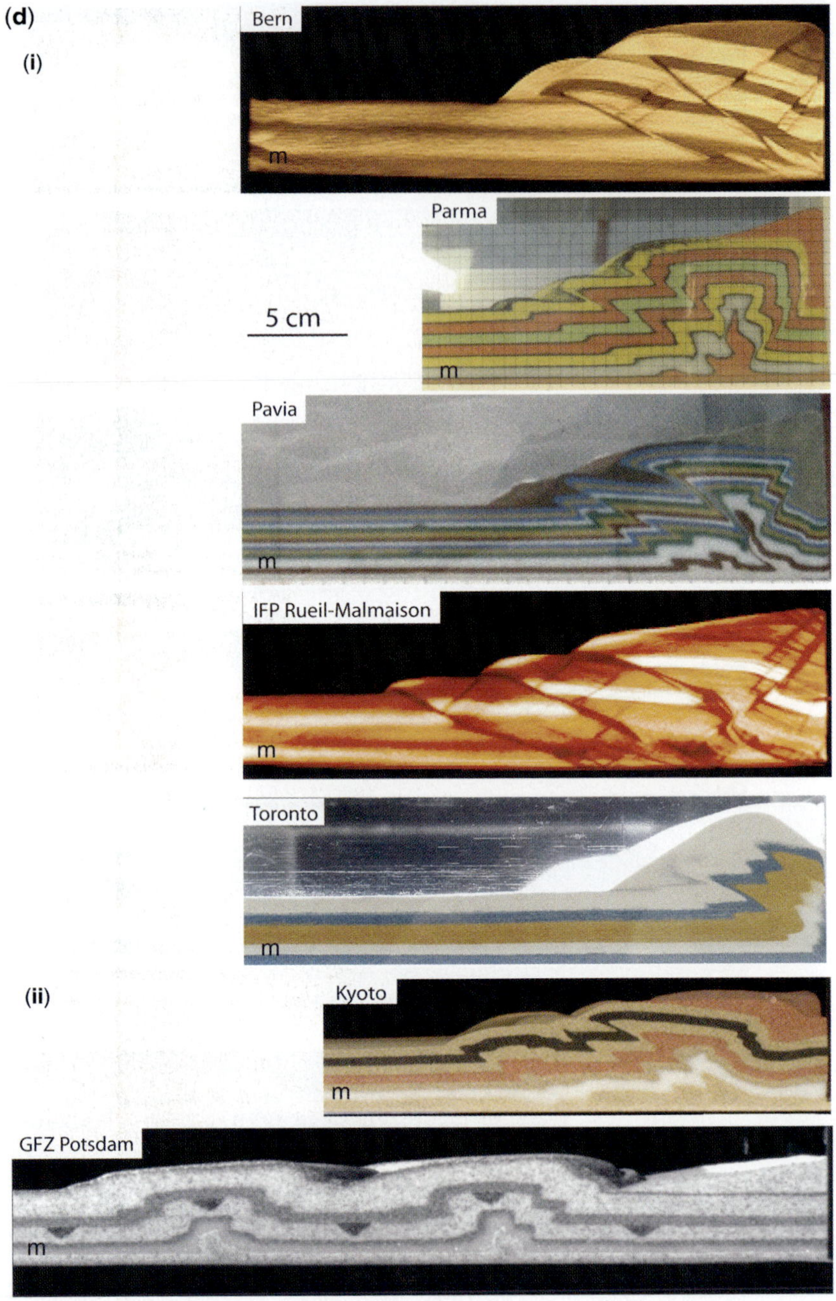

Fig. 2. (**d**) Geometries after 10 cm displacement.

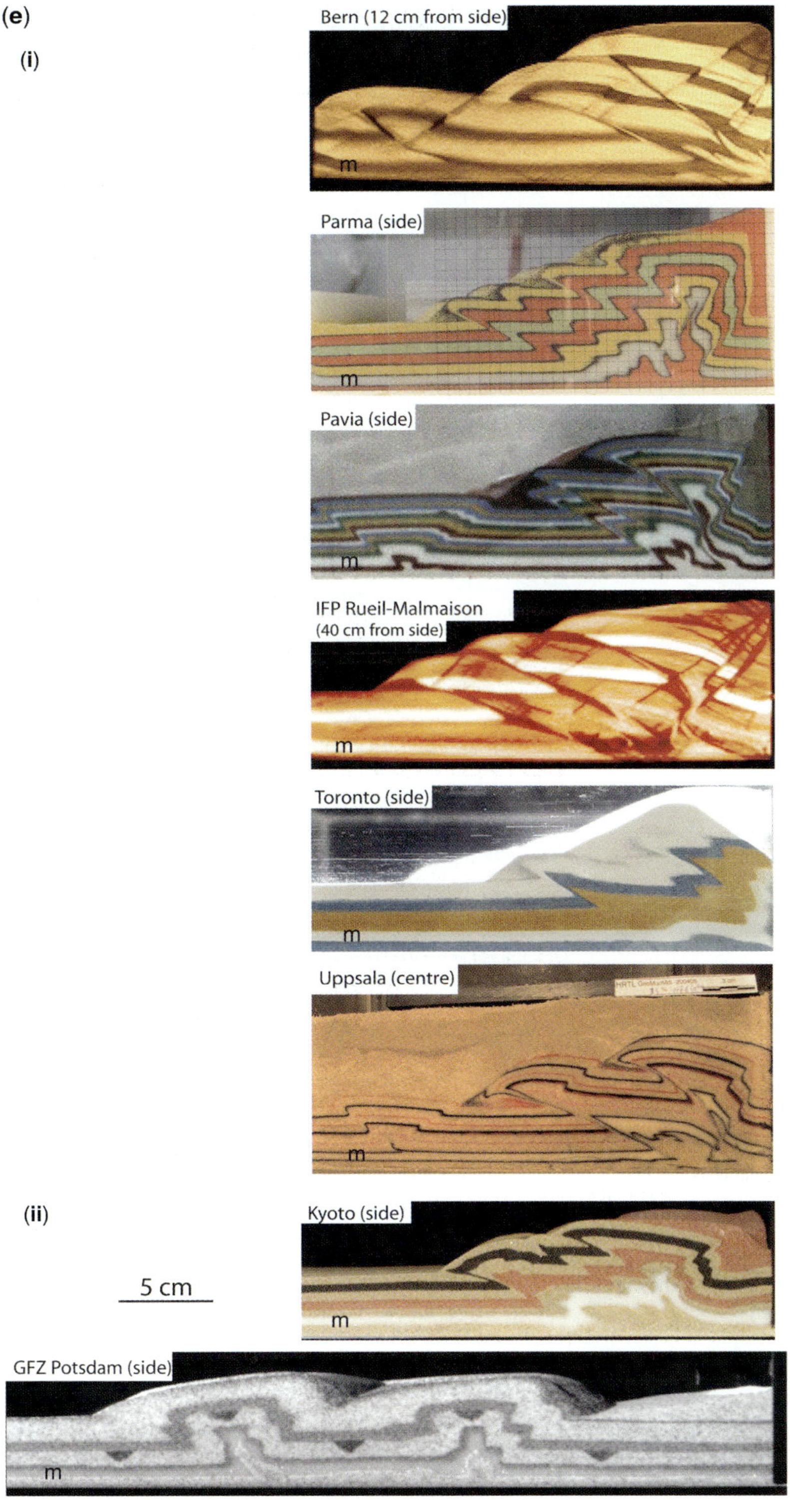

Fig. 2. (e) Geometries after 14 cm displacement. The Uppsala model has been covered with an additional sand layer before cutting of section.

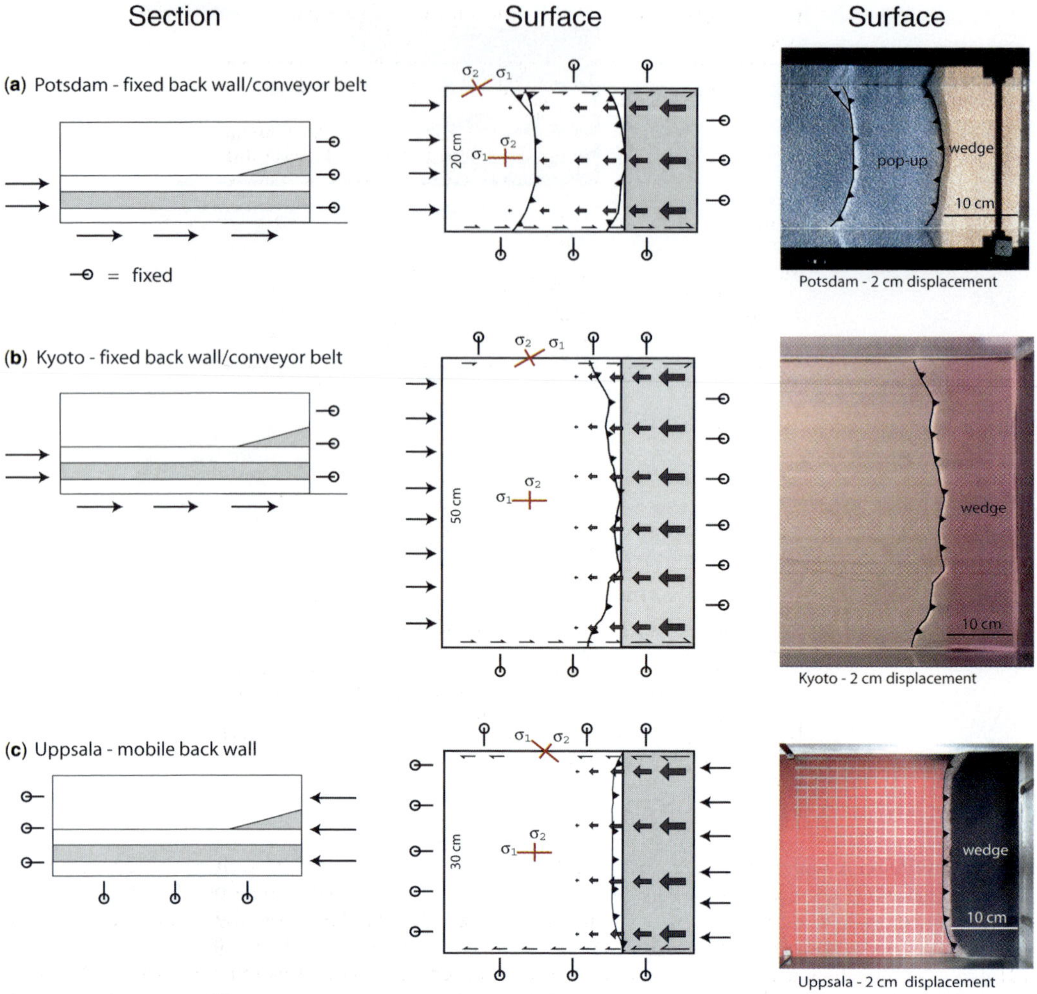

Fig. 3. Left-hand column: Cross sections of fixed and mobile back wall model set-ups. Middle column: Surface views with sense and relative magnitude of horizontal shear stresses acting at the base of the model (filled arrows) and shear stresses acting at depth along the lateral sidewalls (half arrows). The rotation of σ_1 from pure compression in the centre of the model to strike-slip at the sides explains the sense of curvature of the surface traces of the reverse faults. Right-hand column: Surface views of models after 2 cm displacement. (**a**) Fixed back wall model of Potsdam (model width 20 cm). (**b**) Fixed back wall model of Kyoto (model width 50 cm). (**c**) Mobile back wall model of Uppsala. Note the difference in drag of structures along the sidewalls between fixed and mobile back wall models. Grey-shaded area in column in the middle indicates the initial location of the sand wedge.

cut-off points of material layers by the faults. Quantitative analysis used sections along the lateral boundaries (Parma, Pavia and Toronto, initial stage of Uppsala) or through the centre of the model using X-ray CT images (Bern, IFP Rueil-Malmaison). Analysis of the Uppsala model was mainly done by monitoring the surface evolution and by cutting a section through the model at the end of deformation. Thus we could not measure all parameters for this experiment. As in the equivalent numerical models (Buiter *et al.* 2006), we found that the measurements were sensitive to the person measuring due to small differences in interpretation. Therefore, two people measured parameters for all experiments (using cross-sectional images taken at 1 cm increments of displacement) in the same manner and averaged the obtained values. In general the differences between the two measurers were small, with a maximum variation in fault dip angle values of 6°.

Table 3. *Quantification of mobile back wall shortening experiments*

Laboratory	Fault shortening* at 4 cm (cm)	Microbeads layer	
		Activated at (cm)	Detachment length (cm)
Bern	2.85	5	6.6
Parma	n.d.†	11	2.2
Pavia	1.9	no	–
IFP Rueil-Malmaison	2.75	no‡	–
Toronto	1.9	10.8	9.3
Uppsala	n.d.	12.3	12.7

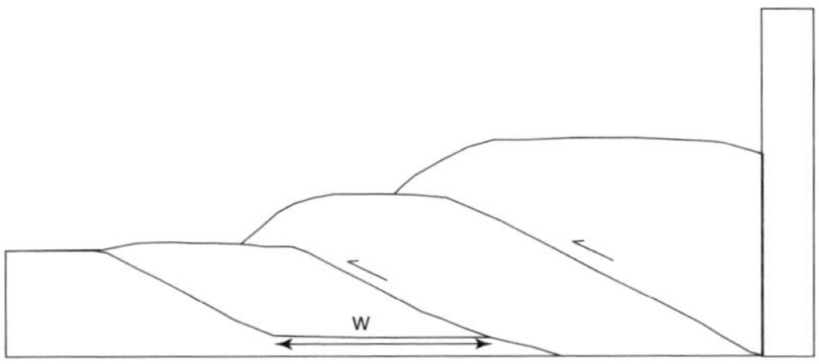

*Fault-controlled horizontal shortening determined by cut-off points between layers and the faults.
†In the model of Parma fault-controlled shortening was not determined as the faults are not narrow shear zones, but rather wide kink-like structures with deflection of layers towards the kink zones.
‡Reactivation of microbeads with a detachment length of 1.8 cm occurred along strike at 9 cm shortening.
Inset at bottom shows how detachment length (w) was measured.

The mobile back wall models share the following similarities: (1) Shortening is accommodated by forward thrusts, which propagate in-sequence towards the foreland, and by back thrusts (Fig. 2). (2) After 2 cm of displacement, an active forward thrust formed in all models (Fig. 2b). (3) The dip angles of this first forward thrust are fairly similar: 28° ± 4° for its lower dip and 29° ± 3° for its upper dip. Upper dip angles are generally slightly steeper (Figs 4b and c). (4) The dip angle of back thrusts is higher than the dip angle of forward thrusts (Fig. 4d).

From Figures 2, 4 and 5 and Table 3 it is also clear that variations in thrust wedge deformation occur among the models: (1) The number of forward thrusts that formed at a particular amount of displacement is variable (maximum difference of two thrusts) (Figs 2 and 4a). (2) Five experiments show the first thrust propagating upward from a point located to the left of the base of the back wall, whereas two experiments (Toronto, Fig. 2b and one model of Parma, Fig. 6a, b) show the forward thrust propagating from the corner at the base of the back wall. The reason for this difference is not entirely clear. A comparison of mobile back wall models whose evolution was monitored through glass sidewalls suggests that the presence of Alkor foil along the sidewalls could promote the formation of forward thrusts emanating from the lower right-hand corner. Cross-sections cut after 14 cm displacement located near or adjacent to the transparent sidewalls are shown in Figure 6 for the two models that applied Alkor foil to the sidewalls (one model of Parma and Toronto). In both models a large number of forward thrusts formed very close to the mobile wall. The absence of back thrusts close to the mobile wall for these models is in agreement with the formation of forward thrusts rooting at the basal right-hand corner, as this does not leave space for the formation of back thrusts propagating from the base. Alternatively, the variation in first thrust configuration could be caused by variation in sand properties. Shear zones may be located close to the mobile wall for sand in which peak strength is reached fast. Variations in friction along the back wall or in the height of the gap below the right-hand wall (however small) may also play a role.

(3) Four experiments (Bern, Parma, Pavia, IFP Rueil-Malmaison) show the formation of a back

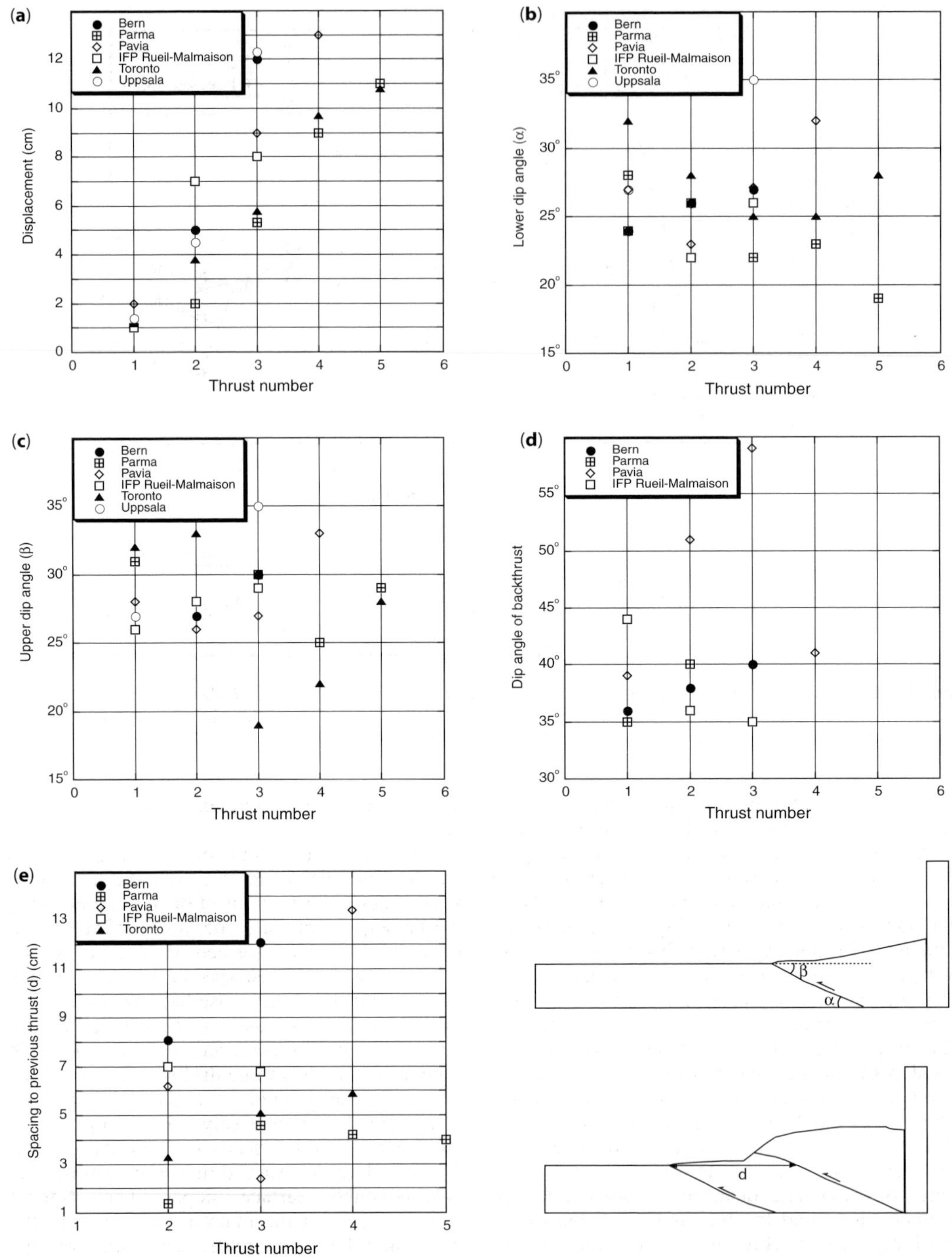

Fig. 4. Quantitative comparison of the six mobile back wall models. All values were measured at the moment a new fault initiated. (**a**) The amount of displacement at which a forward thrust forms. (**b**) Lower dip angle of forward thrusts. (**c**) Upper dip angle of forward thrusts. (**d**) Dip angle of back thrusts. (**e**) Spacing to previously formed forward thrust at the moment of initiation of a new forward thrust. Insets at right-hand lower corner show how lower dip angle (α), upper dip angle (β) and thrust spacing (d) were measured.

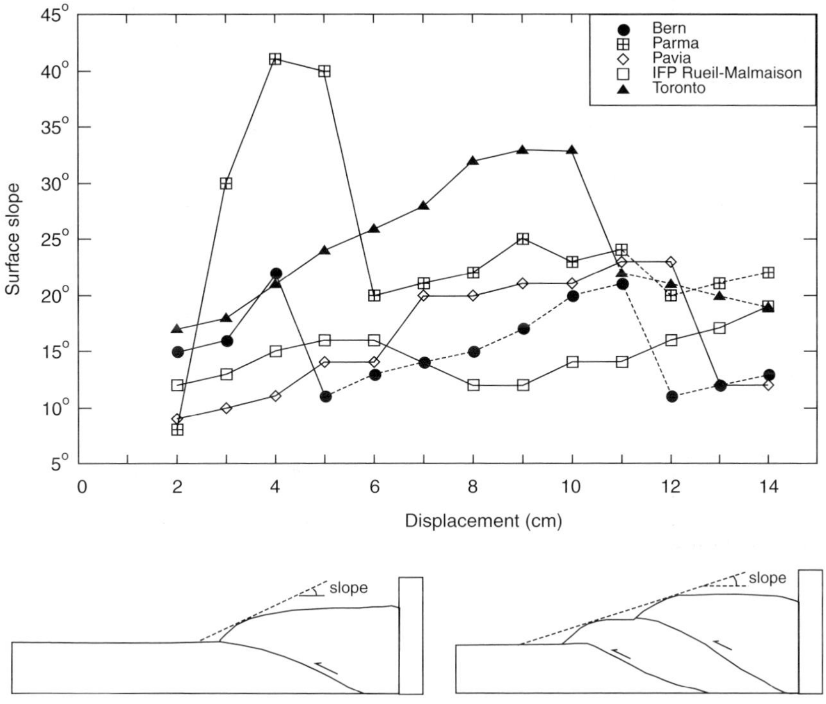

Fig. 5. Surface slope versus displacement for five mobile back wall models. Lines are solid when the microbeads layer has not been activated and dotted after activation of the microbeads layer. The insets show how surface slope was measured for one thrust (left) and for more than one thrust (right).

thrust related to the first thrust. In three of these experiments, the back thrust roots onto the base of the model and connects with the upper right-hand corner of the wedge. In the Bern experiment, however, the first back thrust roots at the top of the microbeads layer and does not connect with the upper right-hand corner of the wedge. In this experiment, CT images reveal that the back thrust forms later than the forward thrust. The back thrust rooting at the top of the microbeads layer suggests sensitivity to small variations in material properties (between sand and microbeads) for the Bern model (Fig. 2). In the Toronto experiment back thrusts did not form until 12 cm of displacement and are then related to the frontal thrust (Fig. 6f). (4) The distance between a newly formed thrust and the previously formed thrust is highly variable (Fig. 4e). (5) The microbeads layer was activated in four experiments (Bern, Parma, Toronto, Uppsala). In one experiment (IFP Rueil-Malmaison) activation of microbeads occurred along strike of the sections shown in Figure 2, based on CT scans. In the Pavia experiment, no activation took place (Table 3). (6) The length of the detachment in the microbeads layer is variable (Table 3).

(7) The surface slope of the wedge is also highly variable (Fig. 5). The high surface slope angles at 3 to 5 cm of displacement in the Parma experiment are the result of a second forward thrust that formed very close to the first one (with a thrust spacing of only slightly over 1 cm; Fig. 4e). Oscillations in surface slope angles occur as new thrusts form. The thrust wedges do not seem to reach a stable state. It may be that more displacement is needed before a stable field is reached (Lohrmann *et al.* 2003). Alternatively, alternating use of the embedded microbeads layer or the base of the model for forward thrust propagation may disturb stable surface slopes. The strength contrast between the microbeads layer and the base may be comparable, depending on the sand and microbeads used. Figure 6 shows no clear effect on surface slope of the first activation of the embedded microbeads layer. (8) Fault-controlled shortening at 4 cm of displacement varies between 48 and 71%, implying that diffuse deformation and deformation taken up by folding ranges from 29 to 52%.

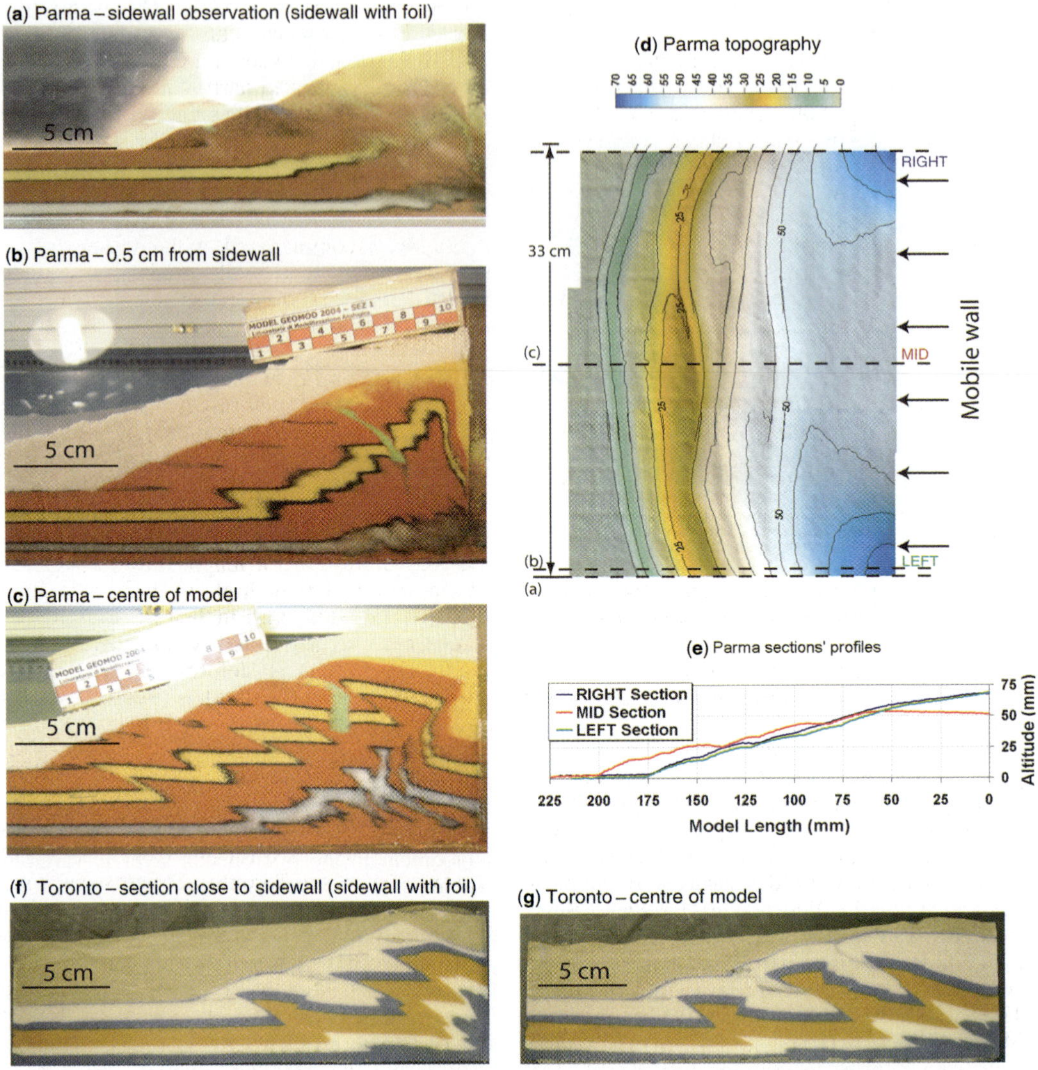

Fig. 6. Influence of lateral shear stresses illustrated in models of Parma (a–e) and Toronto (f and g). Structures are shown after 14 cm displacement. This model of Parma shown is with Alkor foil attached to the sidewalls (and is, therefore, different from the model in Fig. 2). The model of Toronto also has Alkor foil applied to the sidewalls (same model as in Fig. 2). (**a**) Cross section as observed through glass sidewalls. Note that Alkor foil partly obscures model visualization. (**b**) Cross section at 0.5 cm from sidewall. (**c**) Cross section through centre of model. (**d**) Contour map showing topography of deformed model, obtained by laser scanner; scale in cm above model base. (**e**) Topographic profiles 0.5 cm from sidewalls (right and left section) and through centre of model ('mid section'). (**f**) Cross section close to lateral sidewall. (**g**) Cross section through centre of model. Note that in (b), (c), (f) and (g) a sand layer was added on top of the model before cutting.

Effects of experimental set-up, material properties and boundary conditions

The geometry and evolution of the structures in the shortening experiments reflects differences in experimental set-up, material properties and model dimensions.

The surface structures in Figure 3 show that thrusts are convex to the hinterland for fixed back wall models and convex to the foreland for mobile back wall models. Although the absolute magnitude of the basal and lateral shear stresses are not known, the curved geometry along the sidewalls indicates differences in stress trajectories between fixed and mobile back wall models. The sense and relative magnitude of the basal shear stresses at the onset of shortening are the same for both fixed and mobile back wall models. The basal shear stresses are highest near the back wall and decrease away from it as a result of both decreasing wedge load and increasing distance from the back wall. The sense of the lateral shear stresses, however, is different for the fixed and mobile back wall models. The lateral shear stresses in the fixed back wall models are directed towards the fixed back wall, i.e., in a direction opposite to the basal shear stresses. In contrast, in the mobile back wall models the direction of the lateral shear stresses is the same as that of the basal shear stresses. The difference in sense of lateral shear stresses for fixed and mobile back wall models results in the difference in drag of structures near the sidewalls (compare Figs 3a and b with Fig. 3c). In both cases, the drag is in the same direction as the sense of the lateral shear stresses and can be understood from the rotation of the σ_1 (most compressive stress) direction from pure compression in the centre of the models towards strike-slip at the sides (Fig. 3).

In the fixed back wall model of Potsdam, a pop-up structure formed in front of the pre-existing wedge with the back thrust close to its toe (Figs 2b and 3a). The pre-existing wedge itself was not involved in the thrusting. The initial localization of deformation in front of the wedge in the Potsdam experiment (20 cm width) is similar to other fixed back wall model experiments with an initial wedge geometry and comparable initial thicknesses (Gutscher *et al.* 1996, 1998 (30 cm width); Konstantinovskaia & Malavieille 2005 (10 cm width)). In the mobile back wall models, a forward thrust or box fold structure forms close to the mobile wall and deforms the pre-existing wedge (Figs 2 and 3c). The fixed back wall model of Kyoto (Figs 2 and 3b) shows a structural evolution that is more similar to the mobile back wall models, with a forward thrust surfacing close to the toe of the wedge. A main difference between the Potsdam and Kyoto fixed back wall models is the width (measured perpendicular to shortening direction): the Potsdam model is 20 cm wide and 340 cm long, and the Kyoto model is 50 cm wide and 94 cm long. This suggests that model width has an important influence on the overall structural evolution of the thrust wedge. Comparisons of the Kyoto and Potsdam models (Figs 3a and b) suggest that the effect of lateral shear stresses on model structures is more important in narrow models. The width of a model apparatus needs careful attention, therefore, balancing between minimization of edge effects which occur in narrow models and minimization of heterogeneities due to, for example, hand sieving, which tends to be amplified in wide models.

The set-up of our shortening experiment required that at least 14 cm inward displacement could be accommodated without deformation reaching the back wall. Activation of the microbeads layer may cause thrusting to step further out than for a homogeneous sand layer, and a longer initial domain length will consequently be needed in that case. The region in front of the foremost thrust may, depending on the type of sand used, strain-harden upon shortening, and deformation of the wedge may be influenced if the strain-hardening front reaches the back wall. Adam *et al.* (2005) show that the region of diffuse strain accumulation in front of the wedge can have a length of approximately 15 cm (for a 3.5 cm high input layer). Some models in our study may in their late deformation stages have been affected by the length of the model apparatus, as deformation and/or the region of strain-hardening has reached the back wall. The length of the model apparatus should, therefore, also be considered carefully when designing an experiment.

The effects of lateral shear stresses on the shape of the thrust wedge during shortening are illustrated in Figure 6 for two mobile back wall models (Parma and Toronto). The cross-sectional wedge geometries at different locations (Figs 6a–c and 6f–g), the final model topography (Fig. 6d) and topographic profiles along different cross-sections (Fig. 6e) show that the thrust wedge shortens and steepens toward the sidewalls due to the shear stresses along the sidewalls. It is important, therefore, to evaluate the along-strike (perpendicular to the shortening direction) variation of model results. The effects of sidewall properties on model evolution could be tested by repeating the same experiment while varying only the materials of the walls.

As the properties of the granular materials varied significantly, it was not to be expected that all experiments would develop in an identical manner. The set-up of our comparison is, furthermore, not suitable to uniquely distinguish the effects of the differences in material properties from the influence of other variables such as model boundary conditions and dimensions. This can be tested by performing the same experiment in the same apparatus with different granular materials (Lohrmann *et al.* 2003).

Implications of the shortening experiments

The shortening experiments show that differences in material properties, boundary conditions (fixed back wall versus mobile back wall models) and model dimensions may result in significant differences in model results. Despite these differences, a comparison of the mobile back wall models demonstrates similar styles of deformation in which shortening is accommodated by both in-sequence forward propagation of thrusts and, in most models, by back thrusts. We consider this an encouraging result, which will help establish robust features in laboratory tectonic models. The variations in geometry and structural evolution among the mobile back wall models may be attributed to (1) differences in properties of the granular materials, (2) small variations in the exit slot below the mobile wall, (3) differences in the location of observation (at sidewall versus centre of model), (4) the properties of the walls (presence or absence of Alkor foil), (5) width of the box (due to influence of lateral shear stresses), (6) length of the model (for short domain lengths), and (7) set-up technique (for example, small variations may occur when sieving granular layers by hand and levelling of layers after sieving may lead to some compaction).

Our results indicate that care has to be taken when evaluating quantitative results from model experiments and applying them to natural systems. This is especially the case when results are obtained from measurements through transparent sidewalls (where lateral shear stresses are greatest). Although analogue models may be qualitatively used to interpret the mechanics of thrust wedges, limitations apply to quantifiable results such as thrust spacing, surface slope, length of detachment layer and the number of thrusts versus shortening.

Extension experiment

Model set-up

The extension experiment tested the influence of a weak, basal viscous layer on fault localization in overlying brittle materials (Fig. 1c). Five modelling laboratories ran the extension experiment (Bern, Florence, Piscataway, IFP Rueil-Malmaison and Toronto). The initial length (measured parallel to the extension direction) was 20 cm in all models, except for Toronto, where it was 29.2 cm. The width of the model, measured perpendicular to the extension direction, was variable (Table 2). The Florence laboratory ran two experiments with different widths (25 and 90 cm). Alkor foil covered the base and sidewalls of the experimental apparatus. A thin sheet, also covered with Alkor foil, lay on the base of the apparatus. It extended from the centre of the apparatus to a mobile end wall (right-hand side in all figures) and was attached to it. Extension was produced by moving the mobile wall with the attached sheet outward at a constant velocity of 2.5 cm/hour. The edge of the mobile sheet acted as a velocity discontinuity. Total displacement was 5 cm, except for the Piscataway model where it was 3 cm.

A 10 cm long and 5 mm thick slab of viscous PDMS covered the central part of the apparatus. To minimize PDMS sticking to the sidewalls, Vaseline was applied over a height of 5 mm from the right end (mobile) wall to the left end of the PDMS layer. Four models used dry sand and one model (from Piscataway) used wet clay as the brittle analogue material (Table 1). Sand was sieved (from a height of 20 cm with a filling rate of approximately 250 grams/minute) or clay was placed adjacent to and on top of the basal PDMS slab. Multiple sand (or clay) layers had the same material properties (as much as possible), but different colours to help in visualizing deformation. The model was not pre-compacted before the onset of extension. The total height of the model was 3.5 cm. No sedimentation occurred during extension. Two laboratories (Bern and IFP Rueil-Malmaison) monitored the model cross-sectional evolution by X-ray CT imaging through the centre of the model and three laboratories (Florence, Toronto, Piscataway) by monitoring the surface evolution and by cutting a cross-section through the centre after the end of the experiment.

Model results

Figure 7 shows the cross-sectional evolution of the models Bern and IFP Rueil-Malmaison for four stages of displacement (0, 0.5, 1 and 2 cm). Figure 8 illustrates the final cross-sectional geometry of all five models. Surface views at 2 cm displacement are shown in Figure 9. We measured the following parameters for a quantitative comparison of the models: the amount of displacement at which the first faults formed and their fault dip angle (Table 4), the amount of fault-controlled extension (Table 4), the displacement at which faults formed at the edges of the silicone layer (Table 4), and the dip angle and migration of the first-formed normal fault to the right of the velocity discontinuity with increasing extension (Fig. 10). Due to different monitoring procedures not all

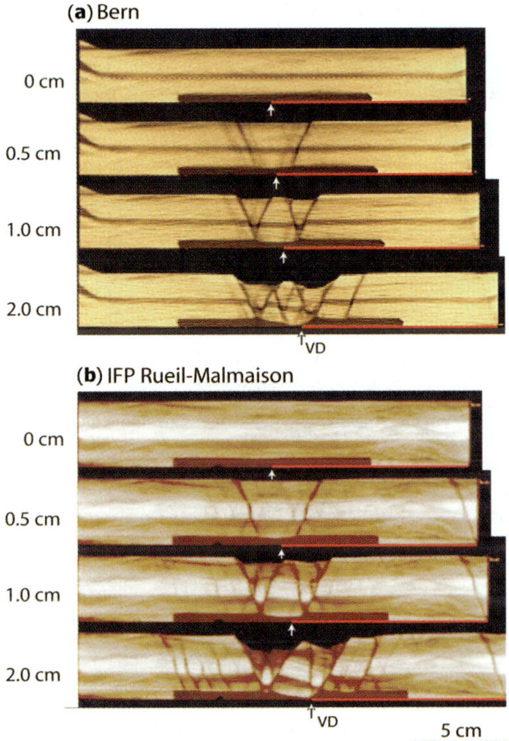

Fig. 7. Cross-sectional geometries of two extension experiments at 0, 0.5, 1 and 2 cm displacement. The horizontal red line indicates the position of basal thin sheet covered with Alkor foil. The tip of the basal sheet is indicated with a white arrow (VD = velocity discontinuity). (**a**) Model of Bern. The cross-section is taken 35 cm from the side (see Fig. 9). (**b**) Model of IFP Rueil-Malmaison. The cross-section is taken through the centre of the model.

parameters could be determined for all laboratories. The values were again measured by two people and then averaged.

Analysis of the qualitative and quantitative aspects of the experimental model results allows the following observations:

(1) After 1 cm of displacement, fault zones are visible in all models. In most models, faults formed near the centre of the model and with additional extension, the faulted zone broadened, with younger faults propagating towards the edges of the silicone layer (Fig. 7). In the narrow model of Florence (25 cm width), the model evolution differed (Fig. 9). The first faults nucleated onto the right-hand edge of the silicone layer and much of the subsequent extension was accommodated along these faults. In this narrow experiment, the velocity discontinuity had no influence on localizing initial deformation. With additional extension, most faulting concentrated in the brittle layers along the right-hand edge of the silicone layer and only minor faulting occurred in the centre and left-hand side of the model. In two other models (Toronto and Florence large-width model), early faulting also nucleated at the right-hand edge of the silicone, but nearly simultaneously with the development of faults near the central velocity discontinuity. Faulting in the central part of the model subsequently accommodated most of the continued extension. All sand-silicone models demonstrate that the rheological change between the viscous silicone layer and the brittle sand cover, combined with friction along the sidewalls, may result in the nucleation of faults along the edges of the silicone layer.

(2) All models show drag structures due to friction along the sidewalls (Fig. 9). The sense of the curvature of the surface views of the normal faults can be understood from the rotation of σ_3 (least compressive stress) from pure extension in the centre of the model towards strike-slip at the sides (see also Vendeville 1987).

(3) The first formed faults have dip angles of 60° to 69° in sand layers and dip angles of 55° to 58° in clay. These dip angles correspond closely to those expected for Mohr-Coulomb faults, for which the fault dip should be $45° + \phi/2$ (Coulomb angle).

(4) The cross-sectional evolution of the Bern and IFP Rueil-Malmaison models, obtained by X-ray CT imaging through the centre of the model, shares many similarities. Extension first leads to the formation of two main fault in the central part of the model on either side of the velocity discontinuity (Fig. 7; panels at 0.5 cm). These faults are initially symmetrically located with respect to the velocity discontinuity. However, the fault extrapolations do not root directly above the velocity discontinuity, but are offset laterally. This may be explained by more distributed thinning of the silicone layer, which smooths out the influence of the velocity discontinuity. Further displacement along the two main normal fault causes further thinning and sideways expulsion of the silicone layer and results in the formation of two conjugate fault sets (Fig. 7; panels at 1 cm). As displacement of the basal sheet continues, the initial symmetry of the first deformation structures is disrupted. The silicone continues to thin, the surface depression becomes wider and asymmetric in shape, and the faulted zone broadens outward.

(5) The dip angle of the first formed normal fault on the right side of the velocity discontinuity that dips away from the mobile wall decreases with continued extension (Fig. 10a). For larger amounts of displacement (5 cm), faults tend to

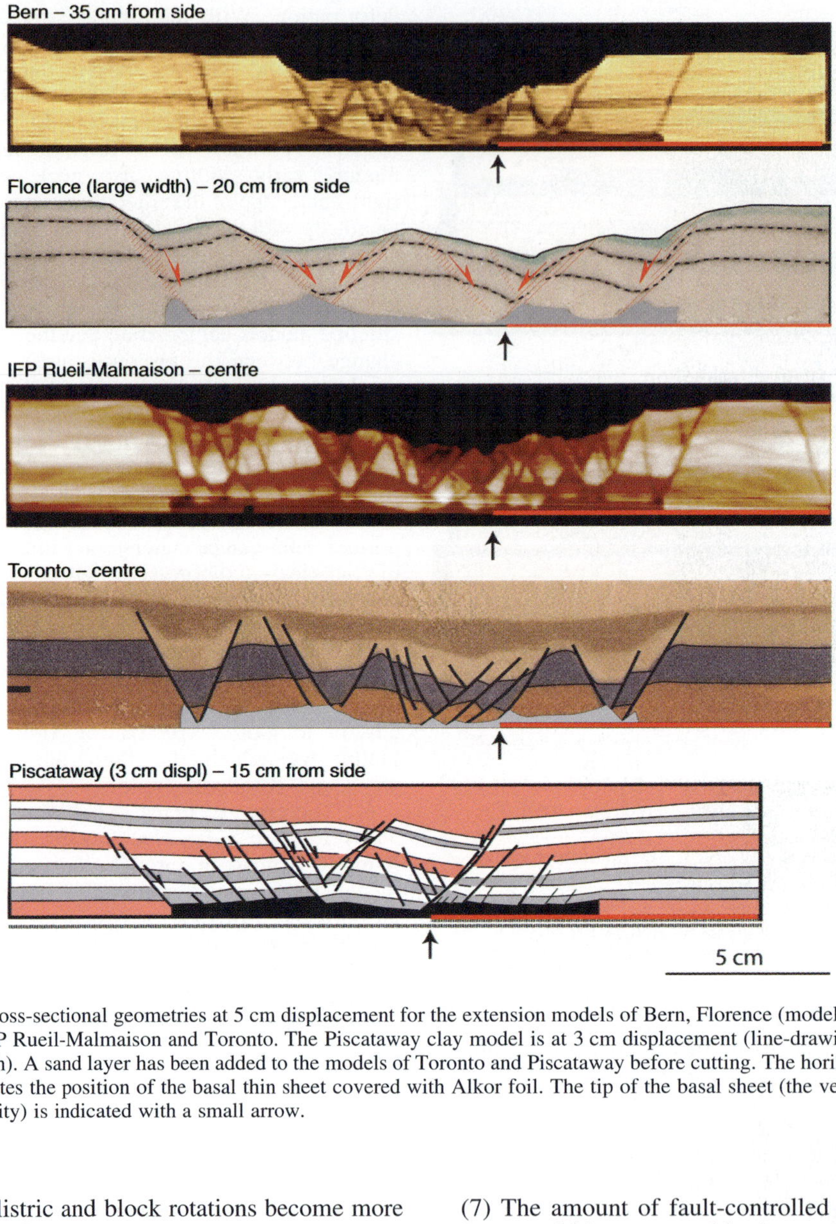

Fig. 8. Cross-sectional geometries at 5 cm displacement for the extension models of Bern, Florence (model of 90 cm width), IFP Rueil-Malmaison and Toronto. The Piscataway clay model is at 3 cm displacement (line-drawing after photograph). A sand layer has been added to the models of Toronto and Piscataway before cutting. The horizontal red line indicates the position of the basal thin sheet covered with Alkor foil. The tip of the basal sheet (the velocity discontinuity) is indicated with a small arrow.

become listric and block rotations become more pronounced.

(6) The first-formed normal fault on the right side of the velocity discontinuity (dipping away from the right end wall) moves to the right as the model is extended (Fig. 10b). In the models of Bern and IFP Rueil-Malmaison fault migration is similar and in both cases slower than the displacement velocity of the mobile wall because continued extensional deformation is also accommodated by the formation of new faults in the block located to the right of the first formed normal fault.

(7) The amount of fault-controlled extension at 2 cm could be measured in three experiments and ranges between 0.8 cm in the silicone-clay model and 0.9 to 0.95 in the silicone-sand models.

Implications of the extension experiments

As in the shortening experiment, we find that the width of the model (perpendicular to the extension direction) may cause differences in model results. Lateral friction seems to have a larger influence on initial fault location and evolution

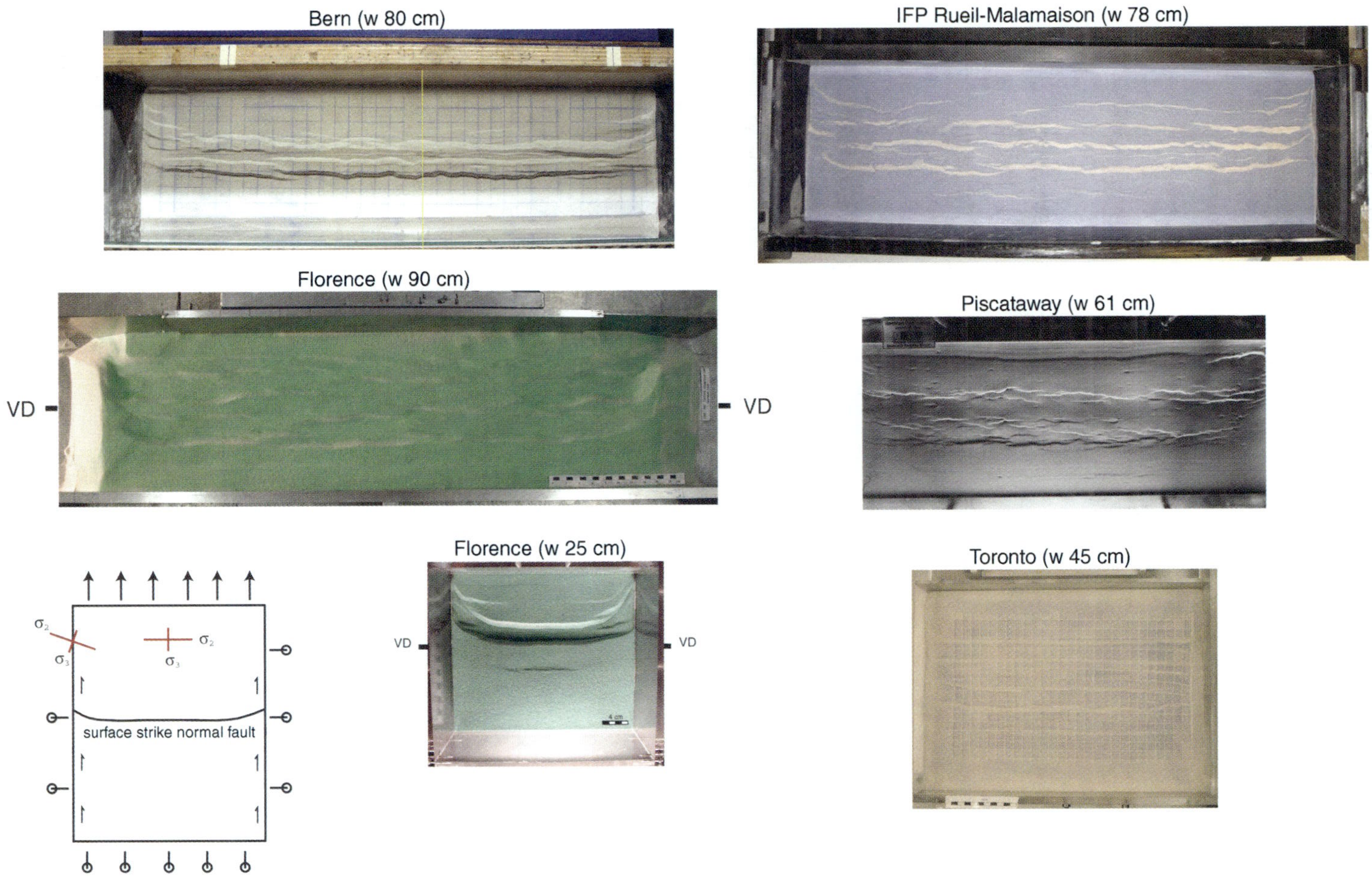

Fig. 9. Surface views of the extension models of Bern, Florence (25 cm and 90 cm wide models), IFP Rueil-Malmaison, Piscataway (57 cm of 61 cm width is shown) and Toronto at 2 cm displacement. The width of the models is indicated with the laboratory names at the top of each panel. The mobile wall is at the top of each panel. Inset shows how the rotation of σ_3 from pure extension in the middle of the model to strike-slip at the sides explains the direction of the drag structures at the sidewalls.

Table 4. *Quantification of extension experiment*

Laboratory	Fault extension* at 2 cm (in cm)	Fault initiation at edges of PDMS (at cm of extension)		First fault on left side of VD dipping toward mobile wall		First fault on right side of VD dipping away from mobile wall	
		Left	Right	Dip angle	At extension (cm)	Dip angle	At extension (cm)
Bern	0.9	never	4–4.5	69°	<0.5	65°	<0.5
Florence[†]	n.d.	1	1	60°[‡]	<0.5	[§]	
Piscataway	0.8	2	never	55°[‖]		58°[‖]	
IFP Rueil-Malmaison	0.95	never	2–2.5	65°	<0.5	60°	<0.5
Toronto	n.d.	1.2	0.6	62°	0.6–1.2	62°	0.6–1.2

*Fault-controlled horizontal extension determined by cut-off points between layers and the faults.
[†]Results shown are for large-width (90 cm) model.
[‡]Fault dip measured at surface.
[§]First two faults that form both dip towards mobile wall.
[‖]Fault dip measured at surface; faults dip shallower towards bottom of model.
VD = velocity discontinuity.

in narrow models than in the wider models. A comparison between the Toronto model (29.2 cm length) and the other models (20 cm length) (Fig. 8) shows no clear indication for influence of the length of the model for this set-up. In wide models the main graben forms initially above the velocity discontinuity, as observed in the equivalent numerical models (Buiter *et al.* 2006). A comparison between these models shows a high level of agreement with respect to initial fault dip and graben location, graben broadening and decrease in fault dip with continued extension. However, variability in structural evolution does occur and includes differences in the number of normal faults that form and the timing with which faulting propagates to the edges of the viscous layer. The differences in evolution between the models may be attributed to (1) differences in material properties, (2) differences in model width, (3) possible small variations of initial silicone slab geometry, (4) variations in base and wall properties, and (5) differences in location of observation. Similar to the shortening experiment, our results indicate that caution is required when quantitative results such as the

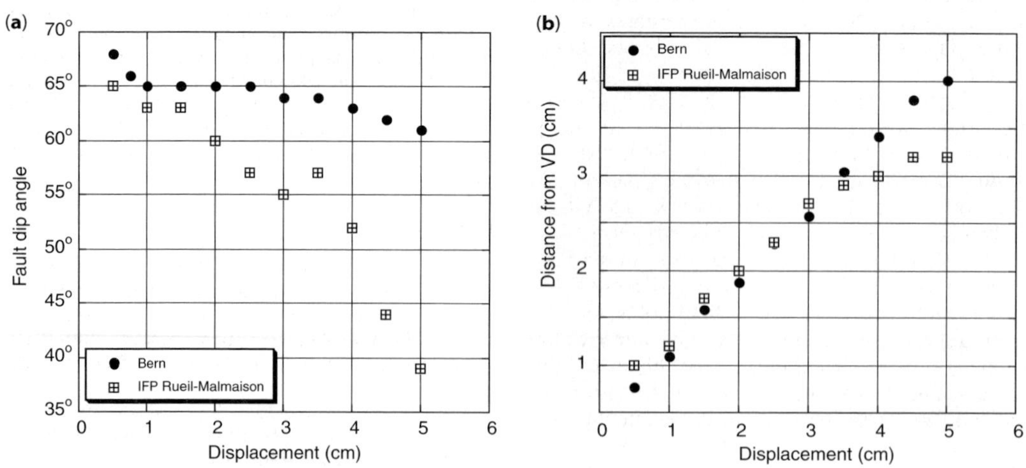

Fig. 10. (**a**) Dip angle versus displacement for the first formed normal fault on the right side of the velocity discontinuity and dipping away from the mobile wall. (**b**) Normal fault migration in the extension experiment. The migration of the first formed normal fault on the right side of the velocity discontinuity and dipping away from the mobile wall is measured relative to the velocity discontinuity (VD).

number and spacing of faults are extrapolated from analogue models to natural systems.

Conclusions and outlook

We used two experimental set-ups to test the reproducibility of model results among different experimental modelling laboratories. We find that all models demonstrate a similar first order evolution. The differences among the models in our study may help to identify main focus points for future comparison studies.

A comparison of eight brittle thrust-wedge experiments shows that model evolution may differ in detail to a considerable extent. These differences may be related to: (1) differences in model set-up (i.e., two laboratories used a conveyor belt below a fixed back wall and six laboratories used a mobile back wall above a fixed horizontal base), (2) differences in model dimensions, (3) location of observation (i.e., at the transparent sidewall or in the centre of the model), (4) material properties of granular materials (angle of internal friction, cohesion and amount of strain softening and strain hardening), (5) mechanical properties along the walls (with or without Alkor foil), (6) small differences in the set-up technique used by individual experimenters ('the human factor'). All mobile back wall experiments show the development of a thrust wedge characterized by in-sequence forward thrusts and minor back thrusting. A large variability exists in quantitative parameters such as fault dip, thrust spacing, number of forward and back thrusts, and surface slope.

The evolution of wide extension models is highly similar. Normal faulting initiates in the brittle layers above the viscous silicone layer, near the basal velocity discontinuity. Fault dip angles tend to shallow with continued extension. Variations between laboratories occur for parameters such as fault dip angle and fault spacing. Reducing the variability in quantifiable parameters such as number of faults and fault spacing will be a challenge for future analogue benchmarks.

The present study is the first ever comparison of analogue experiments between different modelling laboratories and our results may be used as a guide for the design of future comparison studies. Future 'benchmarks' could focus on: (1) a simpler experimental design to minimize the number of differences in experimental set-up among the participating laboratories, (2) use of the same materials in comparison studies between different laboratories (to test the influence of model apparatus and experimenter), (3) measuring the properties of a sample of the same material in different shear testers in order to determine the influence of the measuring method, (4) additional tests of the influence of material properties for different materials and experimental set-ups (where the same experimenter repeats the same experiment in the same laboratory and modelling apparatus), following up on previous studies (e.g., Lohrmann *et al.* 2003; Atmaoui *et al.* 2005; Withjack & Callaway 2000), (5) having different experimenters perform the same experiment using one apparatus (in the same laboratory) to test the 'human factor', (6) testing the variability of quantifiable parameters obtained by one experimenter using one single apparatus and the same materials against the variability of these parameters among laboratories, and (7) testing the effects of the properties of walls and base of the modeling apparatus by performing the same experiment (with the same material and by the same experimenter) for different materials covering the base and walls.

The results of our benchmark study illustrate the importance for analogue experimental studies of a clear specification of modelling apparatus, set-up technique, material properties and boundary conditions, and may help in establishing the degree to which model results may be extrapolated to natural examples.

C. Beaumont, J.-P. Brun, and B. Colletta are thanked for providing constructive feedback on the design of the experiments, J. Adam and S. Ellis for help in determining the list of parameters to quantify, B. Vendeville for insightful discussions on the importance of boundary conditions and experimental procedures, and the participants of the GeoMod2004 pre-conference workshop for helpful discussions. We also thank J. Adam and B. Vendeville for their encouraging and helpful reviews. This study was supported by Swiss National Science Foundation Grant 2000-067952.02 (GS), the Natural Sciences and Engineering Research Council of Canada (ARC), Deutsche Forschungsgemeinschaft as part of SFB 267 (NK), and the Swedish Research Council, VR (HAK and FN).

References

ADAM, J., URAI, J. L. *ET AL.* 2005. Shear localization and strain distribution during tectonic faulting – new insights from granular-flow experiments and high-resolution optical image correlation techniques. *Journal of Structural Geology*, **27**, 283–301.

ATMAOUI, N., KUKOWSKI, N., STÖCKHERT, B. & KÖNIG, D. 2005. Initiation and development of pull-apart basins with riedel shear mechanism: Insights from clay scaled experiments. *International Journal of Earth Sciences*. DOI: 10.1007/S00531-005-0030-1.

BAHROUDI, A., KOYI, H. A. & TALBOT, C. J. 2003. Effect of ductile and frictional décollements on style of extension. *Journal of Structural Geology*, **25**, 1401–1423.

BARNHOORN, A. BYSTRICKY, M., BURLINI, L. & KUNZE, K. 2004. The role of recrystallization on the deformation behaviour of calcite rocks: large strain torsion experiments on Carrara marble. *Journal of Structural Geology*, **26**, 885–903.

BRUN, J.-P. 1999. Narrow rifts versus wide rifts: inferences for the mechanics of rifting from laboratory experiments. *Philosophical Transactions, Royal Society London A*, **357**, 695–712.

BRUN, J.-P., SOKOUTIS, D. & VAN DEN DRIESSCHE, J. 1994. Analogue modelling of detachment fault systems and core complexes. *Geology*, **22**, 319–322.

BUITER, S. J. H., BABEYKO, A. YU. *ET AL.* 2006. The numerical sandbox: Comparison of model results for a shortening and an extension experiment. *In*: BUITER, S. J. H. & SCHREURS, G. (eds) *Analogue and Numerical Modelling of Crustal-Scale Processes.* Geological Society, London, Special Publications, **253**, 29–64.

BYERLEE, J. 1978. Friction of rocks. *Pure and Applied Geophysics*, **116**, 615–626.

CADELL, H. M. 1888. Experimental researches in mountain building. *Transactions of the Royal Society of Edinburgh*, **1**, 339–343.

CASAGRANDE, A. 1932. Research on the Atterberg limits of soils. Public Roads, **13(9)**, 121–136.

COLLETTA, B., BALE P., BALLARD, J. F., LETOUZEY, J. & PINEDO, R. 1991. Computerized X-ray tomography analysis of sandbox models: Examples of thin-skinned thrust systems. *Geology*, **19**, 1063–1067.

COSTA, E., CAMERLENGHI, A. *ET AL.* 2004. Modelling deformation and salt tectonics in the eastern Mediterranean Ridge accretionary wedge. *Geological Society of America Bulletin*, **116**, 880–894.

CRUDEN, A. R., NASSERI, M. B. & PYSKLYWEC, R. 2006. Surface topography and internal strain variation in wide hot orogens from three-dimensional analogue and two-dimensional numerical vise models. *In*: BUITER, S. J. H. & SCHREURS, G. (eds) *Analogue and Numerical Modelling of Crustal–Scale Processes.* Geological Society, London, Special Publications, **253**, 79–104.

DAVIS, D., SUPPE, J. & DAHLEN, F. A. 1983. Mechanics of fold-and-thrust belts and accretionary wedges. *Journal of Geophysical Research*, **88**, B2, 1153–1172.

ELLIS, S., SCHREURS, G. & PANIEN, M. 2004. Comparisons between analogue and numerical models of thrust wedge development. *Journal of Structural Geology*, **26**, 1659–1675.

GARTRELL, A. P. 2001. Crustal rheology and its effect on rift basin styles. *In*: KOYI, H. A. & MANCKTELOW, N. S. (eds) *Tectonic Modeling: A volume in honor of Hans Ramberg*. Geological Society of America Memoir, **193**, 221–233.

GUTSCHER, M.-A., KUKOWSKI, N., MALAVIEILLE, J. & LALLEMAND, S. 1996. Cyclical behaviour of thrust wedges: Insights from high basal friction sandbox experiments. *Geology*, **24**, 135–138.

GUTSCHER, M.-A., KUKOWSKI, N., MALAVIEILLE, J. & LALLEMAND, S. 1998. Material transfer in accretionary wedges from analysis of a systematic series of analog experiments: *Journal of Structural Geology*, **20**, 407–416.

HOTH, S., ADAM, J., KUKOWSKI, N. & ONCKEN, O. 2005. Influence of erosion on the kinematics of bivergent orgens. Results from scaled sandbox simulations. *In*: WILLETT, S.D., HOVIUS, N., BRANDON, M.T. & FISHER, D. (eds) *Tectonics, Climate, and Landscape Evolution*. Special Paper **398**, Penrose Conference Series, The Geological Society of America, Boulder, CO, 201–225, DOI: 10.1130/2006.2398 (12).

HUBBERT, M. K. 1937. Theory of scale models as applied to the study of geological structures. *Geological Society of America Bulletin*, **48**, 1459–1520.

HUBBERT, M. K. 1951. Mechanical basis for certain geological structures. *Geological Society America Bulletin*, **62**, 355–372.

JAEGER, J. C. & COOK, N. G. W. 1979. *Fundamentals of rock mechanics*. John Wiley and Sons, New York, 585pp.

KEEP, M. & MCCLAY, K. R. 1997. Analogue modelling of multiphase rift systems. *Tectonophysics*, **273**, 239–270.

KONSTANTINOVSKAIA, E. & MALAVIEILLE, J. 2005. Erosion and exhumation in accretionary orogens: Experimental and geological approaches. *Geochemistry, Geophysics, Geosystems (G^3)*, **6**, Q02006, DOI 10.1029/2004GC000794.

KOYI, H. A. 1995. Mode of internal deformation in sand wedges. *Journal of Structural Geology*, **17**, 293–300.

KOYI, H. A. 1997. Analogue modelling: From a qualitative to a quantitative technique, a historical outline. *Journal of Petroleum Geology*, **20**, 223–238.

KUKOWSKI, N., LALLEMAND, S. E., MALAVIEILLE, J., GUTSCHER, M.-A. & RESTON, T. J. 2002. Mechanical decoupling and basal duplex formation observed in sandbox experiments with application to the western Mediterranean Ridge accretionary complex. *Marine Geology*, **186**, 29–42.

LALLEMAND, S. E., SCHNÜRLE, P. & MALAVIEILLE, J. 1994. Coulomb theory applied to accretionary and non-accretionary wedges: Possible causes for tectonic erosion and/or frontal accretion. *Journal of Geophysical Research*, **99**, B6, 12033–12055.

LOHRMANN, J., KUKOWSKI, N., ADAM, J. & ONCKEN, O. 2003. The impact of analogue material properties on the geometry, kinematics, and dynamics of convergent sand wedges. *Journal of Structural Geology*, **25**, 1691–1771.

MALAVIEILLE, J. 1984. Modélisation expérimentale des chevauchements imbriqués: application aux chaînes de montagnes. *Bulletin Societé Géologique de France*, **7**, 129–138.

MARONE, C. 1998. Laboratory-derived friction laws and their application to seismic faulting. *Annual Review of Earth and Planetary Sciences*, **26**, 643–696.

MUGNIER, J. L., BABY, P., COLLETTA, B., VINOUR, B., BALE, P. & LETURMY, P. 1997. Thrust geometry controlled by erosion and sedimentation: A view from analogue models. *Geology*, **25**, 427–430.

MULUGETA, G. 1988. Modelling the geometry of Coulomb thrust wedges. *Journal of Structural Geology*, **10**, 847–859.

MULUGETA, G. & KOYI, H. 1987. Three-dimensional geometry and kinematics of experimental piggy-back thrusting. *Geology*, **15**, 1052–1056.

MULUGETA, G. & KOYI, H. 1992. Episodic accretion and strain partitioning in a model sand wedge. *Tectonophysics*, **202**, 319–322.

PANIEN, M. 2004. Analogue modelling experiments of basin inversion using well-characterized granular materials and comparisons with numerical models. Ph.D. thesis. University of Bern, Switzerland.

PANIEN, M., BUITER, S. J. H., SCHREURS, G. & PFIFFNER, O. A. 2006 Inversion of a symmetric basin: Insights from a comparison between analogue and numerical experiments. *In*: BUITER, S. J. H. & SCHREURS, G. (eds) *Analogue and Numerical Modelling of Crustal-Scale Processes*. Geological Society, London, Special Publications, **253**, 253–270.

PYSKLYWEC, R. N. & CRUDEN, A. R. 2004. Coupled crust-mantle dynamics and intraplate tectonics: two-dimensional numerical and three-dimensional analogue modelling. *Geochemistry, Geophysics Geosystems* G^3, **5**(**10**), DOI 10.1029/2004GC000748.

RAMBERG, H. 1981. *Gravity, deformation and the Earth's crust*. Academic Press, London.

SCHELLART, W. P. 2000. Shear test results for cohesion and friction coefficients for different granular materials: scaling implications for their usage in analogue modelling. *Tectonophysics*, **324**, 1–16.

SCHREURS, G., HÄNNI, R. & VOCK, P. 2001. Four-dimensional analysis of analog models: Experiments on transfer zones in fold and thrust belts. In: KOYI, H. A. & MANCKTELOW, N. S. (eds) *Tectonic Modeling: A volume in honor of Hans Ramberg*. Geological Society of America Memoir, **193**, 179–190.

SCHULZE, D. 1994. Entwicklung und Anwendung eines neuartigen Ringschergerätes. *Aufbereitungstechnik*, **35**(**10**), 524–535.

SIMS, D. 1993. The rheology of clay: a modeling material for geologic structures: *EOS, Transactions, American Geophysical Union*, **74**, 569.

STORTI, F. & MCCLAY, K. 1995. Influence of syntectonic sedimentation on thrust wedges in analogue models. *Geology*, **23**, 999–1002.

STORTI, F., SALVINI, F. & MCCLAY, K. 2000. Synchronous and velocity-partitioned thrusting and thrust polarity reversal in experimentally produced doubly-vergent thrust wedges: Implications for natural orogens. *Tectonics*, **19(2)**, 378–396.

TEN GROTENHUIS, S. M., PIAZOLO, S., PAKULA, T., PASSCHIER, C. W. & BONS, P.D. 2002. Are polymers suitable rock analogs? *Tectonophysics*, **350**, 35–47.

TRON, V. & BRUN, J-P. 1991. Experiments on oblique rifting in brittle-ductile systems. *Tectonophysics*, **188**, 71–84.

TURRINI, C., RAVAGLIA, A. & PEROTTI, C. 2001. Compressional structures in a multilayered mechanical stratigraphy: Insights from sandbox modeling with three-dimensional variations in basal geometry and friction. *In*: KOYI, H. A. & MANCKTELOW, N. S. (eds) *Tectonic Modeling: A volume in honor of Hans Ramberg*. Geological Society of America Memoir, **193**, 158–178.

VENDEVILLE, B. C. 1987. *Champs de failles et tectonique en extension: modélisation expérimentale.* Ph.D. thesis, University of Rennes I.

VENDEVILLE, B. C. & JACKSON, M. P. A. 1992*a*. The rise of diapirs during thin-skinned extension. *Marine Petroleum Geology*, **9**, 331–353.

VENDEVILLE, B. C. & JACKSON, M. P. A. 1992*b*. The fall of diapirs during thin-skinned extension. *Marine Petroleum Geology*, **9**, 354–371.

WEIJERMARS, R. 1986. Flow behaviour and physical chemistry of bouncing putties and related polymers in view of tectonic laboratory applications. *Tectonophysics*, **124**, 325–358.

WITHJACK, M. O. & CALLAWAY, S. 2000. Active normal faulting beneath a salt layer; an experimental study of deformation patterns in the cover sequence. *AAPG Bulletin*, **84**, 627–651.

The numerical sandbox: comparison of model results for a shortening and an extension experiment

SUSANNE J. H. BUITER[1], ANDREY YU. BABEYKO[2,8], SUSAN ELLIS[3], TARAS V. GERYA[4], BORIS J. P. KAUS[5], ANTJE KELLNER[2], GUIDO SCHREURS[6] & YASUHIRO YAMADA[7]

[1]*Centre for Geodynamics, Geological Survey of Norway, 7491 Trondheim, Norway (e-mail: susanne.buiter@ngu.no)*

[2]*GeoForschungsZentrum Potsdam, Telegrafenberg, 14473 Potsdam, Germany*

[3]*Institute of Geological and Nuclear Sciences, P.O. Box 30368, Lower Hutt, New Zealand*

[4]*Swiss Federal Institute of Technology, Sonneggstrasse 5, 8092 Zürich, Switzerland*

[5]*Department of Earth Sciences, University of Southern California, Los Angeles, USA*

[6]*Institute of Geological Sciences, University of Bern, Baltzerstrasse 1-3, 3012 Bern, Switzerland*

[7]*Department of Civil and Earth Resources Engineering, Kyoto University, Kyoto 606-5801, Japan*

[8]*Now at Institut für Meteorologie und Geophysik, J.W. Goethe Universität Frankfurt, Germany*

Abstract: We report results of a study comparing numerical models of sandbox-type experiments. Two experimental designs were examined: (1) A brittle shortening experiment in which a thrust wedge is built in material of alternating frictional strength; and (2) an extension experiment in which a weak, basal viscous layer affects normal fault localization and propagation in overlying brittle materials. Eight different numerical codes, both commercial and academic, were tested against each other. Our results show that: (1) The overall evolution of all numerical codes is broadly similar. (2) Shortening is accommodated by in-sequence forward propagation of thrusts. The surface slope of the thrust wedge is within the stable field predicted by critical taper theory. (3) Details of thrust spacing, dip angle and number of thrusts vary between different codes for the shortening experiment. (4) Shear zones initiate at the velocity discontinuity in the extension experiment. The asymmetric evolution of the models is similar for all numerical codes. (5) Resolution affects strain localization and the number of shear zones that develop in strain-softening brittle material. (6) The variability between numerical codes is greater for the shortening than the extension experiment.

Comparison to equivalent analogue experiments shows that the overall dynamic evolution of the numerical and analogue models is similar, in spite of the difficulty of achieving an exact representation of the analogue conditions with a numerical model. We find that the degree of variability between individual numerical results is about the same as between individual analogue models. Differences among and between numerical and analogue results are found in predictions of location, spacing and dip angle of shear zones. Our results show that numerical models using different solution techniques can to first order successfully reproduce structures observed in analogue sandbox experiments. The comparisons serve to highlight robust features in tectonic modelling of thrust wedges and brittle-viscous extension.

Numerical and analogue modelling methods represent two different techniques with which the evolution of geological structures, such as fold-and-thrust belts and sedimentary basins, can be investigated. The underlying assumption with both methods is that their results approximate the development of structures in the real Earth in a reasonable manner. We may then expect that the results of analogue and numerical models look similar when applied to the same

From: BUITER, S. J. H. & SCHREURS, G. (eds) 2006. *Analogue and Numerical Modelling of Crustal-Scale Processes*. Geological Society, London, Special Publications, **253**, 29–64.
0305-8719/06/$15.00

set-up. This expectation motivated us to directly compare results of numerical models obtained with different codes and results of analogue experiments from different laboratories. The aims of our study are (1) to compare results from different numerical codes and (2) to test the similarity of numerical and analogue models, in order to help establish robust features of tectonic models on the scale of the upper crust.

The companion paper (Schreurs *et al.* 2006) presents the results of an analogue comparison study with ten participating modelling laboratories. Two experimental set-ups were tested: (1) a brittle convergent thrust wedge experiment and (2) a brittle-viscous extension experiment. The reproducibility of modelling results between the laboratories was found to be fairly high as shown by the same general trends in the results. Differences between the analogue models were found in, for example, number of faults, fault spacing and dip, and surface slopes. The study highlights the importance of (1) a careful determination of properties of the granular materials and (2) the effects of lateral side boundaries on resulting structures due to friction between model materials and sidewalls. A sandbox needs to be wide enough, therefore, to avoid such boundary effects, and structures should if possible be studied in sections well away from the sidewalls.

We present numerical equivalents of the two analogue experiments of Schreurs *et al.* (2006). The conditions of the analogue modelling apparatus and the analogue material properties are approached as closely as possible. We compare the numerical results to each other through determination of their qualitative (visual) and quantitative (measurements of, for example, surface slope and shear zone dip) similarities and differences. Our study is not a numerical benchmark in its strictest sense as this would require the use of the exact same initial setup, material properties, boundary conditions, mesh size and time step size (see for example the numerical convection benchmarks of Blanckenbach *et al.* 1989; van Keken *et al.* 1997). Instead, we show the variability in results obtained by using different solution methods (finite element method, finite difference method and distinct element method) and allowing variations in mesh and time step size and boundary conditions (e.g., implementation of boundary friction). With our approach, we test the code-independence of tectonic structures predicted by numerical models and our findings are, therefore, of direct relevance for studies in which numerical experiments are used to investigate geological processes. Eight numerical codes were used in our comparison: Abaqus/Standard, LAPEX-2D, I2ELVIS, Microfem, NISA/Static, PFC^{2D}, SloMo and Sopale. A description of these codes is given in the next section.

The combination of our study with the accompanying paper results in a series of numerical and analogue model experiments, which are compared in a qualitative and quantitative manner with each other. Our results can be used to evaluate how closely various numerical solution methods can reproduce analogue model conditions. Issues that should be considered carefully in the 'numerical sandbox' include the suitability of continuum methods (e.g., finite element method) in simulating the behaviour of discontinuous grains, the importance of dilatation and compaction in granular materials, and the representation of velocity discontinuities. Numerical modelling of sandbox experiments poses significant computational challenges requiring that numerical codes are able to (1) calculate large deformations, (2) represent complex boundary conditions, including frictional boundaries and free surfaces, and (3) include a complex rheology involving both viscous and frictional/plastic materials. These challenges reflect directly the state-of-the-art requirements for numerical modelling of large-scale tectonic processes.

The two experimental set-ups were designed to reflect set-ups commonly used in the study of upper-crustal tectonic processes. Convergent thrust wedges have been studied with analytical techniques (Davis *et al.* 1983; Dahlen 1984), analogue experiments (e.g., Mulugeta 1988; Gutscher *et al.* 1998*a*; Storti *et al.* 2000) and numerical models (e.g., Strayer *et al.* 2001; Burbridge & Braun 2002). Similarly, brittle-viscous extension has been investigated with analogue (e.g., Michon & Merle 2000; Bahroudi *et al.* 2003) and numerical (e.g., Behn *et al.* 2002; Wijns *et al.* 2003) methods. A few previous studies have combined numerical and analogue modelling methods for studying a tectonic process on the scale of the crust. Sassi *et al.* (1993) investigated the reactivation of pre-existing faults in a compressive regime and showed how the visualization of the stress field in the numerical results (obtained with a distinct element method) could help in the interpretation of the observed reactivation of some low angle faults. The models calculated with the distinct element method by Saltzer & Pollard (1992) of structures in sedimentary layers overlying basement normal faults show a high resemblance to deformation patterns typically observed in sandbox models. A high level of agreement between analogue and numerical models of brittle and brittle-viscous thrust wedges was obtained by Ellis *et al.* (2004). Their results indicate that the continuum approximation of the finite element method does not necessarily need to be a strong limitation in modelling granular materials. An encouraging agreement between results from finite element and analogue models is also shown by Panien *et al.* (2006); Cruden *et al.* (2006) and Le Pourhiet *et al.* (2006).

Numerical and analogue modelling techniques share the advantages and disadvantages of trying to capture aspects of a geological process in a model. Both methods allow the evolution of structures to be observed, which can be helpful in generating ideas on the origin and development of structures observed in the field. With numerical and analogue models it is possible to vary model parameters easily and determine their relevance for the process under consideration. However, models are also only an approximation of the natural situation and their restrictions should be taken into account when interpreting their results. Advantages of the numerical method are the easy quantification of model results, including the possibility of tracking stresses, strain and strain-rates during the evolution of the model, and the relatively large freedom in choice of material properties (including a relatively easy implementation of temperature-dependent rheologies), boundary conditions and geometries. Some approaches lose accuracy when dealing with large deformation situations (due to remeshing, for example) or large viscosity contrasts. The resolution of 3D models is in general still quite low. Analogue models on the other hand are very suitable for studying the 3D evolution of structures with time. This makes it possible to evaluate the effects of, for example, lateral changes in material properties or oblique extension. The technique is limited in the application of temperature dependent rheologies, phase changes and variations in geometries. Numerical and analogue techniques are partly complementary and the combination of both methods may help establish the robustness of model results (see also Smart & Couzens-Schultz 2001).

Modelling methods

Material properties

The models contain three materials, which have properties resembling those of quartz sand, glass microbeads and polydimethylsiloxane (PDMS). In the analogue comparison, every laboratory used its own granular materials, which typically differ in frictional properties and density (see Schreurs *et al.* (2006) and Table 1). In the numerical experiments fixed values are used to allow the comparison of the numerical results to each other (Table 1). The material behaviour of the numeric brittle materials is characterised by a Coulomb failure-frictional criterion:

$$\sigma_t = \sigma_n \tan(\phi) + C \quad (1)$$

where σ_t denotes shear stress, σ_n normal stress, ϕ the angle of internal friction and C cohesion. Measurements of the properties of analogue granular materials show that deformation initially occurs through a combination of elastic and frictional strain-hardening behaviour until failure at peak strength (ϕ_{peak}), followed by a strain-softening phase until a stable strength

Table 1. *Material properties*

Material	Values in numerical models					Measured analogue properties*				
	Density (kg m^{-3})	C (Pa)	$\phi^{\dagger}_{peak}$	$\phi^{\ddagger}_{stable}$	η (Pa s)	Density (kg m^{-3})	C (Pa)	$\phi^{\dagger}_{peak}$	$\phi^{\ddagger}_{stable}$	η (Pa s)
Sand	1560	10	36°	31°		1350–1740	3–150	33°–45°	28.2°–39°	
Microbeads	1480	10	22°	20°		1400–1600	−25–50	21°–29.2°	20°–26.2°	
PDMS	965				5e4	965				5e4
Side friction		0	19°	19°				20°§	16°–17°§	

*Values from Schreurs *et al.* (2006) for the materials used in the laboratories of University of Bern, IFP Rueil-Malmaison, University of Parma, University of Pavia and University of Toronto. Sand is sieved from 20 cm height with an approximate filling rate of 250 grams/minute. PDMS from Weijermans (1986).

†Angle of internal friction at peak strength.

‡Angle of internal friction at stable strength.

§GFZ sand measured by J. Lohrmann shows an angle of internal friction of 20° softening to 16° (fine-grained sand, 20–200 μm) or 17° (coarser sand, 20–650 μm).

(ϕ_{stable}) is reached (Lohrmann *et al.* 2003; Panien 2004; see also Jaeger & Cook 1979). Five of the eight numeric codes used in our study include elastic material behaviour. For sand, the elastic component is probably relatively small (J. Adam pers. comm.). Most of the numerical models (six of eight) have been run without dilatation and (de)compaction of the materials is, therefore, not automatically taken into account. The strengthening of material in the initial strain-hardening phase seen in analogue models (Adam *et al.* 2005) is in these cases not included. Strain-softening from peak strength to stable strength is assumed to be caused by dilatation of sand at yield and is explicitly prescribed in the numeric models that do not include dilatation. The softening is simulated by a linear decrease in the angle of internal friction over a finite strain interval of 0.5 to 1.0 (finite strain is total accumulated effective strain as measured by the square root of the second invariant of the strain tensor). The cohesion values of the analogue materials were determined using different measuring apparatus (see Schreurs *et al.* 2006) and show a large variation. Cohesion is obtained by extrapolation of the shear stress versus normal stress curve and is, therefore, likely to be an overestimate at low stress values where the stress curve is no longer a straight line, but has a convex-outward shape (Ranalli 1987; Schellart 2000). The boundary friction between sand and the sidewalls and base of the model (which are covered with an 'Alkor' foil in the analogue experiments) is approximately 20°, softening to 16°/17° (for fine sand with grain size 20–200 μm and coarse sand with grain size 20–650 μm, J. Lohrmann, pers. comm.) and a constant value of 19° has been used in the numerical experiments.

PDMS is a Newtonian material with a viscosity of 5×10^4 Pa s (at room temperature and strain rates below $3 \times 10^{-3}\ \text{s}^{-1}$, Weijermars 1986) and low Maxwell times. It is modelled as a linear viscous material: $\tau = 2\eta\dot{\varepsilon}$, for deviatoric stress τ, viscosity η, and strain-rate $\dot{\varepsilon}$.

Scaling

Some of the models are scaled up from the analogue sandbox-scale (cm) to upper-crustal scale (km). This scaling requires that dynamic similarity is observed (Hubbert 1937; see also Ellis *et al.* 2004). Dynamic similarity is maintained when force ratios and length ratios are constant. The models in this study can be scaled completely through the following three relationships: (1) geometric similarity:

$$\frac{l_n}{l_a} = \text{constant} \tag{2}$$

where l_n and l_a are the lengths in the numerical and analogue model, respectively; (2) constant ratio of gravitational force to viscous force between the numerical and the analogue model:

$$\frac{\rho g l^2}{\eta v} = \text{constant} \tag{3}$$

where ρ density, g gravitational acceleration, η viscosity and v velocity; and (3) constant ratio of frictional strength (σ_y) to lithostatic pressure:

$$\frac{\sigma_y}{\rho g l} = \text{constant} \tag{4}$$

The values for ρ, ϕ and g in the numerical models are within the ranges of the analogue values (Table 1).

Description of the numerical codes

In the following, only a brief description of the main features of the numerical codes is given and the reader is referred to the references provided for more detailed information. Specific assumptions concerning the boundary conditions for the experiments are discussed with the description of the model set-up for each experiment.

Abaqus/Standard. Results obtained with Abaqus/Standard are shown for both experiments and were provided by Susan Ellis. Abaqus is a Lagrangian Finite Element code with implicit time integration (Abaqus/Standard User's Manual 2003). Material behaviour is visco-elasto-plastic and materials deform elastically until plastic or viscous yield is reached, after which deformation continues on yield. Tests for the shortening experiment show that material quickly reaches plastic yield once it enters the deforming region. Plastic yielding is modelled using a non-associative plastic flow law with dilatation chosen as zero (to facilitate comparison with the other experiments). Strain softening is included. Boundary friction is implemented through an elastic-stick contact surface formulation. The experiments are scaled such that 1 cm equals 1 km in the numerical model.

I2ELVIS. Taras Gerya provided results from his I2ELVIS code (the successor of I2VIS, Gerya & Yuen 2003) for both experiments. I2ELVIS solves the Stokes equation for creeping flows using the finite difference method on a Eulerian staggered grid combined with a marker-in-cell technique. Material behaviour is visco-elasto-plastic and incompressible (with, therefore, zero dilatation). Elasto-plastic deformation dominates for sand-like material due to the high background viscosity of 10^{12} Pa s, which results in a large visco-elastic relaxation time (10^6 s). In contrast viscous deformation dominates for PDMS-like material due to the low viscosity of 5×10^4 Pa s, which gives small visco-elastic relaxation times (0.05 s). Strain softening is included. Boundary friction is incorporated through frictional contact layers (4 elements wide at the sides and 2.5 elements high at the base). Free surface behaviour is obtained through a low viscosity 'air' layer on top of the model ($\eta = 1$ Pa s and $\rho = 1$ kg m^{-3}). The experiments are not scaled.

LAPEX-2D. Model results for LAPEX-2D were provided by Andrey Babeyko for both experiments. LAPEX-2D solves for balances of mass, momentum and energy through an explicit Lagrangian Finite Difference technique (FLAC-type) (Cundall & Board 1988; Poliakov *et al.* 1993; Babeyko *et al.* 2002). Material and history properties including full stress and strain tensors are tracked and updated by using particles. The material behaviour is visco-elastic with plastic yielding. Dilatation is zero. Strain softening is included. Boundary friction is implemented through explicitly added frictional forces at the boundary nodes (newly added feature for these experiments). The experiments are scaled whereby 1 cm equals 10 km in the numerical model.

Microfem. Susan Ellis also provided model results calculated with Microfem (Fullsack 1995) for both experiments. Microfem is a plane-strain arbitrary Lagrangian Eulerian finite element code. The Stokes equation for incompressible creeping flows is solved on a Eulerian grid on which material properties are tracked using a Lagrangian material grid. A free surface is achieved through vertical stretch or shortening of the Eulerian grid. Material behaviour is either viscous or plastic and incompressible (zero dilatation). Strain softening is included. The version used here is modified from the original code from the Dalhousie Geodynamics group. Frictional contacts are incorporated in a simple manner by allowing nodes to move unconstrained tangential to the boundary once frictional yield is exceeded (Ellis *et al.* 2004). The experiments are scaled such that 1 cm equals 1 km in the numerical models.

NISA II/Static 12.0. Results for the shortening experiment were provided by Antje Kellner. NISA/Static is a Lagrangian Finite Element code (NISA II User Manual 1997). It solves the equations of mechanical equilibrium for visco-elasto-plastic materials, but was used in an elasto-plastic mode for this study. Associated Mohr-Coulomb plasticity is used with a dilatation angle equal to the angle of internal friction. Strain softening is not included. Boundary friction is implemented through the contact element tool: For each frictional boundary two contact surfaces are defined. A contact element combines two contact surfaces facing each other. The experiment is not scaled.

PFC^{2D}. Results for the shortening experiment using the Distinct Element Method PFC^{2D} (PFC^{2D} User's Guide 1999) were provided by Yasuhiro Yamada. The model domain consists of discrete circular elements where only neighbours interact. Linear elasticity (force-displacement law; Cundall & Strack 1979) is incorporated through normal and shear forces at element contacts. The force in the normal direction and a pre-defined friction coefficient determine the inter-element friction in the shear direction. During deformation the size of each element is preserved and only the arrangement in the particle assembly is distorted. Dilatation is incorporated as a change in the inter-element porosity. Strain softening is not included. Boundary friction is incorporated in a similar manner as the inter-element friction. The experiments are scaled whereby 1 cm equals 300 m in the numerical model.

SloMo. Results for the extension experiment were provided by Boris Kaus. SloMo is a plane-strain finite element code (Kaus 2005) which solves the Stokes equations for incompressible materials. The rheology is visco-elasto-plastic with Mohr-Coulomb plasticity. A mixed formulation is employed, with linear, discontinuous, shape-functions for pressure and quadratic shape functions for velocity to avoid spurious pressures (Pelletier *et al.* 1989). Uzawa-type iterations are used to satisfy the incompressibility constraint (Cuvelier *et al.* 1986). In this work, the code has been used in a Lagrangian fashion, with regular remeshing to deal with large deformations (see also Poliakov & Podladchikov 1992). Iterations are performed for plasticity, to ensure that both the force balance and the plastic yield functions are satisfied simultaneously. The interface between sharply varying material properties is tracked by marker-chains, whereas other material properties are stored on the integration points. Strain softening is included. Dilatation is zero. The experiment is not scaled.

Sopale. Model results for Sopale are shown for both experiments and were provided by Susanne Buiter. Sopale is a plane-strain arbitrary Lagrangian Eulerian finite element code (Fullsack 1995) and is the successor of Microfem. A more efficient tracking of material properties (whereby the Lagrangian particles are no longer only constrained to a regular grid and particle injection maintains a high particle density) allows a higher resolution. This version has been modified from the original code to allow horizontal stretch or shortening of the Eulerian grid (in addition to the vertical Eulerian grid modification for free surface behaviour). Strain softening is included. Boundary friction is implemented through the use of frictional contact layers (4 elements wide at the sides and 4 elements high at the base). The experiments are scaled such that 1 cm scales up to 1 km in the numerical model.

Shortening experiment

Model set-up

The shortening experiment follows a more or less classic set-up used to study fold-and-thrust belt evolution with analogue models. Shortening is achieved by moving the right-hand wall inward while the left wall and base are held fixed. The model consists of 'sand' layers with an embedded 0.5 cm thick layer of weaker 'microbeads' (Fig. 1a). Total height is 3.5 cm. The length of the models is variable (Table 2), but since deformation did not reach the left wall in any of the models, all were considered long enough. At the right side an initial wedge 10 cm long with 10° surface slope overlies the model. This wedge slope is in the stable field for a sand wedge (as calculated for this set-up from Dahlen 1984; Zhao *et al.* 1986). The two main problems with representing the analogue set-up numerically lie with the boundary conditions. First, different solutions were applied to simulate boundary friction (contact boundary friction or frictional boundary layers). Second, the mobile wall moving over the base of the sandbox represents a velocity discontinuity, which may lead to high pressures in this corner.

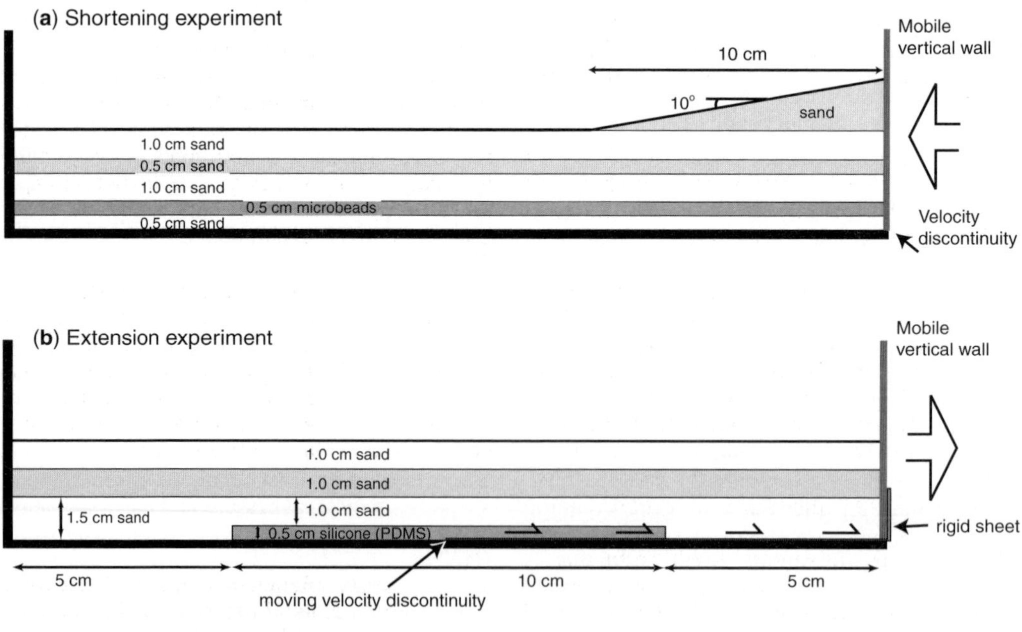

Fig. 1. (**a**) Set-up of the shortening experiment. Horizontal layers of 'sand' (which have the same properties and differ in colour only) with an embedded layer of weaker 'microbeads' are shortened through a mobile wall on the right-hand side which is pushed leftwards. (**b**) Set-up of the extension experiment. A viscous layer (PDMS, 10 × 0.5 cm) lies in the central part of the model on the base. The rest of the model consists of three 'sand' layers (only differing in colour). Extension is achieved by moving the right wall with the attached 10 cm long sheet outwards to the right.

Table 2. *Shortening experiments*

Code	Method*	Nodes†	Track points	Domain length‡	Norm. elem. size§	E‖ (MPa)	ν[1]	η[2] (Pa s)
Abaqus/ Standard	FEM	301 × 36	n.a.	30 km/30 cm	0.1	0.5	0.25	5e6
I2ELVIS	FDM + MIC	900 × 75	393750	45 cm	0.05	3	0.50	1e12
LAPEX-2D	FDM + MIC	351 × 71	400000	400 km/40 cm	0.075	0.1	0.125	4.5e10
Microfem	ALE	201 × 36	7236	28 km/28 cm	0.17	n.a.	n.a.	4e13[3]
NISA/Static	FEM	201 × 14/18	n.a.	50 cm	0.25	0.5–1.0¶	0.25	n.a.
PFC^{2D}/Kyoto	DEM	~18000	n.a.	12 km/40 cm	0.06	n.a.	n.a.	n.a.
Sopale	ALE	401 × 71	253411	40 km/40 cm	0.08	n.a	n.a.	4.6e13

*FEM = Finite Element Method, FDM = Finite Difference Method, ALE = Arbitrary Lagrangian Eulerian, DEM = Distinct Element Method, MIC = Marker-in-Cell.
†Nodes in horizontal × vertical direction, or total number of nodes.
‡First value is for the numerical model, second value is its equivalent scaled down to the sandbox scale. Height is 3.5 (left-hand side) to 5.26 cm (top of wedge at right-hand side).
§Normalized element size is measured as the square root of element area, scaled to sandbox scale (cm).
‖Young's modulus, scaled to sandbox scale.
¶ 0.5 MPa for microbeads, 1.0 MPa for sand.
[1]Poisson ratio.
[2]Background viscosity, scaled to sandbox scale.
[3]Equivalent linear viscosity; a non-linear viscosity was used in the calculations.

The shortening experiment was run with seven numerical codes (Abaqus/Standard, I2ELVIS, LAPEX-2D, Microfem, NISA/Static, PFC^{2D} and Sopale). Five experiments have zero dilatation, while NISA/Static and PFC^{2D} are dilatational. Details of the set-up for each code are given in Table 2. The initial geometry for all numerical models is shown in Figure 2a. The following code-specific assumptions and explanations apply:

Abaqus/Standard. Because Abaqus/Standard uses a Lagrangian formulation and remeshing was not employed, the code cannot handle large deformations and fails to converge at an early step.

I2ELVIS. The rigid mobile wall (with a background viscosity of 10^{14} Pa s and a shear modulus of 10^2–10^4 MPa) is moving inside the computational domain and is, therefore, not simulated through an external boundary condition. The right-hand side frictional boundary layer is limited in vertical extent. During shortening, part of the weak 'air' layer at the top of the model moves spontaneously to behind the mobile wall.

LAPEX-2D. Diffusive erosion (diffusion coefficient of the order 10^{-6} m^2 s^{-1}) is applied at the surface to avoid steep slope angles not supported by the current meshing technique.

Microfem. A small amount of surface smoothing is applied. Tests have shown that this does not significantly affect the dynamic evolution.

NISA II/Static. Because NISA/Static uses a Lagrangian approach without remeshing, the code cannot handle large deformations and fails to converge at an early step.

PFC^{2D}. The density of all materials is 1600 kg m^{-3} and, therefore, differs slightly from the values used in the other experiments (Table 1). The radii size is randomized within 12–24 m (sand) and 5–10 m (microbeads). Shearing tests suggest a shear angle in bulk behaviour of 35° for sand and 25° for microbeads. The basal boundary condition has the same friction angle as sand and is, therefore, higher than the value used in the other experiments.

Sopale. To avoid numerical instabilities associated with an abrupt velocity change the mobile wall velocity is smoothed by extrapolating the applied velocity to 0 over the lowermost 200 m (equals 0.2 cm) of 'sand' (excluding the basal frictional layer). A minimal amount of diffusive erosion (diffusion coefficient 10^{-9} m^2 s^{-1}) is applied at the surface.

Model results

The structural evolution of all models is shown in Figure 2 with plots of geometry and strain-rates after 2, 6, 10 and 14 cm of displacement (unless specified otherwise). Shear zones can readily be identified on the geometry plots, while the strain-rate plots visualize their propagation, moment of initiation and later abandonment.

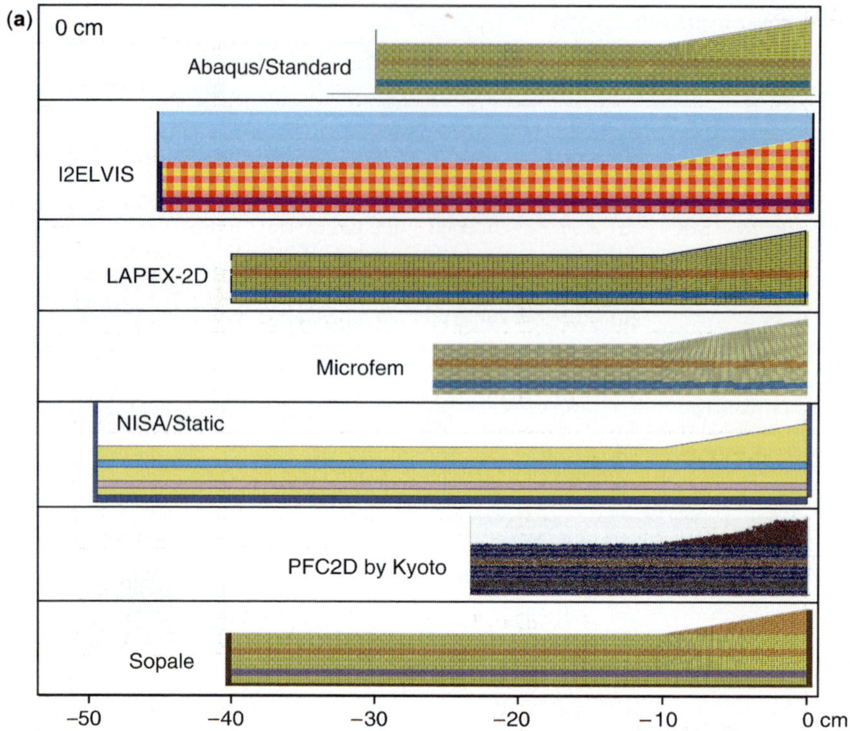

Fig. 2. Results for the shortening experiment. All quantities (including strain-rates) are scaled down to the sandbox scale. (**a**) Initial geometries. The grid of Abaqus/Standard is the calculation grid, while for LAPEX-2D, Microfem and Sopale the shown grid is (a subset of) the tracking grid.

The strain-rate plots also show when the weak embedded layer is utilized and whether shear zones propagate forward along the weak layer or along the base of the model. In continuum numerical models shear displacement is not accommodated along discrete fault planes, but rather along more distributed shear zones. For ease of comparison with the analogue models, we will, however, sometimes refer to 'faults' when describing numerical results.

In order to allow a comparison of all model results in a more quantitative manner we measured the following properties (Table 3, Figs 3 and 4): amount of displacement at which forward thrusts form, their dip angle, the spacing between thrusts, the amount of fault-controlled shortening, surface slope, when the embedded weak layer is activated, and the length of the detachment in this layer. We found, however, that our measurements were influenced by the measurer and subject to small differences in interpretation. The values reported here were, therefore, completely measured by two people, in the same manner, and subsequently averaged. A fault was considered to be initiated when deformation had clearly localized along it as shown by strain-rate maxima and small finite displacement of the deforming material. Dip angles were measured on the finite deformation plots and represent a dip angle for the whole shear zone (not differentiating between upper and lower dip angles). The variation in dip angle values between the two measurers was on average 2° (ranging between 0°–5°). Measurements of surface slope are rather inaccurate in the initial stages of shortening when only one thrust has formed (Fig. 4b), but improve substantially as soon as a second thrust has formed (Fig. 4c). From then on the difference between the values measured by the two measurers did not exceed 3°. Oscillations in surface slope angles occur just before or after the formation of a new thrust, depending on the degree to which a new thrust is incorporated into the wedge. To determine the amount of shortening accommodated by faults, we

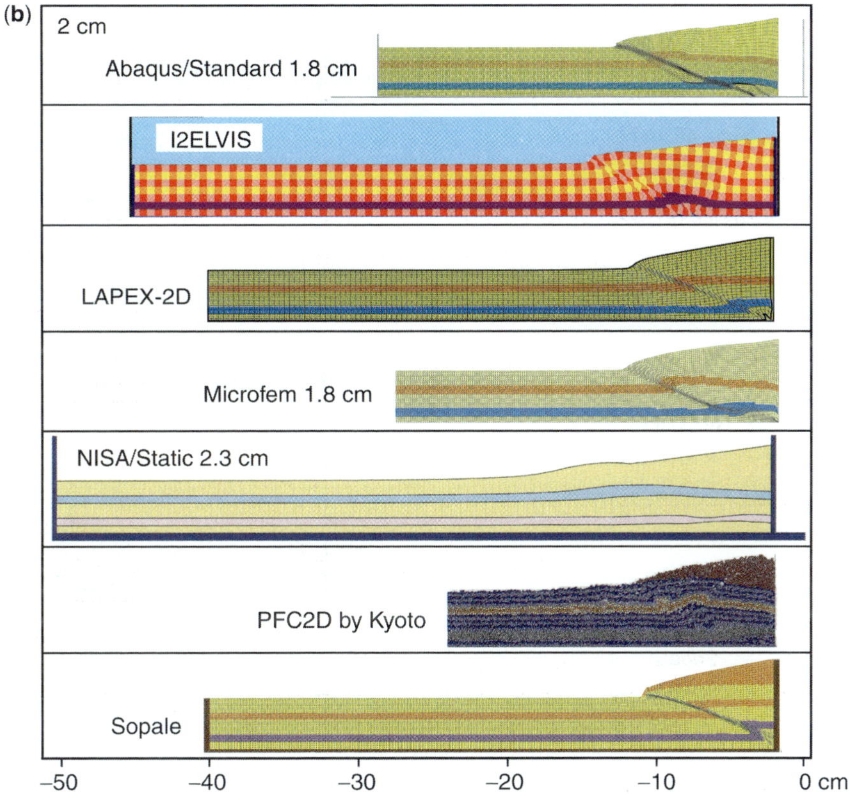

Fig. 2. (**b**) Geometries after 2 cm. Note that Abaqus/Standard and Microfem are shown after 1.8 cm and NISA/Static after 2.3 cm.

measured the offset of layers at a fault by using the cut-off points of the layers and the fault. Folding of layers towards a fault zone was classified as folding-controlled shortening. Fault-controlled shortening was in a number of cases difficult to determine accurately as the cut-off points of layers with faults were smeared out. The ranges in Table 3 reflect the differences between the two measurers.

The numerical models share many similarities: (1) Shortening is accommodated by in-sequence forward propagation of thrusts (Figs 2 and 3a). (2) The first-formed forward thrust roots at the base of the mobile wall in most of the experiments (Figs 2b and c). (3) By 2 cm of displacement an active forward thrust has formed in all models (Fig. 2b). (4) The location where the first-formed forward thrust reaches the surface is influenced by the surface wedge in almost all of the experiments (except maybe NISA/Static) (Fig. 2b). (5) The embedded weak layer accommodates shear displacement in almost all experiments when sufficient shortening has occurred to increase the surface wedge (Table 3) (the strain-rate plots for I2ELVIS show activity of the weak layer, but no finite strain accumulates here). (6) The surface slope remains in the stable field for critical taper theory (calculated from Dahlen (1984) and Zhao *et al.* (1986) for homogeneous sand) (Fig. 4). Activation of the embedded weak layer has no discernable effect on surface slope (Table 3 and Fig. 4). The stable taper field for a cohesive wedge building in 2.5 cm thick material above the weak layer is almost the same as the stable field shown in Figure 4.

Although the manner in which the models accommodate shortening leads to a similar style of deformation, it is also clear from Figures 2, 3 and 4, and Table 3 that variations exist among the models. Differences between the model results are: (1) The number of thrusts that have formed at a specified amount of displacement is variable (maximum two thrusts difference in the displacement range examined here) (Figs 2 and 3a). (2) The dip angle of the

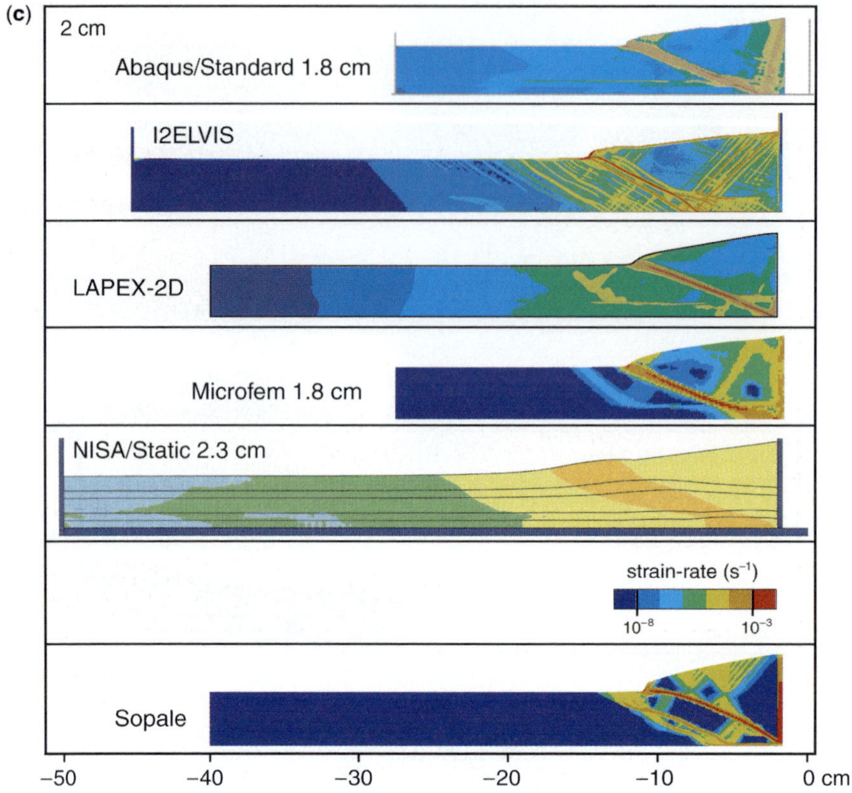

Fig. 2. (c) Strain-rates after 2 cm (Abaqus/Standard and Microfem after 1.8 cm, NISA/Static after 2.3 cm).

forward thrusts shows variations of up to 14°. The dip angle of the first thrusts is most likely influenced by the initial surface wedge. Later thrusts have dip angles between 45° (non-dilatational Roscoe angle) and $45° - \phi/2$ (Coulomb angle) (Vermeer 1990). (3) The distance between a newly formed thrust and the previously formed thrust is highly variable (Fig. 3c). (4) The embedded weak layer is first activated at different amounts of shortening (Table 3). (5) The amount of fault controlled shortening differs highly (Table 3).

Sensitivity analyses

An important advantage of numerical models is the ease with which the sensitivity to model parameters can be investigated. We tested the variability of the numerical results to assumptions regarding the model set-up (mesh resolution and elasticity) and values of material properties (cohesion and basal friction). We found that the evolution of thrusts is influenced by the resolution of the calculation domain, which may provide a partial explanation of differences found between the numerical results. Figures 5a and b show the difference after 10 cm displacement for I2ELVIS caused by increasing the normalized element size (Table 2) from 0.05 to 0.1, thus reducing the resolution. A higher resolution tends to increase the number of shear zones, while the individual shear zones become more focussed. A similar trend can be seen when comparing the result from the two ALE-type codes Microfem (lower resolution) and Sopale (higher resolution). Studies using analogue sandbox experiments found that thrust spacing depends on layer thickness, basal friction and surface slope (Mulugeta 1988; Mulugeta & Koyi 1992; Gutscher *et al.* 1998*b*). However, as pointed out by Mulugeta (1988) changes in material layering may perturb a systematic thrust spacing and it may be, therefore, that the embedded weak ('microbeads') layer in our set-up partly contributes to the observed irregularity in thrust spacing. The sensitivity of shear zone width and spacing is discussed further with the results of the extension

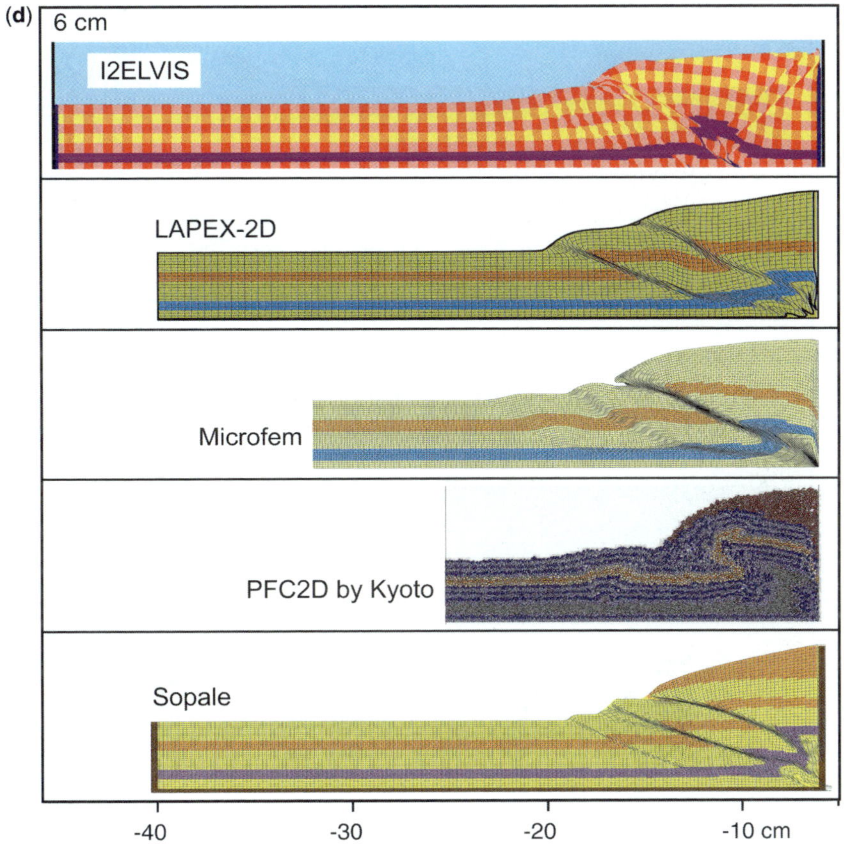

Fig. 2. (**d**) Geometries after 6 cm.

experiment (section on *Localization of shear deformation*).

We have assumed that in sandbox-type experiments the amount of finite strain that can be accommodated by elastic material behaviour is small. Abaqus/Standard, I2ELVIS, NISA/Static and LAPEX-2D include elastic behaviour and their deformation is essentially elasto-plastic in the shortening experiment, while Microfem and Sopale behave essentially rigid-plastic. A general comparison of the large-deformation results shows that the elastic component in material behaviour may lead to slightly more distributed deformation (as strain can also be accommodated between shear zones), but that overall the effect of elasticity is small for our set-up and material properties. In LAPEX-2D, effective elastic strain remains below approximately 2%.

Critical taper theory predicts that the strength of the frictional materials directly influences the taper angle that can be supported. With higher cohesion values the range of stable taper angles increases (from Zhao *et al.* 1986). Ellis *et al.* (2004) found in their thrust wedge experiments that surface slope increases with increasing cohesion, whereby the increase in surface slope mainly occurs at the toe of the material which has been thrust up. Tests with Sopale show that variations in the slope angle of the enveloping surface for cohesion variations from 0 to 100 Pa are limited (2° for displacements of 10 cm or more) without a clear trend towards higher slopes for higher cohesion. Higher cohesion material can support steeper edges of thrust sheets (Figs 5c, d and e). As the higher cohesion value was assigned to 'sand' and 'microbeads' alike, the strength contrast across the 'microbeads' layer is reduced and less shear displacement is consequently accommodated along it.

From Sopale experiments we find that a decrease in the basal friction angle (from 19° to 10°) reduces the surface slope (with around 8°

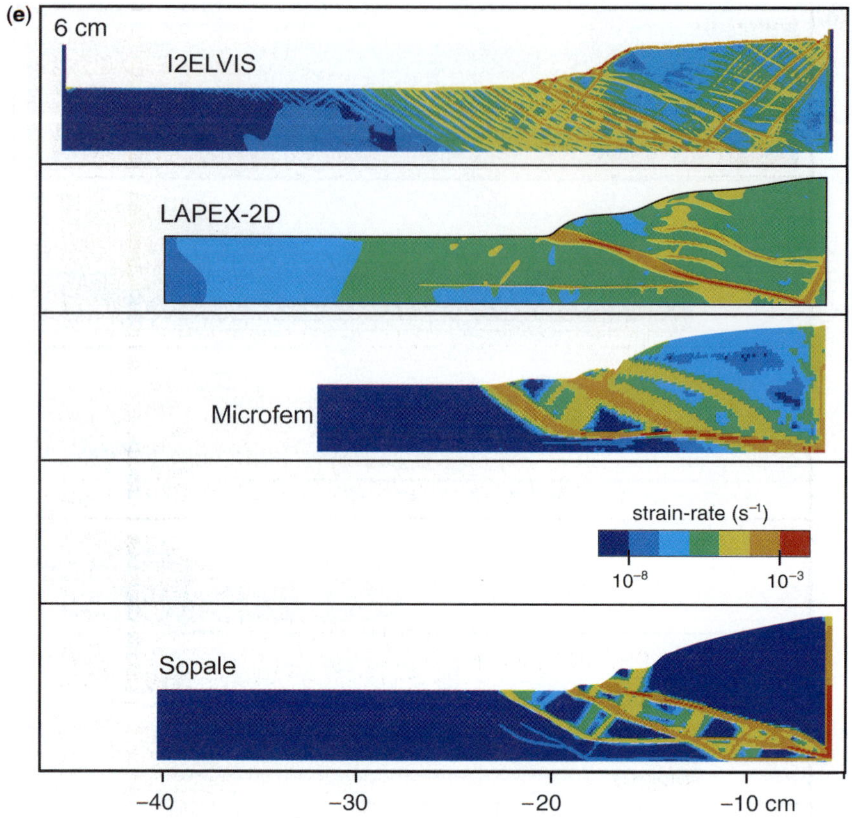

Fig. 2. (**e**) Strain-rates after 6 cm.

for the later shortening stages). This trend is in accordance with critical taper theory, which predicts lower critical taper angles for lower basal friction (Davis *et al.* 1983), and has also been found in other thrust wedge studies (Gutscher *et al.* 1998a; Vanbrabant *et al.* 1999). For low basal friction forward propagation of thrusts occurs faster.

Implications of the numerical experiments

The numerical results for the thrust wedge experiment show similar deformation styles in which shortening is accommodated in a mainly rigid-plastic manner by in-sequence forward propagation of thrusts. Thrust wedge formation and propagation is, therefore, a robust feature of the seven models. However, despite the use of similar initial set-ups, material properties and boundary conditions, variability in structural evolution arises. These differences highlight the possible ranges in structural evolution and measurable quantities (e.g., surface slope) due to the application of different numerical codes, variable resolution and different implementation of boundary conditions. A fascinating unexpected result is the role of measurer bias in determining quantities such as dip and thrust spacing. Our results indicate that thrust spacing and the number of thrusts versus displacement should be considered carefully when evaluating results from numerical simulations and applying these (in a general manner) to natural systems. This implies that although numerical models may be used 'qualitatively' to interpret the mechanics of thrust wedges, limitations exist to the extent to which quantifiable measures such as thrust fault sequences may be compared at the present time. Future code comparisons may be helpful in reducing the differences between the numerical model results. Test experiments should preferably have a simplified set-up (avoiding abrupt changes in velocity boundary conditions which may require numerical smoothing), the same grid resolution, and material properties as similar as possible.

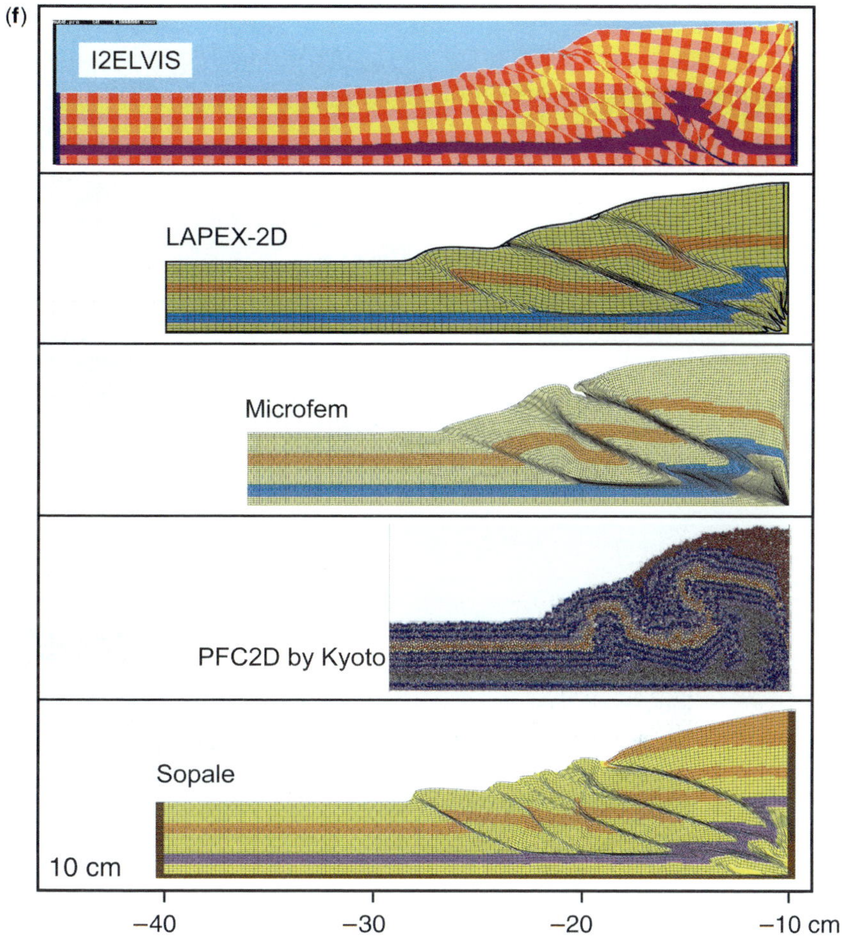

Fig. 2. (**f**) Geometries after 10 cm.

Comparison to analogue results

Figure 6 compares the numerical model geometries at 2 and 14 cm of displacement to results at equivalent stages from the analogue experiments of the laboratories of Bern, IFP, Parma, Pavia and Toronto (Schreurs *et al.* 2006). We have selected those analogue experiments that have the same set-up as our numerical experiments (i.e., no conveyor belt type experiments) and for which cross-sectional information was recorded. The properties of the granular materials used in the five analogue experiments are variable (e.g., sand: ϕ_{peak} between 33.2° and 45°, ϕ_{stable} 29.2°–39°; microbeads: ϕ_{peak} 22.3°–26.4°, ϕ_{stable} 20.6°–26.2°; the exact values can be found in Schreurs *et al.* (2006)) and this probably affected the analogue results. As the analogue granular materials are dilatational, they compact upon shortening. The strain-hardening effect this causes has not been included in most of the numerical experiments. Three analogue experiments were monitored through glass sidewalls (Parma, Pavia, Toronto) while two recorded internal deformation with X-ray CT tomography (Bern, IFP). In the accompanying paper (Schreurs *et al.* 2006) it is shown that edge effects may influence the deformational structures close to the lateral (observation) sidewalls. The extent of edge effects may be evaluated by surface monitoring and cutting of sections after the final stage of the experiment. The quantification of parameters in all analogue and numerical experiments was as much as possible done in the same manner. Ranges of parameters that were measured in the analogue

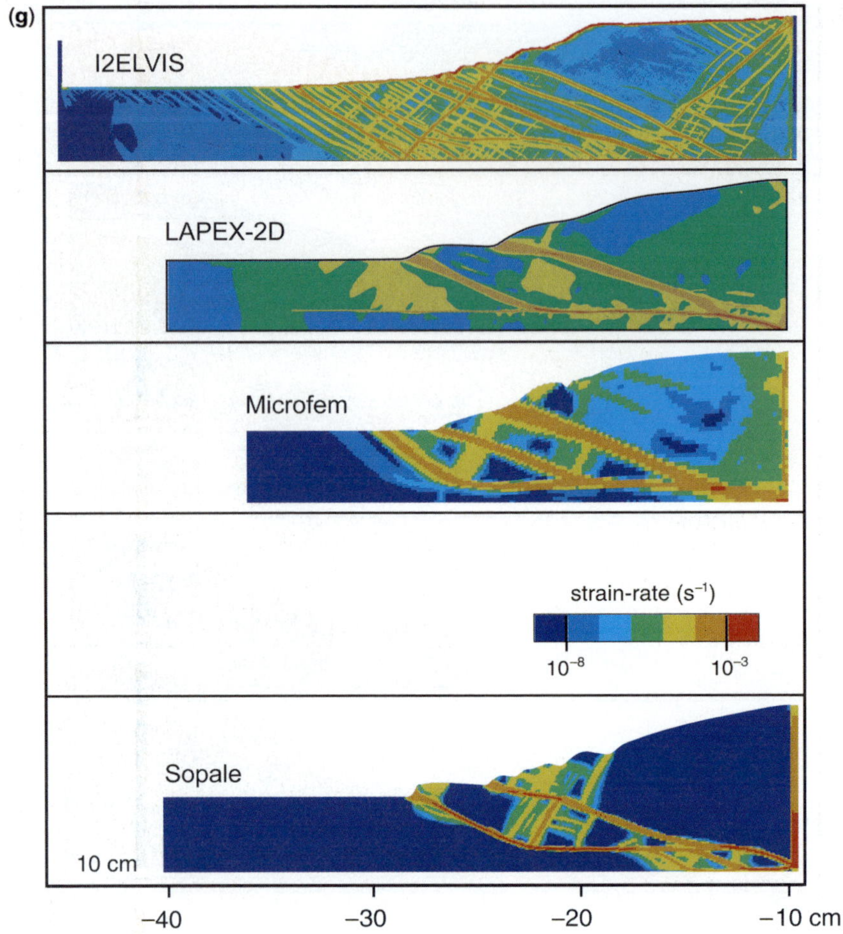

Fig. 2. (**g**) Strain-rates after 10 cm.

experiments (number of thrusts, dip angle, thrust spacing, fault-controlled shortening, microbeads layer activation and surface slope) are indicated in the figures and table which report the numerical values (Figs 3 and 4 and Table 3).

The comparison of our numerical results to the analogue models leads to the following observations: (1) The variability between numerical codes is about the same as the degree of variability between different analogue models. (2) The numerical codes produce the same overall structure and sense of forward-propagation seen in the analogue models. (3) Most of the analogue models show the forward thrust (at 2 cm displacement, Fig. 6a) propagating from a basal point located to the left of the lower right-hand corner, whereas most of the numerical models show the forward thrust propagating from this corner. This may reflect a difficulty in the numerical models of representing the velocity discontinuity at the corner. Panien *et al.* (2006) show that thrust localization near the mobile wall is sensitive to the exact implementation and smoothing of the applied velocity near the basal corner. Alternatively, the actual velocity discontinuity in the analogue models may not be so sharply defined as implied by the set-up (Fig. 1) and just a small exit slot may smooth the impact of the velocity discontinuity. (4) Most of the analogue models show a distinct backthrust at 2 cm displacement which is poorly represented in most of the numerical models (except in the results of I2ELVIS, in which the mobile wall is part of the computational domain and has a 0.05 cm high exit slot at the base) (Fig. 6a). The absence of the backthrust in the numerical models may link to the strong preference for the first forward thrust

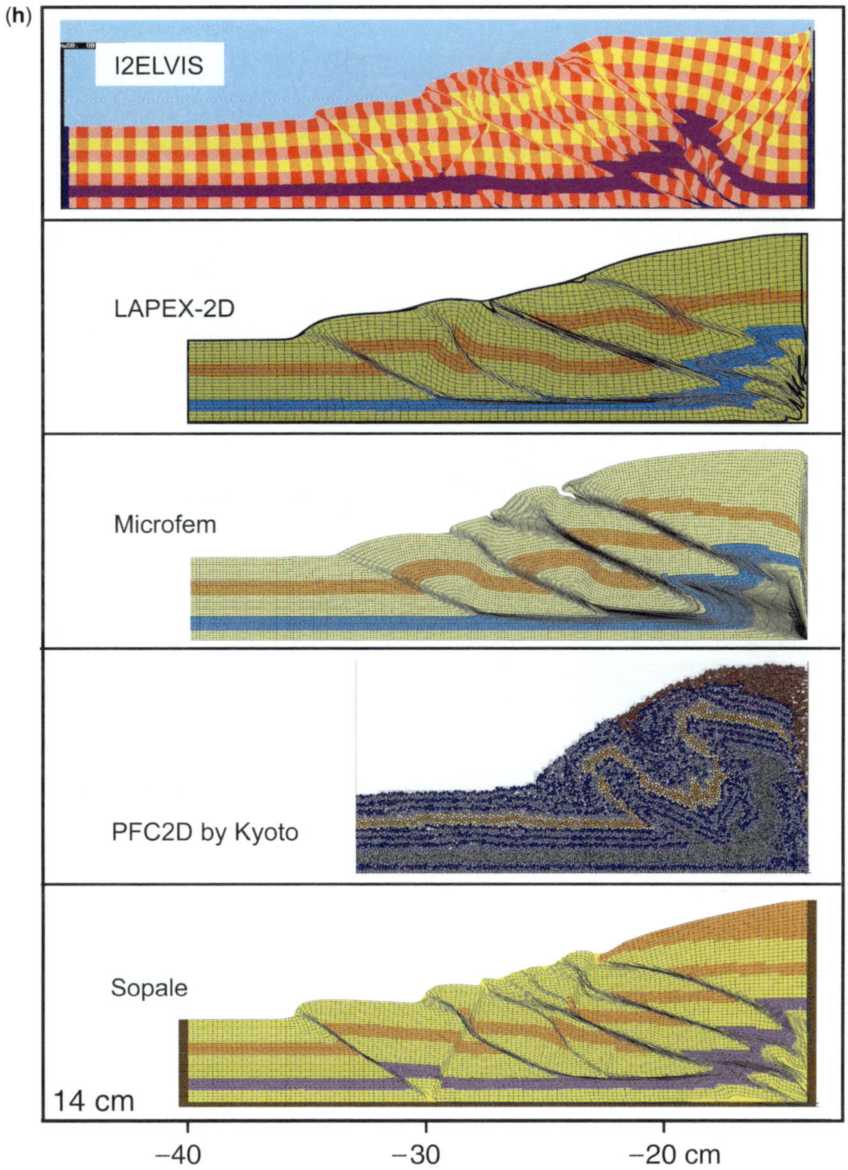

Fig. 2. (**h**) Geometries after 14 cm.

to initiate at the basal right-hand corner, as this configuration does not allow for a backthrust propagating up from the base. Interestingly, in the thrust wedge models of Ellis *et al.* (2004) (which were run with Microfem) a backthrust is visible and the first forward thrust roots to the left of the mobile wall. The exit slot below the mobile wall in these simulations may smooth the corner discontinuity, which suppresses the tendency for the thrust to localize in the basal corner and then allows a backthrust to form. Alternatively, the backthrust in these experiments could be due to the high angle of friction for the backstop. (5) The analogue models show the first thrust intersecting the surface to the left of the imposed wedge, as do the numerical models (Fig. 6a). (6) Localization of deformation in shear zones in the initial stages of shortening occurs slightly earlier in the numerical models than in the analogue models. The analogue

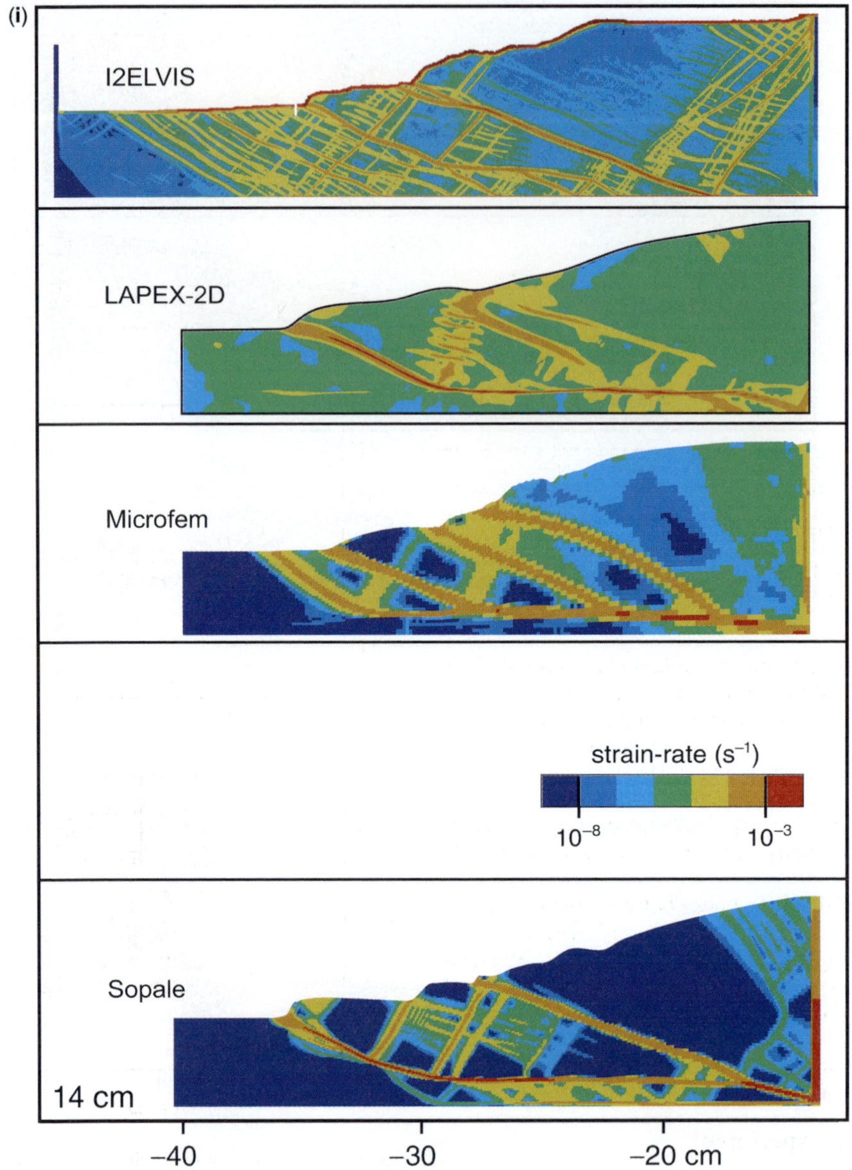

Fig. 2. (**i**) Strain-rates after 14 cm.

models accommodate up to 0.5 cm of shortening through lateral compaction before visible faults form. (7) The variation in the number of thrusts between different numerical codes is approximately the same as the variation between equivalent analogue models (Fig. 3a). (8) The spacing of thrusts is less variable between the numerical codes than between the analogue experiments (Fig. 3c). (9) The dip angles of the forward thrusts are generally slightly lower in the analogue experiments (between 19° and 33°) than in the numerical experiments (between 20° and 39°) (Fig. 3b). The analogue experiments follow the theoretical values for Coulomb dip angles (Vermeer 1990) quite closely, which for the frictional properties of these five analogue experiments range between approximately 23° and 29°. The predicted values for the numeric dip angles lie between 27° (Coulomb dip angle) and 45° (non-dilatational Roscoe dip angle).

Table 3. *Quantification of shortening experiments**

Code	Fault short.[†] at 4 cm	'Microbeads' layer	
		Activated at	Detachm. length
Abaqus/Standard	n.a.	n.a.	
I2ELVIS	2.5–2.6	never	
LAPEX-2D	1.9–2.8	3	7.4
Microfem	1.5–2.3	4	6.2
NISA/Static	n.a.	n.a.	
PFC^{2D}/Kyoto	1.5–1.8	6	8.0
Sopale	2.0–2.8	4	5.0
Analogue ranges[‡]	1.9–2.9	5.0–never	2.2–9.3

*All values referred to sandbox scale in cm. Output stages typically available at 0, 0.5, 1, 2 and then every 2 cm of shortening.
[†]Fault-controlled horizontal shortening determined by cut-off points between layers and the faults.
[‡]Analogue values from experiments of University of Bern, University of Parma, University of Pavia, IFP Rueil-Malmaison and University of Toronto (Schreurs *et al.* 2006).

The first thrusts may be influenced by the surface wedge, leading to lower values for their dip angles. (10) The surface slopes measured for the numerical models plot around the lower ranges of the analogue surface slopes (Fig. 4). This is probably caused by the generally lower strength of the numerical materials in comparison with many of the analogue materials (Table 1).

Most of the differences between the numerical and analogue models may be attributed to the difficulty in representing the exact set-up of the analogue model and to differences in material properties. Despite these differences, the numerical models still produce a similar taper geometry and evolutionary style to the analogue models.

Extension experiment

Model set-up

The extension experiment examines the influence of a weak, basal viscous layer on normal fault localization and propagation in overlying brittle materials. The model is 20 cm wide and 3.5 cm high (Fig. 1b). It consists of three 'sand' layers (only coloured differently) and a 10 cm wide and 0.5 cm high layer of viscous PDMS, which overlies the central basal part of the model. A thin rigid sheet covering the base of the model extends from the middle to the right lateral wall and is attached to it. Extension is achieved by moving the right wall outward to

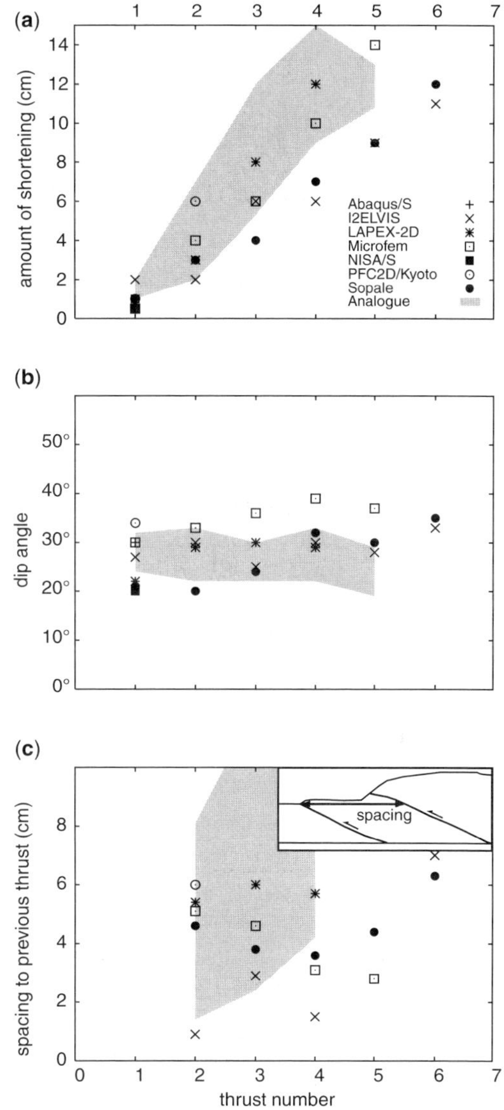

Fig. 3. (**a**) The amount of displacement at which a forward thrust forms shows that all experiments have forward propagation of thrust formation. (**b**) Dip angle of forward thrusts at the moment of their initiation. A dip angle of approximately 19° is expected for the first thrust if its location was entirely controlled by the initial wedge. (**c**) Spacing to the previous forward thrust measured at the moment of initiation of the new forward thrust. Inset shows how thrust spacing is measured. The quantities in (**b**) and (**c**) have been measured at the same stages of shortening as depicted in (**a**). The grey bands denote the range of values measured for the analogue experiments of University of Bern, IFP Rueil-Malmaison, University of Parma, University of Pavia and University of Toronto (see also Fig. 6 and Schreurs *et al.* 2006).

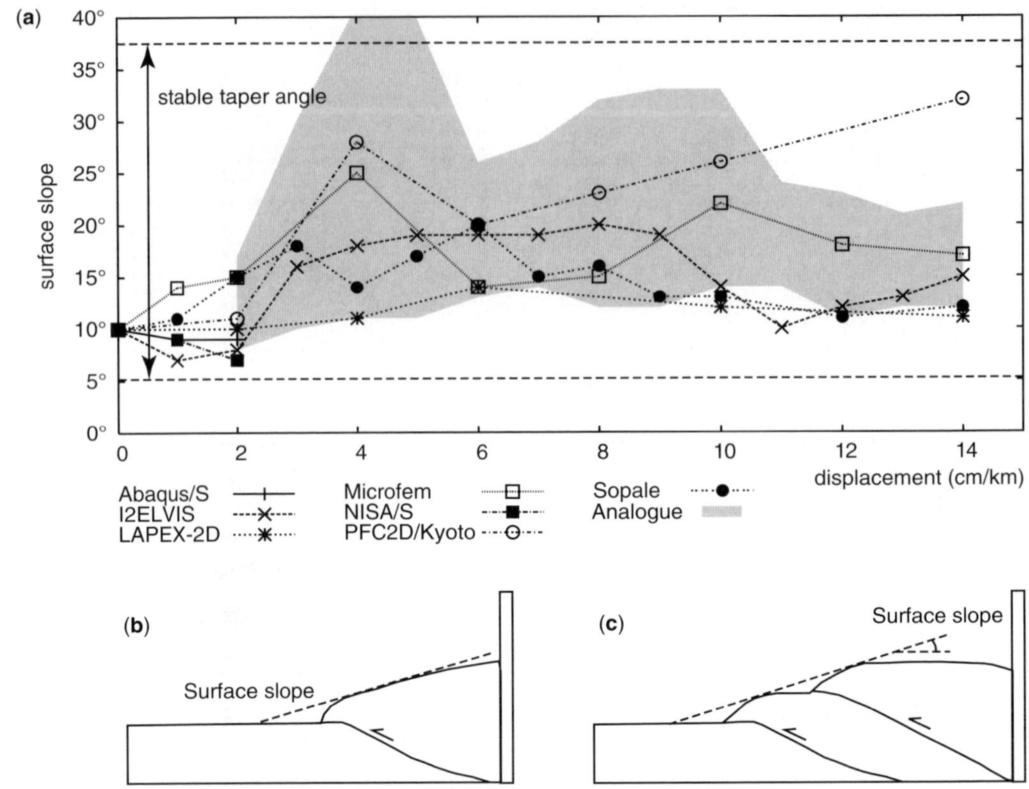

Fig. 4. (**a**) Surface slope versus amount of displacement. Surface slopes as measured in the analogue experiments of University of Bern, IFP Rueil-Malmaison, University of Parma, University of Pavia and University of Toronto (Schreurs *et al.* 2006) are shown as a grey band. (**b**) Schematic figure showing how surface slope has been determined for the early stages of shortening. It is clear that the initial surface slope angles are difficult to determine. (**c**) For two or more thrusts, the surface slope has been determined by drawing the enveloping surface.

the right at a rate of 2.5 cm/hr. This velocity is also transmitted to the right basal half of the model through the basal sheet. The tip of the basal sheet forms a migrating velocity discontinuity. The base of the model, the basal sheet and the sidewalls are again covered with Alkor foil (prescribed boundary friction is 19°). The boundary condition of silicone overlying this foil was treated differently in the models. The extension experiment was run with six numerical codes (Abaqus/Standard, I2ELVIS, LAPEX-2D, Microfem, SloMo and Sopale; Table 4), which were all non-dilatational for this experiment. The initial geometries are shown in Figure 7a. The following code-specific assumptions apply:

Abaqus/Standard. Basal friction is applied along the entire bottom boundary. The change in velocity at the basal velocity discontinuity is accommodated over 4 elements (0.4 km scaling to 0.4 cm). The model diverged after 1.5 cm of displacement.

I2ELVIS. Basal friction is applied along the whole bottom boundary. The rigid mobile wall and a 0.5 mm thin rigid plate (with a background viscosity of 10^{14} Pa s and a shear modulus of 10^2–10^4 MPa) are included into the computational domain. During extension, part of the weak 'air' layer at the top of the model is spontaneously displaced from behind the mobile wall.

LAPEX-2D. Basal points above the moving sheet are assigned the velocity of the sheet, whereas basal points to the left of the moving sheet are held fixed. The change in velocity at the basal velocity discontinuity is accommodated over 4 elements (2.7 km scaling to 0.27 cm).

Microfem. Basal friction is applied along the entire bottom boundary. The change in velocity at the basal velocity discontinuity is smoothed over 3 nodes (0.25 km scaling to 0.25 cm). The domain width is 25 km (25 cm).

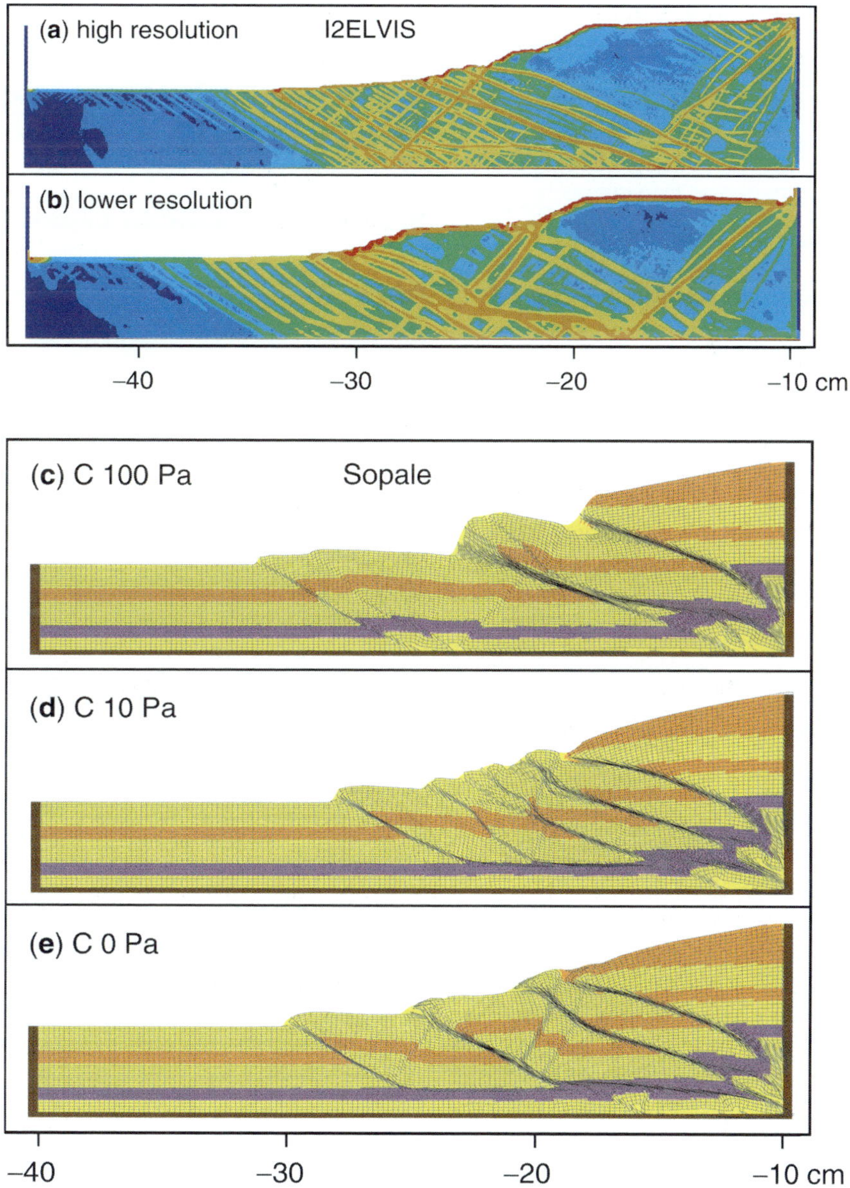

Fig. 5. Sensitivity tests for the shortening experiment. I2ELVIS strain-rates after 10 cm of displacement for (**a**) high resolution (normalized element size 0.05 as in Fig. 2) and (**b**) lower resolution (normalized element size 0.1). Sopale results after 10 cm of displacement for (**c**) a cohesion of 100 Pa (sandbox scale), (**d**) a cohesion of 10 Pa (as in Fig. 2), and (**e**) a cohesion of 0 Pa.

SloMo. Basal points above the moving sheet are assigned the velocity of the sheet, whereas points to the left are held fixed. The basal velocity discontinuity is accommodated by a linear decrease in velocity over 0.5 cm (10 elements).

Sopale. Basal points above the moving sheet are assigned the velocity of the sheet, whereas basal points to the left of the moving sheet are held fixed. The change in velocity at the basal velocity discontinuity is accommodated over 4 elements (0.2 km scaling to 0.2 cm). The vertical sides are held fixed (however, this boundary condition has no effect as deformation does not propagate to the lateral ends of the model).

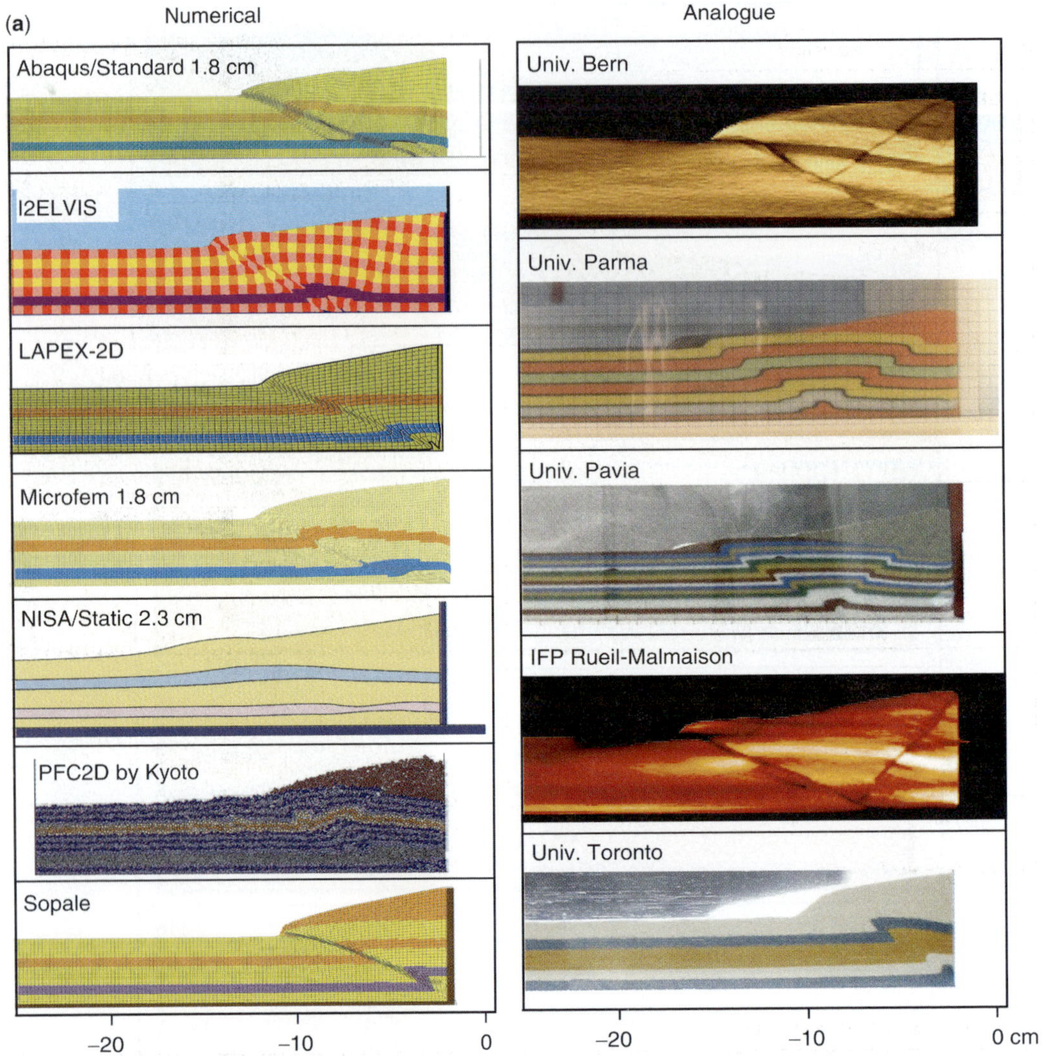

Fig. 6. Comparison of numerical and analogue results for the shortening experiment. (**a**) 2 cm of shortening (note that the model domain has been cropped on the left-hand side). (**b**) 14 cm of shortening.

Model results

The evolution of the six numerical models is shown in Figure 7 with plots of geometry and strain-rate (after 1, 2 and 5 cm of extension) and pressure (after 1 and 5 cm). Measurements of fault dip, location of fault initiation, fault-controlled extension and fault migration are shown in Table 5 and Figure 8. The measurement procedure was the same as for the shortening experiment. The initial set-up of the experiment is symmetric and the very first deformation structures should, therefore, be symmetric. The symmetry is disrupted as the basal sheet is pulled out from underneath the model.

Analysis of the numerical results allows the following observations (Figs 7 and 8, Table 5): (1) Extension first leads to the formation of two conjugate shear zones centred at the basal velocity discontinuity (the tip of the basal sheet) with a dip angle of 45° to 55° (Fig. 7b, Table 5). The dip angles are within the values expected for dynamic pressure-dependent non-dilatational Mohr-Coulomb shear zones, where predictions

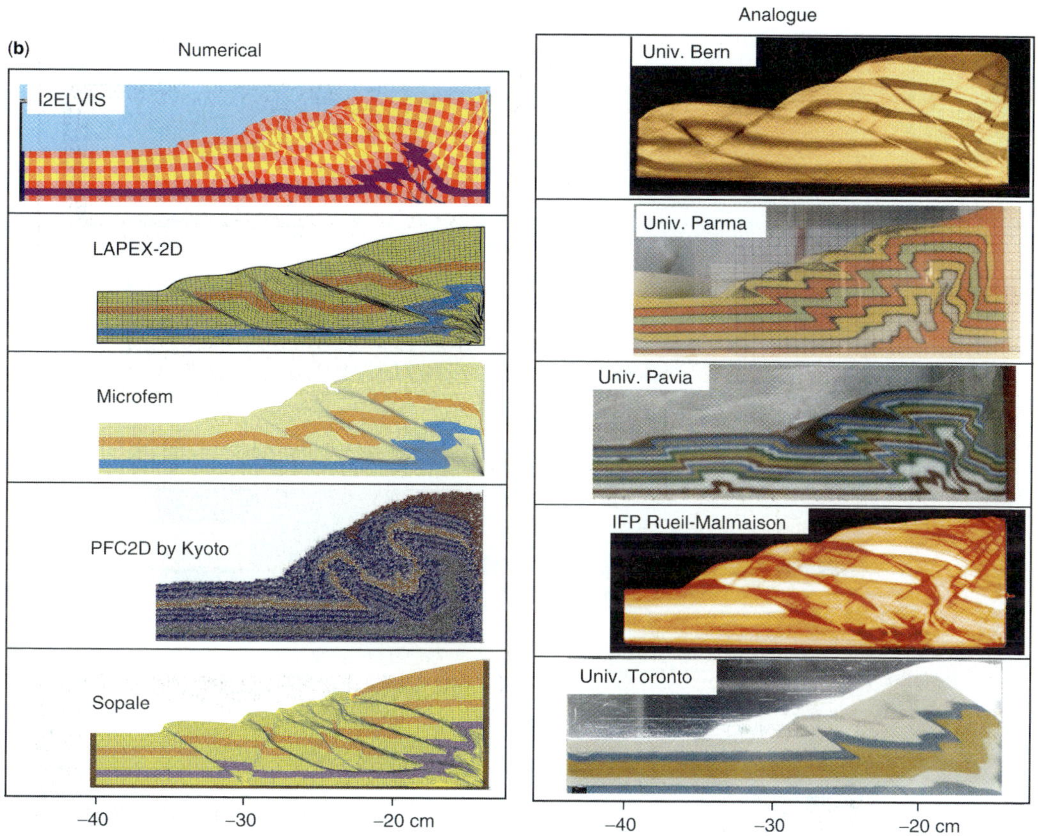

Fig. 6. *Continued.*

Table 4. *Extension experiments*

Code	Method*	Nodes†	Track points	Norm. elem. size‡	E§ (MPa)	ν‖	η¶ (Pa s)	Δt[1]	Veloc.[2]
Abaqus/ Standard	FEM	201 × 36	n.a.	0.1	0.5	0.25	5e6	varies	2.2 cm/yr
I2ELVIS	FDM + MIC	400 × 75	175000	0.05	3	0.50	1e12	/0.008 h	2.5 cm/hr
LAPEX-2D	FDM + MIC	301 × 71	340000	0.0577	0.1	0.125	4.5e10	1 yr/1.4e-3s	1 cm/yr
Microfem	ALE	201 × 61	12261	0.1	n.a.	n.a.	4e13[3]	700 yr/22.2s	2.2 cm/yr
SloMo	FEM	401 × 71	94500	0.05	0.005	0.50	1e9	/5s	2.5 cm/hr
Sopale	ALE	401 × 71	253411	0.05	n.a.	n.a.	4.6e13	1000yr/7.2s	0.5 cm/yr

*FEM = Finite Element Method, FDM = Finite Difference Method, ALE = Arbitrary Lagrangian Eulerian, MIC = Marker-in-Cell.
†Nodes in horizontal × vertical direction.
‡Normalized element size is measured as the square root of element area, scaled to sandbox scale (cm).
§Young's modulus, scaled to sandbox scale.
‖Poisson ratio.
¶Background viscosity (for sand), scaled to sandbox scale. The numerical value for the viscosity of the silicone for the scaled experiments can be derived from the analogue viscosity value (5e4 Pa s), the numerical velocity and the scaling relationships (Equation 3).
[1]Time step used in the calculations. First value is for the numerical model, second value is its equivalent scaled to the sandbox scale.
[2]Velocity on sandbox scale is 2.5 cm/hr for all experiments.
[3]Equivalent linear viscosity; a non-linear viscosity was used in the calculations.

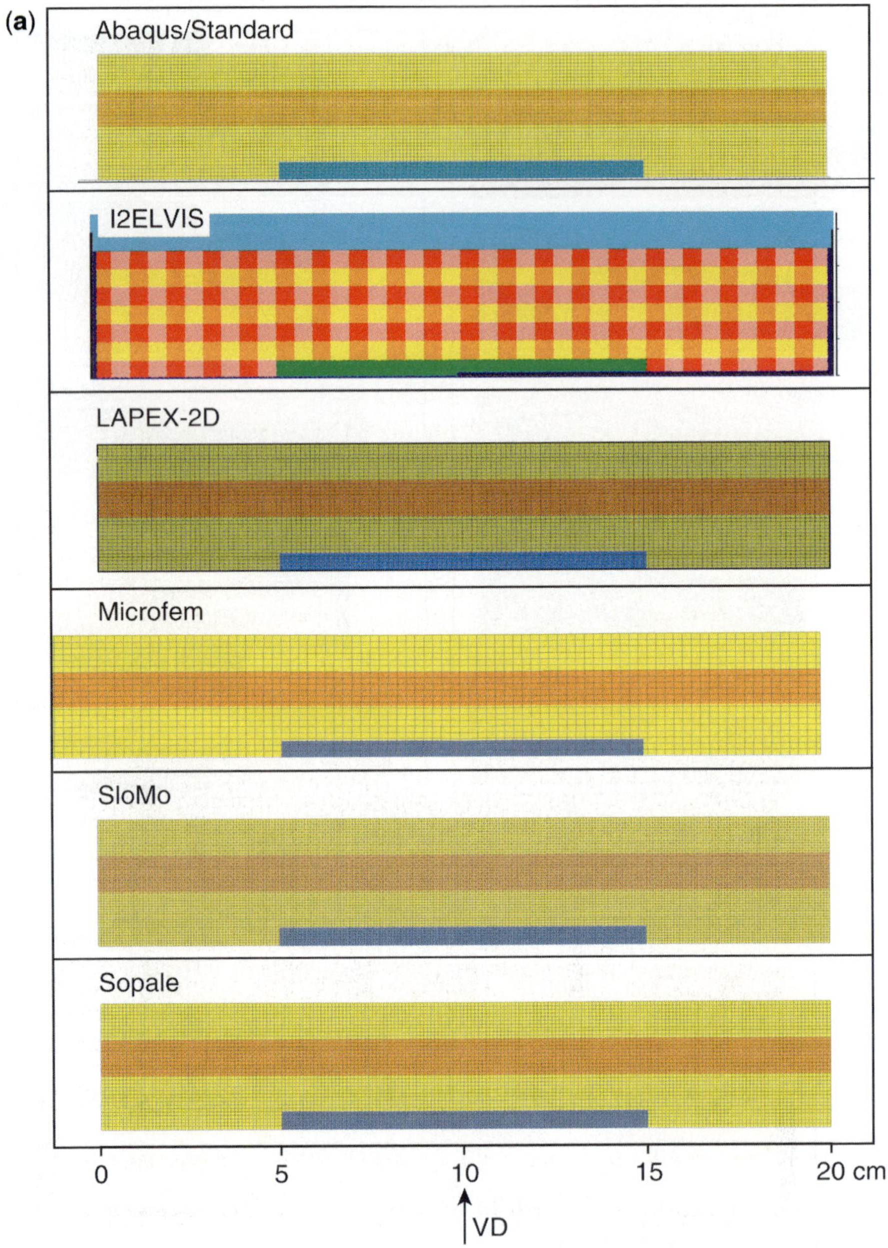

Fig. 7. Results for the extension experiment. All quantities (including strain-rates and pressures) are scaled down to the sandbox scale. (**a**) Initial geometries. The grid of Abaqus/Standard is the calculation grid, while for LAPEX-2D, Microfem, SloMo and Sopale the shown grid is (a subset of) the Lagrangian tracking grid.

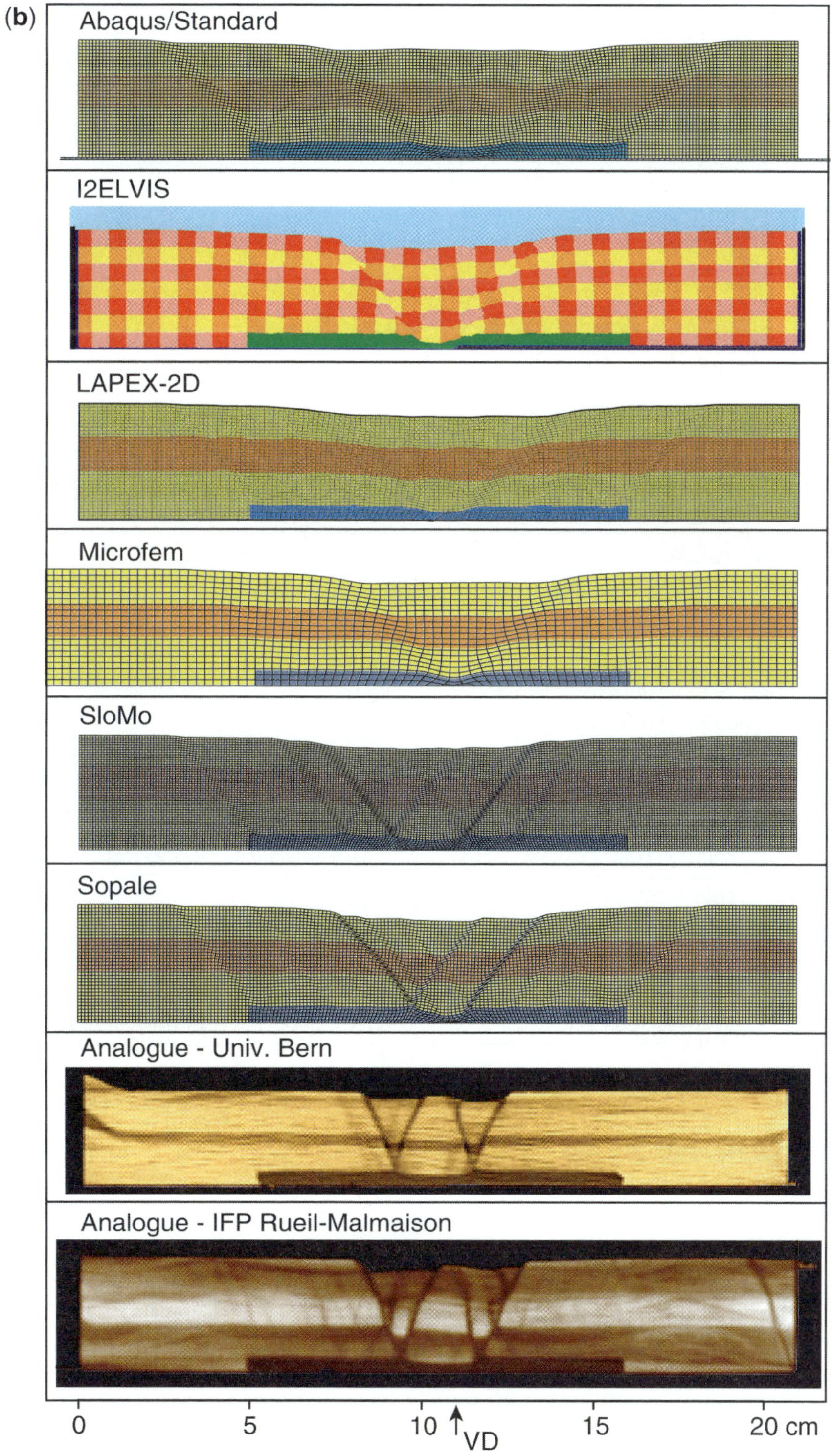

Fig. 7. (**b**) Geometries after 1 cm, including analogue examples of University of Bern and IFP Rueil-Malmaison. VD = velocity discontinuity.

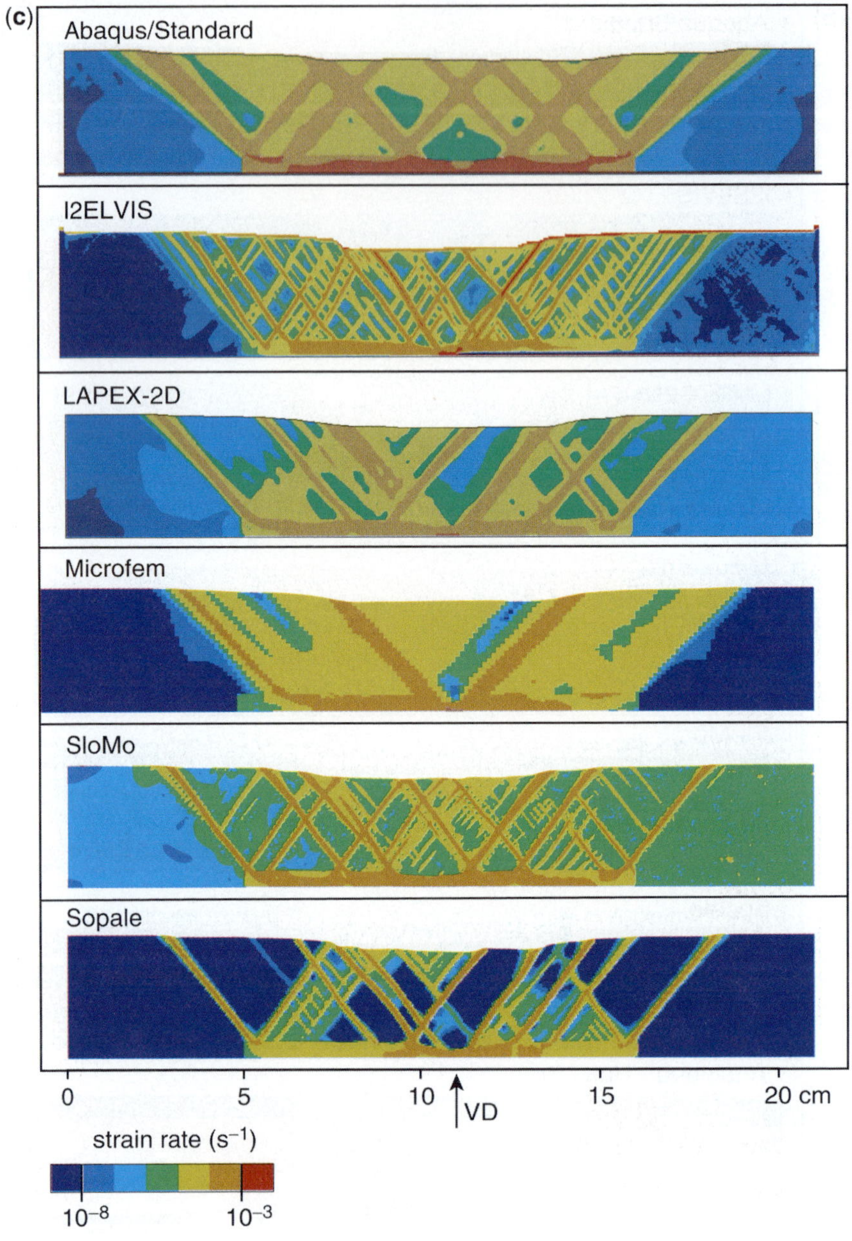

Fig. 7. (**c**) Strain-rates after 1 cm.

for fault inclination range between 45° (Roscoe angle) and $45° + \phi/2$ (Coulomb angle) (Vermeer 1990). The agreement in fault dip angles among the models is reflected in the similarity in dynamic pressures at 1 (and also 5 cm) of extension (Figs 7d and i). (2) The viscous layer distributes deformation quickly and at 1 cm of extension distributed shear zones are visible in all models (Fig. 7c). The left and right side edges of the viscous layer are found quickly (Table 5). (3) With continued extension, the shear zones migrate to the right, the silicone layer thins and the surface depression becomes wider and asymmetric in shape (Figs 7d–g). The structural evolution of all the large-deformation models is very similar, especially in light of the differences in basal boundary condition. This suggests that the basal contact is not

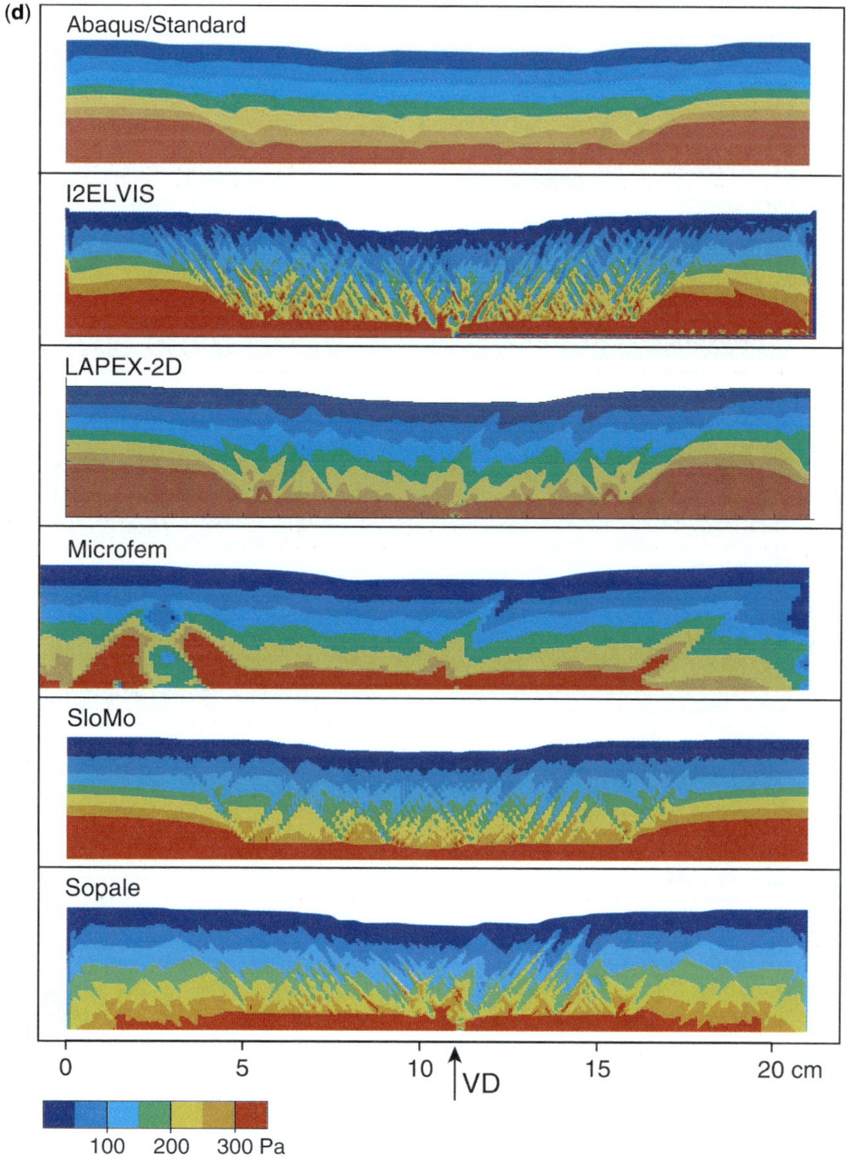

Fig. 7. (**d**) Dynamic pressures after 1 cm.

as important as the interaction between the moving velocity discontinuity and the viscous layer. (4) The evolution in dip angle of the first formed shear zone to the right of the velocity discontinuity is fairly similar for the different models (Fig. 8a). For larger amounts of extension (>5 cm), we find that block rotation becomes more pronounced and fault dip decreases. (5) The first formed shear zone moves to the right as the basal sheet is pulled out (Fig. 8b). A general trend is that fault migration is slower than the velocity with which the basal sheet is pulled because deformation is taken up by the block to the right of the first formed shear zone (e.g., through the formation of new shear zones). The differences between the numerical models are small and the effect on the resulting geometries is small as well. (6) The amount of fault-controlled extension is again highly variable (Table 5), as was observed for the shortening experiment. (7) The dynamic pressure field (Figs 7d and i) is more heterogeneous than the strain-rate field (Figs 7c and h) and visualizes both existing shear zones (currently or previously

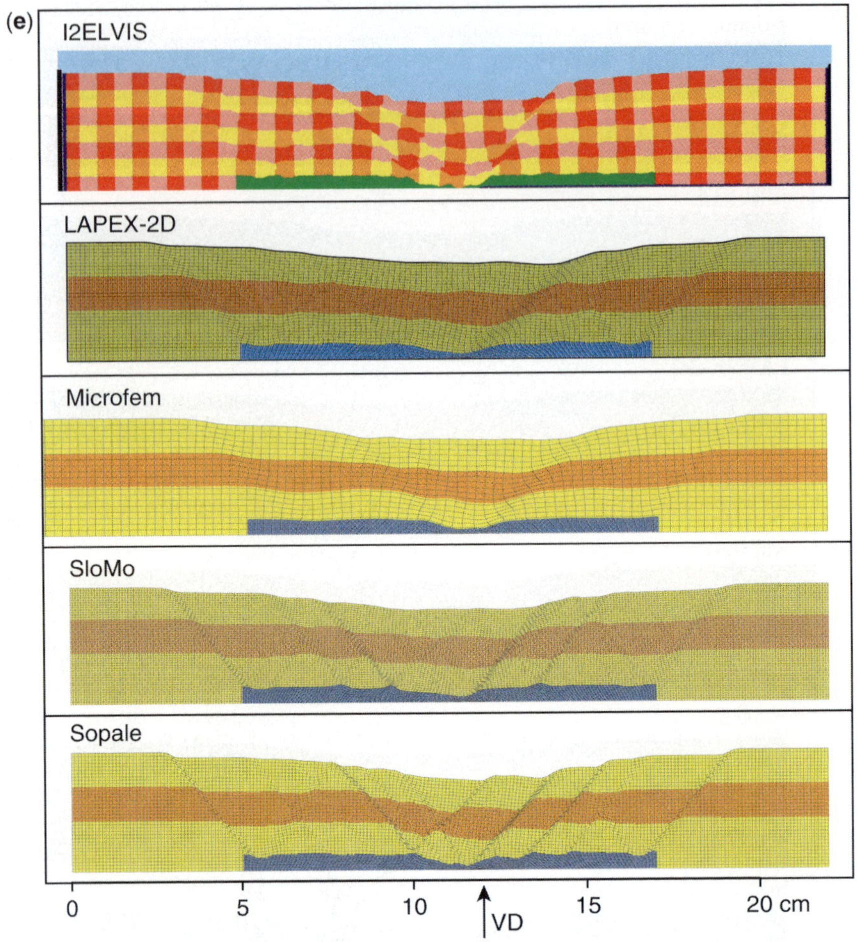

Fig. 7. (e) Geometries after 2 cm.

active) and potential shear zones (which are not necessarily activated). A comparison of the pressure fields for LAPEX-2D and Sopale shows that the seemingly rigid footwall blocks at the sides of the model behave in a different way in the two models. In Sopale the footwalls are at yield, whereas in LAPEX-2D they show an infinitesimal amount of elastic deformation (10^{-5} approximately) and are not at yield. In a purely viscous-plastic code (such as Sopale) any amount of deformation in a high-viscosity limit brings the stresses immediately on yield. This is not the case for a visco-elasto-plastic code where the material has an additional freedom to deform elastically. The dynamic pressure for Abaqus/Standard is less well resolved due to the relatively low resolution of the model, which may perhaps also incorporate a certain degree of pressure smoothing. The pressure plots for Microfem show the effects of the stick-slip friction basal boundary condition. This creates a region of enhanced dynamic pressure in the left footwall block, which is not seen in the results for the other codes. However, the enhanced pressures do not have a large influence on deformation since the average strain-rates in this region are low.

Localization of shear deformation

Comparison of the six experiments shows that the degree of focussing of deformation on shear zones may vary significantly (Table 5, Fig. 7). Results from I2ELVIS and Sopale demonstrate that both the degree of localization and the number of shear zones that form strongly

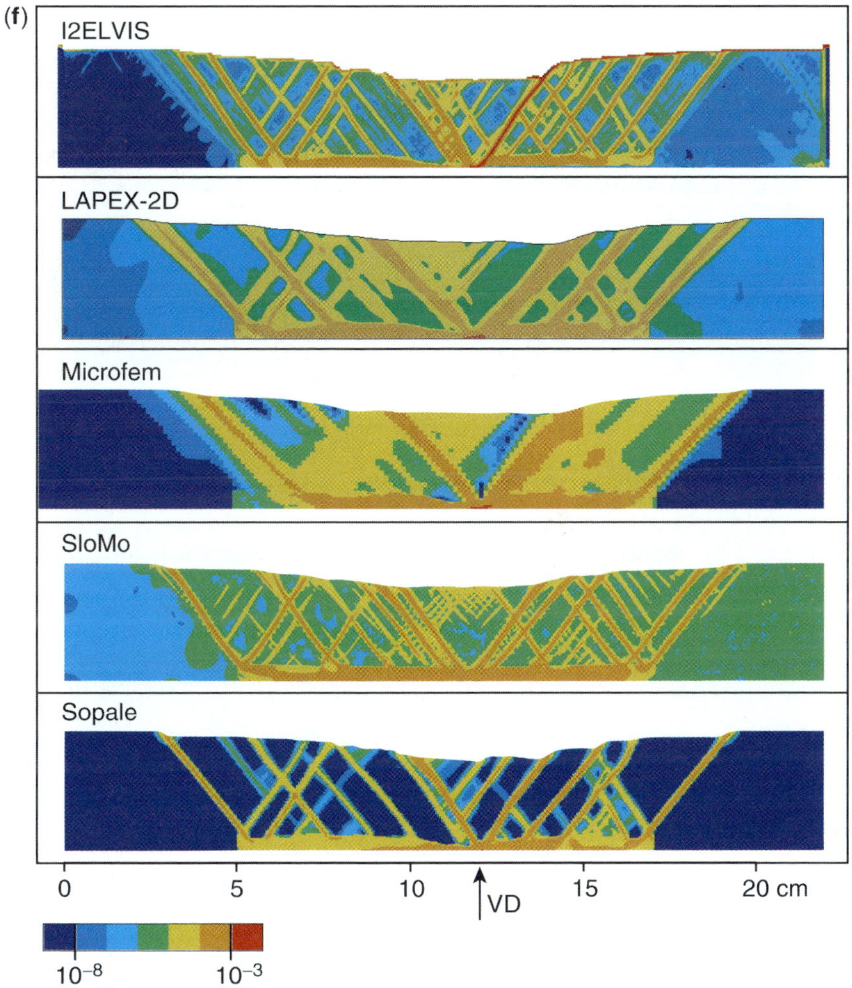

Fig. 7. (**f**) Strain-rates after 2 cm.

increase with increasing resolution of the numerical grid. This can also be seen from a comparison of the results from Microfem (lower resolution ALE code) and Sopale (higher resolution ALE code) (Fig. 7). According to Cundall (1990), the ratio of elastic bulk modulus to lithostatic pressure is an important parameter for fault spacing, whereby an increase in this ratio decreases fault spacing. In nature, the value of this dimensionless ratio is expected to vary over orders of magnitude (for example, between 17 and 4432, using an elastic modulus of 10^{10}–10^{11} Pa, ρ 2300–3000 kg m^{-3} and height 1–20 km) and the variation in this ratio for the four numerical experiments that include elastic material behaviour is of the same order (between 3 and 1867 using the elastic *shear* modulus, ρ 1560 kg m^{-3}, height 3.5 km). However, in the set-up of our experiment, the viscous layer and the velocity discontinuity control the initial location of shear zones and this set-up is, therefore, not suitable for a systematic study of shear zone spacing.

The use of dynamic pressure in Mohr-Coulomb plasticity significantly increases the localization of deformation in shear zones (tests with I2ELVIS and Sopale, Fig. 9). A Von Mises model (no pressure dependence) with an equivalent strength and softening to the Mohr-Coulomb models shows a considerably lower degree of localization (Figs 9a and d). With a depth-dependent pressure (Fig. 9b) localization is still less than in the dynamic pressure case (Figs 9c and e). To be noted here is that the

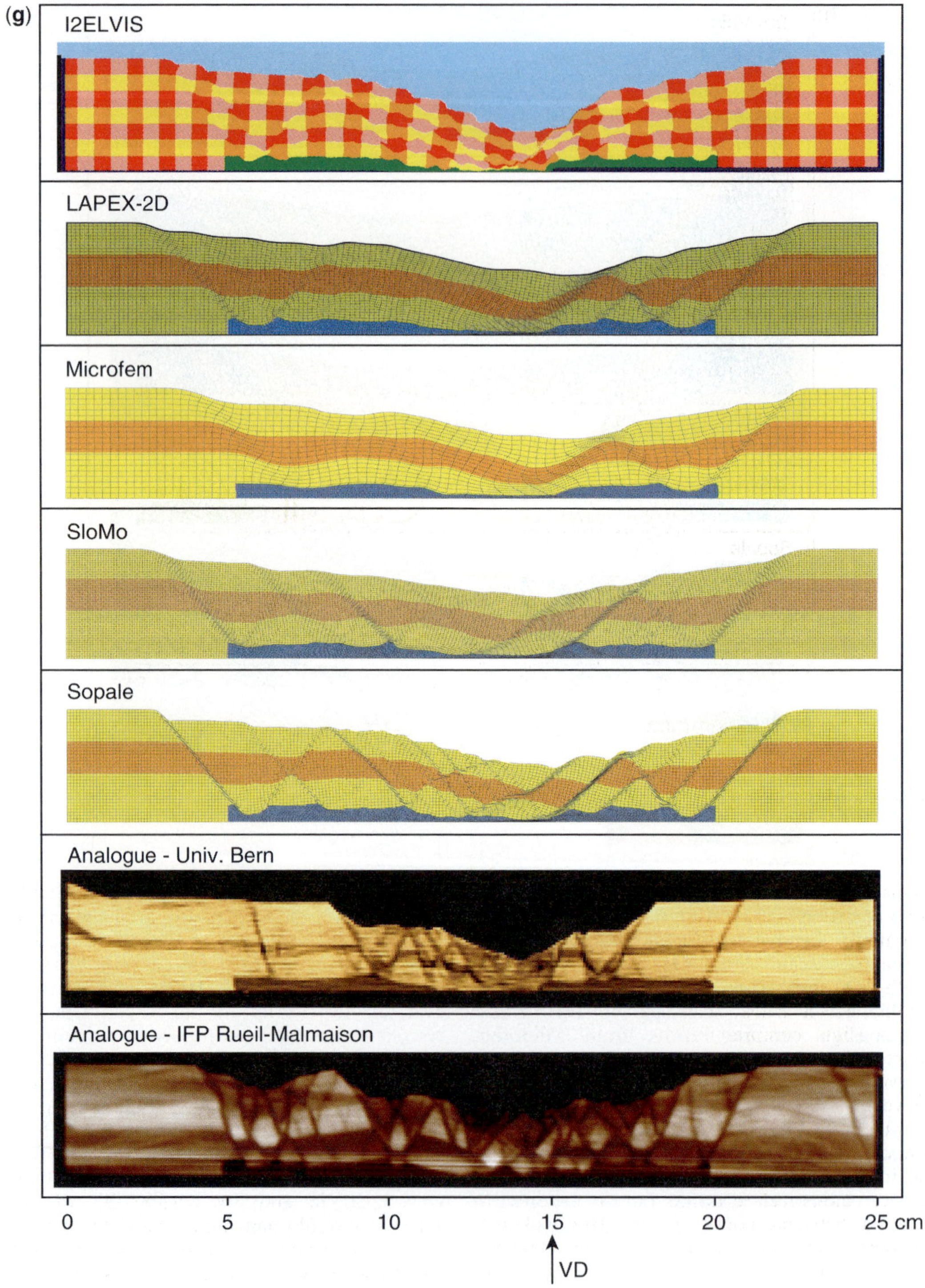

Fig. 7. **(g)** Geometries after 5 cm, including analogue examples of University of Bern and IFP Rueil-Malmaison.

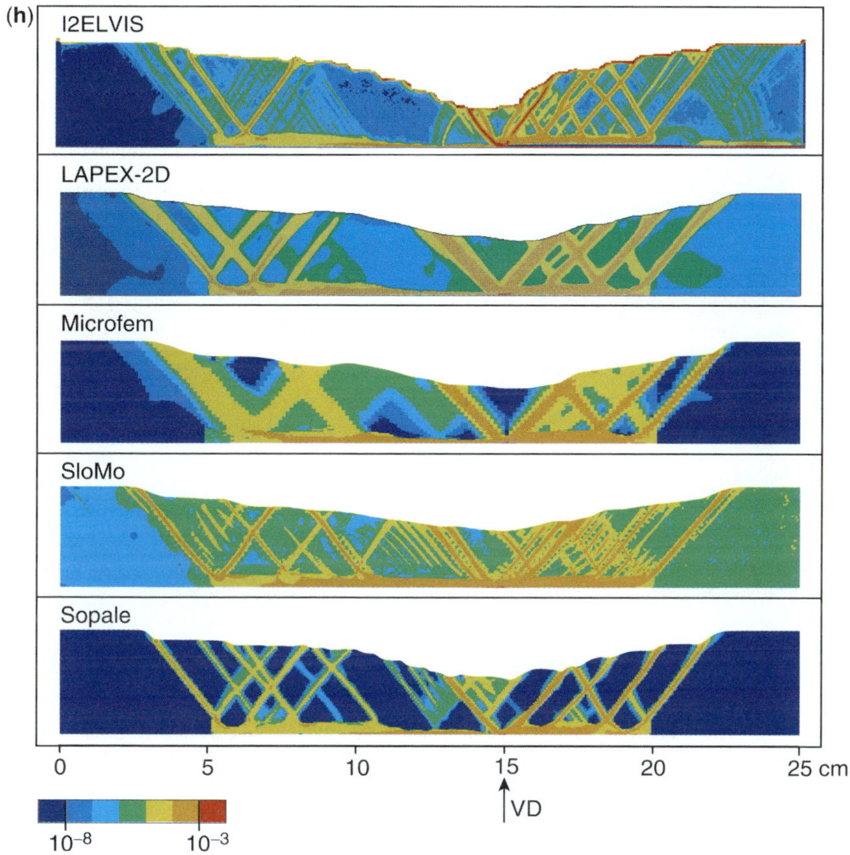

Fig. 7. (**h**) Strain-rates after 5 cm.

dynamic pressure is calculated differently in the two codes compared in Figure 9 (see also Figs 7d and i). Sopale (and also Microfem) calculate dynamic pressure through a penalty formulation with a high but finite penalty ratio (Fullsack 1995) which leads to quasi-incompressibility (or a slight compressibility). In I2ELVIS the purely incompressible continuity equation is solved. Our test calculations with I2ELVIS have also shown that for a marker-in-cell technique, the numerical width of shear zones depends strongly on the interpolation schemes which are used for strain rate, stress and viscosity: numerical diffusion due to the interpolation between nodes of the grid and the markers can cause strong (two- to fivefold) widening of shear zones leading to a decrease in strain-rate contrast. Finite element and finite difference codes have a tendency to reduce shear zone width with decreasing element size, implying that numerical shear zone width is mesh-dependent. Very fine meshes probably underestimate the width of localized shear zones in comparison with observed shear band widths in granular media (see also de Borst & Sluys 1991; Zervos *et al.* 2001).

Sensitivity analyses

In the extensional models presented here, two different boundary conditions were applied at the base. In Abaqus/Standard, I2ELVIS and Microfem, boundary friction was applied along the whole bottom boundary, including along the viscous material. In LAPEX-2D, SloMo and Sopale, grid points on the basal sheet were assigned the velocity with which the sheet was pulled and points to the left of the sheet were held fixed. The high level of agreement between the modelling results of the six experiments shows that any effect of this variation in basal boundary condition is small. The tip of

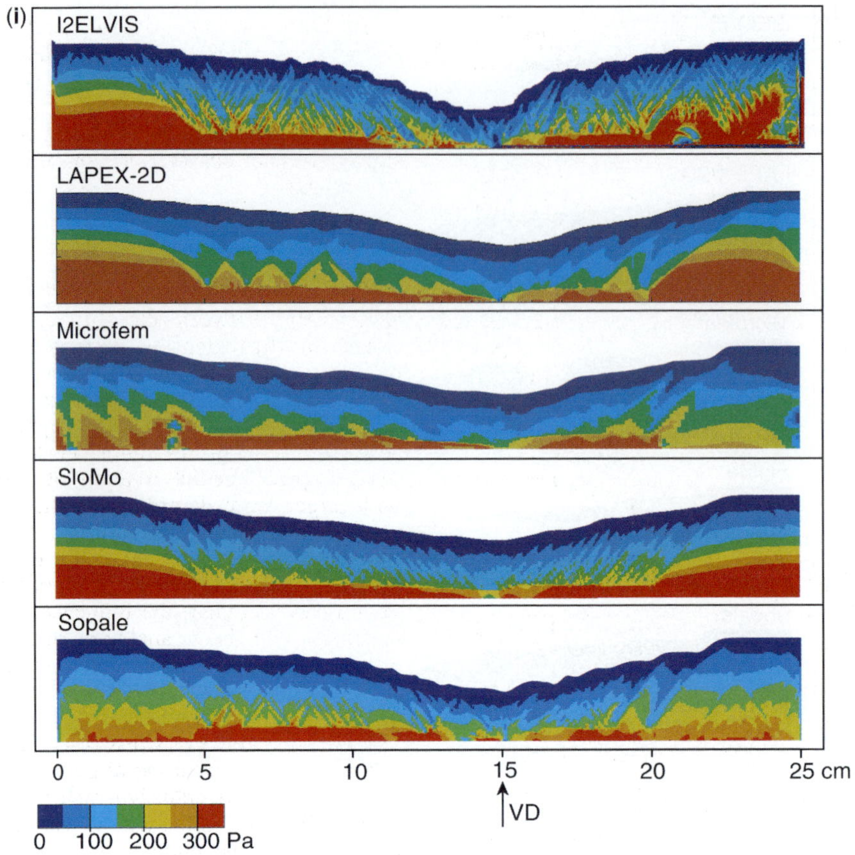

Fig. 7. (i) Dynamic pressures after 5 cm.

the basal sheet serves to initially localize deformation in a symmetric manner in the centre of the model. An experiment with a free slip base without basal sheet (run with Sopale) shows that in this case shear zones first form at the edges of the viscous layer and soon after in the centre of the model.

Experiments with I2ELVIS and Sopale show that a lower extensional velocity (which is equivalent to a lower viscosity for the viscous layer)

Table 5. *Quantification of extension experiments**

Code	Fault ext.† at 2 cm	Fault initiation at ends of PDMS		Fault on left side of VD		Fault on right side of VD	
		Left	Right	dip	when	dip	when
Abaqus/Standard	–	0.5	0.5	45°	0	45°	0
I2ELVIS	0.9–1.2	never	2	51°	1	52°	1
LAPEX-2D	0.7–0.9	0.5	0.5	51°	0.5	53°	0.5
Microfem	0.5–1.2	never	1.5	46°	0.5	48°	0.5
SloMo	0.7–0.8	2	2	53°	1	55°	1
Sopale	1.1–1.3	0.5	0.5	50°	0.5	52°	0.5
Analogue ranges‡	0.9–1.0	never	2–4.5	65°–69°	0.5	60°–65°	0.5

*All values referred to sandbox scale in cm. Output stages typically available at 0, 0.5, 1 and then every cm of extension.
†Fault-controlled horizontal extension determined by cut-off points between layers and the faults.
‡Analogue values from experiments of University of Bern and IFP Rueil-Malmaison (Schreurs *et al.* 2006).

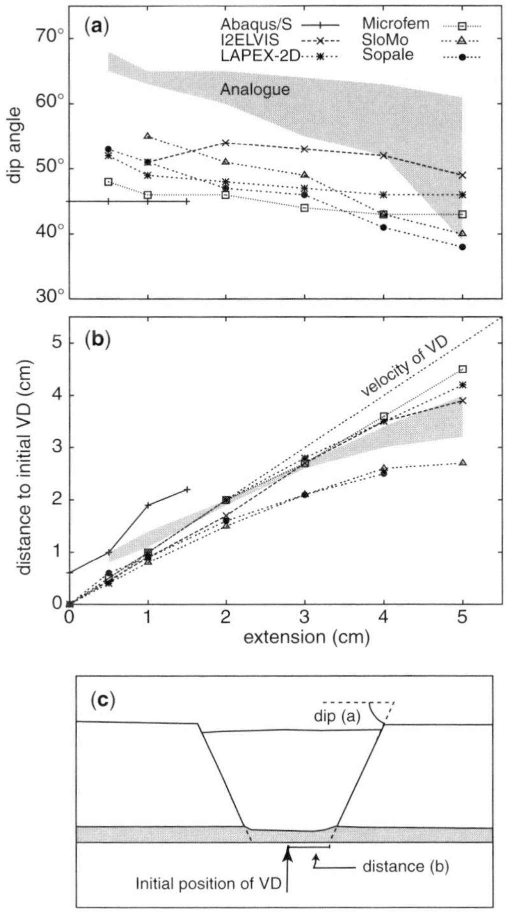

Fig. 8. (**a**) Dip angle of the first-formed normal fault on the right-hand side of the velocity discontinuity (the tip of the basal sheet) versus extension. (**b**) Migration of the same fault. The migration is measured relative to the initial position of the tip of the basal sheet. (**c**) Schematic drawing showing how the quantities in (a) and (b) were measured. The grey bands denote the range of values measured for the analogue experiments of University of Bern and IFP Rueil-Malmaison (see also Fig. 7 and Schreurs *et al.* 2006).

favours a more efficient outward propagation of shear zone development. The surface depression is less deep. A higher cohesion of the 'sand' material leads to locally steeper surface slopes, similar to our observations for the shortening experiment. We found that a reduction or increase in the strain softening range does not affect the results significantly. Tests with I2ELVIS and Sopale also showed that a higher resolution calculation grid (smaller element size) leads to the development of more shear zones which are better focussed, as discussed above. This can also be seen from a comparison of Microfem (lower resolution) to Sopale (higher resolution) results and is again similar to our observation for the shortening experiment.

Implications of the numerical experiments

Comparison between the different numerical codes for the extension experiment shows a high level of agreement. All predict the same location of maximum thinning and show the same general evolution and stepping out of shear zones. High-resolution numerical codes (e.g., Sopale) predict rotation of shear zones with larger finite deformation. The extensional results show much less variation compared to the shortening experiment. Ellis *et al.* (2004) suggest that experiments involving weak viscous layers reduce the influence of boundary conditions on both analogue and numerical results, and instead pass control to rheological strength contrasts within the model domain. In our set-up, the weak viscous basal layer serves to reduce the effect of stress discontinuities due to the basal velocity discontinuity.

Although the overall dynamics and evolution of the different numerical codes are similar, differences in the number of shear zones that develop, the timing at which the ends of the viscous layer are activated and the shear zone dip can be seen. These differences partly appear to reflect differences in mesh resolution. As for the shortening experiment, we find that caution is required in interpretations from numerical results involving number of shear zones and shear zone spacing.

Our numerical experiments may also have important implications for the understanding of pressure distributions in large-scale tectonic processes involving the development of shear zones. As follows from our results, the dynamic pressure field (Figs 7d and i) in a medium subjected to brittle/plastic deformation deviates strongly from the lithostatic distribution often assumed in geosciences. These results are in line with other theoretical studies (e.g., Mancktelow 1995; Petrini & Podladchikov 2000; Burg & Gerya 2005) which predict significant underpressure and overpressure effects in orogenic processes. This, in turn, may have strong consequences for the pressure-temperature evolution of deformed rocks complexes composing the Earth's crust and lithospheric mantle.

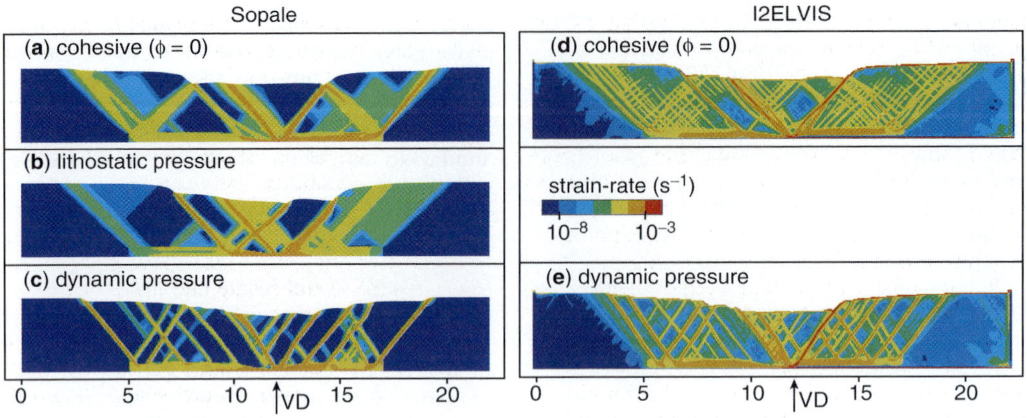

Fig. 9. A comparison of strain-rates for I2ELVIS and Sopale after 2 cm of extension shows the influence of the plasticity description. (**a**) and (**d**) Angle of internal friction $\phi = 0$. Cohesion softens from 170 Pa to 140 Pa (sandbox scale). These values mimic the same strength and strength reduction as for the frictional extension models. (**b**) Sopale for Mohr-Coulomb plasticity with depth-dependent pressure. (**c**) and (**e**) Mohr-Coulomb plasticity with dynamic pressure (standard case).

Comparison to analogue results

Only two analogue experiments documented cross-sectional information at different extension stages and could be used to compare the numerical results against. The analogue results of the University of Bern and IFP Rueil-Malmaison show an overall similar evolution, but differ in the velocity with which normal faults are distributed outwards (Schreurs *et al.* 2006). The geometries of the numerical and analogue experiments after 1 and 5 cm of extension are shown in Figures 7b and g. The dip angle and migration of the first formed shear zone to the right of the velocity discontinuity are compared in Figure 8. In the numerical experiments reported here the mobile wall velocity is directly applied to the right-hand side of the models. However, in the analogue sandbox the role of the mobile wall is likely only to prevent the sand from collapsing and does not provide a driving force. The extension is passed to the model by the moving basal sheet. Because there is a sufficient distance between the right-hand wall and the velocity discontinuity this is, however, not likely to affect the analogue-numerical comparison.

Comparison of the numerical results to the two analogue experiments shows that: (1) The numerical experiments predict the overall dynamics and evolution of the analogue system reasonably well. (2) In both analogue and numerical results, shear zones initiate at the basal velocity discontinuity and with increasing extension new shear zones develop outwards. (3) Both numerical and analogue results show a rotation of shear zones with shallowing dip angles for increasing extension. (4) The velocity with which the first formed normal fault to the right of the velocity discontinuity moves to the right is very similar between the two analogue results and the numerical models (Fig. 8b). (5) The main difference between the analogue and numerical results is the dip angle of the shear zones (Figs 7 and 8a). The dip angles in the analogue experiments are much steeper (between 60° and 69° initially) and follow Coulomb behaviour, with dip angles predicted by $45° + \phi/2$ (63°–68° for the material properties of the two analogue experiments). All numerical extension models assume zero dilatation and for dynamic pressure-dependent Mohr-Coulomb behaviour dip angles between 45° (Roscoe) and $45° + \phi/2$ (Coulomb) are then to be expected (Vermeer 1990) (45°–63° for our material properties). The observed initial dip angles (45°–55°) are in the lower range of the theoretical values. Assigning a non-zero dilatation angle, ψ, to 'sand' will increase the theoretical minimum dip angle value to $45° + \psi/2$. Tests with I2ELVIS show that a non-zero dilatation angle indeed results in higher dip angles for the shear zones.

The numerical sandbox

Our study presents a direct qualitative and quantitative comparison of numerical and analogue model results for two experimental set-ups aimed at simulating upper-crustal tectonic processes. Results from seven numerical codes are compared to results from five analogue laboratories for a brittle thrust wedge experiment, while results from six numerical codes are compared to results from two analogue laboratories for a brittle-viscous extension experiment. The accompanying paper (Schreurs *et al.* 2006) shows additional analogue shortening and extension results, which could not be used in our comparison as either cross-sectional information was not recorded during the experiment or a different material (clay) was used.

A comparison of analogue experiments from different laboratories has to take into account that differences may occur in material properties (whereby properties may even slightly change due to storage conditions or small vibrations), modelling apparatus, model set-up (width and length) and in the set-up technique of individual experimenters. The effects of some variations can be tested by determining the reproducibility of each experiment in the same laboratory, but it is clearly of importance to determine the inter-laboratory reproducibility of analogue model results as well. The accompanying paper (Schreurs *et al.* 2006) shows an encouraging first order agreement of model results of different analogue laboratories, despite variability in model widths and in material properties.

In comparison, numerical experiments have the potential to be much more reproducible, since material properties, model set-up and boundary conditions can be specified exactly. However, set-ups between different experimenters may also vary in subtle ways depending on assumptions used, and even slightly different ways of interpreting and prescribing the initial set-up. The eight numerical codes used in our comparison differ with each other for the following properties and assumptions: (1) resolution, imposed element aspect ratios, and type of element; (2) time step (extension experiment); (3) smoothing of velocity discontinuities; (4) implementation of frictional boundaries; (5) material properties (dilatation and elasticity); (6) scaling; (7) calculation of dynamic pressure; and (8) surface smoothing (shortening experiment). Our results indicate that grid resolution, representation of boundary conditions (boundary friction and velocity discontinuities), calculation of dynamic pressure and dilatation require careful attention in sandbox-type experiments. In contrast, elastic material behaviour, scaling of the experiment and small amounts of surface smoothing appear to play only a minor role.

Our results highlight that different numerical solution methods (finite difference method, finite element method, distinct element method) can successfully be used to reproduce structures seen in analogue sandbox models. The numerical codes can do reasonably well in qualitatively predicting the dynamic evolution of the system in shortening and extension, shear zone (fault) formation, and taper geometry (shortening experiment). The level of variability between the numerical results is about the same as the degree of variability between the analogue results. Numerical codes show differences with each other, and the analogue experiments, for detailed location, dip and spacing of shear zones. As the initial set-up and material properties of analogue models can be relatively well constrained, their results can form a useful test for numerical models. Numerical models may, therefore, benefit from future studies focusing on the differences between analogue and numerical modelling results. Our results indicate that future experiments should preferably choose a simplified set-up that reduces the effects of boundary conditions and especially abrupt changes in boundary velocities, prescribe resolution and time step size, use material behaviour which is as similar as possible and, if possible, quantify results in an unambiguous manner.

Our results emphasize the importance of the following issues for numerical sandbox-type studies: (1) the ability to model large deformation structures; (2) the ability to represent boundary friction, velocity discontinuities and a free surface; (3) the representation of a composite (elasto)-visco-plastic rheology; (4) calculation with a relatively high resolution (important for shear zone evolution, for example); (5) minimization of numerical diffusion; and (6) consensus on the procedure for quantification of the results. These are important for studies in which numerical and analogue models are used in combination, but reflect at the same time requirements to study Earth tectonic processes.

Conclusions

Two experimental set-ups inspired by settings from structural geology were used to compare eight numerical codes to each other. Despite differences in set-up and material behaviour between the individual experiments, the results are useful in showing that: (1) Numerical codes at the present time are capable of modelling the dynamic evolution of compressional and extensional settings, including shear-zone development, forward propagation of thrust wedges, and taper geometry of convergent wedges. The qualitative behaviour of the different codes is similar. (2) A quantitative analysis of the results of these codes shows a higher degree of variability, particularly in number of shear zones that develop and their spacing and dip angle. (3) A comparison between the results of the numerical codes and equivalent analogue experiments shows agreement in the overall evolution of the system and in the localization of deformation onto faults/shear zones. The variability among the numerical results is of the same order as the variations among the analogue results. Differences with the analogue results are also apparent, for example, in dip angles of extensional shear zones, degree of localization, and the development of backthrusts in the shortening experiment. Reducing these differences represents an interesting challenge for numerical modellers.

We hope that the comparisons shown here are a useful guide to the structural geology community, in highlighting the degree to which numerical codes may be used to reproduce experiments in the sandbox and/or processes at the scale of the upper crust.

The figures of the analogue experiments were kindly provided by the laboratories of Bern (GS), IFP Rueil Malmaison (J.-M. Daniel, J.-M. Mengus), Parma (E. Costa, C. Cavozzi), Pavia (A. Ravaglia) and Toronto (D. Boutilier, A. Cruden). J. Lohrmann measured the angle of friction of (GFZ Potsdam) sand on 'Alkor' foil. C. Beaumont, J.-P. Brun and B. Colletta gave us helpful feedback on the design of the experiments. We thank the participants of the GeoMod2004 pre-conference workshop for their insightful discussions and Claudio Faccenna for sharing with us his introductory talk on analogue and numerical modelling which he presented at this workshop. We would like to thank N. Kukowski and the reviewers R. Buck and K. Furlong for their positive feedback, and Y. Podladchikov, S. Hoth and S. Medvedev for stimulating discussions. Microfem and Sopale were developed by P. Fullsack and the Geodynamics Group at Dalhousie University. Simulations by BK have been run at the USC high performance computing facilities. This work was supported by ETH Research Grant TH-12/04-1, RFBR grant #03-05-64633 and the RF President Program 'Leading Scientific School of Russia' (grant #03.5.1645) to TVG and project SFB-267 from the Deutsche Forschungsgemeinschaft for AYuB and AK.

References

ABAQUS/STANDARD USER'S MANUAL 2003. Vol. 1 & 2, version 6.4. Abaqus Inc, Pawtucket, Rhode Island, USA.

ADAM, J. & URAI, J. L. *ET AL.* 2005. Shear localization and strain distribution during tectonic faulting – new insights from granular-flow experiments and high-resolution optical image correlation techniques. *Journal of Structural Geology*, **27**, 283–301.

BABEYKO, A. YU., SOBOLEV, S. V., TRUMBULL, R. B., ONCKEN, O. & LAVIER, L. L. 2002. Numerical models of crustal scale convection and partial melting beneath the Altiplano-Puna Plateau. *Earth and Planetary Science Letters*, **199**, 373–388.

BAHROUDI, A., KOYI, H. A. & TALBOT, C. J. 2003. Effect of ductile and frictional décollements on style of extension. *Journal of Structural Geology*, **25**, 1401–1423.

BEHN, M. D., LIN, J. & ZUBER, M. T. 2002. A continuum mechanics model for normal faulting using a strain-rate softening rheology: implications for thermal and rheological controls on continental and oceanic rifting. *Earth and Planetary Science Letters*, **202**, 725–740.

BLANKENBACH, B. & BUSSE, F. *ET AL.* 1989. A benchmark comparison for mantle convection codes. *Geophysical Journal International*, **98**, 23–38.

BURBIDGE, D. R. & BRAUN, J. 2002. Numerical models of the evolution of accretionary wedges and fold-and-thrust belts using the distinct-element method. *Geophysical Journal International*, **148**, 542–561.

BURG, J.-P. & GERYA, T. V. 2005. Viscous heating and thermal doming in orogenic metamorphism: numerical modelling and geological implications. *Journal of Metamorphic Geology*, **23**, 75–95.

CRUDEN, A. R., NASSERI, M. B. & PYSKLYWEC, R. 2006. Surface topography and internal strain variation in wide hot orogens from three-dimensional analogue and two-dimensional numerical vise models. *In*: BUITER, S. J. H. & SCHREURS, G. (eds) *Analogue and Numerical Modelling of Crustal-Scale Processes*. Geological Society, London, Special Publications, **253**, 79–104.

CUNDALL, P. A. 1990. Numerical modelling of jointed and faulted rock. *In*: ROSSMANITH, A. (ed.) *Mechanics of Jointed and Faulted rocks*. Balkema, Rotterdam, 11–18.

CUNDALL, P. A. & BOARD, M. 1988. A microcomputer program for modelling large-strain plasticity problems. *In*: SWOBODA, G. (ed.) *6th International Conference in Numerical Methods in Geomechanics*. A. A. Balkema, Brookfield, VT, 2101–2108.

CUNDALL, P. A. & STRACK, D. L. 1979. A discrete numerical model for granular assemblies. *Geotechnique*, **29**, 47–65.

CUVELIER, C., SEGAL, A. & VAN STEENHOVEN, A. A. 1986. *Finite Element Methods and Navier-Stokes Equations*. Reidel Publishing Company, Dordrecht, 483pp.

DAHLEN, F. A. 1984. Noncohesive critical Coulomb wedges: an exact solution. *Journal of Geophysical Research*, **89**, 10125-10133.

DAVIS, D., SUPPE, J. & DAHLEN, A. 1983. Mechanics of fold-and-thrust belts and accretionary wedges. *Journal of Geophysical Research*, **88**, 1153–1172.

DE BORST, R. & SLUYS, L. J. 1991. Localization in a Cosserat continuum under static and dynamic loading conditions. *Computer Methods in Applied Mechanics and Engineering*, **90**, 805–827.

ELLIS, S., SCHREURS, G. & PANIEN, M. 2004. Comparisons between analogue and numerical models of thrust wedge development. *Journal of Structural Geology*, **26**, 1659–1675.

FULLSACK, P. 1995. An arbitrary Lagrangian-Eulerian formulation for creeping flows and its application in tectonic models. *Geophysical Journal International*, **120**, 1–23.

GERYA, T. V. & YUEN, D. A. 2003. Characteristics-based marker-in-cell method with conservative finite-difference schemes for modelling geological flows with strongly variable transport properties. *Physics of the Earth and Planetary Interiors*, **140**, 293–318.

GUTSCHER, M.-A., KUKOWSKI, N., MALAVIEILLE, J. & LALLEMAND, S. 1998*a*. Material transer in accretionary wedges from analysis of a systematic series of analog experiments. *Journal of Structural Geology*, **20**, 407–416.

GUTSCHER, M.-A., KUKOWSKI, N., MALAVIEILLE, J. & LALLEMAND, S. 1998*b*. Episodic imbricate thrusting and underthrusting: Analog experiments and mechanical analysis applied to the Alaskan Accretionary Wedge. *Journal of Geophysical Research*, **103**, 10161–10176.

HUBBERT, M. K. 1937. Theory of scale models as applied to the study of geologic structures. *Bulletin of the Geological Society of America*, **48**, 1459–1520.

JAEGER, J. C. & COOK, N. G. W. 1979. *Fundamentals of Rock Mechanics*. John Wiley and Sons, New York, 585pp.

KAUS, B. J. P. 2005. *Modelling approaches to geodynamic processes*. PhD thesis, ETH Zürich.

LE POURHIET, L., MATTIONI, L. & MORETTI, I. 2006. 3D modelling of rifting through a pre-existing stack of nappes in the Gulf of Corinth (Greece): a mixed analogue/numerical approach. *In*: BUITER, S. J. H. & SCHREURS, G. (eds) *Analogue and Numerical Modelling of Crustal-Scale Processes*. Geological Society, London, Special Publications, **253**, 233–252.

LOHRMANN, J., KUKOWSKI, N., ADAM, J. & ONCKEN, O. 2003. The impact of analogue material properties on the geometry, kinematics, and dynamics of convergent sand wedges. *Journal of Structural Geology*, **25**, 1691–1771.

MANCKTELOW, N. S. 1995. Nonlithostatic pressure during sediment subduction and the development and exhumation of high pressure metamorphic rocks. *Journal of Geophysical Research*, **100**, 571–583.

MICHON, L. & MERLE, O. 2000. Crustal structures of the Rhine graben and the Massif Central grabens: An experimental approach. *Tectonics*, **19(5)**, 896–904.

MULUGETA, G. 1988. Modelling the geometry of Coulomb thrust wedges. *Journal of Structural Geology*, **10**, 847–859.

MULUGETA, G. & KOYI, H. 1992. Episodic accretion and strain partitioning in a model sand wedge. *Tectonophysics*, **202**, 319–333.

NISA II™ USER'S MANUAL VERSION 7.0, 1997. Engineering Mechanics Research Corporation, Troy, Michigan, USA.

PANIEN, M. 2004. *Analogue modelling experiments of basin inversion using well-characterized granular materials and comparisons with numerical models*. Ph.D. thesis. University of Bern, Switzerland.

PANIEN, M., BUITER, S. J. H., SCHREURS, G. & PFIFFNER, O. A. 2006. Inversion of a symmetric basin: Insights from a comparison between analogue and numerical experiments. *In*: BUITER, S. J. H. & SCHREURS, G. (eds) *Analogue and Numerical Modelling of Crustal-Scale Processes*. Geological Society, London, Special Publications, **253**, 253–270.

PELLETIER, D., FORTIN, A. & CAMARERO, R. 1989. Are FEM solutions of incompressible flows really incompressible? (Or how simple flows can cause headaches!) *International Journal of Numerical Methods in Fluids*, **9**, 99–112.

PETRINI, K. & PODLADCHIKOV, Y. 2000. Lithospheric pressure-depth relationship in compressive regions of thickened crust. *Journal of Metamorphic Geology*, **18**, 67–77.

PFC^{2D} 1999. *User's Guide* Version 3.0. Itasca Consulting Group, Inc.

POLIAKOV, A. N. & PODLADCHIKOV, Y. Y. 1992. Diapirism and topography. *Geophysical Journal International*, **109**, 553–564.

POLIAKOV, A. N., CUNDALL, P. A., PODLADCHIKOV, Y. Y. & LYAKHOVSKY, V. A. 1993. An explicit inertial method for the simulation of the viscoelastic flow: an evaluation of elastic effects on diapiric flow in two- and three-layers models. *In*: STONE, D. B. & RUNCORN, S. K. (eds) *Flow and Creep in the Solar System: Observations, Modelling and Theory*. Kluwer Academic Publishers, 175–195.

RANALLI, G. 1987. *Rheology of the Earth: deformation and flow processes in geophysics and geodynamics*. Allen & Unwin Inc., Winchester, USA, 366pp.

SALTZER, S. D. & POLLARD, D. D. 1992. Distinct element modelling of structures formed in sedimentary overburden by extensional reactivation of basement normal faults. *Tectonics*, **11(1)**, 165–174.

SASSI, W., COLLETTA, B., BALÉ, P. & PAQUEREAU, T. 1993. Modelling of structural complexity in sedimentary basins: the role of pre-existing faults in thrust tectonics. *Tectonophysics*, **226**, 97–112.

SCHELLART, W. P. 2000. Shear test results for cohesion and friction coefficients for different granular materials: scaling implications for their usage in analogue modelling. *Tectonophysics*, **324**, 1–16.

SCHREURS, G. & BUITER, S. J. H. *ET AL*. 2006. Analogue benchmarks of shortening and extension experiments. *In*: BUITER, S. J. H. & SCHREURS, G. (eds) *Analogue and Numerical Modelling of Crustal-Scale Processes*. Geological Society, London, Special Publications, **253**, 1–28.

SMART, K. J. & COUZENS-SCHULTZ, B. A. 2001. Mechanics of blind thrusting: Comparison of numerical and physical modelling. *The Journal of Geology*, **109**, 771–779.

STORTI, F., SALVINI, F. & MCCLAY, K. 2000. Synchronous and velocity-partitioned thrusting and thrust polarity reversal in experimentally produced, doubly-vergent thrust wedges: Implications for natural orogens. *Tectonics*, **19(2)**, 378–396.

STRAYER, L. M., HUDLESTON, P. J. & LORIG, L. J. 2001. A numerical model of deformation and fluid-flow in an evolving thrust wedge. *Tectonophysics*, **335**, 121–145.

VANBRABANT, Y., JONGMANS, D., HASSANI, R. & BELLINO, D. 1999. An application of two-dimensional finite-element modelling for studying the deformation of the Variscan fold-and-thrust belt (Belgium). *Tectonophysics*, **309**, 141–159.

VAN KEKEN, P. E., KING, S. D., SCHMELING, H., CHRISTENSEN, U. R., NEUMEISTER, D. & DOIN, M.-P. 1997. A comparison of methods for the modelling of thermochemical convection. *Journal of Geophysical Research*, **102**, 22477–22495.

VERMEER, P. A. 1990. The orientation of shear bands in biaxial tests. *Geotechnique*, **40**, 223–236.

WEIJERMARS, R. 1986. Flow behaviour and physical chemistry of bouncing putties and related polymers in view of tectonic laboratory applications. *Tectonophysics*, **124**, 325–358.

WIJNS, C., BOSCHETTI, F. & MORESI, L. 2003. Inverse modelling in geology by interactive evolutionary computation. *Journal of Structural Geology*, **25**, 1615–1621.

ZERVOS, A., PAPANASTASIOU, P. & VARDOULAKIS, I. 2001. A finite element displacement formulation for gradient plasticity. *International Journal for Numerical Methods in Engineering*, **50**, 1369–1388.

ZHAO, W.-L., DAVIS, D. M., DAHLEN, F. A. & SUPPE, J. 1986. Origin of convex accretionary wedges: evidence from Barbados. *Journal of Geophysical Research*, **91**, 10246–10258.

Interaction between normal faults and pre-existing thrust systems in analogue models

GIACOMO CORTI[1], SERENA LUCIA[2], MARCO BONINI[1], FEDERICO SANI[2] & FRANCESCO MAZZARINI[3]

[1]*CNR – Istituto di Geoscienze e Georisorse, Via La Pira, 4-50121 Firenze, Italy (e-mail: cortigi@geo.unifi.it)*

[2]*Dipartimento di Scienze della Terra, Università degli Studi di Firenze, Via La Pira, 4-50121 Firenze, Italy*

[3]*Istituto Nazionale di Geofisica e Vulcanologia, Via della Faggiola, 32-56100 Pisa, Italy*

Abstract: The influence of pre-existing thrusts on the development of later normal faults was investigated using scaled laboratory analogue models. Experiments consisted of a phase of shortening followed by extension at variable angles of obliquity (α) to the shortening direction. Results suggest that the angle α has a major influence on the surface fault pattern and on the interaction between shortening-related structures and later extensional structures. Three different modes of interactions were identified depending upon the extension kinematics. (1) For orthogonal extension ($\alpha = 0^\circ$), shortening-related fold and thrust structures strongly influence the development of normal faults: graben structures nucleate within anticlines and the normal faults reactivate thrusts at depth (*branching at depth mode of interaction*). (2) For highly oblique extension ($\alpha \geq 45^\circ$), shortening-related structures exert no influence on normal faults as extension-related steeply-dipping faults (characterized by an oblique component of movement) displace early thrusts (*no interaction mode*). (3) For intermediate obliquity angles ($\alpha = 15^\circ, 30^\circ$), an *intermediate mode of interaction* characterizes the experiments, where the *no interaction* and *branching at depth* modes coexist in different regions of models. Modelling results can be used to infer regional extension directions as is shown for the Northern Appenines (Italy).

It has been shown that in many cases pre-existing thrust systems play an important role in controlling the architecture of superimposed normal faults. Mechanical theory suggests that the probability of reactivation of pre-existing thrust faults depends on their orientation with respect to the superimposed extensional stress field (e.g., Sibson 1985). The process was investigated in previous analogue models (Krantz 1991; Faccenna *et al.* 1995 and references therein), suggesting that the initial dip of thrust faults has a major control on the development of successive normal faults. Both studies started with a phase of shortening that resulted in progressively steeper thrust faults prior to extension. Model results suggested that three different types of interactions between pre-existing thrusts and successive normal faults may arise (e.g., Faccenna *et al.* 1995): (1) *No interaction* occurs in case of low-dip thrusts ($<32^\circ \pm 1^\circ$) and normal faults form independently, displacing pre-existing thrusts; (2) *branching at depth* of normal faults with thrusts is observed for intermediate dip angles ($32^\circ \pm 1^\circ - 41^\circ \pm 1^\circ$); (3) total reactivation occurs on thrusts dipping at angles higher than $41^\circ \pm 1^\circ$. These three basic mode of interaction between normal faults and thrust systems have been recognized and described in natural systems, such as the Central Apennines (Italy) and the North Sea region (Faccenna *et al.* 1995).

Active extensional structures interacting with early thrust systems have been also recognized in the Northern Apennines (e.g., Finetti *et al.* 2001). In this fold-and-thrust belt, the early compressive structures show a prominent curvature along strike and the successive normal faults display strong variations in terms of trend, arrangement and kinematics in different sectors of the chain (Fig. 1). In particular, where thrusts are orthogonal to the regional NE–SW extension direction (e.g., Mugello Basin; Fig. 1), normal faults are long and linear, display a mainly pure dip-slip kinematics and

From: Buiter, S. J. H. & Schreurs, G. (eds) 2006. *Analogue and Numerical Modelling of Crustal-Scale Processes.* Geological Society, London, Special Publications, **253**, 65–78.
0305-8719/06/$15.00

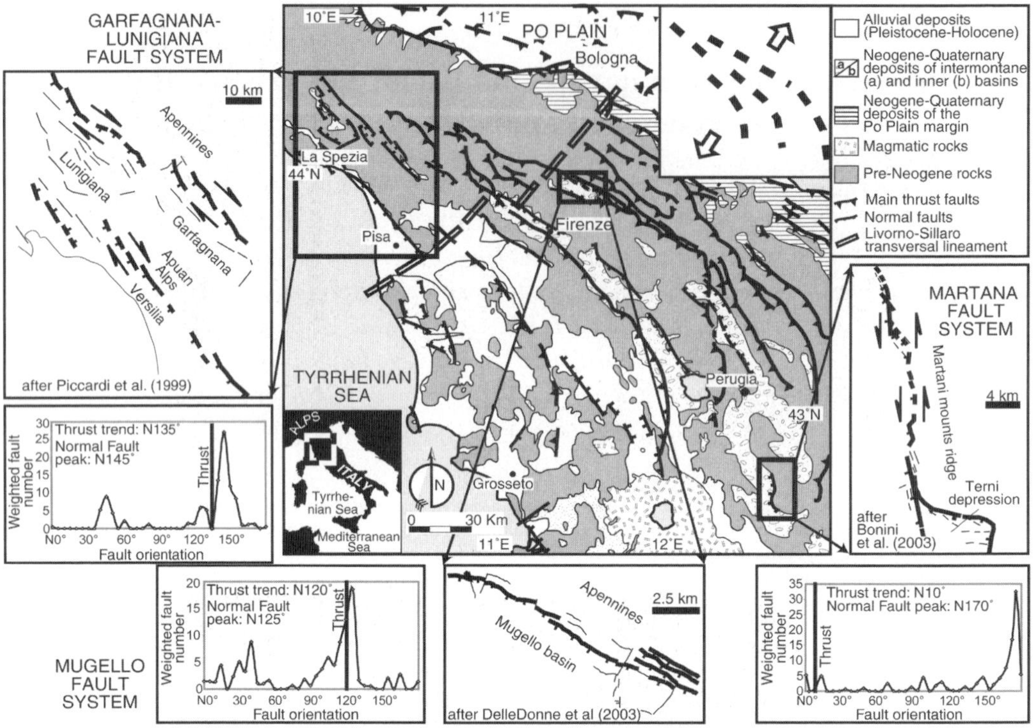

Fig. 1. Schematic structural map of the Northern Apennines (Italy), showing the location of the selected active normal fault systems. Inset shows the location of the investigated area. Graphs display fault orientation, with the weighted fault number plotted against the fault orientation (measured considering the north as the reference direction, i.e., 0°). The weighting factor for each fault is the ratio between the fault length and the minimum length of the whole data set, such that long faults have higher ratio (weight) than short ones. The frequency of the azimuth of a fault directly relates to this ratio; the longer the fault the higher its frequency. The thick black lines in the graphs indicate the thrust direction. The small inset on the map shows the orientation of the main thrust fronts (thick dashed lines) with respect to the roughly NE–SW extension direction. Active normal fault systems in the selected areas are from Piccardi *et al.* (1999), Delle Donne *et al.* (2003) and Bonini *et al.* (2003).

are subparallel to the early compressive structures (Delle Donne *et al.* 2003). Conversely, where thrusts are slightly oblique to the extension (Garfagnana-Lunigiana area; Fig. 1), normal faults are arranged en-echelon, are characterized by an oblique-slip kinematics and display a mean peak in distribution oriented at about 10° with respect to the earlier thrusts (Piccardi *et al.* 1999). In case of high obliquity (e.g., Martani Mounts-Terni Basin area; Fig. 1), normal faults are linear, are characterized by an important (locally dominant) strike-slip component of movement and form at an angle of 20° to the pre-existing structures (Bonini *et al.* 2003). These features suggest that the different orientation of thrusts with respect to the direction of extension may have exerted an important control on the pattern of successive extensional structures. This influence is investigated in this study by means of analogue models. We present a series of experiments on polyphase deformation, consisting of orthogonal shortening followed by extension at variable angles of obliquity.

Analogue modelling

The experiments were performed at the Tectonic Modelling Laboratory of the CNR-IGG located at the Earth Science Department of Florence University (Italy) using the experimental apparatus shown in Figure 2. The models, with initial dimensions of 50 cm × 50 cm × 4 cm, were built over the metal base of the apparatus. Models consisted of a sand-pack overlying a ductile silicone layer. The models were confined on three sides by a mobile L-shaped wall and a fixed wall (Fig. 2). The right-hand frontal end remained unconfined (Fig. 2). Deformation was

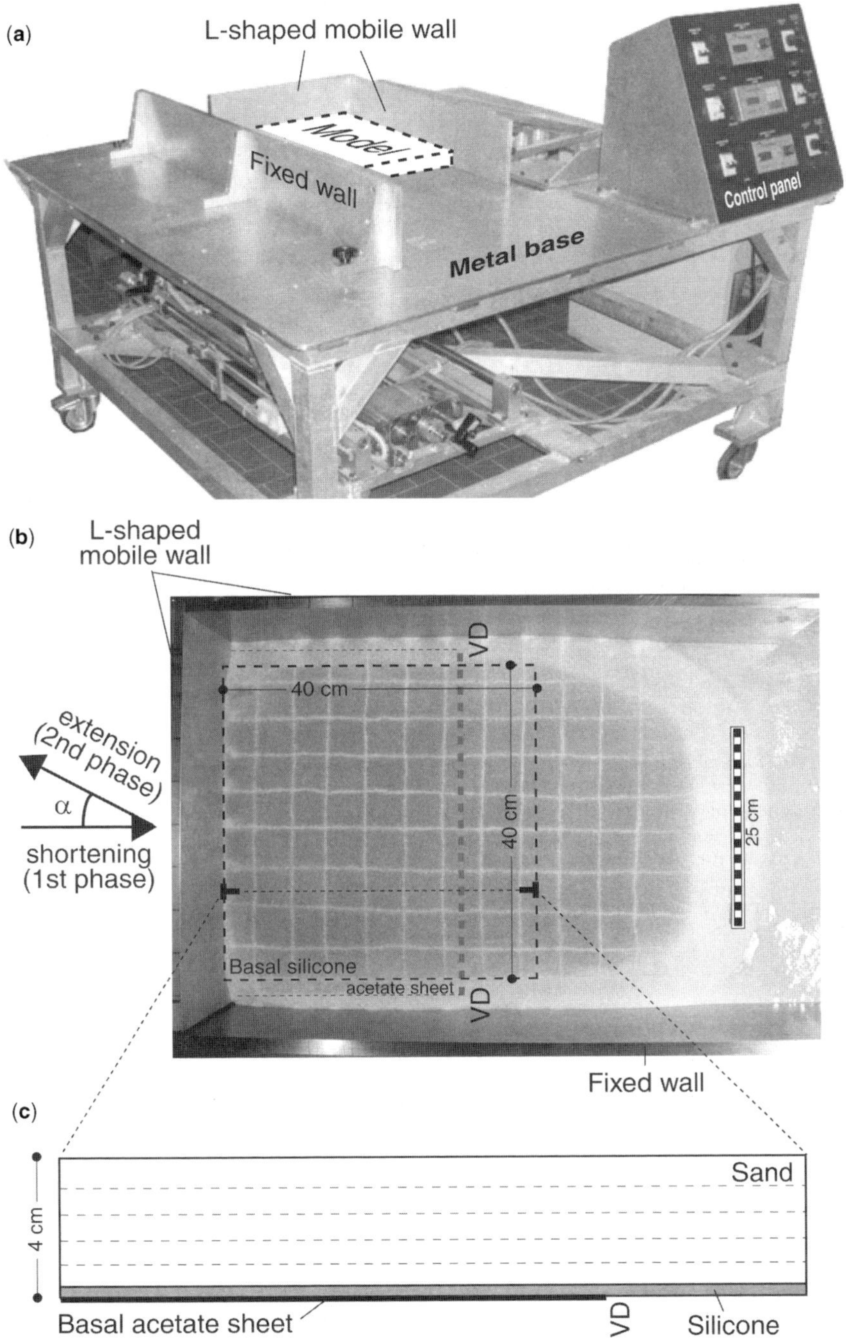

Fig. 2. Experimental set-up. (**a**) Photograph of the experimental apparatus. (**b**) Model top-view, with the two directions of deformation indicated. Angle of obliquity (α) is the angle between the direction of shortening and extension. (**c**) Schematic model cross-section. See text for details.

achieved through displacement of the L-shaped wall, driven by two orthogonal stepper motors controlled by a central unit. In each experiment the shortening direction during the first phase of deformation was parallel to the long axis of the L-shaped wall (Fig. 2). Extension during the second phase was varied in different experiments in order to investigate the influence of the obliquity angle α (angle between extension and shortening direction; Fig. 2b) on the final structural pattern. Obliquity angles of 0°, 15°, 30°, 45° and 60° were investigated (Table 1). Deformation during the extensional phase was localized by a thin acetate sheet (*c.* 0.1 mm thick) at the base of the model and fixed to the L-shaped wall. The thin sheet created a central velocity discontinuity (VD), parallel to the short axis of the mobile wall (Fig. 2). In this set-up, displacement along the VD was transferred by viscous coupling at the base of the sand layer (see Tron & Brun 1991). In nature, the VD simulates a pre-existing discrete fabric at upper crustal levels (e.g., Faccenna *et al.* 1995; Morley 1999); in the case of the Northern Apennines, this may be a basement thrust reactivated during the extensional phase (e.g., Finetti *et al.* 2001).

Shortening occurred at a constant velocity of 40 mm/hr up to 25% bulk shortening, whereas extension took place at 6 mm/hr up to 10% bulk extension (Table 1). In order to improve visualization of surface fault patterns generated by extension, an evenly distributed thin layer (<1 mm) of coloured sand was added on top of the shortened models before the second phase of deformation.

Analogue materials

In the models, a 3.5 cm-thick sand layer was used to simulate the brittle part of the sedimentary cover in nature. We used dry Fontainebleu sand, an almost pure quartz sand (>99% SiO_2) with a rounded grain shape, a grain size <250 μm and a density of *c.* 1550 Kg m^{-3}. Material properties were measured with a ring-shear tester and yielded a cohesion of *c.* 65 Pa and an angle of peak internal friction of *c.* 39°. In order to better visualize and analyse in detail the internal brittle deformation, different layers of coloured sand were introduced as horizontal passive markers within the sand-pack. A 40 cm × 40 cm × 0.5 layer of silicone (Mastic Silicone Rebodissant 29, produced by CRC, France) was placed at the base of the model in order to distribute deformation in the overlying sand-pack, as in the models by Faccenna *et al.* (1995). The silicone is characterized by a density of 1150 kg m^{-3} and a viscosity of 1.6×10^4 Pa s (measured with a coni-cylindrical viscometer at a temperature of *c.* 23°C). The viscous layer simulates a décollement layer (e.g., an anhydritic evaporite horizon) underlying a brittle sedimentary cover, as occurring for example in the Northern Apennines stratigraphic column (Burano formation; e.g., Bonini & Sani 2002). This silicone layer also allowed to reduce the strong stress reorientation associated with a rigid VD (Morley 1999) and thus to avoid its excessive control on fault location (Faccenna *et al.* 1995).

Scaling

Scaling of the models was obtained following the principles of dynamic similitude (e.g., Ramberg 1981). The models were intended to represent a length scale of 10^{-5}, such that 1 cm in the model represents 1 km in nature. Dynamic-kinematic similarity was achieved through dimensionless ratios of forces (e.g., Ramberg 1981; Weijermars & Schmeling 1986). The scaling ratios of the different model and nature parameters are reported in Table 2. The viscosity ratio chosen is 1.6×10^{-15}, implying that silicone (with a viscosity of 1.6×10^4 Pa s) simulates a natural viscosity of 1×10^{19} Pa s, which is a reasonable value for natural evaporitic layers in the Northern Apennines (e.g., Bonini & Sani 2002). The velocity scaling factor implies that deformation occurs at natural rates of

Table 1. *Parameters of model deformation*

Model	Compression velocity (mm/hr)	Bulk compression (%)	Obliquity angle α	Extension velocity (mm/hr)	Bulk extension (%)
COM-EXT-01	40	23	0°	10	10
COM-EXT-02	40	25	15°	13	10
COM-EXT-03	40	25	30°	13	10
COM-EXT-04	40	25	45°	13	10
COM-EXT-05	40	25	60°	13	10

Table 2. *Scaling parameters for brittle and ductile deformation*

Parameter	Model	Nature	Model/Nature
Length, l (m)	0.01	1000	1×10^{-5}
Density brittle layer (kg m^{-3})	1500	2750	0.54
Gravity, g (m s^{-2})	9.81	9.81	1
Coefficient of friction, μ	0.8	*c.* 0.6–0.8	*c.* 1
Stress, σ (Pa)			5.4×10^{-6}
Cohesion, c (Pa)	*c.* 50	1×10^{7}	5.4×10^{-6}
Density ductile layer (kg m^{-3})	1100	*c.* 2400	*c.* 0.5
Viscosity, ν (Pa s)	16000	1×10^{19}	1.6×10^{-15}
Strain rate 1st phase, ε (s^{-1})	2.2×10^{-3}	6.5×10^{-13}	3.4×10^{9}
Strain rate 2nd phase, ε (s^{-1})	3.4×10^{-4}	10^{-13}	3.4×10^{9}
Time 1st phase, t (s)	9×10^{3}	3.1×10^{13}*	2.9×10^{-10}
Time 2nd phase, t (s)	2.3×10^{4}	8×10^{13}†	2.9×10^{-10}
Displacement velocity 1st phase, v (m s^{-1})	1.1×10^{-5}	3.2×10^{-9}‡	3.4×10^{4}
Displacement velocity 2nd phase, v (m s^{-1})	1.7×10^{-6}	5×10^{-11}§	3.4×10^{4}

*corresponding to 1 Ma
†corresponding to 2.5 Ma
‡corresponding to 10 mm/a
§corresponding to 1.5 mm/a

c. 10 mm a^{-1} and *c.* 1.5 mm a^{-1} during the shortening and extensional phase, respectively.

Modelling results

Orthogonal shortening during the first deformation phase resulted in regularly-spaced fore- and back-thrusts parallel to the mobile wall, defining well-developed anticlinal structures (Fig. 3). During shortening, the silicone flowed into the core of these anticlines, giving rise to a bulge in the ductile layer. For low amount of shortening, thrusts localized close to the moving wall, where the most prominent anticlines were observed. Subsequently, thrust structures propagated forward, affecting almost the whole model (Fig. 3). Boundary effects related to the movement of the L-shaped mobile wall and to the lateral confinement of the model resulted, in some cases (e.g., model COM-EXT 01; Fig. 3b), in a slight S-shaped surface strike of structures. However, these effects were restricted to a narrow area (generally <5 cm) and did not alter the geometry of shortening-related structures in the central part of the models.

The dip of thrusts was rather constant in the different experiments, being typically in the range of 32°–35°. Dip of thrusts lower than 25° were occasionally observed and were probably linked to the rotation of secondary structures during progressive displacement along major thrusts (e.g., Krantz 1991).

During the extensional phase, deformation was achieved through activation of the basal VD (Fig. 2). Extension was accommodated by a system of normal faults that nucleated at low bulk extension (*c.* 5–6%; Fig. 3c). Their orientation, arrangement and kinematics, as well as their relation with the early thrusts, was strongly controlled by the angle of obliquity, α (see below).

Orthogonal extension ($\alpha = 0°$)

For orthogonal extension ($\alpha = 0°$; model COM-EXT 01), normal faults developed sub-parallel to the main thrusts and were distributed over a wide area in the left side of the model and mainly close to the moving wall (Fig. 4a). These faults were characterized by a pure dip-slip movement and mainly formed parallel to and within anticlines defined by fore- and back-thrusts. In particular, two major normal faults nucleated within the most prominent anticline, close to the moving wall. Model cross-sections show that normal faults formed above the bulge of the silicone layer, between conjugate couples of fore- and back-thrusts. The main extensional faults connected to and reactivated early back-thrusts at depth (Fig. 4a). The diagram of fault orientation (Fig. 4a) shows a main peak of the normal fault orientation coinciding with the early thrust

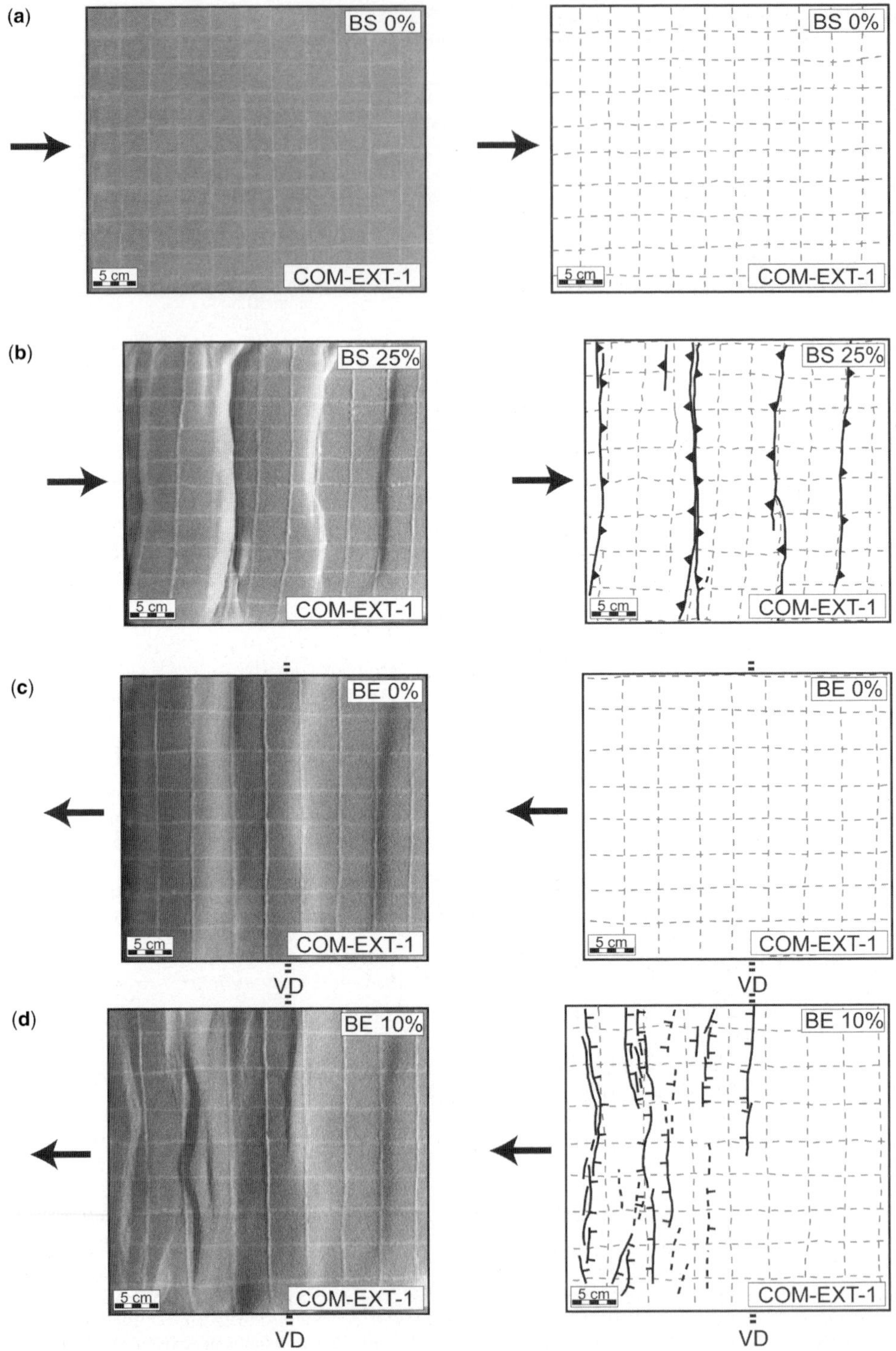

Fig. 3. Deformation of model COM-EXT-01 illustrated as top-view photos (left panels) and line drawings of structures (right panels). Only the portion of the model above the silicone layer is shown. The successive deformation stages are: (**a**) Onset of shortening, (**b**) end of shortening, (**c**) onset of extension, (**d**) end of extension. Note that in (c) the shortened model was covered by a thin layer of coloured sand to allow better observation of extension-related fault patterns.

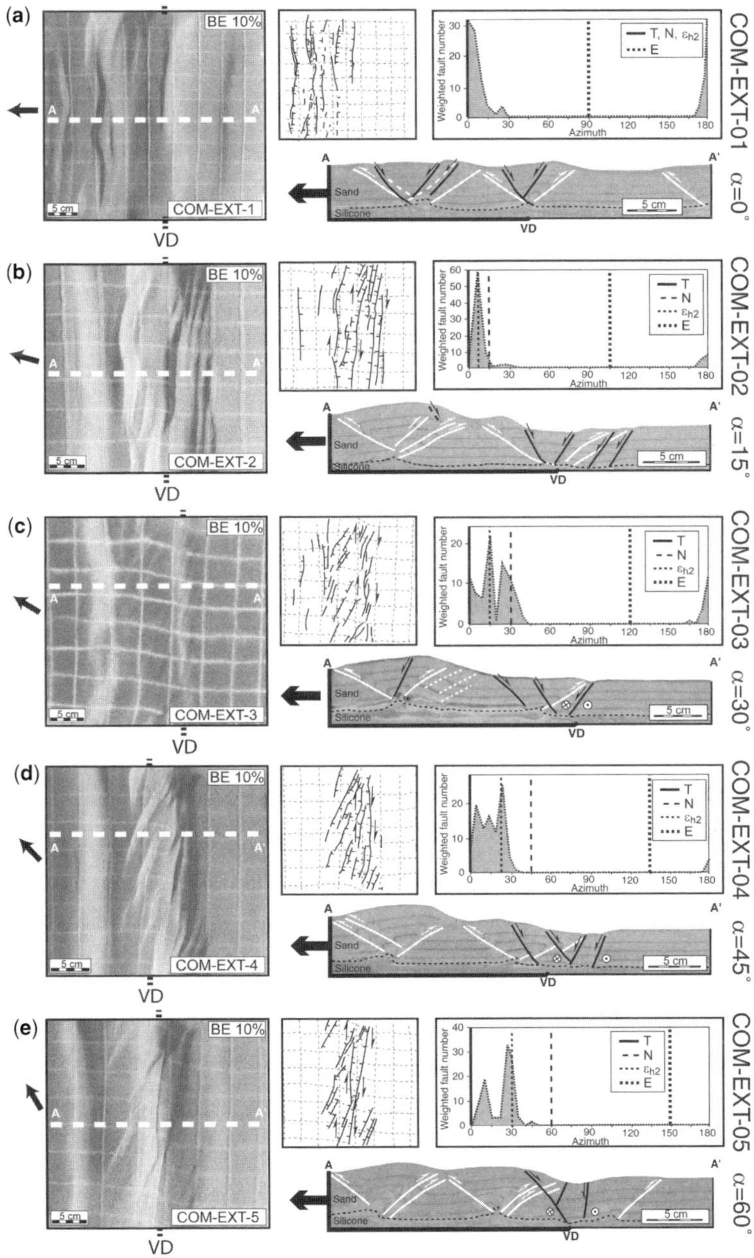

Fig. 4. Results of the experimental series illustrated as top-view photos (left), line drawings of surface deformation (top centre), graphs of fault orientation (top right) and longitudinal cross-sections (bottom centre-right). In the top-view photo, the light is from the right; the trace of the cross-section is indicated with the dashed white line on the model top-view. In the graphs of fault orientation, the azimuth of faults represents the angle (measured clockwise) that extensional structures form with respect to the short side of the mobile wall of the apparatus (i.e., with respect to the VD); the fault number is calculated as explained in Fig. 1. E is the extension direction; N is the perpendicular to the extension direction; T represents the thrust trend, which is parallel to the VD. ε_{h2} indicates the direction of the secondary horizontal principal strain which is orthogonal to the direction of the maximum horizontal extensional strain, ε_{hmax}. In orthogonal extension, since no shear is involved in deformation, ε_{hmax} and the extension direction coincide and consequently N and ε_{h2} are parallel, their azimuths being 0°. During oblique rifting, extension and shear contribute to deformation. Analytical models (Withjack & Jamison 1986) have shown that the angle between ε_{hmax} and the extension direction increases with increasing obliquity angles α and consequently the angle between ε_{h2} and N increases from model COM-EXT 02 to model COM-EXT 05.

trend, indicating that most extensional structures formed parallel to shortening-related structures and perpendicular to the extension direction.

Low obliquity extension ($\alpha = 15°$, $30°$)

Models of low obliquity angles (COM-EXT 02, $\alpha = 15°$; COM-EXT 03, $\alpha = 30°$; Figs. 4b, c) were characterized by more complex patterns of extensional structures, with the majority of normal faults oblique to the early structures and arranged en-echelon. Extensional faults sub-parallel to the thrusts were less common, but still observable in the portion of the model close to the mobile wall (Fig. 4b). Thrust-parallel faults were characterized by almost pure dip-slip kinematics, whereas oblique faults displayed a marked strike-slip component, as inferred from the displacement of the passive grid marker drawn on top of the models (Figs. 4b, c).

In cross-sections, normal faults either cross-cut or reactivate thrusts at depth (Figs. 4b, c). In model COM-EXT 02 ($\alpha = 15°$), the diagram of fault orientation shows a single main peak forming an angle of *c.* 7° to the thrust trend and *c.* 8° to the perpendicular to the direction of extension (Fig. 4b). This peak coincides with the direction of the secondary horizontal strain (ε_{h2}), which is orthogonal to the maximum horizontal extensional strain (ε_{hmax}). Note that, since oblique stretching combines both extension and horizontal shear, ε_{hmax} does not coincide with the extension direction but forms an angle with this direction which is controlled by the obliquity angle α (Withjack & Jamison 1986); analogously, ε_{h2} does not coincide with the normal extension direction (see Fig. 4).

Fault orientation of model COM-EXT 03 ($\alpha = 30°$) is characterized by three main peaks (Fig. 4c). The main peak corresponds to faults trending at *c.* 15° to the early thrusts and to the normal extension direction. Again, this main peak coincides with the orthogonal to the maximum horizontal extensional strain ε_{h2}. A secondary maximum relates to faults forming at an angle of *c.* 25° to the early thrusts, whereas a third peak correlates to thrust-parallel normal faults (Fig. 4c).

High obliquity extension ($\alpha = 45°$, $60°$)

In models COM-EXT-4 ($\alpha = 45°$) and COM-EXT-5 ($\alpha = 60°$) the highly oblique extension was accommodated by the development of oblique faults with a prominent (locally dominant) strike-slip component (Fig. 4d, e). Fault nucleation near the central VD and the area affected by deformation was narrower with respect to previous low-obliquity and orthogonal extension models (see section below). Analysis of model cross-sections reveals that extensional faults displaced the early thrusts and that no reactivation occurred (Fig. 4d, e).

Graphs of fault orientation are characterized by three and two main peaks for model COM-EXT-4 ($\alpha = 45°$) and model COM-EXT-5 ($\alpha = 60°$), respectively. In model COM-EXT-4, the highest peak corresponds to faults developing at angles of *c.* 25° to the early thrust faults and at *c.* 20° to the normal extension direction (Fig. 4e). In model COM-EXT-5, oblique-slip faults develop at *c.* 30° to the pre-existing thrusts and to the normal extension direction (Fig. 4e). Both peaks roughly coincide with the direction orthogonal to the maximum horizontal extensional strain (ε_{h2}).

Parameters of deformation as a function of the obliquity angle α

Figure 5 summarizes the main parameters of model deformation in relation to the varying obliquity angle α. Comparison of the different models suggests that an increase in the angle of obliquity leads to a decrease in width of the deformed zone (Fig. 5a) due to an increase in the dip of the main faults (Fig. 5b) that have a more prominent strike-slip component for high obliquity angles ($\alpha = 45°$, $60°$).

Fault density shows an almost linear decrease with increasing α (Fig. 5c), suggesting that for high-obliquity extension, deformation is accommodated by a few well-developed oblique-slip faults. This is exemplified by model COM-EXT-5, where deformation is mostly accommodated by three main oblique-slip faults extending for almost the entire width of the model (Fig. 4e).

Finally, fault length is characterized by high values for models with $\alpha = 0°$ and 15°; values then decrease, reaching a minimum for models with $\alpha = 30°$ and 45°; then a marked increase is observed for $\alpha = 60°$ (Fig. 5d). This suggests that for orthogonal or low obliquity ($\alpha = 15°$) extension, deformation is taken up by long faults, probably controlled by the pre-existing thrust faults. An increase to intermediate α values leads to patterns of faulting characterized by short, oblique faults. A further increase in α leads to the development of long, oblique slip faults with an important strike-slip component.

Discussion: modes of interaction between pre-existing thrusts and successive normal faults

Our experiments investigated the development of extensional structures following a previous

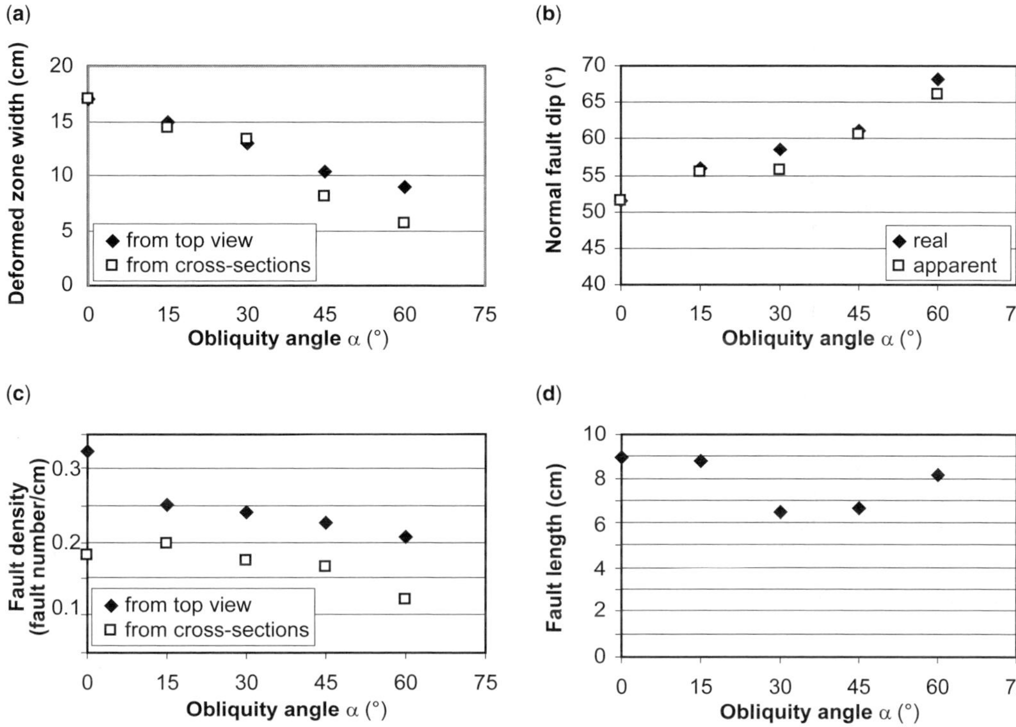

Fig. 5. Quantified parameters as a function of the angle of obliquity (α). All parameters were calculated along three 25 cm-long regularly-spaced reference lines on the model surface and three cross-sections in the central part of the models. (**a**) Deformed zone width. (**b**) Mean dip of normal faults; note that since extensional structures may be oblique with respect to the model cross-section, both real and apparent dip have been reported. (**c**) Density of normal faults (calculated as fault number divided by the length of the reference line or cross-section). (**d**) Tip-to-tip fault length. The discrepancy in the measurements from model top-views or cross-sections is related to the presence of extensional structures characterized by low vertical displacement, which are difficult to recognize in model cross-sections.

contractional phase of deformation, and simulating the reactivation of a deep-seated pre-existing discrete fabric (e.g., a basement thrust simulated by the velocity discontinuity, VD) that is located below an evaporite layer at the base of a brittle sedimentary cover. Extensional reactivation of the discrete anisotropy is transferred through the viscous layer to the overlying brittle sandpack, which deforms by distributed faulting. The kinematics of the pre-existing fabric and in particular the obliquity of extension, which controls the ratio between the extensional and the strike-slip components of motion (e.g., Withjack & Jamison 1986), determine differences in the architecture of extensional structures, and their relations with previous thrusts. Three different main modes of interactions between pre-existing thrusts and later extensional faults are identified (Fig. 6).

(1) For $\alpha = 0^\circ$, early shortening-related structures strongly influence the development of successive normal faults, giving rise to graben structures that nucleate within and dismember previous thrust anticlines (Fig. 6a). Normal faults are parallel to the early thrusts, are long and linear and display a pure dip-slip component. In this case, the master fault of the grabens joins the thrusts at depth, locally inverting the movement on these faults and thus reactivating their deepest portions. The location and geometry of the main normal faults is primarily influenced by both the presence of the silicone bulge between fore- and back-thrusts and the local stress-field influence exerted by the topography related to the presence of the anticlines (e.g., Gartrell 2001). However, reactivation of the lower portion of early thrusts by normal faults suggests that the drop in shear strength due to sand dilatancy along thrusts contributes (although it may not play a determining role) to the localization of subsequent extensional faults, as suggested by the experimental results

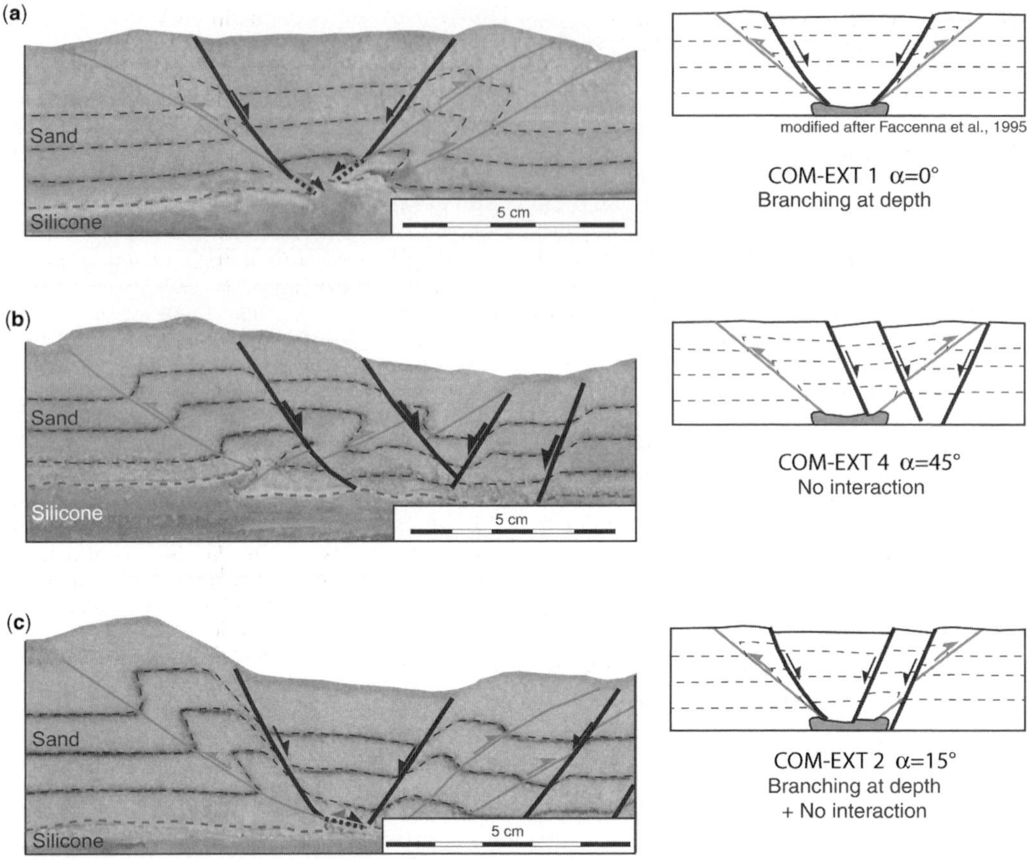

Fig. 6. Modes of interaction between normal faults and pre-existing thrust systems in the different experiments. A schematic representation of fault interaction is also shown. See text for details.

of Faccenna *et al.* (1995). The normal fault geometry is similar to the linked faults of Krantz (1991) and the mode of interaction between thrusts and later faults is equivalent to the *branching at depth mode* obtained by Faccenna *et al.* (1995) for similar initial thrust dips (32°–35°).

(2) For high obliquity angles ($\alpha \geq 45°$), no influence on later extension-related faults is exerted by compressive structures (Fig. 6b), in a similar fashion to the *no interaction mode* of Faccenna *et al.* (1995). Deformation is localized along a few long and steeply-dipping, oblique-slip normal faults that displace thrust faults and nucleate in a narrow zone above the VD. The oblique-slip faults have a large strike-slip component and show no relation with previous anticlinal structures (unlinked faults of Krantz 1991). Although the *no interaction mode* was suggested by Faccenna *et al.* (1995) to characterize models with initial thrust dips $<32° \pm 1°$, in the current models this behaviour is also obtained for slightly higher initial thrust dips in the range of 32°–35°.

(3) For intermediate obliquity angles ($\alpha = 15°$, 30°), an *intermediate mode of interaction* characterizes the experiments. The two beforementioned behaviours (*no interaction + branching at depth*) coexist in different regions of the model (Fig. 6c). Deformation is taken up by short, oblique-slip normal faults, often arranged en-echelon and characterized by an increasing component of strike-slip motion. The majority of faults are oblique to both the thrust (and VD) trend and orthogonal to the direction of extension. Most faults are indeed perpendicular to the maximum extensional horizontal strain (Withjack & Jamison 1986). Increasing the angle of obliquity in the models leads to: (i) the development of faults that are increasingly oblique with respect to the early thrusts, (ii) an almost linear decrease in the width of the

deformed zone, (iii) an increase in fault dip, and (iv) a decrease in fault density. Similar fault patterns were observed in previous sand-box models of oblique (Tron & Brun 1991; Corti *et al.* 2001) and polyphase rifting (Bonini *et al.* 1997; Keep & McClay 1997; Corti *et al.* 2003).

Our model results suggest that the mode of interaction between pre-existing thrusts and newly-formed normal faults is not only influenced by the initial dip of thrusts but also by the kinematics of the extensional phase, which in turn controls fault dip, length and kinematics of later faults. Reactivation of faults along their entire length (as seen in cross-section) was never observed in the models. This supports theoretical predictions from Mohr envelopes for reactivation of pre-existing faults in sand as well as observations by Krantz (1991) and Faccenna *et al.* (1995) that an initial thrust dip higher than *c.* 40° is required in order to have complete fault reactivation.

Relevance of model results for evaluating the regional extension direction in nature: the case of the Northern Apennines

Analogue modelling results have been used in the past to infer the regional extension direction in natural examples (e.g., Withjack & Jamison 1986; Tron & Brun 1991; Dauteuil & Brun 1993; Harris *et al.* 1994; Bonini *et al.* 1997; Clifton *et al.* 2000) mainly on the basis of a statistical analysis of the surface fault pattern. In order to determine the regional extension direction, fault orientations for different angles of obliquity in the models are compared to faults in nature using the thrust trend as a reference direction. The extension direction in nature can be obtained by best fitting the main peaks in the fault orientation in nature with the statistical data from the models.

In the arc-shaped Northern Appenines fold-and-thrust belt extensional faults are superimposed on early thrusts. In different sectors of this chain, the younger normal faults display strong variations in terms of strike, arrangement and kinematics (Figs 1 and 7). These different fault patterns are compared to our analogue models in order to obtain a coherent regional extension direction.

The best fit of graphs of fault orientation suggests that the fault arrangement in the Garfagnana-Lunigiana area (in the northwestern part of the chain) is comparable (though more complex) to low obliquity models (specifically, $\alpha = 15°$). Given the local orientation of the pre-existing thrusts in the area (*c.* N135°), the correspondence in the main peak in fault orientation suggests a N60°-directed local extension (Fig. 7). Eastwards, the main peak in the graph of faults in the Mugello basin (in the central part of the chain; Figs 1 and 7) corresponds to the one characterizing the orthogonal extension model ($\alpha = 0°$), indicating a *c.* N30° direction of extension. Finally, the orientation of faults in the Martani Mounts-Terni Basin area is in good agreement with a high obliquity model ($\alpha = 60°$), indicating a *c.* N30° direction of extension.

Overall, extrapolation of model results to a regional kinematic scenario suggests a compatibility with a roughly NNE to NE-directed extension, consistent with analysis of Quaternary and active deformation in the area (e.g., Piccardi *et al.* 1997, 1999; Bonini *et al.* 2003; Jiménez-Munt *et al.* 2003; Mariucci & Müller 2003). Variations in the extrapolated regional direction of extension in the Garfagnana-Lunigiana area may be related to the influence of rheological anisotropies (as the Apuan Alps metamorphic core complex) that may have locally caused a stress re-orientation, thus contributing to modifications of fault orientations. Both in models and nature normal faults increase their obliquity with early thrusts and their strike-slip component increases with higher angles of obliquity (i.e., passing from the Mugello basin, $\alpha = 0°$, to the Garfagnana-Lunigiana area, $\alpha = 15°$ and to the Martani Mounts-Terni Basin area, $\alpha = 60°$). Fault-slip data collected along the major fault systems are in good agreement with the proposed kinematic scenario, indicating an almost pure dip-slip kinematics in Mugello (Delle Donne *et al.* 2003), a minor dextral component of oblique-slip in Garfagnana-Lunigiana (Bernini & Lasagna 1988; Bernini *et al.* 1991; Piccardi *et al.* 1999) and an important sinistral component of displacement along the Martana Fault System (Bonini *et al.* 2003; Fig. 1).

Conclusions

Laboratory analogue models were used to investigate the influence of pre-existing thrust systems on the development of later normal faults. Experiments consisted of a first phase of shortening, followed by extension at variable angles of obliquity (α). Model results suggest that the trend, arrangement and kinematics of faults are strongly influenced by the obliquity of extension, which also has a major effect on the interaction between normal faults and early thrusts. Three main modes of interactions were

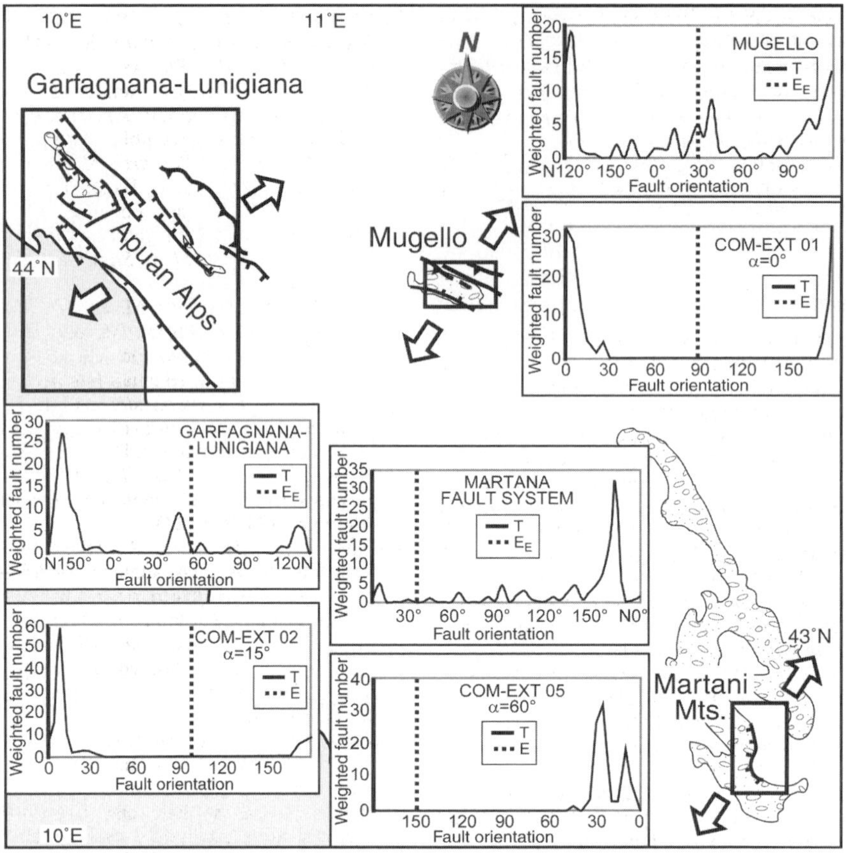

Fig. 7. Extrapolation of model results to nature. White arrows in the selected domains of the Northern Appennines indicate the extension direction as obtained by comparison between the graphs of fault orientation (insets). For this comparison, the reference direction is represented by the direction of early thrusts. Histograms of fault orientation in nature have been shifted in order to make the thrust trend coincide with the one in the model. For the Martana Fault System, the experimental histogram has been reflected horizontally in order to fit the dextral model kinematics with the sinistral deformation inferred in nature. Extension direction in nature was then obtained by best fitting the main peaks in the fault orientation with those of the different experiments. Note that minor peaks at N40°–45° in nature, not present in the graphs of models, correspond to *c.* NE–SW transverse structures reactivating pre-existing discontinuities (e.g., Boccaletti *et al.* 1977; Bemporad *et al.* 1986). E: extension direction; E_E: extension direction extrapolated from models to nature; T: thrust trend. Lithological symbols as in Fig. 1.

identified, depending on the kinematics of extension.

(1) For orthogonal extension ($\alpha = 0°$) normal faults nucleate within previous thrust anticlines reactivating thrust faults at depth (*branching at depth mode of interaction*).
(2) For high obliquity extension ($\alpha \geq 45°$), no relations between extension-related faults and early compressive structures is observed and extension-related steeply-dipping normal faults (characterized by an important and locally dominant strike-slip component) displace thrust faults (*no interaction mode*).
(3) For intermediate obliquity angles ($\alpha = 15°$, $30°$), an *intermediate mode of interaction* characterizes the experiments, as the two behaviours described above (*no interaction + branching at depth*) coexists in different regions of the models.

A comparison of the fault orientations in our analogue models with natural examples of thrust belts that have undergone later extension may yield insights on the regional extension direction, as shown for the Northern Apennines.

We thank reviewers A. Benedicto and L. Harris and editor G. Schreurs for the detailed, constructive comments which helped to improve the manuscript. C. Del Ventisette and D. Montanari are warmly thanked for their support during the experimental work. D. Delle Donne is thanked for providing us data concerning the Mugello basin. We also express our gratitude to M. H. Dickson for improving the English text. Dr. Baldi of the Colorobbia Italia S.p.A. is thanked for providing us the quartz sand used in the experiments.

References

BEMPORAD, S., CONEDERA, C., DAINELLI, P., ERCOLI, A. & FACIBENI, P. 1986. Landsat imagery: a valuable tool for regional and structural geology. *Memorie della Società Geologica Italiana*, **31**, 287–298.

BERNINI, M. & LASAGNA, S. 1988. Rilevamento geologico e analisi strutturale del Bacino dell'Alta Val di Magra tra M. Orsaro e Pontremoli (Appennino Settentrionale). *Atti della Società Toscana di Scienze Naturali, Serie A*, **95**, 139–183.

BERNINI, M., PAPANI, G., DALL'ASTA, M., LASAGNA, S. & HEIDA, P. 1991. The Upper Magra Valley extensional basin: a cross section between Orsaro Mt. and Zeri (Massa province). *Bollettino della Società Geologica Italiana*, **110**, 451–458.

BOCCALETTI, M., COLI, M. & NAPOLEONE, G. 1977. Nuovi allineamenti strutturali da immagini landsat e rapporti con l'attività sismica negli Appennini. *Bollettino della Società Geologica Italiana*, **96**, 679–694.

BONINI, M. & SANI, F. 2002. Extension and compression in the Northern Appennines (Italy) hinterland: evidence from the Late Miocene-Pliocene Siena-Radicofani Basin and relations with basement structures. *Tectonics*, **21(3)**, 10.1029/2001TC900024.

BONINI, M., SOURIOT, T., BOCCALETTI, M. & BRUN, J. P. 1997. Successive orthogonal and oblique extension episodes in a rift zone: laboratory experiments with application to the Ethiopian Rift. *Tectonics*, **16**, 347–362.

BONINI, M., TANINI, C., MORATTI, G., PICCARDI, L. & SANI, F. 2003. Geological and archaeological evidence of active faulting on the Martana Fault (Umbria-Marche Apennines, Italy) and its geodynamic implications. *Journal of Quaternary Science*, **18**, 695–708.

CLIFTON, A. E., SCHLISCHE, R. W., WITHJACK, M. O. & ACKERMANN, R. V. 2000. Influence of rift obliquity on fault-population systematics: results of experimental clay models. *Journal of Structural Geology*, **22**, 1491–1509.

CORTI, G., BONINI, M., INNOCENTI, F., MANETTI, P. & MULUGETA, G. 2001. Centrifuge models simulating magma emplacement during oblique rifting. *Journal of Geodynamics*, **31**, 557–576.

CORTI, G., BONINI, M., CONTICELLI, S., INNOCENTI, F., MANETTI, P. & SOKOUTIS, D. 2003. Analogue modelling of continental extension: a review focused on the relations between the patterns of deformation and the presence of magma. *Earth-Science Reviews*, **63**, 169–247.

DAUTEUIL, O. & BRUN, J. P. 1993. Oblique rifting in a low spreading ridge. *Nature*, **361**, 145–148.

DELLE DONNE, D., PICCARDI, L., SANI, F. & VANNUCCI, G. 2003. Active tectonics of the Mugello Basin (Northern Apennines, Italy). EGS-AGU-EUG Joint Assembly, Nice, France, April 2003, *Geophysical Research Abstracts*, **5**, 06229.

FACCENNA, C., NALPAS, T., BRUN, J. P., DAVY, P. & BOSI, V. 1995. The influence of pre-existing faults on normal fault geometry in nature and in experiments. *Journal of Structural Geology*, **17**, 1139–1149.

FINETTI, I. R., BOCCALETTI, M., BONINI, M., DEL BEN, A., GELETTI, R., PIPAN, M. & SANI, F. 2001. Crustal section based on CROP seismic data across the North Tyrrhenian-Northern Apennines-Adriatic Sea. *Tectonophysics*, **343**, 135–163.

GARTRELL, A. P. 2001.Crustal rheology and its effect on rift basin styles. *In*: KOYI, H. A. & MANCKTELOW, N. S. (eds) *Tectonic Modeling: A Volume in Honor of Hans Ramberg*. Geological Society of America, Memoirs, **193**, 221–233.

HARRIS, L. B., HIGGINS, R., DENTITH, M. C. & MIDDLETON, M. 1994. Analogue modelling of transtensional faulting applied to the structure of the Perth Basin, Western Australia. *In*: PURCELL, P. G. & PURCELL R. R. (eds) *The sedimentary basins of Western Australia*. Proceedings of the Petroleum Exploration Society of Australia Symposium, Perth, 801–810.

JIMÉNEZ-MUNT, I., SABADINI, R. & GARDI, A. 2003. Active deformation in the Mediterranean from Gibraltar to Anatolia inferred from numerical modeling and geodetic and seismological data. *Journal of Geophysical Research*, **108 (B1)**, 2006, DOI 10.1029/2001JB001544.

KEEP, M. & MCKLAY, K. 1997. Analogue modelling of multiphase rift systems. *Tectonophysics*, **273**, 239–270.

KRANTZ, R. W. 1991. Normal fault geometry and fault reactivation in tectonic inversion experiments. *In*: YIELDING, A. M. & FREEMAN, B. (eds) *The Geometry of Normal Faults*. Geological Society, London, Special Publication, **56**, 219–229.

MARIUCCI, M. T. & MÜLLER, B. 2003. The tectonic regime in Italy inferred from borehole breakout data. *Tectonophysics*, **361**, 21–35.

MORLEY, C. K. 1999. How successful are analogue models in addressing the influence of pre-existing fabrics on rift structures? *Journal of Structural Geology*, **21**, 1267–1274.

PICCARDI, L., CORTI, G. & BOCCALETTI, M. 1999. Oblique extension in the Tyrrhenian side of the Northern Apennines (Italy). Proceedings of the XXIV European Geophysical Society General Assembly, The Hague, 19–23 April 1999, *Geophysical Research Abstracts*, **1**, 82.

PICCARDI, L., SANI, F., BONINI, M., BOCCALETTI, M., MORATTI, G. & GUALTIEROTTI, A. 1997. Deformazioni quaternarie nell'Appennino centro-settentrionale: evidenze ed implicazioni. *Il Quaternario*, **10**, 273–280.

RAMBERG, H. 1981. *Gravity, deformation and the Earth's crust*. Academic Press, London.

SIBSON, R. H. 1985. A note on fault reactivation. *Journal of Structural Geology*, **7**, 751–754.

TRON, V. & BRUN, J.-P. 1991. Experiments on oblique rifting in brittle-ductile systems. *Tectonophysics*, **188**, 71–84.

WEIJERMARS, R. & SCHMELING, H. 1986. Scaling of Newtonian and non-Newtonian fluid dynamics without inertia for quantitative modelling of rock flow due to gravity (including the concept of rheological similarity). *Physics of the Earth and Planetaty Interiors*, **43**, 316–330.

WITHJACK, M. O. & JAMISON, W. R. 1986. Deformation produced by oblique rifting. *Tectonophysics*, **126**, 99–124.

Surface topography and internal strain variation in wide hot orogens from three-dimensional analogue and two-dimensional numerical vice models

ALEXANDER R. CRUDEN, MOHAMMAD H. B. NASSERI & RUSSELL PYSKLYWEC

Department of Geology, University of Toronto, 22 Russell Street, Toronto, Ontario M5S 3B1, Canada

Abstract: The post-accretionary deformation of wide, hot orogens is characterized by pure-shear or transpressional shortening of relatively weak lithosphere (the orogen) between converging stronger blocks (the vice). We report on a series of analogue vice models and compare the resulting three-dimensional strain fields and surface topographies to equivalent two-dimensional numerical experiments. In the analogue models a rheologically stratified (frictional/viscous) weak orogenic lithosphere overlying a viscous asthenosphere is squeezed between converging strong lithospheric blocks. Ductile lower crust and mantle in the weak lithosphere is free to flow laterally, parallel to the orogen. The Argand number describes the model dynamics and strongly controls both the orogenic relief and the degree of lower crustal orogen parallel stretching in the analogue models. Cross sections of numerical and analogue experiments display consistent geometries in which upper crustal deformation is characterized by upright folding compared to apparently decoupled horizontal strains in the lower crust. The relative buoyancy and degree of orogen parallel flow in the lower crust of the analogue models has a dramatic influence on three-dimensional strain fields and the kinematics of upper crustal curvilinear shear zones. The analogue and numerical results demonstrate the importance of three-dimensional effects in determining the structure of natural orogens and compare favourably to field and geophysical observations of large hot orogens in the geological record.

Three-dimensional strain partitioning has long been recognized as a key phenomenon in the kinematic and dynamic evolution of convergent orogens (e.g., Oldow *et al.* 1989; Vauchez & Nicolas 1991; Royden 1996). A classic example is the partitioning between orogen-parallel strike-slip fault systems and ductile thickening in the Alpine-Himalayan chain (e.g., Tapponnier *et al.* 1982; Dewey *et al.* 1986; Davy & Cobbold 1988). Recent studies have documented decoupling between upper crustal and mid- to lower-crustal deformation in convergent magmatic arc, collisional orogen and post-orogenic collapse settings (e.g., Axen *et al.* 1998; Garde *et al.* 2002; Klepeis *et al.* 2004). A common expression of such decoupling is the observation of vertical orogen sub-parallel transpression zones, upright folding and cleavage development in the upper crust versus subhorizontal fabric development and lateral flow in the mid to lower crust, as documented by both field and seismic reflection studies in Archean to Mesozoic orogens (e.g., Collins & Vernon 1991; Gapais *et al.* 1992; Jackson & Cruden 1995; Hajnal *et al.* 1996; Moser *et al.* 1996; Royden *et al.* 1997; Axen *et al.* 1998).

The post-accretionary deformation of such wide, hot orogens (e.g., Superior, Grenville, Variscan, Cordilleran, Tibetan) has been attributed to pure-shear or transpressional shortening of relatively weak lithosphere (the orogen) between converging stronger blocks (the vice). In such 'vice models', the orogenic crust is weakened due to the properties of inherited accreted material or thermal/magmatic softening, and deformation is effectively decoupled from mantle dynamics (Ross *et al.* 1995; Ellis *et al.* 1998; Burg *et al.* 2002). The structural and topographic development of the resulting orogen depends on the vertical and lateral rheological structure of the lithosphere, the initial and final widths of the weak zone, surface erosion, and three-dimensional effects (e.g., lateral extrusion; Dewey *et al.* 1986; Ratschbacher *et al.* 1991; Mancktelow & Pavlis 1994; Royden 1996; Seyferth & Henk 2004).

The purpose of this contribution is to investigate the roles of orogen parallel flow and rheological stratification on the vertical and lateral

From: BUITER, S. J. H. & SCHREURS, G. (eds) 2006. *Analogue and Numerical Modelling of Crustal-Scale Processes.* Geological Society, London, Special Publications, **253**, 79–104.
0305-8719/06/$15.00

development of structures in convergent settings. We describe a series of three-dimensional (3D) analogue model experiments and corresponding two-dimensional (2D) numerical models that investigate the structural and topographic response of a model orogenic lithosphere as it is compressed between converging blocks of significantly stronger lithosphere. The analogue experiments inherently permit flow of orogenic material orthogonal to the compression direction. Comparison of these experiments with 2D numerical models, together with detailed structural and topographic data from the model surface and serial sectioning allows us to evaluate the significance of such orogen parallel flow. We will highlight the influence of relative variations in the strength and density of the ductile crust, which dictate the degree of coupling within the crust and body forces, respectively.

Experimental design

The basic design of both our analogue and numerical experiments follows the set up of the original vice models of Ellis *et al.* (1998) and is also similar to recent 3D numerical models of Seyferth and Henk (2004). Due to restrictions imposed by laboratory modelling techniques, our experiments are isothermal, which contrasts with those of Ellis *et al.* (1998) and Seyferth and Henk (2004). However, our focus is to explore the first-order effects of syn-convergent, orogen parallel flow rather than time-dependent phenomena due to variations in thermal and rheological structure. Hence the isothermal nature of our experiments is justified, and the results presented below are complementary to published 2D and 3D numerical thermomechanical models.

The basic configuration of our analogue and numerical experiments is summarized in Figure 1. In the 3D analogue models, a relatively weak model orogenic lithosphere composed of brittle material (upper crust) overlying viscous layers representing ductile lower crust and mantle lithosphere is squeezed between blocks (the vice) of model lithosphere of significantly greater total strength (Fig. 2). Both the vice and orogen float isostatically on a weaker model asthenosphere. Convergence is achieved by the motion of two pistons, which move inward at the same constant rate, driven by a step motor and screw gear assembly (Fig. 3). The vice-piston and vice-orogen boundaries are no-slip, as is the contact between asthenospheric material and the tank walls. The upper boundary in both the analogue and numerical models is a free surface.

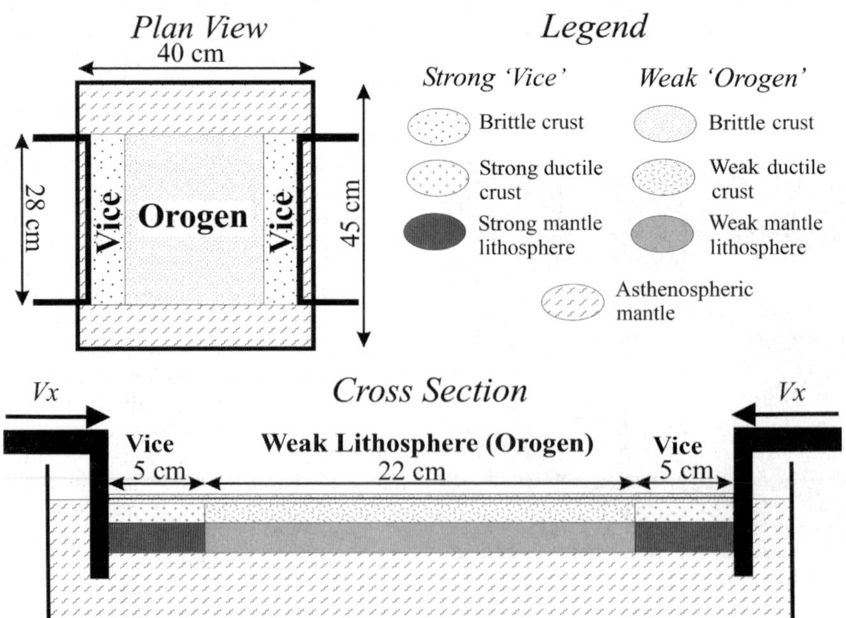

Fig. 1. Vice model set up. Top left is a plan view and bottom is a cross-sectional view of how each analogue experiment is constructed. With the exception of the rigid pistons (black) used to impose orogenic convergence in the analogue models, the cross-section also summarizes the set-up for the numerical experiments.

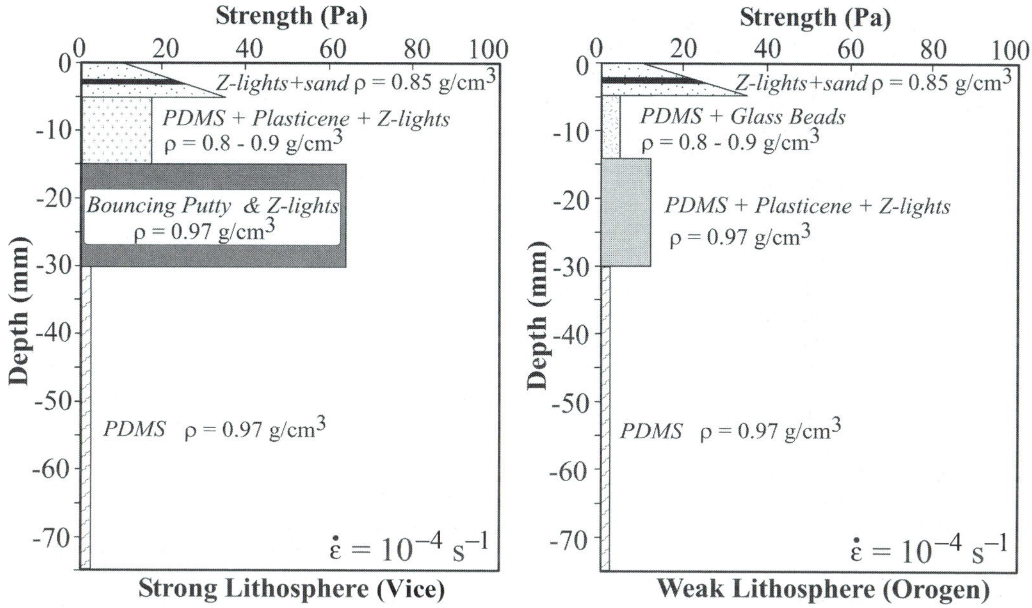

Fig. 2. Representative strength envelopes for the strong lithospheric blocks (vice) and weak orogenic lithosphere employed in the analogue experiments. See Table 1 for a complete summary of the physical properties of materials employed in the experiments.

Orogen parallel flow in the analogue experiments is achieved by constructing the model orogen such that it is unconfined at both ends (Figs 1 and 3). As such, the model orogen is free to flow outwards into a reservoir that is initially filled only with asthenospheric material. A similar approach has been used previously to model radially spreading nappes (Gilbert & Merle 1987; Merle 1989), and indentor (Davy & Cobbold 1988) and escape tectonics (Ratschbacher *et al.* 1991). In nature such unconfined boundaries might be represented by subduction zones, which retreat away from the orogen due to slab roll back (e.g., Schellart & Lister 2004). Lateral flow of model asthenosphere material in our simulated tectonic vice is illustrated in Figure 4. Deformation of a passive marker grid indicates that bulk orogen normal contractional and orogen parallel extensional strain is effectively homogeneous in the area of the models underlain by orogenic lithosphere. More complex strain fields develop adjacent to the lateral walls of the tank and at the corners of the pistons, outside the area of experimental interest. Because asthenospheric material is free to move around and under the pistons (Fig. 4), there is no change of volume or depth of this layer throughout the tank during the experiment.

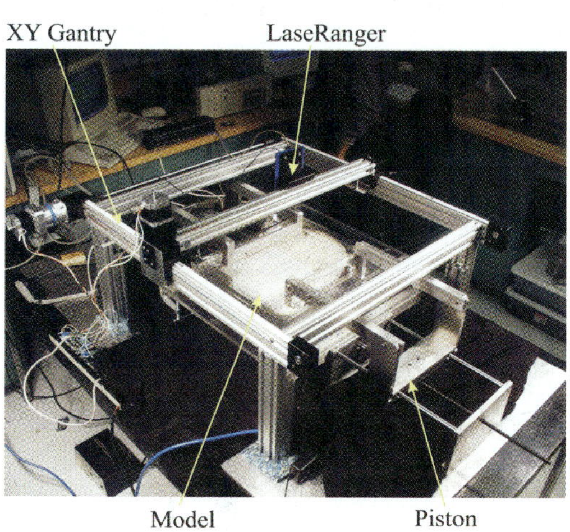

Fig. 3. Oblique view of the experimental apparatus and laser topography scanner. Both the piston and XY-gantry are propelled by computer-controlled stepper motors. Person and control computer in background for scale.

The 2D numerical models are set up with identical cross-sectional geometry and material properties to those of the analogue experiments. Our approach to modelling takes advantage of the relative strengths of laboratory and numerical approaches (e.g., Pysklywec & Cruden 2004).

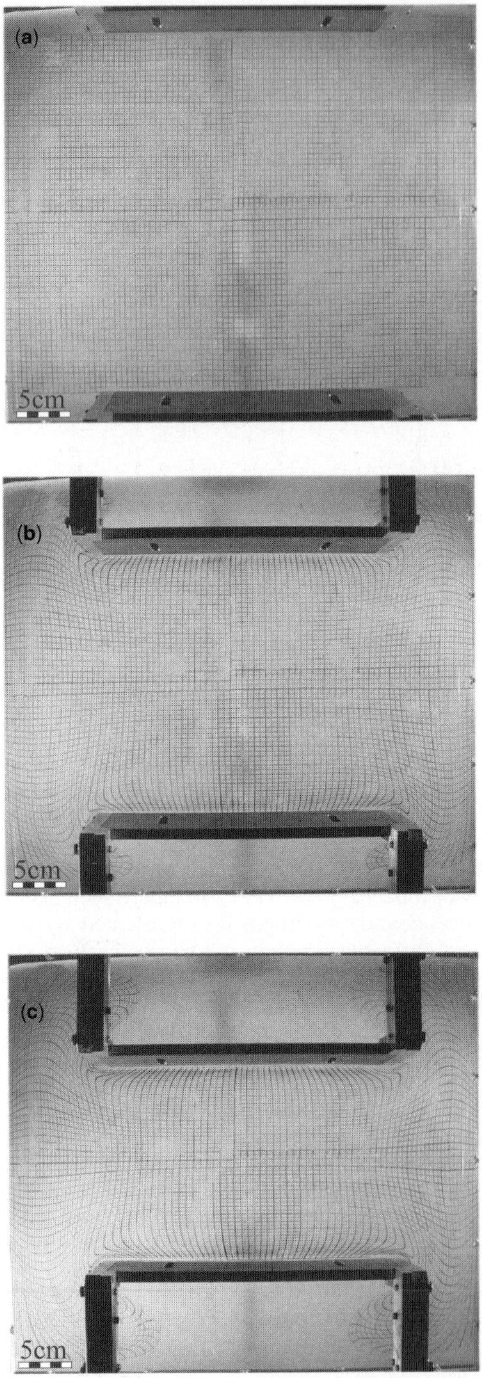

Fig. 4. Deformation of passive marker grid at top of asthenospheric material in response to convergence of pistons. The displacement of the passive grid gives an approximate illustration of the flow of model asthenosphere beneath fully constructed vice models.

In the vice experiments reported here, analogue models provide insights into the contribution of 3D orogen flow on the structural development of the orogen. The 2D numerical experiments act as a control on the laboratory experiments, by allowing for more variations in rheological parameters, and providing quantitative information on the evolving model orogen through time.

Laboratory experiments

Scaling and the general construction of the laboratory experiments (Table 1; Figs 1 and 2) follows the approach for analogue modelling of lithospheric deformation at 1 g described by Davy and Cobbold (1988, 1991). The experiments are approximately scaled to nature by selection of appropriate dimensions (geometric similarity) and laboratory materials (dynamic and kinematic similarity). By necessity the scaling is only approximate because of practical considerations in the laboratory (limitations of model size, time and suitable materials) and uncertainties in the natural system (densities, rheologies, time dependence, etc.).

Materials and scaling

In designing the laboratory experiments we used the physical properties of the asthenospheric mantle material and the length scale of the crust as our principal constraints. For practical purposes we define a length scale based on a model crustal thickness, $l_m = 1.5$ cm, which for a natural continental crustal thickness $l_p = 40$ km gives a length scale ratio $L = l_m/l_p = 3.8 \times 10^{-7}$ (where the subscripts m and p refer to laboratory and natural prototype scales, respectively). As our working fluid to model asthenospheric mantle we use polydimethylsiloxane (PDMS), which is a transparent, high viscosity, high molecular weight silicone polymer that is frequently used in analogue modelling studies (e.g., Weijermars 1986; Pysklywec & Cruden 2004). PDMS has a density $\rho_m = 970\ \mathrm{kg/m^3}$, which for a natural asthenospheric density of $\rho_p = 3100\ \mathrm{kg/m^3}$, sets a density scale of $P = \rho_m/\rho_p = 0.31$. In these experiments at 1 g the gravity scale ratio is $G = g_m/g_p = 1.0$.

The rheological properties of all ductile materials used in this study were measured over a range of strain rates relevant to the experiments with a rotary viscometer (based on a design by Cobbold & Jackson 1992), calibrated against a TA Instruments AR1000 rheometer. PDMS has a slightly non-Newtonian rheology

Table 1. *Parameters for laboratory vice models*

		W (cm)	L (cm)	h (cm)	ρ (kg/m^3)	η_{eff} (Pas)	ϕ	σ_o (Pa)	Material	V_x (cm/hr)	S (cm)	F_t (Pa)	F_b (Pa)	Ar
All Models														
Asthenosphere	Mantle	32	45	4.5	970	2.50×10^4			PS			*c.* 50	*c.* 20	*c.* 0.4
Vice	Lower Crust	5	28	1.0–0.8	900	3.20×10^5			PS + Pl + Gb					
	Mantle Lithosphere	5	28	1.5–2.0	970	$>1.00 \times 10^7$			BP + Zl					
Individual Models														
Experiment #27	Upper Crust	32	28	0.6	850		28°–34°	7–44	S + Zl	1.6	9.6	6.8	25.2	3.7
	Lower Crust	22	28	1	900	5.20×10^4			PS + Zl					
	Mantle Lithosphere	22	28	1.5	970	1.58×10^4			PS + Pl + Zl					
Experiment #29	Upper Crust	32	28	0.6	850		28°–34°	7–44	S + Zl	1.1	7.7	6.3	25.3	4
	Lower Crust	22	28	1	800	4.05×10^4			PS + DF + Gb					
	Mantle Lithosphere	22	28	1.5	970	1.68×10^5			PS + Pl + Zl					
Experiment #30	Upper Crust	32	28	0.4	850		28°–34°	7–44	S + Zl	1	7.5	4.1	16.8	4.1
	Lower Crust	22	28	1	900	4.10×10^4			PS + DF + Gb					
	Mantle Lithosphere	22	28	1.5	970	1.71×10^5			PS + Pl + Zl					
Experiment #32	Upper Crust	32	28	0.4	840		28°–34°	7–44	S + Zl	0.4	11	3.2	16.7	5.2
	Lower Crust	22	28	0.8	900	4.28×10^4			PS + DF + Gb					
	Mantle Lithosphere	22	28	2	970	2.00×10^5			PS + Pl + Zl					
Experiment #33	Upper Crust	32	28	0.4	840		28°–34°	7–44	S + Zl	0.4	11	3.2	16.8	5.2
	Lower Crust	22	28	0.8	800	4.11×10^4			PS + DF + Gb					
	Mantle Lithosphere	22	28	2	970	2.00×10^5			PS + Pl + Zl					
Experiment #34	Upper Crust	32	28	0.45	850		28°–34°	7–44	S + Zl	1.3	8.45	4.8	19	4
	Lower Crust	22	28	0.8	900	1.17×10^4			PS + DF + Gb					
	Mantle Lithosphere	22	28	2	970	1.64×10^5			PS + Pl + Zl					

W = orogen perpendicular width; L = orogen parallel length; h = vertical thickness; ρ = density; η_{eff} = effective viscosity; ϕ = angle of internal friction; σ_o = cohesion (maximum values); Vs = convergence rate; S = amount of shortening; F_t = tectonic force (resistance); F_b = Bouyancy force (gravity collapse); Ar = Argand number = F_b/F_t.
PS = Polydimethylsiloxane (PDMS); Pl = Harbuts plasticene (blue/black); DF = Dow Corning 30,000 cSt Fluid; BP = Dow Corning Bouncing Putty; Gb = 3M Glass Bubbles; S = quartz sand; Zl = 3M Z-light ceramic microspheres.

defined by the flow law:

$$\sigma^n = \eta \dot{\varepsilon} \quad (1)$$

where σ is stress, $\dot{\varepsilon}$ is strain rate, η is a material constant and n is the power law exponent. Because our modelling materials are slightly non-Newtonian ($n = 1.05$ to 1.2) for scaling purposes we define an effective dynamic viscosity $\eta_{\text{eff}} = \sigma/\dot{\varepsilon}$, which for our PDMS (Dow Corning Silastic 4-2901) at a laboratory strain rate of $10^{-5}\,\text{s}^{-1}$ is 2.5×10^4 Pa s. Assuming an effective viscosity of $[\eta_{\text{eff}}]_p = 10^{21}$ Pa s for the natural asthenosphere (Mitrovica & Forte 1997) defines a viscosity scale ratio $M = [\eta_{\text{eff}}]_m/[\eta_{\text{eff}}]_p = 2.5 \times 10^{-17}$. The time scale ratio for the experiments can now be defined as $T = M/[PLG] = t_m/t_p = 2.13 \times 10^{-10}$.

Thicknesses and properties of the crustal and mantle components of the model lithospheres follow from these scaling parameters and are summarized in Table 1 and Figures 1 and 2. Equivalent values of all experimental parameters are given in Table 2.

Variations in the both the density and effective viscosity of ductile mantle lithosphere and lower crust in the experiments are achieved by adding appropriate amounts of Harbut's plasticene (blue and black), low viscosity PDMS (Dow Corning 30,000 cSt Fluid) and solid fillers to PDMS. The effect of blending plasticene with PDMS is to increase both the density and effective viscosity of the mixture. Solid fillers employed in this study are 3M Z-lights ceramic microspheres and 3M Scotchlight Glass Bubbles, which have grain densities of 650 and 320 kg/m^3, respectively. Mixing different proportions of these fillers, which have average diameters *c.* 100 μm, to either pure PDMS or PDMS + Plasticene mixtures decreases the density and increases the effective viscosity of the material. Low viscosity PDMS can be added in small amounts to the above mixtures in order to lower their viscosity. Several different configurations of crustal density and effective viscosity structure were employed in our experiments (Table 1).

Brittle behaviour of the upper crust in the models is simulated by a layer of granular material, whose scaled thickness varies between models from 16 to 11 km. To acquire the desired average upper crustal density we alternated, in the appropriate proportion, layers of Z-light microspheres and quartz sand with bulk densities of 440 and 1570 kg/m^3, respectively. Ring shear tests on both materials in the normal stress range 500 to 2000 Pa determined internal friction angles $\phi = 28°–34°$ and (linearly extrapolated) cohesion values of $\sigma_o = 7–44$ Pa for Z-lights and quartz sand, respectively (J. Lohrmann, GeoForshungsZentrum, Potsdam, unpublished results 2005). Following Schellart (2000) we consider the cohesion values above to be upper bounds on the conditions in our experiments, in which normal stress values vary between 0 and 50 Pa. In this range, the cohesion of dry granular materials approaches 0 Pa (Schellart 2000), hence Z-lights and quartz sand are approximate analogues for modelling the macroscopic, frictional, behaviour of fractured upper crust (Davy & Cobbold 1991; Rossi & Storti 2003).

Model construction

All laboratory experiments were run in a 40 cm (width) × 45 cm (length) × 10 cm (depth) Plexiglas tank (Figs 1 and 3). Models were constructed by sequential addition of asthenospheric and lithospheric layers, with sufficient relaxation time allowed to minimize the development of air bubbles, and to ensure the pre-deformed model achieved isostatic equilibrium. During model construction, lateral flow of ductile material was prohibited by the insertion of vertical sheet metal barriers at the margins of the model orogen. These were removed immediately before the pistons were activated. After removal of the barriers and activation of the pistons, model orogenic material was free to flow at right angles to the compression direction into the reservoir of PDMS. This experimental set-up is designed to maximize the effects of orogen parallel flow driven primarily by piston convergence. Apart from a minor adjustment that occurs when the side barriers are removed, in the absence of piston convergence orogen parallel flow driven by gravity alone would be minimal over the typical duration of our experiments (*ca.* 8 hours).

Observation methods

Progress in all experiments was monitored by time-lapse digital photography from the side and top as well as by linear scanning of the surface topography (Fig. 3). A passive marker grid was imprinted on the top of the model by placing a rectilinear mask over the top surface and sprinkling a thin layer of fine grained white sand through it. Motion of this grid mapped progressive surface deformation in the evolution of the model. In the case of experiments employed to visualize the flow field in the asthenosphere, a passive marker grid was imprinted on the

Table 2. *Scaled natural equivalent parameters in vice models*

		W (km)	L (km)	h (km)	ρ (kg/m^3)	η_{eff} (Pas)	ϕ	σ_o (MPa)	V_x (cm/yr)	S (km)	F_t (MPa)	F_b (MPa)	Ar
All Models													
Asthenosphere	Mantle	853	747	120	3100	1.00×10^{21}					*c.* 420	*c.* 168	*c.* 0.4
Vice	Lower Crust	133	747	21–27	2876	1.28×10^{22}							
	Mantle Lithosphere	133	747	40–53	3100	4.00×10^{23}							
Individual Models													
Experiment #27	Upper Crust	853	747	16	2717		28°–34°	59–286	8.0	256	57	212	3.7
	Lower Crust	587	747	27	2876	2.08×10^{21}							
	Mantle Lithosphere	587	747	40	3100	6.32×10^{21}							
Experiment #29	Upper Crust	853	747	16	2717		28°–34°	59–286	5.5	205	53	213	4
	Lower Crust	587	747	27	2557	1.62×10^{21}							
	Mantle Lithosphere	587	747	40	3100	6.72×10^{21}							
Experiment #30	Upper Crust	853	747	11	2717		28°–34°	59–286	5.0	200	34	141	4.1
	Lower Crust	587	747	27	2876	1.64×10^{21}							
	Mantle Lithosphere	587	747	40	3100	6.84×10^{21}							
Experiment #32	Upper Crust	853	747	11	2685		28°–34°	59–286	2.0	293	27	140	5.2
	Lower Crust	587	747	21	2876	1.71×10^{21}							
	Mantle Lithosphere	587	747	53	3100	8.00×10^{21}							
Experiment #33	Upper Crust	853	747	11	2685		28°–34°	59–286	2.0	293	27	141	5.2
	Lower Crust	587	747	21	2557	1.64×10^{21}							
	Mantle Lithosphere	587	747	53	3100	8.00×10^{21}							
Experiment #34	Upper Crust	853	747	12	2717		28°–34°	59–286	6.5	225	40	160	4
	Lower Crust	587	747	21	2876	4.68×10^{20}							
	Mantle Lithosphere	587	747	53	3100	6.56×10^{21}							

Scaling factors = model/prototype: Length, $L = 3.8 \times 10^7$; Velocity, $V = 1760$; Density, $P = 0.313$; Viscosity, $M = 2.5 \times 10^{-17}$; Stress, $\Sigma = 1.19 \times 10^{-7}$; Time, $T = 2.13 \times 10^{-10}$.

PDMS using a transfer method (Dixon & Summers 1983).

Surface topography measurement was made using a laser triangulation device (Aculux LaseRanger) that is translated linearly over the model surface by an XY-gantry system at a stand-off distance of *c.* 13 cm (Fig. 3). Between 5 (Expt. #27) and 34 (Expt. #34) scan lines were collected at 0.5 to 1 hour intervals during the experiments (the variation was due to improvements made to the speed of data acquisition over the duration of the experimental programme). Individual topographic measurements were made with a spacing of 1 mm along each scan line and the positional accuracy of the XY gantry was $\pm 0.1 \sim$ mm. The distance resolution of the laser device is ± 0.1 mm, but due to gantry vibration and scattering effects of the model surface materials the nominal topographic resolution is $\pm 0.2 \sim$ mm.

The internal structure of models was examined after the end of each experiment. First, the granular layer over one third of the model surface was removed using a vacuum hose, and the exposed brittle ductile interface was photographed. Subsequently, the remaining granular material was saturated with water and the model was placed in a freezer at -50°C (i.e., below the crystallization temperature of PDMS). After 12 hours, serial sections of the model were prepared using a band saw, and photographed. Some viscous relaxation of deformed layer interfaces probably occurs during the initial cooling of the models. However, as shown below, the wavelengths and amplitudes of folded interfaces in the sectioned analogue models are similar to those observed in numerical experiments, indicating that the effects of viscous relaxation are relatively minor.

Numerical experiments

A suite of two-dimensional numerical experiments was conducted as a parallel comparison to the analogue models. The numerical models are scaled to the analogue experiments, and their initial configuration is shown in Figure 1.

The models assume incompressibility of the materials, and the governing equations for the system are:

$$-\nabla p + \nabla \cdot \sigma' + \rho g_z = 0 \qquad (2)$$

$$\nabla \cdot \mathbf{u} = 0 \qquad (3)$$

where $\mathbf{u}$ is the velocity field and g_z is the vertical component of gravity. The full stress tensor $\sigma_{\rm ij}$ has been divided into two components: $\sigma_{\rm ij} = \sigma'_{\rm ij} - p\delta_{\rm ij}$ where σ' is the deviatoric component of the stress tensor and p is pressure ($p = -\frac{1}{3}\sigma_{\rm ii}$ for an incompressible fluid). Viscous or plastic rheologies are imposed for the materials in the numerical model to simulate the behaviour of the analogue materials. For the viscous materials we adopt a power law constitutive relation (Eqn. 1), and for the brittle crust, a Coulomb-type yield criterion of the form $\sigma_{\rm y} = p\sin\phi + \sigma_o$ was used (where $\sigma_{\rm y}$ is the second invariant of the deviatoric stress).

The system of equations is solved using the arbitrary Lagrangian–Eulerian (ALE) finite element method (Hirt *et al.* 1974; Fullsack 1995). The solution space is discretized using a Eulerian resolution of 321×81 nodes and a Lagrangian resolution of 601×151 nodes, evenly distributed through the box. The numerical routine has been benchmarked rigorously with previous studies (Poliakov & Podladchikov 1992; Blankenbach *et al.* 1989; Houseman & Molnar 1997; van Keken *et al.* 1997) to verify its accuracy in computing viscous flow and surface topography.

The top boundary of the box is a free surface and all the other sides have an imposed zero tangential velocity (i.e., no slip). Material is not permitted to penetrate the lower boundary of the box. Deformation in the model is driven by buoyancy forces arising from material density variations and the imposed lithospheric 'vice' contraction. This contraction is simulated as a velocity boundary condition at the sides of the solution space; along the lithospheric portion of each side boundary new lithosphere material is introduced into the box at a horizontal velocity $V_x/2$. An outward flux of asthenospheric material is imposed along the sub-lithospheric portion of the sidewalls of the box in order to balance the mass of injected lithosphere.

Overview and dynamic analysis of experiments

The six laboratory experiments reported here represent a range of orogenic conditions as dictated by rheological and density structure and boundary conditions (convergence rate, V_x). Variations in dynamics and body forces in the experiments can be quantified and direct comparison to nature can be made by considering the Argand number. The Argand number is defined as the dimensionless ratio of buoyancy forces to tectonic forces, $Ar = F_b/F_t$ (England & McKenzie 1982; Houseman & England 1986). For these experiments, the buoyancy force is the static outward force per unit length that results in gravitational collapse of the model lithosphere over the asthenosphere. Following Benes & Davy (1996), the buoyancy force for a three-layer model is given by:

$$F_g = \rho_1 g \frac{h_1}{2} + \rho_1 g \frac{(z_2^2 - z_1^2)}{2} + \rho_1 g \frac{(z_3^2 - z_2^2)}{2} - g \frac{(\rho_1 h_1 + \rho_2 h_2 + \rho_3 h_3)}{2\rho_m} \quad (4)$$

where ρ_1, ρ_2, ρ_3 and h_1, h_2, h_3 are densities and thicknesses of the upper crust, lower crust and lithospheric mantle, respectively; ρ_m is the density of the asthenosphere (PDMS); $z_1 = h_1$, $z_2 = h_1 + h_2$, and $z_3 = h_1 + h_2 + h_3$. Note that F_g is calculated for the start of the experiment and therefore does not consider the buoyancy force caused by crustal thickness changes. F_g is therefore a relative measure of the potential for the model lithosphere to flow laterally over the asthenosphere, and it is resisted by F_t, which is the vertically averaged stress required to deform the model lithosphere at a given strain rate.

The tectonic force in the models is calculated as:

$$F_t = \frac{h_1 \sigma_o + (h_1^2/2\tan\phi) + h_2(\eta_2\,\dot{\varepsilon})^{1/n_2} + h_3(\eta_3\,\dot{\varepsilon})^{1/n_3}}{z_3} \quad (5)$$

where η is the material constant, n is the power law exponent (Eqn 1; subscripts 1, 2, 3 refer to model upper crust, lower crust and lithospheric mantle, respectively) and $\dot{\varepsilon} = V_x/W$ where W is the width of the orogen measured parallel to the shortening direction. Note that because V_x is constant, $\dot{\varepsilon}$ increases with time during each experiment. As with F_g, values of F_t and Ar reported here are for the start of each experiment (Table 1).

F_t in the experimental orogen varies between 3.2 and 6.8 Pa, corresponding to natural values of 41 to 87 MPa (Tables 1 and 2). The main controls on F_t in the model orogen are due to the thickness of the brittle crust, h_1, and the effective viscosity of the lower crust, $[\eta_{eff}]_2$. With the exception of Experiment #34, the model orogenic mantle lithosphere is only marginally stronger than the lower crust. This was mainly predicated by limitations imposed by available analogue materials at the time the experiments were carried out, and the resulting trade-off between maintaining the correct density for the mantle lithosphere layer (important for body force scaling), and its strength. Although the effective viscosity of mantle lithosphere material in the experiments would certainly be too low for modelling of stable continental lithosphere, its relative strength compared to crustal and asthenospheric material is consistent with materials used in other analogue model studies (e.g., Sokoutis *et al.* 2000). The scaled natural effective viscosity of the mantle lithosphere in the experiments (*c.* 7×10^{21} Pa s; Table 2) is also consistent with experimental (e.g., Kohlstedt *et al.* 1995; Hirth & Kohlstedt 1996; Mackwell *et al.* 1998) and geophysics (e.g., Maggi *et al.* 2000) based estimates of the strength of upper mantle rocks in natural orogens, if the geothermal gradient is steep and conditions are relatively wet (e.g., η_{eff} $\sim 10^{21}$ to 10^{22} at strain rates of 10^{-14} to $10^{-15}\,s^{-1}$; Jackson 2002; Toussaint *et al.* 2004).

The strong lithospheric blocks (vice blocks) in the experiments have F_t values that are ~ 10 times greater than the model orogen. The bulk of the strength in these blocks resides in the mantle lithosphere layer (Fig. 2), which is composed of bouncing putty with finite yield strength (Dixon & Summers 1986) mixed with Z-light spheres to achieve the correct density. The effective viscosities of the mantle lithosphere and lower crust in the vices scale up to $>1 \times 10^{23}$ and *c.* 1×10^{22} Pa s, respectively. These values are consistent with a dry upper mantle and granulitic lower crust (Kohlstedt *et al.* 1995; Mackwell *et al.* 1998), as could be expected in stable continental lithosphere. Hence the overall strength of the model orogenic and vice lithospheres and the details of their rheological structure are consistent with a hot orogen with a wet mantle lithosphere and quartz-dominated lower crust (e.g., Toussaint *et al.* 2004) being squeezed between blocks of cold cratonic continental lithosphere (Fig. 2; Table 2).

F_b in the experiments varies between 16.7 and 25.2 Pa, resulting in values of Ar from 3.7 and 5.2 (Fig. 5a; Table 1). Such values indicate that the orogenic lithosphere will tend to collapse if boundary conditions are suitable, and are consistent with conditions employed in previous numerical and analogue model studies of lithospheric convergence and extension (Houseman & England 1986; Benes & Davy 1996). Values of Ar between 1 and 10 are considered to describe the balance of forces in wide, hot orogens such as the Himalaya, and $Ar \geq 3$ is required to develop orogenic plateaux such as the Tibetan plateau (England & McKenzie 1982; Dewey *et al.* 1986; England & Houseman 1986). In the vice blocks, $Ar \ll 1$, which describes the balance of forces expected in stable continental lithosphere, in which gravitational forces are expected to be significantly resisted by the tectonic strength.

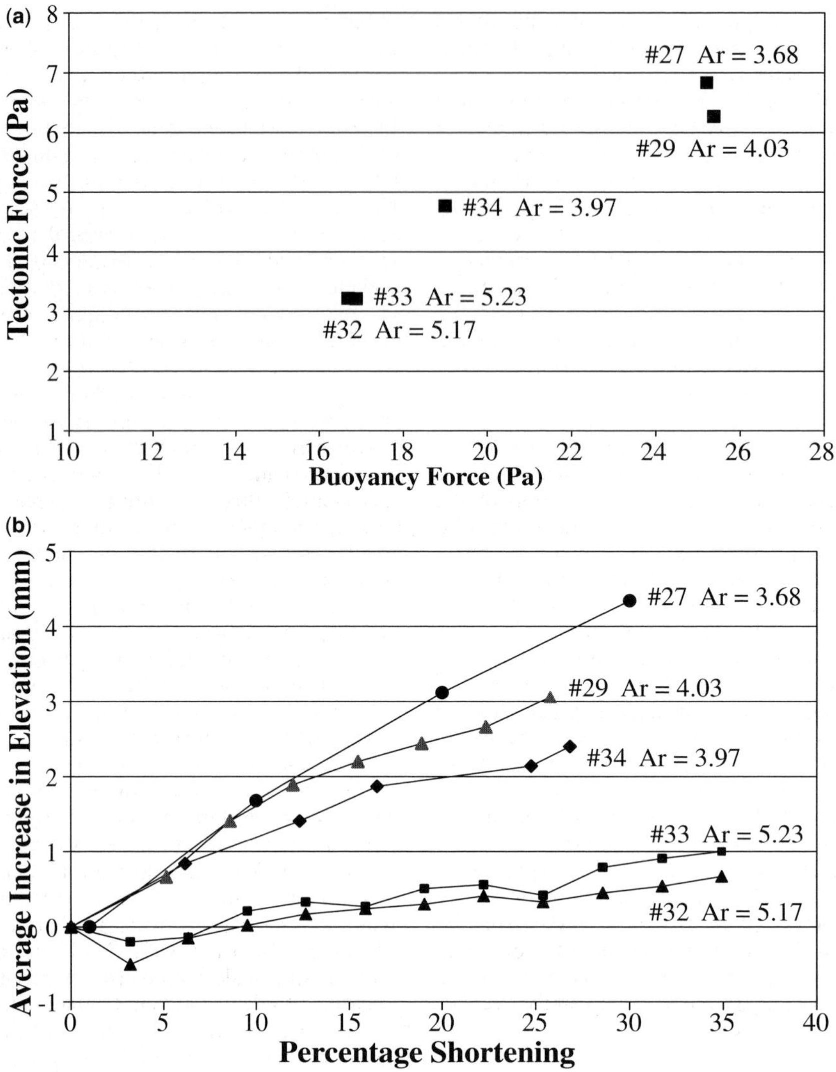

Fig. 5. (**a**) Comparison of tectonic forces (F_t) and buoyancy forces (F_b) in analogue experiments. (**b**) Increase of average surface elevation of orogenic crust in analogue vice models as a function of the amount (%) shortening and Argand number, *Ar*.

Because *Ar* in the experiments is in agreement with values estimated in nature, we can demonstrate approximate dynamic similarity between our vice models and wide, hot orogens. Kinematic similarity is satisfied by imposing model convergence rates of 0.4 to 1.6 cm/hr, which scale up to typical orogenic convergence rates of 2 to 8 cm/a. The model thicknesses and horizontal dimensions scale up to values expected in large orogenic systems (Table 2), which together with total scaled amounts of shortening of 200 to 300 km (25–35% shortening) also satisfy the requirements of geometric similarity.

The effect of *Ar* in the experiments can be evaluated by quantifying how elevation in the model orogen grows with time. Figure 5b plots the average elevation in the area confined between the vice blocks, determined by laser profilometry, as a function of the amount of shortening in each experiment. After an initial drop in elevation due to removal of the barriers at the edge of the model orogen, average elevation increases steadily. As expected, the fastest elevation growth occurs in experiments with smaller Ar values. This is because orogenic thickening outstrips the rate of lateral orogenic collapse allowed by the strength of the model lithosphere. The average height attained in Experiment #27 corresponds to a scaled equivalent mean elevation of *c.* 11500 m, which seems excessive for terrestrial orogens. Experiments #29 and #34 have final scaled average elevations of *c.* 8000 and 6000 m, which is a reasonable fit for Earth, given the absence of erosion in the experiments. In the highest Ar experiments (#33 and #32), final scaled mean elevations are $\leq$2600 m and mean scaled elevations at times that correspond to the end of Experiments #29 and #34, are *c.* 1300 m.

In the presentation of the experimental results below, we will focus on the influence of Ar on the structural and topographic evolution of the model orogen, and examine the role of the relative buoyancy and strength of the lower crust. The attributes of the ductile crust of wide, hot orogens are important because the crust is likely to be weak due to the combined effects of high geothermal gradient, presence of water and quartz-rich composition (e.g., Toussaint *et al.* 2004). These conditions may also lead to partial melting over large areas (e.g., Meissner & Mooney 1998; Rey *et al.* 2001), resulting in a further lowering of the strength of the lower to mid-crust and, if the degree of melting is sufficient, to decrease its bulk density (e.g., Rey 1993; Teyssier & Whitney 2002). Experiments #29, #33 and #34 examine the situation of a weak, non-buoyant ductile crust; in Experiments #29 and #33, the ductile crust is weak and buoyant; and in Experiment #34 the ductile crust is non-buoyant, but significantly weaker than the other models.

Results

Firstly, we consider the influence of lower crustal density on the evolution of the models, using representative experiments having a buoyant (#29) or non-buoyant (#34) lower crust. Secondly, the contribution of Ar on experimental outcomes and the differences between equivalent 3D analogue and 2D numerical experiments are discussed by focussing on results of Experiments #29, #32 and #34.

Buoyant lower crust experiments (#29 and #33)

Surface deformation (analogue only), topography (analogue and numerical) and cross-sectional geometry (numerical only) of Experiment #29 at intermediate (12%) and late amounts (24%) of shortening are shown in Figure 6. Surface deformation in this experiment is characterized by early formation of a conjugate system of shear zones (Fig. 6a) that subsequently evolves into a localized, curvilinear orogen-subparallel zone of thrusting and folding (Figs 6a and 7a). Not evident in Figures 6a and d is the development of upper crustal folds, which are observed in the topographic profiles (Figs 6b and e) and serial sections (Fig. 8a). Overall, the amplitudes and wavelengths of the topographic features associated with these folds are very similar in the analogue and numerical runs. A notable feature in both the analogue and numerical runs of this experiment is the development of significant zones of overthrusting of the model orogen crust over the encroaching vice. A similar degree of overthrusting is also observed in Experiment #33, which also had a buoyant lower crust. Although upper crustal folding in Experiment #33 is more distributed than in Experiment #29, the formation of conjugate sets of upper crustal shear zones is also observed, particularly close to extruding margins of the model orogen.

In comparison to the experiments described below, overthrusting or 'overflow' parallel to the shortening direction at the vice margins, and the development of conjugate upper crustal shear zones within the orogen appear to be common attributes in models with buoyant ductile crust. We attribute the pronounced overflow of buoyant crust and overlying brittle material to be due to a 'water bed' effect. As postulated by Bailey (1999, 2001), a thickened, buoyant and weak ductile crust is gravitationally unstable. Under horizontal plane strain conditions, the resulting net outward force acts opposite to the shortening direction, and provided crustal strength can be exceeded at the orogen margins, overflow will occur (Bailey 2001). The effect of this overflow in our experiments is twofold. Firstly, as seen in the analogue and numerical cross sections (Fig. 8a), outwardly vergent thrust zones develop at the orogen-vice boundary zone. Secondly, upper crustal folding is suppressed because of orogen perpendicular extension caused by the overflow. In the analogue version of Experiment #29, folding in the orogen is confined to a central region, in which the folds are probably amplified by local gravitational instability between the upper and lower crust (Fig. 8a). In the equivalent numerical model, distributed upper crustal folds have significantly lower amplitude than in models with non-buoyant ductile crust. Formation of conjugate upper crustal shear zones in analogue Experiments #29 and #33 is consistent with

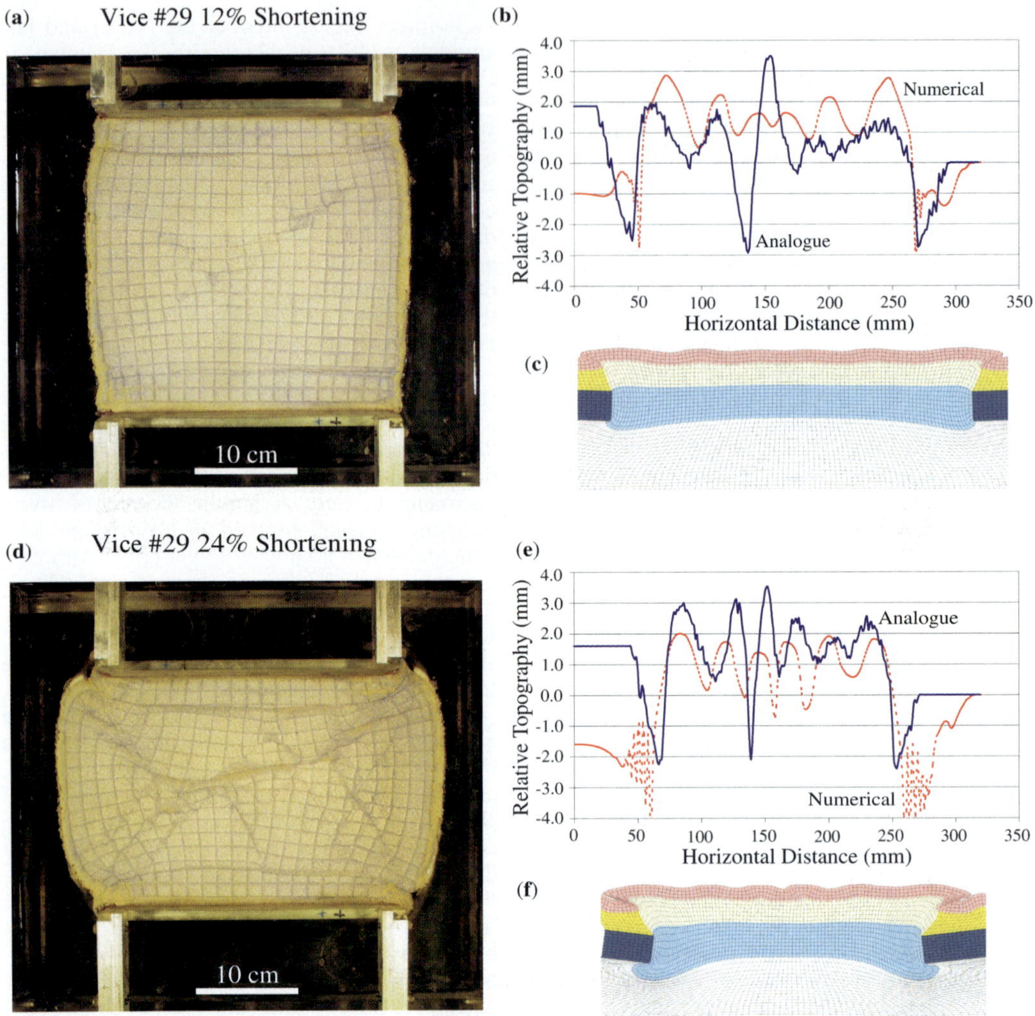

Fig. 6. Results for Experiment #29, with buoyant ductile crust. (**a**) and (**d**) Photographs of top surface after 12% and 24% shortening. (**b**) and (**e**) Comparison of central topographic profiles of analogue and numerical models after 12% and 24% shortening. Relative topography is the difference between the present and original elevation. The analogue and numerical results have been re-levelled for comparison purposes. (**c**) and (**f**) Cross-sections of numerical experiment at 12% and 24% shortening.

orogen parallel flow, driven both by the gravitationally unstable thickened ductile crust, and by orogen-parallel stretching exerted at the base of the model lithosphere by the asthenosphere (Fig. 4).

Due to the very high strength of the mantle lithosphere in the vice blocks, significantly weaker mantle lithosphere material in the model orogen has a tendency to underthrust the vice margins (Figs 6c, f and 8). This effect is observed in all experiments. We consider these features to be artefacts of the experimental boundary conditions and therefore do not ascribe any special tectonic significance to them. Their development does not appear to have a major influence on the formation of structures in the orogenic crust. However, the addition of extra dense material to the base of the vice lithosphere, together with a torque exerted on the strong vice upper mantle layer by the underthrusting of orogenic material, results in a net tilting of the vice blocks towards the interior of the model, as observed in both surface elevation data (Figs 6 and 9) and cross-sections (Fig. 8). Tilting of the

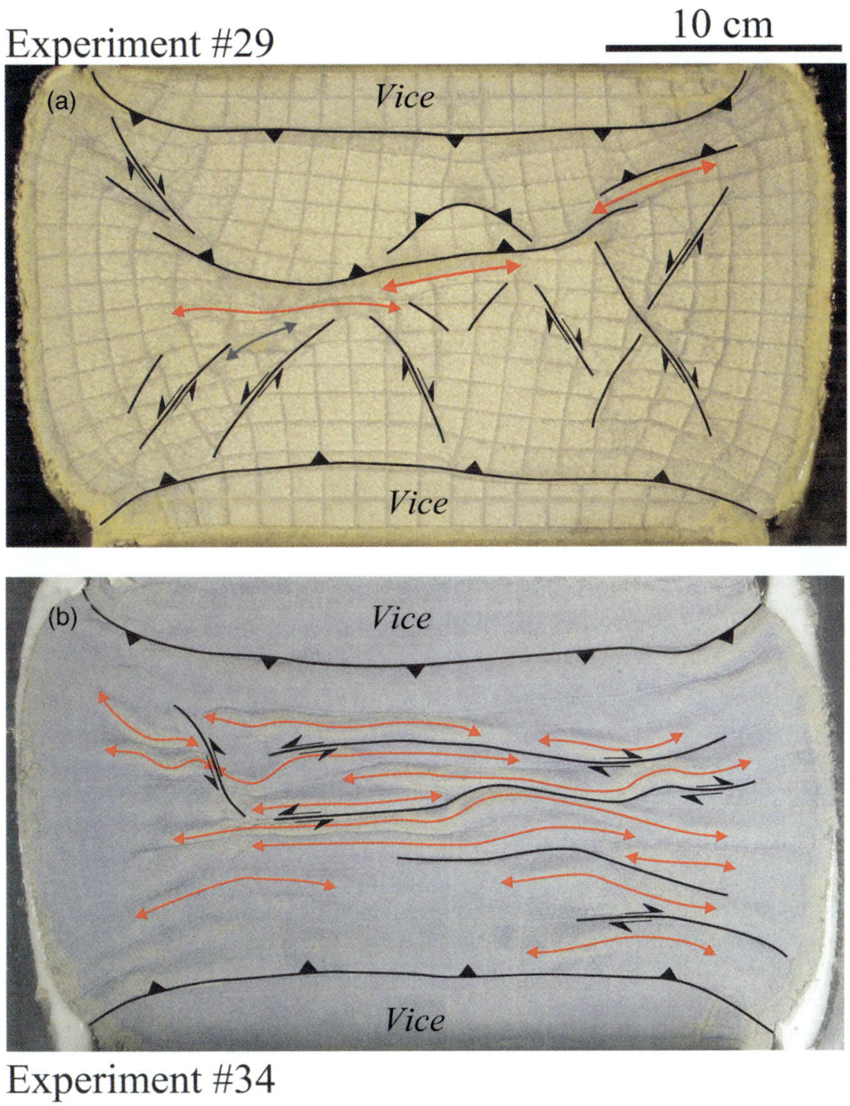

Fig. 7. Structural interpretations of Experiments #29 (**a**) and #34 (**b**), based on surface deformation patterns and serial sections. Red double headed arrows are axial traces of anticlines. Black lines are faults (barbs on hanging wall of thrusts, based on analysis of serial sections; sense of shear on vertical strike slip faults is indicated, based on deflection of marker grid).

vice blocks is more pronounced in the analogue models because of the finite width of the vice blocks over which this torque acts.

Non-buoyant lower crust experiments (#27, #30, #32 and #34)

Results for Experiment #34 are presented in Figures 7 to 9. Although the *Ar* numbers and general topographic growth behaviour are similar (Table 1, Fig. 5), upper crustal deformation in this experiment is markedly different to that in Experiment #29. Namely, the deformation is characterized by the formation of distributed orogen subparallel curvilinear folds from an early stage of the tectonic evolution (Figs 9a, b, c). As shortening progresses the folds lengthen and merge, and at late stages, orogen subparallel shear zones develop,

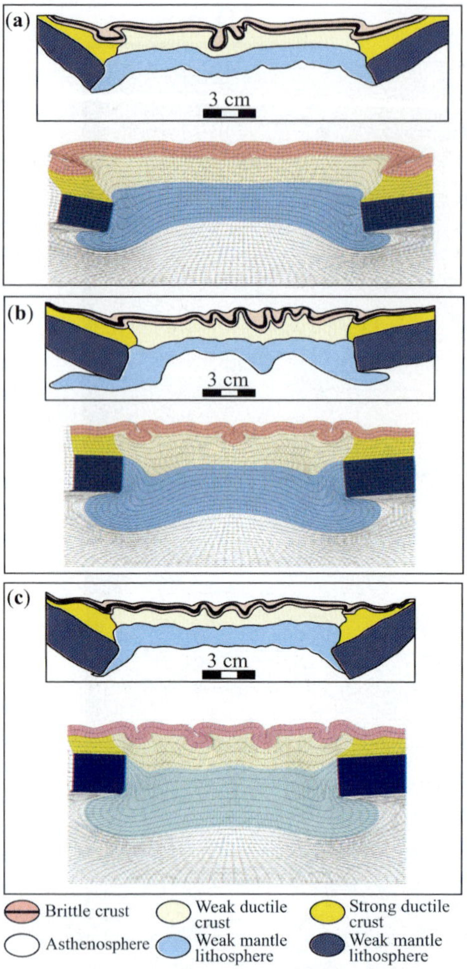

Fig. 8. Central cross-sections for analogue and numerical Experiments #29 (**a**), #34 (**b**) and #32 (**c**) after 32% shortening. Cross-sections of analogue experiments are line drawings traced from photographs of serial sections cut through the centre of the model after freezing.

typically in synclinal cores (Figs 7b and 9d). Conjugate shear zones are generally absent, except at the outer regions of the extruding orogen. The degree of overthrusting of orogenic crust over each vice is significantly less than in models with buoyant crust. Both the amplitude and wavelength of upper crustal folds, as determined from topographic profiles and cross-sections, are quite similar in the analogue and numerical runs.

In experiments with non-buoyant ductile crust, orogen parallel shear zones appear to be favoured in models with thicker brittle upper crust and lower *Ar*, being best developed in Experiment #27 and least well developed in Experiment #32. The geometry of folds at the brittle-ductile boundary also varies according to the effective viscosity of the lower crust. In Experiments #27, #30 and #32 both analogue and numerical runs are characterized by folds at the brittle-ductile interface that are cylindrical in cross-section, with a tendency to develop cuspate geometries in anticlinal cores (Fig. 8c). In Experiment #34, which had a ductile crust an order of magnitude weaker than the other models, anticlinal cores were markedly cuspate, resembling mullion structures in cross-section (Fig. 8b) (e.g., similar to the observations of Sokoutis 1990).

Effects of three-dimensional versus two-dimensional geometry

As noted in the experimental results described above, both the topographic profiles and cross-sections for equivalent analogue and numerical experiments are qualitatively (e.g., fold shapes, vice margin structure; Fig. 8) and quantitatively similar (e.g., close correspondence in amplitude and wavelength of fold-related features on topographic profiles; Figs 6 and 9). However, significant differences between the 3D analogue and 2D numerical experiments are evident when surface deformation patterns are examined (Fig. 7) and in the topographic and internal characteristics of experiments with higher *Ar* numbers (Figs 8c and 10).

The upper crust of the numerical experiments responds to the imposed convergence by folding, and varying amounts of overthrusting at the vice margins. As noted previously, the upper crust of the analogue models is characterized by the development of folds *and* vertical shear zones, either in the form of conjugate arrays (Expts #29 & #33; Fig. 7a) or orogen parallel curvilinear shear zones (Expts #27, #30, #32, #34; Fig. 7b). Thrusts also develop at the vice margins and in some cases in association with pop-up structures formed in the interior of the orogen (e.g., Figs 7a and 8a).

While the lower occurrence of strain localization in the brittle crust of the numerical models can be attributed to choice of model rheology, the development of systems of vertical shear zones in the analogue experiments are due to both strain softening within the granular material *and* orogen-parallel flow. In the case of models with non-buoyant ductile crust, orogenic shortening is accommodated by distributed upper crustal folding and moderate overthrusting

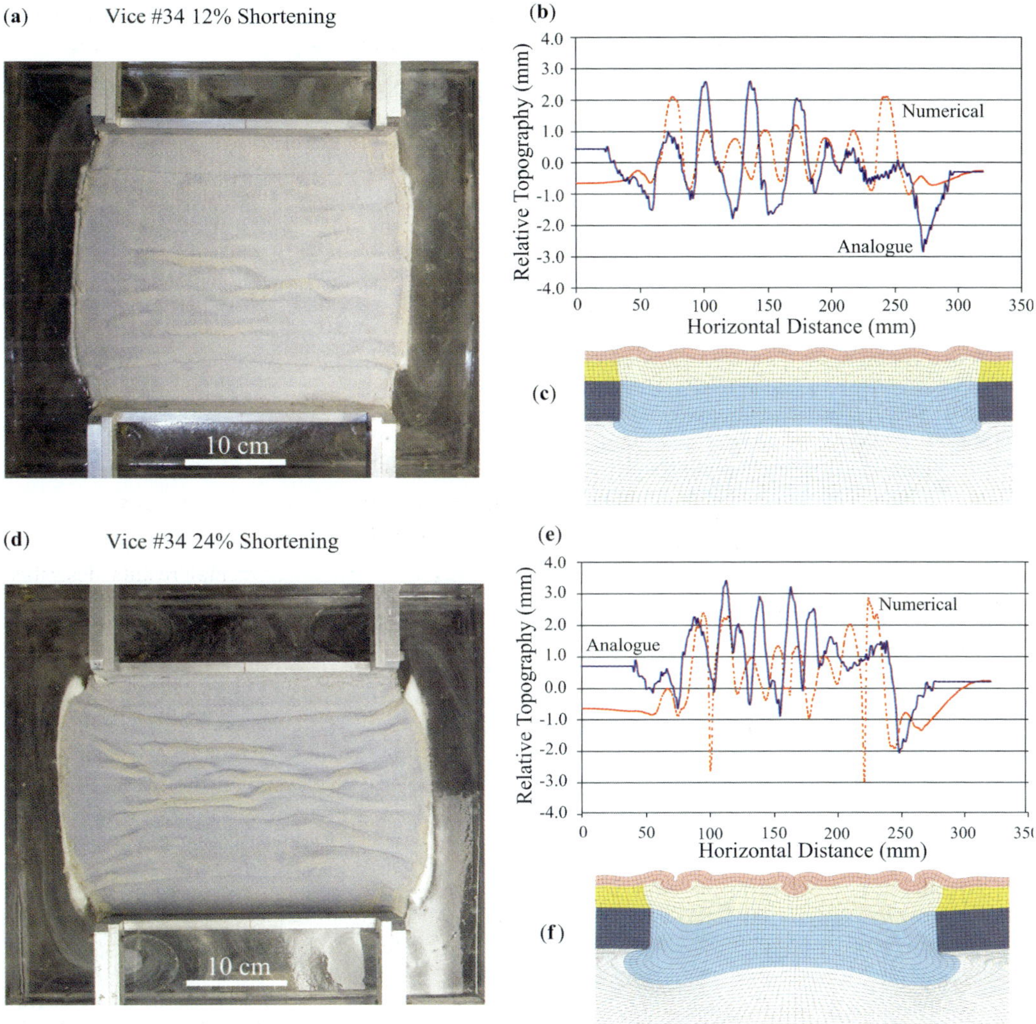

Fig. 9. Results for Experiment #34, with non-buoyant, weak ductile crust. (**a**) and (**d**) Photographs of top surface after 12% and 24% shortening. (**b**) and (**e**) Comparison of central topographic profiles of analogue and numerical models after 12% and 24% shortening. Relative topography is the difference between the present and original elevation. The analogue and numerical results have been re-levelled for comparison purposes. (**c**) and (**f**) Cross-sections of numerical experiment at 12% and 24% shortening.

at vice margins. Orogen-parallel extension in the upper crust of these experiments is taken up by a component of homogeneous stretching, parallel to fold axes and by displacements on orogen subparallel curvilinear shear zones (Fig. 7b). An interesting property of these shear zones is that their sense of shear changes along strike in response to the flow direction in the underlying ductile crust and upper mantle.

In models with buoyant ductile crust, the upper crust responds to convergence by significant overthrusting at vice margins and by localized folding and thrusting in the model interior. Orogen parallel extension in these cases is taken up by relative outward displacements and rotations of large blocks of brittle model upper crust bound by conjugate sets of shear zones, whose acute bisectors are parallel to the convergence direction (Fig. 7a). This style of deformation is attributed to the 3D nature of the 'water bed effect' discussed above. Because the upper crust is lifted up by the buoyant, thickening ductile crust, it is relatively less confined than in experiments with non-buoyant ductile crust. With the exception of a central region of orogen parallel folds and thrusts, this allows the upper crust to break up into relatively undeformed blocks whose motion between conjugate shear zones is controlled by the flow of the underlying ductile crust and upper mantle (Fig. 7a).

In experiments with low to intermediate *Ar* number (#27, #29, #34) we observe a generally close correspondence between analogue and numerical runs in terms of cross-sectional geometry and topographic profiles, when compared to the central part of the models. However, in the analogue models the mean elevation, the amplitude of folds in cross-section and their surface expression decrease outward along the strike of the orogen, introducing significant differences with respect to the numerical results. These effects are due to the increasing importance of orogen-parallel flow on the geometry of upper crustal structure towards the free end of the models. A similar difference between analogue and numerical results is observed even in the central region of models with high *Ar* number (#32, #33). In these experiments both the amplitudes of folds in cross-section and their surface expression in topographic profiles are noticeably less in the analogue models (Figs 8c and 10). This is because in the analogue experiments with high *Ar* number, gravity driven orogen parallel flow can effectively keep up with the slow imposed convergence rate. In 3D this enhanced outward flow has the effect of suppressing the build up of relief and the amplitude of upper crustal folds, even in the central region of the models.

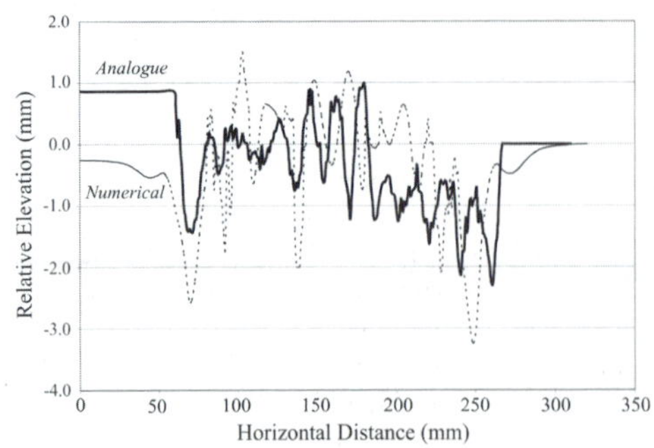

Fig. 10. Relative surface elevation profiles in analogue (central profile; heavy line) and numerical (thin dashed line) runs of Experiment #32. Re-levelled for comparison purposes.

Three-dimensional structure of the brittle-ductile interface

The complex 3D structure of the brittle-ductile interface in the analogue experiments was observed by removal of the upper crustal granular layer with a vacuum hose at the end of each experiment (Fig. 11). Folds at this interface are typically doubly plunging and display an anastomosing pattern in 3D. Fold amplitude and wavelength decreases approaching the free end of the model orogen and the overall strain pattern is similar to that observed in viscous-only experiments (e.g., Fig. 4 and Gilbert & Merle 1987). Significant stretching parallel to fold axes in our experiments is evident from the observation of segmented (boudinaged) fold hinges (Fig. 11). Vertical upper crustal shear zones do not penetrate into the ductile crust, although some shear zones do root into linear zones of very short wavelength, short axial length arrays of en-echelon folds. The brittle-ductile interface therefore represents a decoupling horizon for shear zones, but not for folds, which are clearly continuous across it. However, cross-sections (Fig. 8) show that folding of the brittle ductile interface does not continue to the crust-mantle boundary, indicating that upright folds in the models are disharmonic at the crustal scale.

Fig. 11. Oblique view of the brittle-ductile interface (in foreground) of Experiment #27, after removal of upper granular layer (in background) with a vacuum hose. The width of the model between the vertical pistons is 9.6 cm. Strong vice ductile crust is pink, weak ductile crust is yellow.

Three-dimensional bulk finite strains

Comparison of the analogue and numerical experiments indicates that the centre profile cross-sectional area of the ductile crust and orogenic upper mantle in the analogue models has decreased over the duration of each set of runs. The change in cross-sectional area of these layers is an indication that material has flowed out of the section, parallel to the orogen, and is therefore a measure of orogen-parallel stretch in the models.

Using the coordinate system defined in Figure 12a, we employ the change in cross-sectional area of the ductile crust and orogenic upper mantle, together with known amounts of horizontal shortening, to estimate the bulk finite strain in the central portion of each analogue experiment. The principal strains $(1+e_x)$, $(1+e_y)$ and $(1+e_z)$ are defined to be oriented parallel to the shortening direction (x), the orogen (y) and normal to the model surface (z) and the principal strain ratios in this coordinate system are $R_{xy}=(1+e_x)/(1+e_y)$, $R_{xz}=(1+e_x)/(1+e_z)$ and $R_{yz}=(1+e_y)/(1+e_z)$ (Fig. 12a). The bulk principal strain measured parallel to the shortening direction is given by the known amount of convergence in each experiment, $(1+e_x)=[1+(W-W_o)/W_o]$ where W and W_o are the final and original widths of the orogen. The bulk principal strain measured parallel to the orogen can be estimated from the change in cross-sectional area of the unit in question as $(1+e_y)=A_o/A$ where A_o and A are the initial and final cross-sectional areas measured from digital images of serial sections. Assuming the model materials are incompressible, the vertical bulk principal strain is $(1+e_z)=1/[(1+e_x)(1+e_y)]$, which can be checked against measurements of the change in vertical thickness of the unit from serial sections.

Values of R_{yz} and R_{xz} mantle lithosphere and ductile crust are plotted against Ar in Figures 12b and c. Because of variations between strain ratios estimated for different serial sections, the values presented here are averages determined from bulk strains measured from serial sections in the inner two-thirds of the orogen (i.e., from profiles taken from the centre to two-thirds of the way out to the free edge). Data from the outer one-third has been omitted because strains in this region are highly variable and sensitive to the presence of the unconfined boundary.

The bulk strain ratios in Figure 12 provide a general view of the state of 3D finite strain within the models and therefore can be used to predict fabric elements in the ductile parts of natural vice orogens. Values of R_{yz} increase with increasing Ar, reflecting the greater amount of orogen parallel flow in models with slower shortening rates and larger buoyancy forces (Fig. 12b). For each experiment, R_{yz} is greater in the mantle lithosphere compared to the ductile crust. This reflects the presence of a vertical, orogen-parallel velocity gradient in the models, with the fastest rate of orogen parallel flow occurring at the base of the mantle lithosphere and the slowest at the brittle-ductile interface.

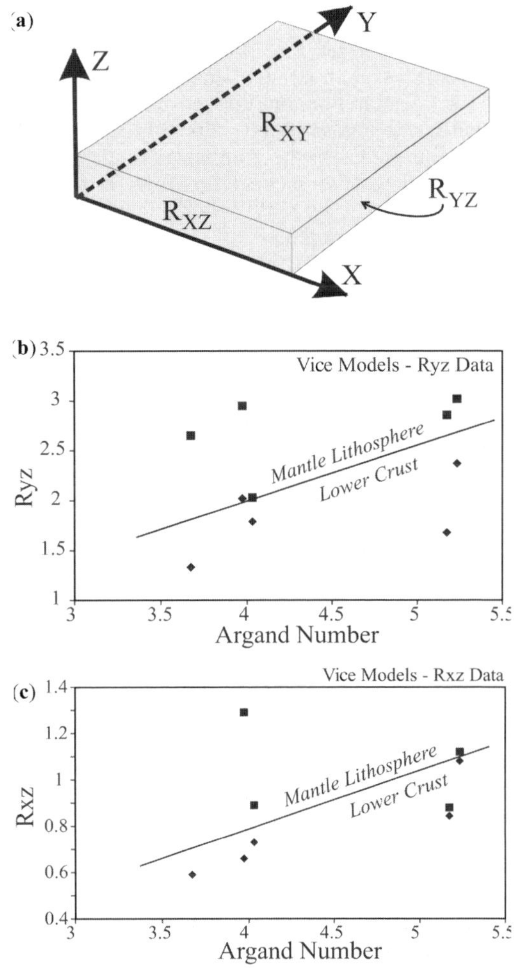

Fig. 12. (**a**) Coordinate system used for determination of bulk strains in analogue models. (**b**) R_{xy} versus Ar number for ductile crust and mantle lithosphere. (**c**) R_{xz} versus Ar number. Squares are bulk strain rations measured for mantle lithosphere, diamonds are ductile crust.

Values of $R_{xz} > 1$ indicate strain ellipsoids with horizontal $(1 + e_x)$ principal strains, while values <1 correspond to strain ellipsoids with vertical $(1 + e_z)$ principal strains. Because of the orogen parallel orientation of $(1 + e_y) > (1 + e_x)$, bulk flattening planes in the ductile crust and mantle lithosphere in the models are therefore either horizontal ($R_{xz} > 1$) or vertical and striking parallel to the orogen ($R_{xz} < 1$). In cases where $R_{xz} = 1$, the bulk strain ellipsoid will be prolate, with a long axis parallel to the orogen. R_{xz} values increase with increasing *Ar*. This is because vertical thinning, $(1 + e_z)$, due to orogen parallel flow, becomes more important with increasing *Ar*. In individual experiments R_{xz} is consistently greater in the mantle lithosphere than in the ductile crust, which again reflects the orogen-parallel vertical velocity gradient.

With exception of Experiment #33 ($Ar = 5.2$), comparison of R_{xz} and R_{yz} values indicates that bulk finite strains in the ductile crust are characterized by vertical orogen-parallel flattening planes with a strong orogen parallel stretch. In Experiment #33, which has the largest Ar number and a buoyant ductile crust, the bulk finite strain is close to prolate orogen parallel stretching in both the ductile crust and mantle lithosphere. Bulk finite strains in the mantle lithosphere in the other experiments vary from close to prolate with strong orogen parallel stretch to general flattening with a horizontal *xy* plane. Bulk plane strains are not observed in any of the analogue models, in contrast to the 2D numerical experiments, which predict, by definition, plane strains.

Local variation in finite strain

The above analysis is useful for predicting the bulk strain symmetry and intensity in the ductile crust and mantle lithosphere of laterally flowing vice orogens. However, examination of analogue model cross-sections (e.g., Fig. 8) and detailed local measurements of local thickness variations in these sections, indicate considerable heterogeneity in finite strain. Although the bulk finite strain in the ductile crust of most models is characterized by a vertical orogen parallel flattening plane, thickness variations indicate that local finite strain in this layer can vary between horizontal flattening and orogen parallel stretching to pure orogen parallel constriction. For example, R_{xz} in Experiment #29 varies locally between 0.35 and 2.6, due entirely to strong vertical thickening (maximum $(1 + e_z) \sim 1.8$) and pronounced vertical thinning (minimum $(1 + e_z) \sim 0.23$) within an individual cross-section. Similar local finite strains are observed in all analogue experiments, indicating that deformation of the ductile crust and mantle lithosphere is characterized by strong across and along-strike variability.

Summary of experimental findings

The major 3D structural characteristics of the vice models, as deduced by analysis of surface deformation patterns, topography data and cross-sections, are summarized in Figure 13. The major finding of the analogue and numerical experiments are as follows:

(1) 3D analogue models highlight the profound effect that syn-convergent orogen parallel flow can have on the development of structures at different levels of the crust (Fig. 13). The significance of these effects is highlighted by comparison to complementary 2D numerical experiments.
(2) Upper crustal deformation is decoupled from lower crustal strain. The boundary between these deformation regimes is transitional and occurs below the brittle-ductile interface in the models.
(3) Finite strain in the ductile crust and mantle lithosphere is characterized by orogen parallel stretching. Strain symmetries vary from horizontal flattening, to orogen parallel constriction, to vertical flattening normal to the shortening direction.
(4) Different upper crustal deformation styles are developed in models with non-buoyant versus buoyant lower crust. Due to the 'water bed' effect, models with buoyant lower crust display marked overflow of the orogenic crust over the bounding vice

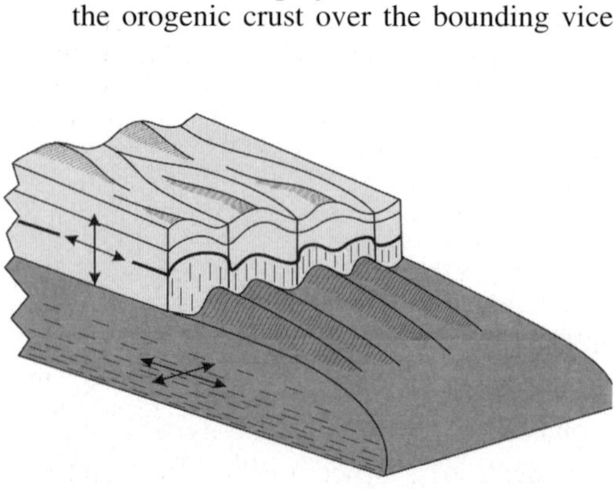

Fig. 13. Interpretation of the first order 3D structural characteristics of the vice experiments. Brittle upper crust is pale grey, ductile crust is dark grey. Arrows indicate orientation of bulk flattening strain (*xy*). Dashes indicate predicted orientation of metamorphic foliations.

blocks, while orogen parallel extension in the upper crust is accommodated by conjugate fracture systems and block rotations. Models with non-buoyant lower crust experience distributed upper crustal folding, and orogen parallel flow in the brittle crust is accommodated by a combination of homogenous stretching and displacements on vertical curvilinear transpression zones, whose kinematics are governed by the polarity of lower crustal flow, rather than the tectonic boundary conditions.

(5) The contribution of gravitational forces to the 3D evolution of the orogen increases with increasing *Ar* number. The *Ar* number controls the mean elevation attained by the orogen and the degree to which orogen-parallel flow will influence the cross-sectional geometry of the orogen. Comparison of analogue and numerical experimental results show that, in the central part of the orogen, amplitudes of folds and their surface expressions are similar in analogue and numerical models with lower *Ar*, while enhanced orogen-parallel flow in analogue models with higher *Ar* results in significant suppression or damping of these structures.

Discussion

Implicit in the vice orogen concept is that the orogenic lithosphere is effectively decoupled from the mantle dynamics responsible for continued plate convergence (Ellis *et al.* 1998). Decoupling of the overriding orogen from the subducting plate appears to be favoured at late stages of collision, provided the orogenic crust becomes sufficiently weak due to the combined effects of thickening and build-up of radiogenic heat (Burg & Ford 1997; Ellis *et al.* 1998). Many large orogens record a similar evolution from initial accretion and collision leading to crustal thickening, followed by gravitationally driven collapse (Dewey *et al.* 1986; Malavieille 1993). In cases where radiogenic heating is great enough, partial melting of the mid to lower crust may occur, leading to enhanced decoupling and the possibility of local buoyancy effects due to the presence of low density felsic melts at depth (Rey 1993; Vanderhaeghe & Teyssier 2001). Hence, the structural and geophysical characteristics of many large orogens are dominated by boundary conditions and mechanical properties that developed at late stages of their history, rather than during the early subduction-related stages of accretion and initial thickening (e.g., Burg *et al.* 2002). The results of the vice model experiments reported above should therefore be evaluated in terms of their potential application to the late-stage structure and dynamics of large hot orogens. Key features of several natural vice orogens are reviewed below in light of the experiments.

The Superior vice: insights from project Lithoprobe

The Superior Province of Canada is thought to have formed by progressive Neoarchean north to south accretion of juvenile volcanic arcs and continental crustal fragments onto a Mesoarchean nucleus (Williams 1990). The large-scale east-west belt structure of the Superior Province (Fig. 14), a variety of syn-accretionary plutonic and sedimentary rocks, together with the preservation of north dipping intra-crustal seismic fabrics and upper-mantle penetrating reflectors, records this *c.* 2720 to 2680 Ma accretionary history (e.g., Card & Ciesielski 1986; Calvert *et al.* 1995; White *et al.* 2003). The post-accretionary history of the Superior province is characterized by widespread granitic intrusion and the development of regional systems of upper crustal, vertical curvilinear shear zones (Fig. 14) (Leclair *et al.* 1993; Card & Poulsen 1998). This late (2660 to 2650 Ma) orogen-scale tectonothermal event coincides with final collision with a continental terrane in the south (Percival *et al.* 1994), thereby putting the entire Superior orogen into the jaws of a north-south vice.

Although generally interpreted in terms of a dextral transpressive regime (e.g., Williams *et al.* 1992), the regional-scale system of curvilinear shear zones have both dextral and sinistral horizontal displacements, depending on their orientation, and typically also record a strong zone normal flattening component (e.g., Wilkinson *et al.* 1999). Seismic reflection profiles indicate that the shear zones root into or are truncated by strong mid-crustal reflectivity fabrics (Jackson & Cruden 1995; Chroukroune *et al.* 1997; White *et al.* 2003). Field observations of tilted crustal sections indicate that mid-crustal reflectivity likely correlates with strong metamorphic fabrics and intrusive layering associated with an episode of east-west extensional flow under high temperature (amphibolite to granulite facies) conditions (Moser *et al.* 1996). U–Pb dating of metamorphic and igneous minerals associated with these fabrics and metamorphic mineral assemblages in upper crustal shear zones indicate that both mid-crustal extensional

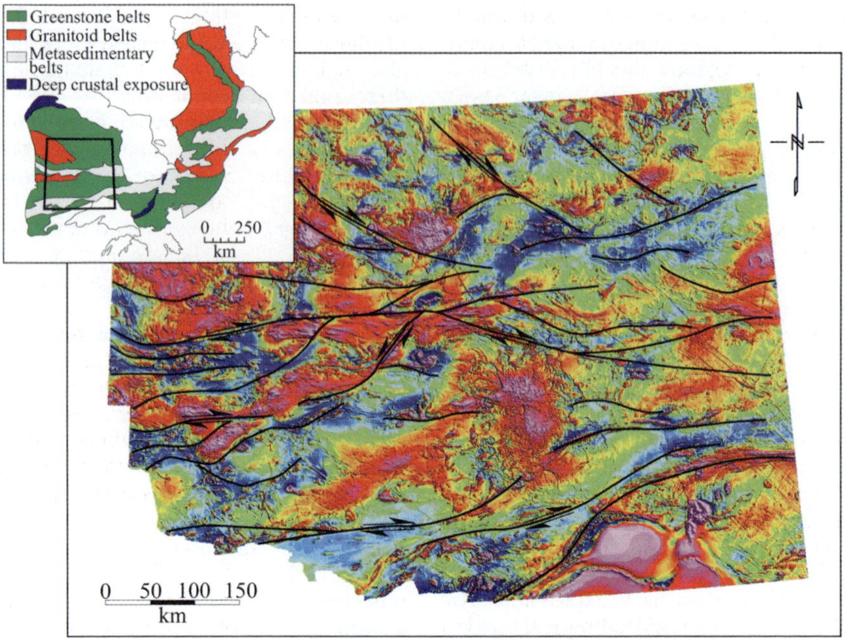

Fig. 14. Total field aeromagnetic map of the western part of the Superior Province (reprocessed after Gupta 1991). Blue tones are low magnetic intensity and red tones are high magnetic intensity. Major curvilinear shear zones, active in the late Archean (*c.* 2680–2650 Ma) are indicated with their sense of horizontal shear, where known. Inset: major subdivisions and location of the Superior Province with respect to the Great Lakes and Hudson's Bay. Box indicates location of the aeromagnetic map.

flow and upper-crustal deformation localized on transpressional shear zones were broadly contemporaneous at *c.* 2660 Ma (Moser *et al.* 1996; Wilkinson *et al.* 1999). Mid-crustal metamorphism continued episodically until *c.* 2600 Ma (Krogh 1993), suggesting that mid to lower crustal flow continued well after upper crustal deformation was locked in.

The geometry, kinematics and timing of upper crustal curvilinear shear zones in the Superior Province and their relationship to mid-crustal, orogen parallel extensional fabrics strongly resembles the characteristics of vice models with non-buoyant lower crust (Fig. 7b). The predominance of low pressure – low to medium temperature metamorphic assemblages within the Superior Province and the local preservation of prehnite-pumpellyite facies rocks indicate that orogenic relief and crustal thickening was likely suppressed. This scenario is most consistent with Experiment #33, which due to a high Ar number of 5.2 only increased in average elevation by *c.* 1 mm (equivalent to *c.* 2600 m) after 35% shortening (Fig. 5b), developed upper crustal curvilinear shear zones and upright folds, and experienced strong bulk mid crustal vertical flattening and orogen-parallel extensional strains (Fig. 12). Continued mid to lower crustal ductile flow in the Superior Province could be accounted for by top-down cooling (and stiffening) of a hot crust during continued orogen-parallel collapse.

Other Archean vice orogens

The vice models also account for the structural features of other Neoarchean orogens. For example, the Yilgarn craton of Australia is also characterized by a system of orogen subparallel curvilinear shear zones that root into mid-crustal horizontal reflectors (e.g., Goleby *et al.* 2002). Structural studies indicate the shear zones are associated with orogen-parallel extension, triggered by weakening due to mid-crustal granite emplacement and subsequent gravitational collapse (i.e., increase in Ar) (Davis & Maidens 2003). The characteristics of Experiments #33 and #34 also best match those of the Yilgarn.

In contrast to the Superior and Yilgarn, late Archean deformation in the Dharwar craton, India, and Zimbabwe craton, Africa, are strongly influenced by buoyancy effects in addition to the formation of upper crustal shear zones and folds (Choukroune *et al.* 1997; Treloar & Blenkinsop 1995). The structural style of the Dharwar craton is characterized by an anastomozing system of orogen sub-parallel shear zones which tend to be associated with synformal keels that wrap around elongate doubly plunging antiforms (domes) cored by granitic gneisses. Bouhallier *et al.* (1995) ascribe this strain pattern to lateral compression of thermally weakened lithosphere and simultaneous growth of gravity instabilities in the crust (diapirism). Such a scenario fits well with Experiment #32, which due to the combination of a buoyant lower crust and high *Ar* number, developed orogen sub-parallel shear zones that nucleated in syncline cores, and elongate anticlines cored and amplified by buoyant lower crustal material. Late Archean deformation of the Zimbabwe craton occurred due to compression caused by collision with the central zone of the Limpopo belt to the south (Treloar & Blenkinsop 1995). In contrast to the examples cited above, this led to the development of conjugate sets of regional NNE-striking sinistral and ESE-striking dextral shear zones that accompanied the intrusion of the Chilimanzi granites (Treloar & Blenkinsop 1995). This pattern is consistent with Experiment #29 (Fig. 7a) and suggests that the Chilimanzi granites were sufficiently voluminous to affect the bulk density of the mid crust. However, southward overflow at the Zimbawe craton–Limpopo belt margin is not observed, perhaps due to the protracted earlier tectonic history of this boundary (McCourt & Vearncombe 1992).

Proterozoic vice orogens

Almost all Proterozoic orogens are wide (*c.* 500–1000 km), have significant restored length (>2000 km) and formed by convergence between older, strong Archean lithospheric blocks (e.g., Trans-Hudson orogen, White *et al.* 1999; Svecofennian orogen, Ehlers *et al.* 1993; East African-Antarctic orogen, Jacobs & Thomas 2004). As in the Archean, these orogens record an early history of accretionary tectonics, and are followed by a protracted latter stage of transpression under high temperature conditions (Gapais *et al.* 2005). Taking the Palaeoproterozoic Trans-Hudson orogen of Canada as an example, this late transpression is expressed by orogen sub-parallel curvilinear shear zones and pervasive metamorphic fabrics that record upper crustal vertical thickening and horizontal extension under ductile conditions. Seismic reflection data indicates that vertical upper crustal shear zones root into strong mid-crustal horizontal reflectivity (Hajnal *et al.* 1996). Seismic and structural studies of the eastern boundary of the Trans-Hudson with the Archean Superior Province indicate an early phase of outward thrusting (overflow), followed by steepening and overturning of the boundary zone (White *et al.* 1999). Although the latter feature is not observed in the vice models, the late orogenic structure of the Trans Hudson and other Proterozoic orogens fits well with experiments with non-buoyant lower crust. Given the development of higher pressure metamorphic assemblages and the longevity of high temperature thermal conditions (Gapais *et al.* 2005), Experiment #34 (Figs 7b and 9), with a lower *Ar* number and therefore greater thickening under more ductile lower crustal conditions, applies best to these orogens.

Phanerozoic vice orogens

Elements of vice tectonics can also be recognized in many Palaeozoic to Recent orogenic systems. The original vice models of Ellis *et al.* (1998) were inspired by structural and geophysical features of the Newfoundland Appalachians, where weak lithosphere of the Dunnage zone is squeezed between the strong vice blocks of the Avalon and Humber zones. The Variscan orogeny of Europe developed in response to convergence between Laurentia and Gondwana in the late Palaeozoic (Vanderhaeghe & Teyssier 2001). This convergence led to the development of a major system of orogen sub-parallel curvilinear shear zones and associated sedimentary basins and well-developed mid-crustal horizontal reflectivity linked to partial melting, granite intrusion and extensional flow (Rey 1993; Malavieille 1994; Vanderhaeghe & Teyssier 2001).

Vice tectonics are also apparent in the Alpine-Himalayan system, particularly in Tibet and Anatolia, which are currently experiencing eastward and westward escape in response to collision between Asia with India and Arabia, respectively (Tapponier *et al.* 1982; Dewey *et al.* 1986). As in the vice orogens cited above, orogen parallel extension in the upper crust is accommodated by displacements on systems of curvilinear shear zones that based on available seismic data in Tibet, root into a mid-crust characterized by horizontal reflectivity and the presence of partial melts (Nelson *et al.* 1996; Vanderhaeghe & Teyssier 2001). Although much recent attention has been focussed on orogen perpendicular channel flow of mid-crustal material facilitated by erosion at the southern side of the Himalaya (e.g., Beaumont *et al.* 2001), the larger scale response of the Tibetan plateau to vice-like convergence between India and Asia is similar to the models presented here and characterized by orogen parallel ductile flow of the lower crust and development of related vertical shear zones in the upper crust (Royden *et al.* 1997).

Likewise, the East Anatolia high plateau appears to be a particularly good example of an active vice orogen. Following an earlier Tertiary accretionary history, this region has experienced vice-like compression between stronger lithospheric blocks represented by the Rhodope-Pontide arc and Arabian plate for the last 11 Ma (Şengör *et al.* 2003). Slab break off at *c.* 11 Ma led to decoupling of the orogenic lithosphere from mantle dynamics as well as crustal weakening and magmatic activity (Keskin 2003). Continued convergence has resulted in the development of the East Anatolia high plateau (mean elevation *c.* 2 km; viz. Experiment #32), significant westward lateral extrusion of the Anatolian block, as well as the development of upper mantle shear wave splitting anomalies that are consistent with orogen parallel flow of the asthenosphere under East Anatolia (Dewey *et al.* 1986; Sandvol *et al.* 2003; Şengör *et al.* 2003).

Implications for fabric studies in orogens

The discussion above focuses on the large-scale surface characteristics and cross-sectional structure of possible vice orogens. Our experiments also provide potential insight on the 3D distribution of strain, fabrics and mesoscopic structures in orogens. For example, field studies find that 'plane strain is more the exception than the rule' in ductilely deformed rocks in orogenic belts (Huddleston 1999) with finite strain measurements indicating a complete spectrum from L-, through L-S and S-tectonites (Ramsay & Huber 1987). The experiments reported here demonstrate that significant variations in finite strain can be achieved by local variation in the degree of vertical thinning or thickening combined with lateral variation in the amount of orogen parallel extension. In particular, they provide insight into the development of orogen parallel stretching lineations observed in the deeper, crystalline parts of some orogens (e.g., Gapais *et al.* 1992; Hatcher 2001) as well as the observation of extension along orogen parallel fold axes (Mancktelow & Pavlis 1994).

A significant feature of all the vice experiments reported here is a transition from upper crustal folding and vertical shear zone development to lower crustal, orogen parallel ductile flow (Fig. 13). The boundary between these regimes in the models is not a discrete detachment; rather it is gradational between the folded brittle-ductile interface and the underlying ductile crust. This is likely an artefact of the model construction: in nature the brittle-ductile transition occurs over a finite thickness of middle crust and field observations indicate that boundaries between 'partitioned' upper and lower crustal structural regimes (Oldow *et al.* 1989; Royden 1996) occur across well-defined detachment structures, often associated with discrete mechanical features such as lithological contacts or magmatic softening horizons (e.g., Collins & Vernon 1991; Axen *et al.* 1998; Klepeis *et al.* 2004).

The common observation of mid to lower crustal seismic reflectivity in orogens is predicted by the orogen parallel flow observed in the models (see also Meissner & Mooney 1998). Because of the weak rheology employed and the vertical velocity gradient observed in the models, orogen parallel stretching is greater in the mantle lithosphere of the experiments than in the ductile crust (Fig. 12b). Strong orogen parallel stretching in the mantle lithosphere and underlying asthenosphere of natural orogens may impart a sufficient degree of anisotropy to cause seismic shear wave splitting anomalies (e.g., Vauchez & Nicolas 1991; Sandvol *et al.* 2003).

Conclusions

Simple vice models provide insight into how different strain regimes can develop at different levels of the crust, in response to boundary conditions, rheological structure and gravitational forces. The experiments provide a framework for the interpretation of a diverse array of structural and geophysical observations in Archean to Recent vice orogens (e.g., upper crustal vs. lower crustal structural regimes; finite strain variation, seismic reflectivity patterns; upper mantle seismic anisotropy). The results also illustrate how in some cases interpretation of local structures in the field (e.g., transpressive shear zones, deep crustal extensional fabrics) might lead to erroneous inferences about palaeotectonic boundary conditions (e.g., oblique convergence; post-orogenic collapse).

The University of Toronto Tectonic Laboratory has been funded by Discovery, Equipment and Lithoprobe Grant programs of the Natural Sciences and Engineering Research Council of Canada (NSERC). We are grateful to D. Boutelier and W. M. Schwerdtner for comments on a draft of the manuscript, and A. Babeyko, S. Ellis and W. Schellart for thorough and constructive reviews, all of which led to substantial improvements, clarifications and corrections. This is Lithoprobe contribution #1412.

References

AXEN, G. J., SELVERSTONE, J., BYRNE, T. & FLETCHER, J. M. 1998. If the strong crust leads, will the weak crust follow? *GSA Today*, **8**, 1–8.

BAILEY, R. C. 1999. Gravity-driven continental overflow and Archaean tectonics. *Nature*, **398**, 413–415.

BAILEY, R. C. 2001. Dynamical analysis of continental overflow. *Journal of Geodynamics* **31**, 293–310.

BEAUMONT, C., JAMIESON, R. A., NGUYEN, M. H. & LEE, B. 2001. Himalayan tectonics explained by extrusion of a low-viscosity crustal channel coupled to focused surface denudation. *Nature*, **414**, 738–742.

BENES, V. & DAVY, P. 1996. Modes of continental lithospheric extension: experimental verification of strain localization processes. *Tectonophysics*, **254**, 69–87.

BLANKENBACH, B. & BUSSE, F. *ET AL.* 1989. A benchmark comparison for mantle convection codes. *Geophysical Journal International*, **98**, 23–38.

BURG, J.-P. & FORD, M. 1997. Orogeny through time: an overview. *In*: BURG, J.-P. & FORD, M. (eds) *Orogeny Through Time*, Geological Society Special Publication, **121**, 1–17.

BURG, J.-P., SOKOUTIS, D. & BONINI, M. 2002. Model-inspired interpretation of seismic structures in the Central Alps: crustal wedging and buckling at mature stage of collision. *Geology*, **30**, 643–646.

BOUHALLIER, H., CHARDON, D. & CHOUKROUNE, P. 1995. Strain patterns in Archaean dome and basin structures: the Dharwar craton (Karnakate, South India). *Earth and Planetary Science Letters*, **135**, 57–75.

CALVERT, A. J., SAWYER, E. W., DAVIS, W. J. & LUDDEN, J. N. 1995. Archaean subduction inferred from a mantle suture in the Superior Province. *Nature*, **375**, 670–674.

CARD, K. D. & CIESIELSKI, A. 1986. DNAG #1 subdivisions of the Superior Province of the Canadian Shield. *Geoscience Canada*, **13**, 5–13.

CARD, K. D. & POULSEN, K. H. 1998. Geology and mineral deposits of the Superior province of the Canadian Shield. *In*: LUCAS, S. B., St. ONGE, M. R. (eds) *Geology of the Precambrian Superior and Grenville Provinces and Precambrian Fossils of North America*, Geological Society of America, **C1**, 13–204.

CHOUKROUNE, P., BOUHALLIER, H. & ARNDT, N. T. 1995. Soft lithosphere during periods of Archaean crustal growth or crustal reworking. *In*: COWARD, M. P. & REIS, A. C. (eds) *Early Precambrian Processes*, Geological Society Special Publication, **95**, 67–86.

CHOUKROUNE, P., LUDDEN, J. N., CHARDON, D., CALVERT, A. J. & BOUHALLIER, H. 1997. Archaean crustal growth and tectonic processes: a comparison of the Superior Province, Canada and the Dharwar Craton, India. *In*: BURG, J.-P. & FORD, M. (eds) *Orogeny Through Time*, Geological Society Special Publication, **121**, 63–98.

COBBOLD, P. R. & JACKSON, M. P. A. 1992. Gum rosin (colophony): a suitable material for thermomechanical modeling of the lithosphere. *Tectonophysics*, **210**, 255–271.

COLLINS, W. J. & VERNON, R. H. 1991. Orogeny associated with anticlockwise P-T-t paths: evidence from low-P, high-T metamorphic terranes in the Arunta Inlier, central Australia. *Geology*, **19**, 835–838.

DAVIS, B. K. & MAIDENS, E. 2003. Archaean orogen-parallel extension: evidence from the northern Eastern Goldfields Province, Yilgarn Craton. *Precambrian Research*, **127**, 229–248.

DAVY, P. & COBBOLD, P. R. 1988. Indentation tectonics in nature and experiment, 1. Experiments scaled for gravity. *Bulletin of the Geological Institute, University of Uppsala, New Series*, **14**, 129–141.

DAVY, P. & COBBOLD, P. R. 1991. Experiments on shortening of a 4-layer model of the continental lithosphere. *Tectonophysics*, **188**, 1–25.

DEWEY, J. F., HEMPTON, M. R., KIDD, W. S. F., SAROGLU, F. & ŞENGÖR, A. M. C. 1986. Shortening of continental lithosphere: the neotectonics of Eastern Anatolia – a young collision zone. *In*: COWARD, M. P. & RIES, A. C. (eds) *Collision Tectonics*, Geological Society Special Publication, **19**, 3–36.

DIXON, J. M. & SUMMERS, J. M. 1983. Patterns of total and incremental strain in subsiding troughs; experimental centrifuged models of inter-diapir synclines. *Canadian Journal of Earth Sciences*, **20**, 1843–1861.

DIXON, J. M. & SUMMERS, J. M. 1986. Another word on the rheology of silicone putty; Bingham. *Journal of Structural Geology*, **8**, 593–595.

EHLERS, C., LINDROOS, A. & SELONEN, O. 1993. The late Svecofennian granite-migmatite zone of southern Finland – a belt of transpressive deformation and granite emplacement. *Precambrian Research*, **64**, 295–309.

ELLIS, S., BEAUMONT, C., JAMIESON, R. A. & QUINLAN, G. 1998. Continental collision including a weak zone: The vise model and its application to the Newfoundland Appalachians. *Canadian Journal of Earth Sciences*, **35**, 1323–1346.

ENGLAND, P. & HOUSEMAN, G. 1986. Finite strain calculations of continental deformation 2. Comparison with the India-Asia collision zone. *Journal of Geophysical Research*, **91** (B3), 3664–3676.

ENGLAND, P. C. & MCKENZIE, D. P. 1982. A thin viscous sheet model for continental deformation. *Geophysical Journal of the Royal Astronomical Society*, **70**, 295–321.

FULLSACK, P. 1995. An arbitrary Lagrangian-Eulerian formation for creeping flows and its application in tectonic models. *Geophysical Journal International*, **120**, 1–23.

GAPAIS, D., PECHER, A., GILBERT, E. & BALLEVRE, M. 1992. Synconvergence spreading of the higher Himalaya crystalline in Ladakh. *Tectonics*, **11**, 1045–1056.

GAPAIS, D., POTREL, A., MACHADO, N. & HALLOT, E. 2005. Kinematics of long-lasting Palaeoproterozoic transpression within the Thompson Nickel Belt (Manitoba, Canada). *Tectonics*. **24**, TC3002, DOI:10.1029/2004TC001700.

GARDE, A., CHADWICK, B., GROCOTT, J., HAMILTON, M. A., MCCAFFREY, K. J. W. & SWAGER, C. P. 2002. Mid-crustal partitioning and attachment during oblique convergence in an arc system, Palaeoproterozoic Ketilidian orogen, southern Greenland. *Journal of the Geological Society, London*, **159**, 247–261.

GILBERT, E. & MERLE, O. 1987. Extrusion and radial spreading beyond a closing channel. *Journal of Structural Geology*, **9**, 481–490.

GOLEBY, B. R., KORSCH, R. J., FOMIN, T., BELL, B., NICOLL, M. G., DRUMMOND, B. J. & OWEN, A. J. 2002. Preliminary 3-D geological model of the Kalgoorlie region, Yilgarn Craton, Western Australia, based on deep seismic-reflection and potential-field data. *Australian Journal of Earth Sciences*, **49**, 917–933.

GUPTA, V. K. 1991. Shaded image of total magnetic field of Ontario, west central sheet. *Ontario Geological Survey*, Map **M2585**, scale 1:1,000,000.

HAJNAL, Z., LUCAS, S. B., WHITE, D. J., LEWRY, J., BEZDAN, S., STAUFFER, M. R. & THOMAS, M. D. 1996. Seismic reflection images of strike-slip faults and linked detachments in the Trans-Hudson Orogen. *Tectonics*, **15**, 427–439.

HATCHER, R. D. Jr. 2001. Rheological partitioning during multiple reactivation of the Palaeozoic Brevard Fault Zone, Southern Appalachians, USA. *In*: HOLDSWORTH, R. E., STRACHAN, R. A., MAGLOUGHLINE, J. F. & KNIPE, R. J. (eds) *The Nature and Tectonic Significance of Fault Zone Weakening*. Geological Society of London, Special Publications, **186**, 257–271.

HIRT, C. W., AMSDEN, A. A. & COOK, J. L. 1974. An arbitrary Lagrangian-Eulerian computing method for all flow speeds. *Journal of Computational Physics*, **14**, 227–253.

HIRTH, G. & KOHLSTEDT, D. L. 1996. Water in the oceanic upper mantle: implications for rheology, melt extraction and the evolution of the lithosphere. *Earth and Planetary Science Letters*, **144**, 93–108.

HOUSEMAN, G. & ENGLAND, P. 1986. Finite strain calculations of continental deformation 1. Method and general results for convergent zones. *Journal of Geophysical Research*, **91**(B3), 3651–3663.

HOUSEMAN, G. A. & MOLNAR, P. 1997. Gravitational (Rayleigh-Taylor) instability of a layer with non-linear viscosity and convective thinning of continental lithosphere. *Geophysical Journal International*, **128**, 125–150.

HUDLESTON, P. J. 1999. Strain compatibility and shear zones: is there a problem? *Journal of Structural Geology*, **21**, 923–932.

JACKSON, J. 2002. Faulting, flow, and the strength of the continental lithosphere. *International Geology Review*, **44**, 39–61.

JACKSON, S. & CRUDEN, A. R. 1995. Formation of the Abitibi greenstone belt by arc-trench migration. *Geology*, **23**, 471–474.

JACOBS, J. & THOMAS, R. J. 2004. Himalyan-type indenter-escape tectonics model for the southern part of the late Neoproterozoic-early Palaeozoic East African-Antarctic orogen. *Geology*, **32**, 721–724.

KESKIN, M. 2003. Magma generation by slab steepening and breakoff beneath a subduction-accretion complex: an alternative model for collision-related volcanism in Eastern Anatolia, Turkey. *Geophysical Research Letters*, **30**, 8046, DOI 10.1029/2003GL018019.

KLEPEIS, K. A., CLARKE, G. L., GEHRELS, G. & VERVOORT, J. 2004. Processes controlling vertical coupling and decoupling between the upper and lower crust of orogens: results from Fiordland, New Zealand. *Journal of Structural Geology*, **26**, 765–791.

KOHLSTEDT, D. L., EVANS, B. & MACKWELL, S. J. 1995. Strength of the lithosphere: constraints imposed by laboratory experiments. *Journal of Geophysical Research*, **100**, 17587–17602.

KROGH, T. E. 1993. High precision U-Pb ages for granulite metamorphism and deformation in the Archean Kapuskasing structural zone, Ontario: implications for structure and development of the lower crust. *Earth and Planetary Science Letters*, **119**, 1–18.

LECLAIR, A. D., ERNST, R. E. & HATTORI, K. 1993. Crustal-scale auriferous shear zones in the central Superior province, Canada. *Geology*, **21**, 399–402.

MACKWELL, S. J., ZIMMERMAN, M. E. & KOHLSTEDT, D. L. 1998. High-temperature deformation of dry diabase with application to tectonics on Venus. *Journal of Geophysical Research*, **103** (B1), 975–984.

MAGGI, A., JACKSON, J. A., MCKENZIE, D. & PRIESTLEY, K. 2000. Earthquake focal depths, effective elastic thickness, and the strength of the continental lithosphere. *Geology*, **28**, 495–498.

MALAVIEILLE, J. 1993. Late orogenic extension in mountain belts: insights from the Basin and Range and the late Paleozoic Variscan belt. *Tectonics*, **12**, 1115–1130.

MANCKTELOW, N. S. & PAVLIS, T. L. 1994. Fold-fault relationships in low-angle detachment systems. *Tectonics*, **13**, 668–685.

MCCOURT, S. & VEARNCOMBE, J. R. 1992. Shear zones of the Limpopo Belt and adjacent granitoid-greestone terranes: implications for late Archaean collision tectonics in southern Africa. *Precambrian Research*, **55**, 553–570.

MEISSNER, R. & MOONEY, W. 1998. Weakness of the lower continental crust: a condition for delamination, uplift and escape. *Tectonophysics*, **296**, 47–60.

MITROVICA, J. X. & FORTE, A. M. 1997. The radial profile of mantle viscosity: results from the joint inversion of convection of post-glacial rebound observables. *Journal of Geophysical Research*, **102**, 2751–2769.

MERLE, O. 1989. Strain models within spreading nappes. *Tectonophysics*, **165**, 57–71.

MOSER, D. E., HEAMAN, L. M., KROGH, T. E. & HANES, J. A. 1996. Intracrustal extension of an Archean orogen revealed using single-grain U-Pb geochronology. *Tectonics*, **15**, 1093–1109.

NELSON, K. D. and ZHAO, W. *ET AL.* 1996. Partially molten middle crust beneath southern Tibet:

synthesis of project INDEPTH results. *Science*, **274**, 1684–1688.

OLDOW, J. S., BALLY, A. W., Avé LALLEMANT, H. G. & LEEMAN, W. P. 1989. Transpression, orogenic float, and lithospheric balance. *Geology*, **18**, 991–994.

PERCIVAL, J. A., STERN, R. A., SKULSKI, T., CARD, K. D., MORTENSEN, J. K. & BÉGIN, N. J. 1994. Minto block, Superior province: missing link in deciphering assembly of the craton at 2.7 Ga. *Geology*, **22**, 839–842.

POLIAKOV, A. & PODLADCHIKOV, Y. 1992. Diapirism and topography. *Geophysical Journal International*, **109**, 553–564.

PYSKLYWEC, R. & CRUDEN, A. R. 2004. Coupled crust-mantle dynamics and intraplate tectonics: two-dimensional numerical and three-dimensional analogue modelling. *Geochemistry, Geophysics, Geosystems (G^3)*, **5**(10), Q10003, DOI 10.1029/2004GC000748.

RAMSAY, J. G. & HUBER, M. I. 1987. *The techniques of modern structural geology. Volume 2. Folds and Fractures.* Academic Press.

RATSCHBACHER, L., MERLE, O., PHILLIPE, D. & COBBOLD, P. 1991. Lateral extrusion in the Eastern Alps: Part I. Boundary conditions and experiments scaled for gravity. *Tectonics*, **10**, 245–256.

REY, P. 1993. Seismic and tectono-metamorphic characters of the lower continental crust in Phanerozoic areas: a consequence of post-thickening extension. *Tectonics*, **12**, 580–590.

REY, P., VANDERHAEGHE, O. & TEYSSIER, C. 2001. Gravitational collapse of continental crust: Definitions, regimes, mechanisms and modes. *Tectonophysics*, **342**, 435–449.

ROSS, G. M., MILKEREIT, B., EATON, D., WHITE, D., KANASEWICH, E. R. & BURIANYK, M. J. A. 1995. Palaeoproterozoic collision orogen beneth the western Canada sedimentary basin imaged by Lithoprobe crust seismic-reflection data. *Geology*, **23**, 195–199.

ROSSI, D. & STORTI, F. 2003. New artificial granular materials for analogue laboratory experiments: aluminium and siliceous microspheres. *Journal of Structural Geology*, **25**, 1893–1899.

ROYDEN, L. 1996. Coupling and decoupling of crust and upper mantle in convergent orogens: implications for strain partitioning in the crust. *Journal of Geophysical Research*, **101**(B8), 17679–17705.

ROYDEN, L., BURCHFIEL, B. C., KING, R. W., WANG, E., CHEN, Z., SHEN, F. & LIU, Y. 1997. Surface deformation and lower crustal flow in eastern Tibet. *Science*, **276**, 788–790.

SANDVOL, E. & TURKELLI, N. *ET AL.* 2003. Shear wave splitting in a young continent-continent collision: An example from Eastern Turkey. *Geophysical Research Letters*, **30**, 8041, DOI 10.1029/2003GL017390.

SCHELLART, W. P. 2000. Shear test results for cohesion and friction coefficients for different granular materials: scaling implications for their usage in analogue modelling. *Tectonophysics*, **324**, 1–16.

SCHELLART, W. P. & LISTER, G. S. 2004. Tectonic models for the formation of arc-shaped convergent zones and backarc basins. *In*: SUSSMAN, A. J. & WEIL, A. B. (eds) *Orogenic curvature: integrating palaeomagnetic and structural analysis*, Geological Society of America Special Paper, **383**, 237–258.

ŞENGÖR, A. M. C., ÖZEREN, S., GENÇ, T. & ZOR, E. 2003. East Anatolia high plateau as a mantle-supported, north-south shortened domal structure. *Geophysical Research Letters*, **30**, 8045, DOI 10.1029/2003GL017858.

SEYFERTH, M. & HENK, A. 2004. Syn-convergent exhumation and lateral extrusion in continental collision zones – insights from three-dimensional numerical models. *Tectonophysics*, **382**, 1–29.

SOKOUTIS, D. 1990. Experimental mullions at single and double interfaces. *Journal of Structural Geology*, **12**, 365–373.

SOKOUTIS, D., BONINI, M., MEDVEDEV, S., BOCCALETTI, M., TALBOT, C. J. & KOYI, H. 2000. Indentation of a continent with a built-in thickness change: experiment and nature. *Tectonophysics*, **320**, 243–270.

TAPPONNIER, P., PELTZER, G., LA DAIN, A. Y., ARMIJO, R. & COBBOLD, P. 1982. Propagating extrusion tectonics in Asia: New insights from simple experiments with plasticene. *Geology*, **10**, 611–616.

TEYSSIER, C. & WHITNEY, D. 2002. Gneiss domes and orogeny. *Geology*, **30**, 1139–1142.

TOUSSAINT, G., BUROV, E. & JOLIVET, L. 2004. Continental plate collision: unstable vs. stable slab dynamics. *Geology*, **32**, 33–36.

TRELOAR, P. J. & BLENKINSOP, T. G. 1995. Archaean deformation patterns in Zimbabwe: true indicators of Tibetan-style crustal extrusion or not? *In*: COWARD, M. P. & REIS, A. C. (eds) *Early Precambrian Processes*, Geological Society Special Publication, **95**, 87–108.

VANDERHAEGHE, O. & TEYSSIER, C. 2001. Partial melting and flow of orogens. *Tectonophysics*, **342**, 451–472.

VAN KEKEN, P. E., KING, S. D., SCHMELING, H., CHRISTENSEN, U. R., NEUMEISTER, D. & DOIN, M. P. 1997. A comparison of methods for the modeling of thermochemical convection. *Journal of Geophysical Research*, **102**, 22477–22496.

VAUCHEZ, A. & NICOLAS, A. 1991. Mountain building: strike-parallel motion and mantle anisotropy. *Tectonophysics*, **185**, 183–201.

WEIJERMARS, R. 1986. Flow behaviour and physical chemistry of bouncing putties and related polymers in view of tectonic laboratory. *Tectonophysics*, **124**, 325–358.

WHITE, D. J., JONES, A. G., LUCAS, S. B. & HAJNAL, Z. 1999. Tectonic evolution of the Superior boundary zones from coincident seismic reflection and magnetotelluric profiles. *Tectonics*, **18**, 430–451.

WHITE, D. J., MUSACCHIO, G., HELMSTAEDT, H. H., HARRAP, R. M., THURSTON, P. C., VAN DER VELDEN, A. & HALL, K. 2003. Images of a lower-crustal oceanic slab: direct evidence for tectonic accretion in the Archean western Superior province. *Geology*, **31**, 997–1000.

WILKINSON, L., CRUDEN, A. & KROGH, T. 1999. Timing and kinematics of post-Timiskaming deformation within the Larder Lake-Cadillac deformation zone, SW Abitibi greenstone belt, Ontario, Canada. *Canadian Journal of Earth Sciences*, **36**, 627–647.

WILLIAMS, H. R. 1990. Subprovince accretion tectonics in the south-central Superior Province. *Canadian Journal of Earth Sciences*, **27**, 570–581.

WILLIAMS, H. R., SCOTT, G. M., THURSTON, P. C., SUTCLIFFE, R. H., BENNET, G., EASTON, R. M. & ARMSTRONG, D. K. 1992. Tectonic evolution of Ontario: Summary and Synthesis. In: *Geology of Ontario*, Special volume **4**, part 2, Ontario Geological Survey, Canada, 1255–1332.

Relative importance of trenchward upper plate motion and friction along the plate interface for the topographic evolution of subduction-related mountain belts

ANDREA HAMPEL & ADRIAN PFIFFNER

Institute of Geological Sciences, University of Bern, Baltzerstr 1, 3012 Bern, Switzerland (e-mail: andrea@geo.unibe.ch)

Abstract: We present finite-element models that investigate the relative importance of both trenchward motion of the upper plate and interplate coupling for the development of topography at convergent margins. Commonly, the role of a trenchward moving continental plate for the growth of topography is neglected in both modelling and field studies. Instead, forces exerted by the downgoing plate on the continental plate as well as interplate coupling are thought to be responsible for the deformation of the upper plate. Our model set-up includes an oceanic plate, which is in contact with a continental plate along a frictional plate interface and driven by slab pull. Both lithospheres have an elasto-visco-plastic rheology. The models demonstrate that friction along the plate interface can only lead to a high topography if the upper plate is moving toward the trench. Without such a trenchward advance, no high topography is generated, as the upper plate subsides owing to the drag exerted by the subducting plate. Increasing the coefficient of friction only amplifies the drag and increases the amount of subsidence. Our findings imply that trenchward motion of the continental plate plays a key role for the development of mountain belts at convergent margins; subduction of an oceanic plate even with high interplate coupling cannot explain the formation of Andean-type orogens.

The growth of topography at convergent margins like the Andean subduction zone has commonly been explained from a point of view that centres on the nature and behaviour of the subducting oceanic plate (Jordan *et al.* 1983; Bott *et al.* 1989; Dewey & Lamb 1992; Shemenda 1993; Willett *et al.* 1993; Gephart 1994; Huang *et al.* 1998; Stern 2002; Lamb & Davis 2003; Nicol & Beavan 2003). In particular, the age and buoyancy of the oceanic plate, its angle of subduction or the ratio between the velocity of plate convergence and of subduction have been drawn upon to explain different styles of deformation in the upper plate (e.g., Molnar & Atwater 1978; England & Wortel 1980; Royden 1993). Furthermore, the material flow in the wedge of mantle between the slab and the overriding plate as well as the viscosity of the mantle wedge have been inferred to modify the topography of the upper plate (Wdowinski *et al.* 1989; Billen & Gurnis 2001). This view assigns a passive role to the overriding plate, which is assumed to react to the forces exerted by the oceanic plate, and neglects the motion of the continental plate.

The use of a framework with an oceanic plate actively moving away from a spreading ridge and sliding down a fixed subduction slot beneath a continent has been criticized by Hamilton (1995, 2003), who emphasizes that oceanic plates subduct because their slabs sink into the mantle owing to their higher density. Hence, the slab pull force acts only in the vertical direction and generally leads to rollback of the slab (e.g., Forsyth & Uyeda 1975; Hamilton 2003). As a consequence, subduction zones (like other plate boundaries) are not stationary and, moreover, downgoing oceanic plates cannot compress their associated upper plates and induce intraplate shortening (Hamilton 1995, 2003). In Jarrard's (1986) comprehensive statistical analysis of the relations between subduction parameters, this is manifested in a poor correlation between the upper plate's strain regime, and the age and other characteristics of the downgoing plate. In contrast, a high correlation exists between the upper plate's strain regime and its absolute motion, which suggests that among the independent subduction parameters the motion of the upper plate is a first-order parameter to control the deformation of the upper plate (Jarrard 1986; Heuret & Lallemand 2005). Likewise, other studies inferred a close relation between the absolute motion

From: Buiter, S. J. H. & Schreurs, G. (eds) 2006. *Analogue and Numerical Modelling of Crustal-Scale Processes.* Geological Society, London, Special Publications, **253**, 105–115.
0305-8719/06/$15.00

of the overriding plate and backarc extension or shortening (Chase 1978; Uyeda & Kanamori 1979; Ruff & Kanamori 1980; Scholz & Campos 1995; Conrad *et al.* 2004).

Another first-order factor discussed as relevant for mountain building in the upper plate is the degree of interplate coupling (e.g., Dmowska *et al.* 1996; Cattin *et al.* 1997; Wang & He 1999; Bevis *et al.* 2001; Husson & Ricard 2004; Yanez & Cembrano 2004). In fact, the topographic profile of the upper plate has been used to *infer* high coefficients of friction and shear stresses along a plate interface, for example for the Andes (Lamb & Davis 2003) or northeast Japan (Huang *et al.* 1997). While Lamb & Davis (2003) and Huang *et al.* (1997) ultimately link the high interplate coupling to the subduction of the oceanic plate, other studies point out the close correlation between high seismic coupling and the trenchward advance of the upper plate (Uyeda & Kanamori 1979; Ruff & Kanamori 1980; Conrad *et al.* 2004). Numerical modelling studies on the influence of friction on topography also reached different conclusions about the role of friction in a subduction system (Cattin *et al.* 1997; Hassani *et al.* 1997; Buiter *et al.* 2001). Hassani *et al.* (1997) constructed finite-element models and assigned an effective elastic thickness to the lower and upper plate. Their upper plate was fixed, while a horizontal velocity and a slab pull force were imposed on the ridgeward and trenchward ends of the oceanic plate, respectively. The models were calculated with friction coefficients of $\mu = 0$, 0.1, 0.2 and 0.5. For all friction coefficients, Hassani *et al.* (1997) found that friction drags the upper plate down with the subducting plate, thereby inhibiting the growth of high topography. These results have been confirmed by Buiter *et al.* (2001) for a friction coefficient of $\mu = 0.1$ in a similar set-up; in another model run with a frictionless plate interface, Buiter *et al.* (2001) observed an upward vertical displacement for a trenchward advancing upper plate. In contrast to Hassani *et al.* (1997) and Buiter *et al.* (2001), Cattin *et al.* (1997) used a model in which the ocean plate subducts with a prescribed velocity and slab curvature beneath an elasto-visco-plastic upper plate, which moves toward the trench at a constant velocity. They concluded that friction coefficients of $\mu \leq 0.2$ are required to create and maintain high topography in the upper plate; only friction coefficients of $\mu > 0.2$ tend to drag the upper plate down. The contrary predictions of the different models with respect to interplate coupling, upper plate motion and the development of topography are not easily resolved, since Cattin *et al.* (1997) incorporate a trenchward moving overriding plate but do not account for slab pull, whereas the models of Hassani *et al.* (1997) and Buiter *et al.* (2001) consider slab pull but in a setting with a moving oceanic and fixed upper plate. The differences in the set-ups and results point, however, to the critical role that the delicate interaction between slab pull, interplate coupling, and motion of the upper plate might play for the deformation and the topographic evolution of the overriding plate.

Our study aims to evaluate the role of two parameters, trenchward upper plate motion and interplate coupling, for the growth of topography in the upper plate. We use finite-element models that include an oceanic plate which is driven by slab pull, and is in contact with a continental plate along a frictional plate interface. Following Cattin *et al.* (1997), Hassani *et al.* (1997), and Buiter *et al.* (2001), we represent the sub-lithospheric mantle by applying a basal boundary condition because this allows to study the interaction of the oceanic and continental lithospheres in a numerically feasible way. Two sets of models, in which the upper plate is fixed or moves toward the trench, will be presented. The models are not intended to represent any specific subduction zone but to shed light on the relative importance of the two parameters, upper plate motion and interplate coupling.

Model set-up

The model consists of a 4000 km-long oceanic plate with a thickness of 80 km and a 2000 km-long continental lithosphere, which is divided into a 30 km-thick crust and a 70 km-thick lithospheric mantle (Fig. 1). Both the oceanic and the continental lithospheres have an elasto-visco-plastic rheology, with a depth-independent Von-Mises yield stress and a temperature-independent linear viscosity. The values of the material properties (*cf.* Turcotte & Schubert 2002; ρ density, E Young's modulus, ν Poisson's ratio, σ_y Von-Mises yield stress, η viscosity) are given in Figure 1. At the bottom of the model linear spring and dashpot elements, which mimic the behaviour of an asthenosphere with a density of $\rho = 3200\ \mathrm{kg/m^3}$ and a viscosity of $\eta = 10^{20}$ Pa s, are applied to account for isostatic effects. In addition, a lithostatic pressure equivalent to the overburden is applied to the bottom of the two plates (oceanic plate: 2.6 GPa, continental plate: 3 GPa). The representation of the sub-lithospheric mantle by boundary conditions means that corner flow in the mantle wedge is neglected.

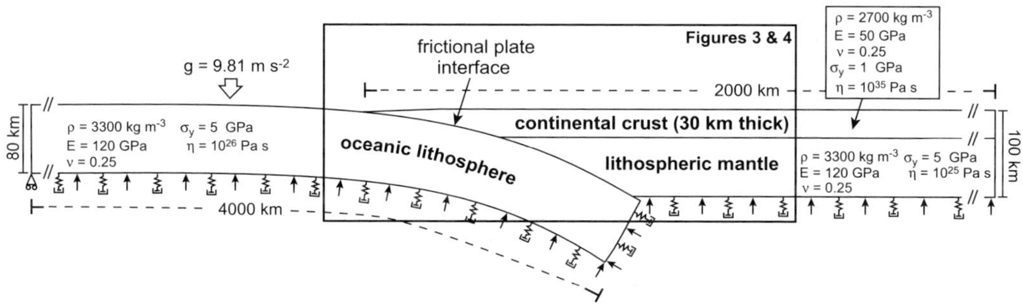

Fig. 1. Model set-up with separate oceanic and continental lithospheres that are in contact along the plate interface. Slip on the plate interface is controlled by a Coulomb criterion. The left end of the oceanic lithosphere is free to move in the horizontal direction and fixed in the vertical direction; the right end of the continental plate may not move in the vertical direction and is either fixed (Figs 2a, 3, 5a and 6a) or moves trenchward (Figs 2b, 4, 5b and 6b). Bending stresses in the oceanic plate are calculated separately and imported as pre-stress field (*cf.* Hibbitt *et al.* 2002). In order to include isostatic effects, lithostatic pressure as well as linear spring and dashpot elements that mimic the asthenosphere are applied at the bottom of both plates. See text for explanation of the material properties.

The plate interface is modelled as the contact zone between the two separate meshes of the oceanic plate and the continental plate. Relative tangential motion of the contacting element surfaces is governed by a Coulomb failure criterion with a critical shear stress ($\tau_{\max}$) and a coefficient of friction (μ) (*cf.* ABAQUS, 2002). In the models presented below, the coefficient of friction is increased stepwise from $\mu = 0$ to $\mu = 0.3$. A higher value of the friction coefficient is unrealistic for subduction systems, as indicated by heatflow data and palaeo-geothermal gradients of high-pressure/low-temperature metamorphic rocks (Molnar & England 1990; Kao & Chen 1991; Tichelaar & Ruff 1993).

The calculations have been performed with the ABAQUS/Standard software (ABAQUS, 2002), which uses a Lagrangian formulation and solves the following equation

$$\vec{\nabla}\bar{\bar{\tau}} + \rho\vec{g} = 0$$

where $\bar{\bar{\tau}}$ is the stress tensor, ρ density, and $\vec{g}$ gravitational acceleration. The elements have an average size of 10×10 km and 5×5 km for the oceanic and continental plate, respectively; no remeshing is applied. In a first step, the bending stresses in the oceanic plate, which is downflexed owing to a vertical force of 8×10^{13} N/m applied to its right edge, are calculated in the absence of the overriding plate (*cf.* Buiter *et al.* 2001; Funiciello *et al.* 2003). In the second step, the continental plate is added and contact is established along the plate interface (this stage of the model is shown in Fig. 1); the system is then allowed to reach isostatic equilibrium again. This procedure avoids the explicit implementation of subduction initiation and allows starting the experiments at a stage at which a slab has already formed. In the third step, the oceanic plate subducts, driven solely by slab pull, which is generated by its own weight and the force applied during the initial phase. No velocity or density anomalies or any additional forces are prescribed on either end of the subducting plate, which is free to move in the horizontal direction. The only boundary condition applied is that the far left edge of the oceanic plate is fixed in the vertical direction. In other words, the subduction velocity of the oceanic plate as well as the curvature of the slab may evolve freely during the main phase of the experiments. The subduction velocity in all models is of the order of a few centimeters per year. The right end of the continental plate is held fixed in the vertical direction. In the first model set presented below, the right edge of the upper plate is fixed in the horizontal direction as well (Figs 2a, 3, 5a and 6a). In the second model set, the boundary condition is such that the upper plate moves toward the trench owing to a displacement of 100 km prescribed at its right edge (Figs 2b, 4, 5b and 6b).

Results

Models with right end of the continental plate fixed

Figure 3 shows the finite vertical displacement fields for the first set of models, in which the right end of the upper plate is not allowed to move in the horizontal direction. All experiments are run until 200 km of relative slip are reached

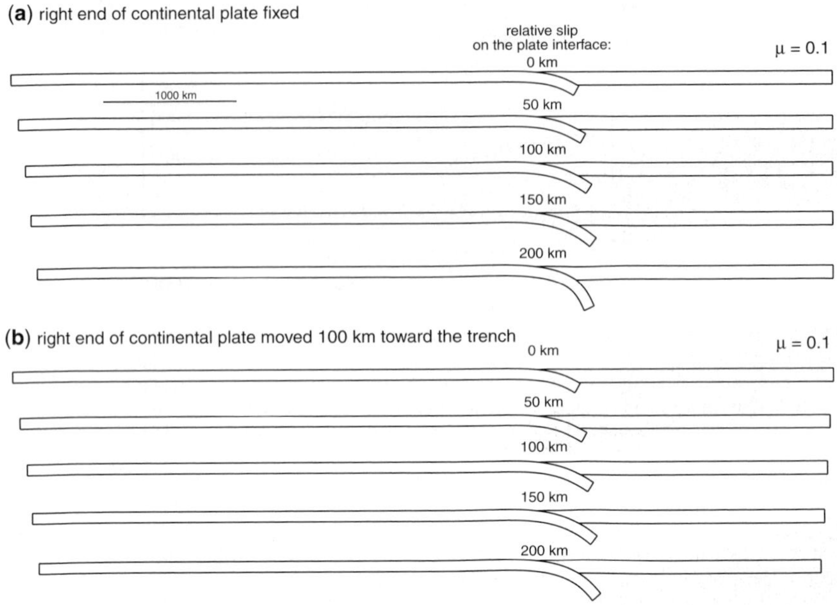

Fig. 2. Different stages of the experiments with a friction coefficient of $\mu = 0.1$ (**a**) without and (**b**) with trenchward motion of the upper plate.

on the plate interface; the slip is distributed uniformly on the plate interface. In order to illustrate the development of the entire model, different stages of the experiments performed with $\mu = 0.1$ are shown in Figure 2. For the frictionless case and for a coefficient of friction of $\mu = 0.05$, a broad zone of upward vertical motion develops landward of the intersection between the base of the continental lithosphere and the top of the subducting plate (region C in Fig. 3a, b). For coefficients of friction of $\mu \geq 0.1$, this zone strongly subsides by up to 9 km (regions C in Fig. 3c–e). In contrast, the frontal part of the overriding plate, within a distance of 30–40 km from the trench, is uplifted by *c.* 2 km regardless of the friction coefficient applied (regions A in Fig. 3). Between the uplifted and strongly subsiding regions A and C the upper plate experiences a moderate amount of subsidence of 1–2 km in all models (regions B).

The strong influence of the friction coefficient on the evolution of the models is also revealed in a comparison of their initial and final topographic profiles (Fig. 5a). For $\mu \leq 0.05$, a broad zone, in which the topography is higher than at the initial stage, has developed, whereas in the case of $\mu \geq 0.1$ the upper plate subsides over a wide region, with the width increasing with higher friction coefficients. Note that in all experiments, including the ones with upper plate motion (presented below), the topography evolves continuously from the initial to the final stage.

With respect to the stress field, the subsiding part of the upper plate is associated with a region of curved trajectories of the principal stresses σ_1 and σ_3 (shown in the upper panel of Fig. 6a for the model with $\mu = 0.1$). The maximum principal stress is vertical near the base of the lithosphere and perpendicular to the plate interface in its lower part. In the lower left corner of the upper plate, the minimum principal stress attains negative values (lower panel of Fig. 6a). This pattern of the principal stresses indicates that the lower left part of the upper plate is dragged down by the subducting plate. At the surface of the subsiding segment, the maximum principal stress has a horizontal orientation and is higher than in neighbouring regions, which is an expression of the upper plate's flexure resulting from the downward drag induced by the oceanic plate.

Models with a trenchward-moving continental plate

In the second set of experiments, the upper plate moves trenchward by 100 km. The finite vertical displacement fields after 200 km of slip are

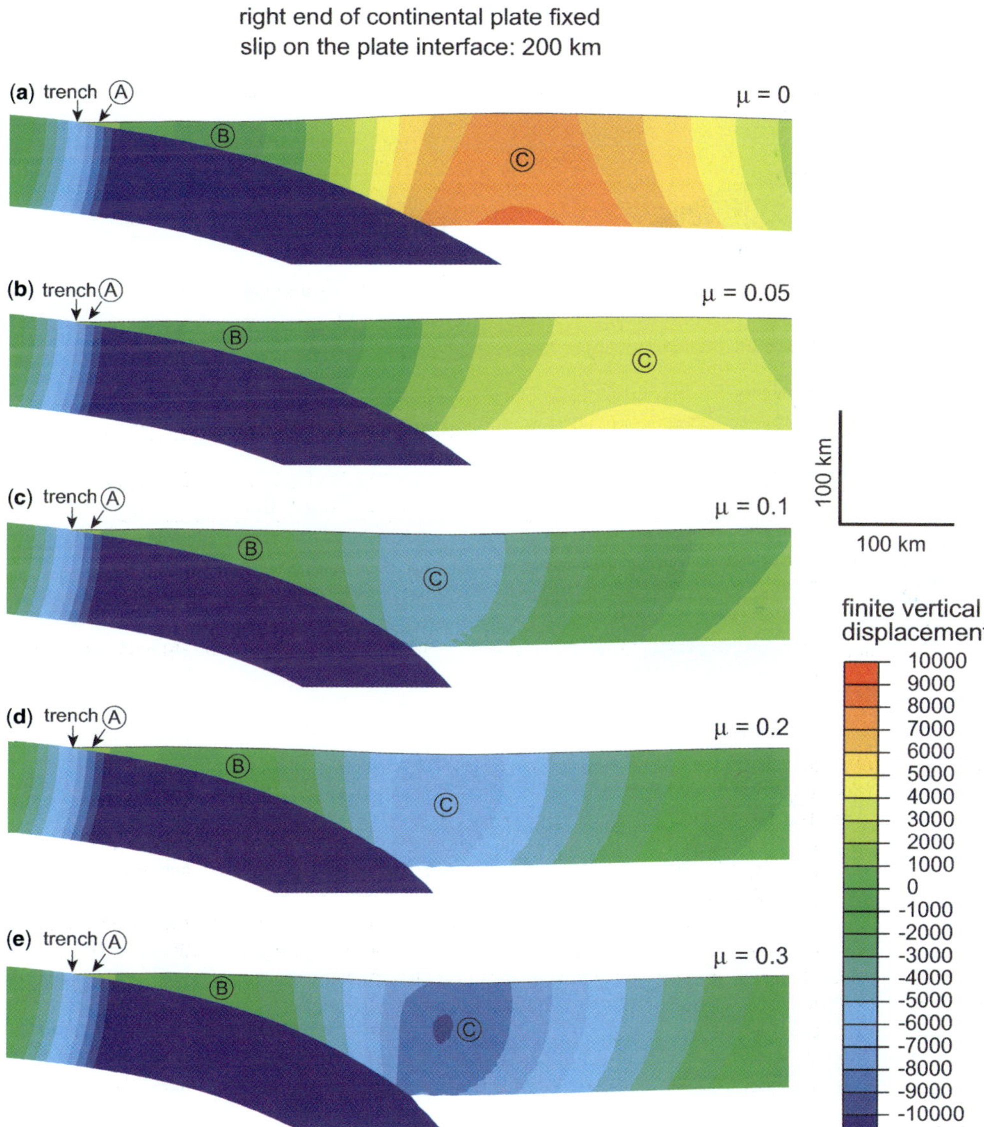

Fig. 3. Finite vertical displacement fields after 200 km of slip on the plate interface for the models with the right end of the continental plate fixed in the horizontal direction. Circles at the bottom of the model mark the spring and dashpot elements. Note that friction coefficients of $\mu \geq 0.1$ lead to strong subsidence of the upper plate (regions C) and inhibit the formation of a mountain belt. The frontal part of the upper plate (A) is uplifted; the adjacent part marked by B experiences moderate subsidence by 1–2 km.

depicted in Figure 4. For all coefficients of friction, the overriding plate experiences uplift, however, the uplift pattern differs for low and high friction values. In the presence of a frictionless plate interface, the distribution of uplift is similar to the one observed in the first model set except that the amount of uplift (Fig. 4a) and the final topography (Fig. 5b) is considerably higher. A maximum of ca. 9 km of uplift is reached slightly landward of the intersection of the base of the continental lithosphere and the top of the subducting oceanic plate (region C in Fig. 4a). For $\mu = 0.05$ and $\mu = 0.1$, the zone of highest uplift broadens and lower maximum

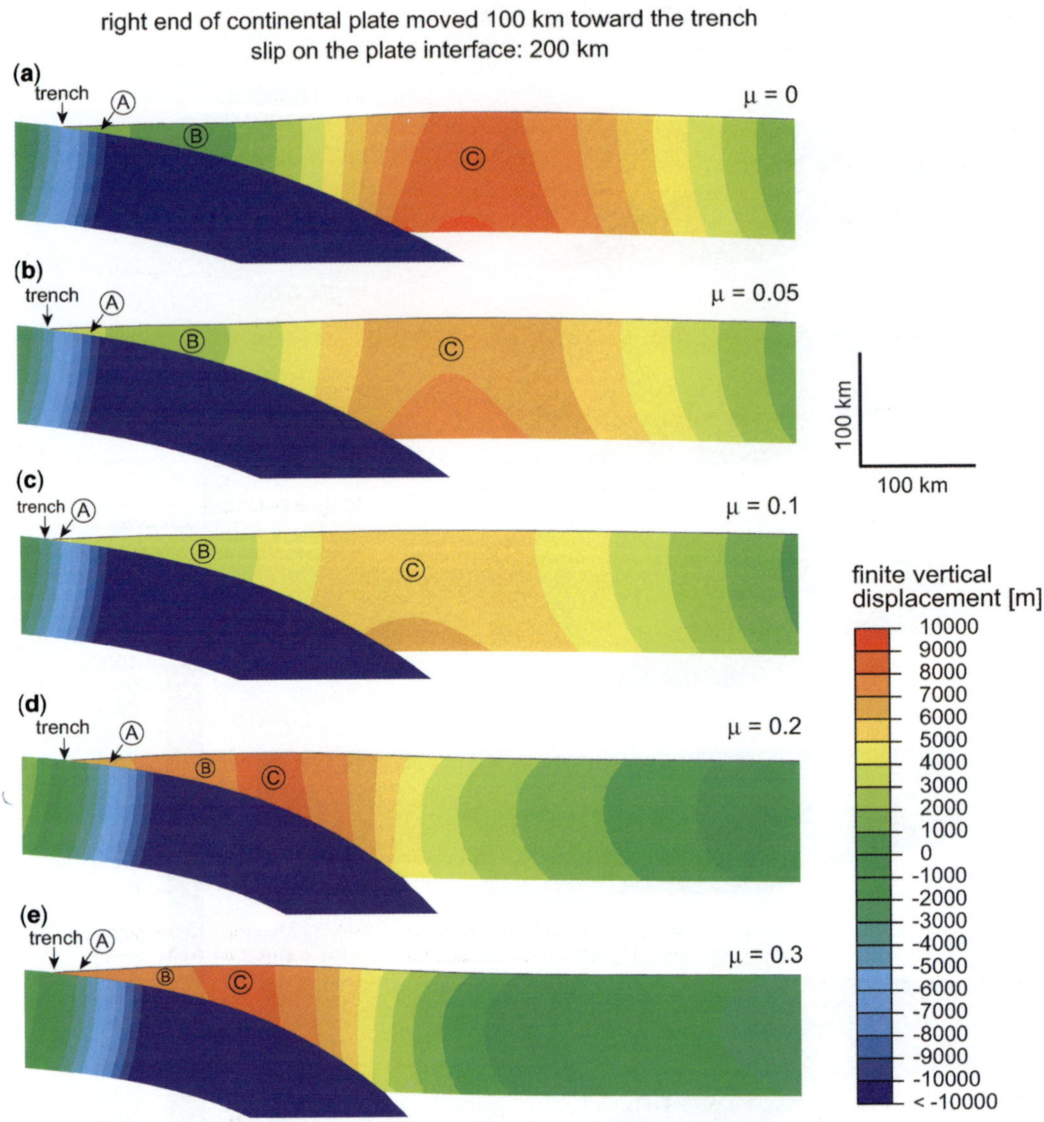

Fig. 4. Finite vertical displacement fields after 200 km of slip on the plate interface for the models with the continental plate advancing 100 km toward the trench. Note that for all coefficients of friction, high topography develops in the upper plate, regardless of the coefficient of friction (regions C). In the models with $\mu \leq 0.1$, the frontal part of the upper plate (A) experiences higher uplift than the adjacent region (B). For friction coefficients of $\mu > 0.1$, uplift in the overriding plate localizes above the plate interface (C in d, e).

values of 5–7 km are observed (regions C in Fig. 4b, c). In the three models with $\mu \leq 0.1$, the frontal part of the upper plate (A) is more strongly uplifted than the adjacent region B. The difference between the maximum and minimum upward displacement observed in regions C and B, decreases from 8 km ($\mu = 0$) to 2 km ($\mu = 0.1$) (Fig. 4a, c). For friction coefficients of $\mu > 0.1$, the uplift pattern changes dramatically: the zone of maximum uplift narrows and shifts toward the trench to a position above the centre of the plate interface, where a maximum vertical displacement of 8–9 km is reached (region C in Fig. 4d, e). The trenchward shift of the region with maximum vertical displacement is associated with a shift

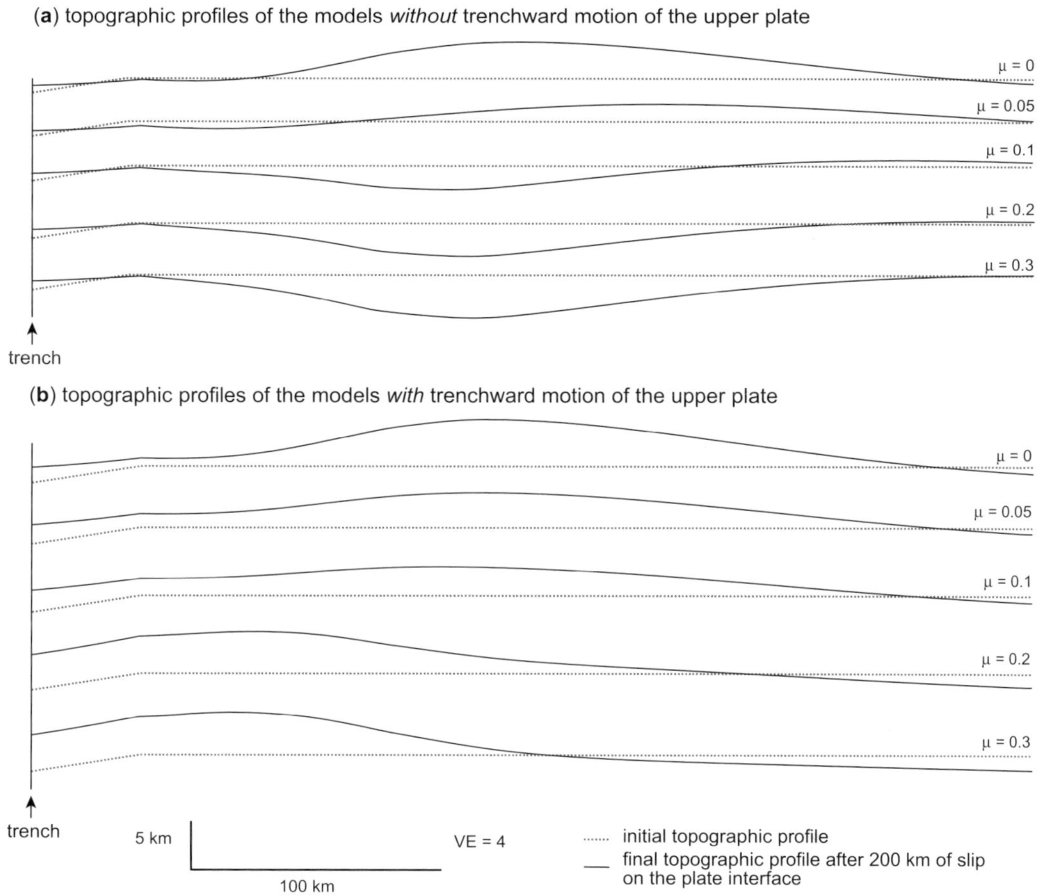

Fig. 5. Topographic profiles of the two model sets (**a**) without and (**b**) with trenchward motion of the upper plate. The dotted grey line marks the initial topography (Fig. 1), while the black line indicates the final topographic profile after 200 km of slip on the plate interface.

of the highest topography toward the trench (Fig. 5b). In the models with $\mu \geq 0.1$, the amount of uplift increases continuously from the trench to the region of maximum uplift (C), i.e., region B experiences stronger uplift than the frontal part of the continental plate (A).

The stress field obtained for the models with upper plate motion differs considerably from the one observed for a stationary upper plate (Fig. 6b). The orientation of the principal stresses indicates that the continental plate is under horizontal compression throughout, with the highest values of the maximum principal stress occurring at the lower left corner of the upper plate. With increasing friction, this region of high compressive stresses becomes larger (not shown in Fig. 6).

Discussion and conclusions

The results of our experiments reveal the crucial role of the upper plate's behaviour for topographic evolution at convergent margins. If the overriding plate remains stationary, the pull of the subducting plate tends to drag it down; this effect is amplified with increasing interplate coupling (Figs 3c–e and 6a). This observation agrees with the modelling results of Hassani *et al.* (1997) and Buiter *et al.* (2001). High topography in the upper plate develops in two cases: (1) if the upper plate remains fixed and the plate interface is frictionless, which is in accordance with Buiter *et al.* (2001), and (2) for $\mu > 0$, if the upper plate advances trenchward. In the latter case, the motion of the overriding plate

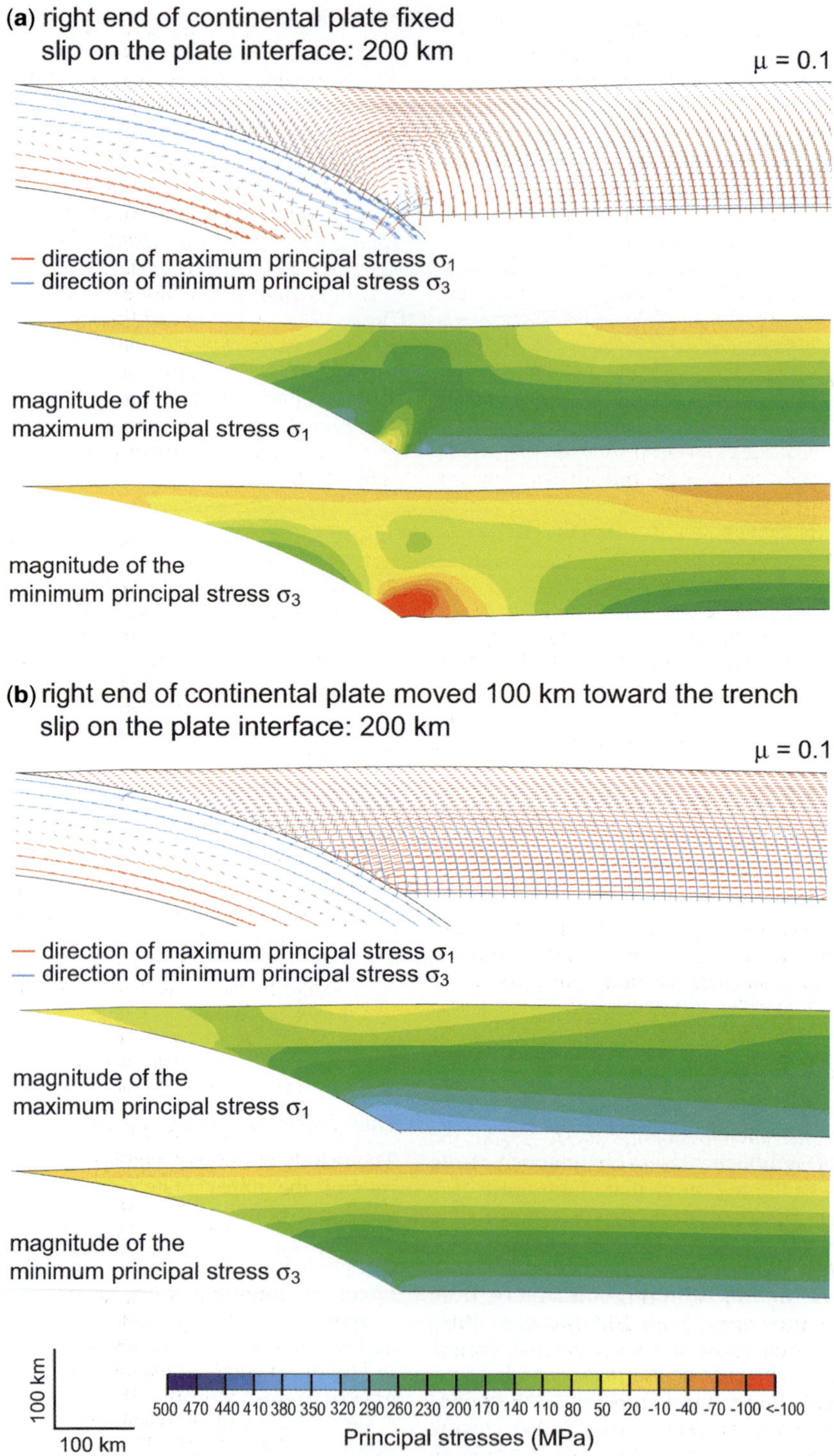

Fig. 6. Orientation and magnitude of the maximum and minimum principal stresses in the experiments with a friction coefficient of $\mu = 0.1$; (**a**) without and (**b**) with trenchward motion of the upper plate.

overcomes the downward drag exerted by the subducting plate, which is a combined effect of slab pull and interplate coupling. Thus the growth of high topography is promoted (Fig. 6b). Our model results imply an active role of the continental plate in the mountain-building process and are in general agreement with Cattin *et al.* (1997). These authors report, however, that friction coefficients of $\mu \leq 0.1$ lead to the development of unrealistically high topography, an effect that does not occur in our experiments (Fig. 4a, b). We interpret this significantly different behaviour as a result from the fact that we account for the pull of the slab, which acts against upward displacement of the overriding plate, whereas Cattin *et al.* (1997) did not include slab pull into their models.

With respect to the relationship between the degree of interplate coupling and a high topography, our experiments demonstrate that it may be misleading to infer a high coefficient of friction simply from the presence of high topography, without accounting for the motion of the upper plate. Such an approach was applied, for example, to the Andes (Lamb & Davis 2003). Although the occurrence of strong subduction earthquakes and the absence of a thick sedimentary cover on the oceanic plate in the central segments of the Andean margin support the conclusion of high interplate coupling, the Andes represent an example where the topography and the intracontinental deformation are not solely related to the motion and nature of the oceanic plate. Instead, geological evidence points to a close relationship between intraplate deformation and upper plate motion. For example, the kinematic analyses of fault populations in the central Andes suggest that temporal changes in the contraction directions reflect the *absolute South American* plate motions during the last 20.5 Ma (Marrett & Strecker 2000) rather than the Nazca–South American relative motion (e.g., Hindle *et al.* 2002; Riller & Oncken 2003), which is dominated by the absolute motion of the Nazca Plate. Faults in the Eastern Cordillera of northwestern Argentina and southern Bolivia indicate WNW–ESE shortening from the Miocene until the Pliocene (Marrett *et al.* 1994; Müller *et al.* 2002). During this time span, from 20.5 Ma to 3.2 Ma, South America moved west–northwestward (Pardo-Casas & Molnar 1987). Since 2–3 Ma, ENE–WSW shortening dominates on the faults (Marrett & Strecker 2000). At 3.2 Ma, South America changed to the west–southwestward direction still maintained today (Pardo-Casas & Molnar 1987; DeMets *et al.* 1990; Norabuena *et al.* 1998). This strongly suggests a causative relation between intraplate contraction and the westward advance of the South American plate, which has been faster than the rollback of the Nazca plate (Marrett & Strecker 2000). South America's absolute motion is primarily caused by the push of the South Atlantic ridge, which leads to a westward motion of both the mid-ocean ridge and South America relative to an approximately stable African Plate (Meijer & Wortel 1992; DeMets *et al.* 1990).

In conclusion, our model results indicate that the motion of the overriding plate plays a key role in the formation of mountain belts at convergent margins. Depending on the behaviour of the overriding plate, friction along the plate interface either leads to subsidence or amplifies crustal shortening and the growth of topography in the upper plate.

We wish to thank the editors, S. Buiter and G. Schreurs, who also co-organized the extremely stimulating GEOMOD 2004 conference in Emmetten at the Lake Lucerne, Switzerland, for their efforts. The thoughtful comments by S. Buiter and by the reviewers M. Gerbault, R. de Franco, and R. Govers improved the manuscript. A. H. thanks R. Hetzel for critical comments on earlier versions of the manuscript. Funding for the project was provided by the Swiss National Science Foundation (grant no. 2000-067952).

References

ABAQUS/Standard User's Manual, 2002. Version 6.3, Hibbitt, Karlson and Sorenson, Inc., Pawtucket, RI, USA.

Bevis, M., Kendrick, E., Smalley, R., Brooks, B., Allmendinger, R. & Isacks, B. 2001. On the strength of interplate coupling and the rate of backarc convergence in the central Andes: an analysis of the interseismic velocity field. *Geochemistry, Geophysics, Geosystems (G^3)*, **2**, 2001GC000198.

Billen, M. I. & Gurnis, M. 2001. A low viscosity wedge in subduction zones. *Earth and Planetary Science Letters*, **193**, 227–236.

Bott, M. H. P., Waghorn, G. D. & Whittaker, A. 1989. Plate boundary forces at subduction zones and trench-arc compression. *Tectonophysics*, **170**, 1–15.

Buiter, S. J. H., Govers, R. & Wortel, M. J. R. 2001. A modelling study of vertical surface displacements at convergent plate margins. *Geophysical Journal International*, **147**, 415–427.

Cattin, R., Lyon-Caen, H. & Chery, J. 1997. Quantification of interplate coupling in subduction zones and forearc topography. *Geophysical Research Letters*, **24**, 13, 1563–1566.

Chase, C. G. 1978. Extension behind islands arcs and motions relative to hot spots. *Journal of Geophysical Research*, **83**, 5385–5387.

CONRAD, C. P., BILEK, S. & LITHGOW-BERTELLONI, C. 2004. Great earthquakes and slab pull: interaction between seismic coupling and plate-slab coupling. *Earth and Planetary Science Letters*, **218**, 109–122.

DEMETS, C., GORDON, R. G., ARGUS, D. F. & STEIN, S. 1990. Current plate motions. *Geophysical Journal International*, **101**, 425–478.

DEWEY, J. F. & LAMB, S. H. 1992. Active tectonics of the Andes. *Tectonophysics*, **205**, 1–3, 79–95.

DMOWSKA, R., ZHENG, G. & RICE, J. R. 1996. Seismicity and deformation at convergent margins due to heterogeneous coupling. *Journal of Geophysical Research*, **101**, B2, 3015–3029.

ENGLAND, P. & WORTEL, R. 1980. Some consequences of the subduction of young slabs. *Earth and Planetary Science Letters*, **47**, 403–415.

FORSYTH, D. & UYEDA, S. 1975. On the relative importance of the driving forces of plate motion. *Geophysical Journal of the Royal Astronomical Society*, **43**, 163–200.

FUNICIELLO, F., MORRA, G., REGENAUER-LIEB, K. & GIARDINI, D. 2003. Dynamics of retreating slabs: 1. Insights from two-dimensional numerical experiments. *Journal of Geophysical Research*, **108**, B4, 2206, DOI 10.1029/2001JB000898.

GEPHART, J. 1994. Topography and subduction geometry in the central Andes: clues to the mechanics of a noncollisional orogen. *Journal of Geophysical Research*, **99**, 12279–12288.

HAMILTON, W. B. 1995. Subduction systems and magmatism. *In*: SMELLIE, J. L. (ed.) *Volcanism Associated with Extension at Consuming Plate Margins*, Geological Society Special Publications, **81**, 3–28.

HAMILTON, W. B. 2003. An alternative Earth. *GSA Today*, **13**, 11, 4–12, DOI 10.1130/1052-5173.

HASSANI, R., JONGMANS, D. & CHÉRY, J. 1997. Study of plate deformation and stress in subduction processes using two-dimensional numerical models. *Journal of Geophysical Research*, **108**, 17951–17965.

HINDLE, D., KLEY, J., KLOSKO, E., STEIN, S., DIXON, T. & NORABUENA, E. 2002. Consistency of geologic and geodetic displacements during Andean orogenesis. *Geophysical Research Letters*, **29**, 8, 2001GL013757.

HUANG, S. S., SACKS, I. S. & SNOKE, J. A. 1998. Compressional deformation of island-arc lithosphere in northeastern Japan resulting from long-term subduction-related tectonic forces: finite-element modeling. *Tectonophysics*, **287**, 1–4, 43–58.

HEURET, A. & LALLEMAND, S. 2005. Plate motions, slab dynamics and back-arc deformation. *Physics of the Earth and Planetary Interiors*, **149**, 31–51.

HUSSON, L. & RICARD, Y. 2004. Stress balance above subduction: application to the Andes. *Earth and Planetary Science Letters*, **222**, 1037–1050.

JARRARD, R. D. 1986. Relations among subduction parameters. *Reviews of Geophysics*, **24**, 217–284.

JORDAN, T. E., ISACKS, B. L., ALLMENDINGER, R. W., BREWER, J. A., RAMOS, V. A. & ANDO, C. J. 1983. Andean tectonics related to geometry of subducted Nazca Plate. *Geological Society of America Bulletin*, **94**, 3, 341–361.

KAO, H. & CHEN, W. P. 1991. Earthquakes along the Ryukyu-Kyushu arc: strain segmentation, lateral compression, and thermomechanical state of the plate interface. *Journal of Geophysical Research*, **96**, 21443–21485.

LAMB, S. & DAVIS, P. 2003. Cenozoic climate change as a possible cause for the rise of the Andes. *Nature*, **425**, 792–797.

MARRETT, R. & STRECKER, M. R. 2000. Response of intracontinental deformation in the central Andes to late Cenozoic reorganization of the South American Plate motions. *Tectonics*, **19**, 3, 452–467.

MARRETT, R., ALLMENDINGER, R. W., ALONSO, R. N. & DRAKE, R. E. 1994. Late Cenozoic tectonic evolution of the Puna Plateau and adjacent foreland, northwestern Argentine Andes. *Journal of South American Earth Sciences*, **7**, 179–207.

MEIJER, P. T. & WORTEL, M. J. R. 1992. The dynamics of motion of the South American Plate. *Journal of Geophysical Research*, **97**, 11915–11931.

MOLNAR, P. & ATWATER, T. 1978. Interarc spreading and cordilleran tectonics as alternates related to the age of subducted oceanic lithosphere. *Earth and Planetary Science Letters*, **41**, 330–340.

MOLNAR, P. & ENGLAND, P. 1990. Temperatures, heat flux, and frictional stress near major thrust faults. *Journal of Geophysical Research*, **95**, 4833–4856.

MÜLLER, J. P., KLEY, J. & JACOBSHAGEN, V. 2002. Structure and Cenozoic kinematics of the Eastern Cordillera, southern Bolivia (21°S). *Tectonics*, **21**, 5, 1037, DOI 10.1029/2001TC001340.

NICOL, A. & BEAVAN, J. 2003. Shortening of an overriding plate and its implications for slip on a subduction thrust, central Hikurangi Margin, New Zealand. *Tectonics*, **22**, 6, 1070.

NORABUENA, E., LEFFLER-GRIFFIN, L. *ET AL.* 1998. Space geodetic observations of Nazca-South American convergence across the central Andes. *Science*, **279**, 358–362.

PARDO-CASAS, F. & MOLNAR, P. 1987. Relative motion of the Nazca (Farallon) and South American Plates since Late Cretaceous time. *Tectonics*, **6**, 233–248.

RILLER, U. & ONCKEN, O. 2003. Growth of the Central Andean Plateau by tectonic segmentation is controlled by the gradient in crustal shortening. *Journal of Geology*, **111**, 367–384.

ROYDEN, L. H. 1993. The tectonic expression of slab pull at continental convergent plate boundaries. *Tectonics*, **12**, 303–325.

RUFF, L. & KANAMORI, H. 1980. Seismicity and the subduction process. *Physics of the Earth and Planetary Interiors*, **23**, 240–252.

SCHOLZ, C. H. & CAMPOS, J. 1995. On the mechanism of seismic decoupling and backarc spreading at subduction zones. *Journal of Geophysical Research*, **100**, B11, 22103–22115.

SHEMENDA, A. 1993. Subduction of the lithosphere and backarc dynamics: insights from physical modeling. *Journal of Geophysical Research*, **98**, B9, 16167–16185.

STERN, R. J. 2002. Subduction zones. *Reviews of Geophysics*, **40**, 4, 1012, DOI 10.1029/2001RG000108.

TICHELAAR, B. W. & RUFF, J. L. 1993. Depth of seismic coupling along subduction zones. *Journal of Geophysical Research*, **98**, 2017–3027.

TURCOTTE, D. L. & SCHUBERT, G. 2002. *Geodynamics*, Cambridge University Press, 2nd edition.

UYEDA, S. & KANAMORI, H. 1979. Backarc opening and the mode of subduction. *Journal of Geophysical Research*, **84**, 1049–1061.

WANG, K. L. & HE, J. H. 1999. Mechanics of low-stress forearcs: Nankai and Cascadia. *Journal of Geophysical Research*, **104**, B7, 15191–15205.

WDOWINSKI, S., O'CONNELL, R. J. & ENGLAND, P. 1989. A continuum model of continental deformation above subduction zones: application to the Andes and the Aegean. *Journal of Geophysical Research*, **94**, 10331– 10346.

WILLETT, S. D., BEAUMONT, C. & FULLSACK, P. 1993. Mechanical model for the tectonics of doubly vergent compressional orogens. *Geology*, **21**, 371–374.

YANEZ, G. & CEMBRANO, J. 2004. Role of viscous plate coupling in the late Tertiary Andean tectonics. *Journal of Geophysical Research*, **109**, B02407, 15191–15205, DOI 10.1029/2003JB002494.

Deformation transfer in viscous detachments: comparison of sandbox models to the South Pyrenean Triangle Zone

HEMIN A. KOYI[1] & MAURA SANS[2]

[1]*Hans Ramberg Tectonic Laboratory, Department of Earth Sciences, Uppsala University, Uppsala, Sweden (e-mail: hemin.koyi@geo.uu.se)*

[2]*Universitat Autonoma Barcelona, Barcelona, Spain*

Abstract: Results of scaled sandbox models, containing three viscous layers located at different geographic and stratigraphic levels simulating three evaporitic units in the South Pyrenean Triangle Zone, and interpreted field data are presented here to explain structural variation and kinematics in shortened areas containing multiple weak horizons acting as detachments. In the Southern Pyrenean Triangle Zone, the Beuda, Cardona and Barbastro thrust fronts have similar geometric features to those developed in the models, suggesting that they could have formed and evolved in a similar way. These deformation fronts are not always perpendicular to the regional shortening direction. Instead, their direction is governed by the initial pinch-out of the viscous horizons. Model results show that triangle zones form when: (1) deformation is transferred to weak horizons located at higher stratigraphic levels, and (2) the deformation front reaches the pinch-out of the weak horizons. Model results also show that the rheology of the detachment horizons controls the geometry of the deformation front. Weak detachments (Cardona Formation, and pure silicone in the models) promote folding and back-vergent structures, and thus formation of triangle zones at the deformation front, irrespective of the location of the thrust front relative to the pinch-out of the viscous detachment. However, over strong (more viscous) detachments (Barbastro and Beuda formations, and impure silicone in the models), folds that eventually evolve to thrusts are dominant. In such cases, backthrusts form only at the pinch-out of the detachment layer. In cases where no viscous detachment is present, no backthrusts form, and therefore the thrust front does not develop a triangle zone geometry. Instead, a foreland-vergent piggyback sequence of thrusts forms. Model results show that the stratigraphic level of a detachment governs size, geometry and spacing of the imbricates formed above it.

The South Pyrenean Triangle Zone (SPTZ, Fig. 1) is an excellent area for studying the geometry of triangle zones because it fulfils two important requisites for formation and preservation of such zones: presence of ductile detachments (Cousenz-Schultz & Whiltscho 1996; Sans *et al.* 1996*a*, *b*) and moderate displacement along faults (Hatcher 1999). A triangle zone is defined as the strata between the basal blind thrust, the hinterland-vergent thrust and the most external of the foreland vergent thrusts to intersect the surface (Gordy *et al.* 1977; Charlesworth *et al.* 1987) irrespective of its internal structure. The presence of ductile detachments is fundamental to the formation of triangle zones. Moreover, the rheology and thickness of the mechanically weak layers control the style of deformation at the deformation front, and therefore at the tip line of the triangle zone (Sans *et al.* 1996*a*). Moderate displacement allows preservation of the triangle zone geometry after the triangle zone is abandoned and deformation progresses towards the foreland. Triangle zones have survived in only a few thrust systems with large displacements. In these cases the hinterland-vergent thrust was inverted. Displacement should therefore be considered a key factor for survival of triangle zones even if all the other attributes are present (Hatcher 1999).

Different aspects of the effect of multiple detachments have been studied using analogue models (e.g., Letouzey *et al.* 1995; Verschuren *et al.* 1996; Bonini 2001). However, in this paper, we present the results of two scaled analogue models prepared to study the formation and evolution of the Southern Pyrenean Triangle Zone, whose stratigraphy includes three evaporitic units located at different levels. In these models, three viscous layers were arranged in a lateral and frontal stepping-up order embedded within layers possessing Coulomb rheology in a similar way to that of the South Pyrenean

From: BUITER, S. J. H. & SCHREURS, G. (eds) 2006. *Analogue and Numerical Modelling of Crustal-Scale Processes.* Geological Society, London, Special Publications, **253**, 117–134.
0305-8719/06/$15.00

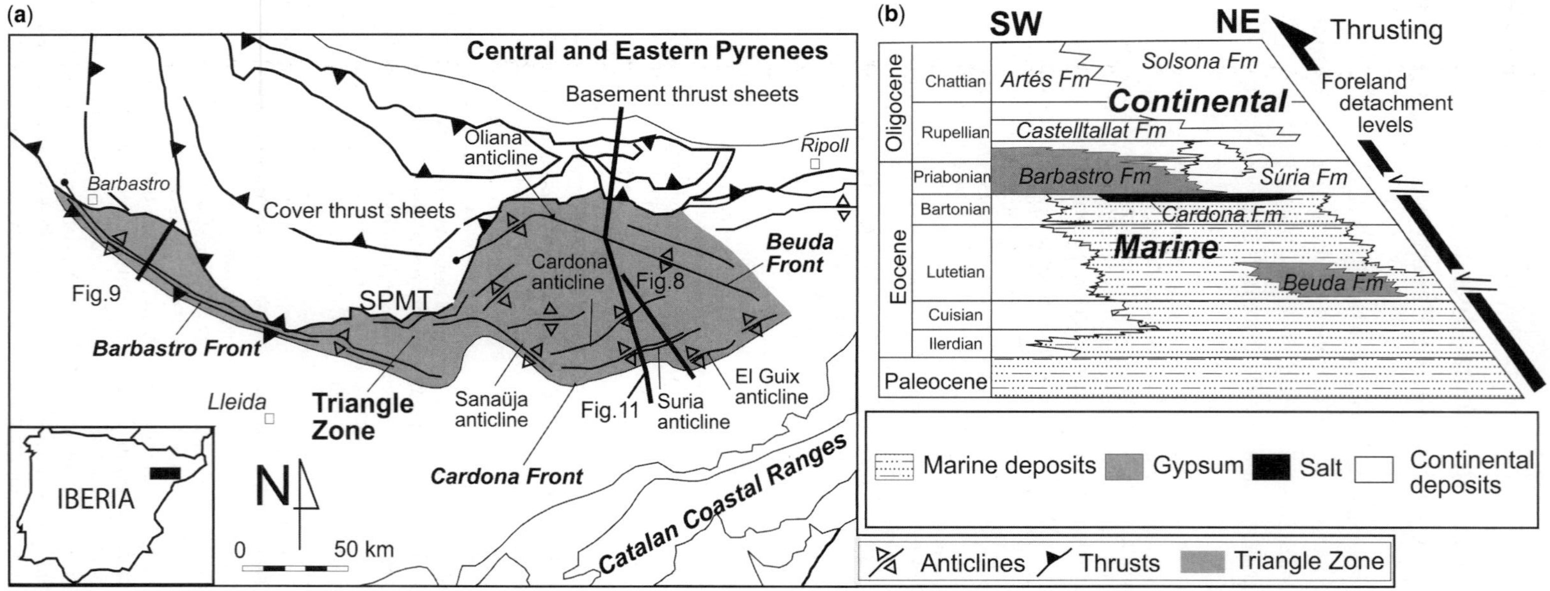

Fig. 1. (**a**) Structural map of the study area (inset shows location). The triangle zone (dark grey) is bounded by the South Pyrenean Main Thrust (SPMT) in the north and the most external structure in the foreland to the south. The triangle zone is internally deformed by anticlines with different orientations indicated in the map. (**b**) Stratigraphy of the Ebro foreland basin. The basin infill changed from marine to continental during thrusting. The three evaporitic levels outlined are the main detachments of the South Pyrenean Thrust System.

Triangle Zone. The purpose of these experiments was to understand the effect of multiple weak horizons in the deformation of a fold-thrust belt, investigate how deformation is transmitted between different detachment levels, study the effect of differences in thickness and rheology of the viscous layers on the geometry of triangle zones, and compare the results with the South Pyrenean Triangle Zone.

South Pyrenean thrust front

The geometry of the South Pyrenean thrust front changes along strike from east to west (Fig. 1). Both in the eastern and the westernmost parts, where there are no evaporitic levels at depth, the front is represented by a foreland-vergent and emergent thrust which places cover rocks on top of the foreland sediments. In contrast, in the central and central eastern parts, where there are evaporitic levels in the stratigraphic section of the foreland sediments, deformation is transferred to the foreland and the thrust front is blind (Vergés *et al.* 1992). Moreover, in this area, the tilted beds of the southern limbs of the antiforms have been passively transported towards the hinterland along a backthrust. The South Pyrenean Triangle Zone is, thus, limited between the South Pyrenean main emergent thrust in the north (hinterland), a basal thrust and the most external of the hinterland vergent structures (Fig. 1) (Sans *et al.* 1996*a*, *b*). The internal structure of the Triangle Zone is characterized by the presence of folds and thrusts with three different orientations. In the northern part the folds are NW–SE, in the central part, they are NE–SW, and in the southern and more western parts they are WNW–ESE (Vergés *et al.* 1992). Each of these areas corresponds, at depth, with a different detachment horizon of the Triangle Zone. In the north the detachment is located in the Beuda formation; in the central part, in the Cardona formation; and in the southwestern part, in the Barbastro formation. These three evaporitic formations are arranged in a lateral and foreland stepping-up geometry (Fig. 1b) which produced a staircase geometry in the basal detachment of the Triangle Zone. Wide flats are located within the evaporitic formations and ramps connect them (Vergés *et al.* 1992; Sans & Vergés 1995; Sans *et al.* 1996*a*, *b*).

Three main thrust wedges have developed along the southern margins of the evaporitic units. The Beuda thrust wedge is located at the southern pinch-out of the Beuda Formation where the basal thrust steps up into the Cardona salt. The Beuda Formation consists of more than 1000 m of alternating shale and anhydrites with 100 m of salt in its middle-upper part (Martínez *et al.* 1989). The intermediate thrust wedge coincides at the surface with the Santa Maria d'Oló, Súria and Sanaüja anticlines and it is related to the southern and southwestern pinch-out of the Cardona formation which consists of a maximum 350 m of salt. At the southern pinch-out of the Cardona formation, a hinterland vergent double anticline has developed (Sans & Vergés 1995; Sans 1999, 2003) representing the southern limit of the Triangle Zone. In the southwestern pinch out, the ramp between the Cardona and the Barbastro Formation coincides at the surface with the Sanaüja anticline. In this anticline, the difference in behaviour between the Cardona Formation (consists mainly of halite) and the Barbastro formation (consists mainly of gypsum and whose thickness increases from east to west, about 1 km thick) can be seen in the different structure under both limbs of the anticline (Sans & Vergés 1995; Sans *et al.* 1996*a*). Whereas in the northeastern limb, the salt of the Cardona Formation flows into the anticline core, the Barbastro gypsums forms a backthrust system that thickens the southwest limb of the anticline. The southernmost thrust wedge is represented at the surface by the Barbastro anticline which formed at the blind tip line of the Barbastro Formation. These details are displayed on seismic data of the area (Gil & Jurado, 1998, fig. 3). The Barbastro anticline consists of a south-vergent antiformal stack of the gypsum levels which, in the southern limb, shows a hinterland-vergent thrust. This thrust may correspond to the reactivation of the previous foreland-vergent upper detachment of the antiformal stack (Sans *et al.* 1996*b*).

Syntectonic sediments related to these three thrust wedges allow dating the formation of the wedges to Late Eocene–Early Oligocene, Late Oligocene and Late Oligocene–Early Miocene, from north to south respectively (Pardo & Villena 1979; Riba *et al.* 1983; Sáez 1987; Burbank *et al.* 1992; Senz & Zamorano 1992; Meigs *et al.* 1996; Vergés & Burbank 1996). From these ages, the three wedges are considered to have been the successive frontal structures of the Triangle Zone during the tectonic evolution of the area (Sans *et al.* 1996*a*). These wedges represent relict thrust fronts abandoned as the basal thrust propagated southwestwards to more external areas.

Modelling materials

Two basic materials have been used: a viscous silicone polymer (SGM36) and dry loose sand. SGM36 is a suitable material to simulate ductile behaviour of evaporites or mudstones, whereas loose sand is appropriate for simulating the brittle behaviour of non-evaporitic sediments in the upper crust (Weijermars 1986; Ballard *et al.* 1987; Koyi *et al.* 1993; Weijermars *et al.* 1993; Bonini 2001). SGM36, manufactured by Dow Corning Ltd, is a Newtonian viscous silicone. We used two kinds of SMG36: pure SGM36, which has an effective viscosity of 5×10^4 Pa s (density 987 kg m^{-3}) and impure SGM36 (mixed with loose sand), which is one order of magnitude more viscous, i.e., 5×10^5 Pa s (density 1100 kg m^{-3}).

The angle of internal friction of rocks in the brittle upper crust ranges between $\phi_n = 31°–40°$ with a cohesive strength of about 50 MPa (Byerlee 1978; Brace & Kohlstedt 1980). Loose sand with the same internal frictional angle, $\phi_m = 36°$ (McClay & Ellis 1987; Mulugeta & Koyi 1987; Koyi *et al.* 1993; Weijermars *et al.* 1993) and a small cohesive strength is a nearly perfect Coulomb material to simulate the brittle behaviour of sedimentary cover (Krantz 1991).

Layers of overburden were constructed to the desired thickness by scraping alternating layers of loose sand (grain size ≥ 35 μm). The bulk density of the loose sand is $\rho_m = 1700$ kg m^{-3} with a cohesive strength of about 140 Pa (Cotton & Koyi 2000).

Model set-up

Two models, referred to as model 1 and model 2 throughout this article, were designed to study the transfer of thin-skinned deformation between detachments at different levels. The models are based on the stratigraphic section shown in Figure 1 to simulate the effect of the three detachment horizons on the evolutionary history of the SPTZ. Each model consisted of three viscous layers (simulating the three detachment levels (Beuda Fm, Cardona Fm and Barbastro Fm in the SPTZ) interbedded with layers of loose sand (Fig. 2). The rheology, thickness and geometrical arrangement of each layer were similar in both experiments, and the minor differences (few millimetres to maximum 0.5 cm) in lateral dimension of the viscous layers between the two models, we believe were of minor significance (Fig. 2). The major difference between both experiments was the amount of bulk shortening: model 1 was shortened up to 15% and model 2 to 22%. Both values are in the range of shortening values achieved in the frontal part of the SPTZ (15–30%).

The brittle-viscous models, which were 30×47 cm, were shortened from one end perpendicular to their width at 2 cm/hr. The models consisted of three viscous layers of SGM36 embedded in sand layers. The viscous layers were smaller in dimension than the model and arranged in a lateral and frontal stepping-up order (Fig. 2). The bottom viscous, layer, which was of impure SGM36 (viscosity *c.* 5×10^5 Pa s), was 0.5 cm thick in both models and simulated the Beuda Formation. This viscous layer was located in the lower right-hand quadrangle of the model, adjacent to the rear wall representing the backstop. A second viscous layer (intermediate viscous layer) of pure SGM36 (viscosity *c.* 5×10^4 Pa s) simulated the Cardona Formation. This viscous layer was 0.5 cm thick in model 1 and 0.8 cm thick in model 2, was placed in the centre of the model and did not overlap with the lower viscous layer. The third and uppermost viscous layer, consisting of impure SGM36, simulated the Barbastro Formation. The top part of the model overlying the uppermost viscous layer consisted of a 1.5 cm thick layer of sand with colour markers at different levels. Initial heights of model 1 and 2 were 4.3 and 4.45 cm, respectively. A passive grid (1.2 cm$\times$1.2 cm) was placed on top of the models and twelve reference points (P1–P12) were printed in three longitudinal rows to monitor the horizontal displacement and vertical growth of the structures during shortening (Fig. 2).

Scaling

Our models are geometrically, kinematically and dynamically scaled to a prototype according to the following parameters. In our models, length ratio is 6.6×10^{-7}, implying that 1 cm in the model simulates 1.5 km in nature.

The models presented here are intended to represent the structural evolution of the SPTZ. Therefore, kinematic similarity was fulfilled by assuming that shortening in model and prototype started after the deposition of all units. However, synkinematic erosion and/or sedimentation, likely to occur in the prototype, were not incorporated in our models.

By choosing modelling materials with relatively similar mechanical properties to that of rocks in the prototype, dynamic scaling was approached. The Mohr–Coulomb criterion, which predicts the rheological behaviour of brittle material, can be used to calculate ratios of the dynamic similarity between models and nature. The Coulomb–Mohr criterion contains

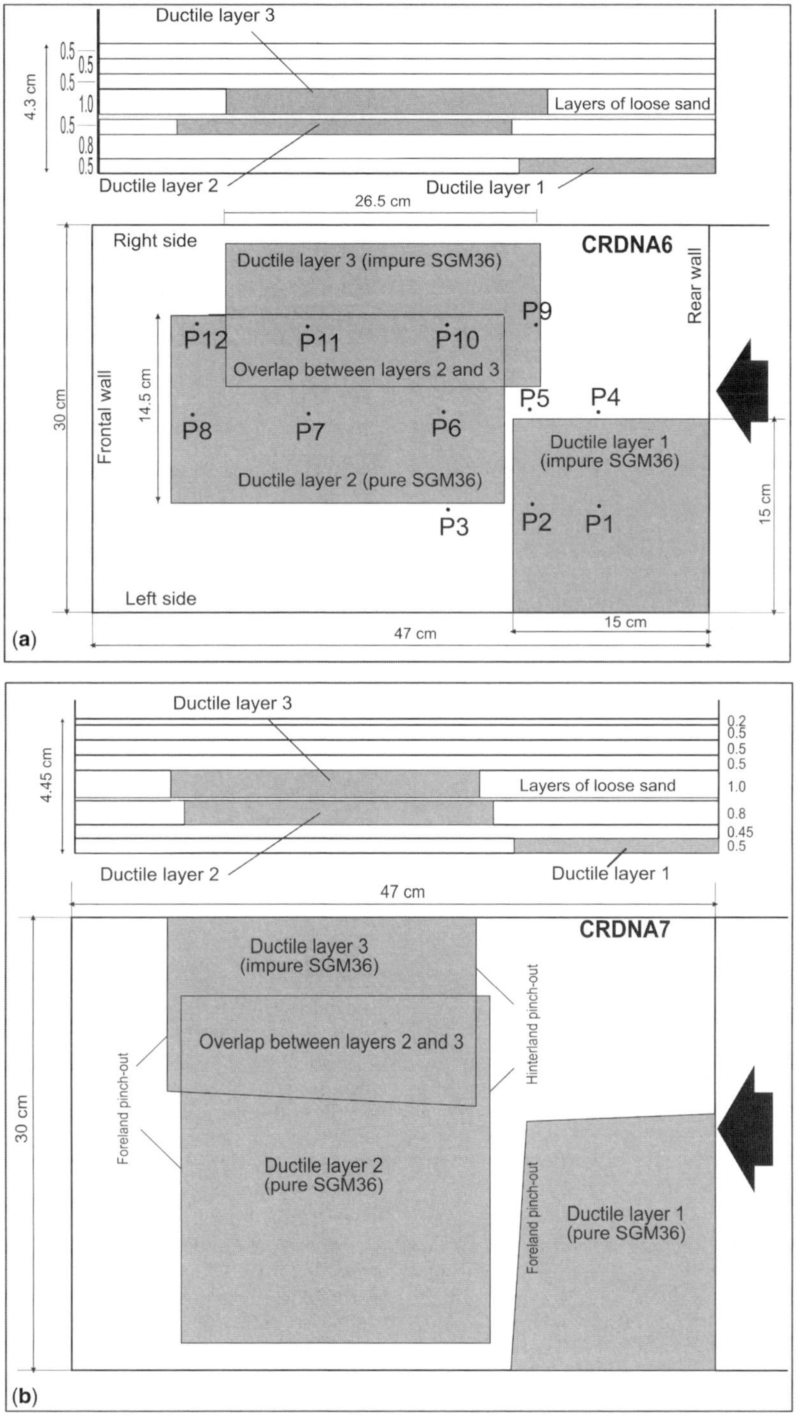

Fig. 2. Initial configuration of (**a**) model 1 (CRDNA6) and (**b**), model 2 (CRDNA7) showing both a schematic profile (top of a and b) and map view (bottom of a and b). The ductile layers are shaded in grey and the sand layers in white. Dimensions are in centimetres. Arrows indicate the shortening direction. In the map view in (a), the location of the marker points (P1 to P12) is indicated. These points were used to monitor deformation in the models.

some intrinsic material properties, for instance cohesive shear strength (C_o) and coefficient of internal friction (μ), which need to be equal in model and nature in order to achieve dynamic similarity:

$$\tau = C_o + \sigma\mu \quad (1)$$

where τ and σ are shear and normal stresses, $\mu = \tan\,\phi$ is the coefficient of internal friction and ϕ is the angle of the internal friction. For rocks in the upper crust, ϕ_n ranges between 31° and 40° (Byerlee 1978; Brace & Kohlstedt 1980) which give a coefficient of internal friction between 0.6 and 0.84. The angle of internal friction of loose sand used in the models is $\phi_m = 36°$ and gives a coefficient of internal friction $\mu_m = 0.73$, comparable to rocks in the brittle upper crust. To scale the model dynamically, the non-dimensional shear strength (τ/σ in model and prototype) also should be equal or within the same order of magnitude (Childs *et al.* 1993; Koyi *et al.* 1993; Weijermars *et al.* 1993). Dividing equation (1) by σ gives:

$$\tau/\sigma = (\mu + C_o)/\sigma \quad (2)$$

Assuming dynamic similarity, the τ/σ ratio for the model (m) and nature (n) should be equal:

$$[(\mu + C_o)/\sigma]_m = [(\mu + C_o)/\sigma]_n \quad (3)$$

Since μ_m and μ_n are almost equal, they can be neglected in equation (3):

$$(C_o/\sigma)_m = (C_o/\sigma)_n \quad (4)$$

Normal stress σ can be defined as

$$\sigma = \rho g l \quad (5)$$

where ρ is density, g is the gravity acceleration and l is the length. Equation (4) can be rewritten as

$$(C_o/\rho g l)_m = (C_o/\rho g l)_n \quad (6)$$

Substituting the values in equation (5), the non-dimensional shear strength values will be 1.9×10^{-1} and 8.9×10^{-1} for nature and model respectively. These values, which are within the same order of magnitude, suggest that dynamic similarity between the models and nature is approached.

For the viscous layers, dynamic similarity was achieved by simulating the ductile behaviour of rock salt (effective viscosity ranging from 10^{17}–10^{18} Pa s) with a Newtonian viscous material (SGM36), whose effective viscosity at room temperature is 5×10^4 Pa s for the pure SGM36 and 5×10^5 Pa s for the impure SGM36, giving a scaling ratio of 5×10^{-14} to 5×10^{-12} for both the pure and impure viscous SGM36 layers, respectively.

Results

After a total bulk shortening of 15% in model 1 and 22% in model 2, the surface geometry of the models showed that in model 1, deformation had not reached the pinch-out of the intermediate viscous layer in the foreland (Figs 3 and 4), whereas in model 2 some structures could be seen as far as the pinch-out of the intermediate and upper viscous layers. At the early stages of shortening, a large foreland-vergent structure developed near the backstop of both models across the width of the model coinciding with the frontal pinch-out of the lower viscous layer and with the rear pinch-out of the intermediate and upper viscous layers. With continuing shortening, structures usually developed in front of the frontal pinch out of the lower viscous layer. These structures were less continuous along strike, and displayed curved geometries close to the lateral pinch-outs of the different viscous layers (Figs 3 and 4). In the purely frictional area, where there was no basal viscous layer in the model stratigraphy, foreland-vergent structures were dominant, whereas in the area where there were one or several viscous layers at depth, both foreland and hinterland-vergent structures were formed (Figs 3 and 4).

Deformation sequence

In the models presented here, it was possible to monitor deformation sequence not only qualitatively through the changes of the surface geometry, but also quantitatively through a continuous monitoring of the temporal and spatial change in the position of the marker points initially 'printed' on the surface of the models (Figs 3, 4 and 5).

In both models, deformation started at the rear part of the model, close to the moving wall, and advanced towards more frontal areas of the model (Figs 3 and 4). Propagation of deformation took place laterally in the direction of tectonic transport as well as across the stratigraphy to the upper and lateral detachments. The first visible fault formed above the frictional area (at 2% bulk shortening), where no viscous layer was present in the stratigraphy (Figs 3 and 4). This foreland-vergent fault was active throughout the whole deformation history, and its position coincided with the rear pinch-out of the upper and intermediate viscous layers. The

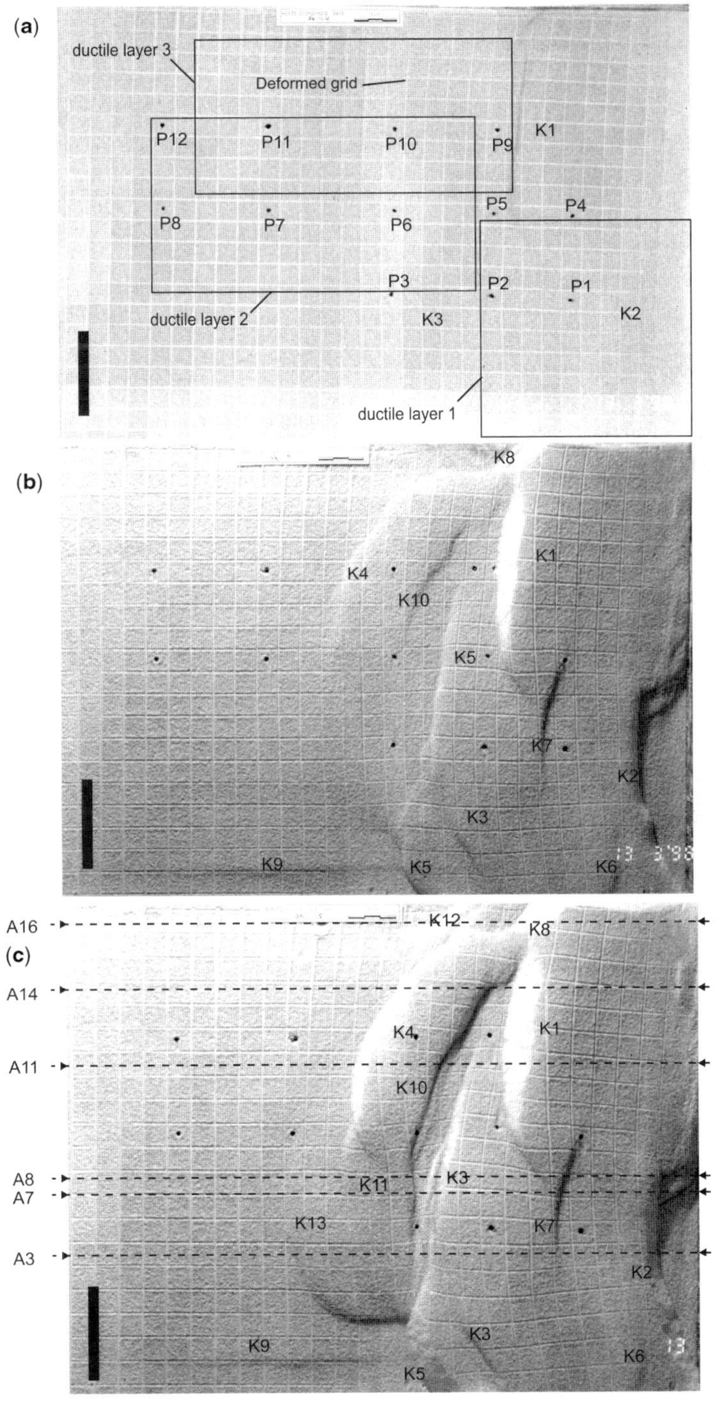

Fig. 3. Top views of model 1 (CRDNA6) during deformation: (**a**) at 4% shortening, (**b**) at 9% shortening, and (**c**) at 15% shortening. The boxes in A indicate the initial distribution of the three ductile layers. Note the irregular geometry and propagation mode of the deformation front and the individual thrusts/kinks (K1, K2, ...) in different domains of the model. Location of sections is shown in (c) (see Fig. 6).

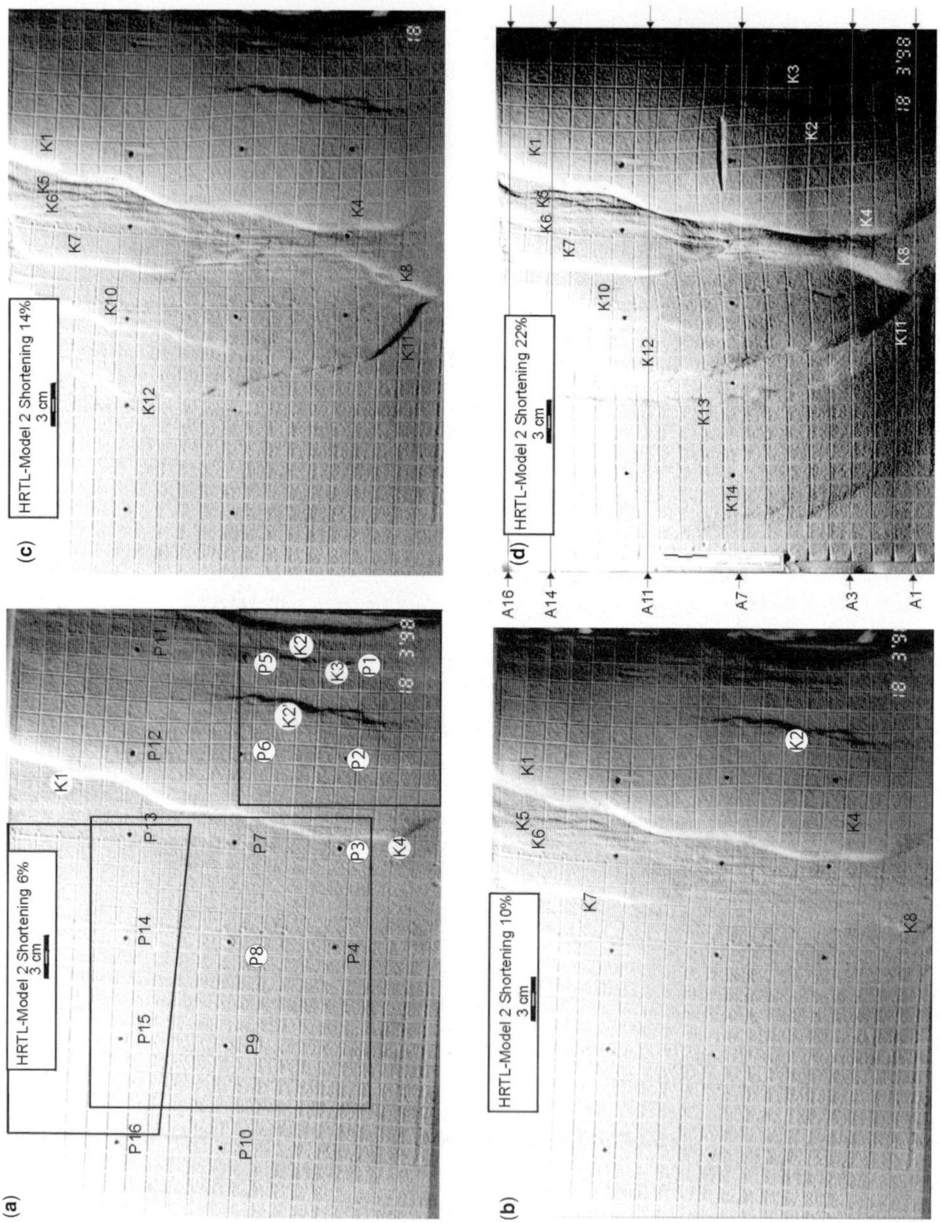

Fig. 4. Top views of model 2 (CRDNA7) during deformation: (**a**) at 6% shortening, (**b**) at 10% shortening, (**c**) at 14% shortening, and (**d**) at 22% shortening. The boxes in (a) indicate the initial distribution of the three ductile layers. Note the irregular geometry and propagation mode of the deformation front and the individual thrusts/kinks (K1, K2, . . .) in different domains of the model. Location of sections is shown in (d) (see Fig. 7).

fault grew along strike by merging with other fault segments that formed later at a similar distance from the backstop in areas where viscous layers were present.

In areas where viscous detachments were present, the first structures formed later (at 4% bulk shortening). Some of these structures, which were verging towards the hinterland of the models, were located close to the backstop, whereas others were foreland-verging and were located at the frontal pinch-out of the lower viscous layer. After the formation of these structures, deformation was transferred from the basal detachments (lower viscous layer and frictional detachment, relative to different parts of the models) to the intermediate and upper detachments (at 5% shortening, Figs 3 and 4). Between 6% and 10% of shortening, the ramp between the lower and the intermediate detachments was accommodating most of the shortening in the central part of the model, whereas in areas of the frictional detachment, new structures were developing towards the foreland (Figs 3 and 4).

After 10% shortening, propagation to the front of the model was achieved in a normal sequence in which deformation propagated progressively more to the frontal areas of the model (Fig. 4, K2, K7 and K10). Only when structures were nucleated above two different detachments did younger structures form in a break-back sequence (e.g., K5 formed behind K4 in Fig. 3). In general, all structures remained active, albeit with variable intensity, throughout the model deformation (Fig. 5). Highest displacements are recorded along the thrusts (ramps) which connected a lower detachment with one at a higher stratigraphic level (e.g., P2–P3 and P5–P6; Figs 3 and 5). Only in the model with 22% bulk shortening did the deformation front reach the foreland pinch-out of the intermediate and upper viscous layers.

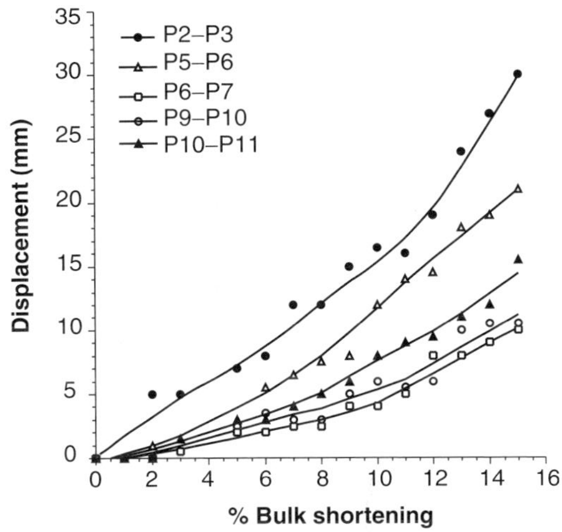

Fig. 5. Graphs of displacement along thrusts monitored through the change in lateral distance between adjacent points in model 1. Here, only five pairs of points are shown (for location of the points see Fig. 3). Note that all thrusts are active throughout model deformation, albeit with variable intensity.

Geometry of structures and their relation to the different viscous layers

In map view and in cross-sections, three different domains could be identified in the models: (1) the external area of the model, where no deformation occurred; (2) the middle part of the model, which is the deformed area located above the intermediate and the upper detachment levels; and (3) the internal area of the model, which corresponds to the deformed area above the basal frictional and lower viscous detachments (lower viscous layer, Figs 3, 4, 6 and 7).

The undeformed external area was larger in model 1 because it was shortened only up to 15% and deformation did not reach the foreland pinch-out of the intermediate and/or upper viscous detachments. In map view, the deformation front, separating the external undeformed area from the intermediate area, was located at variable distances from the backstop; it had advanced furthest in the areas above the intermediate and upper viscous layers that acted as detachment levels (Figs 3 and 4). In model 1, the deformation front was 18 cm away from the backstop above the frictional detachment (Fig. 6a) and above the lower viscous layer (Fig. 6f), whereas it was around 27 cm away from the backstop above the intermediate and upper viscous layers (Figs 3 and 6). However, the major difference in the deformation front between these two areas was not in the position of the deformation front, but in its geometry. Above the intermediate viscous layer (consisting of pure SGM36) the most external structure was hinterland vergent, or backward dominated, irrespective of whether it had formed at the pinch-out of the layer (model 2) or had not reached the pinch-out (model 1). In contrast, above the upper viscous layer (impure SGM36), the frontal structure was foreland-vergent except when it developed at the pinch-out (Figs 7a, b, c and d).

The deformed area above the intermediate and upper viscous layers is characterized by the presence of folds with an initial wavelength of

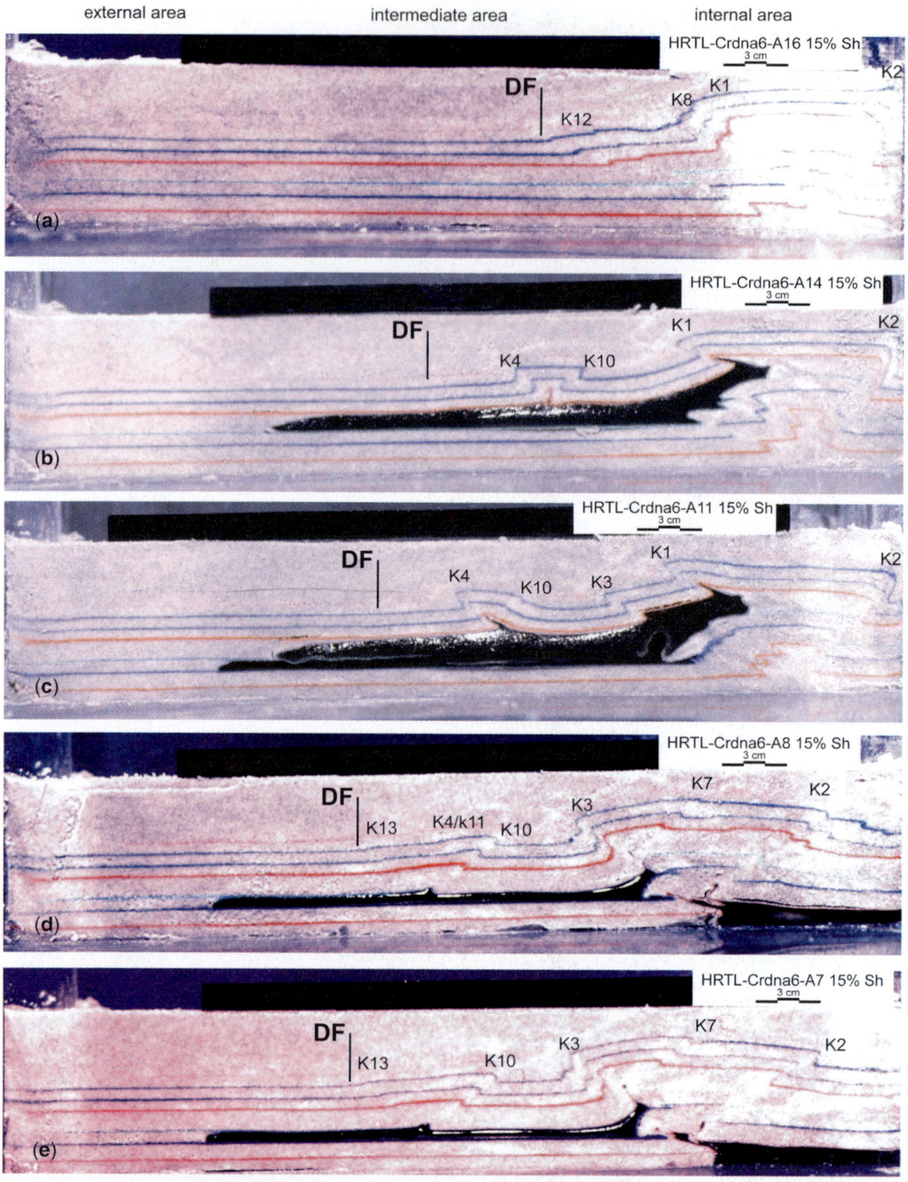

Fig. 6. Sections (**a**)–(**f**) (A3, A7, A8, A11, A14 and A16) of model 1 after 15% shortening. (See Fig. 3 for location.) The lower viscous layer is seen in sections (d), (e) and (f); the intermediate layer in sections (c), (d) and (e); the upper layer is seen in (b) and (c). Section a is taken of an area of the model without any viscous layers. K1, K2, … indicate the sequence of the structures formed during shortening of the model. DF = deformation front.

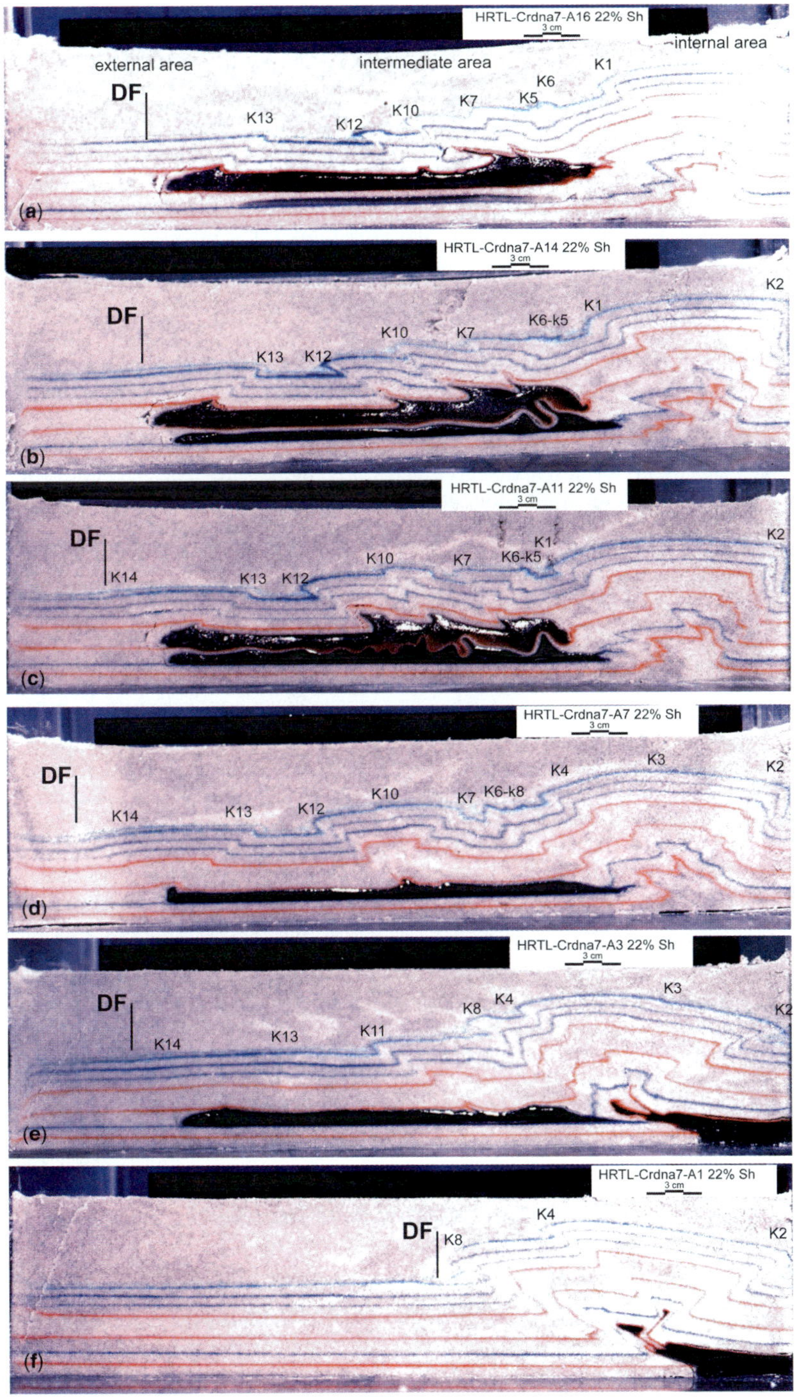

Fig. 7. Sections (**a**)–(**f**) (A1, A3, A7, A11, A14 and A16) of model 2 after 22% shortening. (See locations in Fig. 4.) The lower viscous layer is seen in sections (e) and (f); the intermediate layer in sections (b), (c), (d) and (e); and the upper layer is seen in (a), (b) and (c). K1, K2, ... indicate the sequence of the structures formed during shortening of the model. DF = deformation front.

10–12 cm and an uplift of 1 cm above the intermediate viscous layer; thrust faults have a spacing of 3.5 cm and vertical displacements of 1.25 cm above the upper viscous layer. These structures usually did not form across the entire width of the model; they branched with other structures and were curved along strike coinciding with the lateral pinch-out of the viscous layers (Figs 3 (K13) and 4 (K7)). Above the intermediate viscous layer, where the overburden layers are thicker, the structures have slightly larger wavelengths than above the upper viscous layer (Figs 3 and 6d; K10 and K13) (Figs 4 and 7d; K10 and K13). In cross-section, box folds outlined by a pair of kink-bands formed above the intermediate viscous layer. The kink-bands did not form simultaneously; one of them dominated, giving the general vergence to the fold. Above the intermediate viscous layer, the back-kinks dominate in both models, even though fore-kinks developed in model 2 (Figs 8d and 9d). This is supported by the sequence of formation of the different kinks (Fig. 3, structures K10 and K13). However, the structures behind these kinks may obscure this formation sequence in cross-section. In contrast in both models, foreland vergent structures were dominant above the upper viscous layer, except for the frontal box-fold which verges towards the hinterland (Fig. 6). Above this layer, thrusting, which is the dominant structure in the final stages of model deformation, follows folding.

The internal area of the models is characterized by a large stack of thrusts which extends across the entire width of the models (Figs 3, 4, 6 and 7). Above the frictional areas, the thrust stacks reach the maximum height. A ramp that developed between the lower and the intermediate viscous layers, transferred deformation to the frontal areas (Figs 3e and 4e). In areas where the intermediate detachment level was absent, deformation was not transferred to the front. Instead, the thrust emerged at the surface with a similar geometry to that of the frictional area (Figs 6f and 7f).

In the overburden layers located above the upper viscous layer the dominant deformation mechanism was thrusting, which was mainly foreland-vergent except at the deformation front (Figs 6c, and 7b, c). During shortening of the model, the upper viscous layer thickened in the hanging walls of the thrusts. No offset was seen in the sand layer located between the intermediate and upper viscous layers. This sand layer was folded (with ptygmatic signature) during the flow of the intermediate and upper viscous materials (Figs 6c, and 7b, c). Both the intensity and amplitude of the folds in this layer decreased towards the foreland as the wavelength of the folds increased. The vergence of the folds is not consistent and seems to be controlled by the drag of the upper viscous layer towards the hanging wall of the thrusted units.

Discussion

Comparison of model results with the South Pyrenean Triangle Zone

In the South Pyrenean Triangle Zone, deformation style in each of the three thrust fronts is different and depends on the mechanical behaviour of the overburden units and the detachment layers (salt and/or gypsum) (Sans *et al.* 1996*a*) and on the spatial distribution of the three detachment layers. The frontal wedges, being active during a certain period of time, were abandoned once the sole thrust climbed to a stratigraphically higher detachment level and propagated towards the foreland. The three fronts, despite evolving similarly, have different geometries. Salt-cored anticlines (e.g., Cardona, Suria, el Guix; Figs 1 and 8) formed above the salt detachment, whereas antiformal stacks (e.g., Sanaüja or Barbastro; Figs 1 and 9) formed above

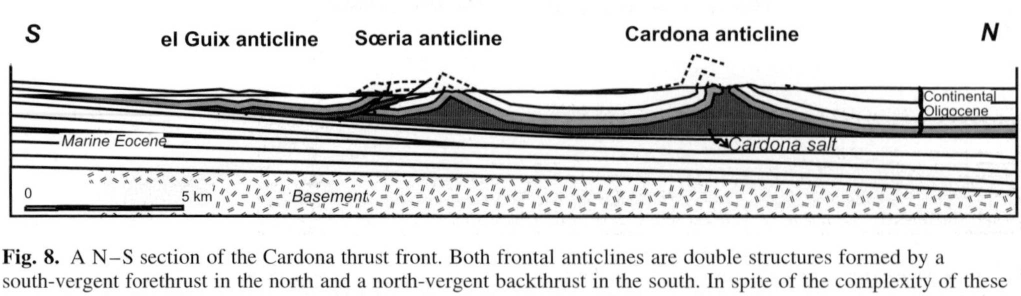

Fig. 8. A N–S section of the Cardona thrust front. Both frontal anticlines are double structures formed by a south-vergent forethrust in the north and a north-vergent backthrust in the south. In spite of the complexity of these structures, the frontalmost structure, which is at the pinch-out of the Cardona salt, is always hinterland-vergent. The Cardona anticline is located in a more internal position and is slightly foreland-vergent. Vertical scale is equal to the horizontal scale.

gypsum and shale detachments (Figs 8 and 9) (Sans *et al.* 1996*a*).

These types of detachments are simulated in the models presented here. Nevertheless, our models do not incorporate the effects of erosion or syntectonic deposition, which have a strong influence on the evolution of the structures in the SPTZ. Mugnier *et al.* (1997) used analogue models to outline the crucial effect of superficial mass-transport phenomena in the tectonic style of thrust belts. Using sandbox models, Storti & McClay (1995) showed that syncontractional sedimentation dramatically changes the shape of the Coulomb thrust wedge, the number of thrust slices, the internal shortening, and the critical taper of the wedge. Rapid, localized sedimentation may influence and even inhibit the development of emergent thrust faults (e.g., Willet *et al.* 1993; Storti & McClay 1995). Erosion may reactive and/or lead to increased displacement along thrusts in the internal areas, whereas sedimentation buries active thrusts and may render them inactive when a deformation front is transferred to the external areas (Koyi *et al.* 2003). Moreover, although the overall shortening in the South Pyrenean foreland was from the north, lateral ramps have formed in the area (Vergés 1993), which together with the irregular disposition of the evaporitic formation with respect to the shortening direction, have resulted in more complex geometries and rotation of the deformation fronts (Sans *et al.* 1996*a*). Similarly, in our models, which are a simplified version of the SPTZ, lateral ramps, discontinuous thrusts and thrusts inclined to the shortening direction have formed (Figs 3 and 4). However, even though the degree of rotation (around a vertical axis) observed in the SPTZ (*c.* 90°), is not entirely duplicated in our models (*c.* 55°), the experiments illustrate the irregular evolution of thrust fronts above multiple detachments. Similar to the natural prototype, lateral ramps formed at the lateral pinch-out of the viscous detachments (Figs 3 and 4).

The Cardona thrust front is characterized by the presence of a 350 m thick salt layer (Cardona Formation). In this area, shortening was in general accommodated by folding, with both fore and back-vergence and development of a counter-regional frontal structure (Fig. 8). This situation was simulated in the models by the intermediate viscous layer (pure SGM36). Structures formed above this layer show, at the end of the experiment, a back-vergent asymmetric box fold (Figs 6d and e). With increasing bulk shortening (model 2), the initially back-vergent structures may increase the displacement along the fore-kink resulting in the back kink being carried upwards and forming box folds with tilted crests towards the foreland (Fig. 7d; K12-K7 and K14-K13). This suggests that, above the intermediate detachment, folds that nucleated at the deformation front with an initial back-vergence that were modified with progressive shortening. As a result, above such type of detachments, subsequent triangle zones formed at the deformation front, regardless of their location relative to the pinch-out of the ductile layer. This type of triangle zones is dominated by folding and could be correlated with type I of triangle zones described by Couzens-Schultz & Wiltschko (1996).

Where the detachment horizon in the South Pyrenean Triangle Zone consists of gypsum (Barbastro and Beuda formations), deformation is mainly dominated by thrusting, with less folding (Sans *et al.* 1996*a*). At the Barbastro front (Fig. 9), back-vergent thrusts are only present at the pinch-out of the evaporitic layer (Barbastro thrust front). This situation was simulated by the upper viscous layer of the impure SGM36, which was one order of magnitude more viscous than the pure SGM36 used for the intermediate viscous layer. Although thrusted

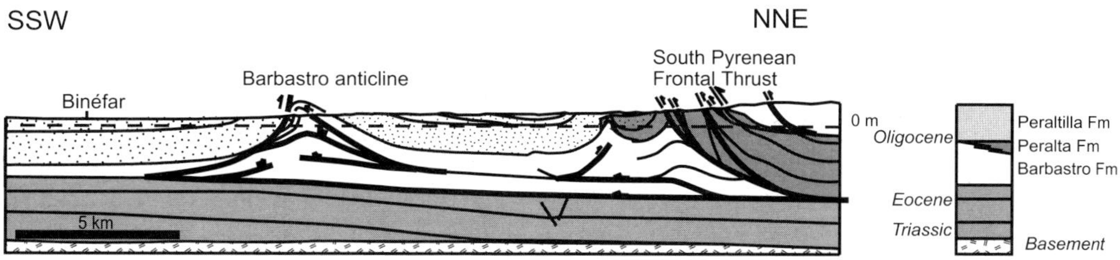

Fig. 9. Balanced cross-section of the Barbastro thrust front constructed at minimum shortening. Gypsum layers of the Barbastro Fm are thrust and piled to form a foreland-vergent antiformal stack. This structure is limited in the front by a hinterland-vergent thrust which defines the thrust wedge geometry of this frontal anticline. Structures more to the north all are foreland-vergent. The stratigraphic column is not to scale. Modified from Sans *et al.* (1996*a*).

folds dominate at the end of the experiment, initially symmetric box folds (Fig. 6b) formed above the upper detachment. As deformation progressed, these folds squeezed the viscous material out of their cores and evolved to foreland-vergent thrusts (Fig. 7a; K7, K10 and K12). However, at the pinch-out of the upper detachment, a back-vergent box fold was formed (Fig. 7a; K13). The asymmetry of the fold was maintained until the end of the experiment (Fig. 7a). As a result, above the upper detachment, triangle zones are formed only at the pinch-out of the viscous layer, and deformation style in the overburden was dominated by thrusting (Fig. 7a).

Finally, when the mechanical stratigraphy of the detachment is complex, with the presence of both rocksalt and gypsum (Cardona and Barbastro formations respectively), the features observed in the field (Sanaüja anticline in the Cardona thrust) indicate that salt and gypsum are folded disharmonically. The evolution of this thrust wedge is sketched in Figure 10 (Sans *et al.* 1996*a*). During the early stages of deformation, as the basal detachment propagated towards the foreland, the Barbastro gypsum, and presumably the Cardona salt, developed flow folds of short wavelength (tens of meters). These folds have generally southwestern vergence in both limbs of the Sanaüja anticline, which indicates that they are not diapiric or second order folds of the anticline (Sans *et al.* 1996*b*). Instead these folds are formed due to the ductile shear in the detachment levels as a result of flow of the Cardona salt during shortening. Further deformation was accommodated by folding and the flow of the Cardona salt from the NE syncline into the anticline. Increasing friction along the basal thrust due to the decrease in thickness of the Cardona Formation to the southwest, led to the development of a counter-regional thrust (northeast-vergent). Later, a foreland-dipping passive roof duplex developed in the forelimb of the anticline. The roof thrust is located a few meters below the upper contact of the gypsum, and also a detachment is present between the Cardona salt and Barbastro gypsum which allows the differential deformation of both lithologies.

Growth of the Sanaüja anticline ended with the complete withdrawal of the Cardona salt formation from the adjacent syncline to the NE. Some segments of the backthrust in the hanging wall of the roof backthrust became inactive due to the growth of the anticline and steepening of the limbs. In the models, this scenario was simulated by two viscous layers (intermediate and upper viscous layers) overlapping each other. Although the details of the evolution of the Sanaüja anticline have not been reproduced in our models, the main geometrical features can be observed. The first similarity is the location of the deformation front and its vergence. In model 2, deformation has reached the pinch-out of the intermediate viscous layer, and both the intermediate and upper viscous layers are deformed. The frontal counter-regional

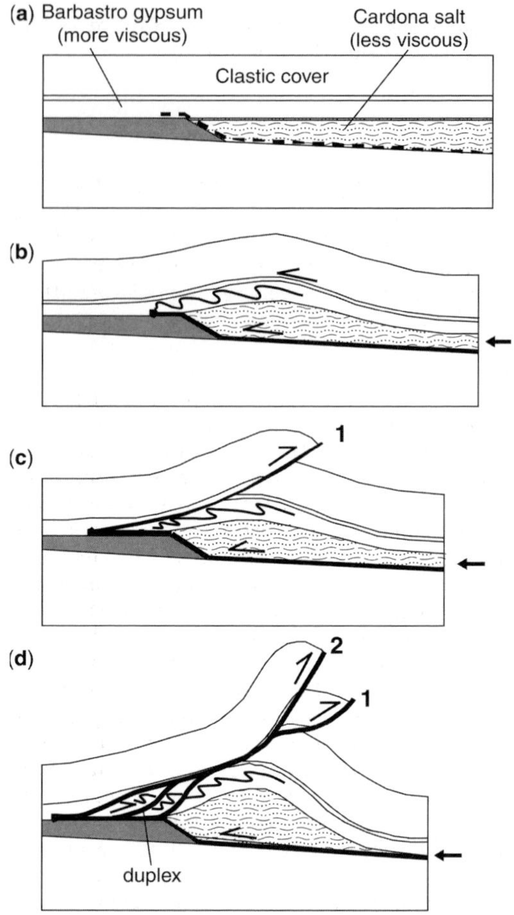

Fig. 10. Evolution of the Sanaüja anticline. (**a**) Initial geometry of the Cardona salt (light grey), the Barbastro gypsum and the clastic cover. (**b**) Cardona salt flowed from the northeast into the growing anticline and folds developed in the Barbastro gypsum related to the shear sense in the detachment. (**c**) A backthrust developed, reactivating previous foreland-vergent structures and separating folded gypsum layers from unfolded gypsum layers. (**d**) Salt flow from the northeastern limb continued at the same time as the development of a hinterland-vergent duplex in the Barbastro Formation. Not to scale.

thrusted fold is the equivalent structure of the Sanaüja anticline (Figs 7b and c).

Deformation of transfer zones

At the early stages of deformation, before any visible structures could be detected at the surface of the model (before 2% bulk shortening), the sand layers accommodated the push from the rear by penetrative deformation, which must have led to internal compaction of the sand layers. This penetrative strain was visible through the square grid, printed initially on top of the model, and which close to the backstop was deformed to rectangles with an aspect ratio of 1.1 at 2% bulk shortening. Also, as deformation proceeded, change in the elevation of the surface started sequentially from more internal to more external areas. Most of the structures grew in amplitude throughout the model shortening.

All three viscous layers acted as effective detachments that allowed deformation to be transmitted vertically from the lower to the upper levels, leaving the sand layers below them undeformed, and horizontally further from the rearwall. Model results show that at the presence of different detachment horizons, deformation is transferred from lower to upper detachment horizons. In this article, two situations are studied: (1) transfer from a lower less viscous detachment to an upper highly ductile detachment without overlap and, (2) transfer from a lower frictional detachment to a higher viscous detachment formed by two partially overlapping viscous layers of different rheologies. Below, we use model results to illustrate these two cases.

In the case where two detachments were present, one in front and at a higher stratigraphic position than the other, deformation was transferred from the lower detachment to the upper one through a ramp (Figs 6d and e, 7e and 11) forming a flat-ramp-flat (FRF) staircase. The basal detachment of the model thus had a flat in the lower viscous detachment, a ramp at its pinch-out and a flat in a higher detachment (intermediate viscous layer). In the lower detachment, a large pop-up formed above the pinch-out of the viscous layer (Figs 6e and f, 7e and f). This structure accommodated shortening until it was transferred to the intermediate detachment. Then the fold was transported above the ramp to the intermediate detachment. Above the ramp, which connected the lower and intermediate detachments, there was mainly a forelandward transport of the anticline (cored by the lower detachment), without any further vertical growth.

The geometry of the models is similar to the geometry of the ramp between the Beuda formation, which is the lower detachment and the Cardona Formation, which is the intermediate detachment in the South Pyrenean Triangle Zone (Figs 6d, e and 11). Above the ramp linking the lower Beuda detachment layer with the upper Cardona detachment, an anticline (Oliana anticline) has formed. This scenario was reproduced above the ramp linking the lower with the intermediate viscous layer, where an anticline formed. However, the models failed to reproduce some other details e.g., the antiformal stack coring the anticline and the geometry of the major backthrust present at the base of the Cardona detachment (Verges 1993), probably because not enough shortening was applied to the models.

In the case where displacement was transferred from a lower frictional detachment to a higher viscous detachment which consisted of two partially overlapping viscous layers of different rheologies (double viscous detachment), deformation was partitioned. Deformation was transmitted through a ramp from the lower frictional detachment to the upper viscous detachment (Figs 6b and c, 7b and c). No deformation is seen in the layers below the double viscous detachment. The sand layers located above the upper detachment were thrusted, with thrust surfaces soling into the upper detachment layer (Fig. 7). In contrast, the sand layer embedded between the upper and intermediate detachments underwent folding, suggesting that deformation between these two viscous layers is partitioned. This is in agreement with field observations reported by Sans & Vergés (1995) and Sans *et al.* (1996*a*), indicating that the top part of the Barbastro gypsum deforms conformably with its overburden unrelated to the Cardona Formation.

In model 1, shortening (15% bulk shortening) was stopped before deformation reached the frontal pinch-out of the intermediate and upper ductile layers. In model 2, shortening (22% bulk shortening) continued until deformation reached the frontal pinch-out. However, the pinch-out was only slightly deformed. It is thus not possible to document how deformation would have proceeded if the model had been shortened further. However, based on the sequence of deformation style above the lower viscous layer at the rear of the model, it is argued here that continued shortening would have been accommodated within the layers above the intermediate and upper viscous layers before propagating to the frictional areas in front of the viscous layers (Fig. 7f).

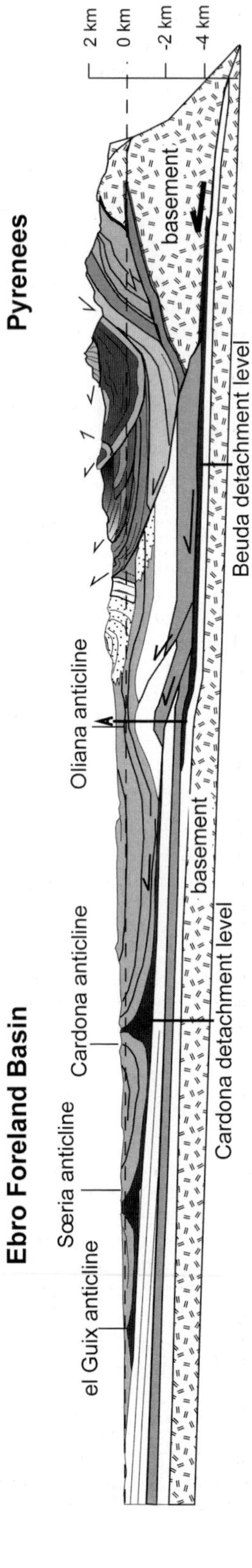

Fig. 11. Regional cross-section through the Beuda and Cardona thrust fronts, showing transfer of deformation from deeper detachments to shallower levels (from Vergés 1993). The Oliana anticline was initiated where the basal thrust ramps from the Beuda Formation (lower detachment) to the Cardona Formation (intermediate detachment) and has been transported over the Cardona flat.

Conclusions

Results of analogue models show that triangle zones form when (1) deformation is transferred to weak horizons located at higher stratigraphic levels, and (2) the deformation front reaches the pinch-out of the weak horizons. In the models, transfer of deformation from lower to upper viscous detachments, which do not overlap, takes place through a connecting ramp. Structures in the lower detachment are highly controlled by the geometry of its foreland pinch-out. In the upper detachment, deformation is accommodated in front of and away from the ramp. Even in cases where a lower viscous detachment is absent, deformation from a frictional detachment is transferred to a viscous, but stratigraphically higher, detachment through a connecting ramp above which an anticline forms.

Model results also show that, in cases where two viscous layers (intermediate and upper layers) are partially overlapping, the two layers may act as one viscous unit above which the upper overburden layers decouple from the lower overburden layers. However, each of the viscous layers accommodate the shortening separately and hence deformation between them is partitioned; overburden layers above the upper viscous layer undergo thrusting, while the layers above the intermediate viscous layer accommodate the shortening by folding.

The models presented here have similar geometric features to those developed in the South Pyrenean Triangle Zone, the Beuda, Cardona and Barbastro thrust fronts, suggesting that they could have formed and evolved in a similar way; the initial distribution of the viscous layers and their pinch-out dictates the position and geometry of the deformation fronts.

J. Vergés kindly provided an electronic version of a regional cross-section of the Pyrenees. Thanks are due to A. Ravaglia and an anonymous reviewer for their thorough review and many insightful comments, which contributed to the improvement of the manuscript. Special thanks to G. Schreurs for useful comments and suggestions. H. A. Koyi acknowledges financial support from the Swedish Research Council (VR).

References

Ballard, J. F., Brun, J-P., Van Den Driessche, J. & Allemand, P. 1987. Propagation des chevauchements au-dessus des zones de décollement: modèles expérimentaux. *Comptes rendus de l'Académie des Sciences*, **305**, Série II, 1249–1253.

Bonini, M. 2001. Passive roof thrusting and foreland-ward fold propagation in scaled brittle-ductile physical models of thrust wedges. *Journal of Geophysical Research*, **106**, 2291–2311.

BRACE, W. F. & KOHLSTEDT, D. L. 1980. Limits on lithosphere stress imposed by laboratory experiments. *Journal of Geophysical Research*, **85**, 6248–6252.

BURBANK, D. W., PUIGDEFÁBREGAS, C. & MUÑOZ, J. A. 1992. The chronology of the Eocene tectonic and stratigraphic development of the eastern Pyrenean Foreland Basin, NE Spain. *Geological Society of America Bulletin*, **104**, 1101–1020.

BYERLEE, J. 1978. Friction of rocks. *Pure Applied Geophysics*, **116**, 615–625.

CHARLESWORTH, H. A. K., JOHNSTON, S. T. & GAGNON, L. G. 1987. Evolution of the triangle zone in the Rocky Mountain Foothills near Coalspur, Central Alberta. *Canadian Journal of Earth Science*, **24**, 1668–1678.

CHILDS, C., EASTON, S. J., VENDEVILLE, B. C., JACKSON, M. P. A., LIN, S. T., WALSH, J. J. & WATTERSON, J. 1993. Kinematic analysis of faults in a physical model of growth faulting above a viscous salt analogue. *Tectonophysics*, **228**, 313–329.

COTTON, J. T. & KOYI, H. A. 2000. Modelling of thrust fronts above ductile and frictional decollements: application to structures in the Salt Range and Potwar Plateau, Pakistan. *Geological Society of America Bulletin*, **112**, 351–363.

COUSENZ-SCHULTZ, B. A. & WILTCHSKO, D. V. 1996. The control of mechanical stratigraphy on the formation of triangle zones. *Bulletin Canadian Petroleum Geology*, **44**, 165–179.

GIL, J. A. & JURADO, M. J. 1998. A 3D geological interpretation and numerical modelling of salt movement in the Barbastro–Balaguer anticline (Southern Pyrenees), *Tectonophysics*, **293**, 141–155.

GORDY, P. L., FREY, F. R. & NORRIS, D. K. 1977. *Geological Guide for the Waterton-Glacier Park Field Conference, Calgary*. Canadian Society of Petroleum Geologists.

HATCHER, R. 1999. Development of triangle zones in the evolution of foreland fold-thrust belts. *American Association of Petroleum Geologists Annual Convention*. Abstracts with program, A56.

KOYI, A. H., JENYON, M. K. & PETERSEN, K. 1993. The effect of basement faulting on diapirism. *Journal of Petroleum Geology*, **16**, 285–312.

KOYI, H. A., HESSAMI, K. & TEIXELL, A. 2000. Epicenter distribution and magnitude of earthquakes in fold-thrust belts: insights from sandbox models. *Geophysical Research Letters*, **27**, 273–276.

KOYI, H. A., SANS, M., TEIXELL, A., COTTON, J. & ZEYEN, H. 2003. The significance of penetrative strain in confractional areas. *In*: MCCLAY, K. R. (ed.) *Thrust Tectonics and Petroleum Systems*. Memoir 8, **2**, American Association of Petroleum Geologists, Tulsa, OK, 207–222.

KRANTZ, R. W. 1991. Measurements of friction coefficients and cohesion for faulting and fault reactivation in laboratory models using sand and sand mixtures. *In*: COBBOLD, P. R. (ed.) Experimental and numerical modelling of continental deformation. *Tectonophysics*, **188**, 203–207.

LETOUZEY, J., COLLETTA, B., VIALLY, R. & CHERMETTE, J. C. 1995. Evolution of salt-related structures in compressional settings. *In*: JACKSON, M. P. A. *ET AL.* (eds) *Salt Tectonics. A Global Prespective*. American Association of Petroleum Geologists Memoir, **65**, 41–60.

MARTÍNEZ, A., VERGÉS, J., CLAVELL, E. & KENNEDY, J. 1989. Stratigraphic framework of the thrust geometry and structural inversion in the southeastern Pyrenees: La Garrotxa area. *Geodinamica Acta*, **3**, 185–194.

MCCLAY, K. R. & ELLIS, P. G. 1987. Geometries of extensional fault systems developed in model experiments. *Geology*, **15**, 341–344.

MEIGS, A. J., VERGÉS, J. & BURBANK, D. W. 1996. Ten million-year history of a thrust sheet. *Geological Society of America Bulletin*, **108**, 1608–1625.

MULUGETA, G. & KOYI, H. 1987. Three-dimensional geometry and kinematics of experimental piggyback thrusting. *Geology*, **15**, 1052–1056.

PARDO, G. & VILLENA, J. 1979. Aportación a la geología de la región de Barbastro. *Acta Geológica Hispánica*, Homenatge a Lluis Solé i Sabarís, **14**, 289–292.

RIBA, O., REGUANT, S. & VILLENA, J. 1983. Ensayo de síntesis estratigráfica y evolutiva de la cuenca terciaria del Ebro. Libro Jubilar J. M. Rios, *Geología de España*, **II**, 131–159.

SÁEZ, A. 1987. *Estratigrafía y sedimentología de las formaciones lacustres del tránsito Eoceno Oligoceno del NE de la Cuenca del Ebro*. Ph.D. thesis, Universitat de Barcelona.

SANS, M. 1999. *From thrust tectonics to diapirism: The role of evaporites in the kinematic evolution of the eastern south-Pyrenean front*. Ph.D. thesis, Universitat de Barcelona.

SANS, M. 2003. From thrust tectonics to diapirism. The role of evaporites in the kinematic evolution of the eastern South Pyrenean front. *Geologica Acta*, **1**, 239–259.

SANS, M. & VERGÉS, J. 1995. Fold development related to contractional salt tectonics: southeastern Pyrenean thrust front, Spain. *In*: JACKSON, M. P. A., ROBERTS, D. G. & SNELSON, S. (eds) *Salt Tectonics: a Global Perspective*. American Associaton of Petroleum Geologists Memoir, **65**, 369–378.

SANS, M., MUÑOZ, J. A. & VERGÉS, J. 1996*a*. Triangle zone and thrust wedge geometries related to evaportic horizons (southern Pyrenees). *Bulletin of Canadian Petroleum Geology*, **44**, 375–384.

SANS, M., SÁNCHEZ, A. & SANTANACH, P. 1996*b*. Internal structure of a detachment horizon in the most exernal part of the Pyrenean fold and thrust belt (northern Spain). *In*: ALSOP, G. I., BLUNDELL, D. J. & DAVISON, I. (eds) *Salt Tectonics*, Geological Society, London, Special Publications, **100**, 65–76.

SENZ, J. G. & ZAMORANO, M. 1992. Evolución tectónica y sedimentaria durante el Priaboniense superior-Mioceno inferior, en el frente de cabalgamientos de las Sierras Marginales occidentales. *Acta Geològica Hispànica*, **27**, 195–209.

STORTI, F. & MCCLAY, K. 1995. Influence of syntectonic sedimentation on thrust wedges in analogue models. *Geology*, **23**, 999–1002.

VERGÉS, J. 1993. *Estudi geològic del vessant sud del Pirineu oriental i central. Evolució cinemàtica en 3D*. Ph.D. thesis, Universitat de Barcelona.

VERGÉS, J. & BURBANK, D. K. 1996. Eocene-Oligocene thrusting and basin configuration in the eastern and central Pyrenees (Spain). *In*: FRIEND, P. F. & DABRIO, C. J. (eds) *Tertiary Basins of Spain*. Cambridge University Press, 120–133.

VERGÉS, J., MUÑOZ, J. A. & MARTÍNEZ, A. 1992. South Pyrenean fold-and-thrust belt: role of foreland evaporitic levels in thrust geometry. *In*: MCCLAY, K. R. (ed.) *Thrust Tectonics*. Chapman & Hall, London, 255–264.

VERSCHUREN, M., NIEUWLAND, D. A. & GAST, J. 1996. Multiple detachment levels in thrust tectonics: sandbox experiments and palinspastic reconstruction. *In*: BUCHANAN, P. G. & NIEUWLAND, D. A. (eds) *Modern Developments in Structural Interpretation, Validation and Modelling*. Geological Society, London, Special Publications, **99**, 227–234.

WEIJERMARS, R. 1986. Finite strain of laminar flows can be visualized in SGM36-polymer. *Naturwissenschaften*, **73**, 33–34.

WEIJERMARS, R., JACKSON, M. P. A. & VENDEVILLE, B. C. 1993. Rheological and tectonic modelling of salt provinces. *Tectonophysics*, **217**, 143–174.

WILLETT, S., BEAUMONT, C. & FULLSACK, P. 1993. Mechanical model for the tectonics of doubly vergent compressional orogens. *Geology*, **21**, 371–374.

Analogue modelling of a reactivated, basement controlled strike-slip zone, Sierra de Albarracín, Spain: application of sandbox modelling to polyphase deformation

S. MERTEN[1], W. G. SMIT[1], D. A. NIEUWLAND[2] & H. E. RONDEEL[1]

[1]*Faculty of Earth and Life Sciences, Vrije Universiteit, de Boelelaan 1085, 1081 HV Amsterdam, The Netherlands (e-mail: sandra.merten@falw.vu.nl)*

[2]*4e Binnenvestgracht 13, 2311 NT Leiden, The Netherlands*

Abstract: This paper presents the results of an analogue modelling study on the reactivation of Riedel shears generated by basement-induced sinistral strike-slip faulting. It is based on a natural example in the Sierra de Albarracín, Iberian Range (Spain). The area has a polyphase deformation history, defined by the Variscan and Alpine orogenies. Late Variscan deformation was concentrated in a wide NW–SE shear zone with accompanying kilometre-scale E–W Riedel shears, which divided the Palaeozoic basement into large fault blocks. Alpine reactivation resulted in differential movements on the Riedel shears, as evidenced by a NW–SE chain of Palaeozoic inliers surrounded by a Mesozoic cover that generally shows minor deformations except near the E–W Riedel shears, where strata locally appear in near-vertical to overturned position.

Sandbox analogue modelling was applied to improve insight into the structural history. It focused on the kinematics of spontaneously developed en echelon Riedel shears, reactivated in a rotated stress field. Sand with a controlled added strength was used to form Riedel shears in a first deformation phase to act as weak zones for a second phase.

The modelling showed that in the first deformation phase large pop-up structures developed between the Riedel shears in a basement-induced sinistral strike-slip zone. Later reactivation in the N060°E and N135°E shortening directions was taken up respectively by sinistral-reverse and dextral-reverse shear along the pre-existing Riedel shears, but only if the sand on one side of the fault zone was allowed to move freely along the other. Scissor faulting along the Riedel shears with their complex 3D-geometry increased the height of the up-squeezed blocks. For experiments with fixed boundaries and no oil-water emulsion layer between the base plate and sand pack, thrusting at the backstop occurred rather than reactivation of the Riedel shears. This approach provided robust insights on the 4D development of the Sierra de Albarracín area.

Although the 3D geometry of strike-slip faults formed in the upper brittle crust is generally well understood, the kinematic and structural processes associated with the reactivation of these faults in rotated stress fields are poorly known. Sandbox analogue modelling offers an opportunity to systematically study the evolution of these faults, as it allows spontaneous growth of faults in three dimensions (Naylor *et al.* 1986). Several authors have described the complex pattern of en echelon curved faults associated with experimental basement-induced strike-slip faulting (e.g., Cloos 1928; Riedel 1929; Wilcox *et al.* 1973; Naylor *et al.* 1986; Richard 1991; Schellart & Nieuwland 2003). Schreurs & Colletta (2002) showed with a series of oblique-slip experiments that the early fault style in transpression experiments depends on the ratio of shear strain rate and shortening rate. For relatively high strain ratios, initial failure in brittle layers is accommodated by steep strike-slip faults; for lower strain ratios pairs of thrust faults develop. Additional deformation is mostly taken up by oblique-slip along favourably oriented faults, creating positive flower structures.

Several authors have carried out analogue sandbox experiments on the reactivation of pre-existing faults. Richard & Krantz (1991) showed that faults produced during initial dip-slip deformation strongly influence the patterns of superimposed strike-slip deformation. Dip-slip faults are reactivated as strike-slip faults at depth; reactivation at the surface is less common. They attributed this behaviour to a drop in cohesion associated with existing

From: BUITER, S. J. H. & SCHREURS, G. (eds) 2006. *Analogue and Numerical Modelling of Crustal-Scale Processes*. Geological Society, London, Special Publications, **253**, 135–152.
0305-8719/06/$15.00

faults, which increases with depth where faults are generated. Reactivation experiments on the influence of pre-existing thrust faults on normal fault geometry were carried out by Facenna *et al.* (1995). They revealed that reactivation in brittle material only occurs when the dip of the pre-existing fault exceeds 40°, because of the drop in friction coefficient, and thus of the shear strength that occurs along the pre-existing fault zone. Buchanan & McClay (1991) and McClay & Buchanan (1992) showed that inversion of extensional fault systems is characterized by high-angle thrust systems that may be convex upwards, steepen downwards and join pre-existing extensional detachments. In these experiments the footwall of the normal fault consisted of a material that was too strong to be faulted in a sandbox model experiment. As a result, the final fault pattern was strongly model driven.

In all the reactivation experiments described above, second phase deformation is induced by movements of a base plate. In this contribution we present the results of sandbox analogue modelling focusing on the geometry and kinematics of spontaneously developed en echelon Riedel shears, reactivated in a rotated stress field. Riedel shears were formed in a first deformation phase induced by strike-slip faulting along a base plate. Sand with a controlled added strength (Mulugeta 1988; Lopes Cardozo *et al.* 2002; Van Mechelen 2004) enabled us to preserve these Riedels as weak zones for a second deformation phase, in which the base plate was fixed.

Mechanically, reactivation processes occur because existing fault zones are weaker than the non-deformed surrounding rocks. Important parameters controlling reactivation are a drop in cohesion between the fault zone and its surrounding rock (Donath & Cronwell 1981), and a drop in friction coefficient. These parameters result in a decrease of shear strength along the pre-existing faults, thus creating the possibility of reactivating faults that are properly oriented with respect to the principal stress directions (Mandl 1988). The boundary conditions for the sandbox models are based on geometries and principal stress directions obtained from a natural example in the Sierra de Albarracín (Spain). The polyphase deformation history of the area provided the opportunity to test different stress fields and compare the results to observed structures.

Geological framework

The Sierra de Albarracín is situated about 200 km east of Madrid in the southeastern part of the Iberian Range (Fig. 1a). Its main structure consists of an extensively deformed Palaeozoic basement unconformably covered by Mesozoic and Tertiary sediments. Several Palaeozoic basement inliers crop out in NW–SE alignment (Fig. 1b). In this paper the main focus is on two of these inliers, the Macizo de Nevera and the Macizo del Tremedal (Fig. 2).

Geological evolution of the Iberian Range

The Iberian Range is a NW–SE trending intraplate mountain range in the central-eastern part of the Iberian Peninsula. Its tectonic evolution is mainly defined by reactivation of Late Variscan strike-slip zones during several phases of Mesozoic extension and Alpine compression (Arche & López-Gómez 1996; Casas-Sainz & Maestro-González 1996). Early Variscan E–W compression led to the build-up of the Variscan orogen and induced N–S folding and thrusting. Late Variscan collapse of the orogen caused the development of conjugate strike-slip faults and of large-scale NW–SE orientated extensional shear zones (Doblas *et al.* 1994). These Late Variscan shear zones controlled the formation of the intracratonic Iberian Basin during Permian and Mesozoic extension (Arche & López-Gómez 1992, 1996; Salas & Casas 1993; Guimerà 1996), when rifting induced vertical movements on the Late Variscan faults, generating volcanic activity (Arthaud & Matte 1977; Alvaro *et al.* 1979) and the formation of local palaeohighs (Salas & Casas 1993). The Serranía de Cuenca Fault and the Molina–Teruel–Espadan Fault (Fig. 1b; SCF and MTEF) acted as basin boundary faults in the southeastern part of the Iberian Basin. These NW–SE normal fault systems are parallel to the fundamental Variscan structures and consist of 50 to 70 km long arcuate sequences, offset by transversal faults trending at high angles (Arche & López-Gómez 1992, 1996). Gravity modelling by Salas & Casas (1993) suggests that the faults go down beyond the Moho directly below the Sierra de Albarracín.

The Alpine orogeny caused the formation of the Iberian Range through inversion of the Iberian Basin (Arche & López-Gómez 1996). Several Alpine phases have been identified, starting with collision between Iberia and Europe during the Palaeogene and followed by collision between Iberia and Africa during the Neogene (Jabaloy *et al.* 2002 and references therein). A local palaeostress analysis of the Macizo de Nevera and the Macizo del Tremedal showed a multiple succession of stress fields. It confirmed that the known regional Alpine stress fields actually dominate this specific area.

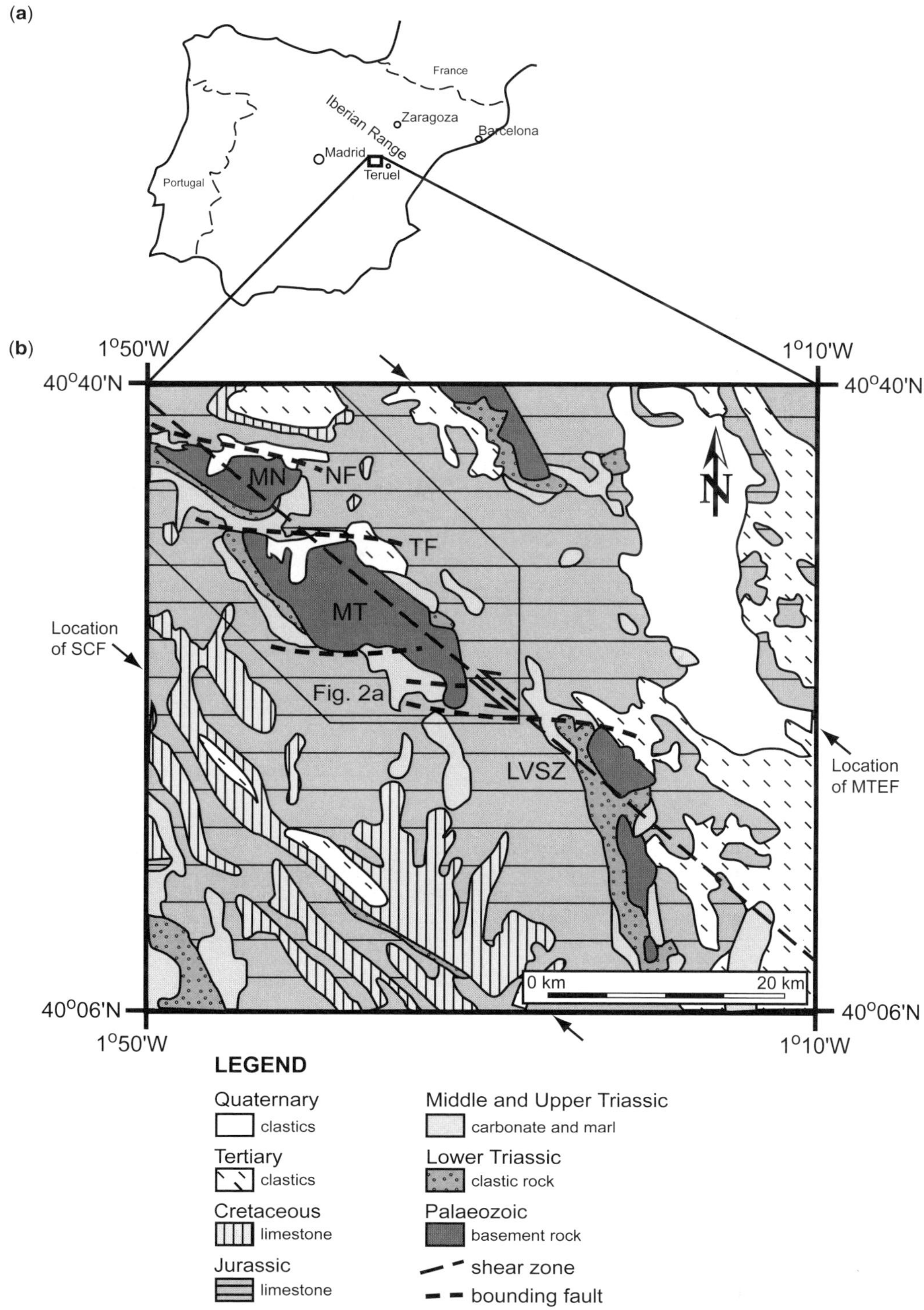

Fig. 1. (**a**) Location map of the Sierra de Albarracín (Spain). Inset shows the location of the study area. (**b**) Main geological features of the NW–SE trending Sierra de Albarracín: MN = Macizo de Nevera; MT = Macizo del Tremedal; LVSZ = Late Variscan shear zone; SCF = Serranía de Cuenca Fault; MTEF = Molina–Teruel–Espadan Fault. Inset shows the location of Fig. 2a.

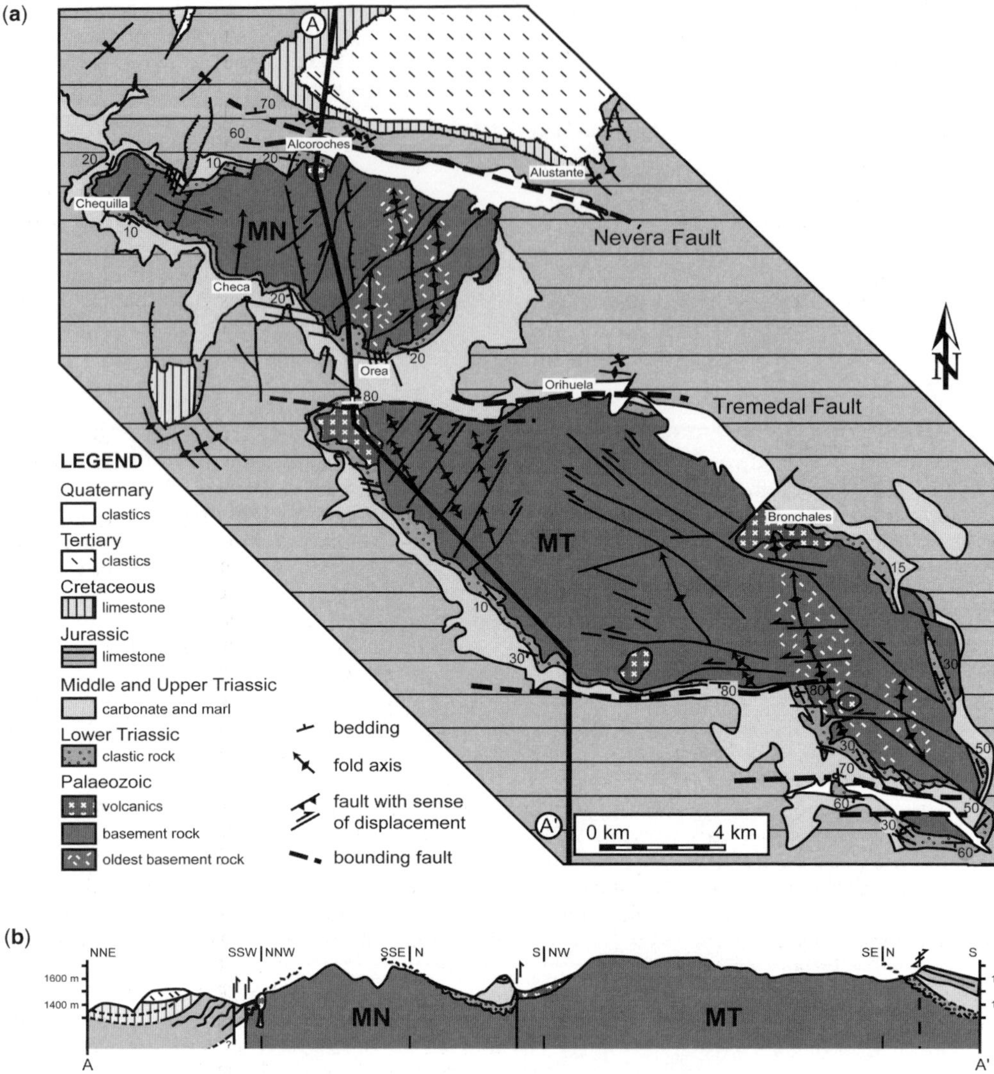

Fig. 2. (**a**) Structural map of the Macizo de Nevera (MN) and the Macizo del Tremedal (MT). The profile line A–A′ indicates the position of (**b**) schematic cross-section through the Macizo de Nevera and the Macizo del Tremedal. Only the main faults that divide the Palaeozoic basement into different blocks are indicated.

De Ruig *et al.* (1991) defined for the Early Eocene a NW–SE compressional stress regime west of the Sierra de Albarracín and a NNE–SSW compressional regime more to the east. From Early Oligocene to Early Miocene times a NE–SW compressional regime, σ_1 oriented at N060°E, reactivated Late Variscan basement faults and was responsible for NW–SE folding of the Mesozoic and Tertiary cover. This regime is considered the most important Alpine stress field in the Iberian Range (Simón-Gómez 1986; Rodriguez-Pascua *et al.* 1994; Muñoz-Martín *et al.* 1998). For the Middle Miocene a NNW–SSE compressional regime is identified, followed by a dominantly NW–SE regime with σ_1 oriented between N130°E and N150°E from the Late Miocene to present activating mainly strike-slip faults (Simón-Gómez 1986; Rodriguez-Pascua *et al.* 1994; De Vicente *et al.* 1996; Muñoz-Martín *et al.* 1998).

Stratigraphy of the Sierra de Albarracín

The folded Palaeozoic rocks of the Sierra de Albarracín consist of a heterogeneous siliciclastic pelitic sequence of Ordovician to Silurian age.

Total thickness of this basement amounts to about 1 to 2 km, with about 400 m of shales at the base (Rondeel *et al.* 1984). It is locally transected and covered by Permian volcanics (Pérez-Arlucea & Sopeña 1983, 1985; Rondeel *et al.* 1984). The absence of Devonian and Carboniferous strata that appear elsewhere in the Iberian Range implies non-deposition and/ or large-scale uplift and erosion.

The Palaeozoic is unconformably overlain by Upper Permian to Lower Triassic continental siliciclastics (Pérez-Arlucea & Sopeña 1983, 1985; Ramos *et al.* 1986; Arche & López-Gómez 1992; Muñoz *et al.* 1992), Middle Triassic dolomites alternating with clay-marl-gypsum levels (Hinkelbein 1969; Pérez-Arlucea & Sopeña 1983, 1985; López-Gómez *et al.* 1993), gypsum-rich Upper Triassic clays and marls (Hinkelbein 1969) and Jurassic limestones. The basal siliciclastics cover a palaeorelief of at least 60 m. They have a strongly variable thickness and are even absent in the eastern part of the Macizo de Nevera where higher units transgressively overlie Variscan rocks (Pérez-Arlucea & Sopeña 1983, 1985; Rondeel *et al.* 1984). These basal siliciclastics are mechanically attached to the basement. Thickness variations within the overlying Triassic are attributed to later décollement excision. The Jurassic limestones behaved as a rigid super-stratum relative to the Triassic (Rondeel *et al.* 1984).

Unconformably covering this sequence are Cretaceous limestones, characterized by relatively strong lateral and temporal shifts in depositional environments due to differential tectonic movements (Vilas *et al.* 1983; Salas & Casas 1993; Van Wees *et al.* 1998). In turn, the Cretaceous is unconformably covered by Tertiary and Quaternary detrital rocks.

Structural style of the Sierra de Albarracín

The complex polyphase deformation history of the Sierra de Albarracín is evidenced by a wide variety of structures (Fig. 2a). In the following section the focus will be on the main faults that divide the Palaeozoic basement into different blocks and are responsible for the present-day structure (Viallard 1983; Pérez-Arlucea & Sopeña 1983, 1985; Ramos *et al.* 1986).

The Variscan fault system of the Macizo de Nevera is predominantly a NE–SW running dextral strike-slip system. This fault system cuts across N–S trending large scale folds with their accompanying axial plane cleavage (Weijermars & Rondeel 1984) and does not continue into the overlying Upper Permian and younger rocks. The situation within the Macizo del Tremedal is strikingly different. In its westernmost domain a great number of dextral NNE–SSW strike-slip faults displace the Variscan fold series. However, in the remainder of the Macizo NW–SE sinistral strike-slip faults are disrupting the folds.

The Variscan fold structure consistently runs NNW–SSE with deviations of no more than 10°, except in two domains in the southern part of the Macizo del Tremedal where E–W faults disrupt the continuity. Here, the folds and accompanying cleavage are sheared in a sinistral sense. The Lower Triassic siliciclastics unconformably cover this Variscan structure. The presence of these faults forms the argument to postulate major sinistral E–W shears of Variscan age to lie buried under the Mesozoic cover at the southern limit of the Macizo del Tremedal (Fig. 2b).

The thickness distribution of the Lower Triassic siliciclastics around the Macizo del Tremedal, being thick and only slightly dipping at western boundaries, but reduced in thickness and strongly inclined (even overturned) at southern boundaries, suggests a reactivation of the shears in vertical motion during and after the Triassic. On the south side of each of the faults delimiting the Macizo del Tremedal to the south, the Lower Triassic consistently shows a continuous change in attitude. Going from E to W the Lower Triassic has a slightly SE-dipping attitude, followed by a steeper S-dipping inclination, a vertical position and an overturned Lower Triassic dipping E to NE, locally even at 30° (Fig. 2a). The vergence of this structure is indicative of westward movement of the material to the south of the faults, thus indicating rejuvenation of the E–W faults in a dextral sense. In this movement the Variscan basement participated with its attached Lower Triassic siliciclastics. Decoupling over the clay-marl-gypsum levels of the Middle and Upper Triassic caused the Jurassic to behave independently as rigid 'floating' blocks (Viallard 1983; Rondeel *et al.* 1984). Amidst these incompetent levels, the dolomitic Middle Triassic folded and faulted passively and decoupled completely.

These observations in the southern part of the Macizo del Tremedal are the basis for the assumption of large E–W shears limiting the Macizos on two sides with sinistral movement in Variscan times and dextral movement in Alpine periods, with intermediate vertical motions. The indications for the occurrence of E–W faults, limiting the Macizos, are numerous. The location of the Tremedal fault is inferred from the non continuity of the Variscan structure between the Macizo de Nevera and the Macizo del Tremedal. In both Macizos, the folds are north plunging and still, the oldest rocks of the

Macizo de Nevera approach the youngest Variscan sediments of the Macizo del Tremedal within a distance of 2 km. A thin zone of vertical to overturned Lower Triassic siliciclastics shows up at the northwestern limit of the Macizo del Tremedal (Fig. 2), and is indicative of its uplift.

The Nevera fault is indicated by the appearance of Palaeozoic basement and strongly inclined Jurassic rocks north of the Macizo de Nevera. Furthermore, small-scale en echelon NE–SW folds within the Jurassic rocks might indicate dextral reactivation of the E–W faults (Fig. 2a). Volcanics occur along all major E–W shears and are probably related to Late Variscan orogenic extension (Doblas *et al.* 1994).

From the presence of the deep-seated Serranía de Cuenca Fault and the Molina–Teruel–Espadan Fault (Salas & Casas 1993) and their strike parallel to the Late Variscan lineaments, we infer a single deep-seated NW–SE Late Variscan shear zone below the Sierra de Albarracín along which major strike-slip faulting occurred during the Variscan orogeny, thus creating the Sierra de Albarracín pop-up structure. It is hypothesized that the major E–W faults bounding the Macizos originated as Riedel shears from this shear zone. The right-stepping Palaeozoic basement inliers are interpreted to represent up-squeezed blocks in an en echelon Riedel shear pattern (Naylor *et al.* 1986) induced by Late Variscan sinistral strike-slip faulting along the shear zone (Fig. 1b).

Later Alpine stress fields caused reverse and dextral reactivation of the E–W faults, evidenced by steeply inclined, overturned and overthrusted Lower Triassic rocks. Of the stress fields described for the Alpine orogeny, we expect the Oligocene-Miocene stress field with a N060°E oriented maximum principal stress (Simón-Gómez 1986; Rodriguez-Pascua *et al.* 1994; Casas-Sainz & Maestro-González 1996; Muñoz-Martín *et al.* 1998) and the Eocene and Late Miocene-present stress field with a N130°E to N150°E oriented maximum principal stress (De Ruig *et al.* 1991; De Vicente *et al.* 1996; Jabaloy *et al.* 2002), later referred to as N060°E and N135°E shortening directions, to be in the most favourable orientation to reactivate the E–W Riedel shears. This would result in sinistral-reverse and dextral-reverse reactivation respectively of the Riedel shears.

Sandbox analogue modelling

Experimental set-up and procedure

Analogue experiments were carried out in a sandbox with a rigid base plate divided in two halves (Table 1 and Fig. 3). This plate was covered by a 20 mm homogeneous pack of dry sand representing the Palaeozoic rocks before Variscan deformation. Coloured marker layers were added to assist structural interpretation of cross-sections at the end of the experiments. To reduce the friction between plate and sand pack, a 2 mm thick lubricant layer of oil-water emulsion (OWE) was spread on top of the base plate for experiments 2 and 4. In these experiments the OWE layer represents the shales at the base of the Palaeozoic rock pile. The ductile, weak elasto-plastic, non-viscous OWE does not flow and forms a localized shear zone. It is an analogue for clays and shales at shallow crustal levels, where the strength is essentially not strain-rate dependent. Silicone putty, commonly used in analogue model experiments to provide detachment levels, is viscous and not ideal for simulating ductile clays and shales. More accurate scaling is presently not possible due to a lack of data on the mechanical properties of décollement shales (Verschuren *et al.* 1996).

The sand was slowly sifted onto the base plate or onto the OWE layer from a container with openings of about 2 mm, followed by careful levelling. Four marker layers were sifted onto the levelled sand pack after every 4 mm of sand added. The horizontal dimensions of the sand pack were 72 cm × 85 cm for experiment 1, 46 cm × 71 cm for experiment 2, and 60 cm × 60 cm for experiments 3 and 4. North Sea sand (Oude Pekela, The Netherlands) was used for experiment 1, while Fontainebleau sand (France) was used for experiments 2, 3 and 4. For North Sea sand, grains are sub-angular with intermediate sphericity and a grain size range of 200–450 μm. The density is about 1.6 g/cm^3 with an internal friction angle of about 40° and a cohesion of 23 Pa. For Fontainebleau sand, grains are sub-rounded with intermediate sphericity and a grain size range of 100–400 μm. The density is about 1.5 g/cm^3 with an internal friction angle of about 40°–45° and a cohesion of 70 Pa (Van Mechelen 2004). Using a scaling factor of 10^{-4}–10^{-5} between cohesion values of natural crustal rock and analogue material (Horsfield 1977), these cohesion values proportionally conform to the cohesion values of natural rocks between 1 and 5 MPa (Byerlee 1978; Van Mechelen 2004).

In the first deformation phase, a sinistral strike-slip displacement was applied to the sand pack by slowly moving one half of the base plate past the other at a constant velocity using a stepping motor (Figs 3a and b). Movement continued until a clear Riedel shear pattern developed in the sand pack.

Table 1. *Experimental parameters*

Experiment number	1	2	3	4
Layering	20 mm sand	2 mm OWE + 20 mm sand	20 mm sand	2 mm OWE + 20 mm sand
Horizontal dimensions (cm)	72 × 85	46 × 71	60 × 60	60 × 60
Phase 1: sinistral strike-slip displacement				
Model set-up	Fig. 3a	Fig. 3a	Fig. 3b	Fig. 3b
Strike-slip direction of base plate	sinistral	sinistral	sinistral	sinistral
Velocity of base plate (mm/hr) (sinistral strike-slip component)	36	18	18	18
Total movement of baseplate (mm) (sinistral strike-slip displacement)	8	18	8	15
Phase 2: shortening oblique to the Riedel shears				
Model set-up	Fig. 3c	Fig. 3c	Fig. 3d	Fig. 3d
Shortening direction	N060°E	N060°E	N135°E	N135°E
Theoretical angle (°) between shortening direction and Riedel shears (E–W faults)	20–25	20–25	45–50	45–50
Shortening velocity (mm/hr)	27	18	18	18
Total amount of shortening (mm)	40	5	23	6
Material properties				
Oil-water emulsion (OWE) (Verschuren et al. 1996)				
Rheology	–	elasto-plastic	–	elasto-plastic
Cohesion (Pa)	–	30–40	–	30–40
Sand (Van Mechelen 2004)	*North Sea*	*Fontainebleau*	*Fontainebleau*	*Fontainebleau*
Rheology	frictional-plastic	frictional-plastic	frictional-plastic	frictional-plastic
Grain size range (μm)	200–450	100–400	100–400	100–400
Grain shape	sub-angular	sub-rounded	sub-rounded	sub-rounded
Sphericity	intermediate	intermediate	intermediate	intermediate
Dry sand (Van Mechelen 2004)				
Density (g/cm^3)	1.59	1.63	1.63	1.63
Angle of internal friction (°)	40	40–45	40–45	40–45
Cohesion (Pa)	23	69	69	69
Damp sand (Mulugeta 1988; Van Mechelen 2004)				
Density (g/cm^3)	1.77	1.69	1.69	1.69
Angle of internal friction (°)	28	28	28	28
Cohesion (Pa)	600–1814	600–2080	600–2080	600–2080

After this first deformation phase, a small amount of water was sprayed onto the sand to strengthen it, to better simulate the Palaeozoic basement rheology. The small amount of water causes a significant increase of the cohesion of the sand up to 600–2000 Pa, while the internal friction angle drops only slightly, to about 28° (Mulugeta 1988; Lopes Cardozo *et al.* 2002; Van Mechelen 2004). These cohesion values proportionally scale to the cohesion values of brittle basement rocks in the upper crust between 10 and 50 MPa (Byerlee 1978; Van Mechelen 2004). Fault zones are dilated zones in the sand and consequently have less grain-to-grain contact points with water bridges. The Riedel shears were therefore preserved as internal weakness zones, thus increasing the difference in strength between the strong sand and the relatively weak inherited faults, analogue to the strong Palaeozoic basement with the relatively weak Variscan faults. Using this technique allows the inherited faults to be reactivated in a

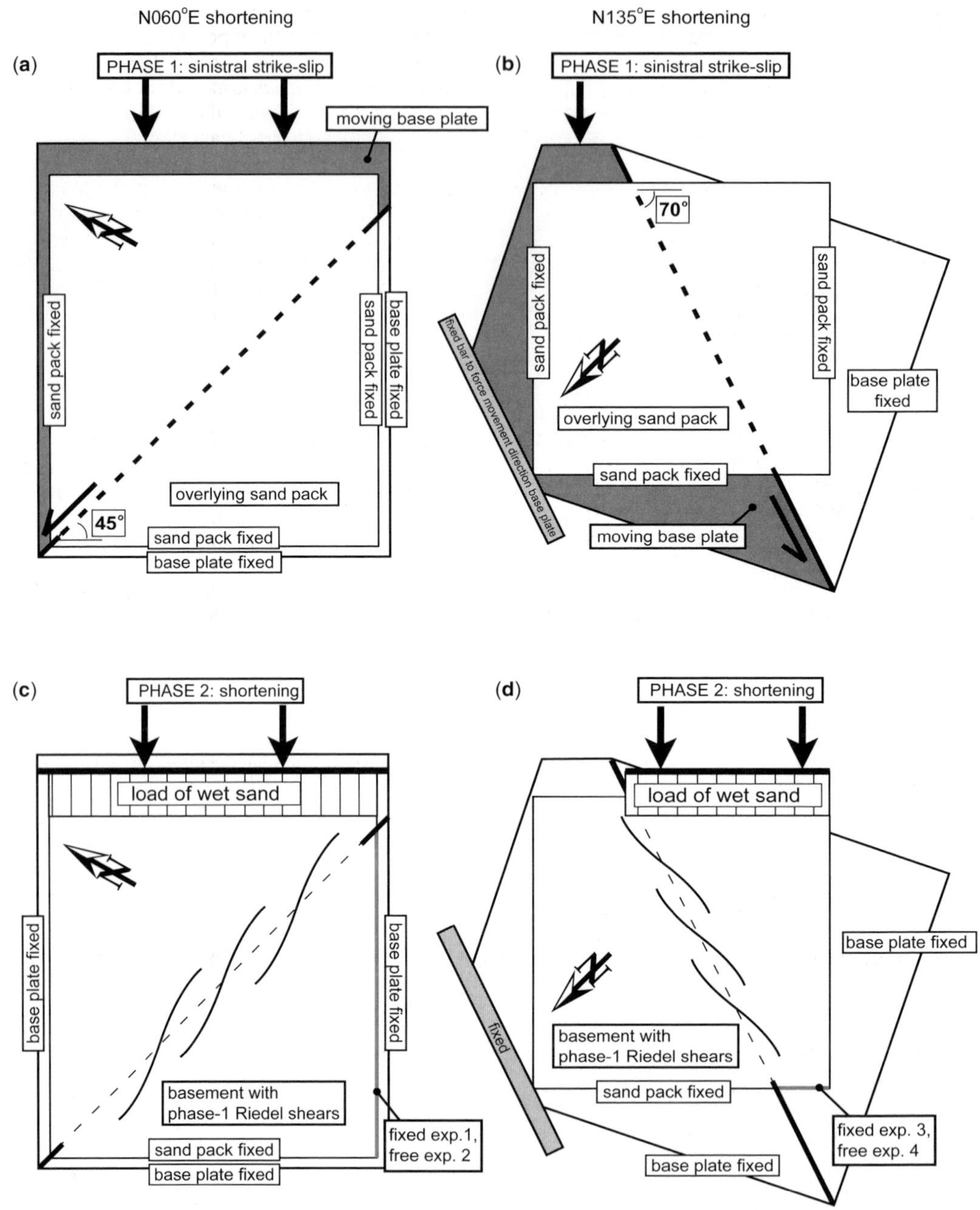

Fig. 3. Experimental set-up. (**a**) and (**b**) Set-up for the first deformation phase. One half of the base plate is laterally driven along the other fixed half using a stepping motor. Basement-induced sinistral strike-slip faulting is continued until a well developed Riedel shear pattern has formed in the overlying sand pack. (**c**) Set-up for the second deformation phase with N060°E shortening applied to the sand pack with the E–W Riedel shears. Both halves of the base plate are fixed. Boundaries of the sand pack are fixed for experiment 1 and free for experiment 2. An extra load of sand was applied at the backstop to force stresses to build up into the Riedel shear zone. (**d**) Set-up for the second deformation phase with N135°E shortening applied to the sand pack with the E–W Riedel shears. Again both halves of the base plate are fixed. Boundaries of the sand pack are fixed for experiment 3 and free for experiment 4.

second deformation phase in a favourable stress field and/or new faults to be formed. This is a critical step in connecting thin-skinned deformation with thick-skinned (deep seated) deformation processes. However, in our models we did not use a complete basement-cover model for the second phase as we started with the simplest approach to study the reactivation of Riedel shears.

In the second deformation phase the base plate was completely fixed and shortening was applied only to the sand pack above the OWE layer, in an attempt to reactivate the Riedel shears. In experiments 1 and 2, the movement direction of the vertical backstop was at 45° to the strike of the base fault, thus making a theoretical angle of about 20°–25° to the Riedel shears (Fig. 3c). In experiments 3 and 4, the movement direction of the vertical backstop was at 70° to the strike of the base fault, thus making a theoretical angle of about 45°–50° to the Riedel shears (Fig. 3d). These two models represented the E–W Riedel shears in second phase N060°E and N135°E shortening directions respectively. A load of sand was applied on top of the sand pack near the backstop, to allow stresses to build up in the Riedel shear zone. In experiments 1 and 3, the boundaries of the model were fixed. In experiments 2 and 4, only the backstop and its opposite side were fixed, thus allowing the sand pack to move freely along the Riedel shears (Figs 3c and d).

The top-views were photographed during the experiments. To study the internal structure of the models, cross-sections were made by vertically slicing the wetted sand pack at the end of the experiment. Before completely wetting the models, a protective additional sand layer was added to preserve the surface topography of the model.

Modelling results

Table 2 and Figures 4 to 7 give an overview of the modelling results. First, the results of the sinistral strike-slip deformation phase will be discussed for all four experiments. Subsequently, the results of the second deformation phase will be discussed with respect to the response of the models to different boundary conditions and to reactivation in the two different stress fields.

Table 2. *Modelling results*

Experimental number	1	2	3	4
Phase 1: sinistral strike-slip displacement				
Development of first Riedel shears (mm of sinistral strike-slip displacement)	6	8	6	8
Termination of first phase after development of Riedel shears along the entire fault zone (mm of sinistral strike-slip displacement)	8	18	8	15
Angles of Riedel shears to the base fault (°)	21–28 thrust component	17–27	20–28	15–26
Average angle of Riedel shears to the base fault (°)	±25	±22	±25	±20
Amount of horizontal displacement along Riedel shears (mm)	4–7	4–8	4–6	4–8
Relief after first phase (mm)	2–3	1–3	2–4	3–5
Phase 2: shortening oblique to the Riedel shears				
Experimental angle (°) between shortening direction and Riedel shears	±20	±23	±45	±40
Reactivation of Riedel shears	only at the backstop	yes	only at the backstop	yes
Displacement along Riedel shears	thrusting at backstop	sinistral-reverse	thrusting at backstop	dextral-reverse
Total amount of shortening (mm)	40	5	23	6
Amount of horizontal displacement along Riedel shears (mm)	–	2–4	–	2–5
Dip of upper section of Riedel shears (°)	35–50	40–50	40–45	40–50
Dip of lower section of Riedel shears (°)	60–80	70–85	75–85	55–75
Total relief after second phase (mm)	2–5	2–5	2–5	4–6

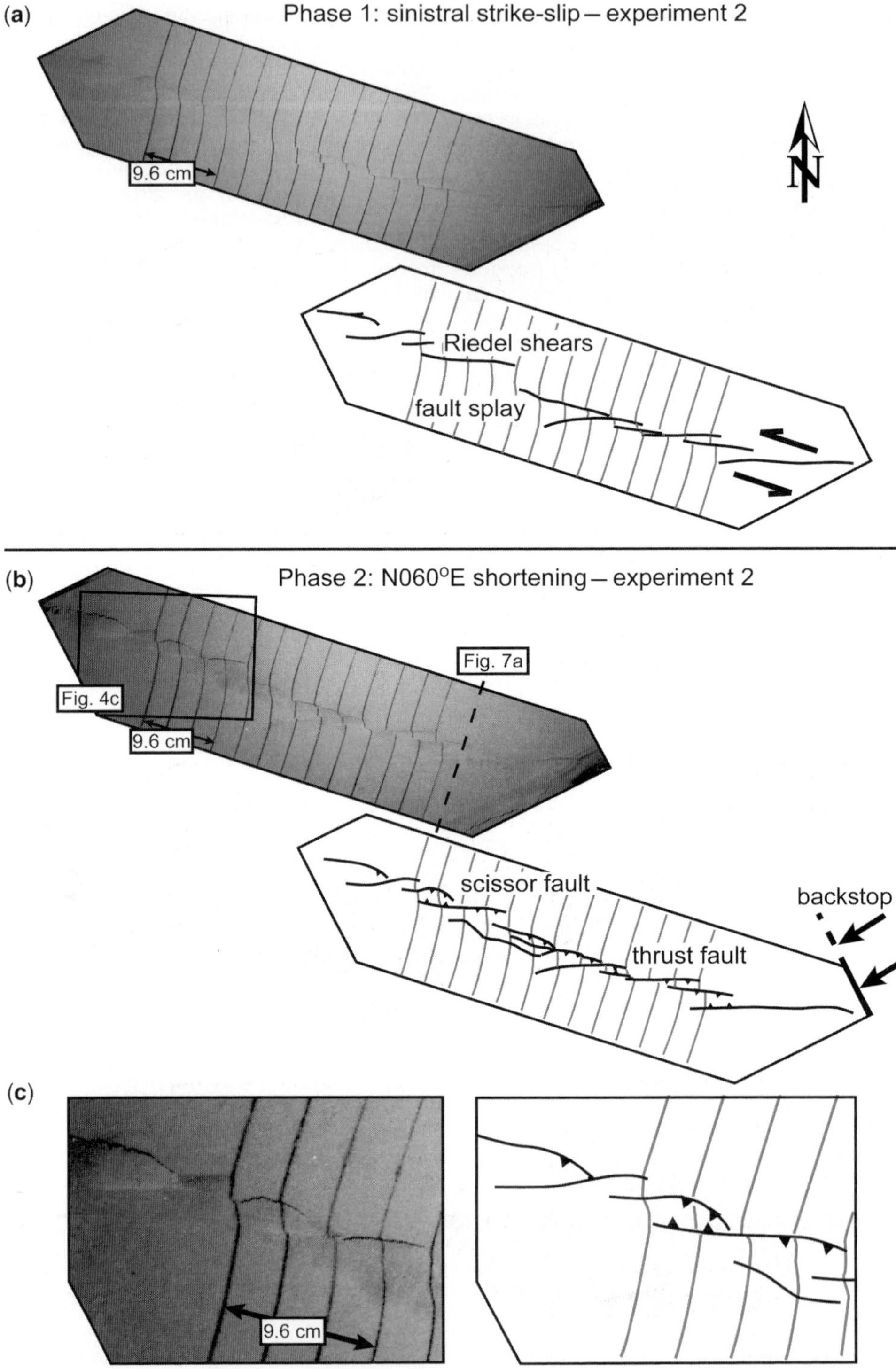

Fig. 4. Top-views and interpretations for reactivation in the N060°E shortening direction. (**a**) Pattern of en echelon Riedel shears induced by sinistral strike-slip faulting at the end of the first deformation phase (experiment 2). (**b**) Riedel shears reactivated by N060°E shortening at the end of the second deformation phase. Location of detailed top view (Fig. 4c) and cross-section (Fig. 7a) is shown. (**c**) Detailed top view of the final model.

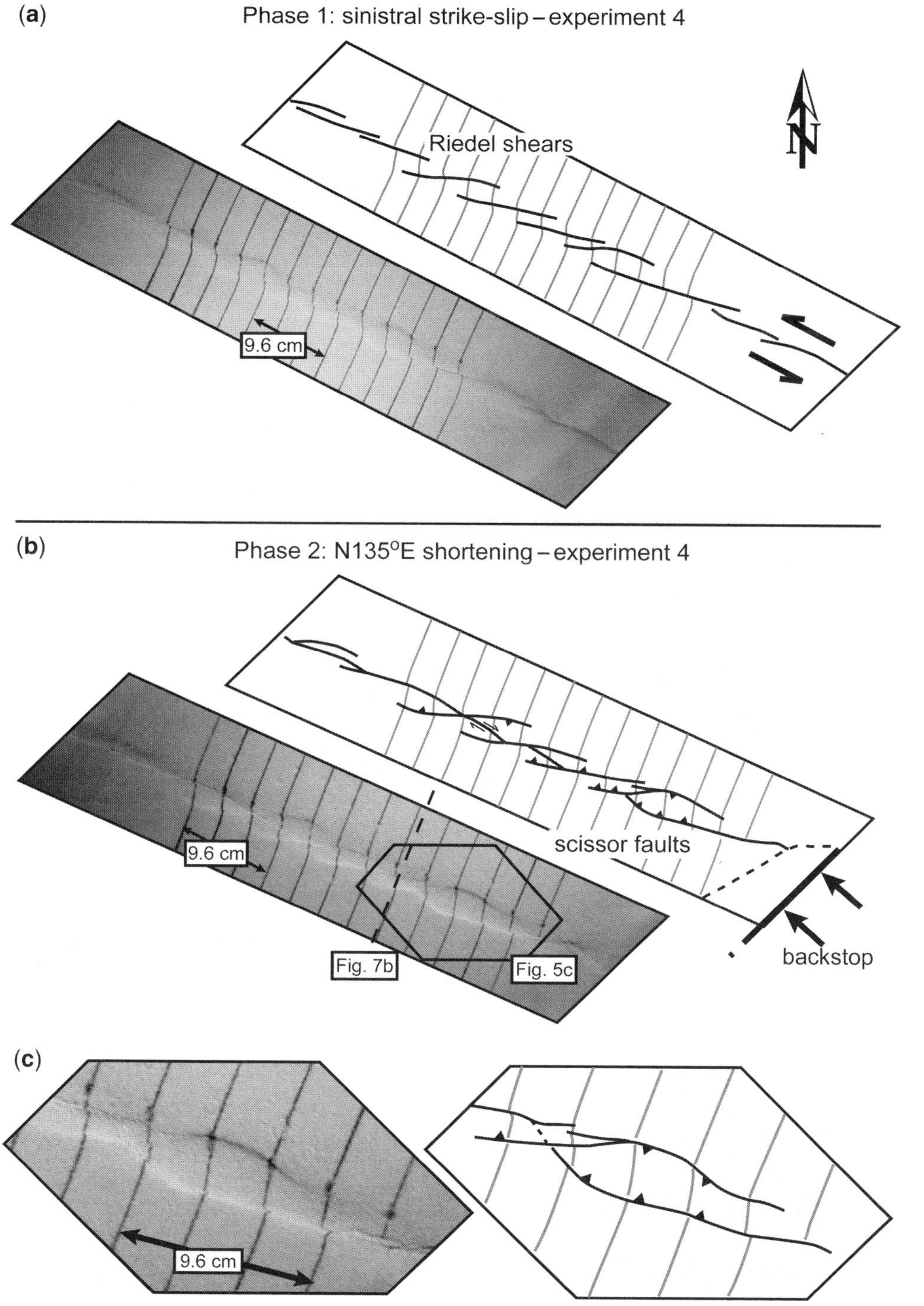

Fig. 5. Top-views and interpretations for reactivation in the N135°E shortening direction. (**a**) Pattern of en echelon Riedel shears induced by sinistral strike-slip faulting at the end of the first deformation phase (experiment 4). (**b**) Riedel shears are reactivated by N135°E shortening at the end of the second deformation phase, mainly by scissor faulting. An irregular fault partly cuts off the Riedel shears. Location of detailed top view (Fig. 5c) and cross-section (Fig. 7b) is shown. (**c**) Detailed top view of the final model.

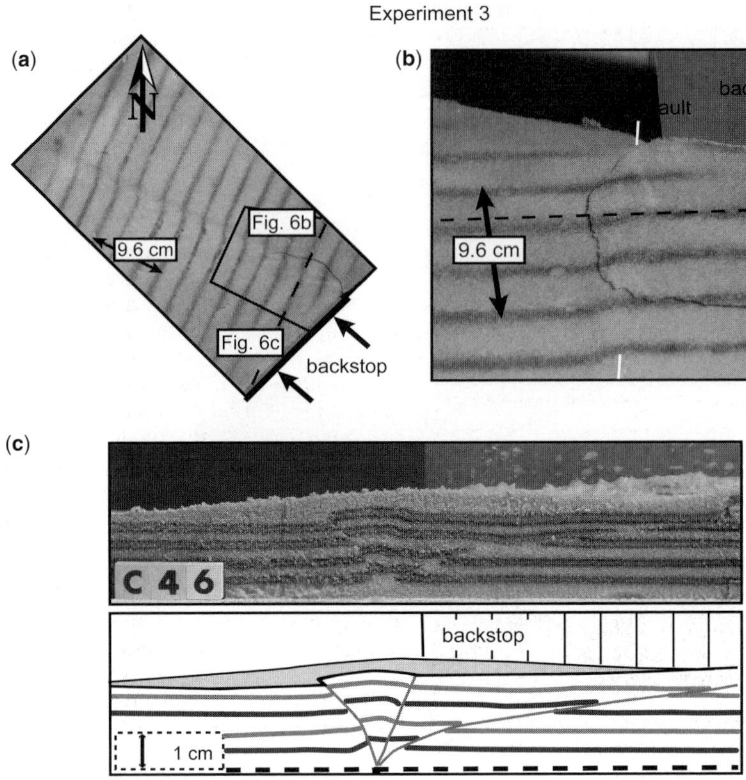

Fig. 6. Final model for second phase shortening in the N135°E direction with fixed boundaries (experiment 3). (**a**) Top view and (**b**) detailed top view showing thrusting in a short zone at the backstop limited to the width of the Riedel shear fault zone, thus only reactivating the Riedel shear close to the backstop. Location of cross-section (Fig. 6c) is indicated by a dashed line. (**c**) Cross-section shows that the newly formed thrust fault is limited by the Riedel shear.

Phase 1: Sinistral strike-slip displacement. In the first deformation phase, sinistral strike-slip faulting resulted in a typical right-stepping synthetic Riedel shear pattern (Figs 4a and 5a) with a zone of higher relief at early stages. Widely spaced Riedel shears started to form after 6 mm of strike-slip displacement in experiments 1 and 3 and a small thrust component developed at the backstop side of the shear zone in experiment 1. At the end of the first phase (8 mm of strike-slip displacement) a well-developed Riedel shear pattern characterized the two models. Riedel shears started to form after 8 mm of strike-slip displacement in experiments 2 and 4. However, a complete Riedel shear pattern above the entire length of the base fault developed only after 18 and 15 mm of strike-slip displacement respectively. At that stage, splay faults had already formed in experiment 2 at the tips of the Riedel shears near the backstop side. The difference in the amount of basal displacement needed to obtain a complete Riedel shear pattern might be due to slightly different model set-ups, such as layering (presence of OWE layer) and boundary conditions. After termination of the first phase, en echelon Riedel shears strike at angles between 20° and 28° to the base fault for experiments 1 and 3, with an average of about 25°. For experiments 2 and 4, Riedel shears strike at slightly lower angles between 15° and 27°, with an average of about 20°–22°.

The amount of horizontal sinistral displacement along the Riedel shears was between 4 and 8 mm, as obtained from the top views. Most Riedels display scissor faulting, with their sense of vertical displacement reversing at their mid-points above the base fault. The region between two Riedel shears is an up-squeezed block. The relief amounted to about 10%–25% of the original thickness of the sand pack (2 to 5 mm). Relief was slightly lower in experiment 2 (5%–15%).

Phase 2: Shortening oblique to the Riedel shears. Second phase deformation depended

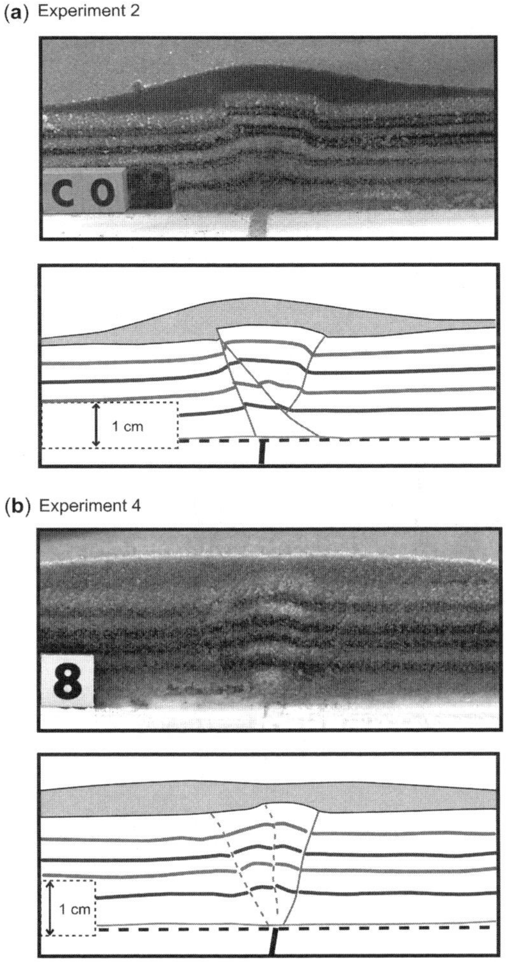

Fig. 7. Cross-sections of the final models for reactivation in the N060°E shortening direction (experiment 2) and N135°E shortening direction (experiment 4). (**a**) The cross-section clearly shows that scissor faulting occurred along the upper section of the Riedel shears. The lower steeper section of the Riedel shears has been cut off by a newly formed, less steep thrust. (**b**) The cross-section shows that clear evidence for second phase deformation cannot be deduced from the cross-sections.

greatly upon the boundary conditions. Riedel shears were not reactivated in the experiments with fixed boundaries and no OWE layer between base plate and sand pack (experiments 1 and 3 with N060°E and N135°E shortening directions respectively). In both these experiments initial shortening was accommodated by compaction (layer parallel shortening) and was followed with ongoing shortening by thrusting close to the backstop and limited to the width of the Riedel shear fault zone, thus only reactivating the Riedel shear near the backstop (Fig. 6). At the end of the second phase, after 40 and 23 mm of shortening in experiments 1 and 3 respectively, relief in the Riedel shear zone had only slightly increased, probably due to compaction of the sand pack.

In experiment 2, Riedel shears were reactivated for shortening at angles of 20°–25° to the Riedel shears (N060°E shortening direction). Reactivation resulted in a sinistral-reverse displacement on the Riedels (Figs 4b and c). Reactivation of Riedel shears near the backstop started after 1 mm of shortening. With ongoing shortening, Riedel shears at the opposite side of the model were reactivated first, and finally Riedel shears in the middle part, after 5 mm of shortening (Fig. 4b). Scissor faulting occurred along most Riedel shears, thus increasing the height of the up-squeezed blocks (Fig. 4c). Riedel shear segments NE of the base fault, in the actively moving part of the sand pack, mostly have a thrust component. The horizontal sinistral displacement along the Riedel shears at the end of the second phase was between 2 and 4 mm, as obtained from the top views. The total relief was about 10%–25% (2–5 mm) of the original thickness of the sand pack.

A cross-section through the pop-up structures displays mostly palm tree geometries with convex upward-shaped Riedel shears (Fig. 7a). Generally, only the upper section of the Riedel shears, which dips between 40° and 50°, appears to have been reactivated. Their steeper, lower part, dipping between 70° and 85°, failed and was cut off by a less steep fault that developed towards the base.

In experiment 4, Riedel shears were reactivated by shortening at angles of 40°–45° (N135°E shortening direction). Reactivation resulted in a dextral-reverse displacement along all shears from the onset of shortening. Ongoing shortening resulted in scissor faulting along the Riedel shears increasing the height of the almond-shaped up-squeezed blocks at the backstop side (Figs 5b and c). At the opposite side, Riedel shears were partly reactivated and partly cut off by an irregular fault roughly parallel to the base fault (Fig. 5b). The experiment was terminated after 6 mm of horizontal shortening. The horizontal dextral displacement along the Riedel shears at the end of the second phase was between 2 and 5 mm, as obtained from the top views. The total relief was about 20%–30% (4–6 mm) of the original thickness of the sand pack.

The cross-section through the pop-up structures displays mostly palm tree geometries with convex upward-shaped Riedel shears (Fig. 7b). Evidence for second phase deformation cannot be deduced from the cross-sections. Syn-tectonic sedimentation might have made such a polyphase analysis possible.

Reactivation of Riedel shears in a rotated stress field

The response of the curved Riedel shears to the second phase stress field is complex, but mechanically understandable. In the models with fixed boundaries (experiments 1 and 3), Riedel shears were not reactivated, probably due to stress build-up at right angles to the shortening direction. The sand on the two sides of the fault zone thus could only escape in a vertical direction. This is comparable to a situation in which the principal stresses in the horizontal plane are relatively large compared to the minimum principal stress, as illustrated with a Mohr diagram (Jaeger & Cook 1968; Sibson 1985; Mandl 1988; Fig. 8a: $\sigma_1 \approx \sigma_2 \gg \sigma_3$). Assuming σ_3 vertical, a plane with $2/3$ C_0 (cohesion) of the intact rock, an internal friction angle (φ) of 30°, and a 20°–40° angle between σ_1 and the strike of the Riedel shears, a new thrust develops rather than pre-existing Riedel shears being reactivated.

In experiments 2 and 4, sinistral-reverse and dextral-reverse reactivation of the Riedel shears occurred respectively. This is attributed to the presence of the OWE layer reducing the basal friction, and to the fact that the sand on one side of the fault zone could freely move along the other side by horizontal movement along the pre-existing Riedel shears. This is comparable to a situation in which the two smaller principal stresses are relatively small compared to the maximum principal stress, as illustrated with a mohr diagram (Fig. 8a; $\sigma_1 \gg \sigma_2 \approx \sigma_3$). For these 'stress' conditions oblique-slip reactivation of the relatively weak Riedel shears may occur.

The upper section of the Riedel shears was clearly reactivated in experiments 2 and 4. In experiment 2 the lower sections of most Riedel shears were cut off by a less steep newly formed thrust and some were reactivated. Merging of strike-slip faults with oblique-slip reverse faults at depth is also reported for transpressional models consisting of a basal shear component and a transverse shortening component (Schreurs & Colletta 2002). However, in the experiments of Schreurs & Colletta (2002) these oblique-slip reverse faults are older faults formed during initial stages of deformation, whereas in our experiments they developed with ongoing deformation from the upper shallower dipping sections of the Riedel shears. In experiment 4, Riedel shears were not cut off by a newly formed thrust towards the base. Instead, with ongoing deformation, an irregular fault developed throughout the fault zone, thus connecting individual Riedel shears.

The relationship between the fault dip and the likelihood of reactivation occurring is (semi) quantitatively analysed in the Mohr diagram of Figure 8b. In order to analyse the reactivation mechanisms for experiment 2, the curved fault planes were divided into two main sections: a steep lower section dipping about 80°, and a shallower upper section dipping 45° (Table 2). The angle between the strike of the Riedel shears and the maximum principal stress is between 20° and 30°. In these experiments with shortening oblique to the Riedel shears, the angle (α) between σ_1 and the intersection of the pre-existing fault plane with the vertically oriented σ_1–σ_3 plane controls whether reactivation occurs. For the steep lower section, α ranges from 62° to 72°, and for the upper section it is between 18° and 26°. Plotting this in the Mohr diagram (Fig. 8b), again assuming σ_3 vertical, a plane with $2/3$ C_0, and an internal friction angle of 30°, reveals that with increasing σ_1 reactivation occurs of fault planes with α between 12° and 48° instead of failure of the matrix rock. For the steep lower section α is too large, thus causing the matrix to fail before reactivation can take place. This explains the cut-off of the lower section of most Riedel shears and a new less steep fault towards the base. For the upper section, a small increase in σ_1 is sufficient to result in the conditions required for reactivation.

In experiment 4 the lower section of the Riedel shears was reactivated. This might possibly be explained in terms of a shallower dipping lower section of the Riedels in combination with a plane with extremely low cohesion. It appears that in this experiment some of the Riedel shears have a shallower dipping lower section compared to the other experiments, even as low as 55° (Table 2). In the case of a plane without cohesion, the range of fault dips that can be reactivated would reach from 3° to 62° (Fig. 8c), thus incorporating faults of the lower section of the Riedel shears observed in this model. A possible explanation for the presence of a relatively weaker plane towards the base is that the drop in cohesion increases with depth (Richard & Krantz 1991), where faults are generated during the first deformation phase. With a maximum fault dip of about 62° for reactivation to occur, this would explain why reactivation of the lower section occurs in experiment 4 (fault dips

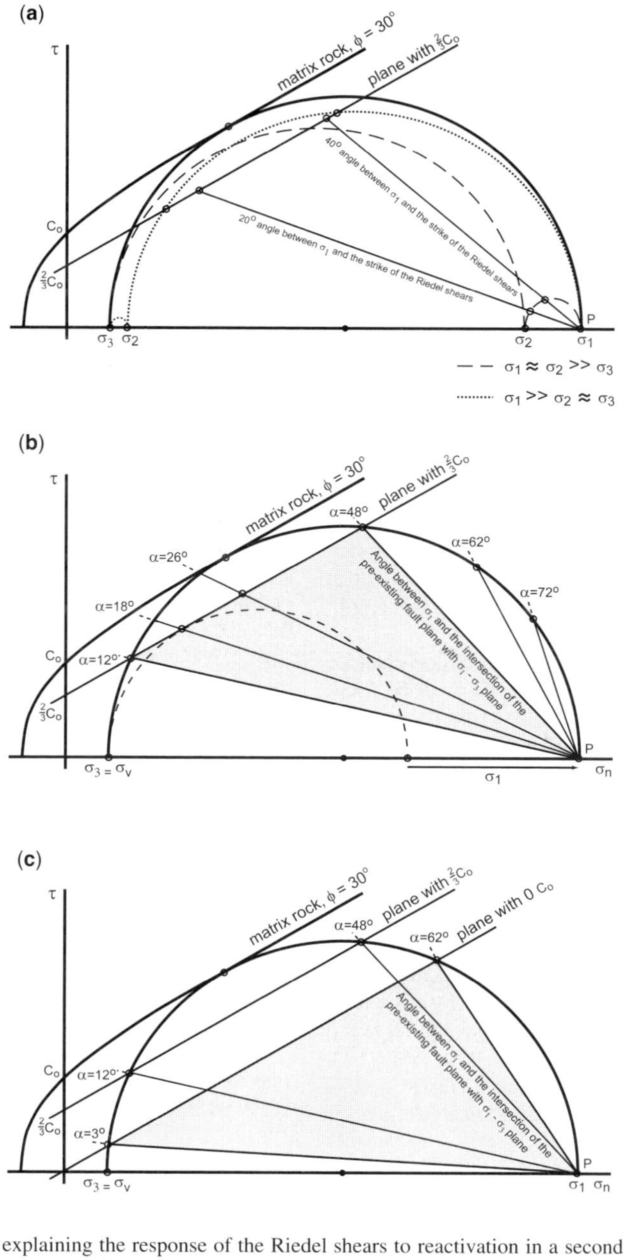

Fig. 8. Mohr diagrams explaining the response of the Riedel shears to reactivation in a second phase stress field. (**a**) Mohr diagram illustrating the effect of the relative size of the principal stresses on the likelihood for strike-slip reactivation to occur. For the condition where $\sigma_1 \approx \sigma_2 \gg \sigma_3$, assuming σ_3 in vertical direction, a plane with $\frac{2}{3}$ C_0 (cohesion) of the intact rock, an internal friction angle (φ) of 30°, and a 20° (experiment 1) to 40° (experiment 3) angle between σ_1 and the strike of the Riedel shears, a new thrust develops rather than strike-slip reactivation occurring (experiments 1 and 3). However, for the condition where $\sigma_1 \gg \sigma_2 \approx \sigma_3$, oblique-slip reactivation of the relatively weak Riedel shears may occur (experiments 2 and 4). (**b**) Mohr diagram illustrating why the steeper lower sections have been cut off and the upper shallower dipping sections were reactivated (experiment 2). Plotting the fault planes, again assuming σ_3 vertical, a plane with $\frac{2}{3}$ C_0, and an internal friction angle of 30°, reveals that with increasing σ_1, reactivation occurs of fault planes with α between 12° and 48° instead of failure of the matrix rock. For the steep lower section, α is too large (62°–72°), thus causing the matrix to fail instead of reactivation of the pre-existing fault planes. (**c**) In the case of a plane without cohesion, the range of fault dips that can be reactivated would reach from 3° to 62°, thus incorporating faults of the lower section of the Riedel shears observed in experiment 4.

between 55° and 75°) and only partly in experiment 2 (fault dips between 70° and 85°). It is however difficult to establish if, and if so why, the lower sections of the Riedel shears developed at lower dip angles in experiment 4.

Discussion and conclusions

Based on structural and palaeostress data two types of models for the polyphase deformation history of the Sierra de Albarracín were tested with sandbox analogue modelling. Both types of models involved two phases of deformation to simulate Late Variscan sinistral shearing followed by shortening in an Alpine stress field with either a N060°E oriented σ_1 (first type of model: experiments 1 and 2) or a N135°E σ_1 (second type of model: experiments 3 and 4). In both types of models reactivation of first-phase Riedels during second phase shortening only occurred in experiment 2 and 4, in which an OWE layer was applied and in which sand on one side of the first phase fault zone could freely move along the other side.

In experiment 2, E–W Riedel shears originating from a Late Variscan sinistral shear zone were reactivated during Alpine N060°E shortening in a sinistral-reverse shear sense, whereas in experiment 4 reactivation of the Late Variscan E–W Riedels during Alpine N135°E shortening occurred in a dextral-reverse shear sense. Both experiments 2 and 4 show that the up-squeezed blocks in the en echelon Riedel shear pattern (Naylor *et al.* 1986) are comparable to the Palaeozoic basement inliers aligned in a NW–SE arrangement and bounded by E–W faults. This specific geometry and the observed sinistral displacement of the oldest Palaeozoic basement rocks can only be obtained by initial Late Variscan sinistral shearing.

For Alpine reactivation, modelling results reveal that the second phase deformation is taken up by oblique-slip along favourably oriented Riedel shears. In the N060°E reactivation model, the steep lower segments of most Riedel shears were cut off by newly formed thrusts, whereas the upper shallower dipping segments were reactivated by sinistral oblique-slip displacement. In the N135°E reactivation model, dextral oblique-reverse slip occurred along the complete Riedel shear segments. In contrast to experiment 2, the lower sections of the Riedel shears in experiment 4 were not cut off by thrust faults. This can be explained by slightly shallower dipping lower sections of the Riedel shears in experiment 4 compared to experiment 2 and an increasing drop in cohesion of pre-existing fault planes with depth (Richard & Krantz 1991).

Vertical displacements are observed along the E–W faults bordering the northern limit of the Macizo de Nevera, but it is unclear whether Alpine reactivation of these faults involved an oblique-slip component and if so, whether reactivation was dextral-reverse or sinistral-reverse. The orientation of these structures would fit both the N060°E and the N135°E reactivation model. The westward vergence of structures within the Lower Triassic south of the E–W faults bordering the Macizo del Tremedal corresponds with the dextral strike-slip movement along pre-existing Riedel shears in the N135°E reactivation model. Thus, the structures observed in the vicinity of the Macizo del Tremedal favour an Alpine N135°E reactivation model to explain the polyphase structure of the Sierra de Albarracín.

It is stressed that second phase reactivation only occurs in experiments 2 and 4, and requires that sand on one side of the fault zone can freely move along the other. This is comparable to a situation in which the maximum principal stress is horizontal and relatively large compared to the other principal stresses. Furthermore, a lubricant OWE layer has to be present between base plate and sand pack, analogous to decoupling of the Palaeozoic rock pile over the shales at its base.

A major advantage of the models is that they explain the present-day structure of the Sierra de Albarracín in terms of a single major Late Variscan megashear zone reactivated during Alpine shortening. This is consistent with the generally accepted idea that the Late Variscan lineaments controlled the evolution of the Iberian Range (e.g., Arche & López-Gómez 1992, 1996).

We thank F. Storti, D. Cunningham and G. Schreurs for constructive and helpful reviews that greatly improved this manuscript. Furthermore, our appreciation goes to our fellow students from the Vrije Universiteit, with whom we had interesting discussions during the Albarracín fieldwork. D. Sokoutis is thanked for the availability of the tectonic analogue modelling laboratory (TecLab) of the Vrije Universiteit. Furthermore, we would like to thank M. Papo and M. Corver for their assistance during the analogue modelling.

References

ALVARO, M., CAPOTE, R. & VEGAS, R. 1979. Un modelo de evolución geotectónica para la cadena Celtibérica. *Acta Geologica Hispanica*, **14**, 172–177.

ARCHE, A. & LÓPEZ-GÓMEZ, J. 1992. Una nueva hipótesis sobre las primeras etapas de la evolución tectosedimentaria de la cuenca pérmico-triásica de SE de la Cordillera Ibérica. *Cuadernos de Geología Ibérica*, **16**, 115–143.

ARCHE, A. & LÓPEZ-GÓMEZ, J. 1996. Origin of the Permian-Triassic Iberian Basin, central-eastern Spain. *Tectonophysics*, **266**, 443–464.

ARTHAUD, F. & MATTE, P. 1977. Late Paleozoic strike-slip faulting in southern Europe and northern Africa: result of a right-lateral shear zone between the Appalachians and the Urals. *Geological Society of America Bulletin*, **88**, 1305–1320.

BUCHANAN, P. G. & MCCLAY, K. R. 1991. Sandbox experiments of inverted listric and planar fault systems. *Tectonophysics*, **188**, 97–115.

BYERLEE, J. 1978. Friction of rocks. *Pure and Applied Geophysics*, **116**, 615–626.

CASAS-SAINZ, A. M. & MAESTRO-GONZÁLEZ, A. 1996. Deflection of a compressional stress field by large-scale basement faults. A case study from the Tertiary Almazán basin (Spain). *Tectonophysics*, **255**, 135–156.

CLOOS, H. 1928. Experimenten zür Inneren Tektonik. *Zentralblatt für Mineralogie, Geologie, Paläeontologie*, **B**, 609–621.

DE RUIG, M. J., SMIT, J., GEEL, T. & KOOI, H. 1991. Effects of the Pyrenean collision on the Paleocene stratigraphic evolution of the southern Iberian margin (southeast Spain). *Geological Society of America Bulletin*, **103**, 1504–1512.

DE VICENTE, G., GINER, J. L., MUÑOZ-MARTÍN, A., GONZALEZ-CASADO, J. M. & LINDO, R. 1996. Determination of present-day stress tensor and neotectonic interval in the Spanish Central System and Madrid Basin, central Spain. *Tectonophysics*, **266**, 405–424.

DOBLAS, M. & LÓPEZ-RUIZ, J. ET AL. 1994. Extensional tectonics in the central Iberian Peninsula during the Variscan to Alpine transition. *Tectonophysics*, **238**, 95–116.

DONATH, F. P. & CRONWELL, R. M. 1981. Probabilistic treatment of faulting in geologic media. *In*: CARTER, N. L., FRIEDMAN, M., LOGAN, J. M. & STRARNS, D. W. (eds) *Mechanical Behaviour of Crustal Rocks*. American Geophysical Union Geophysical Monograms, **24**, 231–241.

FACENNA, C., NALPAS, T., BRUN, J.-P. & DAVY, P. 1995. The influence of pre-existing thrust faults on normal fault geometry in nature and in experiments. *Journal of Structural Geology*, **17**, 1139–1149.

GUIMERÀ, J. 1996. Cenozoic evolution of eastern Iberia: Structural data and dynamic model. *Acta Geologica Hispanica*, **29**, 57–66.

HINKELBEIN, K. 1969. El Triásico y el Jurásico de los alrededores de Albarracín. *Separata de Teruel*, Teruel, **41**.

HORSFIELD, W. T. 1997. An experimental approach to basement-controlled faulting. *Geologie & Mijnbouw*, **56**, 363–370.

JABALOY, A., GALINDO-ZALDÍVAR, F. & GONZÁLEZ-LODEIRO, F. 2002. Palaeostress evolution of the Iberian Peninsula (Late Carboniferous to present-day). *Tectonophysics*, **357**, 159–186.

JAEGER, J. C. & COOK, N. G. 1968. *Elasticity, Fracture and Flow: With Engineering and Geological Applications*. Science Paperbacks, Methuen & Co.

LOPES CARDOZO, G., BADA, G., LANKREIJER, A. C. & NIEUWLAND, D. A. 2002. Analogue modelling of a prograding strike-slip fault: the case study of the Balatonfö fault, western Hungary. *In*: CLOETINGH, S. A. P. L., HORVATH, F., BADA, G. & LANKREIJER, A. C. (eds) *Neotectonics and Surface Processes: the Pannonian Basin and Alpine/Carpathian System*. European Geoscience Union, Stephan Mueller Special Publication Series, **3**, 217–226.

LÓPEZ-GÓMEZ, J., MAS, R. & ARCHE, A. 1993. The evolution of the Middle Triassic (Muschelkalk) carbonate ramp in the SE Iberian Ranges, eastern Spain: sequence stratigraphy, dolomitization processes and dynamic controls. *Sedimentary Geology*, **87**, 165–193.

MANDL, G., 1988. Mechanics of tectonic faulting: models and basic concepts. *In*: H. J. ZWART (ed.) *Developments in Structural Geology*, **1**.

MCCLAY, K. R. & BUCHANAN, P. G. 1992. Thrust faults in inverted extensional basins. *In*: MCCLAY, K. R. (ed.) *Thrust Tectonics*. London Royal Holloway and Bedford New College, Chapman & Hall, 93–104.

MULUGETA, G. 1988. Squeeze box in a centrifuge. *Tectonophysics*, **148**, 323–335.

MUÑOZ, A., RAMOS, A., SÁNCHEZ-MOYA, Y. & SOPEÑA, A. 1992. Evolving fluvial architecture during a marine transgression: Upper Buntsandstein, Triassic, central Spain. *Sedimentary Geology*, **75**, 257–281.

MUÑOZ-MARTÍN, A., CLOETINGH, S., DE VICENTE, G. & ANDEWEG, B. 1998. Finite-element modelling of Tertiary palaeostress fields in the eastern part of the Tajo Basin (central Spain). *Tectonophysics*, **300**, 47–62.

NAYLOR, M. A., MANDL, G. & SIJPESTEIJN, C. H. K. 1986. Fault geometries in basement-induced wrench faulting under different initial stress states. *Journal of Structural Geology*, **8**, 737–752.

PÉREZ-ARLUCEA, M. & SOPEÑA, A. 1983. Estudio estratigrafico y sedimentologico de los materiales Permicos y Triasicos en el noroeste de la Sierra de Albarracín (Provincia de Guadalajara). *Estudios Geológicos*, **39**, 329–343.

PÉREZ-ARLUCEA, M. & SOPEÑA, A. 1985. Estratigrafía del Permico y Triasico en el sector central de la Rama Castellana de la Cordillera Iberica (Provincias de Guadalajara y Teruel). *Estudios Geológicos*, **41**, 207–222.

RAMOS, A., SOPEÑA, A. & PEREZ-ARLUCEA, M. 1986. Evolution of Buntsandstein fluvial sedimentation in the northwest Iberian Ranges (Central Spain). *Journal of Sedimentary Petrology*, **56** (6), 862–875.

RICHARD, P. 1991. Experiments on faulting in a two-layer cover sequence overlying a reactivated basement fault with oblique-slip. *Journal of Structural Geology*, **13**, 459–469.

RICHARD, P. & KRANTZ, R. W. 1991. Experiments on fault reactivation in strike-slip mode. *Tectonophysics*, **188**, 117–131.

RIEDEL, W. 1929. Zur Mechanik geologischer Bruchersheinungen. *Zentralblatt für Mineralogie, Geologie, Paläeontologie*, **B**, 354–368.

RODRIGUEZ-PASCUA, M. A., DE VICENTE, G. & GONZALEZ-CASADO, J. M. 1994. Kinematic and dynamic array of the Alpine deformations in alto Tajo area (Guadalajara). *Cuaderno Laboratorio Xeolóxico de Laxe*, Coruña, **19**, 163–174.

RONDEEL, H. E., WEIJERMARS, R. & VAN DORSSEN, H. G. 1984. Reactivation of early reverse faults associated with oblique strike-slip faulting: A mechanism for crustal shortening (Macizo de Nevera, Sierra de Albarracín, Spain). *Geologie en Mijnbouw*, **63**, 387–398.

SALAS, R. & CASAS, A. 1993. Mesozoic extensional tectonics, stratigraphy and crustal evolution during the Alpine cycle of the eastern Iberian basin. *Tectonophysics*, **228**, 33–55.

SCHELLART, W. P. & NIEUWLAND, D. A. 2003. 3D evolution of a pop-up structure above a double basement strike-slip fault: some insights from analogue modelling. *In*: NIEUWLAND, D. A. (ed.) *New Insights Into Structural Interpretation and Modelling*. Geological Society, London, Special Publications, **212**, 169–179.

SCHREURS, G. & COLLETTA, B. 2002. Analogue modelling of continental transpression. *In*: SCHELLART, W.P. & PASSCHIER, C. (eds) *Analogue Modelling of Large-Scale Tectonic Processes*. Journal of the Virtual Explorer, **6**, 67–78.

SIBSON, R. H. 1985. A note on fault reactivation. *Journal of Structural Geology*, **7** (6), 751–754.

SIMÓN-GÓMEZ, J. L. 1986. Analysis of a gradual change in stress regime (example from the eastern Iberian Chain, Spain). *Tectonophysics*, **124**, 37–53.

VAN MECHELEN, J. L. M. 2004. Strength of moist sand controlled by surface tension for tectonic analogue modelling. *Tectonophysics*, **384**, 275–284.

VAN WEES, J. D., ARCHE, A., BEIJDORFF, C. G., LÓPEZ-GÓMEZ, J. & CLOETINGH, S. A. P. L. 1998. Temporal and spatial variations in tectonic subsidence in the Iberian Basin (eastern Spain): inferences from automated forward modelling of high-resolution stratigraphy (Permian-Mesozoic). *Tectonophysics*, **300**, 285–310.

VERSCHUREN, M., NIEUWLAND, D. & GAST, J. 1996. Multiple detachment levels in thrust tectonics: sandbox experiments and palinspastic reconstruction. *In*: BUCHANAN, P. G. & NIEUWLAND, D. A. (eds) *Modern Developments in Structural Interpretation, Validation and Modelling*. Geological Society, London, Special Publications, **99**, 227–234.

VIALLARD, P. 1983. Le décollement de couverture dans la Chaîne Ibérique méridionale (Espagne): effet de raccourcissements differentials entre substratum et couverture. *Bulletin de la Société Géologique de France*, **7** (3), 379–387.

VILAS, L., ALONSO, A., ARIAS, C., GARCÍA, A., MAS, R., RINCÓN, R. & MELÉNDEZ, N. 1983. The Cretaceous of the Southwestern Iberian Ranges (Spain). *Zitteliana*, **10**, 245–254.

WEIJERMARS, R. & RONDEEL, H. E. 1984. Shear band foliation as an indicator of sense of shear: field observations in central Spain. *Geology*, **12**, 603–606.

WILCOX, R. E., HARDING, T. P. & SEELY, D. R. 1973. Basic wrench tectonics. *Bulletin of the American Association of Petroleum Geologists*, **57**, 74–96.

Lithospheric scale gravitational flow: the impact of body forces on orogenic processes from Archaean to Phanerozoic

PATRICE F. REY[1] & GREGORY HOUSEMAN[2]

[1]*School of Geosciences, The University of Sydney, NSW 2006 and Dept. de Géologie, UMR-CNRS 6524, Université Jean Monnet, St Etienne, F42023, France*

[2]*School of Earth and Environment, University of Leeds, Leeds, LS2 9JT, UK*

Abstract: In the Archaean, the combination of warmer continental geotherm with a lighter sub-continental lithospheric mantle suggests that gravitational forces played a more significant role in continental lithospheric deformation. To test this hypothesis, we compare the evolution of the deformation and the regional state of stress in 'Archaean-like' and 'Phanerozoic-like' lithospheres submitted to the same boundary conditions in a triaxial stress-field with imposed convergence in one direction. For plausible physical parameters, thickening of normal to cold Phanerozoic lithospheres produces relatively weak buoyancy forces, either extensional or compressional. In contrast, for Archaean continental lithospheres, or for anomalously warm Phanerozoic lithospheres, lateral gravitationally-driven flow prevents significant thickening. This conclusion is broadly consistent with: (1) the relative homogeneity of the erosional level now exposed at the surface of Archaean cratons, (2) the sub-aerial conditions that prevailed during the emplacement of up to 20 km of greenstone cover, (3) the relatively rare occurrence in the Archaean record of voluminous detrital sediments, (4) the near absence of significant tectonic, metamorphic and magmatic age gradients across Archaean cratons, (5) the relative homogeneity of strain across large areas, and (6) the ubiquitous presence of crustal-scale strike slip faults in many Late Archaean cratons.

One of the most contentious issues in Archaean geology is the significance of the particular features seen in the surface geology of most Archaean cratons. These involve the characteristic strain pattern of domes and basins; the ubiquity (in particular in Late-Archaean cratons) of strike-slip faults; the relative homogeneity at the craton scale of the finite strain field, the metamorphic facies and the erosion level; together with the particular timing of tectonics, magmatism, and metamorphism that seem to develop craton-wide within a very narrow time window (Binns *et al.* 1976; Choukroune *et al.* 1997; Galer & Mezger 1998; Hamilton 1998; Qiu *et al.* 1999; Rey *et al.* 2003). Based on the same field evidence, these features are interpreted by some authors as evidence of plate-tectonic processes (e.g., Snowden & Bickle 1976; Myers & Watkins 1985; Treloar & Blenkinsop 1995), and by others as evidence of gravitational instabilities (McGregor 1951; Collins 1989; Ramsay 1989; Jelsma *et al.* 1993; Bouhallier *et al.* 1993; Chardon *et al.* 1998) possibly related to plume activity (Choukroune *et al.* 1995; Warren & Ellis 1996). In this three-decade-long ongoing debate, the differences in mechanical behaviour of Phanerozoic and Archaean continental lithospheres have not been properly considered, yet regional finite strain fields depend significantly on the thermo-mechanical properties of the continental lithosphere. Consequently, the particular features observed in Archaean cratons could be related, in part, to contrasting lithospheric mechanical properties rather than to contrasting tectonic processes.

In this paper we aim to illustrate the relatively greater importance of gravitational forces in the tectonic evolution of ancient continental lithospheres. Based on triaxial numerical experiments that compare the evolution of Phanerozoic-like and Archaean-like lithospheres under horizontal convergence, we show that ancient continental lithospheres could have been under an increased stabilizing influence of gravitational forces that promoted homogeneous deformation and therefore impeded the development of strain localization along linear belts. If the buoyancy-derived stress were greater in the Archaean, then lateral escape (strike-slip faults and crustal-scale lateral crustal flow) rather than large-scale thrust and thickening could have become the primary response to tectonic convergence. We also suggest that the secular cooling

From: BUITER, S. J. H. & SCHREURS, G. (eds) 2006. *Analogue and Numerical Modelling of Crustal-Scale Processes.* Geological Society, London, Special Publications, **253**, 153–167.
0305-8719/06/$15.00

of the continental geotherm, combined with the decrease in the buoyancy of the sub-continental lithospheric mantle (SCLM), have changed the thermo-mechanical properties of the continental lithosphere in such a way that the impact of gravitational forces on lithospheric deformation has decreased over geological time.

Argand Ratio: assessing the impact of gravitational forces on modern and ancient continental lithospheres

Scaled models of orogenic systems depend on a few dimensionless parameters that describe the relative importance of different processes in the stress and thermal balance. Amongst these parameters the Argand Ratio characterizes the ability of gravitational forces to intervene in lithospheric deformation (Fig. 1). The Argand Ratio (*AR*) can be defined as the local ratio of the gravitational stress (arising from lateral variation of gravitational potential energy and thus of density) to the averaged lithospheric strength at a nominal strain rate (i.e., the stress-driving deformation). The ratio *AR* may be compared with the Argand Number (*Ar*) of England & McKenzie (1982, 1983), a global parameter by which one can quantify the overall regional balance between gravitational stress and viscous stress in an indentation problem (see also Houseman & England 1986; Buck & Soukoutis 1994; Schmalholz *et al.* 2002, 2005). *AR* is obtained from *Ar* by replacing the nominal buoyant stress factor in *Ar* with the locally variable buoyant stress (relative to a reference column), and similarly replacing the nominal viscous stress scale factor in *Ar* with the locally variable strength of the lithosphere. The resulting Argand Ratio can be used to describe the evolution during deformation of the balance between buoyant stress and the strength of the lithosphere. The Argand Ratio measures the tendency of the lithosphere to deform in response to the variation of gravitational potential energy (ΔGPE), and therefore is a measure of the ability of gravitational forces to intervene in lithospheric deformation.

In modern continental lithospheres, the Argand Ratio may reach values ≥ 1 when large ΔGPE exists and/or when the orogenic system involves a weak lithosphere (i.e., a warmer geotherm). In Tibet as well as in the Basin and Range Province of the western USA, it has been argued that present-day crustal flow is strongly influenced by forces induced by large gradients in gravitational potential energy (England & Molnar 1997; Jones *et al.* 1996; Flesch *et al.* 2000). A warmer geotherm reduces the strength of the continental lithosphere and, as a rule of thumb, it is inferred that a thickened continental crust spreads under its own weight (i.e., Argand Ratio ≥ 1 for a given effective strain rate) when the temperature at the Moho is above *c.* 700°C (e.g., Sonder *et al.* 1987).

In modern continental lithospheres, such a thermal condition is likely to be transient and associated with either crustal thickening or the rise of the lithosphere/asthenosphere interface under continental areas. In the Archaean however, the larger radiogenic heat production, possibly coupled with a stronger basal heat flow, would lead to warmer continental geotherms. For radiogenic-element contents compatible with the present-day average composition of Archaean cratons (Taylor & McLennan 1986) and plausible mantle heat flow, the steady state temperature at the Moho in the Late Archaean could have been close to 700°C (Rey *et al.* 2003). In fact, one can argue that this temperature could have buffered the thickness of the continental crust, since a thicker crust would have spread (Bailey 1999). Hence, the Archaean continental lithosphere would have been less likely to develop large gradients of crustal thickness. In addition, there is solid evidence that the buoyancy of Archaean continental lithosphere was enhanced by a sub-continental lithospheric mantle, more depleted, and therefore less dense, than that of Phanerozoic lithosphere (3310 kg m^{-3} vs 3370 kg m^{-3}, Griffin *et al.* 1998). Upon lithospheric thickening, a more buoyant sub-continental lithospheric mantle would produce relatively greater extensional forces, and therefore a larger Argand Ratio. Figure 1a illustrates the consequences of a warmer geotherm on the integrated strength of the lithosphere (i.e., the denominator of *AR*). Figure 1b illustrates the effect of a more buoyant lithospheric mantle on the local gravitational stress (i.e., the numerator of *AR*). Both effects imply that local gravitational forces might have been more important relative to in-plane stresses arising from distant plate boundaries in the Archaean than in the Phanerozoic.

Models: assumptions and simplifications

Keeping in mind that our objective is to illustrate the difference in lithospheric deformation due to contrasting buoyancy and thermal structure, we assume here that Archaean-like and Phanerozoic-like lithospheres are similar in all but two aspects: the density profile and the temperature profile. There are a number of problems when

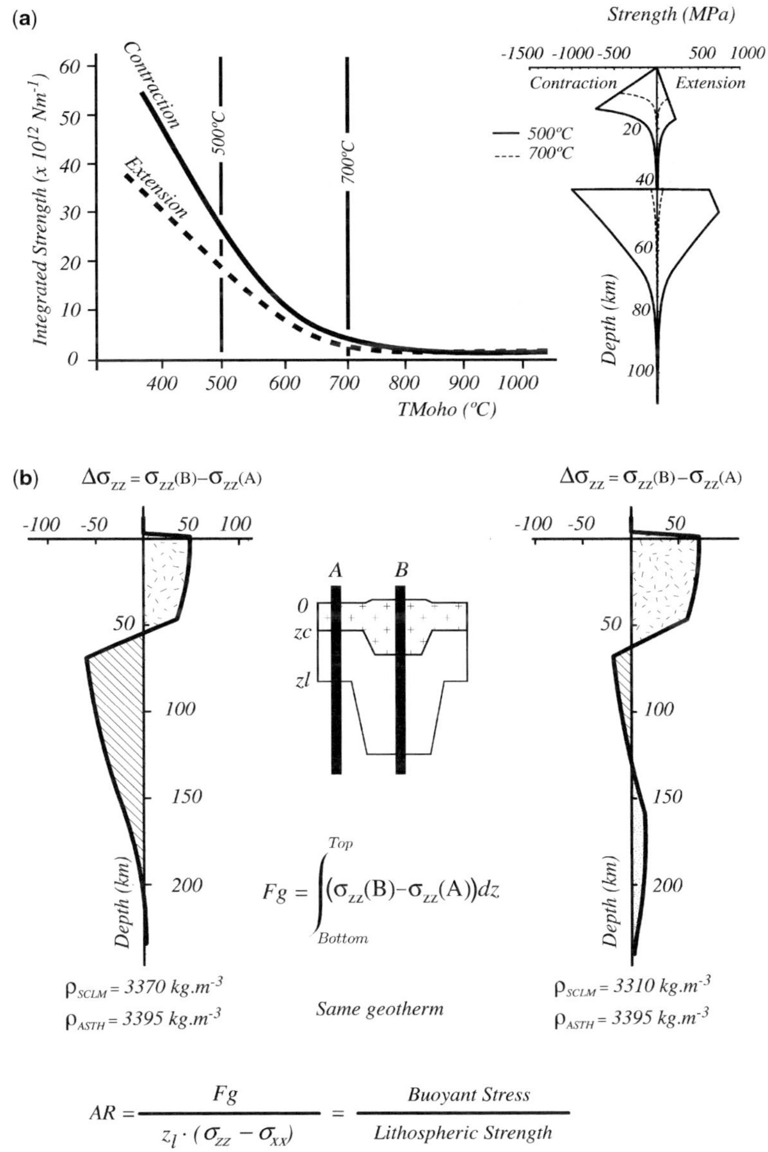

Fig. 1. (**a**) Consequences of a warmer geotherm on the strength of continental lithosphere and therefore on the *AR* denominator, and (**b**) consequences of a more buoyant sub-continental lithospheric mantle on the gravitational force related to homogeneous thickening and therefore on the *AR* numerator. (**a**) The temperature at the Moho is used as a proxy for the continental geotherm. As the temperature at the Moho increases, the integrated strength of the continental lithosphere (integration with depth of the rheological profile) decreases linearly until the Moho reaches *c*. 700°C. Above this temperature, the integrated strength does not change much as the strength of the upper crust, which is now the stronger layer of the lithosphere, is less sensitive to temperature. In inset, the rheological profiles for $T_{\mathrm{Moho}} =$ 500°C and 700°C showing the disappearance of the strong upper mantle in warm continental lithospheres. (**b**) Upon doubling the thickness of a lithospheric column via homogeneous thickening, a gravitational force is created. The gravitational force (Fg) is given by the integration with depth of the difference in the lithostatic pressure ($\Delta\sigma_{zz}$) between the deformed (column B) and the reference lithosphere (column A). Upon thickening, a heavy SCLM generates a deficit in gravitational potential energy (hatched areas in left profile) that balances the excess in gravitational potential energy (stippled areas) in the upper lithosphere. In contrast, a less dense SCLM reduces the gravitational force stored in the lower lithosphere (right profile) but increases the gravitational potential energy stored in the upper lithosphere. Hence, the overall gravitational force created by homogeneous thickening of a lithosphere with a buoyant lithospheric mantle is much larger than that created by the homogeneous thickening of a lithosphere with a denser lithospheric mantle.

trying to define average continental lithospheres representative of Archaean and Phanerozoic eons. Firstly, there is a much larger reduction in crustal radiogenic heat production within the Archaean Eon alone than from the end of the Archaean to now. Secondly, Phanerozoic lithospheres exist in a very large range of thermal regimes. We must account for variation in the properties of both Archaean and Phanerozoic continental lithospheres, for probable systematic differences between the average properties, and for significant overlap in the ranges of properties of the two types of lithosphere. Therefore, we define three Archaean and three Phanerozoic model lithospheres covering a plausible range of thermal regimes (Table 1). Table 2 summarizes the relevant thermo-mechanical parameters.

Density structure

In all models, the continental crust has the same thickness, z_c (40 km) and the same depth-dependent density structure, defined at 0°C. For the crust, the density is assumed to increase from the top of the crust (ρ_a) to the Moho (ρ_b) because of a compositional gradient. In addition, it is dependent on temperature (α_c is the coefficient of thermal expansion).

$$\rho_{\text{crust}}(z) = \left\lfloor \rho_a + \left(\frac{\rho_b - \rho_a}{z_c}\right) \cdot z \right\rfloor \times (1 - \alpha_c \cdot T(z)) \qquad (1)$$

Because of a larger degree of depletion, the density of the sub-continental lithospheric mantle (SCLM) was most likely lower in the Archaean (Jordan 1975, 1997; Griffin *et al.* 1998). Following Griffin *et al.* (1998), the density at 0°C of the SCLM in the Archaean and Phanerozoic models are 3310 kg m^{-3} and 3370 kg m^{-3} respectively. We assume that the asthenospheric mantle acts as an infinite reservoir with constant density through time (ρ_{ao} = 3395 kg m^{-3}). Assumed pressure- and temperature-dependences for the lithospheric mantle density are defined by:

$$\rho_{lm}(z) = \rho_{lmo} \cdot (1 - \alpha_m \cdot T(z) + \chi_m \cdot P) \qquad (2)$$

where χ_m is the coefficient of compressibility, P is the pressure, and ρ_{lmo} is the density of lithospheric mantle at 0°C. For the density profile in the asthenosphere, the density of the asthenosphere ρ_{ao} at 0°C is substituted for ρ_{lmo} in equation (2), with constant temperature equal to that at the base of the lithosphere, and the same value of χ_m.

Continental geotherm

The temperature profile is calculated from a two-layer model with no radiogenic heat production in the SCLM and a depth-independent radiogenic heat production in the crust of thickness z_c. The steady state and transient geotherms are derived from the one-dimensional diffusion-advection equation:

$$\frac{\partial T}{\partial t} = \kappa \cdot \frac{\partial^2 T}{\partial z^2} + \frac{H}{\rho \cdot cp} - v \cdot \frac{\partial T}{\partial z} \qquad (3)$$

where κ thermal diffusivity; H is the radiogenic heat production; ρ is the density; cp is the heat capacity; and v is the velocity of the medium relative to the upper surface. For boundary conditions we assume a constant mantle heat flow entering the base of the lithosphere, and a constant temperature (25°C) at the top of the crust. During deformation, the mantle heat flow remains constant and is applied to the base of the lithosphere, which is defined by the isotherm 1330°C.

For Archaean-like lithospheres, we conservatively choose a crustal radiogenic heat production representative of the Late Archaean. Using the average concentration in radiogenic elements for present-day Archaean cratons

Table 1. *Vertical geometry and thermal parameters for our models of Phanerozoic-like and Archaean-like continental lithospheres*

	z_c (km)	H (10^{-6} Wm^{-3})	Q_o (10^{-3} Wm^{-2})	z_l (km)	T_{Moho} (°C)
Phan160	40	0.48	18.7	160	447
Phan120	40	0.48	24.95	120	546
Phan80	40	0.48	37.4	80	741
Arch160	40	0.99	16.2	160	566
Arch120	40	0.99	21.6	120	652
Arch80	40	0.99	32.3	80	820

Abbreviations: z_c: crustal thickness; H: Crustal radiogenic heat production (depth independent); Q_o: Basal heat flow; z_l: Depth of the isotherm 1330°C, base of the lithosphere; T_{Moho}: Temperature at the Moho.

Table 2. *List of parameter values*

Parameters	Values	Units
Gravity: g	9.81	m s^{-2}
Crust thermal expansion: α_c	3.5×10^{-5}	K^{-1}
Mantle thermal expansion: $\alpha_m = a_0 + a_1T + a_2T^{-2}$	$a_0 = 2.697 \times 10^{-5}$ $a_1 = 1.0192 \times 10^{-8}$ $a_2 = -0.1282$	K^{-1}
Mantle bulk modulus: χ_m	130	GPa
Lithosphere thermal diffusivity: κ	0.97×10^{-6}	m^2 s^{-1}
Crust volumetric heat production: H_c	*cf.* Table 1	Wm^{-3}
Mantle volumetric heat production: H_m	0	Wm^{-3}
Heat capacity: cp	1000	J kg^{-1} K^{-1}
Surface temperature: T_o	293	K
Temperature at the base of the lithosphere: T_l	1603	K
Heat flow at the base of the lithosphere: Q_o	*cf.* Table 1	Wm^{-2}
Crustal thickness: z_c	*cf.* Table 1	km
Lithosphere thickness: z_l	*cf.* Table 1	km
Lithospheric mantle density @ T = 0°C: ρ_{lmo}	Arch: 3310, Phan: 3370	kg m^{-3}
Asthenospheric mantle density @ T = 0°C: ρ_{ao}	3395	kg m^{-3}
Upper crust mass density @ @ T = 0°C: ρ_a	2700	kg m^{-3}
Lower crust mass density @ @ T = 0°C: ρ_b	2900	kg m^{-3}
Water mass density: ρ_w	1030	kg m^{-3}
Ratio pore pressure to overburden stress: λ	0.36	
Universal gas constant: R	8.3144	J mol^{-1} K^{-1}
Crust power law sensitivity: n_c	3	
Crust power law activation enthalpy: Q_c	190000	J mol^{-1}
Crust power law pre-exponent: A	5×10^{-6}	MPa^{-n} s^{-1}
Mantle power law sensitivity: n_m	3	
Mantle power law activation enthalpy: Q_m	520000	J mol^{-1}
Mantle power law pre-exponent: A_m	7×10^4	MPa^{-n} s^{-1}
Mantle Dorn plasticity activation enthalpy: Q_d	540000	J mol^{-1}
Mantle Dorn plasticity law stress threshold: σ_d	8500	MPa
Mantle Dorn plasticity law strain rate: ε_d	3.05×10^{11}	s^{-1}

(Taylor & McLennan 1986) one can work out the crustal radiogenic heat production 2.7 Ga (0.99×10^{-6} Wm^{-3}). For Phanerozoic-like lithospheres, we chose the present-day average crustal radiogenic heat content of Archaean crusts (0.48×10^{-6} Wm^{-3}). We assume for simplicity that the bases of the thermal, mechanical and compositional lithopheres are all coincident. The mantle heat flow controls, z_l the thickness of the lithosphere (i.e., the depth of the isotherm 1330°C). This is highly variable in Phanerozoic lithospheres, and was most likely highly variable in the Archaean too (Lenardic & Moresi 2000). We chose mantle heat flows to get three Archaean-like models (Arch80, Arch120 and Arch160) and three Phanerozoic-like models (Phan80, Phan120 and Phan160) with nominal equilibrium lithospheric thicknesses of 80, 120, and 160 km respectively (Fig. 2, Table 1). Consequently, the Moho temperature for Phanerozoic-like lithosphere with assumed crustal thickness of 40 km ranges from 447°C (Phan160), to 546°C (Phan120), to 741°C (Phan80). Phan80 represents an anomalously warm and thin continental lithosphere, while Phan160 represents a cold and thick lithosphere. The Moho temperature of the Archaean-like models varies from 566°C (Arch160), to 652°C (Arch120), to 820°C (Arch80). Arch80 is clearly an anomalously hot and thin continental lithosphere with a Moho temperature in excess of what would be required for lower crustal flow.

Constitutive equations and rheological profile

We consider here the thin viscous sheet equations in a triaxial situation in which we specify that the horizontal principal stresses, σ_{xx} and σ_{yy}, are externally specified while the vertical stress component, σ_{zz}, is determined simply by gravity. All quantities represent vertical averages through the lithospheric column. Expressed in terms of differential stresses, and

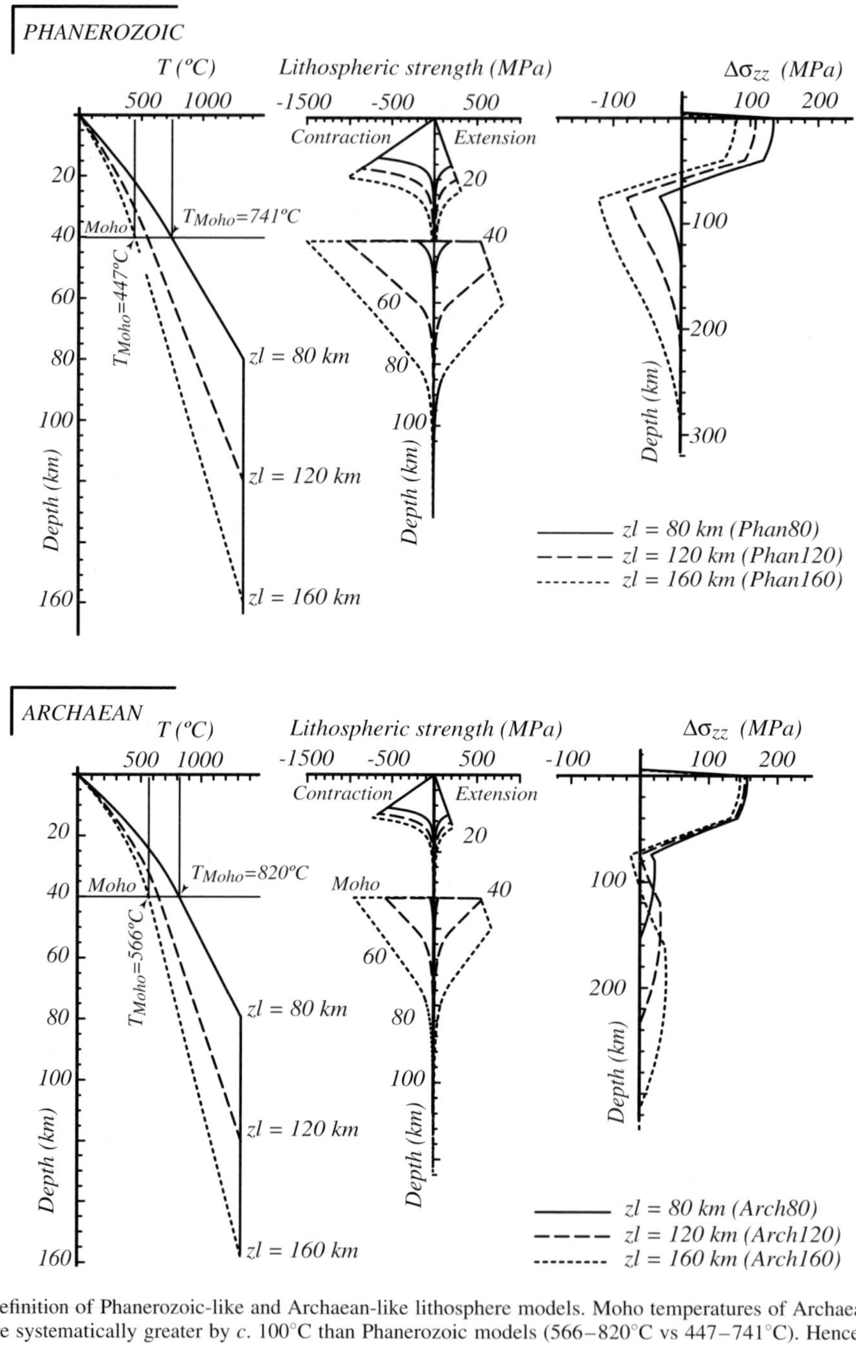

Fig. 2. Definition of Phanerozoic-like and Archaean-like lithosphere models. Moho temperatures of Archaean models are systematically greater by *c.* 100°C than Phanerozoic models (566–820°C vs 447–741°C). Hence, the Archaean models have a reduced strength when compared to the Phanerozoic models. Rheological profiles show that Phan160 and Arch120 have a strong upper mantle whereas Phan80 and Arch80 have a weak upper mantle. The contrasting density structure impacts on the gravitational force, as shown by plotting the $\Delta\sigma_{zz}$ profile for each model, assuming homogeneous doubling of the lithosphere thicknesses. In Archaean models, excesses in gravitational potential energy are stored in both crustal and mantle layers. For Phan80, the $\Delta\sigma_{zz}$ profile is similar to Archaean profiles in the sense that there is a large, unbalanced excess in gravitational potential energy (GPE) in the upper part of the lithosphere. In contrast, for the $\Delta\sigma_{zz}$ profile of Phan160, the deficit in gravitational energy in the lower lithosphere is greater than the excess in GPE in the upper lithosphere.

assuming isotropy for simplicity, the constitutive equations can be expressed as:

$$\sigma_{xx} - \sigma_{yy} = 2\eta \cdot (\dot{\varepsilon}_{xx} - \dot{\varepsilon}_{yy}) \quad (4)$$

$$\sigma_{zz} - \sigma_{yy} = 2\eta \cdot (\dot{\varepsilon}_{zz} - \dot{\varepsilon}_{yy}) \quad (5)$$

$$\sigma_{zz} - \sigma_{xx} = 2\eta \cdot (\dot{\varepsilon}_{zz} - \dot{\varepsilon}_{xx}) \quad (6)$$

where η represents the depth-averaged effective isotropic viscosity, which includes the frictional Byerlee's law relation in the brittle layer and ductile stress laws in the lower layer as summarized below. Noting that the medium is effectively incompressible: $\dot{\varepsilon}_{xx} + \dot{\varepsilon}_{yy} + \dot{\varepsilon}_{zz} = 0$, then the three unknown strain-rates under the prescribed triaxial stress field are given by:

$$\dot{\varepsilon}_{zz} = \frac{1}{6\eta} \cdot (2\sigma_{zz} - \sigma_{xx} - \sigma_{yy}) \quad (7)$$

$$\dot{\varepsilon}_{yy} = \frac{1}{6\eta} \cdot (2\sigma_{yy} - \sigma_{zz} - \sigma_{xx}) \quad (8)$$

$$\dot{\varepsilon}_{xx} = \frac{1}{6\eta} \cdot (2\sigma_{xx} - \sigma_{zz} - \sigma_{yy}) \quad (9)$$

The effective viscosity is obtained by calculating the resistance to deformation at a specified strain-rate (assumed constant with depth) by evaluating the vertically integrated strength of the lithosphere:

$$\textit{Integrated Strength} = \int_{bottom}^{top} (\sigma_1(z) - \sigma_3(z)) \cdot dz \quad (10)$$

We use the standard rheological profile of Brace & Kohlstedt (1980) for the continental lithosphere, in which frictional sliding is the dominant failure mechanism at low temperature and high strain rate (in the upper crust and the upper mantle, Sibson 1974):

$$\sigma_1(z) - \sigma_3(z) = \beta \cdot \rho(z) \cdot g \cdot z \cdot (1 - \lambda) \quad (11)$$

where g is the gravitational acceleration; λ is the ratio of fluid pore pressure to the normal stress; and β is a parameter dependent on the type of faulting, therefore dependent on the tectonic regime. It is given by:

$$\beta = \frac{R - 1}{1 + \delta \cdot (R - 1)} \quad (12)$$

with

$$R = \left(\sqrt{1 + \mu^2} + \mu\right)^{-2}$$

and

$$\delta = \frac{\sigma_{zz} - \sigma_3}{\sigma_1 - \sigma_3} \quad (13)$$

and μ the coefficient of internal friction (*cf.* Houseman & England 1986; Ranalli 1995). As the differential stress ratio δ varies from 0 (when $\sigma_{zz} = \sigma_3$) to 1 (when $\sigma_{zz} = \sigma_1$), β varies from 3 (reverse faults) to 0.75 (normal faults) for $\mu = 0.75$. Intermediate β values describe oblique slip faulting with $3 < \beta < 1.2$ for transpression, and $1.2 < \beta < 0.75$ for transtension. Strike-slip faulting occurs for $\beta = 1.2$ (when $\sigma_{zz} = (\sigma_1 + \sigma_3)/2$). Although different mechanisms (e.g., thrusting and strike-slip faulting) may occur simultaneously in a general triaxial stress-field, our calculation of the effective isotropic viscosity must assume a single value of β, which is determined from the current differential stress ratio δ.

At high temperatures and differential stresses below 200 MPa, the viscous deformation of the crust and the mantle is modelled as power law creep (Evans & Goetze 1979). The creep is thermally activated (Ranalli 1995) and may be written in terms of the second invariant of the strain-rate tensor $\dot{E}^2 = \left(\sum_{i,j} (\dot{\varepsilon}_{ij} \cdot \dot{\varepsilon}_{ij}/2)\right)$ as follows:

$$(\sigma_{ii} - \sigma_{jj}) = \left[\left(\frac{2}{3}\right)^{(n+1)/2n} A^{-1/n} \dot{E}^{(1-n)/n} \times \exp\left(\frac{Q}{n \cdot R \cdot T}\right)\right] (\dot{\varepsilon}_{ii} - \dot{\varepsilon}_{jj}) \quad (14)$$

The numerical factor in (14) is introduced so that A values determined for uniaxial compression experiments can be used in a general triaxial formulation which is independent of the orientation of the axes (Molnar *et al.* 1998).

Finally, at differential stresses larger than 200 MPa we use a Dorn flow law (Evans & Goetze 1979):

$$\sigma_{ii} - \sigma_{jj} = \sigma_d \left[1 - \sqrt{\frac{R \cdot T(z)}{Q_d} \cdot \ln\left(\frac{\dot{\varepsilon}_d}{\dot{\varepsilon}_{ii}}\right)}\right] \quad (15)$$

where σ_d, the so-called Peierls stress, is the yield stress needed to move crystal dislocations at $T = 0$ K. This experimental law limits the strength of the upper mantle to realistic values at high strain rates. We thus define an effective viscosity for the whole system by first integrating the stress difference as a function of depth through the lithosphere (using the different deformation mechanisms that apply in the relevant depth ranges) and dividing by twice the strain-rate difference (compare Eqns 4–6).

Boundary conditions

The model lithospheres are set up in dynamic equilibrium, assuming that the two horizontal principal stress components are initially equal to the vertical principal stress component as determined from the initial density column (each component depth-averaged). The tectonic effect of in-plane stress in the x-direction from a remote plate boundary is then simulated by adding to the stress component, σ_{xx}, an increment that is great enough to drive convergence at an initial strain rate of $5 \times 10^{-15}\,s^{-1}$. Under this tectonic driving force, thin sheet deformation (England & McKenzie 1982, 1983) occurs, and the tectonic stress balance changes with the thickness of the deforming lithosphere. Disregarding erosion and sedimentation, the lithospheric column changes under the action of (1) a triaxial state of stress, (2) local isostasy, and (3) thermal relaxation. These processes are integrated forward in time, using small increments of time whose duration is such that the incremental change in crustal thickness is kept below 250 m and that of the entire lithosphere to under 500 m. Calculation of transient geotherms uses a Crank-Nicholson finite differences scheme with a constant heat flow at the base of the lithosphere and no lateral heat transfer. Deformation strain rates are calculated using the thin sheet approximation, in which the vertically averaged stress-difference is related to the strain rates by the constitutive equations incorporated into Equation (10). The vertical stress-component is always evaluated directly from the current density profile. The two horizontal stress-components are set up as described above: σ_{yy} is constant throughout the experiments and equal to its initial value; σ_{xx} results from maintaining a constant plate boundary force against the vertical section of the deforming lithosphere normal to the x-direction. Therefore, σ_{xx} is inversely proportional to the current thickness of the deforming lithosphere on which the force is applied.

Presentation of the results

The evolution of our six model lithospheres, subject to the above boundary conditions, is presented in Figures 3 to 5. Figure 3 gives the position through time of the three main density interfaces: the surface, the Moho, and the base of the lithosphere (i.e., the isotherm 1330°C). Hence, it illustrates the evolution of the crustal and lithospheric thicknesses during convergence. Figure 4 shows the evolution of the magnitude of the differential stresses $(\sigma_{zz} - \sigma_{xx})$ versus $(\sigma_{zz} - \sigma_{yy})$. In this graph, four domains can be identified, a domain in which extension occurs in both horizontal directions ($(\sigma_{zz} - \sigma_{xx}) > 0$ and $(\sigma_{zz} - \sigma_{yy}) > 0$, therefore $\sigma_{zz} = \sigma_1$); a domain in which convergence occurs in both horizontal directions ($(\sigma_{zz} - \sigma_{xx}) < 0$ and $(\sigma_{zz} - \sigma_{yy}) < 0$, hence $\sigma_{zz} = \sigma_3$); and the two transcurrent domains in which there is a combination of extension in one direction and convergence in the other ($\sigma_{zz} = \sigma_2$).

Figure 4 can also be contoured for the Argand Ratio. We have defined the Argand Ratio as the ratio of the gravitational stress to the depth-averaged strength of the lithosphere at a nominal strain rate. Assuming local isostatic compensation, and disregarding flexural stresses and shear tractions at the base of the lithosphere, the gravitational stress is the depth average of $(\sigma_{zz} - \sigma_{yy})$. It is the difference between the averaged lithostatic stress under the deformed lithosphere and the averaged lithostatic stress determined in the lowland region down to the same depth, i.e., to the base of the deforming lithosphere (Artyushkov 1973; Turcotte 1983; Molnar & Lyon-Caen 1988). The depth-averaged strength at the imposed nominal strain rate is $|\sigma_{zz} - \sigma_{xx}|$, the effective differential stress-driving convergence. Hence, the Argand Ratio is simply:

$$AR = \frac{\sigma_{zz} - \sigma_{yy}}{|\sigma_{zz} - \sigma_{xx}|} \qquad (16)$$

and constant values of AR can therefore be mapped as straight lines onto the $(\sigma_{zz} - \sigma_{yy})$ – $(\sigma_{zz} - \sigma_{xx})$ plane, as shown in Figure 4. In the definition of the Argand Ratio, the absolute value of $(\sigma_{zz} - \sigma_{xx})$ is used so that its sign is consistent with prior use of the Argand Number (Ar), in which the positive Argand Number drives extension or resists convergence (England & McKenzie 1982; Houseman & England 1986). Figure 5 depicts the evolution of the Argand Ratio through time for our six model lithospheres.

The convergence history of all Archaean models and that of the Phanerozoic model with an anomalously warm geotherm (Phan80) unfolds in two stages (Fig. 3). In the first stage, convergence is accommodated by crustal and lithospheric thickening, with a monotonically decreasing rate of thickening. The duration of this first stage is proportional to the thickness of the initial lithosphere (Fig. 3). In the second

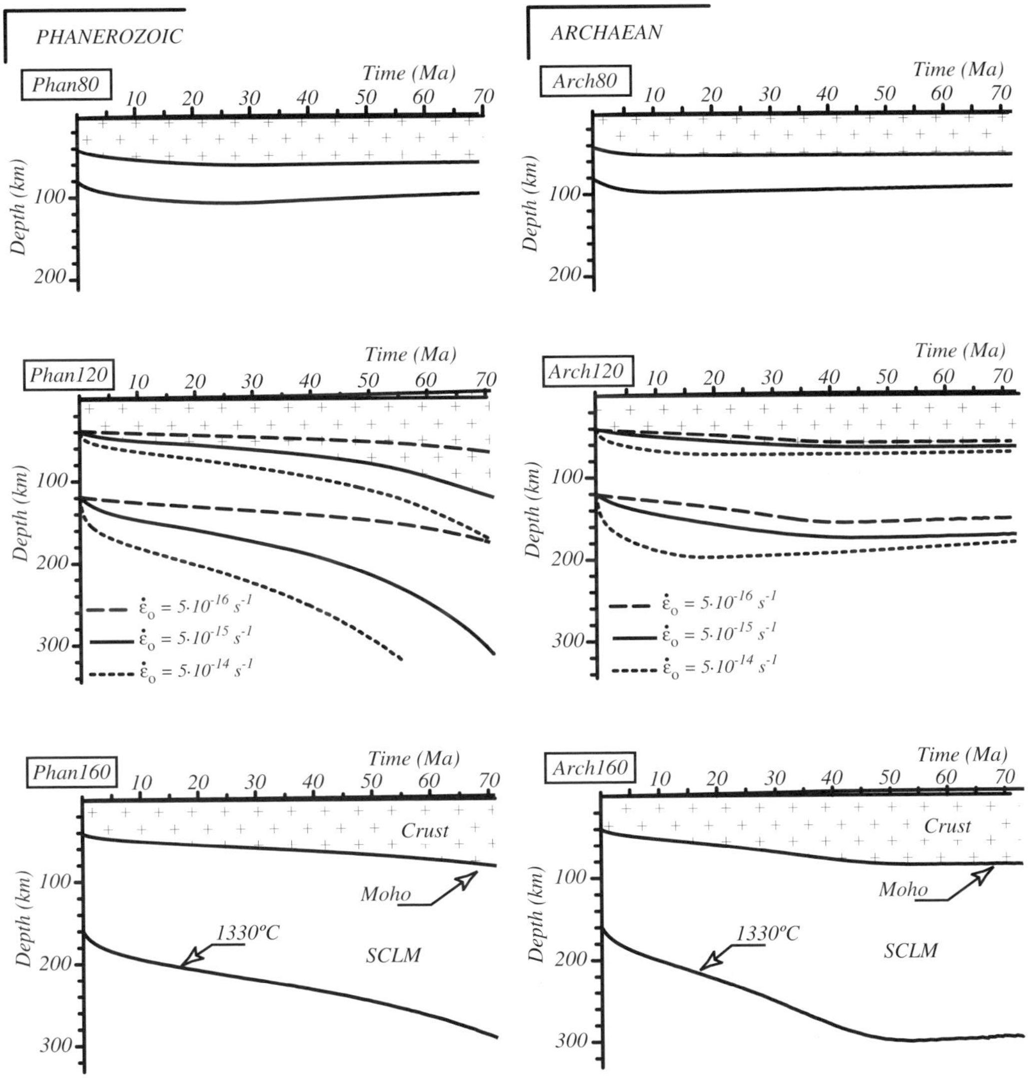

Fig. 3. Evolution of the position of the main density interfaces during convergence. In all Archaean models, the continental lithosphere reaches a plateau stage during which the depths of the density interfaces do not change much. The timescale at which this stage is reached is proportional to the initial thickness of the lithophere. A similar evolution is displayed by Phan80, which in many respects is like the Archaean models. In contrast, the depth of the density interfaces for Phan120 and Phan160 increases at an increasing rate. In these models, the thickening rate increases despite convergence being driven by a decreasing horizontal stress.

stage, convergence continues without significant thickening and even with a slight decrease in the thickness of the SCLM. This near steady-state evolution is maintained for the duration of the convergence. Since convergence is ongoing, shortening in the x-direction is balanced by extension in the direction perpendicular to convergence. Surface deformation at this stage would be primarily in the strike-slip mode. It is interesting to note that during convergence the ratio thickness of the crust to that of the SCLM increases only very slightly (Arch80: 0.5–0.56; Arch120: 0.33–0.363; and Arch160: 0.25–0.269) due to the thermal thinning of the SCLM. Therefore, the two-stage evolution described above cannot be the result of a change in the

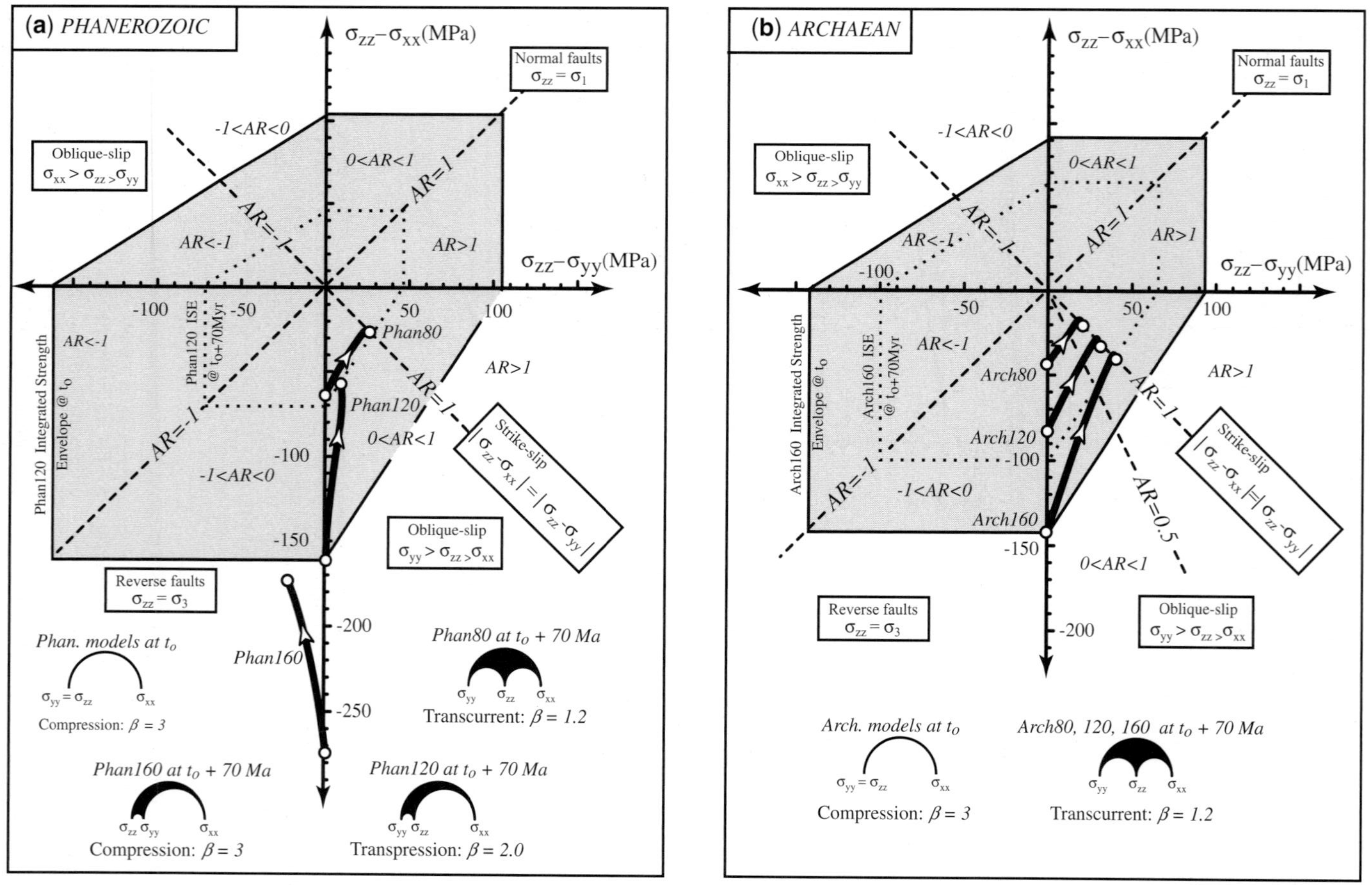

Fig. 4. Evolution of the state of stress during convergence is shown on a graph of $\sigma_{zz} - \sigma_{yy}$ against $\sigma_{zz} - \sigma_{xx}$. (**a**) The Phanerozoic model Phan160 follows a stress path within the domain of convergence in both horizontal directions (least principal stress vertical). For the models Phan120 and Phan80 the state of stress is transpressional. (**b**) The state of stress of all Archaean models becomes transpressional, reaching the transcurrent domain, where $|\sigma_{zz} - \sigma_{yy}| = |\sigma_{zz} - \sigma_{xx}|$. Contours parallel to the boundary of the shaded area define the stress differences required to drive deformation at a specified strain-rate. The Argand Ratio (*AR*) can be mapped into the $\sigma_{zz} - \sigma_{yy}/\sigma_{zz} - \sigma_{xx}$ plane. The stress state of all Archaean models evolves to an increasingly positive Argand Ratio, eventually reaching 1. At that stage, gravitational stress is equal to the stress that drives convergence.

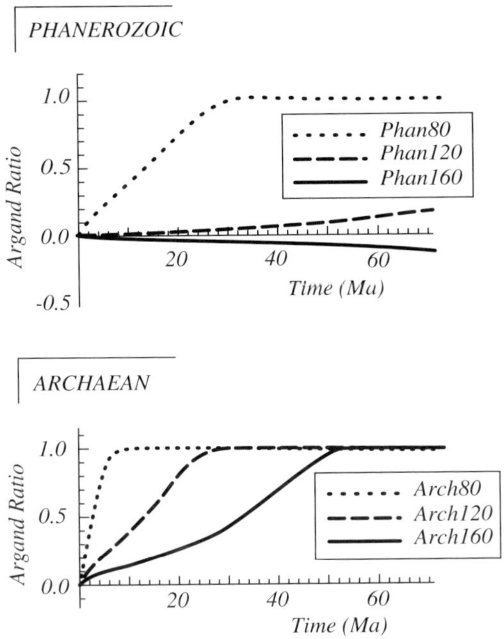

Fig. 5. Evolution of the Argand Ratio during convergence. *AR* for all Archaean models evolves toward 1. At that stage, a steady-state evolution is reached where convergence in one horizontal direction is balanced by extension in the orthogonal horizontal direction. Phanerozoic models show variable evolution of *AR*. For Phan60, the gravitational force enhances the tectonically driven convergence, whereas the gravitational force for Phan80 opposes it.

relative thickness of the crustal and the SCLM layers.

Convergence histories for the Phanerozoic models show large variations, that of model Phan80 being similar to that of Archaean models. Convergence histories for Phan120 and Phan160 models follow a different pattern. As shown in Figure 3, the thickening rate increases monotonically, as demonstrated by the progressive deepening of the Moho and that of the isotherm 1330°C. In these cases, the y-direction spreading is unable to keep pace with the imposed x-direction convergence (Phan120 in Fig. 4) and the increasing gravitational potential energy deficit caused by the thickened lithospheric root actually causes y-direction convergence as well (Phan160 in Fig. 4). Eventually this evolutionary path may lead to the initiation of a mechanical instability as the thickening rate increases exponentially, as suggested by Fleitout and Froidevaux (1982). However, as for the Archaean models, the ratio between the thickness of the crust and that of the SCLM increases very slightly (Phan80: 0.5–0.57; Phan120: 0.33–0.368; and Phan160: 0.25–0.267). The pull of the SCLM is therefore modest.

This contrasting evolution between Archaean and Phanerozoic models is the result of contrasting tectonic regimes that develop in response to an evolving stress balance. Figure 4 illustrates the differences. From its starting point on the biaxial compression axis ($\sigma_{xx} > \sigma_{zz} = \sigma_{yy}$, Fig. 4a), the state of stress for the model Phan160 moves into the domain in which both horizontal axes are convergent along a path where $\sigma_{xx} > \sigma_{yy} > \sigma_{zz}$. Along that path, the parameter $\beta = 3$ and the tectonic regime is in contraction. For such a state of stress, it is inferred that reverse faults and thickening would accommodate the convergence. The state of stress for Phan80 and Phan120 follows a path into the transpression domain where $\sigma_{xx} > \sigma_{zz} > \sigma_{yy}$. However, after 70 Ma of convergence the parameter β for Phan120 is only 2 (oblique-slip faults), whereas it stabilizes at 1.2 (transcurrent tectonic regime with strike-slip faults) for Phan80. The crustal thickening factors f_c increase at an increasing rate for Phan120 and Phan160 but stabilize at *ca.* 1.4 for Phan80. All Archaean lithospheres follow a state of stress history similar to that of model Phan80 (Fig. 4b). Their state of stress moves along a transpression path where $\sigma_{xx} > \sigma_{zz} > \sigma_{yy}$, and then stabilizes in the transcurrent tectonic regime where $(\sigma_{xx} - \sigma_{zz}) = (\sigma_{zz} - \sigma_{yy})$. When the steady-state transcurrent tectonic regime is reached, we infer that strike-slip faults in the upper crust and lateral viscous flow in the lower crust accommodate convergence. This overall lateral flow explains the limited thickening of Archaean lithospheres ($1.4 \leq f_c \leq 2$) and anomalously warm Phanerozoic lithopheres ($f_c \leq 1.4$).

This contrasting evolution is also illustrated by the evolution of the Argand Ratio (Fig. 5). For the Archaean models, and for Phan80, stage 1 of the convergence history leads to a positive Argand Ratio that increases approximately linearly with time (Fig. 5). As the differential stresses have opposite signs they promote opposing flows. Hence, when the Argand Ratio reaches unity (second stage), the tendency of the thickened crust to spread under gravity produces a lateral flow whose rate balances that of convergence, preventing further thickening. In contrast, convergence of the Phanerozoic model Phan160 produces a negative Argand Ratio that decreases at an increasing rate. The gravitational force in this example enhances the tectonic differential stress promoting convergence and thickening. In Phan160, the colder and denser SCLM keel pulls on the deforming lithosphere, enhancing the compressive horizontal stress and promoting thickening. The role of the SCLM in this kind of instability has been proposed in a number of studies (e.g., Fleitout & Froideveaux 1982; Sun & Murrel 1994; Houseman & Molnar 1997, 2001; Conrad 2000), and England & Houseman (1989) noted that the question of whether increased thickening of the lithosphere leads to an increasingly negative or increasingly positive Argand Ratio depends sensitively on assumed density and thermal parameters. Based on the compositional buoyancy of surviving Archaean SCLM, such instabilities would be less likely in Archaean times, and orogeny would more likely lead to convergence in one direction being balanced by extension in the orthogonal direction.

Discussion

Because of the secular cooling of the continental geotherm and the secular change in the composition of the SCLM (Griffin *et al.* 1998), continental thickening in modern and Phanerozoic lithospheres develops with typically lower Argand Ratios compared to those of Archaean times. For moderate to cold thermal regimes, prevalent in Phanerozoic and modern continental lithospheres, AR is $\ll 1$, promoting strain localization necessary for the formation of linear orogens (i.e., mountain belts). During thickening, one could expect that high temperature transient continental geotherms (a response to crustal thickening) would lead to an increasing Argand Ratio. This is because warmer conditions increase the buoyancy of the lithosphere (i.e., increase the AR numerator) while decreasing its integrated strength (i.e., decreasing AR denominator). However, as shown by our experiments Phan120 and Phan160 (Fig. 3) this effect competes with the cooling of the geotherm induced by the downward heat advection driven by thickening. In Phanerozoic and modern continental lithospheres, Argand Ratios ≥ 1, necessary for the development of extensional collapse, require a change in the external factors that affect the orogen. Such changes may be produced either by a decrease of the driving tectonic stresses (reduction of the AR denominator), or by a sudden increase in gravitational potential energy induced by the convective thinning of the thermal boundary layer (England & Houseman 1988, 1989) and the establishment of higher transient continental geotherm (increase of AR numerator). The Basin and Range Province (Jones *et al.* 1996; Flesch *et al.* 2000) and Tibet (England & Molnar 1997) confirm that, in present-day continental lithospheres, gravitational forces indeed play a significant role in the tectonic evolution of anomalously warm and/or buoyant lithospheres. The convergence experiment Phan80 (Fig. 3) illustrates the impact of gravitational force on an initially warm thin continental lithosphere.

Triaxial numerical experiments clearly point toward a more significant role for gravitational forces in lithospheric deformation during the Archaean. The gravitational forces determine the sustainable amount of thickening, the tectonic regime, and the spatial distribution of strain. The reduced ability of continental lithospheres in the Archaean to sustain significant contrast in gravitational potential energy has profound implications on the mode of deformation during both extensional and contractional deformation. In particular, our triaxial tests suggest that convergence involving warm and buoyant lithospheres would result in (1) less topographic relief, (2) more homogeneous lithospheric deformation distributed over broad regions (i.e., less prominent linear belts), and (3) an overall transcurrent tectonic regime where strike-slip faults and crustal-scale gravitational flow are the likely outcomes of sustained convergence. Experiments performed under driving tectonic forces that produce initial strain rates of $5 \times 10^{-16}\ s^{-1}$ and $5 \times 10^{-14}\ s^{-1}$ (Fig. 3) show that these conclusions are not affected by considering different level of tectonic forcing in the Archaean.

Archaean geological records provide plenty of observations consistent with the predictions made above. A low surface elevation and the absence of large gradients of crustal thickness in Archaean cratons are suggested by the rather low and homogeneous erosion level (5 ± 2 km) across most Archaean cratons (Galer & Metzger 1998). In particular, the co-existence of very low-grade greenstones in close proximity with granitic domes suggests that in most cases only a few kilometres of supracrustal rocks have been removed by erosion. Hence, erosion does not reveal any deep orogenic zones, but instead reveals low to medium grade meta-sedimentary and metavolcanic rocks. The evidence of Archaean high-grade terranes such as the Pikwitonei and Kapuskasing domains in the Superior Province, the granulite facies migmatite from the Limpopo belt, and the granulite gneisses of the Karelian craton in central Finland have sometimes been advanced as an argument that regional crustal thicknesses on the order of 60–80 km were created by Archaean orogenies. These rocks, however, record pressures lower than 1100 MPa (*c.* <40 km) and have been exposed to the surface following Proterozoic exhumation (Percival & Peterman 1994; Kamber *et al.* 1995, 1996; Korsman *et al.* 1999), not during Archaean orogenesis involving crustal-scale thrusts. Crustal-scale gravitational flow is also suggested by the maintenance of a sub-aerial to shallow sub-aqueous surface topography during the accumulation of volcano-sedimentary sequences (greenstones), some up to 20 km thick, above mainly felsic basement (Arndt 1997; De Witt & Ashwal 1997). Despite their thickness, greenstone covers failed to thicken the continental crust sufficiently to raise its surface elevation above sea level. The gravitational flow of the lower continental crust during the loading of crust could explain why the accumulation of thick volcano-sedimentary sequences did not translate into significant crustal thickening or significant gradients in crustal thickness.

The triaxial tests presented here suggest that Archaean continental lithospheres reach the plateau stage, and an Argand Ratio close to one, for a smaller amount of thickening. How much smaller depends on the thickness of the SCLM relative to that of the crust. For Arch120 the plateau stage is reached for a crustal thickening factor of 1.5. Therefore, continental convergence in the Archaean, if sustained, should ultimately have led to orogenies being dominated by strike-slip faults and lateral flow. Crustal-scale thrusts of Archaean age appear absent, and have yet to be documented in Archaean cratons. In contrast, craton-scale strike-slip faults of Archaean age are ubiquitous in many Late Archaean cratons such as the Superior Province, the Yilgarn craton, the Dharwar craton, or the Karilian craton. In the Yilgarn craton for instance, they are typically 50–250 km long and 2–10 km in width (Whitaker 2001). These strike-slip faults develop in close temporal relationship with elliptical granitic intrusions whose long axes roughly parallel the trend of strike-slip faults (e.g., Gee 1979). This structural relationship has previously been used to argue for horizontal shortening accommodated by strike-slip faults and horizontal extension (e.g., Treloar & Blenkinsop 1995; Hamilton 1998; Davis & Maiden 2003). The relatively high spatial frequency of these faults (one to three *c.* 300 km long strike-slip fault every 50 km across strike in the Yilgarn craton) contrasts strongly with the few major faults that accommodate lateral escape tectonics in Phanerozoic and modern convergent orogens such as the Himalayas, the European Alps or the European Variscan belt. The relatively homogeneous deformation fields of the Archaean are more likely to have extended over greater areas than in Phaneozoic orogenies. With the possible exception of the Superior Province, many Archaean cratons record regional finite strain fields with similar characteristics over a length scale of hundred of kilometres (Choukroune *et al.* 1997; Hamilton 1998), with little or no metamorphic gradient, and little or no tectono-magmatic age gradient. This strongly contrasts with modern orogens where tectono-metamophic and magmatic events are known to migrate from hinterland to foreland over many tens of millions years.

Conclusion

The steady-state thermal structure of continental lithospheres in the Archaean may have been similar to that of modern and thermally mature orogenic zones. Studies of present-day crustal flow in Tibet and in the Basin and Range Province argue that gravitational forces play a significant role. Both regions have a high Argand Ratio relative to a large anomaly in gravitational potential energy (particularly in Tibet) and a weak lithosphere (particularly in the Basin and Range Province). In ancient continental lithospheres, warmer steady-state thermal conditions combined with a more buoyant SCLM would have ensured the dominant role of the gravitational force in Archaean continental dynamics. Our numerical experiments show that, because of their contrasted thermo-mechanical properties, Archaean-like (warm and buoyant) and Phanerozoic-like (colder and heavier) continental lithospheres evolve along contrasted strain/stress paths when submitted to the same triaxial deformation conditions. In Archaean times, gravitational forces acted as a stabilizing agent by impeding thickening and promoting diffuse deformation over large areas through gravitational flow perpendicular to the direction of convergence. For moderate to cold thermal regimes, prevalent in Phanerozoic and modern times, the continental lithosphere may be too strong to be impacted by gravitational forces. Only when transient high-temperature geotherms are established do gravitational forces intervene in lithospheric deformation. Secular cooling of the continents has increased their strength while the secular increase in density of the sub-continental lithospheric mantle has reduced their buoyancy, both effects diminishing *AR* and therefore the relative importance of gravitational forces in continental dynamics.

Many thanks to Y. Poujdom Djomani, B. Griffin and S. O'Reilly for enlightening discussions on the evolution through time of the density of the SCLM. Thanks to S. Buiter, S. Costa, R. Pysklywec and M. Sandiford for their constructive comments of the first draft of this manuscript. Special thanks are due to K. Stüwe whose thorough review considerably improved its final version. This work was supported by ARC Large Grant N°: A10017138.

References

ARNDT, N. 1997. Why was flood volcanism on submerged continental platforms so common in the Precambrian? *Precambrian Research*, **97**, 155–164.

ARTYUSHKOV, E. V. 1973. Stresses in the lithosphere caused by crustal thickness inhomogeneities. *Journal of Geophysical Research*, **78**, 7675–7708.

BAILEY, R. C. 1999. Gravity-driven continental overflow and Archaean tectonics. *Nature*, **398**, 413–415.

BINNS, R. A., GUNTHORPE, R. J. & GROVES, D. I. 1976. Metamorphic patterns and development of greenstone belts in the eastern Yilgarn Block, Western Australia. *In*: WINDLEY, B. F. (ed.) *The Early History of the Earth*. Wiley, New York, N.Y., 303–316.

BOUHALLIER, H., CHOUKROUNE, P. & BALLEVRE, M. 1993. Diapirism, bulk homogenous shortening and transcurrent shearing in the Archaean Dharwar craton: the Holenarsipur area. *Precambrian Research*, **63**, 43–58.

BRACE, W. F. & KOHLSTEDT, D. L. 1980. Limits on lithospheric stress imposed by laboratory experiments. *Journal of Geophysal Research*, **85**, 6248–6252.

BUCK, W. R. & SOKOUTIS, D. 1984. Analogue model of gravitational collapse and surface extension during continental convergence. *Nature*, **369**, 737–740.

CHARDON, D., CHOUKROUNE, P. & JAYANANDA, M. 1998. Sinking of the Dharwar Basin (South India); implications for Archaean tectonics. *Precambrian Research*, **91**, 15–39.

CHOUKROUNE, P., BOUHALLIER, H. & ARNDT, N. T. 1995. Soft lithosphere during periods of Archaean crustal growth or crustal reworking. *In*: COWARD, M. P. & RIES, A. C. (eds) *Early Precambrian Processes*. Geological Society, London, Special Publications, **95**, 67–86.

CHOUKROUNE, P., LUDDEN, J. N., CHARDON, D., CALVERT, A. J. & BOUHALLIER, H. 1997. Archaean crustal growth and tectonic processes: a comparison of the Superior Province, Canada and the Dharwar Craton, India. *In*: BURG, J. P. & FORD, M. *Orogeny Through Time*. Geological Society, London, Special Publications, **121**, 63–98.

COLLINS, W. J. 1989. Polydiapirism of the Archaean Mount Edgar Batholith, Pilbara Block, Western Australia. *Precambrian Research*, **43**, 41–62.

CONRAD, C. P. 2000. Convective instability of thickening mantle lithosphere, *Geophysics Journal International*, **143**, 52–70.

DAVIS, B. K. & MAIDEN, E. 2003. Archaean orogen-parallel extension: evidence from the northern Eastern Goldfields Province, Yilgarn Craton. *Precambrian Research*, **127**, 229–248.

DE WITT, M. & ASHWALL, L. D. 1997. Greenstone Belts. *Oxford Monographs on Geology and Geophysics*, **35**.

ENGLAND, P. C. & HOUSEMAN, G. A. 1988. The mechanics of the Tibetan plateau. *Philosophical Transaction of the Royal Society of London*, **326**, 301–320.

ENGLAND, P. C. & HOUSEMAN, G. A. 1989. Extension during continental convergence, with application to the Tibetan Plateau. *Journal of Geophysical Research*, **94**, 17561–17579.

ENGLAND, P. C. & MCKENZIE, D. P. 1982. A thin viscous sheet model for continental deformation. *Geophysical Journal of the Royal Astronomical Society*, **70**, 295–321.

ENGLAND, P. C. & MCKENZIE, D. P. 1983. Correction to: A thin viscous sheet model for continental deformation. *Geophysical Journal of the Royal Astronomical Society*, **73**, 523–532.

ENGLAND, PH. & MOLNAR, P. 1997. Active deformation of asia: from kinematics to dynamics. *Science*, **278**, 647–650.

EVANS, B. & GOETZE, C. 1979. The temperature variation of hardness of olivine and its implication for polycrystalline yield stress. *Journal of Geophysical Research*, **84**, 5505–5524.

FLEITOUT, L. & FROIDEVAUX, C. 1982. Tectonics and topography for a lithosphere containing density heterogeneities. *Tectonics*, **1**, 21–56.

FLESCH, L. M., HOLT, W. E., HAINES, A. J. & SHEN-TU, B. 2000. Dynamics of the Pacific-North American plate boundary in the western United States. *Science*, **287**, 834–836.

GALER, S. & METZGER, K. 1998. Metamorphism, denudation and sea level in the Archaean and cooling of the Earth. *Precambrian Research*, **92**, 389–412.

GEE, R. D. 1979. Structure and tectonic style of the Western Australia shield. *Precambrian Research*, **58**, 327–369.

GRIFFIN, W. L., O'REILLY, S. Y., RYAN, C. G., GAUL, O. & IONOV, D. 1998. Secular variation in the composition of the subcontinental lithospheric mantle. *In*: BRAUN, J. ET AL. (eds) *Structure and Evolution of the Australian Continent*, Geodynamics Series, **26**, American Geophysical Union, 1–25.

HAMILTON, W. B. 1998. Archean magmatism and deformation were not products of plate tectonics. *Precambrian Research*, **91**, 143–179.

HOUSEMAN, G. & ENGLAND, P. 1986. Finite strain calculations of continental deformation: 1. Method and general results for convergent zone. *Journal of Geophysical Research*, **91**, 3651–3663.

HOUSEMAN, G. & MOLNAR, P. 1997. Gravitational (Rayleigh-Taylor) instability of a layer with non-linear viscosity and convective thinning of continental lithosphere. *Geophysics Journal International*, **128**, 125–150.

HOUSEMAN, G. & MOLNAR, P. 2001. Mechanisms of lithospheric renewal associated with continental orogeny. *In*: MILLER, J. A., HOLDSWORTH, R. E., BUICK, I. S. & HAND, M. (eds) *Continental Reworking and Reactivation*. Geological Society, London, Special Publications, **184**, 13–37.

JELSMA, H. A., VAN DER BEEK, P. A. & VINYU, M. L. 1993. Tectonic evolution of the Bindura-Shamva greenstone belt (northern Zimbabwe): progressive deformation around diopiric batholiths. *Journal of Structural Geology*, **15**, 163–176.

JONES, C. H., UNRUH, J. R. & SONDER, L. J. 1996. The role of gravitational potential energy in active deformation in the southwestern United States. *Nature*, **381**, 37–41.

KAMBER, B. S., BIINO, G. G., WIJBRANS, J. R., DAVIES, G. R. & VILLA, G. 1996. Archaean granulites of the Limpopo Belt, zone of the Limpopo Belt, Zimbabwe: one slow exhumation or two rapid events? *Tectonics*, **15**, 1414–1430.

KAMBER, B. S., KRAMERS, J. D., NAPIER, R., CLIFF, R. A & ROLLINSON, H. R. 1995. The Triangle Shearzone, Zimbabwe, revisited: new data document an important event at 2.0 Ga in the Limpopo Belt. *Precambrian Research*, **70**, 191–213.

KORSMAN, K., KORJA, T., PAJUNEN, M., VIRRANSALO, P. & GGT/SVEKA WORKING GROUP, 1999. The GGT/SVEKA Transect – structure and evolution of the continental crust in the Palaeoproterozoic Svecofennian Orogen in Finland. *International Geology Review*, **41**, 287–298.

LENARDIC, A. & MORESI, L. 2000. A new class of equilibrium geotherms in the deep thermal lithosphere of continents. *Earth Planetary Science Letters*, **176**, 331–338.

MCGREGOR, A. M. 1951. Some milestones in the Precambrian of Southern Rhodesia. *Transactions of the Geological Society of South Africa*, **54**, 27–70.

MOLNAR, P. & LYON-CAEN, H. 1988. Some simple physical aspects of the support, structure and evolution of mountain belts. *In*: CLARK JR., S. P., BURCHFIELD, B. C. & SUPPE, J. (eds) *Processes in Continental Lithosphere Deformation*. Geological Society of America, Special Papers, **218**, 179–207.

MOLNAR, P., HOUSEMAN, G. A. & CONRAD, C. P. 1998. Rayleigh-Taylor instability and convective thinning of mechanically thickened lithosphere: effects of non-linear viscosity decreasing exponentially with depth and of horizontal shortening of the layer. *Geophysical Journal International*, **133**, 568–584.

MYERS, J. & WATKINS, K. P. 1985. Origin of granite-greenstone patterns, Yilgarn Block, Western Australia. *Geology*, **13**, 778–780.

PERCIVAL, J. A. & PETERMAN, Z. E. 1994. Rb–Sr biotite data from the Kapuskasing uplift and their bearing on the cooling and exhumation history. *Canadian Journal of Earth Sciences*, **31**, 1172–1181.

QIU, Y. M., MCNAUGHTON, N. J., GROVES, D. I. & DALSTRA, H. J. 1999. Ages of internal grantoids in the Southern Cross region, Yigarn Craton, Western Australia, and their crustal evolution and tectonic implications. *Australian Journal of Earth Sciences*, **46**, 971–981.

RAMSAY, J. G. 1989. Emplacement kinematics of a granite diapir: the Chindamora batholith, Zimbabwe. *Journal of Structural Geology*, **11**, 191–209.

RANALLI, G. 1995. *Rheology of the Earth.* Chapman & Hall, London.

REY, P., PHILIPPOT, P. & THÉBAUD, N. 2003. Contribution of mantle plume, crustal thickening, and greenstone blanketing to the 2.75–2.65 Ga global crisis. *Precambrian Research*, **127**, 43–60.

SCHMALHOLZ, S. M., PODLADCHIKOV, Y. Y. & BURG, J. P. 2002. Control of folding by gravity and matrix thickness: implications for large-scale folding. *Journal of Geophysical Research*, **107**, no B1, ETG-1–1, DOI 10.1029/2001JB000355.

SCHMALHOLZ, S. M., PODLADCHIKOV, Y. Y. & JAMVIET, B. 2005. Structural softening of the lithosphere. *Terra Nova*, **17**, 66–72.

SIBSON, R. H. 1974. Frictional constraints on thrust, wrench and normal faults, *Nature*, **249**, 542–544.

SONDER L. J., ENGLAND, P. C., WERNICKE, B. P. & CHRISTIANSEN, R. L. 1987. A physical model for Cenozoic extension of the western North America. *In*: COWARD, M. P., DEWEY, J. F. & HANCOCK, P. L. (eds) *Continental Extensional tectonics*. Geological Society, London, Special Publications, **28**, 187–201.

SNOWDEN, P. A. & BICKLE, M. J. 1976. The Chinamora Batholith; diapiric intrusion of interference fold? *Journal of the Geological Society, London*, **132**, 131–137.

SUN, J. & MURRELL, A. F. 1994. On the growth and collapse of wide orogenic belts. *Geophysical Journal International*, **118**, 255–268.

TAYLOR, S. R. & MCLENNAN, S. M. 1986. The chemical composition of the Archaean crust. *In*: DAWSON, J. B., CARSWELL, D. A., HALL, J. & WEDEPOHL, K. H. (eds) *The Nature of the Lower Continental Crust*. Geological Society, London, Special Publications, **24**, 173–178.

TREOLAR, P. J. & BLENKINSOP, T. G. 1995. Archaean deformation patterns in Zimbabwe: true indicators of Tibetan-style crustal extrusion or not? *In*: COWARD, M. P. & RIES, A. C. (eds) *Early Precambrian Processes*. Geological Society, London, Special Publications, **95**, 87–107.

TURCOTTE, D. L. 1983. Mechanisms of crustal deformation. *Journal of the Geological Society, London*, **140**, 701–724.

WARREN, R. G. & ELLIS, D. J. 1996. Mantle-underplating, granite tectonics, and metamorphic P-T-t paths. *Geology*, **24**, 663–666.

WHITAKER, A. J. 2001. Component and structure of the Yilgarn craton, as interpreted from aeromagnetic data. 4th International Archaean Symposium 2001, Extended Abstracts. *In*: CASSIDY, K. F. ET AL. (eds) 2001. AGSO–*Geoscience Australia, Record*, **37**, 536–538.

Analogue and numerical modelling of accretionary prisms with a décollement in sediments

YASUHIRO YAMADA[1], KEI BABA[2] & TOSHIFUMI MATSUOKA[1]

[1]*Department of Civil and Earth Resources Engineering, Kyoto University, Kyoto 606-8501, Japan (e-mail: yama@electra.kumst.kyoto-u.ac.jp)*

[2]*JAPEX Research Center, Chiba 261-0025, Japan*

Abstract: Active accretionary prisms at subduction margins generally include a horizontal detachment, décollement, within the sedimentary pile. The décollement, and its extension to undeformed regions (i.e., proto-décollement), corresponds to a layer of high fluid pressure. The deformation of the prisms, including such an anomalous layer, can be modelled and examined using analogue experiments and numerical simulations. Both these methods approximate the material under deformation as an assembly of particles (grains). The décollement layer is found to be best modelled by intercalating a layer with smaller internal frictional coefficient than the surrounding materials corresponding to the sediments. Our analogue experiments with dry sand and microglass beads reproduce structural geometry similar to that of interpreted seismic profiles at the toe of the prisms. Thrust faults originate from the horizontal beads layer and propagate upward with a constant angle of about 30°. Each of the fault bends produces a series of minor back thrusts. A particle image velocimetry (PIV) analysis revealed that the fault activity is characterized by intermittent reactivation and segmentation. The numerical simulations based on the distinct element method (DEM) were performed with similar kinematic settings and material properties as the analogue experiments. The numerical simulation results not only reproduce similar geometries as in the analogue experiments, but also show that the particle assembly experiences temporal variations in the deformation velocity and stress field as deformation propagates. This might be related to stick-slip motion of the frictional fault surfaces, which is a common feature of faulting during accretionary processes at subduction margins.

Geological modelling using physical materials in scaled analogue experiments is an extremely useful technique for a detailed examination of the geometry and deformation process of accretionary prisms. Since the mechanics of the upper crust can be approximated by Navier-Coulomb brittle behaviour, granular materials such as dry sand are an appropriate material to model such brittle deformation (McClay 1990) with fulfillment of the scaling theory by Hubbert (1937). Analogue experiments with dry sand (sandbox experiments) have thus been widely applied to simulate a wide variety of geological structures (e.g., Mandl 1988; Cobbold & Castro 1999; Koyi & Mancktelow 2001; Yamada & McClay 2003*a*, 2003*b*, 2004; Yamada *et al.* 2005 and the references therein). Accretionary prism formation, as well as thrust-and-fold belts, has been investigated in detail by a number of analogue models (e.g., Colletta *et al.* 1991; Huiqi *et al.* 1992; Willet 1992; Lallemand *et al.* 1992; Mulugata & Koyi 1992; Gutscher *et al.* 1996; Nieuwland *et al.* 2000; Koyi & Vendeville 2003).

There are two types of model kinematics, either pushing a rigid backstop or pulling a sheet underneath sand, but the results are the same except for the boundary effects of sidewalls (see Schreurs *et al.* 2006). A typical result of these models shows that a foreland propagating (piggyback) sequence of thrusts form a Coulomb wedge with a taper mainly dependent upon the friction of the detachment fault. That is, higher basal friction leads to formation of a steeper slope prism, whereas low basal friction leads to formation of a gentler slope prism. This corresponds with the basic theory of non-cohesive critical Coulomb wedges (Davis *et al.* 1983; Dahlen 1984; Dahlen & Suppe 1984), assuming that an actively accreting wedge attains a critical taper, an internal state of stress on the verge of Coulomb failure throughout. Dahlen (1984) concluded that the surface slope of a critical taper is dependent on the internal and basal friction coefficients, internal and basal fluid pressure ratios, the dip of the basal detachment and the strength of the rock composing the wedge. The theory explains how basal

From: BUITER, S. J. H. & SCHREURS, G. (eds) 2006. *Analogue and Numerical Modelling of Crustal-Scale Processes.* Geological Society, London, Special Publications, **253**, 169–183.
0305-8719/06/$15.00

friction or basal fluid pressure controls the structural styles, including subduction erosion, active accretion, subduction without accretion and normal faulting (Dahlen 1984).

Deformation of granular assemblies can also be simulated numerically using the distinct element method (DEM). The DEM calculates force and velocity in two (normal and shear) directions for all elements, which is repeated for the updated element relations at the next time increment (Cundall & Strack 1979). With significant advances in computing technology, the technique can now handle a large number of elements and be applied to a variety of research fields (e.g., Konietzky 2002; Shimizu *et al.* 2004). Application of the DEM to geological problems requires the geological body to be approximated as an assembly of elements (e.g., Finch *et al.* 2003); thus sandbox experiments are an ideal scenario (e.g., Saltzer 1993; Yamada *et al.* 2004). Numerical models using the DEM are, however, not scale models in the sense of physical models where typical Earth values are used as their input parameters (Strayer & Suppe 2002). Two significant advantages of the DEM are: (1) no need to predefine artificial parameters of discontinuity surfaces, and (2) ability to extract all information from each element during deformation in a quantitative way. The DEM therefore has high potential to be a numerical simulator that can incorporate discontinuity surfaces properly.

This paper shows selected sandbox experiments and DEM simulations for a simplified geological model of accretionary prisms. The results are analysed to investigate the detailed process of fault development.

Accretionary prisms with a décollement in sediments

Recent investigations revealed that the accretion process at subduction margins generally develops a sub-horizontal detachment, décollement, within the sedimentary sequences accumulated in the trench area. This horizon is typically defined by a layer of anomalous high fluid pressure and can generally be easily traced with reverse polarity on seismic profiles (Tobin *et al.* 1994; DiLeonardo *et al.* 2002; Tsuji *et al.* 2004; Bangs *et al.* 2004). The sediments above the décollement are deformed by a series of imbricate thrusts that converge with the décollement surface, whereas those underneath it are subducted without internal deformation.

A typical example can be seen in the Nankai trough located offshore southwest Japan, where the Philippine Sea Plate is subducting underneath the Eurasian Plate at a rate of 4 cm/yr along an azimuth of 310–315° (Fig. 1; Seno *et al.* 1993). Extensive seismic surveys in this area reveal that a décollement is clearly located within the sediments, not at the top horizon of the volcanic basement (Fig. 2a; Bangs *et al.* 2004). The reverse polarity of the décollement suggests that the layer may have extremely high fluid pressure, even at the 'proto-décollement' region (Tsuji *et al.* 2005). The proto-décollement is an extension of the de'collement surface in the undeformed sediments, seaward of the deformation front. This clear identification of the proto-décollement on the profiles suggests that a preferred layer of décollement exists prior to initiation of the actual displacement. Geophysical logging and core analysis at Site 1174 of the Ocean Drilling Project Leg 196 showed that the décollement is located in a hemi-pelagic mudstone sequence at Nankai (Mikada *et al.* 2002).

Such features of décollement development are also observed in Barbados (DiLeonardo *et al.* 2002) and in Cascadia (Tobin *et al.* 1994). The décollement in the Barbados accretionary prism, where the Atlantic Plate is subducting underneath the Caribbean Plate, is located in a radiolarian mudstone of high porosity and low strength (Moore & Klaus 2000). This horizon also forms the proto-décollement. In Cascadia, a large scale accretionary prism develops due to the subduction of the Juan de Fuca Plate underneath the North American Plate, and the décollement (and its proto-décollement) is formed at a boundary horizon between upper turbidites and lower hemi-pelagic mudstones (Westbrook *et al.* 1994). The structural styles

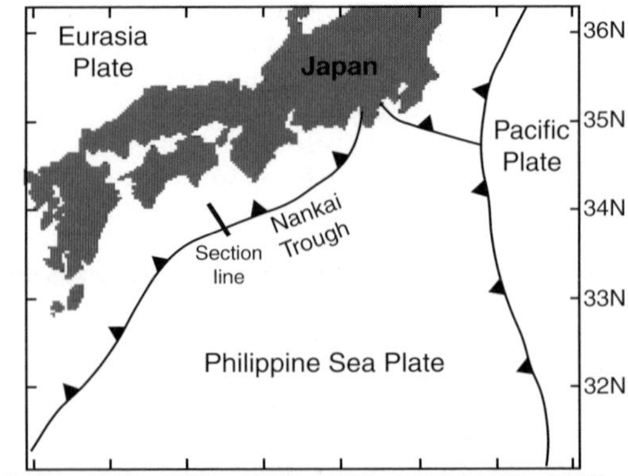

Fig. 1. Location map of the Nankai Trough and the section line.

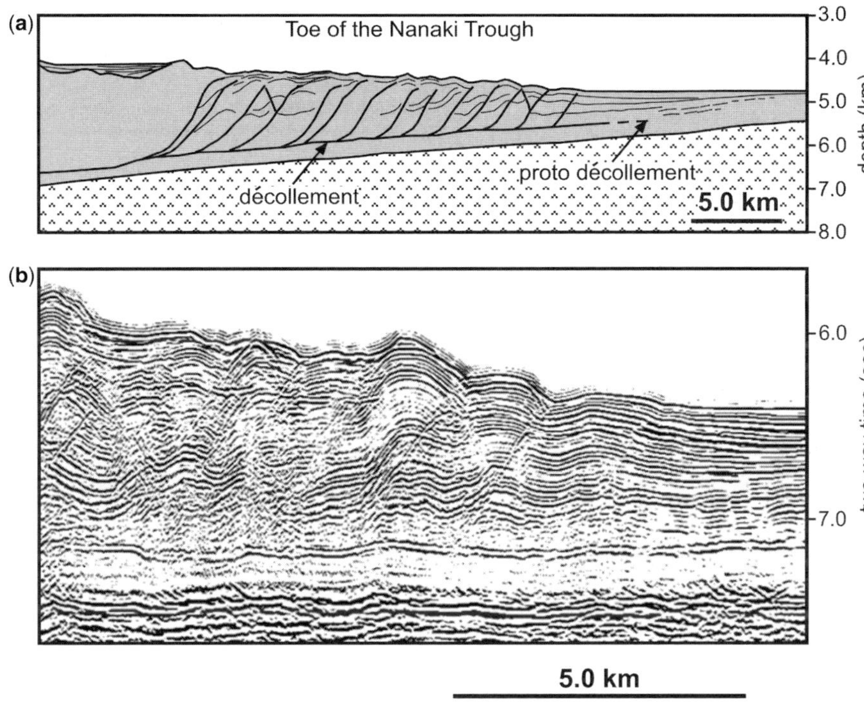

Fig. 2. Section geometry of the Nankai accretionary prism (see Fig. 1 for the location of the section line). (**a**) Schematic geological cross-section (Mikada *et al.* 2002). (**b**) A seismic profile at the toe of the wedge off Muroto Point (Bangs *et al.* 2004). Note that the décollement is not at the uppermost horizon of the volcanic basement (the strong reflector at 7.5 sec), but at the middle (*c.* 7.1 sec) of a thick mudstone sequence (low reflective sediments).

of these prisms are generally similar on seismic profiles.

The experiments and simulations aim to investigate the effects of the décollement within the sediments. Figure 2b is a seismic profile illustrating the characteristic structural style at the toe of the Nankai prism off Muroto Point. The thickness of the sediments above the décollement is *c.* 800 m, fault spacing at the toe is *c.* 1300 m, and the slope angle is *c.* 4°.

Analogue experiments

Modelling set-up

The experimental apparatus is an acrylic box (25 cm × 150 cm × 50 cm) underlain by a flat plastic sheet (Fig. 3). Layers of analogue material are piled on the sheet, which is then pulled by a motor at a constant speed of 1.67×10^{-2} cm sec^{-1} through a slit under the fixed end wall. This slit produces a décollement horizon within the analogue material. The scaling ratio is 4.0×10^{-5}, thus 1 cm in the experiments corresponds to 250 m in nature.

Selection of the modelling materials is an important factor of the analogue experiments. In this study, the effects of the anomalous high fluid pressure along the natural décollement must be considered. The high fluid pressure reduces strength, both in nature and in experiments. Low friction materials thus may be used to approximate the effects of the fluid pressure. In fact, a pioneering work by Cobbold *et al.* (2001) showed that an increase in the inter-granular pore pressure of a layer in their thrust-wedge-type sandbox experiments produced a decreased taper angle and widened fault spacing. These effects are similar to the experiments with a décollement of a smaller internal friction coefficient. Microbeads have been recently regarded as suitable materials for such weaker layers (e.g., Massoli *et al.* 2002; Lohrmann *et al.* 2003; Yamada *et al.* 2005). We used Toyoura dry quartz beach sand and glass microbeads as analogue materials. The average grain sizes are 200 μm and 50 μm respectively. Our shearing tests showed that the microbeads have a much smaller internal friction angle (25°) than Toyoura sand (34°, see Table 1).

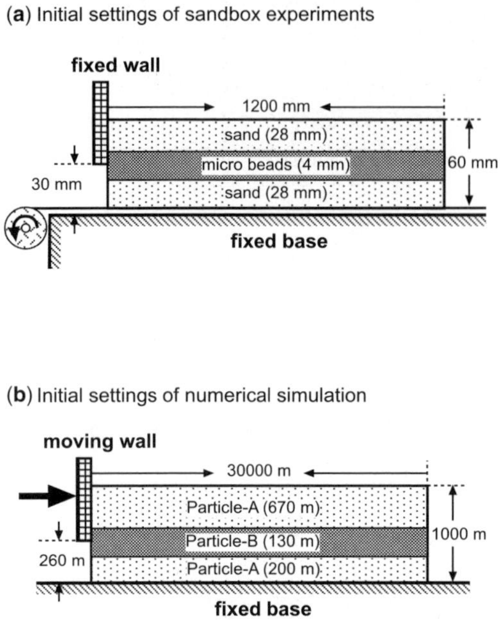

Fig. 3. Set-up for analogue and numerical models. (**a**) The analogue materials are piled on a plastic sheet which is then pulled by an electric motor. Note the slit between the fixed base and the bottom of the fixed end wall. (**b**) The model design for the numerical simulation. This is equivalent to the analogue experiments, including the slit underneath the fixed end-wall, but the polarity of shortening is opposite (i.e., a moving wall instead of a moving sheet).

This paper presents two representative results of sandbox experiments conducted for this study. Experiment I employs pure sand of 6.0 cm thickness, whereas Experiment II includes a layer of microbeads (0.4 cm thickness) midway up the sand pile (6.0 cm in total including the beads layer). The microbeads are white and the sand is naturally light brown, thus they can be easily distinguished. We also used dyed marker sand layers only adjacent to the transparent sidewalls. The marker sands have almost the same physical properties as the undyed sand. These are to maintain the homogeneity of the sand layers. During the experiments, photographs were taken through the sidewall after every 1 cm of displacement of the sheet. The structural style on the free surface suggests that the deformation of the experiments can be approximated as plane strain.

Experimental results

Experiment I. The homogeneous dry sand model reproduced typical thrust-and-fold belt structures including a wedge geometry, and foreland vergent thrusts that propagated upward with a piggyback sequence from a sub-horizontal décollement produced within the sand pack (Fig. 4). The décollement connected to the frontal thrust without forming a bend. The angle of the prism surface (line linking the tips of the thrusts) is *c.* 18–24° and fault spacing near the frontal thrust is *c.* 6.0 cm (corresponds to 1.5 km).

Experiment II. The results of this model, which has a microbeads layer inserted into the sand, showed that the microbeads layer acted perfectly as a horizontal décollement (Fig. 5). Detailed observations revealed that the décollement surface was at the lower boundary of the microbeads layer. The analogue material below the décollement did not deform and was expelled from the experimental rig through the slit. From the décollement, the foreland vergent frontal thrust propagated upward, dipping *c.* 28°–34°. Thus, the active surface of the frontal thrust through the décollement has a bend where the frontal thrust is generated from the décollement. Above the bend, a series of minor back-thrusts was generated in a piggyback sequence, which was only active while the fault underneath was active. The foreland vergent thrusts also formed a piggyback sequence. In Experiment II, the angle of the prism surface (the slope of the line linking the tips of the thrusts) is gentler (*c.* 9–14°), and the fault spacing near the frontal thrust is wider (*c.* 8.3 cm, corresponding to 2.1 km), than in Experiment I (*cf.* Fig. 4).

PIV analysis

In order to analyse the deformation process in detail, time-lapse digital pictures of the

Table 1. *Material properties and grain characteristics of analogue materials*

Material	Density (kg m^{-3})	C (Pa)	ϕ_{stable} (°)	Strain soft. (%)	Ave. grain size (μm)	Grain shape	Composition
Quartz sand	1390	n.d.	34	9	m = 200	Angular	93% SiO_2 4% Al_2O_3
Microbeads	1420	n.d.	25	0	m = 50	Spherical	Na_2O–CaO–SiO_2

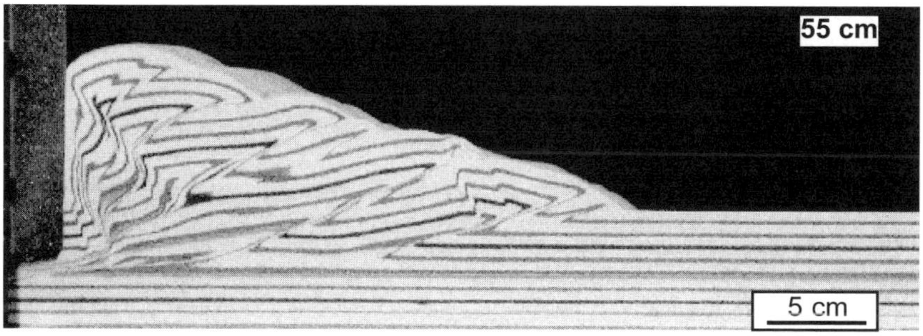

Fig. 4. Result of Experiment I. This pure sand model produces characteristic features of thrust wedges, including a piggyback sequence of thrusting. This picture is taken after 55 cm of shortening and the model has a surface slope of 18°.

experiments were evaluated with an image correlation technique called particle image velocimetry (PIV). This technique is commonly used for dynamic flow analysis and soil mechanics (e.g., Hryciw *et al.* 1997), but has recently also been applied to structural geology (Adam *et al.* 2002; Wolf *et al.* 2003; Adam *et al.* 2005). The method calculates the displacement field of the grains with a theoretical resolution of *c.* 0.5 mm. The displacement accuracy depends on the picture size and the resolution of the digital camera (Wolf *et al.* 2003). PIV data cover the complete range of structural evolution, as well as a major part of the stick-slip field, in experiments with a scaling factor of 10^5–10^6 (Adam *et al.* 2005).

Figure 6a is an example of the photographs that have been taken after every 1 cm of displacement of the plastic sheet and that have been analyzed using PIV. The results are shown in Figure 6b, in which the fixed wall and the basal plate provide axes of a frame of reference for the kinematics of the model. Dark arrows correspond to large displacement, whereas light grey arrows correspond to minor displacement. White areas with no arrows correspond to regions where the software detects no correlation. These displacement distributions identify coherent blocks of active displacement, the discontinuities between such blocks indicating active faulting. The top diagram is the displacement distribution between the images taken after 38 and 39 cm of total displacement, showing that the hanging wall of the frontal thrust is moving to the upper-left relative to the fixed wall, whereas the rest of the regions only show minor displacement in other directions. This suggests that shortening is mainly accommodated by the frontal thrust. The displacement distribution between 39 and 40 cm of total displacement shows a similar pattern to the previous stage described above. However, minor differences can be found in regions of previous thrust sheets, suggesting that these thrusts are reactivated.

The displacement pattern then changes significantly after 40 cm of total displacement. According to the PIV calculations, the frontal area shows a similar displacement distribution between 40 and 42 cm of total displacement as between 39 and 40 cm displacement, but the adjacent area has a large displacement anomaly that drops sharply along the third thrust fault from the frontal thrust. This means that the third thrust fault reactivates between 40 and 42 cm of total displacement. The hanging wall of the thrust fault also shows such a discontinuity in displacement distribution, suggesting that a back thrust also reactivates at this stage. These thrust faults are inactive during the stages from 38 to 40 cm of total displacement.

The PIV results generally confirm the in-sequence thrust process, but each thrust sheet has an intermittent displacement distribution. Adam *et al.* (2005) also reported such temporal activity of faults, characterized by fault reactivation, in their thrust-and-fold type experiments.

Numerical simulations

The distinct element method (DEM) was employed as the numerical simulation technique in this study. The model domain consists of discrete circular elements, and linear elasticity (force-displacement law; Cundall & Strack 1979) is incorporated through normal and shear forces at elemental contacts. The inter-element friction in the shear direction is determined by the normal force and a friction coefficient. The size of each element is preserved during deformation, thus volume change (i.e., dilation) is

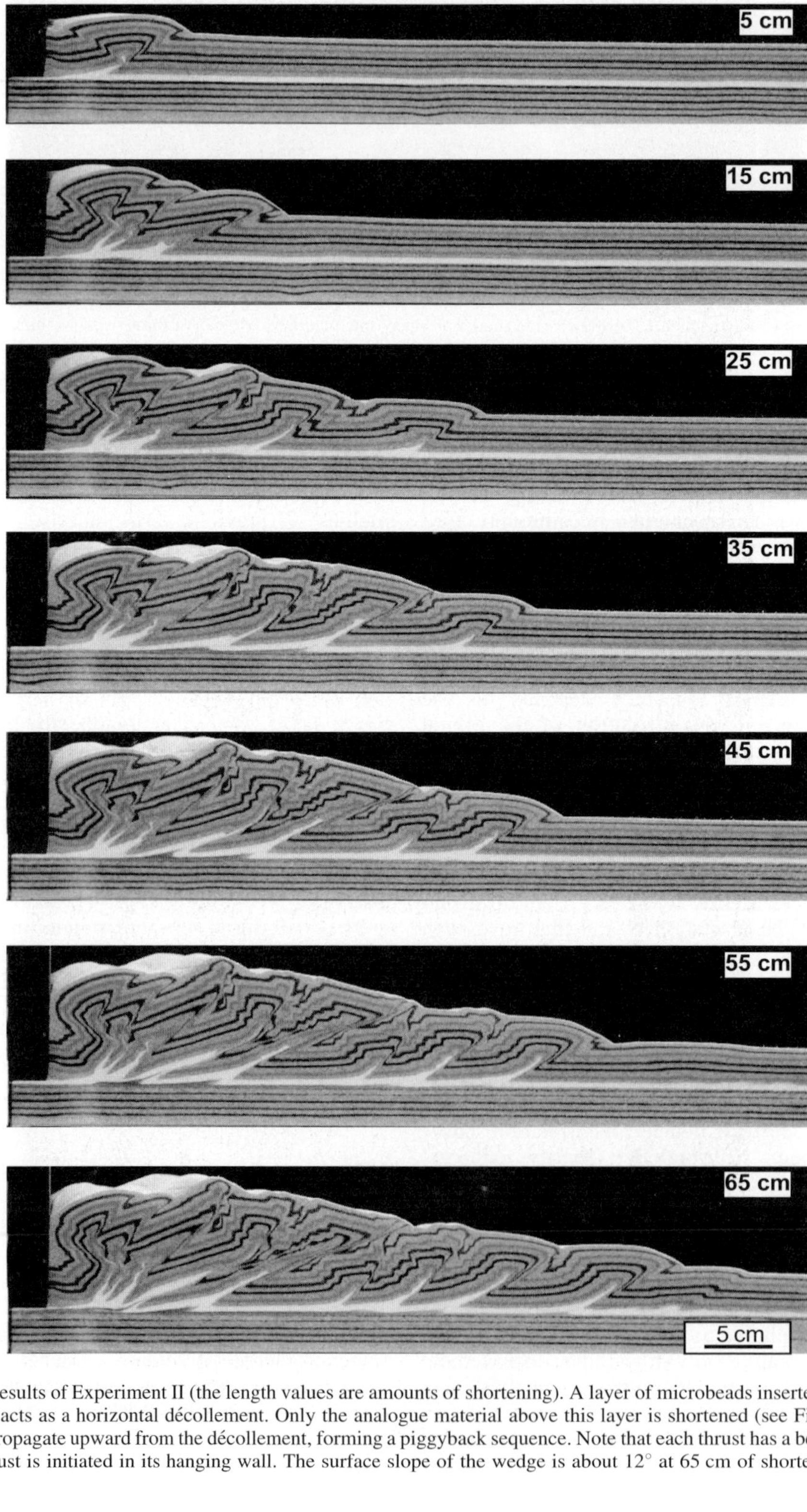

Fig. 5. Results of Experiment II (the length values are amounts of shortening). A layer of microbeads inserted into the sand pile acts as a horizontal décollement. Only the analogue material above this layer is shortened (see Fig. 3a). Thrusts propagate upward from the décollement, forming a piggyback sequence. Note that each thrust has a bend where a backthrust is initiated in its hanging wall. The surface slope of the wedge is about 12° at 65 cm of shortening.

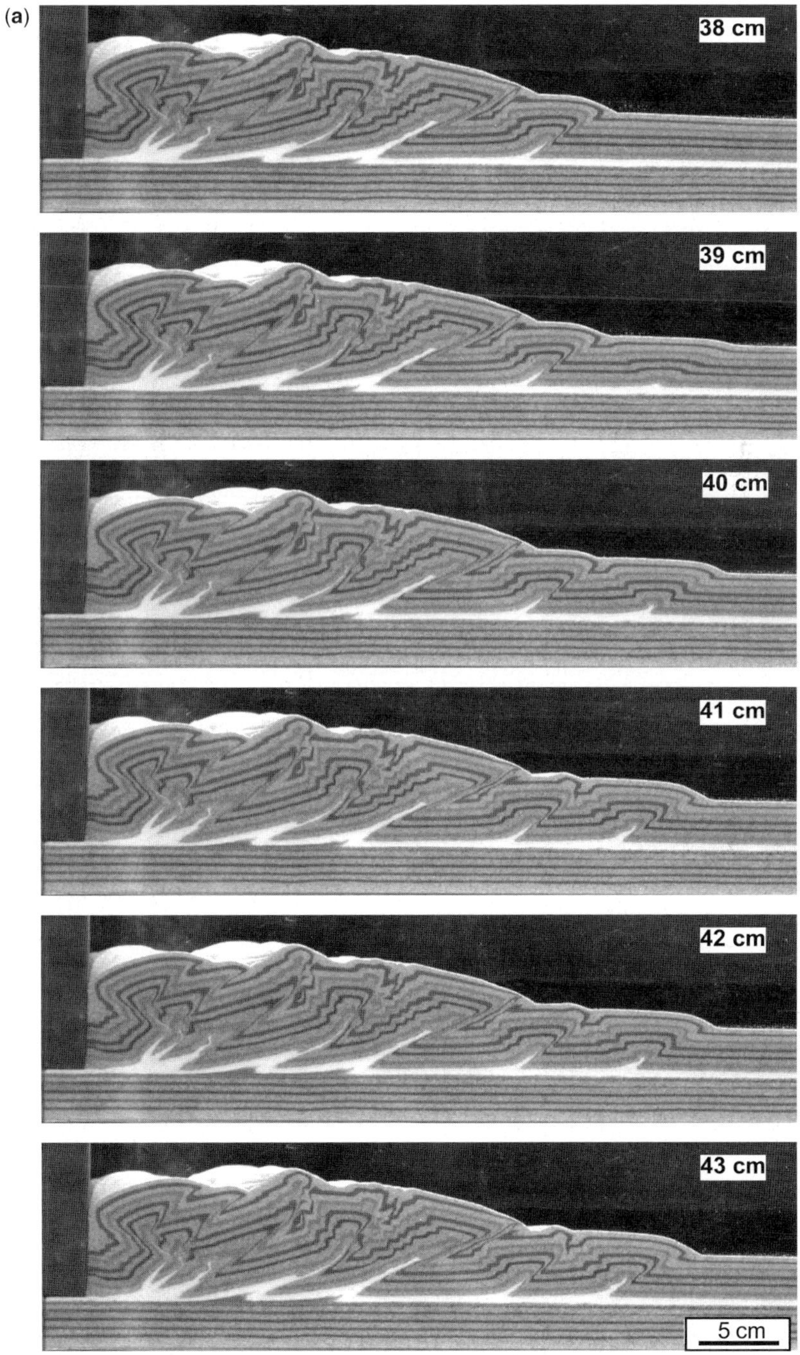

Fig. 6. Particle image velocimetry (PIV) analysis for analogue model II. (**a**) The images analysed are taken at every 1 cm of shortening (the length values are amounts of shortening). (**b**) The results of PIV analysis showing the velocity vectors during shortening. The fixed wall and the basal plate provide axes of a frame of reference for the kinematics of the model. Dark arrows correspond to large displacement, whereas light grey arrows correspond to minor displacement. White areas with no arrow correspond to a region where the software detects no correlation. Note that the velocity patterns suggest sequential fault reactivations.

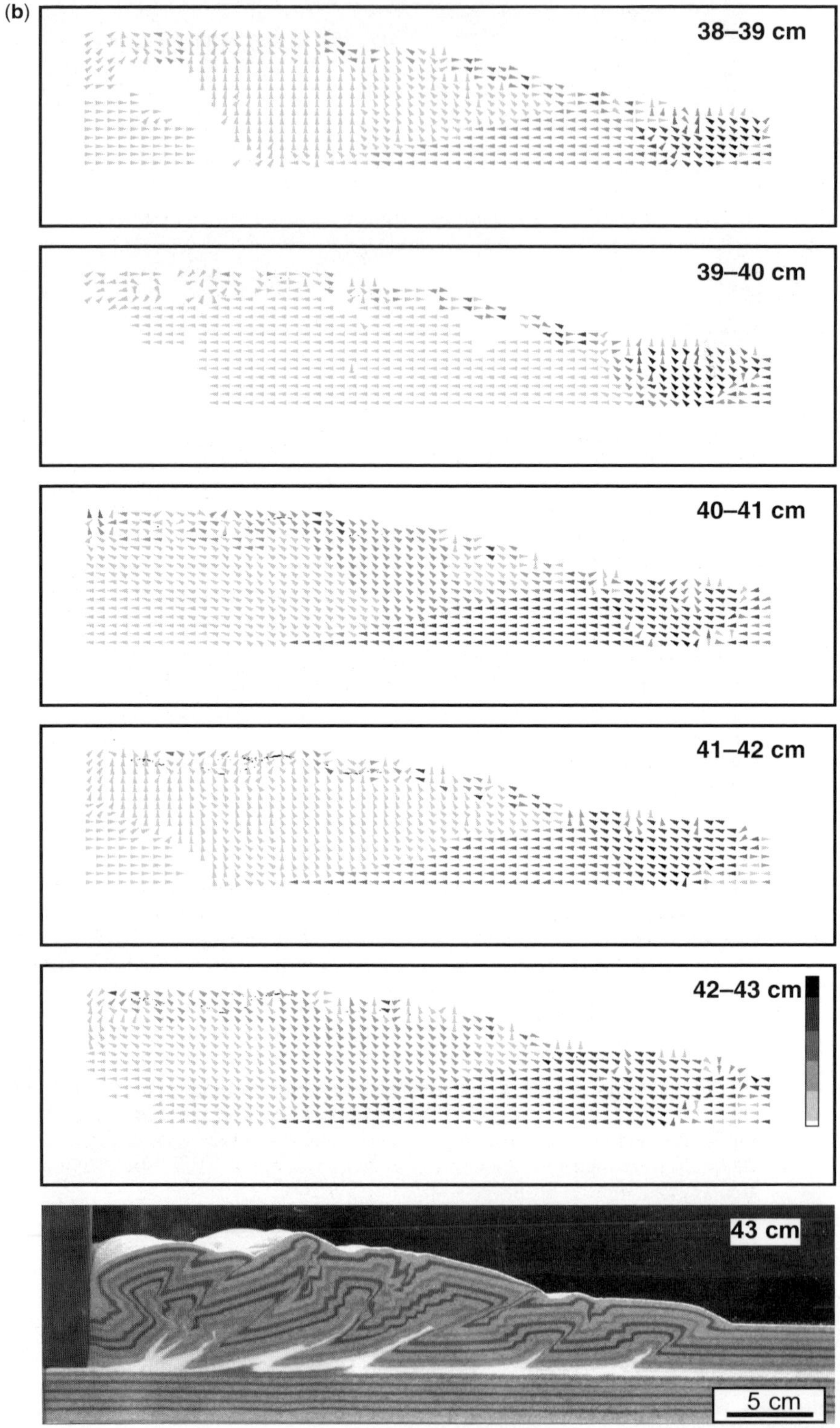

Fig. 6. *Continued.*

accommodated by the inter-element porosity. The method has two steps in each calculation cycle (time-step): the first step is to evaluate interaction forces for every element, and the second step is to move all elements according to numerical integration following Newton's equation of motion for the given external forces. The software employed in this study is PFC 2D developed by the ITASCA Corporation, Minneapolis, USA.

Input parameters

The simulations aim to reproduce the two-dimensional structural style of the accretionary prism with a décollement in the sedimentary sequence. In order to make the initial conditions of the simulations similar to those of the analogue experiments, the input parameters are carefully determined after a series of compression test simulations. Since DEM describes only the parameters related to element contacts, such compression tests are necessary to measure the physical properties of the element assembly. Table 2 shows the physical characteristics of two particle types, which were obtained by bi-axial compression tests with a PFC-2D function called FISH-TANK. Particle A has an internal frictional angle of 35°, while particle B has an internal frictional angle of 25°. These particle types are selected to model the sand and the microbeads respectively in the analogue experiments (*cf.* Table 1). The friction coefficient between the wall and the particles is the same as that of particle A. Initial cohesion and bonding between elements are not introduced, due to the cohesionless nature of the analogue materials. The diameter of the elements has a variety of 20% from the average size to avoid formation of preferential weak planes in the initial arrangement of the element assembly.

The model kinematics is also designed after the sandbox experiments, but the polarity of shortening is opposite (Fig. 3b). In this two-dimensional simulation model, an assembly with a rectangular initial geometry is placed on a fixed base, and a horizontally moving wall pushes one side of the material from the left-hand side. The initial position of the moving wall is determined by the position of the décollement horizon. The displacement rate should be small enough to approximate the deformation as quasi-static. In this study, the rate is 0.9 cm for each calculation cycle. This is why DEM requires a few million time-steps to generate sandbox-type deformations.

The sizes of the elements in this study are larger than the equivalent material sizes of the analogue experiments. The use of the equivalent grain sizes requires the computer to be equipped with an extraordinary size of data storage and memory, and each calculation may also require an extremely long time (about a few months with a current workstation). This is why larger elements were generally used in previous research (e.g., Saltzer 1992; Finch *et al.* 2003; Strayer & Suppe 2002; Yamada *et al.* 2004). In this paper, two representative results of simulations are presented. Simulation I employs a layer of particle A of 670 m thickness without predefined décollement, whereas Simulation II has a layer of particle B (130 m thick) at the middle of a pile of particle A (1000 m in total including the particle B layer). These thicknesses have to be altered from the equivalent values of the analogue experiments, in order to obtain each calculation result within a month.

Simulation results

Simulation I. The overall structural style is a wedge shape that increases in size as the displacement of the wall proceeds (Fig. 7). The simulation reproduces typical foreland vergent thrusts that propagate upward with a piggyback sequence from the basal décollement. The angle of the prism surface (line linking the tips of the thrusts) generally shows a gradual increase toward the deformation front. This value increases up to *c.* 48° adjacent to the frontal thrust after shortening of 6750 m (Fig. 7). The fault spacing around the frontal thrust is *c.* 0.7 km.

Table 2. *Material properties and element characteristics of numerical simulations*

Material	Density (kg m^{-3})	C (Pa)	ϕ_{stable} (°)	Strain soft. (%)	Ave. grain size (m)	Grain shape
Particle A	1600	n.d.	35	n.d.	m = 21	Circular
Particle B	1600	n.d.	25	n.d.	m = 8	Circular

Fig. 7. Results of simulation I consisting of particle A only (the length values are amounts of shortening). The structural style of this particle assembly shows a steep slope angle.

Simulation II. The overall structural style is similar to Simulation I, but the surface slope of the wedge is 21°–22° and the fault spacing around the frontal thrust is *c*. 1.0–1.2 km (Fig. 8). Similar to the sandbox experiment II, these thrusts propagate upward from the décollement layer; thus the frontal thrust surface has a bend where the thrust originates from the décollement. Although backthrusts were also generated at the bend in the analogue experiments, they are unclear in the numerical simulations. Since the DEM calculates velocity and interaction forces of every element for every time-step, these can be extracted to examine the temporal variability of prism evolution. Snapshots of velocity distribution and interaction force, referenced to the fixed base, are shown in Figures 8b and 8c respectively. On the velocity distribution diagrams, a particle group with a same velocity vector behaves as a block and the boundaries of such coherent blocks are the faults. The velocity diagram thus shows the temporal variation of the faulting. For example, the frontal thrust at the 2250 m shortening stage shows a minor displacement rate in the deeper part of the prism with the rate greater in the shallower part, suggesting an interaction of the second thrust from the deformation front. The deeper part of this second thrust is active at the 4500 m shortening stage, but the entire fault surface of this thrust becomes inactive by 6750 m shortening. The deeper part of the frontal thrust at 4500 m shortening becomes active at 6750 m shortening, but the shallower part is inactive. At the 6750 m shortening stage, the frontal thrust is active only at the tip.

In the interaction force diagrams (Fig. 8c), the black lines are directions of maximum gross contact forces, indicating where, how much and in which direction the forces exist in the assembly. The forces initiated from the moving wall that pushes the elements, and can be traced downward to the base (down-to-the-right direction). Immediately above and underneath the décollement layer, the directions of the forces are generally sub-horizontal. This suggests that the layer with low internal friction coefficient acted as a décollement and affected the pattern of the interaction forces. Within the wedge, the interaction forces also show up-to-the-right directions, which may also correspond to the thrust fault activities.

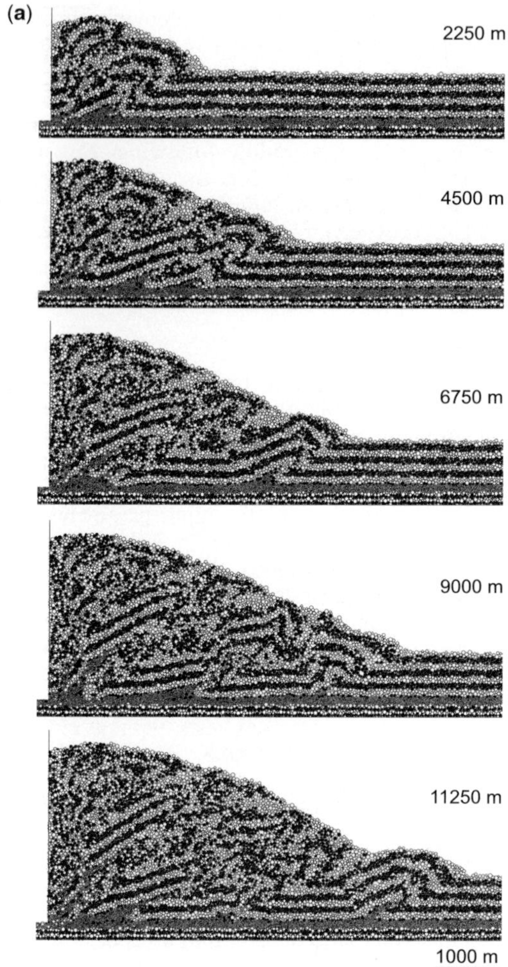

Fig. 8. Results of simulation II (the length values are amounts of shortening). (**a**) Structural style of the particle assembly during deformation. (**b**) Instantaneous velocity distributions of particles showing temporal variations during deformation. (**c**) Instantaneous distribution of maximum gross interaction forces. The curvature suggests that the force is affected by the décollement layer.

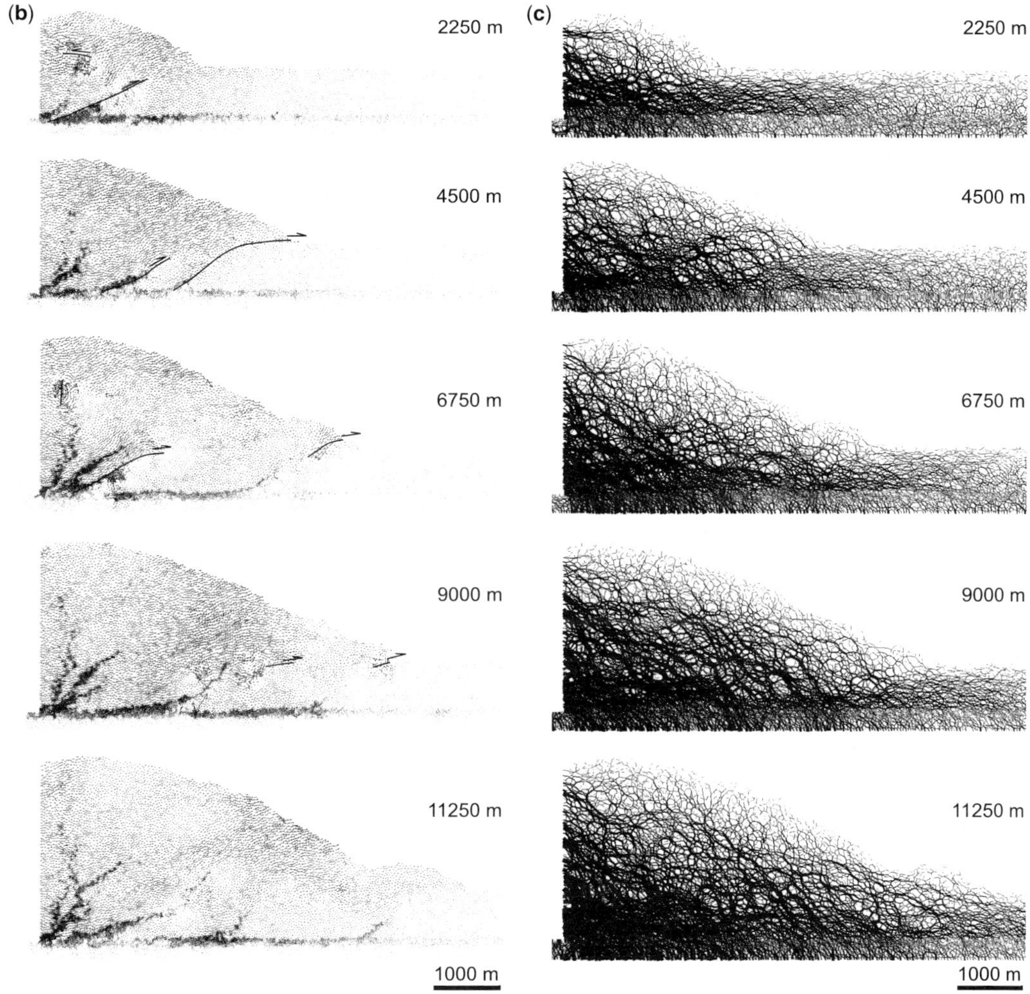

Fig. 8. *Continued.*

Discussion

Effects of décollement and implications for accretionary prisms

In both of the analogue experiments and numerical simulations, inserting the décollement, the layer of a lower frictional coefficient, reduces the slope angle (*c.* 18°–10° in analogue and 45°–23° in numerical; Fig. 9) and increases the fault spacing (*c.* 1.5–2.1 km in analogue and 0.7–1.1 km in numerical). This is in good agreement with the critical taper theory (Davis *et al.* 1983) arguing that the internal frictional coefficient along the décollement has a strong impact on the prism geometry. The internal friction angles of two materials (sand/particle A and microbeads/particle B) are almost the same in the analogue and numerical models. The resultant prism geometries are, however, significantly different. This is possibly caused by the smaller number of particles in the numerical simulations. Theoretically, the thickness of the décollement layer consists of 80 (or more) particles in the analogue models, whereas 16 (or more) are counted in the numerical models. This smaller number of numerical particles could not be enough to maintain the physical property of the layer, and the surrounding particle A could increase the internal friction of the décollement layer.

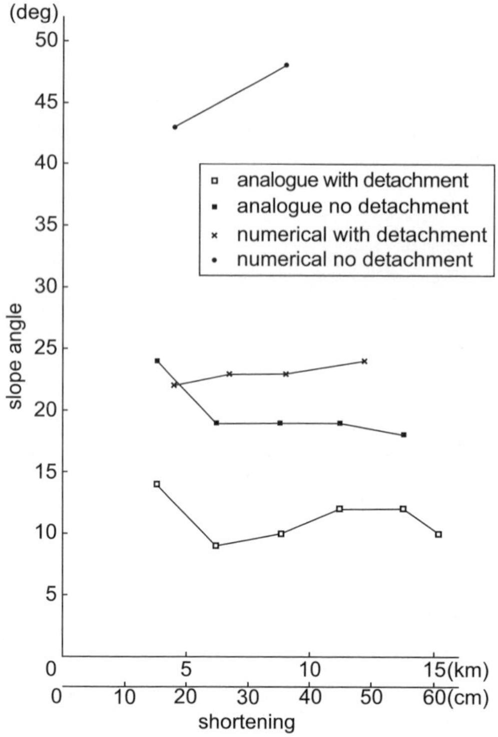

Fig. 9. Slope angles of analogue and numerical models. Inserting a décollement layer reduces the slope angle. Numerical models generally show steeper inclinations.

The effects of the low frictional layer in accretionary wedges can be identified in the prism geometry, the detachment geometry and the geometry of the frontal and backthrusts. The slope angle of the Nankai prism is *c.* 4° around the deformation front where the low friction décollement develops within the sediments, and is increased to *c.* 8° where the décollement is inactive. This may be a clear example showing the relationship between the slope angle and the frictional coefficient of the detachment. The detachment surfaces in the Nankai, Barbados and Cascadia wedges are characteristically flat (Bangs *et al.* 2004; Moore & Klaus 2000; Westbrook *et al.* 1994), similar to the experimental results. This suggests that the décollement geometry around the frontal thrust is strongly controlled by the pre-existing layers that have an anomalously low frictional coefficient. In experiment II, the active surface of the frontal thrust has a bend where the frontal thrust originates from the detachment. The bend produces a kink zone between the flat and the ramp, and compensation faults should be generated in the hanging wall that moves across the kink zone. These faults are observed as backthrusts. In Nankai, the detachment layer also generates a bend in the frontal thrust geometry and backthrusts in the overlaying sediments (*cf.* Fig. 2b).

Faulting and stresses

The analogue and numerical models show variations in the slope angle during prism formation. Nieuwland *et al.* (2000) describe the dynamic equilibrium that is common to the development of critical tapers, the mechanical necessity of out-of-sequence thrusting and the systematically changing angle of a growing critical taper. That is, a new thrust fault lifts up the front of the taper, thereby reducing the angle, then existing thrust faults become reactivated to restore the critical angle. Our observed variations may be produced by a similar effect, and the process of the prism formation would be strongly controlled by sequential fault reactivations.

The directions of the maximum interaction force (Fig. 8a) presumably indicate the direction of the maximum principal stress, σ_1. This direction theoretically makes an angle of 27.5° with the direction of the active shear zone. Figure 10 shows an interpretation of such shear zone directions, based on the calculated stress orientations. It is predicted that the active shear zones would be steepened upward from the décollement and from the moving backstop. The velocity distributions (Fig. 8b) generally show similar directions, thus this prediction may be appropriate. The prediction also suggests that there may be normal faulting under the décollement. The stress measurements at Nankai accord with this extensional stress regime under the décollement (Mikada *et al.* 2002).

Temporal variations due to stick-slip behaviour of frictional faulting

In the numerical results, the velocity and stress distributions show temporal variations during deformation. These distributions are calculated for 0.9 cm of displacement of the moving wall, thus they are regarded as instantaneous values. These variations should be closely related to the deformation behaviour of the particle assembly. Since no breakage of particles is observed during deformation, the shortening should be accommodated by a change of the particle arrangement. Thus, the grain boundaries in the particle assembly behave as displacement surfaces. Since the displacement along these surfaces is controlled by the normal force and frictional

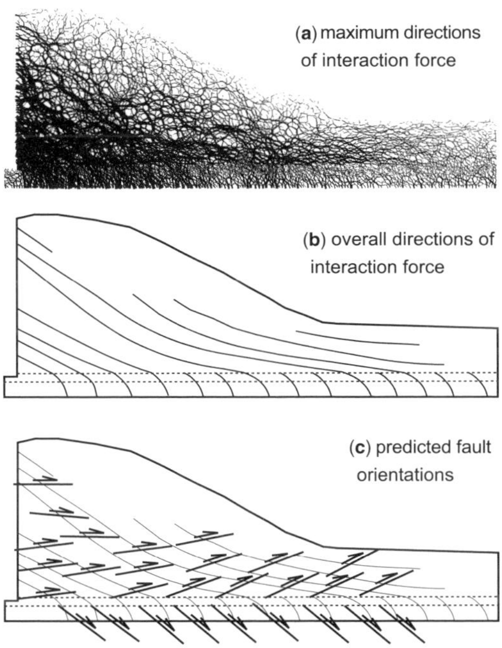

Fig. 10. Schematic diagram of possible active shear zone directions, based on the calculated stress orientations of the numerical model. (**a**) Directions of maximum interaction forces. (**b**) Schematic diagram of the overall direction of the maximum force. (**c**) Predicted active fault orientations. The internal friction angle is approximated as 35° from particle A's property.

coefficient acting on the surface, the displacement may be approximated as frictional gliding. This is grain-size independent and generally intermittent, causing so-called 'stick-slip' motion. Since the stick-slip motion occurs due to cycles of accumulation and release of elastic energy along the gliding surface, the force around the fault surface should be cyclically distorted. According to in-situ stress measurements by Nieuwland *et al.* (1999), low amplitude stress cycling after the initiation of a fault is observed in their thrust-and-fold belt sandbox experiments, interpreted as a stick-slip process in the fault zone. In cases where many faults exist, the distribution of the forces within the geological body will be affected by such distortion due to sequential fault events. The temporal variations in the particle velocity and forces are therefore presumably due to stick-slip behaviour along frictional surfaces.

Such temporal variations can be found in natural structures. Field and seismic observations clearly reveal that geological structures include many faults with a variety of directions even in a region that has been typically formed under a specific tectonic environment. Focal mechanism analysis of aftershocks also always shows such a variation even after one earthquake event (e.g., Shibazaki *et al.* 2002). This suggests that temporal variations of faulting are quite common in natural structures. The temporal variation may also be related to underground fluid flow. Sibson (1990) argues that the underground stress changes the permeability along fault surfaces, thus the variation in the force distributions may affect the fluid flow system in accretionary prisms. The pattern and the paths of the fluid flow in accretionary prisms may be important controls on the accumulation mechanism of methane hydrates. The temporal variations seen in the model results can thus be a key to understanding the heterogeneous distribution of methane hydrates in accretionary prisms (i.e., Baba & Yamada 2004).

Velocity and force variations in natural geological bodies due to stick-slip behaviour of faults are not yet well investigated. The models shown in this paper can be used to understand such natural instability in the geo-environment. This technique might be applied to prevent possible future geo-hazards (e.g., Ueta *et al.* 2000), provided that the models are properly scaled and the magnitude, place and timing of possible earthquakes and related deformations are properly understood from the modelling results.

Conclusions

The analogue (sandbox) and numerical (distinct element method) models used to examine the effect of a décollement layer within sediments show that inserting a lower frictional coefficient layer reduces the slope angle and increases the fault spacing. This agrees with the critical taper theory (Davis *et al.* 1983). The geometric variations observed in the model prism may be explained by fault reactivations. The velocity and force directions extracted from numerical particles are instantaneous distributions and their temporal variations may be due to stick-slip behaviour of frictional gliding along grain boundaries. The active shear zone distributions interpreted from the stress directions of the numerical models predict possible fault curvatures within the prism, and agree with the numerical velocity vectors. The overall geometry of accretionary prisms with a décollement in the sedimentary pile is similar to the models, including the kinked geometry of the sequential thrust surfaces accompanied by backthrusting. The extensional stress regime under the décollement at Nankai is also explained by the numerical stress distributions. This study shows that combinations of analogue and numerical models enable the extraction of more data from the results, that can be examined for detailed deformation characteristics.

This research is partly funded by the methane hydrate research consortium (MH-21) and their permission to publish this paper is gratefully acknowledged. S. Ueda and K. Kaneda assisted the numerical and experimental work respectively. The initial manuscript was greatly improved after constructive criticism by M. Seyferth and D. Nieuwland. The authors also appreciate comments and suggestions from the editors of this volume, G. Schreurs and S. Buiter.

References

ADAM, J. & URAI, J. L. *ET AL.* 2005. Shear localization and strain distribution during tectonic faulting – new insights from granular-flow experiments and high-resolution optical image correlation techniques. *Journal of Structural Geology*, **27**, 283–301.

ADAM, J., LOHRMANN, J., HOTH, S., KUKOWSKI, N. & ONCKEN, O. 2002. Strain variation and partitioning in thrust wedges: High-resolution data from scaled sandbox experiments by 2D-3D PIV analysis. *Bolletino di Geofisica*, **42**, Supplement, 123–125.

BABA, K. & YAMADA, Y. 2004. BSRs and associated reflections as an indicator of gas hydrate and free gas accumulation – An example of accretionary prism and forearc basin system along the Nankai Trough, off central Japan. *Resource Geology*, **54**, 11–24.

BANGS, N. L. B., SHIPLEY, T. H., MOORE, G. F., GULICK, S. P. S. & NAKAMURA, Y. 2004. Evolution of the Nankai trough décollement from the trench into the seismogenic zone: inferences from 3-D seismic reflection imaging. *Geology*, **32**, 273–276.

COBBOLD, P. R. & CASTRO, L. 1999. Fluid pressure and effective stress in sandbox models. *Tectonophysics*, **301**, 1–19.

COBBOLD, P. R., DURAND, S. & MOURGUES, R. 2001. Sandbox modelling of thrust wedges with fluid-assisted detachments. *Tectonophysics*, **334**, 245–258.

COLLETTA, B., LETOUZEY, J., PINEDO, R., BALLARD, J. F. & BALE, P. 1991. Computerized X-ray tomography analysis of sandbox models: examples of thin-skinned thrust systems. *Geology*, **19**, 1063–1067.

CUNDALL, P. A. & STRACK, O. D. L. 1979. A discrete numerical model for granular assemblies. *Geotechnique*, **29**, 47–65.

DAVIS, D., SUPPE, J. & DAHLEN, F. A. 1983. Mechanics of fold-and-thrust belts and accretionary wedges: cohesive Coulomb theory. *Journal of Geophysical Research*, **88**, 1153–1172.

DAHLEN, F. A. 1984. Non-cohesive Critical Coulomb wedges: an exact solution. *Journal of Geophysical Research*, **89**, 10125–10133.

DAHLEN, F. A. & SUPPE, J. 1984. Mechanics of fold and thrust belts and accretionary wedges: cohesive Coulomb theory. *Journal of Geophysical Research*, **89**, 10087–10101.

DILEONARDO, C. G., MOORE, J. C., NISSEN, S. & BANGS, N. 2002. Control of internal structure and fluid migration pathways with the Barbados Ridge Décollement zone by strike-slip faulting: evidence from coherence and 3-D seismic amplitude imaging. *Bulletin of the Geological Society of America*, **114**, 51–63.

FINCH, E., HARDY, S. & GAWTHORPE, R. 2003. Discrete element modelling of contractional fault-propagation folding above rigid basement fault blocks. *Journal of Structural Geology*, **25**, 515–528.

GUTSCHER, M. A., KUKOWSKI, N., MALAVIELLE, J. & LALLEMAND, S. E. 1996. Cyclical behavior of thrust wedges: insight. from high basal friction sandbox experiments. *Geology*, **24**, 135–138.

HRYCIW, R. D., RASCHKE, S. A., GHALIB, A. M., HORNER, D. A. & PETERS, J. F. 1997. Video tracking for experimental validation of discrete element simulation of large discontinuous deformations. *Computers and Geotechnics*, **21**, 235–253.

HUBBERT, M. K. 1937. Theory of scaled models as applied to the study of geological structures. *Bulletin of Geological Society of America*, **48**, 1459–1520.

HUIQI, L., MCCLAY, K. R. & POWELL, D. 1992. Physical models of thrust wedges. *In*: MCCLAY, K. R. (ed.) *Thrust Tectonics*, London, Chapman and Hall, 71–81.

KONIETZKY, H. 2002. *Numerical Modeling in Micromechanics via Particle Methods*. Proceedings of the 1st International PFC Symposium, Gelsenkirchen, Germay, Rotterdam, Balkema Publishers.

KOYI, H. A. & MANCKTELOW, N. S. 2001. *Tectonic Modeling: A Volume in Honor of Hans Ramberg*. Geological Society of America Memoir 193.

KOYI, H. A. & VENDEVILLE, B. C. 2003. The effect of décollement dip on geometry and kinematics of model accretionary wedges. *Journal of Structural Geology*, **25**, 1445–1450.

LALLEMAND, S. E., MALAVIEILLE, J. & CALASSOU, S. 1992. Effects of oceanic ridge subduction on accretionary wedges: experimental modelling and marine observations. *Tectonics*, **11**, 1301–1313.

LOHRMANN, J., KUKOWSKI, N., ADAM, J. & ONCKEN, O. 2003. The impact of analogue material properties on the geometry, kinematics, and dynamics of convergent sand wedges. *Journal of Structural Geology*, **25**, 1691–1711.

MANDL, G. 1988. *Mechanics of Tectonic Faulting*. Elsevier, Amsterdam.

MASSOLI, D., KOYI, H. A., BARCHI, M. R. & ROGLEDI, S. 2002. Analogue modelling of the Po Plain: structural evolution of a fold thrust belt generated by multiple décollement. *Bollettino di Geofisica teorica ed applicata*, **42**, 1/2 supplement, 200–202.

MCCLAY, K. R. 1990. Deformation mechanics in analogue models of extensional fault systems.

In: Knipe, R. J. (ed.) *Deformation Mechanisms, Rheology and Tectonics*, Geological Society of London Special Publications, **54**, 445–453.

Mikada, H., Becker, K. *et al.* 2002. *Proceedings of the Ocean Drilling Project Initial Reports*, **196**. Available from World Wide Web: http://www.odp.tamu.edu/publications/196_IR/196ir.htm.

Moore, J. C. & Klaus, A. 2000. *Proceedings of the Ocean Drilling Project Scientific Results*, **171A**. Available from World Wide Web: http://www.odp.tamu.edu/publications/171A_SR/171asr.htm.

Mulugeta, G. & Koyi, H. A. 1992. Episodic accretion and strain partitioning in a model sand wedge. *Tectonophysics*, **202**, 319–333.

Nieuwland, D. A., Leutscher, J. H. & Gast, J. 2000. Wedge equilibrium in fold-and-thrust belts. Prediction of out-of-sequence thrusting, based on sandbox experiments and natural examples. *Netherlands Journal of Geoscience/Geologie en Mijnbouw*, **79**, 81–91.

Nieuwland, D. A., Urai, J. L & Knoop, M. 1999. Insitu stress measurements in model experiments of tectonic faulting. *In*: Lehner, F. K. & Urai, J. L. (eds) *Aspects of Tectonic Faulting*, Springer Verlag, 151–162.

Saltzer, S. D. 1993. Boundary conditions in sandbox models of crustal extension: an analysis using distinct elements. *Tectonophysics*, **215**, 349–362.

Schreurs, G., Buiter, S. *et al.* 2005. Analogue benchmarking: results of shortening and extension experiments. *In*: Schreurs, G. & Buiter, S. (eds) *Analogue and Numerical Modelling of Crustal Scale Processes*. Geological Society Special Publications, **253**, xx–yy.

Seno, T., Stein, S. & Gripp, A. E. 1993. A model for the motion of the Philippine Sea plate consistent with NUVEL-1 and geological data. *Journal of Geophysical Research*, **98**, 17941–17948.

Shimizu, Y., Hart, R. D. & Cundall, P. 2004. *Numerical Modeling in Micromechanics Via Particle Methods*. Proceedings of the 2nd International PFC Symposium. Leiden, Balkema Publishers.

Shibazaki, B., Yoshida, Y., Nakamura, M., Nakamura, M. & Katao, H. 2002. Rupture nucleations in the 1995 Hyogo-ken Nanbu Earthquake and its large aftershocks. *Geophysical Journal International*, **149**, 572–588.

Sibson, R. H. 1990. Conditions of fault-valve behavior. *In*: Knipe, R. J. & Rutter, E. H. (eds) *Deformation Mechanisms, Rheology and Tectonics*. Geological Society Special Publications, **54**, 15–28.

Strayer, L. N. & Suppe, J. 2002. Out-of-plane motion of a thrust sheet during along-strike propagation of a thrust ramp: a distinct-element approach. *Journal of Structural Geology*, **24**, 637–650.

Tobin, H. J., Moore, J. C. & Moore, G. F. 1994. Fluid pressure in the frontal thrust of the Oregon accretionary prism: experimental constraints. *Geology*, **22**, 979–982.

Tsuji, T., Matsuoka, T. *et al.* 2005. Initiation of plate boundary slip in the Nankai Trough off the Muroto peninsula, southwest Japan. *Geophysical Research Letters*, **32**, L12306.

Ueta, K., Tani, K. & Kato, T. 2000. Computerized X-ray tomography analysis of three-dimensional fault geometries in basement-induced wrench faulting. *Engineering Geology*, **56**, 197–210.

Westbrook, G. K., Carson, B., Musgrave, R. J. & Shipboard Scientists. 1994. *Proceedings of Ocean Drilling Program, Initial Reports*, **146**.

Wolf, H., König, D. & Triantafyllidis, T. 2003. Experimental investigation of shear band patterns in granular material. *Journal of Structural Geology*, **25**, 1229–1240.

Willett, S. D. 1992. Dynamic and kinematic growth and change of a Coulomb wedge. *In*: McClay, K. R. (ed.) *Thrust Tectonics*, London, Chapman and Hall, 19–31.

Yamada, Y. & McClay, K. R. 2003*a*. Application of geometric models to inverted listric fault systems in sandbox experiments. 1:2D hanging wall deformation and section restoration. *Journal of Structural Geology*, **25**, 1551–1560.

Yamada, Y. & McClay, K. R. 2003*b*. Application of geometric models to inverted listric fault systems in sandbox experiments. 2: Insights for possible along strike migration of material during 3D hanging wall deformation. *Journal of Structural Geology*, **25**, 1331–1336.

Yamada, Y. & McClay, K. R. 2004. 3-D analog modeling of inversion thrust structures. *In*: McClay, K. R. (ed.) *Thrust Tectonics and Hydrocarbon Systems*. American Association of Petroleum Geologists (AAPG) Memoir **82**, 276–301.

Yamada, Y., Ueda, S. & Matsuoka, T. 2004. Simulation of accretionary prisms by distinct element method. *In*: Shimizu, Y., Hart, R. & Cundall, P. (eds) *Numerical Modeling in Micromechanics via Particle Methods*. Balkema Publishers, The Netherlands, 187–190.

Yamada, Y., Okamura, H., Tamura, Y. & Tsuneyama, F. 2005. Analogue models of faults associated with salt doming and wrenching: application to offshore UAE. *In*: Sorkhabi, R. & Tsuji, Y. (eds) *Faults, Fluid Flow and Petroleum Traps*. American Association of Petroleum Geologists (AAPG) Memoir **85**, 95–106.

Integrated four-dimensional modelling of sedimentary basin architecture and hydrocarbon migration

S. M. CLARKE[1]*, S. D. BURLEY[1,2], G. D. WILLIAMS[1], A. J. RICHARDS[1]**, D. J. MEREDITH[1] & S. S. EGAN[1]

[1]*Basin Dynamics Research Group, School of Earth Sciences and Geography, Keele University, Staffordshire, ST5 5BG, England, UK (e-mail: smcl@bgs.ac.uk.)*

[2]*BG Exploration and Production India Ltd, Midas Building, Sahar Plaza, M.V. Road, East Andheri, Mumbai, 400 059, India*

**Current address: GLNB, British Geological Survey, Murchison House, West Mains Road, Edinburgh, EH9 3LA, Scotland, UK*

***Current address: Grid Operations Support Centre, CCLRC e-Science Centre, Rutherford Appleton Laboratory, Chilton, Didcot, Oxfordshire, England, UK*

Abstract: Structural geometries, faults and their movement histories, together with the petrophysical properties of flow units, are some of the major controls on hydrocarbon migration pathways within sedimentary basins. Currently, structural restoration, fault-seal analysis and hydrocarbon migration are treated as separate approaches to investigating basin geohistory and petroleum systems. Each of these separate modelling approaches in their own fields is advanced and sophisticated but they are not compatible with each other. Lack of integration produces incorrect palaeogeometries in basin models and inaccurate migration pathways.

A combined structural restoration and fault-seal analysis technique, integrated with fast hydrocarbon migration pathway modelling code based on invasion percolation (IP) methods, is described. These modelling methods are used to develop a 4D basin modelling workflow in which evolving basin geohistories and geometries form an integral part of the analysis of hydrocarbon migration and trapping. By combining structural restoration and 3D fault-seal analysis it is possible to investigate the evolution of structurally complex traps through time. Integration of these techniques with a numerically fast migration pathway modelling technique allows hydrocarbon migration pathways and accumulations to be modelled through the evolution of the basin with time. Additionally, the effects of uncertainties in structural geometry, fault seal or any of the model input parameters can be explored using a risk-driven approach to modelling.

These methods are demonstrated using synthetic, computer generated, 3D models and a well-constrained model of the Moab Fault, Utah, USA. Comparison of modelled structural geometries, fault-seal properties and predicted trapped hydrocarbons with outcrop data is used to validate the integrated modelling approach. The validated techniques are then applied to a seismically derived, 3D model from the southern North Sea, UK, to demonstrate how an integrated, risk-driven approach to modelling allows the effects of uncertainties in the distribution of hydrocarbon accumulations to be investigated.

Basin modelling is an established, powerful technique for predicting the timing of petroleum generation and the distribution of the hydrocarbon phase in the subsurface to reduce exploration risk (Iliffe & Dawson 1996; Welte *et al.* 1997; Burley *et al.* 2000). It is based on decompactional reverse modelling, from an understanding of the present day system, back to an initial depositional point, which is then used as a framework to forward model the basin and characteristics of the petroleum system. Commercial and research basin modelling tools enable many petroleum system processes to be modelled, including subsidence, temperature, compaction, maturation, secondary migration, and hydrocarbon phase (e.g., Mann *et al.* 1997; Wendebourg 2000). As a result, various case histories, 'calibrated' to present day known accumulations, have predicted

From: BUITER, S. J. H. & SCHREURS, G. (eds) 2006. *Analogue and Numerical Modelling of Crustal-Scale Processes.* Geological Society, London, Special Publications, **253**, 185–211.
0305-8719/06/$15.00

hydrocarbon phase and distribution at the basin scale with varying degrees of success (e.g., Schroeder & Sylta 1993; Symington *et al.* 1998; Cowan *et al.* 1999; Johannesen *et al.* 2002).

Remarkably, few of these studies consider the structural evolution of basin geometry with time. Fluid flow conduits, and migration along them, constitute linked 3D networks defined by sediment fill and structural geometries, especially faults, throughout the basin. Accurate prediction of hydrocarbon accumulations therefore requires modelling in 4D, incorporating both 3D geometrical space, and how this geometry changes with time as the basin develops. However, incorporation of basin architecture and its evolution in basin models is at best simplistic. This is because hydrocarbon migration simulators employ an orthogonal grid data framework, suited to fluid flow calculations that conserve volumes, but unable to accommodate complex geometries (Wangen 1997). As a result, palaeogeometries and the effects of structuration on migration and entrapment are incorrect in current basin model realizations if migration pre-dates or is contiguous with structuration. Previous attempts at integrating the effects of faults into migration simulators have incorporated only vertical decompaction, backstripping and vertical shear on simple discontinuities in the orthogonal grid to represent faults and their properties. Methods are being developed to integrate structural restoration into migration simulators without requiring vertical grid alignment (e.g., Mello & Henderson 1997; Welte *et al.* 2000), and most recently, by using sequential time-step output from structurally restored horizons as input into a horizon-based migration simulator (Huggins *et al.* 2005).

The importance of structural restoration in determining basin geohistory for predictive petroleum system modelling is demonstrated in this paper. A new, integrated, modelling methodology is introduced, based on 3D fault restoration and fault-seal analysis software incorporating a hydrocarbon migration simulator employing *invasion percolation* (IP) techniques that enable migration pathways to be investigated through time as the basin geometry and architecture evolve. The fundamental features of the model are illustrated with a computer-generated 3D synthetic dataset, while a well constrained outcrop model of the Moab Fault, Utah, USA is used to demonstrate that accurate 3D restoration of basin geohistory and fault-zone gouge are critical to modelling hydrocarbon migration, and thus petroleum system analysis. This integrated technique also facilities rapid modelling of predicted accumulations, and allows a risk-driven approach to petroleum systems analysis in which multiple realizations can be used to rank the likelihood of migration and accumulation. Uncertainty in aspects of basin architecture, fault seal or petrophysical detail, and their effects on evolving migration pathways, can be assessed and specific accumulations risked accordingly. Application of the technique in a risk-driven scenario is demonstrated using a case study of seismically mapped fault blocks from the southern North Sea, UK.

Each of these models is provided to demonstrate aspects of the 4D modelling approach and to illustrate the effects of accurate restoration on migration pathways and trapping histories. Details of all the models and associated animations are available on the internet at http://www.bdrg.esci.keele.ac.uk/smcresearch.htm.

Restoration, fault seal and migration in basin modelling

Structural restoration and basin modelling have traditionally addressed specialist problems within fields of structural validation and petroleum systems modelling. Appropriate tools for each approach have evolved in isolation from each other, with separate data requirements, assumptions and data structures. Structural modelling tools are based on geometric algorithms that describe rock deformation. By contrast, basin modelling techniques were developed from geochemical prediction of organic matter maturation and the need to calculate migration pathways and trapped volumes. Advances in computing power, visualization and the development of 3D modelling technology now enables the construction of high spatial resolution 3D models that can incorporate both structural restoration and hydrocarbon migration modelling. A more accurate numerical representation of the petroleum system of a basin and its evolution can be achieved by embedding structural modelling into hydrocarbon generation and migration modelling so that migration through palaeogeometries can be investigated. The key to incorporating structural restoration in basin simulators is to represent basin geometry as realistic structural and litho-stratigraphical surfaces (contacts) in 3D space. As faults and horizon discontinuities (such as unconformities) are important architectural components of a basin, these elements should *in principa* be the initial building blocks of the basin model if the basin geometry through time is to be represented.

Structural modelling

Structural validation and modelling tools are based upon geometric algorithms that describe rock deformation, preserving area and/or volumes. In 2D, conservation of volume simplifies to conservation of cross-sectional area, and/or bed length, assuming a section constructed parallel to the mass movement direction, plane strain deformation and a tectonic regime in which bed thickness is unaffected (Dahlstrom 1969; Gibbs 1983). Several 3D approaches have been developed, including strain trajectories (Cobbold 1979), iterative finite element methods (Gratier *et al.* 1991), best-fit unfolding and fault fitting (Rouby *et al.* 2000) and jigsaw fitting (Williams *et al.* 1997). All these techniques enable the restoration of complex faulted and folded structures. New algorithms, such as flexural flow, have been developed that consider out-of-plane movement of material for true 3D modelling (Williams *et al.* 1997).

Handling of complex geometries and deformations in 3D requires a flexible model representation in the form of irregular, triangulated meshes that can change topology (node pattern and connectivity) between modelling steps (Mello & Henderson 1997; Richards 2000). The geohistory of a basin is determined by reverse modelling of the present-day stratigraphical and structural architectures to restore palaeogeometries. Sediment layers are sequentially backstripped and decompacted (e.g., Van Hinte 1978; Bond & Kominz 1984). Differential compaction of layers, with variable thicknesses or discontinuity, has a major impact upon restored horizon palaeogeometries. Different compaction porosity-depth functions also affect restored palaeogeometries (Giles *et al.* 1998), though changes in structural amplitude rather than shape most commonly result.

Backstripping has traditionally been geometric, restoring horizons to an assumed depositional template, whereas latest research models are process-based, using lithospheric models to predict palaeogeometries (Kusznir *et al.* 1995). The simplest backstripping procedure restores the sediment depositional surface to a horizontal datum of a specified bathymetry, usually sea-level (Roberts *et al.* 1998) but can be enhanced by restoring to a palaeobathymetric template, defined from the stratigraphical architecture and biostratigraphy of the layer (Van Hinte 1978). Mechanical models predict basement geometry based upon lithospheric deformation and its isostatic response to basin infilling (Mckenzie 1978; Meredith & Egan 2002), defining evolving horizon geometries and palaeobathymetry.

Structural restoration through time involves a retro-deformational process in which sequential 'snapshots' of the basin evolution are portrayed. Observed or inferred displacement vector fields are combined with geometrical modelling methods that predict hanging wall geometries and define basin architecture with subsequent stages of deformation.

Fault-seal prediction

Faults, their geometries, and their movement histories through time are important controlling elements of hydrocarbon migration and entrapment within sedimentary basins. When faults undergo displacement, their fluid transmissibility may change as a result of (1) juxtaposition of strata, (2) development of a damage zone, and (3) inclusion of material between faulted blocks. In addition, fault displacement modifies the geometry of faulted blocks. These effects have to be combined with 3D structural geometry modelling and incorporated into models of hydrocarbon migration for accurate migration pathways to be predicted.

The sealing properties of a fault can be approximated using 1D and 2D numerical methods (e.g., Allan 1989; Fristad *et al.* 1997) and may be applied to the fault in entirety or to a given stratigraphical sequence and deformational regime. In 3D, the structural geometries of faults and faulted blocks can be represented fully and the sealing properties determined at each spatial point on the fault surface from the interaction of the faulted stratigraphy, the deformation regime, the structural geometry of the fault blocks, and the properties of any fault-zone material (Clarke *et al.* 2005*a*) using 3D derivations of these techniques.

2D fault-surface projections can be constructed to analyse relative cross-fault juxtapositions for a given temporal stage in the deformational history of a fault. However, a fault that is curved along strike will not be represented correctly in the fault-surface section and, if a curved fault is resolved onto a flat plane, stratigraphical cut-off areas will not necessarily be preserved (Harper & Lundin 1997). Furthermore, fault representation as a fault-surface section does not allow investigation of the interaction of multiple faults on sealing capacity, nor analysis of lithology juxtaposition at other temporal stages in the evolution of the fault. These problems can be addressed by mapping properties of the faulted blocks to the footwall and hanging wall of the fault surface directly in 3D (Clarke *et al.* 2005*a*). Furthermore, the interaction of juxtaposition seal between multiple faults is revealed and, by examining relationships at other temporal stages, the evolution of cross-fault juxtaposition can be determined.

Argillaceous smearing is an example of a fault-zone process that can modify transmissibility. This process has attracted much attention due to its particular relevance to mixed arenaceous and argillaceous sequences and the availability of field examples for calibration (e.g., Foxford *et al.* 1996). Many workers have produced 1D and 2D numerical and empirical models to quantify argillaceous smearing and, of these models, the *shale gouge ratio* or SGR (Fristad *et al.* 1997) is commonly employed for the analysis of fault transmissibility using the expression:

$$\mathrm{SGR} = \frac{\sum_{i=1}^{I} (\Delta z f_{cl})_i}{t} \tag{1}$$

where Δz is interval or unit thickness of unit i, f_{cl} is the argillaceous fraction of unit i, t is fault throw, and I is the total number of units that have passed a given point on the fault surface.

In general, an SGR value of *c.* 20% represents the threshold above which smear continuity is sufficient to cause sealing (Gibson 1994; Fristad *et al.* 1997; Yielding 2002), although this value is likely to be affected by case-specific geological factors such as host rock mineralogy and confining pressure. Despite such limitations, the SGR is an accepted technique for the analysis of fault seal potential. Use of the parameter 'throw' in the original published algorithm inherently defines this expression as 1D, although it can be applied repeatedly along the length of a fault to generate a 2D or pseudo-3D analysis. Modern methods project the derivation of throw into a slip-azimuth projection, thus reducing inaccuracy in settings where faults have a significant component of strike-slip. True 3D applications of the SGR, incorporating 3D structural displacement vectors and properties of the faulted sequence, allow fault-zone argillaceous content to be modelled in the full 3D environment and its effects on hydrocarbon migration and accumulation to be investigated in the context of the basin architecture (Clarke *et al.* 2005*a*).

Hydrocarbon migration

Secondary hydrocarbon migration is a multiphase, water-oil-gas process (Schowalter 1979) driven by the balance between buoyancy and capillary forces, although in high permeability sediments pressure gradients, a function of compaction, play a role (Berg 1975). Current petroleum system concepts suggest that hydrocarbons migrate in focused stringers (less than a few metres high, a few tens of metres wide, but up to many kilometres long), at moderate saturations (20–40%; Sylta 2002). Focusing takes place because the permeability fabric of sediments is heterogeneous, so once critical saturations required for the flow condition are established, initial migration pathways are maintained. This 'fingering' dictates that only a small proportion of the basin-fill (in the order of 2%) is invaded by the migrating hydrocarbons, resulting in a highly efficient migration process (Larter *et al.* 1996). Laboratory simulations of hydrocarbon migration confirm that hydrocarbons move vertically under buoyancy until they encounter a permeability barrier, at which point they migrate laterally, or accumulate until the buoyancy pressure of the accumulation exceeds the capillary entry pressure of the confining barrier (Thomas & Clouse 1995).

Three different numerical approaches to modelling these aspects of migration are in widespread use: ray-path, Darcy flow and percolation (Hantschel *et al.* 2000; Carruthers 2003). In all cases, models are constructed from stacked horizon surface grids in which data nodes are vertically aligned. Simple fault geometries and their evolution are incorporated but are handled in a simplistic manner without discrete discontinuities in the grids, and without the complex geometries that can be derived from structural modelling techniques.

Ray-path modelling is a simple approach that allows vertical, buoyancy-driven migration until hydrocarbons impinge on surfaces bounded by low permeability seals, at which point they accumulate beneath the seal in topographical highs. This approach is very effective in reproducing aspects of the migration system, such as focused, high saturation flow in permeable carrier beds. The numerical simplicity of the approach results in very fast simulations, enabling statistical treatment of model results. However, ray-path modelling does not reproduce well migration through low permeability sediments, and is not representative of the whole migration process.

Darcy flow geocellular-based models calculate 3-phase flow as a function of sediment compaction and pressure gradients. Finite element and finite difference approximations are used to model hydrocarbon flow through and between regular or irregular shaped geo-cellular elements, each cell being assigned petrophysical properties. These methods predict the 'state' and direction of flow at each spatial and temporal point within the model in terms of physical and kinetic properties based on defined initial and boundary conditions. This approach honours the physics of matrix Darcy flow but assumes cells are filled with hydrocarbon to a given saturation before migration proceeds. As the numbers of cells in the model is limited by computational power, shortcomings of this approach are dispersion of the migration front at coarse cell dimensions and long simulation times.

Percolation-based models use the principle of gradient invasion percolation, in which petroleum migrates under buoyancy into an opposing network of cells populated with capillarity pressures or permeabilities, pursuing the lowest entry pressure or highest permeability pathway. Computational times are extremely fast and capillary/permeability grids can be defined statistically, allowing enormously large models to be investigated directly from seismic data, and sensitivity studies to be rapidly conducted on model results. This contrasts with the long process times of Darcy flow-derived migration simulation, which renders a risk-driven approach to basin analysis and migration impractical. Over timescales associated with secondary hydrocarbon migration, percolation-based numerical approaches may be the most appropriate technique for modelling petroleum systems (Carruthers 2003).

Modelling hydrocarbon migration through faulted structures

Here we introduce a combined approach to the modelling of sedimentary basin architecture, fault seal properties and hydrocarbon migration, based on percolation theory that enables a risk-driven analysis of hydrocarbon flow through faulted basins (Fig. 1). 3D structural restoration techniques are used to restore models, derive displacement vector fields and subsequently deform basin architecture to provide geometrically valid hanging wall structural geometries. Calculated displacement vectors are incorporated into 3D predictions of fault seal properties. True 3D juxtaposition and shale gouge effects are determined and percolation-based migration techniques are used to predict flow pathways and accumulations within these models. Percolation techniques are used because these provide a realistic representation of hydrocarbon migration and their numerical simplicity allows completion of many realizations (hundreds or thousands) in practical lengths of time. Multiple simulations can be run, having modified one or more input parameters to investigate the sensitivity of the petroleum system to a particular geological scenario. This approach can thus be used to filter low-risk prospects for use in more computationally intensive Darcy flow-based compositional flow simulators.

The structural basin model

To model 3D structure, contacts between lithological and structural units of the sedimentary basin are represented as triangulated surfaces in 3D (Fig. 2a). Details of this methodology are described in Clarke *et al.* (2005*a*). The technique is computationally efficient, and allows structural restoration, as well as calculation and visualization of fault seal properties. However, it does not consider the rock volume between modelled surfaces; a necessary requirement for flow simulation.

To address this issue, we combine the surface model of Clarke *et al.* (2005*a*) with a 3D geocellular model. The rock volume between the surfaces is represented by a lattice in which each cell is populated by a node at its centre (Fig. 2b). Cell size may vary throughout the model, but it is assumed that the cell size at a given point adequately represents the heterogeneity in rock properties within the geological units. Cells adjacent to surfaces representing faults are representative of the fault zone and associated fault rocks. This requires that the cell size adjacent to such surfaces must be of the order of magnitude of the fault zone width and can be approximated from:

$$t_i \approx \frac{d_i}{66} \quad (2)$$

where t_i is the fault zone thickness (in units of d) at point i on the fault surface and d_i is fault displacement at that point (Manzocchi *et al.* 1999).

The triangulated surfaces represent major structural or stratigraphical discontinuities and, as such, divide the geocelluar model into discrete blocks (Fig. 2b). This strong relationship between the surface and geocellular components controls data population of the model. Structural deformation and fault seal properties can be determined using the geometry of the surfaces and integrated with the properties of the geocellular model to control hydrocarbon migration.

Structural restoration and fault seal

For integrated basin modeling, there are two overriding considerations in structural restoration: (1) The algorithm must predict the geometry of the hanging wall with differing degrees of deformation and, (2) the resultant deformation must volume balance. In some scenarios, fault displacement and hydrocarbon migration may occur concurrently. An iterative best-fit algorithm is used to derive the required displacement vector field that best restores the displaced

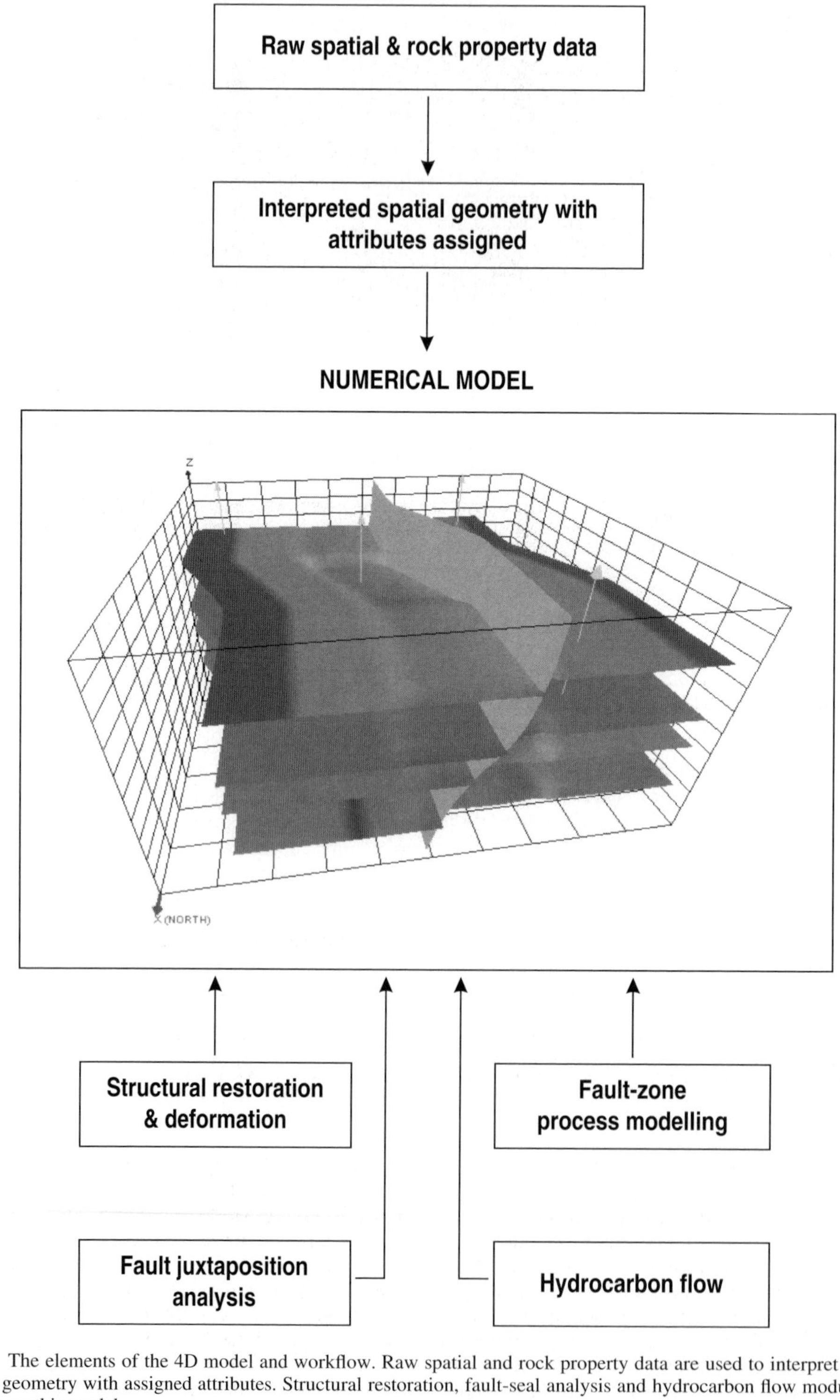

Fig. 1. The elements of the 4D model and workflow. Raw spatial and rock property data are used to interpret spatial geometry with assigned attributes. Structural restoration, fault-seal analysis and hydrocarbon flow modules operate on this model.

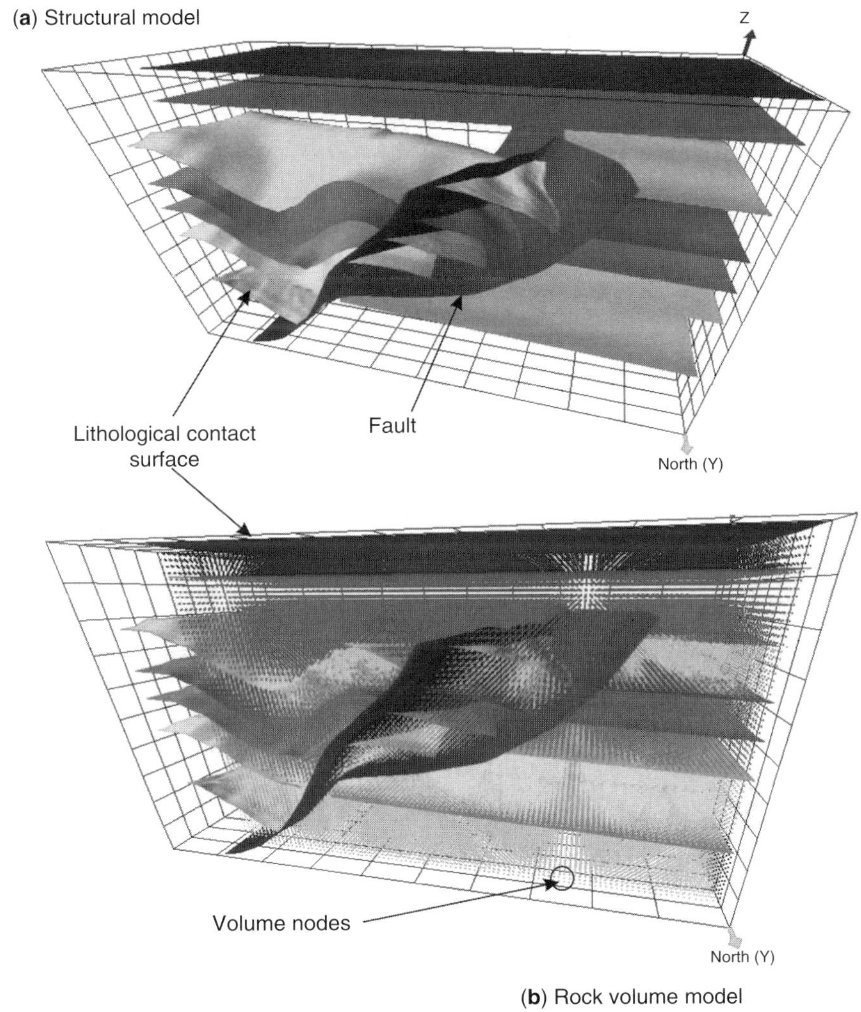

Fig. 2. The numerical model of a kilometre-scale basin. (**a**) Structural, stratigraphical and lithological contacts are represented as triangulated surfaces in 3D space. This modelling strategy allows efficient modelling of structural architecture, restoration and fault seal properties. (**b**) The rock volume between surfaces is represents by a geocellular lattice. Each cell is represented by a node at its centre. Nodes adjacent to faults represent the fault zone. The surfaces divide the rock volume into discrete blocks and constrain data interpolation to the volume.

geometry. First- and second-order derivatives of the map-plane geometry of respective footwall and hanging wall cut-offs are analysed to determine points of significant curvature along the cut-off, termed 'piercing points' (Fig. 3a). Each piercing point along the hanging wall cut-off is matched to a piercing point of the corresponding footwall cut-off in turn, generating a series of possible slip-azimuth vectors (Fig. 3a). The hanging wall cut-off is translated in the direction of each possible slip azimuth in turn and the amount of misfit with the footwall cut-off determined. The slip azimuth that produces the minimum amount of misfit is selected as the most likely restoration slip azimuth (Fig. 3b). The misfit between cut-offs with this slip azimuth is used to define a 'horizontal slip magnitude' gradient (distribution of the horizontal components of displacements in the slip-azimuth direction) for the hanging wall block (Fig. 3c) A 3D solution to the chevron/inclined shear construction (Clarke *et al.* 2005*a*) is used to restore the hanging wall and iteratively refine the slip magnitudes and define true 3D

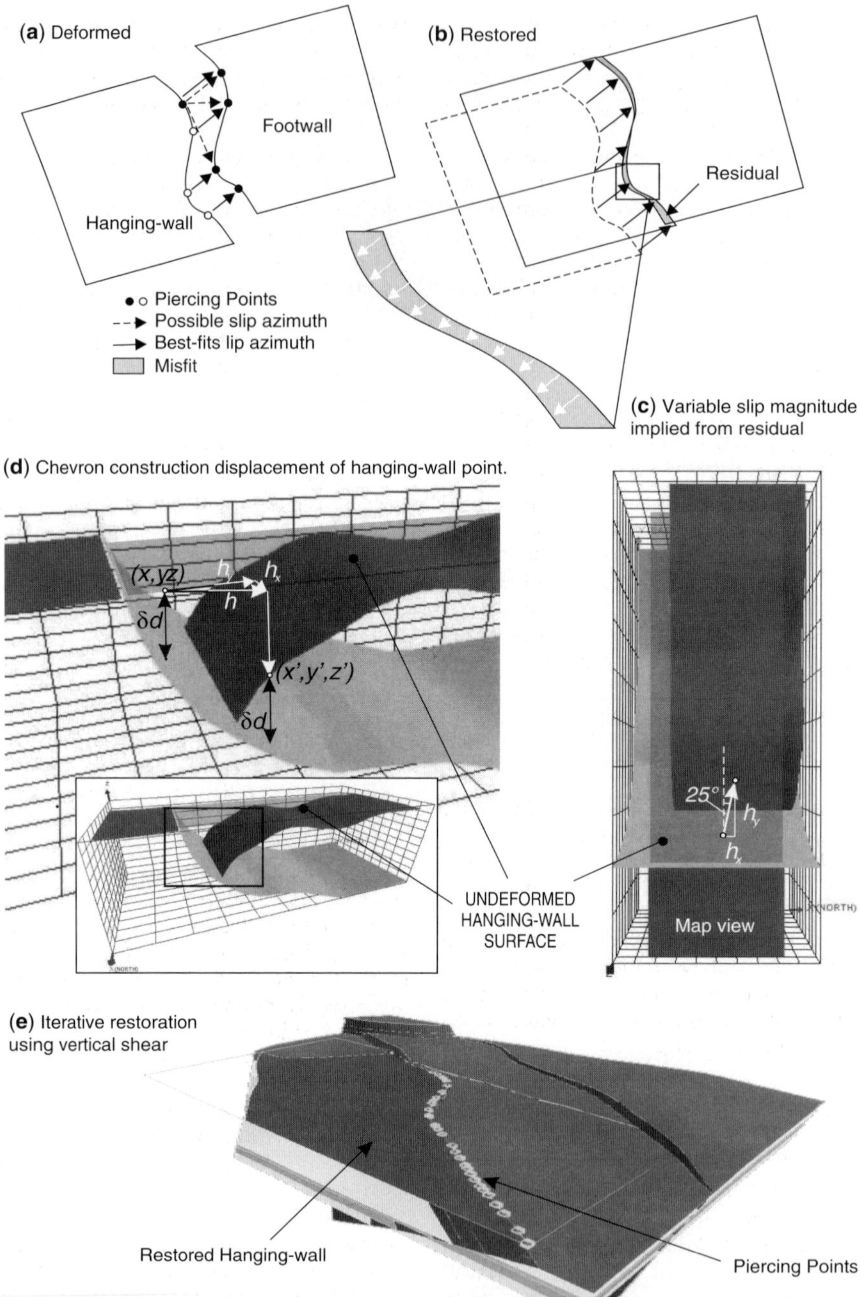

Fig. 3. Automated structural restoration. (**a**) First- and second-order derivatives of footwall and hanging wall cut-offs are used to derive 'piercing points'. The hanging wall cut-off is translated to match piercing points in turn. (**b**) The translation that produces the least misfit is considered the slip azimuth. (**c**) The misfit with this slip azimuth is used to derive the horizontal slip gradient. Iterative restoration using chevron or inclined shear is used to refine these vectors. (**d**) Hanging wall deformation can be achieved using these vectors and the chevron (shown) or inclined shear construction. Each point in the hanging wall is translated in the map plane by Cartesian components (h_x and h_y) of the slip azimuth, and in the vertical plane such that original distances between the surface and the fault (δd) are maintained. (**e**) Hanging wall geometries can be predicted at given snapshots in time.

displacement vectors. Subsequently, the derived best-fit displacement vectors are used to deform hanging wall strata and generate required snapshots of basin evolution (Figs 3d and e). This approach maintains fault-block volumes with deformation. Given the strong relationship between the structural and geocellular components of the model, volume nodes representing the geocellular hanging wall volume are displaced in the same manner and their associated properties translated accordingly. Coherency between the structural and rock volume model is preserved.

Fault properties are modelled directly on the triangulated fault surfaces. Juxtaposition relationships across faults are modelled in 3D by comparing interpolated lithological and petrophysical properties of fault blocks bounded by fault surfaces. Fault-zone gouge is predicted using a 3D application of the SGR (Clarke *et al.* 2005*a*). The argillaceous contents of the faulted stratigraphy are interpolated to the fault surfaces and combined with the restoration-derived displacement vectors to determine gouge for each point on the fault using:

$$\mathrm{SGR} = \frac{\sum_{i=1}^{I} (\Delta a f_{cl})_i}{\sum_{i=1}^{I} \Delta a_j} \qquad (3)$$

where Δa is a the area of a small, defined region i of the fault surface, f_{cl} is the shale fraction of the region and I is the total number of regions, representative of the hanging wall cut-off, that have been displaced past the current fault point. In this way, the true 3D displacement of the hanging wall and the variation in argillaceous content of the footwall stratigraphy over which the hanging wall has moved, are incorporated into the prediction of gouge.

Hydrocarbon migration modelling: invasion percolation

The invasion percolation (IP) approach to fluid flow has received much attention in applied physics (Wilkinson & Willemsen 1983; Meakin *et al.* 1992; Wagner *et al.* 1997; Schwarzer *et al.* 1999) and, more recently, it has been applied to hydrocarbon flow through geological structures (Paterson *et al.* 1996; Meakin *et al.* 2000; King *et al.* 2001; Carruthers 2003).

The technique is an iterative logical selection method that is orders of magnitude faster than numerical calculation (on the scale required for fluid flow simulation). Hence, in general, IP flow modelling is faster than ray-tracing and much faster than Darcy flow-based techniques. However, classical IP methods cannot predict temperature, pressure or associated properties within the flow regime, although advanced methods do facilitate secondary derivation of these properties.

IP theory is based on the concept of displacement of one fluid from a porous medium by another (Wilkinson & Willemsen 1983). The two fluids are immiscible and one fluid is assumed to show a greater affinity for the solid matrix and preferentially spread over the solid surface of the porous medium. This fluid is termed *wetting*; the other fluid is termed *non-wetting*. By considering the porous medium as a network of pores connected by narrower throats, capillary forces will be strongest at the throats and the fluid interface will move fastest through such points. Migration can be represented as a series of discrete ‘jumps’ in which one fluid displaces the other from the smallest available pore on the interface boundary. Accordingly, the principle of IP is that ‘the interface between the two fluids will advance at the point of least resistance’. The general terms *invading fluid* and *defending fluid* are employed to describe the displacing and displaced fluids respectively (Meakin *et al.* 2000).

In an IP model, the porous medium is approximated by a regular, homogeneous, 3D matrix of ‘sites’ (representing the pores) with ‘bonds’ (representing pore throats) between them (Fig. 4a). To introduce heterogeneity, the sites and/or bonds are assigned numerical values representing their *ease of invasion*. At each iterative step, the invading fluid displaces the defending fluid from the *most easily invaded* pore space on the boundary of the invading volume. *Ease of invasion* is defined by the properties of the porous medium and the invading fluid. It is assumed that fluid migration is controlled by capillary forces (Wilkinson & Willemsen 1983). In hydrocarbon migration, the fluids have differing densities and the porous medium is heterogeneous and typically extremely anisotropic.

The effects of density differences between hydrocarbons and water can be incorporated into the invasion percolation algorithm by overprinting the ease of invasion of each site or bond with a depth-dependent factor (i.e., a gravity gradient). This concentrates the flow of the invading fluid into a *finger* or *string* within the porous medium and, with increasing gravity gradient, this string becomes more confined and concentrated (Meakin *et al.* 2000). Large-scale heterogeneities within a porous medium, such as those representing layering or structure, are represented by contrasts of ease of invasion

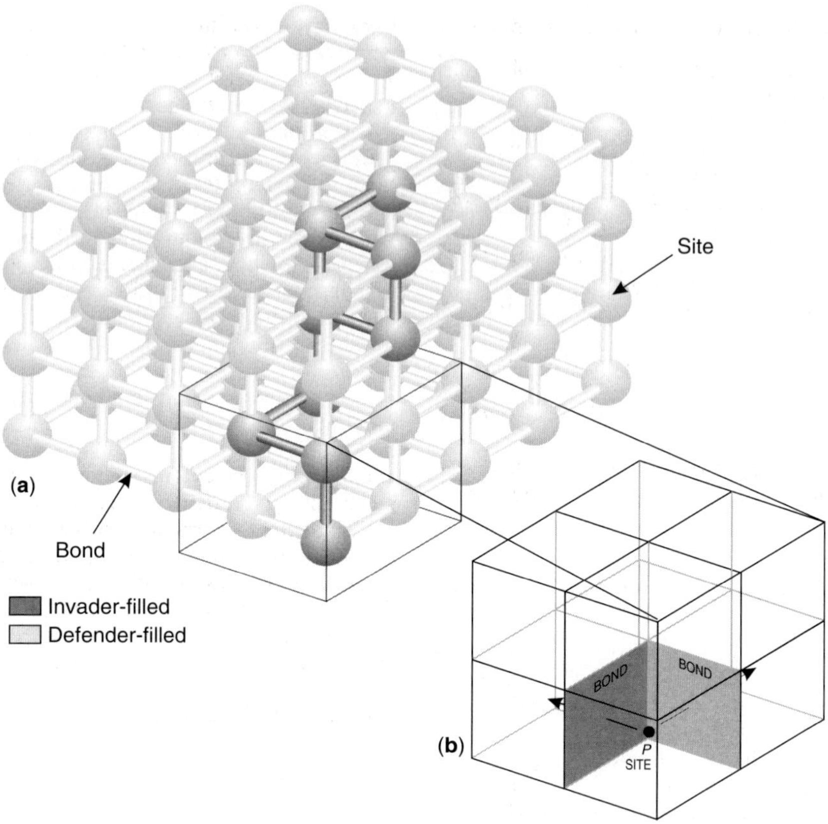

Fig. 4. The principles of invasion percolation. (**a**) Rock volume can be represented as an interconnected lattice of sites (pores) with bonds (pore throats) between them. The invader fluid follows the route of least resistance through the model. (**b**) In practice an exact model of pores of the rock volume is not possible and the lattice is approximated to a grid of cells with cell volumes representing the sites and cell interfaces representing the bonds.

between neighbouring sites or bonds. Such heterogeneities have a controlling influence on the large-scale pattern and pathways of the invading fluid.

In an ideal case, the site/bond matrix is an exact representation of the pores and connecting pore throats of the porous medium that it represents (Fig. 4a) and the value of ease of invasion is simply the size (volume) of each pore and each pore throat. In practice, the porous medium is approximated to a geocellular lattice in which sites and bonds are represented by cell volumes and cell interfaces respectively (Fig. 4b). This approach assumes that the size of the cells adequately reflects the heterogeneity of the porous medium. The low computational load of IP methods has the advantage of allowing many more cells than Darcy flow-based models and therefore a higher resolution model. Permeability (or capillary entry pressure) values are assigned to each cell volume. Values can be deterministically interpolated from available well data, using recognized spatial interpolation algorithms, or probabilistically populated using probability density functions if a comprehensive database is not available. Given the statistical nature of IP, the exact magnitude of permeability for a given cell is not significant. The migration pathways are controlled by the *relative* distribution of permeabilities throughout the model, and relative, dimensionless measures may be used.

In the simulations presented here the two fluids of interest are water and a single hydrocarbon phase. Multiphase flow is not considered since it presents additional theoretical considerations and increases the processing time. Water is always considered the defender and the hydrocarbon phase the invader fluid.

The IP model makes a number of assumptions:

(1) Both fluids are incompressible and homogeneous.
(2) The defending fluid always has a means of escape (withdrawal) when faced with the advance of the invader.
(3) There is an unlimited and constant supply of invading fluid into the model.
(4) Each site is assumed filled to capacity by either phase exclusively.
(5) Geological time is not integrated into the model. The number of iterations over which each realization is simulated is determined by the point at which hydrocarbons migrate out of one edge of the flow domain.

The ease of invasion from one cell to its neighbours is determined from their permeabilities using:

$$T_{AB} = \frac{2}{(1/k_A + 1/k_B)} \quad (4)$$

where k_A is the permeability of cell A, k_B is the permeability of neighbouring cell B, and l is the distance between cell centres (Manzocci *et al.* 1999).

In settings where fault-zone gouge is predicted, the permeability of this material is derived from the predicted SGR (Manzocci *et al.* 1999) and used to populate the volume nodes immediately adjacent to fault surfaces representing the fault zone (Fig. 5). Transmissibility calculations (Eqn. 4) between these nodes and those representing the faulted blocks incorporate both the effects of cross-fault juxtaposition and cross-fault permeability reduction resulting from fault-zone gouge into the transmissibility distribution of the model (Fig. 5).

Populating the model with data

Rock property values are interpolated to the surfaces and to the nodes, representing the geocellular model from high-resolution well data pertaining to wells that penetrate the fault blocks (Fig. 6). The geometry of the surfaces constrains interpolation of values to the geocellular model (Clarke *et al.* 2005*b*). In many simulations, particularly those that are data-poor, rock properties values are assigned from 'standard' ranges or estimated in bulk for a given sequence. IP-driven flow simulation has a significant advantage over more traditional methods in these scenarios. Pathway determination by the easiest route (ease of invasion) relies on relative values rather than absolute quantities. This reduces the error inherent in estimating bulk permeabilities. However, the IP methodology makes the assumption that an 'easiest route' exists. In scenarios where permeability values are estimated in bulk, a given geological unit will have homogeneous permeability, and the easiest route will be controlled by the algorithm rather than by the geology. Homogeneous units are conditioned randomly, with a normal distribution of data with a mean and modal value equal to the homogeneity and a variation over three standard deviations given by:

$$f(x) = \frac{1}{\sigma\sqrt{2\pi}} \exp\left[-\frac{(x-\mu)^2}{2\sigma^2}\right] \quad (5)$$

where σ^2 is variance, σ is standard deviation, μ is mean and $(\mu - 3\sigma) < x < (\mu + 3\sigma)$.

This technique produces a heterogeneous but isotropic unit. On a unit-wide scale, flow pathways will progress evenly in all lateral directions. The value of standard deviation can be employed to express uncertainty in the homogeneity of the unit. Greater heterogeneity can be introduced by incorporating seismic reflection data and attribute volumes into the model to

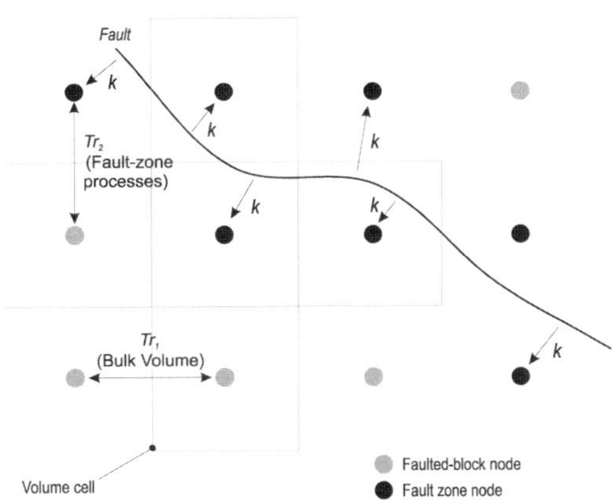

Fig. 5. The surfaces of the model divide up the volume cells (and nodes that represent them) into specific blocks. Those cells that are immediately adjacent to fault surfaces are used to represent the fault zone and populated with permeability values (k) derived from properties of the fault zone (such as gouge) predicted on the fault surfaces themselves. Transmissibility calculations between nodes of fault-block volume (Tr_1) incorporate heterogeneities of the fault blocks whereas transmissibility calculations between these nodes and fault-zone nodes (Tr_2) incorporate the effects of cross-fault juxtaposition and fault-zone properties into the flow model.

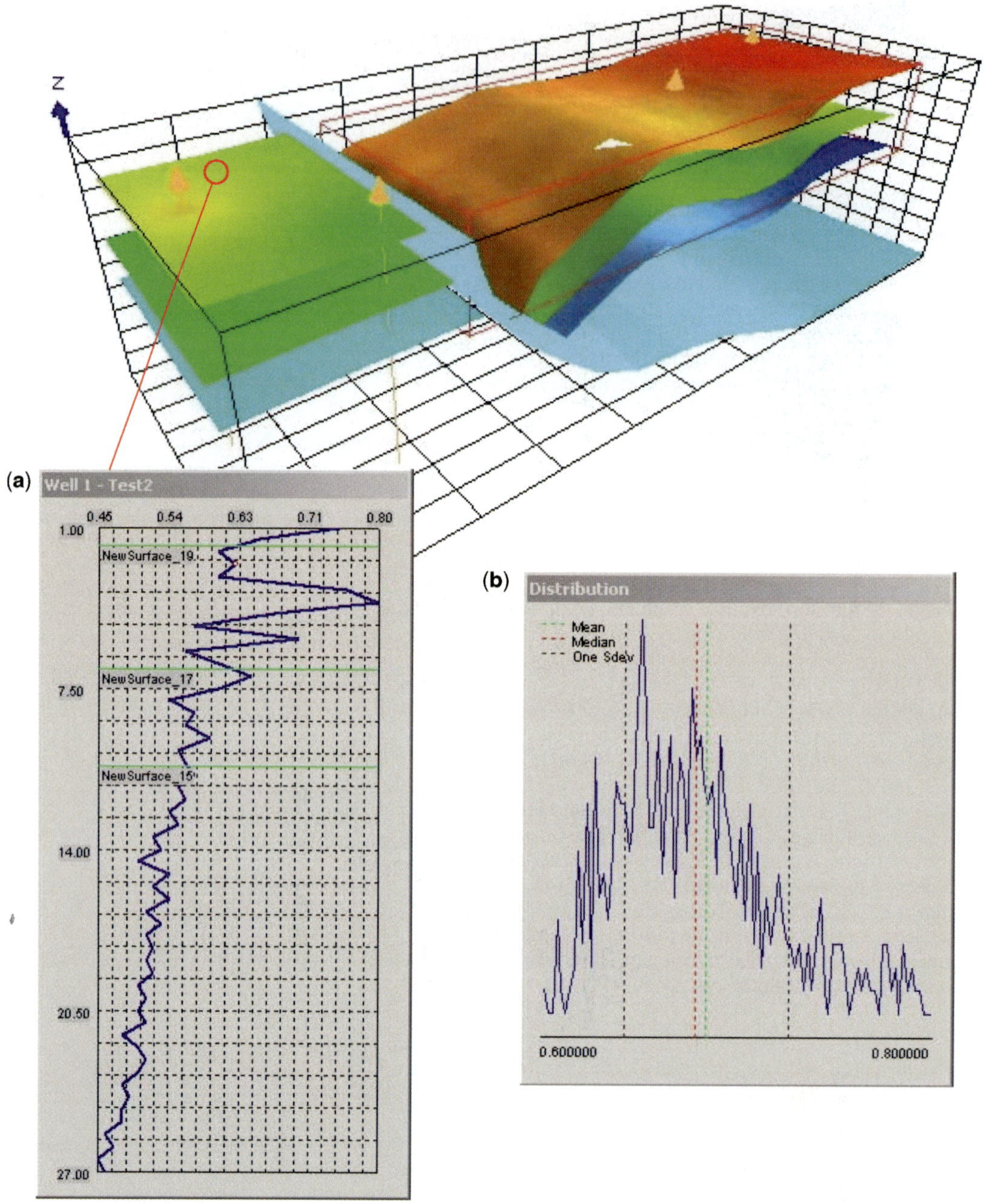

Fig. 6. Populating the model with data. (**a**) Detailed well logs are used to populate data to the modelled surfaces (shown) and volume, where such data are available. (**b**) In the absence of such data, statistical distributions may be used. Data interpolation is constrained by the geometry of the surfaces in 3D. Here, the surface colours show distribution of data interpolated from the wells shown.

'condition' the flow pathways. Finally, volume node permeability values are conditioned further by a depth-dependent buoyancy factor to facilitate upward migration within units.

The combined structural, fault seal and IP-driven hydrocarbon migration modelling approach is demonstrated using a series of computer generated synthetic 3D models (Figs 7 and 8).

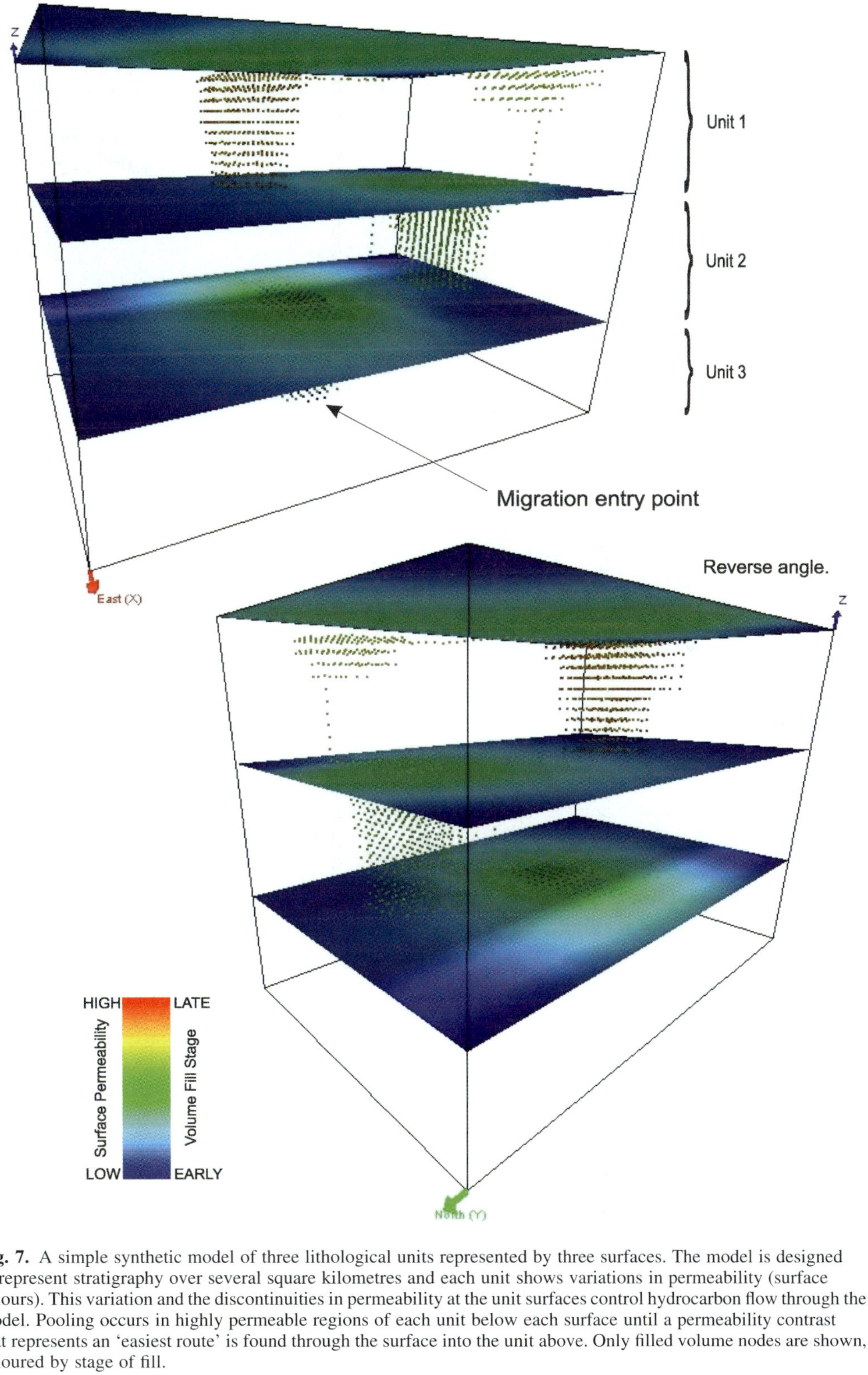

Fig. 7. A simple synthetic model of three lithological units represented by three surfaces. The model is designed to represent stratigraphy over several square kilometres and each unit shows variations in permeability (surface colours). This variation and the discontinuities in permeability at the unit surfaces control hydrocarbon flow through the model. Pooling occurs in highly permeable regions of each unit below each surface until a permeability contrast that represents an 'easiest route' is found through the surface into the unit above. Only filled volume nodes are shown, coloured by stage of fill.

The basic synthetic model contains three lithological units, the contacts between which are represented by three simple surfaces (Fig. 7) that represent major stratigraphical discontinuities. Synthetic values for permeability have been interpolated to the surfaces and to the volume nodes of the model, constrained by the position and geometry of the surfaces (see Fig. 6). Migration pathways from a defined migration entry point at the base of the model are controlled by both the permeability distribution within units and the discontinuities at the surfaces that bound them. Pooling of hydrocarbons takes place beneath each surface until a low permeability pathway is encountered that allows migration through to the overlying unit. Structural complexity is introduced in the form of a major normal fault and associated smaller synthetic fault (Fig. 8). Additionally, seven lithological units with varying permeabilites and shale content are represented by surfaces that bound them. The structural geometry constrains the interpolation of permeability from synthetic wells that penetrate the model to both the surfaces and cells. In Figure 8a migration pathways are predicted through the model based on the juxtaposition of permeabilities across the fault and demonstrate flow pathways up through the stratigraphy as a result of cross-fault migration in regions of juxtaposed high permeability units. Accumulation is predicted in the synthetic fault block and in a highly permeable unit at the top of the model. Figure 8b incorporates fault-zone gouge and demonstrates that the synthetic fault block and highly permeable unit at the top of the model are now sealed off from the flow pathway as a result of this gouge.

Examples of integrated 4D structural modelling with fluid flow simulation

To demonstrate the effectiveness of the integrated methods presented here, the importance of structural geometries in hydrocarbon migration and significance of an integrated approach in risk-driven analysis, two models have been constructed from actual geological data. One model represents part of the Moab Fault, Utah, USA constructed primarily from field data, while the second represents a faulted Rotleigendes prospect from the southern North Sea, constructed from the interpretation of 3D seismic data to produce horizon and fault surfaces.

The model of the Moab Fault (Fig. 9) covers approximately 140 km^2 of the Paradox Basin, southeastern Utah, USA, includes 22 km of the Moab Fault and associated splay faults, and is based on detailed outcrop mapping. The gently folded succession cut by the Moab Fault is represented by contrasting lithological units of clean sands and consolidated shales. The model consists of 14 surfaces in all fault blocks of the Moab Fault and associated splay faults. For clarity, only representative surfaces are shown in Figure 9.

The model from the southern North Sea is an exploration prospect, covering approximately 10 km^2, in which there is significant uncertainty regarding the petrophysical properties of parts of the stratigraphical sequence associated with the main reservoir. Two surfaces are modelled: the top Rotleigendes reservoir and the top of the overlying Zechstein Formation (Fig. 10). One major fault is shown in the model, down-throwing to the west, with two minor faults forming a small antithetic fault block in the hanging wall.

Fault-controlled secondary migration pathways

Field studies of the Moab Fault and associated faulted sequences indicate that the Moab Anticline and footwall closures to the Moab Fault are breached and subsequently exhumed hydrocarbon reservoirs (Garden *et al.* 2001). The widespread iron oxide reduction of sandstones is widely inferred to indicate the passage of a reducing hydrocarbon-related fluid (Chan *et al.* 2000; Garden *et al.* 2001).

1D, 2D and 3D analyses of the present day fault-seal capacity along the Moab Fault demonstrate that juxtaposition of lithologies across the fault and its associated splays provides an effective seal to hydrocarbons in reservoir sequences in both the footwall and hanging wall (Foxford *et al.* 1998; Clarke *et al.* 2005*a*). Furthermore, from study of the argillaceous fraction of the Moab Fault zone, Foxford *et al.* (1998) concluded that much of the fault zone, particularly the area around the Arches Anticline, contains sufficient argillaceous material (>20%) to provide a seal for hydrocarbons. Further north, as displacement decreases, insufficient gouge exists and potential sand-on-sand leak points indicated by juxtaposition are still present, especially in the vicinity of splay faults.

To demonstrate the effect of this complex fault zone geometry on present day and palaeomigration pathways around the Moab Fault, cross-fault juxtaposition and 3D SGR have been included in the model and bulk permeability values have been assigned to each unit,

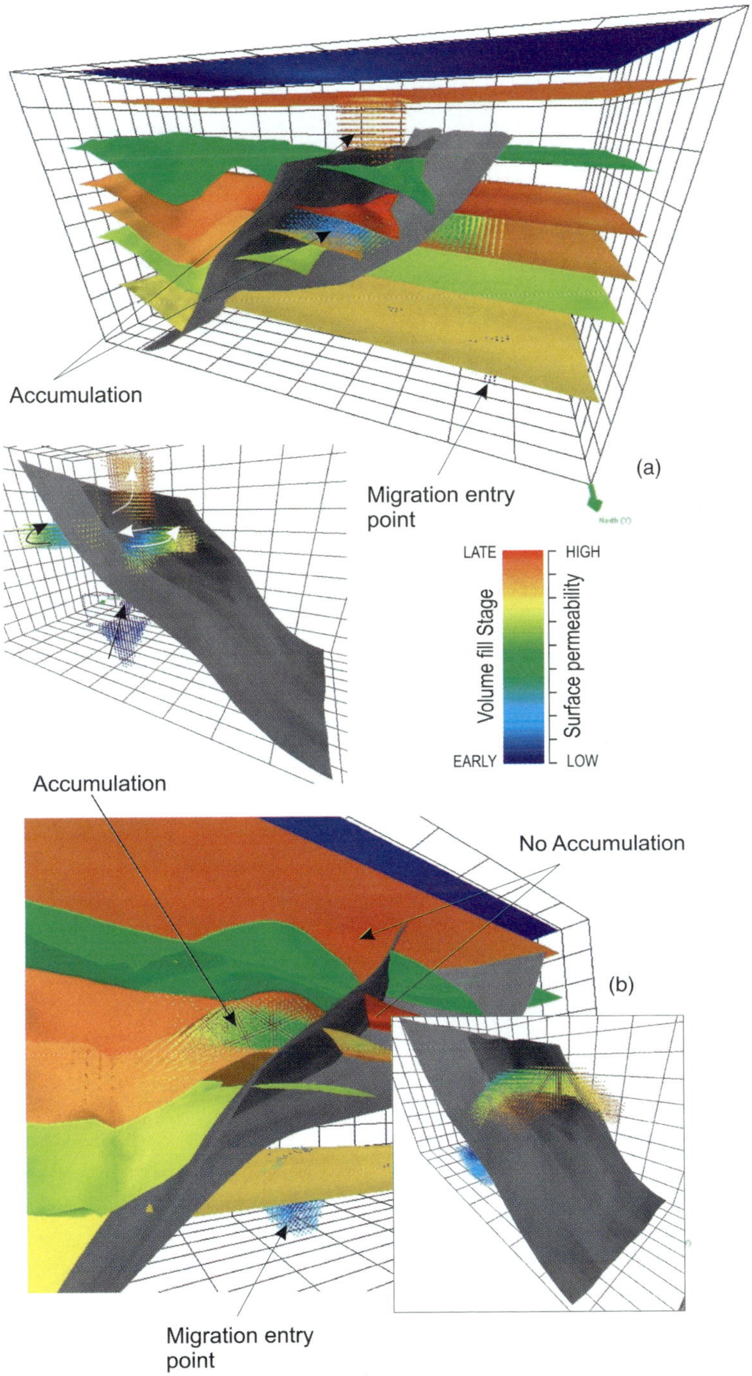

Fig. 8. A simple, synthetic model representing several square kilometres of seven lithological units and two faults. (**a**) Flow pathways are controlled by permeability variations within and between lithological units and juxtapositions across the faults. Migration through the succession is by repeated cross-fault and up-dip flow. Pooling occurs in one unit of the synthetic fault block and in the second highest unit. (**b**) Synthetic shale contents, and restoration-derived displacement vectors are used to calculate 3D shale gouge ratio and the effects of this are included in the flow model. Migration pathways bypass the synthetic fault block trap, which is sealed by fault zone gouge.

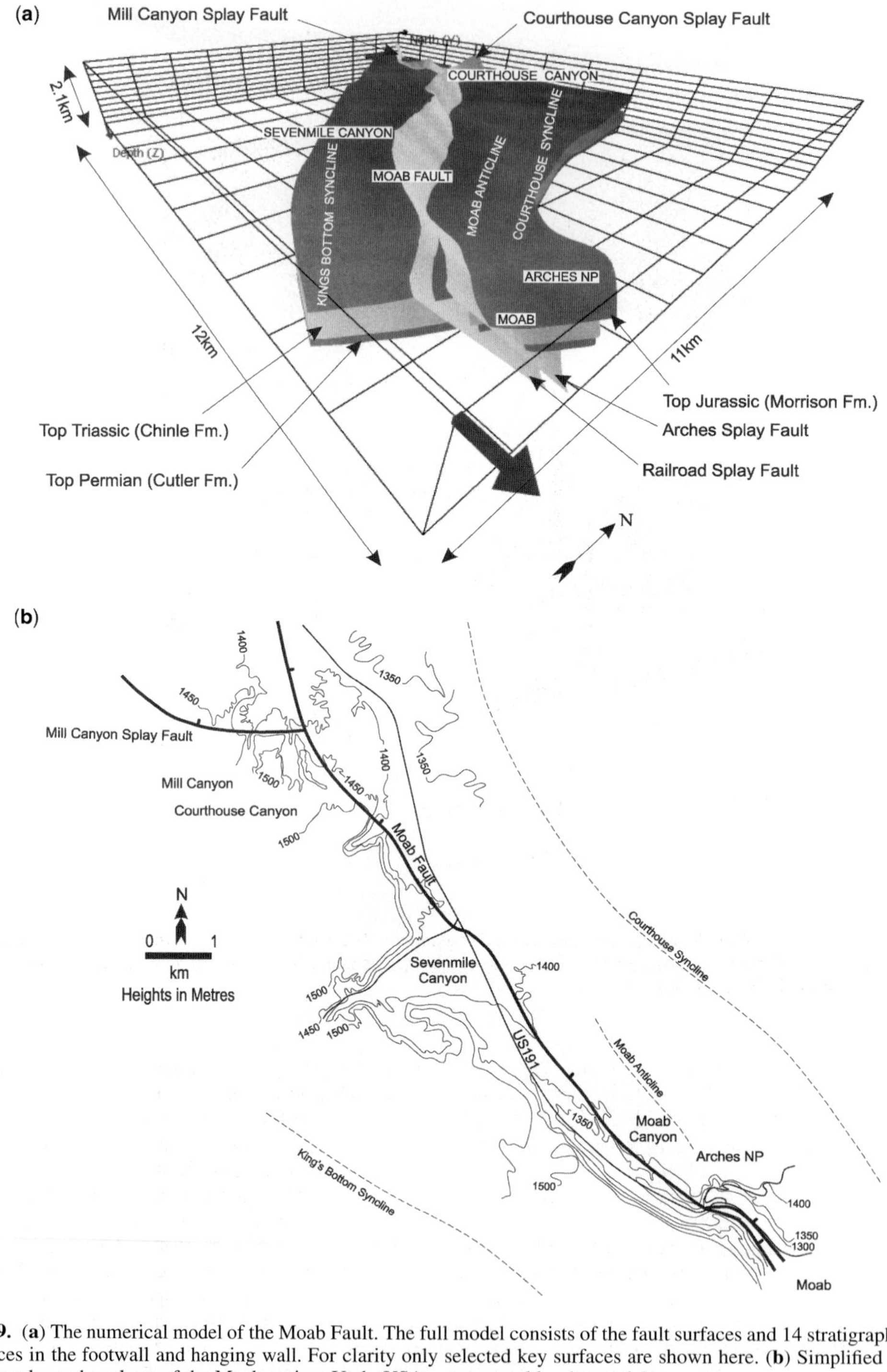

Fig. 9. (**a**) The numerical model of the Moab Fault. The full model consists of the fault surfaces and 14 stratigraphical surfaces in the footwall and hanging wall. For clarity only selected key surfaces are shown here. (**b**) Simplified topography and geology of the Moab region, Utah, USA, represented by the model.

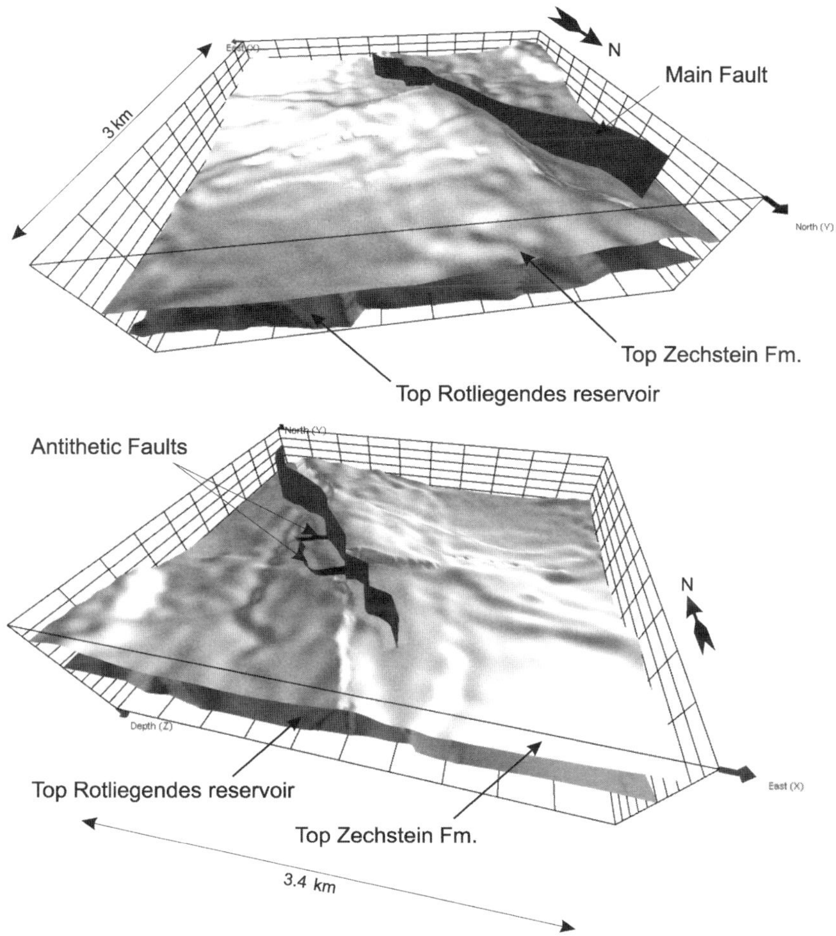

Fig. 10. The numerical model of the southern North Sea example. The model consists of two surfaces, representing the top of the Roteigendes reservoir and the top of the overlying Zechstein (seal) Formation. One major fault and two minor antithetic faults cut the model.

conditioned by normal distributions with standard deviations that reflect their heterogeneity. A series of realizations for both the present day and palaeostructural configurations (at 30 and 70% restoration) are modelled with different hydrocarbon migration entry points for each scenario.

The first series of present day realizations demonstrates migration pathways through the Moab Fault model in the absence of a fault zone, with migration pathways being controlled by cross-fault juxtapositions alone (Fig. 11). In these scenarios, flow pathways from a northeastern migration entry point encounter juxtaposed porous reservoir units in the region of Courthouse Canyon and leak across the fault towards the footwall high (Fig. 11a). Migration windows though the westerly splay are present in the absence of gouge and lead to an accumulation within the northern fault block (Fig. 11b). Choice of alternative migration entry points as an additional variable, results in flow pathways exploiting additional juxtaposed leakage windows. Figure 11c illustrates flow pathways from a migration entry point defined in the hanging wall around Sevenmile Canyon. The Moab Anticline trap spills to the south, and flow pathways develop across the fault to the footwall via juxtaposed reservoir units across the two synthetic splays.

For the second series of migration realizations based on present day structural geometries (Fig. 12), 3D shale gouge is incorporated into the model. Piercing-point restoration of the

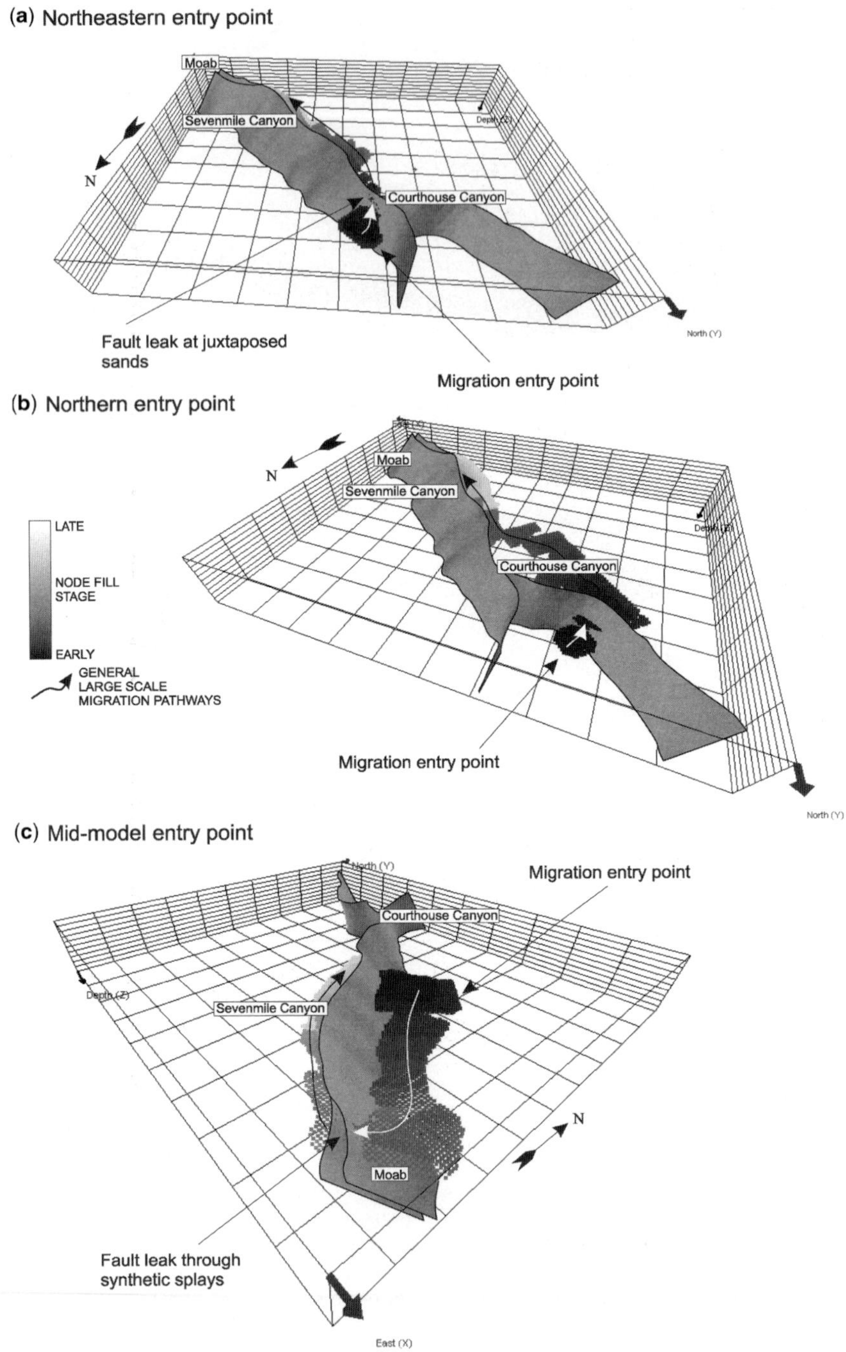

Fig. 11. Predicted migration pathways through the Moab Fault model controlled by cross-fault juxtaposition and present day structural geometries. (**a**) From a northeasterly migration entry point, juxtaposed porous sandstones at Courthouse Canyon provide migration pathways across the fault. (**b**) From a northern migration entry point pathways exist to the footwall across the westerly splay, and (**c**) from a migration entry point in the hanging wall, around Sevenmile Canyon, flow pathways exist to the south and across the fault via juxtaposed porous sandstones at the southern fault splays. Arrows show general trends of migration pathways.

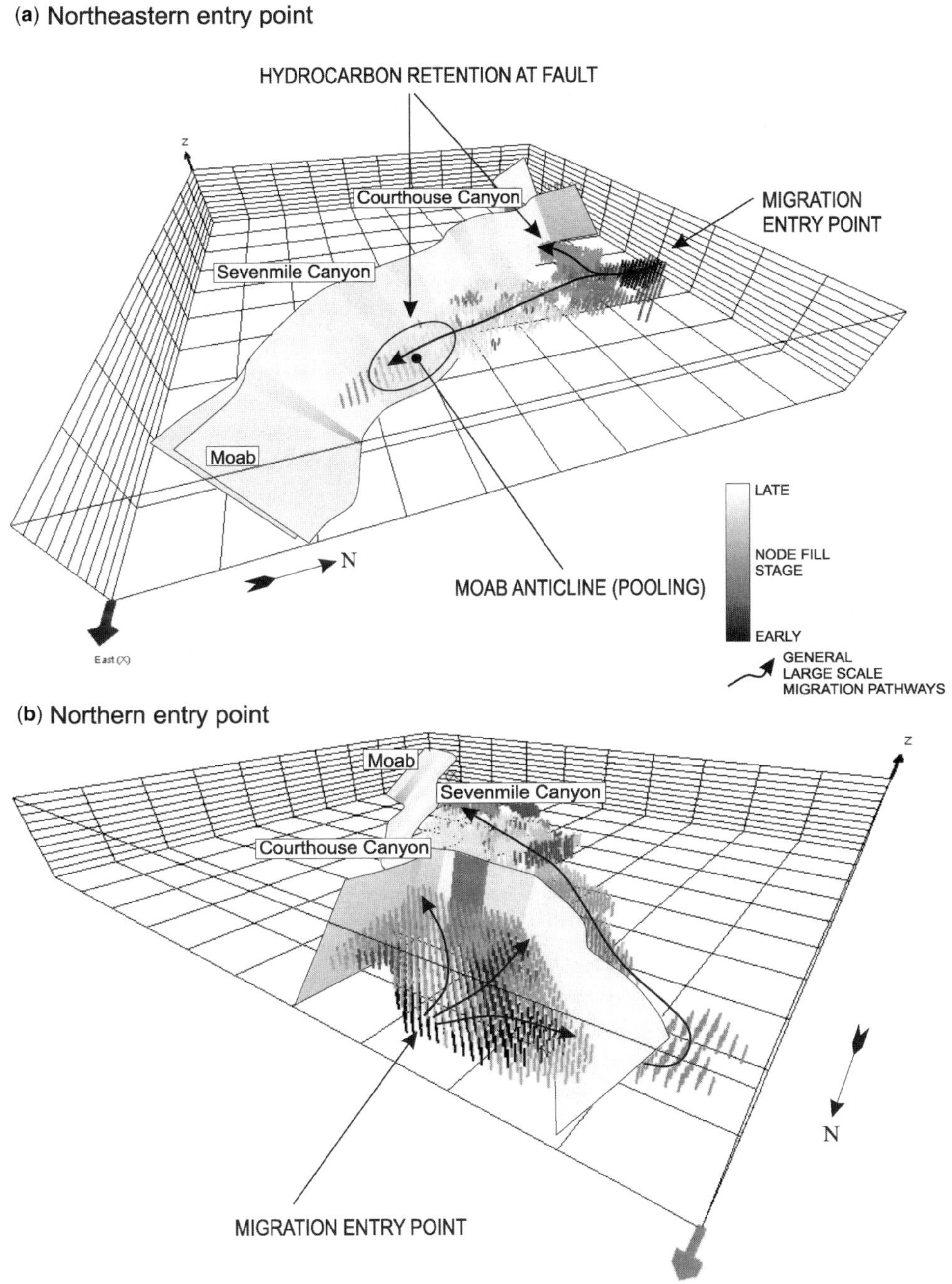

Fig. 12. Predicted migration pathways through the Moab Fault model for present day structural geometries and in the presence of a fault-zone gouge. (**a**) From a northeasterly migration entry point, migration pathways exist up-strata and up-dip to the structural high in the Moab Anticline, sealed at the fault by juxtaposition relationships and argillaceous gouge. (**b**) From a northerly migration entry point, the westerly splay of the Moab Fault still leaks and provides migration pathways to the structural high in the footwall around Sevenmile Canyon. Arrows show general trends of migration pathways.

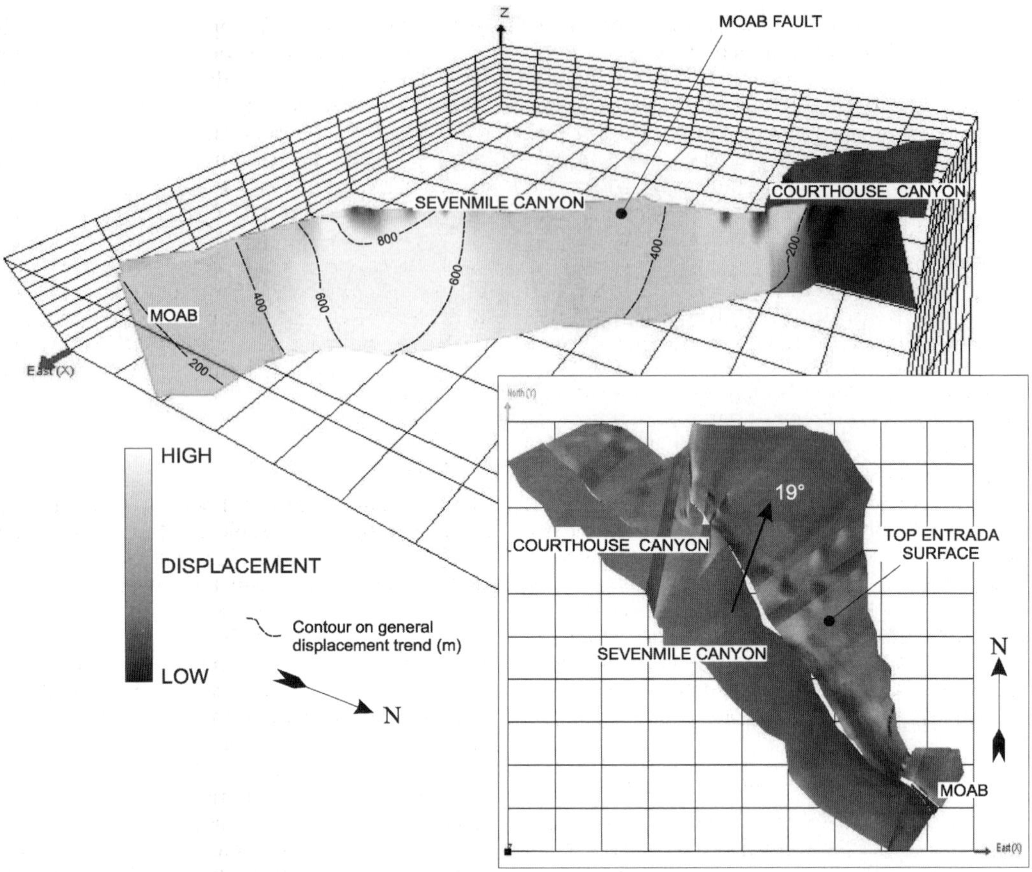

Fig. 13. (**a**) Predicted displacement gradient on the Moab Fault resulting from best-fit piercing point restoration. (**b**) Restored top Jurassic (Entrada Formation) surface using the piercing point-derived displacement gradient and predicted slip azimuth of 19°.

hanging wall (Fig. 13) indicates a slip azimuth of 19° and a displacement gradient that varies from a maximum around the footwall high at Sevenmile Canyon, reducing both northwards and southwards. Using these displacement vectors, the vertical shear chevron construction and the argillaceous contents of the faulted stratigraphy, the 3D distribution of argillaceous material within the fault zone has been predicted. From a northeastern entry point, an accumulation develops in the Moab Anticline, sealed to the east by the Moab Fault (Fig. 12a). This is consistent with field evidence (Garden *et al.* 2001), suggesting that the northeastern splay of the northern section of the Moab Fault, and the main fault around Sevenmile Canyon, have the capacity to seal cross-fault migration from a northeastern source and prevent flow into the footwall. To investigate whether the western splay of the fault has the capacity to seal cross-fault migration in the presence of gouge, another present day realization (Fig. 12b) was initialized from a northern migration entry point, within the fault block bounded by the two splays. Accumulation now takes place within the fault block until a leak point in the westerly splay is encountered. Migration progresses updip of the footwall to accumulate in the footwall high around Sevenmile Canyon, also consistent with field evidence of extensive iron oxide reduction. The presence of fault-zone gouge on the westerly splay is insufficient to cause total sealing of the fault.

In these two simulation scenarios flow pathways are controlled by the combined effects of cross-fault juxtapositions, fault zone properties and the structural geometries of the present day fault architecture. However, migration may have been contemporary with faulting, an uncertainty that can be investigated with structural restoration. Using the derived displacement vectors, the hanging wall has been restored by 30% and 70% and potential flow pathways resulting from the interaction of restored palaeogeometries and cross-fault juxtaposition explored. A 30% restoration (Fig. 14a) is insufficient to juxtapose reservoir units around the footwall high and migration from a hanging wall entry point follows flow pathways defined by regional structural geometry (largely as described in Fig. 11c). However, at 70% restoration (Fig. 14b) juxtaposition of porous reservoir units across the fault in the region of the footwall high creates cross-fault pathways through the fault. Accumulations form in the hanging wall and leak to the footwall, eventually spilling back to the hanging wall to the south. The resulting migration pathways and accumulations are dramatically different from model realizations at full present day deformation, but in this case, not fully consistent with the migration pathways and accumulations inferred from field data (Chan *et al.* 2000; Garden *et al.* 2001). Moreover, the charging histories of the modelled accumulations are reversed from those predicted with both present day geometries and geometries predicted from lesser amounts of restoration.

Fault-controlled field-scale migration

The exploration prospect model from the southern North Sea covers a much smaller area than the Moab Fault model. At this smaller scale, migration pathways are representative of intra-reservoir migration, charging histories and fault block compartmentalization. In this setting, integrated fault seal and hydrocarbon flow modelling can be used to risk specific accumulations within individual fault compartments in the light of uncertainties in the input data.

The geometry of the southern North Sea prospect suggests a potential trap in the footwall of the major fault, sealed by juxtaposition of the Rotliegend reservoir against the overlying Zechstein sequence across the fault. However, there is uncertainty concerning the extent of fracturing (and therefore the permeability) of the Zechstein sequence that is likely to affect the location and volume of hydrocarbon accumulations. Two realizations, integrating structural modelling, fault-seal analysis and hydrocarbon migration pathway prediction have been performed, with the Zechstein defined as either completely impermeable or completely permeable. The sealing capacity of the major fault is different in each scenario and accumulations of different sizes and locations are predicted for the two scenarios. With the unit defined as non-fractured (Fig. 15a), the footwall trap is sealed at the fault by lithological juxtaposition and accumulation takes place in the footwall high. With the unit defined as fractured (Fig. 15b), cross-fault flow results in a smaller footwall trap and additional hanging wall trap within the antithetic fault block. The speed and integrated nature of the approach presented can be exploited to further investigate this problem. In Figure 15c, 200 realizations, each with a different permeability (degree of fracturing) for the Zechstein unit, have been combined. Accumulations are ranked by the number of times each cell fills and therefore the risk associated with each trap location and volume.

Discussion

The realizations presented here of synthetic and actual geological example models demonstrate both the viability and significance of an integrated approach to basin modelling that incorporates structural geometry, fault seal and hydrocarbon migration. In examples using real geological data, it is apparent that migration pathways, trapped hydrocarbon accumulations and their dependence on the 3D sealing properties of faults can become extremely complex, and difficult to interpret. Potential traps are often a result of the interaction of 3D sealing capacity on the fault surface and the structural geometry of the faulted blocks local to the fault surface contemporary with migration. It is not simple to visually determine potential seal and leak points or trapping geometries from an analysis of a fault seal map or fault plane projection alone. The ability to visualize fault geometries in 3D increases the ease of interpretation and is a major advantage of the true 3D approach to fault seal and migration modelling.

Secondly, the inclusion of structural restoration and deformation within an integrated 4D model allows flow pathways to be predicted at temporal stages other than present day. These pathways may be significantly different to those of the present day in terms of fault linkages, leakage points and the geometry of carrier beds. Earlier structural geometries may be critical to both migration pathways and the distribution of trapped hydrocarbons. The ability to model flow pathways through geologically realistic palaeogeometries (at least for normal faults) is a significant advance on current modelling capabilities.

In the models presented here we have employed 3D structural restoration techniques to determine best-fit deformation vectors and predict subsequent hanging wall geometries

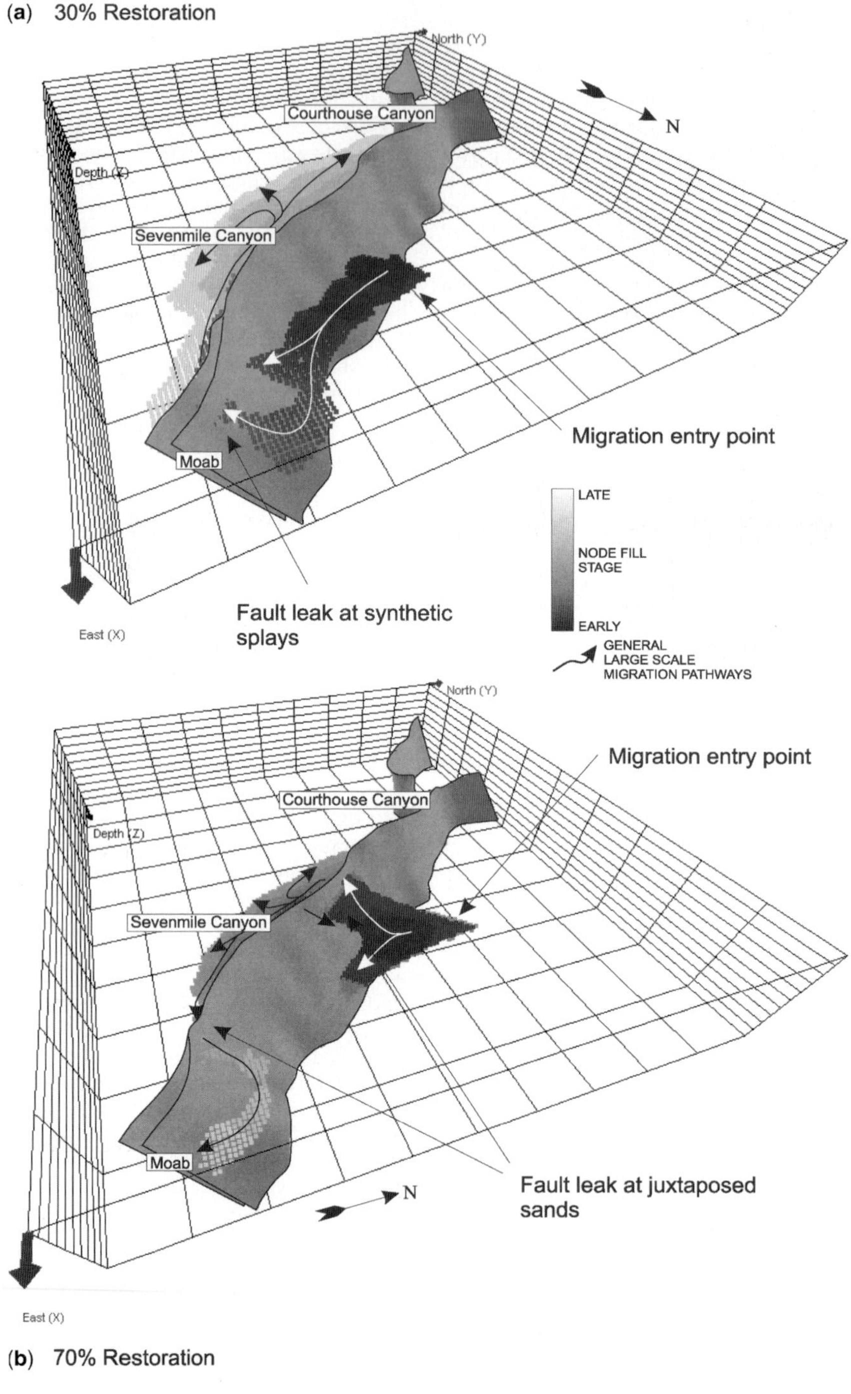

Fig. 14. Migration pathways through the Moab Fault at temporal stages representing (**a**) 30% restoration and (**b**) 70% restoration. At 30% restoration, pathways are similar to those at present day. At 70% restoration sandstones are juxtaposed across the fault and provide migration pathways from hanging wall to footwall and subsequently back to the hanging wall in the south of the model.

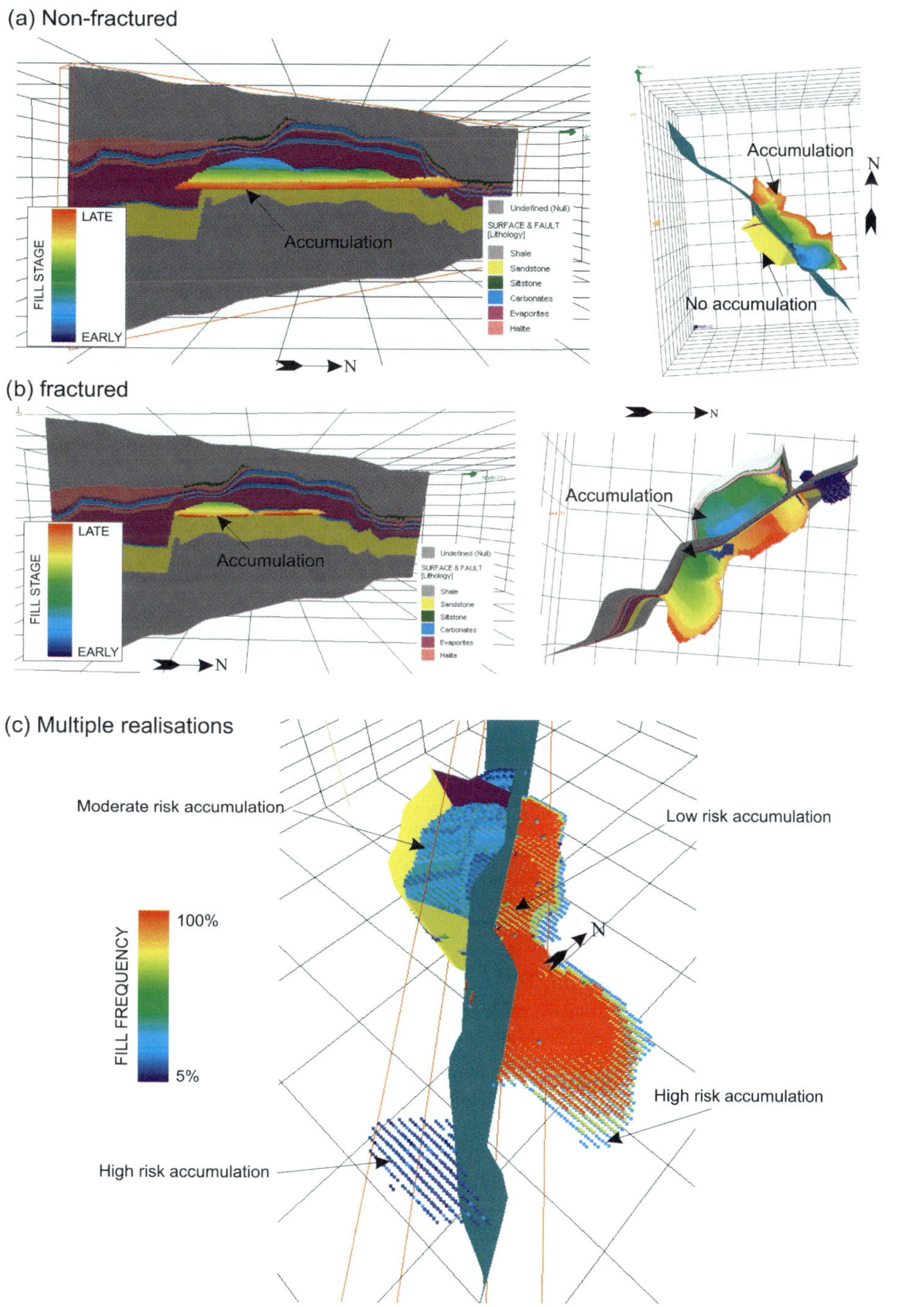

Fig. 15. Migration pathways through the southern North Sea model. (**a**) With the Zechstein defined as completely non-fractured, a large accumulation is predicted in the footwall. (**b**) With the Zechstein defined as completely fractured, cross-fault migration pathways exist to the small antithetic fault block and result in a smaller accumulation in the footwall and an accumulation in the antithetic fault block. (**c**) Comparison of 200 realizations of the southern North Sea model. In each realization the permeability of the Zechstein is slightly different. The resulting accumulations are coloured by their frequency of fill (likelihood of occurrence) and allow risking of the trap volumes and locations.

using fault-bend fold methodologies (chevron/inclined shear constructions). A third key point of the 4D integrated hydrocarbon flow modelling methodology presented is that it is independent of the structural restoration technique employed. Chevron and inclined shear constructions are particularly useful because they generate predictable hanging wall geometries that correlate well with observations of extensional regimes, and they also preserve rock volume with deformation. However, they rely on well-defined interpretations of structural geometry and they do not reproduce internal mechanisms that produce hanging wall deformation nor model subsequent strain effectively (Williams & Chapman 1983). These processes can affect the petrophysical properties of the rock mass and thus subsequent migration pathways. Furthermore, fault-bend-fold methodologies assume a static footwall and do not model isostatic changes in fault block geometry. It follows that the 3D restoration techniques employed here are applicable on the prospect, trap or similarly restricted spatial exploration scale, where isostatically-driven geometrical changes can be largely ignored, but at the larger basin scales (i.e., hundreds of km^2) these effects will be significant and are likely to impact migration pathways (Huggins *et al.* 2006).

Fourthly, by integrating fault seal and structural geology with a numerically fast migration modelling code in 4D, it is possible to test assumptions and uncertainties in the geological data or concept that is being developed. In the Moab Fault model presented here, for example, the presence or absence of fault zone gouge, the timing of migration with respect to deformation, and the assumed migration entry point significantly effect migration pathways. In the southern North Sea prospect, uncertainties related to the properties of the stratigraphical succession have a significant effect on flow and accumulations. The IP-driven approach to hydrocarbon migration through faulted structures is ideal for testing assumptions and uncertainties such as these. The Moab Fault model contains 950,000 volume cells and runs in *c.* 3 min whilst the southern North Sea model contains 5,000,000 cells and runs in *c.* 7 min. These short simulation times render risk-driven modelling approaches to hydrocarbon migration analysis practical. Multiple realizations, each with varied inputs (either deterministic or stochastic) can be used to determine the sensitivity of particular traps to uncertainties in poorly constrained input data. 'What if?' questions can thus be posed to investigate potential simulation outcomes or examine the sensitivity of the model to specific input variables. Comparison of results can be used to risk particular scenarios, and thus exploration prospects. In the southern North Sea model for example, the uncertain permeability of one particular unit has a significant impact on the location and size of predicted accumulations. By performing 200 realizations, each with a different permeability for this unit, it is possible to combine results and risk not only the location but also the potential size of each predicted trap. The combined realizations on this model indicate that a significant proportion of the volume of the footwall trap is very high risk as it depends on the Zechstein being completely impermeable (not fractured in this case); only the topmost fraction is low risk. By contrast, the antithetic fault block trap is moderate risk, as its occurrence and volume is largely independent of the permeable state of the Zechstein unit and the accumulation is only absent if the Zechstein is completely impermeable.

We have demonstrated the integrated 3D application of two aspects of fault seal: juxtaposition and argillaceous gouge. These are by no means the only seal characteristics of faults. Both factors are commonly the dominant sealing mechanisms, but other factors such as diagenetic alteration may contribute to fault seal. These factors have not been simulated in the examples shown here. Additionally, the sealing capacity of argillaceous gouge, predicted from SGR, does not account for entrained blocks within the fault zone and is likely to be affected by case-specific geological factors (e.g., host rock mineralogy and confining pressure). The statistical nature of the IP code implemented here does not facilitate modelling of these factors at present. Furthermore, the methods currently developed may not be appropriate for fault-seal analysis in high fluid pressure settings where mechanisms of seal failure and fluid migration are likely to be different (e.g., hydrocfracturing or seismic pumping/valving; Sibson 1981).

Conclusions

The key to incorporating structural restoration in basin simulators is to represent basin geometry as realistic structural and litho-stratigraphical surfaces in 3D space. As faults and horizon discontinuities (such as unconformities) are the major architectural components of a basin, these elements should form the initial building blocks of the basin model if the basin geometry through time is to be represented. The integrated approach to restoration of basin architecture and hydrocarbon migration simulation presented here provides a fast method of determining flow pathways through structurally complex, exploration prospect scale basins. It integrates 3D structural restoration techniques, 3D fault-seal analysis, the 3D geometry and variations in geology of a sedimentary basin, to predict potential seal and leak points at fault surfaces, highlighting potential trapping geometries in faulted sequences. This approach increases the resolution at which migration pathways can be analysed, and reduces the inherent assumptions in applying these techniques in isolation or with fewer dimensions (Harper & Lundin 1987). The integration of migration pathway prediction into the structural model provides a fast and effective tool for detecting and evaluating such seals, leak points and potential traps. Moreover, migration through restored palaeogeometries can be investigated.

The simplicity and thus speed of the technique makes it ideally suited to a risk-driven approach. Given a 3D model of the basin, structural geometries, fault seal and potential migration pathways and accumulations can be ascertained using this technique. The sensitivity of these potential traps to variations in geological parameters such as the stratigraphy (e.g., variations in sediment type), migration entry point locations, or petrophysical properties (e.g., clay content of sands, entry pressures or permeability) can be determined quickly from multiple realizations of the same model. 'What if?' questions can be posed to investigate potential simulation outcomes or examine the sensitivity of the model to specific input variables. Those potential traps or trap regions that show a high dependence on poorly constrained input parameters are high risk and can be discarded as potential prospects. In this way, the rapid simulation technique and integrated restoration-migration modelling approach presented forms a first pass to filter high-risk prospects prior to time-consuming Darcy flow-based simulation.

The authors wish to thank BG Group for permission to publish, and the Natural Environment Research Council and BG Group for funding this work (NERC grant reference GRC/98/280). S. Buiter, N. Cardozo and P. Nadeau all improved the manuscript with their valuable comments. The basin modelling, structural restoration, 3D fault seal and hydrocarbon migration simulations presented here were performed using software developed at Keele University. In addition to the authorship of this paper, the contributions of many other members of the School of Earth Sciences and Geography at Keele University, particularly P. Greatbatch and D. Wilde, are gratefully acknowledged.

References

ALLAN, U. S. 1989. Model for hydrocarbon migration and entrapment within faulted structures. *AAPG Bulletin*, **73**, 803–811.

BERG, R. R. 1975. Capillary pressures in stratigraphic traps, *AAPG Bulletin*, **59**, 939–956.

BOND, G. C. & KOMINZ, M. A. 1984. Construction of tectonic subsidence curves for the early Paleozoic miogeocline, southern Canadian Rocky Mountains: implications for subsidence mechanisms, age of breakup, and crustal thinning. *Bulletin of the Geological Society of America*, **95**, 155–173.

BURLEY, S. D., CLARKE, S.M. ET AL. 2000. New insights on petroleum migration from the application of 4D basin modelling in oil and gas exploration. *Journal of Geochemistry Exploration*, **70**, 465–470.

CARRUTHERS, D. J. 2003. Modelling of secondary petroleum migration using invasion percolation techniques. *In*: DUPPENBECKER, S. & MARZI, R. (eds) Multi-dimensional basin modelling. *AAPG Discovery Series*, **7**, chpt 3.

CHAN, M. A., PARRY, W. T. & BOWMAN, J. R. 2000. Diagenetic oxides and fault-related fluid flow in Jurassic sandstones, southeastern Utah. *AAPG Bulletin*, **84**, 1281–1310.

CLARKE, S. M., WILLIAMS, G. D. & BURLEY, S. D. 2005*a*. A three-dimensional approach to fault seal analysis: fault block juxtaposition and argillaceous smear modelling. *Basin Research*, **17**, 269–288.

CLARKE, S. M., BURLEY, S. D. & WILLIAMS, G. D. 2005*b*. Dynamic fault seal analysis and flow pathway modelling in three-dimensional basin models. *In*: DORE, A. G. & VINING, B. A. (eds) *Petroleum Geology: North-west Europe and Global Perspectives* (PGC6), Geological Society, London, Miscellaneous Titles.

COBBOLD, P. R. 1979. Removal of finite strain using strain trajectories. *Journal of Structural Geology*, **1**, 67–72.

COWAN, G., BURLEY, S. D. ET AL. 1999. Oil and gas migration in the Sherwood Sandstone of the East Irish Sea Basin. *In*: FLEET, A. J. BOLDY & S. A. R. (eds) *Petroleum Geology of North West Europe*: Proceedings of the 5th Conference of the Geological Society, London.

DAHLSTROM, C. D. 1969. Balanced cross sections. *Canadian Journal of Earth Science*, **6**, 743–757.

FOXFORD, K. A., GARDEN, I. R., GUSCOTT, S. C., BURLEY, S. D., LEWIS, J. J. M., WALSH, J. J. & WATTERSON, J. 1996. The Field Geology of the Moab Fault. *In*: HUFFMAN, A. C., LUND, W. R. & GODWIN, L. H. (eds) *Geology and Resources of the Paradox Basin*. Special Symposium, Utah Geological Association and Four Corners Geological Society Guidebook, **25**.

FOXFORD, K. A., WALSH, J. J., WATTERSON, J., GARDEN, I. R., GUSCOTT, S. C. & BURLEY, S. D. 1998. Structure and Content of the Moab Fault Zone, Utah, USA and its Implications for fault seal prediction. *In:* JONES, G., FISHER, Q. J. & KNIPE, R. J. (eds) *Faulting, Fault Sealing and Fluid Flow in Hydrocarbon Reservoirs*. Geological Society, London, Special Publications, **147**, 87–103.

FRISTAD, T., GROTH, A., YEILDING, G. & FREEMAN, B. 1997. Quantitative fault seal prediction: a case study from Oseberg Sud. *In*: MOLLER, P. &

KOESTLER, A. G. (eds) *Hydrocarbon Seals: Importance for Exploration and Production.* NPF, Special Publications, **7**, 107–124.

GARDEN, R. C., GUSCOTT, S. C., BURLEY, S. D., FOXFORD, K. A., WALSH, J. J. & MARSHALL, J. 2001. An exhumed palaeo-hydrocarbon migration fairway in the Entrada Sandstone of SE Utah, USA. *Geofluids*, **1**, 195–213.

GIBBS, A. D. 1983. Balanced cross sections from seismic sections in areas of extensional tectonics. *Journal of Structural Geology*, **5**, 153–160.

GIBSON, R. G. 1994. Fault zone seals in siliclastic strata of the Columbus Basin, offshore Trinidad. *AAPG Bulletin*, **78**, 1372–1385.

GILES, M. R., INDRELID, S. L. & JAMES, D. M. D. 1998. Compaction – the great unknown in basin modelling. *In*: DUPPENBECKER, S. J. & ILIFFE, J. E. (eds) *Basin Modelling: Practice and Progress*, Geological Society, London, Special Publications, **141**, 15–43.

GRATIER, J. P., GUILLIER, B. & DELORME, A. 1991. Restoration and balance of a folded and faulted surface by best-fitting of finite elements: principle and applications. *Journal of Structural Geology*, **13**, 111–115.

HANTSCHEL, T., KAUERAUF, A. I. & WYGRALA, B. 2000. Finite element analysis and ray trace modelling of petroleum migration. *Marine and Petroleum Geology*, **17**, 815–820.

HARPER, T. R. & LUDIN, E. R. 1997. Fault seal analysis: reducing our dependence on empiricism. *In*: MOLLER, P. & KOESTLER, A. G. (eds) *Hydrocarbon Seals: Importance for Exploration and Production.* NPF, Special Publications, **7**, 149–165.

HUGGINS, P. BURLEY, S. D., SYLTA, O., TOMMERAS, A., Bland, S., KAPE, S. & KUSZNIR, N. 2006. Structural restoration techniques in 3D basin modelling: implications for hydrocarbon migration and accumulation. *Marine and Petroleum Geology*. In press.

ILLFE, J. E. & DAWSON, M. R. 1996. Basin modeling history and predictions. *In*: GLENNIE, K. & HURST, A. (eds) *AD1995: NW Europe's Hydrocarbon Industry*. Geological Society, London, Special Publications, 83–105.

JOHANNESEN, J., HAY, S. J., MILNE, J. K., JEBSEN, C., GUNNESDAL, S. C. & VAYSSAIRE, A. 2002. 3D oil migration modelling of the Jurassic petroleum system of the Statfjord area, Norwegian North Sea. *Petroleum Geoscience*, **8**, 37–50.

KING, P. R., BULDYREV, S. V. *ET AL.* 2001. Predicting oil recovery using percolation theory. *Petroleum Geoscience*, **7**, 105–107.

KUSZNIR, N. J., ROBERTS, A. M. & MORLEY, C. 1995. Forward and reverse modelling of rift basin formation. *In*: LAMBAISE J. (ed.) *The Hydrocarbon Habitat of Rift Basins*, Geological Society, London, Special Publications, **88**, 3–56.

LARTER, S. P., TAYLOR, M., CHEN, B., BOWLER, P., RINGROSE, P. & HORSTAAD, I. 1996. Secondary migration – visualizing the invisible – what can geochemistry do? *In*: GLENNIE, K. & HURST, A. R. *NW Europe's Hydrocarbon Industry*, Geological Society, London, Special Publications, 137–143.

MANN, U., HANTSCHEL, T., SCHAEFER, R. G., KROOS, B., LEYTHAEUSER, D., LITTKE, R. & SACHSENHOFER, R. F. 1997. Petroleum migration: mechanisms, pathways, efficiencies and numerical simulations. *In*: WELTE, D. H., HORSFIELD, B. & BAKER, D. R. (eds) *Petroleum and Basin Evolution.* Springer, 403–509.

MANZOCCHI, T., WALSH, J. J., NELL, P. & YEILDING, G. 1999. Fault transmissibility multipliers for flow simulation models. *Petroleum Geoscience*, **5**, 53–63.

MCKENZIE, D. P. 1978. Some remarks on the development of sedimentary basins. *Earth Planetary Science Letters*, **40**, 25–32.

MEAKIN, P., FEDER, J., FRETTE, V. & JOSSANG, T. 1992. Invasion percolation in a destabilizing gradient. *Physical Reviews* A, **46**(6), 3357–3368.

MEAKIN, P., WAGNER, G., VEDVIK, A., AMUNDSEN, H., FEDER, J. & JOSSANG, T. 2000. Invasion percolation and secondary migration: experiments and simulations. *Marine and Petroleum Geology*, **17**, 777–795.

MELLO, U.T. & HENDERSON, M. E. 1997. Techniques for including large deformations associated with salt and fault motion in basin modeling. *Marine and Petroleum Geology*, **14**, 551–564.

MEREDTIH, D. J. & EGAN, S. S. 2002. The geological and geodynamic evolution of the Eastern Black Sea Basin: insights from 2-D and 3-D tectonic modelling. *Tectonophysics*, **350**, 157–179.

PATERSON, L., PAINTER, S., KNACKSTEDT, M. A. & PINCZEWSKI, W. V. 1996. Patterns of fluid flow in naturally heterogeneous rocks. *Physica A*, **233**, 619–628.

RICHARDS, A. J. 2000. *An integrated approach to three-dimensional modelling of sedimentary basins*. PhD. thesis, Keele University.

ROBERTS, A. M., KUSZNIR, N. J., YIELDING, G. & STYLES, P. 1998. 2D flexural backstripping of extensional basins: the need for a sideways glance. *Petroleum Geoscience*, **4**, 327–338.

ROUBY, D., XIAO, H. & SUPPE, J. 2000. 3D restoration of complexly folded and faulted surfaces using multiple unfolding mechanisms. *AAPG Bulletin*, **84**, 805–829.

SCHOWALTER, T. T. 1979. Mechanics of secondary hydrocarbon migration and entrapment. *AAPG Bulletin*, **63**, 723–760.

SCHROEDER, F. W. & SYLTA, O. 1993. Modelling the hydrocarbon system of the North Viking Grabe. *In*: DORE *ET AL.* (eds) *Basin Modelling, Advances and Applications*, NPF Special Publication, **3**, 469–484.

SCHWARZER, S., HAVLIN, S. & BUNDE, A., 1999. Structural properties of invasion percolation with and without trapping: shortest path and distributions. *Physical Review*, E, **59**, 3262–3269.

SIBSON, R. H. 1981. Fluid flow accompanying faulting: field evidence and models. *In*: SIMPSON, D. W. & RICHARDS, P. G. (eds) *Earthquake Prediction: An Intenational Review*. American Geophysical Union, **4**, 593–603.

Sylta, Ø. 2002. Quantifying secondary migration efficiencies. *Geofluids*, **2**, 285–298.

Symington, W. A., Green, K. E., Huang, J., Pottorf, R. J. & Summa, L. L. 1998. A multidisciplinary approach to modeling secondary migration: a Central North Sea example. *In*: Duppenbecker, S. J. & Illife, J. E. *Basin Modelling, Practice and Progress*. Geological Society, London, Special Publications, **141**, 169–186.

Thomas, M. M. & Clouse, J. A. 1995. Scaled physical model of secondary oil migration. *AAPG Bulletin*, **79**, 19–22.

Van Hinte, J. E. 1978. Geohistory analysis-application of micropalaeontology in exploration geology. *AAPG Bulletin*, **62**, 201–222.

Wagner, G., Meakin, P., Feder, J. & Jossang, T. 1997. Buoyancy driven invasion percolation with migration and fragmentation. *Physica A*, **245**, 217–230.

Wangen, M. 1997. Two-phase oil migration in compacting sedimentary basins modelled by the finite element method. *International Journal of Numerical and Analytical Methods in Geomechanics*, **21**, 91–120.

Welte, D. H., Horsfield, B. & Baker, D. R. 1997. *Petroleum and Basin Evolution – Insights From Petroleum Geochemistry, Geology and Basin Modelling*. Springer Verlag, Berlin, 535pp.

Welte, D. S. H., Hantschel, T., Wygrala, B. P., Weissenburger, K. S. & Carruthers, D. 2000. Aspects of petroleum migration modelling. *Journal of Geochemical Exploration*, **70**, 711–714.

Wendebourg, J. 2000. Modelling mulitcomponent petroleum migration in sedimentary basins. *Journal of Geochemical Exploration*, **70**, 651–656.

Wilkinson, D. & Willemsen, J. F. 1983. Invasion percolation: a new form of percolation theory. *Journal Physics and Mathematics General*, **16**, 3365–3376.

Williams, G. D. & Chapman, T. 1983. Strains developed in the hanging walls of thrusts due to their slip/propagation rate: a dislocation model. *Journal of Structural Geology*, **5**, 563–572.

Williams, G. D, Kane, S. J., Buddin, T. S. & Richards, A. J. 1997. Restoration and balance of complex folded and faulted rock volumes: flexural flattening, jigsaw fitting and decompaction in three dimensions. *Tectonophysics*, **273**, 203–218.

Yielding, G. 2002. Shale Gouge Ratio – calibration by geohistory. *In*: Koestler, A. G. H. & Amsterdam, R. (eds) *Hydrocarbon Seal Quantification*. Elsevier Science, 1–15.

Rifting through a heterogeneous crust: insights from analogue models and application to the Gulf of Corinth

L. MATTIONI[1]*, L. LE POURHIET[1,2] & I. MORETTI[1]

[1]*Institut Français du Pétrole, Rueil Malmaison, France*

[2]*Laboratoire de Tectonique, Université Pierre et Marie Curie, Paris, France*

**Now at Beicip Franlab, Rueil Malmaison, France (luca.mattioni@beicip.fr)*

Abstract: We used analogue models to study the fault evolution produced by extension through a heterogeneous crust. In the experiments, the heterogeneous crust consisted of a gently dipping silicone layer surrounded by brittle material. The viscous silicone level simulates a weak, upper crustal nappe stack that formed during a previous phase of shortening. X-ray scanner facilities allowed us to acquire 3D images of the experimental models at regular time invervals and hence to study the fault pattern development and the location of the main depocenters during rifting. The experimental results show that the inherited weak nappe stack acts as a décollement and localizes deformation. In the early stages of extension a system of conjugate high-angle normal faults initiates close to the upper tip of the gently dipping silicone layer near the free surface and propagates upwards, resulting in an initial symmetrical graben configuration. Further extension results in (1) a progressive asymmetry of the rifted zone, due to migration of its right margin down the nappe, (2) a shift of the main depocentre downward along the décollement, and (3) the simultaneous activity of several normal faults within the rifted zone. When the pre-existing silicone layer is oblique to the extension, the normal faults develop in an en echelon array, with a strike intermediate between the azimuth of the gently dipping silicone layer and the extension direction. The experiments also show how rheological differences between areas with potential intracrustal weak layers and adjacent domains without décollement level can lead to significant differences in fault pattern, dimension and orientation of the rifted zone. Complete asymmetry of a rift and switches in fault dip direction between adjacent domains can be explained by the presence of pre-existing upper crustal heterogeneities.

In regions where extension affects a crust previously deformed by shortening, the recognition of the role exerted by inherited heterogeneities in controlling the subsequent crustal extensional deformation is now being increasingly appreciated (Stewart 1999).

Areas where extension has followed shortening are numerous: in eastern North America, for example, it is well documented that many Palaeozoic shortening-related structures were reactivated during subsequent northwest–southeast directed Triassic to Jurassic rifting (e.g., Olsen & Schlische 1990; Schlische *et al.* 2002, and references therein). In the Mediterranean area, the best examples of pre-existing orogens currently affected by extension are the central Apennines in Italy (e.g., Cooper & Burbi 1986; Lavecchia *et al.* 1987; Ghisetti *et al.* 1993) and the Hellenides in Greece (Jolivet 2001, and references therein).

Data from these and other areas clearly suggest that, among many important parameters, the thickness of weak layers and the inherited pre-rift structural architecture can play a pivotal role on the subsequent development of graben-structures (Stewart 1999).

Overall, the stacking of crustal slices during orogenesis results in a heterogeneous crustal strength profile. The influence of these pre-existing rheological variations on the geometry and kinematics of faulting during subsequent extension represents a key issue for a better understanding of the modes and styles of rifting processes. This problem has already been addressed with a thermomechanical approach by Le Pourhiet *et al.* (2004) using the Gulf of Corinth as an example. In this paper, we used analogue modelling techniques to investigate the influence of inherited heterogeneities on structures that form during rifting. We wish to stress that our modelling does not address fault reactivation, but rather investigates the influence of a pre-existing weak and dipping thrust sheet, embedded within more competent units,

From: BUITER, S. J. H. & SCHREURS, G. (eds) 2006. *Analogue and Numerical Modelling of Crustal-Scale Processes.* Geological Society, London, Special Publications, **253**, 213–231.

on fault development and evolution during rifting. Le Pourhiet *et al.* (2006) discuss the limits of these analogue models, which are especially related to the boundary conditions that can be imposed.

Although the boundary conditions in our analogue models have been chosen to provide insights into the fault pattern and evolution of the Gulf of Corinth, however our study does not intend to reach an ideal fit with the Gulf of Corinth data. In our opinion, the results of our analogue models can be applied to any natural example undergoing extension with a pre-existing upper crustal heterogeneity (a stratigraphic and/or structural one) that is mechanically weaker than the surrounding material.

The Gulf of Corinth (Greece)

In the Gulf of Corinth, a system of roughly east–west normal faults (Moretti *et al.* 2003, and references therein) has been superimposed on the pre-existing structures of the Hellenides thrust belt since Quaternary times (Jolivet 2001) (Fig. 1a). The thrust belt comprises a series of structures striking roughly N150°E, which are therefore oblique to the subsequent north–south oriented extension. The different tectonic units making up the ancient Hellenides belt are best preserved and visible in outcrop in the western part of the Gulf of Corinth.

The Peloponnese, which forms the highlands of the rift to the south, includes uplifted syn-rift sediments. The possible cause of this general uplift may be associated with both the subduction of the Mediterranean plate and the rift opening (Moretti *et al.* 2003, and references therein).

The eastern part of the Gulf of Corinth is characterized by a few major faults that control the current syn-rift sedimentary deposits (Armijo *et al.* 1996) (Figs 1a, b). In its central part, the rift is rather symmetrical (Myriantis 1984; Stefatos *et al.* 2002; Moretti *et al.* 2003) and its marine part is about 30 km wide (Fig. 1a). In contrast, westward, the marine part of the rift is narrower (10 km) and the rift is characterized by several closely spaced normal faults that define a series of small fault-bounded blocks (see Fig. 1a). As shown in Figure 1a, the major faults belonging to the rift system show an average trend of N90°E to N100°E and are disposed en echelon. Furthermore, their age decreases towards the north (Flotté 2002; Micarelli *et al.* 2003, and references therein), but out-of-sequence structures also exist (Causse *et al.* 2004). Earthquakes located beneath the northern shore usually indicate extensional failure on east–west striking planes dipping N20°E–N40°E (Micarelli *et al.* 2003, and references therein). In outcrop these faults are planar, have a high dip angle and possibly root onto a gently dipping (roughly 10°–30° to the north) plane defined by a cluster of microseismicity (Fig. 1c) (Rigo *et al.* 1996).

The inherited thrust sheets of the Hellenides belt also include the Phyllades nappe, which crops out in a tectonic window south of the Gulf of Corinth (Flotté 2002). This nappe is made up mainly of weak phyllosilicates, showing extension lineations towards N30°E (Ghorbal 2002). In the eastern part of the Gulf of Corinth this nappe is absent, or at great depth. In contrast, subsurface data show its presence in all of the western part of the Gulf of Corinth. From a mechanical point of view, the Phyllades nappe represents a major rheological contrast to the surrounding rocks. Recently, Le Pourhiet *et al.* (2004) showed, by means of a thermo-mechanical numerical approach, that the presence of weak and anisotropic phyllosilicates results in an important mechanical weakening factor and influenced faulting during the early stages of the rifting in the Gulf of Corinth. In Le Pourhiet *et al.* (2006), new insights on the generation and development of the Gulf of Corinth have been gained by combining 3D numerical and analogue experiments.

Previous analogue modelling of extension through a heterogeneous crust

Up to now, many sandbox models have been performed to investigate the influence of pre-existing heterogeneities on fault patterns and kinematics during extension (e.g., Brun *et al.* 1985; Vendeville 1987, 1988; Vendeville *et al.* 1987; Tron & Brun 1991; Richard 1991; Nalpas & Brun 1993; Higgins & Harris 1997; Withjack & Callaway 2000; Schreurs *et al.* 2002). More recently, Bahroudi *et al.* (2003) showed that the mechanical characteristics of weak layers play a significant role in controlling the mode and rate of deformation during extension. Experiments where extension has been applied to a brittle-viscous multilayer (Schreurs *et al.* 2002), showed that the shape and the position of a weak silicone level embedded within brittle strata has a major influence on the structural style that develops during extension. All these experiments consider a 'layer cake model', i.e., a model where extension is applied to a brittle sandpack resting above a flat basal layer, usually consisting of silicone. The mechanical aspects of interaction between normal faults and pre-existing thrust faults were studied by

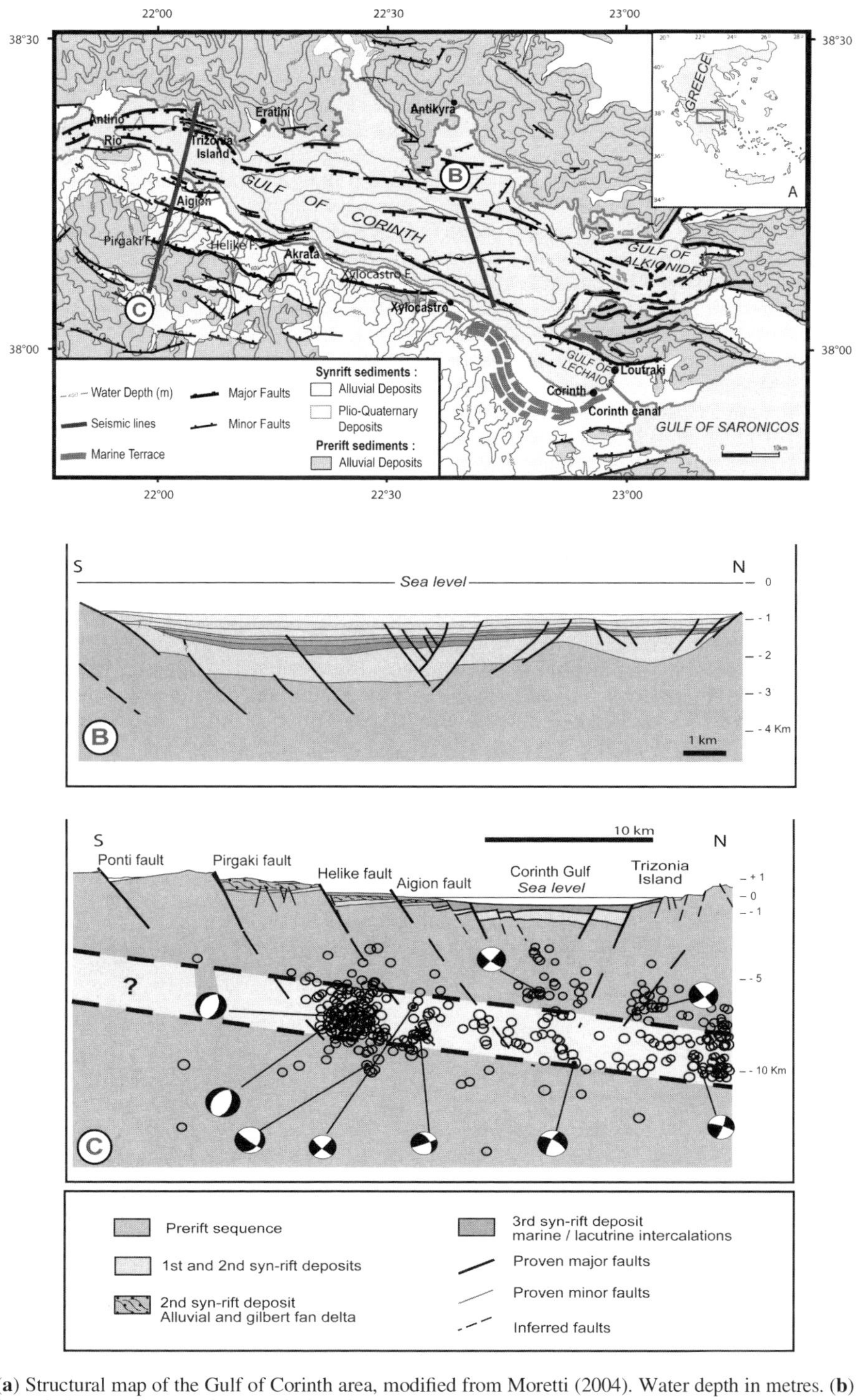

Fig. 1. (**a**) Structural map of the Gulf of Corinth area, modified from Moretti (2004). Water depth in metres. (**b**) Cross-section through the western part of the Gulf of Corinth, based on the interpretation of seismic lines from Hellenic Petroleum. Time-depth conversion is from Clément (2000) and interpretation from Moretti *et al.* (2003). (**c**) Cross-section through the western part of the rift (from Moretti 2004, slightly modified). The inferred seismogenic zone is represented in light green. The focal mechanisms and hypocenter locations of earthquakes are taken from Rigo *et al.* (1996).

Faccenna *et al.* (1995). Their models underwent first, shortening, resulting in thrust faults with variable dips, and then extension. For low-angle thrust faults ($<32^\circ \pm 1^\circ$), no interaction was observed, and normal faults developed independently. For thrust faults with a dip angle of $32^\circ \pm 1^\circ$, the new-formed normal faults branched at depth on the décollement level. Finally, for pre-existing thrust faults whose dip exceeded $41^\circ \pm 1^\circ$, normal reactivation of the thrust planes was observed. In experiments with a sand-pack above a stretching basement, McClay and Ellis (1987) showed that the extension above a rigid sloping basement resulted in faults dipping in the same direction as the basal detachment. More recently, Exadaktylos *et al.* (2003) performed experiments in which the base of the model is gently inclined. Their results showed that extension above a slightly sloping basal detachment leads first to the formation of a symmetrical graben and is followed by the formation of a family of secondary antithetic normal faults which cause an overall asymmetry of the rifted area.

In this context, our experiments are different: the original configuration (before extension) is characterized by the presence of a gently dipping silicone layer embedded within a flat, more competent brittle sand-cake. Our experiments aim to test the control exerted by the weak silicone layer on the style and kinematics of extensional deformation. Further specific parameters investigated in the course of our experiments include (1) the orientation of the embedded silicone layer with respect to the direction of the applied extension, (2) its position within the model, and (3) the role played by sedimentation during extension. The initial burial of the upper tip of the weak thrust sheet as well as its dip are important features for the final geometry; their influence has been discussed in Le Pourhiet *et al.* (2004). In our sandbox experiments, these parameters have been taken as constant.

Analogue models

Experimental procedure

The modelling techniques used for our purposes are similar to those commonly performed for experiments on brittle-ductile systems at the Laboratory of Analogue Modelling at Institut Français du Pétrole (IFP) (e.g., Colletta *et al.* 1991; Schreurs 1992; Mattioni *et al.* 2004). The experimental set-up is shown in Figure 2a and b and consists of a wooden box with two free sides. The other two sides of the box consist of two sliding backstops attached to two base plates, and driven by a double conveyor belt system that moves by means of computer-controlled stepper motors. The conveyor belt system allows the deformation to be localized at the base of the model near its central part (i.e., where the two base plates are in contact) where a velocity discontinuity occurs.

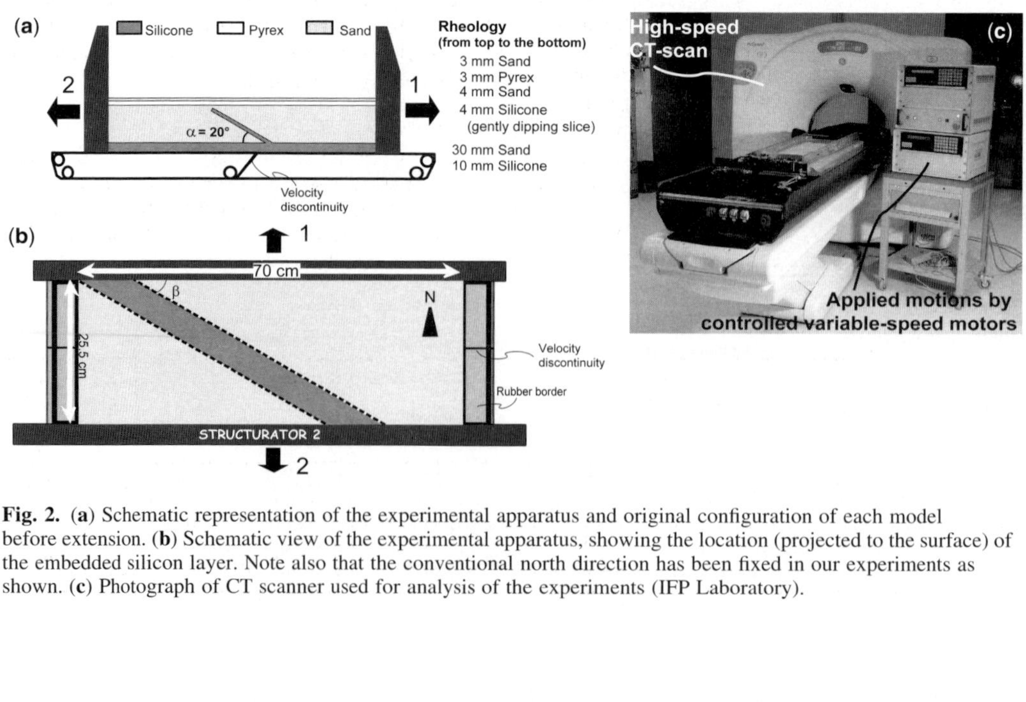

Fig. 2. (**a**) Schematic representation of the experimental apparatus and original configuration of each model before extension. (**b**) Schematic view of the experimental apparatus, showing the location (projected to the surface) of the embedded silicon layer. Note also that the conventional north direction has been fixed in our experiments as shown. (**c**) Photograph of CT scanner used for analysis of the experiments (IFP Laboratory).

During deformation, the central part of the sandbox was scanned by X-ray computerized tomography (Fig. 2c) (Hounsfield 1973; Mandl 1988; Colletta *et al.* 1991). This non-destructive technique generates a series of cross-sectional images through analogue models, thus allowing a detailed analysis of their internal geometry and kinematic evolution (e.g., Colletta *et al.* 1991). Attenuation of x-rays passing through a model depends on the materials used and is a function of density, effective atomic number and thickness. The difference in x-ray attenuation of sand, glass powder and silicone, allows us to visualize a stratified model. Furthermore, fault zones can be imaged in X-ray CT tomography, as they represent zones with lower densities and thus lower attenuation with respect to the unfaulted domains. The images obtained by the CT scanner were processed using specific software and then interpreted, more or less as 3D seismic images.

Materials and scaling

In our experiments, we used granular materials (dry quartz sand and glass powder with an average grain size of *c.* 100 μm) and silicone (polydimethylsiloxane, PDMS). Angles of internal friction of quartz sand and glass powder (about 30° for sand and 39° for glass powder) are similar to those determined by Byerlee (1978) for upper-crustal rocks (40° for normal stresses <2 kbar and 31° for normal stresses between 2 and 20 kbar). Thus, quartz sand and glass powder are considered to be appropriate analogue materials for simulating brittle deformation in the upper crust (Horsfield 1977; Byerlee 1978). This has also been shown by Schellart (2000), who in addition stated that granular materials have negligible cohesion at very low normal stresses (<400 Pa). Lohrmann *et al.* (2003) and Ellis *et al.* (2004) have also pointed out a similarity in the mechanical behaviour of granular sifted materials and the behaviour exhibited by crustal rocks (Barnhoorn *et al.* 2004). In fact, sifted dry quartz sand and corundum sand are characterized by elastic-frictional plastic behaviour with an initial phase of strain-hardening preceding failure, which is followed by strain-softening until a stable dynamic value of strength is reached.

The ductile behaviour of rocks in the lower crust and/or of weak layers within the brittle upper crust is simulated by PDMS. Within the range of strain rates used in this study (lower than 10^{-2} s^{-1}), the silicone behaves as a Newtonian fluid and has a very low yield strength.

In our models, the model-to-prototype ratio for lengths, L^*, was 2.5×10^{-6}, thus implying that 1 cm in the model represents 4 km in nature. To simulate any natural process by analogue modelling, it is necessary to scale the model using the basic principles discussed in detail by Hubbert (1937), Ramberg (1981) and Vendeville *et al.* (1987). Thus, it must be (Table 1):

$$\sigma^* = \rho^* g^* L^* \quad (1)$$

where ρ^* is the ratio for densities and g^* is the ratio for gravity acceleration. As our models were deformed in a natural gravity field, g^* was 1. The material density of materials used in our models was about half that of natural rocks, so that ρ^* was 0.5. Hence, it follows that the stress ratio, σ^*, was 1.25×10^{-6} (Table 1). For the velocity, we assumed that 1 cm/hr in our experiments corresponds to 1 cm/yr in nature. Because in our experiments we varied the velocity of extension from 1 cm/hr to 2 cm/hr, thus V^* changed from 10^4 (for Vmod = 1 cm/hr) to 2×10^4 (for Vmod = 2 cm/hr) (Table 1). Consequently, this implied that:

(1) for $V^* = 10^4$, T^* was 2.5×10^{-10} (i.e., 1 hr in the model corresponds to *c.* 456,000 years in nature), strain rate ratio ε^* ($1/T^*$) was 0.4×10^{10} and viscosity ratio 3×10^{-16}. As the measured viscosity of our own silicone was 2×10^4 Pa s (J. M. Daniel, pers.

Table 1. *Scaling values*

Length ratio L^* = Lmod/Lnat	Gravity ratio g^* = Gmod/Gnat	Density ratio $\rho^* = \rho$mod/ρnat	Stress ratio $\sigma^* = \rho^* g^* L^*$
2×10^{-6}	1	0.5	1.25×10^{-6}

Vmod (cm/hr)	Velocity ratio (V^* = Vmod/Vnat)	Time ratio (T^* = Tmod/Tnat)	Strain rate ratio ($\varepsilon^* = 1/T^*$)	Viscosity ratio ($\eta^* = \eta$mod/ηnat)
1	10^4	2.5×10^{-10}	0.4×10^{10}	3×10^{-16}
2	2×10^4	1.25×10^{-10}	0.8×10^{10}	1.6×10^{-16}

comm.), it implies a natural viscosity of 0.64×10^{20} Pa s.

(2) for $V^* = 2 \times 10^4$, T^* was 1.25×10^{-10} (i.e., 1 hr in the model corresponds to *c.* 913,000 years in nature), strain rate ratio ε^* ($1/T^*$) was 0.8×10^{10} and viscosity ratio 1.6×10^{-16} (thus implying a natural viscosity of 1.2×10^{20} Pa s).

Experimental set-up

The initial configuration of all models consisted of a 1 cm-thick silicone layer directly above the base plates and covered by a 4 cm-thick sandpack (Fig. 2a). Within the sandpack, we placed a gently dipping silicone layer 0.4 cm-thick, which simulated a ductile level embedded within a more mechanically competent series. The angle α of apparent dip of the silicone layer (e.g., the angle measured along a vertical section perpendicular to the velocity discontinuity) was fixed at 20°, which is the dip angle of the seismogenic zone in the western part of the Gulf of Corinth. All the models were subjected to symmetrical extension, with a direction perpendicular to the longitudinal borders of the sandbox. The extension velocity varied between experiments and was either 1 cm/hr or 2 cm/hr (see Table 2).

The boundary conditions and experimental parameters are summarized in Table 2. In the first reference model (experiment 1) the strike of the embedded silicone layer was orthogonal to the direction of the applied extension. Thus, this experiment corresponds to a cylindrical 2D case, discussed in Le Pourhiet *et al.* (2006). In all other models, the strike of the embedded silicone layer was N120°E (Fig. 2b), which corresponds to the expected strike of the weak Phyllades nappe in the Gulf of Corinth. Hence, the angle between the trend of the silicone level and the longitudinal sidewalls of the sandbox (β in Fig. 2b) is 30°. In the experiments we tested the influence of (i) the position of the weak silicone layer within the model (experiments 2 and 3), (ii) its geometry (experiment 4), and (iii) the role of syn-extensional sedimentation (experiment 5) on the resulting fault pattern. With the exception of experiment 1 (called 'reference model' in this paper), all the experiments were divided into two parts: one part with and one part without a dipping silicone layer, referred to as 'heterogeneous' and 'homogeneous' crust respectively. This particular configuration allowed us to compare the two different areas and to check the possible interactions between both areas.

Experimental results

In the description of the experiments, we will call those faults dipping in the same dip direction as the silicone level 'synthetic faults' and all faults dipping in the opposite dip direction 'antithetic faults'. These names are used for convenience and do not mean that the antithetic faults are genetically linked to the synthetic ones. All the structures observed within the models have been orientated with respect to a conventional north direction, fixed as shown in Figure 2b. Thus, in the accompanying cross-sections south is on the left side and north on the right side. A thin layer of glass powder (pyrex) near the surface of each experiment acted as a stratigraphic marker, allowing to identify structures on cross-sections during

Table 2. *Experimental parameters and boundary conditions*

Model number	Thickness of the embedded silicone layer (cm)	Velocity of extension (cm/hr)	Angle β	Apparent dip of the embedded silicone layer (°)	% Bulk extension	Length of the upper border of the silicone layer (cm)	Remarks
1	0.4	1	0	20	16	70	
2	0.4	1	30	20	15.6	24	
3	0.4	2	30	20	16	48	
4	0.4	2	30	20	16	34	Silicone layer with a free lateral termination
5	0.4	2	30	20	16	34	Syn-rift sedimentation Silicone layer with a free lateral termination

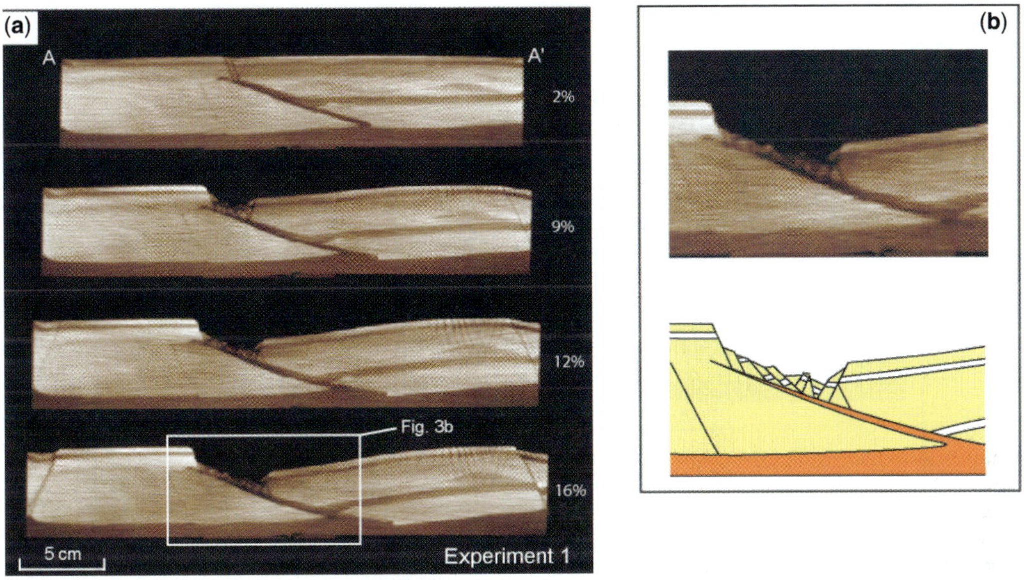

Fig. 3. Experiment 1. (**a**) Vertical section through the model showing fault evolution at four consecutive stages of extension. Location of cross-section is indicated in Fig. 4. The inset at stage 4 displays the location of Fig. 3b. (**b**) Enlargement of rift zone at the end of the experiment.

deformation. Only experiment 1 had a deeper thin layer of glass powder (Fig. 3). A square grid of coloured sand markers was traced on top of each model to monitor the surface evolution.

The reference model (experiment 1)

In this model, a gently dipping embedded silicone layer strikes orthogonal to the extension direction. The initial geometry and boundary conditions therefore correspond to a 2D experiment with a constant east–west strike of the silicone layer. The evolution of this experiment at four consecutive stages of extension is illustrated in vertical sections in Figure 3a.

In the early stages of deformation, a system of conjugate high-angle normal faults initiates roughly at the upper tip of the gently dipping silicone layer near the free surface. The faults propagate upward with increasing displacement along the dipping silicone layer, thus resulting in a more or less symmetrical graben configuration. As deformation goes on, the footwall south of the initial graben does not move, whereas the more northern rift-bounding fault shifts down slightly along the silicone layer, which acts as a décollement level. This leads to a progressive asymmetrical widening of the basin. Increasing deformation results in a progressive large flexure of the entire northern block, with several east–west striking high-angle normal faults that develop far from the basin at the bulge of the flexure. They dip both northward and southward and have limited offset. In contrast, along the southern margin, the footwall of the synthetic rift-bounding fault rift remains essentially undeformed.

With increasing extension the border faults remain the same all along the experiment, but new faults appear within the rift. The main fault with the highest offset is the one bordering the southern side of the rifted zone. Several minor synthetic and antithetic faults develop within the rift and root on the gently dipping silicone layer. Further opening of the rift is accommodated by progressive, anticlockwise tilting of the minor synthetic faults (changing from an average initial dip of 70° to 45° at the end of extension) on the southern side of the rift. In contrast, the antithetic faults in the northern part of the rift maintain their original dip (around 60°) throughout the experiment. After 16% of bulk extension (Fig. 3b), the rift shows a very marked asymmetrical shape, with one main fault at its southern border showing larger displacement with respect to both minor synthetic and antithetic faults cutting through the rift. At the surface, the resultant rift is bordered to the south by the main east–west striking and north dipping fault (Fig. 4). The northern side is more complex and

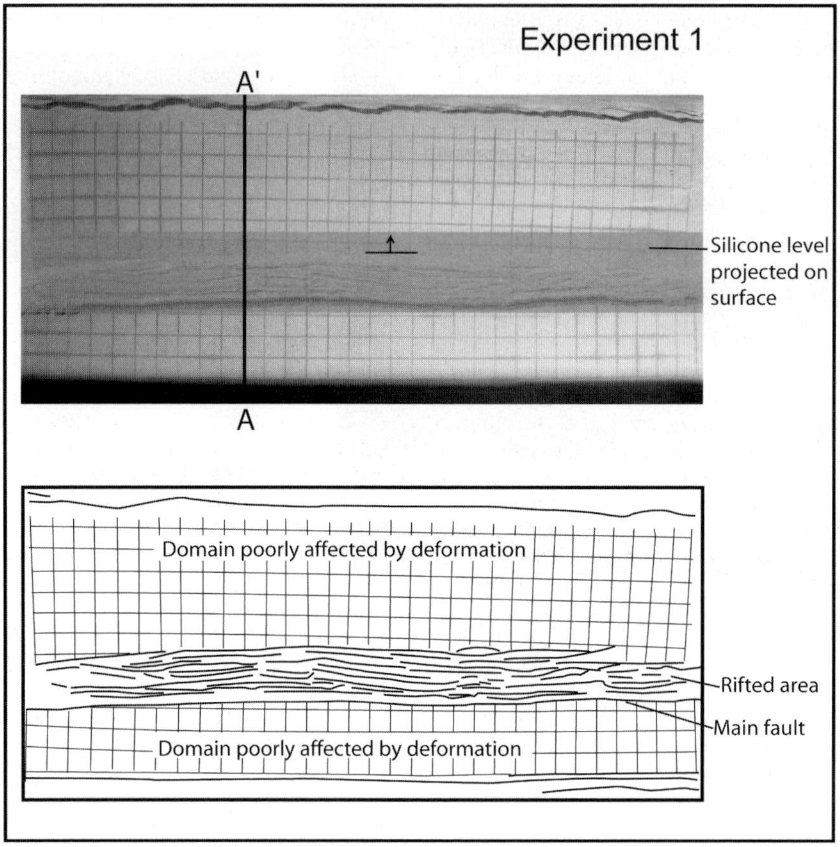

Fig. 4. Experiment 1. Line drawing after surface view at the end of the experiment (16% of bulk extension). The dark grey in the middle of the models displays the location (projected to the surface) of the interbedded gently dipping silicone layer. Arrow indicates its dip direction. Line A–A′ shows the location of the vertical section of Fig. 3.

defined by three east–west oriented and south dipping fault segments that overlap and link during the latest stages of deformation. Within the basin, minor synthetic and antithetic faults have a general east–west strike and are closely spaced. Their length and displacement are variable, and hence the dimension of the resulting fault-bounded blocks varies as well.

The top view allows us to highlight some other important features displayed on Figure 4, which corresponds to the surface fault pattern at the end of the extension. The rifted area is confined to a narrow zone within the model, with two adjacent broad domains showing only little deformation. Another important feature is the progressive migration of the depocentre of the basin in the dip direction of the silicone layer during the course of the experiment. Nevertheless, this does not mean that the active faults are restricted to the border of the depocentre; indeed, simultaneous activity of several normal faults located on both sides of the rift can be noted until the end of the experiment (Fig. 3).

Extension in zones with and without a pre-existing gently dipping silicone layer: a different style in fault genesis and evolution (experiments 2, 3 and 4)

Three analogue experiments were performed in which the dipping silicone layer is only present in the western half of the model. This particular configuration allows us to compare the style of extension in the two areas and to study the structural style in the area of transition. The dipping silicone layer strikes N120°E in all three experiments and is oblique to the north–south oriented extension direction.

In experiments 2 and 3 the initial configuration is the same, with the northwestern and southeastern lateral terminations of the embedded silicone layer being confined by the boundaries of the sandbox. Experiment 4 differs from the previous two as the southeastern lateral termination of the silicone layer ends within the sandpack. The fault evolution above the dipping silicone layer is quite similar for the three experiments, but the abrupt lateral termination of the silicone layer in experiment 4 strongly influences the fault pattern in the transition zone.

Experiments 2 and 3

The initial configuration of these two experiments is the same (see Table 2) and consequently the observed fault evolution is very similar. Thus, for practical reasons, we will discuss only the results of experiment 3. Figures 5 and 6 illustrate the evolution of this experiment, as a cross-section and in map view. Both figures clearly show how the evolution of the normal fault pattern is closely related to the presence or absence of the dipping silicone level at depth. In the eastern half of the model (homogeneous domain), a system of conjugate faults develops in an almost symmetrical way at the early stages of extension (cross-section B–B′, Fig. 5). The normal faults extend down to the top of the basal silicone layer and the initial graben width reflects the depth to the basal silicone layer. With increasing extension, fault offset along existing faults progressively increases and new steep normal faults develop in the footwall of the previously formed normal faults, thus producing a wider graben. Further extension is accomodated by increasing displacement onto the main normal faults and by a progressive tilting of the fault-bounded blocks. Overall, this fault evolution is very similar to the ones previously described in analogue models (e.g., Allemand & Brun 1991; Schreurs *et al.* 2002) and in nature (e.g., the Gulf of Suez, Colletta *et al.* 1988; Moretti 2004).

In the western half of the model (heterogeneous domain), the style and mode of rifting is completely different (section A–A′, Fig. 5). The fault evolution in the vertical north–south sections is similar to the one described for the reference model. In fact, a conjugate system of high-angle normal faults initiates at the tip of the gently dipping silicone layer near the free surface and leads to an initial symmetrical graben. With continuing extension the rifted area widens by migration of its northeastern border down along the weak silicone layer. Progressive widening of the rift is accommodated by the formation of several closely spaced synthetic and antithetic faults that root on the dipping silicone layer. As for the reference model, further extension results in tilting of the synthetic faults cutting through the southern part of the basin, whereas the antithetic faults in the northwestern part of the rift maintain their original dip. At the end of the deformation, the rift becomes asymmetrical in shape (Fig. 5, vertical section A–A′).

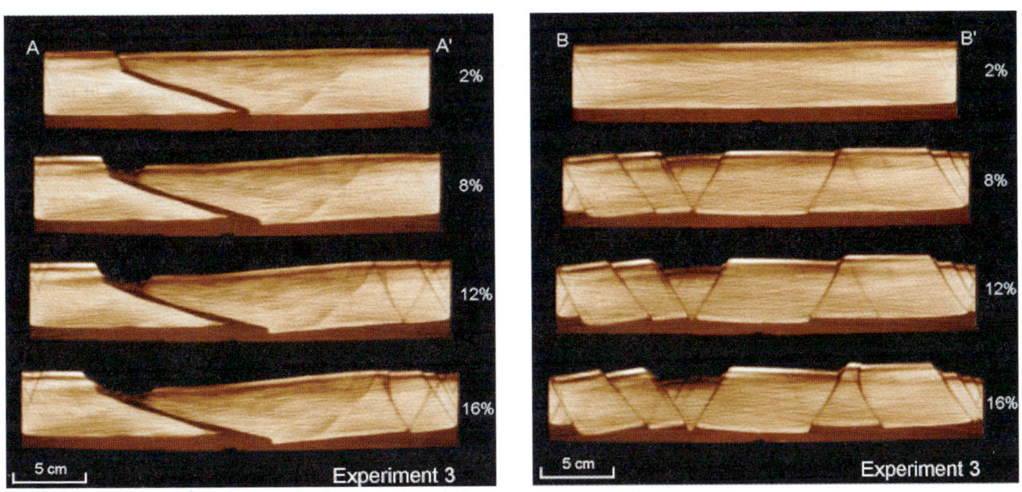

Fig. 5. Experiment 3. Vertical sections showing four consecutive stages of extension. The two cross-sections also display the different style of normal fault evolution within the model for the area where the gently dipping décollement level is present (A–A′) and where it is absent (B–B′). Location of these cross-sections within the model is shown in Fig. 6.

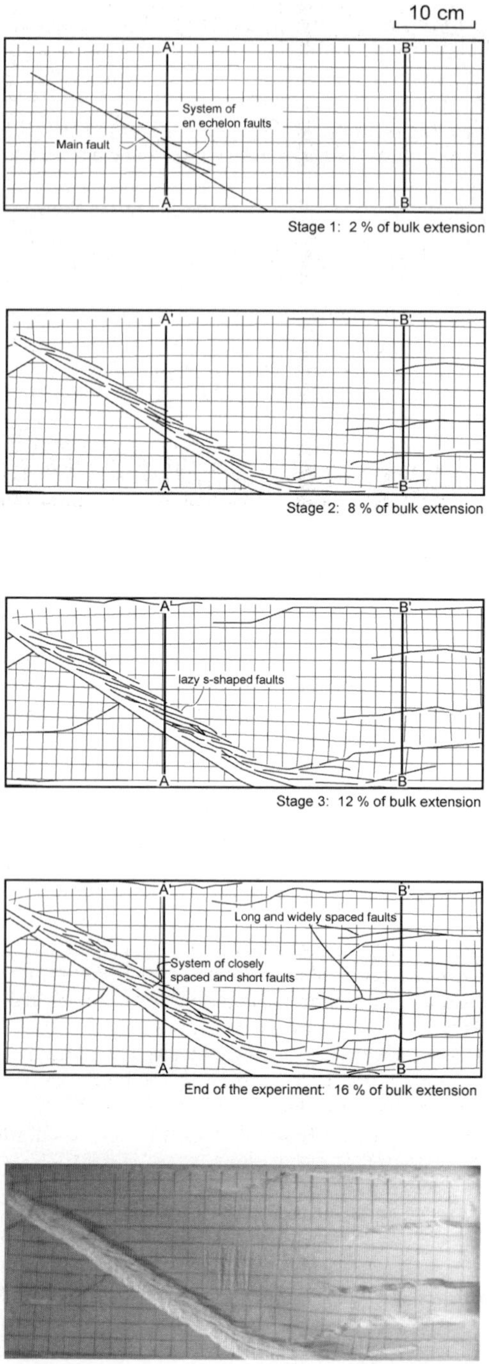

Fig. 6. Experiment 3. Line drawings after surface photographs showing four consecutive stages of extension. Lines A–A′ and B–B′ show the position of vertical sections of Fig. 5. The accompanying photo is representative of the end of the experiment. See text for further explanations.

The surface fault pattern that forms above the dipping silicone layer shows an en echelon pattern of normal faults. This evolution differs from the one described for the reference model and is clearly related to the difference in orientation of the dipping silicone layer with respect to the extension direction. The influence of this obliquity on the surface fault pattern will be discussed in more detail later on.

Experiment 4

In this experiment, one lateral termination of the dipping silicone layer ends within the sandpack. This particular configuration allows us to better investigate the fault pattern in the transition zone between the areas with and without a dipping silicone layer. The experimental results are shown in Figures 7, 8 and 9. The evolution in vertical sections (Figs 7 and 8) is similar to the one described for experiment 3. In surface view (Fig. 9), the transition zone is characterized by the formation of several sub-vertical transfer faults that strike NE–SW, parallel to the lateral termination of the embedded silicone layer. Figure 9 also shows the change in orientation of the main east–west trending normal faults that developed in the western part of the model and propagated towards the embedded silicone layer. Their strike changes from N90°E to N65°E.

Figure 10 illustrates, by means of six serial vertical sections through the final stage of deformation, the drastic lateral changes in the structural style of extension between the homogeneous (i.e., the easternmost and westernmost parts of the model) and heterogeneous central domain. The transition zone between the two domains is marked by several sub-vertical extensional transfer faults, the location and orientation of which are closely linked to the geometry of the embedded silicone layer. These transfer faults have very large offsets, so the deepest part of the rift is located there (see cross-section E–E′ on Fig. 10). In this experiment, the N45°E oriented transfer faults link the N110°E asymmetrical oblique graben above the dipping silicone layer, with the N90°E symmetrical graben eastward. The N45°E transfer faults clearly correspond to the direction of the eastern lateral termination of the silicone level. The transfer faults appear during the early stages of extension, suggesting a strong control of the extremity of the silicone on their location and orientation (Fig. 9). In other words, the lateral tip of the silicone layer acts as a strong localizing factor of deformation. At the end of the experiment, the deformed square grid at the surface shows that no strike-slip displacement occurred along these transfer faults.

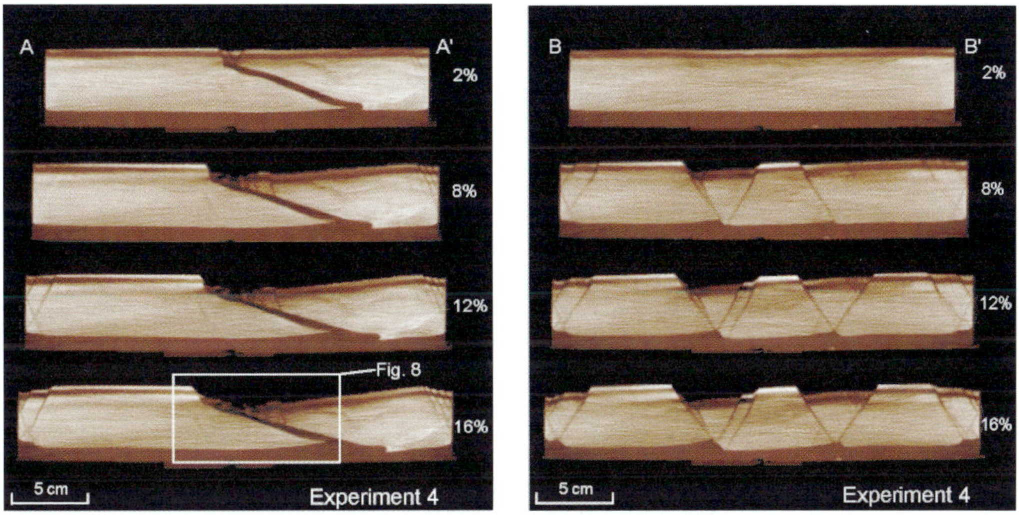

Fig. 7. Experiment 4. Vertical sections showing four consecutive stages of extension. Note the different fault style between areas where the gently dipping décollement level is present (A–A′) and where it is absent (B–B′). Location of cross-sections is shown in Fig. 9. The inset in A–A′ at stage 4 shows location of Fig. 8.

Analysis of the fault pattern at the surface

Figures 6 and 9 show the progressive evolution of structures at the surface of experiments 3 and 4, and illustrate the difference in fault pattern between the homogeneous and heterogeneous domains. In the former domain, faults develop orthogonal to the direction of the applied extension. They are few, straight and widely spaced, so that the resulting rift affects a large area of the model. In the heterogeneous domain, the general trend of the resulting rift is not necessarily orthogonal to the direction of the applied extension; rather, it is roughly N120°E oriented, corresponding to the trend of the underlying dipping silicone layer. Therefore, from the onset of extension a marked asymmetry exists in the orientation and distribution of faults at the two sides of the main rift. The southwestern rift boundary is characterized by one main fault striking N120°E, parallel to the underlying dipping silicone layer. This main fault and the minor synthetic faults that form very close to it, maintain their original trend parallel to that of the underlying silicone layer until the end of the experiment. In contrast, Figures 6 and 9 show that the northeastern border is structurally more complex. It consists of a system of en echelon antithetic faults, organized in a dextral stepping arrangement, that formed during the early stages of extension. Their strike of roughly N105°E is intermediate between the strike of the underlying gently dipping silicone layer and that of the applied extension. As extension continues, these antithetic faults become slightly sigmoidal in shape with their lateral terminations becoming progressively parallel to the strike of the silicone layer.

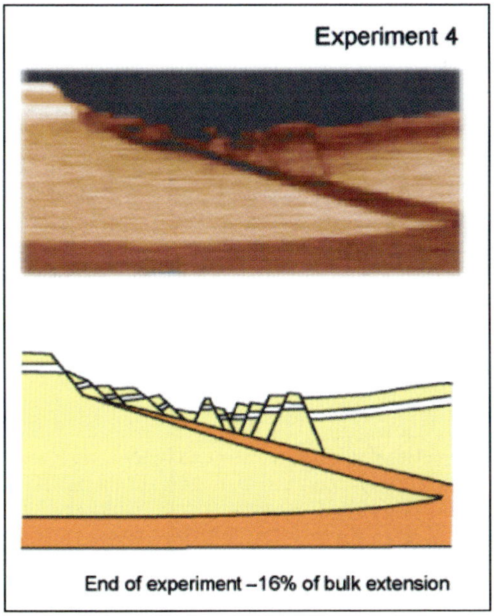

Fig. 8. Experiment 4. Enlargement of rift zone above the décollement level at the end of the experiment. Note the marked asymmetry of the rift with a main fault on the southwestern (left) side of the basin and several minor synthetic and antithetic faults cutting through the basin.

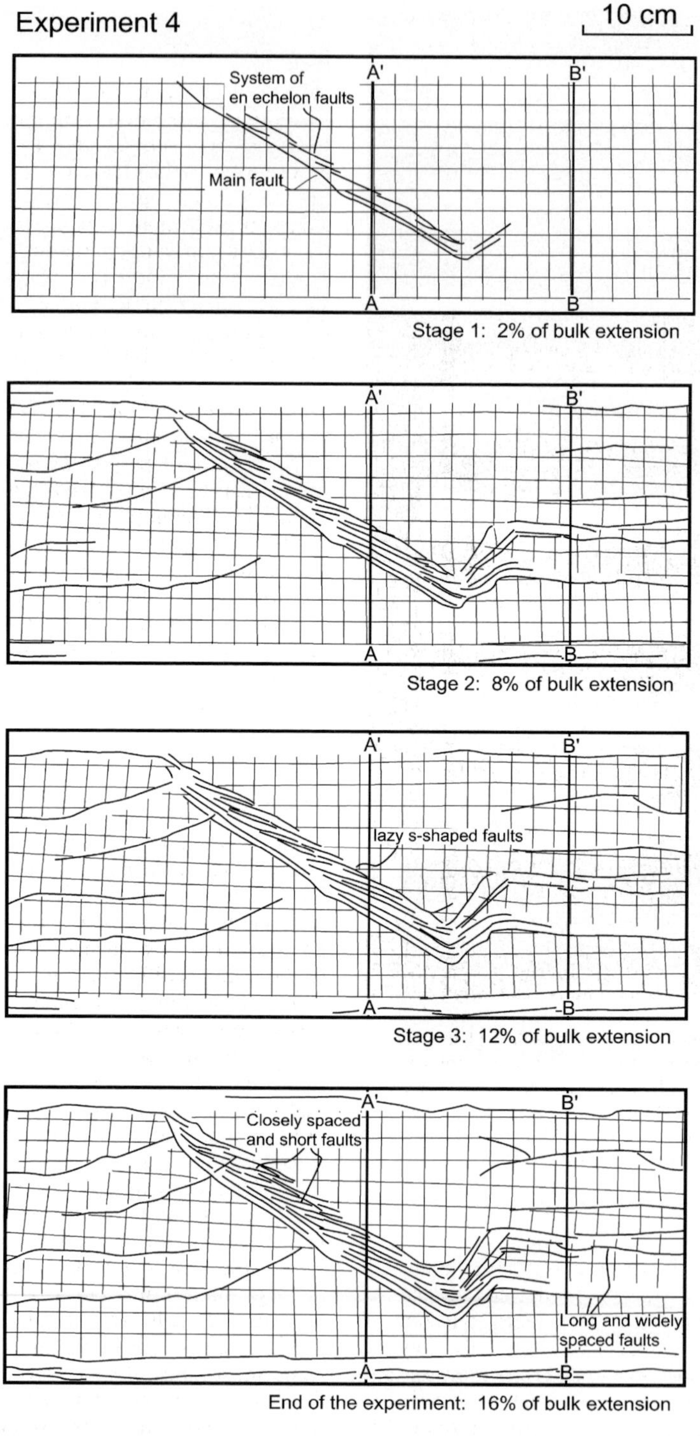

Fig. 9. Experiment 4. Line drawing after surface photographs at four consecutive stages of extension, showing the differences in fault pattern between the area underlain by a gently dipping silicone layer and the adjacent homogeneous areas. Lines A–A′ and B–B′ show the position of vertical sections of Fig. 7.

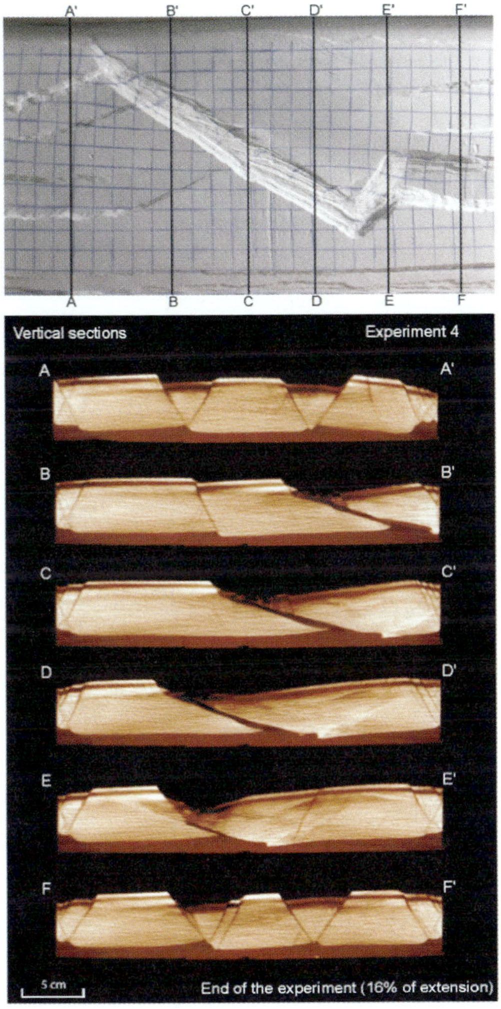

Fig. 10. Experiment 4. Cross-sections through the model at the end of extension, showing the fault pattern across the area where the gently dipping silicone layer is present (B–B′, C–C′ and D–D′) and where it is absent (see A–A′ and F–F′). Cross section E–E′ is representative of the boundary zone between the two areas.

At the end of the deformation (Figs 6 and 9), the surface fault pattern in the heterogeneous domain is clearly different from the one in the adjacent domains. The main rift zone in the heterogeneous domain is very narrow and characterized by several closely spaced and short faults. These faults are not orthogonal to the direction of the applied extension, but the general trend of the rift is roughly parallel to that of the underlying gently dipping silicone layer.

Effect of syn-rift sedimentation (experiment 5)

In this experiment, two cycles of syn-rift sedimentation were simulated: the first one (represented in green in Fig. 11) after 6% and a second one (light green in Fig. 11) after 12% of bulk extension. Figure 11 shows four consecutive stages of deformation through the domain with the gently dipping silicone layer. The initial model configuration is exactly the same as the one for experiment 4 (see Table 2) and the overall fault evolution was rather similar. Nevertheless, several important features are worth noting: for example, at the end of the deformation (Step 4 of Fig. 11), the faults within the rift propagated up to both the first and second cycle syn-rift deposits. These syn-rift deposits are clearly cut and offset, highlighting once again the contemporaneous activity of several faults. Another important feature is the location of the depocenter within the rift with respect to the location of the main fault of the system. The depocenter is located northward, since the graben is deeper in this area. Nevertheless the faults that have the larger offset are along the other margin.

Discussion

Table 3 summarizes the main results from our analogue model experiments. Overall, the experiments show that rheological differences between areas with and without a dipping silicone layer ('heterogeneous' and 'homogeneous' crust, respectively) lead to marked differences in fault evolution.

The following differences between the two areas are evident from the analysis of the analogue experiments:

(1) Extension occurs in a deformation zone that is much wider in the area with homogenous crust than in the area with heterogeneous crust (see Figs 5, 6, 7 and 9).

(2) The extension through the homogenous crust is accommodated by relatively few normal faults in comparison to those developed above the heterogeneous crust. Faults in the homogeneous crust are straight, long, widely spaced and with large displacements, whereas in the heterogeneous crust they are more numerous, short and closely spaced. Furthermore, faults in regions of homogeneous crust sole out into the basal and flat weak silicone level without intersecting each other. In contrast, faults developed in regions of hetereogeneous crust root onto the dipping silicone layer, and some of them intersect each other.

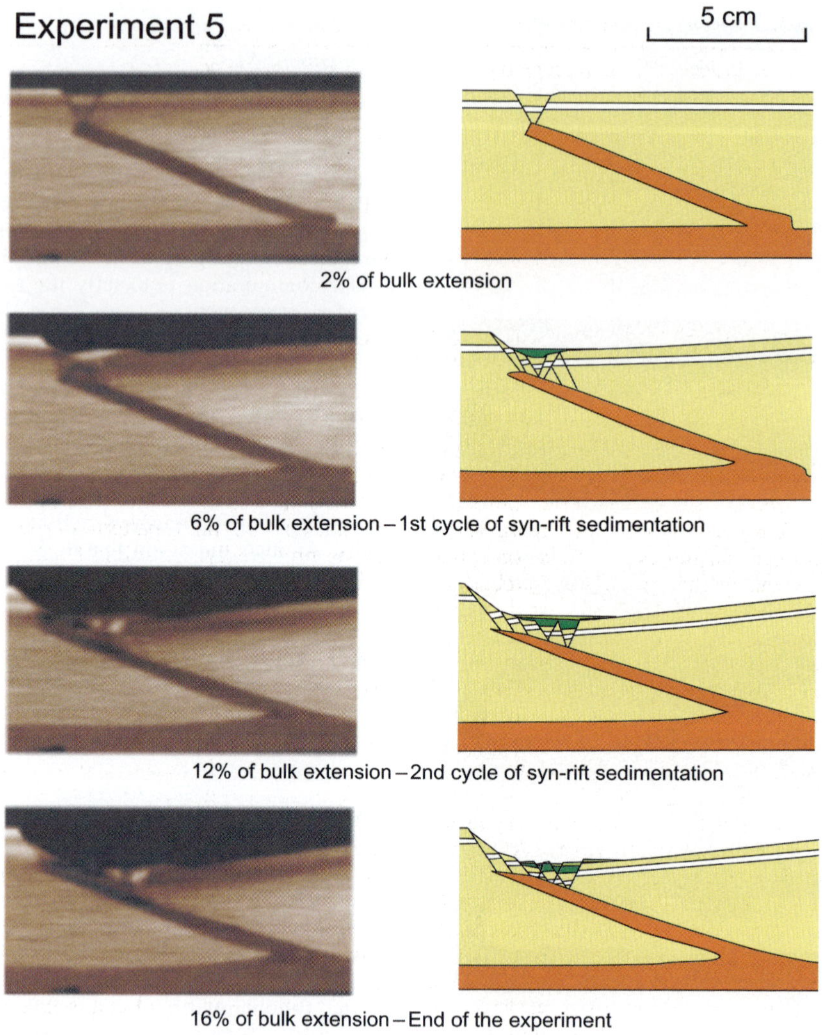

Fig. 11. Experiment 5. Enlargement of the stretched area above the gently dipping silicone level showing the evolution at four successive stages of extension. This experiment differs from the others since two cycles of syn-rift sedimentation have been carried out (the first one marked in green, the second one in light green).

From our models, a first major difference observed between the domains of homogeneous and heterogeneous crust concerns the shape and geometry of the resulting rift. In particular, rifting in regions of heterogeneous crust ultimately results in a marked asymmetry. In fact, from Figures 3b and 8, it may be seen that at the end of the deformation, the resulting basin is characterized by a major fault bordering its southern side and showing larger displacements with respect to those observed along the minor synthetic and antithetic faults cutting through the basin. This particular shape of the rift is not observed in domains where extension affected a homogeneous crust; here in fact, the extensional deformation affects a wider area and the resulting rift is characterized by a symmetrical shape. This symmetry is defined by a system of conjugate faults that develops at the early stages of deformation and extends down to the top of the basal silicone layer. In all experiments, the east–west orientation of faults in the homogeneous domain is consistent with the far-field applied stress system. At the surface, the spacing

Table 3. *Summary of model results*

With an embedded dipping silicone layer ('hetereogenous crust')	Without an embedded dipping silicone layer ('homogeneous crust')
Narrow and asymmetric zone	Wide and symmetric deformation zone
Several normal faults	Few normal faults
Closely spaced, shorter and slightly s-shape normal faults	Widely spaced, longer and straighter normal faults
Irregular horsts and grabens	Regular horsts and grabens
Faults oblique to the direction of extension	Faults perpendicular to the direction of extension
Block rotation	
Simultaneous activity of several faults	

of the faults reflects the brittle layer thickness, as already observed by Vendeville *et al.* (1987).

This particular fault genesis and evolution is markedly different from the one observed within the heterogeneous domain. In the latter domain, fault genesis and evolution is controlled to a large extent by the pre-existing embedded heterogeneity that is mechanically weaker than its surroundings. This heterogeneity results in a strong localization of the deformation in an area located above the silicone layer (Figs 3, 5 and 7). Only at the earliest stages of deformation is the developing rift more or less symmetrical, and is defined by a system of conjugate planar, normal faults. The normal faults extend down to the upper termination of the silicone layer, which acts as a décollement level. As deformation continues, the northernmost fault of this conjugate system progressively shifts down to the ductile level, and the graben gradually becomes asymmetrical. With increasing deformation, the formation of minor synthetic and antithetic faults and tilting of fault-bounded blocks accommodate further widening of the rift. These new faults develop at the hanging wall of the older ones and root on the gently dipping décollement level. Thus the oldest faults are those bordering the two sides of the rift.

The maximum depth and width (measured in sections through the central part of the graben) of the graben above the gently dipping silicone layer have been measured at different phases of bulk extension (Fig. 12). Both maximum graben depth and width increase roughly proportionally with increasing shear displacement along the gently inclined décollement level (see also Exadaktylos *et al.* 2003). The ratio between the maximum graben depth and width at different stages of the extension remains roughly constant, except during the initial phases of extension when this ratio is at its minimum value. This suggests that in the early stages of extension, the rift zone increases mainly in width rather than in depth, whereas with continuing extension, the amount of offset increases and the rift deepens quickly.

The analogue experiments indicate that fault orientation and fault pattern during extension can be profoundly controlled by the orientation and depth of a crustal heterogeneity. In fact, in experiments in which the gently dipping silicone layer strikes obliquely (with $\beta = 30^\circ$, Figs 6 and 9) to the direction of the applied extension, the strike of the normal faults forming above the heterogeneity is not necessarily orthogonal to the direction of extension. Rather, two groups of coeval normal fault orientations can be distinguished; a first group includes faults striking parallel to the direction of the underlying gently dipping silicone layer. These are mainly synthetic faults which form at the southern side of the rift where the silicone level is very shallow. The second group includes the antithetic faults that develop at the northeastern side of the rift, where the silicone layer is at a greater depth. These faults are arranged in a dextral en echelon arrangement with an orientation of N105°, i.e., intermediate between the trend of the underlying gently dipping silicone layer and that of the far-field applied stress regime.

On the other hand, the presence of a gently dipping silicone layer acting as a décollement level and oriented obliquely to the direction of the applied extension seems also to control the fault pattern of faults that formed far from it. Figure 9 shows how major east–west striking faults that developed in regions of homogenous crust deflect abruptly as they approach the embedded silicone layer.

Applicability of the analogue models results to the Gulf of Corinth

Despite the inherent idealizations of analogue models and the fact that they are only partially

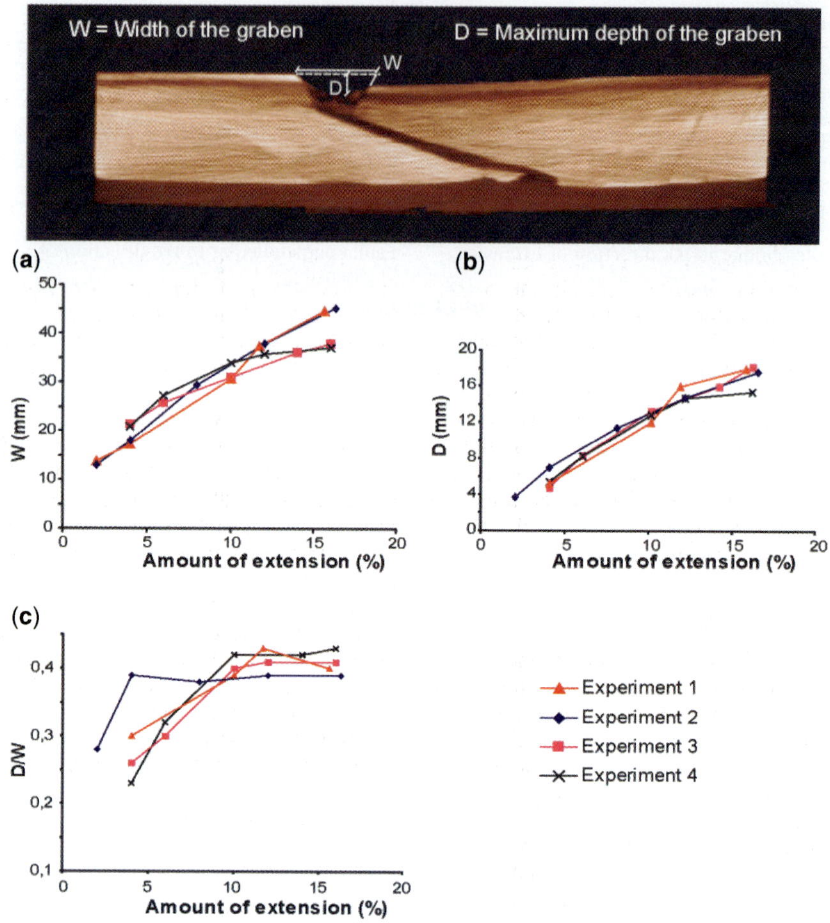

Fig. 12. Diagrams showing (**a**) the graben width (W) against the amount of extension, and (**b**) maximum graben depth at the free surface (D) against the amount of extension. (**c**) ratio D/W (maximal graben depth/graben width) against the amount of extension.

scaled, some important similarities can be observed with respect to the Gulf of Corinth. The west (left)–east (right) changes in fault pattern described in those models with a weak silicone layer oblique to the direction of the applied extension (e.g., experiment 4 in Figs 6 and 9), reflect very well the west–east variations observed in the Gulf of Corinth. There is also good agreement between the eastern homogeneous domain of experiments 3 and 4 (see Figs 6 and 9) and the eastern part of the Gulf of Corinth, where the weak Phyllades nappe is not present, or at least, is at great depth (see Fig. 1a, b). Here, both rifts (model and nature) are rather symmetrical, wide and with few and widely spaced major faults (see also Fig. 1b and cross section B–B′ in Figs 5 and 7). The trend of these faults is roughly east–west, which is clearly the expected direction if one considers the orientation of the present-day north–south extension. The western part of the Gulf of Corinth is characterized by several closely spaced normal faults that define a series of faulted blocks (see Fig. 1a). These faults show an average trend of N90°E–N100°E, which is rougly intermediate between the direction of the present-day extension and the strike of the underlying Phyllades nappe. This structural setting is very similar to the one observed in the western part of experiments 3 and 4 (Figs 6 and 9). From the analogue model results we infer that the en echelon structure observed in the western part of the Gulf of Corinth and the lack of field evidence for major transverse faults can be explained by the obliqueness between the present direction of extension and the dip direction of the crustal heterogeneity presented by the Phyllades nappe.

In the Gulf of Corinth, the present depocenter in the western part is located near the northern shore, southward of the island of Trizonia (Stefatos *et al.* 2002; Moretti *et al.* 2003) (Fig. 1a). The island itself appears to be a horst. Similar features are observed in experiment 4, where horsts exist above the weak thrust sheet and are located either in the central part or near the northern margin (Fig. 8). In the western part of the Gulf of Corinth, the current depocenter is located in front of the Trizonia fault, i.e., near the northern shore of the rift as in experiment 5. However, the results of only one experiment performed with this particular configuration (sediments only deposited in the deepest part of the graben) have to be regarded with caution. Nevertheless, they might be of use for future experimental studies.

Concluding remarks

The role played by inherited heterogeneities on normal fault genesis and evolution has been investigated by means of small-scale analogue model experiments. The experimental results indicate that the presence of structural heterogeneities within the brittle crust can profoundly influence the subsequent fault pattern and the distribution of deformation during a rifting episode. An embedded weak silicone layer simulating an inherited weak nappe behaves as a décollement level during extension, inducing a strong localization of faulting in a narrow area. Rheological differences between areas with intracrustal weak layers and adjacent domains of homogeneous crust can lead to remarkable differences in fault patterns and dimensions of the final rift. In the presence of an embedded gently dipping ductile layer, the resulting rift is completely asymmetrical, with one major border fault showing larger displacements and several other minor synthetic and antithetic faults cutting through the basin. These faults are closely spaced and short and the fault-bounded blocks are small. This fault pattern is markedly different from the one observed in domains of homogenous crust where the applied extension is spread over a wider zone. The resulting rift is symmetrical and the deformation is accommodated by means of a few, widely spaced and long faults. The produced horst-and-graben morphology is characterized by a series of large tilted fault-bounded blocks.

Our results document how the orientation of an interlayered gently dipping ductile layer can control the pattern and orientation of faults rooting on it and of faults that form further away from it. In fact, in experiments where the dipping silicone layer was oblique to the applied extension direction, the faults bordering the northeastern side of the rift above the gently dipping décollement level, form a dextral stepping en echelon arrangement, and their orientation is intermediate between the trend of the underlying gently dipping silicone layer and the far-field applied extension direction. On the other hand, major faults in regions of homogeneous crust also deflect their orientation as they approach the weak dipping silicone layer.

Rift asymmetry and along-strike changes of dominant fault dip in rifts occur frequently in nature (Bosworth 1985; Rosendahl 1987; Colletta *et al.* 1988; Moretti 2004). The analogue models presented here clearly indicate that an upper crustal heterogeneity may induce such an asymmetry.

There is a good agreement between our experiments and the Gulf of Corinth, especially as far as the change in fault pattern from west (presence of a dipping weak layer) to east (absence of a weak layer) is concerned. Our experimental approach may provide useful constraints for the understanding and the interpretation of complex geometries and kinematics in natural zones of extension through a heterogeneous crust.

The authors are grateful to B. Colletta, J. M. Daniel and L. Micarelli for their constructive comments and suggestions. J. M. Mengus and D. Pillot are thanked for their technical assistance in building the models and C. Fichen for her help in obtaining the CT-images. L. M. was partially funded during his stay at IFP by a CNR (Consiglio Nazionale delle Ricerche, Italy) grant (nr 203.22/2002). We thank C. Faccenna, G. Schreurs and an anonymous reviewer for improving the quality of the manuscript with their valuable comments and suggestions.

References

Allemand, P. & Brun J. P. 1991. Width of continental rifts and rheological layering of the lithosphere. *Tectonophyiscs*, **188**, 63–69.

Armijo, R., Meyer, B., King, G. C. P., Rigo, A. & Papanastassiou, A. 1996. Quaternary evolution of the Corinth rift and its implications for the late Cenozoic evolution of the Aegean. *Geophysical Journal International*, **126**, 11–53.

Bahroudi, A., Koyi, H. A. & Talbot, C. J. 2003. Effect of ductile and frictional décollements on style of extension. *Journal of Structural Geology*, **25**, 1401–1423.

Barnhoorn, A., Bystricky, M., Burlini, L. & Kunze, K. 2004. The role of recrystallization on the deformation behaviour of calcite rocks: high strain torsion experiments on Carrara marble. *Journal of Structural Geology*, **26**, 885–903.

Bosworth, W. 1985. Geometry of continental propagating rift. *Nature*, **316**, 625–627.

Brun, J. P., Choukroune, P. & Faugeres, E. 1985. Les discontinuités significatives de l'amincissement crustal: application aux marges passives. *Bulletin de la Société Géologique de France*, **8**, 139–144.

Byerlee, J. 1978. Friction of rocks. *Pure and Applied Geophysics*, **116**, 615–626.

Causse, C., Moretti, I., Ghisetti, F., Eschard, R., Micarelli, L., Ghaleb, B. & Frank, N. 2004. Kinematics of the Corinth Gulf inferred from calcite dating and syntectonic sedimentary characteristics. *Comptes Rendues Académie des Sciences*, **336**, 281–290.

Clement, C. 2000. *Imagerie sismique crustale de la subduction hellénnique et du golfe de Corinthe.* PhD thesis, University of Paris VII.

Colletta, B., Balé, P., Ballard, J. F., Letouzey, J. & Pinedo, R. 1991. Computerized x-ray tomography analysis of sandbox models: examples of thin-skinned thrust systems. *Geology*, **19**, 1063–1067.

Colletta, B., Le Quellec, P., Letouzey, J. & Moretti, I. 1988. Longitudinal evolution of the Suez rift structure (Egypt). *Tectonophysics*, **153**, 221–233.

Cooper, C. & Burbi, L. 1986. Geology of the Central Sibillini Mountains. *Memorie della Società Geologica Italiana*, **35**, 323–347.

Ellis, S., Schreurs, G. & Panien, M. 2004. Comparisons between analogue and numerical models of thrust wedge development. *Journal of Structural Geology*, **26**, 1659–1675.

Exadaktylos, G. E., Vardoulakis, I., Stavropoulou, M. C. & Tsombos, P. 2003. Analogue and numerical modeling of normal fault patterns produced due to slip along a detachment zone. *Tectonophysics*, **376**, 117–134.

Faccenna, C., Nalpas, T., Brun, J. P., Davy, P. & Bosi, V. 1995. The influence of pre-existing thrust faults on normal fault geometry in nature and experiments. *Journal of Structural Geology*, **17**, 1139–1149.

Flotté, N. 2002. *Caractérisation structurale et cinématique d'un rift sur détachement: le rift de Corinthe-Patras, Grèce*. PhD thesis, University of Paris South.

Ghisetti, F., Barchi, M., Bally, A. W., Moretti, I. & Vezzani, L. 1993. Conflicting balanced structural section across the Central Apennines (Italy): problems and implications. *In*: Spencer, A. M. (ed.) *Generation, Accumulation and Production of Europe's Hydrocarbons III*, European Association of Petroleum Geoscience, Special Publication, **3**, 219–231.

Ghorbal, B. 2002. *Quels arguments en faveur d'un detachment Plio-Quaternaire au Sud du Golfe de Corinthe?: Etudes tectono-métamorphiques des Phyllades du Péloponnèse septentrionale*. MA thesis, Univerity of Paris Jussieu.

Higgins, R. I. & Harris, L. B. 1997. The effects of cover composition on extensional faulting above reactivated basement faults: results from analogue modeling. *Journal of Structural Geology*, **19**, 89–98.

Horsfield, W. D. 1977. An experimental approach to basement-controlled faulting. *Geologie en Mijnbouw*, **56**, 363–370.

Hounsfield, G. N. 1973. Computerized transverse axial scanning (tomography). *British Journal of Radiology*, **46**, 1016–1022.

Hubbert, M. K. 1937. Theory of scale models as applied to the study of geologic structures. *Geological Society of America Bulletin*, **48**, 1459–1520.

Jolivet, L. 2001. A comparison of geodetic and finite strain pattern in the Aegean, geodynamic implications. *Earth Planetary Sciences Letters*, **187**, 95–104.

Lavecchia, G., Minelli, G. & Pialli, P. 1987. Contractional and extensional tectonics along the transect Trasimeno Lake-Pesaro (Central Italy). *In*: Boriani, A., Bonafede, M., Piccardo, G. B. & Vai, G. B. (eds) *The Lithosphere in Italy. Advances in Earth Science Research*, Atti Convegno Lincei, **80**, 139–142.

Le Pourhiet, L., Burov, E. & Moretti, I. 2004. Rifting through a stack of inhomogeneous thrusts (the dipping pie concept). *Tectonics*, **23**, TC4005.

Le Pourhiet, L., Mattioni, L. & Moretti, I. 2005. 3D modelling of rifting through a pre-existing stack of nappes in the Gulf of Corinth (Greece): a mixed analogue/numerical approach. *In*: Buiter, S. J. H. & Schreurs, G. (eds) 2006. *Analogue and Numerical Modelling of Crustal-Scale Processes*. Geological Society, London, Special Publications, **253**, 233–252.

Lohrmann, J., Kukowsky, N., Adam, J. & Oncken, O. 2003. The impact of analogue material properties on the geometry, kinematics, and dynamics of convergent sand wedges. *Journal of Structural Geology*, **25**, 1691–1711.

Mandl, G. 1988. *Mechanisms of Tectonic Faulting*. Elsevier, Amsterdam.

Mattioni, L., Le Pourhiet, L. & Moretti, I. 2004. Extension through a heterogeneous crust. The case of Gulf of Corinth (Greece). Part I: Analogue modelling. *Bollettino di Geofisica teorica ed applicata*, **45**, 237–241.

McClay, K. R. & Ellis, P. G. 1987. Geometries of extensional fault system in model experiments. *Geology*, **15**, 341–344.

Micarelli, L., Moretti, I. & Daniel, J. M. 2003. Influence of depth and amount of displacement of the characteristics of normal faults, case study in the Gulf of Corinth – Greece. *Journal of Geodynamics*, **36**, 275–303.

Myrianthis, M. L. 1984. Graben formation and associated seismicity in the Gulf of Korinth (Central Greece). *In*: Dixon, J. E. & Robertson, A. H. F. (eds) *The Geological Evolution of the*

Eastern Mediterranean, Geological Society, London, Special Publications, **17**, 701–707.

MORETTI, I. 2004. Asymmetry of the early rift structures: a comparison between the Gulf of Suez and the Gulf of Corinth. *Studi Geologici Camerti*, Special Issue 2004, 105–111.

MORETTI, I., SAKELLARIOU, D., LYKOUSIS, V. & MICARELLI, L. 2003. The Gulf of Corinth: a half graben? *Journal of Geodynamics*, **36**, 323–340.

NALPAS, T. & BRUN, J. P. 1993. Salt flow and diapirism related to extension at crustal scale. *Tectonophysics*, **228**, 349–362.

OLSEN, P. & SCHLISCHE, R. W. 1990. Transtensional arm of the early Mesozoic Fundy rift basin: penecontemporaneous faulting and sedimentation. *Geology*, **18**, 695–698.

RAMBERG, H. 1981. *Gravity, Deformation and the Earth's Crust*. Academic Press, New York.

RICHARD, P. D. 1991. Experiments on faulting in a two-layer cover sequence overlying a reactivated basement fault with oblique (normal-wrench or reverse-wrench) slip. *Journal of Structural Geology*, **13**, 459–469.

RIGO, A. & LYON CAEN, H. *ET AL.* 1996. A microseismic study of the western part of the Gulf of Corinth (Greece): implications for the large-scale normal faulting mechanisms. *Geophysical Journal International*, **126**, 663–688.

ROSENDAHL, B. 1987. Architecture of continental rifts with special reference to east Africa. *Annual Review of Earth and Planetary Science*, **15**, 445–503.

SCHELLART, W. P. 2000. Shear test results for cohesion and friction coefficients for different granular materials: scaling implications for their usage in analogue modelling. *Tectonophysics*, **324**, 1–16.

SCHLISCHE, R. W., WITHJACK, M. O., EISENSTADT, G. 2002. An experimental study of the secondary deformation produced by oblique-slip normal faulting. *American Association of Petroleum Geologists*, **86**, 885–906.

SCHREURS, G. 1992. Analogue modelling using x-ray computed tomography analysis: experiments on distributed strike-slip shear deformation. *Institut Français du Pétrole Report* 39893.

SCHREURS, G., HÄNNI, R. & VOCK, P. 2002. The influence of brittle-viscous multilayers on faulting during rifting: an analogue modelling approach. *In*: SCHELLART, W. P. & PASSCHIER, C. (eds) *Analogue Modelling of Large-Scale Tectonic Processes*, Journal of Virtual Explorer, **6**, 89–97.

STEFATOS, A., PAPATHEODOROU, G., FERENTINOS, G., LEEDER, M. & COLLIER, R. 2002. Seismic reflection imaging of active offshore faults in the Gulf of Corinth; their seismotectonic significance. *Basin Research*, **14**(4), 487–491.

STEWART, S. A. 1999. Geometry of thin-skinned tectonic systems in relation to detachment layer thickness in sedimentary basins. *Tectonics*, **18**, 719–732.

TRON, V. & BRUN, J. P. 1991. Experiments on oblique rifting in brittle-ductile systems. *Tectonophysics*, **188**, 71–84.

VENDEVILLE, B. 1987. Champs de failles et tectonique en extension: modélisation expérimentale. PhD thesis, University of Rennes, France.

VENDEVILLE, B. 1988. Modèles expérimentaux de fracturation de la couverture contrôlée par des failles normales dans le socle. *Comptes rendus de l'Académie des Sciences*, **307**, 1013–1019.

VENDEVILLE, B. C., COBBOLD, P. R., DAVY, P., BRUN, J. P. & CHOUKROUNE, P. 1987. Physical models of extensional tectonics at various scale. *In*: COWARD, M. DEWEY, J. & HANCOCK, P. L. (eds) *Continental Extensional Tectonics*. Geological Society, London, Special Publications, **28**, 95–107.

WITHJACK, M. O. & CALLAWAY, S. 2000. Active normal faulting beneath a salt layer: an experimental study of deformation in the cover sequence. *American Association of Petroleum Geologists*, **84**, 627–651.

3D modelling of rifting through a pre-existing stack of nappes in the Gulf of Corinth (Greece): a mixed analogue/numerical approach

L. LE POURHIET[1,2*], L. MATTIONI[2**] & I. MORETTI[2]

[1]*Laboratoire de Tectonique, Université Pierre et Marie Curie, Paris, France (e-mail: laetitia@gps.caltech.edu)*

[2]*Institut Français du Pétrole, Rueil Malmaison, France*

**Now at Caltech, Pasadena, CA, USA*

***Now at Beicip Franlab, Rueil Malmaison, France*

Abstract: The Gulf of Corinth is a young (1 Ma) active rift currently extending N00, which displays significant contrasts in structural style along strike. A possible explanation for these variations is the presence of the Phyllades nappe in the basement of the western part of the Gulf. Previous 2D thermo-mechanical models have shown that a strong strength contrast between this metamorphic unit and the rest of the basement can explain the kinematics and the spacing of the faults in the western part. The rift, however, displays a wide variety of 3D features (e.g., en echelon faulting, N30 transverse normal faults) that cannot be taken into account using 2D modelling. To obtain 3D insights into the role of an inherited dipping weakness zone, analogue (sand and PDMS) experiments based on the results of the 2D numerical thermo-mechanical model have been performed. The analogue models show that a 30° discrepancy between the dipping direction of the weak nappe and the direction of extension leads to the formation of en echelon and N30 striking normal faults as observed in the Gulf of Corinth. However, fault spacing and graben width completely misfit both the data and the results of the thermo-mechanical models on which the analogue experiments were based. In order to understand those differences, numerical mechanical benchmarks of the analogue experiments have been run to test different factors (3D lateral displacements, values of the elastic parameters and bottom boundary conditions) that could have affected the dynamics of the analogue model. This approach highlights, for our case study, that the misfits are mostly related to the lack of isostatic compensation at the base of the analogue experiments.

Geological setting

The area of the Gulf of Corinth has been in a continuous extensional regime since the middle Miocene (Jolivet *et al.* 1994; Jolivet & Faccenna 1999). This contribution deals with the last rifting event that has taken place for the last 1 to 1.5 Ma in the area, but as the long-term extension affects the structures, both the long-term and the neotectonic features will be briefly discussed in the following.

In the external zone of the Hellenides, long-term extension has led to the exhumation of metamorphic core complexes along ductile shear zones. In the Peloponnesus, the shear strain associated with these detachments is mainly localized inside the Phyllades nappe (Fig. 1). One of the core complexes crops out at the southern limit of the Gulf of Corinth in the Zarouchla tectonic window (outlined by strips in Fig. 1). The average strike of the stretching lineations affecting this dome structure is parallel to the Miocene direction of extension (i.e., N30 to N50). Hence, at the onset of the neotectonic rifting event, the Phyllades nappe was already retromorphosed from the lowest pressure in the blue schist metamorphic facies to the epimetamorphic facies.

The present day geometry of the shoreline of the Gulf of Corinth shows that the mean topography decreases eastwards. Moho depth follows the same trend (Makris 1978; Tiberi *et al.* 2001). This positive correlation, at the scale of the Gulf, suggests that the base of the crust is isostatically compensated and that the large-scale topographic gradient remains from long-term extension. The widening of the Gulf to the east is also compatible with a clockwise rotation of the Peloponnesus, as inferred from paleomagnetic data (e.g., Duermeijer *et al.* 2000), and reflects the fact that the finite extension has

From: BUITER, S. J. H. & SCHREURS, G. (eds) 2006. *Analogue and Numerical Modelling of Crustal-Scale Processes.* Geological Society, London, Special Publications, **253**, 233–252.
0305-8719/06/$15.00

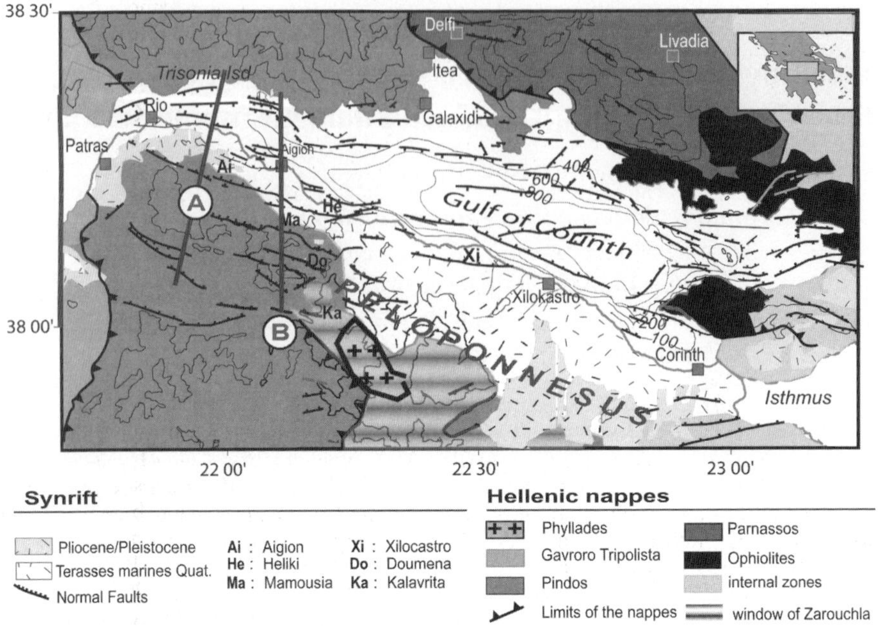

Fig. 1. Structural map of the Gulf of Corinth. Faults after Moretti (2004), limits of the nappes after Aubouin *et al.* (1970). A and B lines correspond, respectively, to the location of the geological cross-section (Moretti 2004) displayed in Fig. 5 and to the micro-seismic cross-section of Rigo (1996) used to constrain the validity of the thermo-mechanical models.

been much greater in the central Aegean domain than in the external part of the Hellenides.

At the onset of rifting, the eastern part of the rift was indeed connected to the Aegean Sea at the level of the Corinth Isthmus and Megara basin further east (Bentham *et al.* 1991). Because the western part of the current rift was still under compression until the Pliocene (Doutsos & Kokkalas 2001), the basin was closed by the Hellenides belt at the level of Patras and, therefore, has not been affected by long-term extension. The sediments from this side of the rift as well as the oldest dated calcite steps found along fault planes are early Pleistocene in age (i.e., <1 Ma; see Causse *et al.* 2004). This timing is compatible with the onset of dextral displacement on the Kefalonia fault (Cocard *et al.* 1999); a strike slip fault located further west. These data suggest a change in the regional kinematics by 1.5 Ma that could explain the discrepancies between geological observations and GPS measurements (Briole *et al.* 1999). GPS data indicates that the current direction of extension is N00 and that the rate of extension increases westward from 0.5 cm/yr at the level of the Corinth Isthmus to 1.5 cm/yr at the level of Patras. This anticlockwise rotation of the Peloponnesus with respect to the northern margin of the Gulf is not compatible with the paleomagnetic data. Therefore, given the short (60 km) length-scale of the Gulf of Corinth, the current rifting phase can be treated separately from the Neogene long-term extension.

From a structural point of view, the Quaternary rift is usually described as an asymmetric graben because the southern shore is covered by uplifted deposits of marine fan deltas and continental Guilbert's fan deltas (Ghisetti *et al.* 2001; Malartre *et al.* 2004), whereas, on the northern shore, the pre-rift basement crops out directly. Although some beach-rocks are locally uplifted, the northern shore is stable or slightly subsiding in general (Lykousis *et al.* 1998).

The Gulf of Corinth is segmented along strike (Ghisetti & Vezzani 2004). In the central-eastern part of the rift, the syn-rift deposits are located in one main graben in which faulting is roughly symmetrical (Stefatos *et al.* 2002). However, the marine terraces uplifted on the southern shore indicate a subsisting asymmetry of the shore (Armijo *et al.* 1996). At the eastern tip, the Gulf is divided into two grabens bordered by faults dipping 45° and branching at the base of the seismogenic zone, which is located much deeper (i.e., 15 to 20 km) than in the west (Jackson *et al.* 1982). These faults have two main strike directions: N90° and N30°. The first ones were responsible for the 1981 earthquake. The last ones were responsible for the closure of the connection between the Gulf and the Aegean Sea during the Quaternary.

In the western part, the basin is elongated N120. The southern shore line reflects the en echelon disposition of the major faults. Their strikes are perpendicular to the current direction of extension (i.e., N90° to N110°, see Moretti *et al.* 2003). N30° normal faults are highlighted by microseismic events (Lyon Caen *et al.* 2004), micro-gravimetric survey (Mrlina 2004*a*, *b*) and seismic reflection data (Naville *et al.* 2004). These normal faults act as transfer between the N90 faults.

On the southern shore, sedimentation has been controlled by numerous and closely spaced faults forming a series of small isolated half grabens. The age of those grabens seems to decrease northward (Flotté 2002; Micarelli *et al.* 2003). Flotté (2002) used anticline deformation data of the syn-rift deposits to constrain kinematic reconstructions of the fault décollement level based on a roll-over model. His models imply that such a level deepens northwards. However, he was not able to propose a conceptual model explaining mechanically that fault spacing reduces from 8 km to 4–2 km in the direction of deepening of the inferred décollement level.

The microseismic and seismic events recorded in the western Gulf of Corinth also outline a northward dipping zone. Moreover, reconstruction of focal mechanisms of the earthquakes located inside this zone indicate very low angle nodal planes (i.e., 10°–30°, Rigo *et al.* 1996; Rietbrock *et al.* 1996) and thus favour the possibility for seismicity along shallow dipping fault(s). Seismologists tend now to agree on the fact that the outcropping major faults are planar up to the base of the seismogenic zone. Although the relation between the low-angle seismic events and the high-angle faults is not yet well understood, the low-angle dipping zone is clearly associated with the presence of fluids, as shown by anomalies in the Vp × Vs (velocity of P-wave × velocity of S-wave) tomography (Latorre *et al.* 2004) and in the magneto-telluric conductivity (Pham *et al.* 2000).

Modelling approach

Based on the observations that (1) the spacing of the normal faults decreases in the same direction as the postulated décollement level deepens and that (2) the N120/N30 direction appears everywhere along this structure, Le Pourhiet *et al.* (2004) proposed that the N120° overall direction of the western part of the Gulf is inherited from the Neogene syn-orogenic extension. Their fully coupled 2D thermo-mechanical models include a weak dipping layer in the pre-rift basement, because:

(1) The southern fault of the zone affected by Quaternary rifting is observed to root at low angle on the Phyllades nappe, in the field (Flotté, 2002).
(2) The Phyllades tectonic unit was originally a pelitic deep basin deposit and was exhumed during the late Miocene (L. Jolivet, pers. comm. for the Gulf of Corinth; Trotet 2000 in the South Peloponesus). During its burial and exhumation, it has been enriched in anisotropic phyllosilicates known to be very weak in the ductile (Ranalli 1987) as in the plastic regime (Bos & Spiers 2002).
(3) Tiberi *et al.* (2001) evidence by means of gravity measurements, a very strong anomaly elongated N120, located at the place where the Phyllades crop out.

Once the presence of the weak dipping nappe in the basement of the Gulf of Corinth is established, thermo-mechanical models have been used to quantify the dynamic influences of the presence of a weak dipping nappe. The weakness of the heterogeneity has been set as a parameter and models have been computed for a wide range of competences. After comparing the kinematics and stress distribution in the model to the fault spacing, earthquakes distribution and fault kinematics in the Gulf of Corinth, it has been found that only a very weak nappe can explain the rotation of P and T axes for the relatively deep earthquakes, the shallow décollement level (inferred from rollover), and the fault spacing in the western part of the Gulf of Corinth.

As a step forward in the understanding of the Gulf of Corinth, this paper and Mattioni *et al.* (2006) aim to verify that the non co-linearity between the current direction of extension and the dipping direction of the weak nappe leads to the formation of en echelon structures similar to those observed in the Gulf of Corinth. 3D analogue models have been designed using the effective rheology (elasto-plastic v. viscous) obtained from the numerical thermo-mechanical computations. Despite this effort made in fitting the effective rheologies, major discrepancies exist in the shape of the graben (width, fault spacing).

Possible causes of these discrepancies fall into three categories: (1) The plane strain assumption in the thermo-mechanical model. (2) The rheological assumptions in the analogue models, e.g., neglecting the elastic effect in the overall model, neglecting plasticity in the weak layer. (3) The basal boundary conditions in the analogue models.

In order to explore the potential impact of these factors, numerical benchmarks of the analogue experiments have been carried out. This part of the study shows how a coupled approach using both 2D numerical modelling and 3D analogue modelling gives better insights into the validity of the models and their relevance for solving geodynamic issues.

After a description of the two modelling tools, the set-up and results of Le Pourhiet *et al.* (2004) are briefly described, as well as a similar thermo-mechanical experiment that will be used as reference for the present study. We then focus on the set-up of the analogue experiments and their results. After comparing the new coupled thermo-mechanical model with the analogue experiments, the paper will seek an explanation for the large discrepancies observed between both approaches, by computing numerical benchmarks of the analogue models and varying the different parameters listed above.

Modelling techniques

Numerical approach

We use Paravoz, a fast Lagrangian analysis of continua (FLAC)-based thermo-mechanical code (Cundall 1989; Poliakov *et al.* 1993). This code solves Newton's second law

$$\rho g_i + \frac{\partial \sigma_{ij}}{\partial x_j} = \rho \frac{\partial V_i}{\partial t} \tag{1}$$

and Fourier's law for heat transfer

$$\frac{DT}{Dt} = \frac{\partial}{\partial x_i}\left(\chi \frac{\partial T}{\partial x_i}\right) + \frac{H_r}{\rho C_p} \tag{2}$$

where ρ = mean density, g = gravity acceleration, σ = stress tensor, x = spatial coordinates, V = velocity, t = time, χ = thermal diffusivity, T = temperature, H_r = radiogenic heat production, C_p = heat capacity. Einstein summation applies for repeated indices.

These partial differential equations are solved using a fully coupled (including thermal stress/shear heating) explicit scheme on a Lagrangian mesh using a mixed (element/nodes) discretization to avoid mesh locking (Cundall & Board 1988). Large strain computations are allowed by remeshing as soon as one of the corners of the triangular mesh reaches a critical minimum angle of 10°. As the remeshing process includes linear interpolation of nodal values (velocity, temperature) and closest neighbour redistribution of elementary values (stresses, physical properties including finite strain), it should be avoided as much as possible for the sake of precision.

Heat advection is implicitly solved considering the rate of deformation of the mesh as indicated by the objective time derivative D in Fourier's equation. The explicit time-marching aspect of this numerical method implies very small time steps and thus enables various non-linear visco-elasto-plastic rheologies to be handled easily. In shear, each element behaves either as compressible an elasto-plastic or Maxwell visco-elastic body, depending on which of these two rheologies produces the lowest value of the second invariant of stress tensor.

Maxwell body behaviour can be written following an incremental rule:

$$\begin{cases} \sigma_{ij} = \sigma_{ij}^{\mathrm{old}} + \dfrac{\partial \tau_{ij}}{\partial t}\Delta t + \delta_{ij}\dfrac{\partial \bar{\sigma}}{\partial t}\Delta t \\ \dfrac{\partial \tau_{ij}}{\partial t} = 2G\left(e_{ij} - \dfrac{\tau_{ij}}{2\mu_{\mathrm{shear}}(T, e^{\mathrm{II}})}\right) \\ \dfrac{\partial \bar{\sigma}}{\partial t} = K\nabla \cdot \mathbf{V} \end{cases} \tag{3}$$

where δ is the Kronecker delta, G and K are respectively the shear and bulk modulus, and e^{II} is the second invariant of the strain rate tensor e, while τ is the deviatoric part of the stress tensor, and $\bar{\sigma}$ is its trace divided by 3. μ_{shear} is the effective dynamic viscosity. It can be set as a constant or may depend on the second invariant of deviatoric strain-rate tensor following the 3D formulation of the power law creep flow rules described in Chen & Morgan 1990.

$$\mu_{\mathrm{shear}}(T, e^{\mathrm{II}}) = \frac{1}{4}(e^{\mathrm{II}})^{((1/n)-1)}\left(\frac{3}{4}A\right)^{-1/n} \times \exp\left(\frac{E}{n\mathrm{RT}}\right) \tag{4}$$

Here, E, A and n are the common non-Newtonian creep parameters obtained from uni-axial creep experiments; T and R are the absolute temperature and the gas constant respectively.

As the time steps are small, a simple elastic predictor/plastic corrector technique is used to compute the non-associated Mohr–Coulomb elasto-plastic flow. This means that when the predicted elastic state of stress reaches the yield criterion f^s, which depends on the internal friction angle Φ of the material,

$$N_\Phi = \frac{1+\sin\Phi}{1-\sin\Phi}$$
$$f^s = \sigma 1 - \sigma 3 N_\Phi + 2C_0\sqrt{N_\Phi} \quad (5)$$
$$f^s \leq 0 \Rightarrow \text{rupture}$$

The plastic strain $\varepsilon^{\text{plas}}$ is assumed to be proportional by a scalar factor λs to the derivative versus stress of plastic potential Qs. Qs depends on the dilatancy angle $\Psi \ll \Phi$ for rocks.

$$N_\Psi = \frac{1+\sin\Psi}{1-\sin\Psi}$$
$$Qs = \sigma 1 - \sigma 3 N_\Psi \quad (6)$$
$$\varepsilon^{\text{plas}} = \lambda s \frac{\partial Qs}{\partial \sigma}$$

The plastic multiplier λs is found assuming that in a static yield regime, stresses stay on the yielding surface for consistency (Eqn. 7) reasons (for details see Vermeer & Deborst 1984).

$$\dot{f}^s = 0 \quad (7)$$

An explanation of the boundary conditions that are used in the numerical models is given in Figure 2.

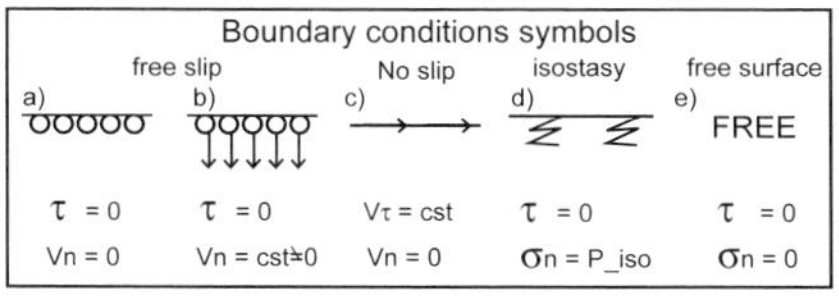

Fig. 2. Type of mechanical boundary conditions: (**a**) and (**b**) Free slip means that the orthogonal velocity is set to a constant (Vn) and the tangential stress (τ) to zero. (**c**) No slip means that the tangential velocity is set to a constant (Vτ) and normal velocity (Vn) is set to zero. (**d**) Local isostasy means that shear stress (τ) is set to zero and normal stress is (σn) calculated with reference to a hypothetic isostatic compensation level at depth. (**e**) Free means that both normal (σn) and shear stress (τ) components are set to zero on the boundary. It constitutes a realistic boundary condition to model the Earth's surface and topography for Lagrangian mesh.

Analogue approach

The analogue experiments are described in further detail in Mattioni *et al.* (2006) and are only briefly described here.

The experiments are analysed by x-ray tomography (Colletta *et al.* 1991), a non-destructive technique that allows the 4D acquisition of the analogue model density (and thus material) distribution.

In the experiments, upper-crustal brittle rocks are simulated with sand and glass powder (Pyrex, usually used together with sand to get a better visualization of the results with the x-ray scanner). Their internal friction angles are 34° and 42°, respectively, while their densities range from 1500 to 1700 kg/m^3, depending on their compaction (note that the sand pile was not pre-compacted in the experiments). As dry granular media are dilatant in shear, the shear bands are easy to image by x-ray tomography as they represent narrow zones of reduced density.

In the present experiments, the Pyrex layer is 3 mm thick and is located 3 mm below the surface at the initial stage, approximately in the middle of the elasto-plastic layer that does not contain the weak dipping heterogeneity.

The ductile behaviour of rocks in the lower crust, and/or of weak layers within the brittle upper crust, is simulated by silicone putty (PDMS) presenting a much lower density than sand (965 kg/m^3) and therefore appearing darker on the cross-sections. PDMS behaves as a nearly Newtonian fluid with a viscosity of 2×10^4 Pa at room temperatures and at strain-rates below 3×10^{-3} s^{-1} (Weijermars 1986).

In order to be considered analogue to Earth, physical experiments have to be properly scaled. This means that the dimensionless physical properties that govern the modelled process must be equal in the analogue experiment and in nature (Ramberg 1981). In this study, the scaling of the physical experiments has followed classical rules for sand silicone experiments, thus neglecting the main governing factor for fault spacing, i.e., elasticity. However, the adjective 'analogue' is used instead of 'physical' to describe the experiments throughout our paper for consistency with other papers in this volume.

In the analogue experiments, extension is applied using diverging plastic sheets which are fixed to smooth diverging walls. Friction along those walls is taken to be negligible compared to the friction in the granular media. In the direction orthogonal to extension, the experiments are bordered by two rubber sheets which are used to simulate free slip boundaries.

Comparing numerical and analogue results

Two main outputs have been chosen: topography and cumulative plastic strain. Although topography stands as a straightforward control parameter, one should note that (1) in the numerical scheme of Paravoz a slight diffusion is applied at the top of the models to reduce the occurrence of remeshing, and (2) as the Pyrex (which has a higher angle of internal friction) is not assigned any geological meaning, it has not been taken into account in the numerical model and will be considered as a possible factor introducing modelling artefact in the analogue approach. The numerical topographies are therefore always slightly smoother than those obtained by analogue modelling.

The second control parameter is cumulative plastic strain. It has been chosen because the dark strips, interpreted as faults in the analogue models, correspond in fact to a dilatation zone associated with plastic shear. For underconsolidated sand (as the case in the present analogue experiments), the dilation is theoretically supposed to be proportional to the plastic shear strain after 2% shear strain (Vermeer 1990). Before this critical strain, the shear bands compact slightly and thus will either not be visible by x-ray tomography or appear with lighter tones. It is thus predictable that some shear bands with low amounts of shear strain or diffuse plastic strain zones predicted by numerical modelling will not be visible through x-ray acquisition in the analogue models.

From 2D thermo-mechanical to 3D analogue models

Thermo-mechanical experiments

All thermo-mechanical experiments were run with Paravoz using a visco-elasto-plastic flow rule including non-Newtonian temperature-dependent viscosity and full thermal coupling. The space was discretized in quadrilateral elements which are initially 200 × 200 m. The mechanical parameters of the runs displayed in this paper are listed in Table 1. The thermal parameters C_p and χ have been set constant for all materials (i.e., 1000 J kg^{-1} K^{-1} and 3 W m^{-1} K^{-1} respectively) as well as H_r, which decreases exponentially with depth according to characteristic decay of $1/e$ for 12 km, taking $H_0 = 10^{-9}$ W kg^{-1} at the surface for all the materials.

Le Pourhiet *et al.* (2004) aimed at getting a better understanding of the impact of a pre-existing weak dipping nappe on the kinematics of faulting during the early stages of rifting. The weak nappe was introduced in the model as a shallow dipping layer (20°) possessing different mechanical properties. The thickness of the nappe (1.5 km), the size of the experiments (60 × 15 km) and the boundary condition were kept constant in the study (Fig. 3a), while two parameters were raised: (1) the contrast in Maxwell relaxation time between the weak dipping nappe and the rest of the pre-rift basement, (2) the pre-rift basement relaxation time. The contrast was obtained by keeping the elastic properties constant as well as the activation energy E and the coefficient n in the creep power law, and by varying the pre-exponential parameter A for the different lithologies. These experiments show that the presence of a weak dipping nappe in the upper crust leads to a local reduction of the integrated strength of the upper crust (Fig. 3b). Consequently, strain localization occurs at a very specific location, the brittle-ductile transition (BDT) of the weak dipping heterogeneity. The parametric study has led to the conclusion that a very weak nappe was indeed needed to avoid the localization on one main decollement level (Fig. 3c) and hence to reproduce the fault spacing, the kinematics and the micro-seismicity depth distribution of the western part of the Gulf of Corinth.

However, to reduce the boundary effects in the analogue models, it was not possible to set the initial geometry of the analogue experiments to the same initial width as in Le Pourhiet *et al.* (2004). Therefore, a new thermo-mechanical run was performed using 102 km instead

Table 1. *Mechanical parameters of the thermo-mechanical experiment*

Material	Density, ρ (kg m^{-3})	A‡ (MPa^{-n})	n‡	E‡ (J mol^{-1})	Young's modulus, E (Pa)	Cohesion (Pa), C_0	Friction angle, Φ (°)
WN*	2600	5×10^3	3	1.9×10^5	7.5×10^{10}	2×10^7	30°
SB†	2600	5×10^{-6}	3	1.9×10^5	7.5×10^{10}	2×10^7	30°

*WN: weak nappe.
†SB: strong basement.
‡Coefficients for non Newtonian creep law are taken from Le Pourhiet *et al.* (2004).

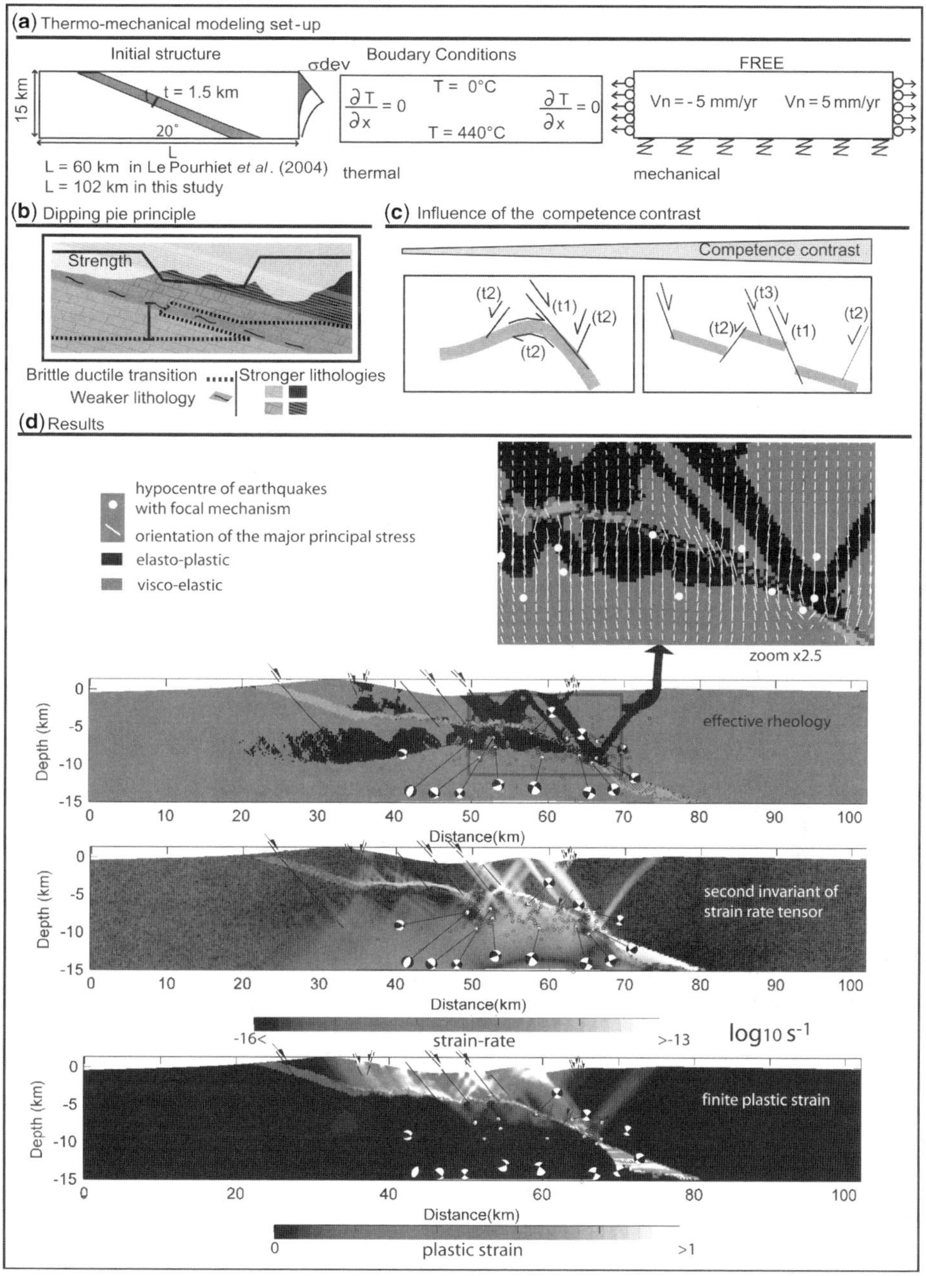

Fig. 3. (**a**) Initial geometry of the thermo-mechanical models; a unique weak nappe is represented in grey. Schematic yield strength envelope for the weak heterogeneity and the strong surrounding are drawn on the right. t is the thickness of the weak nappe. The brittle ductile transition of the weak nappe and background white material are the parameters of the study. The last two boxes display the boundary conditions of the thermo-mechanical models for the thermal and the mechanical parts, respectively. (**b**) The dipping pie conceptualizes a layered media on which stress is not applied parallel to the layers but with a slight angle. It applies to any stack of inhomogeneous units and implies local reduction of the strength (thick black line) of the upper crust associated with lateral variations of the depth of the brittle ductile transition (BDT = dashed line). (**c**) Resulting fault kinematics and distribution as a function of competence contrast. Numbers in brackets indicate the relative timing of faulting. (**d**) Comparison between the results of a thermo-mechanical run and the fault pattern and micro-seismic data from the Gulf of Corinth (Rigo *et al.* 1996).

of 60 km as initial width in the stretching direction, the rest of the parameters remaining unchanged.

The results of the new experiment are shown in Figure 3d. Although the strain rate is now slightly lower, the results obtained with this wider version are similar to those obtained in the previous study in terms of fault spacing and kinematics of faulting, i.e., some of the faults cross the weak nappe and root at the BDT of the upper crust, while the first fault and some out-of-sequence faults root at lower depth on the weak heterogeneity. After 1.5 Ma of extension, there is a good agreement with the fault spacing (i.e., finite plastic strain), the distribution of active faults, the depth distribution of the micro-seismicity, as well as the orientation of the principal stress direction at depth as inferred in the western part of the present day Gulf of Corinth.

Analogue experiments

Set-up. The dimension of the sandbox is determined by the size of the medical scanner (i.e., 0.7×0.255 m). In order to be able to compare numerical and analogue models, the material distribution (i.e., granular or PDMS) in the analogue experiments should reflect the effective rheology (i.e., elasto-plastic or viscous) of the thermo-mechanical experiments. *Sandbox* experiments, unlike numerical experiments, do not enable the temperature dependence of the viscosity to be taken into account. Therefore, an arbitrary limit between the effectively brittle and effectively viscous domain has to be introduced. In order to set this limit, a comparison between the Maxwell relaxation time and the characteristic timescale of our control parameter (i.e., the fault formation and migration relevant for the Gulf of Corinth) has been made. We arbitrarily decided that the part of the thermo-mechanical models having a Maxwell characteristic time five times lower than the time needed to create a new fault, are considered to behave viscously while the rest is considered to behave as an elasto-plastic body. As the faulting timescale is 100 ka, the limit we are seeking corresponds to a Maxwell characteristic time of 20 ka. Taking into account the shear modulus $G = 3 \times 10^{10}$ Pa, this limit corresponds to an effective viscosity of 10^{22} Pa in the thermo-mechanical models.

Because the aim of the study is to get insight into the 3D structure of the Gulf of Corinth, the initial material distribution of the analogue models (Mattioni *et al.* 2006) has been determined using the effective viscosity computed for the best fitting thermo-mechanical model (see top of Fig. 4).

For comparison, we have chosen two analogue experiments presented in Mattioni *et al.* (2006): a 3D cylindrical experiment (PH1), in which the weak nappe dipping direction is in the direction of stretching (i.e., $\beta = 0°$ in Fig. 4), and a non-cylindrical experiment (PH4), in which the dipping direction is not co-linear with the stretching direction (i.e., $\beta = 30°$ in Fig. 4).

The dip of the weak nappe in the analogue experiments depends on its strike, β, and on its apparent dip in the stretching direction, which is set constant, i.e., $\alpha = 20°$. The two analogue experiments have different scaling because they were run with different velocities of extension. As our final goal is to study the Gulf of Corinth, we have arbitrarily chosen that, at nature scale, the velocity of extension is the same but the viscosities of the viscous layers are different. The rest of the scaling parameters are summarised in Table 2.

Results of the analogue experiments. The results of the experiments PH1 and PH4 are described in the companion paper by Mattioni *et al.* (2006) but are summarized here to allow for comparison. In both experiments, the dipping slab of silicone putty (where it exists) allows for the formation of a rift zone rooted above the silicone, which localizes most of the extension in the elasto-plastic domain. In the early stages of deformation, a system of conjugate high-angle normal faults initiates at the upper tip of the dipping layer and propagates to the free surface as the horizontal offset increases at the level of the dipping slab of silicone. As a result, a roughly symmetrical graben is created. As deformation increases, the first synthetic fault remains fixed with respect to the tip of the silicone slab, whereas the rifted zone slightly propagates in the down-dip direction of the silicone level. This leads to a progressive asymmetrical widening of the basin in which the faults rotate passively to a lower angle. As horizontal strain increases, a large downward flexure progressively deforms the hanging wall of the décollement level but no fault seems to form in this area (see cross-section in Fig. 5).

In the non-cylindrical experiment (PH4), the silicone dipping slab is not present everywhere along the strike. This experiment is thus divided into three structural zones (see Mattioni *et al.* 2006), a first one where the slab is present (left side of the surface view experiment in Fig. 5), a second one where the silicone is not present (right side of the surface view experiment in Fig. 5) and a third one which corresponds to the transition zone.

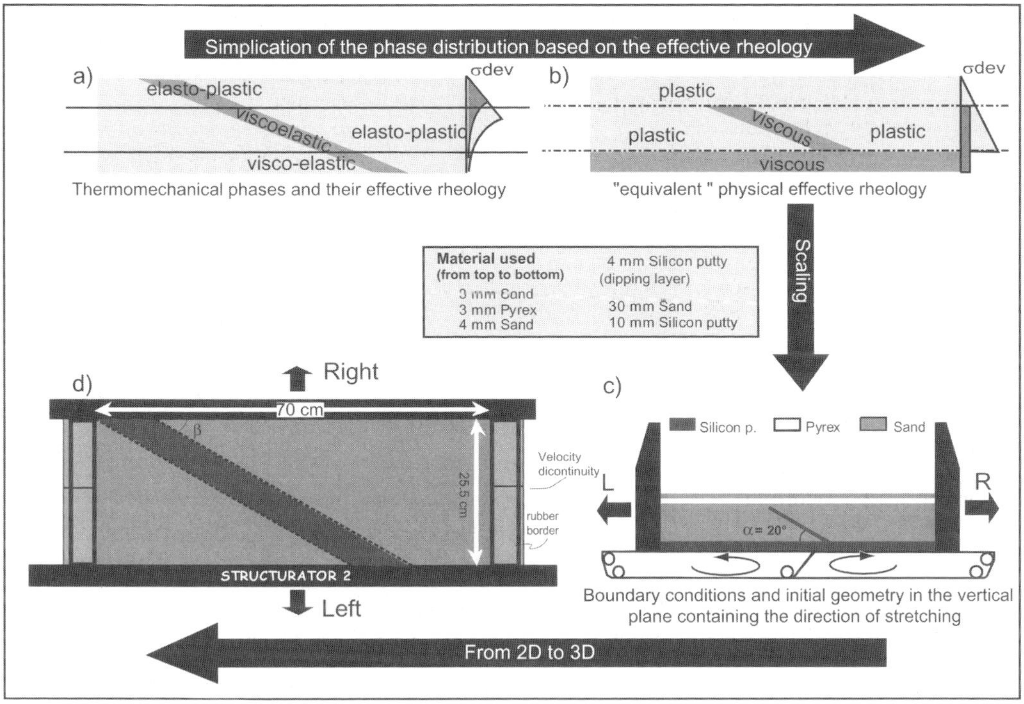

Fig. 4. From thermo-mechanical to analogue modeling: (**a**) Previous thermo-mechanical model and its yield strength envelopes. (**b**) Effective rheology is used to obtain the distribution of viscous and elasto-plastic analogue materials in the analogue experiment. (**c**) Final cross-section of the analogue model in the direction of extension. The apparent dip of the silicone putty in this direction is the same as in the thermo-mechanical model, but the boundary conditions at the base include a velocity discontinuity that is not present in the thermo-mechanical experiment. (**d**) Top view of experiment 4, $\beta = 30^\circ$.

In the first structural zone, the overall trend of the modelled rift is parallel to the strike of the silicone slab. The first fault forming at the tip of the silicone putty maintains its original orientation during the course of the experiments (i.e., parallel to the strike of the underlying viscous layer). The faults forming in the later stage of the experiments are structurally more complex. They nucleate as a system of en echelon faults roughly oriented N105. As the deformation

Table 2. *Scaling parameters for the analogue models PH1 and PH4*

Dimension	Scaling	Nature	Model
L^*	2.5×10^{-6}	4×10^3 m	10^{-2} m
g^*	1	$9.8\ ms^{-2}$	$9.8\ ms^{-2}$
ρ^*	~ 0.5	$2600\ kg/m^3$	$1000-1500\ kg/m^3$
$V^{*\ (1)}$	10^4	1 cm/yr	1 cm/hr
$V^{*\ (4)}$	10^4	1 cm/yr	2 cm/hr
		formula	
σ^*	1.25×10^{-6}	$\rho^* \times g^* \times L^*$	
$t^{*(1)}$	2.50×10^{-10}	L^*/V^*	
$\eta^{*(1)}$	3.13×10^{-16}	$\sigma^* \times T^*$	
$T^{*(4)}$	1.25×10^{-10}	L^*/V^*	
$\eta^{*(4)}$	1.56×10^{-16}	$\sigma^* \times T^*$	

(1) and (4) refer respectively to the analogue experiments PH1 and PH4.

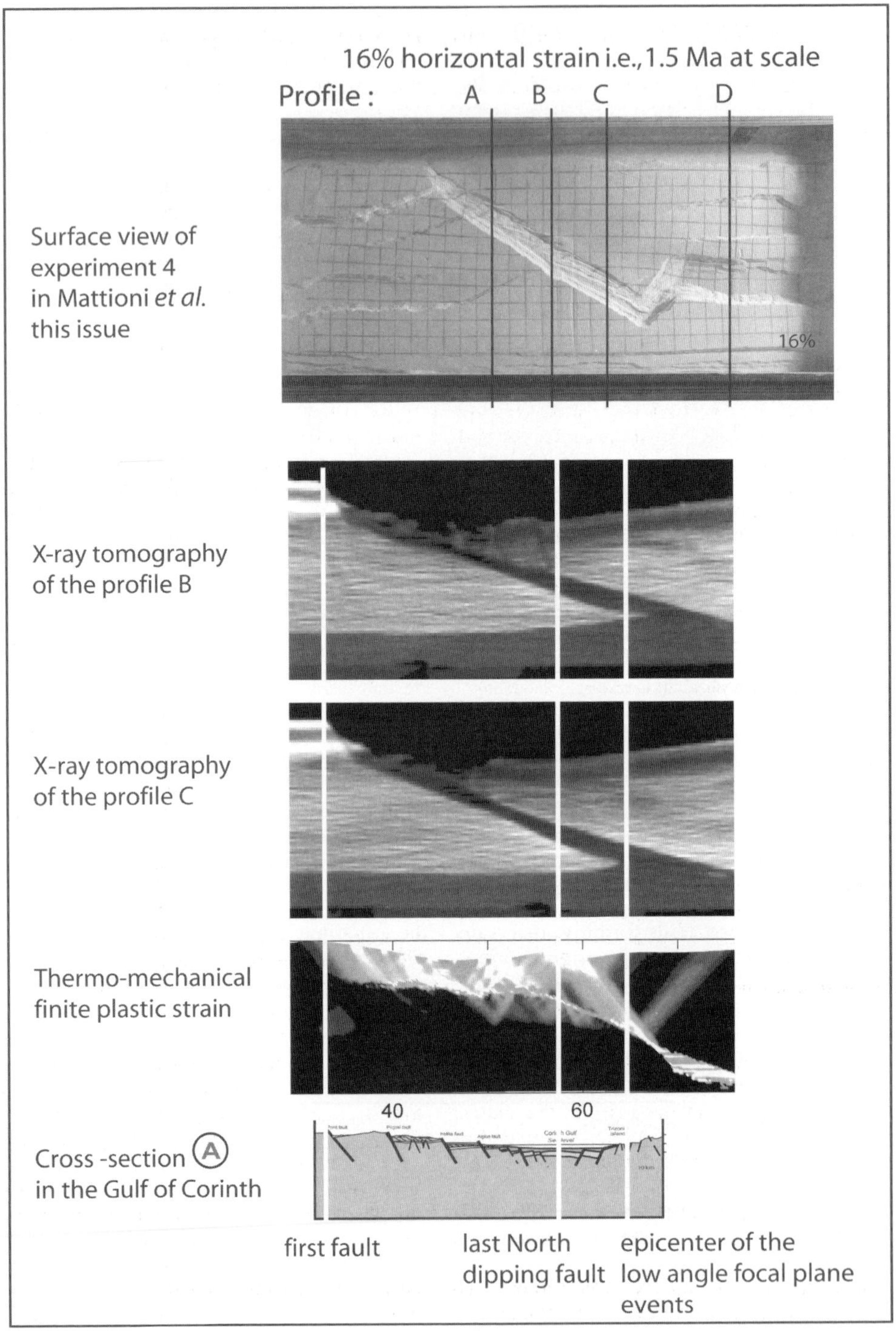

Fig. 5. A comparison between analogue model (PH4), the thermo-mechanical model and a cross-section in the Gulf of Corinth. Top: surface view of experiment PH4 after 16% strain showing the location of the 4 profiles used for comparison. Bottom: All the models are at the same scale, only details of the rifts are displayed. No vertical exaggeration.

progresses, their lateral terminations become progressively parallel to the strike of the viscous layer whereas their central segment remains orthogonal to the direction of extension. This rotation about the vertical axis gives them a roughly sigmoid shape (see surface photography in Fig. 5).

In the second structural zone, rifting is localized in wider grabens striking orthogonal to the direction of extension. Their size reflects the larger thickness of the sand pile. In the third structural zone (i.e., the transition zone), the geometry of the faults depends mostly on the initial geometry of the silicone slab termination (see experiments 2, 3 and 4 in Mattioni *et al.* 2006). However, it is worth noting that in every experiment, the existence of the silicone slab laterally in the model affects the location of the wider grabens in the second structural zone. As soon as the normal faults which border those grabens propagate laterally in the transition zone and further in the first structural zone, their strikes rotate and become perpendicular to the strike of the silicone slab (see surface view in Fig. 5). The strike rotation also occurs for the grabens which nucleate at the rear of the silicone slab (i.e., bottom/left of the surface view in Fig. 5). This rotation does not occur in the cylindrical model and thus cannot be related to border effects. This rotation of the horizontal stress field around the silicone slab suggests that the heterogeneity behaves like an elastic dislocation in the analogue model.

Comparison between thermo-mechanical and analogue experiments (PH 4). Figure 5 shows that the width of the rift as well as the fault spacing obtained in the sandbox experiments is much smaller than in the thermo-mechanical experiment and thus than observed in the Gulf of Corinth. How can those discrepancies be explained?

Some differences are certainly related to the simplifications that had to be done in terms of rheology and boundary conditions for the analogue modelling technique. Rheological assumptions fall into two main groups: (1) the elastic parameters of sand and PDMS are not scaled to the dimension of the model. In other words, elasticity is assumed to be negligible in the problem, and (2) there is no possibility of breaking the silicone putty layer because the transition from brittle to ductile behaviour is abrupt.

Concerning the boundary conditions, two major differences exist between analogue and thermo-mechanical models. Firstly, our 2D numerical models assume plane strain conditions (i.e., no strain occurs in the out-of-plane direction of the numerical model) whereas analogue experiments are 3D. Secondly, bottom boundary conditions differ between the two modelling techniques. The numerical thermo-mechanical models involve isostatic compensation at their base, whereas the analogue models have fixed velocities at their base in both horizontal and vertical direction and display a velocity discontinuity midway between their lateral boundaries where the underlying plastic sheets diverge.

In order to explore the potential effects of contrasting rheology and boundary conditions existing between the two modelling techniques, numerical benchmark tests have been conducted.

Relevance of 2D numerical benchmarks for 3D analogue models

In all the numerical mechanical models, the initial element size is 250 × 250 m and the thermal coupling is disabled because (1) it is not needed to calculate the viscosity (which is set as constant) and (2) thermal stresses or other Boussinesq terms may only have a little effect on the results but may obscure the comparison with the analogue results.

Setting the numerical benchmarks: the REF and REFSLOW models

In the analogue experiments, silicone behaves as a viscous body and sand as an elasto-plastic body. The same rheological laws are used to compute their numerical benchmarks except that for numerical reasons, the weak silicone slab has been replaced by a Maxwell visco-elastic body with a high shear modulus to ensure that it will always behave as an effectively viscous body. Sand and Pyrex have been replaced by a single elasto-plastic body (see Table 3 for the mechanical constants).

The first aim of the simplified numerical experiments is to reproduce the results of the analogue experiments. The second aim is to use them in order to get a better understanding of the discrepancies existing between the analogue and the thermo-mechanical experiments by varying in the numerical benchmarks the parameters that cannot be varied with the sandbox approach.

In order to simulate analogue experiments, numerical boundary conditions have been set as closely as possible to their analogue counterparts except that the sharp velocity discontinuity at the base has been enlarged to a zone of 10 km in order to limit repeated remeshing which leads to numerical diffusion when computing large strains with a Lagrangian strategy. Paravoz is numerically optimized to enter parameters related to lithospheric processes without previous scaling. Hence, the parameters of the benchmarks are similar to the parameters of the sandbox but at nature scale. Some parameters are not used to scale the analogue experiments, (e.g., elastic moduli), but are needed for numerical computations. To compute the reference numerical benchmarks, they are set to their most accepted values in nature.

Because the experiments PH1 and PH4 were made with two different stretching velocities and thus two different viscosities at nature scale, two reference benchmarks had to be set up. Their initial and boundary conditions are described in Figure 6. REFSLOW is similar to the 3D cylindrical experiment PH1, while REF is the closest 2D approximation of a cross-section located at profile B in the analogue model PH4.

Results of the numerical benchmarks

REFSLOW and PH1. The first benchmark test compares the 3D cylindrical analogue experiment PH1 with REFSLOW. Figure 7 displays a comparison between their topographic profiles, and highlights that conceivable differences in amplitude and width exist between the numerical experiments and the analogue experiments. Moreover, the first fault is much steeper in the analogue experiment. This analogue modelling artefact is probably due to the presence of a layer of Pyrex, which possesses a higher friction angle than sand and thus leads to a steeper (Coulombs) fault dip angle.

However, the faulting patterns are quite similar. For example, the number of faults present in the main graben as well as their spacing correlate well between the two models. This shows that the resolution used for the numerical benchmark is sufficient to distinguish between plastic shear bands at the same length-scale as those visualized by x-ray tomography.

Testing the relevance of the 2D plane strain assumption for REF and PH4. The first issue concerning the use of a mixed 2D numerical/3D analogue approach is to check the validity of the plane strain assumption through a series of cross-sections. This is important to validate the results of previous thermo-mechanical models as well as the benchmark approach. Four cross-sections oriented in the direction of extension have been chosen in the analogue experiments. Three of them (A to C) cross the weak nappe and the last profile (D) is located in an area where the weak nappe is not present. Their precise location is indicated on a surface picture of experiment PH4 at the top of Figure 5. For 30° of obliquity between the weak nappe and the direction of extension, analogue experiments clearly show two kinds of 3D structures at the surface: en echelon structures at small length scale, and sand-pile-scaled normal faults propagating from the domain where the weak nappe does not exist towards the weak nappe.

Using topography as a control parameter, profiles A, B and C have been compared to three numerical experiments in which the position of the weak nappe varies relative to the border of the model (Fig. 8). The topographic profiles obtained with these 2D plain strain numerical models are different from each other. As in the test with REFSLOW, the numerical experiments slightly underestimate the amplitudes of topography. However, the variation of width of the faulted zone in the numerical benchmarks is consistent with profiles taken in the direction of extension of the analogue model. This suggests that the variation of width of the rift along the strike of the weak nappe in the analogue experiments is not related to out-of-plane propagating structures but to the mean strain rate affecting the hanging wall of the nappe (i.e., where the rift forms). This rate depends on its position relative to the velocity discontinuity and the diverging wall. The results confirm that the weak nappe controls the depth and the location of localization and show that the topography is largely controlled by the dip of the weak nappe in the direction of extension.

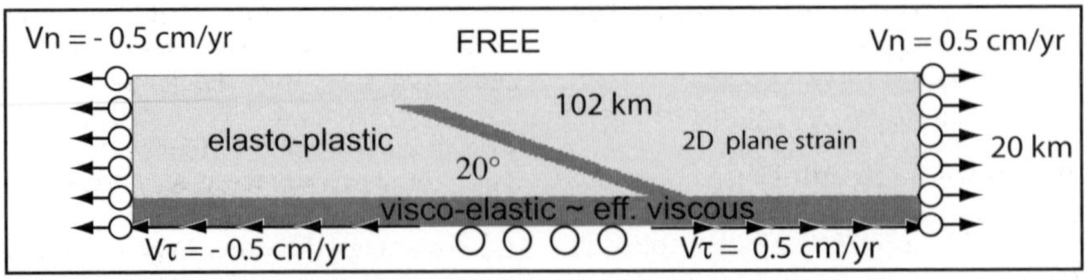

Fig. 6. The distribution of materials in the reference benchmark test, i.e., REF and REFSLOW models, is the same as in the analogue model. The velocity discontinuity at the base has been enlarged to reduce the occurrence of remeshing during computation.

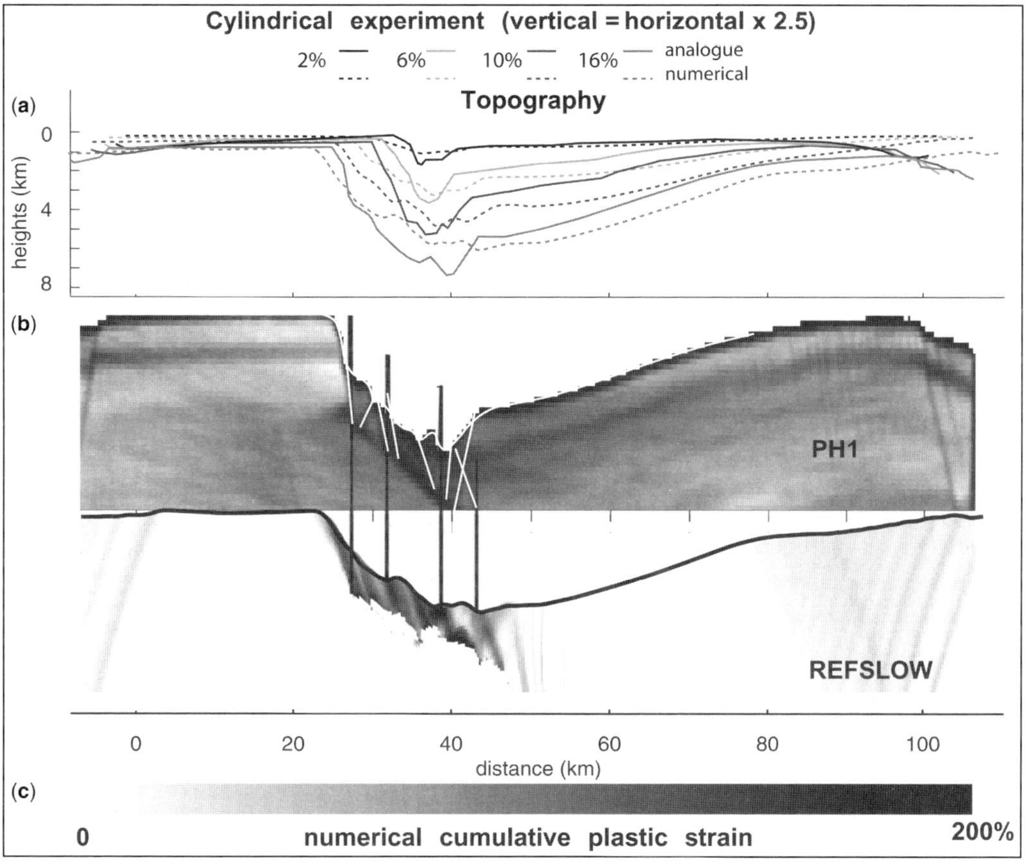

Fig. 7. (**a**): Topographic profiles of the 3D cylindrical experiment PH1 and of the numerical model REFSLOW for different finite stretching. (**b**) Tomographic vertical cross-section of PH1 at 16% stretching with faults and topography interpreted with white lines. (**c**) Cumulative plastic strain at 16% stretching for REFSLOW. The vertical black lines correlate faults between the numerical and the analogue experiments.

A comparison of the x-ray tomographic profile D (at the location where the nappe does not exist) and the cumulative plastic strain in the equivalent numerical model shows that the fault spacing is consistent. However, the location of the grabens differs and suggests that in the analogue model, the location of faults is laterally controlled by the presence of the weak nappe whereas, in the numerical model, it is controlled by boundary effects (Fig. 9).

Testing the effect of rheological assumptions on the benchmarks

Effect of visco-elasto-plasticity in the weak nappe

In the thermo-mechanical experiments, the weak nappe is broken into smaller segments during down dip migration of the subsidence. This phenomenon allows out-of-sequence faults to form at the rear of the major active faults. In the analogue experiments, the weak nappe has a purely viscous behaviour which may prevent the formation of those faults as well as the propagation of faults into the footwall of the weak nappe. To check this possibility, a visco-elasto-plastic (VEP) benchmark test has been computed. By visco-elasto-plastic, we mean that the visco-elastic part of the model has been replaced by a visco-elasto-plastic body with a similar viscosity to that in REF. The comparison between REF and VEP (Fig. 10) shows that a visco-elasto-plastic weak nappe does not affect the width of the faulted zone, but amplifies the effect of the velocity discontinuity by localizing strain rate above it.

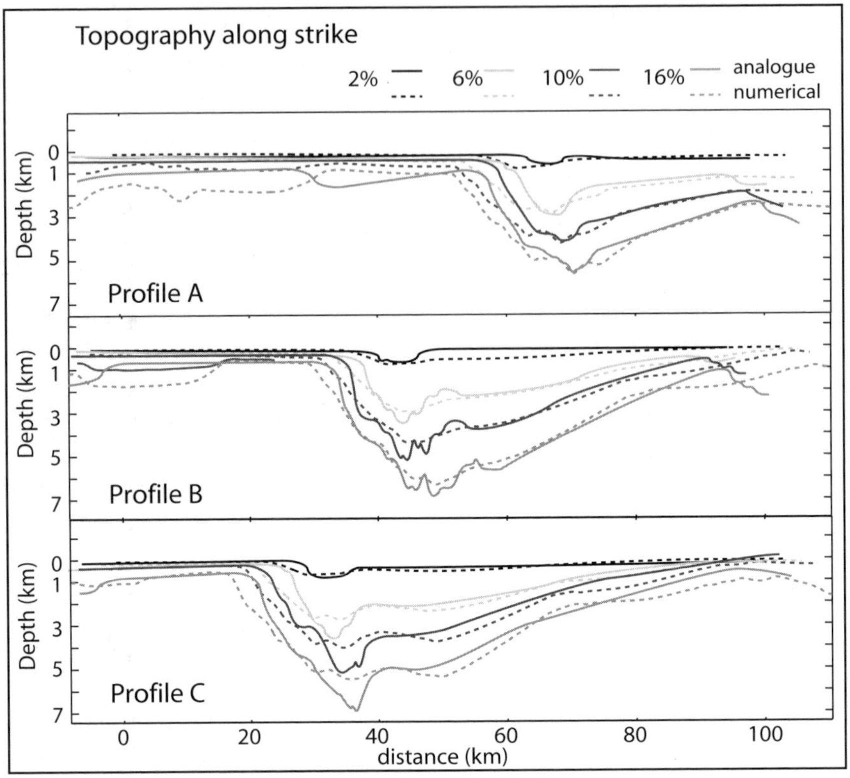

Fig. 8. Analogue model topographies of PH4 experiment are scaled up and shown with dashed lines while their equivalent numerical topographies are represented by plain lines.

Effect of elastic moduli in the elasto-plastic layer

Elastic moduli values are usually not taken into account to scale sandbox experiments. Pa unit parameters are usually scaled following the experimental constrains related to Earth's gravity field, the density of the analogue material, cohesion and dimension of the modelling box.

$$\sigma^* = \rho^* \times L^{*2} \times T^{*-2} \\ = \rho^* \times L^* \times g^* \quad (8)$$

where σ^*, ρ^*, g^* and T^* stand respectively for non-dimensional values of stress, density, gravitational acceleration ($g^* = 1$) and time.

Taking an accepted value of 10^7 Pa (e.g., Vermeer 1990) for the shear modulus of sand and the scaling parameters of experiment 4, the shear modulus G at Earth scale becomes:

$$G^{\text{at scale}} = \frac{G^{\text{sand}}}{\sigma^*} \\ A.N.:G^{\text{at scale}} = 8 \times 10^{12}\,\text{Pa} \quad (9)$$

This value is far higher than values obtained from the PREM model (Poirier 2000) where the highest value is 2.9×10^{11} Pa at the core mantle boundary. Moreover, it differs by two orders of magnitude from the values obtained

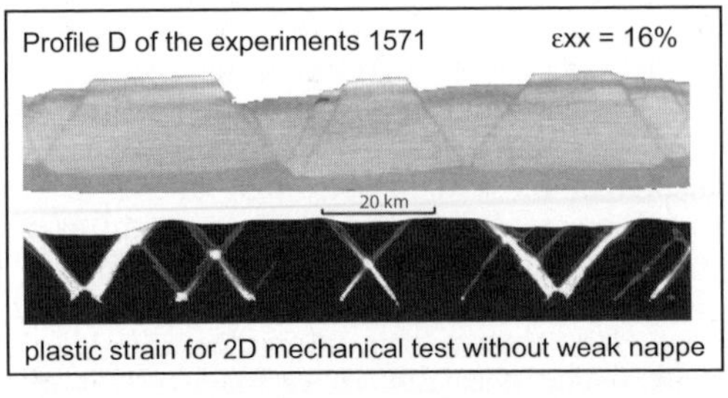

Fig. 9. Cumulative plastic strain without the weak nappe (bottom) is compared with profile D of exp. PH4 (top). In the numerical results, white stripes correspond to localized plastic strain.

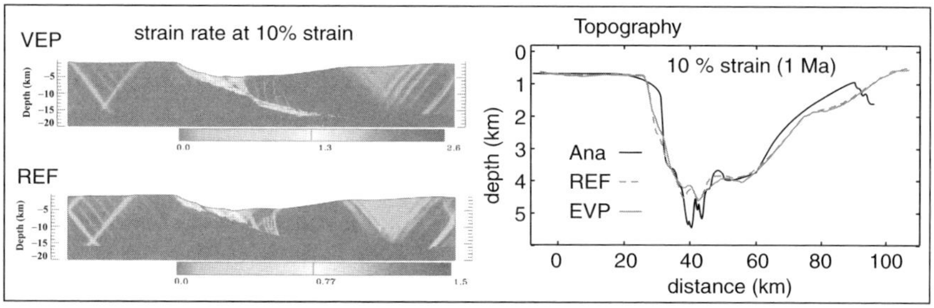

Fig. 10. The effect of a visco-elastic-plastic rheology for the weak nappe. VEP: visco-elasto-plastic; REF: reference model; ANA: experiment PH4 profile B.

for the crust, e.g., 2×10^{10} Pa (Turcotte & Schubert 2002).

However, mechanical codes based on continuum mechanics (e.g., Paravoz) account for an elasto-plastic rheology, and the elastic moduli need to be set. In the REF experiments, elastic moduli are set to values that are relevant for rocks. To test the effects of mis-scaling elastic moduli, we vary their values by one order of magnitude above and below the commonly accepted value for Earth. The resulting topographies for different finite strains are plotted in Figure 11. They suggest that increasing elastic moduli values in the elasto-plastic layer leads to an increase of the topographic amplitudes for a given finite strain.

The misfit between REF and $\times 0.1$ experiments is bigger than between REF and $\times 10$ experiments because the faults (see cumulative plastic strain on Fig. 11) do not propagate down to the basal viscous layer in the model ($\times 0.1$). However, numerical results obtained with an over-estimated elastic coefficient fit much better with the topography of analogue experiments. This means that one should properly scale the elastic properties of sand by adjusting the density or the dimension of the box as much as possible. However, it is very difficult to determine the elastic properties of granular media like sand as they are very sensitive to damage and strain-rate (e.g., Lyakhovsky *et al.* 1993). This misfit highlights a more general problem: is the sand a good analogue for crust? How far do sand and silicone interact elastically? It is beyond the scope of this study to answer those questions because the set-up, dedicated to the Gulf of Corinth, is too complicated to get quantitative insights into these problems and this kind of study should be done by carefully checking the elastic properties of the sand used for the experiments, especially its Poisson coefficient. However, it seems that benchmarks with numerical codes may be helpful in solving this problem.

Testing boundary conditions

Type of boundary conditions

Boundary conditions in the analogue experiments are characterized by a horizontal velocity discontinuity where the underlying sheets diverge and by a vertical velocity set to zero at the base of the models. This differs significantly from the isostatic conditions that were imposed in the thermo-mechanical experiments and leads to the observation that the footwall of the weak nappe hardly rotates in the analogue experiments.

In order to explore the effects of a larger range of boundary conditions, several benchmarks have been run. One of them reproduces the exact boundary conditions of the coupled thermo-mechanical computations (i.e., B4 in Fig. 12) but, as this kind of basement is not available in the analogue experimental device, intermediate boundary conditions available in the lab have been tested. One way to simulate local isostasy is to use a thicker layer of silicone putty at the base of the model. As it is not possible to increase the vertical size of the analogue model, the thickness of the basal layer has been multiplied only by a factor 2. The effect of the basal velocity discontinuity on the results has also been investigated by changing it into free slip boundary conditions in the numerical model.

Results

The strain rates resulting from a wide range of boundary conditions after 10% of bulk extension (Fig. 12) show that the velocity discontinuities, at the base and at the level of the weak nappe, interact to form some parasitic synthetic (almost vertical) faulting at the front of the basin. Such faults are sometimes observed in the analogue results.

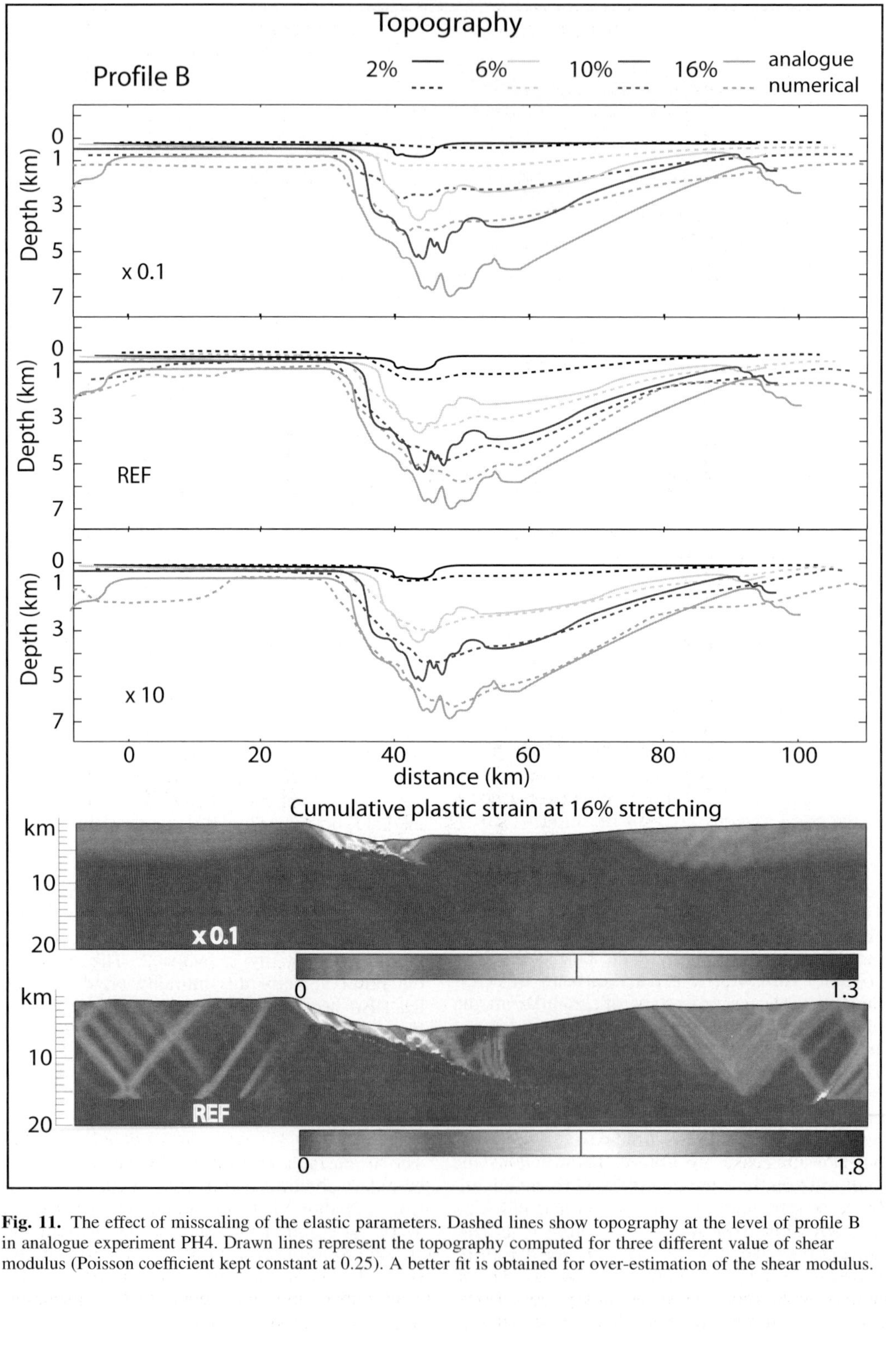

Fig. 11. The effect of misscaling of the elastic parameters. Dashed lines show topography at the level of profile B in analogue experiment PH4. Drawn lines represent the topography computed for three different value of shear modulus (Poisson coefficient kept constant at 0.25). A better fit is obtained for over-estimation of the shear modulus.

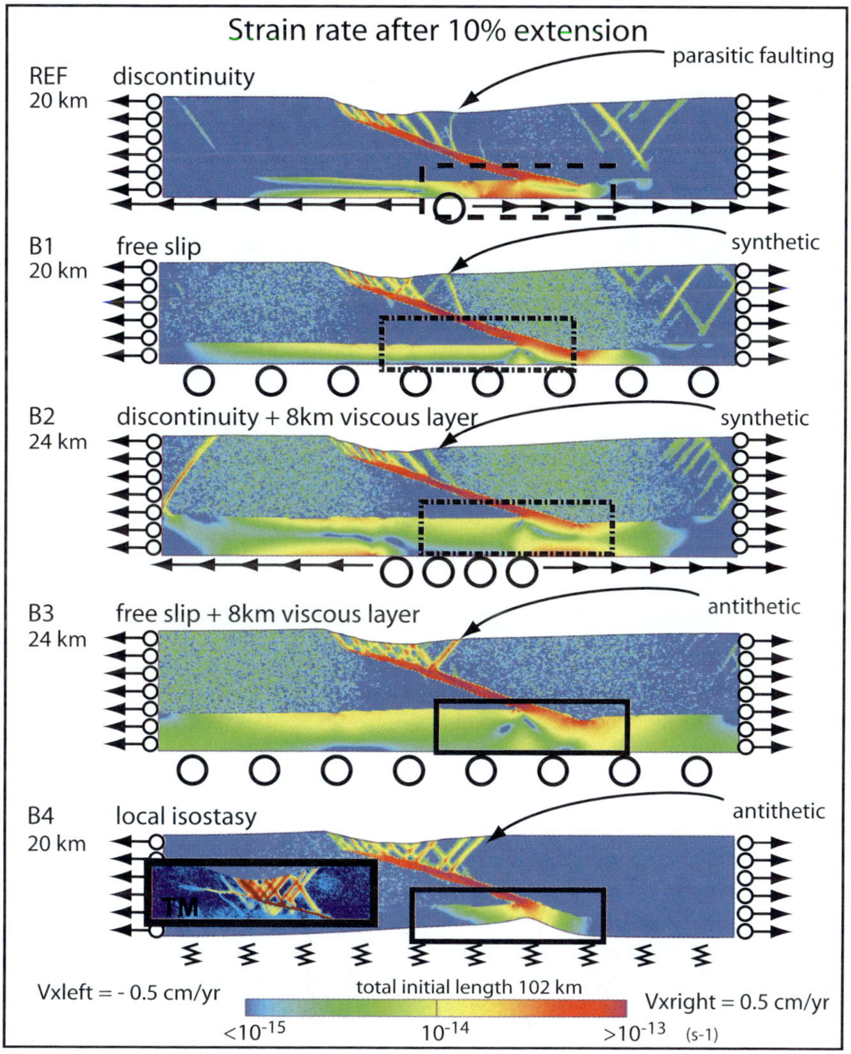

Fig. 12. Strain rate after 10% of bulk extension for a range of boundary conditions at the base: REF: Boundary conditions include a sharp velocity discontinuity in the centre. The base of the model is made of 4 km of viscous body (i.e., 1 cm of silicone putty). B1: Same geometry and vertical velocity to REF model but the base of the model is a free slip surface. B2: Same boundary conditions to REF but 4 km (equivalent to 1 cm of silicone putty) has been added at the base. B3: Same geometry to B2 with the boundary condition of B1. B4: Winkler basement equivalent to the boundary condition of the thermo-mechanical model in Le Pourhiet *et al.* (2004).

For comparison, we also add at the edge of the isostatic case in Figure 12, a detail of strain rate in the thermo-mechanical model for the same amount of bulk extension. It outlines that antithetic faults are active during the formation of the graben. These faults are only present in the B3 and B4 models. These models have their basal boundary conditions closest to those of the thermo-mechanical experiments. However, this detail of the thermo-mechanical results also shows discrepancies with the benchmarks. There are much less active faults in this more realistic experiment. This inactivity of the oldest fault is caused by the presence of faults crossing the footwall which annihilate the shear on the weak nappe and thus relate to the discussed rheological simplifications.

Table 3. *Mechanical properties of the mechanical benchmark test materials*

Properties mat.	Density (kg/m^3)	Shear modulus (Pa)	Poisson ratio	Viscosity (Pa s)	Friction angle	Dilatation angle
Elasto-plastic	2700	3×10^{10}	0.25	no	34°	0°
Visco-elastic		3×10^{11}	0.25	1.25×10^{20}	no	no

Comparing REF and B2 with the other experiments suggests that the removal of the discontinuity at the base permits the antithetic faults to form. However, increasing the thickness of the basal silicone layer is a major factor in widening the rift, because it is the only one that permits the weak layer to flatten by rotation of the footwall (exp. B4). This phenomenon was shown by Le Pourhiet *et al.* (2004) to be one of the effective mechanisms for migration of the basin in the thermo-mechanical models, because upward flexure of the footwall favours faulting (Buck 1988). The annihilation of this flexure in the analogue experiments causes (1) all the faults to remain active, (2) the suppression of the synthetic out-of-sequence faults growth, and (3) the formation of a single basin instead of multiple half-grabens. All dynamic features observed in the thermo-mechanical model thus disappear in favour of a fully kinematically constrained model.

Conclusion on the methodology

Our benchmark tests have shown that for our set-up: (1) The fact that the elastic coefficients are not correctly scaled with nature in the analogue experiments affects the models. However, in this special setting, the boundary conditions at the base and the weak nappe completely control the kinematics and it is difficult to evaluate quantitatively the effect of this mis-scaling. (2) The 2D plane strain approximation used in the thermo-mechanical experiments is valid as long as the weak nappe is present.

Analogue models give insights into expected fault patterns for a real 3D setting which, for software and hardware reasons, cannot be modelled with current numerical tools, but are very limited in terms of available boundary conditions, scaling, and/or acquisition of the results. The sharp rheological contrasts inside the model, as well as the scaling limitation, cause some parasitic phenomena which can easily be isolated using numerical benchmarks of the analogue models. In the present study these benchmark tests have shown that:

(1) The velocity discontinuity at the base of the models interferes with the discontinuity created by the rheological contrast of the weak nappe and leads to the formation of parasitic faults responsible for artefact topographic heights observed in both numerical and analogue models (profile B in Fig. 8).

(2) The vertical dimension of the modelling box, which is constrained by the size of the medical scanner used for the acquisition, does not allow us to put a thick viscous layer at the base of the experiments. The boundary conditions at the base lead to the annihilation of the formation of some of the principal structural features, i.e., faulting in the footwall and migration of the basin by flexural flattening of the weak nappe.

Thus, mixed methodology is a very useful tool because it leads to a better understanding of the features observed in the analogue models. This approach also helps to save time and money for 3D case studies because it enables some analogue experiments to be skipped.

Is the 3D model valid for the Gulf of Corinth?

The 3D analogue and 2D numerical experiments have outlined that:

(1) The results of our previous thermo-mechanical study are only valid if local isostatic compensation occurs at the scale of the upper crust; i.e., the thickness of the elasto-plastic layer should be rather small compared to the thickness of the effectively viscous lower crust. In the case of the Gulf of Corinth, this hypothesis is acceptable, because the elasto-plastic layer extends down to the brittle ductile-transition, i.e., approximately 10–12 km (Rietbrock *et al.* 1996; Rigo *et al.* 1996; Pham *et al.* 2000; Lyon Caen *et al.* 2004), and because the crust is thick enough (40–45 km, according to Makris (1978) and Tiberi *et al.* (2001) to allow for the lower crust to diffuse in short times any displacements at small length scales (5–10 km).

(2) The en echelon structure and the lack of field evidence for major transverse faults can be explained by the discrepancy between the current direction of extension and the dipping direction of the weak heterogeneity.

(3) The seismic events relocated on SW/NE trending faults may not take place on inherited faults as suggested by Lyon Caen *et al.* (2004) but rather on newly-formed transverse normal faults.
(4) The comparison between analogue models and their mechanical numerical benchmarks suggest that the presence of a weak heterogeneity acts as a localizing factor even laterally where the Phyllades nappe does not exist, e.g., in the internal zone of the Hellenides corresponding to the eastern part of the rift.

However, these 3D analogue models and their mechanical benchmarks were not able to reproduce the kinematics (in and out-of-sequence faulting) and the geometry (multiple half-grabens) of the faults in the western part of the Gulf which is obtained with the 2D thermo-mechanical models.

Neither of the two modelling techniques taken separately would have been able to propose an explanation for the 3D dynamics of faulting in the Gulf of Corinth because (1) the boundary conditions imposed by technical limitations of the analogue method do not give full access to the dynamics of the problem, and (2) 3D numerical methods are not efficient enough yet (memory and CPU) to get insights into 3D mechanically consistent shear banding with large strains.

Our study has shown that, by using a mixed approached with a large set of geological data, it is possible to overcome these limitations. The results obtained by this type of approach are, however, case sensitive, because data are needed to test the models at each stage before making new assumptions or approximations. In our study, the dip and the burial depth to the sheet of silicone putty did not change during the analogue experimentation. These two parameters were deduced from systematic numerical modelling results to reduce the number of analogue experiments.

Thanks to E. Burov for helpful discussion and free use of his version of Paravoz (a Podladchickov & Poliakov[ltd] code). J. P. Brun and J. Smit are thanked for their useful comments during the meeting as well as S. Buiter, O. Dauteuil, C. Pascal and an anonymous reviewer for their careful, time-consuming but very useful reviews.

References

ARMIJO, R., MEYER, B. *ET AL.* 1996. Quaternary evolution of the Corinth rift and its implications for the late Cenozoic evolution of the Aegean. *Geophysical Journal International*, **126**, 11–53.

AUBOUIN, J., BONNEAU, M. *ET AL.* 1970. Contribution à la géologie des Hellènides: le Gavrovo, le Pinde et la zone ophiolitique subpélagonienne. *Annales Société Geologique du Nord*, **90**, 277–306.

BENTHAM P., COLLIER, R. E. L. *ET AL.* 1991. Tectono-sedimentary development of an extensional basin: the Neogene Megara Basin, Greece. *Journal of the Geological Society*, **148**, 923–934.

BOS, B. & SPIERS, (2002). Frictional-viscous flow of phyllosilicate-bearing fault rock: Microphysical model and implications for crustal strength profiles. *Journal of Geophysical Research*, **107**, 2028, DOI: 10.1029/2001JB000301.

BRIOLE, P., RIGO, A. *ET AL.* 1999. Active deformation, of the Gulf of Korinthos, Greece: results from repeated GPS surveys between 1990 and 1995. *Journal of Geophysical Research*, **105**, 25605–25625.

BUCK, W. R. 1988: Flexural rotation of normal faults. *Tectonics*, **7**, 959–993.

CHEN, Y. & MORGAN, J. 1990. A non-linear rheology model for mid-ocean ridge axis topography. *Journal of Geophysical Research*, **95**, 17583–17604.

CAUSSE, C., MORETTI, I. *ET AL.* 2004. Kinematics of the Corinth Gulf inferred from calcite dating and syntectonic sedimentary characteristics. *Comptes Rendus Geoscience*, **336**, 281–290.

COCARD, M., KHALE, H. *ET AL.* 1999. New Constraints on the rapid motion of the Aegean region: recent results inferred from GPS measurements (1993) across the West Hellenic Arc, Greece. *Earth and Planetary Science Letters*, **172**, 39–47.

COLLETTA B., BALÉ, P. *ET AL.* 1991. Computerized x-ray tomography analysis of sandbox models: examples of thin-skinned thrust systems. *Geology*, **19**, 1063–1067.

CUNDALL, P. A. 1989. Numerical experiments on localization in frictional materials. *Ingenieur Archiv*, **59**, 148–159.

CUNDALL, P. A. & BOARD, M. 1988. A microcomputer program for modelling large-strain plasticity problems. *In*: SWOBODA, C. (ed.) *Numerical Methods in Geomechanics*, Balkema, Brookfield, 2101–2108.

DUERMEIJER, C. E., NYST, M. *ET AL.* 2000. Neogene evolution of the Aegean arc: palaeomagnetic and geodetic evidence for a rapid and young rotation phase. *Earth and Planetary Science Letters*, **176**, 509–525.

DOUTSOS, T. & KOKKALAS, S. 2001. Stress and deformation patterns in the Aegean region. *Journal of Structural Geology*, **23**, 455–472.

FLOTTÉ, N. 2002. Caractérisation structurale et cinématique d'un rift sur détachement: le rift de Corinthe-Patras, Grèce. PhD thesis, Université Paris Sud.

GHISETTI, F. & VEZZANI, L. 2004. Plio–Pleistocene sedimentation and fault segmentation in the Gulf of Corinth (Greece) controlled by inherited structural fabric. *Comptes Rendus Geoscience*, **336**, 243–249.

GHISETTI, F., VEZZANI, L. *ET AL.* 2001. Tectonic setting and sedimentary evolution of the southwest margin of the Corinth rift (Aigion-Xylocastro area). Public vol. 56207.

JACKSON, J. A., GAGNEPAIN, J. *ET AL.* 1982. Seismicity, normal faulting and the geomorphological development of the Gulf of Corinth (Greece): the Corinth earthquakes of February and March 1981. *Earth and Planetary Science Letters*, **57**, 377–397.

JOLIVET, L. & FACCENNA, C. 1999. The kinematics of back arc basins, examples from Tyrrhenian, Aegean and Japan seas. *In*: MACNIOCAILL, C. & RAYAN, P. D. (eds), *Continental Tectonics*, Geologial Society, London, Special Publications, 21–53.

JOLIVET, L., BRUN, J. P. *ET AL.* 1994. 3D kinematics of extension in the Aegean region from the early Miocene to the present, insights from the ductile crust. *Buletin Société Geologique France*, **165**, 195–209.

LATORRE, D., VIRIEUX, J. *ET AL.* (2004). Converted seismic wave investigation in the Gulf of Corinth from local earthquakes. *Comptes Rendus Geoscience*, **336**(4–5), 259–267.

LE POURHIET, L. 2004. Modélisation thermo-mécanique de l'extension continentale: developpement theorique et applications au golfe de Corinthe (Grèce), Laboratoire de Tectonique, UPMC, 241.

LE POURHIET, L., BUROV, E. B. *ET AL.* 2004. Rifting through a stack of inhomogeneous thrusts (study case in the Gulf of Corinth). *Tectonics*, **23**, TC4005.

LYKOUSIS, V., SAKELLARIOU, D. *ET AL.* 1998. Sequence stratigraphy in the N. margin of the Gulf of Corinth: implication to upper Quaternary basin evolution. *Bulletin of the Geological Society of Greece*, **XXX**, 157–164.

LYON CAEN, H., PAPADIMITRIOU, P. *ET AL.* 2004. On the seismicity of the Corinth Rift laboratory area. *Comptes Rendus Geoscience*, **336**.

LYAKHOVSKY, V., PODLADCHIKOV, Y. *ET AL.* (1993). A rheological model of a fractured solid. *Tectonophysics*, **226**, 187–198.

MAKRIS, J. 1978. the crust and upper mantle of the Aegean from deep seismic soundings. *Tectonophysics*, **46**, 269–284.

MALARTRE, F., FORD, M. *ET AL.* 2004. Preliminary biostratigraphy and 3D geometry of the Vouraikos Gilbert-type fan delta, Gulf of Corinth, Greece. *Comptes Rendus Geoscience*, **336**, 269–280.

MATTIONI, L., LE POURHIET, L. *ET AL.* 2006. Fault pattern produced by extension through a heterogeneous crust: new insights from analogue models. *In*: BRUITER, S. J. H. & SCHREURS, G. (eds), *Analogue and Numerical Modelling of Crustal Processes*. Geological Society, London, Special Publications.

MICARELLI, L., MORETTI, I. *ET AL.* 2003. Influence of depth and amount of displacement of the characteristics of normal faults, case study in the Gulf of Corinth–Greece. *Journal of Geodynamics*, **36**, 275–303.

MORETTI, I. 2004. Asymetry of the early rift structures: a comparison between the Gulf of Suez and the Gulf of Corinth. *In: COST-ACTION 625: Actives Faults: Analysis, Processes and Monitoring*, Camerino, Italy, EDIMOND, 105–111.

MORETTI, I., SAKELLARIOU, D. *ET AL.* 2003. The Gulf of Corinth: a half graben? *Journal of Geodynamics*, **36**, 323–340.

MRLINA, J. 2004*a*. Complex gravimetric investigation of active faults around Aigion, Gulf of Corinth, Greece. *Extended Abstracts, 66th EAGE Conference and Exhibition*, Paris.

MRLINA, J. 2004*b*. Structures and tectonics derived from gravity survey around Aigion, Gulf of Corinth, Greece. *Proceedings 5th International Symposium on Eastern Mediterranean Geology*, Thessaloniki, Greece, **1**, 138–140.

NAVILLE, C., SERBUTOVIEZ, S. *ET AL.* 2004. Pre-drill surface seismic in the vicinity of the AIG-10 well and post-drill VSP. *Comptes Rendus Geoscience* **336**(4–5), 407–414.

PHAM, V. N., BERNARD, P. *ET AL.* 2000. Electrical conductivity and crustal structure beneath the central Hellenides around the Gulf of Corinth (Greece) and their relationship with the seismotectonics. *Geophysical Journal International*, **142**, 948.

POIRIER, J. P. 2000. *Introduction to the Physics of the Earth's interior*. 2nd edn, Cambridge University Press.

POLIAKOV, A. N. B., PODLADCHIKOV, Y. Y. *ET AL.* 1993. Initiation of salt diapirs with frictional overburden: Numerical experiments. *Tectonophysics*, **228**, 199–210.

RAMBERG, H. 1981. *Gravity, Deformation and the Earth's Crust*. Academic Press, New York.

RANALLI, G. 1987. *Rheology of the Earth*: *Deformation and Flow Processes in Geophysics and Geodynamics*, Allen and Unwin, Boston.

RIETBROCK, A., TIBERI, C. *ET AL.* 1996. Seismic slip on a low angle normal fault in the Gulf of Corinth: evidence from high resolution cluster analysis of microearthquakes. *Geophysical Research Letters*, **14**, 1817–1820.

RIGO, A., LYON-CAEN, H. *ET AL.* 1996. A microseismic study in the western part of the Gulf of Corinth (Greece) implications for large-scale normal faulting mechanisms. *Geophysical Journal International*, **126**, 663–688.

STEFATOS, A., PAPATHEODOROU, G. *ET AL.* 2002. Seismic reflection imaging of active offshore faults in the Gulf of Corinth; their seismotectonic significance. *Basin Research*, **14**, 487–502.

TIBERI, C., DIAMENT, M. *ET AL.* 2001. Moho topography beneath the Corinth Rift area (Greece) from inversion of gravity data. *Geophysical Journal International*, **145**, 797–808.

TROTET, F. 2000. Exhumation des roches de haute pression- basse temperature le long d'un transect des Cyclades au Péloponnèse (Grèce). Implications géodynamiques. PhD thesis?, Université Paris XI.

TURCOTTE, D. L. & SCHUBERT, G. 2002. *Geodynamics*; 2nd edn, Cambridge University Press.

VERMEER, P. A. 1990. The orientation of shear bands in biaxial tests. *Geotechnique*, **40**, 223–236.

VERMEER, P. A. & DE BORST, R. (1984). Non-associated plasticity for soil, concrete and rock. *HERON*, **29**(3), 1–64.

WEIJERMARS, R. 1986. Flow behaviour and physical chemistry of bouncing putties and related polymers in view of tectonic laboratory applications. *Tectonophysics*, **124**, 325–358.

Inversion of a symmetric basin: insights from a comparison between analogue and numerical experiments

M. PANIEN[1], S. J. H. BUITER[2], G. SCHREURS[1] & O. A. PFIFFNER[1]

[1]*Institute of Geological Sciences, University of Bern, Baltzerstrasse 1–3, CH-3012 Bern, Switzerland*

[2]*Centre for Geodynamics, Geological Survey of Norway, Trondheim, Norway (e-mail: susanne.buiter@ngu.no)*

Abstract: We use both analogue and numerical experiments to study the inversion by shortening of a symmetric sedimentary basin. The combination of the two modelling techniques uses the strengths of each method to provide insight into basin-inversion processes. The experiments start with a pre-existing basin filled, in part, with weak layers simulating weak sedimentary rocks. Both footwall and hanging wall can deform freely. The physical properties of the materials used in the analogue experiments (sand and microbeads) and the numerical experiments are appropriately scaled to represent upper crustal rocks. We present a systematic study of the effects of basin infill, basin width and basin location and a sensitivity analysis to understand the effects of the boundary conditions. The results of both methods show that the graben fill accommodates most shortening. Weak layers play an important role in localising shortening with limited reactivation of pre-existing (but weakened) faults. In general, forward thrusts and back thrusts nucleate at the lateral contrast of strong and weak materials and cut across the graben-bounding faults. Weak basal detachments are required to transfer shortening to the basin region. The overall evolution of the analogue and numerical models is encouragingly similar.

Analogue and numerical modelling are used in geology to investigate the mechanisms controlling the evolution of basins and orogens. By combining these approaches, it is possible to overcome the weaknesses inherent in each and to gain insight into natural processes (see, for example, Ellis *et al.* 2004). Advantages of numerical methods include: (1) the easy quantification of results, including the possibility of tracking stresses and strains during the evolution of the model, and (2) the relatively large freedom in choice of material properties and boundary conditions. Advantages of analogue approaches include: (1) the low time consumption and expense in model construction, and (2) the relative ease of building 3D models. The two modelling techniques are, therefore, complementary, and comparisons of the modelling results can increase the credibility of the results of each technique (Koyi 1997).

In this study, we present results of analogue and numerical models of inversion of a symmetric sedimentary basin. Basin inversion is the process of shortening a sedimentary basin which was created in extension (Fig. 1). The term 'inversion', introduced by Glennie & Boegner (1981) and used by several authors (Cooper & Williams 1989; Williams *et al.* 1989; Bishop & Buchanan 1995), describes a switch in tectonic mode from extension to shortening. The shortening results in uplift of the basin fill and the reverse reactivation of normal faults (Cooper *et al.* 2001). Many previous analogue experiments on basin inversion focused on hanging wall deformation and have, therefore, used rigid footwalls (e.g., McClay 1989, 1995; Buchanan & McClay 1991; McClay & Buchanan 1992; Keller & McClay 1995; Yamada & McClay 2003). In this type of experiment, inversion results in both the reactivation of the main detachment along the rigid footwall block and the initiation of new thrust faults. Inversion has also been modelled by uplift and rotation of a single rigid 'basement' block (Koopman *et al.* 1987) or by a series of rigid metal plates simulating strong domino fault blocks (Buchanan & McClay 1992; McClay & Buchanan 1991; Mitra 1993; Vially *et al.* 1994; Roure & Colletta 1996). In these cases, the movement of the rigid blocks strongly influences the reactivation of the normal faults.

Experiments where both footwall and hanging wall are able to deform allow the study of the evolution of structures in extension and inversion in a less restricted manner. In the clay models of Mitra & Islam (1994), initial shortening results in

From: BUITER, S. J. H. & SCHREURS, G. (eds) 2006. *Analogue and Numerical Modelling of Crustal-Scale Processes*. Geological Society, London, Special Publications, **253**, 253–270.
0305-8719/06/$15.00

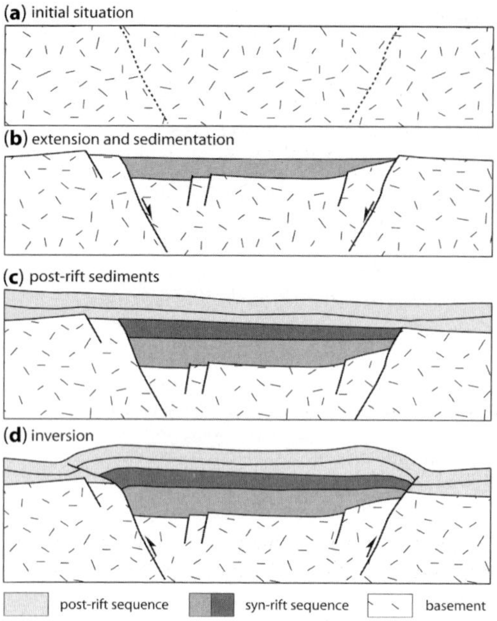

Fig. 1. Cartoon illustrating the formation and inversion of a symmetric sedimentary basin. (**a**) Basement block before extension (future normal faults are indicated). (**b**) During the initial stages of extension weak sediments (such as salt or shales) may be deposited in the shallow basin. (**c**) After extension, the syn-rift sequence is overlain by post-rift sediments. (**d**) Inversion leads to uplift of the basin fill and to reactivation of the major faults that propagate upward into the previously unfaulted post-rift sequence.

the uplift of the graben boundaries and in footwall deformation, but no fault reactivation occurs. The main normal faults in the wet clay models of Eisenstadt and Withjack (1995) are reversely reactivated in the early stages of shortening. During further shortening, low-angle thrust faults cut these reactivated normal faults. Fault reactivation is also observed in the brittle and/or viscous analogue models of basin inversion made by Dubois *et al.* (2002). In their models, Brun and Nalpas (1996) assess the effects of obliquity between the direction of shortening and the graben axis and show that the obliquity has to be less than 45° for graben inversion with fault reactivation to occur.

In our previous numerical experiments (Buiter & Pfiffner 2003) we focused on the inversion of a series of pre-defined half-graben basins. Although these experiments are not scaled equivalents of analogue studies, we could already notice some similarities in the resulting deformational structures to published analogue examples. Nielsen and Hansen (2000) and Hansen and Nielsen (2003) give examples of basin inversion in models on a lithospheric scale, which show the effect of an elastic component in material behaviour.

The focus of our study lies on the comparison of analogue modelling results of inversion of a symmetric basin with results from equivalent two-dimensional numerical models. We built the analogue experiments with granular materials sieved in a sandbox with glass sides and one mobile wall. For the numerical experiments we used a two-dimensional viscous-plastic finite element code. The experiments start with a pre-existing basin, whose bounding faults dip at 60°. We consider a setting in which part of the sedimentary deposits in the basin are weaker than adjacent rocks (Fig. 1). There are examples of inverted basins in nature, where, prior to inversion, less competent sediments are confined largely within the main graben and do not occur outside of it, e.g., in the Broad Fourteenth Basin in the North Sea and in parts of the Atlas foldbelt in northern Africa (Letouzey *et al.* 1995). In our models, therefore, part of the graben fill consists of granular materials which are weaker (microbeads) than the surrounding material (sand). Both footwall and hanging wall can deform freely. The main motivation for a pre-existing basin is to have the same starting geometry for analogue and numerical experiments, which would be difficult to achieve if we also modelled the precursor basin formation phase. We constrain the physical properties of the analogue materials through measurements. This allows us to use the equivalent (but scaled) material properties in the numerical experiments.

We investigate the inversion of symmetric basins with the aim of determining factors controlling the evolution of structures during inversion. We identify such key factors through (1) sensitivity analyses, which study the effect of variations in basal friction in analogue and numerical experiments, the effect of smoothing the velocity applied to the mobile wall and the addition of a strain-hardening phase in the numerical experiments and, (2) systematic studies of geometric parameters in both analogue and numerical experiments, i.e., location and width of the graben, and of the material infilling the basin.

Analogue models

Modelling materials

In our experiments, we used quartz sand (80–200 μm), corundum sand (88–125 μm) and microbeads (70–110 μm) as frictional analogue materials. Table 1 describes the shape of

Table 1. *Grain characteristics and material properties of quartz sand, corundum sand and microbeads*

	Quartz sand	Corundum sand	Microbeads
Grain size	80–200 μm	88–125 μm	70–110 μm
Shape			
–surface texture	irregularly rounded	conchoidal fractures	regular and smooth
–roundness	angular	angular	well-rounded
–form (elongation†)	1.63	2.00	1.04
–sphericity	low	low	high
Density (sieved) in g/cm^3	1.56	1.89	1.48
Coefficient of internal peak friction*	0.71	0.75	0.41
Coefficient of dynamic-stable friction*	0.61	0.63	0.37
Angle of internal peak friction*	35.5°	37°	22°
Angle of dynamic-stable friction*	31°	32°	20.5°
Strain-softening‡	14%	16%	9.7%
Cohesion	21 ± 18 Pa	39 ± 10 Pa	25 ± 4 Pa

*The values for coefficient and angle of friction have been rounded off.
†Elongation is ratio between longest and shortest diameter.
‡Strain softening is the difference between the coefficients of peak strength and dynamic-stable strength divided by peak strength.

the grains of these materials, i.e., their surface texture, roundness, elongation and sphericity. The density strongly depends on the handling technique used to build the models. In our experiments, we sieved the materials from a height of 30 cm. Using this handling technique the density of quartz sand is 1.56 g/cm^3, of corundum sand 1.89 g/cm^3 and of microbeads 1.48 g/cm^3.

We measured the mechanical properties of the analogue materials with a ring-shear tester (Schulze 1994). The results show that granular materials have an elastic frictional-plastic behaviour. An initial strain-hardening phase precedes failure at peak strength and is followed by a transitional phase of strain-softening until stable sliding (dynamic-stable strength) is reached (Lohrmann *et al.* 2003; Ellis *et al.* 2004; Panien 2004). Angles of internal peak friction and dynamic-stable friction are 35.5° and 31° for quartz sand, 37° and 32° for corundum sand, and 22° and 20.5° for microbeads, respectively. All materials have a low cohesion, the value of which is poorly constrained at very low normal stresses.

Analogue model set-up and experimental procedure

We built the models in a sandbox, 40 cm long, 15 cm wide and 15 cm high, with glass side panels, a plastic bottom and a mobile wall. The models consisted of alternating layers of quartz and corundum sand that were sieved in the sandbox to obtain three horizontal layers of 1 cm thick each (Fig. 2). The experiments do not include a phase of extension. A graben structure is created by means of two tilted metallic plates (each 1 mm thick) which are driven at a dip angle of 60° into the layered model. The material between the plates is carefully removed and replaced with brittle layers of alternating strength. The metallic plates are subsequently pulled out, which causes some disturbance of the normal arrangement of the grains along narrow 'fault areas'. The thickness of the disturbed zones is about 2 mm and similar to the initial thickness of faults (as seen in X-ray CT images in analogue experiments using the same granular materials). As the grains are less densely packed in these zones, they represent potential weak zones similar to a fault in a natural environment. To simulate a post-rift sequence, we sieved an additional 0.5 cm thick layer of sand on top of the model. The models are shortened by moving the right vertical wall of the sandbox directly over the base of the model with a hand-driven worm screw.

For ease of description, we refer to parts of the model situated on either side of the graben as 'right or left graben-bounding block', and the disturbed granular zones as 'graben-bounding faults' or 'graben sides'. To track the evolution of the structures, we used marker particles consisting of coloured sand at layer interfaces. These particles do not influence the properties of the surrounding granular material. Our reference model is 27.5 cm long, 15 cm wide and 3.5 cm thick. At the base of this model, the graben is 3.5 cm wide and located at 11 cm from the initial position of the mobile wall (A-ref, Fig. 2). The graben fill consists of a basal layer of weak microbeads overlain by two layers of strong quartz and corundum sand. A basal décollement layer at the base of the sandbox ensures stress transmission during shortening

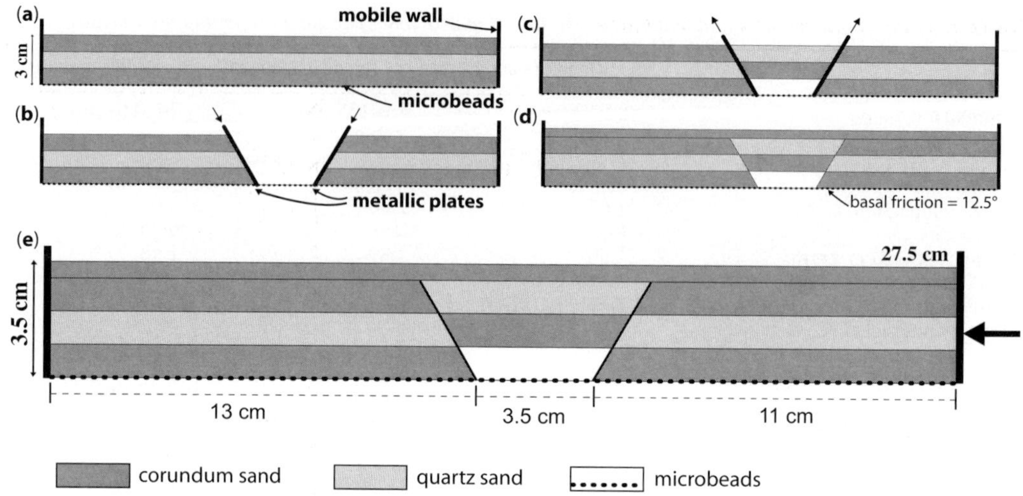

Fig. 2. Set-up of the analogue reference model. (**a**) Three layers of quartz and corundum sand are sieved in the sandbox over a thin layer of microbeads (0.5 mm thick). (**b**) Two tilted metallic plates are then driven at a dip angle of 60° into the model. The layers between the plates are removed and (**c**) replaced with alternating strength brittle layers. The metallic plates are subsequently pulled out and (**d**) an additional 0.5 cm thick layer of corundum sand is sieved on top of the model to simulate a post-rift sequence. (**e**) Analogue reference model showing width and location of graben. The layers in the graben, which are vertically offset from the layers in the left and right graben-bounding blocks, have not experienced any reverse displacement.

as in thin-skinned fold and thrust belts. The décollement layer consists of a thin layer of microbeads (0.5 mm thick) sieved over the plastic base of the sandbox. It reduces the basal friction from originally 28° to 12.5°. We tested the impact of the lateral boundary conditions (the glass side-panels) on the deformation by comparing the reference model to a model with free lateral edges made in a larger sandbox. At the end of shortening, the similarity of the two models shows that the lateral effects are negligible in our models. Table 2 lists all analogue experiments.

Numerical modelling procedure and set-up

We used a two-dimensional plane-strain finite element model (Sopale) (Fullsack, 1995) for the numerical experiments. The model is arbitrary Lagrangian Eulerian (ALE): the equations of mechanical equilibrium are solved on an Eulerian grid to which material properties are assigned using Lagrangian tracking points. The rheology of materials is viscous-plastic. To simulate the brittle behaviour of the analogue granular materials, we use a frictional-plastic Coulomb criterion:

$$(J_2')^{1/2} = p_e \sin(\phi) + C \cos(\phi)$$

where J_2' is the second invariant of the deviatoric stress tensor, p_e the dynamic pressure, C the cohesion, and ϕ the angle of internal friction. The viscosity is chosen high enough to ensure that the materials effectively deform in a rigid-plastic manner. Elastic behaviour is not included. As support for the latter assumption we note that in general, elastic deformation of granular materials such as those used in the analogue experiments seems to be minor (J. Adam, pers. comm.).

We scaled the numerical models such that similarity with the analogue models is as close as possible. One kilometre in the numerical model corresponds to one centimetre in the analogue model. The values for density and angle of internal friction are the same as those of the analogue materials (Table 1). A small amount of strain-softening of the angle of internal friction mimics the strain-softening which has been measured for the analogue materials. Softening occurs in a simple linear manner over a finite strain interval of 0.5 to 1.0 (finite strain is measured as total accumulated effective strain which is defined by the square root of the second invariant of the strain tensor). We scale the cohesion values in the numerical models by requiring that the ratio of frictional strength over gravitational force is equal for the numerical and the analogue models (Hubbert 1937; Ramberg 1981; see also Ellis *et al.* 2004).

Table 2. *List of experiments and related figures*

		Analogue model	Numerical model	Figure
Reference model		A-ref	N-ref	2, 3, 4
Basal friction		A-BF30	N-BF30	5
		A-BF13 (A-ref)	N-BF13	–
			N-BF11 (N-ref)	–
			N-BF5	–
Velocity boundary condition			N-ref	6
			N-VPw	–
			N-VPx	–
			N-VPy	–
			N-VPz	–
Strain-hardening			N-SH	7
Graben width		A-GW2	N-GW2	8
		A-GW3.5 (A-ref)	N-GW3.5 (N-ref)	–
		A-GW5	N-GW5	–
		A-GW6.5	N-GW6.5	–
		A-GW8	N-GW8	
Graben fill	all strong	A-GFas	N-GFas	9a
	all weak	A-GFaw	N-GFaw	9b
	weak at base	A-GF1b (A-ref)	N-GF1b (N-ref)	9c
		A-GF0.5bb		9d
		A-GF0.5bt		9d
	weak in the middle	A-GF1 m	N-GF1m	9c
		A-GF0.5mb		9d
		A-GF0.5mt		9d
	weak at top	A-GF1t	N-GF1t	9c
		A-GF0.5tb		9d
		A-GF0.5tt		9d
Graben location		A-GL9	N-GL9	10
		A-GL11 (A-ref)	N-GL11 (N-ref)	–
		A-GL13	N-GL13	–
		A-GL14	N-GL14	–

A cohesion of 1 Pa in the analogue models is then equivalent to 0.1 MPa in the numerical models. The cohesion is 5 MPa for 'quartz sand' and 'corundum sand' and 2.5 MPa for 'microbeads'. A model with a cohesion of 4 MPa for all materials showed that the structures in the reference model are not sensitive to small variations in cohesion.

The boundary conditions of the numerical model follow the analogue model as closely as possible (Fig. 3). The numerical reference model has 360 × 88 Eulerian elements (width × height, with an average normalized element size, as measured by the square root of element area, of 58.5 m) and a minimum of 280833 Lagrangian tracking points. Thin layers at the sides and base of the model provide boundary friction (4 Eulerian elements wide (312.5 m) or high (300 m), respectively, corresponding to around 3 mm on the sandbox scale). They are Eulerian only and are prevented from changing their overall shape. The basal frictional layer has an angle of internal friction of 11° to simulate the microbeads on plastic/wood friction, whereas the two vertical bars at the left and right side of the model have an angle of internal friction of 28° to simulate sand-wood friction. The left vertical side and the base of the model are held fixed. The right vertical side is moving with a velocity V of 1 cm/yr to the left. We smoothen the velocity at the lower part of the moving side by extrapolating it from V to zero over the height of the microbeads layer (4 elements high corresponding to 50 m in the model and to 0.5 mm on the sandbox scale, note that the vertical element size is variable) and the frictional boundary layer. This avoids numerical instabilities associated with a large velocity contrast over a too small distance. We examine and discuss the effects of this approach further below (in 'Sensitivity Analyses'). The surface of the model is a free surface to which a small amount of diffusive erosion is applied, with a diffusion coefficient of 10^{-8} m^2/s. This simulates the small amount of sand redistribution along

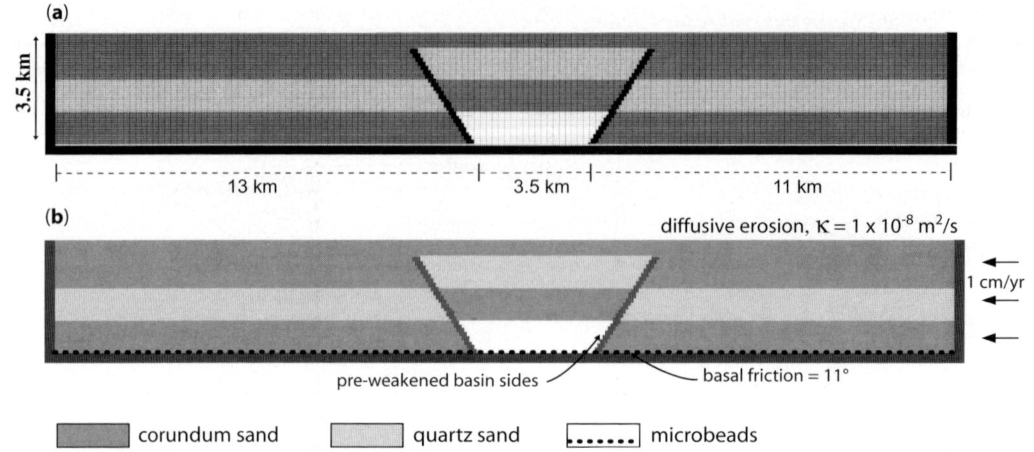

Fig. 3. Set-up of the numerical reference model. (**a**) Dimensions of the model. The grid is a subset of the Lagragian tracking grid (initial tracking grid is 6 times denser). (**b**) Boundary conditions. Thick black lines surrounding the model represent boundary friction layers.

the surface that occurs in the analogue models. We have verified that lower or higher values for the diffusion coefficient (5×10^{-9} to 5×10^{-8}) do not influence the inversion structures we observe in the numerical reference model. Even lower values for the diffusion coefficient cause surface instabilities, where corundum sand slides down when trying to attain its critical taper.

To take into account the disturbance of the graben sides when building the analogue models, we pre-weaken the basin sides of the numerical models by defining bands of quartz and corundum sand with a low (strain-softened) value for the angle of internal friction. The width of these bands is on the order of 3 Eulerian elements (around 240 m, corresponding to 2.4 mm on the sandbox scale). The basin sides and their parallel zone of weakness dip at 60° and do not extend into the post-rift layer.

The basal friction angle is the main exception to the attempted similarity between the numerical and the analogue model, as it needs to have a lower value in the numerical models. When using the friction angle measured in the ring-shear tester for the friction between microbeads and a basal plastic sheet (12.5°), more thrusts localize near the moving wall, as will be discussed later ('Sensitivity Analyses'). The reference model presented here, therefore, has a basal friction angle of 11°. With these parameters the results of numerical models seem to be in good agreement with the analogue experiments.

Caution is required when comparing results of continuum type numerical experiments with those of sandbox experiments. Due to its manual set-up, an analogue model will intrinsically have more heterogeneities than a numerical model (i.e., bedding planes always contain small irregularities). Furthermore, our numerical models do not include elasticity and dilatancy, whereas the analogue model materials do have these properties. As a consequence, shear zones in the numerical model initially develop at angles of around 45°. Table 2 lists all numerical experiments.

Evolution of analogue and numerical reference models

Analogue reference model (A-ref)

During the early stages of shortening, the horizontal displacement produced by the mobile wall propagates leftwards through a detachment at the base of the model. This displacement is accommodated between the mobile wall and the left graben-bounding fault. Stresses transmitted across the model decrease with distance from the mobile wall, resulting in compaction and diffuse deformation of both the right-hand graben-bounding block and graben fill. As a consequence, the width of the graben decreases (with 9% (0.3 cm) after 0.75 cm shortening), the graben fill bulges, the right graben-bounding block thickens and the basin sides steepen. Just prior to 1 cm of shortening, deformation localizes in a narrow zone, leading to the formation of a thrust fault (A in Fig. 4a) that nucleates at the basal transition from strong footwall to weak graben fill and attains a dip of *c.* 30°. This forward thrust accommodates most of the

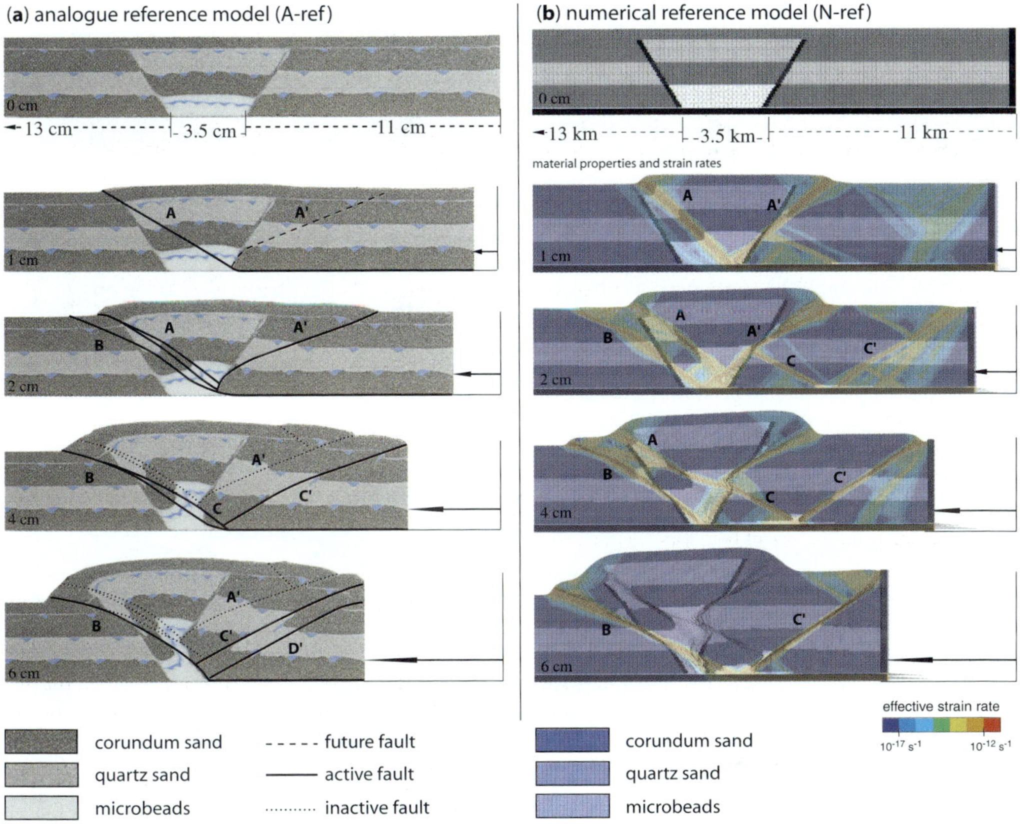

Fig. 4. Analogue and numerical reference models showing the evolution in 5 steps from the initial state until 6 cm of shortening. Note that the left side of the models is not shown completely. Transparent strain-rate plots are superimposed on the numerical models.

bulk shortening and propagates through the graben to the left graben-bounding fault and into the post-rift sediments. It connects more or less with the top of the left basin side. With increasing shortening a conjugate back thrust A′ forms in the weak microbeads. It follows the deeper part of the right graben-bounding weak zone, dipping at 60° (closest to the mobile wall), but propagates upward at a shallower dip into the footwall of the graben and the post-rift sediments. The back thrust A′ displaces the forward thrust A upward. After 2 cm of shortening, thrust A becomes inactive and a new forward thrust B forms at the base of the graben. With further convergence, new conjugate thrust faults (C and C′) nucleate outside the graben, in the hanging wall of thrust B. Forward thrust C cuts the graben-bounding fault and connects with forward thrust B. As back thrust C′ is advected along the forward thrust B, an additional back thrust D′ forms in its footwall. Most displacement takes place along the forward thrust B. The back thrusts are tilted and cease to be active one after the other as they are passively displaced along the forward thrusts C and D. During shortening, the graben fill is progressively uplifted, reaching a stable surface slope dipping at 35°, after which sand slides down along the slope. The stable slope corresponds to the angle of repose of quartz sand and corundum sand (as calculated for our set-up; Zhao *et al.* 1986; Dahlen & Suppe 1988).

Numerical reference model (N-ref)

The dynamics of the numerical model are broadly consistent with those of the analogue model, showing the same general features (Fig. 4b). The formation of strain-softened shear zones, which are comparable to the visible faults in the analogue models, accommodates deformation. A forward thrust A initiates in the right part of the graben at the basal transition from strong footwall block to weak graben fill and propagates through the graben and the post-rift layer. At the same time, a conjugate back thrust A′ forms in the weak microbeads, which follows the right basin-bounding side and then cuts through it. These conjugate thrusts resemble A and A′ in the analogue models and their activity leads to uplift of the basin fill. The initial stages in the numerical model are not characterized by overall thickening of the right graben-bounding block or by bulging of the graben fill as observed in the analogue results. Further shortening leads to the initiation of two new forward thrusts: one (B) in the left part of the model, the other (C) in the right part of the model. The short thrust C connects with the top right corner of the graben microbeads layer and later links up with A. A conjugate back thrust C′ is present in the right-hand graben-bounding block. We find that the conjugate thrusts C and C′ are quite stable features, which are neither related to the piling up of microbeads which are moving up from the basal microbeads layer, nor to the moving wall. The alignment of the former forward thrust A with the newly formed thrust C causes its reactivation after a convergence of 4 km, cutting through the still active back thrust A′. Back thrust C′ progressively displaces thrust AC, which then ceases to be active. After 6 km of shortening, only forward thrust B and back thrust C′ are active. Fewer back thrusts form in the numerical model in comparison with the analogue reference model.

Contrary to the resolution of the analogue models that makes small strains very difficult to observe, strain concentrations in numerical models can easily be observed in strain-rate plots. The strain-rate plots show which shear zones are active at different stages. They also show how deformation is often more distributed before localizing in shear zones (e.g., C′).

Sensitivity analyses

Basal friction

One of the features of the reference analogue model is the presence of a 0.5 mm thick basal layer of weak microbeads. This weak layer reduces the basal friction to 12.5° and allows the transmission of most of the shortening to the graben (A-BF13 in Fig. 5a). In the absence of such a décollement layer, the basal friction is higher (28°, i.e., basal friction between corundum sand and the plastic bottom of the sandbox) and initial deformation occurs closer to the mobile wall (A-BF30 in Fig. 5b). In this case, a

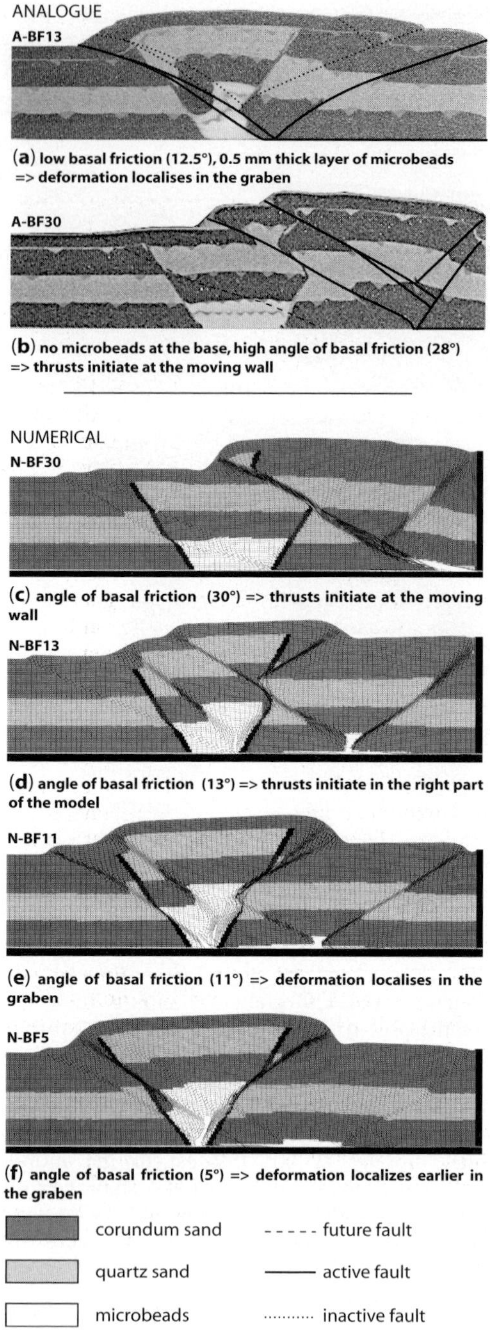

Fig. 5. Sensitivity study on the effect of basal friction. Cross-sections of analogue and numerical experiments after 4 cm and 4 km of shortening, respectively. Low friction favours graben inversion whereas high friction results in stronger effects of the boundary conditions (mobile wall). The left side of the model is not shown completely.

detachment initiates and propagates along the base of the model in the corundum layer. Conjugate thrust faults develop in the right-hand part of the model, with the forward thrust dipping shallower (30°) than the back thrust (between 45° and 55°). As shortening increases, a second and a third in-sequence thrust develop. The numerical model (N-BF30) with a high value of basal friction (30°) shows similar structures (Fig. 5c). The second in-sequence forward thrust forms from the base of the right graben side, whereas in the analogue model (A-BF30, Fig. 5b) the second (and third) thrust is located in the right graben-bounding block.

We first applied a basal friction of 13° in the numerical model (N-BF13, Fig. 5d). After 4 cm of shortening, two sets of conjugate thrusts form in the graben and in the right-hand part of the model. Contrary to the analogue model, the rightmost forward thrust does not link up with the forward thrust situated in the graben, but cuts the graben-bounding fault at the level of the corundum sand layer filling the graben. We tested lower basal frictions to examine the sequence of deformation in the graben and in the right graben-bounding block. With a basal friction of 11° (N-BF11 in Fig. 5e), deformation is initiated in the graben from the start. Thrust faults nucleate at the basal transition from strong footwall to weak graben fill. In this case, the numerical and analogue experiments are similar. Therefore, to obtain comparable results, the basal friction in the numerical models needed to be slightly weaker (11°) than in the analogue models. This small difference is interesting in itself. We speculate that it may either be caused by numerical artefacts, e.g., the way basal friction is implemented, or by a difference in the measured value of the analogue friction in the ring-shear tester and its actual value in the sandbox. In the numerical models it can be seen how microbeads from the basal layer migrate up a shear zone and facilitate further deformation on that shear zone. An even lower value for basal friction (5°) shows how deformation remains mainly focussed on the graben fill during shortening (N-BF5 in Fig. 5f).

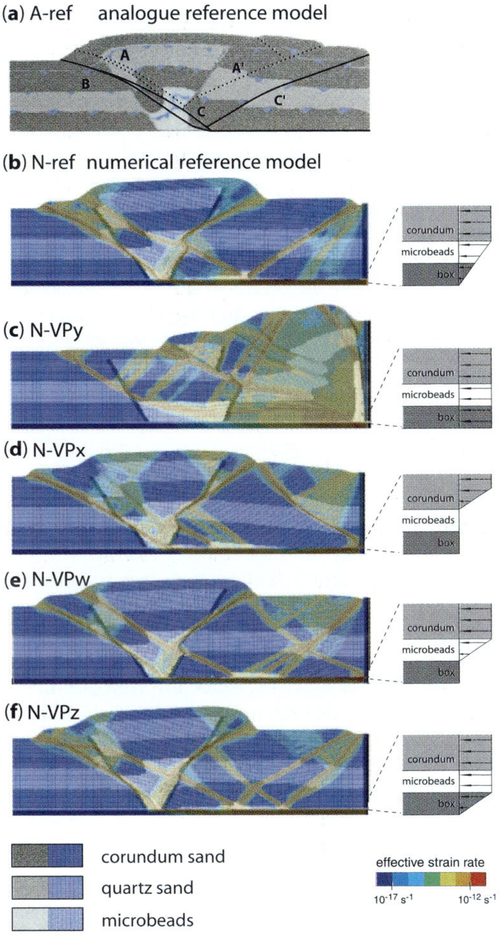

Fig. 6. Analysis of the effect of variations in implementation of the mobile wall velocity profile in the numerical reference model. A-ref, analogue model after 4 cm of shortening for comparison. Numerical models after 4 km of shortening. Sketches of the different applied velocity profiles are given on the right-hand side. 'Box' is short for frictional boundary layer. The left side of the model is not shown completely. Prefixes A and N indicate analogue and numerical models, respectively.

Velocity boundary conditions

We tried to design the boundary conditions of the numerical model such that they follow the analogue model as closely as possible, but the analogue set-up gives a relatively large freedom in the choice of appropriate velocity profile to apply to the mobile wall. In the reference numerical model (Fig. 6b), we applied a velocity V at a uniform rate along the right vertical wall and smoothened it to 0 over the layer of microbeads and the frictional boundary layer at the base. The velocity applied to the side of the frictional boundary layer allows a shear to build up near the side boundary of the model.

Figures 6b–f show model deformation after 4 km of shortening in numerical experiments in which the velocity profile at the right side boundary is varied. When we apply V to the complete right-hand sidewall (including microbeads and frictional boundary layer), thrusts form at the exit slot of the mobile wall and propagate upward to the left (N-VPy model, Fig. 6c). In case only the mobile wall itself is pushed, smoothening the applied velocity to 0 in the layer of corundum sand (model N-Vpx, Fig. 6d), a thrust fault rooting at the basal right corner of the mobile wall is initiated. A similar feature can be seen when we extrapolate the velocity to 0 over the layer of microbeads (N-VPw model, Fig. 6e). This model has difficulties in converging and requires many iterations for every timestep. The N-VPz model (Fig. 6f) is a small variation on the reference model in which the velocity V is applied not only to the wall, but also to the layer of microbeads and then extrapolated to 0 over the frictional boundary layer. After 4 km of shortening, the model is quite similar to the reference model except that two sets of conjugate thrusts are initiated in the right graben-bounding block while only one forms in the reference model.

These tests show that models are very sensitive to the velocity profile applied at the moving wall, and especially to the velocity applied at the interface between the frictional boundary layer and the basal microbeads layer. To prevent the formation of a thrust rooting at the basal corner of the mobile wall, the velocity profile needs to be smoothened and the basal microbeads do not have to be decoupled from the frictional boundary layer.

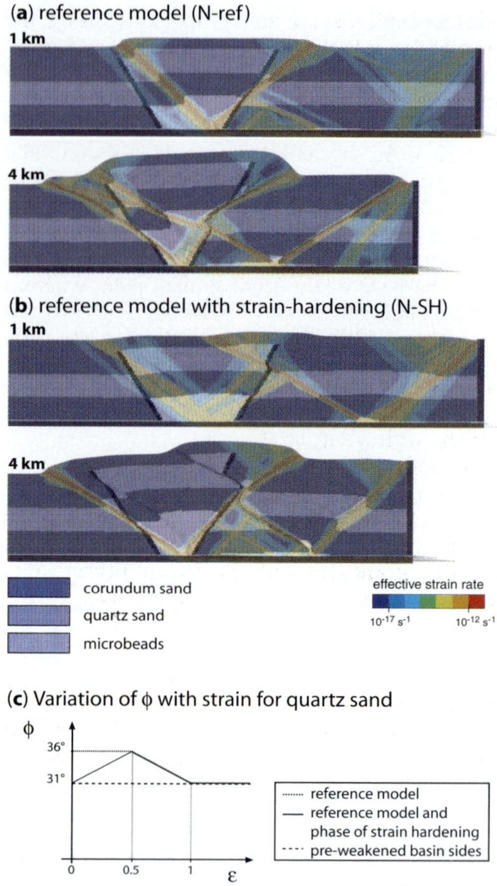

Fig. 7. Effect of a precursory strain-hardening phase in the numerical reference model. N-ref (**a**) and N-SH (**b**) models shown after 1 km and 4 km of shortening. (**c**) Example of the numerical implementation of strain-hardening and strain-softening for quartz sand. ε denotes effective strain. The left side of the model is not shown completely.

Strain-hardening

Shear tests show that the granular materials we use to build the analogue models do not have constant frictional properties. They are characterized by elastic/frictional plastic behaviour, with strain-hardening preceding failure and subsequent strain-softening until a stable value is reached (Lohrmann *et al.* 2003; Ellis *et al.* 2004; Panien 2004). In the N-SH model, we test the effect of a phase of strain-hardening that precedes softening for the numerical reference model (Fig. 7b). Both strain-hardening and softening occur in a simple linear manner, over a finite strain interval of 0–0.5 for hardening and 0.5–1 for softening (Fig. 7c). This maximizes the strain-hardening phase, which in reality might be less in magnitude and/or occur over a smaller strain interval. All microbeads, quartz and corundum sand harden first and then soften. The frictional boundary layers at the sides and the base remain at 28° and 11° respectively, whereas the pre-weakened basin sides remain at their low value for the angle of internal friction. This means that initially there is no strength contrast between the pre-weakened basin sides and the corundum and quartz sand layers. In the N-SH model, deformation first localizes mainly in the right graben-bounding block and later in the basin, while deformation in the reference model initially concentrates in the graben. Deformation is initially more diffuse than in the reference model, as lower internal friction values of quartz and corundum sand layers make the initiation of thrust faults easier.

The introduction of a phase of strain-hardening reduces the initial strength contrast between the layers constituting the model and the weak zones, represented by microbeads and pre-weakened basin sides. The influence of basin sides and microbeads on the deformation of the model is smaller in the first stages of shortening and causes more deformation in the right-hand part of the model in earlier phases than in the reference model.

Effect of geometric parameters

Influence of graben-bounding sides

The basin sides represent potential weak zones that can be partially activated during shortening. In both analogue and numerical reference models, a first forward thrust is initiated at the basal transition from strong footwall block to weak graben fill. It propagates through the graben with a dip angle of about 30° and connects with the basin top in the analogue model. The dip of the line joining the right bottom edge and the left top edge of the graben (called 'reference line' in the following) is around 30° and thus coincides with the forward thrust. We ran additional analogue and numerical experiments with varying graben widths (Figs 8a, b) to determine whether the position of the left graben side influences the orientation of the forward thrust. The dip of the thrusts is measured and compared to that of the reference line in both analogue and numerical experiments (Fig. 8c). In the analogue experiments, thrust faults dip around 30 ± 2° irrespective of the graben width. It follows that the top left basin side of the graben does not influence the forward thrust dip. On the other hand, in the numerical models the graben width influences thrust dip in some cases. In models N-GW5 and N-GW6.5, the dip of the forward thrust is lower (33° and 30°, respectively) than in the other models (±43°). When the basin becomes wider (N-GW8) or narrower (N-GW2 and N-GW3.5) the dip of the forward thrust has a 'stable' value (i.e., thrusts initially develop at angles of 45° in the numerical models). The graben shoulder is either too far away or too close to influence the fault dip.

For both analogue and numerical models, we observe that forward propagation of the thrusts is more clearly developed for the wider graben (GW8), where initially these faults form entirely within the graben.

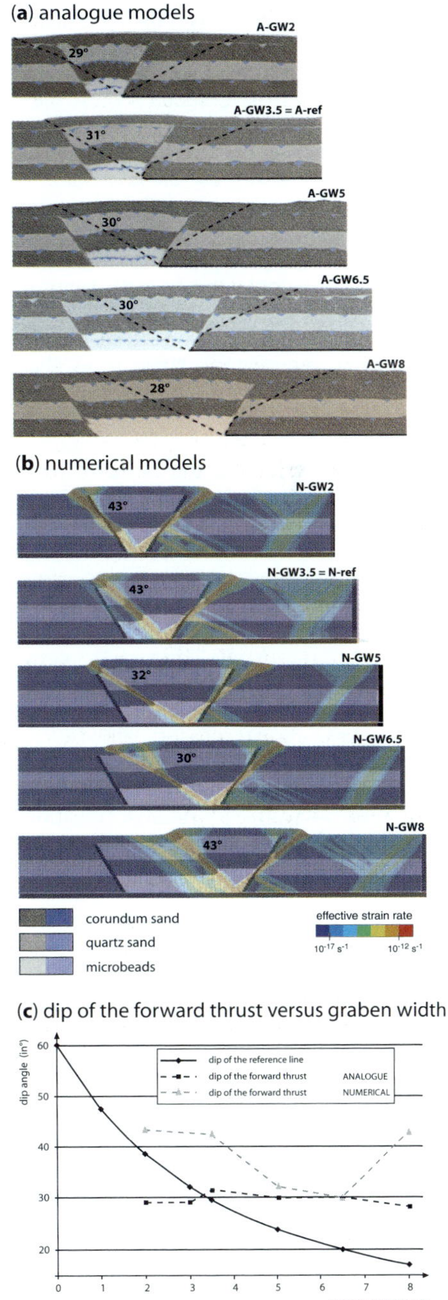

Fig. 8. Influence of the graben-bounding sides. (**a**, **b**) Analogue and numerical experiments for different graben widths. (**c**) The dip of the first forward thrust is measured after 1 cm/km of shortening and compared to that of the reference line (joining right bottom edge of the graben to its top left edge). Graben width is measured along the base of the models and is indicated in the name of the models (e.g., A-GW2 has a 2 cm wide graben). The left side of the model is not shown completely.

Influence of the mechanical stratigraphy of the graben fill

In both analogue and numerical reference experiments, the first forward thrust roots at the base in the right-hand part of the graben. It has a dip of ±30° in the analogue and ±43° in the numerical models. The location of this thrust is linked to the weak zone represented by (1) the graben side and/or by (2) the horizontal transition from strong corundum sand to weak microbeads filling the base of the graben. To test the influence of these two factors, we conducted experiments with varying graben fill. Apart from graben fill, the set-up of these models is the same as for the reference model. In the first two experiments, the graben is solely filled with either strong layers GFas (all strong) made of quartz and corundum sand or with weak microbeads layers GFaw (all weak) (Figs 9a, b). In the other three types of experiments, the position of the weak microbeads layer is at the base (GFb), in the middle (GFm) or at the top (GFt) of the graben. The weak layer is either 0.5 or 1.0 cm thick in the analogue experiments (Figs 9c, d) and 1 km thick in the numerical models. The overall evolution of the models during shortening is similar to the extent that a basal detachment and the formation of conjugate thrust faults accommodate shortening. However, the location of the forward thrusts varies between the models.

In GFas experiments (Fig. 9a), conjugate thrusts form in the right-hand part of the model. During further shortening, activity along the first forward thrust ceases and a second in-sequence forward thrust forms in the analogue model. In the numerical model further shortening is accommodated first by passive transportation of the forward thrust along the back thrust and then by the development of a new in-sequence forward thrust, whose orientation seems related to the top left pre-weakened basin side. In both analogue and numerical models, no faults form at the base of the basin sides. On the other hand, in the GFaw experiments (Fig. 9b), deformation is concentrated within the basin, and conjugate faults are clearly initiated in the graben fill in both analogue and numerical models. In the numerical model, a back thrust develops in the microbeads, parallel to the basin side. The forward thrust is tilted as it is transported by the back thrust during shortening. This leads to bifurcation of the thrust at its base in order to regain a more suitable dip. These two experiments (GFas and GFaw) indicate that the role of the microbeads in the basin is more important in localizing thrusts than the pre-weakened basin sides. Forward and back thrusts develop in the microbeads rather than using the basin side. Moreover, the basin side is not weak enough to influence the initiation of forward thrusts when the graben is only filled with strong materials.

Conjugate thrust faults also form in models where the graben is filled with both weak and strong layers (Figs 9c, d) and in each analogue and numerical experiment the forward thrusts propagate through the weak microbeads layer that partly fills the graben. Although the overall results are basically similar, some details are different in the analogue and numerical experiments. After 4 cm/km of shortening, analogue and numerical models A/N-GF1m and A/N-GF1t differ. In N-GF1m, a new forward thrust forms after 4 km shortening. It partly re-uses the former forward thrust, bifurcates into the microbeads layer at a lower dip, and continues to the surface. In this case, a single set of conjugate thrusts in the analogue model (A-GF1 m) entirely accommodates deformation. In A-GF1t, a second forward thrust forms once the first thrust has been displaced by the back thrust, whereas in the N-GF1t model, only one forward thrust passively transports successive back thrusts and bifurcates at a lower dip into the microbeads layer. Nevertheless, in both analogue and numerical experiments, the weak microbeads layer that partly fills the graben determines the location of the first forward thrust, irrespective of its depth and thickness. The transition from weak microbeads to quartz and corundum sand creates a strong lateral mechanical heterogeneity. During shortening, this heterogeneity causes the initiation of a fault that propagates backward and forward and ultimately connects with the basal detachment and breaks to the surface. This clearly shows that the location of the forward thrust strongly depends on the position of the weak layer inside the graben.

Influence of the graben location

In order to test the effect of the location of the graben in the model, we conducted several experiments in which the base of the graben is 3.5 cm/km wide and located either at 9 cm/km (GL9), 11 cm/km (GL11, reference model), 13 cm/km (GL13) or 14 cm/km (GL14) from the mobile wall.

In the numerical models (Fig. 10b), the deformation of the graben fill is similar. A set of conjugate thrust faults forms at the right-hand side base of the graben. With increasing shortening, the back thrust displaces the forward thrust, leading to the formation of a new forward thrust in the graben. In the N-GL13 and N-GL14 another set of conjugate thrusts is located far away from the graben, almost in the middle of the right graben-bounding block. In the reference model (N-GL11), a set of conjugate thrusts also forms in the right graben-bounding block, but closer to the graben. Moreover, its forward

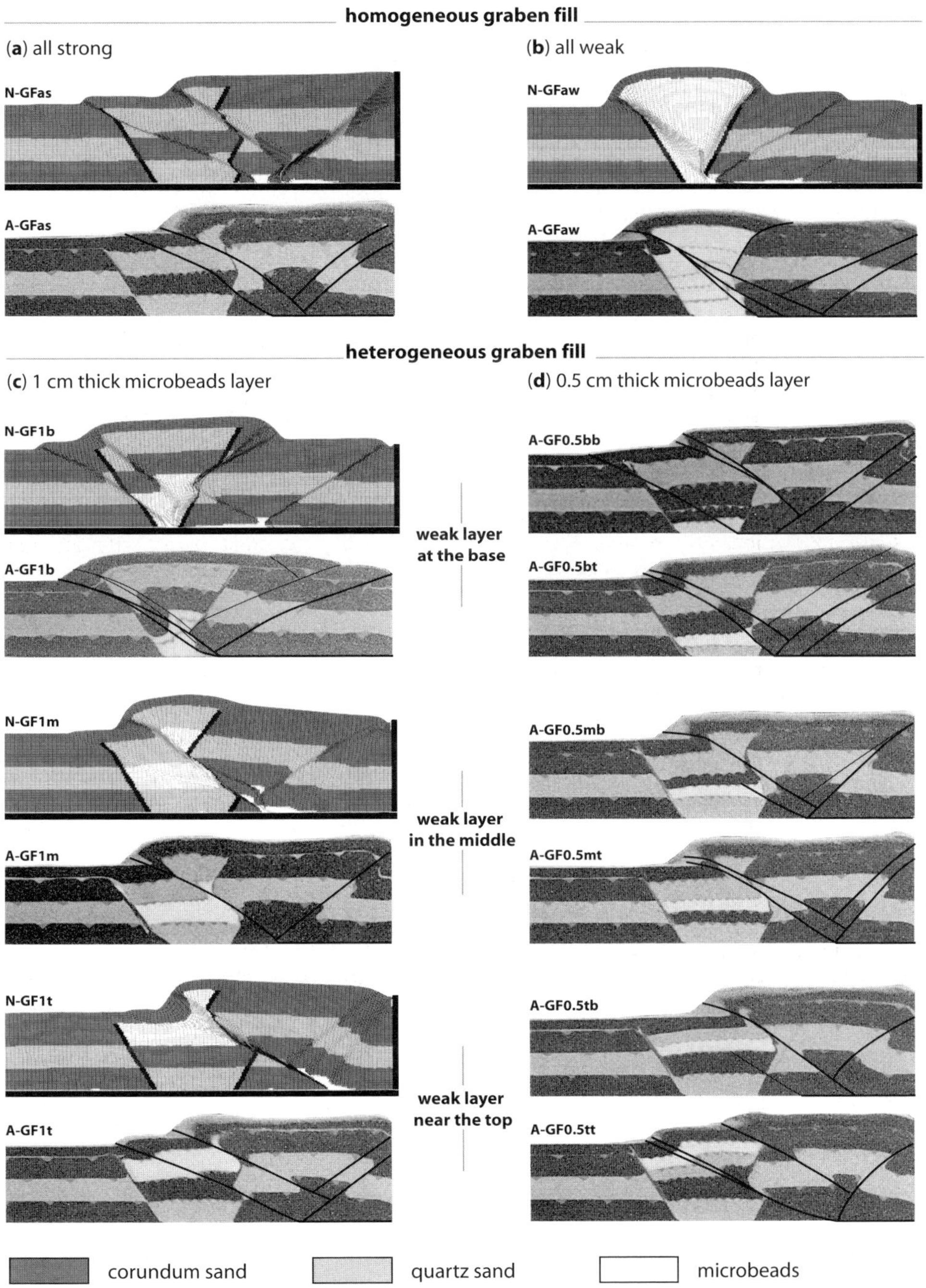

Fig. 9. The influence of varying graben fill on resulting structures during shortening. All experiments shown after 4 cm/km shortening. The left side of the model is not shown completely. Prefixes A and N indicate analogue and numerical models, respectively.

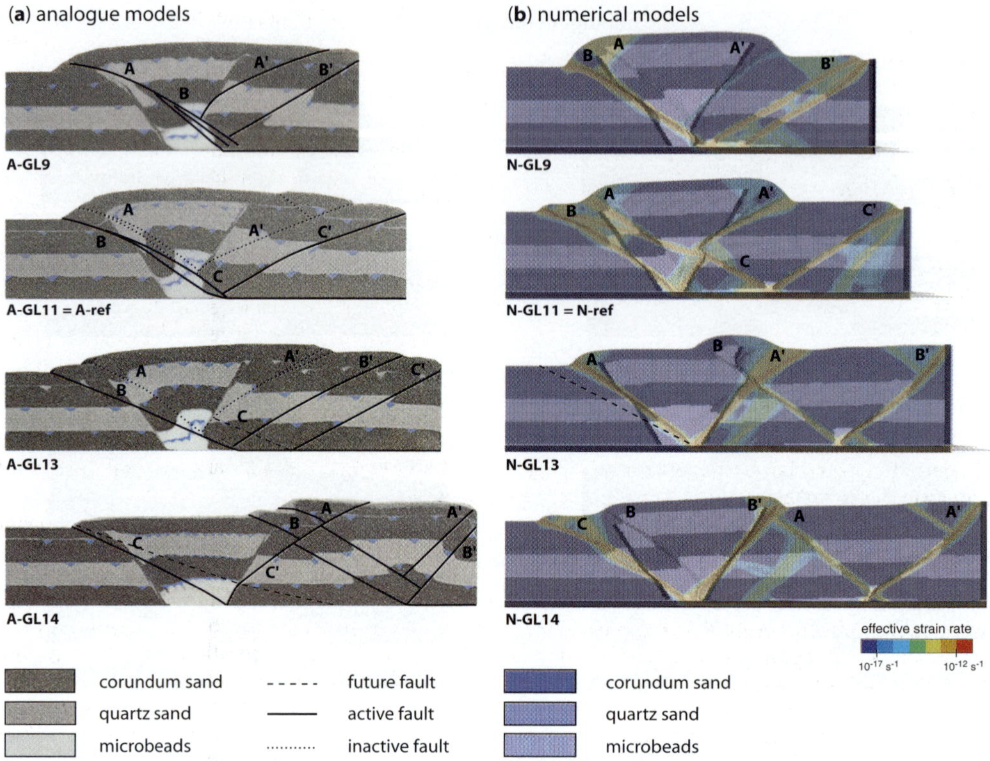

Fig. 10. Test of the influence of graben location. Analogue and numerical experiments are shown after 4 cm/km of shortening. The distance between the mobile wall and the right-bottom edge of the graben is 9 cm/km (GL9), 11 cm/km (GL11 = reference model), 13 cm/km (GL13) or 14 cm/km (GL14). The left side of the model is not shown completely.

thrust aligns with the first forward thrust that is thus reactivated.

In the analogue models A-GL9, A-GL11 and A-GL13 (Fig. 10a), the same mechanism of successive sets of conjugate thrusts accommodates shortening. In the A-GL9 model, the second forward thrust is superimposed on the first one. In the A-GL11 and A-GL13 models, the second set of conjugate thrusts propagates on both sides below the first set, and in the A-GL13 model, a third set of conjugate thrusts appears in the right part of the model. In the A-GL14 model, two sets of conjugate thrusts form in the right graben-bounding block, close to the mobile wall.

In the analogue and numerical GL9, GL11 and GL13 models, deformation localizes first in the graben, prior to localizing in the right part of the model, irrespective of the distance between the graben and the mobile wall. In case of a short right-hand graben-bounding block (GL9), deformation is concentrated in the graben, leading to enhanced bulging and inversion of the graben fill. In contrast, when the right graben-bounding block is longer (GL13), it undergoes internal deformation. As a consequence, less graben fill is expulsed at identical stages of shortening. Moreover, in case of a wider model (A-GL14 and N-GL14), deformation localizes first in the right part of the model.

It results from these experiments that the length of the right graben-bounding block does have an effect on the resulting structures during shortening: the amount of displacement and uplift of the graben varies as a function of its distance from the mobile wall for a certain amount of shortening. Clearly shortening is more effectively transferred to the graben along the basal frictional detachment for short distances. In case of a wider model, deformation takes longer to reach the basin and part of the shortening is taken up by thrusts, which form in the right part of the model. Nevertheless, the inversion of the basin itself is very similar between all these analogue and numerical models.

Discussion

The geometry of the models (i.e., graben width and location), the graben fill, the basal friction, and the velocity boundary condition and strain-hardening phase (for the numerical models) influence to a higher or lesser degree the structures observed in our basin inversion experiments. The effect of these parameters on the structures allows the identification of key factors and points of preferential strain concentration shared by both analogue and numerical experiments.

Identification of key parameters

Our sensitivity study shows that a low value for basal friction is essential for the transmission of deformation through the model. A low basal friction allows the basal detachment to propagate until it reaches a weaker zone (i.e., the weak graben fill). In case of higher basal friction, thrust faults initiate closer to the mobile wall before localizing in the graben of the numerical model or before propagating leftwards in the analogue model. Deformation also strongly depends on the velocity boundary condition applied at the mobile wall. It appears from the numerical tests that the velocity profile needs to be smoothened toward the base of the model to allow the propagation of a basal detachment and avoid thrusts propagating upward from the base of the mobile wall. The addition of a phase of strain-hardening prior to the phase of strain-softening leads to a decrease in the initial strength contrast between the graben-bounding blocks and the weak zones formed by the microbeads layer or the pre-weakened basin sides. Deformation in the reference model initially concentrates in the graben while in the strain-hardening model, deformation localizes first in the right graben-bounding block and later in the graben. The numerical model is not sensitive to small variations in cohesion and amount of surface diffusive erosion.

Analogue and numerical experiments demonstrate that the location of initiation of the forward thrust in the graben is independent of the graben width. Contrary to analogue models, the top left graben-bounding fault in numerical models clearly influences the dip of the forward thrust (N-GW3.5 and N-GW5). For both analogue and numerical models, forward propagation of the thrusts is more clearly developed for the wider graben (GW8), where initially these faults form entirely within the graben. Both analogue and numerical models are sensitive to the mechanical stratigraphy of the graben fill. The experiments demonstrate that the initiation of the first forward thrust strongly depends on the location of the weak layer of microbeads inside the graben. The overall geometry of the analogue and numerical models are similar after 4 cm/km of shortening. The geometry of the graben after inversion hardly depends on graben location in the model. Shortening of the basin is accommodated by the initiation of a set of conjugate thrusts rooted at the base of the right graben-bounding fault, irrespective of the distance of the graben from the moving wall. For large graben-mobile wall distances (GL-14), deformation first localizes in the right part of the model before reaching the graben. Results of analogue and numerical experiments for varying length of the right graben-bounding block are overall in good agreement.

From these observations, it appears that basin fill, basal friction and velocity boundary conditions are key parameters. The weak basin fill is important in localizing shortening in the sedimentary basin, while the basal friction and the velocity boundary condition applied to the mobile wall control the transmission of the deformation to the graben. The pre-weakened basin sides seem to play a less important role than the weak basin fill, but they nevertheless influence the dip of the forward thrust in some numerical models. Basin width and graben location do not control basin inversion structures, but do influence the amount of inversion.

Identification of points of preferential strain concentration

In several experiments, thrust faults seem to initiate at similar locations. This observation seems to indicate that there are points of preferential strain concentration in the models, which are the loci of fault initiation. These 'points of weakness' are shared by several experiments and can be identified thanks to models of varied graben width, infill and location.

In the experiments of varied graben fill, the location for the forward thrust totally depends on the position of the weak microbeads layer inside the graben. The layer of microbeads constitutes the main point of weakness for both analogue and numerical experiments. This is not the case for the pre-weakened basin sides, which are not weak enough to be responsible for the initiation of thrust faults. On the other hand, the top edge of the left basin side may influence the dip of the propagating thrust faults and therefore is considered to be a point of weakness. Its influence is less obvious in the analogue models.

We define two additional points of strain concentration in the numerical experiments, related to the upper corners of the layer of microbeads that form a weak polygon. These points are activated at different stages of shortening. In several numerical experiments, a second and a third forward thrust form in the graben, cutting the graben sides and going through the left and the right top microbeads layer, respectively. The analogue experiments only rarely use these points of weakness.

Comparison of analogue and numerical results

In general the results of the analogue and numerical experiments are in good agreement. We do not observe reactivation of normal faults in the models. The back thrust initiated in the graben propagates parallel and close to the basin sides as the initiation of a thrust in the microbeads is easier than reactivation of the graben-bounding faults. The formation of forward thrusts that initiate at the basal transition from strong footwall to weak graben fill and propagate through the graben and into the post-rift sediments accommodates most of the shortening. The fact that weak sedimentary layers tend to localize deformation during shortening was previously found in numerical models (Buiter & Pfiffner 2003), in analogue models (Brun & Nalpas 1996) and described in natural examples (e.g., Vially *et al.* 1994; Roure & Colletta 1996). Brun and Nalpas (1996) compare their analogue models, in which a ductile layer allows décollement of the cover during inversion, to the Broad Fourteens basin, where the Zechstein salt layer facilitates decoupling of the sedimentary cover from its basement and thrusting of the graben fill onto the basement. Along the North African margin, a layer of Triassic salt constitutes the main décollement level during shortening (Vially *et al.* 1994), whereas in the Alps a weak Triassic layer partly controls the location of thrusts faults (Roure & Colletta 1996).

Despite the overall agreement between analogue and numerical model results, some discrepancies can be found between the two modelling techniques. Analogue models with low basal friction (12.5°) and a weak basal layer of microbeads in the graben show initial bulging of graben fill and cover, and thickening of right graben-bounding block, in contrast to equivalent numerical models. Bulging of graben fill during initial stages of shortening has been described in previous analogue experiments on basin inversion (Panien *et al.* 2005) and in southern England (Chadwick 1993) where the inversion of the Wessex Basin occurred by bulk shortening of the graben-fill and produced a regional upwarp. Chadwick (1993) suggests that bulging takes place by a combination of pure-shear shortening and localized minor reverse faulting in the young, poorly lithified and therefore relatively weak graben-fill sediments and that horizontal shortening at depth is progressively transformed into predominantly vertical motion in the shallower parts of the basin. In the analogue models, diffuse deformation initially accommodates shortening, whereas in the numerical models shortening seems to localize already in early stages along shear zones. In addition, lateral compaction could also play a role in the analogue model. The compaction of the granular material during the initial stages of deformation precedes its decompaction along shear zones during further shortening (Lohrmann *et al.* 2003). Dilation could already have occurred before failure along future shear zones. This dilation could lead to bulging prior to the development of throughgoing shear zones.

Another discrepancy is the influence of the position of the graben-bounding faults on the dip of the first-formed thrust faults in some numerical models, which is not apparent in the analogue models. This discrepancy could imply that the numerical model is more sensitive to lateral heterogeneities, as is the case for the graben-bounding faults.

The overall evolution of the analogue and numerical models is encouragingly similar, which increases faith in the robustness of these models. The combination of the analogue modelling technique, which is relatively simple, inexpensive and not very time consuming, with the numerical technique, in which the results are easily quantified and the input parameters easily modified, is really effective and fruitful as it allows to overcome the weakness inherent in each of the methods and thus to get more insight into the natural processes.

Concluding remarks

Our results of analogue and numerical experiments of inversion of a symmetric basin document a similar structural evolution. The two modelling techniques show that weak layers strongly control the development of thrust zones. The approximations made in the reference numerical experiment (smoothening of the velocity profile toward the base of the model, slightly lower basal friction, absence of strain-hardening phase, pre-weakening of the basin sides) are thus considered to be reasonable. The fact that the basal layer of microbeads needs to be weaker to obtain comparable results and the fact that there are more zones of weakness in the numerical models suggests that numerical models are more sensitive to small parameter changes than analogue models. The comparison of the two modelling techniques brings additional information about the mechanisms of deformation involved in analogue models and can be seen as a validation of numerical codes. The positive results of such a comparison enhance the credibility of modelling results and underlines the accuracy and applicability of both modelling techniques.

The numerical calculations used software (Sopale) developed by P. Fullsack. S. B. thanks C. Beaumont for helpful discussions on the numerical sandbox. A. Liechti is thanked for technical assistance. Funding by the Swiss National Science Foundation (project number 2000-067952.02) is gratefully acknowledged. We thank Martha Withjack and Hemin Koyi for their constructive reviews.

References

BISHOP, D. J. & BUCHANAN, P. G. 1995. Development of structurally inverted basins: a case study from the West Coast, South Island, New Zealand. *In*: BUCHANAN, J. G. B. & BUCHANAN, P. G. (eds) *Basin Inversion*. Geological Society, London, Special Publications, **88**, 549–585.

BRUN, J. P. & NALPAS, T. 1996. Graben inversion in nature and experiments. *Tectonics*, **15**, 677–687.

BUCHANAN, P. G. & MCCLAY, K. R. 1991. Sandbox experiments of inverted listric and planar fault systems. *Tectonophysics*, **188**, 97–115.

BUITER, S. J. H. & PFIFFNER, O. A. 2003. Numerical models of the inversion of half-graben basins. *Tectonics*, **22**, 11–1/11–16.

CHADWICK, R. A. 1993. Aspects of basin inversion in southern Britain. *Journal of the Geological Society*, **150**, 331–322.

COOPER, M. A. & WILLIAMS, G. D. 1989. *Inversion tectonics*. Geological Society, London, Special Publications, **44**.

COOPER, M., WEISSENBERGER, J. *ET AL.* 2001. Basin evolution in Western Newfoundland: new insights from hydrocarbon exploration. *American Association of Petroleum Geologists*, **85**, 393–418.

DAHLEN, F. A. & SUPPE, J. 1988. Mechanics, growth, and erosion of mountain belts. *Geological Society of America Special Paper*, **218**, 161–178.

DUBOIS, A., ODONNE, F., MASSONNAT, G., LEBOURG, T. & FABRE, R. 2002. Analogue modelling of fault reactivation: tectonic inversion and oblique remobilisation of grabens. *Journal of Structural Geology*, **24**, 1741–1752.

EISENSTADT, G. & WITHJACK, M. O. 1995. Estimating inversion: results from clay models *In*: BUCHANAN, J. G. B. & BUCHANAN, P. G. (eds) *Basin Inversion*. Geological Society, London, Special Publications, **88**, 119–136.

ELLIS, S., SCHREURS, G. & PANIEN, M. 2004. Comparisons between analogue and numerical models of thrust wedge development. *Journal of Structural Geology*, **26**, 1659–1675.

FULLSACK, P. 1995. An arbitrary Lagrangian-Eulerian formulation for creeping flows and its application in tectonic models. *Geophysical Journal International*, **120**, 1–23.

GLENNIE, K. W. & BOEGNER, P. L. E. 1981. Sole Pit inversion tectonics. *In*: ILLING, L. V. & HOBSON, G. D. (eds) *Petroleum Geology of the Continental Shelf of North-West Europe; Proceedings of the Second Conference*. London, 110–120.

HANSEN, D. L. & NIELSEN, S. B. 2003. Why rifts invert in compression. *Tectonophysics*, **373**, 5–24.

HUBBERT, M. K. 1937. Theory of scale models as applied to the study of geologic structures. *Bulletin of the Geological Society of America*, **48**, 1459–1520.

KELLER, J. V. A. & MCCLAY, K. R. 1995. 3D sandbox models of positive inversion. *In*: BUCHANAN, J. G. B. & BUCHANAN, P. G. (eds) *Basin Inversion*. Geological Society, London, Special Publications, **88**, 137–146.

KOOPMAN, A., SPEKSNIJDER, A. & HORSFIELD, W. T. 1987. Sandbox model studies of inversion tectonics. *Tectonophysics*, **137**, 379–388.

KOYI, H. 1997. Analogue modelling: from a qualitative to a quantitative technique: a historical outline. *Journal of Petroleum Geology*, **20**, 223–238.

LETOUZEY, J., COLLETTA, B., VIALLY, R. & CHERMETTE, J. C. 1995. Evolution of salt-related structures in compressional settings. *In*: JACKSON, M. P. A., ROBERTS, D. G. & SNELSON, S. (eds) *Salt Tectonics: a Global Perspective*. AAPG Memoir, **65**, 41–60.

LOHRMANN, J., KUKOWSKI, N., ADAM, J. & ONCKEN, O. 2003. The impact of analogue material properties on the geometry, kinematics and dynamics of convergent sand wedges. *Journal of Structural Geology*, **25**, 1691–1711.

MCCLAY, K. R. 1989. Analogue models of inversion tectonics. *In*: COOPER, M. A. & WILLIAMS, G. D. (eds) *Inversion Tectonics*. Geological Society, London, Special Publications, **44**, 41–59.

MCCLAY, K. R. 1995. The geometries and kinematics of inverted fault systems: a review of analogue model studies, *In*: BUCHANAN, J. G. B. & BUCHANAN, P. G. (eds) *Basin Inversion*. Geological Society, London, Special Publications, **88**, 97–118.

MCCLAY, K. R. & BUCHANAN, P. G. 1992. Thrust faults in inverted extensional basins. *In*: MCCLAY, K. R. (ed.) *Thrust Tectonics*. London, 93–104.

MITRA, S. 1993. Geometry and kinematic evolution of inversion structures. *American Association of Petroleum Geologists Bulletin*, **77**, 1159–1191.

MITRA, S. & ISLAM, Q. T. 1994. Experimental (clay) models of inversion structures. *Tectonophysics*, **230**, 211–222.

NIELSEN, S. B. & HANSEN, D. L. 2000. Physical explanation of the formation and evolution of inversion zones and marginal troughs. *Geological Society of America*, **28**, 875–878.

PANIEN, M. 2004. *Analogue modelling experiments of basin inversion using well-characterised granular materials and comparisons with numerical models*. Ph.D. thesis, University of Berne.

PANIEN, M., SCHREURS, G. & PFIFFNER, O. A. 2005. Sandbox experiments on basin inversion: testing the influence of basin orientation and basin fill. *Journal of Structural Geology*, **27**, 433–445.

RAMBERG, H. 1981. The role of gravity in orogenic belts. *In*: MCCLAY, K. R. P. & PRICE, N. J. (eds) *Thrust and Nappe Tectonic*. Geological Society, London, Special Publications, **9**, 125–140.

ROURE, F. & COLLETTA, B. 1996. Cenozoic inversion structures in the foreland of the Pyrenees and Alps. *In*: ZIEGLER, P. A. H. & HORVATH, F. (eds) *Peri-Tethys Memoir 2: Structure and Prospects of Alpine Basins and Forelands*, Paris, Mémoire du Muséum National d'Histoire Naturelle, 173–209.

SCHULZE, D. 1994. Entwicklung und Anwendung eines neuartigen Ringschergerätes. *Aufbereitungs-Technik*, **35**, 524–535.

VIALLY, R., LETOUZEY, J., BENARD, F., HADDADI, N., DESFORGES, G., ASKRI, H. & BOUDJEMA, A. 1994. Basin inversion along the North African margin: the Saharan Atlas (Algeria). *In*: ROURE, F. (ed.) *Peri-Tethyan Platforms*. Paris, Technip, 79–118.

WILLIAMS, G. D., POWELL, C. M. & COOPER, M. A. 1989. Geometry and kinematics of inversion tectonics. *In*: COOPER, M. A. & WILLIAMS, G. D. (eds) *Inversion Tectonics*. Geological Society, London, Special Publications, **44**, 3–15.

YAMADA, Y. & MCCLAY, K. 2003. Application of geometric models to inverted listric fault systems in sandbox experiments. Paper 1: 2D hanging wall deformation and section restoration. *Journal of Structural Geology*, **25**, 1551–1560.

ZHAO, W. L., DAVIS, D. M., DAHLEN, F. A. & SUPPE, J. 1986. Origin of convex accretionary wedges: evidence from Barbados. *Journal of Geophysical Research B*, **91**, 10246–10258.

Lower crustal rheological expression in inverted basins

MIKE SANDIFORD[1], DAVID L. HANSEN[2] & SANDRA N. MCLAREN[3]

[1]*School of Earth Sciences, University of Melbourne, Victoria, 3010, Australia (e-mail: mikes@unimelb.edu.au)*

[2]*Department of Earth Sciences, University of Aarhus, Finlandsgade 6-8, DK-8200, Aarhus, Denmark*

[3]*Research School of Earth Sciences, Australian National University, ACT, Australia*

Abstract: Although lithospheric modelling has provided extraordinary insights into the processes that shape the continental crust, considerable uncertainty surrounds the basic rheology that governs behaviour at geological timescales. In part, this is because it has proved difficult to identify the geological observations that might discriminate, or unify, models of lithospheric rheology. In particular, the relative strength of lower crust and upper mantle remains a contentious aspect of continental lithospheric rheology. We show that various models for lower crustal rheology may produce distinct patterns of inversion in extensional sedimentary basins, consistent with some of the observed natural variability of inversion styles. Inversion of basin interiors, as is common in European Mesozoic basins, is favoured by a lithospheric rheology more sensitive to lateral thermal structure than to changes in the depth of the Moho, consistent with there being little strength contrast between the lower crust and upper mantle in these settings. In contrast, inversion of basin margins, particularly involving basinward verging structures, is consistent with a rheological sensitivity to the depth of Moho as would apply for a lower crust much weaker than the upper mantle. We use an example from central Australia to demonstrate this latter response, together with thermochronologic data that suggests that a relatively weak lower crust in this setting may reflect abnormally high geothermal gradients.

At the continental scale, meaningful predictions about the length-scales of deformation have been derived from models treating the lithosphere in terms of vertically averaged rheology (England & Houseman 1986). While these provide an important framework for understanding the long-wavelength deformation of continental interiors they do not provide much insight into the geometry of deformation at sub-lithospheric scales, where a more explicit treatment of the rheological and compositional stratification of the lithosphere is required (e.g., Beaumont *et al.* 2000). There is general consensus that rheological stratification of the lithosphere arises because of competition between pressure- and temperature-sensitive deformation mechanisms on the one hand, and compositional and mineralogical stratification of the lithosphere on the other (Brace & Kholstedt 1980). However, the distribution of strength remains controversial (Jackson 2002; Afonso & Ranalli 2004), as is the extent to which the strength distribution imposes on the resulting distribution of strain.

The Moho is the most important compositional boundary within the lithosphere, with the associated density contrast providing a fundamental control on the mechanics of crustal deformation. In many rheological models the strength contrast across the Moho (from weak lower crust to strong mantle) is so strongly expressed that the depth (and temperature) of the Moho is the dominant parameter controlling the rheological response of the lithosphere (Sonder & England 1986). This so-called 'jelly sandwich' model has been questioned (Jackson 2002) because it appears at odds with observations that imply that the lower crust is seismically active and capable of sustaining elastic stresses over geological timescales (Maggi *et al.* 2000). Indeed, Jackson (2002) has argued that the lower crust may be as strong as, if not stronger than, the upper mantle beneath, with the implication that the Moho has little relevance to the mechanical behaviour of the continental lithosphere (Fig. 1).

In numerical simulations, we might expect that the 'jelly sandwich' and 'stiff crust' (end-member) rheologies lead to qualitative differences in the styles and patterns of lithospheric deformation, especially for regions characterized by lateral variations in crustal thickness and

From: BUITER, S. J. H. & SCHREURS, G. (eds) 2006. *Analogue and Numerical Modelling of Crustal-Scale Processes.* Geological Society, London, Special Publications, **253**, 271–283.

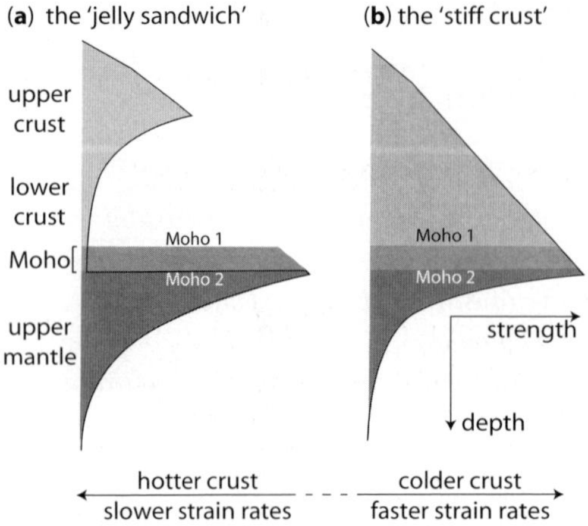

Fig. 1. Schematic illustration of vertical distribution of strength (at depth independent strain rate) in the (**a**) 'jelly sandwich' and (**b**) 'stiff crust' rheological models. An important difference between these models is the greater sensitivity of the 'jelly sandwich' rheology to the depth of the Moho, as indicated by the effect of changing Moho depth (i.e., Moho 1 to Moho 2) on the vertically integrated strength (i.e., the area under a given strength-depth profile). Such hypothetical models are sensitive to the geothermal and strain rate regimes.

thermal regime. The objective of this paper is to explore such qualitative differences by using forward numerical models for the inversion of extensional sedimentary basins. The classic setting showing such lateral structure is old, extended continental crust where the isostatic response to stretching has resulted in the development and filling of sedimentary basins (McKenzie 1978). Here, we consider the way inversion operates on thermally equilibrated extended lithosphere, adopting the assumption that sufficient time has elapsed since extension that all thermal transients associated with the extensional deformation and basin-filling have dissipated.

Many factors have been postulated to influence the style inversion of sedimentary basins, as evidenced by the extensive body of literature dedicated to this very subject (e.g., Cooper & Williams 1989; Buchanan & Buchanan 1995; Turner & Williams 2004). Our objective here is not to provide a comprehensive assessment of all factors that contribute to, and control, inversion, but rather to explore how the two end-member lower crustal rheologies might express themselves in terms of inversion styles. Thus we make a number of simplifying assumptions such as ignoring the anisotropy introduced by basin-forming faulting. This helps to isolate the specific consequences of differing lower crustal rheology and is not meant to imply that these other factors are not equally, if not more, important.

Of course, it is important to evaluate the results of forward models against observations from the real world, and so we begin with some basic observations pertinent to variability in the style of basin inversion in nature. This is followed by an analysis of simple thermo-mechanical models of basin evolution using the end-member rheologies that reproduce some of the essential elements of this natural variation. As summarized by Turner & Williams (2004), a particular problem highlighted in previous analyses of basin inversion has been the mechanical problem of inverting basin interiors where the lithosphere is assumed to have been strengthened by isostatically compensated reductions in Moho depth. This study directly relates to this problem, with our analysis suggesting that, to a large degree, the variation of inversion styles observed in nature may be explicable in terms of variability in lower crustal rheological state.

Styles of inversion in nature

In nature there is a significant variation in the style and distribution of inversion within inverted basins (e.g., Turner & Williams 2004). Here we make some brief remarks about two distinct styles of inversion relevant to our later analysis of numerical modelling results.

The classic style of basin inversion, as evident in numerous examples from Europe (Fig. 2) is characterized by a central zone of basin floor uplift and intense deformation accommodated by reverse faults. The inverted zones coincide with the depocentres of the precursor rift basin and as such the deeper sediments experienced the most uplift. The basin flanks typically subsided and syn-inversion marginal troughs developed. The latter effect reflects the flexural isostatic response to the crustal thickening of the inversion zones (Nielsen & Hansen 2000). In Europe, in the Alpine foreland, the marginal troughs are consistently filled with late Cretaceous chalks, of which only a small fraction is reworked material, signalling the relative simultaneity of the main compressive inversion phases. Here, the time interval between basin formation and inversion is largely set by the phases of rifting, relating to the Mesozoic break-up of Pangea, and the phases of Late Cretaceous and Cenozoic compression relating to the collision of the African and European plates. As this time interval generally compares to the lithospheric thermal time constant, the localization

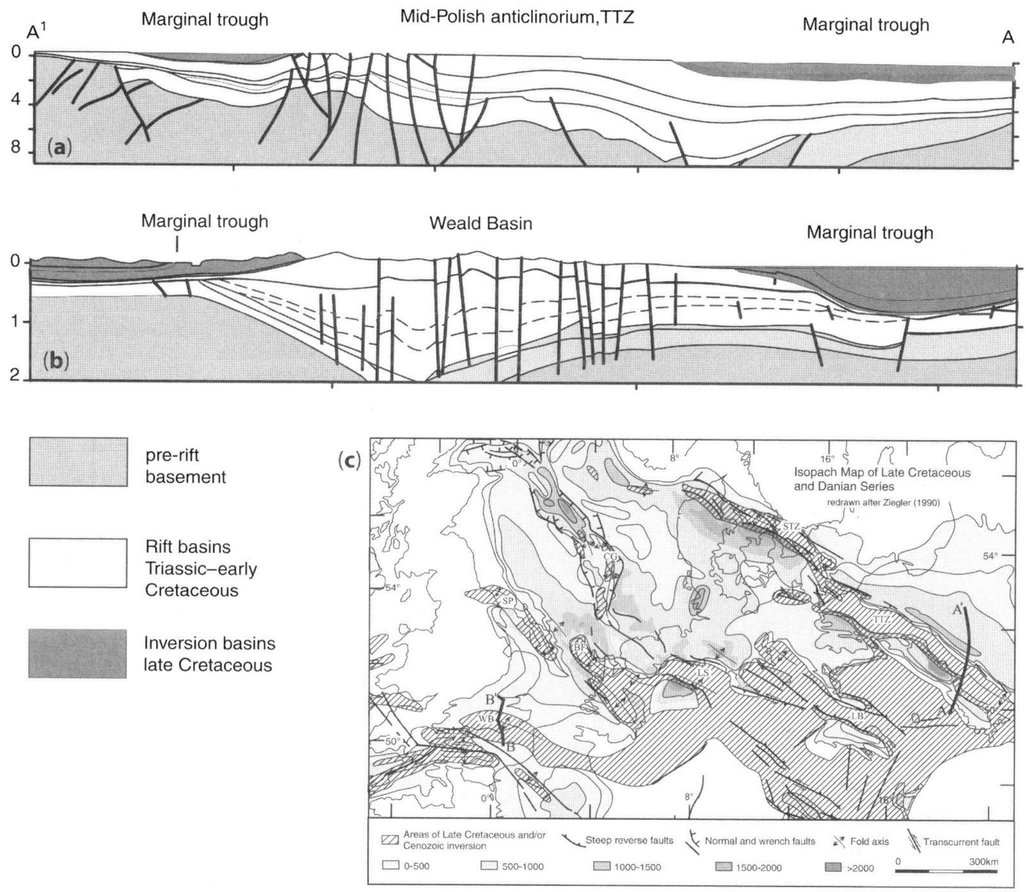

Fig. 2. Cross-section of typical inverted basins (upper panel: Polish trough *c.* 230 km across; lower panel: Weald Basin, SE England, *c.* 160 km across), adapted from Nielsen & Hansen (2000). The locus of maximum inversion intensity typically occurs in the interior of the precursor rift basins. During inversion reworked sediment is shed to marginal troughs that extend across the precursor basin margin towards the foreland. In these cases, inversion occurred at time intervals that preclude localization due to thermal transients associated with the initial rifting. In the case of the Polish Trough, rifting took place from 255 Ma to 152 Ma while the inversion occurred from 85 Ma to 55 Ma. For the Weald Basin, rifting took place from 210 Ma to 130 Ma and inversion from 60 Ma 13. TTZ: Tornquist-Teisseyre zone.

of deformation during inversion cannot be attributed to the thermal perturbations associated with the earlier rifting.

Palaeozoic inverted basins in central Australia show a rather different style of inversion (Fig. 3). Central Australia comprises a mosaic of structurally remnant Neoproterozoic basins initiated at around 800 Ma (e.g., Amadeus, Officer, Georgina and Ngalia basins). These basins are typically several hundred kilometres wide and bounded by basin verging structures that formed at various times through the Palaeozoic, many hundreds of million years after initial basin formation. These inversion structures now expose basement-cored Proterozoic gneissic complexes in their hanging wall (Figs 3 and 4). The distribution of sediment fill and underlying causes controlling the distribution of Neoproterozoic depocentres in the primary basins has been the subject of controversy warranting some discussion of their geodynamic setting. To this end we briefly summarize some salient ideas about these basins. The interested reader can find more comprehensive accounts in Lambeck (1983), Lindsay *et al.* (1987), Korsch and Lindsay (1989), Lindsay and Korsch (1991), Shaw *et al.* (1991), Walter and Gorter (1994), Walter *et al.* (1995) and Sandiford *et al.* (2001).

In one of the earliest accounts (Lambeck 1983) the formation of the central Australia basins was attributed to long high amplitude (*c.*10 km) lithospheric-scale buckling, largely on the basis of the

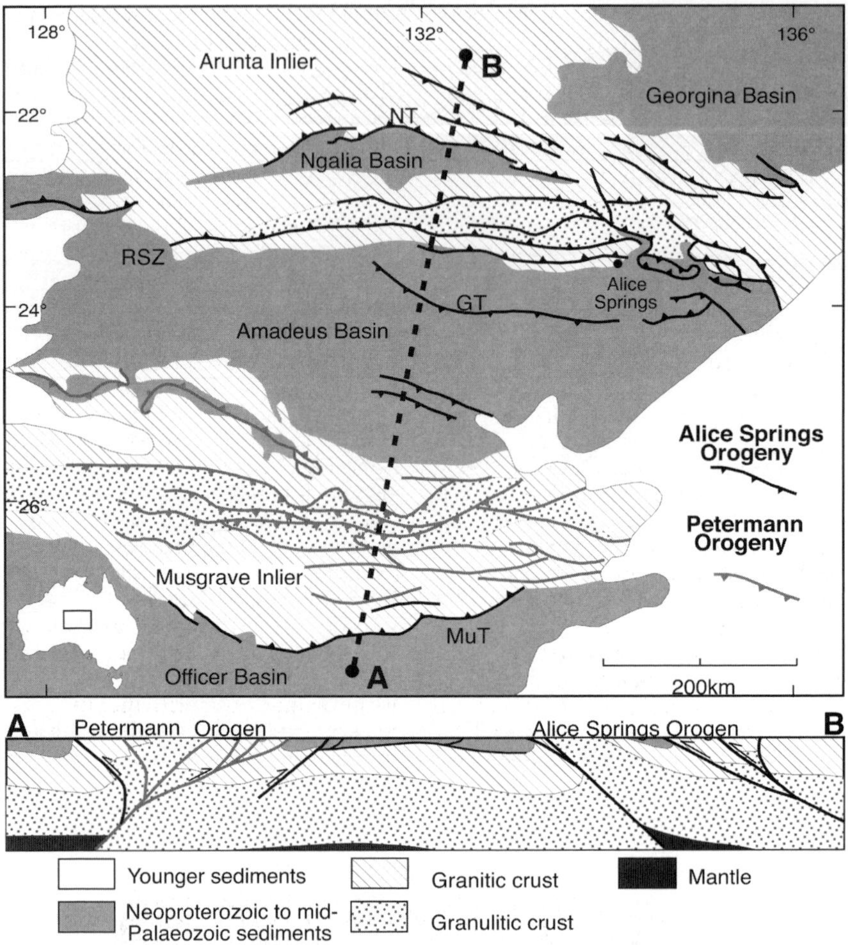

Fig. 3. Geological map of the central Australian region, showing the structurally remnant Neoproterozoic basins (the Officer, Amadeus, Ngalia and Georgina Basins), separated by basement inliers (the Musgrave and Arunta Inliers), adapted from Sandiford *et al.* (2001). Within the basement inliers we can distinguish two distinct types of terrane: (1) gneissic granite terranes that form the peripheral regions of the inliers and which are unconformably overlain by the sediments in the basins; (2) depleted granulite terranes that define the cores of the inliers and which are tectonically juxtaposed with the gneissic granite terranes. This juxtaposition reflects, in part, the strain associated with intraplate orogeny accumulated during the exhumation of the basement inliers from beneath a formerly more or less continuous intracratonic basin. In the Arunta Inlier this intraplate deformation was principally associated with the 450–300 Ma Alice Springs Orogeny, while in the Musgrave Inlier it was associated with the 550–500 Ma Petermann Orogeny. GT: Gardiner Thrust; MuT: Munyari Thrust; NT: Napperby Thrust; RSZ: Redbank Shear Zone.

distinctive gravity pattern of the region (see also Stephenson & Cloetingh 1991). However, subsequent workers (e.g., Teyssier 1985) have demonstrated that the gravity pattern largely reflects late structuring of the basin during inversion (Shaw *et al.* 1991; Sandiford 2002) and therefore does not provide an unequivocal constraint on the earlier basin-forming process. Consequently almost all subsequent analyses have emphasized the role of extensional processes in basin formation (e.g., Lindsay *et al.* 1987; Korsch & Lindsay 1989; Shaw *et al.* 1991; Walter & Gorter 1994; Walter *et al.* 1995; Sandiford *et al.* 2001). The current consensus is that the main rifting associated with basin-initiation stage occurred at >800 Ma. Subsidiary, milder extensional episodes are reflected in changes in the depocentres up until the early Ordovician (e.g., Shaw *et al.* 1991). Relatively new data from sediment provenance studies (Comacho *et al.* 2002) shed important further light on the original extent of these extensional basins. These data demand proximal exposure of basement throughout basin evolution (implying that the present-day distribution of structurally remnant basins mirrors the original pattern of extensional basins). Moreover, from the point of view of our analysis, the findings of Comacho *et al.* (2002) imply that the style of inversion in central Australia must have involved marginal inversion, rather than inversion along a central basin axis as typifies the European Mesozoic basins, with the sediment generated during inversion redistributed towards the original basin centres. Further, the basin-verging asymmetry of the main structures that accommodate basin inversion in central Australia makes them geometrically incompatible with reactivation of plausible primary basin forming, extensional structures that would be expected to dip southwards towards the basin interior (Fig. 4). This is an important inference, since it implies that the existence of basin-forming structures is not likely to be the primary determinant in localizing crustal shortening during inversion in this setting. However, it is necessary to provide a word of caution about this interpretation since the geometry of the primary, basin-forming structures in central Australia is not well understood. This is partly due to the intensity of the subsequent inversion in which large thrust sheets and nappe structures developed as part of a crustal-scale deformation event (e.g., Teyssier 1985; Golbey *et al.* 1989; Sandiford 2002), thereby obscuring earlier formed structures.

One potentially important distinction between the European and Australian basins may relate to the crustal thermal regime at the time of inversion. Heat flows in the Tornquist zone in the Polish Basin (Fig. 2a) are *c.* 65–70 mWm^{-2} (Balling 1995), which are only slightly greater than average continental crustal heat flows. Unfortunately heat flow data from central Australia is sparse, with no available data from the central Australian basins. However, the central Australian crust on which the basin is formed is characterized by unusually high radiogenic element concentrations (Sandiford *et al.* 2001, Fig. 4), as is much of the Australian Proterozoic, where the average heat flow is 85 mWm^{-2} (see McLaren *et al.* 2003 for a full account). Bottom-hole temperatures from exploration wells in the northern half of the Amadeus Basin suggest modern-day thermal gradients of *c.* 25°C/km and locally as high as 35°C/km (Gorter 1984). Using our own unpublished thermal conductivity data (which yield average basin sediment conductivities of *c.* 2.7 W/m/K), these thermal gradients are consistent with heat flows of *c.* 65–95 mWm^{-2}. The existence of

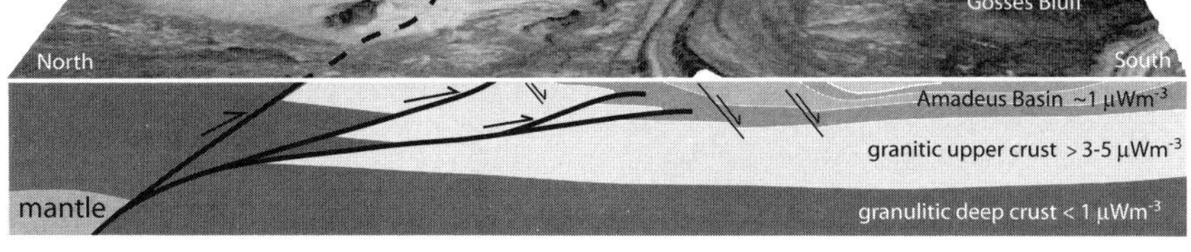

Fig. 4. Airborne radiometric image, data courtesy of Northern Territory Geological Survey, draped over a digital elevation model of the western Macdonnell ranges along the inverted boundary between the Arunta Inlier and the Amadeus basin, with crustal scale cross-section showing the main crustal elements associated with Proterozoic basin formation and Palaeozoic inversion. Note the basin-verging (southward) inversion structures associated with the Redbank shear zone (thick black lines) are geometrically incompatible with the schematically illustrated basin-forming structures (thin black lines). Sample site 01-524 shows the location of the K-feldspars used to constrain thermal regimes during basin inversion. In terms of radioelement concentrations, three distinct elements can be identified (Sandiford *et al.* 2002): (1) the sediments of the Amadeus basin that have heat production rates of *c.* 1 μWm^{-3}, (2) the granites and gneisses that immediately underlie the Amadeus Basin and have heat production rates of >3 μWm^{-3} (locally up to *c.* 10 μWm^{-3}), and (3) the deep crustal mafic granulites that outcrop as isolated monadnocks (e.g., Mount Chapple and Mount Hay) in the desert to the north of the Redbank shear zone with heat production rates <1 μWm^{-3}.

elevated thermal gradients in the central Australian basins prior to inversion is implied by thermochronological data of detrital K-feldspar from a granite boulder in glaciogenic sediments in the lower part of the basin (Fig. 5a). These data (are available online at http://www.geolsoc.org.uk/SUP18240. A hardcopy can be obtained from the Society Library.) show partial resetting of $^{40}Ar/^{39}Ar$ spectra and are believed to reflect the thermal conditions representative of the northern margin of the basin (see Appendix 1 for further elaboration). Importantly, the preservation of pre-basin formation $^{40}Ar/^{39}Ar$ ages (up to 1100 Ma) implies that temperatures *never* exceeded *c.* 230°C during the evolution of this part of the basin (i.e., in the interval 400–800 Ma). Similarly, the fact that the partial resetting of the $^{40}Ar/^{39}Ar$ spectra has demonstrably occurred during the basin-forming interval requires that temperatures must have exceeded 150°C (Fig. 5). Albeit subject to somewhat contentious vagaries of the method, our modelling of the spectra achieves best fits with peak temperatures in the range 170–190°C at the time when inversion commenced (*c.* 400 Ma) followed by cooling to *c.* 330 Ma, coincident with the inversion of the basin at this time (Haines *et al.* 2001). Assuming a surface temperature of about 20°C at the time of inversion then we derive average thermal gradients in the basin of about 28–38°C/km. Using our unpublished conductivity measurements cited above, these data yield heat flows of *c.* 75–100 mWm^{-2} during the late stages of basin evolution, comparable with the range of modern thermal gradients and somewhat higher than in the Tornquist zone in western Europe (Balling 1995).

In terms of potential variations in Moho temperature for the different settings we need to make certain assumptions about the source of the heat flow variation. If we are correct in attributing the differences in heat flow to variations in the heat production of the crust underlying the basin

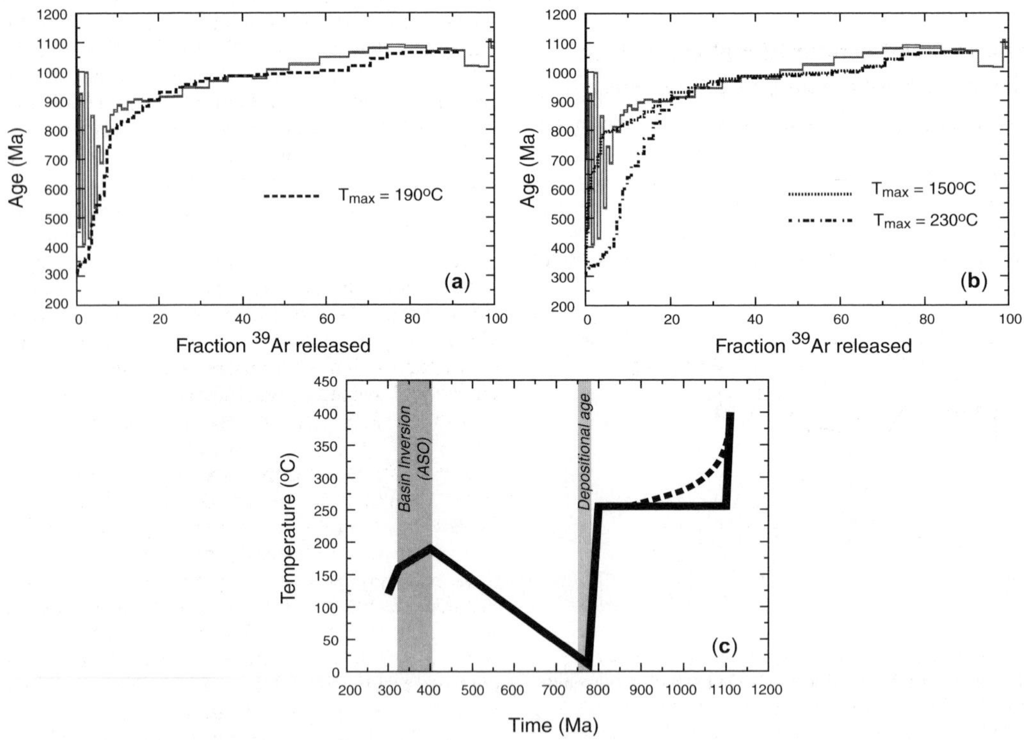

Fig. 5. $^{40}Ar/^{39}Ar$ age spectrum for K-feldspar 01-524 (**a**) laboratory age spectrum with best-fit model age spectrum, (**b**) laboratory age spectrum and model age spectra for post-depositional temperature maxima of 230°C and 150°C. The poor fit to the laboratory data provides robust constraints on this portion of the cooling history. (**c**) Best-fit temperature time history corresponding to a maximum post-depositional temperature of 190°C. Dashed line indicates alternative best-fit history for the higher temperature history. See Fig. 4 for the sample site and Appendix 1 for details of analytical methods and interpretation.

then, following the simple formula introduced by Sandiford and McLaren (2002), the difference in temperature at or beneath the Moho is given by

$$\Delta T_m = \frac{\Delta q_c h}{k} \quad (1)$$

where Δq_c is the difference in the integrated heat production of the crust (as, e.g., reflected in the surface heat flows), h is the mean depth of the heat production and k the thermal conductivity. Assuming a difference in heat production (Δq_c) equivalent to $15\ \mathrm{mWm^{-2}}$ (as suggested by the analysis above), $h = 10$ kms (as is typical of the continents), and $k = 2.7\ \mathrm{Wm^{-1}\ K^{-1}}$ (as indicated by the above quoted conductivity), then we anticipate that temperatures at Moho depths might be as much as 50°C hotter beneath the Amadeus basin compared to the Tornquist zone. The effect of such temperature difference is significant, as in the temperature range 500–600°C, a 10° temperature rise results in approximately 10% reduction in bulk lithospheric strength at a given strain rate (Sonder & England 1986).

Thermal and compositional controls on rheological structuring of rifted crust

At the lithospheric scale the long-term mechanical consequences of extension are potentially influenced by two distinct thermal and compositional effects (Fig. 6, see Sandiford 1999). Firstly, due to density contrasts between basin-filling sediments and extended crust there is a long-term reduction in the depth of the Moho following extension, with attendant Moho cooling (ΔT_m) dependant mainly on the steady-state thermal gradient in the deep crust (typically estimated at 5–10°C/km). The second effect relates to changes in the average crustal thermal gradient (∇T_c) due to the thermal property structure (heat production and thermal conductivity) of the extended crust and associated basin-fill. The attenuation of heat production in the pre-existing crust during extension leads to long-term reductions in ∇T_c, while the burial of this heat production beneath the basin-fill leads to increases in ∇T_c (Sandiford 1999; Hansen & Nielsen 2002; Sandiford *et al.* 2003).

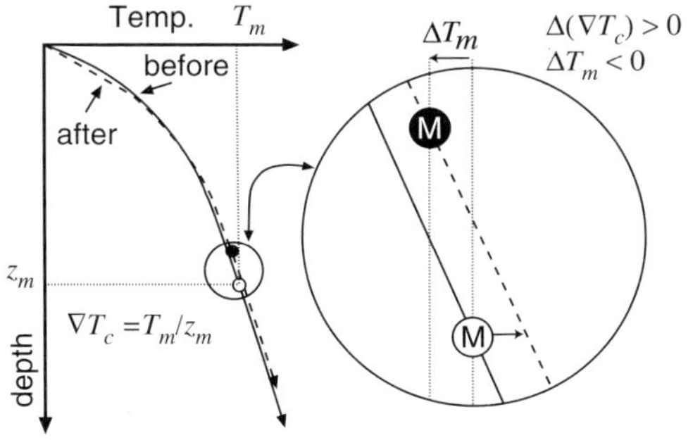

Fig. 6. Schematic illustration of the long-term thermal effects of crustal extension and basin formation (adapted from Sandiford 1999). Long-term Moho temperature changes (ΔT_m) occur primarily due to Moho shallowing while changes in the thermal property structure of the overlying crust (mostly in the basin-fill), lead to changes in the average thermal gradient ($\nabla T_c = T_m/z_m$). See text for further discussion. As is shown in Fig. 7, the characteristic long-term thermal response to basin formation ($\Delta T_m < 0$ and $\Delta(\nabla T_c) > 0$ (see also Hansen & Nielsen 2002).

There is a considerable range in the plausible thermal property structure of continental crust and basin-fill (Sandiford 1999). The large number of parameters that influence the thermal structure of continental crust implies there can be no unique thermal response of the continental crust to extension. Consequently, in order to demonstrate the plausible long-term thermal response, we use a Monte Carlo approach based on the thermal property parameter ranges outlined by Hansen and Nielsen (2002) and Sandiford *et al.* (2003), wherein further details can be found. Figure 7 shows the results of a Monte Carlo simulation (*c.* 4500 individual parameter sets) of thermally-mature extensional basins. The great proportion (*c.* 86%) of modelled parameter sets result in an increase in thermal gradient beneath basins ($\Delta(\nabla T_c) > 0$), with the median value of ($\Delta(\nabla T_c) \sim 1$°C/km) implying that the upper mantle beneath the basin is significantly hotter (by around 30–40°C) than the upper mantle at similar depths outside the basin. Despite this characteristic long-term increase in thermal gradient beneath basin centres, the great majority (76%) of modelled parameter sets result in long-term Moho cooling ($\Delta T_m < 0$) reflecting the significant reduction in Moho depth accompanying the isostatic compensation of low-density sediment fill. Long-term changes in the thermal gradient (∇T_c) beneath basins necessarily result in lateral heat flow, typically directed away from the basin centre because of the increase in the average thermal gradients beneath the basin. For $\Delta(\nabla T_c) > 0$ such lateral heat flow directed away from basin centres results in Moho temperature maxima beneath basin margins (Hansen & Nielsen 2002), where the Moho temperatures may be elevated by up to 20°C, although typically by less than 5°C.

The modelling results summarized in Figure 7 provide an important insight into the thermal structure beneath old, sediment-filled extensional basins. In particular, the lower crust and upper

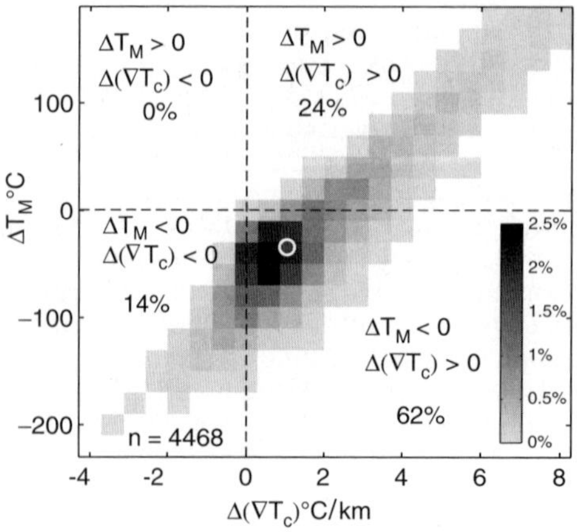

Fig. 7. Monte Carlo simulation of the long-term changes in thermal structure of extended continental crust on which a sedimentary basin has developed. The thermal structure is expressed in terms of the changes in Moho temperature (ΔT_M) and average crustal thermal gradient ($\Delta(\nabla T_c)$). The parameter ranges for the 4468 individual parameter sets are the same as those employed by Hansen & Nielsen (2002). Data are presented as a proportion of the total, binned in 20°C × 0.57°C/km boxes. The median value is illustrated by the circle, with $\Delta T_M = -34$°C and $\Delta(\nabla T_c) = +1$°C/km.

mantle beneath the basin centres is likely to be several tens of degrees hotter than material at identical depths outside the basin, while the Moho is likely to be several tens of degrees cooler beneath the basin. The hottest Moho temperatures are expected at the margin of the stretched crust, where the Moho is deep but close enough to the warm basin interior to 'feel' the effects of lateral heat flow.

Lower crustal rheological expression in inverted basins: thermal-mechanical modelling

Background

The characteristic thermal and compositional structure of old, stretched continental crust (i.e., $\Delta T_m < 0$ and $\Delta(\nabla T_c) > 0$) arguably provides a useful natural experiment for testing models of lower crustal rheology against natural patterns of deformation observed in inverted basins. We use a thermal property set that is close to the median configuration produced by the Monte Carlo simulation (with $\Delta T_m = -34$°C, $\Delta(\nabla T_c) = +1$°C/km), and contrast the style of inversion produced with rheological models appropriate to a 'jelly sandwich' and 'stiff crust' schematically represented in Figure 1. For the 'jelly sandwich' model the lower crust and mantle are characterized by a rheology appropriate to wet anorthosite (Shelton & Tullis 1981) and wet dunite (Chopra & Paterson 1981), respectively, while for the 'stiff crust' model the lower crust and mantle both have the rheological properties of wet dunite (but with densities appropriate to crust and mantle, respectively). In both cases, the upper crust is assumed to have a rheology appropriate to wet quartzite (Paterson & Luan 1990). The response of basins to inversion is modelled using a coupled elasto-plasto-viscous rheology implemented in a thermo-mechanical fully Lagrangian finite element code using 4620 linear triangular elements. Details of the modelling approach are outlined below.

Modelling methodology

A two-dimensional thermo-mechanical continuum model is used to investigate the lithospheric response to compression. The lithospheric profile is initially 300 km long and 120 km deep, consisting of three different mineralogical layers of upper crust (wet quartzite), lower crust (wet anorthosite) and mantle (wet dunite). In the right half of the model the elsewhere 40 km thick crust is thinned to 24 km, with 8 km of sediments resting on top (Fig. 8). The mechanical part of the model is founded on a fully Lagrangian formulation of the principle of virtual work (Green & Zerna 1968; Budiansky 1969; Malvern 1969). The principle of virtual work is complemented by a set of constitutive relations enabling visco-elastic and plastic deformation. The visco-elastic deformation form simulates temperature and time-dependent ductile creep of solid rock mass and is based on a Maxwell relation stating additivity of elastic and viscous strain rates (Jaeger & Cook 1969). While the elastic strain rates satisfy Hooke's law (Ranalli 1995), with Young's modulus at 10^{11} Pa and Poisson's ratio at 0.25, the viscous strain rates satisfy the empirical Dorn equation relating effective strain rates and stress levels for steady state creep (Ranalli 1995). In this case, the associated viscosity, ϑ, may be calculated by

$$\vartheta = B\dot{\varepsilon}^{1/n-1}\exp\left(\frac{Q}{nRT}\right) \qquad (2)$$

where $\dot{\varepsilon}$ is the effective visco-elastic strain rate, T is absolute temperature, and $R = 8.31$ J/(mol K) is the ideal gas constant. n, Q, B, are experimentally derived creep parameters

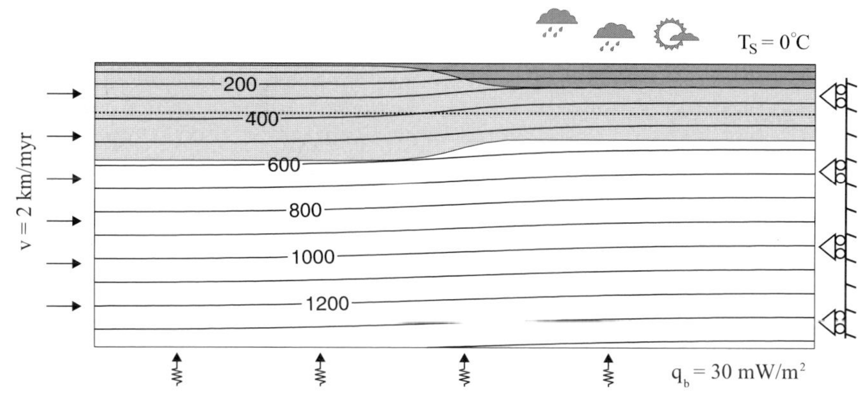

Fig. 8. Schematic of initial model conditions prior to inversion. The modelled domain is initially 300 km wide and 120 km deep. Inversion is imposed on the initial geometry with a thermally mature sedimentary basin above a constant thickness lithosphere. The dashed line indicates the approximate scale of the lithospheric stretching that would be needed to generate the initial basin.

(Chopra & Patterson 1981; Shelton & Tullis 1981) assumed only to depend on mineralogy.

When the yield stress of rocks is reached, the deformation form becomes visco-elastic-plastic, simulating brittle failure. The yield stress is derived from the pressure dependent Drucker-Prager yield function (Khan & Huang 1995):

$$\sqrt{J_2} + (1 - \lambda_v)\alpha(\phi)I_1 = 0 \quad (3)$$

approximating the conditions under which frictional Coulomb type materials fail (Byerlee 1978). Here I_1 is the first invariant of the Cauchy stress tensor and J_2 the second invariant of its deviatoric part. $\alpha(\phi)$ is a function of the angle of internal friction, ϕ (Khan & Huang 1995):

$$\alpha(\phi) = \frac{2\sin(\phi)}{\sqrt{3}(3 - \sin(\phi))}$$

In Equation (3), $\lambda_v = 0.4$ is the pore-fluid factor of near hydrostatic conditions (Sibson 1990). A non-associated plasticity model (Vermeer & De Borst 1984), where incompressible plastic strain rates $\hat{\eta}^{\mathrm{p}}$ are derived from the plastic potential, Ω, is adopted:

$$\hat{\eta}^{\mathrm{p}} = \frac{\partial\Omega}{\partial\sigma} \quad \text{where } \Omega = \sqrt{J_2} + const. \quad (4)$$

The displacement and temperature fields are coupled through the temperature dependence of viscosity (Eqn. 2) and density. The latter is given by

$$\rho = \rho_0(1 - a_v T) \quad (5)$$

where ρ_0 is density at 0°C. $a_v = 3.2 \times 10^{-5}\ \mathrm{K}^{-1}$ is the volumetric expansion coefficient and T is Celsius temperature.

In each model time step the temperature field is found by solving the transient heat transfer equation

$$\rho c\dot{T} = \nabla \cdot (k\nabla T) - \rho c v \cdot \nabla T + A + W \quad (6)$$

where c is specific heat capacity, k is conductivity, A is radiogenic heat production rate, and W is the viscous shear heating rate (Ranalli 1995; Regenauer-Lieb & Yuen 1998). Material advection is represented by the relative velocity between material and grid points. On the mechanical model the temperature field is tracked using the same Lagrangian mesh and hence, here $v = 0$. However, above the mechanical model the thermal mesh is extended to account for the deposition/erosion of sediments and, as such, the thermal mesh follows the topography formed by the changing sediment distribution. The relative velocity then represents the passive flow of material across the model surface (Hansen & Nielsen 2003). The surface temperature is 0°C, whilst at the model base the vertical heat flow is $30\ \mathrm{mW/m^2}$. At the base of the mechanical model pressure conditions are imposed, allowing the entire model to move vertically under the influence of buoyancy forces. At the left vertical axis a constant kinematic boundary condition (2 km/Ma) is imposed, compressing the model profile as the right vertical axis is fixed.

During the entire model evolution surface processes act to re-distribute sediments. These processes are modelled by a diffusion equation, with a source term enabling the transport of sediments to and from the model profile. The transport equation is written

$$\dot{h} = \nabla \cdot (\kappa \nabla h) + \dot{s}(w) \quad (7)$$

$\kappa = 200\ \text{km}^2/\text{Ma}$ is the diffusivity of topography, h, and $\dot{s}(w)$ is a linear function of water depth. In the present study the last term is chosen so that any accommodation space created is immediately filled by sediments. Isochron markers record the chronological evolution of the basin fill by effectively representing stratigraphic boundaries. The porosity of sediment decreases with burial according to $\phi = \phi_0 \exp(z/2\ \text{km})$, where ϕ_0 is surface porosity and z is maximal burial depth. All other parameters are listed in Table 1. The bulk of the sediment, which is present at the post-rift stage of the model start, is included in the thermo-mechanical model, whilst the sediment accumulating during the model evolution is directly included in the thermal model only. However, the redistribution of sediments during the model evolution influences the mechanical model as passive vertical loads.

Modelling results

In the 'jelly sandwich' lithosphere, lateral variations in the strength of the lithosphere are governed primarily by the temperature of the uppermost mantle, immediately below the Moho (Sonder & England 1986). Because vertical thermal gradients are much greater than lateral gradients, variations in Moho depth are therefore crucial, with bulk lithospheric strength varying inversely with crustal thickness. With regard to a characteristic old extensional basin geometry with Moho shallowed beneath the centre of the basin, the lithosphere is strongest beneath the basin centre, and weakest along the outer margin of the basin. Consequently, in-plane compression will localize inversion in the vicinity of the basin margin (Fig. 9a). Deformation occurs in a confined zone that shows distinct localization in a zone propagating up through the crust as a basin verging ramp. Erosion focused on this marginal uplift sheds reworked sediment towards the basin interior, as well as away from the basin margin into the hinterland. Reworked sediment trapped within the basin interior produces a distinctive off-lapping succession progressively restricted towards the internal zones of the underlying rift basin.

In our 'stiff crust' lithosphere, the depth and temperature of the Moho is irrelevant to the rheological response of the lithosphere, and the thermal gradient is the primary control on the bulk lithospheric strength distribution. Because the modelled basin has higher average thermal gradients beneath the basin centre, in-plane compression necessarily localizes deformation beneath basin centres. The numerical simulation (Fig. 9b) produces diffuse strain rate gradients around a broad zone of distributed inversion. Erosion of the inverted domain sheds reworked sediment towards the basin foreland, potentially producing secondary marginal basins, outboard of the primary extensional basin. Structural asymmetry, albeit weak, verges from the basin centre towards the foreland. While not explicitly considered in the models shown here, localization phenomenon due to plastic instabilities should be expected to localize deformation in basin interiors and show a more abrupt transition between the inverted domain and the surrounding crust.

Discussion

Given the extraordinary temperature and mineralogical sensitivity of rheological flow laws pertinent to lithospheric deformation, it seems entirely reasonable that the mechanical response of the lithosphere will vary significantly as a function of its thermal regime and composition. It would therefore be unwise to suggest that the inversion of continental lithosphere can be characterized by the one uniform mode. Indeed, one of the paramount challenges in geodynamics is the understanding of how variations in the mechanical response to external forcing affect the final form and architecture of deformed continental crust. With this in mind, our models for

Table 1. *Rheological and thermal parameters used in numerical models*

Parameter	Symbol [units]	Upper crust	Lower crust	Upper mantle	Sediment
Angle of internal friction	ϕ	30	30	30	30
Thermal conductivity	k [W/m/K]	3.0	2.3	4.0	2.0
Specific heat	c [J/kg/K]	850	900	1000	900
Heat production rate	A [μW/m^3]	1.3	0.3	0.01	1.0
Creep parameter	n	3.10	3.20	4.48	3.10
Creep parameter	B [MPa s$^{1/n}$]	208.0	12.28	0.2628	208.0
Creep parameter	Q [kJ/mol]	135	239	498	135
Density	ρ_0 [kg/m^3]	2800	2900	3400	2300

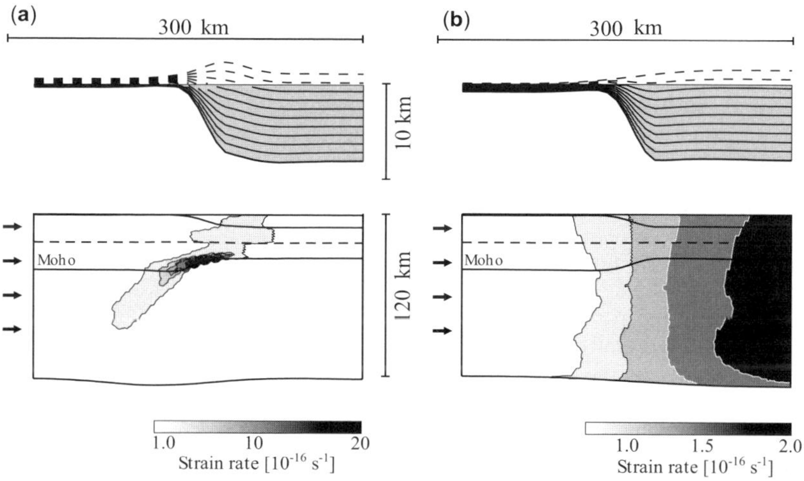

Fig. 9. Models of basin inversion styles predicted for different lower crustal rheology after 20 km of shortening. Upper panels show rift basin stratigraphy after inversion, while lower panels show the mean strain rate during the period of inversion. (**a**) The distribution of deformation for the 'jelly-sandwich' model. The weak lower crust and strong contrast across the Moho make the depth and temperature of the Moho critical determinants to the mechanical response. (**b**) Identical modelling scenario but with a 'stiff crust' rheology (i.e., a relatively strong lower crust, with no contrast in strength across the Moho). Modelling techniques are based on a fully Langrangian elasto-plasto-viscous deformation code, as detailed in the text.

two end-member scenarios of lower crustal rheology provide an intriguing insight into some of the observed variations in the style of inversion in natural inverted basins. Our modelling suggests that the characteristic response of basins in inversion, producing inversion along the central basin axis (Figs 2 and 9b), is consistent with there being little rheological sensitivity to the position of the Moho. For standard thermal regimes, where steady-state Moho temperatures are expected to be no greater than *c*. 500°C, this may therefore reflect the characteristic strong lower crustal rheological mode.

However, it is conceivable that, under different thermal regimes, the rheological contrast across the Moho plays a more prominent role. In support of this notion we can cite the unusual inversion record of Neoproterozoic basins in central Australia (Figs 3 and 4), where Palaeozoic inversion was at least partly localized along the margins of much older (>200 Ma) extensional basins. Here, the inversion structures verge towards basin interiors, similar to Figure 9a, with reworked sediment shed towards the interior of the inverted basins. As such, this central Australian style of inversion is consistent with a more prominent rheological contrast across the Moho. This analysis suggests that the thermal regime provides a principal control on the rheological structure of the lower crust, with the observed natural variation in inversion styles reflecting, at least in part, variations in the rheological state of the lower crust.

The model results we have presented show how thermal and compositional structuring of the lithosphere can lead to distinctive patterns of strain localization at the lithospheric scale, without appealing to strain-dependent weakening mechanisms. Many studies have shown that strain-dependent weakening is of fundamental importance to the behaviour of the continental lithosphere, and an important challenge is to establish the relative importance of the various localization mechanisms. With regard to this, our analysis suggests that the rheological structuring of the lithosphere associated with changes in Moho temperature of order 10°C is comparable to, if not more significant than, the rheological structuring associated with strain-dependent weakening processes during basin formation.

Appendix 1. $^{40}Ar/^{39}Ar$ thermochronology

K-feldspar 01-524 was irradiated for 480 hours in the Australian Nuclear Science and Technology Organization HIFAR reactor, NSW. The sample was subject to a series of 43 steps between 450°C and 1450°C. The gas released was analyzed using a VG Isotech MM3600 gas source mass spectrometer with a sensitivity of 3.5×10^{-17} mol/mV (This data is available online at http://www.geolsoc.org.uk/SUP18240. A hardcopy can be obtained from the Society Library.). The resultant $^{40}Ar/^{39}Ar$ age spectrum was interpreted using the multiple-diffusion-domain method assuming that argon release is by a thermally activated diffusion process (Lovera *et al.* 1989; Richter *et al.* 1991). We computed a distribution of activation energies and domain length scales (volume fraction and effective size) using eight slab-shaped domains (This data is available online at http://www.geolsoc.org.uk/SUP18240. A hardcopy can be obtained from the Society Library.). The geological setting explicitly requires a reheating history associated with basin formation and we are able to calculate the best-fit temperature-time paths by inputting trial thermal histories and minimizing the differences between the laboratory and modelled-age spectra by iteration. Elsewhere (McLaren *et al.* 2002) we have shown that K-feldspar MDD modelling can be used to recover robust thermal histories from Australian Neoproterozoic basins, and that paper contains a rigorous discussion of our approach and the problems that need to be addressed to demonstrate that this approach yields meaningful results. The $^{40}Ar/^{39}Ar$ age spectrum of the sample (from estimated palaeodepths of 5.1 ± 0.2 kms in the Areyonga formation; see location in Fig. 4) shows a distinctive monotonically increasing age gradient, which suggests that the sample started to accumulate argon around 1100 Ma, with final closure to argon loss occurring in the Palaeozoic (Fig. 5). The preservation of Neoproterozoic ages provides an important constraint on the post-depositional temperatures in the basin. Modelling produces good fits with slow cooling from 1100 Ma until at least 900 Ma (as indicated by the broad range of ages in this interval). A period of rapid cooling at around 900–800 Ma is required to generate the sedimentary detritus by the inferred depositional age of the sediment (*c.* 780 Ma, Lindsay 1989). While excess argon in the early released gas precludes a robust reconstruction of the reheating history in the depositional interval 780–400 Ma, we are able to constrain the range of temperature experienced during this interval. The age spectrum requires a period of slow reheating from 780 Ma to between 170 and 190°C at 400 Ma at *c.* 0.5°C/Ma at broadly consistent with the depositional history of the overlying sedimentary pile (e.g., Haines *et al.* 2001).

This work was supported by ARC grants DP0209157 and F10020050 and by the ACcESS MNRF. Constructive comments from L. Gemmer, R. Stephenson and E. Eide, and S. Buiter have helped our preparation of the manuscript. Irradiation of K-feldspar 01-524 was undertaken by the Australian Nuclear Science and Technology Organization through the Australian Institute of Nuclear Science and Engineering.

References

AFONSO, J. C. & RANALLI, G. 2004. Crustal and mantle strengths in continental lithosphere: is the jelly sandwich model obsolete? *Tectonophysics*, **394**, 221–232.

BALLING, N. 1995. Heat flow and thermal structure of the lithosphere across the Baltic Shield and northern Tornquist Zone. *Tectonophysics*, **244**, 13–50.

BEAUMONT, C., MUNOZ, J. A., HAMILTON, J. & FULLSACK, P. 2000. Factors controlling the Alpine evolution of the central Pyrenees inferred from a comparison of observations and geodynamical models. *Journal of Geophysical Research*, **105**, 8121–8145.

BRACE, W. F. & KOHLSTEDT, D. L. 1980. Limits on lithospheric strength imposed by laboratory experiments. *Journal of Geophysical Research*, **85**, 6248–6252.

BUCHANAN, J. G. & BUCHANAN, P. G. 1995. *Basin Inversion*. Geological Society, London, Special Publications, **88**.

BUDIANSKY, B. 1969. Remarks on theories of solid and structural mechanics. *In*: LAVRENT'EV, M. (ed.) *Problems of Hydrodynamics and Continuum Mechanics*. SIAM, Philadelphia, 77–83.

BYERLEE, J. D. 1978. Friction of rocks. *Pure and Applied Geophysics*, **116**, 615–626.

CAMACHO, A., HENSEN, B. J. & ARMSTRONG, R. 2002. Isotopic test of a thermally driven intraplate orogenic model, Australia. *Geology*, **30**, 887–890.

CHOPRA, P. N. & PATERSON, M. S. 1981. The experimental deformation of dunite. *Tectonophysics*, **78**, 453–473.

COOPER, M. A. & WILLIAMS, G. D. 1989. *Inversion Tectonics*. Geological Society, London, Special Publications, **44**.

ENGLAND, P. & HOUSEMAN, G. 1986. Finite strain calculations of continental deformation II: comparison with the India-Asia collision zone. *Journal of Geophysical Research*, **91**, 3664–3676.

GOLEBY, B. R., SHAW, R. D., WRIGHT, C., KENNETT, B. L. N. & LAMBECK, K. 1989. Geophysical evidence for 'thick-skinned' crustal deformation in central Australia. *Nature*, **337**, 325–330.

GORTER, J. D. 1984. Source potential of the Horn Valley Siltstone, Amadeus basin. *Australian Petroleum Exploration Association Journal*, **24**, 66–90.

GREEN, A. E. & ZERNA, W. 1968. *Theoretical Elasticity*. Oxford University Press, New York.

HAINES, P., HAND, M. & SANDIFORD, M. 2001. Palaeozoic syn-orogenic sedimentation in central and northern Australia: a review of distribution and timing with implications for the evolution of intracontinental orogens. *Australian Journal of Earth Sciences*, **48**, 911–928.

HANSEN, D. L. & NIELSEN, S. B. 2002. Does thermal weakening explain basin inversion, stochastic modelling of the thermal structure of sedimentary basins. *Earth and Planetary Science Letters*, **198**, 112–127.

HANSEN, D. L. & NIELSEN, S. B. 2003. Why rifts invert in compression. *Tectonophysics*, **373**, 5–24.

JACKSON, J. A. 2002. Strength of the continental lithosphere: time to abandon the jelly sandwich? *GSA Today*, **12**, 1–4.

JAEGER, J. C. & COOK, N. G. W. 1969. *Fundamentals of Rock Mechanics*. Chapman and Hall Ltd., London.

KHAN, A. S. & HUANG, A. 1995. *Continuum Theory of Plasticity*. John Wiley & Sons Inc., New York.

KORSCH, R. J. & LINDSAY, J. F. 1989. Relationships between deformation and basin evolution in the intracratonic Amadeus Basin. *Tectonophysics*, **158**, 5–22.

LAMBECK, K. 1983. Structure and evolution of the intracratonic basins of central Australia, *Journal of the Royal Astronomical Society*, **74**, 843–866.

LINDSAY, J. F. 1989. Depositional controls on glacial facies associations in a basinal setting, late Proterozoic, Amadeus Basin, central Australia. *Palaeogeography, Palaeoclimatology, Palaeoecology*, **73**, 205–232.

LINDSAY, J. F. & KORSCH, R. J. 1991. The evolution of the Amadeus Basin, central Australia. *In*: KORSCH, R. J. & KENNARD, J. M. (eds) *Geological and Geophysical Studies in the Amadeus Basin, Central Australia*. Bureau of Mineral Resources, Geology and Geophysics Australia Bulletin, **236**, 7–32.

LINDSAY, J. F., KORSCH, R. J. & WILFORD, J. R. 1987. Timing the breakup of a Proterozoic supercontinent: evidence from Australian intracratonic basins. *Geology*, **15**, 1061–1064.

LOVERA, O. M., RICHTER, F. M. & HARRISON, T. M. 1989. The $^{40}Ar/^{39}Ar$ thermochronometry for slowly cooled samples having a distribution of diffusion domain sizes. *Journal of Geophysical Research*, **94**, 17917–17935.

MAGGI, A., JACKSON, J. A., MCKENZIE, D. & PRIESTLEY, K. 2000. Earthquake focal depths, effective elastic thickness, and the strength of the continental lithosphere. *Geology*, **28**, 495–598.

MALVERN, L. E. 1969. *Introduction to the Mechanics of a Continuous Medium*. Prentice-Hall, Inc., New Jersey.

MCKENZIE, D. 1978. Some remarks on the development of sedimentary basins. *Earth and Planetary Science Letters*, **40**, 25–32.

MCLAREN, S., DUNLAP, J., SANDIFORD, M. & MCDOUGALL, I. 2002. The thermochronological record of extraordinary heat production at Mount Painter, South Australia: implications for tectonic reactivation of continental interiors. *Tectonics*, 10.1029/2000TC001275.

MCLAREN, S., SANDIFORD, M., HAND, M., NEUMANN, N., WYBORN, L. & BASTRAKOVA, I. 2003. *The Hot Southern Continent: Heat Flow and Heat Production in Australian Proterozoic Terranes*. Geological Society, Australia, Special Publications, **22**, 157–167.

NIELSEN, S. B. & HANSEN, D. L. 2000. Physical explanation of the formation and evolution of inversion zones and marginal troughs. *Geology*, **28**, 875–878.

PATERSON, M. S. & LUAN, F. C. 1990. Quartize rheology under geological conditions. *In*: KNIPE, R. & RUTTER, E. (eds) *Deformation Mechanisms, Rheology and Tectonics*. Geological Society, London, Special Publications, **54**, 299–307.

RANALLI, G. 1995. *Rheology of the Earth*. Chapman & Hall, London.

REGENAUER-LIEB, K. & YUEN, D. 1998. Rapid conversion of elastic energy into shear heating during incipient necking of the lithosphere. *Geophysical Research Letters*, **25**, 2737–2740.

RICHTER, F. M., LOVERA, O. M., HARRISON, T. M. & COPELAND, P. 1991. Tibetan tectonics from $^{40}Ar/^{39}Ar$ analysis of a single K-feldspar samples. *Earth and Planetary Science Letters*, **105**, 266–278.

SANDIFORD, M. 1999. Mechanics of basin inversion. *Tectonophysics*, **305**, 100–120.

SANDIFORD, M. 2002. Low thermal Peclet number intraplate orogeny in central Australia. *Earth and Planetary Science Letters*, **201**, 309–320.

SANDIFORD, M. & MCLAREN, S. 2002. Tectonic feedback and the ordering of heat producing elements within the continental lithosphere. *Earth and Planetary Science Letters*, **204**, 133–150.

SANDIFORD, M., FREDERIKSEN, S. & BRAUN, J. 2003. The long-term thermal consequences of rifting: implications for basin reactivation. *Basin Research*, **15**, 23–24.

SANDIFORD, M., HAND, M. & MCLAREN, S. 2001. Tectonic feedback, intraplate orogeny and the geochemical structure of the crust: a central Australian perspective. *In*: MILLER, J., HOLDSWORTH, R., BUICK, I. & HAND, M. (eds) *Continental Reactivation and Reworking*. Geological Society, London, Special Publications, **184**, 195–218.

SHAW, R. D., ETHERIDGE, M. A. & LAMBECK, K. 1991. Development of the late Proterozoic to mid-Palaeozoic intracratonic Amadeus Basin in central Australia: a key to understanding tectonic forces in plate interiors. *Tectonics*, **10**, 688–721.

SHELTON, G. & TULLIS, J. 1981. Experimental flow laws for crustal rocks. *EOS (Transactions), American Geophysical Union*, **62**, 396.

SIBSON, R. 1990. Conditions for fault-value behaviour. *In*: KNIPE, R. J. & RUTTER, E. H. (eds) *Deformation Mechanisms, Rheology and Tectonics*. Geological Society, London, Special Publications, **54**, 143–165.

SONDER, L. & ENGLAND, P. 1986. Vertical averages of rheology of the continental lithosphere; relation to thin sheet parameters. *Earth and Planetary Science Letters*, **77**, 81–90.

STEPHENSON, R. A. & CLOETINGH, S. 1991. Some examples and mechanical aspects of continental lithosphere folding. *Tectonophysics*, **188**, 27–27.

TEYSSIER, C. 1985. A crustal thrust system in an intracratonic environment. *Journal of Structural Geology*, **7**, 689–700.

TURNER, J. P. & WILLIAMS, G. A. 2004. Sedimentary basin inversion and intraplate shortening. *Tectonophysics*, **65**, 277–304.

VERMEER, P. A. & DE BORST, R. 1984. Non-associated plasticity for soils, concrete and rock. *Heron*, **29**, 1–62.

WALTER, M. R. & GORTER, J. D. 1994. The Neoproterozoic Centralian Superbasin in Western Australia: the Savory and Officer Basins. *In*: PURCELL, P. G. & PURCELL, R. R. (eds) *The Sedimentary Basins of Western Australia*. Proceedings Petroleum Exploration Society Australia Symposium, Perth, 851–864.

WALTER, M. R., VEEVERS, J. J., CALVER, C. R. & GREY, K. 1995. Neoproterozoic stratigraphy of the Centralian Superbasin, Australia. *Precambrian Research*, **73**, 173–195.

Geometric and experimental models of extensional fault-bend folds

MARTHA O. WITHJACK & ROY W. SCHLISCHE

Department of Geological Sciences, Rutgers University, Piscataway, New Jersey, 08854, USA (e-mail: drmeow3@yahoo.com)

Abstract: We use geometric and experimental models to study the development of extensional fault-bend folds. The geometric models show that fault shape, fault displacement, and patterns of aggradation/erosion profoundly affect the distribution of growth beds, the magnitude and direction of dip of pregrowth and growth beds, and the location and dip of the outer limit of folding in pregrowth and growth beds. Complex structural and stratigraphic patterns develop if the rate of aggradation/erosion relative to the rate of fault displacement changes through time. The experimental models (with dry sand and wet clay) show that several deformational styles can accommodate extensional fault-bend folding. In sand models, a few, relatively major, secondary antithetic normal faults accommodate most hanging wall deformation. Pregrowth layers, although faulted, remain flat. The effective shear direction parallels the antithetic normal faults, and the shear angle is about 60°–65°. In clay models, numerous, relatively minor, secondary normal faults (antithetic and synthetic) and cataclastic flow accommodate most hanging wall deformation. The deformed pregrowth and growth layers dip gently toward the main fault. The effective shear angle (35°–50°) is considerably less than the dip of the antithetic normal faults. In the sand models and geometric models with a large shear angle (60°), more displacement occurs on the main normal fault and the hanging wall collapses in a relatively narrow zone. In the clay models and geometric models with a small shear angle (35°), less displacement occurs on the main normal fault. Instead, the hanging wall stretches substantially and collapses in a relatively wide zone.

Normal faults have a great variety of shapes in cross-sectional view. Some are listric (i.e., concave upward), either dipping (Fig. 1a) or flattening at depth (Figs 1b–d). Others steepen with depth (Fig. 1e). Still others have complex ramp-flat-ramp geometries (Fig. 1f). Movement on nonplanar normal faults causes the hanging wall to deform, typically by extensional fault-bend folding. Several researchers have used experimental and geometric models to study the development of extensional fault-bend folds. Experimental studies with wet clay (e.g., H. Cloos 1928, 1930; E. Cloos 1968; Withjack *et al.* 1995) and dry sand (e.g., McClay & Ellis 1987*a*, *b*; Ellis & McClay 1988; McClay & Scott 1991) have simulated the secondary faulting and folding associated with movement on nonplanar normal faults. Geometric studies (e.g., White *et al.* 1986; Dula 1991; White & Yielding 1991; Kerr & White 1992; Xiao & Suppe 1992; Withjack & Peterson 1993) have examined how fault shape, fault displacement and depositional patterns affect hanging wall geometries and tested the applicability of geometric modelling.

Each modelling approach has strengths and weaknesses. With geometric models, it is easy to vary fault shape, fault displacement and depositional/erosional patterns. It is necessary, however, to prescribe a hanging wall deformation mechanism, typically layer-parallel shear in compressional settings or inclined simple shear in extensional settings (Fig. 2). With experimental models, it is difficult to vary fault shape without imposing unrealistic boundary conditions (e.g., Withjack *et al.* 1995; Hauge & Gray 1996). The hanging wall, however, can deform by a range of deformation mechanisms that can vary spatially and temporally. In this paper, we use these complementary modelling approaches to better understand the development of extensional fault-bend folds. Specifically, we use the geometric models to document how fault shape, fault displacement, depositional and erosional patterns, and multiple episodes of faulting affect the geometry of extensional fault-bends folds. We use experimental models with dry sand and wet clay to study the hanging wall deformation that accommodates extensional fault-bend folding.

From: BUITER, S. J. H. & SCHREURS, G. (eds) 2006. *Analogue and Numerical Modelling of Crustal-Scale Processes.* Geological Society, London, Special Publications, **253**, 285–305.
0305-8719/06/$15.00

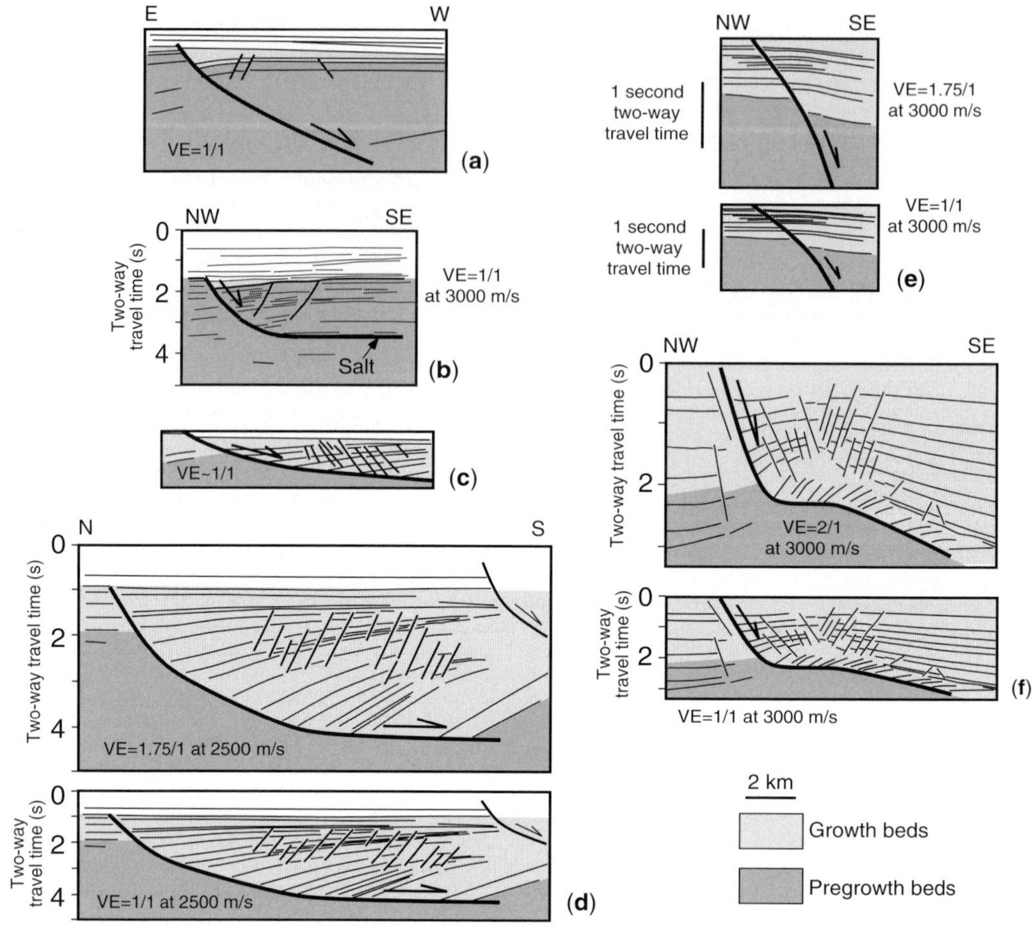

Fig. 1. Examples of normal faults and extensional fault-bend folds. (**a**) Line drawing of seismic line from offshore Norway (converted to depth) showing listric normal dipping at depth, rollover fold and secondary normal faults (Yielding *et al.* 1991). (**b**) Line drawing of migrated seismic line from offshore Norway showing listric normal fault flattening at depth in salt, rollover fold and antithetic secondary normal faults (after Withjack & Peterson 1990). (**c**) Line drawing of seismic line from Gulf of Oman with listric normal fault flattening at depth, rollover fold and synthetic secondary normal faults (after Wernicke & Burchfiel 1982; Hauge & Gray 1996). The line is displayed with approximately no vertical exaggeration. (**d**) Line drawing of migrated seismic line from Gulf of Mexico with listric normal fault flattening at depth, rollover fold and antithetic secondary normal faults. (**e**) Line drawing of migrated seismic line from offshore Australia (Northwest Shelf) showing normal fault that steepens with depth and associated hanging wall fold. (**f**) Line drawing of migrated seismic line from Gulf of Mexico showing normal fault (Brazos fault) with ramp-flat-ramp geometry, hanging wall folding, and complex pattern of secondary normal faulting. The line drawings in panels d–f are displayed with (top) and without (bottom) vertical exaggeration.

Geometric modelling

Movement on a nonplanar fault surface creates a potential gap between the footwall and hanging wall (Fig. 2a). In our geometric models, we assume that deformation of the hanging wall by inclined simple shear closes this gap, forming an extensional fault-bend fold. Numerous case studies have shown that inclined simple shear is a reasonable approximation of the bulk deformation in the hanging walls of nonplanar normal faults (e.g., White & Yielding 1991; Kerr & White 1992; White 1992; Xiao & Suppe 1992; Withjack & Peterson 1993; Withjack *et al.* 1995; Hauge & Gray 1996). For simplicity, we assume that the inclined shear direction dips 70° toward the main normal fault. We also assume that the footwall remains rigid, that

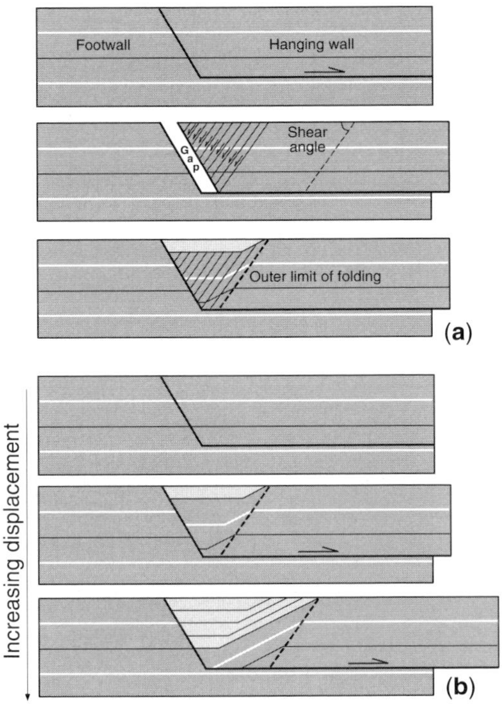

Fig. 2. Geometric modelling technique. (**a**) Movement on a nonplanar normal fault creates a potential gap between the footwall and hanging wall. Deformation of the hanging wall by inclined shear closes the gap and forms an extensional fault-bend fold. A growth bed fills the depression above the fold. (**b**) As fault displacement increases, fault-bend folding continues and new growth beds fill the resultant depressions above the fold. The pregrowth and growth beds deform as fault displacement increases.

compaction is negligible, and that growth beds completely fill any depressions that develop during faulting up to a prescribed regional datum (Fig. 2b). Although these factors certainly affect the geometry of extensional fault-bend folds (e.g., White & Yielding 1991; Xiao & Suppe 1992), the focus of this paper is to systematically study the effects of fault shape, fault displacement, and patterns of deposition and erosion for one or more episodes of normal faulting.

Experimental modelling

In our experimental models, either dry sand or wet clay represents the sedimentary overburden. The dry sand is composed of quartz grains with diameters of less than 0.5 mm (Withjack & Callaway 2000). Its density is about 1600 kg m^{-3}. It has a negligible cohesion, and a coefficient of internal friction of about 0.6. Localized cataclastic faulting (e.g., Rutter 1986) is the primary deformation style, even when strains are small. Thus, the dry sand has a negligible ductility as defined by Rutter (1986). The wet clay is composed predominantly of kaolinite and water (*c*. 40% by weight) (Withjack & Callaway 2000). Its density is about 1600 kg m^{-3}. It has a low cohesion (*c*. 50 Pa) and a coefficient of internal friction of about 0.5. Distributed cataclasis (e.g., Rutter 1986) is the primary deformation style in the wet clay when strains are small. With increasing strain, however, deformation becomes localized and cataclastic faulting is the primary deformation style. Thus, the wet clay has a greater ductility than that of the dry sand. The cohesion and coefficient of internal friction of the dry sand and wet clay are appropriate to ensure dynamic similarity between the models and natural prototypes (see Appendix). The sand and clay, however, simulate different types of rocks. The dry sand, with its negligible cohesion and ductility, represents rock that deforms primarily by localized cataclastic faulting (e.g., granite that deforms into large fault blocks with little internal deformation). The wet clay, with its greater cohesion and ductility, represents rock that deforms by distributed cataclasis and localized cataclastic faulting (e.g., interbedded sandstone and shale that deforms into fault blocks with considerable internal deformation).

Our experimental models have two different set-ups. One set-up has two overlapping metal plates beneath a 4 cm-thick layer of dry sand or wet clay (Fig. 3a). During the experiments, the lower plate slides outward at 4 cm hr^{-1}. In response, a normal fault propagates upward from the edge of the fixed, upper plate through the overlying sand or clay layer. The second set-up, originally described in Withjack and Callaway (2000), has two juxtaposed metal blocks beneath a 4 cm-thick layer of dry sand or wet clay (Fig. 3b). The 45°-dipping boundary between the metal blocks represents a dipping normal fault. During the experiments, the hanging wall block slides down the normal fault at 0.25 cm hr^{-1}. In response, a normal fault propagates upward from the edge of the footwall block through the overlying sand or clay layer. During the experiments, we fill subsiding areas with either dry sand or wet clay to simulate deposition during deformation. Also, we photograph the top surface of the models to document the surface deformation through time. After the experiments, we vertically slice the models, creating serial cross sections that

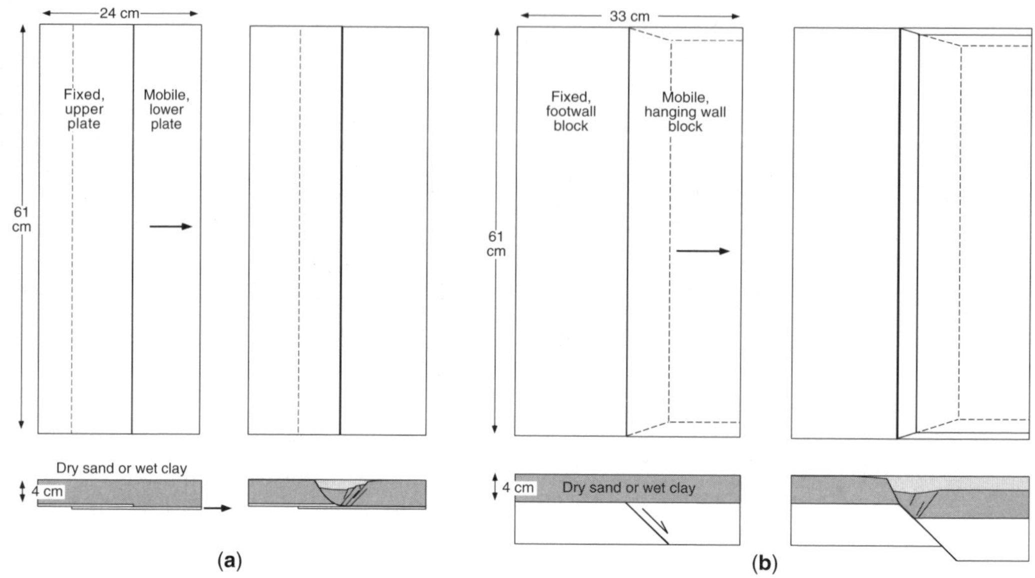

Fig. 3. Experimental modelling technique. (**a**) Experimental set-up with two overlapping metal plates beneath a layer of dry sand or wet clay. During the experiments, the lower plate slides away from the upper plate. (**b**) Experimental setup with two juxtaposed metal blocks beneath a layer of dry sand or wet clay. The 45°-dipping boundary between the blocks represents a dipping normal fault. During the experiments, the hanging wall block slides downward.

record the final deformation throughout the models. Colored (but mechanically identical) sub-layers within the sand and clay help identify faults and folds.

Results of geometric modelling

Fault shape

Our geometric models, like those of White *et al.* (1986), Dula (1991), White and Yielding (1991), Kerr and White (1992), Xiao and Suppe (1992) and Withjack and Peterson (1993), show that fault shape profoundly affects the shape of an extensional fault-bend fold.

If a normal fault flattens/detaches at depth (Fig. 4, left column), then growth beds have a limited lateral distribution, provided no aggradation occurs during faulting. Far from the fault, pregrowth beds in the hanging wall rise to the level of their footwall counterparts. If the upper fault segment is planar (Fig. 4a), then a triangular panel of undeformed, flat-lying beds exists near the fault. Adjacent to this triangular panel, the deformed pregrowth and growth beds dip uniformly toward the normal fault. If the upper fault segment is listric (Fig. 4b), then pregrowth and growth beds dip toward the fault. The dip of the growth beds systematically increases with depth, creating a fanning geometry.

If a normal fault dips at depth (Fig. 4, right column), then growth beds have a wide lateral distribution. All hanging wall beds are below the level of their footwall counterparts. The outer limit of the folding in the growth beds parallels the lowest fault segment, provided no aggradation or erosion occurs during faulting. If the upper fault segment is planar, then a triangular panel of undeformed, flat-lying beds exists near the fault. If the planar segment dips more steeply than the lower fault segment (Fig. 4c), then the tilted pregrowth and growth beds dip uniformly toward the main normal fault. If the planar segment dips more gently than the lower fault segment (Fig. 4d), then the tilted pregrowth and growth beds dip uniformly away from the main normal fault. If the upper fault segment is concave-upward (Fig. 4e), then the pregrowth and growth beds near the fault dip toward the fault. The dip of the growth beds systematically increases with depth, creating a fanning geometry. If the upper fault segment is convex-upward (Fig. 4f), then the pregrowth and growth beds near the fault dip away from the fault. Dips systematically increase

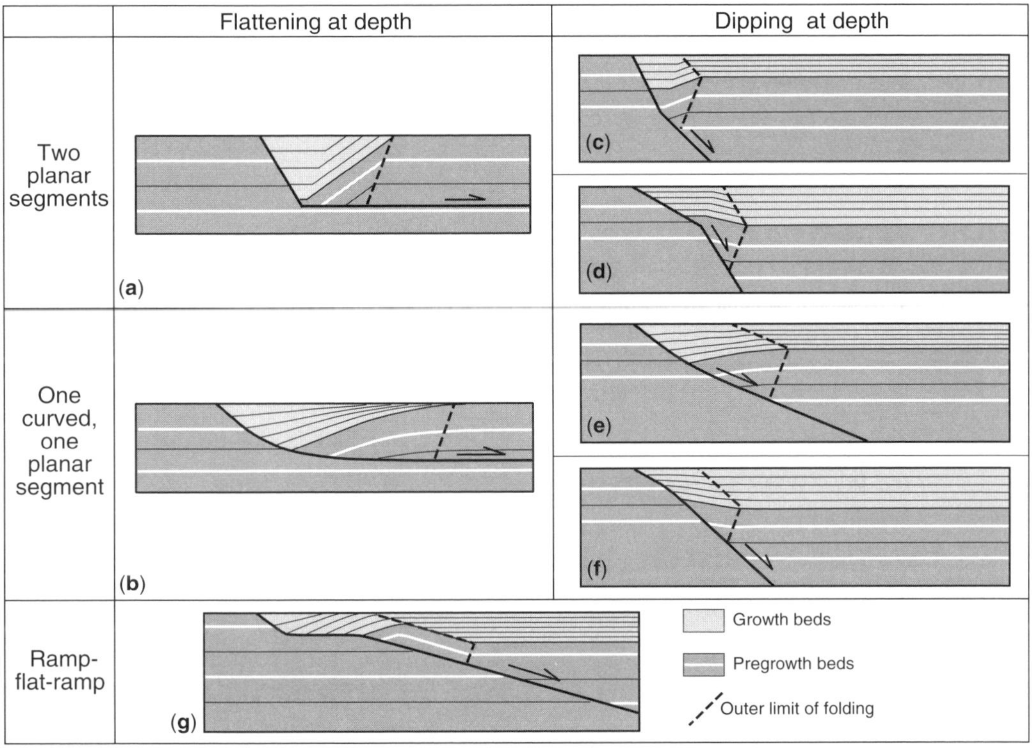

Fig. 4. Geometric models showing the effect of fault shape on the geometry of an extensional fault-bend fold. No aggradation or erosion occurs during faulting. Fault shape affects the lateral distribution of growth beds and the direction and magnitude of bedding dip. See text for further discussion.

with depth. This latter type of extensional fault-bend fold resembles a fault-propagation fold or a compaction-related fold.

If a normal fault has a ramp-flat-ramp geometry (Fig. 4g), then an anticline develops in the pregrowth beds. Growth beds near the fault dip and thicken toward the fault. Growth beds far from the fault are flat-lying. The outer limit of the folding in the growth beds parallels the lower ramp, provided no aggradation or erosion occurs during faulting.

Fault displacement and fold evolution

The geometry of an extensional fault-bend fold changes through time as fault displacement increases. If a fault has two planar segments (Fig. 2b), then fault displacement affects the distribution, but not the dip, of the deformed pregrowth and growth beds. As displacement increases, more of the pregrowth and growth beds deform. The dip of the deformed pregrowth and growth beds, however, remains constant. For a listric normal fault that flattens at depth (Fig. 5a), the dip of the pregrowth beds increases as fault displacement increases. Growth beds develop a pronounced fanning geometry, and pregrowth and growth beds become thinner and longer. When fault displacements are very large, the oldest growth beds are substantially thinned and lengthened. Their fanning geometry becomes subtle, making it difficult to distinguish between growth and pregrowth layers.

For a normal fault with a ramp-flat-ramp geometry (Fig. 5b), a wide anticline initially forms in the pregrowth beds above the flat. Near the fault, growth beds dip and thicken toward the fault. Far from the fault, growth beds are flat-lying. No deposition occurs above the central part of the flat across the crest of the anticline. As fault displacement increases, the anticline narrows and moves down the lower ramp. Growth beds continue to accumulate near and far from the fault, but not above the central part of the flat. With additional fault displacement, the site of no deposition moves down the lower ramp, producing a dipping, diachronous unconformity. Correlative beds are offset across

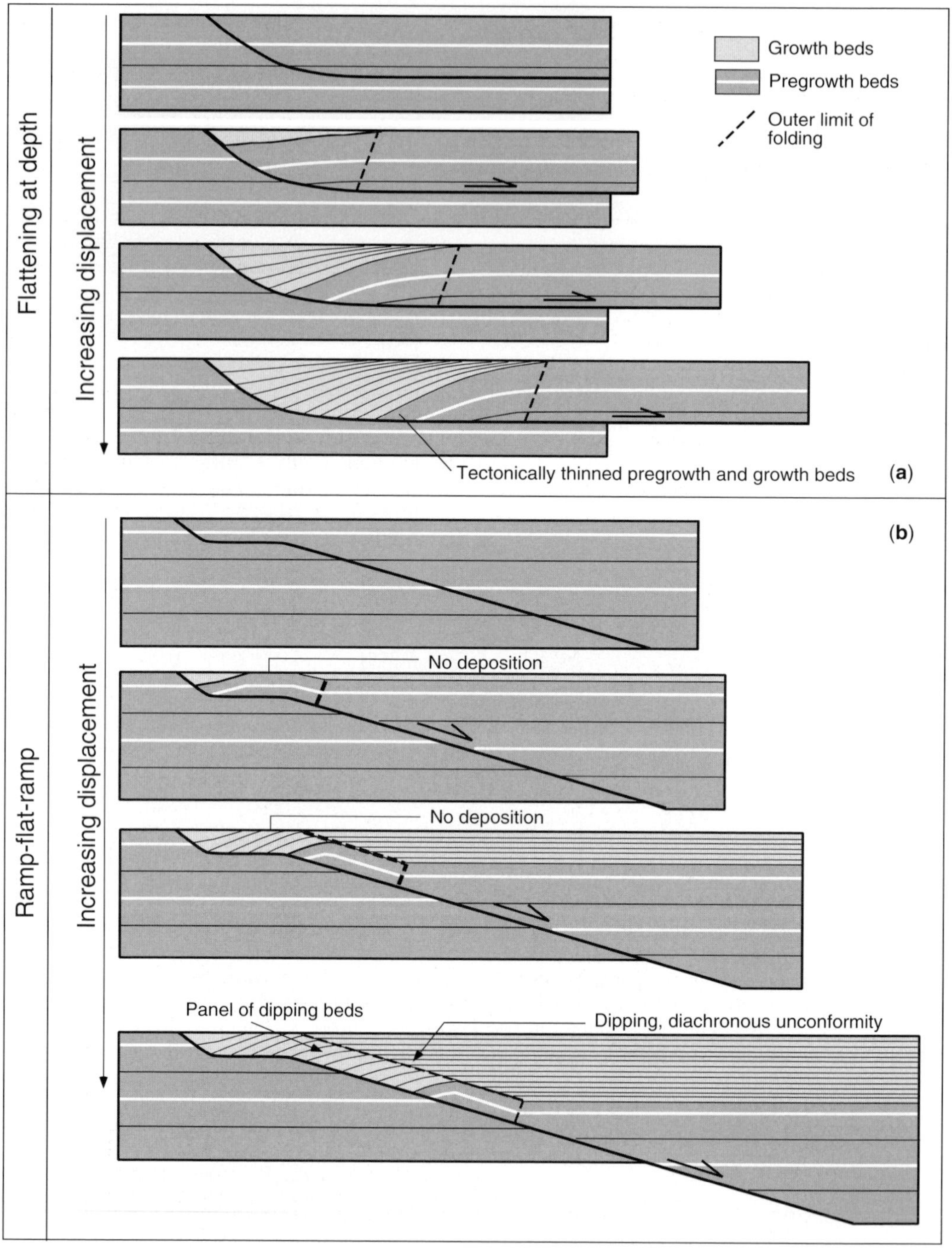

Fig. 5. Geometric models showing the effect of fault displacement on the geometry and evolution of an extensional fault-bend fold. No aggradation or erosion occurs during faulting. (**a**) Listric fault that flattens at depth. (**b**) Fault with ramp-flat-ramp geometry. See text for further discussion.

the unconformity. Thus, the unconformity resembles a normal fault that parallels the lower ramp. A panel of dipping growth beds develops directly above the lower ramp.

Aggradation and erosion

In many extensional settings, regional subsidence, regional uplift, and fluctuations in sediment supply or sea level lead to aggradation or erosion during normal faulting. Our geometric models (Figs 6 and 7) support the conclusion of Xiao and Suppe (1992) that the rate of aggradation relative to the rate of fault displacement profoundly affects the shape of an extensional fault-bend fold. Additionally, our geometric models show that the rate of erosion relative to the rate of fault displacement also profoundly affects the shape of an extensional fault-bend fold.

For a listric fault that flattens at depth (Fig. 6a), growth beds have a wide lateral distribution with aggradation and a very limited lateral distribution with erosion. With erosion, growth beds have a more subtle fanning geometry and can, in fact, closely resemble pregrowth beds. For a fault that dips at depth (Fig. 6b), the dip of the outer limit of the folding in the growth beds depends on the amount of aggradation or erosion that occurs during faulting. As mentioned previously, the outer limit of the folding in the growth beds parallels the lower fault segment, if no aggradation or erosion occurs (Fig. 4). With aggradation, the outer limit of the folding dips more steeply than the lower fault segment. The greater the amount of aggradation, the steeper the dip. With erosion, the outer limit of the folding dips more gently than the lower fault segment. The greater the amount of erosion, the gentler the dip. For a normal fault with a ramp-flat-ramp geometry (Fig. 7), a panel of dipping growth beds develops above the lower ramp with aggradation and with

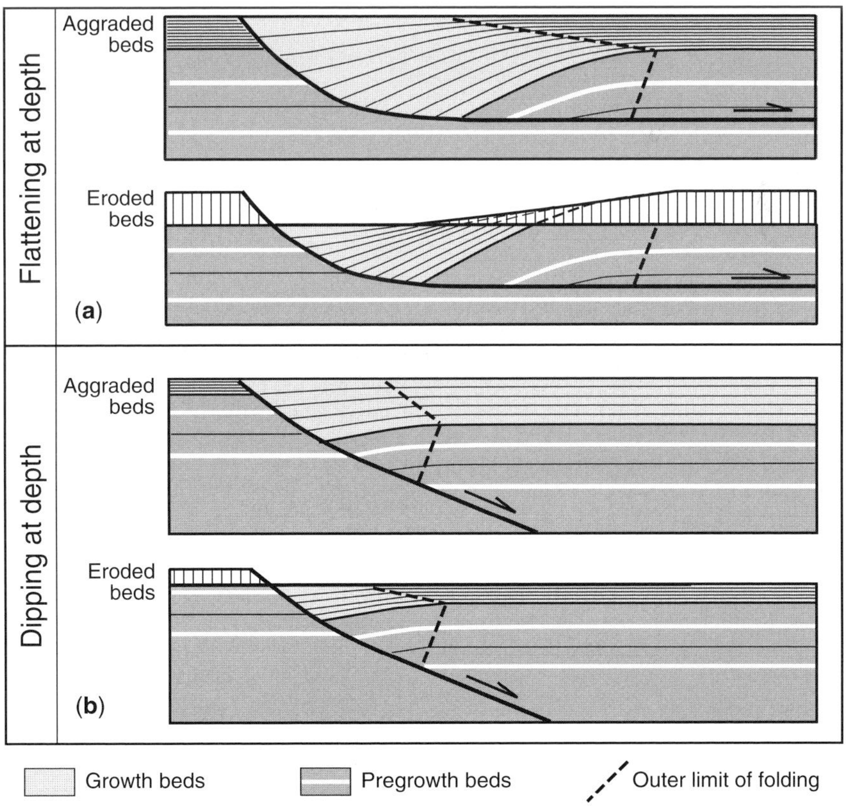

Fig. 6. Geometric models showing the effect of aggradation and erosion on the geometry of an extensional fault-bend fold. (**a**) Fault that flattens at depth. (**b**) Fault that dips at depth. The occurrence of aggradation/erosion affects the lateral distribution of growth beds, the dip of the growth beds, and the outer limit of the folding. See text for further discussion.

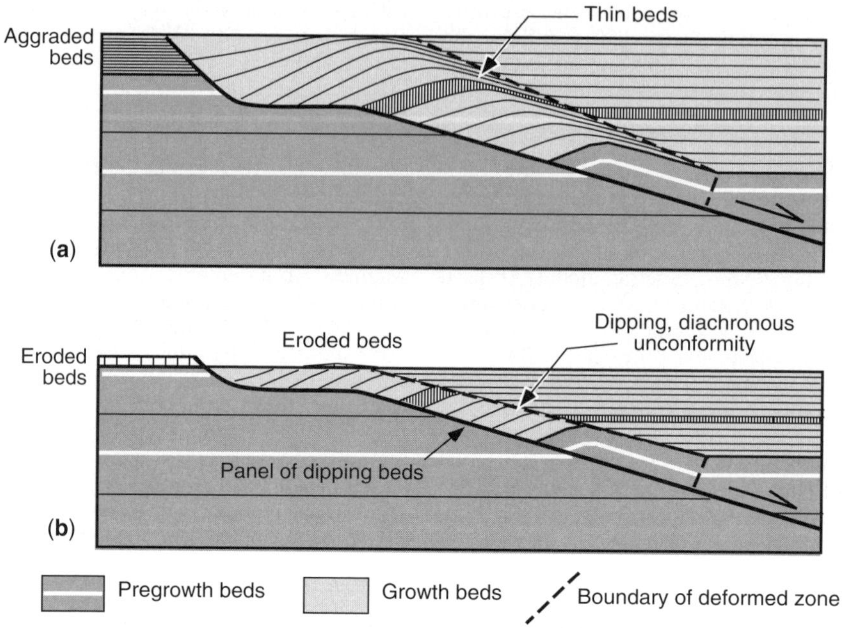

Fig. 7. Geometric models of faults with ramp-flat-ramp geometry showing the effect of (**a**) aggradation and (**b**) erosion on the geometry of an extensional fault-bend fold. The occurrence of aggradation/erosion affects the lateral distribution of growth beds, the dip of the growth beds, and the outer limit of the folding. See text for further discussion.

erosion. With aggradation, the panel widens upward, and thin growth beds dipping away from the fault overlie the panel (Fig. 7a). With erosion, the panel narrows upward, and a dipping, diachronous unconformity overlies the panel. The unconformity dips more gently than the lower ramp (Fig. 7b).

Multiple episodes of faulting

Complex structural and stratigraphic patterns develop with multiple episodes of normal faulting if the rate of aggradation/erosion relative to the rate of fault displacement changes through time (Fig. 8). In Figures 8a–c, the normal fault is listric and flattens at depth. In Figure 8a, aggradation occurs during the second episode of faulting but not during the first episode. The distribution and geometry of the growth beds associated with each faulting episode differ substantially. The outer limit of the folding is not planar. It parallels the inclined shear direction in the pregrowth beds, it flattens above the growth beds associated with the first faulting episode, and it dips away from the fault in the growth beds associated with the second faulting episode. In Figure 8b, widespread deposition occurs between episodes of normal faulting. The outer limit of the folding parallels the inclined shear direction in the pregrowth beds. It flattens above the growth beds associated with the first faulting episode, and it parallels the inclined shear direction in the beds deposited between the faulting episodes. In Figure 8c, erosion occurs between the episodes of normal faulting. The growth beds associated with the first faulting episode have a very subtle fanning geometry. In fact, they closely resemble the underlying pregrowth beds. In Figure 8d, the normal fault has a ramp-flat-ramp geometry. If significant aggradation occurs during the second episode of faulting but not during the first episode, then the distribution and geometry of the growth beds associated with each faulting episode differ substantially. The outer limit of the folding is not planar. It parallels the lower ramp in the growth beds associated with the first faulting episode, and it dips toward the fault in the growth beds associated with the second faulting episode.

Results of experimental modelling

As discussed below, the sand and clay models provide information about the styles of deformation associated with extensional fault-bend folding (Figs 9–11).

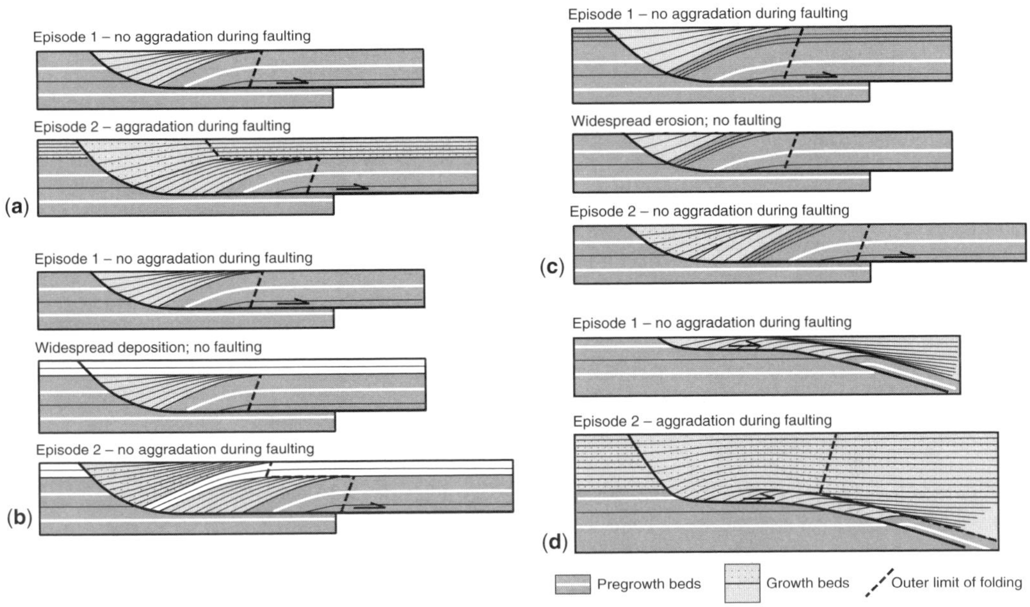

Fig. 8. Geometric models showing the effect of multiple episodes of faulting on the geometry of an extensional fault-bend fold. The rates of aggradation/erosion relative to the rates of fault displacement vary during the faulting episodes. See text for further discussion.

Models with overlapping plates: horizontal detachment at depth

In the sand model with the overlapping plates (Fig. 9a), a fault propagates upward from the edge of the fixed upper plate. The fault is planar and dips steeply, about 60°–65°. This steep fault and the horizontal mobile plate, together represent a normal fault that flattens/detaches at depth. A fault bend separates the steep and horizontal segments. As the deformation progresses, a series of steeply dipping antithetic faults develops in the hanging wall. Each antithetic fault forms near the fault bend, moves with the mobile plate past the fault bend, and becomes inactive. Thus, the antithetic faults become progressively older the farther from the fault bend. The antithetic faults have relatively large displacements and accommodate most of the hanging wall deformation, accounting for about 40% of the imposed displacement of the mobile plate (Fig. 9c). The upper segment of the main normal fault accounts for another 45% of the imposed displacement. The remainder (15%) is internal deformation within the fault blocks. Kautz and Sclater (1988) determined similar amounts of internal deformation in their extensional sand models. The effective shear direction in the sand model (based on the outer limit of the folding) is subparallel to the antithetic normal faults. Thus, the inclined shear angle in the sand model is about 60°–65°, agreeing with the value obtained by Xiao and Suppe (1992) for a similar sand model published by McClay and Ellis (1987*a*).

In the clay models with the overlapping plates (Figs 9b and 11a–d), a fault propagates upward from the edge of the fixed upper plate. Unlike the sand models, fault-propagation folding and secondary faulting precede the development of the through-going normal fault. The fault is listric, steeply dipping (*c.* 75°) near the surface and gently dipping (*c.* 35°) near the base of the clay layer. This steep fault and the horizontal mobile plate, together, represent a normal fault that flattens/detaches at depth. A fault bend separates the steep and horizontal segments. As displacement increases, an upward-widening zone of deformation develops, emanating from the fault bend (Fig. 11a). Deformation includes secondary normal faulting (antithetic and synthetic) and fault-bend folding (layers dip toward the main normal fault). The deformation zone moves with the mobile plate, passing the fault bend. Eventually, it becomes inactive and a new deformation zone replaces it (Figs 11b, c). Thus, deformation becomes progressively older

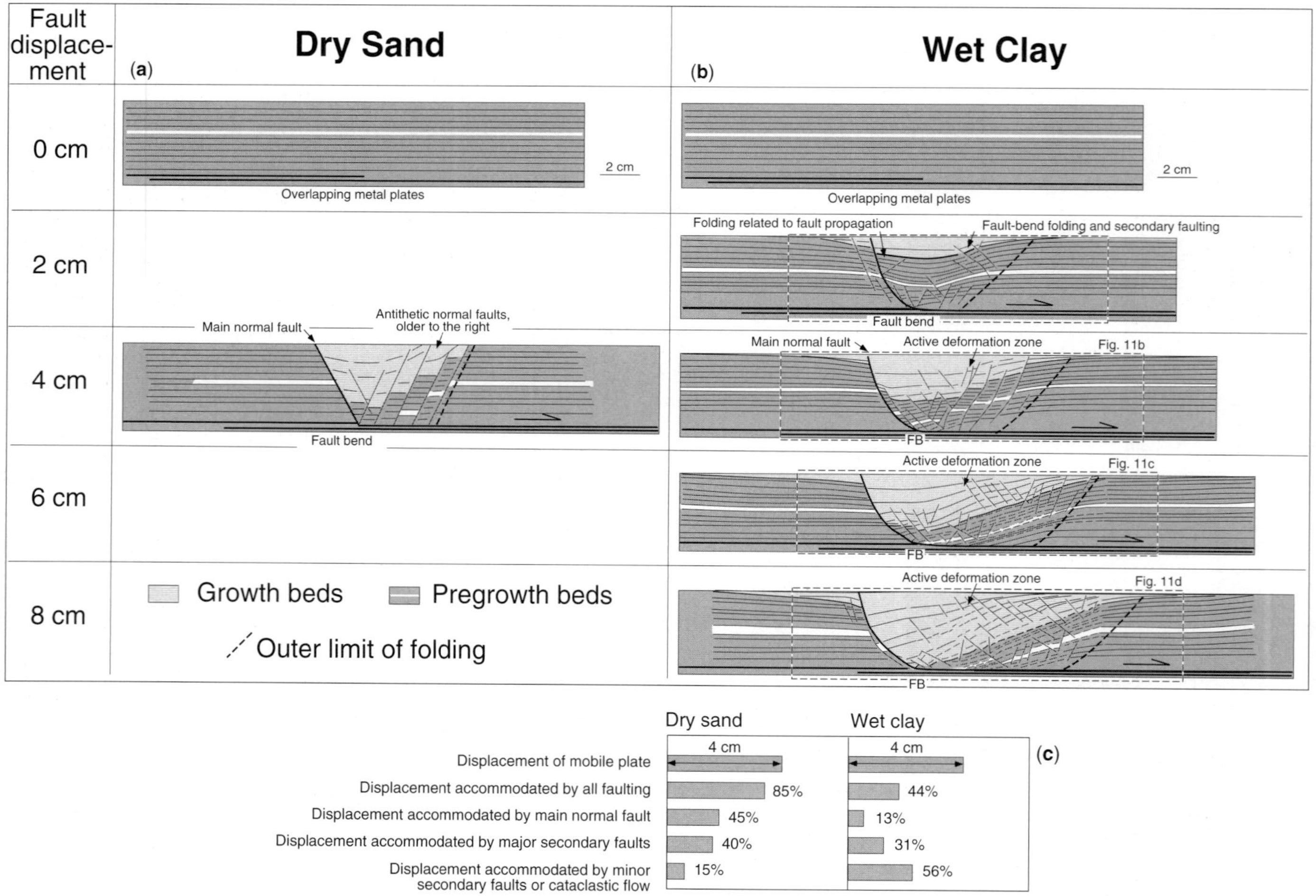

Fig. 9. Line drawings of cross sections through (**a**) experimental model with dry sand and (**b**) four experimental models with wet clay. The sand or clay layer overlies overlapping metal plates. During the experiments, the lower plate moves to the right by a prescribed displacement. The experimental models simulate fault-bend folding associated with a normal fault that flattens at depth. The boxes show the locations of the photographs in Fig. 11. FB is fault bend (**c**) Accommodation of imposed displacement (4 cm) in the sand and clay models.

the farther from the fault bend. By 8 cm of displacement of the mobile plate (Fig. 11d), most pregrowth and growth layers are faulted and dip gently (*c.* 20°) toward the main fault. Fault-bend folding rotates many of the secondary faults. Thus, synthetic faults dip more gently and antithetic faults dip more steeply through time. The pregrowth beds and the oldest growth beds have thinned and lengthened significantly. In fact, many of the oldest growth beds lack an obvious fanning geometry and resemble pregrowth beds (Fig. 11d).

In the clay models, cataclastic faulting, cataclastic flow and rigid-body rotation accommodate the fault-bend folding. For example, in the model with 4 cm of displacement of the mobile plate, the major secondary faults account for about 31% of the imposed displacement (Fig. 9c). The upper segment of the main normal fault accounts for another 13% of the imposed displacement. Minor faulting, cataclastic flow, and rigid-body rotation account for 56% of the imposed displacement. Kautz and Sclater (1988) determined similar amounts of internal deformation within their extensional clay models. The effective shear angle is about 35°–50° (based on the average dip of the outer limit of the folding), considerably less than the dip of the antithetic faults. White and Yielding (1991) calculated a similar value (49°) for a published clay model by E. Cloos (1968), and Withjack *et al.* (1995) obtained similar values (50°–55°) for their clay-model experiments.

Models with juxtaposed blocks: dipping fault at depth

In the sand model with the juxtaposed blocks (Withjack & Callaway 2000), a normal fault propagates upward from the edge of the fixed block (Fig. 10a). It is planar, and dips steeply, about 65°. This steep fault segment and the 45°-dipping boundary between the juxtaposed blocks represent a normal fault that dips at depth. A steeply dipping antithetic fault forms near the fault bend between the fault segments. It moves downward with the mobile block as fault displacement increases. Pregrowth and growth layers near the main fault dip gently toward the fault. The effective shear direction in the sand model (based on the dip of the outer limit of the folding) is subparallel to the antithetic normal fault. Thus, the inclined shear angle is about 65°.

In the clay model with the juxtaposed blocks (Withjack & Callaway 2000), a normal fault propagates upward from the edge of the fixed block (Figs 10b and 11e). Fault-propagation folding and secondary faulting precede the development of the through-going normal fault. The fault is listric, steeply dipping (*c.* 90°) near the surface and gently dipping (*c.* 45°) near the base of the clay layer. This fault segment and the 45°-dipping boundary between the juxtaposed blocks represent a normal fault that dips at depth. An upward-widening zone of deformation develops in the hanging wall of the main normal fault. The zone, emanating from the fault bend, moves downward with the mobile block as fault displacement increases. Deformation within the zone includes secondary faulting (mostly antithetic normal faults) and fault-bend folding with pregrowth and growth layers dipping gently toward the main fault. The effective shear angle (based on the average dip of the outer limit of the folding in the pregrowth layers) is about 50°.

Discussion and applications

Comparisons of geometric and experimental models

The style of hanging wall deformation varies considerably in the experimental models, reflecting the differences in the cohesion and ductility of the dry sand and wet clay. In the sand models, a few, relatively large, secondary antithetic normal faults accommodate most of the hanging wall deformation. Pregrowth layers, although faulted, are flat-lying. The effective shear direction parallels the antithetic normal faults, and the shear angle is about 60°–65°. In the clay models, secondary faulting (antithetic and synthetic) and cataclastic flow accommodate most of the hanging wall deformation. The deformed pregrowth and growth layers dip gently toward the main normal fault. The effective shear angle (35°–50°) is considerably less than the dip of the antithetic normal faults.

The style of deformation in the sand models differs from that in the geometric models in several ways. Deformation is focused on a few, relatively large, antithetic normal faults in the sand models, not uniformly distributed as it is in the geometric models. Pregrowth layers are nearly flat-lying in the sand models, not dipping at they are in the geometric models. Although the distributed style of deformation in the clay models is similar to that in the geometric models, the clay models also differ from the geometric models in several ways. Fault-propagation folding precedes the development of through-going normal faults in

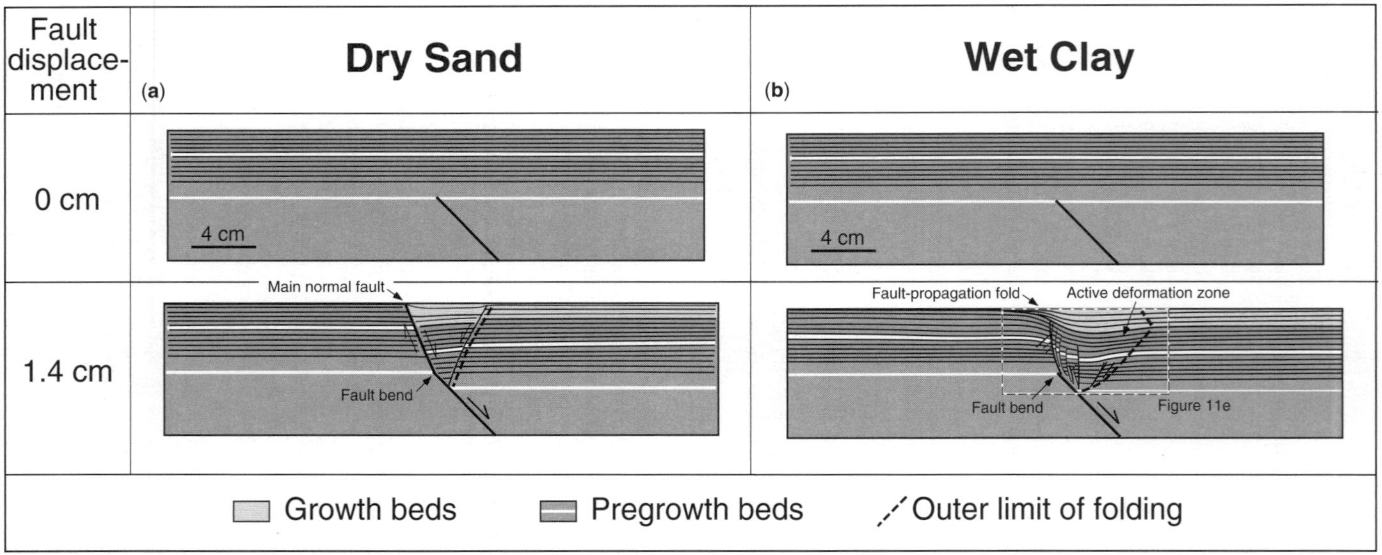

Fig. 10. Line drawings of cross sections through experimental models with dry sand and wet clay (after Withjack & Callaway 2000). The sand or clay layer overlies juxtaposed metal blocks. During the experiments, the hanging wall block moves downward by a prescribed displacement. The experimental models simulate fault-bend folding associated with a normal fault that dips at depth. The box shows the location of the photograph in Fig. 11e.

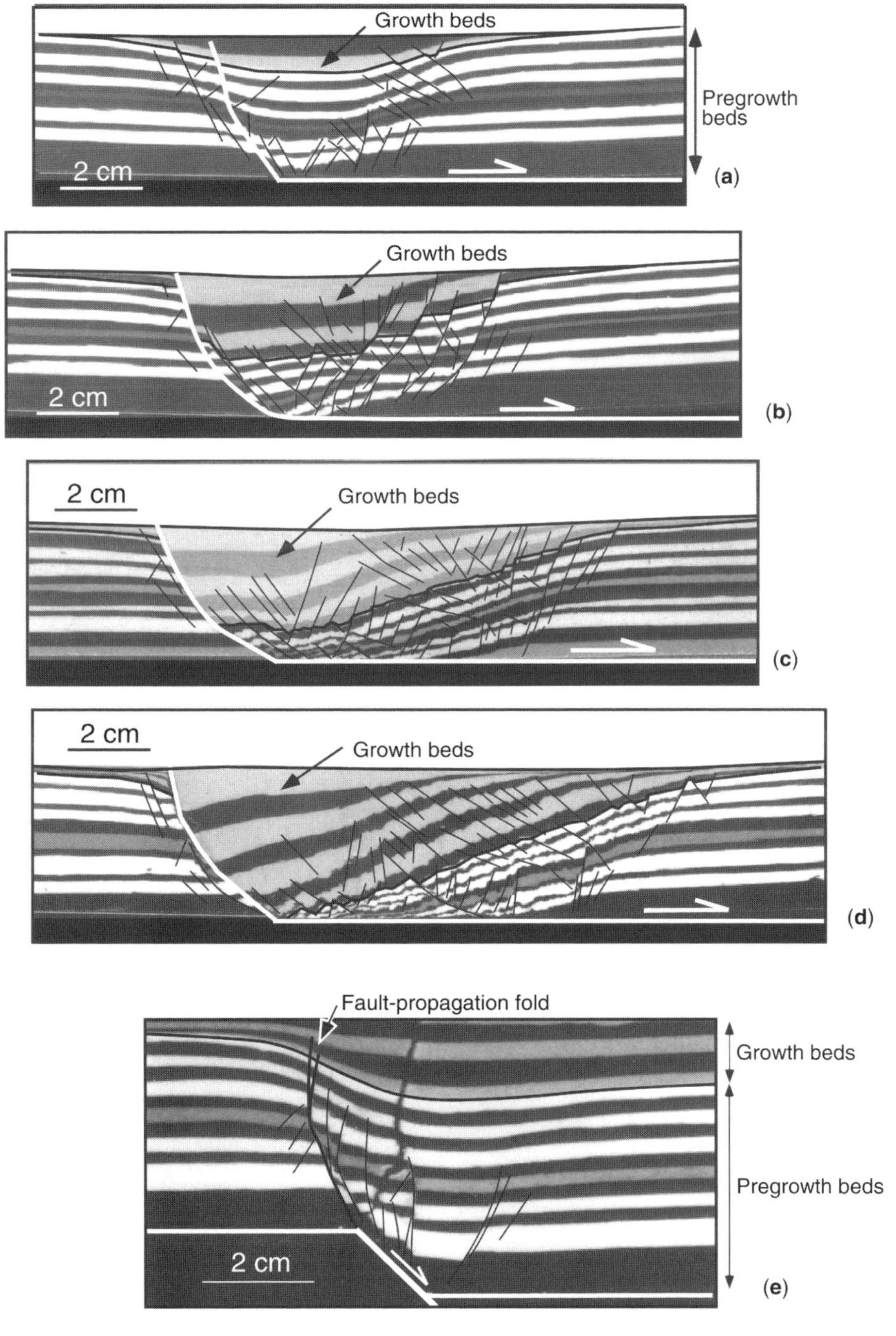

Fig. 11. Photographs of cross sections through five clay models. Major secondary faults are shown in black. (**a**) Cross section through model with 2 cm displacement of mobile plate. (**b**) Cross section through model with 4 cm displacement of mobile plate. (**c**) Cross section through model with 6 cm displacement of mobile plate. (**d**) Cross section through model with 8 cm displacement of mobile plate. (**e**) Cross section through model with 1.4 cm of displacement on mobile block.

the clay models. Thus, not all deformation in the clay models is related to fault-bend folding. Also, deformation zones in the pregrowth layers widen upward in the clay models. In geometric models with inclined shear, the width of the deformation zones in the pregrowth layers remains constant with depth (Withjack *et al.* 1995).

Despite differences in the styles of deformation, the geometric and experimental modelling approaches yield similar results for the overall shape and evolution of an extensional fault-bend fold. Specifically, in models with a normal fault that flattens at depth, growth beds have a limited lateral distribution. In models with a normal fault that dips at depth, growth beds have a wide lateral distribution. Both the geometric and experimental models show that rollover folds develop in the hanging walls of normal faults whose dip decreases with depth. The geometric and clay models with a listric, upper fault segment show that the dip of the growth beds increases with depth, creating a fanning geometry. If fault displacements are large, then pregrowth and growth beds thin and lengthen significantly during extensional fault-bend folding.

Figure 12 compares sand and clay models (with 4 cm of displacement of the mobile plate) with similar geometric models with shear angles of 60° and 35°, respectively. The distribution of deformation in the experimental models closely resembles that in the geometric models. In the sand model and the geometric model with the 60° shear angle, more displacement occurs on the upper segment of the main normal fault than in the clay model and the geometric model with the 35° shear angle. The hanging wall collapses in a relatively narrow zone in the sand model and the geometric model with the 60° shear angle. The hanging wall stretches substantially and collapses in a relatively wide zone in the clay model and the geometric model with the 35° shear angle.

Figure 13 compares a geometric model with a shear angle of 35° with the cross section from the clay model with 8 cm of displacement of the mobile plate. The clay model closely resembles the geometric model. Specifically, the location of the hanging wall cutoffs, the dip of the deformed hanging wall beds (*c.* 20°), and the displacement distribution for the pregrowth and growth beds are similar in both models. In the clay model and the geometric model, the hanging wall stretches considerably, requiring less displacement on the upper segment of the main normal fault. In the clay model, the youngest growth beds near the main normal fault dip and thicken toward the fault. In the geometric model, the growth beds near the main normal fault are flat-lying. This difference reflects the listric shape of the upper fault segment in the clay model and the planar shape of the upper fault segment in the geometric model.

Geological examples

Observed hanging wall folds in many natural examples resemble those in the geometric and experimental models. Rollover folds develop in the hanging wall of normal faults that flatten at depth (Figs 1b, c and d). Growth beds dip and thicken toward the main fault, and the oldest growth beds commonly have subtle fanning geometries. Normal faults with convex-up bends (Fig. 1e) have hanging wall beds that dip away from the main fault. Normal faults with ramp-flat-ramp geometries (Fig. 1f) have complex hanging wall deformation. Beds dip toward the fault near the main fault, and a dipping unconformity that parallels the lower ramp commonly develops above the ramp. The secondary normal faults in many natural examples resemble those in the experimental models. Many secondary normal faults are antithetic to the main fault (Figs 1a, b and d). Some, however, are synthetic to the main normal fault (Fig. 1c), like those in the clay models (Figs 9b and 11). Most secondary normal faults have relatively small displacements (Figs 1, 14 and 15), like those in the clay models (Figs 9b and 11). Calculated inclined shear angles for geological examples of extensional fault-bend folds (Hauge & Gray 1996) range from about 45°–70°, similar to those in the experimental models.

The Blackberry fault from the Gulf of Mexico (Fig. 14) has up to 10 km of normal displacement. It is listric, flattening within shales. Two small fault bends are present at shallow levels. Secondary normal faults in the hanging wall are antithetic to the Blackberry fault and have relatively small displacements. As predicted by the geometric and experimental models, a large rollover fold affects its hanging wall. Growth beds dip and thicken toward the fault. Second-order fault-bend folding occurs near the small fault bends. This folding, with beds dipping away from the Blackberry fault, resembles fault-propagation folding. It is, however, associated with beds passing through a convex-upward fault bend. Similar second-order fault-bend folds develop in our clay models (Fig. 14d). Xiao and Suppe (1992) have reported similar folds elsewhere in the Gulf of Mexico.

The Brazos fault from the Gulf of Mexico (Fig. 15) has two major fault bends, a northwest

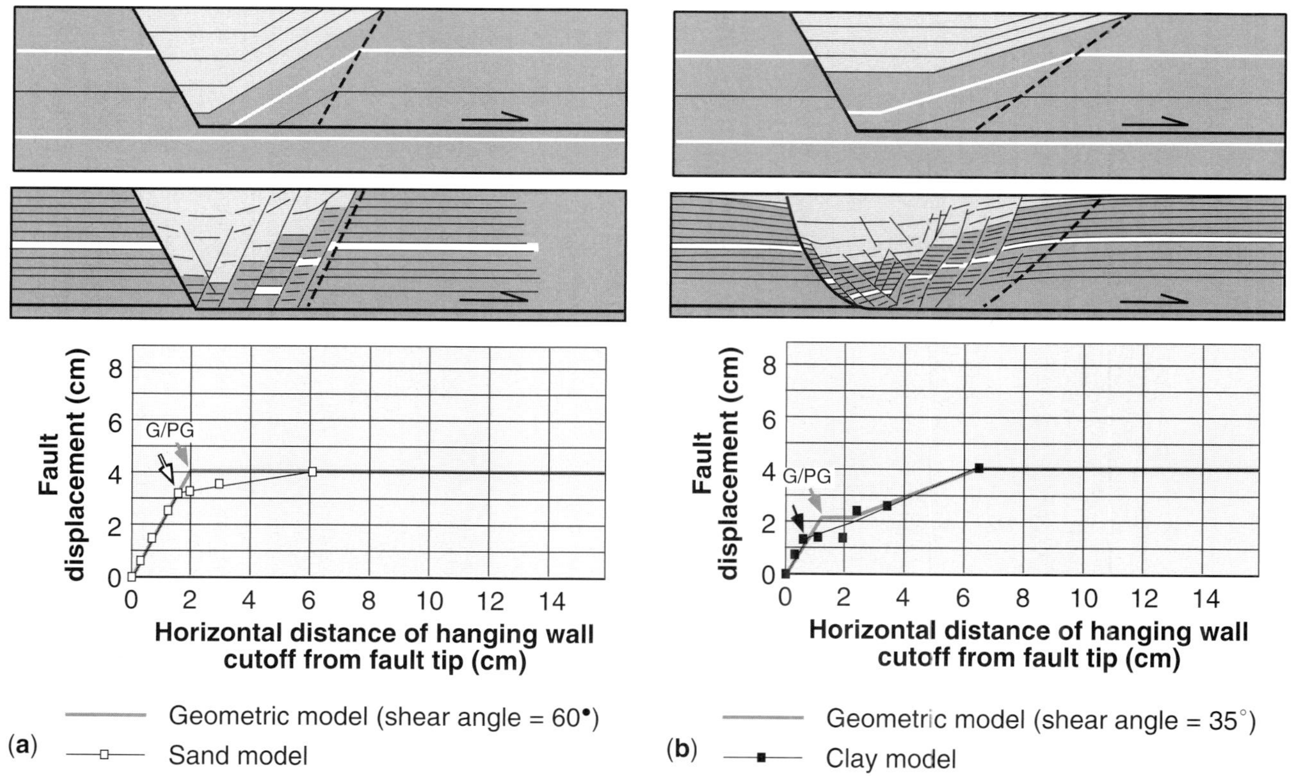

Fig. 12. (**a**) Line drawing of sand model (4 cm displacement of the mobile plate) and geometric model with 4 units of displacement and inclined shear angle of 60°. Plot of fault displacement versus horizontal distance of hanging wall cutoffs from fault tip for sand model and geometric model. (**b**) Line drawing of clay model (4 cm displacement of the mobile plate) and geometric model with 4 units of displacement and inclined shear angle of 35°. Plot of fault displacement versus horizontal distance of hanging wall cutoffs from fault tip for clay model and geometric model. G/PG indicates the boundary between the growth and pregrowth beds.

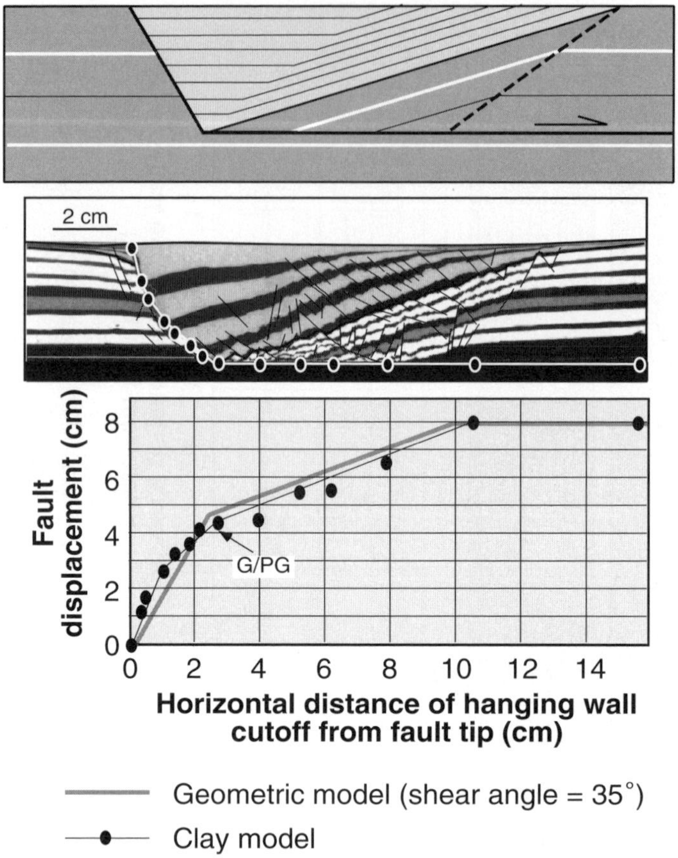

Fig. 13. Geometric model of fault-bend fold with 8 units of displacement (top). Inclined shear angle is 35°. Photograph of cross section of clay model after 8 cm of displacement (centre). Circles show locations of hanging wall cutoffs for key growth and pregrowth beds. Plot of fault displacement versus horizontal distance of hanging wall cutoffs from fault tip for clay model and geometric model (bottom). G/PG indicates the boundary between the growth and pregrowth beds.

concave-up bend and a southeast convex-up bend. The geometry of the fault changes considerably along strike (Vogler & Robinson 1987). On parts of the Brazos fault, the segment between the fault bends dips gently to the southeast (Figs 15a–c). On other parts of the Brazos fault, the segment between the fault bends is nearly horizontal (Fig. 15d). The hanging wall deformation closely resembles the deformation in the geometric models with the ramp-flat-ramp geometry (Figs 4g, 5b, 7 and 8d). Near the fault, growth beds dip toward the fault. Farther from the fault, growth beds dip gently away from the fault. A panel of dipping growth beds develops directly above the lower ramp. The geometry of the extensional fault-bend fold varies depending on the dip of the central segment. Where the segment is flat (Fig. 15d), the growth beds directly above the panel of dipping growth beds are very thin. In fact, Vogler and Robinson (1987) report that a dipping unconformity locally overlies the panel of dipping growth beds. Where the segment is dipping (Fig. 15c), the growth beds directly above the panel of dipping growth beds are slightly thicker. The abrupt change in the style of the fault-bend folding with depth (i.e., tight folding at depth; broad folding at shallow levels) probably reflects significant aggradation during the latter stages of faulting and folding (Fig. 8d).

The low-angle, east-dipping Vicksburg fault underlies the Slick Ranch field in southern Texas (Fig. 16) (Erxleben & Carnahan 1983). The deformation in the hanging wall of the Vicksburg fault resembles that in the hanging

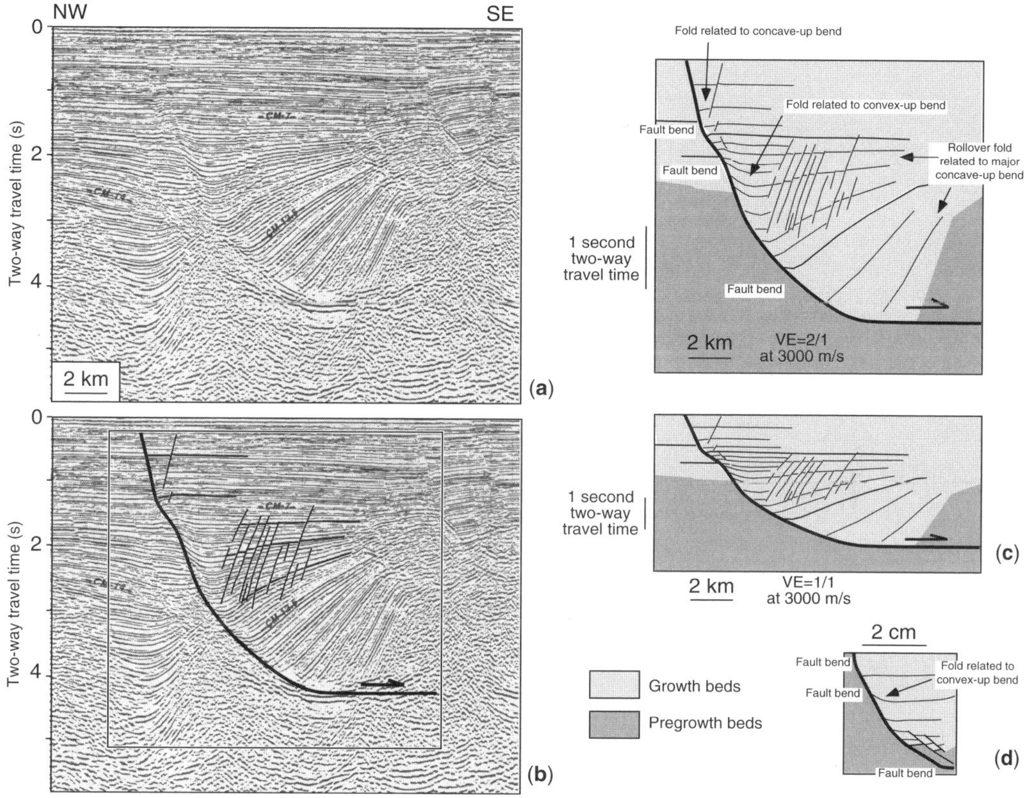

Fig. 14. Blackberry normal fault, offshore Texas, Gulf of Mexico. (**a**) Time-migrated seismic line (after Bally *et al.* 1991). CM-7 and CM-12.5 represent strata that are about 7 Ma and 12.5 Ma, respectively. (**b**) Interpreted seismic data. Black box shows location of line drawing. (**c**) Interpreted line drawing with (top) and without (bottom) vertical exaggeration. Second-order fault-bend folding occurs near small fault bends. (**d**) Second-order fault-bend fold in clay model with overlapping metal plates and 6 cm displacement.

wall of the Brazos fault in terms of geometry and dimensions (Fig. 15e). A panel of dipping growth beds overlies the low-angle fault. A dipping unconformity overlies the panel of dipping growth beds. As in the case of the Brazos fault, this deformation probably reflects movement on a fault with a ramp-flat-ramp geometry.

Conclusions

We have used geometric and experimental models to better understand the development of extensional fault-bend folds. The geometric models show that:

(1) The shape of a normal fault controls the distribution of growth beds, the magnitude and direction of dip of the pregrowth and growth beds, the outer limit of folding in the pregrowth and growth beds, and the level of hanging wall beds relative to their footwall counterparts (Fig. 4).

(2) If a fault has two planar segments, then fault displacement controls the distribution, but not the dip, of the deformed pregrowth and growth beds (Fig. 2b). If a fault has a curved segment, then fault displacement controls the dip, thickness, and distribution of the deformed pregrowth and growth beds (Fig. 5).

(3) Aggradation and erosion control the distribution of growth beds, the dip of the pregrowth and growth beds, and the outer limit of folding in the pregrowth and growth beds (Figs 6 and 7). The outer limit of the folding in the growth beds parallels the lower fault segment, if no aggradation or erosion occur (Fig. 5). With aggradation, the outer limit of the folding dips more steeply than the lower fault segment. With erosion, the outer limit of the folding dips more gently than the lower fault segment.

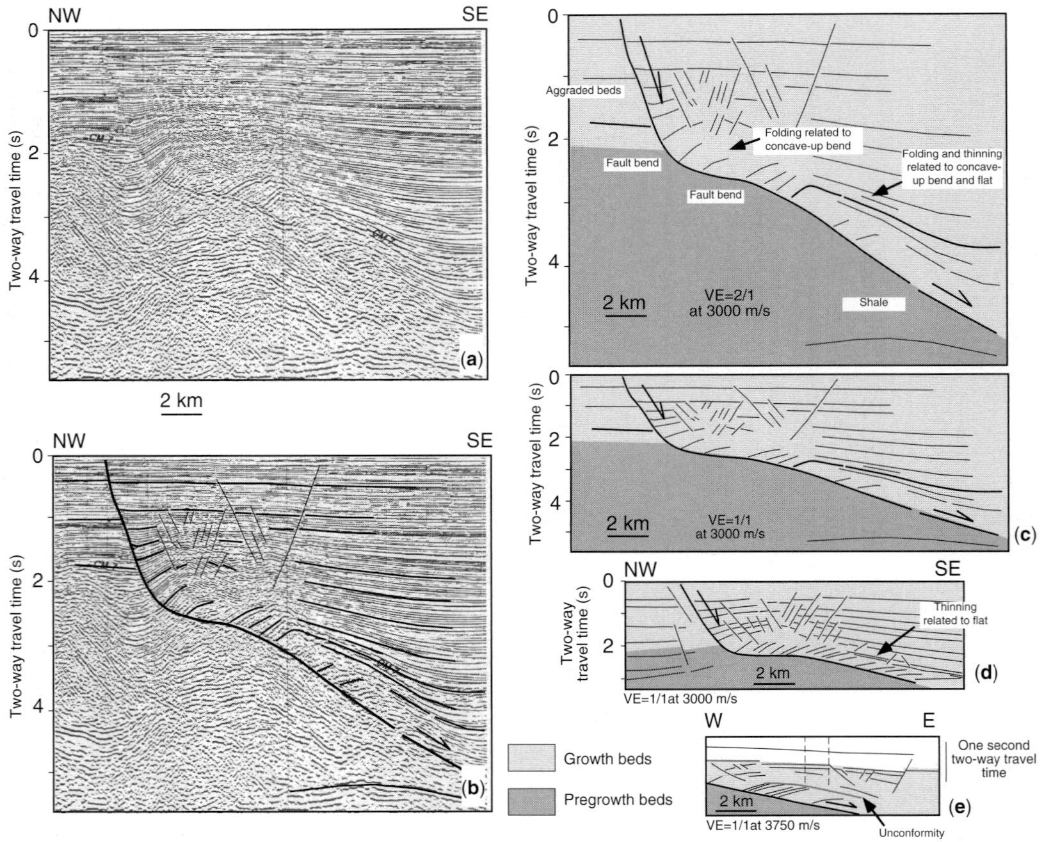

Fig. 15. (**a**) Time-migrated seismic line of Brazos fault, offshore Texas, Gulf of Mexico (after Bally *et al.* 1991). CM-7 represents strata that are about 7 Ma. (**b**) Interpreted seismic data. (**c**) Interpreted line drawing with (top) and without (bottom) vertical exaggeration. This part of the Brazos fault has two major fault bends. The intervening fault segment dips gently to the southeast. Beds near the fault dip toward the fault. Beds slightly farther from the fault dip away from the fault. Antithetic and synthetic normal faults affect the hanging wall. (**d**) Interpreted line drawing of seismic line from different part of the Brazos fault. This part also has two major fault bends. The intervening fault segment is flat-lying. (**e**) Vicksburg fault near Slick Ranch field, southern Texas. Hanging wall deformation resembles that of Brazos fault in terms of geometry and dimension. Vertical dashed lines show well locations. See also Fig. 16.

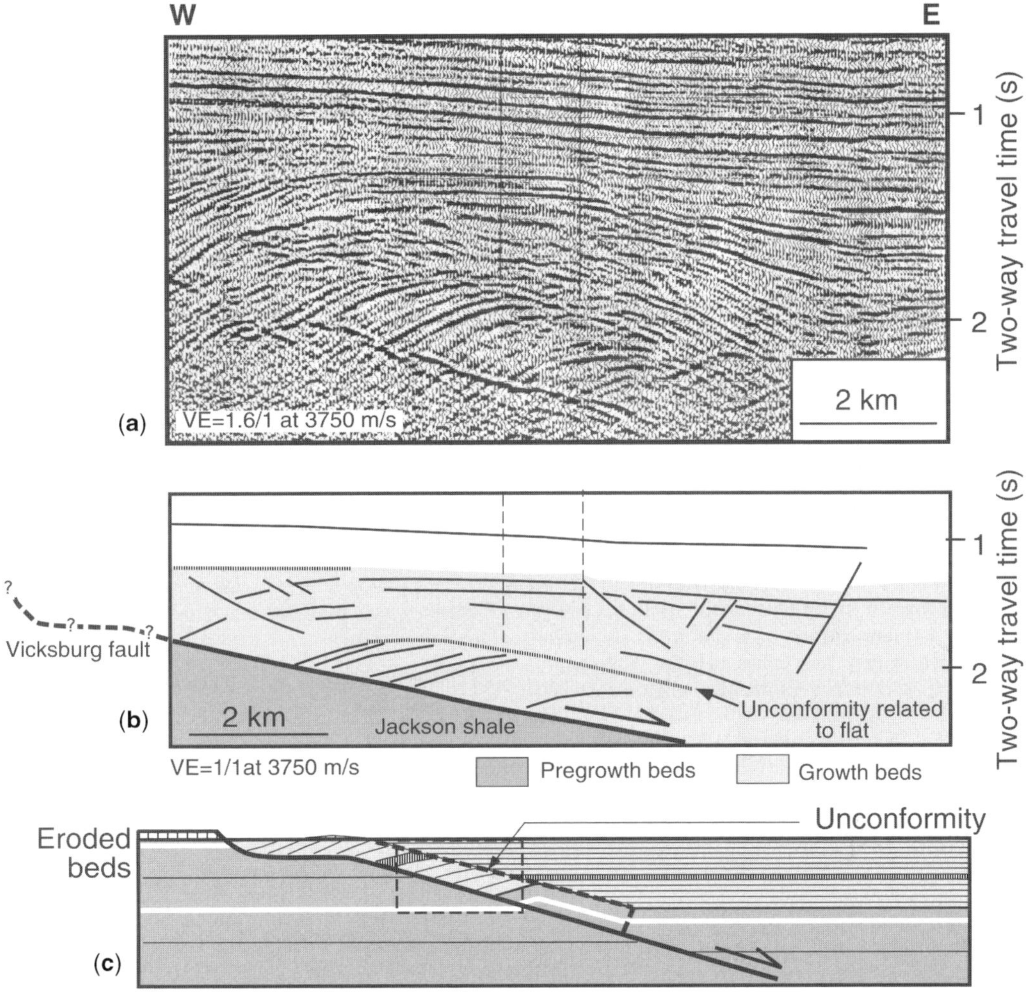

Fig. 16. Vicksburg fault near Slick Ranch field, southern Texas. Vertical dashed lines show well locations. (**a**) Uninterpreted seismic line with vertical exaggeration (after Erxleben & Carnahan 1983). (**b**) Interpreted line drawing of time-migrated seismic line displayed with approximately no vertical exaggeration (after Erxleben & Carnahan 1983). An unconformity separates a lower panel of growth beds dipping toward the fault from an upper panel dipping away from the fault. (**c**) Geometric model of ramp-flat-ramp fault showing unconformity similar to that in (b).

(4) Complex structural and stratigraphic patterns develop with multiple episodes of normal faulting if the rate of aggradation/erosion relative to the rate of fault displacement changes through time (Fig. 8). The outer limit of the folding dips more gently for growth packages associated with episodes of normal faulting without aggradation and more steeply for growth packages associated with episodes of normal faulting with aggradation. If erosion occurs between episodes of normal faulting, then growth beds associated with the first faulting episode can have a subtle fanning geometry and closely resemble pregrowth beds.

The experimental models indicate that the style of hanging wall deformation can vary considerably.

(1) In sand models, a few, relatively large secondary antithetic normal faults accommodate most hanging wall deformation. Pregrowth layers, although faulted, remain flat. The effective shear direction parallels the antithetic normal faults, and the shear angle is about $60°–65°$.

(2) In clay models, secondary faulting (antithetic and synthetic) and cataclastic flow accommodate most of the hanging wall deformation. The deformed pregrowth and growth layers dip gently toward the main normal fault. The

effective shear angle (35°–50°) is considerably less than the dip of the antithetic normal faults.

(3) The sand and clay models resemble geometric models with large (*c.* 60°) and small (*c.* 35°) shear angles, respectively. In the sand models and the geometric models with large shear angles, more displacement occurs on the main normal fault and the hanging wall collapses in a relatively narrow zone. In the clay models and the geometric models with small shear angles, less displacement occurs on the main normal fault. Instead, the hanging wall stretches substantially and collapses in a relatively wide zone.

Observed hanging wall folds in nature resemble those in the geometric and experimental models. Rollover folds develop in the hanging wall of normal faults that flatten at depth. Normal faults with convex-up bends have hanging wall beds that dip away from the main fault. Normal faults with ramp-flat-ramp geometries have complex hanging wall deformation; beds dip toward the fault near the main fault, and a dipping unconformity that parallels the lower ramp commonly develops above the ramp. Observed secondary normal faults in nature resemble those in the experimental models, especially the clay models. Many secondary normal faults are antithetic to the main fault, but some are synthetic to the main normal fault. Most secondary normal faults have relatively small displacements.

Appendix: scaling

The strength of most upper crustal rocks increases with depth, obeying a Mohr-Coulomb criterion of failure (e.g., Byerlee 1978). According to this criterion,

$$\tau = c + \mu \cdot \sigma_n, \quad (1)$$

where τ and σ_n are, respectively, the shear and normal stresses on a potential fault surface, c is the cohesion, and μ is the coefficient of internal friction. This empirical criterion of failure describes the initiation of new faults rather than the frictional reactivation of existing faults. For most sedimentary rocks, μ ranges from about 0.55–0.85 (e.g., Handin 1966; Byerlee 1978). For intact sedimentary rocks, c is about 10–20 MPa (Handin 1966), whereas for highly fractured sedimentary rocks, c is significantly less (e.g., Byerlee 1978; Brace & Kohlstedt 1980). To ensure dynamic similarity between the models and natural prototypes, two conditions must be satisfied (e.g., Hubbert 1937; Weijermars *et al.* 1993; Vendeville *et al.* 1995). First, the modelling materials and the rocks in nature must have similar coefficients of internal friction. This condition is satisfied with either dry sand or wet clay as the modelling material. Second,

$$c^* = \rho^* \cdot g^* \cdot l^*, \quad (2)$$

where c^*, ρ^*, g^* and l^* are model-to-natural prototype ratios for cohesion, density, gravity and length, respectively. In our models, the values of ρ^* and g^* are about 0.7 and 1, respectively, and l^* is about 10^{-5} (i.e., 1 cm in the models equals about 1 km in nature). Thus, to ensure dynamic similarity between the models and natural prototypes, the cohesion of rock must be about 10^5 greater than that of the modelling materials. This condition is satisfied with either dry sand or wet clay as the modelling material.

We thank E. Peterson for his assistance with the geometric modelling, and S. Callaway, S. Dixon, G. Eisenstadt and D. Sims for their help with the experimental models. We also wish to thank G. Corti, T. Dooley and B. Vendeville for their helpful reviews of the manuscript. The research was partially supported by NSF grant EAR-0408878.

References

Bally, A., Withjack, M., Meisling, K. & Fisher, D. 1991. *Seismic expression of structural styles.* Geological Society of America, Short Course Notes, 1991 National Meeting.

Brace, W. F. & Kohlstedt, D. L. 1980. Limits on lithospheric stress imposed by laboratory experiments. *Journal of Geophysical Research*, **85**, 6248–6252.

Byerlee, J. 1978. Friction of rocks. *Pure Applied Geophysics*, **116**, 615–626.

Cloos, E. 1968. Experimental analysis of Gulf Coast fracture patterns. *American Association of Petroleum Geologists Bulletin*, **52**, 420–444.

Cloos, H. 1928. Experimente zur inneren tektonik. *Centralblatt für Mineralogie*, **B**, 609–621.

Cloos, H. 1930. Kunstliche gebirge, II. *Natur und Museum*, **60**, 258–269.

Dula, W. F. 1991. Geometric models of listric normal faults and rollover folds. *American Association of Petroleum Geologists Bulletin*, **75**, 1609–1625.

Ellis, P. G. & McClay, K. R. 1988. Listric extensional fault systems – results of analogue model experiments. *Journal of Basin Research*, **1**, 55–70.

Erxleben, A. W. & Carnahan, G. 1983. Slick Ranch area, Starr County, Texas. *In*: Bally, A. W. (ed.) *Seismic Expression of Structural Styles.* American Association of Petroleum Geologists, Studies in Geology, **15**, 2.3.1, 22–26.

HANDIN, J. 1966. Strength and ductility. *In*: CLARK, S. P. Jr. (ed.) *Handbook of Physical Constants*. Geological Society of America, **Memoir 97**, 223–289.

HAUGE, T. A. & GRAY, G. G. 1996. A critique of techniques for modelling normal-fault and rollover geometries. *In*: BUCHANAN, P. G. & NIEUWLAND, D. A. (eds) *Modern Developments in Structural Interpretation, Validation and Modelling*. Geological Society, London, Special Publications, **99**, 89–97.

HUBBERT, M. K. 1937. Theory of scale models as applied to the study of geologic structures. *Geological Society of America Bulletin*, **48**, 1459–1520.

KAUTZ, S. A. & SCLATER, J. G. 1988. Internal deformation in clay models of extension by block faulting. *Tectonics*, **7**, 823–832.

KERR, H. G. & WHITE, N. 1992. Laboratory testing of an automatic method for determining normal fault geometry at depth. *Journal of Structural Geology*, **14**, 873–885.

MCCLAY, K. R. & ELLIS, P. G. 1987*a*. Analogue models of extensional fault geometries. *In*: COWARD, M. P. DEWEY, J. F. & HANCOCK, P. L. (eds) *Continental Extensional Tectonics*. Geological Society, London, Special Publications, **28**, 109–125.

MCCLAY, K. R. & ELLIS, P. G. 1987*b*. Geometries of extensional fault systems developed in model experiments. *Geology*, **15**, 341–344.

MCCLAY, K. R. & SCOTT, A. D. 1991. Experimental models of hangingwall deformation in ramp-flat listric extensional fault systems. *Tectonophysics*, **188**, 85–96.

RUTTER, E. H. 1986. On the nomenclature of mode of failure transition in rocks. *Tectonophysics*, **122**, 381–387.

VENDEVILLE, B., HONGXING, G. & JACKSON, M. P. A. 1995. Scale models of salt tectonics during basement-involved extension. *Petroleum Geoscience*, **1**, 179–183.

VOGLER, H. A. & ROBINSON, B. A. 1987. Exploration for deep geopressured gas–Corsair trend, offshore Texas. *American Association of Petroleum Geologists Bulletin*, **71**, 777–787.

WEIJERMARS, R., JACKSON, M. P. A. & VENDEVILLE, B. 1993. Rheological and tectonic modeling of salt provinces. *Tectonophysics*, **217**, 143–174.

WERNICKE, B. P. & BURCHFIEL, B. C. 1982. Modes of extensional tectonics. *Journal of Structural Geology*, **4**, 105–115.

WHITE, N. 1992. A method for automatically determining normal fault geometry at depth. *Journal of Geophysical Research*, **97**, 1715–1733.

WHITE, N. J. & YIELDING, G. 1991. Calculating normal fault geometries at depth: theory and examples. *In*: ROBERTS, A. M., YIELDING, G. & FREEMAN, B. (eds) *The Geometry of Normal Faults*. Geological Society, London, Special Publications, **56**, 251–260.

WHITE, N. J., JACKSON, J. A. & MCKENZIE, D. P. 1986. The relationship between the geometry of normal faults and that of the sedimentary layers in their hanging walls. *Journal of Structural Geology*, **8**, 897–909.

WITHJACK, M. O. & CALLAWAY, J. S. 2000. Active normal faulting beneath a salt layer – an experimental study of deformation in the cover sequence. *American Association of Petroleum Geologists Bulletin*, **84**, 627–651.

WITHJACK, M. O. & PETERSON, E. T. 1993. Prediction of normal-fault geometries – a sensitivity analysis. *American Association of Petroleum Geologists Bulletin*, **77**, 1860–1873.

WITHJACK, M. O., ISLAM, Q. & LAPOINTE, P. 1995. Normal faults and their hanging-wall deformation – an experimental study. *American Association of Petroleum Geologists Bulletin*, **79**, 1–18.

XIAO, H. & SUPPE, J. 1992. Origin of rollover. *American Association of Petroleum Geologists Bulletin*, **76**, 509–529.

YIELDING, G., BADLEY, M. E. & FREEMAN, B., 1991, Seismic reflections from normal faults in the northern North Sea. *In*: ROBERTS, A. M., YIELDING, G. & FREEMAN, B. (eds) *The Geometry of Normal Faults*. Geological Society, London, Special Publications, **56**, 79–89.

Recent advances and current problems in modelling surface processes and their interaction with crustal deformation

JEAN BRAUN

Géosciences Rennes, Université de Rennes 1, Rennes, France
(e-mail: Jean.Braun@univ-rennes1.fr)

Abstract: I present a brief summary of recent advances in the field of computational geomorphology and various attempts to couple numerical models of landscape evolution to models of crustal/lithospheric deformation. The most commonly used formulations for the various physical processes at play during surface erosion, transport and deposition are presented, as well as an outline of how they have been incorporated in a variety of numerical schemes. I also explain how the coupling between erosion and tectonics has been performed under various simplifying assumptions. Determining the rate constants for each of the proposed landforming mechanisms remains a difficult challenge that has recently been helped by the advent of new low temperature thermochronometers and exposure dating by cosmogenic radionuclides. I demonstrate how the information contained in the relationship between age and elevation can be used to provide constraints on the 'age' of a landscape, as well as how important rate information can be extracted from various datasets by using simple modelling techniques. This paper demonstrates why the field of computational geomorphology needs to harmonize the various parameterizations (often the legacy of empirical relationships derived from observations at the human scale), quantitative estimates of the value of the numerous rate parameters and improvement of the numerical techniques.

In recent years, much work has been devoted to understanding the coupling between tectonics, erosion and climate (Molnar & England 1990; Koons 1994; Beaumont *et al.* 1999; see also Burbank & Pinter (1999) for an introduction to this subject). We know that tectonic movements create topography at the Earth's surface by which atmospheric circulation can be perturbed, thereby creating a first-order interaction between tectonics and climate. This perturbation often takes the form of enhanced precipitation on the windward side of a mountain belt which, in turn, can be responsible for enhanced rates of erosion. This feedback is less obvious to demonstrate but may be responsible for much of the morphological variety that is observed among active and ancient mountain belts.

The Southern Alps of New Zealand are often presented as a classical example of a tectonic feature which has perturbed atmospheric circulation: it rains on average ten times more on the west coast of the South Island of New Zealand than on the east coast. It is also often taken as an example of an area where the feedback has been extreme: rocks are preferentially exhumed from a much larger depth along the western front of the orogen (Wellman 1979; Koons 1990; Beaumont *et al.* 1996). Thus the first-order features of an orogen, such as the surface distribution of metamorphic rocks, or the geometry and location of large structures, may be dictated by patterns in the local climate.

At the larger scale, it has been postulated that the Tibetan Plateau is the result of ongoing continental shortening and thickening, unbalanced by erosion, principally due to the lack of rainfall in the region (Royden 1996; Beaumont *et al.* 2001). Once a proto-plateau formed, potentially as early as 35 Ma, rainfall was even more perturbed, and dry conditions established themselves on the plateau and most of its flanks, limiting erosion to the Himalayan front, the only plateau boundary facing an open body of water.

At an even larger scale, Molnar and England (1990) postulated that the apparent increase in global erosion rates since the beginning of the Cenozoic is the product of a strong feedback between tectonics and climate. According to this hypothesis, higher erosion rates resulting from the colder climatic conditions experienced by the Earth over the past 65 Ma leads to local relief production which, by flexural isostasy, results in mountain peak uplift and further perturbs the climate; this perturbation would ultimately lead to more efficient erosion in regions of high elevation.

From: BUITER, S. J. H. & SCHREURS, G. (eds) 2006. *Analogue and Numerical Modelling of Crustal-Scale Processes*. Geological Society, London, Special Publications, **253**, 307–325.
0305-8719/06/$15.00

Numerical modelling of the coupling between tectonics, erosion and climate

The links between tectonics, climate and erosion have been demonstrated by many numerical models of the coupled lithosphere–hydrosphere system in which varying erosional conditions have led to clear modifications in tectonic regime. In particular, in active orogenic settings, it has been demonstrated that the large-scale morphology of a mountain belt can be modified by perturbing the dominant wind direction (Willett *et al.* 1993; Willett 1999). As an example, I show the results of a series of computations in which a thermo-mechanical model of the continental lithosphere has been coupled to a simple, one-dimensional model of surface erosion, transport and deposition (Fig. 1). The crust is assumed to be shortened by the subduction of the underlying lithospheric mantle. Rock rheology is a combination of pressure-dependent brittle failure and temperature-dependent creep. The heat transport equation is also solved and the resulting temperature used to calculate the viscosity of rocks. The results show that the behaviour of the system is strongly controlled by the efficiency of the surface processes. It has been assumed that, as it evolves, the orogen is abruptly subjected to an increase in erosion rate, potentially corresponding to a change in climatic

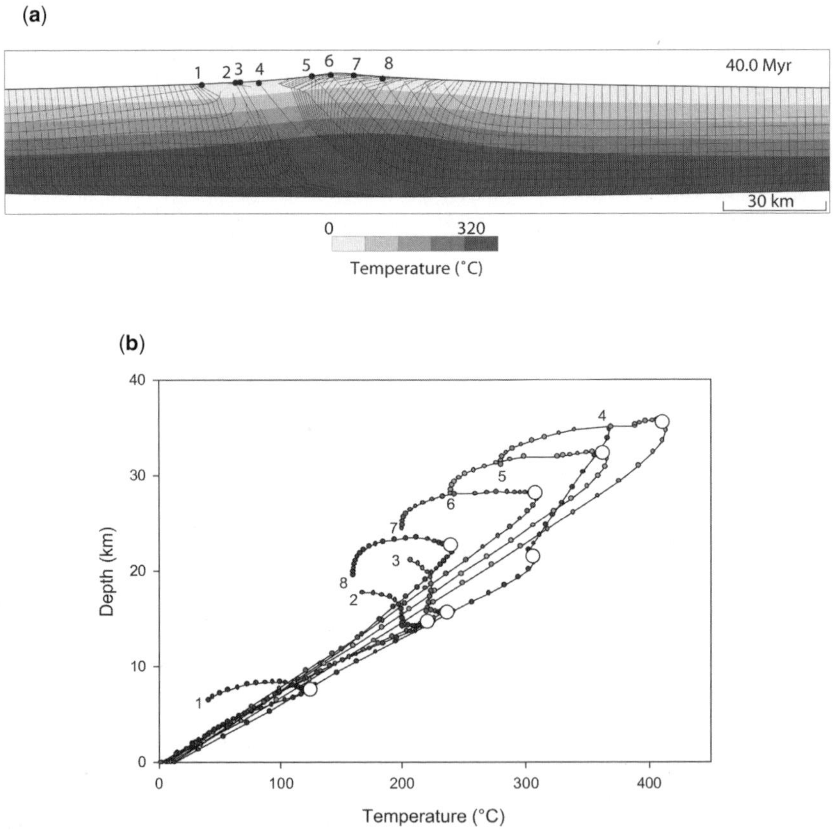

Fig. 1. Results of a numerical model of crustal deformation coupled to surface erosion; see Braun and Sambridge (1994) for details of the numerical method. Deformation is driven by an imposed velocity discontinuity at the base of the crust; a uniform heat flow is imposed at the base of the model. Erosion rate is assumed proportional to (1) local slope and drained area (fluvial erosion) and (2) curvature of the topography (hillslope processes). At 20 Ma in the evolution of the model, fluvial and diffusive erosion parameters are increased by a factor 2 to represent an increase in erosion rate. (**a**) Temperature and geometry of an initially rectangular grid; the dark circles are Lagrangian markers advected towards the surface by the computed velocity field. (**b**) Depth-temperature-time paths of the 8 particles tracked in panel (**a**). The small circles are time markers at 1 Ma intervals in the evolution of the model; the large white circle corresponds to the 20 Ma time step.

conditions or to its emergence above sea level. When erosion increases (Fig. 1a), the orogen is rapidly transformed; it evolves from a state of lateral growth to a phase of rapid and localized exhumation, steepening of the structures controlling the deformation, and the exhumation and, subsequently, narrowing of the mountain belt.

In another series of numerical experiments in which a tectonic model was coupled to a surface processes model through a simplified representation of orographic precipitation, Beaumont *et al.* (1992) and later Batt and Braun (1999) illustrated the complete feedback between local climate and tectonics (Fig. 2). During the early stages of development of a mountain belt, erosion is relatively uniform as little or no orographic control exists on precipitation. As the mountain grows, the local wind circulation is affected by the rising topography and precipitation becomes focused on the windward side of the orogen, leading to an asymmetric erosion pattern across the orogen. The 'wet' side of the orogen is characterized by much steeper topography than the 'dry' side. Further growth of the orogen enhances the orographic effect and the asymmetry of the resulting topography.

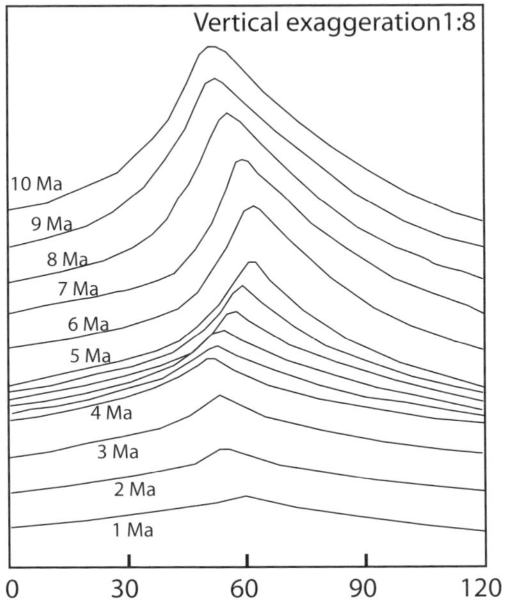

Fig. 2. Simulated evolution of the topography across an active mountain belt, from the time of onset of collision until 10 Ma in the evolution of the orogen. Model set-up is similar to that shown in Fig. 1 but precipitation is orographically controlled. Dominant wind direction is from left to right. As topography grows, orographic control on precipitation leads to asymmetric erosion patterns which leads to an asymmetric topography and a further enhancement of the orographic control on precipitation. Modified from Batt and Braun (1999).

Through the results of numerical simulations, Beaumont *et al.* (2001, 2004) and Jamieson *et al.* (2004) have shown that the dynamic evolution of the Himalayan orogen and the adjacent Tibetan Plateau are causatively linked and that the strong rainfall along the Himalayan front is the main driving mechanism for the exhumation of a mobile lower crust that formed in response to thickening of the Plateau (Fig. 3). As shown by Royden (1996) and Clark and Royden (2000), it is the difference in potential energy between the Plateau and the adjacent Indian continent that provides the pressure gradient necessary to mobilize the lower crust, but it is the focused erosion along the Himalayas that causes it to be exhumed so rapidly, and, in doing so, creates the tallest mountain range at the Earth's surface (Beaumont *et al.* 2001; Hodges *et al.* 2001).

This coupling between erosion and tectonics arises from the very strong density gradient that exists at the interface between the lithosphere and the overlying hydrosphere. Indeed, small vertical deflections of that interface can create very large horizontal pressure (stress) gradients in the underlying lithosphere that can strongly affect the force balance and enhance or inhibit deformation (Fig. 4). Recently, it has been postulated that the pressure change caused by the rapid incision of a deep river valley may be sufficient to cause localization of the deformation in the area just beneath the valley. This scenario has been formulated by Zeitler *et al.* (2001*a*, *b*) and Koons *et al.* (2002) to explain the apparent 'pinning' of the Himalayan front at its two syntaxes.

Finally, it must also be noted that the recent technological and algorithmic developments that have made possible the coupling between models of lithospheric deformation and models of surface erosion and transport, have also made it possible to greatly improve the prediction capabilities of these models and their ability to reproduce geological observations/data. Indeed, it is erosion that is responsible for the exhumation of rocks towards the surface where they are observed and collected by geologists. An important consequence of this development has been the further realization that tectonic processes, combining with spatially variable erosional processes, are not only responsible for the vertical movement of rocks towards the surface but also for very large horizontal transport. As they 'travel' through an orogen, the temperature and pressure history of exhumed rocks will be strongly affected by this lateral transport component.

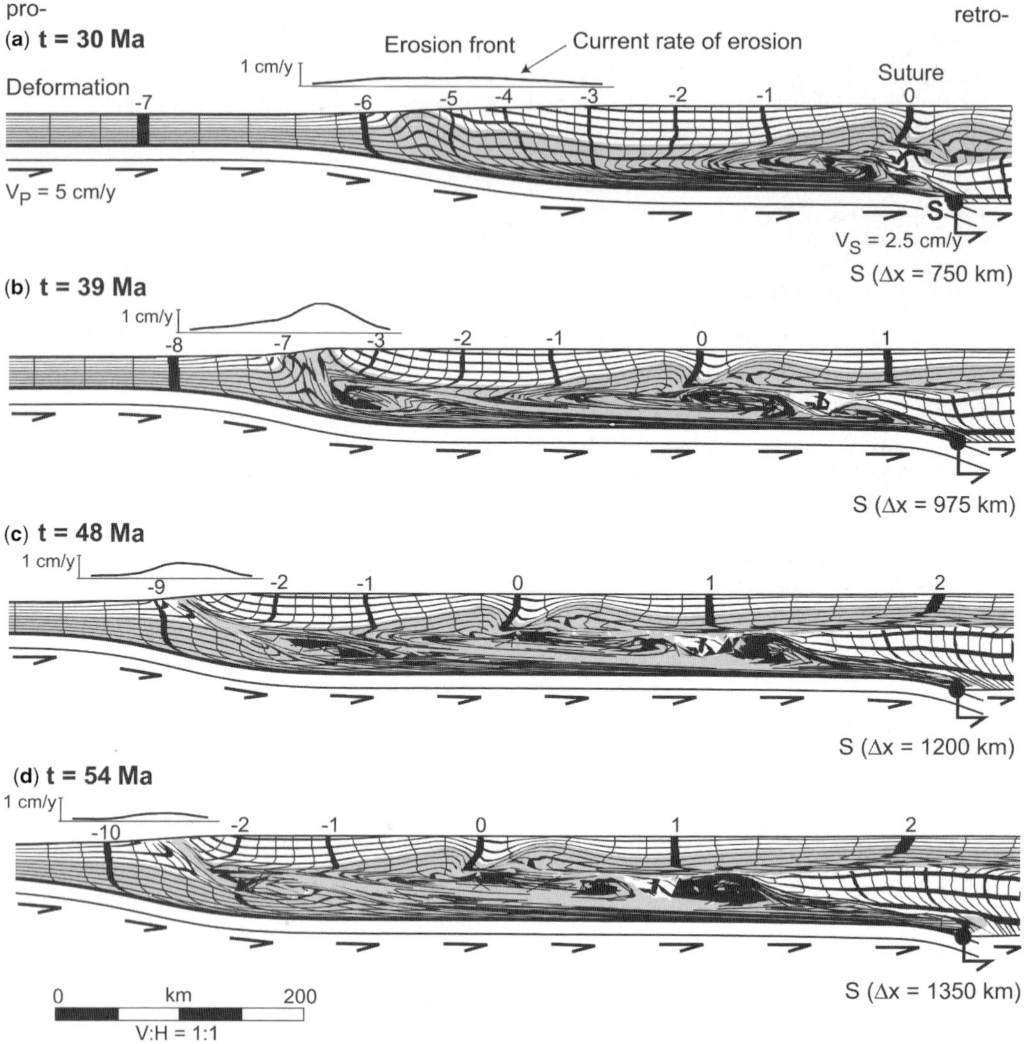

Fig. 3. Numerical model of the tectonic development of the Tibetan Plateau and Himalayan front. Convergence is driven by imposed subduction of the pro- (Indian) side of the orogen beneath the retro- (Eurasian) side. Extrusion of the lower crust is driven by a horizontal pressure gradient caused by the elevated topography of the Tibetan Plateau with respect to the Indian continent and made possible by efficient erosion along the Himalayan front. The four panels correspond to four different stages of evolution of the model showing the formation of a mobile lower crust and a focusing of the erosion along the Himalayan 'erosional front'. Total convergence is (**a**) 1500 km, (**b**) 1950 km, (**c**) 2400 km and (**d**) 2700 km, respectively. Modified from Beaumont *et al.* (2004).

The results presented in Figure 1 illustrate this point: a series of rock particles originating from a range of locations, both in the horizontal and vertical dimensions, are brought together towards their final surface position during the tectonic event. Each particle also experiences a clear change in its pressure and temperature evolution at the time of increase in erosional conditions at the surface (the large white circles in Fig. 1b), but the amplitude and sign of this pressure/temperature change will be different in various parts of the orogen, therefore affecting each rock particle differently.

Parameterization of landform evolution

To quantify the coupling between tectonics, erosion and climate, many numerical models of

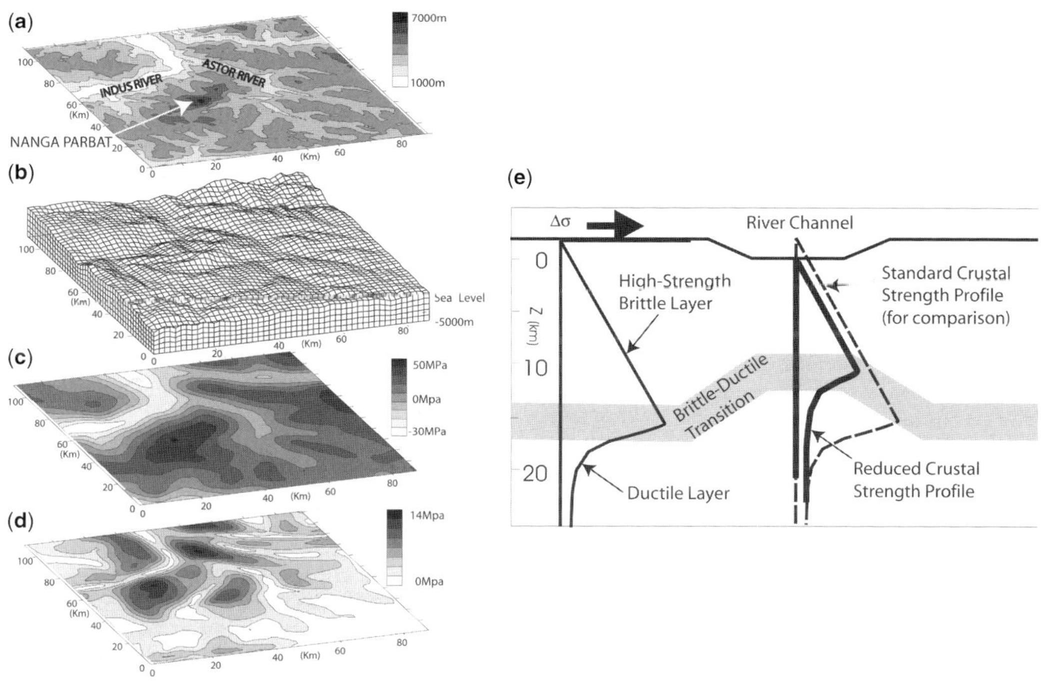

Fig. 4. Large gradient in surface topography (**a**) and (**b**) caused by the incision of a river channel may lead to perturbation in the underlying state of stress (**c**) and (**d**). (**c**) Topographic contribution to the vertical normal stress as represented by the departure of the vertical normal stress (σ_{zz}) at 5 km below sea level from that predicted by a constant 2 km elevation over the region. (**d**) Vertical shear stress component $(\tau_{xz}^2 + \tau_{yz}^2)^{1/2}$ generated by topography at 5 km below sea level. As shown in the right panel (**e**), two factors contribute to strength reduction beneath a deeply incised river channel: the erosional unloading and the thermal perturbation caused by rapid advection of heat towards the surface. Modified from Koons *et al.* (2002).

surface processes, often called Surface Processes Models (SPMs) or Landscape Evolution Models (LEMs), have been developed in recent years. It is not possible here to cite all the models. Instead, I will try to summarize the most common algorithms that have been used, often based on an extrapolation of observations made at the human scale to the scale of tectonic processes, both in space and time. Apart from a few examples (Stock & Montgomery 1999; van der Beek & Bishop 2003; Tomkin *et al.* 2003), this extrapolation is often not tested and the first-order match between model predictions and large-scale features of natural landscapes is taken as proof of the appropriateness of the extrapolation.

Landform evolution is determined by a wide range of processes that have been parameters in a variety of ways (Dietrich *et al.* 2003). In general, one tries to express the rate of change of topography, h, as a function of parameters and quantities that can either be computed from the shape of the landscape (local slope, drained area, etc.) or determined empirically. In most LEMs, the landscape is divided into small cells in which the elevation is commonly assumed to be uniform or vary linearly; a notable exception is the continuum approach developed by Simpson and Schlunegger (2003). One then introduces a rule for local erosion/deposition and, combining it with a form of the continuity equation, one obtains a law that encompasses erosion, sedimentation and transport.

Stream-power law

Water channels (rivers or streams) are often regarded as the controlling elements of many land surfaces, where the interaction between the hydrosphere and the lithosphere is focused. Stream incision is commonly parameterized by the *stream-power law*:

$$\frac{\partial h}{\partial t} = K_f S^n A^m \quad (1)$$

in which incision rate is assumed to be proportional to the product of the local slope, $S = \|\overrightarrow{\text{grad}}\, h\|$, and water discharge. Drainage area, A, is commonly used as a proxy for discharge (this is appropriate if one assumes that precipitation, evaporation and infiltration are uniform over the catchment), K_f is a constant that depends on local precipitation, rock type and other factors such as vegetation, n and m are power-law exponents. The value of n is assumed to vary between $\frac{2}{3}$ and $\frac{5}{3}$ (Whipple & Tucker 1999), but a commonly used value is 1. The value of m is less well known. If one assumes steady-state, the ratio m/n can be determined by the downstream rate of increase of channel slope with drainage area (Whipple & Tucker 1999). This ratio is often referred to as the *concavity* of a river profile and has a value that is typically around $\frac{1}{2}$ (Whipple 2001).

River incision may be limited by the mechanical strength of the substratum, in which case a threshold, C_0, needs to be introduced (Tucker & Slingerland 1994)

$$\frac{\partial h}{\partial t} = K_f S^n A^m - C_0 \tag{2}$$

Note, however, that if a threshold is used, the stochastic nature of many quantities, such as precipitation or bedrock resistance to incision, will play an important role in determining the time-integrated behaviour of the system (Tucker 2004).

Transport-limited incision

One can also assume that it is not the rate of incision that controls the rate of landform evolution but the rate at which material can be transported away from its source by the channel. To model this behaviour, one introduces a quantity called the *carrying capacity* of the river, Q_{eq}, that is the flux of sediment (volume per unit time) that a river can transport for a given slope, S, and catchment area, A:

$$Q_{eq} = K_t S^{n'} A^{m'} \tag{3}$$

where K_t is a proportionality constant. n' and m' are poorly known and commonly assumed to be similar to n and m in Equation 1. Assuming that a river is always carrying sediment at capacity, the rate of evolution of the surface is given by the continuity equation (Willgoose *et al.* 1991):

$$\frac{\partial h}{\partial t} = -\frac{1}{W}\frac{\partial Q_{eq}}{\partial x} \tag{4}$$

where x is the downstream direction and W is the width of the river channel.

Sediments can have two effects in a river channel: when they are in motion within the water column, they can be regarded as tools to abrade the substratum (Sklar & Dietrich 1998); when they are at the bottom of the channel, they can protect it. This latter role of sediments has been introduced in an alternate formulation for river incision that also incorporates transport and deposition/sedimentation (Kooi & Beaumont 1994) in which the rate of incision/deposition is assumed to be proportional to the disequilibrium between carrying capacity and local sediment flux, Q_{sed}:

$$\frac{\partial h}{\partial t} = -\frac{1}{WL_f}(Q_{eq} - Q_{sed}) \tag{5}$$

where L_f is an erosion/deposition length scale required for disequilibrium to be reduced by a factor of (e) when Q_{eq} is constant (Kooi & Beaumont 1994). Sediment flux can be calculated by integrating the erosion/deposition flux over the upstream part of the catchment.

Hill-slope processes

Stream incision leads to a local increase in surface slope along the sides of the valley carved by the stream. The other type of processes that will control the long-term evolution of the landscape is thus the transport along hill/valley slopes. In rapidly uplifting, tectonically active regions, landslide is the process dominating mass transport (Hovius *et al.* 1997), in lower relief, tectonically inactive regions, soil creep controls hill slope transport (Anhert 1967).

On soil-mantled hill-slopes, the rate of landform evolution is controlled by the rate at which soil is produced, transported and/or deposited:

$$\frac{\partial h}{\partial t} = -\frac{\partial Q_S}{\partial x} + (\kappa - 1)P \tag{6}$$

where Q_S is the soil flux, x is the downhill direction, κ is the ratio of bedrock to soil density, and P is the rate of soil production (Braun *et al.* 2001).

Soil production is often regarded as proportional to soil thickness, as the chemical alteration of bedrock to form soil is dependent on the availability and thus penetration of water into the soil. Based on cosmogenic isotope measurements at the bedrock/saporolite interface at a few sites worldwide, Heimsath *et al.* (2000, 2001) have proposed a soil production function,

P, that varies as the negative exponential of soil thickness:

$$P = P_0 e^{-\xi/\xi_0} \tag{7}$$

where ξ is soil thickness, P_0 is the production rate at zero soil thickness and ξ_0 is the production e-folding depth, that is, the soil thickness necessary to reduce production by a factor e.

Soil transport is a complex physical process that is likely to involve more than one mechanism. These commonly include simple soil creep, for which soil flux, Q_S, is linearly related to local slope, S (Anhert 1967):

$$Q_S = -K_d S \tag{8}$$

Martin (2000) suggests a range of values for K_d, the linear diffusion coefficient, that are suited to the long-term evolution of soil-mantled hill slopes. Because soil flux is the product of soil thickness by vertically-integrated soil velocity, this commonly used law implies that mean soil velocity is inversely proportional to soil thickness. Furthermore, a linear transport law predicts that, at steady state, all hill slopes should be convex, not straight as is often observed in moderately active tectonic settings (Roering *et al.* 1999). To resolve these problems, depth-dependent soil transport has been used and is represented by:

$$Q_S = -K_v \xi^{m''} S^{n''} \tag{9}$$

where K_v is a proportionality constant that varies with soil type, climate and vegetation. If soil flow conforms to Manning's equation for liquid flow (Olson 1973), m'' and n'' are 1.67 and 0.5, respectively which is also compatible with *in situ* measurements (Clarke *et al.* 1999). Finally, soil may be transported by overland flow, in which case soil flux can be expressed as (Moore & Burch 1986):

$$Q_S = -K_w A^k S^p \tag{10}$$

The exponents k and p have values between 1 and 2 (Prosser & Rustomji 2000). The proportionality constant K_w is poorly known and depends mostly on soil type and vegetation (Prosser & Rustomji 2000).

In situations where river incision is fast (mm/yr or more), due, for example, to a very large rate of tectonic uplift and/or base level drop, important slopes develop and landsliding can play a dominant role in the transport of material from hill slopes to river channels (Hovius *et al.* 1997, 2000; Montgomery & Brandon 2002). Proper parameters of the conditions for slope instability and of the volume and rate of transport of individual landslides is difficult to include in landform evolution models, principally due to their lack of spatial resolution. Indeed, critical slope angles for landsliding are of the order of 30° or more, which would require a spatial resolution of the order of a few tens of meters, well below what is commonly used (typically 100 m to 1 km) in most large-scale LEMs.

One can either assume an artificially low critical angle for soil movement by sliding (Champel *et al.* 2002) or that a simple linear transport equation is adequate to represent all hill-slope processes, including landsliding (Kooi & Beaumont 1994). This latter approach is, however, not consistent with the observation that at high uplift rate, erosion rates adjust primarily through changes in the frequency of landsliding rather than by increased hill-slope steepness (Montgomery & Brandon 2002). This implies that a non-linear form of the diffusion equation (Martin 2000) or a threshold must be introduced (Roering *et al.* 2000; Densmore *et al.* 1998).

Glacial erosion

In the recent geological past, the Earth's climate has been characterized by fluctuations on a time scale of a few tens to a hundred thousand years during which conditions in most mountainous regions of the world have oscillated between fluviatile and glaciated. In fact, over the past 3 million years, mountains in temperate to cold regions of the globe have experienced glaciated conditions more often than not. It is thus unfortunate that the effect of glacial erosion on landform evolution in tectonically active areas has been neglected in most landscape evolution models (Braun *et al.* 1999).

Assuming that abrasion, glacial plucking or quarrying are the dominant mechanisms responsible for glacial erosion, the rate of surface lowering by glaciers can be shown to be linearly related to ice sliding velocity, u_s (Hallett 1979), through the following relationship:

$$\frac{\partial h}{\partial t} = -K_g u_s \tag{11}$$

where K_g is a proportionality constant that depends on rock type. The sliding velocity must be computed from the solution of the ice dynamics equation which may be obtained by combining the force balance equation (Newton's first and second laws), mass conservation and a rheological law that relates ice deformation to local stress conditions. In cases where variations in ice thickness take place over length scales that are much greater than the thickness of the ice, the thin-ice approximation can be used to compute the basal sliding velocity (Hutter 1981). A simple climatological model must also be introduced to estimate where ice accumulation and ablation takes place. Finally the temperature distribution at the base of the ice must also be estimated to determine where the ice is sliding on the bedrock or whether it is frozen (Braun *et al.* 1999). Such an approach has been used to simulate the effect of varying climatic conditions, and thus landforming mechanisms, on the evolution of a high relief, tectonically active mountain range in two dimensions (Braun *et al.* 1999), or the formation of high-relief steps and overdeepenings in longitudinal glacial valley profiles generated at the confluence of tributary glaciers (MacGregor *et al.* 2000).

Chemical weathering

In regions where bedrock lithology is dominated by calcareous rocks, chemical erosion must be taken into account. A large-scale representation of the processes of karst formation and their relationship to surface runoff, and the evolution of the surface landform is still to be developed. Combining a simple law for chemical weathering and the consequent downstream evolution of carbon dioxide pressure in a flowing river, Kauffman and Braun (2001) have shown that incision rate in a karst-dominated limestone landform may simply be related to discharge, or drainage area.

Numerical models of landform evolution

Many landform evolution models exist. Of interest to us are those aimed at studying the evolution of landforms at the time scale of orogenic processes and at the length scale of a small to medium size orogen, or at least a large catchment. These include SIBERIA (Willgoose *et al.* 1991), PRECIPITON (Chase 1992), DRAINAL (Beaumont *et al.* 1992), the detachment-limited model of Howard (1994), GOLEM (Tucker & Slingerland 1994), CASCADE (Braun & Sambridge 1997), CAESAR (Coulthard *et al.* 1997), ICE-CASCADE (Braun *et al.* 1999), ZSCAPE Densmore *et al.* 1998), LAPSUS (Schoorl *et al.* 2000), CHILD (Tucker *et al.* 2001), HEXSOIL (Braun *et al.* 2001) and EROS (Crave & Davy 2001). A good description of these various numerical methods up to 2001 is given in Coulthard (2001). In broad terms, they differ by the type of spatial discretization used to represent the landscape (rectangular, hexagonal, irregular), whether the channels are treated on a separate network, and the combination of equations/processes used.

To illustrate the purpose of a typical landform evolution model, I show in Figure 5 the results of computations made by Tomkin (2000) using the ICE-CASCADE model which combines fluvial, hill-slope and glacial erosion laws. These computations (Tomkin & Braun 2002) confirmed the suggestion of Whipple *et al.* (1999) that glaciations do not increase catchment-scale relief unless temperature is sufficiently low for ice to be frozen to and protect mountain peaks. In another study, Braun *et al.* (1999) used ICE-CASCADE to demonstrate that the rate of mountain erosion is, in most cases, more efficient under cyclic climatic conditions during which the dominant landforming process switches between fluvial and glacial erosion, with such a frequency (100 ka) that the shape of the landscape is continuously out of balance with the dominant process. During ice-free periods, near-vertical glacial valley walls are gravitationally unstable and prone to landslides (Hovius *et al.* 1997). In actively uplifting regions, this process combines with others to rapidly transform glacial landforms into graded, V-shaped fluvial valleys. Once the ice returns, these valleys are filled with fast-flowing glaciers and are rapidly reshaped into glacial troughs (Harbor 1992). Braun *et al.* (1999) demonstrated that this continuous imbalance leads to a more efficient rate of erosion than if the conditions were stable in one of the two states. This process is thought to occur in the South Island of New Zealand, with a partial or complete reshaping of the landform on a time scale of a few tens of thousands of years (Whitehouse 1987).

It is worth mentioning the recent development of laboratory set-ups in which landforming processes are studied using analog materials (Crave *et al.* 2000; Lague *et al.* 2003). Scaling of the geomorphological laws to the size of the experimental apparatus is difficult, especially due to the finite size of the material grains. Despite these limitations, the method is useful in studying the scaling properties of channelized systems.

Coupling with tectonics

The most commonly used, zeroth-order approach to 'couple' tectonics to a landscape evolution model is through the definition of a purely kinematic uplift function to which the landscape

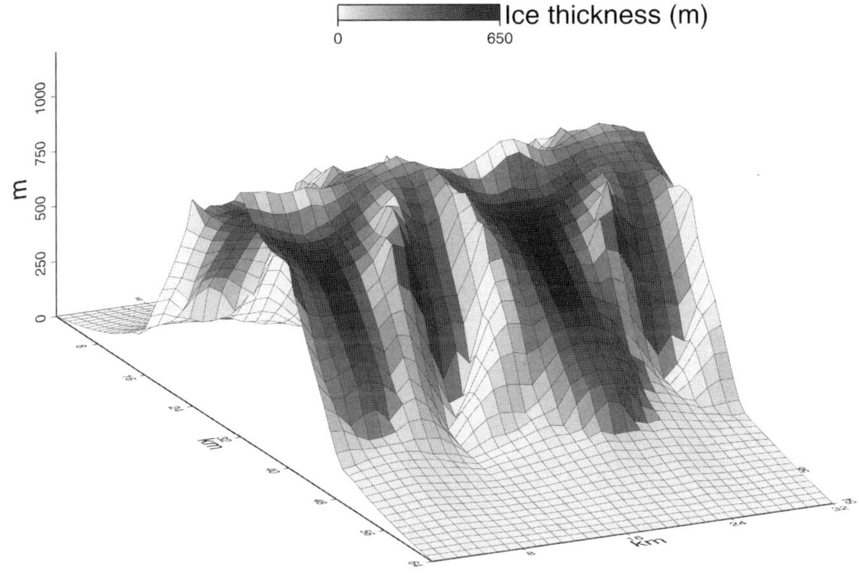

Fig. 5. Ice draped over topography. Ice thickness and landform computed by ICE-CASCADE (Tomkin *et al.* 2003) assuming a simple mass balance between precipitation, ablation and transport, that ice deforms as a power-law viscous fluid and that glacial erosion is linearly proportional to ice basal sliding velocity. Note that where the ice is frozen to the landscape, no erosion takes place.

evolution model reacts. There is no feedback from the erosional system onto the tectonic forcing.

The first-order coupling is to calculate the isostatic rebound associated with the redistribution of mass caused at the Earth's surface by erosional and depositional processes. In practice, the change in surface topography over a time step of the LEM is converted into an equivalent surface load (positive where deposition took place, negative where erosion took place) that is then applied to an isostatic foundation, under the assumption of local or regional/flexural isostasy. In most cases where flexural isostasy is considered, a simple linear, thin elastic plate foundation is used (Beaumont *et al.* 1999; Braun & Sambridge 1997; Braun *et al.* 1999). Under this strategy there is a feedback between surface processes and the tectonic forcing but only on its amplitude, not its geometry.

Second-order coupling, i.e., involving a feedback that will affect the distribution of forces within the underlying lithosphere, has been achieved in many different ways. The simplest is to use the predictions of the LEM to determine the parameters (angles) of the analytical solution to the critical wedge problem (Dahlen 1984). This approach was originally used by Johnson and Beaumont (1995) and, more recently, further developed by Hilley and Strecker (2004), Whipple and Meade (2004) and Roe *et al.* (2005).

Coupling to fully dynamic models of tectonic deformation is performed under the assumption of two dimensionality, either by using a plane-form, thin-sheet model of lithospheric/crustal deformation to which an LEM can be directly connected (Ellis 1995; Braun & Shaw 2001; Simpson 2004), or by using a 2D (vertical cross-section) tectonic model which requires a strike average of the results of the LEM (Beaumont *et al.* 1992; Stolar *et al.* 2005) or coupling to a one dimensional LEM (Batt & Braun 1999). Examples of these approaches (thin-sheet approximation and coupling with a 1D LEM) are shown in Figure 6.

Deformation of the crust/lithosphere during a tectonic event leads to uplift but also horizontal deformation. A proper coupling between an LEM and a tectonic model must incorporate both effects which, unfortunately, may lead to numerical problems. Indeed, most LEMs are based on a regular discretization of space which prevents them from incorporating a horizontal advection term of the landform unless some sort of spatial interpolation of the landform is made (Densmore *et al.* 2004), or the coupling between tectonics and erosion is made at strategically chosen time intervals at which parts of the landscape are moving by an amount exactly equal to

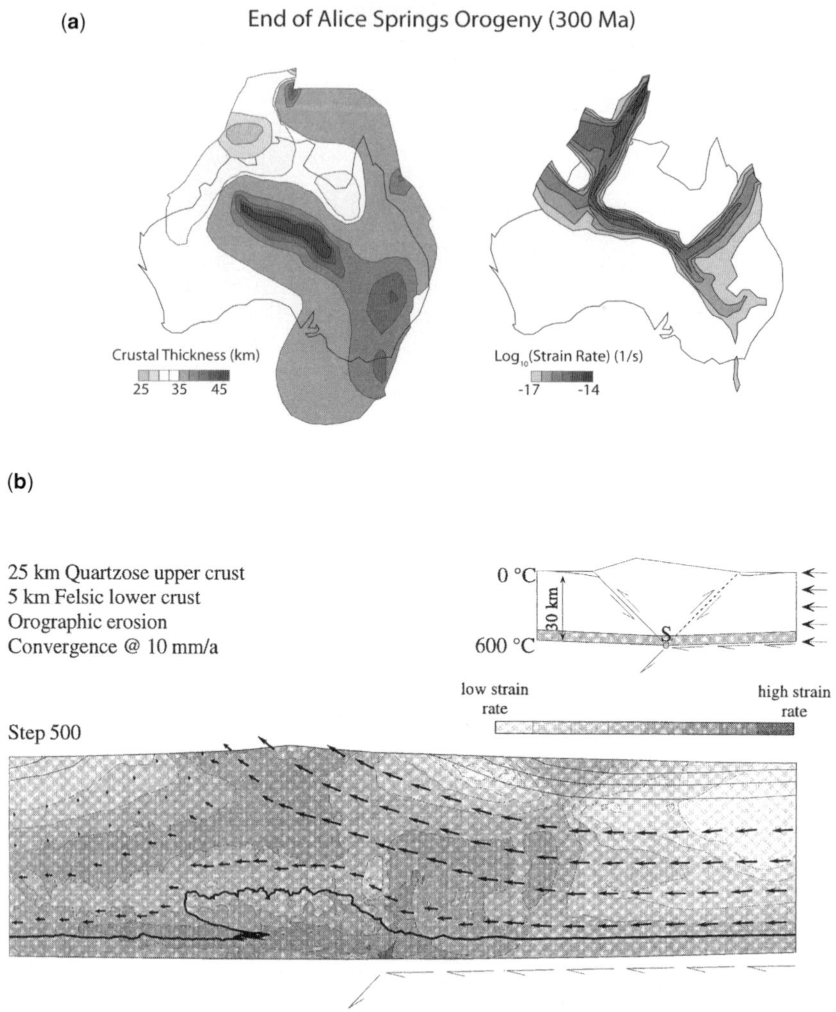

Fig. 6. Results of fully coupled erosion-tectonic models. (**a**) A model for the Palaeozoic deformation of the Australian continent based on thin sheet approximation and coupled to a LEM; the feedback from erosion to tectonics is through variations in pressure and temperature within the lithosphere that lead to variations in strength (from Braun & Shaw 2001); at this scale, it is assumed that surface mass transport is linearly related to slope; variations in crustal thickness are the result of the time-integrated deformation of the lithosphere whereas strain rate is a measure of instantaneous deformation; the predictions of the model are shown near the end of the Palaeozoic Alice Springs Orogeny. (**b**) A sophisticated 2D vertical tectonic model in which the solution of the equation of force balance in a visco-elasto-plastic material is used to predict horizontal and vertical tectonic transport; it has been linked to a 1D LEM in which river incision and hill slope diffusion have been incorporated; an orographically-controlled precipitation function can also be activated to include the feedback from tectonics unto local climate; at this stage in the evolution of the model total convergence is 50 km, i.e., approximately twice the thickness of the crust (from Batt & Braun 1999).

the spatial discretization (Garcia-Castellanos 2002). In models where an irregular discretization of space is used, this problem can easily be overcome (Braun & Sambridge 1997; Willett *et al.* 2001).

Finally, it is worth mentioning here a recent attempt at coupling a numerical model of landscape evolution with an analogue model of crustal deformation (Persson *et al.* 2004). In this later approach, a 'sandbox' analog model is used to represent the internal deformation of the Earth's crust under the forcing of imposed displacements along its side or bottom boundaries. The geometry of the deformed surface is carefully measured and used to determine an uplift function that is then introduced in a landscape evolution model. The LEM is used to compute the evolution of the surface over an interval of time equivalent to the scaled time step of the laboratory experiment by a combination of fluvial and hill-slope processes. The resulting change in surface topography is imposed onto the surface of the sandbox model by carefully removing sand grains with a small vacuum pump.

Constraining the rate of landscape evolution

One of the major issues facing LEMs concerns their prediction capability: we do not know the value of the rate parameters entering the geomorphological laws (e.g., K_f, K_t, K_d, K_v, K_w and K_g) with enough accuracy to ensure that the model predictions can be trusted in absolute terms. In Table 1, I give a range of values derived by selected authors for some of these rate constants, demonstrating how poorly constrained they commonly are. In fact, we still lack a basic understanding of the long-term processes at play during the carving of a landform to state with confidence the time it takes to achieve it. Commonly, parameter values are chosen from empirical considerations, from rate measurements at the human scale or in such a way that the LEMs predictions compare well to first-order rate observations, like catchment integrated erosion rates (Summerfield & Hulton 1994). These approaches require an extrapolation in space and time that is often difficult to justify.

Several recent studies (Stock & Montgomery 1999; van der Beek & Bishop 2003; Tomkin *et al.* 2003) have attempted to provide quantitative constraints on the rate of landscape evolution on geological time scales with the use of a variety of geochronological methods. Notably, Stock and Montgomery (1999) and, more recently, van der Beek and Bishop (2003), used well dated basaltic flows to define the initial geometry of the present-day landform in southeastern Australia (Stock & Montgomery 1999; van der Beek & Bishop 2003), Hawaii, Japan and California (Stock & Montgomery 1999) and, from it, a first-order estimate of the rate at which the system evolved. van der Beek and Bishop (2003) provided estimates of the fluvial incision constants in agreement with a previous study in the same area (van der Beek & Braun 1998). Similarly, Tomkin *et al.* (2003) combined a study of fluvial terraces with low temperature thermochronological data to obtain a good estimate of incision rates along the Clearwater River in the northwestern USA. They provided some constraints on the value of fluvial incision parameters but also concluded that many of the parameterizations were incapable of describing the evolution of the landform within realistic bounds.

The advent of the low-temperature apatite (U-Th)/He chronometer has driven a series of studies to determine the age of a variety of landforms (House *et al.* 1998, 2001; Reiners *et al.* 2003*a*, *b*). Good summaries of how the thermochronometer can be used to constrain the rate of landform evolution can be found in Braun (2002*b*, 2005) and Ehlers and Farley (2003). The use of cosmogenic isotopes has also flourished to determine current, instantaneous rate of erosion (Cerling & Craig 1994). In the following sections I provide two examples of how these dating methods have been used to derive quantitative constraints on the rate of landform evolution.

Low-T thermochronology

The perturbation caused by a finite amplitude topography on the thermal structure of the

Table 1. *Values of the constants appearing in the equations*

Parameter	Value	Reference
K_f	$10^{-7} - 10^{-2}$ m$^{0.2}$/yr (for $m = 0.3$–0.5 and $n = 1$)	Stock & Montgomery 1999
l_f	5–30 km	van der Beek & Bishop 2003
K_d	$10^{-4} - 1$ m^2/yr	Martin 2000; Braun *et al.* 2001
K_v	$1 - 3 \times 10^{-3}$ m$^{2-m''}$/yr	Braun *et al.* 2001
K_w	$1 - 3 \times 10^{-2}$ m^{2-2k}/yr	Braun *et al.* 2001
K_g	0.01 – ? (poorly constrained)	Tomkin & Braun 2002
P_0	50–200 m/Ma	Heimsath *et al.* 1999, 2000; Dietrich *et al.* 1995
ξ_f	45–50 cm	Heimsath *et al.* 1999, 2000

Values from a limited number of sources, demonstrating how variable and/or poorly constrained these values are.

underlying crust is a function of the wavelength of the topography. More precisely, the perturbation decreases exponentially with depth over a depth range that is equal to the wavelength of the topography (Turcotte & Schubert 1982). Braun (2002*a*) demonstrated that by considering the relationship between age and elevation as a function of wavelength, one can extract independent information on the mean rate of rock exhumation and the rate of landform evolution.

For most thermochronological systems, the 'age' of a rock corresponds to the time in the past when the rock cooled through the so-called 'closure temperature'. This temperature is different for each dating method; for example, it ranges from 65° to 85°C for (U-Th)/He dating in apatite. Considering an age-elevation profile at short wavelength, one can assume that, because the depth of the isotherm corresponding to the closure temperature is constant across the profile, the variation in age is directly proportional to the rate of rock exhumation and cooling through the closure temperature isotherm. Conversely, at long wavelength, isotherms should perfectly follow the shape of the landscape and there should be no variation of age with elevation, unless the shape of the landform has changed since the rocks cooled through the closure temperature isotherm. Any variation of age with elevation at long wavelength can thus be interpreted as an indication that the relief has increased (in the case of a positive age-elevation slope) or decreased (in the case of a negative slope) (Braun 2002*a*).

Accurate estimates of both mean exhumation rate and rate of landform evolution can be derived from appropriately sampled age-elevation profiles by performing a spectral analysis on the datasets. A so-called 'gain function' can be calculated as the complex ratio between the age and elevation spectra. At short wavelength, the inverse of the gain function value is directly proportional to the mean exhumation rate. The ratio of short to long wavelength gain values provides an estimate of the change in relief between the time corresponding to the mean age along the profile and today (Braun 2002*a*).

This method has been used to derive estimates of the amount of relief loss in the Sierra Nevada since the end of the Laramide Orogeny (Braun 2002*a*), based on (U-Th)/He datasets collected in the area by House *et al.* (1998). The results clearly demonstrated that the larger features of the landscape, including the 30 km-wide Kings Canyon, are indeed very old (60–70 Ma) and that, since the end of the Laramide Orogeny, relief amplitude has decreased by a factor of approximately 2. Plotting the age and elevation datasets along the profile (Fig. 7) clearly demonstrates that there is a strong positive correlation between the two at short wavelength and a clear anti-correlation at longer wavelength. The method was also used to determine the rate of relief evolution in the Dabie Shan of Eastern China (Braun & Robert 2005), confirming that the present-day relief in the area is also relatively old and results from the isostatically-compensated, erosional decay of a late Cretaceous mountain belt.

Braun and van der Beek (2004) also used low-temperature thermochronological data from a transect across the marginal escarpment of south-eastern Australia to demonstrate that this geomorphic feature formed very rapidly to reach its present-day position within 15 Ma of the onset of rifting, some 90 Ma ago and has since remained relatively stable. Braun and van der Beek (2004) also demonstrated that the existing thermochronological data cannot be used to differentiate between competing scenarios for the evolution of the escarpment.

More recently, Herman and Braun (2005) and Herman *et al.* (2005) have used (U-Th)/He dating in zircon and apatite to determine the rate of landform evolution in an extremely active tectonic environment, along the west coast of the South Island, New Zealand. They demonstrated that the high relief glaciated landscape that forms on the east side of the main divide is tectonically advected towards the west coast where it is

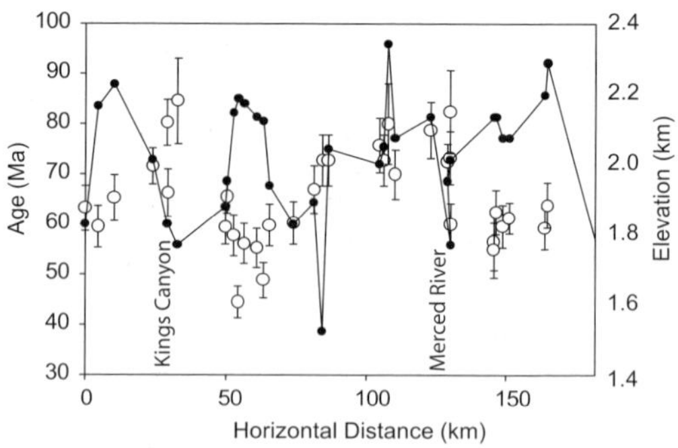

Fig. 7. Age and elevation along a linear profile in the Sierra Nevada, California (from House *et al.* 1998) clearly showing a positive correlation between (U-Th)/He apatite age (white circles) and elevation (black circles) along the narrow features of the landform and a negative correlation across the broader features. This data can be interpreted to demonstrate that the large-scale features of the landform are very old and that relief amplitude has decreased by a factor 2 over the past 60 Ma (Braun 2002*a*).

rapidly transformed into its present-day form. The very young ages and their relationship to elevation suggest that the reshaping of the landform from long, linear, flat-bottom glacial troughs into V-shaped graded river valleys is a relatively rapid process that takes place on a scale of less than 100 thousand years.

These various studies have further demonstrated that there exists a wide range of rates at which landforming processes operate and that the main factor controlling this rate is clearly related to tectonic uplift. Indeed, in regions of low tectonic activity, such as in the Sierra Nevada (House *et al.* 1998; Braun 2002*a*), the Dabie Shan (Reiners *et al.* 2003*b*; Braun & Robert 2005) or along the escarpments of southeastern Australia (Persano *et al.* 2002; Braun & van der Beek 2004) or southern Africa (Brown *et al.* 2002; van der Beek *et al.* 2002), very old landforms, i.e., 10 to 100 Ma old, are preserved. Despite their age, they can be characterized by very large local relief – several thousand meters in the Sierra Nevada. At the other end of the spectrum, in regions of high tectonic uplift, erosion processes can rejuvenate a landform on a much smaller time scale (Herman *et al.* 2005), i.e., of the order of 1 Ma or less. This implies that erosional systems must be characterized by a threshold beyond which their efficiency is greatly enhanced. This threshold could take the form of a mean critical slope beyond which the frequency of landslides increases (Montgomery & Brandon 2002).

Cosmogenic isotopes

Cosmogenic isotopes (^{10}Be, ^{26}Al in quartz, ^{36}Cl in calcite, Kfeldspar and whole rock samples, and ^{3}He and ^{21}Ne in olivine and feldspar) are formed by high energy cosmic ray reactions in the uppermost 1–2 m of the crust (Cerling & Craig 1994). Isotope production rates depend strongly on depth, hence isotope accumulation reflects a sample's shielding history in the uppermost crust. Where a sample's exposure geometry was altered instantaneously by a geomorphic event (a landslide, fault rupture, etc.), cosmogenic isotopes provide age information. Where a geomorphic process continuously alters a sample's depth, the process rate can be determined from cosmogenic isotope data via a model of the balance between local erosion rate and *in situ* isotope production (Cerling & Craig 1994; Bierman 1994).

In situ cosmogenic isotope concentrations have thus been used to determine the rate of soil production (Heimsath *et al.* 2000). Combining such data with morphometric measurements at two locations (southwestern California and southeastern Australia), Braun *et al.* (2001) have provided constraints on the various parameters entering the laws controlling soil transport. In their study, Braun *et al.* (2001) used a two-dimensional finite volume model to predict the relationship between soil thickness and topographic surface curvature, using the cosmogenic data to constrain a production function, in which production decreases exponentially with soil thickness (Heimsath *et al.* 2000). The results were compared to measurements of soil thickness and surface curvature to provide direct constraints on the efficiency of each of the mechanisms assumed to be efficient at transporting soil on soil-mantled hill slopes, on the residence time of the soil on the landscape, and on the response time of the system to rapidly changing environmental conditions.

Cosmogenic isotopes have been used to constrain the rate of landform evolution in a range of settings, including, for example, glaciated landscapes (Stroeven *et al.* 2002). They are commonly used to date fluvial terraces (Hancock *et al.* 1999). See Duller (2000) for a short review of the method and its application to geomorphology.

Current and outstanding problems

Parameterization and scaling

The most pressing problem in the development and use of landscape evolution models is the formulation of a widely accepted parameterization of the basic processes at play during erosion, transport and deposition. Unlike tectonic models that are based on the solution of Newton's laws of force balance, there is no consensus in the field of geomorphology about the fundamental equations that should be used to represent, for example, the physical processes of fluvial erosion. Should we use the stream power law? Should we include a carrying capacity term or introduce a threshold to represent the strength of the bedrock? What form of the soil transport law should we use? How do we parameterize landsliding, chemical erosion, or eolian transport? Note that it is not just the value of the various parameters that is unknown, it is also the form of the equations to solve that remains uncertain.

Potentially, this lack of simple or uniformally accepted parameterization is a reflection of the wide range of processes involved in the formation of landforms. Which is most important certainly depends on environmental conditions including climate, vegetation, tectonic setting, etc. It might therefore be utopian to believe that a single parameterization could ever represent all processes efficiently and/or accurately. On the other hand, this state of affairs may also be the result of how the field of quantitative or predictive geomorphology has evolved from its observational and empirical beginnings. Most of the laws we use have been constructed to represent physical processes at the human scale. The extrapolation to longer scales has been done with little justification. Thankfully, work is already under way (e.g., see Davy & Crave 2000; or Martin & Church 2004) to understand how this scaling may be done with more rigour, which, ultimately, may lead to the formulation of equations that have only little resemblance to the 'classical' geomorphic laws but are more appropriate to describe the system over long timescales. Of particular interest in the formulation of these more fundamental parameterizations, are the effects of infrequent but significant (large) events such as floods or mega-size landslides on the long-term evolution of the landform (Lague *et al.* 2003).

Thermochronology: sampling strategy

Geochronology, and in particular thermochronology, are the most widely used tools to determine the rate of Earth processes. The use of these techniques was, until recently, confined to the field of tectonics. The development of accurate low-temperature thermochronometers has opened the door to a broader use of these techniques and, potentially, will permit the quantification of the rate of landform evolution. Unfortunately, as shown by Braun (2002*a*) and Braun and van der Beek (2004), sampling strategies developed to study tectonic processes are usually incompatible with those necessary to document erosional processes. Typically, to constrain tectonic systems, rocks are collected along transects that run perpendicular to the main controlling structure, whereas to estimate the rate of river incision in an active tectonic setting, rocks need to be collected along a profile parallel to the main controlling structure to ensure not only that the mean exhumation rate is uniform along the transect, but also to sample the relief at the most meaningful wavelengths. To date, very few datasets exist that satisfy this requirement.

Future work

More efficient algorithms

Despite ever-increasing computer power, there is a need to continue developing efficient algorithms to solve the complex geomorphic equations for the following two reasons: (1) to test the validity of scaled laws and the importance of critical but rare events requires a very fine spatial and temporal discretization; (2) to determine the sensitivity of LEMs behaviour to variations in relatively poorly constrained parameters requires that large numbers of model runs be performed. Ideally, in cases where thermochronological and morphometric datasets exist and are used to constrain the value of model parameters, a proper inversion procedure should be used (van der Beek & Braun 1998; van der Beek & Bishop 2003; Tomkin *et al.* 2003; Braun & van der Beek 2004); it commonly requires that thousands to tens of thousands of LEM runs be executed, in order to adequately sample parameter space in all directions.

The most challenging part of these computations is to calculate the geometry of catchments and compute sediment flux by integration of upstream erosion/sedimentation rate. Ranking of the discrete topographic heights is the most commonly used algorithm (Beaumont *et al.* 1992), but it becomes rather inefficient for very large problems. Others have opted for a stochastic approach that does not require the explicit computation of catchment geometry (Chase 1992; Crave & Davy 2001) but cannot be used to estimate upstream sediment flux. With the increase availability of parallel computers, there is a need to develop a parallelized version of a dynamic network routing algorithm to compute drainage networks and the geometry of catchments.

3D tectonic models

The study of the coupling between erosion and tectonics has been limited so far by the simplifying assumption of two dimensionality in the vertical or horizontal direction. To fully comprehend the complex interactions between surface processes and tectonics requires the solution of the force balance equation in three dimensions. Surface processes are strongly affected by subtle variations in the distribution of uplift. Long-range interaction mechanisms, such as channel incision and fluvial transport, may lead to efficient mass redistribution on the landscape, which will produce important pressure gradients in the underlying lithosphere. Thus the feedback or coupling between erosion and tectonics cannot, in principle, be averaged in any particular direction.

Because most thrust faults have a gentle dip, horizontal advection must play as important a role as tectonic uplift in regions of continental convergence. Horizontal advection of landforms by tectonic movement across faults or drainage divides has so far been neglected or treated empirically. The accuracy of the methods used needs first to be established. Most would agree that, in nature, the processes involved are discrete: a drainage divide may be redefined during a landslide and drainage reorganization may take place during a single slip event on a fault. The behaviour of these systems will therefore depend on the stochastic nature of these processes, which needs to be better defined.

Define target areas where the efficiency of the coupling can be estimated

Finally, one must be reminded that the demonstration that a coupling between erosion and tectonics exists has so far been limited to the results of computer modelling of the system. As amply demonstrated in a collection of jointly published papers (Reiners *et al.* 2003*a*; Burbank *et al.* 2003; Dadson *et al.* 2003), our efforts to document this coupling in nature have been rather inconclusive. The most pressing development to improve our understanding of the lithosphere–hydrosphere system might be the definition of a natural situation where the efficiency of the mechanisms we invoke can be measured or estimated independently. For example, a good situation to document might be a long compressional orogen in which one of the two driving mechanisms (erosion or tectonics) is clearly uniform and constant and where the other varies along strike. The two key elements in the selection of such a natural laboratory will be (1) the ability to discern between causes and effects and (2) the ability to describe a system that is not at steady-state.

The author wishes to thank Y. Martin, D. Stolar and an anonymous reviewer for very constructive comments on an earlier version of this manuscript.

References

ANHERT, F. 1967. The role of the equilibrium concept in the interpretation of landforms of fluvial erosion and deposition. *In*: MACAR, P. (ed.) *L'Evolution des versants*. Université de Liège, Belgium, 22–41.

BATT, G. E. & BRAUN, J. 1999. The tectonic evolution of the Southern Alps, New Zealand: insights from fully thermally coupled dynamical modelling. *Geophysical Journal International*, **136**, 403–420.

BEAUMONT, C., FULLSACK, P. & HAMILTON, J. 1992. Erosional control of active compressional orogens. *In*: MCCLAY, K. R. (ed.) *Thrust Tectonics*. Chapman and Hall, New York, 1–18.

BEAUMONT, C., JAMIESON, R., NGUYEN, M. & LEE, B. 2001. Himalayan tectonics explained by extrusion of a low-viscosity crustal channel coupled to focused surface denudation. *Nature*, **414**, 738–742.

BEAUMONT, C., JAMIESON, R., NGUYEN, M. & MEDVEDEV, S. 2004. Crustal channel flows: 1. Numerical models with applications to the tectonics of the Himalayan-Tibetan orogen. *Journal of Geophysical Research*, **109**(B06406): DOI 10.1029/2003JB002809.

BEAUMONT, C., KAMP, P., HAMILTON, J. & FULLSACK, P. 1996. The continental collision zone, South Island, New Zealand: comparison of geodynamical models and observations. *Journal of Geophysical Research*, **101**, 3333–3359.

BEAUMONT, C., KOOI, H. & WILLETT, S. 1999. Coupled tectonic-surface process models with applications to rifted margins and collisional orogens. *In*: SUMMERFIELD, M. (ed.) *Geomorphology and Global Tectonics*. John Wiley and Sons Ltd, New York, 29–55.

BIERMAN, P. 1994. Using in situ produced cosmogenic isotopes to estimate rates of landscape evolution: a review from the geomorphic perspective. *Journal of Geophysical Research*, **99**, 13885–13896.

BRAUN, J. 2002*a*. Estimating exhumation rate and relief evolution by spectral analysis of age-elevation datasets. *Terra Nova*, **14**, 210–214.

BRAUN, J. 2002*b*. Quantifying the effect of recent relief changes on age-elevation relationships. *Earth and Planetary Sciences Letters*, **200**, 331–343.

BRAUN, J. 2005. *Thermochronology*, MSA-RiMG, in press.

BRAUN, J. & ROBERT, X. 2005. Constraints on the rate of post-orogenic erosional decay from thermochronological data: example from the Dabie Shan, China. *Earth Surface Processes and Landforms*, **30**, 1203–1225.

BRAUN, J. & SAMBRIDGE, M. 1994. Dynamical Lagrangian remeshing (DLR): a new algorithm for solving large strain deformation problems and its application to fault-propagation folding. *Earth and Planetary Sciences Letters*, **124**, 211–220.

BRAUN, J. & SAMBRIDGE, M. 1997. Modelling landscape evolution on geological time scales: a new method based on irregular spatial discretization. *Basin Research*, **9**, 27–52.

BRAUN, J. & SHAW, R. 2001. A thin-plate model of Palaeozoic deformation of the Australian lithosphere: implications for understanding the dynamics of intracratonic deformation. *In*: MILLER, J., HOLDSWORTH, R., BUICK, I. & HAND, M. (eds) *Continental Reactivation and Reworking*. Geological Society, London, Special Publications **184**, 165–193.

BRAUN, J. & VAN DER BEEK, P. 2004. Evolution of passive margin escarpments: what can we learn from low-temperature thermochronology? *Journal of Geophysical Research*, **109**, F04009, DOI 10.1029/2004JF000147.

BRAUN, J., HEIMSATH, A. & CHAPPELL, J. 2001. Sediment transport mechanisms on soil-mantled hillslopes. *Geology*, **29**, 683–686.

BRAUN, J., ZWARTZ, D. & TOMKIN, J. 1999. A new surface processes model combining glacial and fluvial erosion. *Annals of Glaciology*, **28**, 282–290.

BROWN, R., SUMMERFIELD, M. & GLEADOW, A. 2002. Denudational history along a transect across the Drakensberg Escarpment of southern Africa derived from apatite fission track thermochronology. *Journal of Geophysical Research*, **107**, 2350, DOI 10.1029/2001JB000745.

BURBANK, D. & PINTER, N. 1999. Landscape evolution: the interactions of tectonics and surface processes. *Basin Research*, **11**, 1–6.

BURBANK, D., BLYTHE, A. *ET AL.* 2003. Decoupling of erosion and precipitation in the Himalayas. *Nature*, **426**, 652–655.

CERLING, T. & CRAIG, H. 1994. Geomorphology and in-situ cosmogenic isotopes. *Annual Review of Earth and Planetary Sciences*, **22**, 273–317.

CHAMPEL, B., VAN DER BEEK, P., MUGNIER, J. & LETURMY, P. 2002. Uplift and lateral propagation of fault-related folds in the Siwalik foothills, Nepal Himalaya: rates, mechanisms, and geomorphic signature. *Journal of Geophysical Research*, **107**, 2111, DOI 10.1029/2001JB000578.

CHASE, C. 1992. Fluvial landscuplting and the fractal dimension of topography. *Geomorphology*, **5**, 39–57.

CLARK, M. & ROYDEN, L. 2000. Topographic ooze: Building the eastern margin of Tibet by lower crustal flow. *Geology*, **28**, 703–706.

CLARKE, M., WILLIAMS, M. & STOKES, T. 1999. Soil creep: problems raised by a 23 year study in Australia. *Earth Surface Processes and Landforms*, **24**, 151–175.

COULTHARD, T. J. 2001. Landscape evolution models: a software review. *Hydrological Processes (HP Today)*, **15**, 165–173.

COULTHARD, T., KIRKBY, M. & MACKLIN, M. G. 1997. Modelling hydraulic, sediment transport and slope processes, at a catchment scale, using a cellular automaton approach. *In*: PASCOE, R. (ed.) *Proceedings of the Second Annual Conference: GeoComputation 97*. University of Otago, Dunedin, New Zealand, 309–318.

CRAVE, A. & DAVY, P. 2001. A stochastic 'precipiton' model for simulating erosion/sedimentation dynamics. *Computers and Geosciences*, **27**, 815–827.

CRAVE, A., LAGUE, D., DAVY, P., KERMARREC, J.-J., SOKOUTIS, D., BODET, L. & COMPAGNON, R. 2000. Analogue modelling of relief dynamics. *Physics and Chemistry of the Earth*, **6–7**, 549–553.

DADSON, S., HOVIUS, N. *ET AL.* 2003. Links between erosion, runoff variability and seismicity in the Taiwan orogen. *Nature*, **426**, 648–651.

DAHLEN, F. 1984. Noncohesive critical Coulomb wedges: an exact solution. *Journal of Geophysical Research*, **89**, 10125–10133.

DAVY, P. & CRAVE, A. 2000. Upscaling local-scale transport processes in large-scale relief dynamics. *Physics and Chemistry of the Earth*, **25**, 533–541.

DENSMORE, A., DAWERS, N., GUPTA, S., GUIDON, R. & GOLDIN, T. 2004. Footwall topographic development during continental extension. *Journal of Geophysical Research*, **109**, F03001, DOI 10.1029/2003JF000115.

DENSMORE, A., ELLIS, M. & ANDERSON, R. 1998. Landsliding and the evolution of normal-fault-bounded mountains. *Journal of Geophysical Research*, **103**, 15203–15219.

DIETRICH, W., BELLUGI, D., SKLAR, L., STOCK, J., HEIMSATH, A. & ROERING, J. 2003. Geomorphic transport laws for predicting landscape form and dynamics. *In*: WILCOCK, P. & IVERSON, R. (eds) *Prediction in Geomorphology. Geophysical Monograph*, American Geophysical Union, Washington, DC, **135**, DOI 10.1029/135GM09.

DIETRICH, W., REISS, R., HSU, M.-L. & MONTGOMERY, D. 1995. A process-based model for colluvial soil depth and shallow landsliding using digital elevation data. *Hydrological Processes*, **9**, 383–400.

DULLER, G. 2000. Dating methods: geochronology and landscape evolution. *Progress in Physical Geography*, **24**, 111–116.

EHLERS, T. & FARLEY, K. 2003. Apatite (U-Th)/He thermochronometry: methods and applications to problems in tectonic and surface processes. *Earth and Planetary Sciences Letters*, **206**, 1–14.

ELLIS, S. 1995. *Continental convergence: length scales, aspect ratios, and styles of crustal deformation*. PhD thesis, Dalhousie University, HALIFAX, NS, Canada.

GARCIA-CASTELLANOS, D. 2002. Interplay between lithospheric flexure and river transport in foreland basins. *Basin Research*, **14**, 89–104.

HALLETT, B. 1979. A theoretical model of glacial abrasion. *Journal of Glaciology*, **89**, 39–50.

HANCOCK, G., ANDERSON, R., CHADWICK, O. & FINKEL, R. 1999. Dating fluvial terraces with ^{10}Be and ^{26}Al profiles: application to the Wind River, Wyoming. *Geomorphology*, **27**, 41–60.

HARBOR, J. 1992. Numerical modelling of the development of U-shaped valleys by glacial erosion. *Geological Society of America Bulletin*, **104**, 1364–1375.

HEIMSATH, A., CHAPPELL, J., DIETRICH, W., NISHIIZUMI, K. & FINKEL, R. 2000. Soil production on a retreating escarpment in southeastern Australia. *Geology*, **28(9)**, 787–790.

HEIMSATH, A., CHAPPELL, J., DIETRICH, W., NISHIIZUMI, K. & FINKEL, R. 2001. Late Quaternary erosion in southeastern Australia: a field example using cosmogenic nuclides. *Quaternary International*, **83(5)**, 169–185.

HEIMSATH, A., DIETRICH, W., NISHIIZUMI, K. & FINKEL, R. 1999. Cosmogenic nuclides, topography, and the spatial variation of soil depth. *Geomorphology*, **27**, 151–172.

HERMAN, F. & BRAUN, J. 2005. Fluvial response to horizontal shortening and glaciations: a study in the Southern Alps of New Zealand. *Journal of Geophysical Research*, in press.

HERMAN, F., BRAUN, J. & DUNLAP, W. 2005. Low temperature themochronology to constrain the tectono-morphic evolution of the Southern Alps of New Zealand. *Journal of Geophysical Research*, in press.

HILLEY, G. & STRECKER, M. 2004. Steady-state erosion of critical Coulomb wedges with application to Taiwan and the Himalaya. *Journal of Geophysical Research*, **109**, B01411-10.1029/2002JB002284.

HODGES, K., HURTADO, J. & WHIPPLE, K. 2001. Southward extrusion of Tibetan crust and its effect on Himalayan tectonics. *Tectonics*, **20**, 799–809.

HOUSE, M., WERNICKE, B. & FARLEY, K. 1998. Dating topography of the Sierra Nevada, California, using apatite (U-Th)/He ages. *Nature*, **396**, 66–69.

HOUSE, M., WERNICKE, B. & FARLEY, K. 2001. Paleogeomorphology of the Sierra Nevada, California, from (U-Th)/He ages in apatite. *American Journal of Science*, **301**, 77–102.

HOVIUS, N., STARK, C. & ALLEN, P. 1997. Sediment flux from a mountain belt derived from landslide mapping. *Geology*, **25(3)**, 231–234.

HOVIUS, N., STARK, C., HAO-TSU, C. & JIUM-CHUAN, L. 2000. Supply and removal of sediment in a landslide-dominated mountain belt: Central Range, Taiwan. *Journal of Geology*, **108**, 73–89.

HOWARD, A. 1994. A detachment–limited model of drainage basin evolution. *Water Resources Research*, **30(7)**, 2261–2285.

HUTTER, K. 1981. The effect of longitudinal stress on the shear stress of an ice sheet: in defence of using stretched co-ordinates. *Journal of Geology*, **27**, 39–56.

JAMIESON, R., BEAUMONT, C., MEDVEDEV, S. & NGUYEN, M. 2004. Crustal channel flows: 1. Numerical models with implications for metamorphism in the Himalayan-Tibetan orogen. *Journal of Geophysical Research*, **109** (B06407): DOI 10.1029/2003JB002811.

JOHNSON, D. & BEAUMONT, C. 1995. Preliminary results from a planform kinematic model of orogen evolution, surface processes, and the development of clastic foreland basin stratigraphy. *In*: DOROBEK, S. & ROSS, G. (eds) *SEPM Special Publication on Stratigraphic Evolution of Foreland Basins*, 3–24.

KAUFFMAN, G. & BRAUN, J. 2001. Modelling karst denudation on a synthetic landscape. *Terra Nova*, **36**, 1381–1391.

KOOI, H. & BEAUMONT, C. 1994. Escarpment evolution on high-elevation rifted margins: insights derived from a surface processes model that combines diffusion, advection and reaction. *Journal of Geophysical Research*, **99**, 12191–12209.

KOONS, P. 1990. Two-sided orogen: collision and erosion from the sandbox to the Southern Alps, New Zealand. *Geology*, **18**, 679–682.

KOONS, P. 1994. Three-dimensional critical wedges: Tectonics and topography in oblique collisional orogens. *Journal of Geophysical Research*, **99**, 12301–12315.

KOONS, P., ZEITLER, P., CHAMBERLAIN, C., CRAW, D. & MELZER, A. 2002. Mechanical links between erosion and metamorphism in Nanga Parbat, Pakistan Himalaya. *American Journal of Science*, **302**, 749–773.

LAGUE, D., CRAVE, A. & DAVY, P. 2003. Laboratory experiments simulating the geomorphic response to tectonic uplift. *Journal of Geophysical Research*, **108**, 10.1029/2002JB001785.

MACGREGOR, K., ANDERSON, R., ANDERSON, S. & WADDINGTON, E. 2000. Numerical simulations of glacial-valley longitudinal profile evolution. *Geology*, **28**, 1031–1034.

MARTIN, Y. 2000. Modelling hillslope evolution: linear and nonlinear transport relations. *Geomorphology*, **34**, 1–21.

MARTIN, Y. & CHURCH, M. 2004. Numerical modelling of landscape evolution: geomorphological perspectives. *Progress in Physical Geography*, **28**, 317–339.

MOLNAR, P. & ENGLAND, P. 1990. Late Cenozoic uplift of mountain ranges and global climate change: chicken and egg? *Nature*, **346**, 29–34.

MONTGOMERY, D. & BRANDON, M. 2002. Topographic control on erosion rates in tectonically active mountain ranges. *Earth and Planetary Sciences Letters*, **201**, 481–489.

MOORE, I. & BURCH, G. 1986. Sediment transport capacity of sheet and rill flow: application of unit stream power theory. *Water Resources Research*, **22**, 1350–1360.

OLSON, R. 1973. *Essentials of Engineering Fluid Mechanics*. 3rd edn, Feefer and Simons International University Editions, New York.

PERSANO, C., STUART, F., BISHOP, P. & BARFORD, D. 2002. Apatite (U-Th)/He age constraints on the development of the Great Escarpment on the southeastern Australian passive margin. *Earth and Planetary Sciences Letters*, **200**, 79–90.

PERSSON, K., GARCIA-CASTELLANOS, D. & SOKOUTIS, D. 2004. River transport effects on compressional belts: first results from an integrated analogue-numerical model. *Journal of Geophysical Research*, **109**, DOI 10.1029/2002JB002274.

PROSSER, I. & RUSTOMJI, P. 2000. Sediment transport capacity relations for overland flow. *Progress in Physical Geography*, **24**, 179–193.

REINERS, P., EHLERS, T., MITCHELL, S. & MONTGOMERY, D. 2003*a*. Coupled spatial variations in precipitation and long-term erosion rates across the Washington Cascades. *Nature*, **426**, 645–647.

REINERS, P., ZHOU, Z., EHLERS, T., XU, C., BRANDON, M., DONELICK, R. & NICOLESCU, S. 2003*b*. Post-orogenic evolution of the Dabie Shan, eastern China, from (U-Th)/He and fission-track thermochronology. *American Journal of Science*, **303**, 489–518.

ROE, G., STOLAR, D. & WILLETT, S. 2005. The sensitivity of a critical wedge orogen to climatic and tectonic forcing. *In*: WILLETT, S., HOVIUS, N., BRANDON, M. & FISHER, D. (eds) *Tectonics, Climate, and Landscape Evolution*, Geological Society of America, Special Publication.

ROERING, J., KIRCHNER, J. & DIETRICH, W. 1999. Evidence for a non-linear, diffusive sediment transport on hillslopes and implications for landscape evolution. *Water Resources Research*, **35(3)**, 853–870.

ROERING, J., KIRCHNER, L., SKLAR, J. & DIETRICH, W. 2000. Hillslope evolution by nonlinear creep and landsliding: an experimental study. *Geology*, **29(2)**, 143–146.

ROYDEN, L. 1996. Coupling and decoupling of crust and mantle in convergent orogens: implications for strain partitioning in the crust. *Journal of Geophysical Research*, **101**, 17679–17705.

SCHOORL, J., SONNEVELD, M. & VELDKAMP, A. 2000. Three-dimensional landscape process modelling: the effect of DEM resolution. *Earth Surface Processes and Landforms*, **25**, 1025–1034.

SIMPSON, G. 2004. A dynamic model to investigate coupling between erosion, deposition, and three-dimensional (thin-plate) deformation. *Journal of Geophysical Research*, **109**, F02006. DOI 10.1029/2003JF000078.

SIMPSON, G. & SCHLUNEGGER, F. 2003. Topographic evolution and morphology of surfaces evolving in response to coupled fluvial and hillslope sediment transport. *Journal of Geophysical Research*, **108**, F02300, DOI 10.1029/2002JB002162.

SKLAR, L. & DIETRICH, W. 1998. River longitudinal profiles and bedrock incision models: stream power and the influence of sediment supply. *In*: TINKLER, K. & WOHL, E. (eds) *Rivers over Rock: Fluvial Processes in Bedrock Channels*. Geophysical Monograph, American Geophysical Union, Washington, D.C., 237–260.

STOCK, J. & MONTGOMERY, D. 1999. Geologic constraints on bedrock river incision using the stream power law. *Journal of Geophysical Research*, **104**, 4983–4993.

STOLAR, D., ROE, G. & WILLETT, S. 2005. Evolution of a critical orogen under various forcing scenarios: findings from a numerical sandbox. *In*: WILLETT, S., HOVIUS, N., BRANDON, M. & FISHER, D. (eds) *Tectonics, Climate, and Landscape Evolution*. Geological Society of America, Special Publication.

STROEVEN, A., FABEL, D., HATTESTRAND, C. & HARBOR, J. 2002. A relict landscape in the centre of Fennoscandian glaciation: cosmogenic radionuclide evidence of tors preserved through multiple glacial cycles. *Geomorphology*, **44**, 145–154.

SUMMERFIELD, M. & HULTON, N. 1994. Natural controls of fluvial denudation rates in major world drainage basins. *Journal of Geophysical Research*, **99**, 13871–13883.

TOMKIN, J. 2000. *Landforming processes in glaciated orogens: a numerical study*. PhD thesis, the Australian National University, Canberra, ACT, Australia.

TOMKIN, J. & BRAUN, J. 2002. The effect glaciation has on the relief of a fast growing orogen: a numerical modelling study. *American Journal of Science*, **302**, 169–190.

TOMKIN, J., BRANDON, M., PAZZAGLIA, F., BARBOUR, J. & WILLETT, S. 2003. Quantitative testing of bedrock incision models, Clearwater River, WA. *Journal of Geophysical Research*, **108**, 2308, DOI 10.1029/2001JB000862.

TUCKER, G. 2004. Drainage basin sensitivity to tectonic and climatic forcing: implications of a stochastic model for the role of entrainment and erosion thresholds. *Earth Surface Processes and Landforms*, **29**, 185–205.

TUCKER, G. & SLINGERLAND, R. 1994. Erosional dynamics, flexural isostacy, and long-lived escarpments: a numerical modeling study. *Journal of Geophysical Research*, **99**, 12229–12243.

TUCKER, G., LANCASTER, S., GASPARINI, N. & BRAS, R. 2001. The Channel-Hillslope Integrated Landscape Development (CHILD) Model. *In*: HARMON, R. & DOE III, W. W. (eds) *Landscape Erosion and Evolution Modeling*. Kluwer Academic/Plenum Publishers, 349–388.

TURCOTTE, D. & SCHUBERT, G. 1982. *Geodynamics: Applications of Continuum Physics to Geological Problems*. 1st edn, John Wiley and Sons, New York.

VAN DER BEEK, P. & BISHOP, P. 2003. Cenozoic river profile development in the Upper Lachlan catchment (SE Australia) as a test of quantitative fluvial incision models. *Journal of Geophysical Research*, **1108**, 2309, DOI 10.1029/2002JB002125.

VAN DER BEEK, P. & BRAUN, J. 1998. Numerical modelling of landscape evolution on geological time scales: a parameter analysis and comparison with the southeastern highlands of Australia. *Basin Research*, **10**, 49–68.

VAN DER BEEK, P., SUMMERFIELD, M., BRAUN, J., BROWN, R. & FLEMING, A. 2002. Modelling post-breakup landscape development and denudational history across the southeast African (Drakensberg Escarpment) margin. *Journal of Geophysical Research*, **107**, 2351, DOI 10.1029/2001JB000744.

WELLMAN, H. 1979. An uplift map for the South Island of New Zealand, and a model for uplift of the Southern Alps. *In*: WALCOTT, R. I. & CRESSWELL, M. M. (eds) *The Origin of the Southern Alps*. Bulletin, Royal Society of New Zealand, 13–20.

WHIPPLE, K. 2001. Geomorphic limits to climate-induced increases in topographic relief. *American Journal of Science*, **301**, 313–325.

WHIPPLE, K. & MEADE, B. 2004. Controls on the strength of coupling among climate, erosion, and deformation in two-sided, frictional orogenic wedges at steady state. *Journal of Geophysical Research*, **109**, F01011, DOI 10.1029/2003JF000019.

WHIPPLE, K. & TUCKER, G. 1999. Dynamics of the stream-power incision model: implications for height limits of mountain ranges, landscape response timescales and research needs. *Journal of Geophysical Research*, **104**, 17661–17674.

WHIPPLE, K., KIRBY, E. & BROCKLEHURST, S. 1999. Geomorphic limits to climate-induced increases in topographic relief. *Nature*, **401**, 39–43.

WHITEHOUSE, I. 1987. Geomorphology of a compressional plate boundary: Southern Alps, New Zealand. *International Geomorphology*, **1**, 897–924.

WILLETT, S. 1999. Orogeny and orography: the effects of erosion on the structure of mountain belts. *Journal of Geophysical Research*, **104**, 28957–28981.

WILLETT, S., BEAUMONT, C. & FULLSACK, P. 1993. Mechanical model for the tectonics of doubly-vergent compressional orogens. *Geology*, **21**, 371–374.

WILLETT, S., SLINGERLAND, R. & HOVIUS, N. 2001. Uplift, shortening, and steady state topography in active mountain belts. *American Journal of Science*, **301**, 455–485.

WILLGOOSE, G., BRAS, R. & RODRIGUEZ-ITURBE, I. 1991. A physically based coupled network growth and hillslope evolution model: 1. Theory. *Water Resources Research*, **27**, 1671–1684.

ZEITLER, P. & MELTZER, A. *ET AL.* 2001*a*. Erosion, Himalayan geodynamics, and the Geomorphology of Metamorphism. *GSA Today*, **11**, 4–9.

ZEITLER, P. K. & KOONS, P. O. *ET AL.* 2001*b*. Crustal reworking at Nanga Parbat, Pakistan: evidence for erosional focusing of crustal strain. *Tectonics*, **20**, 712–728.

Macroscale dynamics of experimental landscapes

STEPHANE BONNET & ALAIN CRAVE

Géosciences Rennes, UMR 6118 CNRS/Université de Rennes 1, 35042 Rennes cedex, France
(e-mail: sbonnet@univ-rennes1.fr)

Abstract: We review results from laboratory-scale modelling of erosion and relief dynamics under variable uplift and rainfall rates. Under constant values of these forcing parameters an experimental landscape typically evolves towards a steady-state between uplift and erosion, and we show how the geometry of the steady-state landscape adjusts to the rates of uplift and rainfall. The comparison between these laboratory-scale landscapes and the natural ones is not straightforward because contrary to analogue modelling in tectonics, natural conditions of relief evolution cannot be downscaled to the laboratory without any scale distortions. Laboratory-scale modelling in geomorphology is therefore only experimental, not analogue. Despite these limitations, experimental models may be used to provide physical tests for numerical models and they give insights into first-order behaviours and directions for future research.

The recognition of the importance of erosion in geodynamics (Beaumont *et al.* 1991, 2001; Avouac & Burov 1996; Horton 1999; Willett 1999; Montgomery *et al.* 2001; Zeitler *et al.* 2001; Finlayson *et al.* 2002; Wobus *et al.* 2003; Simpson 2004; Thiede *et al.* 2004; Vannay *et al.* 2004) has led to a need for a better understanding of topographic evolution at geological timescales. However the changing form of the landscape with time is a long-standing question (Davis 1859; Penck 1953; Hack 1960) that is difficult to assess because of the timescale of the processes involved. Numerous recent theoretical advances have been made regarding the physical description of erosional processes (e.g., Howard 1994) and their incorporation into numerical models of erosion and landscape dynamics at geological timescales. Recent experimental models in geomorphology have been primarly built to provide physical tests for numerical models of erosion (Hasbargen & Paola 2000; Hancock & Willgoose 2001; Lague *et al.* 2003). Experimental dynamics have been successfully reproduced by some numerical models (e.g., EROS code: Crave & Davy 2001) given the use of appropriate parameters of the erosion law (e.g., Lague *et al.* 2003). Experimental modelling in geomorphology has been widely developed for the study of erosional/depositional processes such as hillslope erosion and landsliding (Densmore *et al.* 1997; Roering *et al.* 2001; Gabet 2003), bedrock abrasion (Sklar & Dietrich 2001), sediment transport (Paola *et al.* 1992), or channel dynamics (Shepherd & Schumm 1974; Holland & Pickup 1976; Ashmore 1982; Gardner, 1983; Schumm *et al.* 1987; Meunier & Métivier 2000). Some other experiments have also focused on drainage network growth (Schumm *et al.* 1987; Pelletier, 2003) and on the dynamics of landscape (Crave *et al.* 2000; Hasbargen & Paola 2000; Hancock & Willgoose 2001; Bonnet & Crave 2003; Lague *et al.* 2003; Babault *et al.* 2005). We use here a selection of experiments to illustrate how experimental landscapes adjust to the applied uplift and rainfall rates. We discuss the meaning of the experiments regarding scaling problems and how they give insights into controls on, and dynamics of, relief generation.

Experimental modelling in geomorphology

Principle

The basic principle of modelling in geomorphology is very simple, and consists of surface erosion and sediment transport by runoff of water. However, differences between models arise from many aspects: the material; the size of the model; the device used to generate rainfall, which controls the rate of rainfall and/or the size of raindrops; the mechanism of applying uplift or base-level change; and the measurement of the experimental results; by photographs, digitization of the topography, and/or measurement of the eroded flux. Because of this diversity, model experiments performed in different laboratories (e.g., University

From: Buiter, S. J. H. & Schreurs, G. (eds) 2006. *Analogue and Numerical Modelling of Crustal-Scale Processes*. Geological Society, London, Special Publications, **253**, 327–339.
0305-8719/06/$15.00

of Minnesota: Hasbargen & Paola 2000; University of Rennes1/CNRS: Crave *et al.* 2000; University of Newcastle: Hancock & Willgoose 2001), even if simple in principle, are very difficult to compare with each other. For example, the experiments developed at Rennes are the only ones where erosion may act to compensate an applied uplift of the model. The basic data generated by these experiments consist of topographic measurements through time, whereas in other experiments, for example, by Hasbargen and Paola (2000), the data consist of flux measurements (weight of eroded material through time).

We only detail herein results of experiments made in Rennes over the last six years. In a first step, an innovative experimental design was built and successfully used in different studies (Crave *et al.* 2000; Bonnet & Crave 2003; Lague *et al.* 2003; Crave *et al.* 2005). However the use of this system revealed some defects that have led to different improvements and the development of a new device (Fig. 1) that is presently in use (Babault *et al.* 2005). These experimental set-ups were designed to study the dynamics of small-scale systems under simple conditions of erosion, uplift and rainfall rates. Rainfall and uplift rates are totally independent and set by the operator. No feedback loops exist between the surface uplift of the model and the rainfall rate (orographic effect) or between the pattern of erosion of the model and the uplift rate (isostatic effect). These simplifications of the natural conditions are not solely due to technical problems but are also deliberate, mainly for two reasons. Firstly, the complexity of geomorphological processes is still poorly understood so we prefer to minimize the parameters involved, instead of adding hypothetical analogue processes in the experiments. Secondly, simple case studies are necessary to define a baseline of experimental observations to help the interpretation of results from more complex experimental set-ups.

Whatever the simplifications of the geomorphological processes, the experimental device offers the possibility of studying a large open field of questions on relief dynamics that are not yet physically explored. The main assumption of this set-up is to reduce all the processes of erosion to the sole action of runoff. Other erosion processes like eolian transport or slope instability are certainly important in sculpting topography, but we believe that runoff is the major erosion vector at the global scale and is responsible for the drainage organization of most continental surfaces.

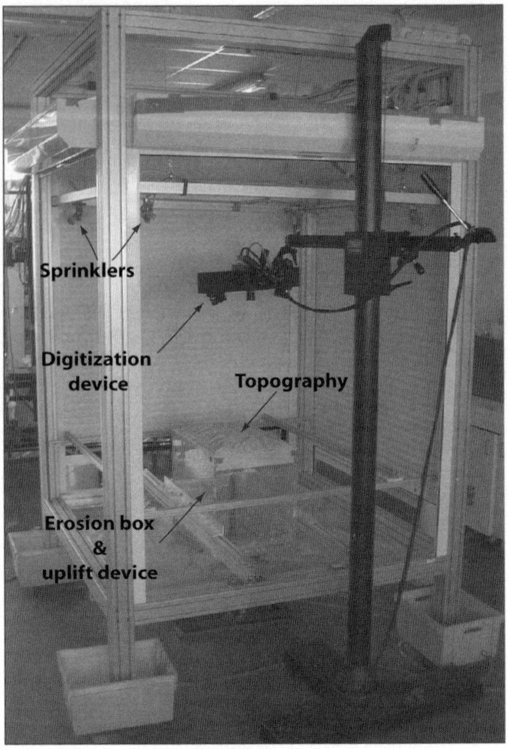

Fig. 1. Experimental device in use at Geosciences Rennes since 2003. Experiments are run within a rainfall simulator (size: 2 m × 2 m) whose four walls consist of roll shutters. Four sprinklers are positioned in the top corners. The erosion box is filled with the material to be eroded. Base movements are driven by a screw and a computer-controlled stepping motor. During an experiment, the base moves upward at a constant rate and pushes the material relative to the top of the erosion box at a rate defined as the uplift rate.

Experimental set-up

Eroded material. The main material used is a paste made by mixing a granular silica powder (median grain size of between 10 and 20 μm) with water. Some experiments have also been made with loess, as in the study by Lague *et al.* (2003). Adding water during the preparation of the material leads to pore saturation so infiltration is negligible during the subsequent erosion of the model. Silica has been chosen to avoid any chemical reaction with water. Because of the small grain size, small shear stresses are needed for particle transportation, and cohesion allows for the development of a fine network of topographic incisions with steep slopes. The resistance of the material to erosion is partially due to the angular shape of the silica grains we use. The addition of rounded grains of silica with the same size as the angular ones increases the erosion rates (Fig. 2) because of the decrease of material resistance to erosion.

Erosion box and uplift device. Tectonic uplift is one of the main parameters that controls erosion and relief dynamics at the Earth's surface at geological timescales, so the incorporation of an uplift device is essential for modelling purpose. We built an erosion box (Fig. 1) that we use for experiments where erosion acts to compensate an uplift applied to the model. It consists of a plexiglas box that is filled with the material to be eroded. It is open at its top side while its base can move upwards or downwards within the box. The movements of the base are driven by a screw and a computer-controlled stepping motor. During an experiment, the base moves upward at a constant rate. It pushes the silica paste relative to the top of the erosion box uniformly at a rate defined as the uplift rate, usually between 1 and 50 mm/hr. The four walls of the erosion box act as the constant-elevation base level for the eroding system. Erosion boxes of several different sizes have been used: 14×20 cm (Bonnet & Crave 2003; Crave *et al.* 2005), 20×30 cm (Lague *et al.* 2003) and 40×60 cm (Babault *et al.* 2005). The influence of the size of the erosion box on the experimental topography is discussed below.

Rainfall simulator. Experiments are run within a simulator that produces rain, whose kinetic energy is low enough to avoid the detachment effect of raindrop impacts (splash effect). In a first step (Crave *et al.* 2000; Bonnet & Crave 2003; Lague *et al.* 2003; Crave *et al.* 2005), we produced an artificial fog by atomization of a high-pressure water flow through as many as 44 sprinklers. With this system the raindrop size was less than 10 μm and the possible rates of rainfall varied between 50 to 350 mm/hr depending on the number of active sprinklers, their orientation and configuration, and the water pressure. However the spatial variability was often high (around 20%) because the high pressure induced a 3D circulation of the fog. The variability of rainfall rate in time was also high, between 5 and 10%. In a second step (Babault *et al.* 2005) we have built a new rainfall simulator in which rainfall is produced by using four sprinklers that deliver a mixture of water and pressurized air (Fig. 1). This yields less spatial and temporal variability of rainfall (around 5% and 3% respectively) over the same range of rainfall rates.

Before each experimental run, we measure the spatial variability of rainfall rate by collecting water in pans that are regularly spaced within the rainfall simulator at the erosion box location. During the run, it is not possible to measure the rainfall rate at the erosion box location so we collect rainfall in four pans located at the corners of the erosion box to avoid any perturbation of rainfall above the model.

Digitization of the topography

The acquisition of high-resolution digital elevation models (DEMs) of the experimental topography is of first importance in experimental geomorphology. Firstly, as explained below, during an experimental run we follow the steps of development of a relief by computing its mean elevation from the produced DEMs. To track the development of topography during an experiment, we need a topographic acquisition facility that allows for the fast acquisition and computation of surfaces that are possibly rugged. The resulting DEMs must have a high resolution (less than 1 mm) in order to characterize the geometry of the landscape.

Two main devices have been used in our experiments. The first consisted of five telemetric lasers mounted on a rack whose displacements on the horizontal axis were controlled precisely by two stepping motors. This allowed the digitization of the topography with a vertical accuracy of 40 μm and a horizontal accuracy of 500 μm. The second device (Fig. 1) is an industrial stereo-photography device, the Advanced TOpometric Sensor (ATOS) developed by the gOm Company. This device allows the quick digitization of complex topography with near vertical slopes with an accuracy of 30 μm. By using these two systems we produce high-resolution DEMs (usually 0.5–1 mm square-grid DEMs Fig. 3).

The production of the DEMs requires the momentary cessation of the experiment because of the opacity within the rainfall simulator and of the vulnerability of the measurement devices to rainfall. Consequently we regularly stop the uplift and rainfall devices, usually for 10 to 30 minutes of erosion, to digitize the model topography and generate a DEM. The interruption of erosion for an interval of less than 30 minutes does not significantly modify the surface of the topography (due to drying of the surface and formation of a solid crust) and its subsequent evolution. The digitization of a small model (area around 300 cm^2) took around 5 minutes with the first device and less than 5 minutes with the second one. However for larger models (area > 2000 cm^2) the acquisition time using the first device is too long, more than

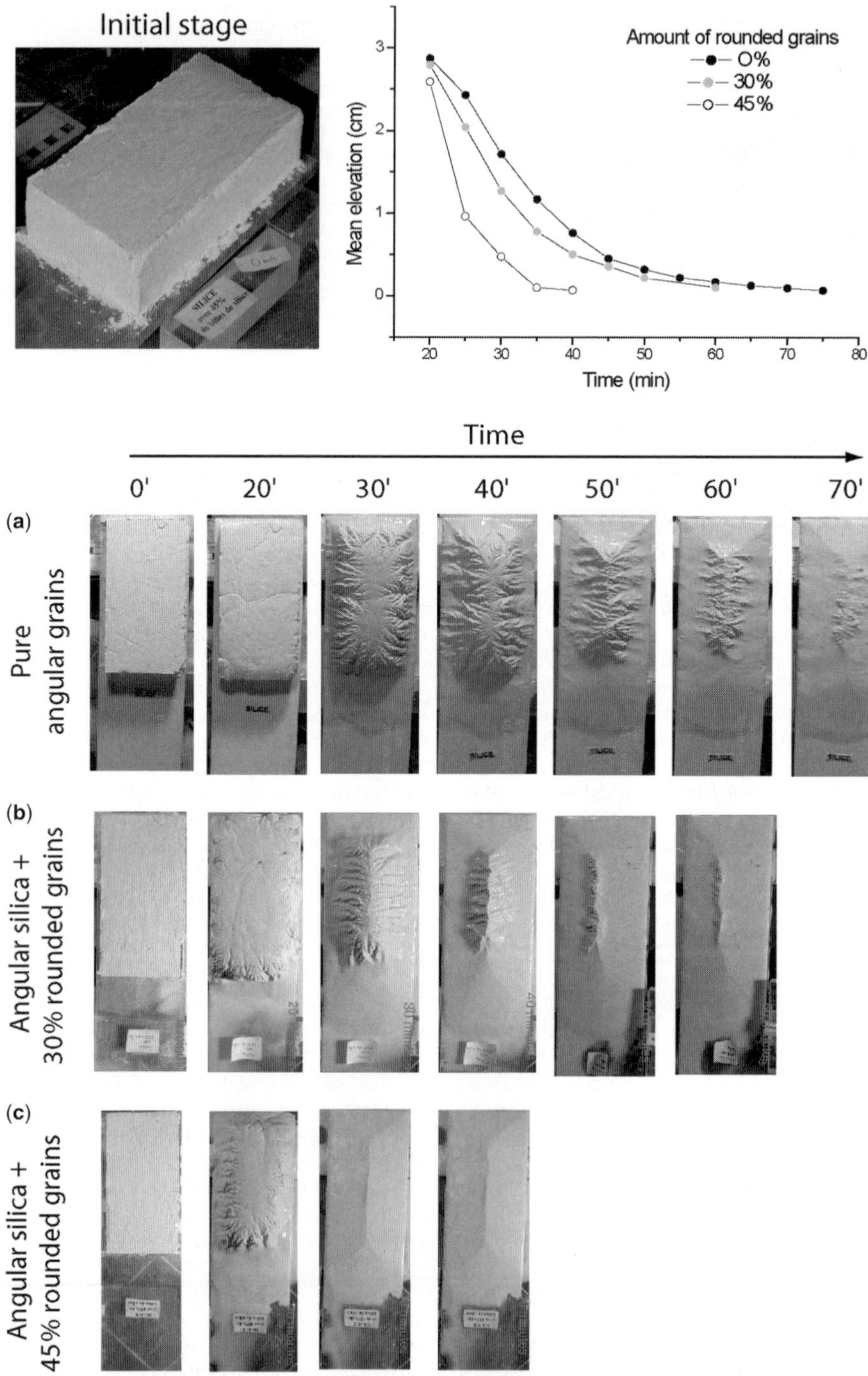

Fig. 2. Erosional decay of silica blocks (initial size 20 × 10 cm; height 3 cm) made of pure angular grains of silica or of a mixture of angular and rounded grains of silica (30 and 45% of rounded grains are shown). No uplift is applied during erosion. Rainfall rate is 60 mm/hr. The graph shows the decrease of the mean elevation of the blocks during erosion and illustrates that the addition of rounded grains accelerates the rate of elevation decay. The addition of rounded grains then decreases the resistance to erosion of the material as illustrated qualitatively by the sequence of photos. (**a**) 0% of rounded grains. (**b**) 30% of rounded grains. (**c**) 45% of rounded grains.

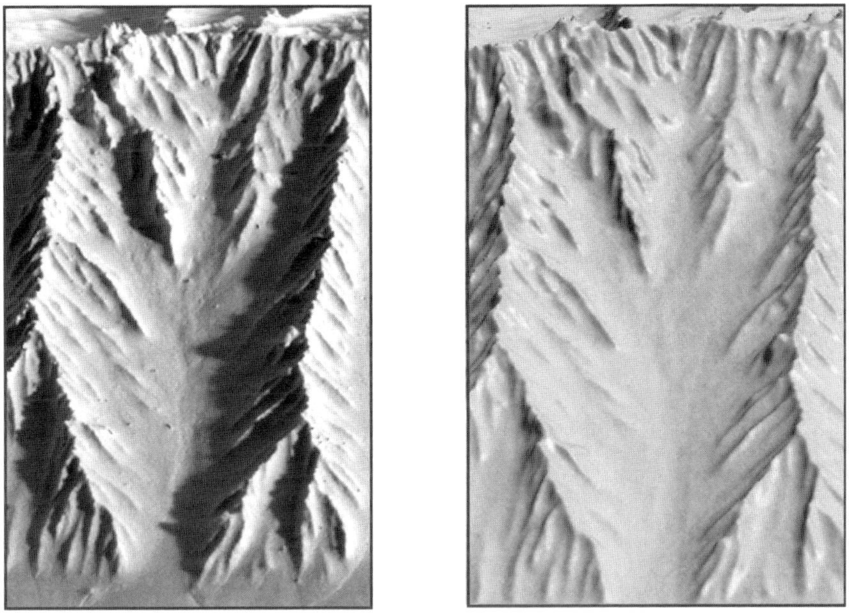

Fig. 3. Photograph (left) and 3D view of the corresponding DEM (right; 0.5 mm square grid) of an experimental drainage basin. The width of the view is around 5 cm.

1 hour, whereas it takes around 10–15 minutes using the ATOS. During an experimental run, we follow the relief dynamics by computing the mean and maximum elevations of the topography at each step of the digitization (Fig. 4). The difference between two successive DEMs is used to measure the output sediment flux and to elaborate denudation maps of the landscape.

Results

Typical evolution of an experimental landscape

Most of the experiments begin with the erosion box filled with the silica paste up to the top of the erosion box so the initial stage of a run corresponds to the uplift of a plateau (Fig. 4). An experiment begins when we start the uplift and rainfall devices. For constant uplift and rainfall rates a topography typically involves a growth phase and a steady-state phase (Fig. 4). When the initial plateau begins to be uplifted, some topographic incisions form along the four borders of the model (Fig. 4). As uplift continues, incisions propagate inward until there is complete dissection of the initial plateau. This growth phase is characterized by an increase in both the mean and maximum elevations (Fig. 4). In a second step, both the mean and maximum elevations stabilize with time (Fig. 4), while the applied uplift is still active. This indicates that erosion fully balances uplift at the scale of the model, that is, the output eroded flux equals the input uplift flux (flux steady-state: Willet & Bradon 2002). The repetition of some experiments shows that for a given rainfall and uplift rate the elevation curves are highly reproducible, with a variability of the stabilization value of the mean elevation of less than 10%.

A constant mean elevation with time is one criterion that can be used to define a steady state of the topography (e.g., Willett & Brandon 2002). At topographic steady state (Willett & Brandon 2002) the elevation of all the points of the topography does not change with time and hillcrests erode at the same rate as the valley floors. It should be noticed that a constant mean elevation is not a sufficient criterion to infer a topographic steady state. The shape of the relief could vary while its mean elevation remained constant, if, for example, uplift of summits is compensated by valley lowering. This is the reason why we also follow the maximum elevation during an experiment and we use the stability of this criterion together with the mean elevation to infer that an experimental relief approachs a topographic steady state (Fig. 4). Note that as also observed in other experimental models by Hasbargen and Paola (2000), our experimental

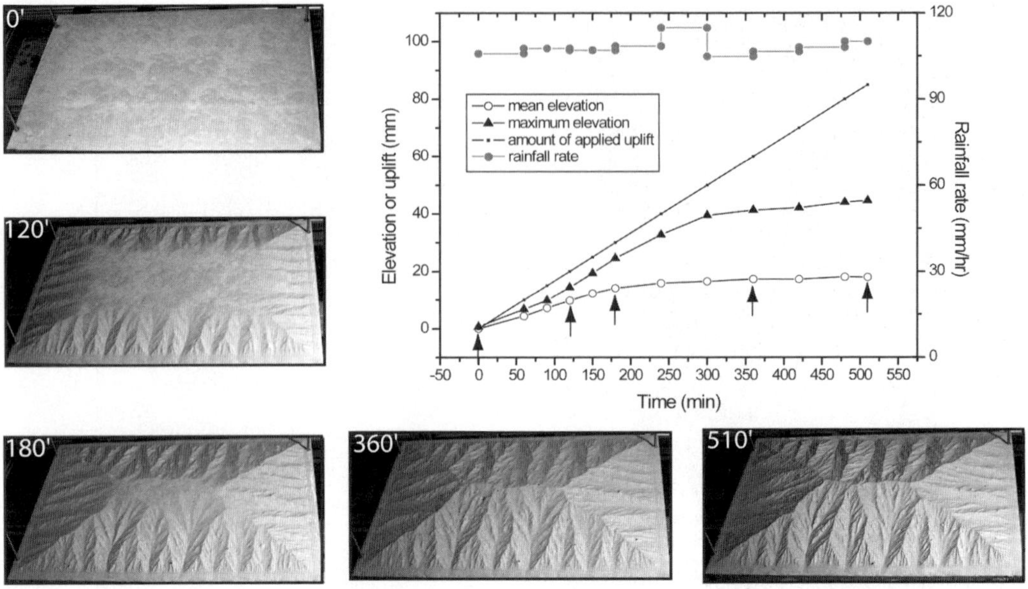

Fig. 4. Evolution of mean and maximum elevation of an experiment under constant rates of uplift (10 mm/hr) and rainfall (110 mm/hr) up to steady state, calculated from high-resolution DEMs (0.5 mm square grid) taken at different time intervals. Photos show the topography during the growth phase (photos taken at 0, 120 and 180 min) and during steady state (photos taken at 360 and 510 min), and correspond to arrows on the time-elevation plot. The size of the model is 40 × 60 cm.

landscapes never reach a pure topographic steady state. We usually observe a spatial migration of ridges and rivers at the local scale (less than 1 cm) that we assume to come from small uncontrolled spatial variations in the rainfall rate. The steady-state landscapes we characterized are then mostly at flux steady state.

Influence of the rainfall rate

We report here results from experiments made under the same uplift rate of 20 mm/hr but different rainfall rates, ranging from 100 to 300 mm/hr. Figure 5 shows the dependency between mean elevation of the steady-state topographies and the rainfall rate. Steady-state elevations show a clear inverse dependency with the rainfall rate (Fig. 5): the higher the rainfall rate, the lower the steady-state topography (Bonnet & Crave 2003). Figure 6 shows three experimental landscapes at steady state under different rainfall rates which illustrate this outcome. Note that these three landscapes experience the same uplift rate (20 mm/hr) so they all denude at the same rate of 20 mm/hr. It is commonly accepted that the erosion rate at any point in a landscape is set by the product of the local slope and the water discharge coming from upstream (Howard & Kerby 1983), this latter being approximated by the product of the rainfall rate and the drainage area. Steady state then corresponds to a combination of low rainfall plus high slope or high rainfall plus low slope. Therefore, low mean

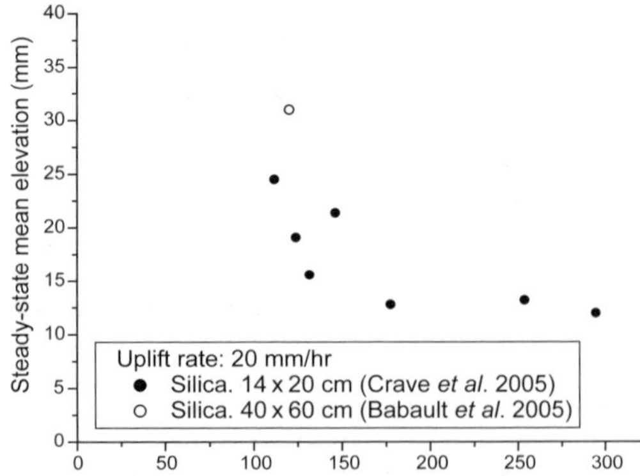

Fig. 5. Mean elevation of experimental landscapes at steady state against rainfall rate. The uplift rate is 20 mm/hr in all experiments.

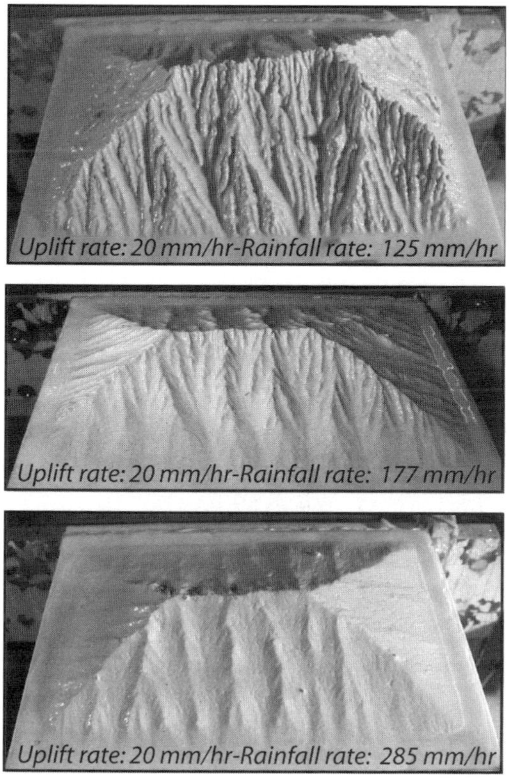

Fig. 6. Photos of steady-state topographies (model size: 14×20 cm) under a similar uplift rate of 20 mm/hr but different rainfall rates of 125 (upper), 177 (middle) and 285 (lower) mm/hr. Note that because erosion and uplift rates are balanced, denudation rates are the same in all three experiments.

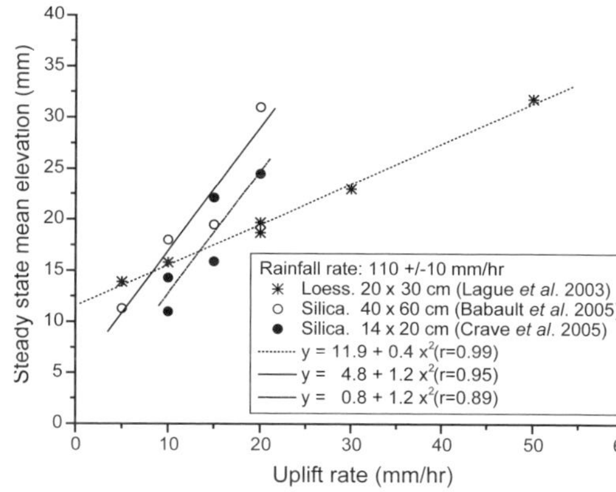

Fig. 7. Mean elevation of experimental landscapes at steady state against uplift rate. The rainfall rate is around 110 mm/hr in all experiments. Experiments made with loess are shown by stars, those with silica are shown by circles. Note that the system varies between 14×20 cm (solid circles) and 40×60 cm (open circles) for the latter set of experiments.

elevation and slopes do not imply a low uplift rate. Montgomery *et al.* (2001) argued that climate exerts a strong control on the large-scale morphology of the Andes. They suggested that the low elevations of the northern Andes could result from high precipitation rather than lower rates of crustal shortening and rock uplift. This phenomenon is plausible in view of our results (Bonnet & Crave 2003).

Given the present set of data we cannot properly fit the trend of mean elevations against rainfall rate with a well-defined function (see Crave *et al.* 2005 for a full discussion). From theoretical considerations (Whipple & Tucker 1999; Lague *et al.* 2003) this function should be a power law whose exponent would depend on the exponent on the drainage area of the transport law, but its determination is presently limited by the range of possible rainfall rates within the rainfall simulator.

Influence of the uplift rate

We report results from experiments made under the same rainfall rate of 110 ± 10 mm/hr but different uplift rates. Figure 7 shows the dependency between the mean elevation at steady state and the uplift rate for three sets of experiments: models made of loess with a size of 20×30 cm (Lague *et al.* 2003), and models made of silica with a size of 14×20 cm (Crave *et al.* 2005) and 40×60 cm (Babault *et al.* 2005). Each individual data set shows a positive linear correlation between the mean elevation at steady state and the uplift rate (Fig. 7). It should be noticed that the slope of the linear fits clearly depends on the material used in the experiments (Fig. 7). As all the other parameters are similar, this could be directly linked to their difference in erodibility. Also, each data set shows a non-zero intercept of the linear fit that implies a positive elevation of the landscape in the absence of uplift. This value corresponds to the non-zero mean elevation reached by a relaxing landscape (i.e., one that evolves without uplift) after a long time period of erosion, and is interpreted as the signature of an erosion threshold (Lague *et al.* 2003).

Influence of the model size

Experiments with silica have been made with two sizes of the erosion box that span almost a tenfold difference in the eroded area: 280 cm^2

(14 × 20 cm: Bonnet & Crave 2003; Crave *et al.* 2005) against 2400 cm^2 (40 × 60 cm: Babault *et al.* 2005). As erosion depends on the aggregation of the water flux coming from upstream, our aim in increasing the experiment size was to explore the possibility of more complex erosional dynamics in larger experiments compared to the smaller ones. Surprisingly, their steady state mean elevation versus uplift rate trends are very similar in terms of the slope of the linear fits, the only difference being a small offset between the two trends (Fig. 7). Because of the small number of experiments and the small range of uplift rates that has been investigated, one can ask if these results are significant or not. Analysis of the relationships between local slopes and their corresponding drainage area (Fig. 8) allows us to go deeper into this problem.

It is well known that slope and drainage area correlate (Hack 1957) and these correlations have been used to characterize the local shapes

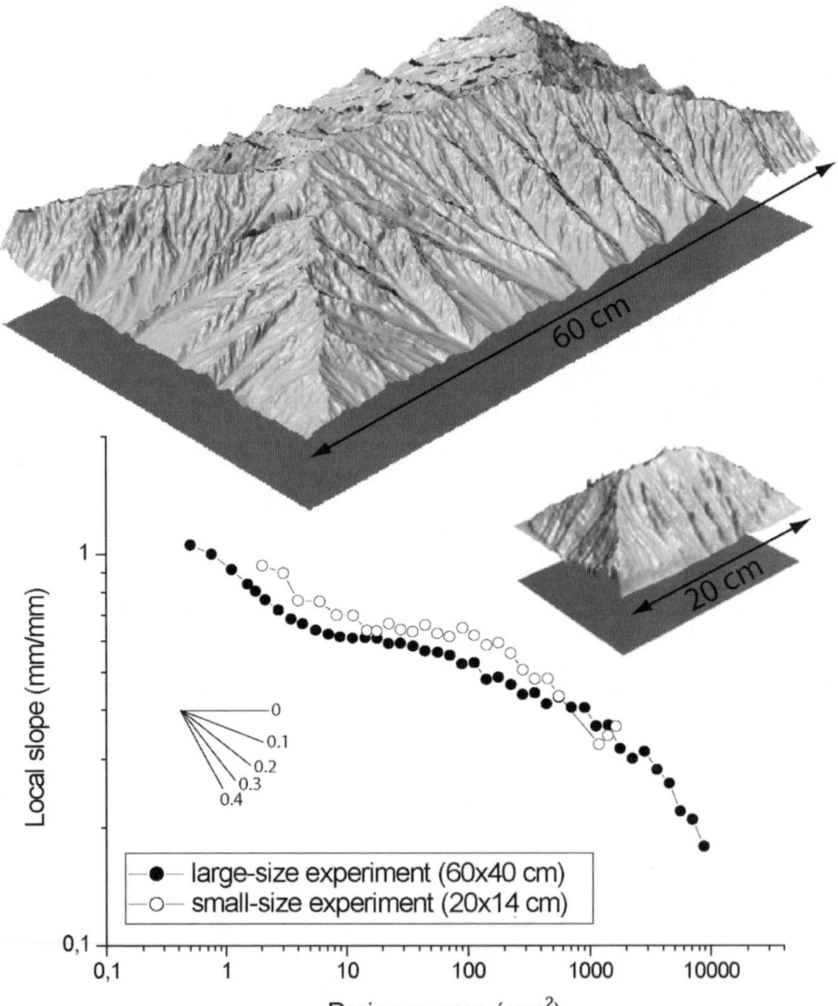

Fig. 8. Plot of local slope against drainage area for two experiments made with silica and at steady state under the same uplift (20 mm/hr) and rainfall (110 mm/hr) rates. The two experiments differ by their size (14 × 20 cm and 40 × 60 cm). The data are the logarithm bin averages of raw data for the entire surfaces. Note that the offset between the two curves is not relevant with regard to standard deviation (not shown for visibility). Also shown are 3D views of the corresponding DEMs (0.5 mm square grid) represented at the same scale.

of the landscapes with regard to erosional processes (e.g., Montgomery & Foufoula-Georgiou 1993). Note that in this kind of analysis, drainage area is used as a surrogate for water discharge. A log-log plot of the local slope against the drainage area allows the definition of linear trends whose slope parameter is a measure of the concavity of the topography (concavity exponent: Snyder *et al.* 2000), which could depend on the erosional processes. For example a river typically shows a negative trend that indicates a decrease of its local slope with increasing drainage area (i.e., a concave-up shape) with a concavity exponent that is typically between 0.3 and 0.7 (Whipple & Tucker 1999). The determination of the concavity exponent of a landscape over a given range of drainage areas could help to define the prevailing erosional process over this range, by the sole analysis of the topography. Note that even if widely used, intepretation of slope-area data must be done with great caution as it has been recently demonstrated that slope-area dependence can be entirely unrelated to erosional processes (Schorghofer & Rotham 2002).

Figure 8 shows the slope/area relationships of two topographies that developed under the same conditions of uplift and rainfall rates but whose size is different. The two curves have a very similar shape and show three main domains of slope reduction for increasing drainage area: a first domain with a concavity exponent around 0.2–0.3, a second domain with an exponent less than 0.1, and a third domain with an exponent of 0.3–0.4 (see Crave *et al.* 2005 for a full discussion of the significance of these domains). The visual inspection of the experimental landscapes confirms the existence of three topographic domains characterized by different shapes that we believe to result from different action of the running water, possibly because of different hydraulic regimes. The identification of these three domains over roughly the same ranges of drainage area on both landscape would indicate that despite the difference in system size, the two models evolved under the same set of erosional processes, as defined using slope/area relationships. It would imply that these two kind of models probably behave the same way regarding external parameters and that the similarity between the two trends defined on Figure 7 would probably not be an artefact. It appears that a tenfold increase in the size of the experiments is not sufficient to change the erosional dynamics of the experimental landscapes.

Discussion and conclusions

Scaling and analogy

The systems we study are deliberately oversimplified and do not take into account many processes that may play a role in natural landscapes (for example vegetation, weathering, etc.). Our objective is to understand first how individual external parameters influence a simple physical system. Nevertheless, a major unsolved problem lies in the scaling of the experiments. Unlike some laboratory experiments in tectonics (e.g., Davy & Cobbold 1991), experiments in geomorphology are not analogous to natural systems. The main reason is that it is quite impossible to downscale natural conditions to the laboratory without any scale distortions, so the ratios of significant dimensions and forces are not equal to those in nature. Processes involved in the formation of natural landscapes span a very wide range of timescales, from hydraulical (hours) to geological (millions of years), and the reduction of the geomorphic systems to the laboratory does not preserve their ratio. Similarly the ratio between the size of the experimental drainage basins and the grain size is much smaller than in natural conditions and its preservation would need the use of materials with a granulometry at least ten to a hundred fold smaller than the one we use. These two examples illustrate some of the reasons that make modelling in geomorphology experimental instead of analogue (Schumm *et al.* 1987).

Despite these scaling problems one of the most striking feature of experimental reliefs is that they often look very much like natural landscapes, and they obey some fundamental geometric laws deduced from natural systems, such as Hack's law or slope/area relationships (Hack 1957). As explained above, the study of the local shape of a topography through analysis of its slope/area relationship may give information about the processes responsible for the sculpting of its relief (Howard & Kerby 1983; Montgomery & Foufoula-Georgiou 1993). The detailed analysis of slope/area relationships on experimental landscapes (Lague *et al.* 2003; Crave *et al.* 2005) reveals that their concavity exponent is never higher than 0.4 whatever the material used, the size of the model or the topographic domain considered (Fig. 8). This has important implications because in nature, rivers are often characterized by a high concavity, with values of the slope/area exponent near 0.3–0.7 (Whipple & Tucker 1999). Thus the erosional processes within our experiments never form topographic incisions with concave shapes as observed for natural rivers. Consequently our experimental reliefs cannot be viewed as systems whose dynamics is equivalent to fluvial-driven landscapes. In contrast, large areas of the experimental landscapes are characterized by a low value of the slope/area exponent, near 0.1 (central domain on Fig. 8), comparable to those seen in debris-flow dominated colluvial channels (Montgomery & Foufoula-Georgiou 1993). Consequently, rather than simulating all the landscape components of a mountain range, our experiments can be better viewed as an equivalent of the landmass that is between the main streams of the landscape, including debris flow-dominated channels. In this scheme, the four borders of the erosion box can be viewed as streams at equilibrium in which all the material eroded from the landscape is transferred outside the system.

Relief dynamics: insights from experimental models and research needs

The development of numerical models of erosion and relief dynamics has stimulated a drive for tools and criteria to test their validity. Some geometric criteria obeyed by natural landscapes, like Horton's laws (Horton 1945), have been used but it has been demonstrated that these relations are independent of the origin or structure of landscapes (Kirchner 1993). Their use for testing numerical models is thus limited and experimental models of relief dynamics have been built first to provide some physical tests for numerical models (Hasbargen & Paola 2000; Hancock & Willgoose 2001; Lague *et al.* 2003). The SIBERIA (Willgoose & Bras 1991) and EROS (Crave & Davy 2001) codes have for example successfully reproduced the form and the macroscale dynamics of some experimental landscapes given an appropriate erosion law (Hancock & Willgoose 2001; Lague *et al.* 2003).

As pointed out by Hooke (quoted in Schumm *et al.* 1987, p. 5) it can be logically assumed that a process which produces a characteristic behaviour in experiments has the same effect on natural systems. If correct, this implies that despite scaling problems some macroscale experimental dynamics are probably partly relevant to natural systems evolution, so modelling can be used to emphasize general concepts of landscape behaviour. Our experiments demonstrated, for example, a positive correlation between the mean elevation of steady-state landscapes and the uplift rate (Fig. 7), in agreement with theoretical formulations (e.g., Whipple & Tucker 1999) and inferences from natural systems (Snyder *et al.* 2000; Lague & Davy 2003), as long as landsliding is not the main process of hillslope erosion in response to river incision (Schmidt & Montgomery 1995; Burbank *et al.* 1996). Experimental landscape response to uplift is therefore partly analogous to natural systems and the main problem lies in the representativeness of the linear relationship observed in experiments.

From a detailed analysis of experimental steady-state landscapes, Lague *et al.* (2003) demonstrated that this linear relationship results from a linear dependency between the erosion rate and the local slope of the experimental landscapes. No consensus exists about the nature of this relationship in natural systems. Some studies support, for example, a linear relationship for hillslope erosion (Small *et al.* 1999; Lague & Davy 2003), whereas others argue for a non-linear one (Roering *et al.* 1999), and the same discrepancy exists about erosion at the drainage or continental scale (Ahnert 1970; Pinet & Souriau 1988; Montgomery & Brandon 2002). The lack of knowledge of natural systems makes the representativeness of the experimental landscape dynamics with regard to uplift difficult to fully assess. Note that the question of the nature of the relationship between erosion rate and local slope is crucial with respect to the tectonic/erosion problem because analytical developments predict that the degree of non-linearity of this relationship controls the response time of the topography with respect to the uplift rate (Whipple & Tucker 1999). One important outcome is the prediction that the timescale to steady state would be independent of the uplift rate in the case of linearity between erosion rate and slope (Whipple & Tucker 1999), which is effectively observed in our experiments (Lague *et al.* 2003; Babault *et al.* 2005).

Experimental models also give insights into general concepts of landscape behaviour relative to climate through rainfall. Gabet *et al.* (2004) observed a decrease in the mean hillslope angles with increasing rainfall rate in the Himalayas, in agreement with the decrease in the mean elevation with increasing rainfall rate we observed in the experiments (Fig. 5) and with theoretical formulations (Willgoose *et al.* 1991; Tucker & Bras 1998). As in our experiments, Gabet *et al.* (2004) cannot propose a fit to their data, but of interest for our purpose here is the qualitative similarity of behaviour between this natural system and our experiments. As a consequence of the inverse correlation between mean steady-state elevation and the rainfall rate, Bonnet and Crave (2003) demonstrated experimentally a shift from high to low rainfall rate induces surface uplift of the landscape, even with constant uplift rate. This phenomenon is defined as a climatically induced surface uplift (Bonnet & Crave 2003), as opposed to a tectonically induced surface uplift, and corresponds to a specific dynamic of denudation, so that it is theoretically possible to differentiate between the climatic or tectonic causes of surface uplift from records of output sediment fluxes (Bonnet & Crave 2003). If we assume that the experimental landscape response to rainfall variations is qualitatively relevant to natural systems, this raises the question of the sensibility of landforms (in terms of magnitude and timescale of response) to climatic perturbations (Whipple 2001), with important issues regarding the origin of continental surface uplifts.

One of the primary values of experimental modelling in geomorphology is to generate hypotheses which can be tested by reference to natural lanscapes (Schumm *et al.* 1987; Bonnet & Crave 2003; Babault *et al.* 2005). Even under constant external forcing, experimental models are also more dynamic than numerical models would suggest (Hasbargen & Paola 2000), so they can potentially improve our understanding of landscape sculpting by erosional processes. Recent field studies have demonstrated a discrepancy between measured erosion rates and some theoretical parameters of erosion efficiency such as stream power (Dadson *et al.* 2003; Reiners *et al.* 2003), which implies an inadequancy of the current theory to describe and predict erosion (Molnar 2003). Experimental modelling could also be a valuable tool for improving the theory of erosion, given the possibility of directly evaluating the erosion law of the models.

We thank A. Densmore for a constructive review that greatly helped improve this paper. We also thank P. Davy, D. Lague, J. Van den Driessche and J. Babault for helpful discussions and J.-J. Kermarrec, whose technical assistance made this work possible. This study was funded by CNRS-INSU.

References

AHNERT, J. 1980. Contemporary uplift and erosion of the southern Alps, New Zealand: summary. *Geological Society of America Bulletin*, **91**, 2–4.

ASHMORE, P. E. 1982. Laboratory modelling of gravel braided stream morphology. *Earth Surface Processes and Landforms*, **7**, 201–225.

AVOUAC, J.-P. & BUROV, E. B. 1996. Erosion as a driving mechanism of intracontinental mountain growth. *Journal of Geophysical Research*, **101**, 17747–17769.

BABAULT, J., BONNET, S., CRAVE, A. & VAN DEN DRIESSCHE, J. 2005. Influence of Piedmont sedimentation on erosion dynamics of an uplifting landscape: an experimental approach. *Geology*, **33**, 301–304.

BEAUMONT, C., FULSACK, J. & HAMILTON, J. 1991. Erosional control of active compressional orogens. *In*: MCCLAY, K. R. (ed.) *Thrust Tectonics*. Chapman and Hall, New York, 1–18.

BEAUMONT, C., JAMIESON, R. A., NGUYEN, M. H. & LEE, B. 2001. Himalayan tectonics explained by extrusion of low-viscosity crustal channel coupled to focused surface denudation. *Nature*, **414**, 738–742.

BONNET, S. & CRAVE, A. 2003. Landscape response to climate change: insights from experimental modelling and implications for tectonic versus climatic uplift of topography. *Geology*, **31**, 123–126.

BURBANK, D. W. *ET AL.* 1996. Bedrock incision, rock uplift, and threshold hillslopes in the northwestern Himalaya. *Nature*, **379**, 505–510.

CRAVE, A. & DAVY, P. 2001. A stochastic 'precipiton' model for simulating erosion/sedimentation dynamics. *Computers & Geosciences*, **28**, 815–827.

CRAVE, A., BONNET S. & LAGUE, D. 2005. Influence of rainfall rate on laboratory-scale reliefs. *Journal of Geophysical Research*, under review.

CRAVE, A., LAGUE, D., DAVY, P., KERMARREC, J.-J., SOKOUTIS, D., BODET, L. & COMPAGNON, R. 2000. Analogue modelling of relief dynamics. *Physics and Chemistry of the Earth*, **25**, 549–553.

DADSON, S. J. *ET AL.* 2003. Links between erosion, runoff variability and seismicity in the Taiwan orogen. *Nature*, **426**, 648–651.

DAVIS, W. M. 1899. The geographical cycle. *Geographical Journal*, **14**, 481–504.

DAVY, P. & COBBOLD, P. R. 1991. Experiments on shortening of a 4-layer model of continental lithosphere. *Tectonophysics*, **188**, 1–25.

DENSMORE, A. L., ANDERSON, R. S., MCADOO, B. G. & ELLIS, M. A. 1997. Hillslope evolution by bedrock landslides. *Science*, **275**, 369–372.

FINLAYSON, D. P., MONTGOMERY, D. R. & HALLET, B. 2002. Spatial coincidence of rapid inferred erosion with young metamorphic massifs in the Himalayas. *Geology*, **30**, 219–222.

GABET, E. J. 2003. Sediment transport by dry ravel. *Journal of Geophysical Research*, **108**, 2049, DOI 10.1029/2001JB001686.

GABET, E. J., PRATT-SITAULA, B. A. & BURBANK, D. W. 2004. Climatic controls on hillslope angle and relief in the Himalayas. *Geology*, **32**, 629–632.

GARDNER, T. W. 1983. Experimental study of knickpoint and longitudinal profile evolution in cohesive, homogeneous material. *Geological Society of America Bulletin*, **94**, 664–672.

HACK, J. T. 1957. Studies of longitudinal stream profiles in Virginia and Maryland. *United States Geological Survey Professional Paper*, **294B**, 45–94.

HACK, J. T. 1960. Interpretation of of erosional topography in humid temperate region. *American Journal of Science*, **258**, 80–97.

HANCOCK, G. & WILLGOOSE, G. 2001. Use of a landscape simulator in the validation of the Siberia catchment evolution model: declining equilibrium landforms. *Water Resource Research*, **37**, 1981–1992.

HASBARGEN, E. L. & PAOLA, C. 2000. Landscape instability in an experimental drainage basin. *Geology*, **28**, 1067–1070.

HOLLAND, W. N. & PICKUP, G. 1976. Flume study of knickpoint development in stratified sediment. *Geological Society of America Bulletin*, **87**, 76–82.

HORTON, B. K. 1999. Erosional control on the geometry and kinematics of thrust belt development in the central Andes. *Tectonics*, **18**, 1292–1304.

HORTON, R. E. 1945. Erosional development of streams and their drainage basins: hydrophysical approach to quantitative morphology. *Geological Society of America Bulletin*, **56**, 275–370.

HOWARD, A. D. 1994. A detachment-limited model of drainage basin evolution. *Water Resource Research*, **30**, 2261–2285.

HOWARD, A. D. & KERBY, G. 1983. Channel changes in badlands. *Geological Society of America Bulletin*, **94**, 739–752.

KIRCHNER, J. W. 1993. Statistical invevitability of Horton's laws and the apparent randomness of stream channel networks. *Geology*, **21**, 591–594.

LAGUE, D. & DAVY, P. 2003. Constraints on the long-term colluvial erosion law from the analyzis of slope-area relationships at various tectonic uplift rates in the Siwaliks Hills (Nepal). *Journal of Geophysical Research*, **108**, 2129, DOI 10.1029/2002JB001893.

LAGUE, D., CRAVE, A. & DAVY, P. 2003. Laboratory experiments simulating the geomorphic response to tectonic uplift. *Journal of Geophysical Research*, **108**, 2008, DOI 0.1029/2002JB001785.

MOLNAR, P. 2003. Nature, nurture and landscape. *Nature*, **426**, 613–614.

MEUNIER, P. & MÉTIVIER, F. 2000. Permanence des flux de masse d'une rivière en tresses expérimentales. *Compte Rendu de l'Académie des Sciences*, **331**, 105–110.

MONTGOMERY, D. R. & BRANDON, M. T. 2002. Topographic controls on erosion rates in tectonically active mountain ranges. *Earth and Planetary Science Letters*, **201**, 481–489.

MONTGOMERY, D. R. & FOUFOULA-GEORGIOU, E. 1993. Channel network source representation using digital elevation models. *Water Resource Research*, **29**, 3925–3934.

MONTGOMERY, D. R., BALCO, G. & WILLET, S. D. 2001. Climate, tectonics, and the morphology of the Andes. *Geology*, **29**, 579–582.

PELLETIER, J. D. 2003. Drainage basin evolution in the rainfall erosion facility: dependence on initial conditions. *Geomorphology*, **53**, 183–196.

PAOLA, C., PARKER, G., SEAL, R., SINHA, S. K., SOUTHARD, J. B. & WILCOCK, P. R. 1992. Downstream fining by selective deposition in a laboratory flume. *Science*, **258**, 1757–1760.

PENCK, W. 1953. *Morphological Analysis of Land Forms*. Macmillan, London.

PINET, P. & SOURIAU, M. 1988. Continental erosion and large-scale relief. *Tectonics*, **7**, 563–582.

REINERS, P. W., EHLERS, T. A., MITCHELL, S. G. & MONTGOMERY, D. R. 2003. Coupled spatial variations in precipitation and long-term erosion rates across the Washington Cascades. *Nature*, **426**, 645–647.

ROERING, J. J., KIRCHNER, J. W. & DIETRICH, W. E. 1999. Evidence for nonlinear, diffusive sediment transport on hillslopes and implications for landscape morphology. *Water Resource Research*, **35**, 853–870.

ROERING, J. J., KIRCHNER, J. W., SKLAR, L. S. & DIETRICH, W. E. 2001. Hillslope evolution by non linear creep and landsliding: an experimental study. *Geology*, **29**, 143–146.

SCHMIDT, K. M. & MONTGOMERY, D. R. 1995. Limits to relief. *Science*, **270**, 617–620.

SCHORGHOFER, N. & ROTHMAN, D. H. 2002. Acausal relations between topographic slope and drainage area. *Geophysical Research Letters*, **29**, 1–5.

SCHUMM, S. A., MOSLEY, M. P. & WEAVER, W. E. 1987. *Experimental Fluvial Geomorphology*. John Wiley, New York.

SHEPHERD, R. G. & SCHUMM, S. A. 1974. Experimental study of river incision. *Geological Society of America Bulletin*, **85**, 257–268.

SIMPSON, G. 2004. Dynamic interactions between erosion, deposition, and three-dimensional deformation in compressional fold belt setting. *Journal of Geophysical Research*, **109**, F03007, DOI 10.1029/2003JF000111.

SKLAR, L. S. & DIETRICH, W. E. 2001. Sediment and rock strength controls on river incision into bedrock. *Geology*, **29**, 1087–1090.

SMALL, E. E., ANDERSON, R. S., HANCOCK, G. S. & FINKEL, R. S. 1999. Estimates of regolith production from ^{10}Be and ^{26}Al: evidence for steady state alpine hillslopes. *Geomorphology*, **27**, 131–150.

SNYDER, N. P., WHIPPLE, K. X., TUCKER, G. E. & MERRITTS, D. J. 2000. Landscape response to tectonic forcing: DEM analysis of stream profiles in the Mendocino triple junction region, northern California. *Geological Society of America Bulletin*, **112**, 1250–1263.

THIEDE, R. C., BOOKHAGEN, B., ARROWSMITH, J. R., SOBEL, E. R. & STRECKER, M. R. 2004. Climatic control on rapid exhumation along the Southern Himalayan front. *Earth and Planetary Science Letters*, **222**, 791–806.

TUCKER, G. E. & BRAS, R. L. 1998. Hillslope processes, drainage density, and landscape morphology. *Water Resources Research*, **34**, 2751–2764.

VANNAY, J.-C., GRASEMANN, B., RAHN, M., FRANK, W., CARTER, A., BAUDRAZ, V. & COSCA, M. 2004. Miocene to Holocene exhumation crustal wedges in the NW Himalaya: evidence for tectonic extrusion coupled to fluvial erosion. *Tectonics*, **23**, TC1014, DOI 10.1029/2002TC001429.

WHIPPLE, K. X. 2001. Fluvial landscape response time: How plausible is steady-state denudation? *American Journal of Science*, **301**, 313–325.

WHIPPLE, K. X. & TUCKER, G. E. 1999. Dynamics of the stream-power incision model: implications for the height limits of mountain ranges, landscape response timescales, and research needs. *Journal of Geophysical Research*, **104**, 17761–17674.

WILLET, S. D. 1999. Orogeny and orography: The effects of erosion on the structure of mountain belts. *Journal of Geophysical Research*, **104**, 28957–28981.

WILLET, S. D. & BRANDON, M. T. 2002. On steady states in mountain belts. *Geology*, **30**, 175–178.

WILLGOOSE, G. R. & BRAS, R. L. 1991. Results from a new model of river basin evolution. *Earth Surface Processes & Landforms*, **16**, 237–254.

WILLGOOSE, G. R., BRAS, R. L. & RODRIGUEZ-ITURBE I. 1991. A physically based coupled network growth and hillslope evolution model, 1: theory. *Water Resource Research*, **27**, 1671–1684.

WOBUS, C. W., HODGES, K. V. & WHIPPLE, K. X. 2003. Has focused denudation sustained active thrusting at the Himalayan topographic front? *Geology*, **31**, 861–864.

ZEITLER, P. K. & MELTZER, A. M. *ET AL.* 2001. Erosion, Himalayan geodynamics and the geomorphology of metamorphism. *GSA Today*, **11**, 4–9.

Numerical modelling of erosion processes in the Himalayas of Nepal: effects of spatial variations of rock strength and precipitation

V. GODARD[1], J. LAVÉ[2] & R. CATTIN[1]

[1]*Laboratoire de Géologie, Ecole Normale Supérieure, 24, rue Lhomond, 75005 Paris, France (e-mail: godard@geologie.ens.fr)*

[2]*Laboratoire de Géodynamique des Chaînes Alpines, CNRS, 1381, rue de la Piscine, 38400 St-Martin d'Hères, France*

Abstract: The interplay between tectonics and erosion has a predominant control on the evolution of the morphology of mountain belts. Here we investigate the modalities of deformation in Central Nepal on a *c.* 100 ka time scale in response to tectonic and external forcings, through the use of a finite-element thermomechanical model coupled with an integrative denudation formulation that accounts for fluvial incision and hillslope landsliding. We study the complex coupling existing between tectonics and erosion, with special emphasis on the influences of rock strength and rainfall distributions. Our results underline the key role played by lithologic variations in the elevation of both rivers and mean topography. We show that the location of the Main Frontal Thrust is mainly controlled by the low erodability of the unconsolidated sandstone in the Siwaliks Hills. As previously suspected (Burbank *et al.* 2003), our simulations demonstrate that the pattern of uplift in Nepal is mainly dependent on both erodability and fault geometry, rather than on rainfall distribution.

It has long been understood that the landscape of mountain belts is the result of the balance between tectonic uplift and surface processes, as modulated by denudation and sedimentation (Molnar 2003). Outside glaciated regions, river incision into bedrock has been recognized as a primary agent in both landscape evolution and large-scale surface mass transport (e.g., Burbank *et al.* 1996; Willett 1999; Whipple & Tucker 1999; Lavé & Avouac 2001). Erosion influences tectonics by modifying the mass distribution in the orogen, which controls the isostatic response and the strain localization. On the other hand, tectonic uplift or subsidence is related to the distribution of erosion and sedimentation zones.

Over the last fifteen years many numerical models have been developed to study feedbacks and interactions between tectonics and denudation in mountainous areas. However, in most of these models, transport of mass at the surface is either reduced to a diffusion law (Avouac & Burov 1996; Cattin & Avouac 2000), to a linear function of the local slope (Beaumont *et al.* 2001), or to a simple fluvial incision law where the mean topographic profile is represented by a river profile (Willett 1999). Those approaches do not take into account the dual nature of denudation in mountain ranges, which combines river bedrock incision and hillslope landsliding.

Here we explore how patterns of precipitation and rock strength affect both river and topographic profiles across the range of the Himalayas and southern Tibet. Following the approach of Godard *et al.* (2004) we have coupled a 2D thermomechanical model, which includes localization of deformation through frictional faults, to a surface process model based on a detachment-limited fluvial incision law. In contrast with Willett (1999), the mean topographic profile is computed using both the calculated trans-Himalayan river profile and an implicit description of the tributaries and hillslopes (Lavé 2005). Our purpose here is to evaluate the importance of rainfall and rock strength distributions in controlling the modalities of landscape denudation. We investigate the response of the mechanical model to forcings corresponding to variations of those two parameters. We first review the geodynamic and hydrological setting of the study area that encompasses the Himalayas of Nepal from the Gangetic Plain to the southern edge of the Tibetan Plateau. Next, we describe the modelling approach and the surface processes formulation introduced as a boundary condition at the surface of the model. Then, we present the behaviour of a reference

From: BUITER, S. J. H. & SCHREURS, G. (eds) 2006. *Analogue and Numerical Modelling of Crustal-Scale Processes.* Geological Society, London, Special Publications, **253**, 341–358.
0305-8719/06/$15.00

model which closely mimics the main features of the Himalayan–Tibetan system. Starting from this reference model, the sensitivity of the results to rock strength and rainfall distribution is investigated.

Geodynamical setting

Himalayan tectonics

The Himalaya results from the collision between India and the southern edge of Eurasia. Since *c.* 20 Ma, tectonic and erosive processes have built an impressive landscape characterized by a steep topographic front from the 5000 m elevation Tibetan Plateau down to the Gangetic plain. Four major morphotectonic domains are usually recognized from north to south: the low-relief South Tibetan plateau, the High Himalaya (HH) with deep gorges and *c.* 8000 m summits, the mountainous to hilly relief of the Lesser Himalaya (LH), and the frontal low elevation relief of the Siwalik Hills (Fig. 1). The Himalayan range results from the successive activation of major thrust zones: the main central thrust (MCT) and the main boundary thrust (MBT) (Le Fort 1986). These two faults are presumed to branch upward from a major mid-crustal décollement, the main Himalayan thrust (MHT) (Schelling & Arita 1991; Zhao *et al.* 1993; Pandey *et al.* 1995; Lavé & Avouac 2001). Coeval with thrusting, the Indo-Gangetic foredeep formed in front of the rising Himalaya and accumulated several kilometres of Cenozoic molasse deposits, material eroded from the areas of high relief. Thin-skinned thrust-faulting has incorporated these sediments into the hanging walls of several thrusts, now expressed as the Siwaliks Hills. These frontal faults branch from the MHT, which roots at a depth of 30–40 km beneath the South Tibet (Zhao *et al.* 1993), and displays a ramp-and-flat geometry beneath the HH and LH domains (Schelling & Arita 1991; Pandey *et al.* 1995; Lavé & Avouac 2001). The long-term shortening rate across the range is *c.* 20 mm yr^{-1} (Lyon Caen & Molnar 1985; Armijo *et al.* 1986) and is similar to present convergence imaged by GPS (Bilham *et al.* 1997). In central Nepal, geomorphic evidence

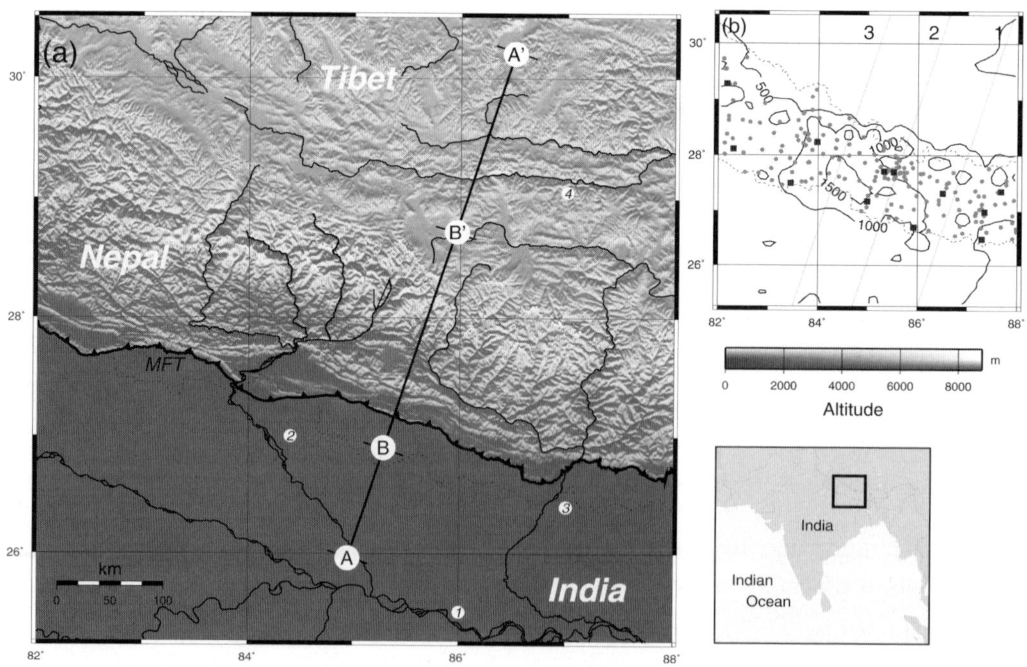

Fig. 1. Situation. (**a**) Topographic map of the study region (GTOPO30 DEM), showing the principal hydrographic features, the position of the main frontal thrust, and the cross sections AA′ (Fig. 3) used in the modelling and BB′ presented on Fig. 8. 1 – Ganga, 2 – Narayani, 3 – Sapt Kosi, 4 – Tsangpo. (**b**) Rainfall map derived from NOAA (2005) estimates. The position of the 3 cross-sections of Fig. 2 is indicated. Light grey dots and dark grey squares represent the rain gauges of GAME (2005) and NMFD (2005) networks, respectively.

suggests in addition that most of the present shortening across the Himalayas is transferred toward the main frontal thrust (MFT) (Lavé & Avouac 2000).

Precipitation, hydrology and erosive processes

Precipitation in Nepal is mainly controlled by the barrier of the Himalayas, with a brutal condensation against the HH of the wet air coming from the Indian Ocean during the monsoon. Whereas a marked rain shadow develops on the northern flank of the HH, the prominent fluvial network of the southern flank, fed by intense rainfall, is deeply entrenched in the topography and actively contributes to the denudation of the orogen. Important lateral variations of the distribution of rainfall exist in Nepal (Figs 1 and 2), but, despite the amplitude of change, the consistent main patterns are the increasing precipitation from the Gangetic plain to the HH, the orographic barrier of the HH, and the low precipitation on the Tibetan Plateau. Several major north-south rivers drain the Himalayas of Nepal from southern Tibet down to the Indo-Gangetic plain. In Central and East Nepal, across the HH, those trans-Himalayan rivers are on average *c.* 50 km apart before joining, in the southern part of the LH, two major rivers systems, the Narayani and

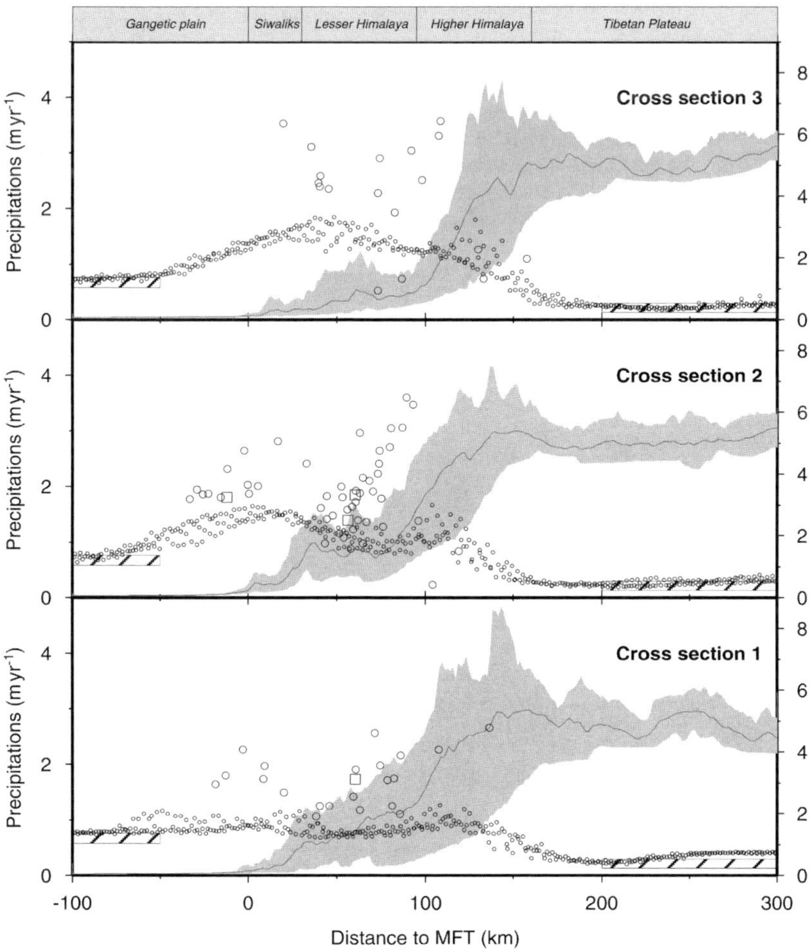

Fig. 2. Projection of the precipitation data on the three swathe profiles presented in Fig. 1, with corresponding mean elevation from GTOPO30 DEM (grey area indicates minimal and maximal values). Dots derive from NOAA dataset and circles and squares refers to measurements from GAME (2005) and NMFD (2005) gauge networks, respectively. Dashed boxes in Tibet and India represent compilation of data from GAME (2005) and IITM (2005), respectively. All swathes are 100 km wide.

Sapt Kosi basins. Both rivers are tributaries of the Ganga (Fig. 1). Trans-Himalayan rivers and fluvial terraces profiles in Nepal (Lavé & Avouac 2000, 2001), and cosmogenic data in India (Vance *et al.* 2003) suggest that erosion is maximal across the Siwaliks and the HH, lower in the LH, and minimal in South Tibet.

Modelling approach

Geometry of the system

Our model is based on a 700 km long N18°E cross-section perpendicular to the range, from the Gangetic Plain to the Tibetan Plateau (see Fig. 1 for the location). The principal geometric characteristics of our model are similar to Cattin and Avouac's model (2000) and our previous work (Godard *et al.* 2004) and are presented in Figure 3. The geometry of the Moho beneath South Tibet is derived from INDEPTH seismic profile (Zhao *et al.* 1993) and from gravity data (Cattin *et al.* 2001). The initial elevation for both river and mean topography mimics the main features of the present day topography.

Rheology

We use a 2D Lagrangian finite element code (Hassani *et al.* 1997) that allows us to solve the constitutive rheological equations expressing the relationships between stress and strain tensors ($\underline{\underline{\sigma}}$ and $\underline{\underline{\varepsilon}}$, respectively). The behaviour of the materials is considered elasto-visco-plastic. The elasticity is expressed by Hooke's law (see specification of parameters in Table 1),

$$\varepsilon_{ij} = \frac{1+\nu}{E}\sigma_{ij} - \frac{\nu}{E}\mathrm{trace}(\underline{\underline{\sigma}})\delta_{ij} \tag{1}$$

while the non-Newtonian viscous behaviour, dependent on temperature T, is controlled by the following relationship between deviatoric stress and strain rate $\dot{\varepsilon}$,

$$\dot{\varepsilon} = \gamma_0(\sigma_1 - \sigma_3)^n e^{E_a/RT} \tag{2}$$

where σ_1 and σ_3 are the maximum and minimum principal stresses, respectively. The limit between the visco-elastic and the plastic domains is defined by a Drucker-Prager failure

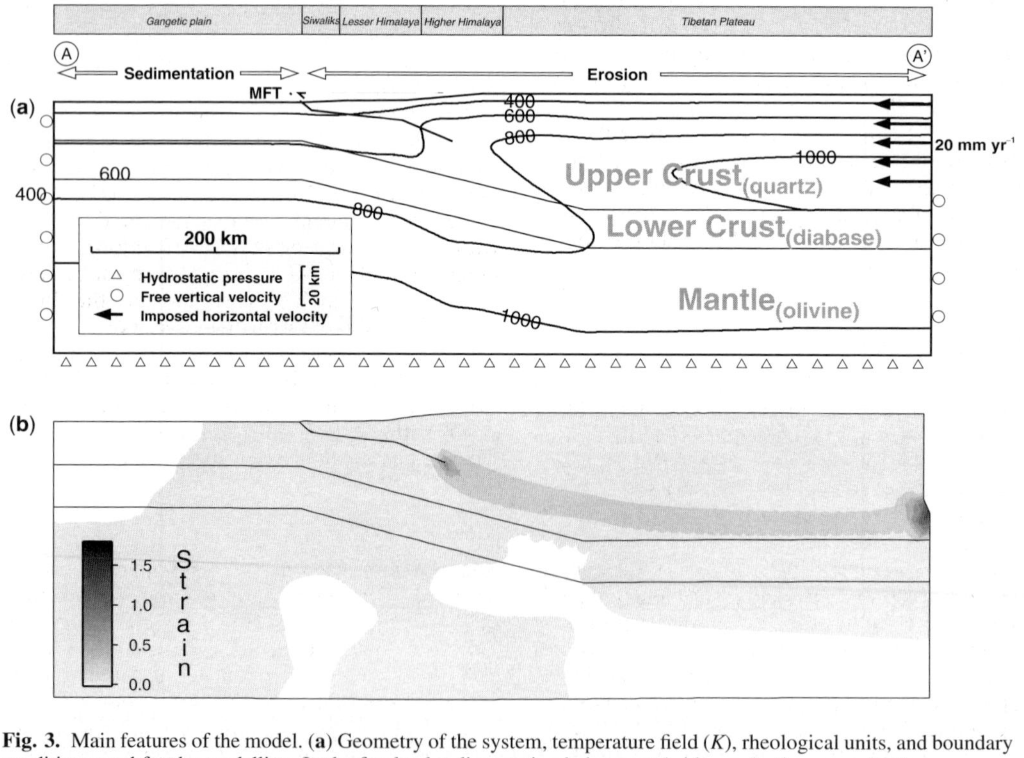

Fig. 3. Main features of the model. (**a**) Geometry of the system, temperature field (K), rheological units, and boundary conditions used for the modelling. In the foreland sedimentation balances subsidence. In the range, landscape evolution is controlled by the fluvial incision law. (**b**) Second invariant of the deviatoric strain tensor at the end of the simulation. Horizontal and vertical dimensions of model are 700 km and 130 km, respectively.

Table 1. *Rheological parameters used for the mechanical model*

	Quartzite	Diabase	Olivine
Density, ρ (kg m^{-3})	2900	2900	3300
Young modulus, E (Gpa)	20	20	70
Poisson's ratio, ν	0.25	0.25	0.25
Cohesion, c (Mpa)	10	10	10
Internal friction angle, ϕ	30°	30°	30°
Standard fluidity, γ_0 (Pa^{-n} s^{-1})	6.03×10^{-24}	6.31×10^{-20}	7.00×10^{-14}
Power law exponent, n	2.72	3.05	3.0
Activation energy, E_a (kJ mol^{-1})	134	276	510

Parameters from Carter & Tsenn 1987; Tsenn & Carter 1987; Kirby & Kronenberg 1987. Universal gas constant $R = 8.314$ J mol^{-1} K^{-1}.

criterion, according to,

$$\frac{1}{2}(\sigma_1 - \sigma_3) = \left(c(\cot\phi) + \frac{1}{2}(\sigma_1 + \sigma_3) \right) \times \sin\phi \qquad (3)$$

Godard *et al.* (2004) have shown that a composite quartz-diabase rheology associated with a fluvial incision law is required to explain both erosion rate and topographic profile in the Himalayas. Three layers are distinguished for the lithosphere: the upper and lower crusts, and the upper mantle. Each layer has specific mechanical properties. We use the empirical rheological equations and laboratory-derived material properties for quartz, diabase and dry olivine (Table 1). The temperature field from Henry *et al.* (1997) (Fig. 3) is used to compute the viscosity of the material. Considering the typical duration (*c.* 100 ka) of our simulations, we do not solve the heat equation. This temperature field is thus considered stable, relative to the mesh, over the run.

Boundary conditions

The boundary conditions applied to the system are constrained by geodynamical data. Most of the observations available in this area (see Fig. 8 caption), including GPS measurements (Bilham *et al.* 1997; Larson *et al.* 1999) estimate a shortening rate of 20 mm yr^{-1}. Recently, Chen *et al.* (2004) suggested velocities closer to 13 mm yr^{-1}. A variation of this magnitude in convergence rate does not modify significantly our conclusions. We apply, therefore, a 20 mm yr^{-1} horizontal velocity on the northern vertical face to a depth of 40 km, which is the depth of the high temperature and low viscosity zone under Tibet, used as a decoupling level in the crust. Vertical velocities on the other vertical faces are left free. The structure is supported by hydrostatic pressure at its base, which allows isostatic compensation and thus generates a coupling between rock-uplift and denudation.

A fault with a simple Coulomb friction law is introduced and follows the ramp-and-flat geometry proposed for the MHT. Due to the duration of our simulations we consider long-term deformation, which integrates a large number of seismic cycles, and slip on a low friction MHT ($\mu = 0.01$) is considered as continuous (Cattin & Avouac 2000). This low frictional surface connects to a mid-crustal high temperature zone that allows the localization of a narrow ductile deformation zone corresponding to the continuation of the MHT (Fig. 3), imaged by the INDEPTH experiment (Zhao *et al.* 1993).

The main purpose of this study is to focus on the influence of the surface processes applied to the topography as a boundary condition. We distinguish two domains in term of surface processes: the foreland south of the MFT ($x < 0$), with active sedimentation, and the mountain range north of the MFT, dominated by active erosional processes. The first domain, the Indo-Gangetic basin is classically described as a low elevation over-filled basin (Lyon-Caen & Molnar 1985): we thus assume in the following a constant *c.* 0 m elevation for it. The characteristics of the fluvial incision formulation used in the range are described further down.

Numerical method

The approach used here for simulating tectonic processes employs the dynamic relaxation method (Underwood 1983) coupled with the finite element method. The dynamic relaxation (DR) method is an explicit iterative method for the static solution of mechanical problems. The method can easily take into account large deformations of materials with non-linear behaviour. It is based on the fact that the static solution is the steady-state part of the transient response for a temporal load,

$$M\ddot{q} + C\dot{q} + Kq = F_e \quad (4)$$

with q, $\dot{q}$ and $\ddot{q}$ the displacement, velocity and acceleration vectors, respectively. C is the damping matrix, K the stiffness matrix, and F_e the external forces applied on the system, respectively. M is the fictitious mass matrix, which can be chosen to be non-singular. The DR algorithm evaluates in each time step the unbalanced forces, that is, the acceleration $\ddot{q}$ due to the disequilibrium between external and internal forces,

$$\ddot{q} = M^{-1}(-C\dot{q} - Kq + F_e) \quad (5)$$

The quantity $C\dot{q}$ represents a numerical viscosity introduced for stabilization purposes, and Kq is the internal forces computed from the integration of the constitutive material laws (Eqns 1, 2 and 3).

The time integration scheme computes velocity and displacement using an explicit (forward in time) finite difference method. The explicit algorithm is conditionally stable, the condition being

$$\Delta t \leq \frac{2}{\omega_{\max}} \quad (6)$$

where $\lambda_{\max} = \omega_{\max}^2$ is the highest eigenvalue of the matrix $M^{-1}K$.

Time scenario and stability

The total duration of the simulations is *c.* 320 ka, with an effective simulation duration (Fig. 4) of 80 ka and a time step of 1.5 yr. The elements are triangular with a typical size (diameter of circumscribing circle) of 3 km.

The convergence of the algorithm is associated with the minimization of unbalanced forces (i.e., acceleration); the different processes acting on the system are thus introduced gradually (Fig. 4). The simulation is divided into two steps: (1) a stabilization period where tectonics and denudation are introduced

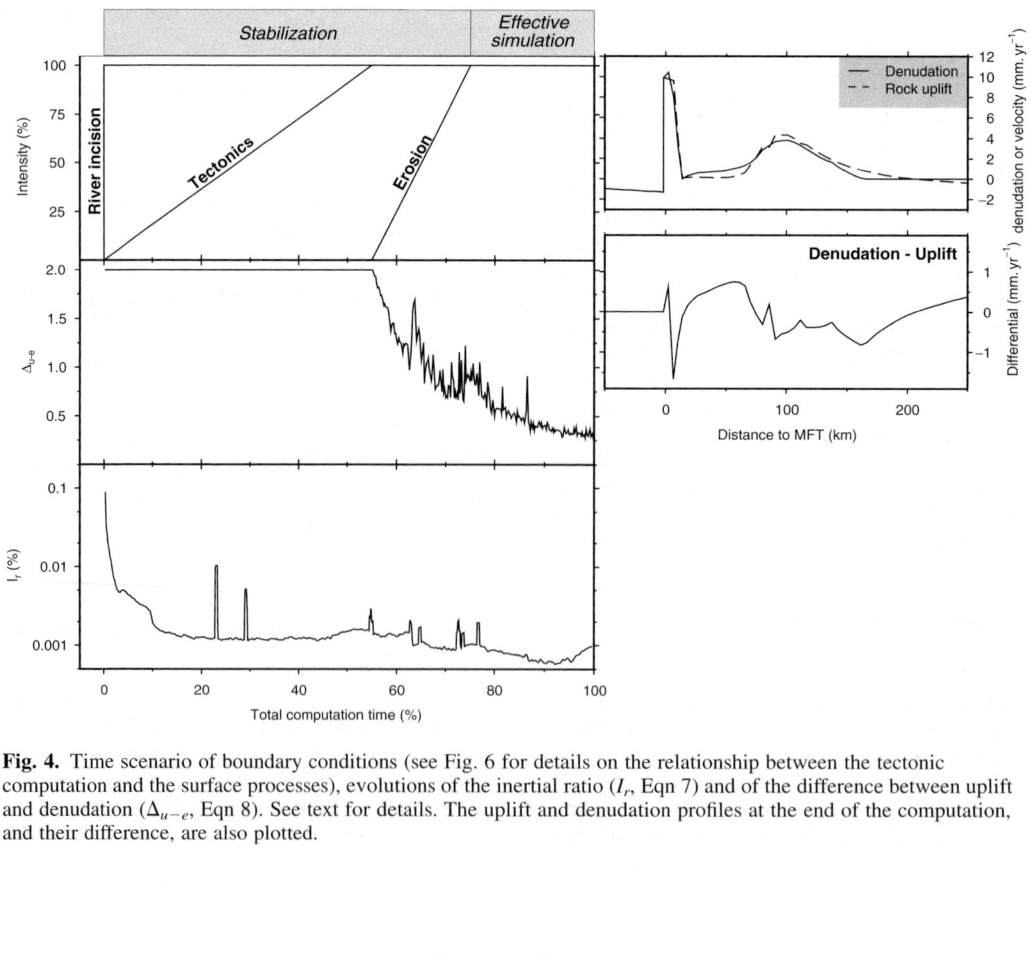

Fig. 4. Time scenario of boundary conditions (see Fig. 6 for details on the relationship between the tectonic computation and the surface processes), evolutions of the inertial ratio (I_r, Eqn 7) and of the difference between uplift and denudation (Δ_{u-e}, Eqn 8). See text for details. The uplift and denudation profiles at the end of the computation, and their difference, are also plotted.

progressively, (2) an effective simulation period, when all the processes act at full intensity. The initial river profile is the same as the initial mean topographic profile. Starting fluvial incision before topographic denudation is thus necessary to promote the stabilization of the river (i.e., convergence toward the steady-state profile) and to allow a significant entrenchment to exist at the beginning of the effective simulation period, thereby initiating topographic denudation. To evaluate the stability of the computation, we use the inertial ratio, I_r computed as,

$$I_r = \frac{\|\vec{F_e} + \vec{F_i} + \vec{F_c}\|}{\|\vec{F_e}\| + \|\vec{F_i}\| + \|\vec{F_c}\|} \quad (7)$$

where $\|\vec{F_e}\|$, $\|\vec{F_i}\|$ and $\|\vec{F_c}\|$ are respectively the external, internal and reaction forces acting on the system. The mechanical steady state is defined as an average equilibrium between uplift and denudation rates, which is computed as the average, at each point along the profile, of the normalized difference between denudation, e_i, and rock-uplift, u_i,

$$\Delta_{u-e} = \frac{\sum_{i=1}^{N_p} |u_i - e_i|}{\sum_{i=1}^{N_p} (u_i + e_i)/2} \quad (8)$$

The time limit of the simulations is imposed by distortions of the mesh in high erosion areas (Siwaliks foothills).

Figure 4 shows that the inertial ratio I_r is lower than 0.01% during the effective simulation period, which attests to the overall numerical stability of the model. The slight increase of the ratio at the end of the run is associated with the reduction of the size of the elements due to erosive processes. During the effective simulation period, Δ_{u-e} converges toward a constant value of *c.* 0.3 which indicates a slight degree of disequilibrium. Part of the mismatch between uplift and erosion corresponds to the contribution of horizontal velocity to uplift. Our main concern is to reach a relative equilibrium that allows inter-model comparison.

Surface processes

Fluvial incision

Recent studies have underlined the key role played by fluvial incision in driving unglaciated landscape denudation (Whipple & Tucker 1999). Whereas different functional forms have been proposed to model fluvial incision (Whipple & Tucker 2002), in an attempt to develop a simple approach, we have used a detachment-limited relation that provides satisfying first-order results in the Subhimalaya (Lavé & Avouac 2001). This approach states that the bedrock incision rate of a river, at abscissa x, is proportional to the fluvial shear stress τ_x in excess of some threshold τ_c:

$$\left(\frac{\partial h_{\mathrm{riv}}}{\partial t}\right)_x = K_x(\tau_x - \tau_c) \quad (9)$$

with $\tau_x = \rho \cdot g \cdot H_x \cdot S_x$, ρ the density of water, S_x the river slope, H_x the flow depth and K_x an erodability coefficient that primarily depends on rock strength with respect to abrasion (Lavé & Avouac 2001), but which could mask hidden dependencies on sediment flux and flood distribution. For the sake of simplicity, we will ignore the two latter variables. The shear stress τ represents the force exerted by the flowing water on an area unit of the channel bottom, and is derived from energy expenditure equation. Assuming Manning's equation to describe mean water flow velocity, and employing a scaling law between river width and discharge, it can be shown (Howard *et al.* 1994) that shear stress can reduce to an expression of channel slope and drainage area. The central Himalaya, like many other mountain ranges, displays a relative cylindrical structure and linear drainage systems perpendicular to the range axis. Along these linear drainage systems, 50 km apart, the drainage area is linearly related to the cross-range distance (Fig. 5). Shear stress can then be expressed by:

$$\tau_x = k_1(\bar{P}_x - P_r)^{\gamma}(L(X - x))^{\beta}\left(\frac{S_x}{s_0}\right)^{\alpha} \quad (10)$$

where k_1 is a coefficient that depends on the river network geometry and sediment size, L the width of the watershed, s_0 the average sinuosity of the main stream (Fig. 5) and α, β, γ are constant exponents. $\bar{P}$ is the average precipitation on the watershed and P_r some threshold runoff. X is the abscissa of the drainage divide. The values of these parameters are given in Table 2.

Landscape denudation by fluvial incision and landslides

Many numerical models (e.g., Willett 1999) compute the denudation rates applied to topography directly from Equation 9. However, despite their prominent role, the main rivers do not account for the mean topography, which is the pertinent variable for the upper boundary

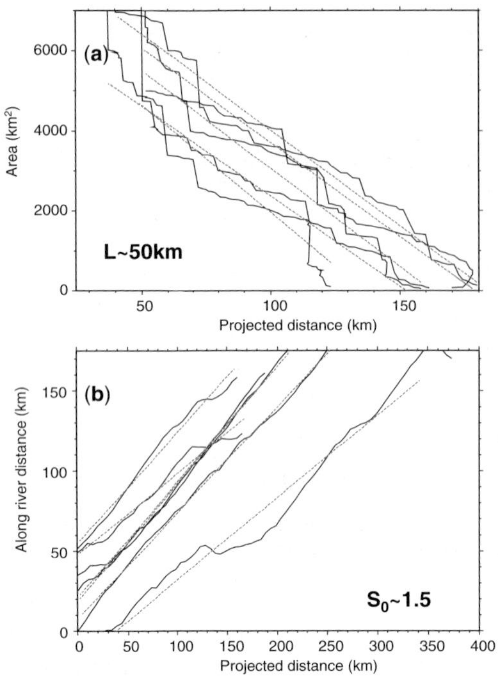

Fig. 5. Two examples of parameter determination for the fluvial incision formalism. (**a**) Evolution of the area of the watershed of some major Trans-Himalayan rivers as a function of projected distance (abscissa on a N18° trending line) illustrating the existence of a linear relationship between those two parameters. The slope of the fit (dashed line) gives the width L of the watershed, ranging from 45 to 53 km. (**b**) Evaluation of the sinuosity for some major Trans-Himalayan rivers through the relation between projected distance (N18°) and along river distance (measured along the river path). The slope of the fit (dashed line) gives the inverse of sinuosity $1/s_0$, sinuosity ranges from 1.4 to 1.9. Fluvial network and watersheds extracted from a 90 m DEM.

Table 2. *Denudation law parameters*

Parameter	Value
Hillslope angle of repose, θ_c (degrees)	40
Himalayan valley width, L (km)	50
Sinuosity, s_0	1.5
Slope exponent, α	0.7
Area exponent, β	0.27
Precipitation exponent, γ	0.33
Critical shear stress, τ_c (Pa)	30
k_1 (Pa m$^{-0.87}$ s$^{0.33}$)	5677
k_2 (Pa m$^{-1.03}$ s$^{0.33}$)	2400
Threshold runoff, P_r (m)	0.2

From Lavé & Avouac (2001) or Himalayan river measurements (e.g., Fig. 5).

condition of mechanical modelling. The elevation profile of the trans-Himalayan rivers in fact represents the base level for the network of tributaries that are draining the whole topography, from their sources at the base of the hillslopes to their confluence with the trunk stream. At a given abscissa, the mean elevation of the topography $\bar{h}$ is therefore the sum of three contributions: (1) the elevation of the main river h_{riv}, (2) the fluvial relief associated to the tributaries Δh_{trib} that we assume to be controlled by the same incision law as the main river, and (3) the mean relief of the hillslope from the fluvial network to the crest, Δh_{hill}. In active orogens, hillslopes are dominated by landslides (Hovius *et al.* 1997); we thus assume that they display a critical slope angle of repose θ_c and that they react instantaneously to any local base-level drop. A new formulation to integrate the fluvial relief associated with the tributary network (Lavé 2005) enables computation at each time step of the changes of the elevation of the trunk stream from Equation 10 and the changes in mean topography according to

$$\left(\frac{\partial \bar{h}}{\partial t}\right)_x = \left(\frac{\partial h_{\text{riv}}}{\partial t}\right)_x + K_x(k_2(P_x - P_r)^{\gamma} \times \Delta h^{\alpha}_{\text{trib}_x} - \tau_c) \quad (11)$$

with,

$$\Delta h_{\text{trib}} = \Delta h_{\text{total}} - \Delta h_{\text{hill}} = \Delta h_{\text{total}} - \frac{\Delta l}{2}\tan\theta_c \quad (12)$$

where k_2 is a coefficient which depends on the tributary network geometry and Δl is the horizontal distance between the crest and the base of the hillslope.

The solving scheme is based on the interplay between two profiles: the mean elevation profile and the river profile (Fig. 6). The river profile is advected with the tectonic velocity field computed by the mechanical solver and the incision rates computed from Equation 10. The mean topographic profile constitutes the upper boundary of the mechanical model. Feedback between tectonics and erosion is achieved by this dynamic interface: (1) the mean topography is advected with the tectonic velocity field and the denudation rates from Equation 11, and (2) the internal stress field is affected by mass removal from the mean topography by denudation processes.

The above formulation does not satisfy, strictly speaking, the mass conservation equation and could be viewed as incomplete. However, in

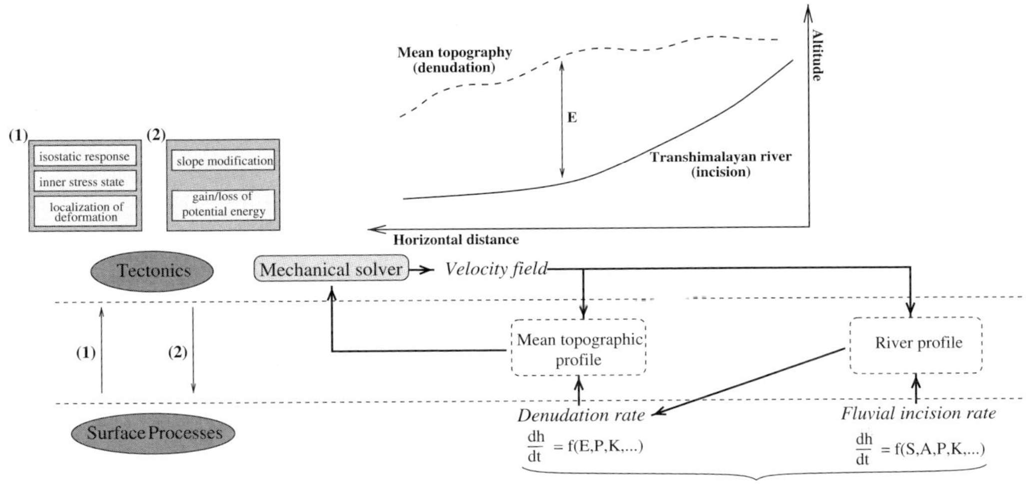

Fig. 6. Relationship between the mechanical solver and the surface processes computation. S: slope of the river, A: area of the watershed, P: precipitation over the watershed, K: erodability, E: entrenchment.

a detachment-limited system, the required hydraulic energy to incise bedrock is considered to be much higher than the energy required to transport the eroded sediments toward the foreland basin. In the above formalism, the sediments are systematically and efficiently evacuated from the range and do not play any explicit role in erosion processes, so mass conservation is not strictly required.

Controlling parameters in fluvial incision and landscape denudation

The threshold τ_c of the incision law is close to the threshold of pebble motion in rivers and, in steep mountains, much lower than average shear stress. Neglecting this threshold term in Equations 9 and 11 provides two simple expressions for river incision and landscape denudation, as the products of a local stream or regional topographic gradient (Δh and S respectively) and of a coefficient which depends on several factors. The denudation term depends on local precipitation, local erodability and valley width L (embedded in k_2). The main trans-Himalayan river incision depends not only on local erodability and valley width, but also on regional precipitation as averaged over the length of its contributing area. For a mountain range subjected to uniform erodability conditions and precipitation, the study of coupling between erosion and tectonics can be reduced to an investigation of the role of a single erosion parameter, $E^* = f(P, K)$. Depending on the precipitation exponent γ, the sensitivity of the system to precipitation can be viewed as weak relative to the sensitivity to abrasion rock strength. For a precipitation exponent $\gamma = 0.33$ and according to Equations 9 and 11, increasing the precipitation by a factor 8 is equivalent to increasing the erodability by a factor 2.

However, the Himalayas display large cross-range variations in precipitation and rock types. Because of these variations, and because tributary denudation and main river incision depend on local and integrated precipitation respectively, the above non-dimensional analysis and reliance on a single erosive parameter loose their validity. To investigate these complexities in the interaction between landscape denudation, main river incision and tectonics, we consider different case studies in which the erodability and precipitation profiles are varied independently. To illustrate lateral (Fig. 1) and cross-range variations in precipitation, a series of profiles representing the lateral amplitude variation of rainfalls in Nepal will first be tested.

Rock types across the Himalayan range display strong variation, from poorly cemented sandstones in the Siwalik Hills, to schists, quartzite and sandstones in the LH, or gneissic units in the HH, or sedimentary units in the South Tibetan series (Fig. 7). A recent study (Attal & Lavé 2006) has shown that these lithologic variations can lead to differences up to 3 orders of magnitude in pebble abrasion rates. Similarly, the bedrock erodability of the Siwalik units represents an erodability coefficient 15 times larger than the LH units: $K_{\mathrm{siw}} = 1.05 \times 10^{-1}$ mm yr^{-1} Pa^{-1} and $K_{\mathrm{LH}} = 0.76 \times 10^{-2}$ mm yr^{-1} Pa^{-1}, respectively

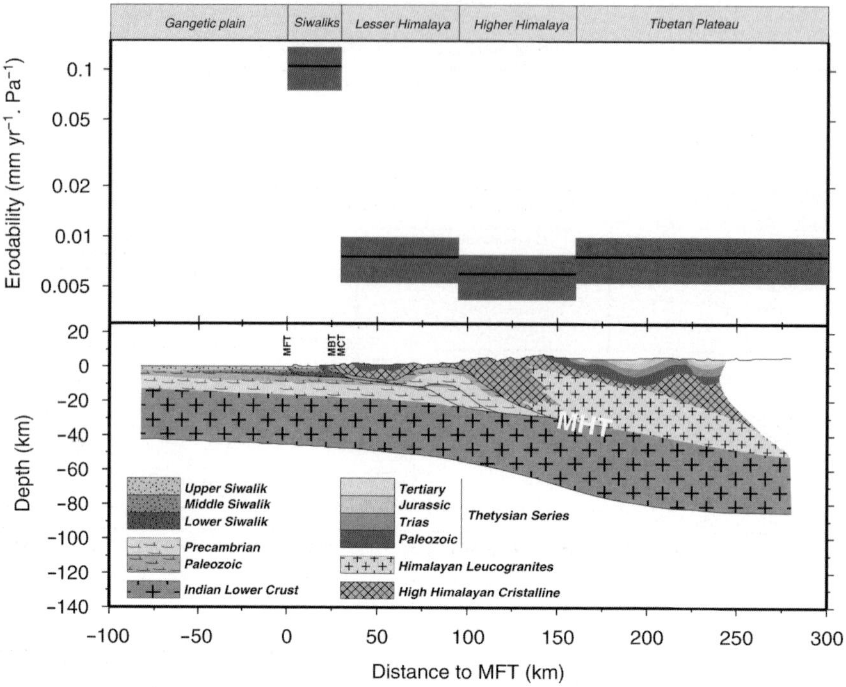

Fig. 7. Erodability values determinated across Central Nepal (Lavé & Avouac 2001) and corresponding geological cross-section modified from Lavé & Avouac (2001). MBT: Main Boundary Thrust; MCT: Main Central Thrust.

(Lavé & Avouac 2001). In order to evaluate the sensitivity of our model to this parameter, we test different kinds of profiles. First we consider that erodability in the Siwalik Hills is K_{siw} while the erodability of the Lesser Himalaya (K_{LH}) is applied to the entire range north of the frontal foothills. Then, we consider homogeneous profiles across the range, where no variation of erodability is considered, with erodability ranging from 0.5×10^{-2} mm yr^{-1} Pa^{-1} to 2×10^{-2} mm yr^{-1} Pa^{-1}. Finally, considering that the erodability in the HH has been evaluated by Lavé and Avouac (2001) such that K_{HH} *c.* 0.8 K_{LH}, we introduce this variation to evaluate the influence of slight changes in erodability.

Reference model

From the observations of Figures 2 and 7, two reference profiles are defined for both precipitation and erodability. These two profiles are used as an input for the modelling. The results obtained with this model (Fig. 8, hereafter refered to as the reference model) are in good agreement with the available data sets in the study area (Fig. 1).

The horizontal velocity profile (relative to India) shows that the 20 mm yr^{-1} of convergence applied as a boundary condition are transferred to the MFT. This is related to the low-friction coefficient of the fault, which allows free slip on it. The calculated convergence rate across the range is in agreement with the data, including (1) the horizontal shortening deduced from the uplift of fluvial terraces in the Siwalik foothills (Lavé & Avouac 2001), (2) the progradation of the Gangetic Plain sediments associated with the flexure of the Indian plate (Lyon-Caen & Molnar 1985), and (3) South Tibet Quaternary grabens extension (Armijo *et al.* 1986). Due to gravitational collapse of the range, the horizontal velocity is slightly higher than 20 mm yr^{-1} in the Lesser Himalaya.

This gravitational collapse also accounts for the offset of the calculated topographic profile, which is shifted southward relative to the present-day topography. However, it still preserves both the clear slope transition between the HH and the LH, and the respective average values of slopes in those areas. The present-day topography is used as initial profile for both the mean topography and the main trans-Himalayan river. To allow a significant entrenchment to be created before the beginning of the effective simulation period, the incision of the main

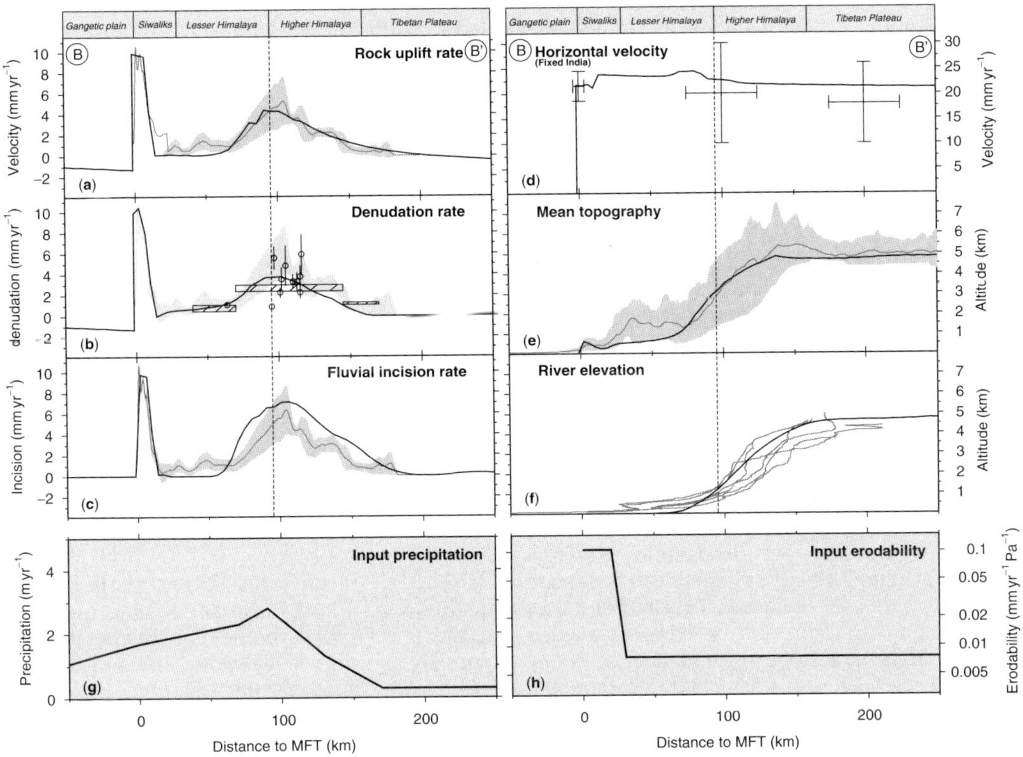

Fig. 8. Reference model. Thick black lines represents model outputs averaged on the last 1% of computation time. The data used to constrain the model outputs are listed below, see text for details. (**a**) Uplift profile, data from Lavé and Avouac (2001): mean value (solid thin line) and 1σ confidence interval (grey area). (**b**) Denudation profile, fission track data from Burbank *et al.* (2003) and averaged cosmogenic denudation rates from Vance *et al.* (2003) (dashed boxes). (**c**) Fluvial incision profile, data from Lavé & Avouac (2001): mean value (solid thin line) and 1σ confidence interval (grey area). (**d**) Horizontal velocity profile, control points from Lavé and Avouac (2000), Lyon-Caen & Molnar (1985) and Armijo *et al.* (1986) (north to south). (**e**) Topographic profile, data derived from GTOPO30 DEM mean (solid thin line) and extreme values (grey area). (**f**) River profile, trans-Himalayan river profiles from Lavé and Avouac (2001). (**g**) and (**h**) present rainfall and erodability profiles used as an input in the modelling. Note that time parameterization is different than that used by Godard *et al.* (2004).

trans-Himalayan river starts at the beginning of the simulation, while the denudation of topography begins later (Fig. 4). Even though the river profile does not reach a strict steady state, it tends to stabilize at the end of the run (Fig. 9). The modelled river profile displays a shape which is in the range of that observed for the main rivers crossing the range.

Fluvial incision rates across the range were calculated by Lavé and Avouac (2001) from a detachment-limited fluvial incision law calibrated in the Siwaliks and the LH. The incision rates on the main trans-Himalayan river computed from Equation 10 give values slightly higher than the data. However, the wavelength and position of the extrema in the Siwaliks and at the LH-HH boundary are consistent with observations.

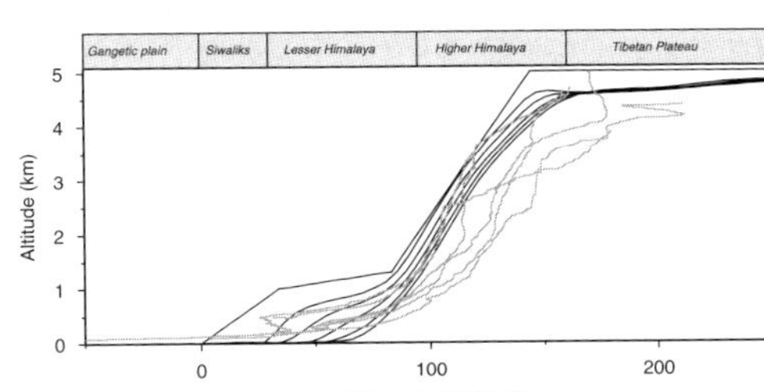

Fig. 9. Evolution of the river profile during the reference run (black lines). The time interval between each profile is *c.* 60 ka. The grey lines represent observed trans-Himalayan river profiles.

Denudation is compared with exhumation rates deduced from apatite fission-track ages in the HH (Burbank *et al.* 2003) and cosmogenic dating of river sediments, which provide an estimation of short-term denudation rate on the watershed (Vance *et al.* 2002). An estimation of uplift rates across the range was deduced from the fluvial incision profile (Lavé & Avouac 2001). This data set indicates that denudation and uplift reach a maximum in the Siwaliks and at the boundary between the Lesser and Higher Himalaya and decrease to a minimum in the LH and South Tibet. Computed denudation and vertical velocity profiles from the model display similar patterns, which are consistent in amplitude and wavelength with what is observed in Central Nepal.

A maximum for uplift, denudation and incision can be observed in the Siwaliks; its presence is mainly due to the high erodability value assigned to this area. The localization of another maximum at the LH-HH boundary can be associated with three mechanisms: (1) presence of a high precipitation zone in the HH, (2) localization of high river slopes in this area, implying high incision rates and, as a result, high entrenchment and denudation, and (3) existence of a ramp for the MHT below the HH. The relative importance of those three contributions is examined in the two next sections.

Sensitivity to precipitation patterns

Influence of rainfall amplitude

The precipitation (collected by the hydrographic network) directly influences the incision on the main trans-Himalayan river according to Equations 10 and 11. The response of the system was tested for various rainfall profiles (Fig. 10), which differ in the amplitude of precipitation in the high range. The erodability profile is the same as for the reference model (Fig. 8).

The increase of precipitation at the LH-HH transition leads to higher denudation rates in this area, whereas the changes in rainfall have no significant effects on denudation in the southern part of the LH. This can be related to the fact that the rock-uplift is close to 0 in the LH, which reduces the sensitivity of the river to the variations of discharge. Fluvial incision of the trans-Himalayan rivers is much less affected than the local incision along tributaries (inferred from the denudation profile) and is, therefore, less than the topographic denudation. This is because discharge along the trans-Himalayan rivers reflects the integration of precipitation along the whole profile and because precipitation is kept constant in South Tibet. Furthermore the variability of incision in the HH leads to important variations of entrenchment (i.e., difference of elevation between the topographic and river profiles, Δh_{total} in Eqn 12), which propagates to the denudation rates in this area.

The near steady-state topographic profile is relatively insensitive to variations in precipitation at the timescale of our study. The increase in denudation near the LH-HH boundary is thus correlated to an increase in the uplift in this area. This highlights a level of coupling between tectonics and erosion, where increased precipitation induces more denudation and more uplift as an isostatic response. As a counterbalance, the uplift, by increasing topographic gradients and tributary relief, will favour an increase in landscape denudation.

Constant precipitation rates over the range

In order to quantify the effects of the rainfall distribution, the behaviour of the model is tested for a set of profiles where precipitation rates are constant between the MFT and southern Tibet. This constant precipitation rate ranges from 0.3 m yr^{-1} to 3.5 m yr^{-1}. On the Tibetan Plateau the precipitation rate falls to 0.3 m yr^{-1} for all the profiles (Fig. 11), to take into account the orographic barrier of the HH. The results reveal that the model is relatively insensitive to the existence of a maximum in precipitation at the LH-HH transition.

When comparing simulations with constant rainfall over the range to what is obtained with the reference model, it appears that the existence of a high precipitation zone between the HH and the LH does not have a noticeable consequence, in our model, on the position of the maximum for denudation, incision or uplift. However the amplitude of this maximum depends directly on the amount of precipitation provided to the system.

The uplift pattern stays unchanged whatever the rainfall profiles, with a maximum at the transition between the HH and the LH. Part of this is due to the rather low value of the precipitation exponent ($\gamma = 0.33$) in the detachment limited incision law (Eqns 10 and 11). In area/slope analysis (Whipple & Tucker 2002), the discharge/area or precipitation exponents cannot be measured, but only estimated relative to the slope exponent. Some studies along rivers draining regions with variable uplift rates (Tomkin *et al.* 2003; Van der Beek & Bishop 2003) have, however, proposed absolute values for the discharge/area exponent, but with large uncertainties. According to Lavé & Avouac's study (2001) in the Siwaliks, and based also on

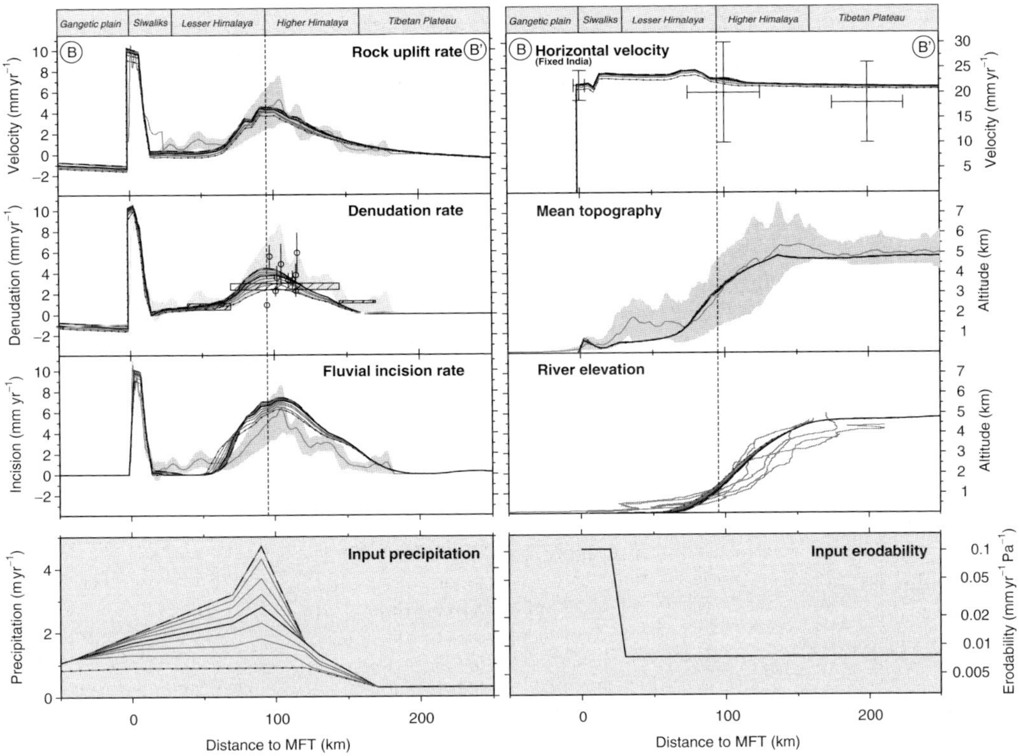

Fig. 10. Results of the simulations with various rainfall profiles. Dashed and dotted lines refer to the maximum and minimum rainfall profiles respectively. The solid bold line is the reference model presented in Fig. 8. The data sets used to compare with the model outputs are the same as in Fig. 8.

a larger data set of worldwide rivers, Lavé and Attal (2003) suggest that the shear stress exponent in the detachment limited incision law is ranging between 0.7 and 1.3. Consequently, the precipitation exponent would range between 0.23 and 0.43. Even though more work is still required to better assess the physical expression of bedrock incision laws, we hypothesize that the sensitivity of incision to precipitation is relatively low. In contrast with previous studies (Thiede *et al.* 2004), it brings the idea that the maximum for rock uplift at x $\sim$ 100 km, observed both in the model outputs and the data from Lavé and Avouac (2001), is mainly related, in our model, to the ramp-flat geometry of the MHT rather than to the location of the peak of precipitation.

Sensitivity to erodability

Constant erodability

A set of profiles with constant erodability from the MFT to the Tibetan Plateau, ranging from 2.5×10^{-3} mm yr^{-1} Pa^{-1} to 2×10^{-2} mm yr^{-1} Pa^{-1} were tested (Fig. 12). Due to the linear relationship between erodability and both incision and denudation rates (Eqns 10 and 11), a slight increase in erodability has important implications on the global behaviour of the model.

Increasing erodability leads to higher incision and denudation rates and a coincident augmentation in the uplift rate. The sensitivity is maximum in the HH, where the use of an erodability of 2×10^{-2} mm yr^{-1} Pa^{-1} leads to denudation rates up to 10 mm yr^{-1} in the HH, which far exceed the values of the denudation rate estimated in this area. Erodabilities in the $0.5–1 \times 10^{-2}$ mm yr^{-1} Pa^{-1} range gives values for denudation and uplift in much better agreement with the observations.

Variations of erodability strongly influence both the amplitude and the location of the maximum of fluvial incision rate. This can be related to the regressive incision of the main trans-Himalayan river in response to increasing erodability, which offsets both the maximum for

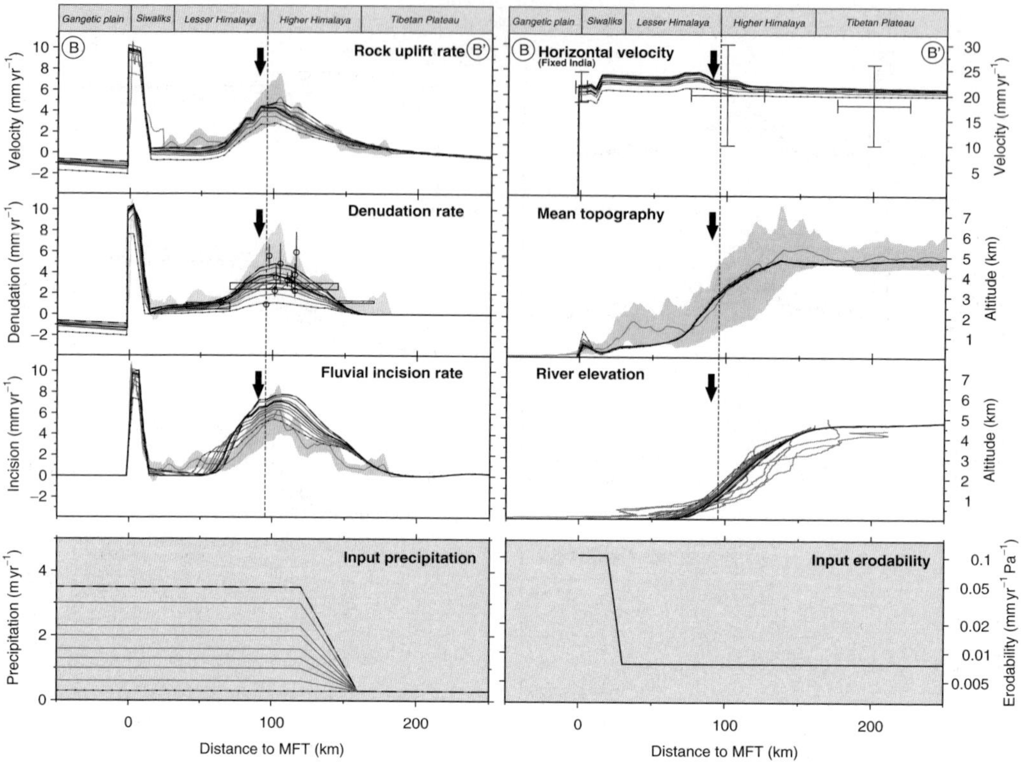

Fig. 11. Results of simulations where the precipitation rate is assumed to be constant from the MFT to southern Tibet, ranging from 0.3 m yr^{-1} to 3.5 m yr^{-1}. Dashed and dotted lines refer to the maximum and minimum rainfall profiles, respectively. The solid bold line is the reference model presented in Fig. 8. The black arrow indicates the maximum of the reference rainfall profile (Fig. 8). The data sets used to compare with the model outputs are the same as in Fig. 8.

the slope of the river and the maximum in incision northward, in accordance with Equation 10. The topographic profile exhibits a lower sensitivity to erodability than does the river profile. More surprisingly, horizontal velocities are significantly affected by the variations of erodability. The rate of convergence in the frontal foothills, like the uplift rates, decreases with increasing rock strength (i.e., with erodability decrease). This phenomenon appears to reflect the role of erosion in stripping crustal material and in favouring the full transfer of the convergence toward the frontal structures.

Changes of erodability along the transect

The models presented in the previous section assume relatively low erodabilities in the Siwaliks foothills, in comparison with the value calculated by Lavé and Avouac (2001) of 1.05×10^{-1} mm yr^{-1} Pa^{-1}. This leads to the development of an unrealistic *c.* 2000 m-high topographic front that cannot be counterbalanced by erosion (the run duration does not permit the equilibrium topography to be reached so this value represents a minimum value). The introduction of a high erodability zone in the Siwalik associated with the unconsolidated sandstones of the foothills, is thus required, first, to obtain a reasonable topographic front in this area, and second, to be able to localize deformation and a very high uplift rate without propagating the deformation southward.

Finally, a more refined erodability profile (dash-dot line, Fig. 12), which imposes slightly lower erodability at the LH-HH boundary in comparison with the southern LH (6×10^{-3} mm yr^{-1} Pa^{-1} instead of 7.6×10^{-3} mm yr^{-1} Pa^{-1} in the reference model), is tested. This low variation of erodability does not have noticeable effects on horizontal shortening and topography. However, the calculated river profile is significantly less entrenched than in the reference case, and it induces a decrease of 1–2 mm yr^{-1} for incision, denudation and uplift rates in the HH. One noteworthy point is

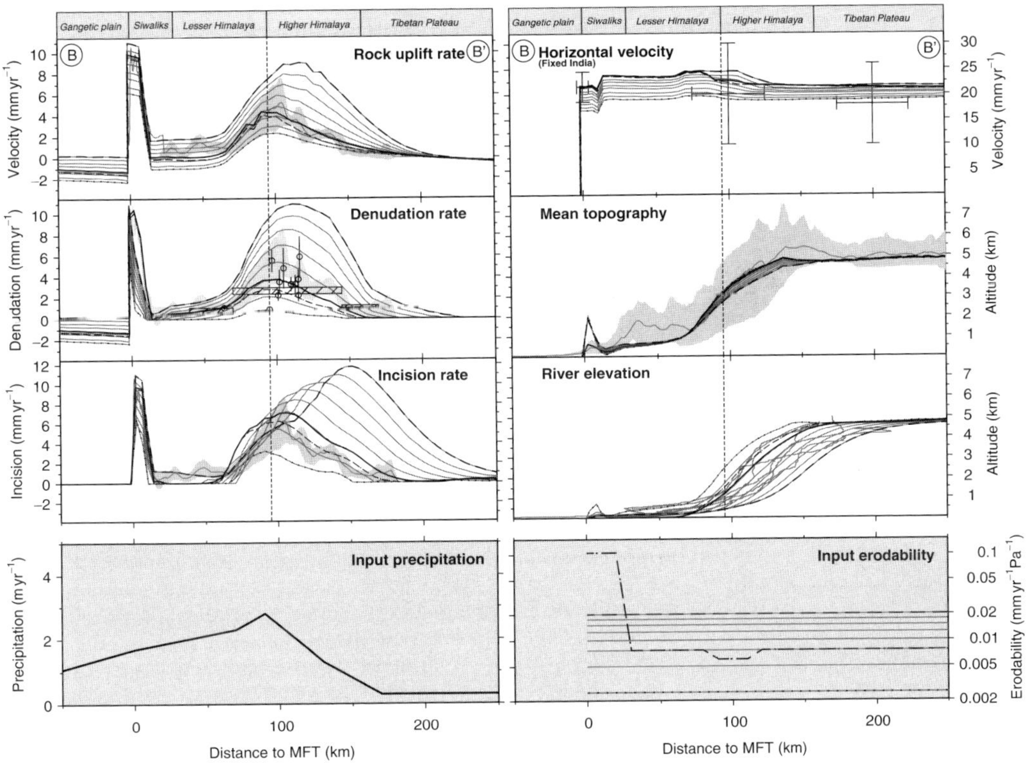

Fig. 12. Results of the simulations for various erodability profiles. Dashed and dotted lines refer to the maximum and minimum erodability profiles, respectively. The solid thick line is the reference model presented in Fig. 8. Dot dashed black line represents the case where a distinction of erodability is made between the different units of the range. The data sets used to compare with the model outputs are the same as in Fig. 8.

that the area affected by this modification (in comparison to the reference profiles) is wider and extends farther north than the defined low erodability area (*c.* 60 km against *c.* 30 km). This is interpreted as a response to an upstream adjustment of the main river near-equilibrium profile, which tends to be less steep. It leads to less entrenchment and, as a consequence, to a relative diminution of denudation and rock-uplift in an area wider than just the low erodability zone.

Discussion and conclusions

Godard *et al.* (2004) have shown that the choice of the erosion law in the modelling of an orogenic system can strongly influence the results of the models and the associated interpretations. Using a fluvial incision formulation, instead of, for example, a classical diffusion law, brings some significant improvements. In particular, an incision-driven model allows the use of external parameters in modelling, measurable on the field, such as precipitation and erodability. The goal of this study was to test the behaviour of the system in response to variations of those two quantities in terms of both amplitude and spatial distribution.

Erodability is a relatively poorly constrained parameter, and even if its importance in orogenic evolution has been recognized (Schlunegger & Simpson 2002) in most modelling, it is considered as uniform. The results presented in this study clearly reveal that rock strength has a first-order control on the evolution of an orogenic wedge and that lithologic considerations really matter when introducing a denudation formulation in a mechanical model. Furthermore, this study demonstrates the predominant role of the high erodability of the Siwalik sediment in the localization of the deformation front and in the regulation of topography.

Moreover, it appears that small variations in erodability (for example between the HH and the LH), may have small, but significant, consequences on the amplitude of denudation and uplift. As significant uncertainties are often associated with the quantification of the erodability coefficient, it appears that systematic study of the relationship between lithology and erodability is required to develop realistic coupled modelling of tectonics and surface processes. This perspective is certainly influenced by the way denudation processes are formulated in this study and is perhaps particularly relevant to our area of interest. Nevertheless the erodability parameter may still have a predominant control on the modalities of landscape denudation, as observed with the almost linear relationship existing between this parameter and the total sediment flux (Fig. 13).

Due to the values of the parameters used in the fluvial incision formulation, our modelling is less sensitive to rainfall than to erodability (exponent $\gamma = 0.33$ for precipitation, linear dependency for erodability). Precipitation is far better documented than erodability, but the spatial and temporal variabilities of both parameters are extremely important. The approach used in this study was to consider only first-order long-wavelength features characterizing the spatial distribution of precipitation, as, for example, the rain shadow between the HH and Tibetan Plateau (Fig. 2). The amplitude of precipitation directly controls the amplitude of denudation and uplift (Figs 10 and 11). However, the spatial distribution of rainfall seems to have a limited impact on the behaviour of the system, given that the presence or absence of a localized high-precipitation area at the LH-HH boundary does not induce noticeable modifications of the uplift pattern in this area, although it clearly affects the magnitude of denudation. The important variable to take into account is the global volume of water provided to the system by precipitation (Fig. 14). The relative distribution of that precipitation in the range has second order effects and does not modify the global amount of sediment eroded from the system (Fig. 14). Short-wavelength variations exist (Fig. 1) but the results presented in this study show that their relative influence on the tectonics of the system is limited. Our study demonstrates that the existence of localized high denudation and uplift rates can be related to lithology (high erodability in the Siwaliks) or structural and morphological features (ramp-flat geometry of the MHT, high slopes in the HH), rather than to a particular high precipitation area, as suggested by Thiede *et al.* (2004). In a mechanically coupled system, the tectonic behaviour is mostly sensitive to the global amount of precipitation and not to small-scale variations: in the case of the Himalayas, looking for spatial correlation between denudation and precipitation at a wavelength smaller than 20–40 km is not likely to be justified.

Such phenomena and the general low sensitivity of the chosen fluvial incision law to precipitation could explain the recent data along the Marsyandi river in Central Nepal (Burbank *et al.* 2003). This study displays a zone across the HH with uniform values of the FT closure

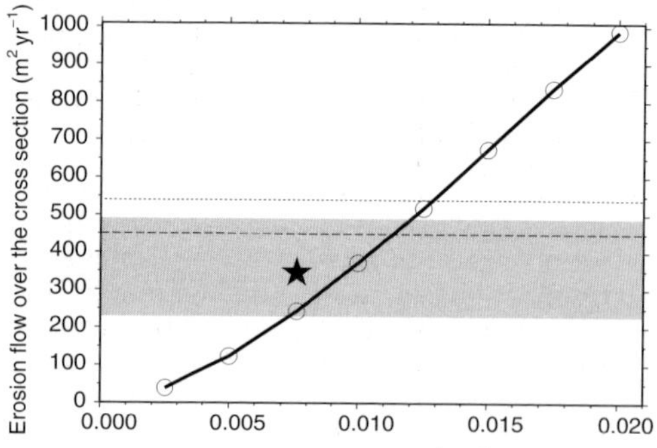

Fig. 13. Erodability versus outcoming flux of sediments for the constant erodability simulations (Fig. 12). The grey area, the dashed and dotted bold lines represent the erosion flux estimated by Métivier *et al.* (1999), Pinet and Souriau (1988) and Summerfield and Hutton (1994), respectively. The position of the reference model (Fig. 8) is indicated by the black star. The mean erodability associated with the reference model takes into account the variations of this parameter along the studied profile.

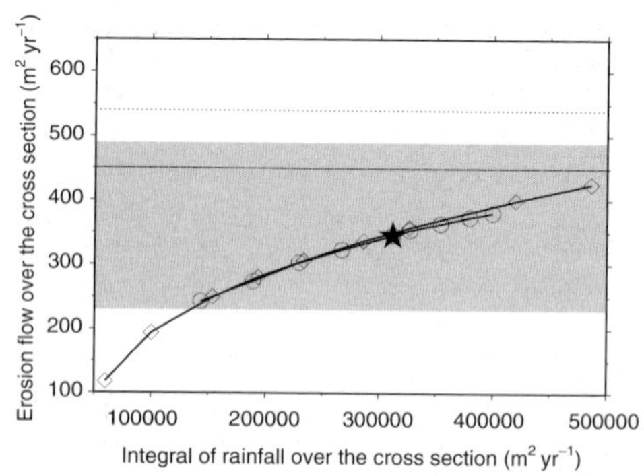

Fig. 14. Integrated precipitations over the cross-section versus outcoming flux of sediments (integration of the denudation profile). Circles and diamonds represent outputs for runs from Figs 10 and 11, respectively (variable precipitation and constant erodability). The black star indicates the position of the reference model (Fig. 8). The data are the same as in Fig. 13.

ages in apatite, and probably of the denudation rates, despite a three to four fold drop of the rainfall profile across the HH. These ages would primarily reflect the underlying uniform tectonic uplift associated with the mid-crustal ramp, and the lower sensitivity of the trans-Himalayan rivers to the local precipitation. This illustrates the difficulties or, perhaps, the systematic bias in trying to correlate local rainfall with local denudation to demonstrate or invalidate coupling between tectonics and climate.

Our assumption that applied present-day observed rainfall amplitudes to the timescale of the Quaternary in our simulations could be questioned. In particular, the influence of short to middle term evolution of climatic variables (e.g., monsoon, glaciations) on the tectonics of mountain ranges is still unclear and is worth investigating.

Most of the illustrations of this paper were prepared using GMT (Wessel & Smith 1991). We are grateful to J. Chéry and R. Hassani for providing the finite element code and Y. Dramais for corrections on the manuscript. Constructive reviews by F. Schlunegger, D. W. Burbank and one anonymous reviewer contributed to improve the manuscript. This work is supported by a RELIEF grant from the Centre National de la Recherche Scientifique.

References

ARMIJO, R., TAPPONIER, P., MERCIER, J. L. & TONGLIN, H. 1986. Quaternary extension in southern Tibet. *Journal of Geophysical Research*, **91**, 13803–13872.

ATTAL, M. & LAVÉ, J. 2006. Changes of bedload characteristics along the Marsyandi River (central Nepal): Implications for understanding hillslope sediment supply, sediment load evolution along fluvial networks and denudation in active orogenic belts. *In*: WILLETT, S. D., HOVIUS, N., BRANDON, M. T. & FISHER, D. (eds), *Tectonics Climate, and Landscape Evolution*, Geological Society of America Special Paper, **398**, p. 143–171, DOI: 10.1130/2006.2398(09).

AVOUAC, J. P. & BUROV, E. B. 1996. Erosion as a driving mechanism of intracontinental mountain growth. *Journal of Geophysical Research*, **101**, 17747–17769.

BEAUMONT, C., JAMIESON, R. A., NGUYEN, M. H. & LEE, B. 2001. Himalayan tectonics explained by extrusion of a low-viscosity crustal channel coupled to focused surface denudation. *Nature*, **414**, 738–742.

BILHAM, R. K., LARSON, K., FREYMULLER, J. & PROJECT IDYLHIM MEMBERS, 1997. Indo-Asian convergence rates in the Nepal Himalayas. *Nature*, **386**, 61–64.

BURBANK, D. W., LELAND, J., FIELDING, E., ANDERSON, R. S., BROZOVIC, N., REID, M. R. & DUNCAN, C. 1996. Bedrock incision, rock uplift and threshold hillslopes in the northwestern Himalayas. *Nature*, **379**, 505–510.

BURBANK, D. W., BLYTHE, A. E. *ET AL.* 2003. Decoupling of erosion and precipitation in the Himalayas. *Nature*, **426**, 652–655.

CARTER, N. L. & TSENN, M. C. 1987. Flow properties of continental lithosphere. *Tectonophysics*, **136**, 27–63.

CATTIN, R. & AVOUAC, J. P. 2000. Modelling moutain building and the seismic cycle in the Himalaya of Nepal. *Journal of Geophysical Research*, **105**, 13389–13407.

CATTIN, R., MARTELET, G., HENRY, P., AVOUAC, J. P., DIAMENT, M. & SHAKYA, T. R. 2001. Gravity anomalies, crustal structure and thermo-mechanical support of the Himalayas of Central Nepal. *Geophysical Journal International*, **147**, 381–392.

CHEN, Q., FREYMUELLER, J. T., YANG, Z., XU, C., JIAN, W., WANG, Q. & LIU, J. 2004. Spatially variable extension in southern Tibet based on GPS measurements. *Journal of Geophysical Research*, **109**, DOI 10.1029/2002JB002350.

GAME DATA CENTRE. 2005. World Wide Web address: ftp://ftpprd.ncep.noaa.gov/pub/cpc/fews/S.Asia/

GODARD, V., CATTIN, R. & LAVÉ, J. 2004. Numerical modelling of mountain building: interplay between erosion law and crustal rheology. *Geophysical Research Letters*, **31**, DOI 10.1029/2004GL021006.

HASSANI, R., JONGMANS, D. & CHÉRY, J. 1997. Study of plate deformation and stress in subduction processes using two-dimensional numerical models. *Journal of Geophysical Research*, **102**, 17951–17965.

HENRY, P., LE PICHON, X. & GOFFÉ, B. 1997. Kinematic, thermal and petrological model of the Himalayas: constraints related to metamorphism within the underthrust Indian crust and topographic elevation. *Tectonophysics*, **273**, 31–56.

HOVIUS, N., STARK, C. P. & ALLEN, P. A. 1997. Sediment flux from a mountain belt derived from landslide mapping. *Geology*, **25**, 231–234.

HOWARD, A. D., DIETRICH, W. E. & SEIDL, M. A. 1994. Modelling fluvial erosion on regional to continental scales. *Journal of Geophysical Research*, **99**, 13971–13986.

INDIAN INSTITUTE OF TROPICAL METEOROLOGY (IITM). 2005. World Wide Web address: http://www.tropmet.res.in/

KIRBY, S. H. & KRONENBERG, A. K. 1987. Rheology of the lithosphere: Selected topics. *Reviews of Geophysics*, **25**, 1219–1244.

LARSON, K. M., BURGMANN, R., BILHAM, R. & FREYMULLER, J. 1999. Kinematics of the India-Eurasia collision zone from GPS measurements. *Journal of Geophysical Research*, **104**, 1077–1094, DOI 10.1029/1998JB900043.

LAVÉ, J. 2005. Analytic solution of the mean elevation of a watershed dominated by fluvial incision and hillslopes landslides. *Geophysical Research Letters*, **32**, L11403, DOI 10.1029/2005GL022482.

Lavé, J. & Attal, M. 2003. Landscape evolution due to river incision in active mountains. *Geophysical Research Abstracts*, **5**.

Lavé, J. & Avouac, J. P. 2000. Active folding of fluvial terraces across the Siwaliks Hills, Himalayas of central Nepal. *Journal of Geophysical Research*, **105**, 5735–5770.

Lavé, J. & Avouac, J. P. 2001. Fluvial incision and tectonic uplift across the Himalayas of central Nepal. *Journal of Geophysical Research*, **106**, 26561–26591.

Le Fort, P. 1986. Metamorphism and magmatism during the Himalayan collision. *In*: Coward, M. P. & Ries, A. C. (eds) *Collision Tectonics*. Geological Society Special Publications, **19**, 159–172.

Lyon-Caen, H. & Molnar, P. 1985. Gravity anomalies, flexure of the Indian plate, and the structure support and evolution of the Himalaya and Ganga basin. *Tectonics*, **4**, 513–538.

Métivier, F., Gaudemer, Y., Tapponier, P. & Klein, M. 1999. Mass accumulation rates in Asia during the Cenozoic. *Geophysical Journal International*, **137**, 280–318.

Molnar, P. 2003. Nature, nurture and landscape. *Nature*, **426**, 612–614.

Nepalese Meteorological Forecasting Divison (NMFD). 2005. World Wide Web address: http://www.mfd.gov.np.table.php

NOAA Climate Prediction Centre. 2005. World Wide Web address: http://hydro//iis.u-tokyo.ac.jp/GAME-T.GAIN-T/routine/Nepal

Pandey, M. R., Tandukar, R. P., Avouac, J. P., Lavé, J. & Massot, J. P. 1995. Interseismic strain accumulation on the Himalayan crustal ramp (Nepal). *Geophysical Research Letters*, **22**, 751–754.

Pinet, P. & Souriau, M. 1988. Continental erosion and large scale relief. *Tectonics*, **7**, 563–582.

Schelling, D. & Arita, K. 1991. Thrust tectonics, crustal shortening and the structure of the Far Eastern Nepal Himalaya. *Tectonics*, **10**, 851–862.

Schlunegger, F. & Simpson, G. 2002. Possible erosional control on lateral growth of the European Central Alps. *Geology*, **30**, 907–910.

Summerfield, M. A. & Hutton, N. J. 1994. Natural control of fluvial denudation rates in major world drainage basins. *Journal of Geophysical Research*, **99**, 13871–13883.

Thiede, R. C., Bookhagen, B., Arrowsmith, J. R., Sobel, E. R. & Strecker, M. R. 2004. Climatic control on rapid exhumation along the Southern Himalayan Front. *Earth and Planetary Science Letters*, **222**, 791–806.

Tomkin, J. H., Brandon, M. T., Pazzaglia, F. J., Barbour, J. R. & Willett, S. D. 2003. Quantitative testing of bedrock incision models for the Clearwater River, NW Washington State. *Journal of Geophysical Research*, **108**, 2308, DOI 10.1029/2001JB000862.

Tsenn, M. C. & Carter, N. L. 1987. Upper limits of power law creep of rocks. *Tectonophysics*, **136**, 1–26.

Underwood, P. 1983. *Dynamic Relaxation. Computational Methods for Transient Analysis*. Elsevier, Amsterdam, 245–265.

Vance, D., Bickle, M., Ivy-Ochs, S. & Kubik, P. W. 2003. Erosion and exhumation in the Himalaya from cosmogenic isotope inventories of river sediments. *Earth and Planetary Science Letters*, **206**, 273–288.

Van der Beek, P. A. & Bishop, P. 2003. Cenozoic river profile development in the Upper Lachlan catchment (SE Australia) as a test of quantitative fluvial incision models. *Journal of Geophysical Research*, **108**, 2309, DOI 10.1029/2002JB002125.

Wessel, P. & Smith, W. H. F. 1991. Free software helps map and display data. *EOS Transactions AGU*, **72**, 441, 445–446.

Whipple, K. X. & Tucker, G. E. 1999. Dynamics of the stream-power river incision model: implications for height limits of mountain ranges, landscape response timescales, and research needs. *Journal of Geophysical Research*, **104**, 17661–17674.

Whipple, K. X. & Tucker, G. E. 2002. Topographic outcomes predicted by stream erosion models: sensitivity analysis and intermodel comparison. *Journal of Geophysical Research*, **107**, 2179, DOI 10.1029/2001JB000162.

Willett, S. D. 1999. Orogeny and orography: The effects of erosion on the structure of mountain belts. *Journal of Geophysical Research*, **104**, 28957–28981.

Zhao, W., Nelson, K. D. & Project INDEPTH Team. 1993. Deep seismic-reflection evidence continental underthrusting beneath southern Tibet. *Nature*, **366**, 557–559.

Effects of compaction processes on stresses, faults, and fluid flow in sedimentary basins: examples from the Norwegian margin

KNUT BJØRLYKKE

Department of Geosciences, Box 1047, University of Oslo, 0316 Oslo, Norway (e-mail: knut.bjorlykke@geo.uio.no)

Abstract: Modelling of sediment compaction requires that the rate limiting processes are understood. The compaction of uncemented sediments at relatively shallow burial depths should be modelled as a function of effective stress following soil mechanical principles and using experimental compaction data for calibration. In siliceous rocks chemical compaction is dominant at depths greater than 2–3 km (80–100°C). Chemical compaction should be modelled as a function of the temperature history and the mineralogical and textural composition of the sediments. The rate of chemical compaction for siliceous sediments is to a large extent a function of the quartz cementation, which is an exponential function of temperature, while the effective stress plays a minor role. In the case of carbonate sediments the kinetics of precipitation of cement is much faster and the effective stress is more important than temperature.

The magnitude and distribution of effective *in situ* stresses is a complex function of external tectonic stresses, gravitational forces and fluid pressures. Sediments undergo mechanical compaction when subjected to high effective stress and are much more compressible than basement rocks. Chemical compaction also results in a reduction in rock volume and this has a strong feedback on the *in situ* stresses. If the horizontal stress is greater than the vertical stress, both mechanical compaction and chemical compaction will also occur in the horizontal direction, thus relaxing *in situ* stresses unless there is significant basin shortening. Calculations show that relatively large *in situ* stress anomalies (10 MPa) may be relaxed in 5–10 ka by chemical compaction during basin subsidence. Chemical compaction may also continue during uplift; it is fundamentally different from mechanical compaction and must be modelled separately.

Faults may be conduits or barriers to fluid flow in sedimentary basins. In basins like the North Sea they tend to be barriers (Fisher *et al.* 2003). The formation of faults and their properties with respect to fluid flow depends on the nature and magnitude of stress and on the physical properties of the rocks responding to the imposed stress. Compaction processes determine the rocks' mechanical properties at the time of deformation and, therefore, the properties of the fault zone. The permeabilities of faults are subsequently modified by mechanical stress and cementation along the fault zone.

The compaction of sediments in sedimentary basins involves very complex mechanical and chemical processes. Modelling sediment compaction and fluid flow requires that we can identify the rate-limiting processes and then attempt to quantify these. Compaction processes cause a reduction in fluid volume, which is defined as the porosity, and an increase in rock density. Mechanical compaction is a function of the effective stress (σ_e):

$$\sigma_e = \sigma_t - P_f \quad (1)$$

where σ_t is the total stress which may the weight of the sedimentary overburden and P_f is the pressure in the fluid phase (pore pressure). The volume change of the solid phase is negligible due to the low compressibility of the mineral phases. The change in rock volume is then almost exclusively due to the change in fluid content, which is the porosity (φ), and the fluids must be expelled for compaction to continue. Equation (1), which is the Terzagi equation, has been modified to include a factor α_B, the Biot constant (Biot 1955) which modifies the effect of the pore pressure:

$$\sigma_e = \sigma_t - \alpha_B P_f \quad (2)$$

For sandstones and other relatively permeable sediments the Biot constant is close to 1 but it is

From: BUITER, S. J. H. & SCHREURS, G. (eds) 2006. *Analogue and Numerical Modelling of Crustal-Scale Processes*. Geological Society, London, Special Publications, **253**, 359–379.
0305-8719/06/$15.00

difficult to determine this constant in fine grained rocks like shales where it may be lower. (For discussions see Lothe *et al.* 2004.) Compaction of sediments can be modelled by the rules of soil mechanics (Lambe & Whitman 1979) as long as there is no chemical process involved such as dissolution and precipitation of minerals. This however requires accurate input about the effective stresses as a function of time and the sediment composition, determining the mechanical compressibility.

Chemical compaction involves dissolution and precipitation of solids (mostly minerals) and these processes are controlled by the thermodynamic stability of the mineral phases and by the kinetics of the reactions. The degree to which the system is open or closed with respect to solid transport is critical for basin modelling. On a 1–10 m scale the chemical composition of the sediments may be assumed to remain nearly constant during burial so that cementation equals dissolution, except near the surface or on large permeable fractures. The amount of solids dissolved in the pore water is determined by the solubility of the minerals and is in most cases very small. Precipitation or dissolution due to advection is a function of the fluid flux and the solubility/temperature gradient. Calculations show, however, that in silicates transport of solids in solution by compaction-driven fluid flow is in most cases insignificant due to the low solubility of silicate minerals and the low fluid fluxes (Bjørlykke 1994). The volume of cement (V_c) precipitated is (Bjørlykke 1994):

$$V_c = Ft\nabla T\alpha_T\rho^{-1} \quad (3)$$

Here F is the vertical component of the fluid flux, t is time, ∇T is geothermal gradient, α_T is the solubility/temperature gradient of a mineral phase and ρ is the mineral density. Modelling shows that it requires about 10^9 volumes of water to precipitate one volume of quartz, assuming normal geothermal gradients (Bjørlykke 1994). Also in the case of carbonates the fluid transport is limited in compacting sedimentary basins due to low solubility gradients which are retrograde with respect to temperature (Wood 1994). The solubility gradients imply that if quartz is precipitated by advective flow, 30 to 100 times more calcite will dissolve and the presence of calcite is therefore evidence of limited quartz precipitation (Bjørlykke & Egeberg 1993). The distribution of cement is therefore in most cases controlled by local diffusion rather than by advective transport, except near the surface and along permeable faults. Pore water is normally in equilibrium with calcite which is nearly always present, at least in marine sediments. The volume of the solid phase may, however, be reduced by mineral dehydration and by diffusion over shorter distances. The generation of petroleum also represents a phase transformation of solids (kerogen) to fluids.

Sedimentary rocks in subsiding sedimentary basins differ in very important aspects from metamorphic and igneous basement rocks. Sedimentary rocks with significant porosity (>5%) may shrink significantly by chemical compaction, while this is not the case for well-cemented sedimentary rocks and metamorphic rocks.

By studying well exposed outcrops the geometrical distribution of faults and fractures can be mapped in great detail (e.g., Antonellini & Aydin 1995). In sedimentary basins the composition of faults can only be studied when sampled by cores, but the subsidence and temperature history is mostly well constrained by well logs. In many cases the rocks are presently at their maximum burial depth compared to their previous burial history. The response to external stress depends on the timing of the deformation in relation to compaction and diagenesis during subsidence. It is very important to distinguish between deformation occurring during subsidence when sediments may be normally consolidated, and during uplift when the rocks are mostly over-consolidated and brittle (Figs 1 and 2). In subsiding sedimentary basins mechanical compaction will normally occur under increasing effective stress. The sediments are then normally consolidated and have ductile properties. This means that these sediments

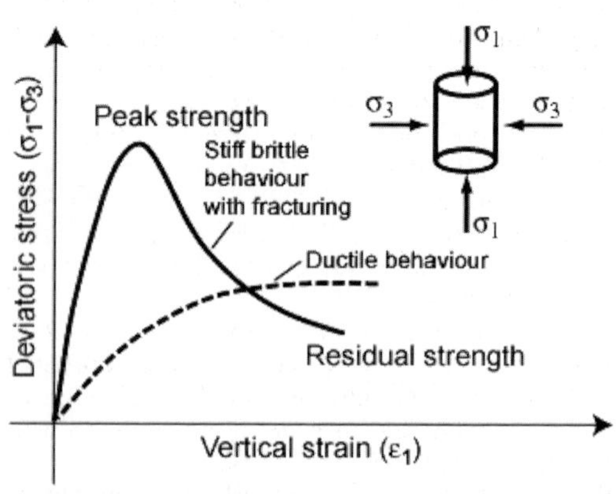

Fig. 1. Principal difference between brittle and ductile response to deviatoric stresses. Normally consolidated rocks will tend to show a ductile response while over-consolidated rocks are brittle. Modified from Bjørlykke and Høeg (1997).

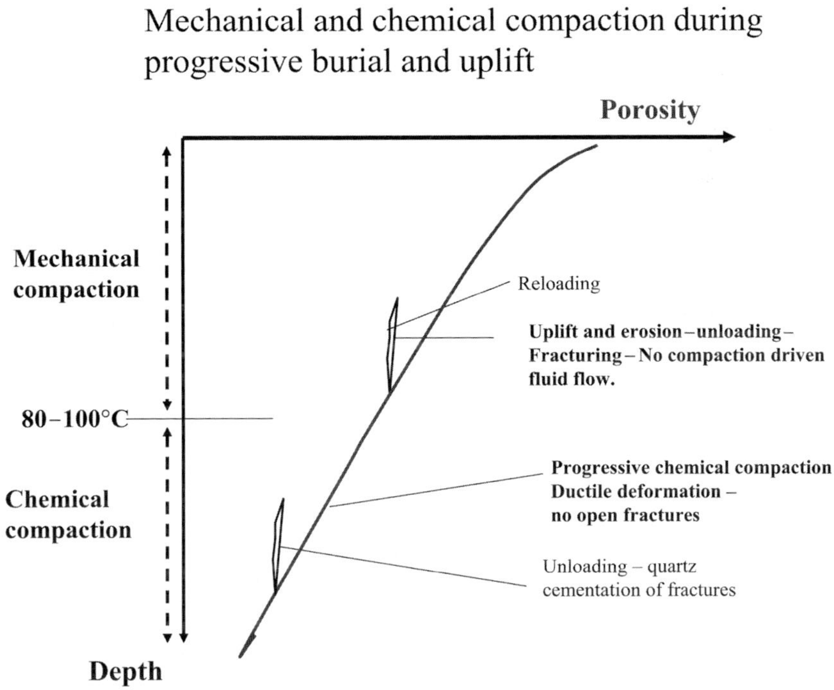

Fig. 2. Rock properties as a function of progressive burial and uplift (exhumation). Reduced overburden stress causes extension and brittle behaviour. No compaction-driven flow then occurs, because there is no porosity loss and lower than hydrostatic pressures may develop due to the extension because of a slight increase in porosity. Chemical compaction due to quartz cementation may, however, continue during exhumation for temperatures >80–100°C, because it is driven by temperature rather than effective stress.

have no peak strength (Fig. 1), are ductile and do not deform by brittle fracturing. They then fall on the solid line for compaction during progressive burial and increasing stress (Fig. 2) (Bjørlykke & Høeg 1997; Bjørlykke 1999). Open fractures can then not exist. Uplift and erosion will, however, reduce the effective vertical stress and cause the sediments to be over-consolidated and brittle. Open fractures are then likely to develop, due to high differential stresses and extension. A build up of overpressure may also reduce the effective stress despite deeper burial and cause sediments to become over-consolidated.

When discussing fluid flow in sedimentary basins we must always consider the source of the fluids and maintain a mass balance. In subsiding sedimentary basins water is expelled upwards at an average rate which is constrained by the porosity loss of the underlying sedimentary sequences. This porosity reduction may be due to mechanical compaction caused by increased effective stress or chemical compaction due to dissolution and precipitation of minerals. In addition water is released by dehydration of minerals, mostly clay and evaporite minerals. Although the average water flow is upwards in relation to the subsiding sedimentary layers, it can be shown to be downwards in relation to the sea floor (Bjørlykke 1994). The actual flow rates are controlled by the distribution of permeabilities in the sedimentary sequences. During uplift (exhumation) there is very little compaction of the sedimentary sequence and if extension and fracturing occurs the pore volume increases, resulting in lower than hydrostatic pressure and a slight downwards flow of meteoric water.

The distribution of stress is critical for both the generation of faults and the properties of the fault zones, and it is important to understand the processes controlling the distribution and magnitude of stresses in sedimentary basins. The present paper will discuss the effect of mechanical and chemical compaction on stresses in sedimentary basins and how this affects the properties of fault zones with respect to fluid flow. The North Sea and Haltenbanken basins of offshore Norway provide good field datasets for studying the effects of compaction processes during subsidence. These principles will apply also to other basins that have been subjected to recent subsidence and where the sedimentary layers are presently at their maximum depth and stress. Examples are basins in a passive margin setting which are formed during rifting and subject to post-rift cooling and subsidence.

The principles discussed in this study are of direct importance for basin modelling studies. Modelling of sediment compaction, distribution of stresses and fluid flow should for shallow depths and cold temperatures be based on soil and rock mechanical principles. For the deeper parts (<80°C–100°C) the principles of chemical compaction apply in siliceous sediments where the rate of compaction is an exponential function of temperature integrated over time (Walderhaug *et al.* 2001; Bjørkum & Nadeau 1998; Bjørkum *et al.* 1998*a*). Failure to distinguish between these two very different processes in basin modelling may produce rather erroneous results. Mechanical compaction will stop or continue at a very slow rate due to creep when the effective stress is constant, while chemical compaction may continue at constant or even much reduced stress. Chemical compaction will also continue at reduced temperature during uplift, but at a slower rate if the temperature is above 80°C in siliceous sediments. The rate of chemical compaction can be modelled as an exponential function of temperature, grain size and other textural parameters controlling the surface area available for quartz cementation, following Walderhaug (1996), Perez *et al.* (1999), Walderhaug *et al.* (2001) and Lander *et al.* (2003). The principles involved in such modelling are relatively well understood, but the main difficulty is to estimate the input parameters for the modelling, particularly the textural parameters such as grain size and the presence of clay coatings.

Distribution of stresses during mechanical compaction

We must distinguish between stresses generated by gravitational forces internally within the basins and stresses imposed by external tectonic forces. In a sedimentary basin the total vertical stress (σ_v) is a function of the weight of the overburden. The effective stress (σ_{ve}) transmitted to the solid phase (mostly minerals) is $\sigma_{ve} = \sigma_v - p_p$, where p_p is the pressure in the fluid phase.

Mechanical compaction of sand as a function of increasing effective stress can be tested in the laboratory. The mechanical compressibility is a function of the composition and sorting of the sand. Well-sorted coarse grained sands are significantly more compressible than fine grained sand, due to grain crushing (Chuhan *et al.* 2002, 2003). At an effective stress of 20–30 MPa (corresponding to 2–3 km of overburden and hydrostatic fluid pressure) coarse grained sand is reduced to 26–30% porosity while fine grained sand has 36–38% porosity. The porosity prior to uniaxial loading was in both cases 42% (Chuhan *et al.* 2002, 2003).The rate of compaction in mudstones varies greatly and smectitic mudstones compact much less and at much lower rates than other types of mudstones. This results in much lower log velocities (Storvoll *et al.* 2005).

In a uniform horizontal sedimentary sequence the horizontal stress should be expected to be the same in all directions. If the horizontal stress is higher than the vertical stress, compaction will also occur in the horizontal direction. A gradual tectonic shortening of the basin is then required to maintain differential stresses. On passive margins horizontal stresses are produced by the ridge push effect (Grollimund *et al.* 2001; Gölke & Coblentz 1996), but these stresses will mainly be transmitted through the basement which is much stiffer (higher modulus) than the overlying sedimentary sequence (Kjeldstad *et al.* 2003). The elevation of basement rocks on a margin mainland will also generate a horizontal stress component similar to ridge push, but this is also mainly transmitted through the basement rather than through the sedimentary sequence (Fig. 3). Across mid Norway and the northern North Sea, some reverse faults can be related to compression in the underlying basement (Dore *et al.* 2002; Faleide *et al.* 2002; Dore & Lundin 1996). Most of the Cretaceous and Cenozoic sedimentary sequence in the North Sea and Haltenbanken (offshore Norway) show, however, no evidence of horizontal compression. The small shortening of the basin indicated by the reverse faults is taken up by compaction of the overlying sedimentary sequence. The late tertiary sequence was probably deposited after the compression which may have been of Miocene age (Dore & Lundin 1996). In the North Sea and Haltenbanken brittle faulting is mostly associated with late Jurassic rifting when much of the basin was uplifted (Gabrielsen *et al.* 2001). Few of the faults extend far into the Cretaceous sequence which was deposited during continued subsidence (Fig. 4).

In this passive margin setting basement faults may still be reactivated because they represent zones of weakness which have lower shear strength than the rest of the basement. In the

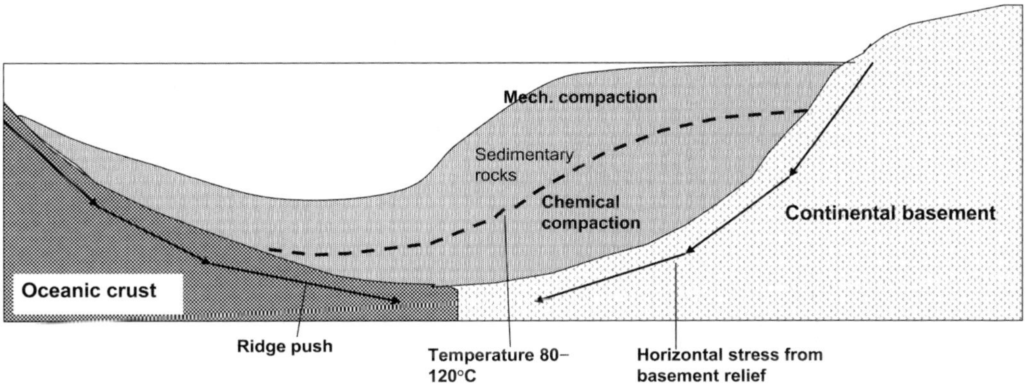

Fig. 3. Schematic presentation of a sedimentary basin in a passive margin setting. The horizontal stress component, resulting from the topography of the spreading ridges (ridge push) and the relief of the continental basement, is transmitted mostly through the basement rocks. Only a small part of this stress is transmitted though the sediments in a subsiding basin where the sedimentary rocks are undergoing mechanical and chemical compaction because of their higher compressibility both mechanically and chemically.

late Mesozoic and Tertiary sequence offshore Norway few faults have formed (Fig. 4) and they are in most cases not reactivation of older faults. Faults in Jurassic reservoirs in the North Sea and Haltenbanken nearly always have lower porosity and permeability than the rock matrix and are often preferentially cemented by quartz cement or carbonate cement (Fisher & Knipe 2001; Fisher *et al.* 2003). They have higher shear strength than the rock matrix and are not zones of weakness which are likely to be reactivated. Based on very extensive studies of faults undertaken at the University of Leeds, Quentin Fisher states: ‘One of the best pieces of evidence that indicates fault reactivation has not lead to leakage is that despite all of the numerous analyses we have conducted on faults from within cores of the reservoirs on Haltenbanken and the North Sea (including those from Visund) – we have never found faults that have shown evidence of being formed at depths greater than 500 m below the seafloor or containing extra-formational cements precipitated by localized fluid flow. This is unlike examples where there is clear seismic evidence of late-stage fault reactivation, where there are often numerous examples of such features’ (Q. Fisher, pers. comm. 2005). There are also few evidences of late faulting on seismic sections (Fig. 4).

In many of the over-pressured fields in the northern North Sea and the Central Graben (offshore Norway) the present day overpressures agree well with the measured fracture pressure (Moss *et al.* 2003). This is certainly also true for the Gullfaks Field (Heum 1996) which is leaking and draining large parts of the Tampen area. Leakage below fracture pressure has been attributed to high horizontal stresses caused by glacial loading (Wiprut & Zoback 2000; Grollimund & Zoback 2003). There is, however, no evidence for glacio-tectonic deformation on seismic sections (Faleide *et al.* 2002). High horizontal effective stress (up to 60 MPa) at about 3 km depth has been inferred from Haltenbanken based on bore hole breakouts (Norgaard Bolås & Hermandrud 2002), but this would have resulted in significant horizontal compaction and grain breakage. This could not have been possible without a shortening of the basement (and the basin) corresponding to the horizontal compaction.

Seabed topography and dipping sedimentary layers of different density will produce gravity-driven differences in the horizontal stresses (Bjørlykke & Høeg 1997). The overall bathymetric relief of the continental slope west of Haltenbanken exceeds 1 km, producing a lateral variation in vertical effective stress of >10 MPa at the same depth. When the horizontal stress is a function of the vertical stress this sea floor topography will set up a corresponding horizontal stress. The slope will produce extension near the top and compression near the base.

When modelling compaction and stresses in sedimentary basins, normally consolidated and over-consolidated sedimentary rocks must be treated differently since they have different compressibilities (Fig. 2). Normally consolidated sediments fall on the continuous stress/strain curve. At hydrostatic pressures the effective stress increases with depth as a function of the sediment density. When sediments are subjected

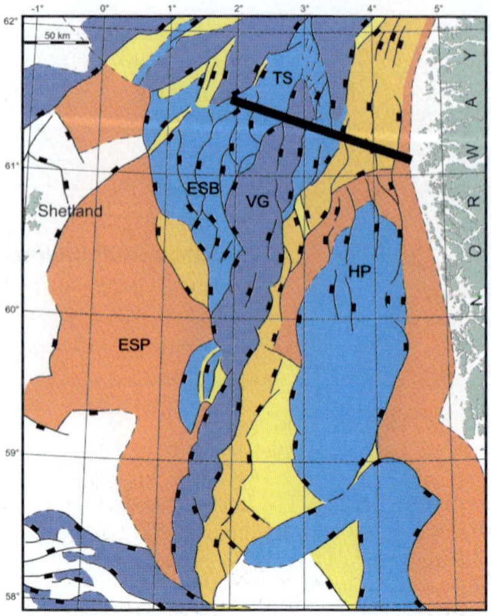

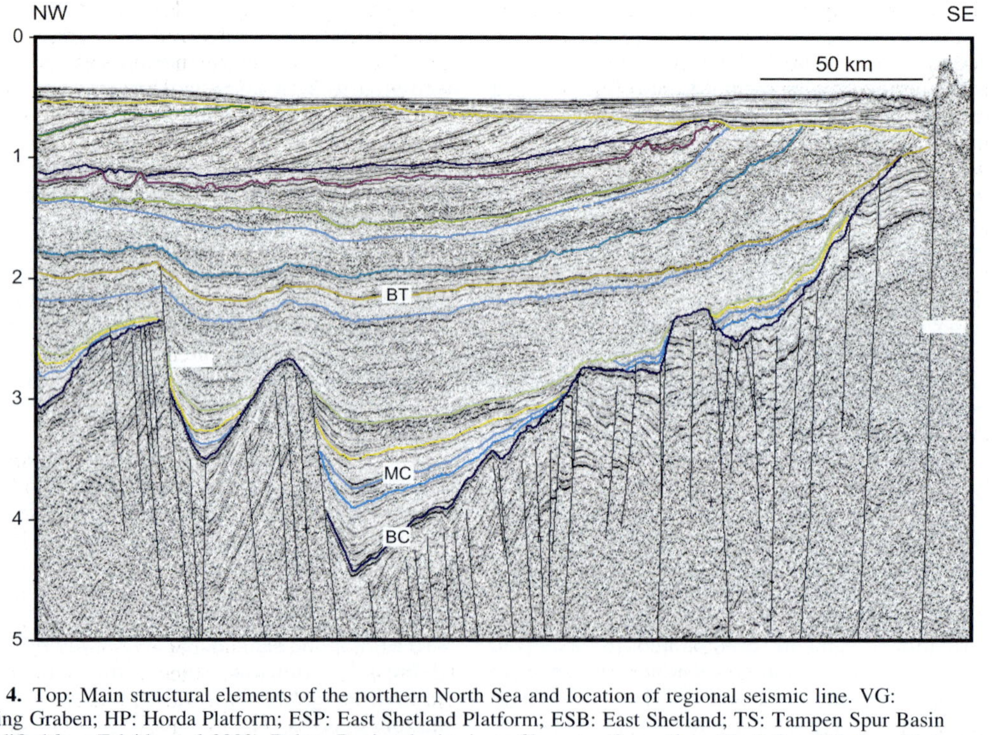

Fig. 4. Top: Main structural elements of the northern North Sea and location of regional seismic line. VG: Viking Graben; HP: Horda Platform; ESP: East Shetland Platform; ESB: East Shetland; TS: Tampen Spur Basin (modified from Faleide *et al.* 2002). Below: Regional seismic profile across the northern North Sea. BT: Base Tertiary; MC: Mid Cretaceous; BC: Base Cretaceous (Modified from Faleide *et al.* 2002).

to increasing effective stress during burial they are normally consolidated and behave in a ductile manner (Fig. 2). After uplift and exhumation the stress is reduced and sediments become mechanically over-consolidated. If they are then subjected to reloading, they have a much lower compressibility (Fig. 2). Where the sediments have earlier been subjected to higher effective stresses before they are over-consolidated, very little compaction will take place. When reloading over-consolidated sedimentary rocks, compaction is partly elastic and also due to closing of open extensional fractures. The stress/strain curve follows the hysteresis loop on the unloading-reloading curve (Fig. 2) and the compressibility is much lower than for normally consolidated sediments. In sedimentary basins, renewed sedimentation after a period of erosion will cause reloading, and normal compaction will resume once the previous maximum effective stress has been exceeded (Fig. 2). In the North Sea brittle faulting occurred during rifting associated with uplift and extension (Færseth 1996), but there is little evidence for that in the Cretaceous and younger sediments. Cementation of the grain framework (e.g., by carbonate or quartz cement) increases the rocks strength and makes the rocks over-consolidated (Fig. 5). Build up of overpressure may also reduce the effective stress and cause over-consolidation.

Erosion reduces the overburden and thus the vertical stress, causing elastic expansion of the sedimentary layers in the vertical direction. Horizontal expansion, however, is usually limited in the horizontal dimensions and the horizontal stresses are, therefore, normally not relaxed to the same extent, resulting in high ratios of horizontal to vertical stress. When sedimentary rocks are normally consolidated they will yield (compact) in response to any increase in the effective stress, following the normal compaction curve (Fig. 2). In extensional stress regimes the highest effective stress will be in the vertical direction due to the weight of the overburden. The horizontal stress is then a function of the vertical stress. Should the horizontal stress be higher than the vertical stress, compaction would result in shortening in the horizontal direction as long as the horizontal stresses are greater than the vertical stresses. If there is no shortening of the underlying basement during continued subsidence, to maintain the high horizontal stress it must, therefore, be lower than the vertical stress. When the vertical gravity-generated stress is the maximum effective stress (σ_{ve}), the horizontal effective stress $\sigma_{he} = K_0 \cdot \sigma_{ve}$ related to the Poisson's ratio (ν). For isotropic uniaxial elastic deformation $K_0 = \nu/(1-\nu)$ (Jaeger & Cook 1979).

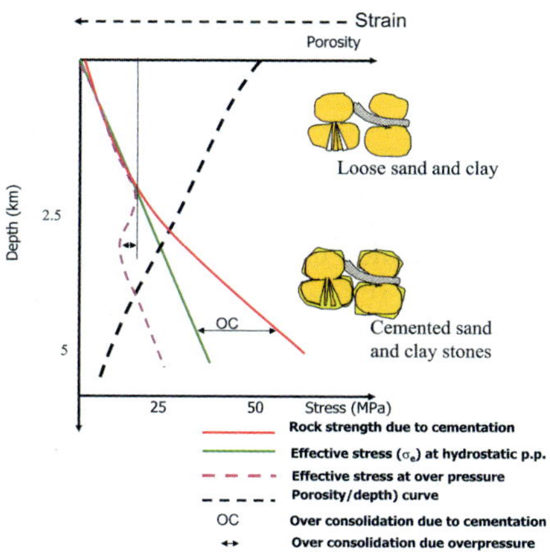

Fig. 5. Changes in stress and rock properties of a sediment layer as a function of increasing burial. Overconsolidation may be produced by overpressure or by cementation. A build up of overpressure may reduce the effective stress so that the rock becomes overconsolidated. At hydrostatic pore pressure the effective stress will increase as a function of depth causing the sedimentary rocks to be normally consolidated. However the rocks may become 'pseudo-over consolidated' due to cementation at shallow depth by carbonate and at greater depth (>80–100°C) by quartz cement.

In the North Sea, the ratio between the total horizontal and vertical stresses (σ_h/σ_v) measured by leak off tests (LOT) may typically be between 0.8–0.95, approaching 1.0 (Moss *et al.* 2003). The relation between the effective horizontal and vertical stresses, particularly in over-pressured rocks may be significantly lower. The values measured by leak-off tests (LOT) may be too high because of the drop in pressure across low permeability mud cakes next to the formation (Raaen *et al.* 2001). Even when taking this into account the stress ratios (σ_h/σ_v) in basins like the North Sea are relatively high. The may be the result of mechanical creep as well as chemical compaction, which will tend to make the stress distribution more isotropic and the ratio between the horizontal and vertical directions closer to 1.0.

Chemical compaction

Chemical compaction involves dissolution at grain contacts which are load bearing and precipitation of cements in open pores where there is no effective stress, a process resulting in reduced rock volume. Strain rates during chemical compaction are controlled by the rate of dissolution of grains and precipitation of cements. Quartz dissolution and cementation is one of the main processes in siliceous sediments, but dissolution and precipitation of clay minerals and feldspars may also play a role. In the deeper parts of sedimentary basins the compaction of siliceous sandstones is mainly chemical and controlled by temperature (Bjørkum 1996; Walderhaug 1996; Walderhaug *et al.* 2001). In sandstones and other siliceous sediments the precipitation of quartz is assumed to be the rate-limiting step and this is a function of temperature rather than stress (Bjørkum 1996; Oelkers *et al.* 1996; Walderhaug 1996; Walderhaug *et al.* 2001). The rate of porosity loss and vertical shortening as a function of time and temperature has been modelled by the authors listed above. The rate of compaction is an exponential function of temperature integrated over time and of the grain surface area available for quartz cementation. The rate of quartz cementation per unit area (available grain surface) (r) can be calculated from (Walderhaug *et al.* 2001):

$$r = Ae^{E_a/RT} \quad (4)$$

Here A is a constant with units of mol cm^{-2} s^{-1}, E_a is the activation energy in J mol^{-1}, R is the universal gas constant and T is temperature in degrees Kelvin. The rate of compaction as function of temperature and time can then be modelled (Fig. 6).

Clastic mica and also clay minerals have a catalytic effect on the dissolution of quartz (Bjørkum 1996) determining where dissolution occurs, but the precipitation stage may still be rate limiting. The quartz dissolution rates in pure water increase by a factor of about 10^5 from 20°C to 100°C (Palandri & Kharaka 2004) and surface controlled precipitation may speed up by a factor up to 10^7 for the same temperature interval (Giles 1997). The precipitation of quartz cement increases the degree of consolidation and rock strength of the sediments in a way which is nearly independent of stress and this is often referred to as pseudo-overconsolidation (Fig. 5).

Compaction of siliceous mudstones is mostly mechanical at temperatures below about 80°C. At higher temperatures (80–110°C) smectite become thermally unstable and opal A and Opal CT is dissolved and the silica reprecipitated

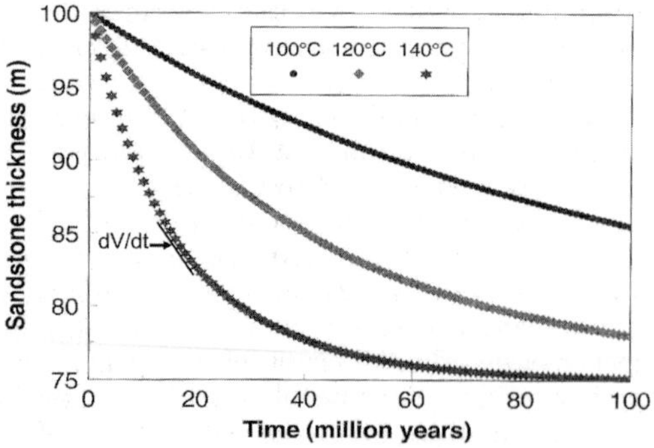

Fig. 6. Results from numerical modelling of chemical compaction of sandstones due to quartz cementation at constant temperatures. Modified from Walderhaug *et al.* (2001). The rate of compaction ($d\varphi/dt$ or dV/dt) at a given time is the derivative of this compaction curve and is an expression of the rate of rock volume reduction.

as quartz. Both mechanical and chemical compaction of mudstones varies greatly with the mineral composition and particularly with the smectite content. Chemical compaction of mudstones is more difficult to model than sandstones because of the uncertainty about textural parameters, but here also quartz cementation is one of the main controls in addition to the stability of the clay minerals. Carbonate cement in shales strongly influences the compaction behaviour of mudstones. They become overconsolidated and much less compressible than uncemented mudstones even at relatively shallow depth (Nygård *et al.* 2004).

In the case of carbonates compaction is more sensitive to stress and to a lesser degree to temperature, compared with siliceous sediments (Giles 1997). The kinetics of carbonate mineral dissolution and precipitation is relatively fast, also at low temperatures. For calcite the activation energy for dissolution and precipitation is much lower than for quartz (Palandri & Kharaka 2004). Increasing effective stress at grain contacts or along stylolites increases the solubility of carbonate minerals like calcite. This provides a thermodynamic drive for chemical compaction. The dissolution step is, therefore, probably rate-limiting for what is often referred to as pressure solution in carbonates, but here also clay plays an important role in the dissolution process. Clay minerals also increase the rate of dissolution and they become concentrated along stylolites where much of the dissolution occurs.

Chemical compaction may also involve the dissolution of hydrous minerals like smectite and kaolinite and precipitation of less hydrous minerals like illite, thus reducing the volume of the solid phase (Fig. 2). The principles of chemical compaction due to diagenetic processes apply also to progressive metamorphism where the increased density of less hydrous minerals accounts for most of the loss of rock volume. This could also potentially reduce *in situ* stresses, but the boundary conditions in terms of tectonic stress may be very different than for sedimentary basins on passive margins.

In summary, chemical compaction is driven by thermodynamics and kinetics, while mechanical compaction is a result of effective stress and rock strength.

Fluid flow in sedimentary basins

In a sedimentary basin water will always be the continuous phase on a large scale and hence control the overall fluid potential. The pressure in the hydrocarbon phase will be controlled by the thickness of the hydrocarbon columns, and the difference in pressure is balanced against the water pressure by the capillary forces.

The flow of water can be modelled using the simple Darcy equation (Verruijt 1982):

$$F = k\frac{\nabla P}{\mu} \qquad (5)$$

The fluid flux (F), however, can in most cases not be modelled from these parameters because the pressure gradient (∇P) and the distribution of permeability, k, are not sufficiently well known within a sedimentary basin. The viscosity (μ) of the fluids is better constrained. To predict the fluid flow in a sedimentary basin the three-dimensional distributions of permeability must be known over large distances and this is generally not possible. The flow in permeable sandstones in the deeper parts of a sedimentary basin depends only to a small degree on the permeability of the sandstones but mainly on the permeability of the surrounding tight shales. The compaction driven fluid flux (F) in the Darcy equation is a function of the rate of compaction and the viscosity, which may be relatively constant over a limited period of geological time. The pressure gradient (∇P) is then a function of the permeability (k).

It is important to note that in the case of chemical compaction in siliceous sedimentary rocks the rate of compaction and fluid flow is a function of temperature and nearly independent of stress and degree of overpressure. The actual flow rate is a function of the distribution of permeabilities. The main factor controlling the degree of overpressure is the minimum permeability along a pathway all the way to the surface. The distribution of pressure is also very difficult to predict because it depends on the distribution of permeabilities in three dimensions. From the Halten Terrace, offshore Mid-Norway, the presence of high overpressure can be shown to depend on the lateral drainage rather than on the permeability of the overburden (Olstad *et al.* 1997). Also in the northern North Sea the pressure compartments are controlled by faults restricting the lateral drainage (Moss *et al.* 2003) which implies that the overpressure is not primarily a function of the permeability of the overlying shale (seal) but of the drainage in three dimensions.

If all input parameters in the Darcy equation were known on a local and regional scale the pressure distribution would already be known and modelling of fluid fluxes would become circular. The main constraint on fluid flow in sedimentary basins is the source of the fluids. Both the meteoric water flowing into the basin and the water flux released by compaction can be quantified or at least constrained. The compaction-driven flux is a direct function of the reduction in porosity in the underlying sediments and the water released by mineral reactions. The contribution from each layer to the compaction-driven water flux is the derivative of the porosity/depth curve, $d\varphi/dz$ (Fig. 2), but this curve is different for each lithology. The upward flow of water is always lower than the sedimentation (subsidence) rate except where there is a very high degree of focussing (Bjørlykke 1993). Assuming a given porosity/depth curve the compaction-driven flux can be calculated as a function of the sedimentation rate.

The effect of faults on fluid flow in sedimentary basins

A clear distinction must be made between flow perpendicular to faults in sandstones and flow along faults. Faults in sandstone reservoirs may be important barriers for fluid flow and the permeability across the fault is controlled both by the mechanical shearing and the subsequent diagenetic reactions. The type of fracture depends both on the magnitude and distribution of effective stresses and the shear strength of the rocks (Fig. 7). Shear deformation in soft (uncemented) sediments at shallow depth is likely to cause compaction (negative dilation, Fig. 7a) while well-cemented sedimentary rocks and metamorphic rocks will tend to develop dilation and increased permeability (Fig. 7b).

Shear deformation in uncemented sand depends on the normal stress and on the composition and size of the grains. This can be tested experimentally. For quartz-rich sand, significant grain crushing starts at 5–10 MPa for coarse to medium grade, while fine-grained sand requires normal stresses exceeding 10 MPa, corresponding to at least 1 km of overburden, for significant grain crushing to occur (Kjeldstad 2002) (Fig. 8). Shear deformation of relatively well-sorted sand produces a dense, interlocking grain fabric and strain hardening (Wong et al. 1997) and reduced permeability even before grain crushing. Because of strain hardening the zone of shear deformation will tend to shift, resulting in a broad zone of shear deformation without a well-defined shear plane. This type of shear deformation in clean sand is not likely to introduce clay smear from adjacent clay layers because the offset will be small.

Diagenetic reactions, particularly quartz cementation, cause a reduction in porosity and

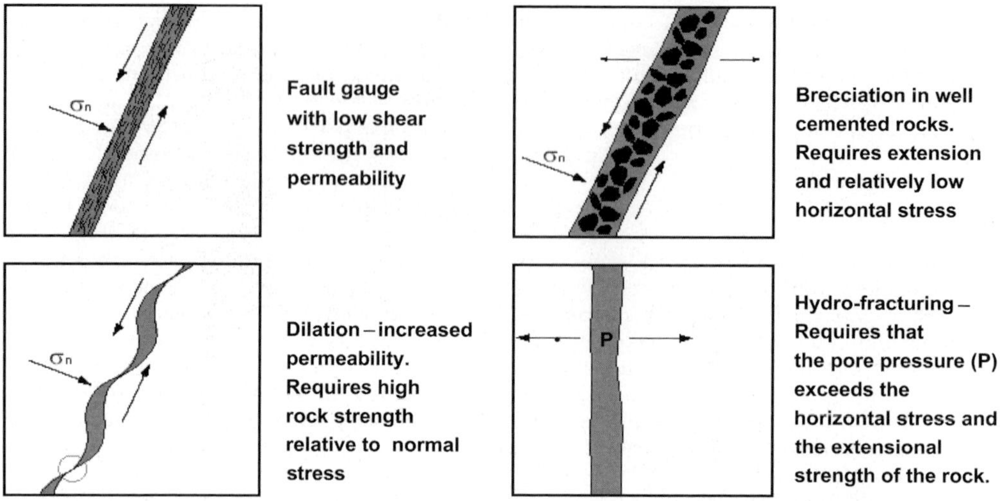

Fig. 7. Different types of fractures. In soft sediments (low shear strength) a fault gauge with negative dilation may develop at relatively low normal stresses (6a). In well cemented rocks and metamorphic rocks shear deformation may result in brecciation (6c) and dilation (6b), causing increased permeability. When the pore pressure exceeds the minimum stress, normally the horizontal stress, hydro-fracturing occur (6d).

permeability as a function of time and temperature during burial. This process can be modelled numerically (Walderhaug 1996; Walderhaug *et al.* 2001) and such modelling is done on a commercial basis in most major oil companies. The porosity in the shear zone at the onset of quartz cementation is one of the most critical factors determining the final porosity and permeability (Fisher *et al.* 2003). If the shearing occurs after the precipitation of clay coating, grain crushing will expose fresh quartz surfaces, promoting higher rates of quartz cementation within the fault plane (Fisher & Knipe 2001; Chuhan *et al.* 2002; Hesthammer *et al.* 2002; Fisher *et al.* 2003). Diagenetic healing of faults may change their permeability and capacity to serve as pressure barriers through time as shown by examples from the Haltenbanken area (Olstad *et al.* 1997; Karlsen *et al.* 2004). Small amounts of clay, like mica and kaolinite, present in the sand will be concentrated along the fault plane and may later develop into stylolites which have a strong effect on the flow properties, reducing the permeability and increasing the capillary entry pressure. The effective permeability of shear zones is difficult to predict when the fault zone is broad and diffuse and contains numerous discontinuous shear zones.

In the Haltenbanken basin and also the North Sea basin, faulting occurred during Late Jurassic rifting (Corfield & Sharp 2002) (Fig. 9). Most of the faults seem to end near the base Cretacous reflector. Even so the permeability is relatively low due to dense grain packing (Fisher & Knipe 2001). This implies that the faults intersecting Middle Jurassic reservoir rocks were active at a time when the reservoir sand was buried at a depth of only a few hundred metres. Faults from the Kristin Field at Haltenbanken show shear bands which have probably formed during strain hardening in relatively clean sand. This has resulted in a dense packing of small and coarse grains giving relatively low porosity and permeability compared to the matrix (Fig. 10). Here the shearing has occurred before the precipitation of grain-coating chlorite and illite. The grain coating is therefore well preserved and consequently there is little quartz cementation in the shear zones (Fig. 11). At burial depths less than 600–700 m the vertical stress may have been 5–6 MPa and the horizontal stress would have been less than 5 MPa. This is insufficient horizontal stress to cause grain crushing in medium to fine grained sand (Kjeldstad *et al.* 2003) (Fig. 7).

In metamorphic rocks, fractures account for nearly all the porosity and changes in the apertures of faults and fractures provide a mechanism for seismic pumping, as described by Sibson (1981). In sedimentary basins, however, most of the fluids are stored in the pores, even in well-cemented sandstones, and the effects of fractures are very different. Well cemented sedimentary rocks and basement rocks may cause dilation when sheared (Fig. 7), while in poorly cemented weak rocks shearing is likely to

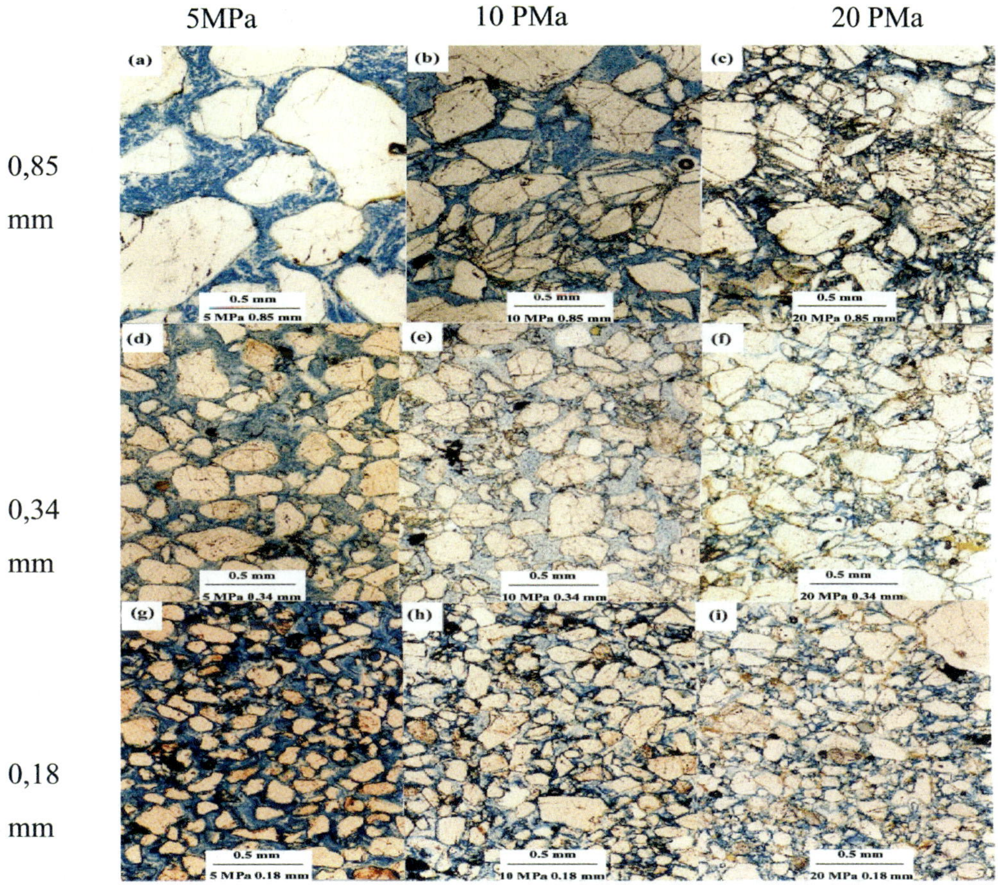

Fig. 8. Experimental shearing of sand with different grain sizes (from Kjeldstad 2002). Coarse grained sand (0.85 mm) becomes more fractured than fine grained (0.18 mm) sand subjected to the same stress. At normal stresses below 10 MPa there is little grain crushing in fine to medium grained sand. At 20 MPa there is significant grain crushing in coarse grained sand.

cause compaction or negative dilation, as shown by experimental shearing (Fig. 8).

Extensional faulting due to rifting is often associated with uplift and erosion of the margin. This will stop mechanical compaction because of the reduced effective stress and it will reduce the rate of chemical compaction by quartz cementation. There will therefore be little porosity reduction or no compaction-driven fluid flow to fill the fractures. The increase in porosity caused by fracturing may lower the pore pressure below hydrostatic levels and provide a potentiometric gradient for a drawdown of meteoric water from above (Bjørlykke 1999).

Permeable faults and fractures are rarely filled with cement precipitated from advective water flow along the fractures. Quartz cementation requires temperatures above 70–100°C and much higher fluid fluxes than can realistically be obtained, while calcite will dissolve when upward-flowing water is cooled (Bjørlykke & Egeberg 1993). Modelling of fluid flow in sedimentary basins shows that precipitation of one unit volume of quartz by upwards flow of cooling water will cause dissolution of 30–100 unit volumes of calcite, depending on the pH and the geothermal gradient (Bjørlykke & Egeberg 1993). This means that the presence of calcite precludes significant introduction of quartz by advection. Fractures are most likely to be filled by crystal growth sourced by diffusion from the adjacent rocks (Egeberg & Saigal 1991; Hesthammer *et al.* 2002). Fractured grains provide good nucleation sites for quartz overgrowth, and fault zones, therefore, often have more quartz cement than the matrix

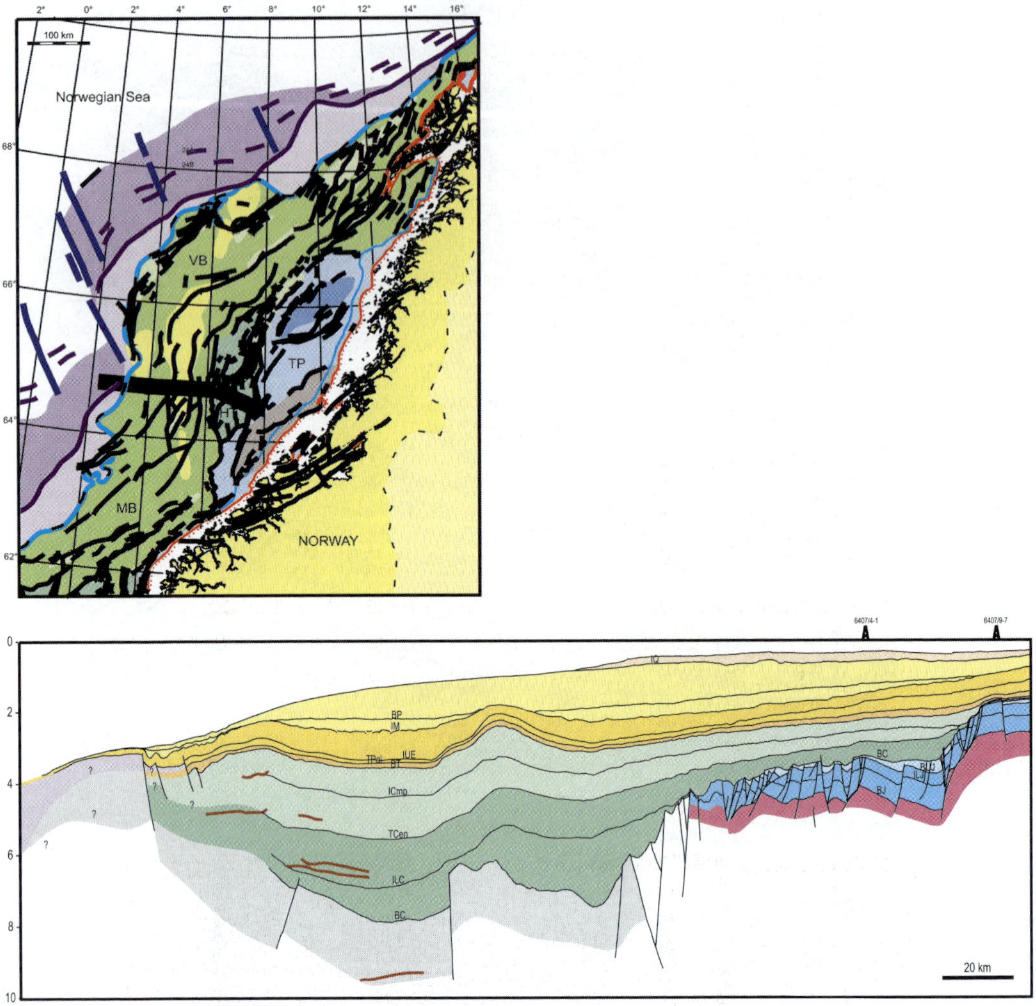

Fig. 9. Main structural elements of the continental shelf and margin off mid-Norway (top) and interpreted regional seismic profile (below). MB: Møre Basin; VB: Vøring Basin; HT: Halten Terrace; TP: Trøndelag Platform. The thick solid line in the top figure indicates the location of the regional profile. (modified from Blystad *et al.* 1995).

(Fisher & Knipe 2001). They are therefore unlikely to be reactivated, because the fault plane has acquired higher shear strength than the matrix. Reactivation of basement faults and differential compaction and stress along juxtaposed lithologies may, however, produce new faults close to older cemented faults. In this case failure is not due to reduced shear strength along pre-existing faults. The rate of cementation of fractures depends on the porosity and diffusion constant in the rock matrix adjacent to the fracture.

Fractures in well-cemented sedimentary rocks and in metamorphic rocks may remain open longer than in porous sandstones. In these rocks, faulting produces a plane of weakness (reduced shear strength) and when subjected to repeated shear stresses the shearing tends to follow existing shear planes. Clay-rich sediments may develop a clay smear along fault planes which results in strain softening, and these faults may therefore be reactivated. During burial, however, the shear strength of clayey sediments may also change, due to increased compaction or diagenetic alteration. Studies in the North Sea faults show that burial depth and temperature have a major effect on the permeability of faults (Fisher & Knipe 2001).

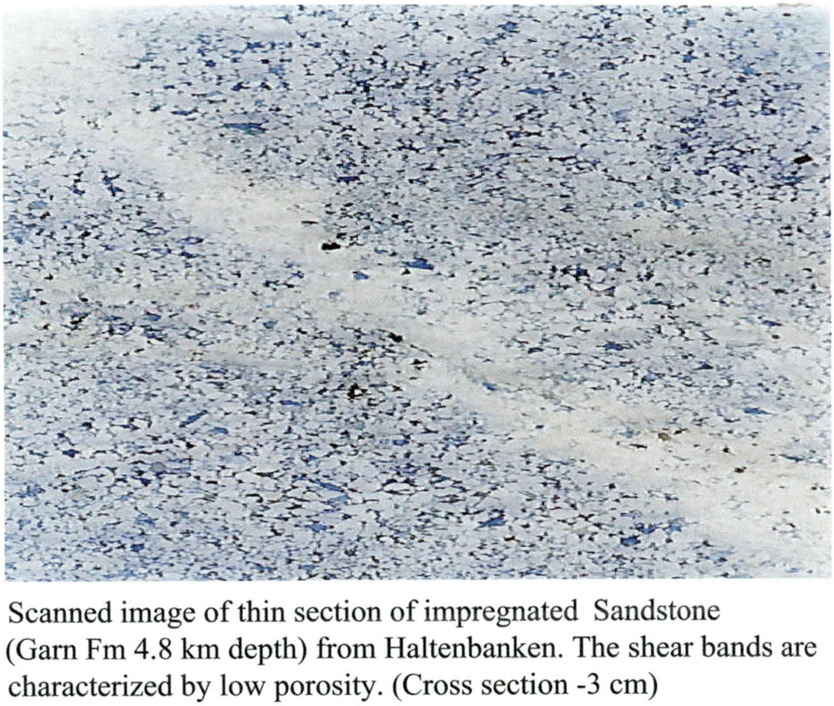

Fig. 10. Scanned image of thin section (3 cm), impregnated with blue epoxy. From the Middle Jurassic Garn Formation, Kristin Field, 5 km depth. Note that the shear zone produced by strain hardening has low porosity.

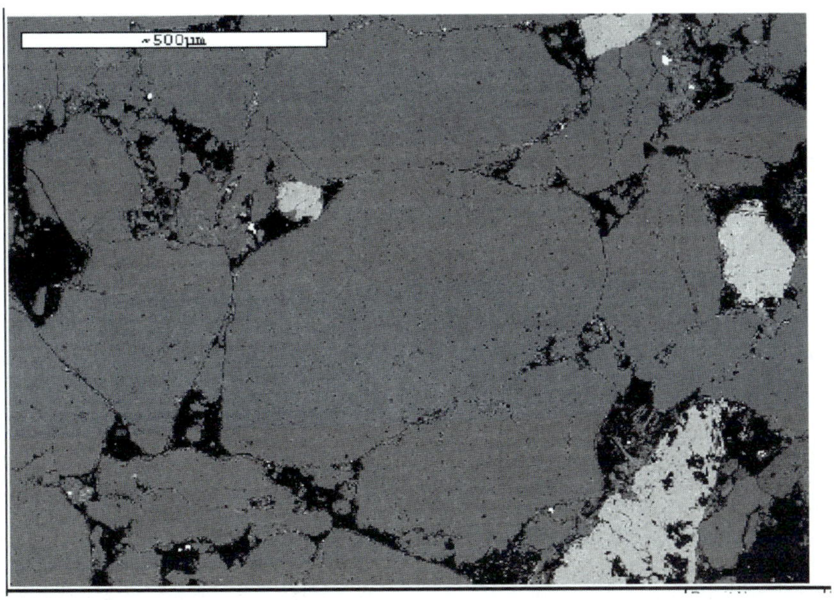

Fig. 11. SEM picture of shear zone in the Kristin Field produced by dense packing of grains resulting in strain hardening, but very little grain breakage. The grains were subsequently coated with chlorite coating which has prevented quartz overgrowth, preserving much of the remaining porosity.

Recrystallization of clay, such as the illitization of smectite, and precipitation of quartz along the fault plane, can increase the shear strength of the fault gouge so that it becomes stronger than the matrix. Carbonate cementation along the fault plane may also increase the shear strength.

Even if faults have sufficient permeability not to be pressure barriers for water over geological time, they may be sealing for oil and gas in water-wet reservoirs. During production the sealing capacity becomes even higher than in a static setting and at the Snorrre Field in the northern North Sea sealing capacities up to 15 bar (1.5 MPa) for hydrocarbon flow across some faults (Sverdrup *et al.* 2003).

Cap rock integrity and leakage

Cap rock may leak due to intergranular flow when the capillary entry pressure for the hydrocarbons is exceeded, or due to hydro-fracturing. The North Sea Basin provides good examples of over-pressuring and hydro-fracturing (Gaarenstroom *et al.* 1993; Leonard 1993; Heum 1996; Isaksen 2004). Shales are in most cases impermeable for oil and fracturing is, therefore, required before leakage can take place. The permeability in mudstones may be low at relatively shallow depth due to carbonate cementation, but clay mineral reaction and quartz cementation are important in the making of a tight seal. Fracturing occurs when the pore pressure (P_p) exceeds the least principal stress. In subsiding sedimentary basins this is normally the horizontal stress (σ_h). The requirement for hydrofracturing is that $P_p > \sigma_h + \tau$. The tensional strength (τ) is assumed to be small compared to σ_h.

The formation or reactivation of brittle faults and fractures may cause leakage at pressures below the normal fracture pressure (Sibson 2003). Chemical compaction involving the cementation of faults and fractures reduces the probability of fault reactivation. Even if high overpressure makes cap rocks brittle in a mechanical sense, chemical compaction will produce a ductile response to low strain rates as discussed above. Furthermore, shear failure will not necessarily cause leakage since it may also result in negative dilation if the normal stress is high compared to the rock strength.

The thickness of the hydrocarbon column determines the pressure difference between the hydrocarbon phase and the water phase, but this difference is accommodated by the capillary pressure. If the cap rock is water-wet, fracturing of the cap rock will occur when the pressure in the water phase exceeds the fracture pressure, because it is the water phase which is critical for fracturing in a water-wet cap rock. Thus if the capillary entry pressure for oil or gas of the cap rock is not exceeded, the thickness of the hydrocarbon column has little effect on the risk for leakage by hydro-fracturing (Bjørkum *et al.* 1998*b*).

The effects of uplift and erosion on pore pressure have been modelled (Luo & Vaseur 1995) based on assumptions about mechanical compaction, but these models do not take the effects of chemical compaction into account. In the case of hydrostatic pressure erosion will cause an overpressure which takes time to bleed off (Luo & Vaseur 1995). Parts of basins which are under high overpressure may upon uplift and leakage continue to compact due to increasing effective stress, and behave as normally consolidated rocks in a ductile manner for a period when the effective stress increases during deformation.

High overpressures occur most commonly in pressure compartments with limited lateral drainage. If the pore pressure approaches the fracture pressure, the strength of the cap rock cannot prevent fracturing. The compaction-driven water flux in siliceous sediments at temperatures above about 80–100°C is primarily controlled by temperature and time (Walderhaug 1996). The underlying sediments will therefore continue to compact and provide water flux even if the effective stress is reduced.

Calculation of relaxation of tectonic stresses by mechanical and chemical compaction

During progressive burial in sedimentary basins, sediments are usually subjected to increasing effective stresses and, if there is limited cementation (chemical compaction), are normally consolidated. The sedimentary rocks will then compact and yield to any increase in the effective stress, potentially in both the vertical and horizontal directions. High lateral stress resulting from ridge push will primarily be transmitted through oceanic crust and continental basement (Fig. 3). Also, well cemented sedimentary rocks with low porosity and negligible potential for further compaction could transmit stress over long distances.

High horizontal (tectonic) stresses in subsiding sedimentary rocks will gradually be relaxed by the compaction process unless there is additional shortening of the basement across the basin. For normally consolidated sediments, increases in the effective stress will cause compaction and this then reduces the stress if there is no basin shortening. The vertical stress σ_v is driven by the weight of the overburden and is usually the direction of maximum stress.

Without significant shortening of the underlying basement rocks, imposed horizontal stresses, e.g., from ridge push, will be relaxed by the compaction process. If there was a net shortening of the basin and the sedimentary layers, the sediments would still be normally consolidated with respect to the horizontal stress, and compact horizontally. Experimental compaction (Chuhan *et al.* 2002, 2003) shows that well sorted medium to fine grained sands with an initial porosity of about 42% have been compacted to 30–37% at 20 MPa effective stress (σ_{ve}), corresponding to about 2 km burial depth where there is little quartz cement. In the North Sea, reservoirs at 2–2.5 km depth have maximum porosities between 30 and 35% (Ramm & Bjørlykke 1994). A 20 MPa increase in effective stress (σ_{ve}) results in a porosity reduction of 5–12% (0.05–0.12). This is equal to a compression modulus of $B = \sigma_{ve}/\varepsilon_v = 400$–150 MPa.

If the width of the basin remains constant, nearly all the compaction (strain) is in the vertical direction. When subjected to increased horizontal stress exceeding the vertical stress, the basement must be shortened to compensate for the strain in the horizontal direction. Horizontal stresses exceeding the vertical stress are therefore only possible in over-consolidated sediments, because normally consolidated sediments would compact mechanically and produce a significant inelastic horizontal strain which could also involve significant grain crushing (Figs 2 and 8). The maximum horizontal stress is therefore governed by the degree of over-consolidation reducing mechanical compaction. For chemical compaction the precipitation of quartz is assumed to be the rate-limiting step and is controlled by kinetics (temperature) (Bjørkum 1996; Walderhaug *et al.* 2001). Even though the dissolution at grain contacts requires some stress, the rate of dissolution is mainly a function of the rate of precipitation and therefore nearly independent of the effective stress. The compaction processes involve a reduction in rock volume (shrinkage) and this will relax *in situ* stresses. This is a very slow process which may determine the *in situ* stress in the rocks for slow strain rates. The rate of deformation is controlled by the temperature and the grain surface area available for precipitation (Walderhaug 1996). In well-cemented sandstones with little porosity the potential for chemical compaction is limited by the small surface area for precipitation of cement.

Even if the rate of dissolution is controlled by the rate of precipitation in the open pores, it requires some effective stress and is probably also favoured by the presence of clay and mica (Bjørkum 1996). Dissolution at grain contacts and also along stylolites results in a shortening in the vertical direction, but may also result in a shortening in the horizontal direction when subjected to tectonic stress with a horizontal component. Thus chemical compaction will result in ductile properties with an approximately isotropic distribution of stress for low strain rates.

Only a small fraction of the total strain ($\Delta\varepsilon$) is elastic and reversible (Fig. 12) and chemical compaction corresponding to the elastic strain (ε_{el}) is required to relax a tectonic stress, assuming no additional basin shortening. Thus the change in rock volume due to chemical compaction ($\Delta V/V$) will cause a reduction in stress ($\Delta\sigma$). The rate of stress relaxation ($\Delta\sigma$) can be calculated as a function of the chemical compaction and Young's modulus (E) for elastic deformation:

$$E = \frac{\Delta\sigma}{\Delta\varepsilon_{el}} \tag{6}$$

The volumetric strain $\Delta\varepsilon$ is the change in rock volume $\Delta V/V$ due to compaction and can in the case of sandstones be modelled as a function of temperature and sediment composition (Walderhaug 1996) or estimated from porosity/depth curves ($d\varphi/dz$) and changes in

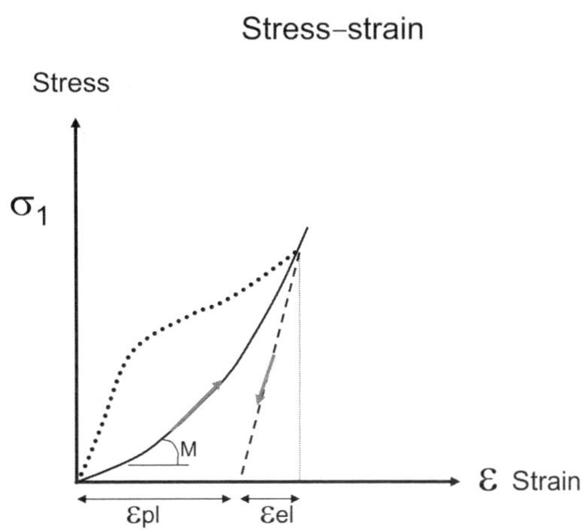

Fig. 12. Stress–strain diagram illustrating that the elastic strain which can be measured after unloading is only a small fraction of the total strain. Chemical compaction corresponding to the elastic strain will however relax *in situ* stresses. M is the modulus which is the ratio between the stress (σ) and the strain (ε).

the rock volume of the solid phase (dV_s/dz):

$$\frac{dV}{dz} = \frac{dV_s}{dz} + \frac{dV_\varphi}{dz} \tag{7}$$

Based on the sedimentation (subsidence) rate (dz/dt) the rate of compaction as a function of time can be calculated:

$$\frac{dV}{dt} = \frac{dV}{dz} \times \frac{dz}{dt} \tag{8}$$

Integrating the volumetric strain (dV/dt) as a function of time and temperature we obtain the total strain ($\Delta\varepsilon$) over a certain or time or depth interval depending on the subsidence rate. Even if this deformation is ductile, the total strain will relax *in situ* stress according to Equation (6) for elastic deformation (Fig. 12). If the elastic Young's modulus E for quartz is 90 GPa (9.10^{10} Pa) (Karner *et al.* 2003), a strain ($\Delta\varepsilon$) of 10^{-4} would cause a reduction in stress ($\Delta\sigma$) of 9 MPa (9.10^6 Pa). The time (t) required to obtain a strain ($\Delta\varepsilon$) of 10^{-4} can then be calculated from the rate of porosity loss dV_φ/dz and the subsidence rate (dz/dt). If the density and volume (dV_s/dz) of solid phase (minerals) remains constant, the reduction in volume dV/dz is equal to the porosity loss dV_φ/dz (Eqn. 7). Mineral reactions involving dehydration such as the dissolution of smectite and kaolinite and the precipitation of illite increase the density of the solid phase and represent compaction in addition to the porosity loss.

The porosity gradient in the North Sea basin is typically 1–2%/100 m (Ramm & Bjørlykke 1994). If the sedimentation rate is 10^{-4} m/yr, the rate of volume loss (dV/dt) is $1–2 \times 10^{-8}$/yr. In porous sediments the change in bulk volume is mostly due to porosity loss (dV_ϕ), and the change in solid phase (dV_s) is small except due to mineral dehydration. The time (t) required to obtain a strain of 10^{-4} is then 5000–10,000 years (Fig. 13) and this will relax the *in situ* stress correspondingly. This means that excess horizontal stresses ($\Delta\sigma = 10$–20 MPa) in basins like the North Sea and Haltenbanken may be relaxed by chemical compaction in 5000–20,000 years. For cemented sandstones the value of E may be lower than for pure quartz and the time required for stress relaxation would be correspondingly longer.

Compaction of mudstones is more complex than the compaction of sandstones. If there is no carbonate cement, mudstones compact mechanically at temperatures up to about 80°C. At higher temperatures chemical compaction may occur due to dissolution and precipitation of clay minerals, mainly dissolution of smectite and precipitation of illite and quartz (80–100°C), and dissolution of kaolinite and K-feldspar and precipitation of illite and quartz (130°C) (Bjørlykke *et al.* 1995). In addition, quartz cementation sourced by amorphous silica and also dissolution of clastic quartz may be important for the compaction process in mudstones turning into shales. The compaction of mudstones is therefore to a large extent controlled by temperature.

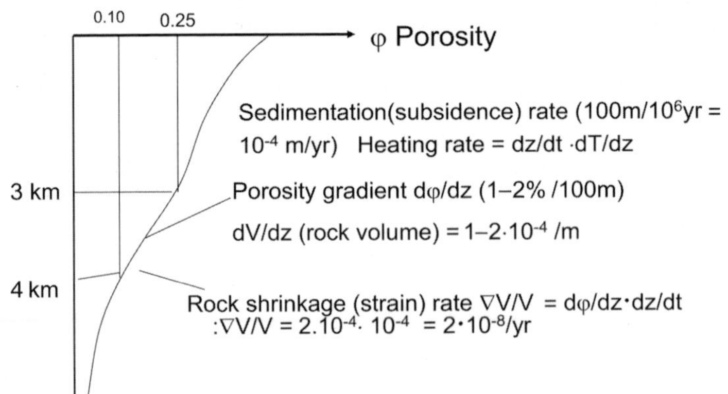

Fig. 13. Calculations showing the rate of stress relaxation due to chemical compaction. If the porosity gradient and the sedimentation (subsidence) rate are known, the rate of volume shrinkage dV/dt can be calculated. In 10^4 years the total strain ($\Delta\varepsilon$) will be 10^{-4}, assuming an annual strain of 10^{-8}. Using the elastic modulus (E) for quartz (90 GPa) the stress will be relaxed by 9 MPa in 10000 years. See discussions in the text.

Carbonate-cemented mudstones and sandstones reach a high stiffness and already have a low compressibility at shallow depth and thus have a smaller potential to relax tectonic stress.

If it is assumed that there is no shortening of the basin due to tectonics, the horizontal stress cannot exceed the vertical stress in subsiding sedimentary basins like the North Sea and Haltenbanken. Mechanical compaction requires continued subsidence and increase in effective stress to produce the shrinkage in rock volume which relaxes the differential stresses. Chemical compaction in siliceous sediments will, however, continue without progressive subsidence or heating and will also occur during uplift and cooling, as long as the temperature is above 80–100°C and the porosity is high enough to accommodate it. The shrinkage of the rock volume resulting from chemical compaction would relax tectonically-produced *in situ* lateral stresses also during uplift.

We may therefore conclude that as long as sedimentary rocks are subsiding and being subjected to compaction, horizontal stresses cannot be higher than the vertical stress in the situation where there is no shortening of the basement. Even if there should be some basement shortening, horizontal stresses would still be reduced by horizontal compaction. The *in situ* stresses would then be a function of the rate of horizontal shortening, i.e., due to tectonic stress, the mechanical compressibility of the rocks and the rate of chemical compaction. The minimum horizontal stress (σ_{vmin}) in both the northern North Sea and Haltenbanken, as measured by the fracture pressure, is close to the overburden stress (σ_v) (Moss *et al.* 2003). If the maximum horizontal stress cannot exceed the vertical stress it follows that $(\sigma_{hmax}/(\sigma_{hmin})$ must be close to 1.0.

Horizontal stress anomalies may, however, be produced by a component of the gravitational force due to sloping sea floors or dipping sedimentary layers of different density. In sedimentary basins subjected to uplift and erosion, reduction of the vertical stress due to unloading may lead to a situation where the horizontal stresses exceed the vertical stress if the former are not reduced to the same extent due to lack of expansion in the horizontal direction for extensive layered rocks. While mechanical compaction will stop, chemical compaction will continue and reduce the horizontal stress after uplift if the temperatures still are high enough for significant quartz cementation (>80–90°C).

Chemical compaction may, as shown above, relax differential stress produced tectonically and reduce the build-up of *in situ* stresses which can cause brittle deformation and earthquakes. Focal depths are reported to be deep in the southern North Sea, northern North Sea and offshore Mid-Norway, while they are shallow in onshore western Norway, onshore mid-Norway and in Finnmark, Northern Norway (Hicks *et al.* 2000). It is clear that the brittle basement rocks of onshore Norway have shallow epicentres compared to areas where there are 5–10 km or more of sedimentary rocks. This is probably because basements rocks have very limited potential for both mechanical and chemical compaction and external stress may then result in brittle failure.

Stress orientations based on bore hole breakouts in the North Sea and Haltenbanken have been interpreted to be due to plate tectonics stress (ridge push) (Lindholm *et al.* 1995; Hicks *et al.* 2000). The ridge push stress is transmitted through oceanic crust into continental crust and it is difficult to see how much of this stress should be transmitted into the more compressible sediments. The sediments in the uppermost 2–3 km burial depth undergo compaction, mostly due to mechanical compaction, and that can also occur in the horizontal directions. The deeper part of the basin will be dominated by chemical compaction which also can have a horizontal component if there are high horizontal stresses (Fig. 13). Even if there should be some shortening in the basement, high compressibility in the overlying sedimentary rocks may prevent the build up of high *in situ* stresses. The orientation of stresses indicated by earthquakes, bore hole breakouts and over-coring shows that the stress anomalies are not a direct result of the ridge push effect (Byrkjeland *et al.* 2000).

The basement rock underlying the North Sea and Haltenbanken consists of metamorphic rocks, mostly formed at depths exceeding 20–30 km, and subsidence of up to 10 km will not result in significant compaction and reduction of *in situ* stresses. Shear stress may therefore build up to reach failure while the stress in the sedimentary sequence above may be relaxed by chemical compaction. Well-cemented sedimentary rocks with low porosity have limited potential for further compaction and will be able to sustain stresses nearly to the same degree as the basement.

Implications for basin modelling

Modelling of sediment compaction is an important part of basin analyses because it affects the sediment density and rock properties and, therefore, the distribution of stress. This study has shown that sediments compact by very different processes and any modelling must first identify the critical and rate-limiting processes.

Mechanical compaction of sedimentary rocks is in principle well understood as described in rock and soil mechanics. The main problem is to predict the highly variable physical properties of sedimentary rocks and the processes responsible for the stresses. It is important to distinguish between external tectonic stress and gravitational stress generated within the basins. Mechanical compaction is dominant in uncemented sand and in mudstones without significant carbonate cement.

At depths greater than 2–3 km (>80–100°C) both sandstones and mudstones tend to become mechanically 'pseudo over-consolidated' due to cementation, and further compaction is controlled mostly by chemical processes. Carbonate cements in sandstones and mudstones at shallow depth drastically reduce the mechanical compressibility upon further subsidence. Chemical compaction follows completely different principles compared to mechanical compaction. In the case of siliceous sediments dissolution and precipitation of quartz exerts the main control on the rate of compaction. As discussed above, chemical compaction can be modelled as an exponential function of temperature and time (Walderhaug 1996; Bjørkum & Nadeau 1998; Walderhaug *et al.* 2001; Lander *et al.* 2003). The main problem is to predict the initial sediment composition, including parameters such as grain size and grain coatings. Using the observed amounts of quartz cement in sandstones the burial (temperature) history can be modelled using, for example, the EXEMPLAR model (Perez *et al.* 1999). When modelling compaction of mudstones the thermodynamic stability of the clay minerals, such as smectite and kaolinite, are important because the solid phase is dissolved and reprecipitated as less hydrous minerals. In carbonate rocks the kinetics of mineral reactions are much faster than in siliceous rocks, implying that dissolution at grain contacts (pressure solution) rather than precipitation may be rate limiting. Compaction of carbonate rocks is therefore more sensitive to effective stress and not very sensitive to temperature.

Both mechanical compaction and chemical compaction are sensitive to the primary rock composition. Modelling of sediment compaction and fluid flow on a regional scale in a sedimentary basin requires knowledge of the distribution of sedimentary facies and also of provenance. Basin modelling must therefore be linked to facies analyses and sequence stratigraphy. The properties of shear zones depend not only on the stress distribution, but also substantially on the timing relative to mechanical and chemical compaction processes (diagenesis). Studies of outcrops may be a good basis for mapping out fault zones, but they provide very poor constraints on the rock properties at the time of deformation. In subsiding basins the burial history relative to deformation episodes is much better constrained, thus providing a better basis for realistic basin modelling in combination with experimental compaction data.

Hydraulic leakage by pressure and stress simulations have been modelled using examples offshore Mid-Norway (Lothe *et al.* 2004). Simulations of the build up of pore pressures are, however, very difficult because of the poorly defined permeability distributions in the shales (seals). Even if the temperature history of the sediments can be relatively well constrained, the rate of chemical compaction in shales and sandstone, and thereby the fluid flux, will also to a large degree depend on mineralogical and textural parameters which are poorly known. This is also the case for the degree of focusing of the fluids and lateral drainage. The distribution of stresses may influence the fracture pressure, but the rate of chemical compaction and the build up of pore pressure by chemical compaction is nearly independent of the effective stresses. It is very difficult to model the build up of pore pressures and the timing for hydrofracturing because of the poorly constrained variables and the variable sediment composition. In particular the distribution of permeabilities in three dimensions will vary greatly.

Conclusions

Realistic modelling of sediment compaction and rock properties requires that the rate-limiting processes are identified and then quantified. The properties of faults and fractures with respect to fluid flow must be predicted based on rock and soil mechanical principles and on chemical processes including mineral precipitation and dissolution. These are very different processes that must be modelled separately. Mechanical compaction should be based on the principles of soil mechanics when the sediments are mostly uncemented. In calcareous sediments chemical compaction may start at relatively shallow depth. In siliceous sediments quartz dissolution and cementation is controlling much of the chemical compaction and this starts at 2–3 km (>80–100°C) and must be modelled mainly as a function of temperature and sediment composition.

Faults rarely serve as conduits for fluid flow during continued subsidence and compaction in sedimentary basins. When fracture pressure is reached, leakage will occur by hydro-fracturing independently of the faults. During uplift and extension fractures tend to be open, but there is little compaction-driven fluid flow along fractures because the sediments are not compacting. The horizontal stresses must be lower than the vertical stress in a basin that is subsiding without significant shortening.

Compaction in siliceous sedimentary rocks can be modelled as a function of the rate of quartz cementation which is controlled by temperature and time. Chemical compaction causing reduced porosity and reduced rock volume will then occur both during subsidence and uplift as long as the temperature is high enough for quartz cementation to continue. Shrinkage of the rock volume represents a strain which will also reduce the elastic *in situ* stress in the horizontal direction, which may then occur both during subsidence and uplift as long as the chemical compaction continues. In the case of well-cemented sedimentary sandstones and shales the porosity may be too low for significant continued chemical compaction and this is also usually the case for metamorphic rocks. High horizontal stress may then be preserved during exhumation and reduction of the vertical stress, resulting in high ratios between horizontal and vertical stress. During subsidence of reservoir rocks the *in situ* lateral stresses at deeper burial are to a large extent controlled by chemical compaction. During petroleum production, however, increased effective stress due to reduced fluid pressure will cause a purely mechanical response at least in siliceous sedimentary rocks, changing the ratio between the horizontal and the vertical stresses.

The principles involved in sediment compaction are relatively well known and may be modelled as functions of effective stresses, temperature and time. Such models are however very sensitive to variations in the primary sediment composition, and basin modelling therefore requires input from facies analyses and studies of sediment provenance.

The author would like to thank K. Høeg, J. Jahren and A. Read for comments on the manuscript and J. I. Faleide and P. Aagaard for discussions. Thanks are also due to the reviewers K. Heffer and A. Aydin. The editor S. Buiter has helped to improve the manuscript and to make it more accessible to readers from other disciplines. This contribution was based on projects funded by NFR and Statoil (Vista).

References

ANTONELLINI, M. & AYDIN, A. 1995. Effect of faulting on fluid flow in porous sandstones: geometry and spatial distribution. *American Association of Petroleum Geologists Bulletin*, **79**, 642–671.

BIOT, M. A. 1955. Theory of elasticity and consolidation for a porous anisotropic solid. *Journal of Applied Physics*, **26**, 182–185.

BJØRKUM, P. A. 1996. How important is pressure in causing dissolution of quartz in sandstones? *Journal of Sedimentary Research*, **66**, 147–154.

BJØRKUM, P. A. & NADEAU, P. H. 1998. Temperature controlled porosity/permeability reduction fluid migration, and petroleum exploration in sedimentary basins. *Australian Petroleum Production and Exploration Association Journal*, **38**, 453–465.

BJØRKUM, P. A., OELKERS, E. H., NADEAU, P. H., WALDERHAUG, O. & MURPHY, W. M. 1998*a*. Porosity prediction in quartzose sandstones as a function of time, temperature, depth, stylolite frequency, and hydrocarbon saturation. *American Association of Petroleum Geologists Bulletin*, **82**, 637–648.

BJØRKUM, P. A., WALDERHAUG, O. & NADEAU, P. H., 1998*b*. Physical constraints on hydrocarbon leakage and trapping revisited. *Petroleum Geoscience*, **4**, 237–239.

BJØRLYKKE, K. 1993. Fluid flow in sedimentary basin. *Sedimentary Geology*, **86**, 137–158.

BJØRLYKKE, K. 1994. Fluid flow and diagenesis in sedimentary basins. *In*: PARNELL, J. (ed.) *Geofluid: Origin, Migration and Evolution of Fluids in Sedimentary Basins*. Geological Society, London, Special Publications, **78**, 127–140.

BJØRLYKKE, K. 1999. An overview of factors controlling rates of compaction and fluid flow in sedimentary basins. *In*: JAMTVEIT, B. & MEAKIN, P. (eds) *Growth, Dissolution and Pattern Formation in Geosystems*. Kluwer Akademic Publishers, 381–404.

BJØRLYKKE, K. & EGEBERG, K. 1993. Quartz cementation in sedimentary basins. *American Association of Petroleum Geologists Bulletin*, **77**, 1538–1548.

BJØRLYKKE, K. & HØEG, K. 1997. Effects of burial diagenesis on stresses, compaction and fluid flow in sedimentary basins. *Marine and Petroleum Geology*, **14**, 267–276.

BJØRLYKKE, K., AAGAARD, P., EGEBERG, P. K. & SIMMONS, S. P. 1995. Geochemical constraints from formation water analyses from the North Sea and Gulf Coast Basin on quartz, feldspar and illite precipitation in reservoir rocks. *In*: CUBITT, J. M. & ENGLAND, W. A. (eds) *The Geochemistry of Reservoirs*. Geological Society, London, Special Publications, **86**, 33–50.

BLYSTAD, P., BREKKE, H., FÆRSETH, R. B., LARSEN, B. T., SKOGSEID, J. & TØRUDBAKKEN, B. 1995. Structural elements of the Norwegian continental shelf: Part II. The Norwegian Sea region. *Norwegian Petroleum Directorate Bulletin*, **8**.

BYRKJELAND, U., BUNGUM, H. & ELDHOLM, O. 2000. Seismotectonics of the Norwegian continental margin. *Journal of Geophysical Research*, **105**, 6221–6236.

CHUHAN, F. A., KJELDSTAD, A., BJØRLYKKE, K. & HØEG, K. 2002. Porosity loss in sand by grain crushing. Experimental evidence and relevance to reservoir quality. *Marine and Petroleum Geology*, **19**, 39–53.

CHUHAN, F. A., KJELDSTAD, A., BJØRLYKKE, K. & HØEG, K. 2003 Experimental compression of loose sands simulating porosity reduction in petroleum reservoirs during burial. *Canadian Geotechnical Journal*, **40**, 995–10011.

CORFIELD, S. & SHARP, I. R. 2000. Structural style and stratigraphic architecture of fault propagation folding in extensional settings: a seismic example from the Smorbukk area, Halten Terrace, Mid-Norway. *Basin Research*, **12**, 329–341.

DORE, A. G. & LUNDIN, E. R. 1996. Cenozoic compressional structures on the NE Atlantic margin; nature, origin and potential significance for hydrocarbon exploration. *Petroleum Geoscience*, **2**, 299–311.

Dore, A. G., CARTWRIGHT, J. A., STOCKER, M. S., TURNER, J. P. & WHITE, N. 2002. *Exhumation of the North Atlantic margin; timing, mechanism and implications for petroleum exploration*, Geological Society, London, Special Publications, **196**.

EGEBERG, P. K. & SAIGAL, G. 1991. North Sea chalk diagenesis – cementation of chalks and healing of fractures. *Chemical Geology*, **92**, 339–354.

FALEIDE, J. I., KYRKEBØ, R., KJENNERUD, T., GABRIELSEN, R. H., JORDT, H., FANAVOLL, S. & BJERKE, M. D. 2002. Tectonic impact on sedimentary processes during Cenozoic evolution of the northern North Sea and surrounding areas. *In*: DORE, A. G., CARTWRIGHT, J. A., STOKER, M. S., TURNER, J. P. & WHITE, N. (eds) *Exhumation and Implications for Petroleum Exploration.* Geological Society, London, Special Publications, **196**, 235–269.

FÆRSETH, R. B. 1996. Interaction of Permo-Triassic and Jurassic extensional fault-blocks during the development of the northern North Sea. *Journal of the Geological Society*, London, **153**, 931–944.

FISHER, Q. J. & KNIPE, R. J. 2001. The permeability of faults within siliclastic petroleum reservoirs of the North Sea and Norwegian Continental Shelf. *Marine and Petroleum Geology*, **18**, 1063–1081.

FISHER, Q. J., CASEY, M., HARRIS, X. & KNIPE, R. J. 2003. Fluid-flow properties of faults in sandstones: the importance of temperature history. *Geology*, **31**, 965–968.

GAARENSTROOM, L. R., TROMP, A. J., DE JONG, M. C. & BRANDENBURG 1993. Over-pressures in the Central North Sea. Implications for trap integrity and drilling safety. *In*: PARKER, J. R. (ed.) *Petroleum Geology of Northwest Europe*. Proceedings from the 4th Conference, Geological Society, London, 1305–1313.

GABRIELSEN, R. H., KYRKJEBØ, R., FALEIDE, J. I., FELDSKAR, W. & KJENNERUD, T. 2001. The Cretaceous post-rift development in the northern North Sea. *Petroleum Geoscience*, **7**, 137–154.

GILES, M. R. 1997. *Diagenesis: A Quantitative Perspective*. Kluwer.

GROLLIMUND, B. & ZOBACK, M. D. 2003. Impact of glacially induced stress changes on fault-seal integrity offshore Norway. *American Association of Petroleum Geologists Bulletin*, **87**, 493–506.

GROLLIMUND, B., ZOBACK, M. D., WRIPUT, D. J. & ARNESEN, L. 2001. Stress orientation, pore pressure and least principal stress in the Norwegian sector of the North Sea. *Petroleum Geoscience*, **7**, 173–180.

GÖLKE, M. & COBLENTZ, D. 1996. Origins of the European regional stress field. *Tectonophysics*, **266**, 11–24.

HESTHAMMER, J., BJORKUM, P. A. & WATTS, L. 2002. The effect of temperature on sealing capacity of faults in sandstone reservoirs: Example from the Gullfaks Field. *American Association of Petroleum Geologists Bulletin*, **86**, 1733–1751.

HEUM, O. R. 1996. A fluid dynamic classification of hydrocarbon entrapment. *Petroleum Geoscience*, **2**, 145–158.

HICKS, E. C., BUNGUM, H. & LINDHOLM, C. D. 2000. Stress inversion of earthquake focal mechanism solutions from onshore and offshore Norway. *Norsk Geologisk Tidskrift*, **80**, 235–250.

ISAKSEN, G. H. 2004. Central North Sea hydrocarbon systems: generation, migration, entrapment and thermal degradation of oil and gas. *American Association of Petroleum Geologists Bulletin*, **8**, 1545–1572.

JAEGER, J. C. & COOK, N. G. W. 1979. *Fundamentals of Rock Mechanics*. Chapman and Hall, London.

KARLSEN, D. A., SKEIE, J. E. *ET AL.* 2004. Petroleum migration, faults and over pressure. Part II. Case history. The Haltenbanken Petroleum Province, offshore Norway. *In*: CUBITT, J. M., ENGLAND, W. A. & LARTER, S (eds) *Understanding of Petroleum Reservoirs; Towards an Integrated Reservoir Engineering Approach.* Geological Society, London, Special Publications, **237**, 305–372.

KARNER, S. L., CHESTER, F. M., KRONENBERG, A. K. & CHESTER, J. S. 2003. Subcritical compaction of granular sand. *Tectonophysics*, **377**, 357–381.

KJELDSTAD, A. 2002. *Compaction and shear deformation in sedimentary basins*. Ph.D. thesis, University of Oslo.

KJELDSTAD, A. J., SKOGSEID, J., LANGTANGEN, H. P., BJØRLYKKE, K. & HØEG, K. 2003. Differential loading by prograding sedimentary wedges on continental margins: an arch forming mechanism. *Journal of Geophysical Research*, **108**, 2036, DOI 10.1029/2001 JB001145.

LAMBE, T. W. & WHITMAN, R. V. 1979. *Soil Mechanics, SI Version*. John Wiley, New York.

LANDER, R. H., HELSET, H. M., MATHEWS, J. C. & BONELL, L. M. 2003. Can sandsteons diagenesis induce overpressure? *In*: DUEPPENBECKER, S. J. & MARZI, R. W. (eds) *Multidimensional Basin Modelling. American Association Petroleum Geologists of Discovery Series*, 352–353.

LEONARD, R. C. 1993. Distribution of sub-surface pressure in the Norwegian Central Graben and applications for exploration. *In*: PARKER J. R. (ed.) *Petroleum Geology of Northwest Europe. Proceedings of the 4th Conference.* The Geological Society, 1295–13003.

LINDHOLM, C. D., BUNGUM, H., BRATLI, R. K., AADNØY, B. S., DAHL, N. T., RUDBAKKEN, B. & ATAKAN, K. 1995. Crustal stress in the northern North Sea as inferred from borehole breakouts and earthquake focal mechanisms. *Terra Nova*, **7**, 51–59.

LOTHE, A. E., BORGE, H. & GABRIELSEN, R. H. 2004. Modelling of hydraulic leakage by pressure and stress simulations and implications for Biot's constant: an example from the Halten Terrace, offshore Mid Norway. *Petroleum Geoscience*, **10**, 199–213.

LUO, X. & VASSEUR, G. 1995. Modelling of pore pressure evolution associated with sedimentation and uplift in sedimentary basins. *Basin Research*, **7**, 35–52.

MOSS, B., BARSON, D., RAKHUT, K., DENNIS, H. & SWARBRICK, R. 2003. Formation pressure and formation water. *In*: EVANS, D., GRAHAM, C., ARMOUR, A. & BATHURST, P. (eds) *The Millennium Atlas: Petroleum Geology of the Central and Northern North Sea*. Geological Society, London, 317–329.

NORDGAARD BOLÅS, H. M. & HERMANDRUD, C. 2002. Rock stress in sedimentary basins – implication for trap integrity. *In*: KOESTLER, A. G. & HUNSDALE, R. (eds) *Hydrocarbon Seal Quantification* Norwegian *Petroleum Society*, Special Publication, **11**, 17–35.

NYGÅRD, R., GUTTIERREZ, M., HØEG, K. & BJØRLYKKE, K. 2004. Influence of burial history on microstructure and compaction behaviour of Kimmeridge Clay. *Petroleum Geoscience*, **10**, 269–270.

OELKERS, E. H., BJØRKUM, P. A. & MURPHY, W. M. 1996. A petrographic and computational investigation of quartz cementation and porosity reduction in North Sea sandstones: *American Journal of Science*, **296**, 420–452.

OLSTAD, R., BJØRLYKKE, K. & KARLSEN, D. K. 1997. Pore water flow and petroleum migration in the Smørbukk Field area, offshore Norway. *In*: MØLLER-PEDERSEN, P. & KOESTLER, A. G. (eds) *Hydrocarbon Seals – Importance for Exploration and Production. Norwegian Petroleum Society*, Special Publication, **7**, 201–216.

PALANDRI, J. L. & KHARAKA, Y. K. U. 2004. A compilation of rate parameters of water-mineral interaction kinetics to geochemical modelling. *U.S. Geological Survey Report*, 2004–1068.

PEREZ, R. J., CHATELLIER, J. I. & LANDER, R. H. 1999. Use of quartz cementation kinetic modeling to constrain burial histories; examples from the Maracaibo Basin, Venezuela. *Revista Latino-Americana de Geoquimica Organica*, **5**, 39–46.

RAAEN, A. M., SKOMEDAL, E., KJØRHOLT, H., MARKESTAD, P. & ØKLAND, D. 2001. Stress determination from hydraulic fracturing tests: the system stiffness approach. *International Journal of Rock Mechanics and Mining Sciences*, **38**, 529–541.

RAMM, M. & BJØRLYKKE, K. 1994. Porosity/depth trends in reservoir sandstones: assessing the quantitative effects of varying pore – pressure, temperature history and mineralogy. Norwegian Shelf data. *Clay Minerals*, **29**, 475–490.

SIBSON, R. H. 1981. Fluid flow accompanying faulting. Field evidence and models. *In*: SIMPSON, D. W. & RICHARDS, P. G. (eds) *Earthquake Prediction: An International Review*. American Geophysical Union, Maurice Ewing Series, **4**, 593–603.

SIBSON, R. H. 2003. Brittle-failure controls on maximum sustainable overpressure in different tectonic regimes. *American Association of Petroleum Geologists Bulletin*, **87**, 901–908.

STORVOLL, V., BJØRLYKKE, K. & MONDOL, N. H. 2005. Velocity-depth trends in Mesozoic and Cenozoic sediments from the Norwegian Shelf. *American Association of Petroleum Geologists Bulletin*, **89**, 359–381.

SVERDRUP, E., HELGESEN, J. & VOLD, J. 2003. Sealing properties of faults and their influence on water-alternating-gas injection efficiency in the Snorre field, northern North Sea. *American Association of Petroleum Geologists Bulletin*, **87**, 1437–1458.

VERRUIJT, A. 1982. *Ground Water Flow*. Macmillan Press, London.

WALDERHAUG, O. 1996. Kinetic modelling of quartz cementation and porosity loss in deeply buried sandstone reservoirs. *American Association of Petroleum Geologists Bulletin*, **80**, 731–745.

WALDERHAUG, O., BJØRKUM, P. A., NADEAU, P. H. & LANGNES, O. 2001. Quantitative modelling of basin subsidence caused by temperature-driven silica dissolution and precipitation. *Petroleum Geoscience*, **7**, 107–113.

WIPRUT, D. & ZOBACK, M. D. 2000. Fault reactivation and fluid flow along a previously dormant normal fault in the northern North Sea. *Geology*, **28**, 595–598.

WONG, T.-F., DAVID, C. & ZHU, W. 1997. The transition from brittle faulting to cataclastic flow in porous sandstones. Mechanical deformation. *Journal of Geophysical Research*, **102**, 3009–3025.

WOOD, J. R. 1994. Chemical Diagenesis. *In*: WILSON, M. D. (ed.) *Reservoir Quality Assessment and Prediction in Clastic Rocks. SEPM Short Course*, **30**, 119–136.

Multiple faults in ductile simple shear: analogue models of flanking structure systems

ULRIKE EXNER[1,*], BERNHARD GRASEMANN[2] & NEIL S. MANCKTELOW[1]

[1]*Geologisches Institut, ETH-Zentrum, CH-8092 Zürich, Switzerland*

[2]*Institut für Geologische Wissenschaften, University of Vienna, A-1090 Vienna, Austria*

[*]*Current address: University of Vienna, Althanstrasse 14, 1090 Vienna, Austria*

(e-mail: ulrike.exner@univie.ac.at)

Abstract: Rotational behaviour and deformation around multiple faults was investigated in analogue experiments using a linear viscous matrix material under simple shear boundary conditions. Previous analogue and numerical studies have shown that, for single faults, characteristic deformation geometries are produced in initially straight marker lines parallel to the shear zone boundary (flanking structures). Observations from several natural shear zones suggest that not only single faults, but often several parallel or conjugate fault planes are subjected to progressive shear resulting in distinctive deflection geometries. If the distance between faults is on the order of their length, or less, then the perturbation flow fields interfere and coalescence, and finite deflection structures develop that are distinctly different from those around single fractures. In particular, coeval contractional and extensional geometries may develop across conjugate faults, although for bulk simple shear the total length of marker lines parallel to the shear zone boundary cannot change. This advises caution in inferring shear-zone parallel contraction or extension from secondary slip surfaces. In contrast to single flanking structures, conjugate flanking structure systems occurring in natural shear zones are reliable shear sense indicators due to their triclinic symmetry.

Planar discontinuities, such as fractures, veins or dykes, may be subjected to progressive deformation in ductile shear zones. Depending on the flow geometry of the shear zone and the initial orientation of the discontinuity with respect to the shear zone boundary (SZB), various deflection geometries of passive, initially planar, markers may develop. The resulting structures have been collectively called flanking structures (Passchier 2001). Several studies (Grasemann & Stüwe 2001; Grasemann *et al.* 2003; Exner *et al.* 2004; Wiesmayr & Grasemann 2005) have systematically investigated the formation of such structures around single, isolated discontinuities, termed cross-cutting elements (CE; Fig. 1), embedded in linear viscous materials under isochoric plane strain boundary conditions. Flanking structures are the geometric consequence of a perturbation of the homogeneous flow field induced by planar discontinuities, recorded by planar markers in the surrounding material. This effect is referred to as perturbation strain (e.g., Cobbold 1975; Mancktelow 1991).

In all orientations other than parallel to the eigenvector of the flow, CEs co-rotate with respect to the bulk shear sense in simple shear. Nevertheless, the sense of movement along the CE with respect to the bulk shear sense, the type of offset of a marker line, as well as the sense of deflection of the marker line in the immediate vicinity of the CE, may vary. Furthermore, the deflection of marker lines changes along the length of a CE, and the characteristic geometric features are ideally observed along the central marker line (CML; Fig. 1). In simple shear, six different geometries of flanking structures are possible at small strains (Fig. 2; Grasemann *et al.* 2003). These geometries can be characterized in terms of the offset of the CML along the CE relative to the bulk shear (s = synthetic or co-shearing, a = antithetic or counter-shearing), the drag of the CML against the CE (normal or reverse, *cf.* Hamblin 1965), and whether there is locally extension or contraction along the CML across the CE. This classification requires knowledge of the bulk shear sense, which is given for experimental models and may be independently established in field examples.

In this study, we apply the definition of flanking structures as introduced by Grasemann & Stüwe (2001), where a pinned slip surface (or an elliptical

From: BUITER, S. J. H. & SCHREURS, G. (eds) 2006. *Analogue and Numerical Modelling of Crustal-Scale Processes*. Geological Society, London, Special Publications, **253**, 381–395.

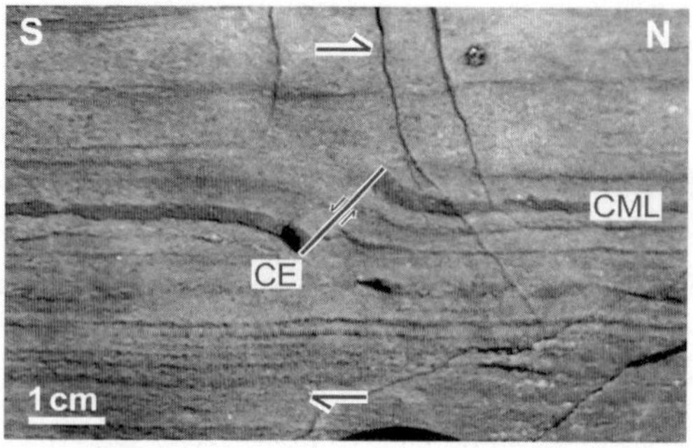

Fig. 1. Natural example of a flanking fold with extensional, antithetic offset and reverse drag in mylonitic marble with known shear sense (arrows) on Naxos, Greece (N37°11′23.6″, E25°30′55.1″). The central marker line (CML) is located in the approximate symmetry centre of the structure; at a distance to the structure, marker lines above and below the tips of the cross-cutting element (CE) are straight and continuous.

low-viscosity inclusion) embedded in a linear viscous matrix accounts for the formation of a confined flanking structure with monoclinic symmetry. Shear bands are a particularly common, but special example, that develop at low angles with the SZB (0–45°), showing extensional, synthetic offset of the CML (Fig. 2). At medium angles (45–135°), flanking folds with antithetic displacement along the CE are formed, having either contractional or extensional offset of the CML. Above 135°, flanking folds with synthetic, but now contractional, offset can be observed.

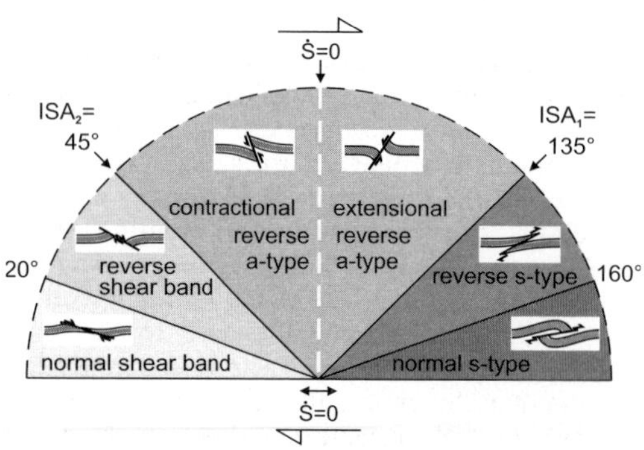

Fig. 2. Instantaneous structure development of flanking structures in dextral simple shear, depending on the initial orientation of the CE to the shear zone boundary (after Exner *et al.* 2004). Instantaneous stretching axes (ISA) delimit sectors of synthetic and antithetic sense of shear along the CE; lines of zero stretch ($\dot{S} = 0$) are at 0° and 90° to the SZB in simple shear, separating contractional and extensional structures within the syn/antithetic sectors.

With progressive deformation, it is also possible for one type of flanking structures to evolve into another (Exner *et al.* 2004).

Observations from natural shear zones suggest that not only single discontinuities, but also sets of parallel or conjugate features are commonly affected by subsequent ductile deformation. Multiple (sub)parallel shear bands in mylonite zones are a well known example, but other combinations of multiple flanking structures are, although quite common in natural rocks, completely unstudied. We want to introduce the term *flanking structure system*, defining a complex structure resulting from the interference of perturbation strains in a set of two or more CEs. Perturbation strains of CEs overlap when the distance between the CEs roughly equals the length of the CE (Pollard & Segall 1987).

In this study we present a series of analogue experiments, which were designed to examine the systematics of deformation around conjugate or parallel planar features within a homogeneous ductile matrix. A major advantage of analogue experiments over numerical models is that it is relatively simple to introduce more than one fault surface. Moreover, the amount of finite shear strain is unlimited in the ring shear apparatus employed in this study, permitting the observation of structure development to large strains. Although the basic principles of flanking structures can easily be applied to a series of parallel or conjugate discontinuities, there are some major differences in structural evolution and finite geometries, especially if the perturbation strains induced by different CEs in the host rock interfere.

The aim of this paper is to investigate analogue models of parallel and, in more detail, conjugate flanking structure systems by means of finite perturbation fields and to compare the results with natural examples. The study has important consequences for: (1) the coexistence and simultaneous activity of contractional and extensional faults; (2) the development of triclinic flanking structure systems; and (3) the correct interpretation of shear sense derived from flanking structure systems.

Experimental procedure

The analogue experiments were conducted in a ring shear apparatus (Arbaret *et al.* 2001). This analogue rig consists of two concentric vertical cylinders rotating in opposite directions at the same velocity. The volume between the two cylinders is filled with a linear viscous polymer, PDMS (polydimethylsiloxane SGM 36, manufactured by Dow Corning), resting on a lubricant glycerine layer. Isochoric plane strain cylindrical shear flow is produced in the analogue material between the counter-rotating cylinders, with increasing strain rate from the outer to the inner cylinder.

The setup of the analogue experiments closely follows the procedure described in Exner *et al.* (2004). At the centre of the model shear zone, 'faults' were cut into the PDMS and filled with soap or silicone oil. A passive marker grid was imprinted on top of the model surface using partially fixed photocopies (Dixon & Summers 1985).

The distance between the CEs was chosen to accord with that observed in natural examples, avoiding direct connection of the fault surfaces, yet allowing the perturbations around the faults to produce combined geometries instead of two completely independent structures. Whether interference occurs depends on the spacing and initial orientation of the faults relative to their length. This relationship was also investigated in the current experiments.

Results of analogue experiments

Pinned (non-propagating) slip surfaces embedded in a linear viscous matrix produce flanking structures if subjected to further deformation (e.g., Grasemann & Stüwe 2001). The main feature of flanking structures, apart from a more or less significant offset of initially straight markers along the CE, is the deflection of these markers as they approach the CE. This deflection, especially the phenomenon of reverse drag, was first described by Hamblin (1965), and extended by, e.g., Reches and Eidelman (1995) and Grasemann *et al.* (2003). In contrast to all previous interpretations, the main conclusion of the most recent work is that fault drag is not produced by the frictional resistance along the CE, but by the perturbation field induced by slip along the CE (Passchier *et al.* 2005). The slip along the CE causes a localized strain that differs from the imposed bulk strain, and this difference must be compensated in the immediate vicinity of the fault. This results in the development of four open perturbation loops (Mancktelow 1991), which represent an inhomogeneous vector field with highest magnitudes adjacent to the fault, decreasing to zero at some distance into the matrix (Passchier *et al.* 2005).

The finite perturbation fields presented in this study are constructed by subtracting the heterogeneously deformed finite grid from the deformed grid not influenced by the presence of a CE. Under ideal simple shear this part, which is not influenced by the CE, would be the homogeneous part of the finite deformation, sometimes called the background strain (Weijermars 1992). However, because in our experiments the strain rate is a function of the radius of the rig, our background strain is in fact also heterogeneous. An intuitive approach for investigating perturbation fields is that these vector fields represent the deflections of straight marker lines in the vicinity of heterogeneities (e.g., fault drag along a CE), because in a homogeneous strain field, initially straight passive marker lines remain straight. Note that these finite perturbation fields are the same as the 'finite vector displacement field of the heterogeneous component of deformation' of Mancktelow (1991, Fig. 5), but different from the instantaneous perturbation velocity fields discussed by Passchier *et al.* (2005).

The present analogue study was designed to investigate (1) the shape and variation of the perturbation loops, (2) the rotational behaviour of multiple CEs in simple shear, and (3) the deflection geometries of planar makers around multiple conjugate or parallel CEs resulting from the combination of perturbation flow and fault rotation.

Mechanisms

Generally, there are two different mechanisms responsible for the finite geometries of markers around planar discontinuities: (1) the existence of an instantaneous perturbation velocity field, and (2) the progressive rotation of the CEs with respect to the SZB, which causes the perturbation velocity field to change with time.

Perturbation displacement fields. Perturbation flow accounts for the deflection of markers around the CEs. Following the concept of Passchier *et al.* (2005), perturbation flow can be visualized as a vector field. Vector fields can generally be divided into a series of open or (in special cases) closed loops separated by lines where the polarity of flow changes, and points of no perturbation flow.

For single CEs, there exist four instantaneous perturbation loops, which are variably developed depending on the orientation of the CE. When the CE is parallel to one of the instantaneous stretching axes (ISA), the polarity of the loops is inverted. CE orientations symmetrical about the ISA have identical instantaneous perturbation velocity fields except for a reversal in polarity (Passchier *et al.* 2005, Fig. 7c,g).

Parallel flanking structure systems exhibit a coalescence of the intermediate perturbation loops. Figure 3 shows the perturbation field

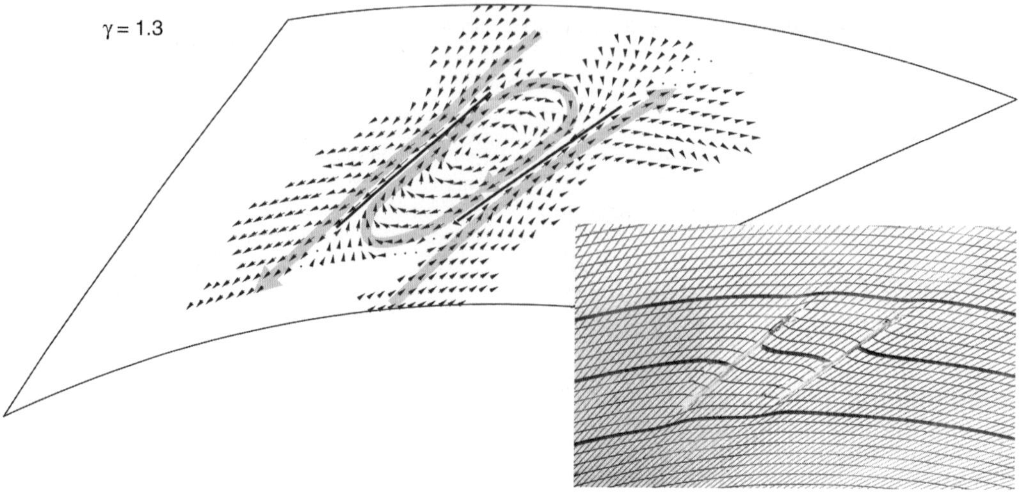

Fig. 3. Finite perturbation displacement field and image of analogue experiment of a parallel flanking structure system with initial orientation of the CEs at 90° to the SZB at $\gamma = 1.3$. A concentric perturbation loop accounts for the clockwise rotation and reverse drag of the CML between the CEs.

around two parallel reverse a-type flanking fold geometries, where the intermediate region is defined by a clockwise closed circular loop, which is explicitly separated from the two external, anti-clockwise open loops. Because the CEs have the same orientation within the flow field, they also have the same rotation sense and rate. Therefore, the whole finite structure shows a monoclinic symmetry, even when several CEs with the same orientation are arranged en echelon. A natural example of parallel flanking structure systems are asymmetric boudinage structures, which have been classified as Type I forward rotating and Type II backward-rotating shear geometries (Swanson 1992), depending on the orientations of the CEs and thus the sense of the coalescent perturbation loops. A complete geometric description of boudin geometries was discussed in detail by Goscombe *et al.* (2004).

Conjugate flanking structure systems are controlled by a more complex arrangement of perturbation loops. Here, the loops are no longer monoclinic. Although the configuration for separate isolated CEs with the same orientation consists of four independent loops around each CE, in combined systems the neighbouring cells between the two CEs immediately merge to form a common vector field, where boundaries between distinct loops cannot be distinguished (see Fig. 7).

Perturbation loops are characterized by a flow field deviating significantly from the bulk flow field. Hence, lines initially already parallel to the bulk fabric attractor (in simple shear also a line of no finite elongation) may nonetheless experience rotation away from this orientation and consequently, significant extension and/or contraction during progressive deformation. Passive markers record the finite sum of all previous (instantaneous) perturbation fields. Their heterogeneous deformation is a direct consequence of the finite perturbation field, which may deviate strongly from the instantaneous perturbation field. Deformed marker lines, both in experiment and nature, provide a visualization of the finite effects of perturbation flow and thus a practical basis for comparison and classification.

Fault rotation. It is important to note that rotation of a fault (or CE) is not a prerequisite for producing drag effects in adjacent marker layers. Faults that are stable or meta-stable in orientation may also produce flanking structures (*cf.* Fig. 7a of Passchier *et al.* 2005). Nevertheless, the sense and rate of rotation is critical in controlling the progressive deformation and finite geometry of flanking structures (Exner *et al.* 2004), especially if they rotate into a different field of instantaneous flanking structure development. This effect is particularly important in conjugate flanking structure systems.

A comparison between the rotation rate of passive markers and CEs (i.e., active slip surfaces) shows that, within experimental error, they have the same rotational behaviour. Kocher and Mancktelow (2004) came to the same conclusion when comparing analytical solutions

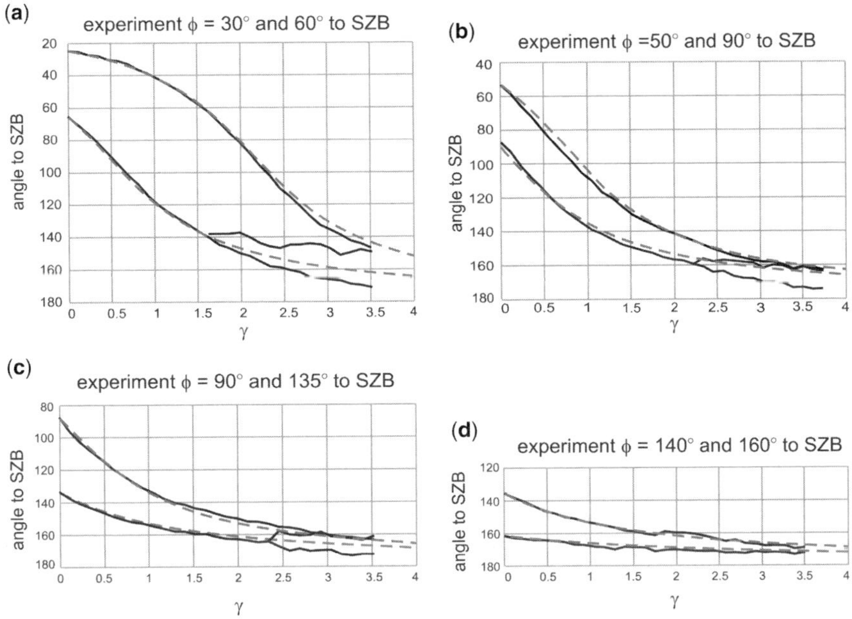

Fig. 4. Rotation of CEs measured from experiments (continuous lines) and theoretical rates of passive markers (dashed lines) in simple shear calculated with Eqn (2). Corresponding to Fig. 7, experiments with initial angles of CEs at (**a**) 30° and 60°, (**b**) 50° and 90°, (**c**) 90° and 135° and (**d**) 140° and 160° are shown. Note the excellent match of the model with theory, even though the model set-up clearly differs from the geometrical constraints of the theory.

and finite element models of rotating ellipses with theoretical curves for passive marker line rotation. This is quite general: any elongate inclusion (R > *c.* 10) will effectively rotate as a passive line whether the inclusion is weaker or stronger than the matrix (e.g., see the analytical solutions of Schmid & Podladchikov 2003). In Figure 4a–d, experimental data for rotating CEs are compared to the corresponding theoretical curves for purely passive markers in simple shear. Following Ramsay (1967), the rotation rate $\dot{\omega}$ of passive markers can be calculated using

$$\dot{\omega} = (\dot{\gamma}/2)(1 - \cos 2\alpha) \tag{1}$$

Alternatively, the orientation of passive markers can be determined using the trigonometric relationship

$$\cot \alpha' = \cot \alpha + \gamma \tag{2}$$

Any difference between the experimentally observed values for CEs and the theoretical values for a passive line can be attributed to the inherent heterogeneity in strain that occurs in the ring shear machine. The shear strain rate (and therefore rotation rate) is greater toward the inner cylinder, so that initially straight CEs (and marker lines) become curved at higher shear strain values.

An important corollary of these experimental results is that, even though CEs induce a complex perturbation flow field, their rotational behaviour is not affected. Even when the perturbation fields of neighbouring CEs interfere, the rotation rates are unaffected and they still behave like passive marker lines. This finding is consistent with observations from analogue experiments on rigid particles in simple shear, whose rotational behaviour is not changed unless the distance between them is shorter than their own length (Ildefonse *et al.* 1992). It follows that the only stable orientation of CEs in simple shear is the eigenvector, or fabric attractor (Passchier 1997), parallel to the shear direction (Fig. 5).

Geometry of passive marker lines

Deflection and offset of passive marker lines around multiple CEs is controlled by (1) the bulk flow geometry of the shear zone, (2) the initial orientation of the CEs with respect to the SZB, (3) the initial angle of the CEs with

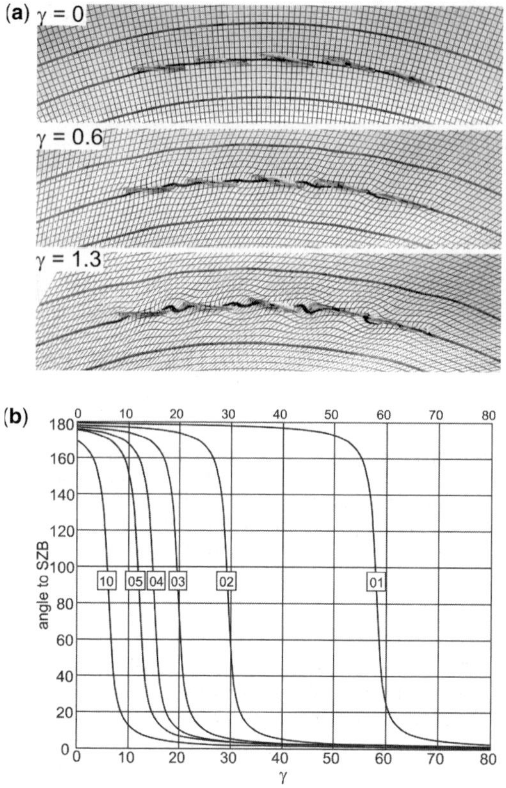

Fig. 5. (**a**) Analogue experiment with 5 CEs initially oriented at *c.* 10° to SZB at $\gamma = 0$, 0.6 and 1.3. Development of pronounced flanking structures can be observed, whereas the rotation rate of the CEs is rather insignificant. (**b**) Rotational behaviour of passive markers at low angles (between 1° and 10°) to SZB. Small variations in the initial orientation result in major differences of the value of shear strain, at which the markers start to rotate significantly into the shear direction.

respect to each other, and (4) the distance between the CEs relative to their length.

Conjugate flanking structure systems. For simple shear flow considered here, all initially horizontal marker lines (i.e., parallel to the SZB) remain straight and parallel and maintain the same spacing in the far field of flanking structures around single discontinuities. In contrast, marker lines in the area between closely spaced conjugate CEs are subjected to a combination of two (unequal or even opposed) perturbation fields. The mutual interference between these different vector fields results in a combined perturbation flow field that may involve the coalescence of loops or even the formation of new loops. Thus, due to the small distance between the CEs (relative to their length), initially SZB-parallel markers in the space between the two CEs are not only deflected, but may be translated (either upwards or downwards) and/or rotated with respect to their initial position.

Figure 6 shows a selection of five possible combinations of initial conjugate CE orientations. The amount of finite shear strain is small for all structures (between 0.6 and 1.3), allowing a comparison at a similar stage in their evolution. Corresponding to the concept of flanking structures around single CEs, the drag and sense of offset along the two CEs depends on the initial orientation of the CE with respect to the SZB. More specifically, special combinations of CE orientations with respect to the SZB result in special rotation and drag geometries of the marker lines between the CEs. Interestingly, instantaneous geometries may develop that are quite unexpected for simple shear flow. For example, not only co- and counter-rotation of the inter-CE marker lines can be observed, but also apparently extensional (Fig. 6b) and contractional (Fig. 6d) geometries. Such geometries are only possible if the distance between the CEs is small enough to allow the two perturbation fields to interfere. Note that some of the model results have striking similarities with large-scale fault bounded structures like horst and graben (Fig. 6b), domino- or bookshelf-type halfgraben (Fig. 6c), extruding-wedge geometries (Fig. 6d) or pop-up structures. Therefore, we speculate that the concept of interfering perturbation fields also work on scales which are several orders of magnitude larger than our models.

Finite conjugate flanking structure systems generally have triclinic symmetry, even if the initial orientation of the conjugate CEs was monoclinic. This is an effect of the induced simple shear flow forcing differently oriented CEs to rotate with different rates (Fig. 4). As an important consequence, one CE might be in an orientation of synthetic (or co-) shearing, while the other CE might be in an orientation of antithetic (or counter-) shearing. From a detailed investigation of natural examples on all scales, it can be seen that triclinic conjugate flanking structure systems are more the rule than the exception.

As an example, Figure 7a shows the perturbation displacement field around two CEs initially oriented at an angle of 30° and 60° to the SZB, respectively, at $\gamma = 0.6$. As can be seen in the inset photograph of the analogue experiment, the CEs have already rotated to higher angles, and the initially straight (or more precisely, concentric) marker lines show a distinct offset and deflection. The finite perturbation

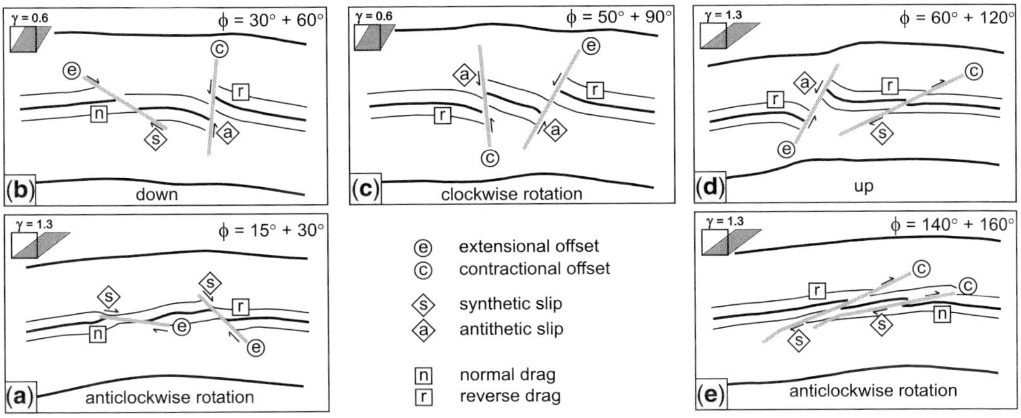

Fig. 6. Five possible combinations of conjugate flanking structure systems. Various initial conjugate angles result in different geometries. Note the variation in rotation/relative position of CML between the CEs, resulting from different combinations of syn/antithetic slip, extensional/contractional offset and normal/reverse drag. Some of the model results resemble tectonic structures, which occur in nature at much larger scales compared to the size of our model: e.g., (**a**) high and low angle normal faults, (**b**) graben, (**c**) half-graben, (**d**) extruding wedge, (**e**) duplex.

field has developed two distinct external perturbation loops, whereas the intermediate loops form a common, more or less homogeneously oriented vector field. Although it is not immediately evident in the photograph of the experiment, the perturbation arrows show a larger offset in the internal area between the CEs, especially for the left (less steeply inclined) CE.

At larger shear strain ($\gamma = 1.3$), the picture becomes more complex (Fig. 7b). Now, only the external perturbation loop on the right side of the conjugate CEs displays a simple concentric vector field. The central and left perturbation loops are connected by a zone of short, upwards pointing arrows. Here, the vectors transform from a downwards-directed field into a left-directed movement. Moreover, the perturbation direction along the right margin of the left CE is separated into two diverging, oppositely directed loops, approximately in the centre of the CE. Consequently, marker lines above this divergence point show a normal, below it a reverse, drag. The finite perturbation field is to some extent counter-intuitive, but this is a consequence of the procedure by which it is determined. Tracing the relative movement of marker points follows the finite displacement paths and represents an integration of instantaneous perturbation flow fields that change as the individual CEs rotate during progressive deformation. Thus, areas with low initial perturbation will show the finite effects of a reversal in the instantaneous sense of movement much earlier than areas with initially high perturbation.

Figure 7c–f presents two experiments with initial angles of the conjugate CEs at 50° and 90° (Fig. 7c,d) or 90° and 135° (Fig. 7e,f) to the SZB, respectively. Both experiments are shown at two different stages ($\gamma = 0.6$ and 1.3), to allow comparison of the different evolution of the perturbation loops, and especially of the direction of the perturbation displacement fields between the CEs with different initial angles. It can be seen that the perturbation loops change their orientation, size and also magnitude, while the CEs rotate at different rates according to their orientation with respect to the SZB (Fig. 4).

Parallel flanking structure systems. For parallel flanking structure systems, the midpoint of the CML between two CEs always maintains the same position relative to the SZB. This reflects the monoclinic symmetry of both the initial geometry and the imposed simple shear boundary conditions. Even for complex progressive histories, where the sense of rotation (counterclockwise/clockwise) and offset direction (antithetic vs. synthetic) of the CML changes with time, the symmetry is maintained (e.g., Figs 3 and 5). In contrast, flanking structure systems where the CEs were initially conjugate have an initial triclinic symmetry, which is also maintained during deformation (Fig. 7).

A special type of parallel flanking structure system is represented by parallel shear bands, SCC[1] or ECC fabrics. They are frequently observed in anisotropic rocks (micaschists or

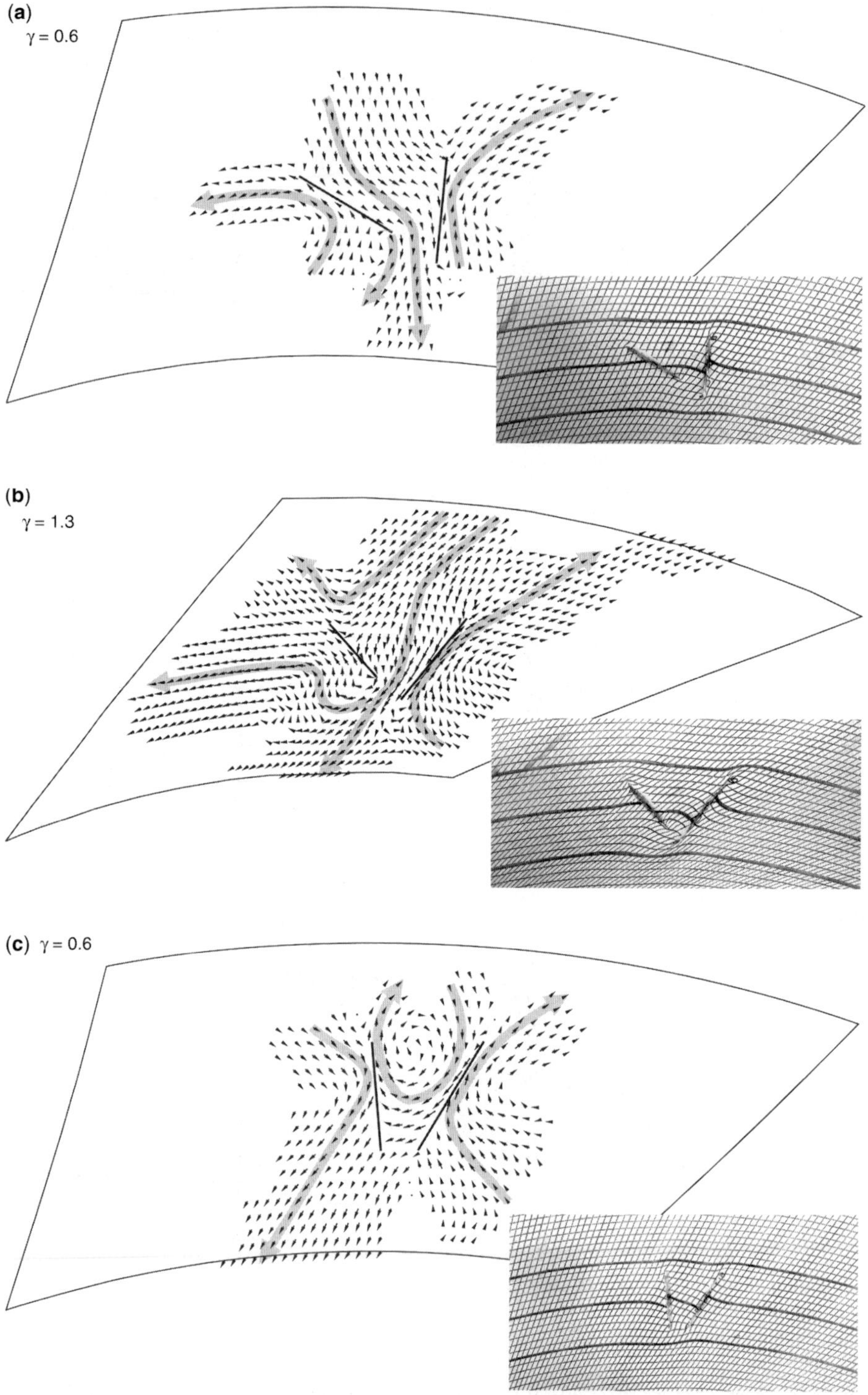

Fig. 7. Perturbation displacement fields of selected experiments of conjugate flanking structure systems. Initial orientations of the CEs are at 30° and 60° for (**a**) and (**b**), at 50° and 90° for (**c**) and (**d**) and at 90° and 135° for (**e**) and (**f**). Each experiment is shown at γ of 0.6 and 1.3. The large grey arrows drawn in the background should help to highlight trajectories of the perturbation flow field. Note that this is not identical to a particle flow field. Images of analogue experiments are provided for comparison of the perturbation fields with the actual geometry of the marker lines.

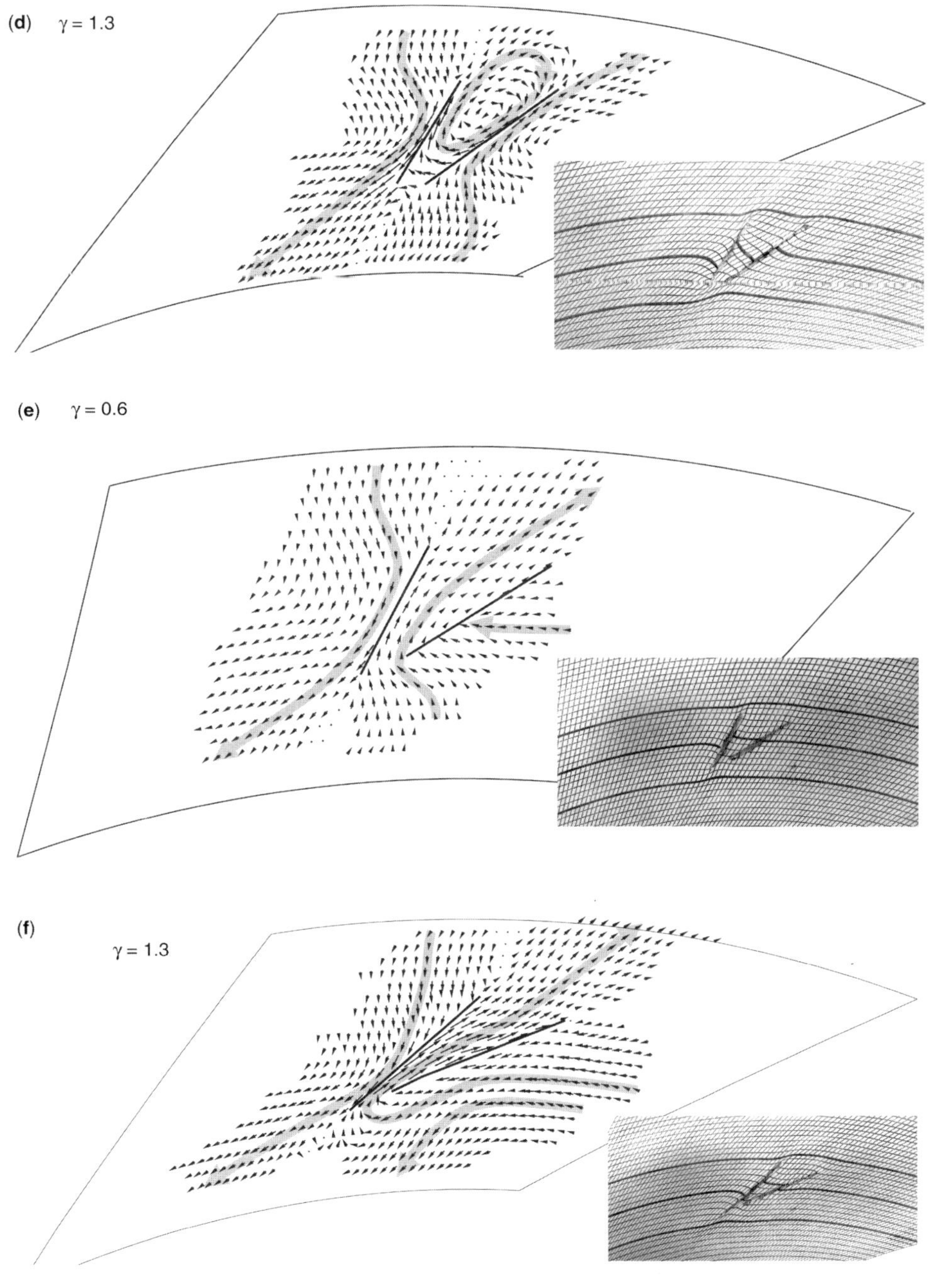

Fig. 7. *Continued.*

paragneisses). Even though the mechanism of formation of the CEs is not investigated in the present study, the geometric result of synthetic shearing along moderately inclined secondary CEs is identical to the one observed in our analogue experiments. For example, in an experiment with five CEs initially oriented at 10° to the SZB (Fig. 5a), the fault orientation changes very little up to (very) large γ. For an even smaller initial angle to the SZB, CEs retain their 'stable' position even longer, up to $\gamma \sim 50$ for an angle of 1° to the SZB (Fig. 5b).

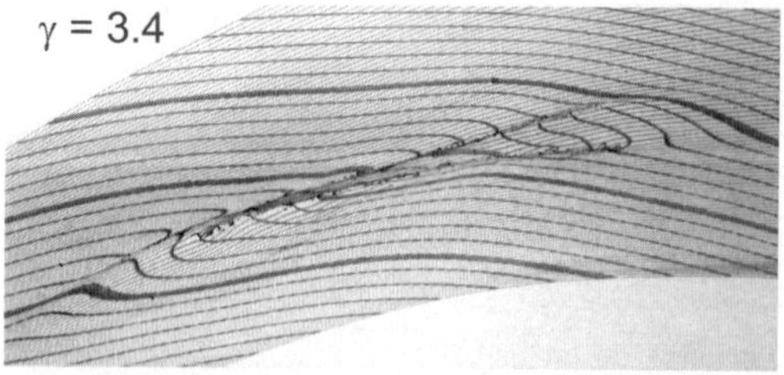

Fig. 8. Analogue experiment with initial orientations of CEs at 90° and 135°, at $\gamma = 3.4$. With increasing strain, some markers develop tight folds along the CE.

Discussion

As outlined above, parallel flanking structure systems maintain their monoclinic symmetry during progressive deformation. En echelon arrangement of parallel flanking structure systems, including natural examples, have been extensively studied in several papers on boudinage in layered and foliated rocks (e.g., Platt & Vissers 1980; Swanson 1992; Goscombe & Passchier 2003). Figures 9a and b show two photographs of natural parallel flanking structure systems with CEs at low and high angles to the SZB, respectively. It can be seen that the finite geometries of CMLs are strongly dependent on the distance between the CEs. Recent geometric studies of parallel flanking structure systems in high-grade rocks have proposed several different mechanisms for the phenomenon of 'back-rotation' of the foliation between parallel shear zones or faults (e.g., Harris 2003). We suggest that most of the effects reported are a direct consequence of the perturbation strains induced by these CEs and that the concept of flanking structure systems is also applicable to the description of marker rotation between shear zones (Blenkinsop 2004). In this study, we focus on the discussion of conjugate flanking structure systems, which (1) may develop complex triclinic geometries, (2) can show contemporaneous extensional and contractional faults, and (3) may allow determination of the bulk shear sense. Figure 9 comprises natural examples of flanking structures with monoclinic and triclinic symmetry from Naxos, Greece, where a strong non-coaxial component of flow is well documented (e.g., Urai *et al.* 1990), thus allowing qualitative comparison with our simple shear analogue experiments.

Flanking structure systems with triclinic symmetry

The concept of an ideal isolated fault ellipse (Walsh & Watterson 1989), which rotates during progressive deformation to produce a single monoclinic flanking structure, is probably more the exception than the rule in nature, although beautiful examples have been described (e.g., Passchier 2001; Grasemann *et al.* 2003). The simple analogue models of two conjugate CEs deforming in plane strain, as presented here, provide some important insights into the development of interfering perturbations caused by CEs, but natural three-dimensional examples of conjugate fault systems may be considerably more complex.

Figure 6 shows five of our modelled triclinic flanking structure systems, which represent a subset of the wide range of structures that could develop from a set of two CEs with differing initial orientations. Here, we focus on the possible instantaneous structures that may form in simple shear from a combination of the initial orientation of the CEs relative to the six fields of instantaneous flanking structure development (Fig. 1). Considering two CE orientations with six deformation sectors gives 15 different combinations. Including the possibility that both CEs could be also located in the same field with only slightly different orientations results in $15 + 6 = 21$ different instantaneous conjugate flanking structure systems. In fact, the number of possible finite structures is even greater, because during progressive deformation the CEs will rotate with different rates into the next field or even the field after that (Exner *et al.* 2004). The range of potential structures is further increased in general shear compared to simple shear, because CEs may not only co-, but also counter-rotate (Grasemann *et al.* 2003; Wiesmayr & Grasemann 2005). The natural examples in Figure 9 highlight this potential complexity in real systems.

Coeval extension and contraction in conjugate flanking structure systems

In the following, we focus on contemporaneous contractional and extensional CEs and their transient development. It should be emphasized that these geometries develop despite the fact that all structures were modelled under simple shear boundary conditions, where in the far field flow there is no extension or contraction parallel to the shear zone.

In simple shear, the instantaneous stretching axes (ISA) and principal stress directions are oriented at 45° and 135° to the SZB. By definition, these directions experience no instantaneous shear and separate fields of synthetic (or co-) and antithetic (or counter-) shearing. In the fields <45° and >135°, CEs show synthetic shearing, whereas CEs oriented between 45° and

(a)

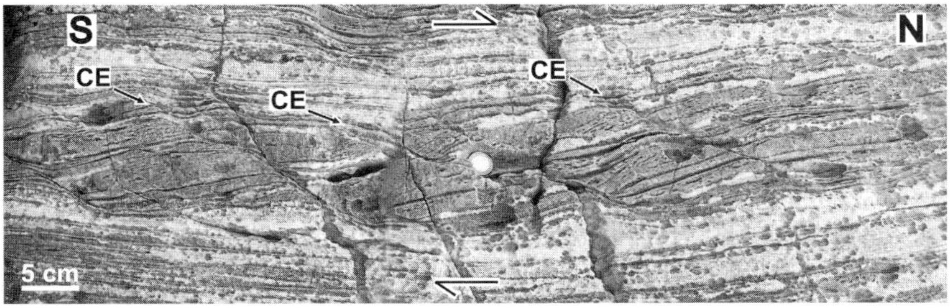

(b)

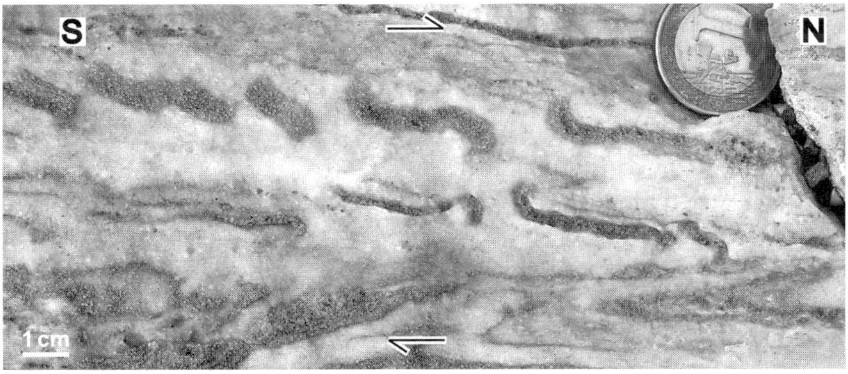

(c)

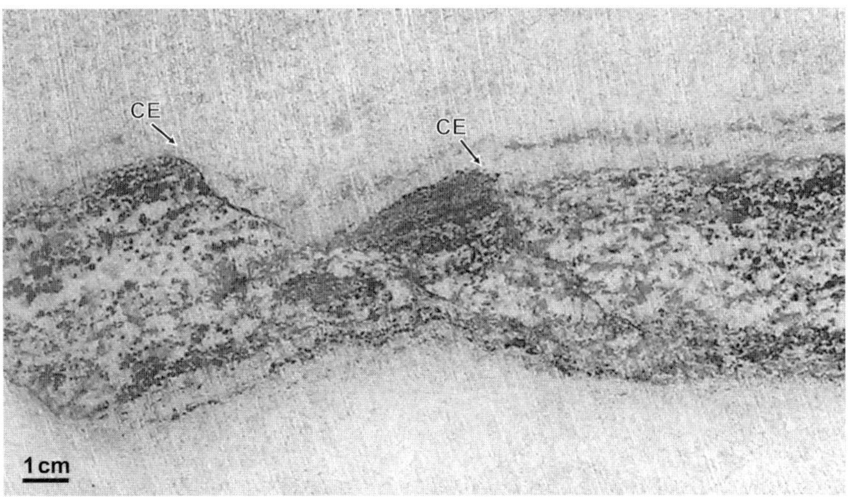

Fig. 9. Natural examples of flanking structure systems. (**a**) Parallel flanking structure system at low angle to SZB, showing synthetic, extensional offset of markers in a calcite marble with dolomitic layers from Naxos, Greece, along a vertical section with a top to the north sense of shear in the core of the Naxos dome (N37°11′26.2″, E025°30′48.8″); (**b**) parallel flanking structure system with CEs at high angles to the layering, taken from a similar lithology at the same location as (a); (**c**) amphibolite layer in a block from a marble quarry in Kinidaros (Naxos, Greece), with a low-angle conjugate flanking structure system producing a triclinic geometry of the marker layers; (**d**) conjugate flanking structure system from the same location as (a), combining a contractional with an extensional CE, resulting in a clockwise rotation of the CML between the CEs; (**e**) conjugate flanking structure system with several CEs at high angles to the layering, on the southern coast of Naxos, Greece, along a horizontal section at the western limb of the Naxos dome with bulk dextral shear sense (N36°58′22.9″, E25°24′21.6″); (**f**) at the same location as the structure in (e), two CEs oriented at high angles to the SZB form a conjugate flanking structure system. Drag effects of the marker lines between them vary significantly, from convex in the central parts of the structure to sigmoidal shapes in the upper right area, strongly resembling the geometry of markers in the experiment shown in Fig. 8.

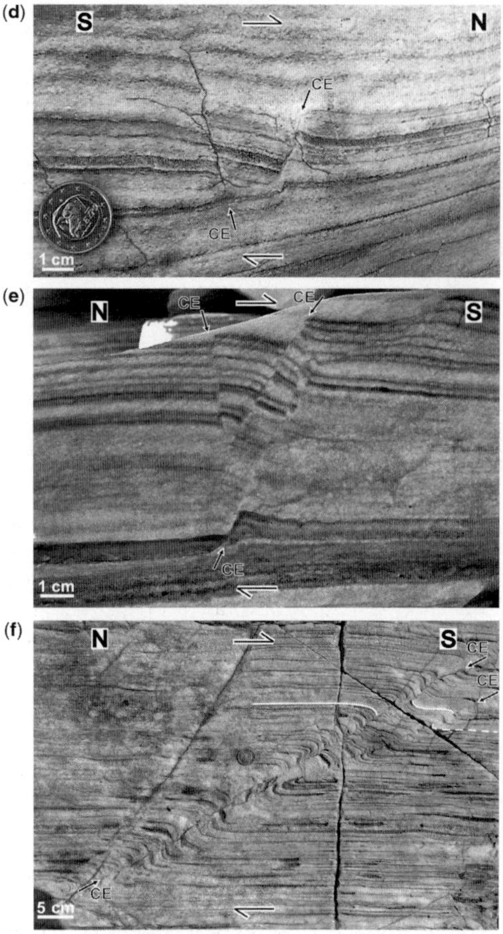

Fig. 9. *Continued.*

135° show antithetic shearing. Furthermore, the lines of zero instantaneous stretch oriented at 0 and 90° to the SZB separate alternate quadrants of instantaneous extension and contraction. Depending on the initial orientation of the conjugate CEs and their progressive rotation and shearing, complicated conjugate flanking structure systems with extensional and contractional offsets may coexist.

An amphibolite layer in a marble from Naxos, Greece (Fig. 9c), is cut by a low-angle CE with extensional offset and reverse drag, which is pronounced on the right side of the structure. On closer inspection, a second CE at a slightly steeper angle to the layering can be identified at the right side of the prominent CE. Hence, the triclinic symmetry of the structure can be attributed to the conjugate flanking structure system with two slightly oblique CEs, comparable to Figure 6a.

Another conjugate flanking structure system developed in a layered marble from Naxos, Greece (Fig. 9e). Several CEs with variable lengths and angles are developed, each featuring a different drag, amount of slip and, in some cases, even a different sense of offset. In this case, a sequential initiation of the CEs may also be possible.

A conjugate flanking structure system also from Naxos, Greece, with CEs oriented at high angles to the SZB is presented in Figure 9f. Variable drag effects identical to those in Figure 8 can be observed along the two CEs, which merge into a single CE toward the bottom left. These examples suggest caution in using secondary slip surfaces in shear zones to imply overall shear zone parallel extension or contraction in natural examples. Contraction of a shear zone parallel marker across a thrust, which repeats the marker in a section normal to the shear direction, does not necessarily imply contraction parallel to the shear zone in the far field. For example s-type flanking structures can develop in our analogue models while there is no extension or contraction parallel to the SZB (*cf.* Fig. 1; Exner *et al.* 2004). Similarly, extension of a shear zone parallel marker across a normal fault (e.g., shear bands, C′ fabric) does not necessarily represent an extension of the whole shear zone in the far field. Thus, the expression 'extensional crenulation cleavage' (Platt 1979), used to describe synthetic, low-angle, parallel flanking structure systems occurring as a dominant fabric element, can be misleading, by suggesting an extensional component parallel to the shear zone boundary. In fact, these kind of parallel flanking structure systems are possible not only under simple shear boundary conditions, but also in transtensive flow where there is even a component of contraction parallel to the SZB (Wiesmayr & Grasemann 2005).

Shear sense derived from conjugate flanking structure systems

Finite element modelling and natural examples suggest that single flanking structures may be ambiguous shear sense indicators, because mirror symmetries exist for all structures (Grasemann *et al.* 2003). Similarly, parallel flanking structure systems, which maintain their monoclinic symmetry, are also ambiguous. However, the analogue models presented in Figure 6 show that simple shear conjugate flanking structure systems are very reliable shear sense indicators due to their triclinic symmetry.

For instance, the natural example in Figure 9d demonstrates how a triclinic flanking structure system can be used as kinematic indicator. The marker lines are cut by two conjugate CEs: the left one shows a thrust geometry, while a normal offset is recognizable in the right one. Although the normal fault could be kinematically compatible with top-to-south shear, an anticlockwise rotation of the left CE would require extensional offset, which is incompatible with the observed contractional offset. Thus, overall this geometry clearly indicates bulk top-to-north shear sense.

Strain estimates from conjugate flanking structure systems

Comparing geometries of the experimental conjugate flanking structure systems with natural examples allows a tentative estimate of the amount of shear strain that was accommodated by the specific structure, and thus the approximate degree of rotation of the CE. Figure 9d provides a natural example of a conjugate flanking structure, which represents the combination of a contractional and an extensional reverse a-type flanking fold. The comparable analogue experiment (Fig. 7c) suggests a shear strain of $\gamma \sim 1$. In the experiment, the CEs rotated about 30°–35° from their initial positions at *c.* 50° and 90° to the SZB. Reverse numerical modelling of specific natural conjugate flanking structure systems has potential for constraining both the bulk strain and flow type during their development. The analogue experiments suggest that this constraint may be tighter than for single flanking structures, which are already capable of providing quantitative results in well-defined natural examples (Kocher & Mancktelow 2004).

Application to large-scale structures

Even though our simple physical models do not account for rheological contrasts, isostasy, flexural loading/unloading or gravity, there are nevertheless clear similarities between the small-scale models and some natural large-scale structures, suggesting that the basic geometric effects induced by planar discontinuities are valid at all scales. Deflections of the regional foliation or bedding planes near low- and high-angle faults are well known from natural examples. Low-angle normal faults are normally described in association with a regional-scale normal drag, whereas reverse drag is frequently associated with high-angle normal faults (e.g., Hamblin 1965; Tearpock & Bischke 2003). The geometric similarities between such natural regional scale faults and our model results are remarkable (e.g., compare Fig. 6b and c to the Cinderhill Fault in Figs 7 and 8 of Barnett *et al.* 1987).

The geometry of the set of conjugate normal faults in Figure 6b, with the resulting normal offset of the central part of the structure, is directly comparable to graben structures in rift systems (e.g., Hills 1963) and to the flank uplift and subsidence observed at rifted continental margins (Sclater & Célérier 1987). Similarly, Figure 6c would commonly be interpreted as a steep normal fault with a roll-over antiform, which is crosscut by a conjugate reverse fault acting as accommodation structure in the hanging wall (e.g., McClay & Scott 1991). However, our experiments make it clear that this geometry can also be explained as a conjugate flanking structure system.

Furthermore, some model results show extruding wedge geometries (e.g., Fig. 6d), with a thrust fault at the base and a normal fault at the top. In our models, there is large offset along the faults and decreasing strain towards the centre of the wedge. There is also a background regional strain that results in a rotation of the bounding faults, particularly the upper, more steeply inclined normal faults. In contrast, current models of extruding wedges assume that the bounding blocks are basically rigid and the limiting faults are therefore fixed in their orientation, and explain the observed geometry with wedge-normal shortening and general shear extrusion along stretching faults (e.g., Grujic *et al.* 1996; Vannay & Grasemann 2001). Nevertheless, fault analyses, e.g., in the Buthan Himalayas (Wiesmayr *et al.* 2002) indicate progressive rotation of the normal faults at the upper margin of the extruding wedge (i.e. the Higher Himalaya Crystalline), which is in line with the observations from our analogue experiments (e.g., Figs 7e, f and 8). Finally, the structure shown in Figure 6e is geometrically similar to natural imbricate thrust systems (e.g., Butler 1982), in spite of the fact that there is no shear zone parallel contraction in our analogue models.

Although there may be fundamental differences between the mechanics of our models and the natural examples, the effects of flow perturbations caused by variable slip along faults are scale independent and these effects can also be significant for large-scale structures. It may be argued that linear viscous material and brittle rocks in the upper crust have quite different rheologies. Nevertheless, the resulting perturbations in the velocity field are a purely geometric effect, which can also develop in brittle and/or anisotropic materials. This is confirmed by the analogue scale models of Brun *et al.* (1994) and Koyi and Skelton (2001), which consist of a brittle upper layer above a viscous lower layer. Both these layers develop drag geometries that are very similar to those of our experiments. These analogue experiments were scaled to model large-scale geodynamic processes, involving both gravity and a free upper surface, yet still show the same perturbation effects.

Conclusions

Perturbations of the homogeneous flow field around discontinuities cause flanking structures. Intersection of perturbations preconditions the development of conjugate and parallel flanking structure systems. Due to the perturbation of the homogeneous bulk simple shear flow field, lines initially parallel to the fabric attractor (in simple shear a line of no finite elongation) may (1) rotate in any direction away from this orientation, and (2) may experience elongation or contraction during progressive deformation. The sense and rate of rotation of faults is the same as for passive marker lines. Perturbation fields around conjugate (and parallel) flanking structure systems are markedly different from the simpler patterns developed around single discontinuities. During progressive deformation and coeval rotation of the faults, geometry, size and direction of the perturbation cells may change significantly, thus resulting in a complex evolution of the geometry of initially planar markers. Various combinations of fault orientations can produce geometric results that would not intuitively be attributed to simple shear boundary conditions: extensional horst and graben-like and contractional extruding wedge geometries can develop simultaneously in simple shear without layer-parallel deformation. The amount of background strain and fault rotation necessary to produce significant offset and deflection of marker layers is remarkably low ($\gamma \sim 1$). Conjugate flanking structure systems are suitable as shear sense indicators, because in simple shear they produce characteristic geometries depending on their orientation relative to the shear zone boundary. Geometries from analogue models can be used as a rough estimate of the amount of shear strain in natural examples. The only relevant scale factor of flanking structures is the length of the CE, which controls the size of the perturbation loops and thus the area and amount of deflection of passive markers.

The authors gratefully acknowledge financial support from ETH project 0-020998-02 and Austrian Science Foundation FWF project P-15668-Geo. Collaborative field work with G. Wiesmayr and T. Kocher and comments on an earlier version of the manuscript by J. P. Burg are highly appreciated. Reviews by F. Odonne and P. Bons and editorial comments by G. Schreurs also helped to improve the quality of the manuscript.

References

ARBARET, L., MANCKTELOW, N. S. & BURG, J.-P. 2001. Effect of shape and orientation on rigid particle rotation and matrix deformation in simple shear flow. *Journal of Structural Geology*, **23**, 113–125.

BARNETT, J. A. M., MORTIMER, J., RIPPON, J. H., WALSH, J. J. & WATTERSON, J. 1987. Displacement geometry in the volume containing a single normal fault. *American Association of Petroleum Geologists Bulletin*, **71**, 925–937.

BLENKINSOP, T. 2004. Comment on 'Folding in high-grade rocks due to back-rotation between shear zones' by Lyal B. Harris. *Journal of Structural Geology*, **26**, 601–602.

BRUN, J. P., SOKOUTIS, D. & VAN DEN DRIESSCHE, J. 1994. Analogue modeling of detachment fault systems and core complex. *Geology*, **22**, 319–322.

BUTLER, R. W. H. 1982. The terminology of structures in thrust belts. *Journal of Structural Geology*, **4**, 239–245.

COBBOLD, P. R. 1975. Fold propagation in single embedded layers. *Tectonophysics*, **27**, 333–351.

DIXON, J. M. & SUMMERS, J. M. 1985. Recent developments in centrifuge modelling of tectonic processes: equipment, model construction techniques and rheology of model materials. *Journal of Structural Geology*, **7**, 83–102.

EXNER, U., MANCKTELOW, N. S. & GRASEMANN, B. 2004. Progressive development of s-type flanking folds in simple shear. *Journal of Structural Geology*, **26**, 2191–2201.

GOSCOMBE, B. D. & PASSCHIER, C. W. 2003. Asymmetric boudins as shear sense indicators – an assessment from field data. *Journal of Structural Geology*, **25**, 575–589.

GOSCOMBE, B. D., PASSCHIER, C. W. & HAND, M. 2004. Boudinage classification: end-member boudin types and modified boudin structures. *Journal of Structural Geology*, **26**, 739–763.

GRASEMANN, B. & STÜWE, K. 2001. The development of flanking folds during simple shear and their use as kinematic indicators. *Journal of Structural Geology*, **23**, 715–724.

GRASEMANN, B., STÜWE, K. & VANNAY, J.-C. 2003. Sense and non-sense of shear in flanking structures. *Journal of Structural Geology*, **25**, 19–34.

GRUJIC, D., CASEY, M., DAVIDSON, C., HOLLISTER, L. S., KÜNDIG, R., PAVLIS, T. & SCHMID, S. 1996. Ductile extrusion of the Higher Himalayan Crystalline in Bhutan: evidence from quartz microfabrics. *Tectonophysics*, **260**, 21–43.

HAMBLIN, W. K. 1965. Origin of 'reverse drag' on the downthrown side of normal faults. *Geological Society of America Bulletin*, **76**, 1145–1164.

HARRIS, L. B. 2003. Folding in high-grade rocks due to back-rotation between shear zones. *Journal of Structural Geology*, **25**, 223–240.

HILLS, E. S. 1963. *Elements of Structural Geology*. Methuen and Co., London.

ILDEFONSE, B., SOKOUTIS, D. & MANCKTELOW, N. S. 1992. Mechanical interactions between rigid particles in a deforming ductile matirx. Analogue experiments in simple shear flow. *Journal of Structural Geology*, **14**, 1253–1266.

KOCHER, T. & MANCKTELOW, N. S. 2004. Numerical modelling of progressive evolution of flanking structures. *Bollettino di Geofisica*, **45**, 304.

KOYI, H. A. & SKELTON, A. 2001. Centrifuge modelling of the evolution of low-angle detachment faults from high-angle normal faults. *Journal of Structural Geology*, **23**, 1179–1185.

MANCKTELOW, N. S. 1991. The analysis of progressive deformation form an inscribed grid. *Journal of Structural Geology*, **13**, 853–864.

MCCLAY, K. R. & SCOTT, A. D. 1991. Experimental-models of hanging wall deformation in ramp flat listric extensional fault systems. *Tectonophysics*, **188**, 85–96.

PASSCHIER, C. W. 1997. The fabric attractor. *Journal of Structural Geology*, **19**, 113–127.

PASSCHIER, C. W. 2001. Flanking structures. *Journal of Structural Geology*, **23**, 951–962.

PASSCHIER, C. W., MANCKTELOW, N. S. & GRASEMANN, B. 2005. Flow perturbations: a tool to study and characterize heterogeneous deformation. *Journal of Structural Geology*, **27**, 1011–1026.

PLATT, J. P. 1979. Extensional crenulation cleavage. *Journal of Structural Geology*, **1**, 95.

PLATT, J. P. & VISSERS, R. L. M. 1980. Extensional structures in anisotropic rocks. *Journal of Structural Geology*, **2**, 397–410.

POLLARD, D. D. & SEGALL, P. 1987. Theoretical displacements and stresses near fractures in rocks. *In*: ATKINSON, B. K. (ed.) *Fracture Mechanics of Rock*. Academic Press, London, 277–349.

RAMSAY, J. G. 1967. *Folding and Fracturing of Rocks*. New York, McGraw-Hill.

RECHES, Z. & EIDELMAN, A. 1995. Drag along faults. *Tectonophysics*, **247**, 145–156.

SCHMID, D. W. & PODLADCHIKOV, Y. Y. 2003. Analytical solutions for deformable elliptical inclusions in general shear. *Geophysical Journal International*, **155**, 269–288.

SCLATER, J. G. & CÉLÉRIER, B. 1987. Extensional models for the formation of sedimentary basins and continental margins. *Norsk Geologisk Tidsskrift*, **67**, 253–267.

SWANSON, M. T. 1992. Late Acadian – Alleghenian transpressional deformation: evidence from asymmetric boudinage in the Casco Bay area, coastal Maine. *Journal of Structural Geology*, **14**, 323–341.

TEARPOCK, D. J. & BISCHKE, R. E. 2003. *Applied Subsurface Geological Mapping*. Prentice Hall, New Jersey.

URAI, J. L., SCHUILING, R. D. & JANSEN, J. B. H. 1990. Alpine deformation on Naxos (Greece). *In*: KNIPE, R. J. & RUTTER, E. H. (eds) *Deformation Mechanisms, Rheology and Tectonics*. Geological Society, London, Special Publications, **54**, 509–522.

VANNAY, J.-C. & GRASEMANN, B. 2001. Himalayan inverted metamorphism and syn-convergence extension as a consequence of a general shear extrusion. *Geological Magazine*, **138**, 253–276.

WALSH, J. J. & WATTERSON, J. 1989. Displacement gradients on fault surfaces. *Journal of Structural Geology*, **11**, 307–316.

WEIJERMARS, R. 1992. Progressive deformation in anisotropic rocks. *Journal of Structural Geology*, **14**, 723–742.

WIESMAYR, G. & GRASEMANN, B. 2005. Sense and non-sense of shear in flanking structures with layer-parallel shortening: implications for fault-related folds. *Journal of Structural Geology*, **27**, 249–264.

WIESMAYR, G., EDWARDS, M. A., MEYER, M., KIDD, W. S. F. LEBER, D., HÄUSLER, H. & WANGDA, D. 2002. Evidence for steady fault-accommodated strain in the High Himalaya; progressive fault rotation of the southern Tibet detachment system in NW Bhutan. *In*: De MEER, S., DRURY, M. R., DE BRESSER, J. H. P. & PENNOCK, G. M. (eds) *Deformation Mechanisms, Rheology and Tectonics: Current Status and Future Perspectives*. Special Publications **200**, Geological Society, London, 371–386.

Using an elastic dislocation model to investigate static Coulomb stress change scenarios for earthquake ruptures in the eastern Marmara Sea region, Turkey

JORDAN R. MULLER[1,2], ATILLA AYDIN[1] & TIM J. WRIGHT[3]

[1]*Department of Geological and Environmental Sciences, Stanford University, 450 Serra Mall, Bldg. 320, Stanford, CA, 94305, USA*

[2]*Present address: NASA Goddard Space Flight Center, Geodynamics Branch, Mail Code 921, Greenbelt, MD, 20771, USA (e-mail: jmuller@core2.gsfc.nasa.gov)*

[3]*Department of Earth Sciences, Oxford University, Parks Road, Oxford OX1 3PR, UK*

Abstract: Using an elastic dislocation model, we incorporate a historical earthquake catalog, mapped Marmara Sea fault traces, and fault slip distributions for the 1999 Izmit earthquake inferred from InSAR and GPS data to determine various stress change scenarios crucial for evaluating future earthquake potential in the eastern Marmara Sea. We have tested six plausible past rupture configurations arising from the uncertainty in the location of the western termination of 1999 Izmit earthquake rupture and the location of the 1963 Yalova earthquake rupture. Coulomb stresses calculated are increased on the Princes' Islands, Çinarcik, and Armutlu fault segments in each case. In four of the six plausible configurations of previous ruptures, the Çinarcik fault receives the greatest average stress change. In one other configuration, the average stress increase on the Princes' Islands fault is greatest. In another, the stress changes on the Çinarcik and Princes' Islands faults are comparable. Moreover, we show that rupture initiating on either the Princes' Islands or Armutlu faults would be favoured to propagate onto the Central Marmara, or Imrali fault, respectively, based on the favourable geometries of the respective fault intersections. Rupture initiating on the Çinarcik fault, however would be limited to a much shorter length based on its mapped western termination. Therefore, while the earthquake-induced stress changes may, in most cases, be greatest on the Çinarcik fault, an earthquake initiating on this fault segment may produce a shorter cumulative rupture compared to rupture initiated on the two other major eastern Marmara Sea fault segments. These results are encouraging for the use of geomechanical modelling tools in addressing uncertainties inherent in most geological and geophysical data applied to earthquake-related problems.

The cascading sequence of large earthquakes progressing along the North Anatolian fault from 1912 to 1999 has led scientists to an intuitive, yet troubling, conclusion: the next large earthquake is most likely to rupture along the northern portion of the Sea of Marmara and within tens of kilometres of Istanbul, a city of 14 million inhabitants (Fig. 1) (Le Pichon *et al.* 1999; Hubbert-Ferrari *et al.* 2000; Parsons *et al.* 2000). To address this threat, earth scientists aim to provide geologically constrained probabilities of future rupture scenarios (e.g., Stein *et al.* 1997). Following the 1999 Izmit earthquake, Parsons *et al.* (2000) combined earthquake records over the past several centuries, earthquake recurrence intervals and computed stress changes associated with recent earthquakes, to estimate that there is a $62 \pm 15\%$ probability for strong shaking to occur in Istanbul before 2030. The contribution of the computed static stress changes to this probability estimate is large – without the stress changes associated with the 1999 Izmit earthquake, the probability is reduced by *c.* 15%. Realistic future rupture scenarios and effective incorporation of earthquake interaction in probability studies therefore require accurate calculation of stress changes on a well-constrained Marmara Sea fault geometry.

Several recent studies have calculated stress changes on vertical faults with uniform or smoothly varying strike directions throughout the Sea of Marmara due to large earthquakes within the 20th century (Nalbant *et al.* 1998; Hubert-Ferrari *et al.* 2000; Parsons *et al.* 2000; Pinar *et al.* 2001; Cakir *et al.* 2003). This

From: BUITER, S. J. H. & SCHREURS, G. (eds) 2006. *Analogue and Numerical Modelling of Crustal-Scale Processes.* Geological Society, London, Special Publications, **253**, 397–414.
0305-8719/06/$15.00

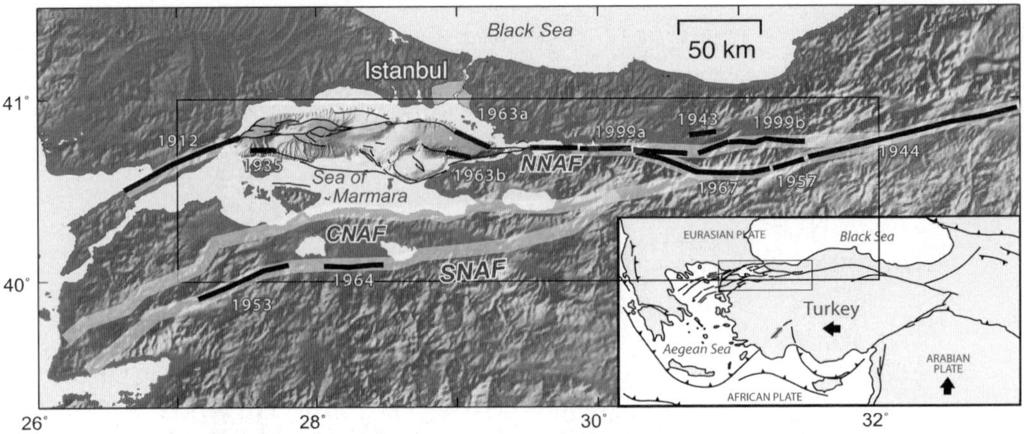

Fig. 1. Seismotectonic setting of the Marmara Sea and surrounding region. Surface traces of $M_s > 6.4$ earthquakes since 1900 are plotted in black. The ruptures 1963a and 1963b refer to the same event on September 18 but show the different rupture locations interpreted in other publications. Thick light gray lines show the locations of the northern (NNAF), central (CNAF) and southern (SNAF) strands of the NAF. Marmara Sea fault geometry is modified from Armijo *et al.* (2002). Inset figure shows the regional setting of the larger figure. Large rectangle in larger figure shows boundaries of Fig. 2a.

'optimally oriented fault' approach is only accurate to the degree that the local fault orientation in the Sea of Marmara matches the assumed regional fault orientation in the model. This approach therefore introduces inaccuracies in regions such as the eastern Sea of Marmara, where faults that bound the Çinarcik basin dip 60° or shallower and deviate in strike from the assumed plate boundary by up to 30°. Others have computed stress changes resolved onto individual Sea of Marmara fault segments (Hubert-Ferrari *et al.* 2000; Pinar *et al.* 2001). In general, the calculated stress increases range from 0.01 to 0.5 MPa depending on the proximity of the fault segment to the nearby modelled earthquake ruptures. Not surprisingly, the greatest stress increases are located in the eastern Sea of Marmara region, to the west of the presumed termination of the 1999 Izmit earthquake. In this region, the northern North Anatolian fault (NNAF) splits into three strands (Fig. 2), and Pinar *et al.* (2001) suggest that the Yalova-Hersek fault segment (0.45 MPa) was subjected to a greater stress change due to the Izmit earthquake than either the Princes' Islands (0.18 MPa) or Çinarcik-Yalova (0.09 MPa) fault segments.

The recent earthquakes in this region, however, have ambiguous rupture configurations, making it difficult to constrain a single stress-change scenario. The locations of both the western termination of the 1999 Izmit earthquake rupture and the rupture of the 1963 eastern Marmara earthquake (commonly referred to as the 'Yalova' earthquake) are debated within the literature. Various authors have suggested that the 1999 Izmit earthquake has either terminated east of the Hersek Delta within the Bay of Izmit, or propagated anywhere between 12 and 33 km further west (Fig. 2b). The 1963 earthquake has been thought to occur on either the NW-trending Princes' Islands fault (Taymaz *et al.* 1991; Parsons *et al.* 2000) or the WNW-trending southern Çinarcik fault (Jackson & McKenzie 1988; Nalbant *et al.* 1998). The stress changes that are calculated within the eastern Sea of Marmara will depend significantly on the configurations of these ruptures.

Using geomechanical models, we aim to evaluate how future rupture scenarios, in terms of Coulomb stress changes, depend on the locations of these recent eastern Sea of Marmara earthquake ruptures. We argue that future rupture is most favoured along either the southern Çinarcik fault or the Princes' Islands fault, depending on the location of earlier ruptures. The difference between these two scenarios has important implications for the potential hazard posed to the region by a future earthquake due to the proximity of the faults to Istanbul. The rupture lengths that may develop for an earthquake initiating on either fault are likely to be different. The Çinarcik fault is mapped to have an abrupt western termination whereas the Princes' Island fault may more effectively transfer slip as it links with the

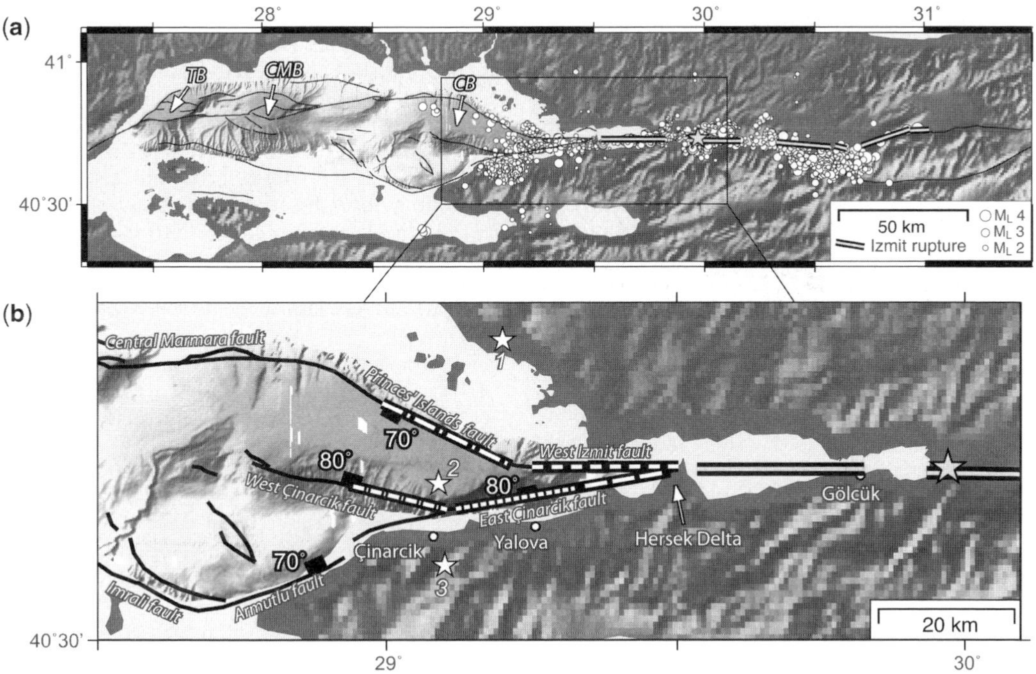

Fig. 2. (**a**) Location of major 1999 Izmit earthquake rupture segments (white bounded by black) and epicentre (star), Izmit earthquake aftershocks (Özalaybey *et al.* 2002), and faults within the Sea of Marmara (thin black) (Armijo *et al.* 2002). TB: Tekirdag basin; CMB: Central Marmara basin; CB: Çinarcik basin. (**b**) Configuration of eastern Sea of Marmara faults and the three different westernmost Izmit earthquake rupture segment geometries (white, dashed) and two different 1963 earthquake rupture locations (white, dot-dashed) tested in our study. Small stars indicate epicentral locations of 1963 Yalova earthquake proposed by (1) Taymaz *et al.* (1991), (2) Jackson and McKenzie (1988), and (3) Ambraseys and Finkel (1988).

Central Marmara fault that continues tens of kilometres further west.

1963 earthquake location

The size of the $M_s = 6.2–6.4$ 1963 earthquake (Table 1) suggests that the fault segment ruptured in this earthquake released at least several decades of stress accumulation; therefore, the local distribution of stress change on the eastern Sea of Marmara fault segments depends significantly on the location of this event. Seismological studies have determined epicentral locations up to tens of kilometres apart (Fig. 2b) (Ambraseys & Finkel 1988; Jackson & McKenzie 1988; Taymaz *et al.* 1991). Relative to the locations of the Princes' Islands and Çinarcik faults (Fig. 2b), Taymaz *et al.* (1991) determine the event to be to north of both faults, Jackson & McKenzie (1988) place it between the faults, and Ambraseys & Finkel (1988) provide a location south of both faults. Based on the ambiguity of these locations, previous stress change studies have modelled the 1963 rupture either on the SW-dipping Princes' Islands fault (Parsons *et al.* 2000) or

Table 1. *Location of the 1963 earthquake in seismological and stress modelling studies*

Long (°)	Lat (°)	Strike (°)	Dip (°)	Rake (°)	Reference
29.1	40.7	268	70	235	Jackson & McKenzie (1988)
29.1	40.6	–	–	–	Ambraseys (1988)
Princes' Islands fault		296	~60	–	Parsons *et al.* (2000)
29.2	40.9	304	~56	−82	Taymaz *et al.* (1991)
Çinarcik fault		304	~56	−82	Nalbant *et al.* (1998)

the north-dipping southern Çinarcik fault (Nalbant *et al.* 1998; Hubert-Ferrari *et al.* 2000).

Taymaz *et al.* (1991) provide the most comprehensive source determination and compute almost pure normal slip on south- and north-dipping nodal planes that approximately matches the orientations of the Princes' Islands and southern Çinarcik faults, respectively. However, their resolved epicentral location is not near to the major faults in the region (Fig. 2a), it is *c.* 15 km northeast of the Princes' Islands fault and *c.* 25 northeast of the southern Çinarcik fault. Damage from the earthquake was localized on the southern coast of the Sea of Marmara, leading Nalbant *et al.* (1998) to attribute the earthquake to rupture of the southern Çinarcik fault, and the north-dipping nodal plane of Taymaz *et al.* (1991). Parsons *et al.* (2000) have attributed the event to rupture of the Princes' Island fault, presumably based on its closer proximity to the epicentre location, and the southwest-dipping nodal plane of Taymaz *et al.* (1991). The combination of epicentral location, nodal plane solutions and damage reports does not clearly define which fault ruptured in 1963. Therefore, we test and evaluate separate scenarios of rupture associated with the 1963 event on both the Princes' Islands and the southern Çinarcik faults.

Possible locations for the western termination of the 1999 Izmit rupture

Previous slip inversions

Since most of the Izmit surface rupture west of the epicentre occurred underwater within the Bay of Izmit, we rely upon a variety of geophysical data sets in constraining the rupture location. These include GPS and InSAR geodetic data, seismic waveform data, and aftershock locations. Inverting for Izmit earthquake fault slip, Reilinger *et al.* (2000) show that GPS geodetic measurements do not account for slip in excess of 0.6 m more than 10 to 15 km west of the Hersek Delta along the Yalova segment of the Izmit rupture. Synthetic aperture radar interferometry (InSAR) indicates that the best-fitting length of rupture beyond the Hersek Delta is 12 km, with good fits to the interferograms for fault lengths between 9 and 18 km (Wright *et al.* 2001*b*). At fault lengths beyond 12 km west of the Hersek Delta, the increase in misfit is reduced, allowing for the possibility that rupture extended further than 12 or even 18 km west of the Hersek Delta. However, the small increases in misfit may be attributed to the lack of InSAR data coverage west of the Hersek Delta, reduced data coverage as the Gulf of Izmit widens, and a lack of coherence on the southern shore of the Gulf of Izmit in the interferograms (Wright *et al.* 2001*b*).

Seismic waveform data reveals different interpretations of the western rupture termination depending on the data set and processing method. Yagi and Kikuchi (2000) use near-field strong motion data and teleseismic body wave data to invert for the slip distribution, and show that subsurface slip extended no further than 25 km west of the epicentre near Gölcük (15 km east of the Hersek Delta). Gülen *et al.* (2002) invert teleseismic body waves from 25 seismic stations, and when they include the latter phases of the seismograms (60–100s) in their inversion they find a rupture sub-event west of the Hersek delta corresponding to 1.3 m of slip approximately 10 km west of the Hersek Delta. They propose that this sub-event corresponds to rupture of the Yalova segment and is responsible for the lineation of surface cracking nearby onshore, and the alignment of aftershocks extending along the coast and into the Marmara Sea. Aydin and Kalafat (2002) argued for a similar southwest extension of the rupture based on the field observation of earthquake-related faults and increased damage in the Yalova region. Based primarily on a lineation of aftershocks, Gülen *et al.* (2002) also suggest the possibility that the Izmit rupture propagated west along a more northerly fault segment extending from the Hersek Delta to the southern portion of the Princes' Islands fault. However, this rupture scenario is hard to constrain because it is well within the Sea of Marmara, placing it further from coastal sites and geodetic measurements.

Based on the ground motion from five near-fault accelerometers, Bouchon *et al.* (2002) compute fault slip that sharply decreases from a local surface slip maximum of 2 m near the Hersek Delta on a fault that trends east–west into the Marmara Sea. Also using near-fault strong motion waveforms, Sekiguchi and Iwata (2002) suggest that slip extended west of the Hersek Delta for at least 5 km. Several authors use joint inversion of different types of geodetic data (Feigl *et al.* 2002), or combined geodetic and seismic data (Delouis *et al.* 2002) to constrain the western rupture termination. Similar to the Bouchon *et al.* (2002) results, the solutions of both Delouis *et al.* (2002) and Feigl *et al.* (2002) include less than a metre of slip about 50 km west of the epicentre that decays to zero about 15 km west of the Hersek Delta.

In summary, the various slip inversion studies suggest that the western termination of the 1999 Izmit earthquake rupture is located 10–20 km west of the Hersek Delta (50–60 km west of the epicentre), although the latitudinal position is not well constrained because the inversion models presuppose a fault trend.

Aftershock location studies

The distribution of aftershocks following the Izmit earthquake extend approximately 75 km west of the epicentre (Karabulut *et al.* 2002; Özalaybey *et al.* 2002) and are clustered in three general regions within and west of the Bay of Izmit (Fig. 2a). One cluster is a linear trend of aftershocks that extend about 35 km west of Hersek with an uninterrupted east–west trend parallel to the axis of Izmit Bay. The vast majority of the hypocentres occurred at 5–17 km depth (Karabulut *et al.* 2002; Özalaybey *et al.* 2002) and have predominantly strike-slip focal mechanisms (Karabulut *et al.* 2002).

Further west, the second cluster branches to the NW and extends to within 10 km south of the Princes Islands. Coseismic slip in this region is not confirmed or ruled out by the geodetic or strong motion inversions and may be suggested by teleseismic body wave inversion (Gülen *et al.* 2002). Focal mechanisms of three $M_L \geq 4.0$ events that occurred along this lineation show right-lateral strike-slip mechanisms along WNW-trending fault planes (Özalaybey *et al.* 2002), while many smaller events show normal faulting mechanisms (Karabulut *et al.* 2002).

In contrast, the third cluster of aftershocks, mostly onshore and near the city of Yalova, show 45°–60° north-dipping normal faulting mechanisms (Özalaybey *et al.* 2002) that cluster along a N50°-dipping zone of seismicity. The number of aftershocks within the Yalova cluster significantly increases two days after the mainshock. This delayed onset, together with the dip-slip focal mechanisms, indicate that this cluster represents triggered seismicity and is not likely to be part of the mainshock rupture; however, the localized cluster suggests a local stress concentration that could be related to nearby termination of the Izmit rupture.

Pinar *et al.* (2001) concluded that the Izmit rupture terminated east of 29.6°E. They argue that this rupture termination location, as opposed to one further west at 29.1°E, places a greater number of aftershocks within regions of increased Coulomb stresses. Several observations complicate this approach. First, the vertical faults with an optimally oriented strike direction along which Coulomb stress were calculated in their model is less accurate in view of the variation of the focal mechanism nodal planes in this region. Second, in addition to regions beyond the lateral fault tips, aftershocks can be triggered along the upper and lower tiplines of ruptures. In map view, the aftershocks along the upper or lower tiplines would plot along the main trace of the rupture, resulting in a rather continuous lineation of aftershocks along the fault and well past its tips. This makes it difficult to clearly discern where within this linear cluster the lateral terminations of the rupture are located. Indeed, there are aftershocks all along the rupture trace for both locations of western termination that Pinar *et al.* (2001) evaluate, suggesting that aftershocks are a poor constraint for determining the lateral limits of this rupture.

Review of the geophysical data and other investigations suggests four possible western termination locations for the Izmit rupture (Fig. 2b): (1) Rupture continuing 15 km WSW of the Hersek Delta along the southern strand of the NNAF within the Bay of Izmit, as suggested by slip inversions from seismic and geodetic data. (2) Rupture continuing 20 km west of the Hersek Delta along the northern strand of the NNAF towards the Princes' Islands fault as suggested by aftershocks and slip inversions by Gülen *et al.* (2002). (3) Rupture continuing 30 km WSW of the Hersek Delta along the southern strand of the NNAF as suggested by aftershock clustering and strong shaking in Çinarcik and Yalova (Özalaybey *et al.* 2002). (4) Rupture terminating just east of the Hersek Delta as suggested by Pinar *et al.* (2001). We note that the available data cannot rule out the possibility that several of these faults strands ruptured simultaneously during the earthquake; however, we choose to investigate single-strand rupture geometries which should produce the end-member magnitudes of possible stress changes on the eastern Marmara faults.

Investigated Izmit earthquake rupture scenarios

To evaluate the stress changes that each of these different rupture scenarios would produce on faults in the eastern Sea of Marmara, we invert InSAR (Wright *et al.* 2001*a*, *b*) and GPS (Reilinger *et al.* 2000) ground surface displacement data to solve for the Izmit earthquake slip distributions for each of the four different

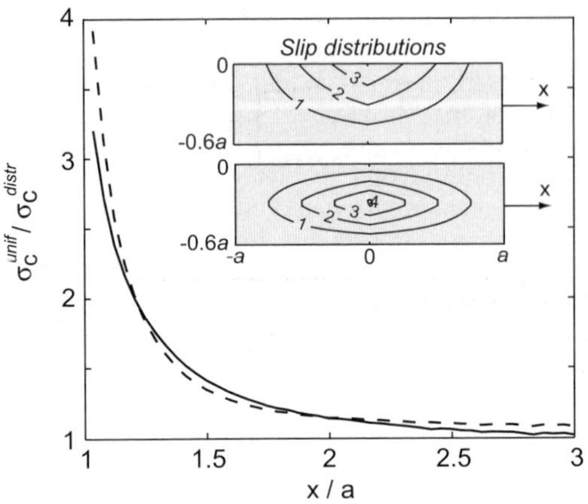

Fig. 3. Comparison of stress change magnitudes as a function of distance from the fault edge for uniform versus distributed slip. Uniform slip produces Coulomb stress changes that are several times greater than stress changes from distributions with either maximum slip at the surface (solid) or at depth (dashed). Faults (grey shaded) with slip distributions are shown in inset with the average slip in both cases (1 m) matching the case of uniform slip.

fault configurations. Solving for the distributed slip, rather than simply applying a uniform slip magnitude to an entire fault surface, is important because faults with uniform slip produce stress changes up to several times greater than that of tapered slip within a half-length of the ruptured fault (Fig. 3). Using well-constrained distributed slip solutions will be particularly important for the faults in the eastern Sea of Marmara because the NNAF splits into three different strands within a half-length of the major fault segments of the Izmit rupture (Fig. 2a).

We solve for a distribution of Izmit earthquake fault slip that provides the best fit to ground surface displacements from InSAR interferometry (Wright *et al.* 2001*a*) and GPS data (Reilinger *et al.* 2000) in a least squares sense. We use an ERS-2 ascending interferogram computed from satellite passes on 13 August and 17 September 1999. The western edge of the interferogram is located about 25 km west of the Hersek Delta. The computed fault slip values will therefore be less well constrained, approaching or exceeding the western edge of the interferogram. The GPS data were acquired within the first two months of the earthquake and were recorded at 51 sites (Reilinger *et al.* 2000).

The fault segments used in our inversion are rectangular and we fix their geometry (Fig. 4, Table 2). Surface displacements for each slip solution are calculated, with each fault patch modelled as a rectangular dislocation in a homogeneous, isotropic elastic medium. We assume pure right-lateral fault slip on the fault segments and determine its distribution using a standard linear least squares inversion procedure (e.g., Jónsson *et al.* 2002; Wright *et al.* 2004). We use a smoothing parameter, based on the finite difference approximation to the Laplacian operator, to constrain the variation in slip between each slip patch and its neighbours. The relative weight of the smoothing is chosen by plotting the slip roughness against data misfit: the value selected gives the smoothest solution that still has a satisfactory fit to the data. A positivity constraint is also applied to ensure that slip within each entire fault segment is right-lateral. A zero-slip boundary condition is prescribed along the lateral and bottom edges of the fault. Following the method of Wright *et al.* (2004), we weighted the GPS and InSAR data using weight matrices, **W**, the Cholesky decompositions of the respective data variance-covariance matrices, Σ ($W^TW = \Sigma^{-1}$). Because we include off-diagonal terms for the variance-covariance matrix of the InSAR data, to account for the high degree of spatial correlation, the InSAR data do not swamp the GPS data. The final slip solution is that which provides the minimum misfit between the observed geodetic surface displacements and those computed using the model slip solution (Wright *et al.* 2004).

The misfits between the modelled and observed InSAR and GPS surface displacements indicate that the scenario with no faulting west of the Hersek Delta (scenario 4) cannot reproduce observed displacements as well as cases with slip extending further west (Fig. 5, Table 3), so this scenario is not considered for use in the stress change calculations. The multi-segment slip distributions for the three cases where slip extends west of the Hersek Delta, projected onto a single fault only for display, are shown in Figure 4. For the most part, they are very similar, with a slip maximum of 6.7–6.8 m located west of the epicentre, along the eastern portion of the Bay of Izmit. Slip decreases to less than 3 m within 10 km of the east Hersek Delta and just west of Lake Sapanca. These three cases, however, all fit the data equally well. This is likely to be due to the proximity of the western edge of the InSAR interferogram and the lack of data within the Bay of Izmit. This means that the three scenarios with rupture extending west of the Hersek Delta are equally justified based on the available InSAR and GPS data. This uncertainty motivates our modelling approach and requires

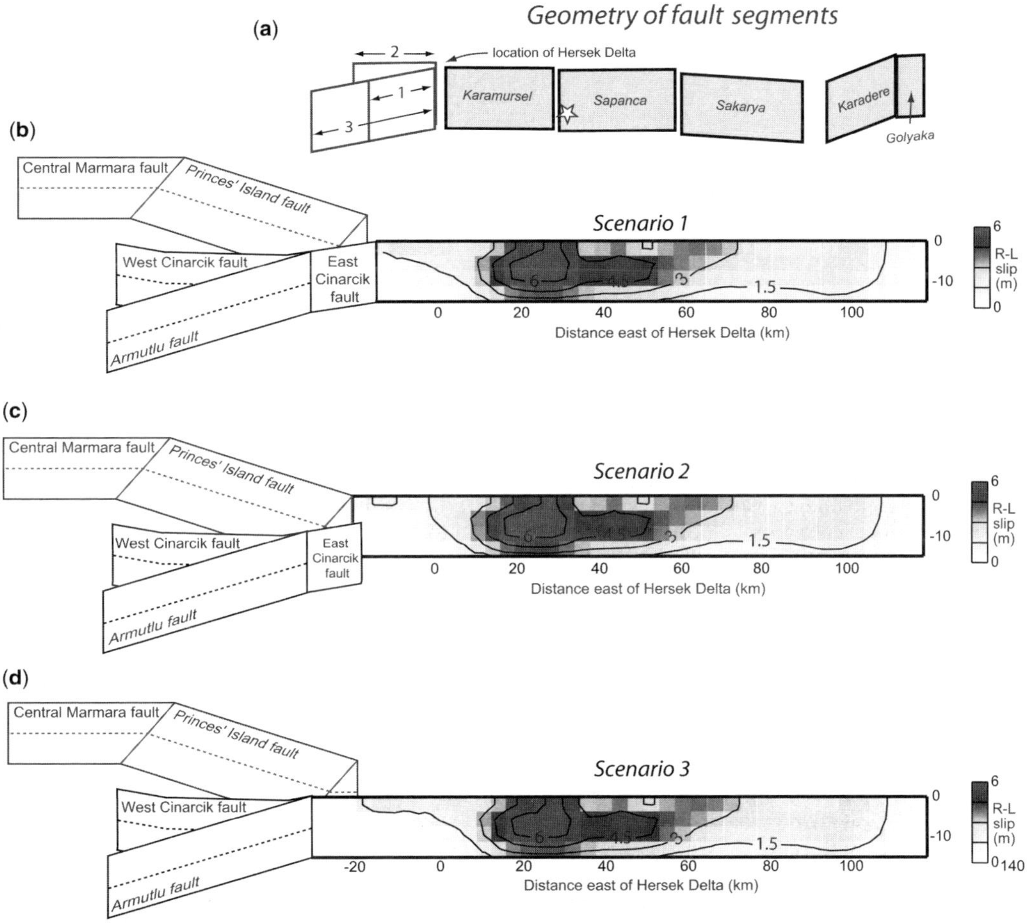

Fig. 4. (**a**) 3-D geometry of Izmit earthquake rupture segments used for slip inversion showing the three different western termination geometries where slip extends west of the Hersek Delta. Segment parameters are given in Table 2. The named grey segments with black outline are the same for all inversions. The star indicates the location of the Izmit earthquake hypocentre. The three different westernmost segments outlined in red and numerically labelled correspond to the segments included for inversion in scenarios 1, 2, and 3. The inversion smoothes slip across the discontinuous segment boundaries. (**b**) Computed slip distribution for scenario 1 (fault extends 15 km southwest of Hersek Delta, along the Bay of Izmit coast). Slip values are projected onto a planar surface for presentation only. (**c**) Slip distribution for scenario 2 (fault extends 20 km west towards Princes' Islands fault). (**d**) Slip distribution for scenario 3 (fault extends 30 km southwest along Yalova/Çinarcik coast to intersect Armutlu fault).

us to test these three different Izmit rupture configurations.

Stress changes after 1999

To gain an accurate measure of the Coulomb stress change in the eastern Marmara Sea for each scenario, we resolve the shear and normal stress on the fault surfaces at a depth of 7.5 km. This depth is half of the approximate 15 km thickness of the seismogenic crust based on the lower depth of aftershocks following the 1999 Izmit earthquake (Karabulut *et al.* 2002; Özalaybey *et al.* 2002). The surface traces of the model faults (Fig. 2b) are those of Armijo *et al.* (2002) and the dips and dip directions are based on seismic reflection data (Parke *et al.* 1999; Okay *et al.* 2000; Le Pichon *et al.* 2001) and the modelling analysis of different proposed Marmara Sea fault geometries (Muller & Aydin 2005).

The Coulomb stress change is defined as

$$\sigma_c = |\Delta\tau| + \mu'\Delta\sigma$$

Table 2. *Fault segment parameters*

Izmit rupture segment	Strike (°)	Dip (°)	Rake (°)	Length (km)	Depth (km)	Marmara Sea fault segment	Strike (°)	Dip (°)
Scenario 1						West Izmit	270	90
Westernmost segment	77	90	180	17.0	15	Princes' Islands	116	70
Scenario 2						East Çinarcik	260	80
Westernmost segment	88	90	180	21.7	15	West Çinarcik	284	80
Scenario 3						Armutlu	245	70
Westernmost segment	77	90	180	33.1	15			
Karamursel	89	90	180	28.4	15			
Sapanca	88	90	180	30.6	15			
Sakarya	92	90	180	32.0	15			
Karadere	64	90	180	19.8	15			
Golyaka	84	90	180	7.2	15			

where $\Delta\tau$ is the change in the shear traction on the fault surface due to a neighbouring earthquake, $\Delta\sigma$ is the change in the normal traction (compression is negative), and μ' is the effective coefficient of friction. It is necessary to keep track of the sign of τ (traction inducing right-lateral slip being positive) because shear tractions of different sign, but equal magnitude, on adjacent areas of a fault surface could produce a uniform positive Coulomb stress change, while this situation would not physically lead to failure in any given direction. Such a scenario is unlikely to occur, but should be considered in the eastern Marmara Sea region where perturbations from dip-slip and strike-slip earthquakes in close proximity are modelled.

The effective coefficient of friction μ' is used to relate pore pressure to the normal stress acting on the fault surface and can be defined as $\mu' = \mu(1 - B)$ where Skempton's coefficient B

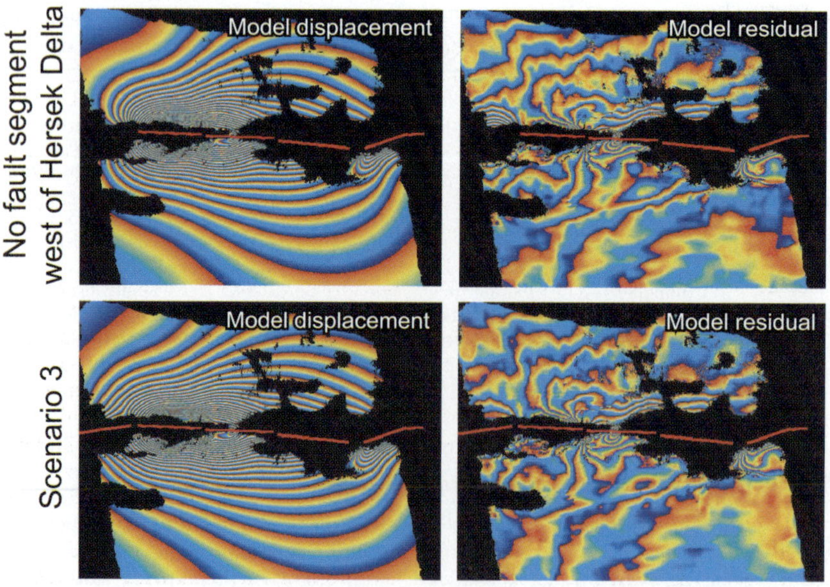

Fig. 5. Computed satellite line-of-sight surface displacement interferograms (left column) and residual interferograms (right column) for endmember Izmit fault geometries. Residuals are calculated by subtracting the model interferogram from the actual interferogram. In upper panels, there is no Izmit rupture segment west of the Hersek Delta. In the lower panels, the fault geometry corresponds to scenario 3 in Fig. 4. Note the concentration of fringes indicating large residuals in the western end of the Izmit rupture when there is no fault segment west of the Hersek Delta (upper right panel).

Table 3. *Root-mean-squared misfits between modelled and observed InSAR and GPS data*

Scenario	Western termination of Izmit rupture	InSAR residual (cm)	GPS residual (cm)
1	15 km west of the Hersek Delta, along southern coast of Bay of Izmit	3.29	4.59
2	20 km west of the Hersek Delta, at intersection with Princes' Islands fault	3.30	4.50
3	30 km west of the Hersek Delta, along southern coast of Bay of Izmit	3.27	4.60
4	Hersek Delta	3.45	4.99

theoretically ranges from 0 for fully drained conditions to 1 for fully saturated conditions (Harris 1998). However, μ' is not physically well-constrained because we commonly do not know the appropriate value of B to employ for the fault rocks in the seismogenic zone and because pore pressures can change with time (Scholz 2002). Studies of stress changes produced by earthquakes suggest that μ' can range from 0 to 0.75, with values of $\mu' = 0.4-0.5$ appropriate in earthquake-triggering investigations (e.g., Harris *et al.* 1995; Harris & Simpson 1996; Stein *et al.* 1997). We set $\mu' = 0.4$, for direct comparison with other Coulomb stress change studies along the NAF (e.g., Stein *et al.* 1997; Hubert-Ferrari *et al.* 2000; Pinar *et al.* 2001).

Modelling method

We compute stress changes using Poly3D (Thomas 1993), a boundary element code that provides quasi-static stress changes at any point in an isotropic and homogeneous linear elastic half-space due to slip on dislocation surfaces. The boundary element method is computationally efficient for dislocation modelling because only the dislocations, and not the surrounding body, require discretization. The dislocation, or fault, surfaces embedded in the model half-space consist of multiple planar, triangular elements. Each triangular element is composed of six superposed angular dislocations (Comninou & Dunders 1975) arranged in such a way that a discontinuity in displacement exists only in a desired triangular region (Jeyakumaran *et al.* 1992). The full derivation of the solution for the stress at any point within the elastic half-space due to prescribed displacement discontinuity boundary conditions on one, or many, triangular elements is given in Thomas (1993, appendix A). The displacement discontinuity is constant on each element, but multiple elements can be used to model an arbitrary number of mechanically interacting fractures or faults with non-uniform slip distributions (Crouch & Starfield 1983). Unlike rectangular fault elements used in many dislocation models (e.g., those based on the solution of Okada 1985), the triangular fault elements in Poly3D allow for three-dimensional curved fault surfaces and tiplines without introducing artificial gaps and overlaps among elements (Resor 2004). In our models, the average element size is approximately 7 km^2. The stresses calculated within our model are static in that time-dependent, or dynamic, stresses that are associated with the propagation of seismic waves during earthquake rupture are not considered.

The slip distributions computed from the InSAR and GPS inversions (Fig. 4) are applied as the displacement discontinuity boundary conditions on the Izmit fault segments in our six test cases. We test permutations of the three different western terminations of the 1999 Izmit rupture and the two possible locations of the 1963 Yalova earthquake, for a total of six different eastern Marmara rupture scenarios (Fig. 6a–f, col. 1).

In addition to the 1963 Yalova and 1999 Izmit earthquakes, we include in our models all $M_s \geq 6.4$ earthquakes in the Sea of Marmara region since 1900 (Fig. 1; Table 4). Besides the 1963 Yalova and 1999 Izmit events, these earthquakes contribute stress changes of the order of 0.01–0.02 MPa to the faults in the eastern region of the Sea of Marmara, which represent *c.* 5–10% of the total stress changes when the 1963 Yalova and 1999 Izmit earthquakes are included in the model. For all earthquakes besides the 1999 Izmit event, a constant displacement discontinuity is applied on the entire fault surface, with slip magnitude rake and direction given in Table 4.

Stress increases associated with secular plate motion, commonly incorporated in stress change calculations, are hard to accurately constrain in the Marmara Sea region. Due to the lack of GPS survey measurements close to the fault strands, it is unclear how the multiple upper crustal fault strands in this region connect with deep plate motion beneath the locked seismogenic portion of the lithosphere. Studies suggest different locking depths and

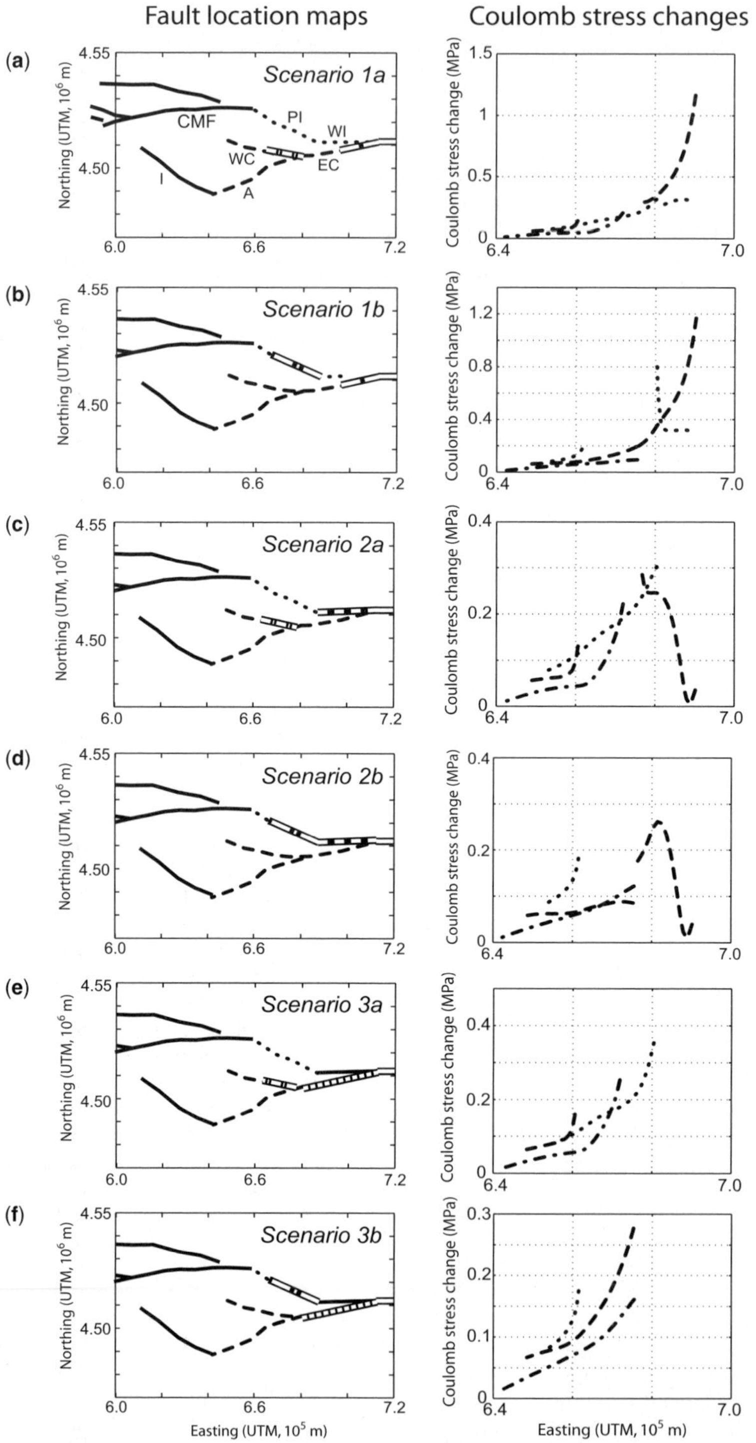

Fig. 6. Rupture scenarios (left column) and Coulomb stress changes (right column) for the eastern Marmara Sea faults. WI: West Izmit fault (dotted); PI: Princes' Islands fault (dotted); EC: East Çinarcik fault (dashed); WC: West Çinarcik fault (dashed); A: Armutlu fault (dot-dashed); I: Imrali fault (solid); CMF: Central Marmara fault (solid). Western segment of Izmit earthquake rupture is shown as thick dashed white line corresponding to Fig. 2b. 1963 earthquake rupture is shown as thick dot-dashed white line corresponding to Fig. 2b.

Table 4. *Earthquakes since 1900 included in the stress change model*

Date	M_s	M_o (10^{19} Nm)	Area of model fault[6] (10^8 m^2)	Dip of model fault (°)	Slip[7] (m)	Rake on model fault[8] (°)
Jan. 19, 1912	7.4	12.6[1]	11.7	90	3.60	180
Feb. 4, 1935	6.4	0.4[1]	3.4	60N	0.39	−90
June 20, 1943	6.4	0.4[1]	2.4	90	0.54	180
Feb. 1, 1944	7.3	8.9[1]	19.4	90	1.53	180
March 18, 1953	7.2	6.3[1]	9.0	90	2.35	180
May 26, 1957	7.0	3.2[1]	3.0	90	3.57	180
Sept. 18, 1963	6.4	0.1[2]	3.0[4]	70S[4]	0.10[4]	90
			2.1[5]	80N[5]	0.15[5]	−90
Oct. 6, 1964	6.9	0.4[2]	7.8	60N	0.18	−90
July 22, 1967	7.1	11.0[3]	1.2	90	2.96	180
Nov. 12, 1999	7.2	6.3[1]	7.8	60N	2.69	180

[1]Scalar seismic moment computed by assuming $M_w = M_s$ and using the relationship $M_o = 10^{(1.5^*M_w+9)}$ (Shearer 1999).
[2]Scalar seismic moment from Taymaz *et al.* (1991).
[3]Scalar seismic moment from Pinar *et al.* (1996).
[4]Rupture on the Princes' Islands fault (e.g., Parsons *et al.* 2000).
[5]Rupture on the Çinarcik fault (e.g., Nalbant *et al.* 1998).
[6]Faults have lengths shown in Fig. 1, dips in column 5, and extend to 15 km depth.
[7]Average slip, $u = M_o/(\mu A)$, where $\mu = 30$ GPa, and A is fault area (Shearer 1999).
[8]Rake assumed to be pure dip-slip or strike-slip.

plate boundary locations in the Marmara Sea (Hubert-Ferrari *et al.* 2000; Meade *et al.* 2002). In addition, due to uncertainties in the rupture location, it is unclear if the 1719, 1754, or 1766 event was the last major earthquake in this region (Hubert-Ferrari *et al.* 2000; Parsons *et al.* 2000). Regardless of this, based on these dates and the slip rate of the NNAF it is clear that the eastern Marmara Sea is in the late stages of an interseismic cycle and the secular stress accumulation should be large. Using the most conservative penultimate earthquake date of 1766 and estimates of secular stress loading in these faults for a conservative locking depth of 15 km, the Princes' Islands and Çinarcik faults will have accumulated at least 3–4 MPa of tectonic stress (Muller & Aydin 2005). Since this accumulated stress is an order of magnitude greater than the stress changes due to nearby earthquakes, the location, orientation, and locking depth of deep plate boundary slip are important factors in determining the absolute stress conditions in the eastern Marmara faults. However, we are focusing solely on the stress changes due to recent earthquakes because (1) the true location of the plate boundary structure beneath the eastern Marmara Sea, and hence the amount of stress transfers onto the shallow eastern Marmara faults, is not precisely resolvable, and (2) the additional stress perturbation from preceding earthquakes is commonly the trigger for subsequent earthquakes along the North Anatolian fault (e.g., Stein *et al.* 1997).

Stress changes

The results show that, in all cases, all of the Marmara faults receive increases in Coulomb stress due to the combined effects of all earthquakes since 1999. However, the fault segment receiving the greatest stress change is not the same for each scenario (Fig. 6, col. 2, Table 5). In four of the six scenarios, the Çinarcik fault receives the greatest stress change (Figs. 6a, b, d, f). In one scenario, the Çinarcik and Princes' Islands faults receive approximately equal stress increases (Fig. 6c), and in another the Princes' Islands fault receives the greatest increase (Fig. 6e). In all tests, the maximum stress increase on the Armutlu fault is less than those on the Çinarcik and Princes' Islands faults, but of the same order of magnitude. The direction of the maximum shear stress vector along the faults, in all cases, never deviates by more than ±20° from pure right-lateral shear.

Scenario 1a and 1b

If the Izmit earthquake ruptured 15 km SW of the Hersek Delta along the southern coast of the Bay of Izmit, and the 1963 earthquake ruptured along the Çinarcik fault (scenario 1a, Fig. 6a), the East Çinarcik segment receives the greatest stress change, with a maximum of over 1 MPa. This magnitude of Coulomb stress change is greater than the average increase of 0.4 MPa calculated at the epicentres of large earthquakes triggered during the 1939–1992 earthquake sequence along the NAF (Stein

Table 5. *Maximum and average Coulomb stress changes on eastern Marmara Sea faults*

Scenario	1a		1b		2a		2b		3a		3b	
Izmit rupture western termination	15 km west of Hersek Delta, along coast		15 km west of Hersek Delta, along coast		20 km west of Hersek Delta, to Princes' Islands fault		20 km west of Hersek Delta, to Princes' Islands fault		30 km west of Hersek Delta, along coast		30 m west of Hersek Delta, along coast	
Fault ruptured in 1963 event	Çinarcik		Princes' Islands		Çinarcik		Princes' Islands		Çinarcik		Princes' Islands	
	Coulomb stress change (MPa)											
Fault segment	Max.	Av.	Max.	Av.	Max.	Av.	Max.	Av.	Max.	Av.	Max.	Av.
West Izmit	0.32	0.29	0.81	0.38	–	–	–	–	–	–	–	–
Princes' Islands	0.32	0.16	0.17	0.11	0.30	0.16	0.18	0.11	0.35	0.16	0.18	0.11
East Çinarcik	1.16	0.54	1.17	0.50	0.29	0.17	0.26	0.16	–	–	–	–
West Çinarcik	0.14	0.07	0.19	0.10	0.13	0.07	0.09	0.07	0.16	0.08	0.28	0.13
Amutlu	0.21	0.06	0.09	0.06	0.22	0.06	0.12	0.06	0.25	0.07	0.16	0.07

et al. 1997). The next largest average increases occur on the West Izmit and Princes' Islands fault segments. Based on these results, it is difficult to assess which future rupture scenario is more likely due solely to stress changes from previous earthquakes. To produce an earthquake rupture of at least 40 km, westward propagating rupture on the East Çinarcik fault would have to either propagate through the portion of the fault assumed to have ruptured in 1963, or it would have to propagate onto the Armutlu fault. Therefore, throughgoing rupture along the West Izmit and Princes' Islands faults or initial slip on the East Çinarcik segment that branches onto the Armutlu fault may be more plausible rupture scenarios than rupture west along the Çinarcik fault.

If instead the 1963 earthquake occurred on the Princes' Islands fault (scenario 1b, Fig. 6b), the East Çinarcik fault receives the greatest stress change. In this case, there is no impediment to throughgoing rupture on the Çinarcik fault because the fault hasn't ruptured recently, so complete rupture of the Çinarcik fault is possible. In this case, although the West Izmit fault segment has a 30% greater average stress increase than that in scenario 1a, the location of the 1963 rupture would be likely to impede throughgoing rupture along this fault segment.

Scenario 2a and 2b

If the Izmit earthquake ruptured 20 km directly west of the Hersek Delta to the intersection with the Princes' Islands fault segment, and the 1963 earthquake was on the Çinarcik fault, the East Çinarcik and Princes' Islands faults receive comparable maximum and average Coulomb stress increases (scenario 2a, Fig. 6c). Following earlier arguments, while the stress changes equally advance the likelihood of rupture initiation on either segment, throughgoing rupture seems more likely on the Princes' Islands fault because, in this case, it is not assumed to have ruptured in 1963. Branching rupture onto the Armutlu fault is again plausible, but the average stress increase on the Princes' Islands fault is at least twice as large as that on the Armutlu fault.

If the 1963 earthquake ruptured the Princes' Islands fault (scenario 2b, Fig. 6d), the East Çinarcik fault again receives the greatest stress changes. The stress changes on the West Çinarcik and Armutlu faults are comparable, which suggests that rupture initiating on the East Çinarcik fault could be equally likely to propagate onto either segment.

Scenario 3a and 3b

If the Izmit earthquake ruptured 30 km west of the Hersek Delta along the southern coast of the Bay of Izmit, and the 1963 earthquake ruptured along the Çinarcik fault (scenario 3a, Fig. 6e), the Princes' Islands fault receives the greatest stress change, with a maximum of 0.3 MPa. Unlike what we found in this case, the alignment of the western part of the Izmit rupture with the Armutlu fault could intuitively suggest that the Armutlu fault should receive the greatest Coulomb stress changes due to the large shear stress concentrations ahead of the Izmit rupture tip. However, it must be noted that although the rupture extends 30 km west of the Hersek Delta, the amount of slip resolved in the slip inversion is less than 20 cm within several kilometres of the western edge of the Izmit rupture. This, combined with the reduced normal stresses acting on the Princes' Islands fault due to its location within the extensional deformation quadrant of the Izmit rupture, are the likely causes for the greatest Coulomb stress increases being located on the Princes' Islands fault. This results contradicts the singular prediction by Pinar *et al.* (2001) that the Yalova-Hersek fault segment received the greatest stress change after the Izmit earthquake. Following our stress change result, if the Princes' Islands fault ruptured in the 1963 earthquake (scenario 3b, Fig. 6f), thereby releasing stresses, stress changes must increase the likelihood of initiation and continued westward rupture along the Çinarcik fault.

Future westward rupture propagation

Having analysed which of the faults in the eastern Marmara Sea have an increased likelihood of rupture for each scenario, we now focus on the potential for the initial rupture to extend further west within the Marmara Sea. We consider geometry of the faults mapped further west and calculate Coulomb stress changes produced by the scenarios of rupture in Figure 6 that most favour failure of each of the three different major eastern Marmara Sea fault segments (Princes' Islands, Çinarcik and Armutlu faults). In the future failure scenario for each eastern Marmara fault, we apply uniform slip from 0 to 15 km depth with a magnitude derived from an empirical relationship between surface rupture length and moment magnitude (Table 6) (Wells & Coppersmith 1994).

The potential rupture length of an earthquake rupture initiating on the Princes' Islands fault segment would depend on the ability of the rupture to continue onto the vertical, nearly east–west trending Central Marmara fault (Muller & Aydin 2004). Coulomb stress changes on the Central Marmara fault, calculated for either a strike-slip or normal-slip earthquake on the Princes' Islands fault (see Table 6), are greatly increased near the intersection of this fault with the Princes' Islands fault and are positive for the entire fault segment (Fig. 7a). For this hypothetical scenario, the uniform slip on the Princes' Islands fault exaggerates the magnitudes of the stress change close to the fault intersection, therefore, we consider the results to be qualitative. Nevertheless, it is clear that the relative orientations of the Princes' Islands fault and the Central Marmara fault favour throughgoing rupture from the standpoint of static stress change.

The length of rupture associated with a Çinarcik fault earthquake would strongly depend on the accuracy of the fault configuration; specifically, the connection or lack thereof between this fault and the Central Marmara fault further west (Parke *et al.* 1999; Okay *et al.* 2000; Le Pichon *et al.* 2001; Armijo *et al.* 2002). Strike-slip earthquake ruptures have not been known to jump between fault segments

Table 6. *Rupture parameters for the eastern Marmara faults used in potential future rupture scenarios*

Fault	Area* (km^2)	$M_w^{\dagger}$	$M_o^{\ddagger}$ (1019 Nm)	Av. slip[§] (m)	Rake (°)
Princes' Islands (SS)	507	6.74[a]	1.29	0.85	180
Princes' Islands (NS)	507	6.69[b]	1.08	0.71	−90
Çinarcik	1230	7.13[a]	4.95	1.34	180
Armutlu	588	6.80[a]	1.58	0.90	180

*Faults extend from surface to 15 km depth along surface traces in Fig. 7. Fault dips are given in Fig. 2b.
†a Calculated from empirical relationship for strike-slip earthquakes $M_w = 3.98 + 1.02{*}\log(A)$, where A is rupture area (Wells & Coppersmith 1994).
†b Calculated from empirical relationship for normal-slip earthquakes $M_w = 3.93 + 1.02{*}\log(A)$ (Wells & Coppersmith 1994).
‡Scalar seismic moment, $M_o = 10^{(1.5{*}M_w+9)}$ (Shearer 1999).
§Average fault slip, $u = M_o/(\mu A)$ (Shearer 1999).

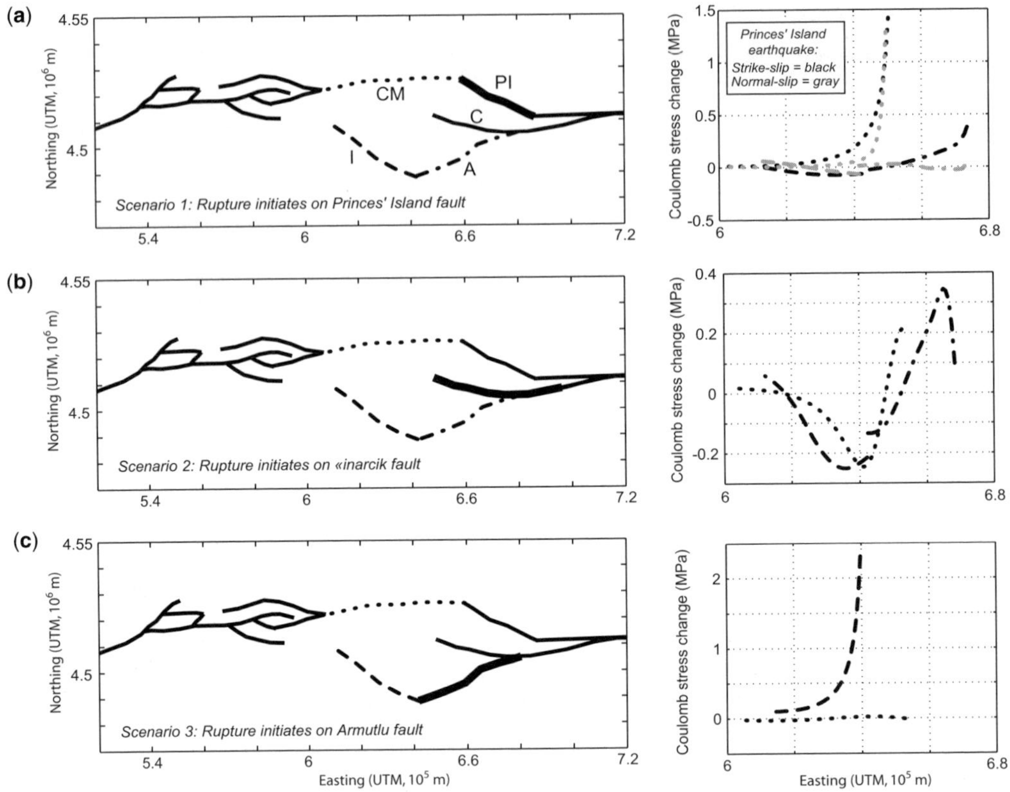

Fig. 7. Coulomb stress changes (right column) on faults in the central Marmara Sea due to potential initial eastern Marmara ruptures (left column). PI: Princes' Islands fault (solid); C: Çinarcik fault (solid); A: Armutlu fault (dot-dashed); I: Imrali fault (dashed); CM: Central Marmara fault (dotted). (**a**) A case in which the initial future rupture (thick black line) occurs on the Princes' Islands fault. Stress plot on right shows stress changes for either a strike-slip or dip-slip Princes' Islands earthquake. Black patterned lines show results for the strike-slip mechanism and gray patterned lines for a normal dip-slip mechanism (Table 6). (**b**) A case in which the initial future rupture occurs on the Çinarcik fault. (**c**) A case in which the initial future rupture occurs on the Armutlu fault.

separated by more than 5–10 km (Harris & Day 1993), so it would be unlikely to propagate further west. Rupture of this Çinarcik fault geometry would also decrease the Coulomb stresses acting on the Central Marmara and Imrali fault further west (Fig. 7b; Table 7), perhaps impeding or delaying the continued migration of slip through the central and western portions of the NNAF within the Marmara Sea.

In many of our stress change scenarios, the Armutlu fault receives stress changes greater than 0.2 MPa, even though in no case is it the segment with the greatest stress increase. The scenario in which Izmit rupture has propagated along the coast past Yalova (Fig. 6e) shows the greatest possibility for future failure along the Armutlu fault. Despite the large change in fault strike, the Coulomb stress is significantly increased along the Imrali fault (Fig. 7c). This is primarily because the fault is located in what would be the extensional quadrant of the Armutlu rupture segment. Transition of slip between these fault segments would produce a rupture with a combined length exceeding 80 km.

Discussion

By considering the uncertainties in the rupture geometries of the 1963 Yalova earthquake and the western termination of the 1999 Izmit earthquake, our analysis represents a systematic investigation of the Coulomb stress changes for the most plausible scenarios involving the eastern Marmara Sea faults. Assuming that static stress changes do provide a measure of the future failure tendency for large earthquakes (e.g., King *et al.* 1994; Stein *et al.* 1997), our

Table 7. *Maximum and average Coulomb stress changes on Marmara Sea faults west of potential eastern Marmara Sea ruptures*

Scenario Future rupture segment	1 Princes' Islands fault (strike-slip offset)		1 Princes' Islands fault (normal-slip offset)		2 Çinarcik fault		3 Armutlu fault	
	Coulomb stress change (MPa)							
Fault segment	Max.	Av.	Max.	Av.	Max.	Av.	Max.	Av.
Central Marmara	1.42	0.15	1.28	0.06	0.21	−0.05	0.02	−0.01
Armutlu	0.38	0.08	0.02	0.00	0.34	0.08	–	–
Imrali	0.01	−0.05	0.06	0.00	0.06	−0.13	2.31	0.40

results show that predicting future rupture scenarios depends on accurately constraining the recent earthquake configurations and slip distributions (Muller *et al.* 2003). For the majority of our stress change scenarios, the Çinarcik fault receives a greater stress increase than the Princes' Islands fault segment. In an analysis of the stress changes from the Izmit earthquake alone, Pinar *et al.* (2001) reach the same conclusion. This result does not support the notion that combined failure along the Princes' Islands and Central Marmara faults is the only, or most likely future rupture scenario (e.g., Le Pichon *et al.* 1999) or the suggestion that the stress changes on the Princes' Islands and Çinarcik fault are essentially identical (e.g., Hubert-Ferrari *et al.* 2000). Our results instead show that several future scenarios are possible and that, in the case of the Çinarcik fault rupture, a future earthquake may not be as hazardous because of the limited length of such a rupture if the fault is not connected to other Marmara Sea fault strands further west.

One way to evaluate the effects of different stress change magnitudes on different faults is to compare the magnitude of the stress change with the long-term stress loading rate on each fault due to plate motion. GPS measurements (Reilinger *et al.* 1997) and elastic modelling (Flerit *et al.* 2003; Muller & Aydin 2005) of slip along the plate boundary structures in NW Turkey suggests that 30 mm/yr of right-lateral slip on the NAF east of the Marmara Sea and 24 mm/yr on the NNAF and 6 mm/yr on the central strand of the NAF (CNAF) within the Marmara Sea region are appropriate long-term slip rates. These slip rates applied to plate boundary with a locking depth of 15 km (Muller & Aydin 2005) provide approximate stressing rates of *c.* 1.5×10^{-2} MPa/yr at 7.5 km depth on the faults of interest. The average stress changes from scenarios 1a and 1b have therefore advanced the rupture time by approximately 30 years for the East Çinarcik fault. In the other scenarios, the averages stress changes shorten the time until rupture on the Princes' Islands and West Çinarcik fault by approximately a decade relative to their long-term stressing rate. These time advances toward rupture are comparable to the 25 year average advance calculated by Stein *et al.* (1997) for the sequence of large earthquakes along the eastern and central portions of the NAF from 1939–1992; therefore, the progression of rupture into the Marmara Sea seems likely within the next several decades.

Several factors that we have not included in our model could affect the future rupture scenario in the eastern Marmara Sea. First, the location of the 1894 earthquake that caused widespread damage in the Bay of Izmit between the cities of Yalova and Sapanca is not very well constrained. From extensive review of historical damage reports, Ambraseys & Finkel (1991) suggest that the rupture occurred in the Bay of Izmit and further east. Hubert-Ferrari *et al.* (2000) support this location and suggest the event ruptured the mountainfront fault between the 1967 Mudurnu Valley earthquake and the 1999 Izmit earthquake. Parsons *et al.* (2000) suggest, however, that the rupture occurred within the Bay of Izmit and extended further west along the Armutlu fault. The latter interpretation would have relieved some of the long-term stress accumulation in the eastern Marmara Sea, whereas the earlier interpretation would have added an increased stress perturbation. The locations of damage reported in this earthquake, including Istanbul, are similar to those reported in the Izmit earthquake; therefore, it does not appear that rupture extending far along the Armutlu fault is necessary to explain the damage pattern. We therefore hesitate to rule out the Armutlu fault as a potential candidate for future rupture based on the 1894 rupture location interpreted by Parsons *et al.* (2000).

Second, the partitioning of loading from deep slip along the plate boundary onto the shallower eastern Marmara Sea faults is not well-constrained. GPS data is able to show that slip below the locked shallow faults is several times greater along the NNAF than along the CNAF (Straub *et al.* 1997), but within the shallow NNAF strands themselves, it is unclear if certain faults are being loaded with a greater stressing rate. A shallow fault preferentially aligned above the deep plate boundary shear zone would be likely to fail more frequently than faults further from the plate boundary.

In this paper we do not investigate the stress changes resolved on the fault geometry in the western Marmara Sea. In focusing on the eastern Marmara Sea region, we are by no means implying that the next large earthquake in the Marmara Sea could not occur further west, in the region of increased Coulomb stress east of the 1912 Ganos earthquake. We recognize that the stress change scenario in this region will be affected by uncertainty in the location of the eastern termination of the Ganos earthquake. This $M_s = 7.4$ earthquake ruptured only 40 km on land before entering the Gulf of Saros to the west and the Marmara Sea to the east, suggesting that a significant portion of its rupture occurred offshore to the east or west. Armijo *et al.* (2003) suggest that a set of surface breaks extending 60 km east in to the Marmara Sea from the onshore rupture, as observed from a submarine remote-operated vehicle, may be associated with the 1912 rupture. Evaluation of earthquake-induced stress changes in the western Marmara Sea region would therefore need to consider the ambiguities in the 1912 rupture geometry as well as the complex fault geometries bounding the Tekirdag and Central Marmara basins (Fig. 2a). To this end, the methodology in this study can be used to assess each potential scenario involving the past rupture locations and fault configurations.

Conclusions

Using a geomechanical modelling tool, we find that earthquake-induced stress changes of several tenths of an MPa load faults in the eastern Marmara Sea for each possible different location of the 1963 Yalova earthquake and western termination of the 1999 Izmit earthquake. In four of the six permutations of these past rupture locations that we test, a portion of the Çinarcik fault receives the greatest average stress change. In one case, when Izmit earthquake slip extending up to 30 km west of the Hersek Delta and the 1963 rupture occurred on the Çinarcik fault, the Princes' Islands fault receives the greatest stress increase. For the case where the Izmit rupture propagated straight west and the 1963 event was on the Çinarcik fault, the stress increases on the Princes' Islands and portions of the Çinarcik fault are comparable. The magnitudes of the stress increases due to the recent perturbations relative to the long-term tectonic stressing rate, suggest that the time before future rupture is advanced by approximately 10–30 years.

Based on the mapped fault configuration extending west from the initial failure of these faults, an earthquake along the Princes' Islands fault is likely to be the most hazardous due to its proximity to Istanbul and because its intersection geometry with the Central Marmara fault favours continuous propagation of westward rupture. This scenario would result in a cumulative rupture length of at least 80 km before the rupture intersected the Central Marmara basin. In contrast, rupture of the Çinarcik fault would not be likely to exceed 50–60 km, based on its isolated western termination and therefore, although it may represent the scenario most likely in terms of stress change, it may present slightly less hazard to Istanbul. Earthquake-induced average stress changes on the Armutlu fault, while positive, do not exceed those on the other eastern Marmara faults for any tests. If, however, rupture were to initiate on this segment, then static stress changes would favour the transition of westward rupture onto the NW-trending Imrali fault segment, thereby producing a combined rupture length exceeding 80 km.

Our results show that identification of the fault segment receiving the greatest earthquake-induced stress increase can vary depending on minor changes in local rupture geometry and slip distribution, thus highlighting the need for accurate surface mapping or active fault and earthquake rupture geometries as well as quality geodetic or seismological data for modelling earthquake slip distributions. Given that uncertainties in these parameters are unavoidable, geomechanical modelling provides a unique tool to test the significance of each case and its resulting realization.

Izmit earthquake aftershock data was generously provided by T. Taymaz and S. Özalaybey. Several figures were plotted with GMT software (Wessel & Smith 1998). The Stanford Rock Fracture Project provided partial financial support of this work. T. Wright is funded by a Royal Society University Research Fellowship. The authors thank reviewers K. Johnson and M. Belardinelli and editor S. Buiter for constructive criticism that improved this manuscript.

References

AMBRASEYS, N. & FINKEL, C. 1988. The Anatolian earthquake of 17 August 1668. *In*: LEE, W. H. K., MEYERS, H. & SHIMIZAKI, K. (eds) *Historical Seismograms and Earthquakes of the World*. Academic Press, 173–180.

AMBRASEYS, N. N. & FINKEL, C. F. 1991. Long-term seismicity of Istanbul and of the Marmara Sea region. *Terra Nova*, **3**, 527–539.

ARMIJO, R., MEYER, B., NAVARRO, S., KING, G. & BARKA, A. 2002. Asymmetric slip partitioning in the Sea of Marmara pull-apart: a clue to propagation processes of the North Anatolian Fault? *Terra Nova*, **14(2)**, 80–86.

ARMIJO, R. & THE MARMARASCARPS CRUISE PARTY 2003. Young earthquake breaks in the Sea of Marmara floor: a possible underwater extension of the 1912 earthquake rupture? *In*: *International Workshop on the North Anatolian, East Anatolian, and Dead Sea Fault Systems*, Middle East Technical University, 27.

AYDIN, A. & KALAFAT, D. 2002. Surface ruptures of the August 17 and November 12, 1999, Izmit and Düzce earthquakes in NW Anatolia, Turkey: their tectonic and kinematic significance and the associated damage. *Bulletin of the Seismological Society of America*, **92**, 95–106.

BOUCHON, M., TOKSÖZ, M. N., KARABULUT, H., BOUIN, M., DIETRICH, M., AKTAR, M. & EDIE, M. 2002. Space and time evolution of rupture and faulting during the 1999 Izmit earthquake. *Bulletin of the Seismological Society of America*, **92**, 256–266.

ÇAKIR, Z., BARKA, A. & EVREN, E. 2003. Coulomb stress interactions and the 1999 Marmara earthquakes. *Turkish Journal of Earth Sciences*, **12**, 91–103.

COMNINOU, M. A. & DUNDERS, J. 1975. The angular dislocation in a half-space. *Journal of Elasticity*, **5**, 203–216.

CROUCH, S. L. & STARFIELD, A. M. 1983. *Boundary Element Methods in Solid Mechanics*. Allen and Unwin, London.

DELOUIS, B., GIARDINI, D., LUNDGREN, P. & SALICHON, J. 2002. Joint inversion of InSAR, GPS, teleseismic, and strong-motion data for the spatial and temporal distribution of earthquake slip: application to the 1999 Izmit mainshock. *Bulletin of the Seismological Society of America*, **92**, 278–299.

FEIGL, K. L., SARTI, F. *ET AL.* 2002. Estimating slip distribution for the Izmit mainshock from coseismic GPS, ERS-1, RADARSAT, and SPOT measurements. *Bulletin of the Seismological Society of America*, **92**, 138–160.

FLERIT, F., ARMIJO, R., KING, G. C. P., MEYER, B. & BARKA, A. 2003. Slip partitioning in the Sea of Marmara pull-apart determined from GPS velocity vectors. *Geophysical Journal International*, **154**, 1–7.

GÜLEN, L., PINAR, A., KALAFAT, D., ÖZEL, N., HORASAN, G., YILMAZER, M. & ISIKARA, A. M. 2002. Surface fault breaks, aftershock distribution, and rupture process of the 17 August 1999 Izmit, Turkey, earthquake. *Bulletin of the Seismological Society of America*, **92**, 230–244.

HARRIS, R. A. 1998. Introduction to special section: stress triggers, stress shadows, and implications for seismic hazard. *Journal of Geophysical Research*, **103**, 24347–24358.

HARRIS, R. A. & DAY, S. M. 1993. Dynamics of fault interaction: parallel strike-slip faults. *Journal of Geophysical Research*, **98**, 4461–4472.

HARRIS, R. & SIMPSON, R. W. 1996. In the shadow of 1857 – the effect of the great Ft. Tejon earthquake on subsequent earthquakes in southern California. *Geophysical Research Letters*, **23**, 229–232.

HARRIS, R. A., SIMPSON, R. W. & REASENBERG, P. A. 1995. Influence of static stress changes on earthquake locations in southern California. *Nature*, **375**, 221–224.

HUBERT-FERRARI, A., BARKA, A. *ET AL.* 2000. Seismic hazard in the Marmara Sea region following the 17 August 1999 Izmit earthquake. *Nature*, **404 (6775)**, 269–273.

JACKSON, J. & MCKENZIE, D. P. 1988. The relationship between plate motions and seismic moment tensors, and the rates of active deformation in the Mediterranean and Middle East. *Geophysical Journal of the Royal Astronomical Society*, **93**, 45–73.

JEYAKUMARAN, M., RUDNICKI, J. W. & KEER, L. M. 1992. Modeling slip zones with triangular dislocation elements. *Bulletin of the Seismological Society of America*, **82**, 2153–2169.

JÓNSSON, S., ZEBKER, H., SEGALL, P. & AMELUNG, F. 2002. Fault slip distribution of the 1999 Mw 7.1 Hector Mine earthquake, California, estimated from satellite radar and GPS measurements, *Bulletin of the Seismological Society of America*, **92**, 1377–1389.

KARABULUT, H., BOUIN, M., BOUCHON, M., DIETRICH, M., CORNOU, C. & AKTAR, M. 2002. The seismicity in the eastern Marmara Sea after the 17 August 1999 Izmit earthquake. *Bulletin of the Seismological Society of America*, **92**, 387–393.

KING, G. C. P., STEIN, R. S. & LIN, J. 1994. Static stress changes and the triggering of earthquakes. *Bulletin of the Seismological Society of America*, **84(3)**, 935–953.

LE PICHON, X., SENGÖR, A. M. C. *ET AL.* 2001. The active Main Marmara Fault. *Earth and Planetary Science Letters*, **192**, 595–616.

LE PICHON, X., TAYMAZ, T. & SENGOR, C. 1999. The Marmara fault and the future Istanbul earthquake. *In*: KARACA, M. & URAL, D. N. (eds) *ITU-IAHS International conference on the Kocaele Earthquake 17 August 1999*, Istanbul Technical University, 41–54.

MEADE, B. J., HAGER, B. H. *ET AL.* 2002. Estimates of seismic potential in the Marmara Sea region from block models of secular deformation constrained by global positioning system measurements. *Bulletin of the Seismological Society of America*, **92**, 208–215.

MULLER, J. & AYDIN, A. 2004. Rupture progression along discontinuous oblique fault sets: implications for the Karadere rupture segment of the 1999 Izmit earthquake, and future rupture in the Marmara Sea. *Tectonophysics*, **391**, 283–302.

MULLER, J. R. & AYDIN, A. 2005. Using mechanical modelling to constrain fault geometries proposed for the northern Marmara Sea. *Journal of Geophysical Research*, **110**, B03407, DOI: 10.1029/2004JB003226.

MULLER, J., AYDIN, A. & MAERTEN, F. 2003. Investigating the transition between the 1967 Mudurnu Valley and 1999 Izmit earthquakes along the North Anatolian fault with static stress changes. *Geophysical Journal International*, **154**, 471–482.

NALBANT, S. S., HUBERT, A. & KING, G. C. P. 1998. Stress coupling between earthquakes in northwest Turkey and the north Aegean Sea. *Journal of Geophysical Research*, **103**, 24469–24486.

OKADA, Y. 1985. Surface deformation due to shear and tensile faults in a half-space. *Bulletin of the Seismological Society of America*, **75**, 1135–1154.

OKAY, A. I., KASLILAR-OZCAN, A., IMREN, C., BOZTEPE-GUNEY, A., DEMIRBAG, E. & KUSCU, I. 2000. Active faults and evolving strike-slip basins in the Marmara Sea, Northwest Turkey; a multichannel seismic reflection study. *Tectonophysics*, **321(2)**, 189–218.

ÖZALAYBEY, S., ERGIN, M., AKTAR, M., TAPIRDAMAZ, C., BICMEN, F. & YÖRÜK, A. 2002. The 1999 Izmit earthquake sequence in Turkey: seismological and tectonic aspects. *Bulletin of the Seismological Society of America*, **92**, 376–386.

PARKE, J. R., MINSHULL, T. A. *ET AL.* 1999. Active faults in the Sea of Marmara, Western Turkey, imaged by seismic reflection profiles. *Terra Nova*, **11**, 223–227.

PARSONS, T., TODA, S., STEIN, R. S., BARKA, A. & DIETERICH, J. H. 2000. Heightened odds of large earthquakes near Istanbul: an interaction-based probability calculation. *Science*, **288(5466)**, 661–665.

PINAR, A., HONKURA, Y. & KIKUCHI, M. 1996. A rupture model for the 1967 Mudurnu Valley, Turkey earthquake and its implications for seismotectonics in the western part of the North Anatolian fault zone. *Geophysical Research Letters*, **23(1)**, 29–32.

PINAR, A., HONKURA, Y. & KUGE, K. 2001. Seismic activity triggered by the 1999 Izmit earthquake and its implications for the assessment of future seismic risk. *Geophysical Journal International*, **146**, F1–F7.

REILINGER, R. E., ERGINTAV, S. *ET AL.* 2000. Coseismic and postseismic fault slip for the 17 August 1999, M = 7.5, Izmit, Turkey earthquake. *Science*, **289**, 1519–1524.

REILINGER, R. E. & MCCLUSKY, S. C. 1997. Global positioning system measurement of the present-day crustal movements in the Arabia-Africa-Eurasia plate-collision zone. *Journal of Geophysical Research*, **102**, 9983–9999.

RESOR, P. 2004. *Deformation associated with continental normal faults.* PhD thesis, Stanford University.

SCHOLZ, C. H. 2002. *The Mechanics of Earthquakes and Faulting*. Cambridge University Press, New York.

SEKIGUCHI, H. & IWATA, T. 2002. Rupture process of the 1999 Kocaeli, Turkey, earthquake estimated from strong-motion waveforms. *Bulletin of the Seismological Society of America*, **92**, 300–311.

SHEARER, P. M. 1999. *Introduction to Seismology*. Cambridge University Press, New York.

STEIN, R. S., BARKA, A. A. & DIETERICH, J. H. 1997. Progressive failure on the North Anatolian fault since 1939 by earthquake stress triggering. *Geophysical Journal International*, **128**, 594–604.

STRAUB, C., KAHLE, H.-G. & SCHINDLER, C. 1997. GPS and geological estimates of the tectonic activity in the Marmara Sea region, NW Anatolia. *Journal of Geophysical Research*, **102(12)**, 27587–27601.

TAYMAZ, T., JACKSON, J. & MCKENZIE, D. 1991. Active tectonics of the north and central Aegean Sea. *Geophysical Journal International*, **106**, 433–490.

THOMAS, A. L. 1993. *Poly3D: a three-dimensional, polygonal element, displacement discontinuity boundary element computer program with applications to fractures, faults, and cavities in the earth's crust.* MS Thesis, Stanford University.

WELLS, D. L. & COPPERSMITH, K. J. 1994. New empirical relationships among magnitude, rupture length, rupture width, rupture area, and surface displacement. *Bulletin of the Seismological Society of America*, **84(4)**, 974–1002.

WESSEL, P. & SMITH, W. H. F. 1998. New, improved version of generic mapping tools released. *EOS*, **79(47)**, 579.

WRIGHT, T., FIELDING, E. & PARSONS, B. 2001*a*. Triggered slip: observations of the 17 August 1999 Izmit (Turkey) earthquake using radar interferometry. *Geophysical Research Letters*, **28(6)**, 1079–1082.

WRIGHT, T., PARSONS, B. & FIELDING, E. 2001*b*. The 1999 Turkish earthquakes: source parameters from InSAR and observations of triggered slip. *In*: TAYMAZ, T. (ed.) *Symposia on Seismotectonics of the North-Western Anatolia–Aegean and Recent Turkish* Earthquakes, Istanbul Technical University, 54–71.

WRIGHT, T., LU, Z. & WICKS, C. 2004. Constraining the slip distribution and fault geometry of the Mw 7.9, 3 November 2002, Denali fault earthquake with interferometric synthetic aperture radar and global positioning system data. *Bulletin of the Seismological Society of America*, **94(6B)**, S175–189.

YAGI, Y. & KIKUCHI, M. 2000. Source rupture process of the Kocaeli, Turkey, earthquake of August 17, 1999, obtained by joint inversion of near-field data and teleseismic data. *Geophysical Research Letters*, **27(13)**, 1969–1972.

Oil reservoirs in foreland basins charged by thrustbelt source rocks: insights from numerical stress modelling and geometric balancing in the West Carpathians

MICHAL NEMČOK[1] & ANDREAS HENK[2]

[1]*Energy & Geoscience Institute, University of Utah, 423 Wakara Way, Suite 300, Salt Lake City, UT 84108, USA*

[2]*Geologisches Institut, Universität Freiburg, Albertstrasse 23b, D-79104 Freiburg, Germany (e-mail: henk@uni-freiburg.de)*

Abstract: The West Carpathian thrustbelt advanced northeastwards over the European Platform. Its thrust sheets comprise sediments of the Early Cretaceous rifts that evolved on a passive margin of the European Platform, the Late Cretaceous–Paleocene basins formed by rift inversion, and the Eocene-Oligocene flexural basin. Geochemical analyses established a clear link between pooled oils in the foreland and the Oligocene Menilite Formation inside the thrustbelt. In order to understand the driving forces for this oil migration scenario, finite-element models of fault-propagation and fold-bend folds are used to study the mean stress distribution in the thrust sheets and the foreland. Mean stress has a profound control on the pore fluid pressure through the relationship affected by sediment porosity, and sediment skeleton and fluid compressibilities. Modelling results suggest that only fault-propagation folds are capable of generating foreland-directed mean stress gradients as they are characterized by a large foreland area of decreased mean stress, by coupled increased/decreased mean stress areas on advancing/receding sides of the ramp tip, and an overall mean stress decrease inside the thrust sheet in the direction towards the foreland. This interpretation is in accordance with the dominant fold-and-thrust style in the Western Carpathians inferred from balanced cross-section restoration. It shows that frontal fault-propagation folding was active during the late Oligocene–Early Miocene, providing an effective tectonic driving force for hydrocarbon migration from source rocks inside the thrustbelt towards reservoirs in the foreland.

The West Carpathian thrustbelt (Fig. 1a) formed during the Tertiary by a NE- and eastward accretion and advance over the European passive margin and adjacent oceanic crust (e.g., Royden & Báldi 1988). It developed in front of the advancing Carpathian microplates (e.g., Balla 1984; Kovács *et al.* 1989; Csontos *et al.* 1992). The development was driven by subduction roll-back (e.g., Royden *et al.* 1982) with some influence from the eastward lateral mass extrusion from the Eastern Alps (Neubauer & Genser 1990; Ratschbacher *et al.* 1991). Subduction placed the remnant Carpathian Flysch Basin between the Carpathian microplates and the European Platform and floored by oceanic and thinned continental crust (e.g., Nemčok *et al.* 1998). The thrustbelt itself comprises sediments deposited in Early Cretaceous rifts evolved on the European margin (e.g., Książkiewicz 1977), in Late Cretaceous–Paleocene basins formed by an inversion of Early Cretaceous rifts (e.g., Suk *et al.* 1984 and references therein; Roca *et al.* 1995), and in Eocene–Oligocene foreland basins (Winkler & Ślączka, 1992; Roure *et al.* 1993). The European Platform underneath and in front of the thrustbelt shows Palaeozoic–Triassic sediments related to combined effects of Palaeozoic rifting, late Palaeozoic contraction and Mesozoic rifting. Jurassic sediments evolved in Mesozoic rifts, Upper Cretaceous sediments related to contraction and, finally, Lower-Middle Miocene molasse sediments of the flexural foreland basin.

Apart from fields in the thrustbelt, there are about 67 producing hydrocarbon fields in various stratigraphies of the foreland close to the thrustbelt (Karnkowski 1999). Of these, 13 contain oil, allowing oil/source rock correlations. Oils in these fields are very similar to thrustbelt oils, having a light to medium character, with densities of 810–860 kg m^{-3} and paraffin content of 2.32–9.37% (Weil *et al.* 1998). Based on total

From: BUITER, S. J. H. & SCHREURS, G. (eds) 2006. *Analogue and Numerical Modelling of Crustal-Scale Processes.* Geological Society, London, Special Publications, **253**, 415–428.

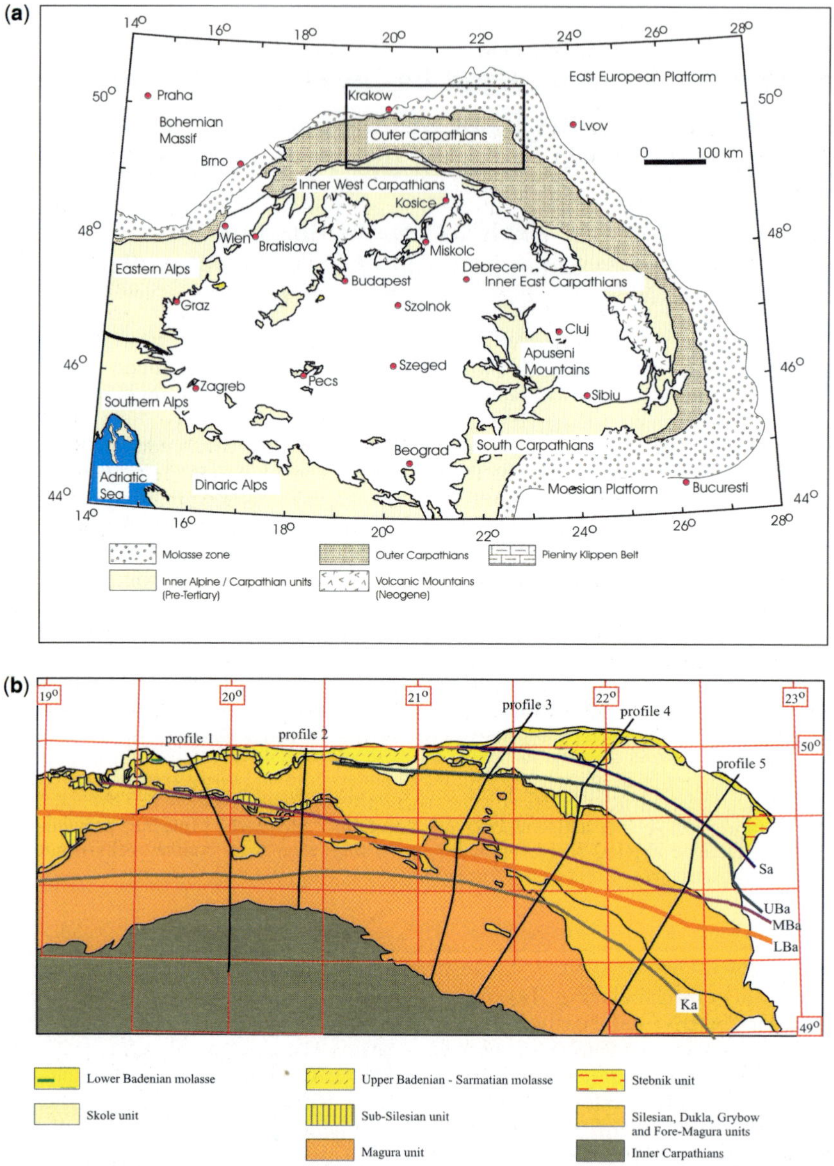

Fig. 1. (**a**) Map of the Carpathians with major tectonic units (modified after Krs & Roth 1977; Krs & Potfaj 1991; Krs *et al.* 1993; Koráb *et al.* 1981; Royden & Báldi 1988; Sandulescu 1988; Márton & Márton 1989; Túnyi & Kováč 1991; Patrascu *et al.* 1994). Rectangle indicates location of figure (b). (**b**) Portion of the West Carpathians with location of balanced cross-sections (after Nemčok *et al.* 1999). Coloured lines indicate southernmost extent of the various autochthonous molasse sediments below the accretionary wedge (Ka: Karpatian, Lba: Lower Badenian, Mba: Middle Badenian, UBa: Upper Badenian, Sa: Sarmatian).

organic carbon content and other source rock characteristics, initial source rock candidates for described fields in the orogenic foreland are: Lower Carboniferous and Miocene formations of the foreland, and Lower Cretaceous and Oligocene thrustbelt formations (Bessereau *et al.* 1996; Ślączka 1996; PUGzG Geopol 1998). Geochemical data and angiosperm-derived biological markers from oils of the autochthonous foreland (Grobla, Partynia-Podborze and Tarnów fields; Ten Haven *et al.* 1993; Bessereau *et al.* 1996) indicate that the source rock has Late

Cretaceous-Tertiary age (Moldowan *et al.* 1991). This excludes Lower Carboniferous sediments of the foreland and Lower Cretaceous thrustbelt sediments from being a source rock for fields in the autochthonous foreland. Miocene sediments of the foreland are not mature enough to be considered (Kotarba *et al.* 1987). A combination of the various arguments outlined above allows determination of the source rock for the foreland reservoirs. The only candidate matching all constraints is the Oligocene thrustbelt formation, the Menilite shale (Bessereau *et al.* 1996), indicating that source rock for oils in the autochthonous foreland was located inside the thrustbelt. The age of oil migration is Late Oligocene and younger (Ten Haven *et al.* 1993; Bessereau *et al.* 1996).

How was it possible for oils to migrate from the thrustbelt into its foreland? Which migration driving forces could have an effect on this special scenario? This paper tries to address these questions by studying thrustbelt structures by balancing and simulating migration controls by finite-element modelling. While several authors have focused on different numerical techniques to model fold-and-thrust structures (e.g., Sassi & Faure 1996; Salvini *et al.* 2001; Burbidge & Braun 2002; Kwon & Mitra 2004), we attempt a coupled geometric-mechanical approach to document the incidence of thrusting and related palaeostress evolution as a driving mechanism for fluid transfer out of a thrustbelt.

Methods

Balancing

In order to determine the geometries and kinematics of the thrust sheets in the Western Carpathians at the time of oil migration, we have constructed five balanced cross-sections from the West European Platform to the Inner Carpathians (Fig. 1b). Details for cross-section construction are partly described in Nemčok *et al.* (1999, 2000, 2001) and further information is given below. Data constraints for balancing were provided by:

(1) magnetotelluric data (e.g., Ryłko & Tomas, 1995) that helped to determine the basement top;
(2) reflection seismic profiles (e.g., profiles 5-3-73K, 5-1-78K, 5A-1-78K; Nemčok *et al.* 1999, 2000) that imaged thrust sheet geometries;
(3) bore holes that provided thicknesses of thrust sheets and thicknesses of involved sediments; and
(4) our own outcrop data and data from available geological maps that provided thrust sheet geometries and thicknesses of sediments.

Dip domain and kink band analyses of planar data were made manually, in order to exclude local complexities such as small-scale folding. Small-scale complexities were filtered out from large-scale geometries of thrust sheets, constructed without either small-scale faulting or complex deformation zones, which were smaller than the resolution of regional balanced cross-sections. Obtained fold geometries cleaned from small-scale complexities comprise fault-bend folds (Suppe 1983), fault-propagation folds (Suppe & Medwedeff 1984) and their evolutional combinations (e.g., Mitra 1990), providing an exact understanding of their mechanisms and interactions. Cross-sections were constructed using the program GeoSec 2D® (Paradigm Geophysical Ltd., Houston, USA). Surface level and well profiles provided the best constraints. The construction downward was less precisely controlled by the wedge base line interpreted from seismics and magnetotellurics. Pin lines for the Magura and Silesian Units were located in their fronts. Volume preservation was simulated by the area preservation within the cross-section. Deformed cross-sections were validated by restoration to undeformed state, made by the same software, using the flexural slip algorithm.

Numerical modelling

In order to understand oil migration driven by mean stress gradients, we study a set of 2D finite-element models of fault-propagation and fault-bend folds. Model geometries describing different stages of fold development were made by forward balancing from undeformed to most mature folding stages using programs GeoSec 2D® and Move 2D® (Midland Valley Exploration Ltd., Glasgow, UK) (Fig. 2). Rock mechanics were added by transferring the various balancing stages into the finite-element program ANSYS® (Ansys Inc., Houston, USA), analyzing for stress field perturbations and strain localization during fold growth.

Modelled profiles comprise 2 km thick pre-tectonic sediments composed of three lithologic layers. The upper one is 620 m thick, the middle one is 710 m thick and the lower one is 670 m thick (Fig. 2). Ramp angles for fault-propagation and fault-bend folds are 28° and 26°, respectively. The pre-tectonic section of the siliciclastic scenario is composed of a sandstone–shale–sandstonesuccessionmimicking

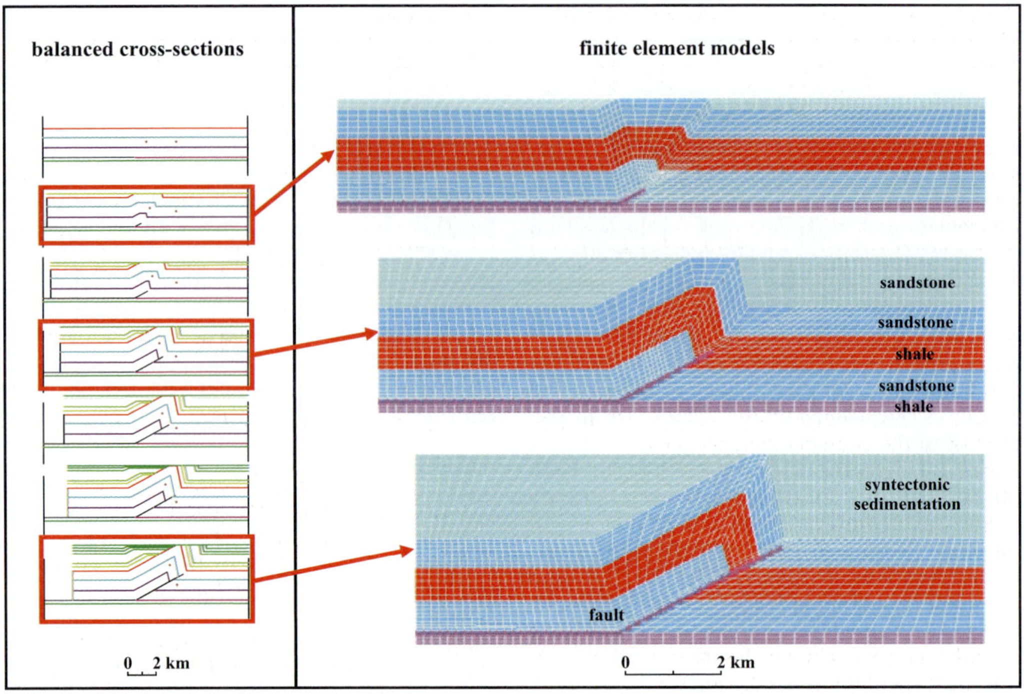

Fig. 2. Finite element models based on balanced sections representing different growth stages of fault-propagation folds. Thickness of the pre-tectonic thrust sheet is 2 km. Vertical scale equals horizontal scale.

the conditions in the West Carpathian thrust belt. For comparison, a more brittle dolomite–limestone–dolomite succession is also studied. Both scenarios experienced either 50% erosion of the evolving folds or deposition keeping up with the growth structure during simulated shortening, in order to test both continental and marine settings. Syn-tectonic sediments are either shale or sandstone.

Each of the finite-element models describes a cross-section through the folded sedimentary sequence, which is broken by planes of weakness, i.e., faults. ANSYS® uses a Lagrangian formulation to simulate plane strain deformation using four-node isotropic elements to represent the individual pre- and syn-tectonic sediment layers. Their mechanical behaviour in the elastic domain is described by the generalized Hooke's law, relating strains to stresses via Young's modulus and Poisson's ratio (e.g., Mandl 1988). Brittle failure is governed by the Coulomb criterion according to Jaeger and Cook (1979):

$$\tau_{\text{crit}} = c + \mu\sigma_{\text{n}}, \quad (1)$$

where τ_{crit} is the critical shear stress, c is the cohesion, μ is the coefficient of internal friction and σ_{n} is the stress normal to the shear fracture. The various physical properties of the lithologies used in the numerical models are summarized in Table 1.

Faults are described by so-called contact elements. This approach handles large differential movements between parts of the model, but does not describe cracking, i.e., the fault propagation itself. Contact elements are defined at opposing sides of the pre-assigned fault and stiffness values similar to the Young's moduli of the rocks in contact are used to enforce compatibility between adjacent fault surfaces. Contact elements are also capable of describing frictional sliding, so that the influence of different friction coefficients can be studied. Depending on the restoration stage, up to 2100 planar elements and 900 contact elements were used to represent the model geometry.

Boundary conditions for the fault-propagation fold models are a fixed base and right side, while the top is a free surface. Displacement constraints are applied to the nodes on the left side of the hanging wall to simulate compression. Boundary conditions for the fault-bend fold models differ in one only aspect: a lithostatic pressure boundary condition is applied to the

Table 1. *Physical properties of rocks used in the finite-element models*

Rock	Angle of internal friction (degree)	Coefficient of internal friction	Cohesion (MPa)	Young's modulus (GPa)	Poisson's ratio	Density (kgm^{-3})
Sandstone	28.8	0.55	27.2	18.30	0.38	2180
Shale	21.0	0.38	0.69	5.52	0.25	2470
Limestone	26.4	0.50	26.8	77.40	0.26	2720
Dolomite	39.0	0.81	35.9	106.00	0.29	2840

Values are taken from Clark (1966); Lama & Vutukuri (1978); Čermák & Rybach (1982); Ibrmajer *et al.* (1989); Cloetingh *et al.* (1995).

part of the left model side representing the syntectonic sediments. The displacement boundary condition is 2.3 mm a^{-1}. Models are run until excessive mesh distortion and/or mesh penetration at the contact elements cause an instable numerical solution. This typically occurs after 50–60 m of total convergence.

Stresses in our models were built up by block motions, i.e., movement of the hanging wall relative to the footwall. Special focus is on the mean stress distribution and stress field orientation during the various restoration stages. However, as only an approximate description of the real sediment rheology is used, we do not attempt to predict absolute values but concentrate on the relative patterns of mean stress in relation to different thrust geometries and their kinematics. Mean stress has a profound control on the forces driving and opposing fluid migration. Driving forces are (e.g., Vandenbroucke 1993):

$$\text{buoyant force} = (\rho_w - \rho_o)g \sin(\phi)\, l \quad (2)$$

where ρ_w and ρ_o are water and oil densities, g is the acceleration of gravity, φ is the angle between migration and horizontal directions, l is the oil column length; and (e.g., Mandl 1988)

$$\text{hydraulic head} = h_p + h_e \quad (3)$$

where h_p and h_e are pressure and elevation heads. h_e represents the potential energy of a fluid due to its elevation above a reference level, while

$$h_p = \Delta p/(\rho g) \quad (4)$$

where Δp is pore fluid pressure difference and ρ is fluid density.

Opposing forces are (e.g., Winter 1987):

$$\text{capillary pressure} = 2\gamma \cos(\theta)/r \quad (5)$$

where γ is the interfacial tension between oil and water, θ is the wettability expressed as the contact angle of the oil–water interface against the rock surface, and r is the average radius of the interconnected pore space; and (e.g., Mandl 1988)

$$\text{frictional resistivity force} = -\eta\, u/k \quad (6)$$

where η is the fluid viscosity, u is the filtration velocity and k is the permeability.

Loading of the thrust sheet by tectonic stress generated by block motions affects fluid migration forces in various ways. In the case of the buoyant force, the shortening would result in a growth of the fold structure, increasing the angle between migration and horizontal directions. This effect is not directly coupled to mean stress and, therefore, remains effective even after the end of the thrustbelt shortening. We call this control passive because it is effective even after shortening.

In the case of the hydraulic head, the shortening affects both elevation and pressure heads. The elevation head changes with growth of the fold structure. Remaining changed after the end of shortening in the thrustbelt, the elevation head also qualifies for being a passive control. The pressure head is coupled with mean stress development via its effect on the pore fluid pressure (e.g., Shi & Wang 1986; Ge & Garven 1992):

$$\Delta p = (a_n/[a_n + nb(1-n))]\Delta\sigma_m \quad (7)$$

where Δp is the change in pore fluid pressure, a_n is the porosity compressibility, n is the porosity, b is the fluid compressibility, and $\Delta\sigma_m$ is the change in mean stress. Because mean stress perturbations develop in each stress build-up cycle, they qualify for an active control, as they vanish after the end of thrustbelt shortening.

In the case of the capillary pressure, the mean stress controls the average radius of the interconnected pore space via its effect on the pore space (e.g., Zimmerman *et al.* 1986). Because the pore space decreases both elastically, being able to return to its earlier volume after loading release, and plastically, undergoing permanent pore space changes, this control has passive and active components.

In the case of the frictional resistivity, the permeability and filtration velocity reacts to the mean stress (Vandenbroucke 1993; Brown *et al.* 1994; R. Sigal 1998, pers. comm.) both temporally and permanently; this control has both passive and active components.

Results

Balancing results

All balanced cross-sections are pinned on the European Platform and end at the Pieniny Klippen Belt (Fig. 1b), except profile 1, which goes further to the Inner Carpathians. As profiles 1 and 2 have already been discussed in detail in Nemčok *et al.* (1999, 2000, 2001), this paper focuses on the description of the remaining profiles 3 to 5 (Fig. 3). The general style of deformation and the dominant fold-and-thrust style are similar in all five sections.

The frontal half of the thrustbelt along profile 3 accreted small volumes of the Neogene molasse sediments (Fig. 3a). The youngest of them are of Badenian-Lower Sarmatian age. The frontal half of the wedge comprises a Silesian sediment section accreted in 4.3–18.6 km wide thrust sheets. This width, increased in comparison with profile 4 and roughly comparable to profile 5, is caused by the dramatic thickness increase of the Cretaceous portion of the sedimentary section in the unit defined as the Skole Unit, which caused a strength increase of the accreted sedimentary package. The maximum thickness of the package is 3.6 km. Thrust sheets were formed by fault-propagation folding, as indicated not only by our balancing but also by available reflection seismics. They were thrust over incompetent formations of the autochthonous molasse. Two preserved ramp tips are located inside the Upper Cretaceous–Paleocene inversion-related section, one at the base of the Oligocene–Lower Miocene syn-tectonic sediments. Syn-tectonic sediments show large thickness variations, which are due to erosion of shortened structures and the existence of complex topography during their deposition.

The frontal third of the accretionary wedge along profile 4 incorporates a small volume of the Neogene molasse, Lower Sarmatian being the youngest (Fig. 3b). Thrust sheet widths are affected by thickness changes of the pre-Eocene portion of the sedimentary section. The thick section, which has a thickness of about

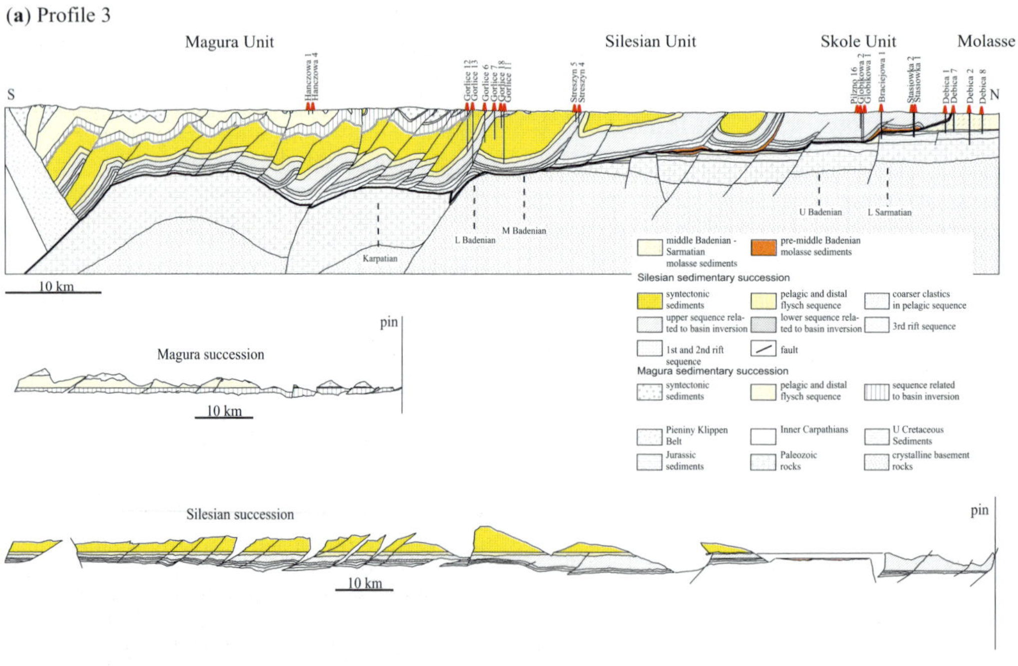

Fig. 3. Balanced deformed and restored cross sections through the Western Carpathians (see Fig. 1b for location). Vertical scale equals horizontal scale. The thick dashed lines indicate the southernmost extent of autochthonous molasse sediments of indicated age below the accretionary wedge. The thrustbelt incorporates sediments of two main units; Magura and Silesian Units.

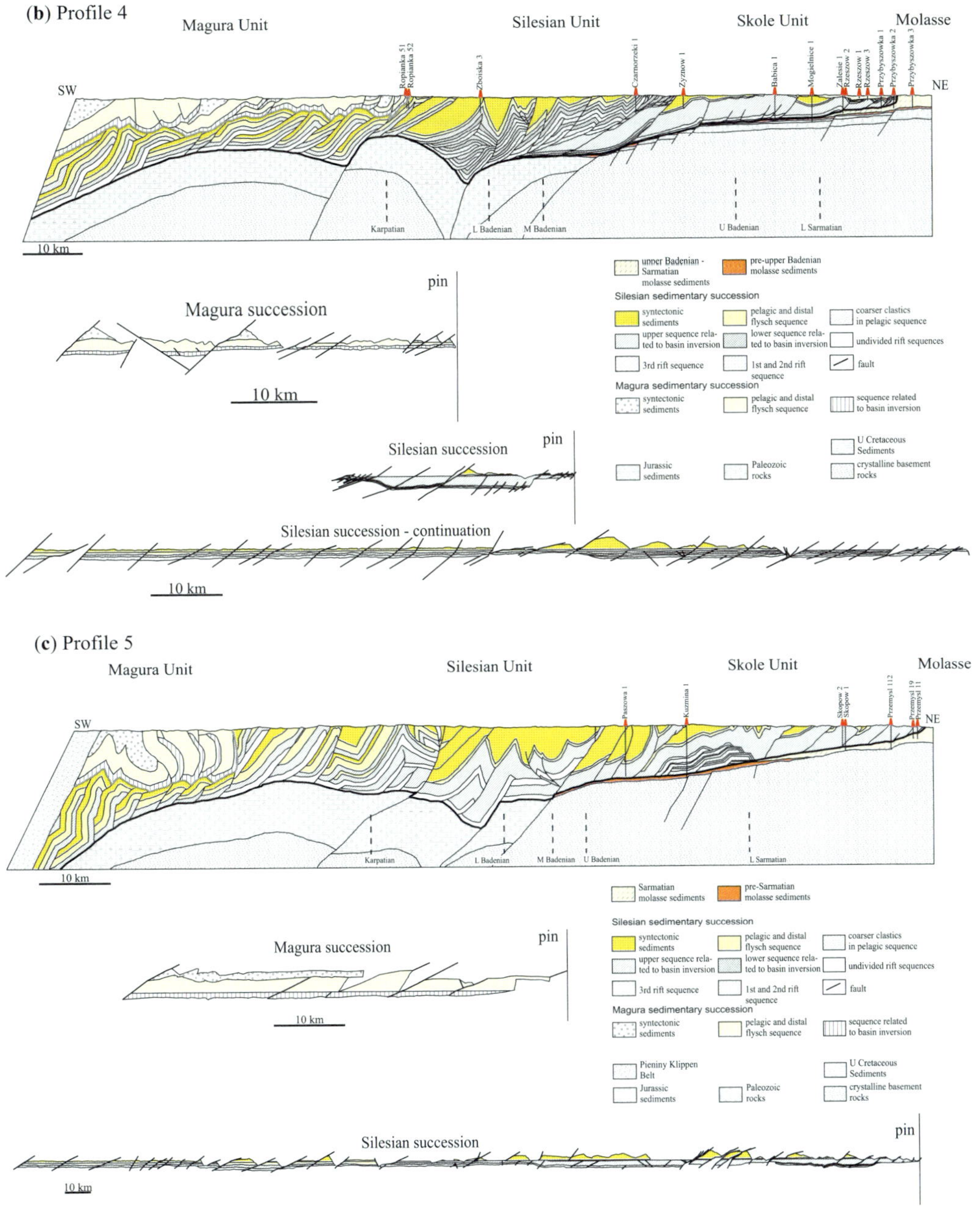

Fig. 3. *Continued.*

1.92 km, forms thrust sheets 5.5–7.1 km wide. The thin section, located at both southern and northern sides of the thick section, which has a thickness of about 0.5 km, forms thrust sheets only 0.55–1.65 km wide. Most of these thrust sheets are formed by fault-propagation folding, thrust over incompetent Neogene molasse sediments. Four ramp tips are located in the middle of the upper sequence related to basin inversion. One ramp tip is located inside the Eocene pelagic and distal flysch sequence. The frontal six ramps propagated upward into Neogene molasse sediments deposited in a piggyback basin carried on top of the frontal thrust sheets. This is the only profile with preserved geometries of frontal thrust sheets. Profiles 3 and 5 have their frontal structures deeply eroded.

The frontal half of the thrustbelt along profile 5 comprises a Silesian sediment section imbricated in 2.1–26.4 km wide thrust sheets (Fig. 3c). This increased width in comparison with profile 4 and roughly comparable to profile 3, is caused by the thickness increase of the Cretaceous sedimentary section in the Skole Unit, which increased its strength. Its maximum thickness, comparable to profile 3, is 3.6 km. The frontal part of the wedge was thrust over less competent middle Badenian–Sarmatian facies of the autochthonous molasse. Due to the deep erosional level, only three of the frontal thrust sheets show the evidence of fault-propagation folding. Two preserved ramp tips are located at the base of the Eocene section. Variable thickness of the thrust sheets controls the thrust sheet widths. Thrust sheets with a width of 7.1–26.4 km have a thickness close to its maximum value, containing both Lower and Upper Cretaceous sections. The thickness of 2.1–8.6 km wide thrust sheets ranges between 1.4 and 2.1 km. These thrust sheets do not contain the Lower Cretaceous section and the thickness of the Upper Cretaceous section is frequently reduced.

Rear portions of the described profiles are not composed of simple fault-propagation fold structures. The rear half of the thrustbelt along profile 3 has a complex structure, which includes thick-skin tectonics, buried Silesian duplexes and the overlying Magura Unit (Fig. 3a). The rear two-thirds of the thrustbelt along profile 4 includes thick-skin tectonics, two levels of buried duplexes in the Czarnorzeki area, an antiformal stack and two triangle zones with back-thrusting in the Zboiska area and the Magura Unit above Silesian buried duplexes (Fig. 3b). The rear portion of the wedge along profile 5 contains thick-skin tectonics, several triangle zones and back-thrusts in the Silesian section (Fig. 3c).

Results of our balancing demonstrate that fault-propagation folding developed large frontal portions of the West Carpathians. This observation is in accordance with observations made from our profiles 1 and 2 (Nemčok *et al.* 1999, 2000, 2001).

Finite-element modelling results

Analysis of the finite-element modelling results focused on the mean stress distribution in relation to different fold-and-thrust geometries and balancing stages, respectively, in order to assess the driving forces for fluid migration. Figures show the mean stress distribution for various growth stages of a fault-propagation fold with syn-tectonic sedimentation (Fig. 4) and with syn-tectonic erosion (Fig. 5). Similarly, mean stresses for two growth stages of a fault-bend fold are shown in Figure 6.

Fault-propagation fold scenarios with syn-tectonic deposition (Fig. 4) indicate the following features of the mean stress perturbation:

(1) Maximum mean stress anomaly in the advancing side of the propagating ramp tip, whereas minimum mean stress anomaly exists in the receding side of the propagating ramp tip.
(2) Minimum mean stress anomaly in the anticline during early stages of the stress build-up.
(3) Maximum mean stress anomaly in the hanging wall in the bend area above the flat/ramp transition. The whole lower beam unit (sandstone) above the flat experiences a mean stress increase, which diminishes in direction toward the ramp tip.
(4) Medium mean stress anomaly in the anticline.
(5) Distinct minimum mean stress anomaly in the footwall in front of the upper portion of the ramp.
(6) The footwall underneath the flat is characterized by mean stress increase during all stages of the stress build-up.
(7) The footwall in front of the lowermost portion of the ramp progressively develops mean stress increase by the end of the fold growth during last stages of the stress build-up.

Fault-propagation fold scenarios with syn-tectonic erosion (Fig. 5) show the following mean stress distribution: (1) Only slightly increased mean stress in the advancing side of the propagating ramp tip and a minimum mean stress anomaly in the receding side of the propagating ramp tip during the last stages of the fold growth and last stages of the stress build-up. (2) A broad minimum mean stress anomaly in the upper parts of the anticline and in the foreland in front of the upper half of the ramp. (3) Slightly increased mean stress in the hanging wall in the bend area above the flat/ramp transition.

Fault-bend fold scenarios are depicted in Figure 6 for the dolomite–limestone–dolomite succession with syn-tectonic shale deposition, which shows a more variable mean stress pattern than the corresponding siliciclastic

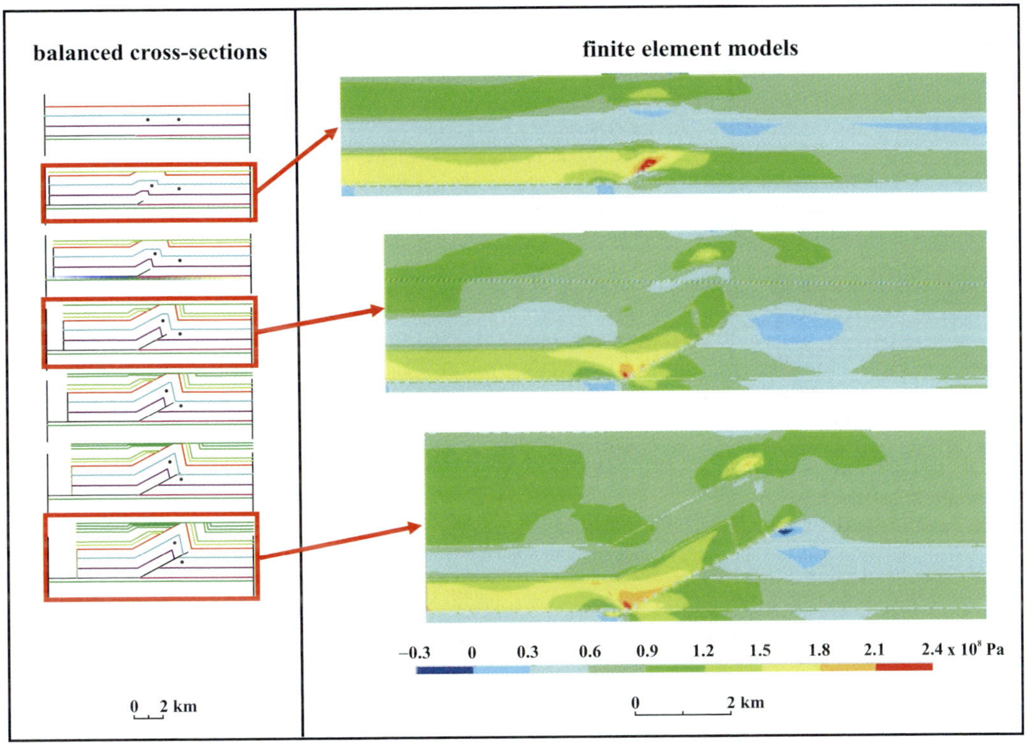

Fig. 4. Finite-element models of mean stress perturbations during various development stages of the fault-propagation fold scenario with syn-tectonic sedimentation. Models show the siliciclastic scenario represented by an upper sandstone, a middle shale and a lower sandstone layer of the pre-tectonic section. The syn-tectonic sediment keeping up with growing structure is sandstone. Colour scale is in Pa.

scenario. They indicate the following features for the mean stress perturbation: (1) Maximum mean stress anomaly is developed in the foreland in front of the lowermost portion of the ramp. It develops during the last stress build-up stages, but only in carbonate scenarios, which are 'more brittle' than siliciclastic ones. (2) A broad minimum mean stress anomaly in the anticline, which is developed in both carbonate and siliciclastic scenarios. Carbonate scenarios generate

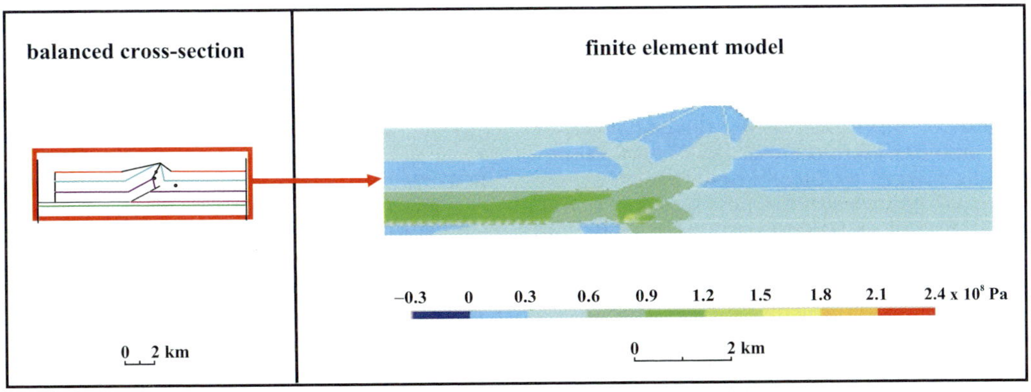

Fig. 5. Mean stress distribution of fault-propagation fold scenario with syn-tectonic erosion. Colour scale is in Pa. A corresponding scenario with syn-tectonic sedimentation is shown in Fig. 4 middle.

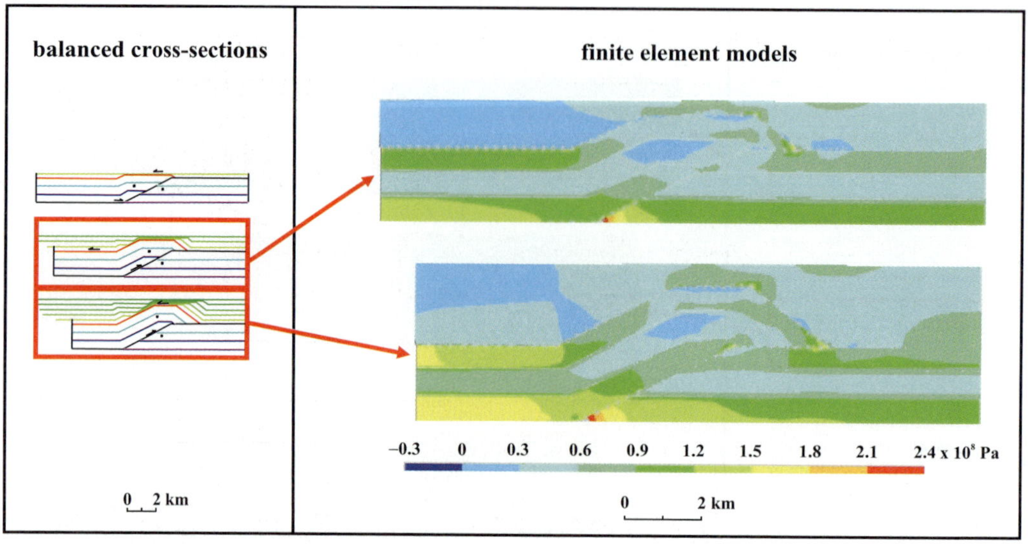

Fig. 6. Finite-element models of mean stress during mature development stages of the fault-bend fold type. Models show the carbonate scenario represented by upper dolomite, middle limestone and lower dolomite layers of the pre-tectonic section. The syn-tectonic sediment keeping up with growing structure is shale. Colour scale is in Pa.

stronger anomalies. (3) A slight minimum mean stress anomaly in the foreland in front of the central portion of the ramp.

During growth of the fault-bend fold, continuing block movements build up a new stress cycle ultimately leading to failure and rapid stress release. A very early stage of this build-up is captured in Figure 7b. This figure shows that several areas of the profile do not 'feel' loaded by tectonic stress yet. In these areas the maximum principal stress σ_1 is still oriented vertically and has not yet changed into sub-horizontally oriented compression. An advanced stage of the build-up is shown in Figure 7c. This is the time when the whole profile affected by block movements experiences loading by sub-horizontally and sub-vertically oriented maximum and minimum principal compressional stresses σ_1 and σ_3, typical for thrusting regime. Several areas of the profile experience various stress perturbations caused by complexities added by the frictional resistance of fault planes, flexure, internal deformation of the thrust sheet and uneven distribution of gravity forces. Stress perturbations are not only characterized by changes in orientations of principal stresses but also by changes in their magnitudes. The last stress build-up stage, before the failure of the model, is captured in Figure 7d. The stress regime is similar to that of Figure 7c but stress perturbations are more pronounced.

Discussion

Our finite element modelling results indicate that each fold type has its own typical stress distribution during each stress build-up and growth stage. The straining, which controls the stress transfer, differs among various fold types (compare Figs 4 and 6), because their different geometries and fault constellations also control what portions of the stress generated by the driving force would be spent on overcoming the frictional resistance of pre-existing faults, internal deformation of the thrust sheet, overcoming gravity forces, and deformation of the foreland.

Mean stress in various portions of fault-bend fold models is affected by the complex stress perturbations described earlier. The whole anticline is the place where the mean stress decreases below the regional level (Fig. 6). This observed mean stress decrease, together with an anticline being the morphologically highest feature of the thrust sheet, causes a match of favourable elevation and pressure terms of the hydraulic gradient, which would have a strong tendency to drive fluids from broader hanging wall to hanging wall anticline in all tested fault-bend scenarios. The drive will be strongest in the carbonate case with sealing syn-tectonic shale. Figure 6 indicates that practically the whole anticline serves as a

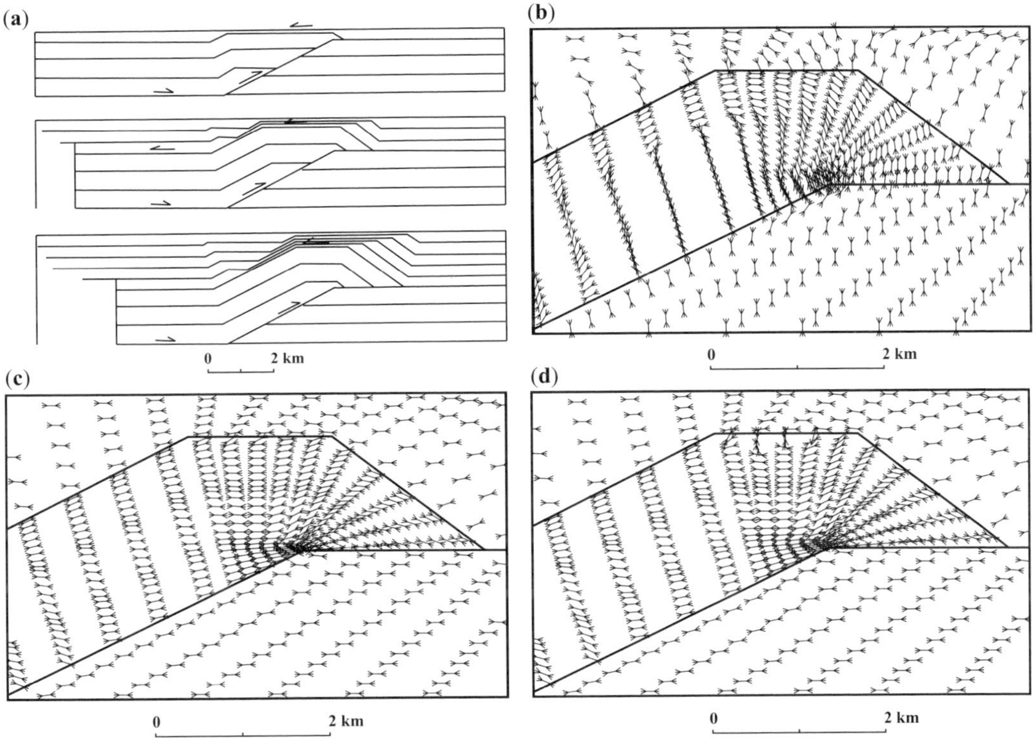

Fig. 7. (**a**) Forward model of the fault-bend fold (passive roof duplex). Roof thrust is formed by syn-tectonic sandstone, the deposition rate of which was faster than uplift of the growth structure during simulated shortening. (**b–d**) Finite-element models capturing the orientation of the maximum principal stress during a stress build-up cycle within the passive-roof duplex in its mature growth stage shown in Fig. 7a bottom. Arrows indicate orientation of the largest principal stress σ_1 after 4.6 m (b), 26.3 m and 39.1 m (d) of horizontal convergence.

fluid sink. A comparison of Figure 6 with Figure 4 indicates that the fault-bend fold is not capable of generating large-scale mean stress perturbations and can be defined as 'low energy' structure.

The fault-propagation fold is characterized by: coupled mean stress increase and decrease on hanging wall and footwall sides of the ramp tip; mean stress decrease in the initial anticline followed by increase during later anticline growth; progressive mean stress increase in the broader hanging wall; and progressive mean stress decrease in the foreland.

Observed mean stress features indicate hydraulic gradients, which can potentially drive fluids (Fig. 4): from hanging wall to footwall by tip-tapping mechanism in the ramp tip region; from the advancing to the receding side of the ramp tip; thus acting as a source-sink mechanism between anticline and foreland.

These modelling results strongly suggest that only the fault-propagation folds can generate mean stress perturbations, which would drive fluids into their foreland if any other fluid-migration driving and opposing forces were ignored. The Carpathian data indicate that the mean-stress controlled fluid drive into the foreland dominated over the fluid drive into growing anticlines created by fault-propagation folding. The fluid drive into anticlines would be controlled by (1) a buoyant force enhanced by the shortening-controlled angle between migration and horizontal directions; and (2) the elevation head enhanced by shortening.

The additional enhancement of the foreland-ward fluid drive was provided by active portions of the mean stress control over the pore volume gradients (i.e., capillary pressure gradients) and permeability gradients, which provided an additional fluid pumping mechanism from more compacted to less compacted, or less permeable to more permeable, areas. In doing this, fault-propagation folds are dynamically different from any other fold types, as discussed earlier.

Separation of various fluid-drive controls into passive and active provides an understanding of which controls remain effective or vanish after the end of the thrustbelt shortening. Fault-propagation folds are capable of modifying both pressure and elevation portions of the hydraulic head when they are active. They can modify the elevation head when they are both active or passive. The pressure head controlled by mean stress perturbations in active fault-propagation folds creates pore fluid pressure gradients that act in addition to the elevation head effect. The changing structural geometry forms elevation controls. When folds are passive, the hydraulic head is controlled only by the structural architecture developed earlier during the active thrusting. Therefore, the answer whether folds were active during hydrocarbon migration becomes very important, because that is the only time when fault-propagation folds can pump fluids into their forelands. When they are passive they become accumulators just like any other fold types. Mean stress perturbations in the modelled folds just make the oil accumulation in the anticline stronger when folds are active. Other active controls affected by mean stress perturbations in growing fault-propagation folds over fluid migration forces are effects on pore space, permeability and filtration velocity.

Conclusions

The combination of geochemical results (Ten Haven *et al.* 1993; Bessereau *et al.* 1996) with our balancing data indicate that oil migration from the Oligocene Menilite source shale inside the West Carpathian thrustbelt to fields in its autochthonous foreland took place during the Late Oligocene–Early Miocene. This was the time when the thrustbelt underwent shortening and its frontal structures adjacent to the autochthonous foreland developed by fault-propagation folding. The source-reservoir correlation and the dominant tectonic style inferred from geometric balancing are in accordance with the predictions made by the numerical simulations. Only the fault-propagation folds are characterized by a large foreland area of decreased mean stress, by a coupled increased/decreased mean stress areas on advancing/receding sides of the ramp tip and the overall mean stress decrease inside the thrust sheet in the foreland direction. These mean stresses and related hydraulic gradients provided an effective fluid pumping mechanism from frontal fault-propagation folds into their footwalls. It can operate when the influence of the pressure head on the fluid migration dominates over the influence of elevation head and buoyant force, and becomes enhanced by advantageous distribution of capillary pressure and frictional resistivity gradients that further enhance fluid flow from the thrust sheet into its foreland.

Numerical modelling portion of the study was made within the project 'Systematics of hydrocarbon exploration and production in thrustbelts' sponsored by Elf, Enterprise, Occidental, Repsol YPF and Texaco. The balancing portion of the study was sponsored by Amoco. B. Welker and C. Christensen helped with figures. S. Buiter, F. Roure and B. Sperner are thanked for their helpful comments on an earlier version of the manuscript.

References

BALLA, Z. 1984. The Carpathian loop and the Pannonian basin: a kinematic analysis. *Geophysical Transactions*, **30**(**4**), 313–353.

BESSEREAU, G., ROURE, F., KOTARBA, A., KUSMIEREK, J. & STRZETELSKI, W. 1996. Structure and hydrocarbon habitat of the Polish Carpathians. *In*: ZIEGLER, P. A. & HORVATH, F. (eds) Peri-Tethys Memoir 2: Structure and prospects of Alpine basins and forelands. *Memoires du Museum National & Histoire Naturelle*, **170**, 343–373.

BROWN, K. M., BEKINS, B., CLENNELL, B., DEWHURST, D. & WESTBROOK, G. 1994. Heterogeneous hydrofracture development and accretionary fault dynamics. *Geology*, **22**, 259–262.

BURBIDGE, D. R. & BRAUN, J. 2002. Numerical models of the evolution of accretionary wedges and fold-and-thrust belts using the distinct element method. *Geophysical Journal International*, **148**, 542–561.

ČERMÁK, V. & RYBACH, L. 1982. Thermal conductivity and specific heat of minerals and rocks. *In*: G. Angenheister (ed.) *Landolt-Bornstein Numerical Data and Functional Relationships in Science and Technology*. New Series, 16, Springer Verlag, Berlin.

CLARK, S. P. Jr. 1966. *Handbook of Physical Constants*. GSA Memoir, New York, **97**.

CLOETINGH, S. A. P. L., ZOETEMEIJER, R. & VAN WEES, J. D. 1995. *Tectonics I. Tectonics and basin formation in convergent settings; thermo-mechanical evolution of the lithosphere and basin evolution in compressive tectonic regimes*. Short course, Vrije University, Amsterdam.

CSONTOS, L., NAGYMAROSY, A., HORVÁTH, F. & KOVÁČ, M. 1992. Tertiary evolution of the intra-carpathian area: a model. *Tectonophysics*, **208**, 221–241.

GE, S. & GARVEN, G. 1992. Hydromechanical modelling of tectonically driven groundwater flow with application to the Arkoma foreland basin. *Journal of Geophysical Research*, **97**, 9119–9144.

IBRMAJER, J., SUK, M. *ET AL.* 1989. *Geophyiscal Image of the CSSR* (in Czech). ÚÚG Praha.

JAEGER, J. C. & COOK, N. G. W. 1979. *Fundamentals of Rock Mechanics*. Chapman & Hall, London.

KARNKOWSKI, P. 1999. *Oil and Gas Deposits in Poland*. The Geosynoptics Society 'GEOS', Cracow.

KORÁB, T., KRS, M., KRSOVÁ, M. & PAGÁČ, P. 1981. Palaeomagnetic investigations of Albian (?) – Paleocene to Lower Oligocene sediments from the Dukla unit, East Slovakian Flysch, Czechoslovakia. *Západné Karpaty, Séria Geológia*, **7**, 127–149.

KOTARBA, M., SZAFRAN, S. & ESPITALIE, J. 1987. A study of organic matter and natural gases of the Miocene in the Polish part of the Carpathian Foredeep. *Chemical Geology*, **64**, 197–207.

KOVÁCS, S., CSASZAR, G., GALACZ, A., HAAS, J., NAGY, E. &VOROS, A. 1989. The Tisza superunit was originally part of the northern (European) margin of Tethys. *In*: RAKÚS, M., DERCOURT, J. & NAIRN, A. E. M. (eds) *Evolution of the Northern Margin of Tethys*. Memoires de la Société Geologique de France, Nouvelle Serie, Paris, 81–100.

KRS, M. & POTFAJ, M. 1991. Palaeo-magnetic investigations of the flysch belt in the Orava region, Magura unit, Czechoslovak Western Carpathians. *Geologické Práce*, Správy, **92**, 125–151.

KRS, M. & ROTH, Z. 1977. A paleomagnetic study of Cenomanian–Lower Turonian sediments in the Moravskoslezké Beskydy Mts (in Czech). *Věstník ÚÚG*, **52(6)**, 323–332.

KRS, M., KRSOVÁ, M. & CHVOJKA, R. 1993. Palaeomagnetic investigations in the Biele Karpaty mountains, flysch belt of the West Carpathians. *Geologica Carpathica*, **45(1)**, 35–43.

KSIĄŻKIEWICZ, M. 1977. Plate motion hypothesis and Carpathian development (in Polish). *Annales de la Société Geologique de Pologne*, **47**, 321–353.

KWON, S. & MITRA, G. 2004. Three-dimensional finite element modelling of a thin-skinned fold-thrust belt wedge; Provo Salient, Sevier Belt, Utah. *Geology*, **32**, 561–564.

LAMA, R. D. & VUTUKURI, V. S. 1978. *Handbook on Mechanical Properties of Rocks*. Transtech Publications, Bay Village, 2.

MANDL, G. 1988. *Mechanics of Tectonic Faulting. Models and Basic Concepts*. Elsevier, Amsterdam.

MÁRTON, E. & MÁRTON, P. 1989. A compilation of paleomagnetic results from Hungary. *Geophysical Transactions*, **35**, 117–133.

MITRA, S. 1990. Fault-propagation folds: geometry, kinematic evolution, and hydrocarbon traps. *American Association of Petroleum Geologists Bulletin*, **74**, 921–945.

MOLDOWAN, J. M., FAGO, F. J., HUIZINGA, B. J. & JACOBSON, S. R. 1991. Analysis of oleanane and its occurrence in upper Cretaceous rocks. *In*: MANNING, D. A. C. (ed.) *Advances and Applications in Energy and the Natural Environment. Organic Geochemistry*. Manchester University Press, Manchester, 195–197.

NEMČOK, M., COWARD, M. P., SERCOMBE, W. J. & KLECKER, R. A. 1999. Structure of the West Carpathian accretionary wedge: insights from cross-section construction and sandbox validation. *Physics and Chemistry of the Earth*, **24**, 659–665.

NEMČOK, M., NEMČOK, J. *ET AL*. 2000. Results of 2D balancing along 20° and 21° longitude and pseudo-3D in the Smilno tectonic window: implications for shortening mechanisms of the West Carpathian accretionary wedge. *Geologica Carpathica*, **51**, 281–300.

NEMČOK, M., NEMČOK, J. *ET AL*. 2001. Reconstruction of Cretaceous rifts incorporated in the Outer West Carpathian wedge by balancing. *In*: CLOETINGH, S., NEMČOK, M., NEUBAUER, F., HORVÁTH, F. & SEIFERT, P. (eds) *The Hydrocarbon Potential of the Carpathian–Pannonian Region*. Marine and Petroleum Geology, **18**, 39–64.

NEMČOK, M., POSPÍŠIL, L., LEXA, J. & DONELICK, R. A. 1998. Tertiary subduction and slab break-off model of the Carpathian–Pannonian region. *Tectonophysics*, **295**, 307–340.

NEUBAUER, F. & GENSER, J. 1990. Architektur und Kinematik der östlichen Zentralalpen: eine Übersicht. *Mitteilungen des Naturwissenschaftlichen Vereines für Steiermark*, **120**, 203–219.

PATRASCU, S., PANAIOTU, C., SECLAMAN, M. & PANAIOTU, C. E. 1994. Timing of rotational motion of Apuseni Mountains (Romania): paleomagnetic data from Tertiary magmatic rocks. *Tectonophysics*, **233**, 163–176.

PUGZG GEOPOL 1998. *Poland Petroleum Opportunities*. Geopol, Warsaw.

RATSCHBACHER, L., FRISCH, W., LINZER, H. G. & MERLE, O. 1991. Lateral extrusion in the Eastern Alps, part 2: structural analysis. *Tectonics*, **10(2)**, 257–271.

ROCA, E., BESSEREAU, G., JAWOR, E., KOTARBA, M. & ROURE, F. 1995. Pre-Neogene evolution of the Western Carpathians: constraints from the Bochnia–Tatra Mountains section (Polish Western Carpathians). *Tectonics*, **14**, 855–873.

ROURE, F., ROCA, E. & SASSI, W. 1993. The Neogene evolution of the outer Carpathian flysch units (Poland, Ukraine and Romania): kinematics of a foreland/fold-and-thrust belt system. *Sedimentary Geology*, **86**, 177–201.

ROYDEN, L. H. & BÁLDI, T. 1988. Early Cenozoic tectonics and paleogeography of the Pannonian and surrounding regions. *In*: ROYDEN, L. H. & HORVÁTH, F. (eds) *The Pannonian Basin. A Study in Basin Evolution*. American Association of Petroleum Geologists Memoir, Tulsa, 1–16.

ROYDEN, L. H., HORVÁTH, F. & BURCHFIELD, B. C. 1982. Transform faulting, extension and subduction in the Carpathian-Pannonian region. *Geological Society of America Bulletin*, **73**, 717–725.

RYLKO, W. & TOMAS, A. 1995. Morphology of the consolidated basement of the Polish Carpathians in the light of magnetotelluric data. *Geological Quarterly*, **39**, 1–16.

SALVINI, F., STORTI, F. & MCCLAY, K. 2001. Self-determing numerical modelling of compressional fault-bend folding. *Geology*, **29**, 839–842.

SANDULESCU, M. 1988. Cenozoic tectonic history of the Carpathians. *In*: ROYDEN, L. H., HORVÁTH, F. (eds) *The Pannonian Basin. A Study in Basin Evolution*. (American Association of Petroleum Geologist Memoir), **45**, 17–26.

SASSI, W. & FAURE, J. L. 1996. Role of faults and layer interfaces on the spatial variation of stress regimes in basins; inferences from numerical modelling. *Tectonophysics*, **266**, 101–119.

SHI, Y. & WANG, C. Y. 1986. Pore pressure generation in sedimentary basins: overloading versus aquathermal. *Journal of Geophysical Research*, **91**, 2153–2162.

SLACZKA, A. 1996. Oil and gas in the Northern Carpathians. *In*: WESELY, G. & LIEBL, W. (eds) *Oil and Gas in Alpidic Thrustbelts and Basins of Central and Eastern Europe*. EAGE Special Publication, **5**, 187–195.

SUK, M., BLÍŽKOVSKÝ, M. *ET AL*. 1984. *Geological History of the Territory of the Czech Socialist Republic*. Czech Geological Survey, Prague.

SUPPE, J. 1983. Geometry and kinematics of fault-bend folding. *American Journal of Science*, **283**, 684–721.

SUPPE, J. & MEDWEDEFF, D. A. 1984. Fault-propagation folding. *GSA Bulletin*, Abstracts with Programs, **16**, 670.

TEN HAVEN, H. L., LAFARGUE, E. & KOTARBA, M. 1993. Oil/oil and oil/source rock correlations in the Carpathian Foredeep and overthrust, south-east Poland. *Organic Geochemistry*, **20(7)**, 935–959.

TÚNYI, I. & M. KOVÁČ. 1991. Paleomagnetic investigation of the Neogene sediments from the Little Carpathians. *Contributions of Geophysical Institute of Slovak Academy of Sciences*, **21**, 125–146.

VANDENBROUCKE, M. 1993. Migration of Hydrocarbons. *In*: BORDENAVE, M. L. (ed.) *Applied Petroleum Geochemistry*. Paris, Technip, 125–148.

WEIL, W., GORECKI, W., KRUCEK, J. & KARNKOWSKI, P. 1998. Discoveries of petroleum deposits in the years 1987–96 and current reserves of oil and gas in Poland. *Oil and Gas News*, **8**, 19–33.

WINKLER, W. & SLACZKA, A. 1992. Sediment dispersal and provenance in the Silesian, Dukla and Magura flysch nappes (Outer Carpathians, Poland). *Geologische Rundschau*, **81**, 371–382.

WINTER, A. 1987. Percolative aspects of hydrocarbon migration. *In*: DOLIGEZ, B. (ed.) *Migration of Hydrocarbons in Sedimentary Basins*. Editions Technip, Paris, 237–255.

ZIMMERMAN, R. W., SOMERTON, W. H. & KING, M. S. 1986. Compressibility of porous rocks. *Journal of Geophysical Research*, **91**, 12765–12777.

Relation between effective friction and fault slip rate across the Northern San Andreas fault system

ANN-SOPHIE PROVOST & JEAN CHÉRY

Laboratoire Dynamique de la Lithosphère, Université Montpellier 2, cc. 060, Place E. Bataillon, 34095 Montpellier cedex 5, France (e-mail: jean@dstu.univ-montp2.fr)

Abstract: The San Andreas Fault system is a complex tectonic ensemble that accommodates most of the relative plate motion between the Pacific and the North American plates. The structure and rheological properties of the faults vary along the plate boundary and lead to the distribution of deformation that we observe today. In order to learn more about the mechanical behaviour of such a fault system, a model of the northern California fault system is built, constrained by heat flow data, GPS and palaeoseismological measurements of slip rates (on the San Andreas, the Maacama and Bartlett Springfaults), and stress orientations. Our basic assumption is that the upper crust has a high frictional strength and that major faults represent weak zones with a lower effective friction. Several combinations of effective fault frictions on the three major faults of the system in the model are tested. We find that slight variations of the effective friction angle on one of the three active strands lead to an important redistribution of slip rates through the system. If present in nature, this fault behaviour could explain why fault slip rates vary in time, as suggested by slip rate variations over geological scales in intracontinental fault systems.

The San Andreas Fault system, slipping at a rate of 39 mm/yr, accommodates 80% of the relative plate motion between the Pacific plate and the North American plate (Freymueller *et al.* 1999; Argus & Gordon 2001). The distribution of the deformation varies largely along the fault zone. In central California south of 36.5° latitude, 90% of the differential motion between the Sierra Nevada and the Pacific plate is concentrated on a single strand of the San Andreas Fault, with a geodetic slip rate of 34 mm/yr (Lisowski *et al.* 1991). The fault system north of San Francisco is more complex than in the central region (Fig. 1). Slip rate is distributed over three major fault zones, including the San Andreas Fault to the west, the Maacama fault just to the east of the San Andreas, and farther to the east the Bartlett Springs fault (Castillo & Ellsworth 1993). The two latter zones are the northward continuations of the Hayward and the Calaveras faults in the San Francisco Bay area, and dip at 60° to 70° toward the NE. Here we attempt to quantify the slip distribution over the faults in order to understand fault interactions.

Because the kinematic boundary conditions of the system remain almost the same from central to northern California, it has been proposed that the broader distribution of deformation over different fault strands in Northern California should be related to differences in fault zone properties between Northern and Central California (Chéry *et al.* 2001). Indeed, a variation of the effective fault friction on a fault results in a change in the slip velocity of the fault system. The deformation is also dependent on the thermal state of the lithosphere, which is believed to strongly influence the lithospheric rheology (Kirby 1985). We quantify the distribution of the long-term deformation on the fault system (i.e., averaged over several seismic cycles) by means of numerical modelling and we sort out the factors that determine the activity of the three major fault zones of northern California.

Both geodetical and geological observations of the deformation have been made in the region. Based on GPS measurements corrected for elastic loading (Freymueller *et al.* 1999) the long-term slip rate is 17.4, 13.9 and 8.2 mm/yr for the San Andreas, Maacama and Bartlett Spring faults respectively, at latitude 39°N. On a longer time scale, palae oseismological studies are consistent with slip rates of 20–25 mm/yr for the San Andreas fault (Prentice 1989; Niemi & Hall 1992), *c.* 6.4–10.4 mm/yr for the Rogers Creek fault just south of the Maacama fault (Schwarz *et al.* 1992), and

From: BUITER, S. J. H. & SCHREURS, G. (eds) 2006. *Analogue and Numerical Modelling of Crustal-Scale Processes.* Geological Society, London, Special Publications, **253**, 429–436.
0305-8719/06/$15.00

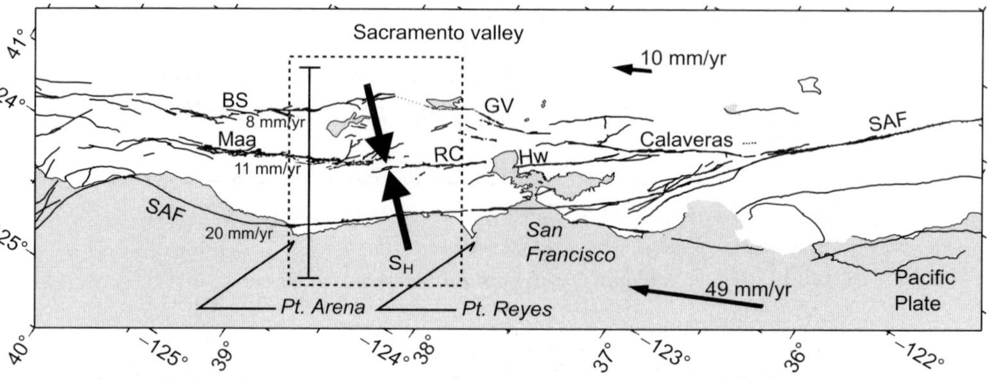

Fig. 1. Map of northern California fault system (Jennings 1992). The vectors show the relative plate motion of the Pacific plate (49 mm/yr) and the Sacramento valley (10 mm/yr) relative to North America. The estimated slip rates on the three major faults of the system are; 20, 11 and 8 mm/yr for the San Andreas, the Maacama and the Bartlett Spring faults respectively (see text for details). The maximum horizontal compression S_H is found to be at *c.* 60° of the strike of major faults in this area (Provost & Houston 2003) as indicated by the facing arrows. The area studied is delimited by the line across the fault system. The zone of interest for modelling corresponds to the dashed rectangle. BS: Bartlett Spring fault; GV: Green Valley fault; Hw: Hayward fault; Maa: Maacama fault; Pt. Arena: Point Arena; RC: Rogers Creek fault; SAF: San Andreas fault.

c. 5 mm/yr for the Green Valley fault just south of the Bartlett Spring fault (Frizell & Brown 1976). Although short-term (geodetical) and long-term (geological) fault slip rate estimates may not necessarily coincide (Friedrich *et al.* 2003), we develop our analysis with combining these slip rates, which allows us to build a simplified deformation field across the fault system. We define an area close to Point Arena (latitude 39°), extending 180 km inland and stack our data on a cross-section perpendicular to the average fault system orientation (Fig. 1). Assuming that most of the strain is accommodated by a three-fault system, we choose the following slip rates: 20, 11, and 8 mm/yr for the San Andreas, Maacama and Bartlett Spring faults respectively and we use this information to constrain our model.

Mechanical model

As mentioned earlier, the San Andreas Fault system has a different mechanical behaviour in northern and central California. The central segment of the San Andreas fault displays a maximum horizontal compression S_H oriented at very high angles (*c.* 84°) to the fault (Mount & Suppe 1987; Zoback *et al.* 1987; Provost & Houston 2001), while the compression is more oblique (60°) in the northern San Andreas fault system (Provost & Houston 2003). The distribution of the deformation in northern California has been explained as an increase of the effective fault friction with respect to the central California segment (Chéry *et al.* 2001). Alternatively, it has been proposed that the viscosity of the lower crust has a dominant effect on the distribution of the deformation in the upper crust (Roy 1998). Also, the surface geological fault slip rate could be entirely controlled by the averaged velocity of the lower lithosphere underneath these crustal blocks (Bourne *et al.* 1998). In this view the influence of the brittle crust would be negligible. With a thermomechanical model we investigate the effect of the effective fault friction and of the lateral variation of the temperature field on the slip rate of a multistrand fault system.

We build our model using a 3D Lagrangian finite element code, ADELI3D (Chéry *et al.* 2001; Provost *et al.* 2003). We solve a system of equations formed by the momentum equation, the constitutive law of the medium, and static and kinematical boundary conditions. The numerical experiments are conducted with meshes having 9918 tetrahedral elements corresponding to a typical element size of *c.* 2 km. The entire model is subject to gravitational forces (g in Fig. 2) and a density of 2800 kg/m^3 is assumed. Geometry and boundary conditions of our model are shown in Figure 2. The model is 180 km long (x) perpendicular to the strike of the faults, and 30 km deep (z). We assume here no variation of deformation and stress along strike (y). Because this hypothesis is enforced by means of boundary conditions on

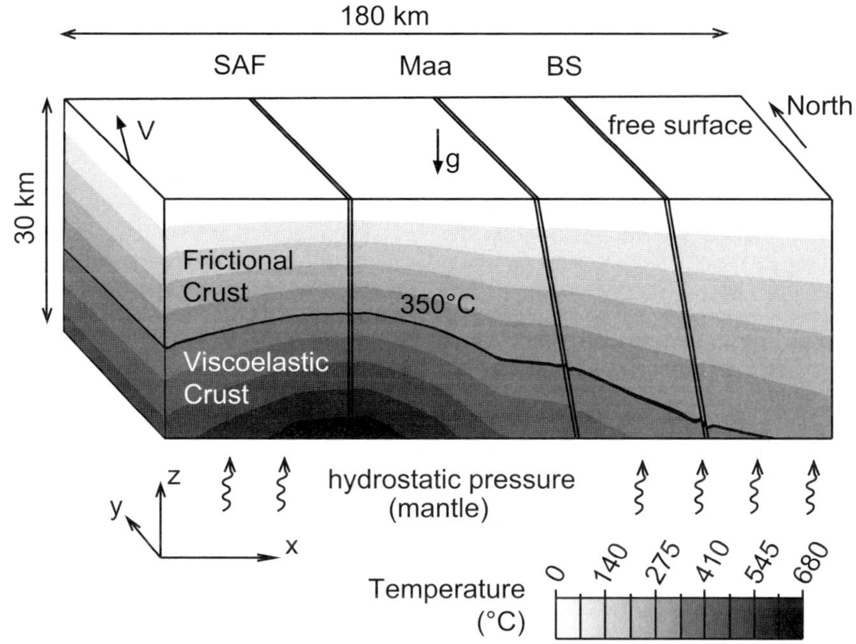

Fig. 2. Geometry and boundary conditions of the 3D mechanical model used in this analysis. The north is along the y-axis. The model is 30 km deep (z-axis) and 180 km long (x-axis). On the model The dip of the Maacama and Bartlett Spring faults is 65°. The vector V indicates the velocity (Vx; Vy) imposed on the western part of the model, and g is the gravitational force applied on it. Abbreviations are as in Fig. 1.

x-z planes, the mesh width in the y-direction is arbitrary. Our resulting model is therefore between a 2D and a 3D model, the non-zero components of the strain tensor ε being ε_{xx}, ε_{xy}, ε_{xz}, ε_{yz}, ε_{zz}.

The velocity of the eastern part of the model (Sacramento Valley) is fixed to zero, and we impose a velocity on the western part of the model (Pacific plate) of 39 mm/yr along strike (Vy) and of 3.5 mm/yr perpendicular to it (Vx) in order to simulate the transpressive stress state occurring in central and northern California. However, this perpendicular value could be somewhat inaccurate and rely on geological (Thompson 1999) and geodetical (Gordon & Argus 1998; Argus & Gordon 2001) estimates. Because the fault zone corresponds to a hot lithospheric region (based on seismic tomography), the uppermost mantle viscosity is probably low (Furlong *et al.* 1989) and we assume that the lithospheric strength is located mainly in the crust. Therefore, we incorporate the buoyancy effect of the underlying mantle as a hydrostatic pressure applied at the base of the crust. As temperature measurements in wells have shown (Lachenbruch & Sass 1980), the heat flow is much higher close to the San Andreas Fault (*c.* 80 mW/m^2) than farther east in the Sacramento Valley (*c.* 30 mW/m^2).

We solve explicitly the 1D conductive heat flow equation in order to derivate a steady-state temperature field in agreement with these observations (Fig. 2). To do so, we use standard values of thermal parameters for the continental crust (conductivity of 3 Wm^{-1} K^{-1}, radiogenic crustal heating of 1 μW/m^3) and we adjust the mantle heat flow at the base of the model. As a result, the San Andreas Fault area corresponds to a hot geotherm where the 350°C isotherm is at 12 km depth. Cooler crustal temperatures occur towards the east and the 350°C temperature is found at 30 km depth in the Bartlett Spring area.

The crustal rheology is modelled using a viscoelastoplastic law according to pressure and temperature variations within the crust. Elastic parameters correspond to a Young's modulus of 10^{11} Pa and to a Poisson's coefficient of 0.25 in the whole medium. At low P-T, we assume a elastic-frictional behaviour for the crust for which we use a Drucker-Prager model (Leroy & Ortiz 1989). We adopt an effective friction angle ϕ of 15°. Because the Drucker-Prager model involves the intermediate stress, this friction angle is not directly comparable to the one involved in the Mohr-Coulomb criterion, which depends on minimum and maximum principal stresses only. However, the value of ϕ of 15° would be compatible with a Mohr-Coulomb friction coefficient of 0.51 and a hydrostatic pore pressure for the surrounding crust. We set the cohesion to a low value of 10^6 Pa for both the crust and the faults. Assuming that faults are weaker than the surrounding crust (Bird & Kong 1994), we define active faults with a distinct material behaviour and we use lower effective friction angles for the fault zones (from 3° to 12° which correspond to effective friction coefficients ranging from 0.1 to 0.44). At high P-T a viscoelastic behaviour is assumed and a temperature-dependent, linear Maxwell model is used in order to reproduce the ductile behaviour of quartzo-feldspathic rocks (Kirby 1985). The thermal dependence is controlled by the fluidity $\gamma = \gamma_0 \exp(-E/RT)$, γ_0 is a pre-exponential constant, E the activation energy, R the gas constant ($R = 8.32$), and T the temperature in K. The viscous parameters are determined by assuming an effective viscosity of 10^{23} Pa s at 350 °C and of 10^{20} Pa s at 650 °C, leading to $\gamma_0 = 8.49 \times 10^{-15}$ Pa s and E = 110 KJ/mol.

Results

We run five tests in which we vary the angle of friction, ϕ, on the San Andreas (SAF), the Maacama (Maa) and the Bartlett Spring (BS) faults, as well as the temperature field distribution. Our first tests are run with the temperature field described as in Figure 2 and with the same effective friction angle on the three faults. We start with a very low friction on the fault system, with ϕ (SAF, Maa, BS) = (3°,3°,3°), increase it up to ϕ = (8°,8°,8°) and then to ϕ = (12°,12°,12°). We list in Table 1 the velocities obtained for each of these three tests, across the faults and in between them (off-fault regions), as well as the percentage of deformation absorbed on-fault and off-fault, and the average orientation of the maximum horizontal compression S_H with respect to the strike of the faults. These results lead to a dramatic variation of the distribution of the deformation across the fault system with effective friction angle changes.

Case 1 with ϕ = (3°,3°,3°) shows that the distribution of the deformation is mostly on-fault (98%) with a slip rate on the San Andreas fault of 28.8 mm/yr that is much larger than the slip rate on the two other faults: 6.2 and 3.3 mm/yr on the Maacama and Bartlett Spring faults respectively. These slip rates are not in agreement with the observed ones of 20, 11, and 8 mm/yr respectively. When the angle of friction ϕ increases by equal amounts on the three faults, the slip rate on the San Andreas fault remains largest, but some deformation appears off-fault, 30% for *case 2* with ϕ = (8°,8°,8°) and up to 75% for *case 3* where ϕ = (12°,12°,12°). The latter case is extreme, because almost all the deformation imposed on the model is accommodated by the San Andreas Fault and its surroundings, leaving the eastern and cooler regions almost not affected by the deformation. One can notice that for these three cases, even though the effective friction angle ϕ is the same on the three faults, the slip rate on the San Andreas fault is always much larger than those found for the Maacama and Bartlett Spring faults. This is caused by the influence of the temperature field on the mechanical behaviour of the crust: the lower crust in the vicinity of the San Andreas Fault is warmer than for the other faults at the same depth, thus less viscous than in the eastern regions of the model. Consequently, the viscous lithosphere surrounding the San Andreas Fault at depth is more compliant than the one surrounding the other faults and can accumulate more strain.

Because the slip rates computed by our previous three tests are not in agreement with the estimated ones, we run several tests in which we applied different angles of friction to the faults. In Table 1 we list the results for a model (*case 4*) that reasonably satisfies the geological and geodetical slip rates. Starting from *case 1*

Table 1. *Differential velocities (mm/yr) across the faults and off-faults for cases 1 to 5*

Case number	ϕ SAF, Maa, BS	V_{off1}	V_{SAF}	V_{off2}	V_{Maa}	V_{off3}	V_{BS}	V_{off4}	$V_{on\text{-}faults}$ (%)	$V_{off\text{-}faults}$ (%)	S_H (SAF)	S_H (far-field)
1	3, 3, 3	0.2	28.8	0.2	6.2	0.1	3.3	0.1	98.5	1.5	61°	75°
2	8, 8, 8	6.9	25.0	1.7	1.7	0.3	0.3	0.2	69.3	30.7	52°	56°
3	12, 12, 12	16.7	8.7	0.7	0.7	0.8	0.1	0.3	24.4	75.6	48°	49°
4	4, 3, 3	0.3	18.1	10.8	10.8	0.1	9.3	0.2	98.0	2.0	62°	76°
5	4, 3, 3	0.2	3.6	17.3	17.3	0.1	17.3	0.4	97.8	2.1	59°	69°

with $\phi = (3°,3°,3°)$, a slight increase of ϕ on the San Andreas fault to 4° leads to a drastically reduced slip rate down to 18.1 mm/yr, inducing a slip rate increase on the Maacama and the Bartlett Spring faults to 10.8 and 9.3 mm/yr respectively. Profiles of the velocity along strike (in the y-direction) are shown in Figure 3a. The black line shows the velocity at the surface while the dashed line represents the velocity at 25 km depth. Surface deformation in this case occurs mainly on faults, while at depth only warmer regions (under the San Andreas Fault) allow the deformation to spread off-fault. This broadening of the deformation is clearly visible on the cross-section of Figure 3b showing the strain rate for *case 4*. Between 15 and 30 km depth, under the San Andreas Fault, a wide zone (*c.* 20 km) is deforming viscously at strain rates above $1.3 \times 10^{-15}\ s^{-1}$. Such strain rates are visible only below 20 km depth

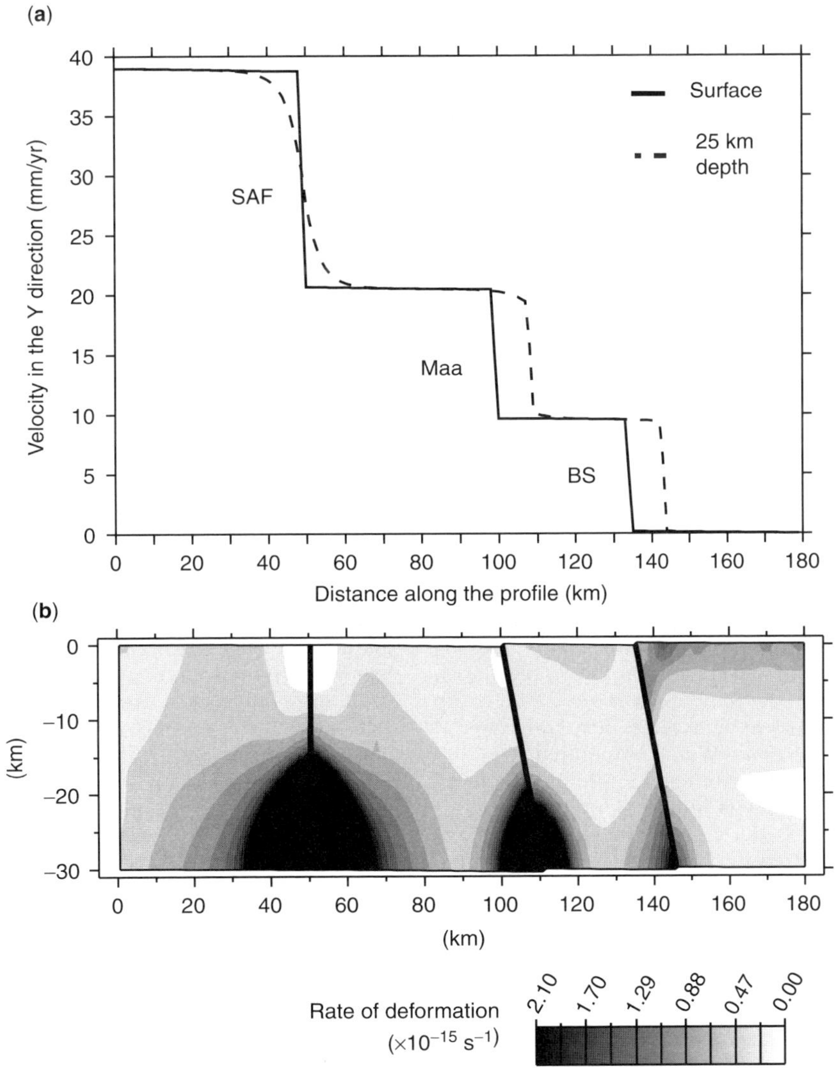

Fig. 3. Velocity profiles and distribution of the deformation across the model for the case $\phi = (4°, 3°, 3°)$. (**a**) Profiles across the fault system of the velocity along strike (y-direction) at the surface (black line) and at 25 km depth (dashed line). The shift of the dashed line at the location of the Maacama and Bartlett Spring faults is due to their dipping planes. (**b**) Distribution of strain rate deduced from the model. Here the strain rate is expressed as the quadratic invariant of the strain rate tensor. Abbreviations are as in Fig. 1.

over a 10 km wide zone at Maacama and correspond to a very limited area under the Bartlett Spring fault.

In order to clarify the role of temperature on the fault slip rate, we present a last test (*case 5*) with $\phi = (4°,3°,3°)$ as for *case 4* but having a constant surface heat flow of 80 mW/m^2, corresponding to a 350 °C isotherm at 15 km. This run leads to very different slip rate values with respect to *case 4*. The San Andreas Fault slips at a low rate of 3.6 mm/yr whereas the Maacama and Bartlett Spring faults slip rates increase up to 17.3 mm/yr for both. These values are obviously in agreement with field observations of recent fault activity. Clearly, it should be possible to obtain a better agreement of measured slip rate using a constant heat flow model with *slightly* decaying the friction angle of the San Andreas Fault. Although we do not perform this exercise here, we point out that a trade-off exists between the a priori thermal model and the set of friction angles to be used. Similarly, the optimal friction angle set may also depend on the viscous flow laws. The exploration of this effect is beyond the scope of this paper. However, we found through numerical experiments that the influence of the flow law remains small for values typical of the continental crust (Kirby 1985).

Discussion

As already mentioned, the average orientation of the maximum horizontal compression S_H is at about 60° from the strike of major faults in Northern California (Provost & Houston 2003), with a little variation (10°) between orientations very close to and far from the main faults. Because our modelling involves a velocity boundary condition between the two vertical sides, S_H orientations are not initially set but depend on the solution of the momentum equation. We obtain similar values (56° to 76°) for *cases 2, 4* and *5* (i.e., low friction cases). S_H appears to strongly depend on the effective friction. For large ϕ values (12° as for *case 3*), S_H is oriented at smaller angles to the faults (49°). Therefore, our best kinematical model (*case 4*) with $\phi = (4°,3°,3°)$ also satisfies the seismological observations. Interestingly, we remark that even if S_H is oriented at smaller angles to the faults in this region (compared to Central California where S_H is *c.* 84°), this does not necessarily imply that the fault zone's frictional strength is higher. This finding is somewhat different than the one obtained for a one-strand strike-slip fault model (Chéry *et al.* 2001). For such a tectonic setting, these authors demonstrate that a decrease of the effective fault friction angle always corresponds to an increase in the value for S_H.

We find that the mechanical rules that hold for a single strike-slip fault (Rice 1992) may not be directly applicable to a multistrand fault system. Our analysis shows that a fault system composed of more than one strand does not behave as one single fault. Indeed, communication between the faults balances the overall distribution of the deformation in the system. We have shown that the slip rate can vary largely from one fault to the other if rheological or thermal parameters are changed. Perhaps the most striking feature of our model is that a variation as small as 1° of the effective friction angle on one fault profoundly influences the slip distribution of the whole fault system. Because a multi-fault system (such as the San Andreas Fault in northern California) has more degrees of freedom than a single-strand fault (the San Andreas Fault in central California), we suggest that it is thus more sensitive to the variation of the friction angle on one fault, resulting in the variation of the fault velocities. This behaviour is due to the coupling between frictional and viscous deformation in the fault zone. The strength of the fault plane (here the y-z plane) corresponds to the sum of a frictional stress, which is rate independent, and a viscous stress, which increases with the strain rate. For a constant slip rate, a decrease of the effective friction on one fault of the system brings the system out of equilibrium due to a decrease of the strength of the fault. As a quasi-static equilibrium is required for a slowly evolving geological system (as is our model), the fault must adapt by increasing its viscous strength and therefore its slip rate.

On a global scale, plate tectonics display a good agreement of differential plate motion over geodetical and geological time scales (Sella *et al.* 2002). On a smaller spatial scale, it has often been postulated that fault slip rates slowly evolve over geological times. That would explain why geodetic strain rates may coincide with geological estimates. This view appears correct for the Central San Andreas Fault where geological and geodetical slip rates are coincident (Sieh & Jahns 1984; Lisowski *et al.* 1991). However, a growing geological data archive for California also shows that fault slip rates vary largely over various time scales (0.01–10 Ma). For the northern San Andreas Fault system, a spatial and temporal slip distribution analysis leads to a fault model in which slip rates vary on short geological time periods (Wakabayashi 1999). This author suggests that the fastest slipping group of faults has shifted several times during the northward migration of the Mendocino triple junction. The western faults were more active than the eastern ones from 18 to 16 Ma and from 12 to 8 Ma, but from 16 to 12 Ma this trend was reversed. In southern California, the slip rates of San Andreas and the San Jacinto faults have also evolved largely during the Quaternary (Kendrick *et al.* 2002; Bennett *et al.* 2004). Other evidence for rapid fault slip rate variation has recently been discovered in other continental zones such as the Basin and Range in the western United States (Wallace 1987; Friedrich *et al.* 2003) and in Mongolia (Ritz *et al.* 2003).

Because our modelling indicates that a small change of effective friction strongly affects the slip rate of a group of faults, this behaviour may provide a mechanical basis for rapid temporal changes. As a result, comparing Quaternary and geodetically determined slip rates could be not as straightforward as we postulated in the introduction of this paper. Indeed, we demonstrate that a variation of 1° of the friction angle drastically affects the slip rate on a fault embedded in a multistrand fault system and one can compute that the equivalent shear stress change on the fault plane is *c.* 5 MPa for a seismogenic depth of 10 km. This change is a small fraction of the total differential stress that can be accommodated by crustal rocks (*c.* 100 MPa). Therefore, 5% change of the differential stress could lead to a slip rate variation of 30%, as shown by the comparison between cases 1 and 4.

As most of large intracontinental faults are earthquake prone, it is plausible that effective friction variations on fault planes in nature are related to earthquake mechanics and seismic cycle. Because our model is based on a steady-state approximation, we cannot directly model the effect of repeated earthquakes on the averaged fault friction. However, one can remark that the stress change we mention (5 MPa) is close to the mean stress drop of 3 MPa associated with continental earthquakes (Hanks 1977). Combined with the fact that earthquakes do not regularly occur on large faults (Grant & Sieh 1994), the occurrence of a few earthquakes clustered in time may be an indication that the maximum stress sustainable by the fault is dropping by a few MPa. Consequently, the interseismic fault slip rate could increase dramatically, while other faults slow down according to constant plate boundary conditions. This feedback process may lead low slip rate faults to cease their activity and heal, while the accelerating fault accommodates most of the differential plate motion, as in central California. As relative plate motion slowly evolves over a few million years, the remaining active fault should slip with a constant velocity for this corresponding time period.

We thank M. Nyst, F. Pollitz and S. Buiter for their constructive reviews. This work has been supported by the research program ACI 'prevention des catastrophes naturelles' of the French ministry of Research.

References

ARGUS, D. F. & GORDON, D. 2001. Present tectonic motion across the Coast Ranges and San Andreas fault system in central California. *Geological Society of America Bulletin*, **113(12)**, 1580–1592.

BENNETT, R., FRIEDRICH, A. M. & FURLONG, K. P. 2004. Codependent histories of the San Andreas and San Jacinto fault zones from inversion of fault displacement rates. *Geology*, **32**, 961–964.

BIRD, P. & KONG, X. 1994. Computer simulations of California tectonics confirm very low strength of major faults. *Geological Society of America Bulletin*, **106**, 159–174.

BOURNE, S. J., ENGLAND, P. C. & PARSONS, B. 1998. The motion of crustal blocks driven by flow of the lower lithosphere: implications for slip rates of faults in the south island of New Zealand and Southern California. *Nature*, **391**, 655–659.

CASTILLO, D. A. & ELLSWORTH, W. L. 1993. Seismotectonics of the San Andreas Fault System between Point Arena and Cape Mendocino in northern California: implication for the development and evolution of a young transform. *Journal of Geophysical Research*, **98(B4)**, 6543–6560.

CHÉRY, J., ZOBACK, M. D. & HASSANI, R. 2001. An integrated mechanical model of the San Andreas Fault in central and northern California. *Journal of Geophysical Research*, **106(B10)**, 22051–22066.

FREYMUELLER, J. T., MURRAY, M. H., SEGALL, P. & CASTILLO, D. 1999. Kinematics of the Pacific-North America plate boundary zone, northern California. *Journal of Geophysical Research*, **104(B4)**, 7419–7441.

FRIZELI, V. A. J. & BROWN, R. D. Jr. 1976. *Map showing recently active breaks along the Green Valley fault, Napa and Sonoma Counties*, California, M(200) MF no. 743, US Geological Survey.

FRIEDRICH, A. M., WERNICKE, B. P., NIEMI, N. A., BENNETT, R. A. & DAVIS, J. L. 2003. Comparison of geodetic and geologic data from the Wasatch region, UTAH, and implications for the spectral character of Earth deformation at periods of 10 to 10 million years. *Journal of Geophysical Research*, **108(B4)**, DOI 10.10292001JB000682.

FURLONG, K. P., HUGO, W. D. & ZANDT, G. 1989. Geometry and evolution of the San Andreas fault zone in northern California. *Journal of Geophysical Research*, **94(B3)**, 3100–3110.

JENNINGS, C. W. 1992. *Preliminary fault activity map of California*. California Division of Mines and Geology, Sacramento.

GORDON, R. G. & ARGUS, D. F. 1998. Present tectonic motion across the Coast Ranges and San Andreas Fault System in central California. *EOS Transaction AGU*, **79(95)**, S204.

GRANT, L. B. & SIEH, K. 1994. Paleoseismic evidence of clustered earthquakes on the San Andreas fault in the Carrizo plain, California. *Journal of Geophysical Research*, **99**, 6819–6841.

HANKS, T. 1977. Earthquake stress drop, ambient tectonic stress, and stress that drives plate motions. *Pageoph*, **115**, 441–458.

KENDRICK, K. J., MORTON, D. M., WELLS, S. G. & SIMPSON, R. W. 2002. Spatial and temporal deformation along the Northern San Jacinto Fault, Southern California: Implications for slip rates. *Bulletin of the Seismological Society of America*, **92**, 2782–2802.

KIRBY, S. H. 1985. Rocks mechanics observations pertinent to the rheology of the lithosphere and the localization of strain along shear zones. *Tectonophysics*, **119**, 1–27.

LACHENBRUCH, A. H. & SASS, J. H. 1980. Heat flow and energetics of the San Andreas fault zone. *Journal of Geophysical Research*, **85(B11)**, 6185–6222.

LEROY, Y. & ORTIZ, M. 1989. Finite element analysis of strain localization in frictional materials. *International Journal for Numerical and Analysis Methods in Geomechanics*, **13**, 53–74.

LISOWSKI, M., SAVAGE, J. C. & PRESCOTT, W. H. 1991. The velocity field along the San Andreas fault in central and southern California. *Journal of Geophysical Research*, **96(5)**, 8369–8389.

MOUNT, V. S. & SUPPE, J. 1987. State of stress near the San Andreas fault: implications for wrench tectonics. *Geology*, **15**, 1143–1146.

NIEMI, T. M. & HALL, N. T. 1992. Late Holocence slip rate and recurrence of great earthquakes on the San Andreas fault in northern California. *Geology*, **20**, 195–198.

PRENTICE, C. 1989. *Earthquake geology of the northern San Andreas fault near Point Arena, California*. Ph.D. thesis, California Institute of Technology Pasadena.

PROVOST, A.-S. & HOUSTON, H. 2001. Orientation of the stress field surrounding the creeping section of the San Andreas fault: evidence for a narrow mechanically-weak fault zone. *Journal of Geophysical Research*, **106(B6)**, 11373–11386.

PROVOST, A.-S. & HOUSTON, H. 2003. Stress orientations in northern and central California: evidence for the evolution of frictional strength along the San Andreas Plate boundary system. *Journal of Geophysical Research*, **108(B3)**, 2175, DOI 10.1029/2001JB001123.

PROVOST, A.-S., CHÉRY, J. & HASSANI, R. 2003. 3D mechanical modeling of the GPS velocity field along the North Anatolian fault. *Earth and Planetary Science Letters*, **209**, 361–377.

RICE, J. R. 1992. Fault stress states, pore pressure distributions, and the weakness of the San Andreas fault. *In*: B. EVANS and T.-F. WONG (eds) *Fault Mechanics and Transport Properties of Rock*, Academic Press, London, 475–503.

RITZ, J.-F., BOURLES, D., BROWN, E. T. *ET AL*. 2003. Analysing the uplift rate along the Gurvan Bulag thrust fault (Gobi-Altag, Mongolia) during the past 130,000 yrs using surface-exposure 10Be datings. *Journal of Geophysical Research*, **108**(B3), 2162, DOI: 10.1029/2001JB000553, 2003.

ROY, M. 1998. Evolution of fault system at a strike-slip plate boundary: a viscoelastic model. *Geophysical Research Letters*, **25(15)**, 2881–2884.

SCHWARZ, D. P., PANTOSTI, D., HECKER, S., OKUMURA, K., BUDDING, K. E. & POWERS, T. 1992. Late Holocence behavior and seismogenic potential of the Rogers Creek fault zone. *The second Conference on Earthquake Hazards in the Eastern San Francisco Bay Area*, 393–398.

SELLA, G. F., DIXON, T. H. & MAO, A. 2002. REVEL: A model for recent plate velocities from space geodesy. *Journal of Geophysical Research*, **107(B4)**, ETG 11-1,11-32.

SIEH, K. & JAHNS, R. H. 1984. Holocence activity of the San Andreas Fault at Wallace Creek, California. *Geological Society of America Bulletin*, **95(8)**, 883–896.

THOMPSON, G. A. 1999. California coast ranges tectonics in the light of geophysical constraints: A tribute to Ben Page. *International Geology Review*, **41**, 383–390.

WAKABAYASHI, J. 1999. Distribution of displacement on and evolution of a young transform fault system: the northern San Andreas fault system, California. *Tectonics*, **18(6)**, 1245–1274.

WALLACE, R. E. 1987. Grouping and migration of surface faulting and variations in slip rates on faults in the Great Basin province. *Bulletin of the Seismological Society of America*, **77**, 868–876.

ZOBACK, M. D., ZOBACK, M. L. *ET AL*. 1987. New evidence on the state of stress of the San Andreas fault system. *Science*, **238**, 1105–1111.

Index

Page numbers in italic, e.g. *45*, refer to figures. Page numbers in bold, e.g. **35**, signify entries in tables.

DEMOCRACY'S GUARDIANS

Democracy's Guardians

A History of the German Federal Constitutional Court
1951–2001

JUSTIN COLLINGS
Brigham Young University

OXFORD
UNIVERSITY PRESS

OXFORD
UNIVERSITY PRESS

Great Clarendon Street, Oxford, OX2 6DP,
United Kingdom

Oxford University Press is a department of the University of Oxford.
It furthers the University's objective of excellence in research, scholarship,
and education by publishing worldwide. Oxford is a registered trade mark of
Oxford University Press in the UK and in certain other countries

First Edition published in 2015

Published in the United States of America by Oxford University Press
198 Madison Avenue, New York, NY 10016, United States of America

British Library Cataloguing in Publication Data
Data available

Library of Congress Control Number: 2015945189

ISBN 978–0–19–875337–7

For Lia

'... because you are to me the chief woman in the world—the throned lady whose colours I carry between my heart and my armor.'

—Daniel Deronda

Acknowledgements

This book began as a dissertation submitted to the history department of Yale University. Perhaps someone, somewhere, sometime has had as fine a dissertation committee as I, but I rather doubt it. Adam Tooze, my advisor, has been an inexhaustible fount of encouragement, energy, and ideas. For nearly a decade, Bruce Ackerman has been a consummate teacher and mentor, forever pushing me to do better and to think bigger. Jay Winter is an exquisite gentleman, a magnificent teacher, and a warm human being—a friend and guide since my first day of graduate school. Dieter Grimm has been wondrously kind to an American interloper writing about the Court he once served so ably. All four are giants in their fields—powerful thinkers, lucid writers, captivating teachers, committed citizens. They have been models for me as well as mentors. I hope, in my scholarly life, to repay them by emulation.

This project was interrupted for a year while I worked as a law clerk for the Honorable Guido Calabresi, another supreme mentor and one of the truly great ones of the earth. My co-clerks, Farah Peterson and Brian Soucek, blessed me with their wisdom, friendship, and encouragement. This book, and indeed my whole life, is richer for that experience.

Some of the historical material in Chapter 1 appeared previously in German in an essay called 'Gerhard Leibholz und der Status des Bundesverfassungsgerichts. Karriere eines Berichts und seines Berichterstatters,' in a collection edited by Anna-Bettina Kaiser and entitled *Der Parteienstaat. Zum Staatsverständnis von Gerhard Leibholz* (Nomos 2013). I thank the Nomos publishing house for permission to reproduce that material here.

I wish to thank the gracious and gifted archivists and librarians of the Staatsbibliothek zu Berlin, the Bundesarchiv in Koblenz, the Friedrich-Ebert-Stiftung in Bonn, the Hans-Seidel-Stiftung in Munich, and the Landesarchiv Baden-Württemberg in Stuttgart. The Fox International Fellowship, under the aegis of the MacMillan Center at Yale, generously sponsored my research in Germany during the 2011–12 academic year. I also wish to thank Frau Marianne Leibholz for permission to examine the personal papers of her father, Justice Gerhard Leibholz. Galen Fletcher, Kory Staheli, Dianne Davenport, Dennis Sears, and other librarians at BYU Law School have been skilled and supportive. Adam Ott, Janet Lawrence, Alexandra Thomas Sandvik, Amy West, and Whitney Nelson provided invaluable help as I prepared the manuscript for publication. Several colleagues at BYU read parts of this book and gave helpful feedback. Donald Kommers, the dean of German constitutional studies, graciously read and responded to the entire manuscript. Mitchel Lasser, my friend and former teacher, deserves recognition both for inspiring me to become a comparative constitutionalist and for helping to get this book published.

Alex Flach and Elinor Shields of Oxford University Press have been gracious and helpful at every step of the publication process. Ambiga Jayakumar and others have done exceptional copy-editorial and production work.

My largest thanks go to my family. Our daughters, Julia, Elisabeth, and Katharine, endured this project patiently from beginning to end. Julia and Eli even put up with a year in German schools while I did the bulk of the research and writing. Abdiel and Gwyndelin joined our family while my dissertation was becoming a book. I've been asked many times how I survived law school, graduate school, the job market, and my first year of teaching with so many small children. In fact, I could never have survived without them. On my darkest days they gave me a hero's welcome when I came home. They have stirred my soul with their childlike belief that their daddy could do anything—even write the history of the German Constitutional Court.

I have neither the space nor the power to thank adequately my wife, Lia. I hope my life can be my gratitude. For now, she will have to settle for the dedication of this book, which, like its author, would be nothing at all without her.

Contents

Table of Cases

FEDERAL CONSTITUTIONAL COURT

FEDERAL CONSTITUTIONAL COURT CHAMBER DECISIONS

FEDERAL COURT OF JUSTICE

EUROPEAN COURT OF JUSTICE

UNITED STATES SUPREME COURT

Prologue

The View from 1949

My ladies and gentlemen! What exactly is the aim of the work which we have undertaken here? What does 'Parliamentary Council' mean? What does 'Basic Law' mean? When in a sovereign state the People summons a constitution-giving national assembly, its commission is clear and need not be discussed further: It has a constitution to craft.... If we were to work under such conditions, we wouldn't need to ask the question, 'What exactly are we about?'

—Carlo Schmid (1948)[1]

[T]he Basic Law was fortunate, not only in the circumstances that attended its inauguration, but also in the preconditions of its genesis. What? I hear someone say: a defeated, decimated, and ruined land, divided in four by foreign powers, morally and physically desolated; a hungry and freezing populace, millions of homeless refugees and expellees, everyone occupied with bare survival; a people from whom every thought of political engagement had been beaten out—these are felicitous preconditions for the formation of a democratic constitution? Now, no one would maintain that the Germans of 1948/1949 were happy. They were as unhappy as it is possible to be. But precisely the deepest misfortune can be the most felicitous precondition for a new beginning.

—Sebastian Haffner (1976)[2]

I. *De Profundis*

A spectre haunted the Council's deliberations—the spectre of the collapse of the Weimar Republic. The Parliamentary Council had convened on 1 September 1948 in the stately Koenig Museum in Bonn. Its purpose was to craft an interim constitution for Germany's Western occupation zone. The Council consisted of seventy delegates, appointed by Germany's Western states, roughly one delegate for every 750,000 citizens.[3] The Council, which included four women, was dominated by jurists and civil servants. Several delegates had suffered greatly during the Nazi era—some lost jobs, some fled abroad, five were interned in concentration camps. A handful enjoyed successful or influential careers during the Third Reich. Few delegates were young; their average age was fifty-six. Nearly all had come of political age before or during the Weimar era. They were haunted by the Weimar Republic's failure; tormented by the calamities and atrocities that followed. The Weimar constitution, with its putatively fatal and fateful flaws, was the demon that drove the Council's deliberations.[4]

[1] 'Speech of Delegate Carlo Schmid to the Parliamentary Council, 8 September 1948' in *Parlamentarischer Rat. Stenographische Berichte über die Plenarsitzung* (Bonner Universitäts-Druckerei 1948/1949) 70.

[2] Sebastian Haffner, 'Der Erfolg des Grundgesetzes' in Sebastian Haffner, *Im Schatten der Geschichte. Historisch-politische Variationen aus zwanzig Jahren* (Deutsche Verlags-Anstalt 1985) 191.

[3] The five delegates from West Berlin were not voting members of the Council.

[4] For reflections on the spectre of Weimar in German political discourse, see Christoph Gusy (ed), *Weimars lange Schatten—'Weimar' als Argument nach 1945* (Nomos 2003).

The Council was not a constitutional convention or constituent assembly in any traditional sense. It was convened, not at the behest of the German people but at the command of their allied occupiers. Its members were chosen, not by popular election but through appointment by the parliaments of the several states—states that had been reconstituted shortly after the War. The constitution the Council promulgated on 23 May 1949 was never submitted for popular ratification of any kind.

It was a constitution, moreover, that dared not speak the name. As one writer has observed, the very designation 'constitution' (*Verfassung*) struck the delegates as 'too lofty, too solemn, too definitive' for the document they produced.[5] Instead, the Council christened its handiwork a 'Fundamental Law' or, as we shall call it in this work, a 'Basic Law' (*Grundgesetz*), one that by the terms of its own preamble was preliminary, provisional, transitional, and temporary.[6] The Basic Law, the preamble declared, would apply during a 'transitional period' (*Übergangszeit*) until the 'entire German people' could, 'in free self-determination, complete the unity and freedom of Germany'.[7] The point was repeated, emphatically, in the Basic Law's final provision. 'This Basic Law', Article 146 stipulated, 'shall cease to apply on the day when a constitution [*Verfassung*] established by free choice of the German people enters force'.[8]

But for all the disclaimers, the Basic Law read, in substance and in style, like a full-fledged constitution. The preamble spoke of 'the German people' acting with a 'consciousness of its responsibility before God and men' and 'by virtue of its constitution-giving authority'. It was a provisional constitution, perhaps, but it was a constitution all the same, complete with a catalogue of fundamental rights and a thorough allocation of powers between branches and levels of government.

The Basic Law both created a West German state and confirmed the division of Germany. There was little doubt that, once the Western Allied powers—Britain, France, and the United States—had overseen the creation of a new state in the Western occupation zone, the Soviet Union would follow suit in the East. Western and Soviet policy concerning Germany had long been on a collision course. The 'German Question' became the first concrete controversy of an incipient Cold War.

Only forty months separated the convening of the Parliamentary Council from the subterranean suicide of Adolf Hitler and the subsequent surrender of the German Reich. As early as January 1943, in Casablanca, the Allies had resolved to press relentlessly for unconditional surrender. And this, in fact, is what they achieved. Hitler himself had wished it so. Already in the mid-1920s he had written in *Mein Kampf* that it was Germany's destiny to become 'either World Power or nothing at all'.[9] *Alles oder nichts*—all or nothing—had been the *Führer*'s mantra. When Germany's bid for world conquest failed, national dissolution was, in Hitler's view, the natural consequence. In his final hours, Hitler raged that the German people had proven unworthy of him—unworthy even of survival. The Germany people had 'proven themselves the weaker', and should share, Hitler thought, in his own *Götterdämmerung*.[10]

Germany's allied conquerors—with several caveats and without the frenzied prism of Hitler's radical, Manichean cosmic Darwinism—agreed. 'Germany', observed General Charles de Gaulle, 'driven to fanaticism by its dream of mastery, so conducted the war

[5] Christoph Schönberger, 'Anmerkungen zu Karlsruhe' in Matthias Jestaedt and others, *Das Entgrenzte Gericht. Eine kritische Bilanz nach sechzig Jahre Bundesverfassungsgericht* (Suhrkamp 2011) 11.

[6] Throughout this book, references to specific articles of the Basic Law will use the abbreviation 'GG' for *Grungesetz*—i.e. 'Article 3(1) GG' refers to Article 3, section 1 of the Basic Law.

[7] Preamble, GG. [8] Article 146 GG. [9] Adolf Hitler, *Mein Kampf*, 37th edn (Eher 1933) 742.

[10] In Albert Speer, *Erinnerungen* (Ullstein 1969) 446.

that the contest became—materially, politically, and morally—a total conquest. The victory must therefore be a total victory. This has occurred. With regard to state, power, and doctrine, the German Reich is destroyed'.[11] Unconditional surrender gave the Allies the right to reorganize the German state. As one German historian has written, 'Germany had become the plaything [*Spielball*] of the Allies'.[12]

On the question what to do with this plaything, the Allies were deeply divided, and grew more divided as time went on. All agreed, however, that Germany must be punished. Germany was responsible—and, in Europe, solely responsible—for the bloodiest conflict in world history, a war that had killed some sixty-six million people. More than five million Germans had died, including 500,000 civilians. But there could be no question that the Germans were more sinning than sinned against—worlds without end, more sinning. Germany had murdered more Jews than it had lost soldiers. It became clearer with each Allied 'liberation' of a German concentration camp that the German state had commanded—and individual Germans had committed—crimes of horrific, sickening, and unprecedented scope. But even before this became obviously known, the Allies were resolved that Germany should pay for the cost of the War in treasure and blood. The Allies were resolved that any recurrence of expansionist German aggression should be rendered forever impossible. At the Tehran Conference in the final weeks of 1943, the Allied 'Big Three'—Roosevelt, Churchill, and Stalin—agreed that after the War Germany must be stripped of all capacity to train soldiers or produce armaments. It should have no military capabilities at all. Roosevelt thought the word 'Reich' should be stripped from the German language.[13]

At War's end it was clear that the peace would be Carthaginian. The Germans must beat their swords into ploughshares—literally, some thought. Henry Morgenthau, the American treasury secretary, advocated the wholesale de-industrialization of Germany—its sweeping reconversion into an agricultural land.[14] This was impracticable, as other Allied leaders realized. Germany was too densely populated and too urban to return to some pre-industrial bucolic idyll. What's more, the Allies feared the political consequences of avulsive economic disruption. They remembered the radicalization that flourished in the instability that followed the previous War. Still, they were resolved to treat the Germans roughly. The American military governor received orders to exercise uncompromising severity. Germany's Eastern borders were rolled back—a move that forced the westward flight of millions of German nationals—and its reduced territory divided into four occupation zones, each to be overseen by one of the victorious Allies. The Allies' occupation policy was to be guided by four 'D's': demilitarization, denazification, decentralization, and democratization.

Of these, denazification was the most controversial and the most unevenly applied.[15] Implementation by the Americans was in some respects the most militant. In early 1947, the

[11] Charles de Gaulle, *Discours de Guerre (Mai 1944—September 1945)*, vol 3 (Egloff 1945) 214.

[12] Manfred Görtemaker, *Geschichte der Bundesrepublik Deutschland* (CH Beck 1999) 19.

[13] See Keith Sainsbury, *The Turning Point: Roosevelt, Stalin, Churchill, and Chiang-Kai-Shek, 1943: The Moscow, Cairo, and Teheran Conferences* (Oxford University Press 1986); Keith Eubank, *Summit at Teheran* (William Morrow & Co. 1985).

[14] See John Dietrich, *The Morgenthau Plan: Soviet Influence on American Postwar Diplomacy*, 2nd edn (Algora Publishing 2013) 4.

[15] For recent general accounts of denazification, see Francis Graham-Dixon, *The Allied Occupation of Germany: The Refugee Crisis, Denazification, and the Path to Reconstruction* (IB Tauris 2013); Fred Taylor, *Exorcising Hitler: The Occupation and Denazification of Germany* (Bloomsbury Press 2011). On denazification in the Eastern Occupation Zone, see Timothy Vogt, *Denazification in Soviet-Occupied Germany: Brandenburg, 1945–1948* (Harvard University Press 2000).

former American president, Herbert Hoover, reported to the sitting American president, Harry S. Truman, that in the American and British occupation zones some 90,000 former Nazis were held in internment camps (Hoover called them 'concentration camps') and 1,900,000 others were under sanctions by which they were allowed to engage only in manual labour.[16] Denazification in the Western zone would end only in February 1950, two years later than in the Soviet zone. By then, millions of Germans had filled out questionnaires regarding their activities during the Nazi era and their views concerning Nazi ideology. Thousands were required to view grisly footage of the liberated death camps. The process was bitterly resented by the German populace. Its educative impact was hotly contested.

Denazification coincided with the spectacular trials of German leaders and Nazi operatives for war crimes and crimes against humanity. Of these the most sensational were the trials held in Nuremberg between November 1945 and October 1946. The trials were conducted by extraordinary courts. The criminal jurisdiction of ordinary German courts was limited to simple crimes perpetrated by Germans against Germans (or against stateless persons). German courts would acquire jurisdiction over alleged Nazi crimes only in 1955.[17]

In 1946, the philosopher Karl Jaspers published *The Question of German Guilt*, a seminal essay in which he probed the problem of Germans' culpability for Nazi crimes.[18] Jaspers distinguished between criminal guilt, which involved breaking positive law and was punished by the courts; moral guilt, which involved personal failure to meet moral duties and must be processed by each individual; political guilt, which implicated citizens in the crimes of a murderous regime and was being dealt with by the War's victors; and metaphysical guilt, which involved ruptures in the basic solidarity of human interaction and could be judged only by God.[19] Jaspers' taxonomy would exert enormous influence on subsequent efforts to come to terms with the German past.[20] But its nuances would have been lost on the vast majority of Jaspers' immediate contemporaries, even if they had taken the time to read his essay. Most Germans in the immediate postwar era were more concerned with provisions than with penance, in more desperate search of food and fuel than of pardon and absolution.

Life in the occupation zones was grim. Allied bombing campaigns had wrought massive destruction of dwellings. An epidemic of homelessness was compounded by the arrival of thousands of 'displaced persons' from the East. Millions of Germans were living in the War's residual rubble, waging a daily battle against hunger and cold. Not until 1948 would the average German in the Western zone reach the daily caloric intake requirements set forth by the League of Nations in 1936.[21] Malnutrition and under-nutrition heightened the risk of the spread of disease. Existence was particularly precarious for refugees and expellees from the East. To many observers it seemed a kind of divine visitation on a nation that had ground countless slave labourers to death by starvation and famine, and that had systematically murdered millions of political dissidents, homosexuals, Roma, and Jews.

Some observers feared that these bleak conditions would foment radicalization. But for most the battle for bare subsistence rendered all ideologies superfluous.[22] In any case, Allied officials and soldiers were on hand to nip signs of nascent radicalism in the bud.

[16] Herbert Hoover, *The President's Economic Mission to Germany and Austria: Report no. 1—German Agricultural and Food Requirements* (1947) 2.

[17] Görtemaker (n 12) 26–27.

[18] Karl Jaspers, *Die Schuldfrage. Ein Beitrag zur deutschen Frage* (Artemis-Verlag 1946).

[19] ibid 10–14.

[20] See Jan-Werner Müller, *German Intellectuals, Unification and National Identity* (Yale University Press 2000) 27–29.

[21] Görtemaker (n 12) 29. [22] ibid 30.

Under these circumstances, it is hardly surprising that there was little popular political engagement during the founding moments of postwar West German politics. It was not a season of rebirth, but a time for old hands to collect shattered fragments. Several new parties formed in the early postwar months, nearly all of them led by Weimar veterans. The Christian Democratic Union (CDU) was founded 17 June 1945 in Cologne on the 'unshakable foundation of Christianity and Western [*abendländisch*] culture'.[23] Among the party's founders was Cologne's sixty-nine-year-old mayor, Konrad Adenauer. The German People's Party (DVP) formed three months later under the leadership of Reinhold Maier (age fifty-six) and Theodor Heuss (age sixty-one). Thomas Dehler, who helped found the liberal Free Democratic Party in December, was a comparatively youthful forty-eight. Almost immediately after the War, the Leftist Social Democratic Party (SPD) reemerged from its Nazi-era ban under the direction of Kurt Schumacher, who had been interned during the War in various concentration camps.

The creation or reestablishment of political parties coincided with the reorganization, shortly after the War, of the individual West German states (*Länder*). It was for influence within these reorganized states that the parties first contended. The contending factions sought control of municipal councils and state parliaments. In time, however, events prompted the Western occupying powers to call for the creation of a West German state. At that point the parties became national; they vied for influence—but also collaborated as statesmen—in the shaping of a new nation.

The critical year was 1948. In March, discussions among the four Allied powers regarding a common currency reform failed definitively. Nazi policies for financing the War had ruined the Mark irrevocably, but as yet the Mark had no replacement. The principal coin of the realm in the immediate postwar era was the cigarette. In July, failure among the four Allies to unite on the currency question led the Western Allies to introduce a new currency, the Deutsch-Mark, within the Western occupation zone. Efforts to extend the new currency to West Berlin provoked a Soviet blockade—and a prolonged crisis in Berlin.

The Soviet gambit backfired. It fostered solidarity among the Western powers and forged, among West Germans, an emotional bond with West Berlin. Berlin became a symbol of Western resilience in the face of Soviet aggression. The instantly famous 'airlift' by which the Western powers—primarily the United States—provisioned the Western half of Germany's fallen capital was a logistical miracle and symbolic triumph. Between June 1948 and May 1949 more than 270,000 flights delivered more than 1.8 million tons of goods at a cost of more than $200 million. In Germany's Western zone, the drama helped reconcile the victors and vanquished of the recent War.[24]

II. The Parliamentary Council

As a steady stream of flights roared into Berlin—sometimes as frequently as one every few minutes—politicians in Frankfurt and Bonn were quietly creating a West German constitution. In the early months of 1948, British and American leaders ventilated the idea of forging a tight-knit group of Western democracies in which a new West German state would play an appropriate role. As the year progressed, the Western Allies—despite French

[23] 'Kölner Leitsätze' in Ossip K. Flechtheim (ed), *Dokumente zur parteipolitischen Entwicklung in Deutschland seit 1945*, vol 2 (Wendler & Co. 1969) 31.

[24] Accounts of the airlift are plentiful; many are panegyrical. For a sampling, see Richard Reeves, *Daring Young Men: The Heroism and Triumph of the Berlin Airlift, June 1948–May 1949* (Simon & Schuster 2010); Roger G. Miller, *To Save a City: The Berlin Airlift, 1948-1949* (Texas A&M University Press 2000).

reservations—increasingly pushed for the promulgation of a West German constitution and the creation of a West German state. In July, the minister presidents of the several Western German states met in Frankfurt to confer about the project. Some state premiers worried that the fact and nature of Allied occupation were incompatible with the promulgation of a free constitution and the founding of a new state. They also feared that enacting a West German 'constitution' would cement German division. They preferred, as noted, to call the document a 'Basic Law'. Ultimately, the Allies yielded on nomenclature, and the states forged ahead. In September 1948, the states' delegates convened as a 'Parliamentary Council' in Bonn, the birthplace of Beethoven on the banks of the Rhine.

Bonn was a fateful, if not an obvious choice. Frankfurt, home of the ill-starred 'Paul's Church' constitution of 1848/49, was the clear symbolic option. There would have been something poignantly appropriate in the Council's convening on the site and at the centennial of Germany's first attempt at a liberal democratic constitution. But the minister presidents were shy about signals of solemnity and shows of permanence. They wished to wear their symbolism lightly. Besides, Frankfurt was the headquarters of the Allied occupation, and the premiers wanted to give the councillors some distance from their overseers. Bonn was a compromise choice. Most minister presidents had proposed something closer to home—sites near their own state capitals. Reinhold Maier, minister president of Württemberg-Baden, for instance, suggested Karlsruhe, a sleepy city near the Southwest Franco-German border. In time, Karlsruhe would receive the consolation prize of housing the new country's preeminent courts of law.

Before the Parliamentary Council convened in Bonn at the beginning of September, a preliminary convention was held for thirteen days in August on the Herrenchiemsee Island in lower Bavaria. The Herrenchiemsee convention consisted of a handful of delegates—ostensibly civil servants rather than politicians—tasked by the minister presidents to issue nonbinding recommendations to the Parliamentary Council. The convention represented the last chance for the minister presidents to influence the formation of the Basic Law. In fact, that influence was enormous. Many of the Herrenchiemsee convention's most striking proposals—a strong federal government, a weak head of state, an aversion to plebiscites, a ban on amendments that would overturn the constitution's democratic premises—made their way, in one form or another, into the Basic Law itself.

In order to create a home for the Parliamentary Council, Bonn's Pedagogical Academy—a Spartan outfit near the centre of town—was hastily converted into a parliamentary building. The Academy's ceremonial hall was made into an assembly room. A classroom was placed at the disposal of each of the Council's partisan factions. No contemporary foresaw that the building would serve as a seat of government for five decades. On 1 September 1948, the ceremonial opening of the Council was staged nearby in the more imposing Koenig Museum. Here too adjustments had to be made. The hall that housed the festivities was ordinarily a showplace of taxidermy. Many animals were hurriedly removed, and others swiftly covered, before the show began.

As shows go, the Council's opening was a sombre one, featuring sumptuous music by Beethoven and Bach and sober speeches by Karl Arnold, minister president of the host state of North Rhine Westphalia, and Christian Stock, premier of Hesse and chair of the minister presidents' conference. Stock was at particular pains to deny that the Council's convening sealed Germany's division. 'If it is said that today in Bonn is completed the division of the German people', Stock intoned, 'then I hereby declare to the German people: We do not divide, we lead together and unite'.[25] Strikingly, no representative of the Allied powers

[25] In Theodor Eschenburg, *Geschichte der Bundesrepublik Deutschland. Jahre der Besatzung 1945–1949* (Deutsche Verlags-Anstalt 1983) 485.

spoke during the service—a fact that seemed to confirm Stock's proud remark that the assembly represented the first in postwar German history instigated, not by the Diktat of the conquerors, but 'according to agreements between the military governors and the minister presidents'.[26] In the months that followed, Allied influence over the Council's deliberations wavered between slender and non-existent. The architects of the West German constitution were West Germans, a fact that gave credence to later insistence that the Basic Law was not an *oktroyierte* constitution (one dictated 'from above'), but a product, in some sense, of regional self-determination.[27]

The Council's first order of business was to choose its leaders. Here, naturally, each party wished to place its own chieftains in important posts. The two principal factions—the CDU/CSU faction[28] and the SPD faction—each had twenty-seven delegates. This might have led to an impasse in choosing the Council's president, but for the SPD's calculation that the president would prove less influential than the chair of the Council's central committee (*Hauptausschuss*). The calculation was not unreasonable. For one thing, the president's role was, in an important sense, nonpartisan. He would govern the Council's deliberations but (it was supposed) have little impact on the substance of its proposals. What's more, in the German tradition parliamentary presidents rarely went on to become leading politicians. The presidency was viewed as a terminal position—all the more so in this case since the CDU/CSU presidential candidate, Konrad Adenauer, was seventy-two years old when the Council convened. The SPD gladly yielded the presidency to the Union parties and Adenauer in exchange for the central committee chairmanship, which went to Carlo Schmid. As it happened, the Social Democrats were spectacularly wrong about Adenauer—both about his influence within the Council and about his future political career.

Adenauer and Schmid were a study in contrasts. The septuagenarian Adenauer was a sober figure with a cold and craggy visage that somehow sharpened the impact of his frequent humorous interventions. He could deadpan to deadly effect. Adenauer was a forceful politician whose very lack of glamour and flashiness enhanced his aura of statesmanlike gravitas. Schmid was twenty years Adenauer's junior and as ebullient as Adenauer was composed. Schmid was a brilliant academic lawyer and a man of immense cultivation—urbane, witty, and warm, with a genial round face and unruly greying hair. He was a multi-faceted figure who combined multiple roles with grace and panache—a scholar, a politician, and a man of letters. (In idle hours he translated Macchiavelli and Baudelaire.) Schmid disarmed opponents and delighted the press with his immense personal charm and inexhaustible supply of sparkling one-liners. One journalist found Schmid's very vanity—which she termed 'childlike'—endearing.[29] Whereas Adenauer's outlook was firmly anchored in his native Rhineland, Schmid—who was born to a French mother in the south of France—took a resolutely cosmopolitan view of the world. Schmid saw himself, and was seen by others, as a European, a 'Westerner', a citizen of the world.

Both men exerted enormous influence within the Parliamentary Council. Adenauer was no mere figurehead; he played a pivotal role in the Council's debates. He also presided with dignity and skill, exuding a sobriety and poise that earned him international attention and led, in September 1949, to his becoming the first federal chancellor of the infant Federal Republic (as the new West German state was to be called). Schmid, for his part, insisted

[26] ibid.
[27] See Udo di Fabio, 'Vom Recht, Recht zu Sprechen' (2011) 61 *Aus Politik und Zeitgeschichte* 3, 4.
[28] That is, the Christian Democrats and their Bavarian sister party, the Christian Socialist Union.
[29] In Görtemaker (n 12) 63.

during the Council's earliest discussions that the delegates were not assembled to erect a new state. They were to craft an 'administrative statute' (*Verwaltungsstatut*)—an interim measure that would govern the Western territories until the Allies reached a more permanent resolution to the German Question. The real 'constitution', Schmid suggested, would be the Allies' occupation statute. But Schmid threw himself into the Council's labours, and those labours soon carried him away. The work enthralled him. The constitutionalist in Schmid could not resist the allure of constitution making. The student of Machiavelli and Plutarch could not refuse the chance to play Numa or Solon. With Schmid leading the way, the Council proceeded, over long laborious months, to frame a full-blown constitution, to charter a brand new state. Through the autumn, winter, and spring of 1948/49, Carlo Schmid stood at the centre of the West German political constellation. And he shone brilliantly. Schmid was West Germany's James Madison. Perhaps more than any other delegate to the Parliamentary Council, Schmid deserves to be called 'the father of the Basic Law'. But it was Adenauer who would become the father of the Federal Republic. After the Parliamentary Council had finished its work, Schmid's star would never shine so brightly again; Adenauer's star had only begun to rise.

On 8 May 1949—four years to the day after the unconditional surrender of the German Reich—the Parliamentary Council agreed, by a vote of 53:12, to adopt a Basic Law constituting the new Federal Republic of Germany. The negative votes came from the two communist delegates (who remained seated while their peers stood to vote); the two delegates from the German Party; the two representatives of the Centre Party; and six of the eight delegates from the Bavarian Christian Socialist Party (CSU). The CSU holdouts, who insisted that the proposal did too little to honour states' rights, were taken firmly to task by a fellow Bavarian, the fiery FDP delegate Thomas Dehler, one of three members of the Council's redaction committee. The Council had tried repeatedly, Dehler said, to meet Bavarian objections halfway. 'The thanks', he complained, 'was that again and again they said no'.[30] His choler rising, Dehler accused the CSU delegates of irresponsibility, duplicity, and even sabotage. Dehler's vehemence caused a stir, and Adenauer had to call for order. The Bavarians sought revenge four months later by opposing (unsuccessfully) Dehler's appointment as the Federal Republic's first minister of justice—a role in which we shall encounter Dehler and his trenchant tongue later in these pages.

Dehler may have been the most outspoken, but he was not the most influential FDP delegate in the Council. That honour went to Dehler's friend and party comrade Hermann Höpker-Aschoff, a courtly Berlin politician and former Prussian minister of finance. Höpker-Aschoff, who stood directly in front of Dehler for the Council's final vote, played such an able and important role in shaping the Basic Law's tax and finance provisions that he was later dubbed 'father of the financial constitution'.[31] The sixty-six-year-old Höpker-Aschoff had been a member of the Reichstag during the Weimar era and hoped for high political office in the neophyte Federal Republic. Instead he had to settle, firmly against his will, with becoming the first chief justice of the Federal Constitutional Court.

The new Court was one of the Basic Law's many innovations, most of which responded in one fashion or another to the collapse of the Weimar Republic or the crimes and calamities of the Third Reich. For example, the Basic Law circumscribed strictly the role of the federal president. The Weimar Reich president had been a powerful figure, popularly

30 In Eschenburg (n 25) 509.

31 See Frank Spieker, *Hermann Höpker-Aschoff. Vater der Finanzverfassung* (Duncker & Humblot 2004). For Höpker-Aschoff's own assessment of the provisions he helped craft, see Hermann Höpker-Aschoff, 'Das Finanz- und Steuersystem des Bonner Grundgesetzes' (1949) 75 *Archiv des öffentlichen Rechts* 306.

elected and able, in case of parliamentary impasse or impotence, to dissolve the Reichstag and govern by emergency decree. As the Weimar Reichstag descended into instability and chaos, the Reich president made increasing use of his power to issue emergency decrees. The president's office became something of a plebiscitary dictatorship. Worse still, it was Reich President Paul von Hindenburg who, in January 1933, summoned Adolf Hitler to the chancellorship and asked the Nazi leader to form a government. The president of the Federal Republic, by contrast, would be a largely ceremonial figure. He would be elected by Parliament, not by popular vote, and he would wield no independent power to dissolve Parliament.[32]

Parliament itself was considerably strengthened. The Council's delegates were tormented by the memory of Weimar's chaotic carousel of shifting coalitions and collapsing governments, of immobile parliaments repeatedly dissolved. Under the Basic Law, only Parliament had power to appoint or dismiss a government. And Parliament could not dissolve the government merely by denying its confidence to the head of government, the federal chancellor. It could do so only by voting affirmatively to appoint a new chancellor. It could topple the sitting government, in other words, only by appointing its successor. Such a 'constructive vote of no confidence' was one of the Basic Law's major innovations. The framers wished for strong, stable leadership, and they sought to provide it in an energetic federal government under the leadership of a vigorous federal chancellor. But the chancellor was to be appointed, not by plebiscite or by the president, but by Parliament. Here as elsewhere the Basic Law shunned plebiscitary elements.[33]

It also shunned centralized power. This was particularly so in the constitution's federalist provisions, which reserved crucial realms of governance and administration to the states. The Basic Law also divided Parliament into two chambers: the Bundestag, comprised of members chosen in countrywide elections; and the Bundesrat, comprised of representatives of the several states—members, indeed, of individual state governments.

The framers sought further to protect Parliament from the poisonous influences that had undermined parliamentary government in the twilight days of the Weimar Republic. It was conventional wisdom among the delegates that the Weimar Republic was a democracy incapable of defending itself against democracy's enemies. Drawing on the putative 'lessons of history', the delegates of the Parliamentary Council sought to establish a 'militant democracy' (*wehrhafte Demokratie*), a notion elaborated during the 1930s by the émigré political theorist Karl Loewenstein.[34]

The Basic Law was militant in various ways. It provided, first of all, that anyone who abused basic rights to attack the Basic Law's 'free democratic basic order' (*freiheitliche demokratische Grundordnung*) would forfeit those rights.[35] What's more, a political party

[32] See Konrad Adenauer, *Erinnerungen 1945–1953* (Deutsche Verlags-Anstalt 1965) 153 ('A general principle on our part was that we must draw the necessary conclusions from the errors of the Weimar Republic. The position of the future Federal President must not be endowed with the authority that the Reich President of the Weimar Republic possessed'.).

[33] On the influence of 'Weimar arguments' in the decision against plebiscites and referenda, see Andreas Wirsching, 'Konstruktion und Erosion: Weimarer Argumente gegen Volksbegehren und Volksentscheid' in *Weimars lange Schatten* (n 4) 335. As one member of the Parliamentary Council put it, 'The plebiscite, the popular initiative ... is, in the time of mass politics [*Vermassung*] and uprootedness, and in large-scale democracies, a blessing to every demagogue'. In ibid.

[34] See Karl Loewenstein, 'Militant Democracy and Fundamental Rights I' (1937) 31 *Am Pol Sc Rev* 417; Karl Loewenstein, 'Militant Democracy and Fundamental Rights II' (1937) 31 *Am Pol Sc Rev* 638; Karl Loewenstein, 'Legislative Control of Political Extremism in European Democracies I' (1938) 38 *Colum L Rev* 591; Karl Loewenstein, 'Legislative Control of Political Extremism in European Democracies II' (1938) 38 *Colum L Rev* 725; see also Max Lerner, *It Is Later Than You Think: The Need for a Militant Democracy* (Viking Press 1938).

[35] Article 18 GG (1949).

that sought to undermine or abolish the free democratic basic order was 'anti-constitutional' (*verfassungswidrig*) and could be banned.[36] These constitutional shields against radicalism were buttressed further by election law provisions that closed the doors of Parliament to any party that failed to secure at least 5 per cent of the votes in a federal election.[37]

'Free democratic basic order' was an imposing phrase—doomed, alas, to infelicitous translation—that appeared three times in the text of the Basic Law. In one of its most dramatic provisions, the Basic Law sought to render the Republic's democratic essence immutable and unamendable. Article 79 announced that the Basic Law's text could be amended by a two-thirds vote in both chambers of Parliament. But an amendment 'through which the division of the federation in states, the fundamental involvement of the states in legislation, or the principles laid down in Articles 1 and 20 are affected is impermissible'.[38] Article 1 declared that human dignity was sacrosanct and inviolable, and that the fundamental rights catalogued in Articles 2 through 19 were directly binding on all legislative, executive, and judicial authorities. Article 20 proclaimed that the new Federal Republic was 'a democratic and social federal state'. Article 79's 'eternity clause' shielded these provisions against future alteration.

The Basic Law's treatment of fundamental rights marked another profound break from the Weimar tradition. The Weimar constitution's fundamental rights provisions appeared only in the latter portions of the document, and Article 48 explicitly empowered the Reich president, in a national emergency, to set several of these provisions aside. Furthermore, Weimar-era constitutionalists sharply disputed whether the constitution's 'social rights' provisions were enforceable as positive law. Most public lawyers differentiated between these and classical, negative rights, insisting that only the latter were judicially enforceable; the former had a merely hortatory, programmatic character. Defensive rights were applied by the administrative courts, but the Weimar Republic had no exclusively constitutional tribunal.

The Basic Law changed all this. The Bonn framers placed the Basic Law's fundamental rights catalogue at the very beginning of the document and gave fundamental rights central significance. As noted, the constitution's first article proclaimed that '[t]he following fundamental rights bind legislature, executive, and judiciary as directly valid law'.[39]

The fundamental rights then listed included many of the classical, 'negative' rights of the liberal tradition. The catalogue protected citizens against state interference in religious belief and practice,[40] speech and expression,[41] research and teaching,[42] art and scholarship,[43] associations and assemblies,[44] domestic travel and postal communications,[45] homes and property,[46] choice of work and petitions to public authority.[47] The Basic Law also guaranteed to all people the right to life and bodily integrity;[48] equality before the law; and protection against discrimination on the basis of gender, ancestry, race, language, religious belief or political ideology.[49] In an elliptical phrase, each person was promised 'the right to the free development of his personality', but only so long as he didn't interfere with the rights of others, the constitutional order, or the moral law.[50]

[36] Article 21(2) GG.

[37] For the 1949 Bundestag elections, this so-called 'five-percent hurdle' or '*Sperrklausel*' applied separately within each state. From 1953 onwards, it applied generally nationwide.

[38] Article 79(3) GG. [39] Article 1(3) GG. [40] Article 4(1)–(2) GG. [41] Article 5(1) GG.

[42] Article 5(3) GG. [43] Article 5(3) GG. [44] Article 9(1) GG; Article 8(1) GG.

[45] Article 11(1) GG; Article 10 GG. [46] Articles 13–14 GG.

[47] Article 12(1) GG; Article 17 GG. [48] Article 2(2) GG. [49] Article 3(1), (3) GG.

[50] Article 2(1) GG.

The members of the Parliamentary Council agreed early in their deliberations to limit the Basic Law's rights catalogue to such classical rights. But this agreement was in some particulars abandoned.[51] The Catholic Church exerted a strong influence on the CDU/CSU delegates, who pushed successfully for provisions placing 'marriage and family' under the 'special protection of the state', and recognizing that the 'care and upbringing of children' was the 'natural right of parents'.[52] Social Democrat delegates, for their part, successfully included a provision guaranteeing equal legal and social status to 'non-marital children'.[53] And champions of gender equality, led by the stalwart Elisabeth Selbert (SPD), successfully inserted the concise phrase, 'Men and women have equal rights' (*Männer und Frauen sind gleichberechtigt*).[54]

How were these rights to be enforced? Article 19, the last article in the Basic Law's fundamental rights catalogue, declared that 'the path of the law' (*der Rechtsweg*) was open to any whose rights were violated by public authority.[55] This was a gesture toward the courts, and more specifically to a court created by the Basic Law itself—a court that would, in the fullness of time, do more than perhaps any other public institution to shape the legal, constitutional, political, and even the moral culture of the Federal Republic.

III. The Federal Constitutional Court

The Basic Law referred no fewer than twenty times to the Federal Constitutional Court (*Bundesverfassungsgericht*). The document called for the Court's creation, and endowed it with extraordinary powers. The Court would consider appeals from various Bundestag decisions dealing with electoral law and various Bundesrat decisions dealing with administrative matters.[56] It would decide on the accuracy of parliamentary allegations that the federal president or judges of the ordinary courts had engaged in unconstitutional behavior.[57] Within the Basic Law's militant democracy, the Court would determine to what extent persons who abused fundamental rights would forfeit those rights,[58] and it would decide whether (on appeal from Parliament or the federal government) to ban allegedly anti-constitutional political parties.[59] Most portentous of all, the Court would wield exclusive jurisdiction over all controversies arising between branches and levels of government, and all challenges to the constitutionality of state or federal law.[60] The Court would police the boundaries of constitutional authority between the chambers of Parliament; between Parliament and the federal government; between the federation and the states; and between individual states.

The Court's constitutional jurisdiction would be exclusive; and its jurisdiction would be exclusively constitutional. The Parliamentary Council embraced thus the so-called 'Austrian model' of judicial review (exemplified by the ill-starred Austrian Constitutional Court of the 1920s) over the 'American model' (exemplified by the United States Supreme Court). West Germany's Federal Constitutional Court would not be a supreme appellate court of general jurisdiction, but a specialized body empowered to decide *all* constitutional controversies and *only* constitutional controversies. If, in the course of a legal proceeding, an 'ordinary' court concluded that a law crucial to its decision was of doubtful constitutionality, the Basic Law directed that court to adjourn the proceedings, refer the constitutional

[51] Eschenburg (n 25) 502. [52] Article 6(1)–(2) GG. [53] Article 6(5) GG.
[54] Article 3(2) GG. [55] Article 19(4) GG. [56] Article 41(2); Article 84(4).
[57] Article 61(1) GG; Article 98(2) GG. [58] Article 18 GG. [59] Article 21(2) GG.
[60] Article 93 GG; Article 100 GG.

question to the Federal Constitutional Court, accept the Court's answer as binding, and resolve the case on that basis.[61]

The framers established or anticipated two features of the Court's jurisdiction that would profoundly shape the history of the Federal Republic. The first was the power of abstract judicial review (*Normenkontrolle*), by which the Court could—on appeal from a qualified parliamentary minority, the federal government, or a state government—review the constitutionality of federal legislation directly. This prerogative dispensed with an American-style 'case or controversy' requirement. In a very short time, it would give opposition parties a remarkable weapon in opposing majority initiatives. To this was later added the right of individual complaint, by which private citizens, having exhausted other legal remedies, could appeal to the Court to redress infringements of their constitutional rights. Such complaints, though subject to standing requirements, would give ordinary citizens access to the country's supreme constitutional tribunal. In years to come, West German citizens filed individual complaints by the tens of thousands. The procedure would become a central source of the Court's public trust and popular esteem.

Taken together, these provisions amounted to a staggering conferral of judicial authority. Rarely in world history had a court wielded such wide competence. If the Basic Law exuded the aura of a full-fledged constitution, the Constitutional Court bore all the trappings of an authorized interpreter and enforcer. The Court would give teeth, it was thought, to the Basic Law's most innovative provisions.

With minor qualifications, the military governors of the British, French, and American occupation zones approved the proposed Basic Law on 12 May 1949. The legislatures of the several states were swift to follow suit. Not then, nor ever since, was the document submitted for popular ratification. This ran counter to the wishes of the Allies, but reflected the framers' fear that a negative vote in Bavaria would thwart the entire project. The Basic Law officially entered into force just over two weeks after the Council concluded its deliberations. Forever after, it would carry the date of 23 May 1949, which entered history as the birthdate of the Federal Republic.

Almost immediately the Basic Law encountered sharp criticism from powerful figures within the legal elite. One of the most formidable early attacks came from Werner Weber, a disciple of Carl Schmitt and a prominent professor of public law. Like many other German academics around the time of the Basic Law's passage, Weber was returning to the academy after being dismissed at War's end for alleged complicity with the Nazi regime. Weber, who lost his academic chair in 1945 because of his membership in the Nazi party and in the SA, assumed a professorship in Göttingen during the summer of 1949. His inaugural lecture, delivered on 18 June 1949, was a blistering attack on the newborn Basic Law.

The central accusation of Weber's polemic was that the framers had drawn exaggerated lessons—and the wrong lessons—from the demise of the Weimar Republic. 'Like the ghostly apparition of one unhappily departed after a life of failure', Weber began, 'the Weimar constitution pervaded and oppressed the Bonn deliberations'.[62] Instead of crafting a charter drawing on the many virtues of the Weimar constitution—which Weber called the last gasp of the great European Liberal–Democratic constitutional movement of the nineteenth century[63]—the Council had devoted its energies to exorcising suppositious Weimar demons.[64] The result, Weber complained, was a document that gave a nominal show of humility—calling itself a mere 'Basic Law'—but that in substance paraded 'in the

[61] See Article 100(1) GG.
[62] Werner Weber, *Weimarer Verfassung und Bonner Grundgesetz* (Fleischer 1949).
[63] ibid 6–7.
[64] ibid 5.

character of a perfect constitution with all the attendant pretentions'.[65] What's more, the new charter lacked democratic legitimacy. It was jurists' law (*Juristengesetz*), not a people's law (*Volksgesetz*)—and even jurists would have trouble deciphering it.[66]

Weber's democratic critique ran deep. Seldom, he said, had a Western constitution arisen with so little public scrutiny. 'The great democratic power of public opinion remained unused and unaddressed'.[67] Weber laughed at the preamble's invocation of 'the German people' and its 'constitution-giving authority', a phrase to which Weber appended a parenthetical '(sic!)'.[68] 'The People', Weber sneered, had nothing to do with creating the Basic Law and would have little to do under the Basic Law, which provided 'no institutional possibility for the expression of public opinion as a whole, unmediated, in its relationship, for instance, to a statesman or to a given constitutional-political fact'.[69] There was, under the Basic Law, only one thing left for the people to do: elect the Bundestag.[70] The real power in the new state lay elsewhere. For the rest, the people's influence was 'fully and without exception mediated by political parties'.[71] Parties would, 'in all organs and functions, monopolise the power of political decisions'.[72]

Worse still, political parties would themselves be mediated and constrained by courts. Weber railed at the 'unheard-of proliferation of elements of the judicial state [*justizstaatlicher Elemente*]'.[73] Already worried at the disempowerment of the president, Weber was appalled that the power of suspending fundamental rights had been transferred to the Constitutional Court.[74] That Court, Weber feared, would be well-nigh omnipotent. 'There is hardly an essential process of constitutional life', he marveled, 'that cannot become embroiled in a proceeding before the Federal Constitutional Court'.[75] By creating such a Court, the Basic Law's framers had thrown open the gates to 'the double danger of a juridification of politics and the politicisation of justice'—and this 'despite various warnings and terrifying precedents from earlier times'.[76]

It was not entirely clear which 'precedents' Weber had in mind. But the source of the 'warnings' was obvious. In connection with constitutional judicial review, the terms 'juridification of politics' and 'politicisation of justice' stemmed from the Weimar pen of Weber's mentor, Carl Schmitt.[77] And it was in a thoroughly Schmittian vein that Weber closed his lecture. 'The perfected judicial-state ideal system of the Bonn Basic Law' had naively ignored the relentless and implacable realities of the political. It was an 'old truth', Weber noted, 'that responsible action and political daring, not judicial pronouncements, determine the destiny of nations'.[78] No judicial fiat could provide 'work, bread, clothing, housing', and protection against 'the enemy'. The Basic Law's ultimate failure was that it viewed life 'in reaction to the immediate past', and with insufficient attention to 'the future of the German people'.[79]

It was a ferocious philippic. Carl Schmitt himself, who in his diary had mocked the Basic Law's framers as 'the poor little men in Bonn' and said that he read the finished constitution 'with the amusement of an all-knowing old man', expressed astonishment at 'the vehemence with which Werner Weber criticises the Opus'.[80] More moderate voices in the legal academy answered Weber by suggesting that public law scholars, rather than polemicize against the new state, should seek creative solutions to its problems.[81] Yet moderates, too,

[65] ibid 10–11. [66] ibid 12. [67] ibid 9. [68] ibid 11. [69] ibid 16. [70] ibid.
[71] ibid 17. [72] ibid [73] ibid 25. [74] ibid 23–24. [75] ibid 25. [76] ibid 26.

[77] Carl Schmitt, 'Das Reichsgericht als Hüter der Verfassung' in Carl Schmitt, *Verfassungsrechtliche Aufsätze aus den Jahren 1924–1954. Materialien zu einer Verfassungslehre* (Duncker & Humblot 1958) 63.

[78] Weber (n 62) 28. [79] Weber (n 62) 29.

[80] In Michael Stolleis, *Geschichte des öffentlichen Rechts in Deutschland 1945–1949*, vol 4 (CH Beck 2012) 131.

[81] See, e.g., Georg Strickrodt, 'Die notwendige Position der Staatsrechtslehre' (1952) 7 *JuristenZeitung* 385.

predicted an enduring transfer of power to the judiciary.[82] But neither Weber nor his detractors appreciated how enduring and extensive that transfer would prove to be. Whether fearful or favourable, critical or commendatory, all contemporary assessments of the Basic Law's durability and of the Court's future impact proved understated. No one foresaw what was coming, or how swiftly it would come.

[82] See, e.g., Hans-Peter Ipsen, *Über das Grundgesetz. Rede gehalten anläßlich des neuen Amtsjahres des Rektors der Universität Hamburg am 17. November 1949* (Selbstverlag der Universität 1949).

Introduction

There is hardly a single essential development in our constitutional life that cannot become embroiled in a proceeding before the Federal Constitutional Court.... Thus are the gates thrown open, despite various warnings and terrifying precedents from earlier times, to the double danger of a juridification of politics and a politicization of justice.

—Werner Weber (1949)[1]

Tell me your attitude toward constitutional justice, and I will tell you your conception of the constitution itself.

—Werner Kägi (1945)[2]

In the Beginning

Six years after the self-immolation of Adolf Hitler and the total surrender of his millennial Reich, there was established in the sleepy south-west city of Karlsruhe one of the most powerful and influential fundamental rights tribunals in the history of the world. No one present at the creation of the German Federal Constitutional Court saw it coming. Among the scores of constitutional tribunals formed across the globe since the end of the Second World War, the German Court has been, by many estimates, the most successful. By nearly all accounts, it has become the most powerful.[3] Now in its seventh decade, the self-proclaimed 'Guardian of the Constitution' (*Hüter der Verfassung*) sits secure at the centre of the German legal and political order. At home and abroad, the Court is regarded as a bastion of fundamental rights and an emblem of liberal–democratic stability. Few crucial questions in German politics escape the Court's review. Tocqueville observed long ago that in Jacksonian America, most political questions sooner or later became judicial questions.[4] In contemporary Germany, nearly all major political questions eventually become constitutional questions.

Frequently they begin as constitutional questions. Political controversy in Germany often assumes a constitutional colouring from the outset. Parliamentary debate in Berlin is conducted with an eye towards constitutional review in Karlsruhe. The terms of policy discussion are shaped by canons of Constitutional Court jurisprudence. Across the gamut of state activity, the Court supplies the standards of what is permissible and what is proscribed, of what is excessive and what is required.[5]

These standards are derived, ostensibly, from the constitution itself. But they increasingly draw upon the thick tradition of the Court's own precedents—precedents cited not by name or date, but merely by page number and volume in the Court's official reports—a practice that seems to strip precedents of historical context and endow them with an aura

[1] Werner Weber, *Weimarer Verfassung und Bonner Grundgesetz* (Fleischer 1949) 25–26.

[2] Werner Kägi, *Die Verfassung als rechtliche Grundordnung des Staates* (Polygraph 1945) 147.

[3] In every comparative sense, writes one leading scholar, the Federal Constitutional Court is 'an extraordinarily powerful and impactful court'. Matthias Jestaedt, 'Phänomen Bundesverfassungsgericht. Was das Gericht zu dem macht, was es ist' in Matthias Jestaedt and others, *Das Entgrenzte Gericht. Eine kritische Bilanz nach sechzig Jarhen Bundesverfassungsgericht* (Surkamp 2011) 149.

[4] Alexis de Tocqueville, *De la démocratie en Amérique*, vol 1 (first published 1835, Laffont 1986) 258.

[5] On the Court's role as a 'standard-setter', see Oliver Lepsius, 'Die maβstabsetzende Gewalt' in *Das Entgrenzte Gericht* (n 3) 159.

of timeless principle—chapter and verse in an ancient code. The Court has been elaborating constitutional norms, both textual and doctrinal, for more than sixty years. In the process, it has become a judicial lawmaker par excellence. In most domains of domestic governance, and even in spheres of foreign policy, the roads of German rule lead one and all to Karlsruhe.

Naturally, the Court's accumulation and exercise of political power has provoked significant criticism; but such criticism has been consistently overwhelmed by remarkable levels of popular esteem and public support. The Constitutional Court has always been and remains one of the most trusted institutions of the German state. Popular acceptance of the Court has consistently surpassed—often it has far surpassed—approval of the Federal Republic's other constitutional organs: the two Houses of Parliament (Bundestag and Bundesrat), the federal government, and the federal president.[6] Support for the Court has even increased at times when the Court's immense power might have seemed most obvious and most ominous.

Such was the case in autumn 2012. Americans had spent months speculating breathlessly how their own Supreme Court might vote on a constitutional challenge to President Barack Obama's historic healthcare reform; but the rest of the world turned its attention to a different judge in a different court in a different case. That judge was Andreas Voβkuhle; that court the German Federal Constitutional Court; and that case a constitutional challenge to the Bundestag's approval of the European Stability Mechanism (ESM).

Voβkuhle professes not to love the limelight. A tall man with a rounded face, complacent eyes, bookish glasses, and a benevolent smile, Voβkuhle has the look of a man content to spend his days in patient, pleasant academic plodding. He is indeed an academic, but his academic career was meteoric rather than plodding. He finished his second dissertation at age thirty-four; became a full professor at Freiburg University the following year; and in 2008, at age forty-four, became the youngest rector in the university's history. His tenure as rector, however, was brief. In May 2008, the Bundesrat made Voβkuhle vice president of the Constitutional Court, with the understanding that he would later become—as in March 2010 he did become—the youngest president, or chief justice, in the Court's history. In February 2012, Angela Merkel, the chancellor, offered Voβkuhle the position of federal president after the embattled Christian Wulff resigned. Voβkuhle mulled the matter over, but ultimately declined. As chief justice of the Constitutional Court, quipped the monthly magazine *Cicero*, Voβkuhle was already the country's 'real head of state'.[7]

In the months following his refusal to become federal president, Voβkuhle became the most visible chief justice in the Court's history. He assumed centre stage, not only of German politics, but of European politics—even of global politics. What was at stake in the constitutional challenge to the ESM was nothing less than the survival of the Euro on one hand and, on the other, the identity and self-understanding of postwar Germany. Voβkuhle had gone on record insisting that European integration must not come at the cost of democracy and the rule of law. German citizens must not 'wake up one morning to discover that those whom they have elected have nothing left to decide'.[8] There were limits, Voβkuhle stressed, to how much 'Europe' Germany's constitution could allow. If Germany wanted

[6] On how the Court crowned itself a 'constitutional organ', and not merely a court of law, see Chapter 1.

[7] The cover of *Cicero*'s April 2012 issue read, 'Andreas Voβkuhle: The President of the Federal Constitutional Court is Germany's true head of state'.

[8] In 'Der Kapitän' *Der Spiegel* (Hamburg, 10 September 2012) 28. Because this book refers to hundreds of newspaper articles, the city of publication will not be given in each instance. The sites of publication for those papers cited most frequently are as follows: *Der Spiegel* (Hamburg); *Die Zeit* (Hamburg); *Frankfurter Allgemeine Zeitung* (Frankfurt); *Süddeutsche Zeitung* (Munich); *Die Welt* (Hamburg and Berlin); *Frankfurter Rundschau* (Frankfurt); *Christ und Welt* (Bonn); *Bulletin des Presse- und Informationsdienstes der Bundesregierung* (Bonn); *Stuttgarter Zeitung* (Stuttgart), and *Badische neuste Nachrichten* (Karlsruhe).

to transgress those limits, it would need 'to give itself a new constitution', presumably by plebiscite.[9]

The ESM pressed the limits of European integration and pressed them hard. Germany's leading weekly, *Der Spiegel*, reported that if the plans went forward, Voßkuhle might be remembered 'not only [as] the youngest president of the Constitutional Court of a sovereign Federal Republic, but also [as] the last'.[10] The ESM case seemed to pit Germany's sovereignty against Europe's survival. And the power to decide that question lay in the hands of eight German judges.

An astonishing share of the German citizenry was pleased that this was so. As the Court considered the constitutional challenge to the proposed ESM, a banner article in the *Frankfurter Allgemeine Zeitung*, Germany's leading daily, praised the Court as the nation's 'bulwark'.[11] 'In a time of great insecurity', the article summarized, 'the Federal Constitutional Court is becoming an institution that citizens trust to preserve the political order and national interests. The Court's prestige [*Ansehen*] is greater than ever before'.[12] Contemporaneous surveys suggested that 82 per cent of Germans were grateful the Court could annul political decisions. A vast majority thought the Court's power was growing, but only 5 per cent thought it too powerful.[13] While popular trust in many state institutions was waning, trust in the Court was growing.[14] At the same time, German politicians continued to worry—as they have always worried—that the Court might dash some treasured legislative or diplomatic achievement. And, at that troubled hour, politicians round the globe feared that the Court might someday explode the Euro.[15] That fear, though the Court later approved the ESM without qualification,[16] lingers on.

This is a book about how the German Constitutional Court came to possess such extraordinary power, and why so many Germans are grateful that it does. How the Court became so mighty and why so few resist its strength are questions, I contend, that can be answered only by telling the Court's story and by situating that story in a broader historical frame. What follows, then, is a narrative history—one meant to be read from beginning to end. It is also a public history—a political and societal, rather than a doctrinal and jurisprudential, history of the Constitutional Court. It is the story of the Court's interventions in postwar (West) German history, and of the (West) German public's engagement with their Constitutional Court. This public orientation is reflected both in the book's periodization and in the sources on which it relies. Although the book is naturally organized around the Court's major decisions (the only way in which a court of law can intervene in history), it also examines popular and political reactions to those decisions, drawing heavily on newspaper accounts of major judgments and material from the archives of individual

[9] ibid. [10] ibid.

[11] 'Bundesverfassungsgericht. Das Bollwerk' *Frankfurter Allgemeine Zeitung* (hereinafter FAZ) (Frankfurt, 21 August 2012) http://www.faz.net/aktuell/politik/inland/bundesverfassungsgericht-das-bollwerk-11863396.html> accessed 21 September 2013.

[12] ibid. [13] ibid.

[14] That growth has continued. In May 2014, when Germany celebrated the sixty-fifth anniversary of its postwar constitution, an Allensbach survey found that 92 per cent of Germans trusted their constitution and 86 per cent trusted their Constitutional Court. 'Deutsche schätzen das Grundgesetz' *FAZ* (Frankfurt, 21 May 2014) <http://www.faz.net/aktuell/politik/inland/umfrage-zum-verfassungstag-deutsche-schaetzen-das-grundgesetz-12950505.html> accessed 25 May 2014.

[15] See Quentin Peel, 'Germany's Judgment Day' *Financial Times* (London, 7 August 2012) http://www.ft.com/cms/s/0/78df7420-dfa5-11e1-9bb7-00144feab49a.html>. The Court's judgment upholding the ESM gave rise to a global sigh of relief, as well as deep misgivings. See 'Germany and the Euro: Yes, but ...' *The Economist* (London, 15 September 2012); Bruce Ackerman and Miguel Maduro, 'Broken Bond' *Foreign Policy* (Washington, 17 September 2012).

[16] See 135 BVerfGE 317 (2014).

politicians and judges. The account also draws on the rich scholarly literature devoted, since the Court's founding, to evaluating its decisions. But it adduces scholarly commentary less as an exercise in intellectual history than as a means of capturing historical mood.

In a similar way, the narrative's divisions are linked to major events in the political history of the Federal Republic, rather than to events internal to the Court itself.[17] The one partial exception comes in the book's first chapter. Chapter 1 begins with the Court's founding in 1951 and ends in 1959, just after the Court's landmark early fundamental rights decisions and coinciding with Gebhard Müller's appointment as chief justice. But 1959 also marked the beginning of the declining fortunes of Chancellor Konrad Adenauer after a decade of steadily mounting strength. Chapter 2 covers the period from 1959 until the advent in 1969 of postwar West Germany's first socialist-led government under Chancellor Willy Brandt. Chapter 3 chronicles the Court's activity during the 'Social–Liberal era', from 1969 to 1982. Chapter 4 takes the story from the Christian Democrats' return to power in 1982 through German reunification in 1990. Chapter 5 deals with the 'reunification decade', closing with the global convulsions that attended the terrorist attacks on the United States in September 2001, and with the Court's golden jubilee, celebrated that same month.

Each chapter highlights one of the book's major themes—one explanation of the public's acceptance of the Court's power. Those themes are liberalization, democratization, historicization, integration, and mediation. Obviously, it would be worse than artificial to relegate each theme to a given time period. Naturally, each theme recurs throughout the book. But the comparative prominence of each theme has shifted across time, and the narrative seeks to underscore those shifts. Unifying all the themes is a central thesis of this book, which is that the Court's remarkable rise and the public's remarkable trust in the Court have been the product, not of some persistent national sequacity and deference to authority, but of the Court's perceived role in redeeming the country from its authoritarian past and restoring it to the community of liberal democracies. Accounting for the Court's popularity and power requires more than a nuanced historical account of the Court's meteoric rise in the postwar Federal Republic. It requires situating the Court within the broader sweep of German history. The Court's history is the product of its pre-history; its success the product of Germany's failures.

I. A Special Path?

The critical dates are 1848 and 1933.[18] The former marked a pivotal crossroads in what some historians have controversially called the German *Sonderweg*, or special path. In its most pointed form, the *Sonderweg* thesis suggested that while other Western nations—primarily Great Britain and France—had progressed naturally down an evolutionary course

[17] The common American practice of apportioning Supreme Court history and naming its eras for the tenures of chief justices has even less to recommend it in the German context than in the American. For one thing, the tenure of the chief justice is often brief. Under the current rules, no justice may serve longer than twelve years or beyond her sixty-eighth birthday. Only two chief justices—Gebhard Müller (1959–71) and Ernst Benda (1971–83)—have served full twelve-year terms. What's more, although the chief justice plays an important administrative and symbolic role, he or she (there has been one she) casts only one vote and in only one of the Court's two Senates, or chambers. The chief justice has little additional influence within her own Senate and none at all in the other. With rare exceptions, the Court's chief justices have not been its most influential jurists. The great justices have often been powerful academics, unfit for administrative duties. Assigning epochal significance to the appointment of a new chief justice would convey a false impression.

[18] Revisionist historians have argued powerfully for a shift in emphasis in the chronology of German historiography from 1933, when the Nazis seized power, to 1941, when they launched the Final Solution. See Helmut Walser Smith, *The Continuities of German History: Nation, Religion, and Race Across the Long Nineteenth Century* (Cambridge University Press 2008).

culminating in liberal–democratic modernity, Germany had gone its own way, describing a disastrous career of illiberalism, authoritarianism, military aggression, dictatorship, total warfare, and the unimaginable horrors of the Holocaust. Germany was the 'belated nation' (*Verspätete Nation*), united not in 1848 as the culmination of a liberal–national constitutional moment, but in 1871 under the imperial auspices of Prussia and its indomitable prime minister, Prince Otto von Bismarck. Bismarck was utterly contemptuous of the failed effort to frame a united German constitution in 1848. He scorned the National Assembly that met in Frankfurt to craft a united German constitution as a utopian 'parliament of professors' (*Professorenparlament*)—a convocation of idealists and dreamers who mired themselves in abstract orations on first principles and frittered the momentous occasion in endless disquisitions on the fine points of fundamental rights. It was an assembly given to words, not deeds. What Germany needed, and what Bismarck himself was only too willing to supply, was a forceful statesman who could seize the historical moment with the power of decisive action.

A. The turning point at which Germany failed to turn?

No such action was forthcoming in 1848/49. When the National Assembly's emissaries presented the draft constitution to Prussia's King Frederick William IV with a request that he become the united Reich's first Kaiser, that monarch declined the offer with princely disdain. Frederick William was already King of Prussia by divine right and God's grace. He neither needed nor desired the commission of a constituent body that would make him Kaiser of all Germany. In April 1849, the National Assembly issued its constitution all the same, but it was a dead letter—merely declaratory and wholly ignored. In retrospect, the British historian, AJP Taylor, famously (and controversially) scored 1848/49 as a time when 'German history reached its turning-point and failed to turn'.[19] The first great effort to fashion for Germany a liberal, democratic constitution with a comprehensive catalogue of binding, judicially enforceable rights had failed spectacularly. Seven decades would transpire before the next attempt.

Meanwhile, bitterly disappointed liberals consoled themselves as they could. Until the debacle of 1848/49, German liberals—who stemmed, by and large, from the professional and mercantile middle class—had championed liberal values and political participation as an inseparable unity, a coherent and indivisible ideological whole. The rule of law and popular self-government were, for most liberals, intimate partners in a common project. After 1848, they parted paths. Liberals who had hoped for a constitutional democracy settled for a state based on the rule of law—for a *Rechtsstaat*, in the evocative, untranslatable German phrase. In the latter half of the nineteenth century, middle-class liberals gave up their quest for political participation and self-government against the princes, and sought instead to secure and protect fundamental liberties in the private and economic sphere. Denied a liberal democracy, the middle classes pursued a liberal *Rechtsstaat*. With time, they made a virtue of necessity—savouring the stability of the rule of law while revelling in the cultural and diplomatic triumphs of the German Reich. Having settled for the rule of law when they could not also have democracy, many Germans came to prefer the rule of law over democracy. The preference had a long shelf life. Its influence endures today. After the First World War, the preference deepened.

[19] AJP Taylor, *The Course of German History: A Survey of the Development of German History since 1815*, 2nd edn (Methuen 1961) 69.

B. Descent into barbarism

Germany's first national experiment with democracy was a tempestuous one. The Weimar Republic was rocked by a carousel procession of governments rising and falling, of coalitions forming and collapsing, of parliamentary deadlock and dissolution. The first German Republic had twenty governments in fourteen years. Carl Schmitt, one of Germany's most powerful legal minds, delivered a devastating critique of parliamentary democracy in his 1923 essay, *Die geistesgeschichtliche Lage des heutigen Parlamentarismus* ('The Intellectual–Historical Position of Contemporary Parliamentarism'). Modern parliamentary governance, in Schmitt's view, was a sham, its pretence of representative legitimacy and rational deliberation a farce. Members of Parliament were operatives of political parties, not representatives of the people. Modern parliaments did not deliberate and decide; they postured in public and made deals behind closed doors. They were incapable of the vigorous decision-making essential to a stable modern state.

To many German minds, the ten years that followed the publication of Schmitt's strictures gave colour to his critique. Narrow partisan bickering and endless regime changes fostered deadlock and disruption against a backdrop of world economic depression. The early 1930s were a grim and fearful time. With each passing election the electoral fortunes and parliamentary presence of Adolf Hitler's National Socialists grew dramatically. In July 1932, the Nazi vote rose to 37.3 per cent—nearly sixteen percentage points better than the second-place Social Democrats (SPD). In January 1933, Reich President Paul von Hindenburg asked Hitler to assume the chancellorship and form a government.

Immediately the new regime set about dismantling the *Rechtsstaat* and replacing it with its opposite. The Third Reich was an Un-*rechtsstaat*—an arbitrary dictatorship marked by the wholesale suspension of fundamental rights, the systematic recasting of German law in the image of Nazi ideology, the methodical persecution of ethnic minorities and political opponents, and, in the regime's final years, aggressive warfare and industrial genocide on a staggering, sickening scale. The consequences for Germany—politically, economically, socially, morally—were devastating.

In the aftermath of total war and unconditional surrender, most Germans were more focused on their own immediate needs and their own recent sufferings than on the victims of their fallen government or their own complicity in that victimization. The German people on the morning after the apocalypse were more interested in stability than introspection, in search more of repose than of redemption. As the political scientist Wilhelm Hennis, then a soldier returning from active duty, later recalled: 'for working through the past [*Vergangenheitsbewältigung*] my generation had neither inclination nor time'.[20] Having endured and sometimes embraced a season of total politicization, Germans were weary of politics and wary of politicians, mistrustful of ideologies and antagonistic toward political parties. Most were more eager for the return of basic legal protections and elementary economic survival than for a revival of democratic participation and popular self-government. Their ancestors pursued the *Rechtsstaat* on its own when they couldn't combine it with democracy. Now many longed for the *Rechtsstaat*'s return, but cared little for democracy or were even hostile to it. This was a posture that endured beyond the founding of the Federal Republic.

[20] Wilhelm Hennis, *Politikwissenschaftliche Abhandlungen: Regieren im modernen Staat*, vol 1 (Mohr Siebeck 1999) 398.

II. Liberalization, Democratization, and Legitimacy

The political culture in which the Federal Constitutional Court took root, then, was a pre-democratic culture marked by a long German tradition that prized the rule of law over democracy.[21] Independent of that tradition, the rule of law was regarded by many in the Federal Republic's founding generation as a necessary precondition of democracy, not a limitation or constraint. This perspective redounded to the benefit of a neophyte Constitutional Court that many perceived as a harbinger and defender of democracy, not as a threat to it.[22] Modern students of the Court have hailed it as a 'midwife of the second German democracy'.[23] 'It became', one commentator has written, 'the strict but benevolent guardian of an immature democracy that didn't quite trust itself'.[24]

A. Guardians of the constitution

The German word for 'guardian' in the statement just quoted is *Vormund*, a term used to designate the legal guardian of a minor child. But there is another German term, central to the Court's history and self-understanding, which also translates as 'guardian': *Hüter*. In a 1952 manifesto the Court proclaimed itself *Hüter der Verfassung*—Guardian of the constitution.

It was a redolent phrase. During the 1920s and 1930s German legal theorists fiercely debated who should stand as 'guardian' of the embattled Weimar constitution. 'Who will protect the constitution against its enemies?' was the pointed query of Carl Schmitt. Schmitt's answer was the Reich President, along with the public administration—the civil service.[25] Hans Kelsen, Schmitt's most formidable antagonist, championed constitutional judicial review.[26] In the short run, Schmitt seemed to prevail. In the long run, the cup went to Kelsen. As parliamentary governance in the Weimar Republic dwindled into chaos and instability, the Reich President—who enjoyed the prestige of plebiscitary appointment—ruled the land by emergency decree. Ultimately, as noted, it was Reich President von Hindenburg who invited the Nazis to assume power. This was the fateful precedent that led the Federal Republic's founders to eschew plebiscites and weaken the presidency. On the other hand, the framers erected a powerful Constitutional Court to police the constitution's boundaries and enforce its basic rights.

B. Fundamental rights as a precondition for democracy

No one foresaw quite how powerful that court would become. In the Court's early years, its power was accepted, and even applauded, largely because the Court's role in enforcing fundamental rights and guarding constitutional structures was seen as a precondition to the

[21] See Christoph Schönberger, 'Anmerkungen zu Karlsruhe' in *Das Entgrenzte Gericht* (n 3) 43. See also Brun-Otto Bryde, 'Die Rolle der Verfassungsgerichtsbarkeit in Umbruchsituationen' in Joachim Jens Hesse and others (eds), *Verfassungsrecht und Verfassungspolitik in Umbruchsituationen* (Nomos 1999) 197–210; Ludger Helms, 'Ursprünge und Wandlungen der Verfassungsgerichtsbarkeit in den konsolidierten liberalen Demokratien' (2006) 53 *Zeitschrift für Politik* 50, 68.

[22] I am indebted to Dieter Grimm for this insight. [23] Schönberger (n 21) 27. [24] ibid 43.

[25] Carl Schmitt, *Hüter der Verfassung* (JCB Mohr 1931).

[26] See Hans Kelsen, 'Wesen und Entwicklung der Staatsgerichtsbarkeit' (1929) 5 *Veröffentlichung der Vereinigung der Deutschen Staatsrechtslehrer* 30; Hans Kelsen, *Wer soll der Hüter der Verfassung sein?* (JCB Mohr 1931). Kelsen had some practical experience with judicial review as a justice of the short-lived Austrian Constitutional Court.

successful establishment of a second German democracy. Liberalization, many thought, must precede democratization.

In its first two decades, the Court contributed to both. It vindicated the legitimacy of political and parliamentary opposition,[27] invigorated the position of political parties, and fortified rights of speech and expression. It exalted fundamental rights as the crown jewels of the constitutional order. In doing so, it reconceived such rights as objective values arrayed in a systematic order that pervaded all fields and fora of positive law.[28] The Court constitutionalized—and thereby liberalized—all of West German law. In the process, it helped liberalize West German society.[29] And, by helping to fulfil preconditions of democratic governance, the Court hastened the democratization of West German political culture.

The Court benefited institutionally from having been planted in pre-democratic soil.[30] In the jaded, suspicious environment of postwar West German politics, and in the legitimacy vacuum left by the discredited and departed elites of the fallen regime, the Karlsruhe justices emerged as neutral, non-partisan experts—stable, trustworthy guardians of a new and fledgling order.[31] One of the Court's greatest early contributions was to help democratize the culture that surrounded it. Ironically, it often did so by constraining democratically elected actors. But the net gain to democracy was positive, and the net gain to liberal values and fundamental rights was enormous.

Such, at least, is the claim advanced in the first two chapters of this book. It was not a linear progression, nor an uninterrupted symphony of sweetness and light. The Court moved forward by fits and starts. Some landmark early judgments struck contemporary observers and later critics as illiberal and undemocratic. The Court banned two political parties as unconstitutional, even as it vindicated the constitutional status of other parties. The Court outlawed regional plebiscites, even as it freed the channels of democratic discourse in the public sphere. Even so, by 1970, it was possible for a leading public law scholar to note that '[t]he jurisprudence of the Federal Constitutional Court has, as we all know, fostered the development of a true democratic understanding in the Federal Republic—not least in the sphere of public opinion. There are decisions of the Federal Constitutional Court that, as harbingers of the free *Rechtsstaat* and of democracy, were years ahead of public opinion'.[32]

C. Democracy and the rule of law

In its early jurisprudence, then, the Court bolstered both democracy and the rule of law—and it helped reunite the two in the West German mind. What's more, the democracy- and liberty-enforcing strain in the Court's decisions helped counterbalance to the constitution's (and the Court's own) protection of the democracy principle through repressive

[27] See Klaus Stüwe, 'Der "Gang nach Karlsruhe". Die Opposition im Bundestag als Antragstellerin vor dem Bundesverfassungsgericht' (1997) 28 *Zeitschrift für Parlamentsfragen* 545; Klaus Stüwe, *Die Opposition im Bundestag und das Bundesverfassungsgericht* (Nomos 1997).

[28] See 6 BVerfGE 32 (1957); 7 BVerfGE 198 (1958).

[29] See Ulrich Herbert, 'Integration der jungen Republik durch Verfassungsrecht?' in Thomas Henne and Arne Riedlinger (eds), *Das Lüth-Urteil aus (rechts-) historischer Sicht: die Konflikte um Veit Harlan und die Grundrechtsjudikatur des Bundesverfassungsgerichts* (Berliner Wissenschafts-Verlag 2005) 85; Brun-Otto Bryde, 'Der Beitrag des Bundesverfassungsgerichts zur Demokratisierung der Bundesrepublik' in Robert Christian van Ooyen and Martin H.W. Möllers (eds), *Das Bundesverfassungsgericht im politischen System* (Verlag für Sozialwissenschaften 2006) 321–31; Schönberger (n 21) 27.

[30] Schönberger (n 21) 43. [31] ibid 44.

[32] Wilhelm Karl Geck, 'Diskussionsbemerkung' in 29 Veröffentlichung der Vereinigung der Deutschen Staatsrechtslehrer 113 (1970).

means—means that were in any case relativized and softened by the Court's steady application of the proportionality principle.[33]

If, in all of this, the Court was ahead of its time, it also operated largely outside of the public's awareness. The Court won great popular acclaim for certain celebrated decisions, such as its invalidation (in 1961) of Chancellor Konrad Adenauer's plans for a second public television network—a network manifestly calculated to serve the sitting government.[34] But for the most part, and particularly with regard to the landmark doctrinal decisions that did so much to expand the Court's jurisdiction and enhance the scope of its review, the public paid little attention to the Court during the long era of Christian Democrat rule from the founding of the Federal Republic to the accession of Willy Brandt's Social–Liberal coalition in 1969.[35] The Court's early accumulation and exercise of power was not so much embraced by the public as it was opaque to the public.

As late as 1974, no respondent to an Allensbach survey thought the Court exerted too great an influence on West German politics.[36] This might suggest more ignorance than endorsement; but what the public knew of the Court it largely liked. And what it did not know, it would learn to appreciate later on. The Court's early work—its reinforcement of democratic values and institutions, its explicit anti-Nazism, its receptivity to citizen efforts to vindicate fundamental rights—provided a reference from which appreciative observers could defend the Court when it later became embattled.

III. History and Legitimacy

With the advent of a reformist government in 1969, the Court was placed, for the first time, in a conservative posture. Hitherto the Court had been ahead of the times; the times now caught up. Willy Brandt, the new chancellor, came to power on the strength of the slimmest of parliamentary majorities. But he spoke as one commissioned by 'We the People' to proclaim a new dispensation. In his inaugural message to Parliament, Brandt famously vowed to 'dare more democracy'. The Federal Republic, he announced, had barely begun its experiment in democratic self-government. Hitherto the country had thought as a child and spoken as a child. It had now arrived at democratic maturity. It was time to put away childish things. The Federal Republic was emerging from its adolescence. Brandt's ambitions for its adulthood knew no bounds.

Brandt's proclamation of a new constitutional moment put the Constitutional Court in an awkward position. The promise of a newer, bolder, more daring democracy endangered the Court's self-understanding: Mature individuals don't need guardians.

In the years that followed, the government's commitment to sweeping reform and the Court's devotion to objective constitutional values clashed with mounting frequency and intensity. The Court demolished or defanged prized initiatives, and government officials cried foul. The resultant crisis was the severest the Court had faced since its founding. Various factors—some external to the Court, some not—allowed the Court to weather the storm. For one thing, the reformist energy of the Social–Liberal government steadily declined—particularly after the world oil crisis of 1973, which sobered reformers' sanguinity; and Brandt's resignation in turmoil in 1974, which left them without a leader. The

[33] See Michael Stolleis, *Geschichte des öffentlichen Rechts in Deutschland*, vol 4 (CH Beck 2012) 325.

[34] See Gary S. Schaal, Sabine Friedel, and Andreas Endler, *Die Karlsruher Republik. Der Beitrag des Bundesverfassungsgerichts zur Entwicklung der Demokratie und zur Integration der bundesdeutschen Gesellschaft* (Stiftung Mitarbeit 2000) 92.

[35] ibid 87. [36] ibid 100.

Court's crisis continued to mount even after these events, but the flow of dramatic reforms into the Court's docket steadily ebbed. In addition, the Court itself displayed renewed deference and caution in its 1979 approval of a major labour law reform.[37]

But beyond the partial backtracking of both sides, there was another ironic factor at play. The Court's ongoing clash with the federal government brought unprecedented attention to the Court as an institution—to its structure, its administration, its jurisprudence, and its history.

The last element named was particularly significant. Fierce criticism of the Court's political interventions led some observers to highlight the Court's historical contributions. One might quibble, said the Court's defenders, with a given decision on a highly politicized issue; but there could be no gainsaying the Court's historic role in nurturing a liberal-democratic *Rechtsstaat* in the early Federal Republic. The Court had protected fundamental rights robustly, and had secured the essential preconditions of democracy. It had helped West Germany rise from the shadows of Nazism and assume its place morally—not just economically—among the world's liberal democracies. This was reason enough, the argument ran, for citizens to swallow the odd, unpalatable decision.

The Court's intermediate crisis of the mid-1970s—chronicled in Chapter 3—marked a turning point in its history. The Court's survival of that crisis, and swift subsequent ascent to unprecedented levels of popular regard, is a product and illustration of another central thesis of this book—namely, that a Court whose power was accepted initially because it had helped ensure the preconditions of liberal democracy came to be regarded as itself a precondition of liberal democracy. The observation that the Court had helped to democratize a pre-democratic political culture led, in the fullness of time, to a conviction that the Court was a constitutive element of the democratic political culture it helped create. The key to the Court's success was the public's embrace of the following syllogism:

No democracy without fundamental rights (major premise);
No fundamental rights without constitutional justice (minor premise);
No democracy without the Constitutional Court (conclusion).

This is the secret of the Court's staggering popularity in the face of the obvious restraints it places on democratic self-governance and political freedom of action. It is the secret of the Court's survival of one crisis in the mid-1970s and of another in the mid-1990s. The syllogism may offend as a matter of logic.[38] But, as a good lawyer once said, the life of the law has not been logic; it has been experience.

IV. Constitutional Justice and Societal Integration

As noted, the Court not only survived its mid-1970s crisis, but rose swiftly thereafter to unmatched levels of trust and esteem.[39] In 1982, a new coalition came to power under the leadership of Helmut Kohl, a Christian Democrat. The Court had fewer occasions to clash with a conservative than with a reformist government. But the Court in the early Kohl era

[37] See 50 BVerfGE 290 (1979).

[38] For a sharp critique of the notion that fundamental rights protection is a sufficient source of the Court's legitimacy, see Christoph Möllers, 'Legalität, Legitimität und Legitimation des Bundesverfassungsgerichts' in *Das entgrenzte Gericht* (n 3) 344–50.

[39] As of 1984, nearly 70 per cent of West Germans reported that they trusted the Court. See Hans Vorländer and André Brodocz, 'Das Vertrauen in das Bundesverfassungsgericht. Ergebnisse einer repräsentativen Bevölkerungsumfrage' in Hans Vorländer (ed), *Die Deutungsmacht der Verfassungsgerichtsbarkeit* (Verlag für Sozialwissenschaften 2006) 264.

also benefited, as discussed in Chapter 4, from its traditional roles as a motor of societal integration and check on partisan political power.

Rudolf Smend, a prominent Weimar-era legal theorist and a kind of patron saint to the early Constitutional Court, argued as early as the 1920s that constitutions and constitutionalism should foster political and societal integration.[40] The notion was not, in its original formulation, a liberal one (Smend harboured an early admiration for Italian fascism), nor was it a reflection on the role of constitutional courts.[41] But the integration thesis, applied to the Constitutional Court and its jurisprudence, has had tremendous staying power.[42] From very early in the Court's history, observers have contended that the Court plays or should play an integrative role in the society and politics of the Federal Republic.

What exactly is meant by integration has been as elusive as it has been controversial.[43] Conceptually, the notion is vague. It can mean all things to all observers. In practice, it can lure the Court afield from the tangible business of deciding concrete cases. But there can be no question that the Court has often acted as arbiter of social and political conflict. Many citizens now *expect* the Court to exert an integrative influence. The Court has been most popular when its integrative role has been most pronounced. One such moment fell in the early- and mid-1980s, when the Court's approval ratings reached historic highs.

The Court's integrative role was on prominent display in several decisions responding to petitions from the environmentalist–pacifist Green Party, the first new party to enter Parliament since the late 1940s. The Greens were not uniformly successful in their trips to Karlsruhe, but their repeated complaints and occasional victories helped link the new party to the constitutional order of the Basic Law. At the same time, the Court won spectacular popular affection in 1983, when it annulled a federal census law that only the Greens had opposed.

Throughout the 1980s, the Court also benefited from its contrast with traditional politicians and established political parties. Even before the Kohl era began, many observers complained of the supreme power of political parties. Many saw the so-called 'Party State' (*Parteienstaat*) as symptomatic of a legitimacy crisis in parliamentary democracy.[44] In the years following the Court's *Census* judgment, the Federal Republic was racked by a sequence of parliamentary scandals that affected all major parties. Traditional German suspicion of parties and politicians rekindled. One of the era's watchwords was *Politikverdrossenheit*—roughly (and weakly) 'disenchantment with politics'.[45] Gloom about the country's politics fostered gratitude for the constitutional guardians who kept

[40] See Rudolf Smend, 'Verfassung und Verfassungsrecht' (1928) in Rudolf Smend, *Staatsrechtliche Abhandlungen und andere Aufsätze*, 2nd edn (Duncker and Humblot 1968) 89.

[41] See Christoph Möllers, Der Staat als Argument (CH Beck 2011) (especially chapter 4).

[42] For a comment on the theme from the Court's then chief justice, see Jutta Limbach, 'Die Integrationskraft des Bundesverfassungsgerichts' in Jutta Limbach, *Im Namen des Volkes: Macht und Verantwortung der Richter* (Deutsche Verlags-Anstalt 1999) 157. See also Gary Schaal, *Integration durch Verfassung und Verfassungsrechtsprechung? Über den Zusammenhang von Demokratie, Verfassung, und Integration* (Duncker and Humblot 2000); Ulrich Haltern, 'Integration als Mythos. Zur Überforderung des Bundesverfassungsgerichts' (1997) 45 *Jahrbuch des öffentlichen Rechts* 31; Ulrich Herbert, 'Integration der jungen Republik durch Verfassungsrecht?' in Thomas Henne and Arne Riedlinger (eds), *Das Lüth-Urteil aus (rechts-)historischer Sicht: die Konflikte um Veit Harlan und die Grundrechtsjudikatur des Bundesverfassungsgerichts* (Berliner Wissenschafts-Verlag 2005) 85.

[43] See Ulrich R. Haltern, 'Integration als Mythos—zur Überforderung des Bundesverfassungsgerichts' (1997) 45 *Jahrbuch des öffentlichen Rechts der Gegenwart* 31.

[44] See Michael Stolleis, 'Parteienstaatlichkeit—Krisensymptome des demokratischen Verfassungsstaats?' (1986) 44 *Veröffentlichungen der Vereinigung der Deutschen Staatsrechtslehrer* 41.

[45] See, e.g., Adolf M. Birke (ed), *Politikverdrossenheit: der Parteienstaat in der historischen und gegenwärtigen Diskussion: ein deutsch-britischer Vergleich* (KG Saur 1995); Jürgen Rüttgers, *Dinosaurier der Demorkatie: Wege aus Parteienkrise und Politikverdrossenheit* (Hoffmann und Campe 1993).

politicians in check. The public sometimes felt closer to the justices, whom they did not elect, than to Members of Parliament, whom they did elect. Richard von Weizsäcker, the highly regarded federal president, spoke for many when he described the Court as a 'devoutly desired supra-partisan oasis'.[46]

The Court's popularity in the 1980s was also bolstered by its muscular enforcement of rights of public protest. The early 1980s witnessed the hotly controversial placement by the North Atlantic Treaty Organization (NATO) of short-range nuclear missiles on West German soil. The placement unleashed protests that raised vexing questions about limits on the rights of protest and assembly. The Court answered many of these questions in a manner that expanded or reinforced protester rights. This was a major contribution to democratic culture in a country whose constitution provided few formal opportunities for the *vox populi* to make itself heard.

V. Mediating the Conflicts of German Reunification

The enormous protest movements in West Germany during the early 1980s were calm compared to the storms unleashed at decade's end to the East of the inner-German border. The fall of the Berlin Wall, the collapse of the communist regime in East Germany, and the reunification of Germany's divided halves was the central constitutional event of postwar German history and the defining public experience of nearly every German who lived through it. But although reunification answered one incarnation of the 'German Question' with an exclamation point, it also raised a new and insistent version. How would the world respond to a powerful, united Germany? How would united Germany respond to itself?

These questions had multiple ramifications, many of them constitutional. In sundry shapes and various forms, the aftermath and implications of reunification made their way before the Constitutional Court.[47] The reunification era tested and fused two of the Court's traditional functions: correcting the German *Sonderweg* and integrating German society. German division was both a by-product and a phase in Germany's putative special path, and reunification posed special problems of societal integration. The Court played a central role in stabilizing liberal constitutional democracy in united Germany and in mediating conflicts between the country's sections. This dual role is the focus of Chapter 5.

The political structure of reunification exacerbated its integrative challenge. Germany's sections united, not through the combined promulgation of a new constitution, but through the East German states' accession to an enlarged Federal Republic. This accession was effected, moreover, not by popular referendum but by state legislatures. In March 1990, the East German citizens who ousted the Socialist Unity Party of Germany (*Sozialistische Einheitspartei Deutschlands*, or SED) regime and toppled the Berlin Wall cast a clear vote for reunification, but they were otherwise excluded from deliberations about their constitutional future. In some of the most stirring political images of the twentieth century, East German protesters had chanted 'We are the People!', and later 'We are *one* People!' But 'We the People' played little role in shaping the constitutional structure of German reunification.

[46] Richard von Weizsäcker, *Richard von Weizsäcker im Gespräch mit Gunter Hofmann und Werner A. Perger* (Eichborn 1992) 158–59.

[47] See Hans Meyer, 'Die Wiedervereinigung und ihre Folgen vor dem Forum des Bundesverfassungsgerichts' in Peter Badura and Horst Dreier (eds), *Festschrift 50 Jahre Bundesverfassungsgericht*, vol 1 (Mohr Siebeck 2001) 83–129.

Some of those who called most forcefully for meaningful East German self-determination were current and former justices of the Constitutional Court. There was a certain irony in such extra-judicial advocacy. The self-proclaimed guardians of a constitution never submitted for popular approval and thoroughly hostile to plebiscitary initiatives emerged as paladins of direct democracy and popular self-determination. But the justices appreciated keenly the vulnerable democratic legitimacy of the constitution they enforced. Perhaps some hoped the Court's own legitimacy would be bolstered by a ringing popular affirmation of a united German constitution, new or old.[48] In any case, in the years following reunification, the justices emerged as something of a mediation tribunal for East–West conflicts. This is not to say that East German petitioners uniformly prevailed in their appeals to Karlsruhe. But the Federal Republic's newest citizens won significant (if sometimes partial) victories before the Court that they could never have achieved through ordinary politics. In an atmosphere in which many Easterners feared, with some justification, that their interests would be trampled by superior Western political heft, the Court played an important mediating role.

Within the 'old' states of the Federal Republic, however, the Court's post-unification jurisprudence was exquisitely divisive. A trio of searingly controversial decisions delivered in the mid-1990s plunged the Court into an unparalleled institutional crisis. The three decisions overturned the conviction of sit-in protesters for criminal coercion; remanded libel convictions for use of the phrase 'soldiers are murderers'; and nixed a Bavarian statute requiring that a crucifix be affixed to the wall of every public schoolroom. The judgments unleashed a firestorm of criticism unlike any the Court had seen. Longtime Court-watchers descried a transition from a traditional criticism of individual judgments (*Urteilsschelte*) to a categorical attack on the Court as an institution (*Institutionenschelte*).[49] Prominent politicians led protest marches and breathed defiance at the *Crucifix* judgment. For the first time in its history, the Court faced the prospect that its decisions would be not only criticized but ignored. The Court's approval ratings plummeted.[50]

In retrospect, the most telling memorials of the Court's end-of-century crisis are not the intemperate criticisms levelled at the Court but the full-throated defences—many from non-lawyers—sounded on its behalf. As with earlier crises, the Court's defenders adduced its history. They cited the Court's accomplishments in shielding fundamental rights, advancing liberalism, and securing democracy. They praised the Court as a guardian of individual rights, a bulwark of the *Rechtsstaat*, a constraint on self-interested politicians, and a crucial catalyst of the end of Germany's *Sonderweg*. Discontent with individual judgments notwithstanding, they argued, Germans should take pride in their Constitutional Court and feel gratitude for its achievements.

VI. Coda: Constitutional Patriotism

In 1979, the political theorist Dolf Sternberger called on citizens of the Federal Republic to rally round their state in the name of 'constitutional patriotism' (*Verfassungs-patriotismus*), a phrase Sternberger himself coined.[51] Seven years later, Jürgen Habermas adduced the

[48] Perhaps a similar sensitivity was at work in the controversial rhetoric about democracy in the Court's cautious approval, in 1993, of the Maastricht Treaty creating the European Union. 89 BVerfGE 155 (1993).

[49] Rainer Wahl, 'Quo Vadis—Bundesverfassungsgericht? Zur Lage von Verfassungsgerichtsbarkeit, Verfassung und Staatsdenken' in Bernd Guggenberger and Thomas Würtenberger (eds), *Hüter der Verfassung oder Lenker der Politik? Das Bundesverfassungsgericht im Widerstreit* (Nomos 1998) 83–84.

[50] See Vorländer and Brodocz (n 39) 264.

[51] Dolf Sternberger, 'Verfassungspatriotismus' *FAZ* (23 May 1979).

phrase on a more abstract level in his argument that, after Auschwitz, the only brand of patriotism available to West Germans was a constitutional patriotism rooted in the universal values of Western liberalism.[52] Anchoring the Federal Republic in an international community united by such values was, Habermas contended, the proudest achievement of his generation.

The concept of constitutional patriotism has been assailed as conceptually incoherent and practically implausible.[53] How many Germans, one might reasonably ask, have felt stirrings akin to those described (or prescribed) by Sternberger or Habermas? On the other hand, Germans *have* been grateful for the Federal Republic's stability and success, and they have ascribed some of that success story to the constitution under which it has unfolded.[54] Moreover, Court and constitution are inseparable in many citizens' minds. As a journalist in the *Süddeutsche Zeitung* wrote during the mid-1990s crisis, vituperative attacks on the Court affected 'necessarily the Basic Law as well'. For, 'as nearly fifty years have shown, both vital forces depend on one another. The battle for Karlsruhe, therefore, is also a battle for the constitution'.[55] When that crisis passed, the *longue-durée* view of the Court prevailed. By century's end, the Court featured prominently in celebratory accounts (composed for the country's fiftieth birthday in 1999) of the Federal Republic's 'success story'. Similar panegyrics sounded when the Court honoured its own golden anniversary two years later. By the dawn of the twenty-first century, the Federal Republic's capital had moved from Bonn to Berlin. But Germany remained—and, despite new challenges, it still remains—the Karlsruhe Republic.[56]

[52] Jürgen Habermas, 'Eine Art Schadensabwicklung' *Die Zeit* (11 July 1986) 40.

[53] For an excellent critical essay on the concept, see Jan-Werner Müller, *Constitutional Patriotism* (Princeton University Press 2007).

[54] See Dieter Grimm, 'Identität und Wandel' (2009) 37 *Leviathan* 603, 614; Dieter Grimm, 'Verfassungspatriotismus nach der Wiedervereinigung' in Hauke Bunkhorst and Peter Niesen (eds), *Das Recht der Republik* (Suhrkamp Verlag 1998) 307; Stolleis, *Geschichte des öffentlichen Rechts* (n 33) 378; Schönberger (n 21) 44–47.

[55] Helmut Kerscher, 'Die Kesselschlacht um Karlsruhe' *Süddeutsche Zeitung* (2 March 1996) 4.

[56] See Schaal, Friedel, and Endler (n 34); Gerhard Casper, 'The "Karlsruhe Republic"—Keynote Address at the State Ceremony Celebrating the 50th Anniversary of the Federal Constitutional Court' (2001) 2 German Law Journal, http://www.germanlawjournal.com/index.php?pageID=11&artID=111 last accessed 26 May 2015.

1

Consolidation, 1951–1959

We have known since Marbury v. Madison *what a dynamic constitutional court can mean, and the nine men under Earl Warren have proved it anew. There is little sense, therefore, in discussing the matter. You have the authority and the normative strength of the factual on your side. This is more than enough.*

—Hugo Marx to Gerhard Leibholz[1]

Constitutional questions are not originally questions of law, but questions of power.

—Ferdinand Lassalle[2]

Introduction

The Federal Constitutional Court opened in a solemn ceremony in Karlsruhe, held on 28 September 1951. In his inaugural remarks, Chancellor Konrad Adenauer suggested, in a towering understatement, that the new tribunal had its work cut out for it. The Court, Adenauer noted, was to defend and enforce the constitution of a republic that had been framed—to put the matter mildly—under inauspicious circumstances. It was to do so without the guidance of tradition and within a legal system sorely burdened by its recent past.[3] Even so, the chancellor called on the Court to stand as a pillar of the Federal Republic and supreme guardian of its fundamental law.[4] The Court's responsibility was solemn; its challenges were enormous. Adenauer had barely begun to enumerate them.

The Court was to enforce a democratic constitution in a country where popular democratic conviction ran thin. The political culture of the Federal Republic was pre-democratic, if not downright anti-democratic. This presented the Court with both a challenge and an opportunity. The challenge was to implement liberal and democratic values in a political and legal culture that was skeptical of liberalism and democracy. The opportunity was to establish constitutional judicial review without having to defend its democratic legitimacy. In the German legal tradition, the principles of *Demokratie* and *Rechtsstaat*—of democracy and the rule of law—had parted paths after the failed revolution of 1848. More recently, the Nazis had obliterated both principles. The Court's commission was to restore them; its supreme achievement was to reunite them. But the rule of law came first. As the Court, in its earliest years, worked to restore the rule of law, citizens began to view the Court as a necessary precondition—and not as an institutional threat—to democracy.

In the process the Court helped the society and political culture of the Federal Republic to become both more liberal and more democratic. The Court did this by developing and deploying doctrinal precepts and interpretive methodologies that required all West German law—and, later, all West German politics—to respect fundamental rights. The Court characterized such rights as objective values, and pronounced that those values must condition

[1] Hugo Marx to Gerhard Leibholz (undated) Bundesarchiv Koblenz (hereinafter BAK) N 306/4 (Nachlass Hugo Marx).

[2] Ferdinand Lassalle, *Über Verfassungswesen* (Meyer and Zeller 1863) 59.

[3] In 'Eröffnung des Bundesverfassungsgerichts' (1951) 4 *Neue Juristische Wochenschrift* 791 (hereinafter *NJW*).

[4] ibid.

and constrain all positive law and all state power. The Court's jurisprudence transformed the Basic Law into a totalizing constitution. The Court itself became that constitution's ubiquitous and watchful guardian.

This jurisprudence brought the Court into conflict with its surrounding political and societal milieu. Both politics and society were undergoing the birth pangs of democracy. The process was slow and painful, its progress often difficult to detect.

Many citizens of the early Federal Republic were indifferent toward their new state; others were actively hostile. A survey conducted around the time of the Court's founding asked West Germans when things had gone best for their country. Forty-five per cent pointed to the pre-1914 Wilhelmine Empire; 40 per cent to the peacetime years of National Socialism. Only 7 per cent chose the Weimar Republic; only 2 per cent the present.[5] Many citizens of the infant Republic rejected it out of hand as the sham imposition of an occupying power. (A small minority decried its continuities with the Third Reich.) Many still viewed members of the wartime resistance as traitors.[6]

Six years after the collapse of the Nazi regime, its sympathizers and self-styled successors were on the political march. In the 1951 state elections, far-Right parties uniformly won between 7 and 11 per cent of the vote. The strongest of these parties was the Socialist Reich Party (SRP), which at its peak boasted 40,000 members. Its ranks were alarmingly young; its message alarmingly attractive to former officers and soldiers—disillusioned drifters with no professional prospects in a land without a military. The party evoked memories of the Weimar-era *Stahlhelm* (steel helmet) organizations. Many feared it would expand exponentially, and that its implacable hatred of the new Republic would appeal to refugees expelled from areas east of the Oder-Neiβe boundary.[7] Before Parliament even began to staff the Constitutional Court, the Adenauer government considered asking the Court to ban the ascendant party.

The SRP was toxic to the political mainstream, but so were efforts to make Germans reckon with their recent past. Four months before the Court's opening ceremonies, the Bundestag formally proclaimed the end of denazification, an object of popular loathing whose demise all major political parties advanced and applauded.[8] A parliamentary statute of 11 May 1951 smoothed the path for the near-universal rehabilitation and reintegration of Nazi-era civil servants. The law fulfilled the mandate of Article 131 GG, which required Parliament to clarify the legal standing of civil servants who lost their posts after the unconditional surrender of May 1945 but had not yet returned to public employment.

Quickly christened the '131 Law', it provided financial compensation, and allowed reentry, to those it affected—the '131-ers'. Some 430,000 former officials claimed the law's benefits.[9] By 1953—when the Constitutional Court sustained the law's constitutionality in the most provocative manner imaginable—some 30 per cent of federal administrative offices were filled by 131-ers. In the foreign and interior ministries, the rate was even higher.[10] The 131-ers were a variegated group, including former officials from the state of Prussia, the military, and the East. More than a fourth of the 200,000 or so 'civil 131-ers', however, had lost their jobs after denazification trials.[11] After the 131 Law was passed,

[5] Elisabeth Noelle and Erich Peter (eds), *Jahrbuch der Öffentlichen Meinung 1947–1955* (Allensbach Verlag für Demoskopie 1956) 126.

[6] ibid 138.

[7] Eckart Conze, *Die Suche nach Sicherheit. eine Geschichte der Bundesrepublik Deutschland von 1949 bis in die Gegenwart* (Siedler 2009) 148.

[8] See Norbert Frei, *Vergangenheitspolitik. Die Anfänge der Bundesrepublik und die NS-Vergangenheit* (CH Beck 1996) 54–68.

[9] ibid 70. [10] ibid 85. [11] See Conze (n 7) 155–56.

persons once dismissed for their Nazi ties flooded the Federal Republic's civil service. Former Nazis even filled cabinet posts. Restoration and rehabilitation reached other areas of national life as well. In theatres throughout West Germany, filmgoers flocked to see the first postwar film of Nazism's star director, Veit Harlan—later the plaintiff–appellee in the Court's most famous case.[12]

Prominent among the rehabilitated 131-ers were judges and law professors. Some of these became leading skeptics of constitutional judicial review. Even before the Court's founding, the Association of Public Law Professors expressed its anxiety—and latent hostility—in the very theme of its 1950 meeting: 'The *Limits* of Constitutional Justice', the conference was called.[13] Almost to a man—and they were all men—public law professors agreed that the Court's authority should not involve political questions.

Fears that the Court would spawn an unholy mingling of law and politics found popular resonance as well. The Court was a week old when the Catholic-conservative *Rheinischer Merkur* wondered whether constitutional justice would produce 'Jurisprudence or Politics', whether the Court was capable of forging a '*legal* tradition', and whether its judges were jurist enough for their jobs. The *Merkur* grumbled that the justices were merely robed politicians dabbling in a kind of juristic dilettantism—politicians who might amass frightening power in an insalubrious 'jurists' monopoly'.[14]

Worries that political power might concentrate in Karlsruhe were compounded by signs of political instability in Bonn. The Federal Republic's first nation-wide elections in August 1949 had given a narrow Bundestag majority to a coalition headed by the Christian Democratic Union (CDU)—with its Bavarian sister, the Christian Socialist Union (CSU)—flanked by the liberal Free Democratic Party (FDP) and the nationalist German Party (DP). The coalition placed Konrad Adenauer, the seventy-three-year-old former mayor of Cologne and president of the Parliamentary Council, at the head of the first federal government. Adenauer filled many cabinet positions with Christian Democratic comrades, but also gave five posts to leaders from the CDU's coalition partners. Among these was Thomas Dehler, Bavarian chairman of the FDP, whom Adenauer named minister of justice.

By autumn 1951, when the Constitutional Court opened its doors, the Adenauer government was already embattled. Only a third or so of the citizenry approved of Adenauer's policies and performance. The Bundestag was a battleground of fiery debate between government and opposition. The most contentious arena of all was foreign policy. Adenauer, whose craggy countenance seemed to embody the ancient landscape of his native Rhine valley, was a visceral man of the West. In Adenauer's cultural cartography the Rhineland was Christian Europe's heart, the Cologne Cathedral its crown. To the East lay the Asiatic steppes (beginning roughly in Braunschweig) and the Siberian wilderness (coterminous with the late kingdom of Prussia). Adenauer's foreign policy was a relentless impulse westward. Its watchwords were *Westintegration* (Western integration) and *Westbindung* (binding with the West); its overarching goal was to bind the Federal Republic to its Western neighbours within a coalition of liberal democracies. Adenauer envisioned a Western defence alliance that included an armed West Germany. To this prospect the Social Democrats were

[12] On Harlan, see Thomas Harlan, *Veit* (Rowohlt 2011).

[13] *Veröffentlichung der Vereinigung der Deutschen Staatsrechtslehrer*, vol 9 (1952) (emphasis added). The conference was held 20/21 October 1950. See also Thomas Henne, '"Von 0 auf Lüth in 6 ½ Jahren": Zu den prägenden Faktoren der Grundsatzentscheidung' in Thomas Henne and Arne Riedlinger (eds), *Das Lüth-Urteil aus (rechts-) historischer Sicht: die Konflikte um Veit Harlan und die Grundrechtsjudikatur des Bundesverfassungsgerichts* (Berliner Wissenschafts-Verlag 2005) 200.

[14] 'Rechtsprechung oder Politik', *Rheinischer Merkur* (14 September 1951) 2 (emphasis added).

passionately opposed. The conflict ultimately came before the Constitutional Court, where it occasioned the Court's first and greatest crisis.

The brightest spot on the national landscape was the economy. Three years after the postwar currency reform, the Deutschmark (DM) was gaining strength; a year after the outbreak of the Korean War, a war boom was beginning to be felt. Marshall-Plan aid supplied an additional fillip. In retrospect, the decade's first years marked the beginning of an unprecedented period of economic growth—an 'economic miracle' (*Wirtschaftswunder*), as even contemporaries dubbed it. West German GDP rocketed from 97 billion DM in 1950 to 216 billion DM in 1957. Average growth between 1950 and 1955 exceeded 9 per cent; in 1955, it topped 12 per cent. As of September 1951, however, most of this was hidden from view. Growth in 1950 was halting. Ludwig Erhard, the economic minister, would later be lionized as the 'Prophet of the Free Market' and 'Hero of the Economic Miracle'. But in early 1951, he was one of the least popular politicians in the country. The SPD opposition clamoured to end the market economy. Nightmarish memories of Weimar resurfaced. When the Court opened its doors, the economic miracle had begun, but no one knew it. The future seemed anything but certain.

In the face of such uncertainty, West Germans grew nostalgic for old absolutes and longed for continuity with quieter times. One source of continuity was religion; both of Germany's major denominations enjoyed something of a postwar boom. Another was the family. The ideal of a breadwinning father, a homemaking mother, and cherubic children living in unity and security under a single roof experienced a popular renaissance, though in the real world such arrangements were rare. Millions of men had been killed in the War and thousands more remained Soviet prisoners. The War brought many women, mothers among them, into the workplace. After the War, social pressures pushed many of these women to leave their jobs and make room for returning soldiers. The Basic Law trumpeted that men and women were equal before the law,[15] but many politicians, preachers, and public law scholars insisted that this precept was limited by the imperatives of the traditional family.[16] The constitution gave Parliament a grace period in which to update laws that flatly offended the principle of gender equality, but Parliament had not yet done so. This matter, too, would wend its way to Karlsruhe.

So would many of the major societal and political issues of the early Federal Republic. Some of these controversies placed the Court at the centre of the public's attention. On the whole, the Court's handling of them secured the public's regard. In the most dramatic cases, the Court often deferred to government policy. Never in the 1950s did the Court directly invalidate an initiative of the Adenauer government. Indeed, the Court issued its most aggressive early judgments—banning the Socialist Reich Party in 1952, the German Communist Party in 1956, and local referenda on atomic armaments in 1958—at the federal government's instigation.

But never, on the other hand, did the Court surrender an ounce of independence or an inch of jurisdiction. The early Court mingled pragmatic caution with steely defiance, procedural self-assertion with substantive restraint. The Court was flexible in matters of government policy but adamantine on questions of institutional authority. The Court assumed enormous, but always latent, political power and conspicuously eschewed its partisan exercise. The Court made haste slowly, yielding frequently to government

[15] Article 3(2) GG.

[16] See the account in Christine Franzius, *Bonner Grundgesetz und Familienrecht. Die Diskussion um die Gleichberechtigung von Mann und Frau in der westdeutschen Zivilrechtslehre der Nachkriegszeit 1945–1957* (Vittorio Klostermann 2005).

demands while reserving the right, in future cases, not to yield. When the Court did assert authority it generally did so to the detriment only of unsympathetic (or in any case politically powerless) litigants. And it did so largely outside the public gaze. This was especially true of the Court's early fundamental rights landmarks—in hindsight the most momentous matters the Court has ever decided. Many groused that the Court was meddling in politics, but few complained that it was undermining democracy. Instead, by establishing itself as a safeguard of the rule of law, the Court emerged as a herald of democracy—the guarantor of the one and the precursor of the other. At the same time, the Court consolidated power, the future exercise of which would redefine both politics and democracy.

I. Institutional Independence

A. Choosing justices

'*Wir haben ein Verfassungsgericht*', intoned the 11 September 1951 issue of the *Frankfurter Allgemeine Zeitung*—'We have a constitutional court'.[17] Days earlier, the Federal Constitutional Court had issued its first major decision. The decision temporarily forbade a proposed plebiscite on the creation of the Southwest German state of Baden-Württemberg. The referendum could not go forward, the justices held, until after the Second Senate had time to rule on the merger's constitutionality. The Senate approved the union, to the dismay of the Baden patriots who opposed it, six weeks later.[18] The controversy over the Southwest state placed the Court at the centre of public attention for the first time. More attention, and greater controversy, followed in swift succession.

Nearly two and a half years passed between the promulgation of the Basic Law and the creation of the Constitutional Court. The framers left many details concerning the Court's organization to future parliaments, and a divided Bundestag took eighteen months to craft the Court's statutory charter, the Federal Constitutional Court Act (FCCA), of 12 March 1951.[19] This law called for half of the justices on each Senate to be appointed by the Bundestag, half by the Bundesrat.

It took Parliament nearly six months to staff the Court.[20] Nominally, each party in Parliament would propose a number of candidates commensurate with the party's parliamentary strength. In fact, sub-committee members from the coalition parties nominated together. As a result, candidates were classed as government nominees or opposition nominees. These designations had portentous consequences for early perceptions of the Court in the political and public sphere. The Adenauer government also played a significant role in the nominating process through the ministry of justice, led by Thomas Dehler, the justice minister, and his confidante, Willi Geiger. Dehler thought the government should work to make the Court non-partisan. Instead of nominating only candidates politically

[17] Hans Baumgarten, 'Wir haben ein Verfassungsgericht' *Frankurter Allgemeine Zeitung* (hereinafter *FAZ*) (11 September 1951).

[18] 1 BVerfGE 1 (1951); 1 BVerfGE 14 (1951).

[19] *Das Bundesverfassungsgerichtsgesetz* ('BVerfGG' in subsequent footnote citations). For contemporary comments on the Act, see Eduard Dreher, 'Glanz und Elend der Staatsgerichtsbarkeit. Zum Gesetz über das Bundesverfassungsgericht vom 12. März 1951' (1951) 4 *NJW* 377; Willi Geiger, *Gesetz über das Bundesverfassungsgericht vom 12. März 1951. Kommentar* (Vahlen 1952). For the standard scholarly study, see Reinhard Schiffers, *Grundlegung der Verfassungsgerichtsbarkeit—Das Gesetz über das Bundesverfassungsgericht vom 12. März 1951* (Droste 1984).

[20] See Richard Ley, 'Die Erstbesetzung des Bundesverfassungsgerichts' (1982) 13 *Zeitschrift für Parlamentsfragen* 521.

close to the government, the coalition parties should seek judges 'enjoying a general public regard'.[21] Dehler would later think this effort had backfired.

The Court's original twenty-four justices—twelve in each Senate—varied widely in outlook and experience. As a group, they were most striking for their lack of judicial experience and their anti-Nazi credentials. This outcome was no accident. Conservative members of Parliament, apprehensive about the Court's potential interference in political affairs, wanted to populate the Court with justices whose broad experience in public affairs would attune them to the political implications of sensitive cases. Social Democrats, on the other hand, mistrusted the traditional judiciary, whose hostility to the Left during the Weimar era was surpassed only by its fealty to the *Führer* during the Third Reich.[22] Only six of the twenty-four founding justices came to Karlsruhe directly from another judicial post. Only half of the justices had ever worked as judges in any capacity—and some of these only briefly and decades earlier.[23] As one student of the selection process concluded, most of the early justices lacked a 'typically judicial outlook [*Lebensgefühl*]' and were 'foreign to the traditions of this caste'.[24]

Distance from the traditional judiciary meant distance from the Nazi-era judiciary. Unlike other official circles, the original Court, by and large, was free from the stain of Nazi collaboration. More than two-thirds of the new justices had suffered personally under Nazi misrule—eight were forced out of office; five had career paths slowed; four resigned voluntarily.[25] Four justices left Germany and spent the Nazi era in exile.

Some of the justices had suffered very personally indeed. Gerhard Leibholz, the moral and intellectual leader of the early Court, was a leading Weimar-era jurist. Leibholz lost his professorship in Göttingen because of his Jewish ancestry in 1935. In 1938, he emigrated—first to Great Britain, where he was briefly interred as an 'enemy alien', later to the United States. Leibholz was accompanied in exile by his wife, Sabine Bonhoeffer, twin sister of Dietrich Bonhoeffer, the martyred resistance hero. Leibholz returned to Göttingen after the War, but to a chair, not in law, but in political science—a discipline dismissed by German traditionalists as a dubious American import.[26]

Rudolf Katz, the Court's first vice president, was also Jewish. Shortly after the Nazi seizure of power, Katz fled to the United States, where he taught at the Institute of Public Administration at Columbia and edited the *Neue Volkszeitung*, the sole SPD press organ in exile. Erna Scheffler, the only woman on the original Court, lost her judgeship after the Nazis passed the Nuremberg racial laws. Later she was disbarred as well. In 1934, the Nazis dissolved her marriage to Georg Scheffler: Hitler's millenarian Reich refused to recognize the marriage of an Aryan to a 'half-Jewess'. Scheffler spent the War working as a bookkeeper

[21] Thomas Dehler to Walter Strauβ (26 May 1951) BAK B 136/4436.

[22] See Ingo Müller, *Furchtbare Juristen: die unbewältigte Vergangenheit unserer Justiz* (Knaur 1987). See also *Der Unrechts-Staat: Recht und Justiz im Nationalsozialismus* (edited by *Kritische Justiz*) (Nomos 1983); Lothar Gruchmann, *Justiz im Dritten Reich 1930–1944: Anpassung und Unterwerfung in der Ära Gürtner* (Oldenbourg 1987).

[23] Werner Billing, *Das Problem der Richterwahl zum Bundesverfassungsgericht: Ein Beitrag zum Thema 'Politik und Verassungsgerichtsbarkeit'* (Duncker and Humblot Gmbh 1969) 181. The FCCA required that eight justices be chosen from the ranks of the federal judiciary, Parliament met this requirement by proposing four nominees for midnight appointment to the Federal Court of Justice. This gave them the required status as federal judges, though they had never actually sat on the Federal Court of Justice.

[24] ibid 186.

[25] See Udo Wengst, *Staatsaufbau und Regierungspraxis, 1948–1953: zur Geschichte der Verfassungsorgane der Bundesrepublik* (Droste 1984) 243.

[26] On Leibholz, see Anna-Bettina Kaiser (ed), *Der Parteienstaat. Zum Staatsverständnis von Gerhard Leibholz* (Nomos 2013); Peter Unruh, 'Erinnerung an Gerhard Leibholz (1901–1982)—Staatsrechtler zwischen den Zeiten' (2001) 126 *Archiv des öffentlichen Rechts* 61; Manfred H. Wiegandt, *Norm und Wirklichkeit: Gerhard Leibholz (1901–1982)—Leben, Werk, und Richteramt* (Nomos 1995).

in the business of a female friend, who shared ration cards with her. Beginning in January 1945, she hid in a summer home outside Berlin. After the War, she and Georg Scheffler 're-married'.[27] With their appointment to the Federal Constitutional Court, Leibholz, Katz, and Scheffler became the highest ranking Jews in the West German state.

Justice Erwin Stein worked in various positions as a government lawyer and judge in Hessen until 1933, when he was forced resign because his wife, Hedwig, was Jewish. Stein spent the War working as a private lawyer in Offenbach-am-Main. Fearing deportation to extermination camps in the East, Hedwig Stein committed suicide in 1943.[28] Martin Drath, an SPD stalwart since the mid-1920s and former student and assistant of Hermann Heller, a leading Weimar-era jurist, spent the Nazi era working in a humble position as a clerk and accountant. Wilhelm Ellinghaus lost his job as a senior regional civil servant. Julius Federer left the justice ministry for reasons of conscience after the Nazi seizure of power and joined the ministry of the Catholic Church. Ernst Friesenhahn's promising academic career was scuttled by his open opposition to the regime.[29] Gerhard Heiland lost his position as deputy police chief because of his open democratic convictions.

Many years after the Court's founding, Justice Willi Geiger reflected that those involved in selecting the first justices—Geiger himself among them—had been animated by a spirit of redress or compensation (*Wiedergutmachung*) when they appointed members of groups persecuted by the Nazis. For these individuals, Geiger noted, high office in other branches of government was largely foreclosed.[30] Geiger himself had openly supported the Nazi regime.[31]

For the Court's first president, or chief justice, Hermann Höpker-Aschoff, elevation to the bench came as a bitter disappointment, rendered doubly bitter by the manner in which it came about. The Adenauer government played an aggressive role in choosing the chief justice—rather to the irritation of parliamentarians who thought the chancellor and justice minister were overstepping. After a series of preferred candidates declined the job because of illness or disinclination, the cabinet's choice fell, in the summer of 1951, on Höpker-Aschoff. The cabinet minutes give no explanation of the choice, but it seems likely that Adenauer, who initially took little personal interest in the Court, wished to appease his FDP coalition partners by naming one of their own as chief justice. FDP leaders, including the justice minister, Thomas Dehler, had long been calling for greater representation at the highest levels of government. Dehler and Höpker-Aschoff also happened to be personal friends, then living in the same apartment complex in Bonn.

[27] On Scheffler, see Christian Waldhoff, 'Erna Scheffler—erste Richterin des Bundesverfassungsgerichts' (2008) 56 *Jahrbuch des öffentlichen Rechts der Gegenwart* 261.

[28] On Stein, see Andreas Hedwig and Gerhard Menk (eds), *Erwin Stein (1903–1992). Politisches Wirken und Ideale eines hessischen Nachkriegspolitikers* (Hessisches Staatsarchiv 2004); Richard Ley, 'Nachruf Erwin Stein' (1992) 45 *NJW* 3217.

[29] Friesenhahn, a student of Carl Schmitt, had joined the SA in 1933 and applied to join the NSDAP. In 1934, however, he left the SA and increasingly distanced himself from Schmitt when his former teacher openly advocated the regime.

[30] See Wengst (n 25) 244 (citing a personal conversation with Geiger).

[31] Geiger was an officer of the SA and a member of the NSDAP. His dissertation on the National Socialist press law of 4 October 1933 warmly praised the law's categorical ban on Jewish journalists: 'With one blow, this provision abolished the over-powerful, Volk-noxious, and culture-destroying influence of the Jewish race in the sphere of the press'. Willi Geiger, *Die Rechtsstellung des Schriftstellers nach dem Gesetz vom 4. Oktober 1933* (Kichler 1940) 40. During the Third Reich, Geiger secured at least five death sentences as a state prosecutor before the Sondergericht in Bamberg. In 1943, he was conscripted into the Wehrmacht and led a company on the Eastern Front. It is unclear how much was known of Geiger's past at the time of his appointment to the Constitutional Court. His 1947 denazification proceeding classed him as '*nicht betroffen*' (not affected)—effectively an acquittal.

Höpker-Aschoff, a Berlin politician with angular features, penetrating eyes, and an aquiline nose (there is something fiercely aviary in the surviving pictures), was enormously disinclined to take the job. He chafed at leaving an incipient academic career that was steadily growing on him and he balked at the idea of leaving the capital for provincial Karlsruhe. Worst of all, he cringed at the thought of abandoning all hope of a postwar political career. At length, however, the Prussian in Höpker-Aschoff prevailed. After some internal agonizing and a brief delay, he reported to the justice ministry that, urged by a sense of duty, he was willing to answer his country's call.

But his country, by that point, was reluctant to call him. Rumours of Höpker-Aschoff's candidacy produced loud alarms from Catholics, who were appalled by his liberalism, and from federalists, who were alarmed by his centralism. Adenauer now realized that if he pushed through Höpker-Aschoff's appointment, he might alienate two crucial, often overlapping, bastions of his electoral strength: the Catholic Church and the CDU's Bavarian sister party, the Christian Socialist Union (CSU). Unwilling to risk so much for the Constitutional Court, Adenauer asked Höpker-Aschoff to withdraw his candidacy. In an awkward letter of explanation, Adenauer regretted that Höpker-Aschoff's wartime involvement with *Haupttreuhandel Ost*—the agency responsible for distributing property confiscated in occupied Poland—would provoke discussions 'that will be truly disagreeable and would be injurious to your position'.[32] Wartime activities that didn't dissuade the government from choosing Höpker-Aschoff in the first place now furnished the pretext for dismissing him on unrelated political grounds. Höpker-Aschoff, deeply wounded, fired a frosty reply:

> I did not seek the office of president of the Federal Constitutional Court. I declared myself willing after two cabinet ministers, commissioned by the federal government, informed me that the government had unanimously determined to recommend me for this office, and after both ministers adjured me to accept it. The decision was difficult for me, for it was clear to me what I must give up: political activity, my beloved teaching at the University of Bonn, and my posts on the board of directors of two significant German firms. In making my decision, however, I assumed that no reservations would be raised against my nomination and that general trust would be placed in me. Since this is obviously not the case, I am constrained, to my sorrow, to inform you that I am no longer ready to accept this office.[33]

By the beginning of September, however, the political landscape had altered. Efforts to find a willing candidate who pleased both the government and the parliamentary selection committee had repeatedly failed. A ruling on the constitutionality of the Southwest state plebiscite had become urgent. Pressure mounted to find a chief justice and launch the new Court. Moreover, and decisively for Höpker-Aschoff's candidacy, Adenauer had sparked a row with the FDP leadership by holding talks in Bürgenstock with national trade union leaders on the question of workers' participation in industrial governance.[34] The chancellor sought to soothe the strain by returning to Höpker-Aschoff who, still bristling at Adenauer's earlier volte-face, ultimately accepted. Wounded pride yielded to duty; the self submitted to the state. But the insult was not forgotten.

Catholics and Federalists were appalled, as were many Unionists who thought the ruling party should pick one of its own to the head the highest court. But carp as detractors might,

[32] In Thomas Aders, *Die Utopie vom Staat über den Parteien: biographische Annäherungen an Hermann Höpker-Aschoff (1883–1954)* (Peter Lang 1994) 304.

[33] ibid 304–05. [34] See Wengst (n 25) 239; Aders (n 32) 307.

the government stood by its choice. On 7 September 1951, the Bundestag made Hermann Höpker-Aschoff the first chief justice of the Federal Constitutional Court.

B. The 'status question'

The Court was now staffed with its two dozen justices. On 9 September 1951, it opened its doors. What role the Court would play in the postwar political order remained to be seen. Because the Court was empowered to police the constitutionality of government action, the SPD entertained high hopes that the Court would serve as an instrument of opposition. The governing coalition feared the Court for the same reason. The opposition championed the Court's institutional and budgetary independence; the coalition called for strong ties between the Court and the federal government. In the end, the opposition view won out. The Court, as if in grateful payment, repeatedly affirmed the constitutional legitimacy—and democratic necessity—of parliamentary opposition. This would prove to be one of the early Court's greatest contributions to the democratization of West German politics.

At its inception, the Constitutional Court lacked both a building and a budget of its own. Like the other federal courts, it was placed under the supervisory authority of the ministry of justice. The justice ministry controlled the Court's internal administration, including salary and staffing. SPD politicians hotly opposed this state of affairs; cabinet ministers vigorously defended it. Beneath the tumult of this general clash was a personal duel between two men: Adolf Arndt, the leading legal mind of the opposition, and Thomas Dehler, the federal minister of justice.[35]

Fiery and stubborn, volatile and intractable, Dehler was a man of strong views sharply expressed. He was a mason and a liberal, a polemist and a firebrand. He was also a man of great courage and conviction. Dehler first met and befriended Hermann Höpker-Aschoff and Theodor Heuss, the first president of the Federal Republic, during the Weimar era when the three men were fellow partisans of the German Democratic Party. In 1925 Dehler married Irma Frank, a Jew, to whom he remained fiercely loyal throughout the Nazi era. Peers within the Nazi-dominated bar pressed Dehler to renounce his marriage and forsake his Jewish clients. He did neither. Dehler also refused to give the Hitler greeting and, in his legal practice, he represented the regime's opponents. In the mid-1930s, he joined the Robinson-Strassman Group, one of the longest-surviving Nazi-era opposition groups. In 1938, Dehler was briefly imprisoned by the Gestapo. Because of his Jewish ties, he was deported in 1944 to a forced labour camp. After the War, Dehler worked briefly as a denazification prosecutor in Bavaria, but he soon concluded that denazification had gone far enough. Perhaps he thought it had gone too far.

Dehler's justice ministry was marked by a policy of amnesty and amnesia. 'I have never doubted', Dehler wrote, 'that it is not only desirable, but in the interest of the internal pacification of our people it is indispensable, that a thick line be drawn over the past'.[36] As justice minister, Dehler knowingly hired several former Nazis.[37] One of the brightest of them, and one of Dehler's closest counsellors, was Willi Geiger, the future justice.[38] Dehler was also accused of blinking at the recent past in the matter of judicial appointments. On 3 January 1951, an article in the *Frankfurter Rundschau* sharply criticized the justice ministry's

[35] On Arndt, see Dieter Gosewinkel, *Adolf Arndt. Die Wiederbegründung des Rechtsstaats aus dem Geist der Sozialdemokratie (1945–1961)* (Verlag J.H.W. Dietz Nachf 1991); on Dehler, see Udo Wengst, *Thomas Dehler, 1897–1967: Eine politische Biographie* (Oldenbourg 1997).

[36] In Wengst, *Thomas Dehler* (n 35) 158.

[37] ibid 142–43; see also Frei, *Vergangenheitspolitik* (n 8) 31.

[38] Wengst, *Thomas Dehler* (n 35) 142.

appointments to the ordinary courts of the federal judiciary. The article noted that of the fifty or so federal judges appointed during Dehler's tenure, nearly half had been members of the Nazi party. The article also claimed that Dehler had privately pronounced himself 'against, in principle, Jewish judges and such judges … as have even a suspected connection to one of the opposition parties'. On 12 January 1951, the *Rundschau* printed Dehler's passionate rebuttal. Seared into the justice minister's memory were the names of the article's SPD signatories. Foremost among them was Adolf Arndt.

Seven years Dehler's junior, Arndt came of age politically in the tempestuous Berlin of the Weimar Republic. He became a judge in 1932 but resigned his post in 1933 so as, he recalled, 'not to cooperate with them [the National Socialists]'. In 1933, he wrote a letter to Nazi authorities, petitioning for admission to the bar and describing his judicial career as compatible with National Socialist ideals. The letter, preserved in the archives of the Federal Ministry of Justice, later fell into the hands of Thomas Dehler. Arndt spent the Nazi era working as an attorney in the Berlin office of Fritz Schönberg, representing business clients and victims of political persecution. During the War, Arndt was classed by the authorities as a 'half Jew'. In 1943, he became a forced labourer in the military production plant *Organisation Todt* ('organization Death'). He was interned in the summer of 1944, escaped incognito in January 1945, and fled with his family to Westphalia in February 1945.

In 1949, Arndt began a twenty-year parliamentary career. He quickly distinguished himself as the SPD's most gifted lawyer and one of its best debaters. Colleagues revered his immense knowledge; opponents feared his withering tongue. Journalists quickly christened the endlessly quotable Arndt the 'crown jurist' of social democracy. Arndt abominated the title. It smacked of monarchy, he thought, and it echoed the nickname once given to Carl Schmitt, quondam 'crown jurist' of the NSDAP.

As German patriots, Dehler and Arndt both dreamed of restoring the rule of law in the aftermath of Nazi misrule. But they had very different visions of how to do so. Dehler's hopes centred on his own ministry. He envisioned the justice ministry emerging as a full-fledged 'law ministry' (*Rechtsministerium*) commissioned to 'safeguard the legality and constitutionality of all acts of state authority and, so far as the competences of the federal government reach, ensure that its measures are in harmony with these principles'.[39] In Dehler's view, the justice ministry would stand as 'Guardian of the Constitution' (*Hüter der Verfassung*).[40]

Arndt ascribed the same role and title to the Federal Constitutional Court. Within his own party, on the floor of Parliament, and in the press, Arndt was an indefatigable champion of the Karlsruhe tribunal.[41] He was a militant advocate of the Court's independence. Less than three weeks after the Court opened its doors, Arndt stood on the floor of the Bundestag and called for the Court's budgetary and administrative independence from the justice ministry.[42]

This Dehler was most unwilling to concede. The justice minister viewed the new Court with more anxiety than anticipation. Dehler never doubted the Court would play an important political role. This would have its healthy side, he acknowledged. In January 1951 Dehler characterized the Court as an 'outlet' for 'political tensions', a rule-of-law forum where 'controversies can be worked out'.[43] But Dehler was also disconcerted by the Court's 'extraordinary plethora of power'.[44] Just before the FCCA was passed, Dehler confided to

[39] ibid 143. [40] ibid. [41] See Gosewinkel (n 35) 471.
[42] *Verhandlungen des Deutschen Bundestages. Stenographische Berichte*, 165. Sitzung (27 September 1951) 6753.
[43] In Wengst, *Thomas Dehler* (n 35) 151. [44] ibid.

Robert Lehr, the interior minister, that he thought the Basic Law had granted the Court 'vastly excessive' powers, overburdening it 'with decisions ... hardly reconcilable with its character as a court'. Scrupulous care must be taken, Dehler cautioned, to keep the Court from making 'political decisions'. Otherwise, the public might see the justices 'less as the [country's] highest court, than as a political organ'.[45] Dehler cast this caution in the passive voice. But it was very clear by whom such 'scrupulous care' must be taken. Dehler's own ministry of justice would keep the Court at bay.

In the months that followed, Dehler's anxiety increased. He rarely saw eye to eye with the parliamentary committees charged with choosing the justices and was largely unhappy with their choices.[46] It was some consolation to have his friend Hermann Höpker-Aschoff presiding over the First Senate and his confidante Willi Geiger sitting on the Second. Such correspondents in key places, Dehler hoped, would facilitate his supervision of the Court's work.

Dehler soon moved to justify and secure this supervisory role. Already needled by Arndt's call for the Court's separation from the justice ministry, Dehler was further provoked in October 1951, when Justice Kurt Zweigert, without consulting the justice minister, submitted his resignation from the Court directly to the federal president, Theodor Heuss, and the president of the Bundestag, Hermann Ehlers (CDU). Zweigert, Dehler fumed, had gone over the justice minister's head. Dehler asked justice ministry lawyers to prepare two memos, one addressing the 'Relationship Between the Federal Minister of Justice and the Federal Constitutional Court', and another addressing the 'Budget of the Federal Constitutional Court'.[47] Both memos proclaimed the Court's administrative and budgetary subordination to the justice ministry.

This unilateral assertion of authority roused the justices to a united response. At the behest of his peers, Justice Gerhard Leibholz prepared a circular that the Court later expanded into an official memo, which they submitted on 27 June 1952 to the federal president, the chairs of both chambers of Parliament, and the federal government.[48] Soon styled the 'Status Memo' (*Status-Denkschrift*), the document breathed defiance to Dehler's claim of supervisory prerogative. It was the nine-month-old Court's declaration of institutional independence.

The memo's message was straightforward: the Court's current institutional position was untenable. More pointedly, it was unconstitutional. The legal basis of the Court's argument was the first sentence of its founding charter: 'The Federal Constitutional Court is, vis-à-vis all other constitutional organs, a self-sufficient and independent court of justice'.[49] The FCCA's reference to all *other* constitutional organs' clearly indicated that the Court itself was a constitutional organ, coequal and coordinate with the 'other constitutional organs' from which the FCCA proclaimed the Court independent. As Leibholz wrote in his report, the Court 'must be addressed, not only as the head of judicial authority and in this capacity as guardian and guarantor of the constitution, but also as a constitutional organ, endowed with supreme authority, which politically as well as for purposes of constitutional law must

[45] ibid 152. [46] ibid 154–55.

[47] See 'Justizverwaltungsverhältnis zwischen Bundesjustizminister und Bundesverfassungsgericht; Haushalt des Bundesverfassungsgerichts' in *Rechtsstellung des Bundesverfassungsgerichts*, BAK, B 141/83. See also Frank Spieker, *Hermann Höpker-Aschoff: Vater der Finanzverfassung* (Duncker and Humblot 2004) 209.

[48] Both Leibholz's report and the Court's memo were later published. See 'Bericht des Berichterstatters an das Plenum des Bundesverfassungsgerichts zur 'Status-Frage' (1957) 6 *Jahrbuch des öffentlichen Rechts* 120; 'Denkschrift des Bundesverfassungsgerichts—Die Stellung des Bundesverfassungsgerichts' (1957) 6 *Jahrbuch des öffentlichen Rechts* 144.

[49] BVerfGG § 1 (1951).

be placed beside the Bundestag, the Bundesrat, the federal government, and the federal president'.[50]

From this lofty status, Leibholz derived three concrete demands, which the Court should present to its coordinate organs: first, remove the Court from the jurisdiction of the justice ministry and place its administration under the sole authority of the chief justice; second, give the Court its own budget, rather than a mere chapter within the budget of the justice ministry; and third, establish for the justices an official rank comparable to that of a cabinet minister or member of Parliament.

While the Court was still debating Leibholz's report and its recommendations, Dehler got wind of its content. The justice minister was apoplectic. He fired off a personal letter to Adenauer, in which he roared that the justices had 'fallen prey to an unparalleled hubris'.[51]

Other politicians, however, even within the chancellor's own party, were sympathetic to the Court's demands. After the Court circulated its memo on 27 June 1952, Heinrich von Brentano, a CDU parliamentarian, wrote to Adenauer urging that the government take a comprehensive position on the question as soon as possible.[52] In light of the myriad cases pending before the Court—including a handful of petitions from SPD parliamentarians—Brentano thought it politic to yield 'at least to the justified requests' set forth in the Court's memo. It would be rash, he warned, to make enemies in Karlsruhe.[53]

The government dug in. Dehler documented his own position but didn't publish it. Instead, he hoped to exploit divisions within the Court itself. Two of the justices disputed their colleagues' conclusions. One of them, Willi Geiger, took the extraordinary step of publishing unofficially what amounted to a dissenting opinion.[54] The other abstaining justice, more circumspect in his difference of opinion, was Hermann Höpker-Aschoff. Dehler hoped to make Höpker-Aschoff's resistance to the memo's 'demands' a rallying point of opposition. Dehler coaxed and cajoled his former comrade to oppose the memo publicly. Höpker-Aschoff hummed and hawed, but finally consented to write a position paper of his own—but only after his holiday. The chief justice's paper didn't appear until 13 October 1952.[55]

It was not the blistering refutation Dehler had hoped for. The chief justice merely distanced himself from the strongest statements in the Status Memo. Höpker-Aschoff argued that the Court was indeed a constitutional organ, but not in the strong sense urged by his colleagues. He saw in Section 1 of the FCCA no more than a guarantee of independence and self-sufficiency similar to that enjoyed by the other federal courts. He embraced the Memo's conclusions only on matters of secondary importance. On the budgetary question he remained undecided.

More forceful than Höpker-Aschoff's personal statement was the other justices' response to it. On 29 October, Vice President Rudolf Katz, joined by nineteen other justices,[56] wrote to Dehler directly. Katz rejected Höpker-Aschoff's arguments and repeated the demands of the original Memo.[57] The constitution, Katz insisted, required 'the Federal Constitutional Court's general independence, as a supreme federal organ, from *every*' other governmental

[50] 'Bericht des Berichterstatters' (n 48) 127.
[51] In Wengst, *Thomas Dehler* (n 35) 317.
[52] ibid 318.
[53] ibid.
[54] This move was all the more remarkable since the FCCA did not allow individual justices to publish dissents in the Court's formal judgments. See Willi Geiger, 'Ergänzende Bemerkungen zur Stellung des Bundesverfassungsgerichts' (1957) 6 *Jahrbuch des öffentlichen Rechts* 161.
[55] See Hermann Höpker-Aschoff, 'Schreiben des Präsidenten des Bundesverfassungsgerichts' (1957) 6 *Jahrbuch des öffentlichen Rechts* 44.
[56] At this point two seats on the Court were empty. Only Geiger didn't sign Katz's rebuttal.
[57] See Rudolf Katz, 'Schreiben des Vizepräsidenten des Bundesverfassungsgerichts an den Bundesminister der Justiz' (1957) 6 *Jahrbuch des öffentlichen Rechts* 156.

institution.[58] At this point Dehler again set out a position of his own, but again decided not to publish it.[59] Instead, after considerable delay, he commissioned Richard Thoma, a prominent conservative law professor (whom the government viewed as a potential justice)[60] to answer the Court's Memo.[61] Thoma did so at moderate length and some level of abstraction, completing his advisory opinion in late March 1953 and publishing it shortly thereafter.[62]

Thoma's paper found a warm reception both with his patrons in the government and within the conservative professoriate. Walter Jellinek wrote that he rejoiced over his colleague's memo, 'in the first place, because [it] shows you in top form, but also because you have really given it to the arrogant justices of the Constitutional Court'.[63] Speaking for the Court, Leibholz responded to Thoma's paper; but Thoma let the matter rest there, explaining that he valued his personal friendship with Leibholz too much to risk 'a genuine jurists' duel'.[64]

Meanwhile the Court received unexpected aid from Parliament. During its October 1952 budget debates, the Bundesrat proposed 'a special individual plan for the Constitutional Court within the federal budget'.[65] The Bundesrat renewed this proposal two months later.[66] In a letter to Adenauer, Rheinhold Maier (the Bundesrat's president) explained the motion in terms reminiscent of the Status Memo. Similar language now appeared over the signature of Hermann Höpker-Aschoff. Speaking for the Court on 16 January 1953, the chief justice stressed that 'the Federal Constitutional Court is not directly subordinate to a federal minister, but is rather a self-sufficient supreme federal organ'.[67] Though Höpker-Aschoff's personal view of the matter was unchanged,[68] in his official capacity he now joined his colleagues in a united call for institutional independence. Events of the previous month, not directly related to the status question but with Thomas Dehler at the centre of them, helped prompt the chief's apparent change of heart.

In March 1953, Social Democrats in the Bundestag, led by Adolf Arndt, proposed to grant the Court an independent budget.[69] The Bundestag promptly referred the proposal to its budget committee. While the committee deliberated, Dehler wrote a memo to his cabinet

[58] ibid.

[59] See 'Stellungnahme des Bundesministeriums der Justiz' (December 1952) BAK N 1194/8 (Nachlass Richard Thoma).

[60] See 'Vorschlagsliste der Bundesregierung' BAK B 136/4436.

[61] Mosler to Richard Thoma (29 December 1952) BAK N 1194/10 (Nachlass Richard Thoma) (asking for an advisory opinion in Dehler's behalf). See also Thomas Dehler to Richard Thoma (12 January 1953) BAK N 1194/10 (Nachlass Richard Thoma) (offering 2,000 DM for the solicited memo); Richard Thoma to Thomas Dehler (14 January 1953) BAK N 1194/10 (Nachlass Richard Thoma) (accepting the job on Dehler's terms). On Thoma, see Hans-Diether Rath, *Positivismus und Demokratie: Richard Thoma 1874–1957* (Duncker and Humbolt 1981).

[62] See Richard Thoma, 'Rechtsgutachten betreffend die Stellung des Bundesverfassungsgerichts: Auf Ersuchen des Herrn Bundesministers der Justiz' (1957) 6 *Jahrbuch des öffentlichen Rechts* 161.

[63] Walter Jellinek to Richard Thoma (29 June 1953) BA, N 1194/10 (Nachlass Richard Thoma). Another leading jurist, Ulrich Scheuner, wrote to Thoma, praising his memo and criticizing what he termed the 'Begriffsjurisprudenz', or conceptual jurisprudence, of the Court's own position. Ulrich Scheuner to Richard Thoma (21 June 1953) BAK N 1194/10 (Nachlass Richard Thoma).

[64] See Richard Thoma to Gerhard Leibholz (22 June 1953) BAK N 1194/10 (Nachlass Richard Thoma). Leibholz accepted the truce. See Gerhard Leibholz to Richard Thoma (25 June 1953) BAK N 1194/10 (Nachlass Richard Thoma). By this point, Thoma had also been alienated by Dehler's intemperate public railing against the Court. See Richard Thoma to Justice Ernst Friesenhahn (23 June 1953) BAK N 1194/10 (Nachlass Richard Thoma).

[65] *Stenographischer Bericht über die 94. Sitzung des Bundesrates* (24 October 1952) 495.

[66] *Stenographischer Bericht über die 94. Sitzung des Bundesrates* (19 December 1952) 602.

[67] See Hermann Höpker-Aschoff, 'Einzelplan des Bundesverfassungsgerichts im Haushaltsplan' (1957) 6 *Jahrbuch des öffentlichen Rechts* 159.

[68] As late as 25 June 1953, Leibholz wrote to Richard Thoma that Höpker-Aschoff and Geiger remained opposed to the position the Court had taken. Gerhard Leibholz to Richard Thoma (25 June 1953) BAK N 1194/10 (Nachlass Richard Thoma).

[69] See BAK B 141/84 116.

colleagues in which he embraced Thoma's arguments and rejected the Court's demands.[70] The cabinet adopted Dehler's position. The Court stood its ground. On 3 June, the Court answered Thoma's memo, summarizing the Status Memo and repeating its calls for reform.[71] The impasse was finally broken on 14 July, when the Bundestag amended the Federal Civil Servants Law to allow the chief justice to appoint civil servants on the same terms as the presidents of the Bundesrat and the Bundestag. At the same time, the Bundestag Budget Committee embraced the SPD proposal and granted the Court its own budget.

Unwilling to concede defeat, Dehler solicited memoranda from several prominent law professors on the Court's status.[72] But the theoretical question was moot; memoranda could not repeal statutes. Whatever the professors might say, the Court was now, de jure and de facto, a sovereign constitutional organ, coordinate in rank with the other branches of the federal government. Dehler continued to howl against the tide,[73] but he soon lost his voice. In autumn 1953, Adenauer declined to renew Dehler's appointment as justice minister. Höpker-Aschoff and others had threatened to resign if Dehler remained. Dehler learned by sad experience the wisdom of Brentano's warning. In a matter not directly related to the 'Status' dispute, the justice minister had roused enmity in Karlsruhe. It cost him his office, and thereby hangs a tale.

II. Judicial Politics

By refusing to renew Dehler's appointment after the 1953 elections, Adenauer proved less loyal to his justice minister than Dehler had been to him. Especially in the fractious field of foreign policy, Dehler had toed the Adenauerian line. That line, as noted, pointed aggressively westward. The chancellor wished to integrate the Federal Republic within an alliance—including a military alliance—of Western European democracies and the United States. Dehler made Adenauer's vision of Western integration his own. 'I will take this path with you', he promised the chancellor in January 1952.[74]

A. Crisis and solidarity: the European Defence Community Treaty

That path, as it happened, ran through Karlsruhe. Days after Dehler's letter to Adenauer, 144 members of the Bundestag petitioned the Constitutional Court to enjoin the centrepiece of Adenauer's foreign policy—a treaty creating the European Defence Community (ECD)—as unconstitutional. The petition was prophylactic. The treaty had not yet been signed. The parliamentary complainants were represented by two among their number: Bernhard Reismann from the Centre Party, and Adolf Arndt of the SPD.

The challenged treaty grew out of the so-called Pleven Plan, a proposal put forth in 1950 by René Pleven, the French prime minister, as an alternative to West German accession to NATO. The Plan sought to protect Western Europe against Soviet encroachments. The EDC

[70] Thomas Dehler to the Staatssekretär des Bundeskanzleramtes (13 April 1953) BAK B 122/2168.

[71] See Das Bundesverfassungsgericht, 'Bemerkungen des Bundesverfassungsgerichts zu dem Rechtsgutachten betr. Die Stellung des Bundesverfassungsgerichts von Geh. Hofrat Dr. Thomas vom 3. Juni 1953' (1957) 6 *Jahrbuch des öffentlichen Rechts* 194.

[72] See letters from Thomas Dehler to Professors Kaufmann, Thoma, Scheuner, and Mosler (12/13 June 1953) BAK B 141/85.

[73] As late as 17 June 1953, Dehler wrote to Adenauer that he considered his criticisms of the Court over the previous months to be 'justified and necessary'. Thomas Dehler to Konrad Adenauer (17 June 1953) BAK B 136/4436.

[74] In Wengst, *Thomas Dehler* (n 35) 203.

envisioned a merger of the national armies of its six member states—France, Italy, West Germany, and the three Benelux nations—with command vested at the supranational level.

The SPD's vehement opposition to the proposal was not the product of unbending pacifism. After the outbreak of the Korean War, the SPD had no objection to rearmament in principle. The crucial question for the opposition was *how* the Federal Republic was to rearm—under what circumstances and at whose behest. The SPD's hostility to the proposed treaty had a nationalist impulse. Full restoration of (West) German sovereignty, the party maintained, must precede international defence alliances. Any such alliance, moreover, must not postpone German reunification indefinitely. In the SPD's eyes, the EDC Treaty failed on both counts.

Policy objections nurtured constitutional scruples. The opposition argued that the government lacked constitutional authority to sign the treaty and that the treaty violated substantive constitutional provisions. The theory underlying these objections rejected the existence of implied powers and insisted that enumerated powers be construed strictly. Initially, the government viewed the constitutional challenge with more contempt than concern. In February 1952, an insouciant Adenauer dismissed the petition as 'utterly hopeless and superfluous'.[75] In months to come, however, the chancellor's confidence began to crumble.

Adolf had already made aggressive use of constitutional challenges to oppose the government's foreign policy. He had challenged the constitutionality of the so-called Petersburg Agreement (which partly restored West German sovereignty at the price of international control of the Ruhr) and the German–French Economic Agreement (a trade and reparations deal that focused on the Saarland). In separate judgments dated 29 July 1952, the Court rejected both petitions.[76] The very next day, the First Senate rejected as inadmissible the SPD's prophylactic petition against the EDC Treaty.[77] The challenge, the Senate held, was premature. Only after the treaty had been ratified in Parliament could the Court assess its constitutionality.

This was a disappointment to the opposition, but also a partial victory. The Court's holding implied that treaties become law, not when the executive signs them, but when Parliament approves them. By the time the Court ruled on this procedural point, the treaties had already been signed—in Paris on 26 May, in Bonn the day after. And yet the Court held that the treaty was not yet law. The Court's decision meant delay, not defeat. The SPD could renew its challenge after the treaty was affirmed in Parliament.

However the opposition might console itself, the government was jubilant. 'The federal government', crowed an official bulletin of 31 July, 'finds particular satisfaction in the fact that its point of view in the procedural question has now been vindicated by the Federal Constitutional Court. It now awaits with confidence the Federal Constitutional Court's resolution, in response to the request of Federal President Heuss, of the contested substantive issue'.[78]

The presidential petition mentioned in the bulletin had been submitted on 10 June. The FCCA allowed the federal president to solicit from the Court an advisory opinion on the constitutionality of a pending law. Presumably at Adenaur's urging, Heuss had done so with respect to the EDC Treaty. Heuss presented his request as a statesmanlike bow to the letter

[75] *Verhandlungen des Deutschen Bundestags. Stenographische Berichte*, 190. Sitzung (7 February 1952) 8106.

[76] 1 BVerfGE 351 (1951) (Petersberg Agreement); 1 BVerfGE 372 (1951) (economic agreement).

[77] 1 BVerfGE 396 (1952).

[78] See *Bulletin des Presse- und Informationsamtes der Bundesregierung* (31 July 1952).

of the constitution; opposition politicians and journalists perceived it as a tactical manipulation of the Constitutional Court.

In choosing the Court's first members, coalition and opposition made contradictory calculations about the relative importance of the Court's two Senates. Adolf Arndt, the SPD's leading negotiator, was convinced that the First Senate, with jurisdiction over abstract review of legislation and questions of fundamental rights, would be the more important chamber politically. Willi Geiger, the government's leading counsellor, was sure the Second Senate, charged with mediating disputes between branches and levels of government and with policing the boundaries of state power, would wield greater political weight. The opposition parties worked to place their favoured candidates on the First Senate; the coalition parties to place theirs on the Second.[79]

Though few of the justices were openly affiliated with a political party, observers calculated that seven of the eleven justices on the First Senate favoured the opposition, while seven of the eleven judges on the Second Senate supported the government.[80] Wags saddled both Senates with Stendhalian epithets—the First Senate, ostensibly opposition-friendly, was branded the 'Red' Senate; the Second Senate, ostensibly government-friendly, the 'Black'.

As an appeal for abstract review (*Normenkontrol*), the opposition's challenge to the EDC Treaty would go before the First ('Red') Senate. Government sympathizers accused the opposition of gaming the Court's procedures to place the petition before 'their' Senate. Whether or not the accusation was justified, the government responded in kind. Under the FCCA, a presidential request for an advisory opinion went before a plenum of both Senates. Those in the know assumed that the plenum would consist of eleven 'Black' justices, ten 'Red' justices, and one neutral.[81] Presumably an advisory opinion from the plenum would bind each individual Senate in future judgments on the merits. What's more, the Court gave temporal precedence to requests for advisory opinions. The government hoped the plenum would validate the treaty and preempt review by the hostile First Senate.

Though Dehler had toyed with the idea of requesting an advisory opinion as early as February, he grew skeptical when, towards the end of May, Adenauer announced his intention to embrace it.[82] Rumours of Dehler's private seething about the Court—its 'unparalleled hybris' and general unfitness[83]—had reached the justices, and Dehler may have thought it wise to avoid further provocation.

But it was not in Dehler's nature to avoid provocation for long. When Adenauer floated the plan again on 6 June, Dehler raised no objection. With the cabinet agreed, the government approached Heuss. The President demurred. Heuss, a grey eminence of the young Republic and founder of the postwar FDP, took seriously his constitutional obligation of strict neutrality, an obligation that made him reluctant to intervene in a partisan dispute.[84] What's more, Heuss was a liberal of the old school, hostile to law's intrusion into politics and skeptical, therefore, of judicial review.[85] Heuss didn't want matters of state decided by judges. For this reason, he initially balked at Adenauer's request. Perhaps for the same reason, he ultimately acquiesced. Better, perhaps, to have all of the red robes in Karlsruhe

[79] See Wengst (n 25) 241. [80] ibid 242.

[81] Two seats, those of Kurt Zweigert and Carl Leusser, had been vacated, but Parliament, convulsed by the controversy over the ECD, could not agree on replacements.

[82] See Wengst (n 25) 206; Arnulf Baring, *Außenpolitik in Adenauers Kanzlerdemokratie* (Oldenbourg 1969) 226. As early as February 1952, Adenauer's advisors had raised the possibility of putting the case before the Second Senate or the Plenum. See, e.g., Hans Globke to Konrad Adenauer (26 February 1952) BAK B 136/989.

[83] See Wengst (n 25) 206.

[84] On Heuss, see Thomas Hertfelder and Christiane Ketterle, *Theodor Heuss: Publizist, Politiker, Präsident* (Stiftung Bundespräsident-Theodor-Heuss-Haus 2003).

[85] See Baring (n 82) 224.

sustain government policy than to have half of them scuttle it. In any case, on 10 June 1952, Heuss asked the Court for an advisory opinion. He wished to know, he explained, whether he could, in good conscience, approve the treaty. Dehler and his assistants helped write the petition.[86]

The opposition was indignant. Adolf Arndt, who begged Heuss in vain not to submit the request, now called on Heuss to resign.[87] Arndt growled privately that he no longer considered Heuss president. Some of the justices, however, greeted the request with relief. Höpker-Aschoff beamed when he learned of it.[88] Responsibility for the decision would now be shared by the entire Court. Perhaps the plenum format would silence talk of a Red Senate and a Black. The First Senate proposed that the abstract review petition be set temporarily aside, so the plenum could focus on Heuss's request for an advisory opinion. The government consented eagerly;[89] the opposition firmly declined.[90] Faced with the split, the Court decided first to address whether the opposition complaint was even admissible, and then to consider Heuss's request. His ruse apparently foiled, Adenauer groaned that 'the entire construction of the Federal Constitutional Court now proves itself a failure'.[91]

The failure seemed less calamitous six weeks later, when the First Senate declared the prophylactic petition inadmissible. The 'Red' Senate had ruled against the SPD, thus challenging the predictive power of partisan calculations. Such unpredictability would prove a two-edged sword to a government that, on the basis of partisan calculations, had placed the substantive question of the ECD's constitutionality before the Court's plenum. Meanwhile the stakes continued to rise. On 4 August, Heuss expanded his request, asking the Court to rule on the constitutionality of the so-called 'Germany Treaty'—a transitional treaty hastening the end of occupation—as well.[92]

The opposition urged the Court to reject the entirety of Heuss's request as inadmissible in its entirety.[93] Arndt argued that, if the Court couldn't assess the constitutionality of pending legislation in response to an abstract review petition, it couldn't do so in response to a request for an advisory opinion either.[94] Unimpressed by the analogy, the Court ruled on 8 September that Heuss's request was admissible. Justice ministry lawyers went further, arguing that the plenum's response to Heuss's request would bind both Senates in subsequent decisions on the merits.[95] The opposition continued to insist that the request was inadmissible, but agreed nonetheless to take part in the proceedings.

Those proceedings drew international attention. In the theatre of West German politics, they assumed centre stage. The Court itself kept a low profile, conducting oral hearings and deliberating internally under its traditional terms of confidentiality. The leading parties to the case, by contrast, erupted in partisan polemics. In the middle of the fiery ring stood Thomas Dehler and Adolf Arndt.

Enmity between the two men began when the Bundestag debated the Petersberg Agreement in November 1949 and escalated over the following months as Arndt emptied

[86] See Chef des Bundespräsidialamtes to Otto Lenz (7 June 1952) BAK B 136/990.

[87] Wengst (n 25) 207. [88] See Baring (n 82) 227.

[89] See Thomas Dehler to Hermann Höpker-Aschoff (17 June 1952) BAK B 136/990.

[90] See Hans Globke to Konrad Adenauer (18 June 1952) BAK B 136/990. Globke indicates that Arndt and Reismann had sent the Court a telegram earlier that day. See also Adolf Arndt and Bernhard Reismann to Bundesverfassungsgericht, First Senate (19 June 1952) BAK B 136/990 (a further explanation for the previous day's refusal). The government knew the opposition would decline the proposal before sending its own acceptance. See Hans Globke to Konrad Adenauer (16 June 1952) BAK B 136/990.

[91] In Wengst (n 25) 207. [92] *Der Kampf um den Wehrbeitrag*, vol 2 of 2 (Isaar 1958) 4.

[93] See the exchange of letters between Arndt and Höpker-Aschoff in ibid 829–35.

[94] See Adolf Arndt to Hermann Höpker-Aschoff (14 August 1952), in ibid 830–31.

[95] See the statements of Walter Strauβ (of the federal justice ministry) and Erich Kaufmann (of the chancellery) during the oral arguments on 18 July 1952 in *Der Kampf um den Wehrbeitrag*, vol 1 of 2 (n 92) 390, 427, 433.

a quiver of venomous barbs in Dehler's direction. In a 1950 newspaper article Arndt called Dehler 'the grave-digger of justice'. A month later he quipped in the Bundestag that 'when Herr Minister Dehler speaks, it is always a national misfortune'.[96] Dehler sued, unsuccessfully, for libel. He longed for the day when Arndt's parliamentary immunity would expire.[97] As the simultaneous controversies over the EDC Treaty and the status of the Constitutional Court unfolded in 1952, Dehler saw in Arndt the mastermind behind an insidious conspiracy to make the Court a repository of political power and a tool of socialist aims. Rumours that the Court would rule against the treaties suggested to Dehler's apprehensive mind that Arndt was succeeding.

In November, Dehler engaged in a mutual exchange of abuse with Arndt's SPD colleague, Otto-Heinrich Greve. Arndt wrote afterwards to Wilhelm Laforet, chairman of the Bundestag legal committee, that 'already once in the plenum of the Bundestag, I have unfortunately had to say that a federal minister of justice, who wishes to make political lies punishable by law, must first behave accordingly himself'.[98] The following day, at an FDP convention in Bad-Ems, Dehler delivered a speech that, even by his standards, was a monument of ill-consideration. 'You know', Dehler warned his hearers, 'that the barriers of the Constitutional Court apply to our treaties as well'. 'I would hope', he continued, 'that [in Karlsruhe] the spirit of socialism has no effect. I would hope that the highest German court makes decisions not of political will, but of law'.[99] Dehler's meaning was unmistakable, his words both admonitory and accusing. He insinuated openly that the Court leaned leftward, and that a ruling against the EDC Treaty would confirm that suspicion. Dehler's words were understood—by press and public, in Karlsruhe and in Bonn—as a thinly veiled threat.[100]

The SPD responded on 25 November with a censure motion in the Bundestag. Dehler's speech, they said, was calculated 'to raise doubts concerning the lawfulness and nonpartisanship of the Federal Constitutional Court'. Adenauer pled with Dehler to say nothing in response to the motion and to remain 'as calm as possible'. 'The statement in Ems', the chancellor wrote, 'was not very fortunate'.[101] While Adenauer was laconic, Heuss was indignant. Dehler's remarks about the Court, Heuss wrote, were 'impossible'. The insult to 'the judges themselves' was 'simply too injurious'. After reading the full report of the speech, Heuss sent Dehler an even sharper missive. The second message suggested that the personal bond between Dehler and Heuss had been severed.[102]

Dehler thus found himself with his party fraying internally; his friends turning against him; his chancellor upbraiding him; the Bundestag deliberating whether to censure him; and his justice ministry certain (he thought) to lose the most important legal case in its history. His political world was crumbling, and lurking behind the debacle he saw everywhere the figure of Adolf Arndt.

It was under these circumstances that Dehler learned of Arndt's 20 November letter to Laforet. Dehler responded with a bare-fisted personal attack. He wrote to Laforet on 2 December, explaining that for some time he had had in his possession a copy of a letter, dated 9 April 1933, in which Arndt petitioned National Socialist authorities for a promotion. In the letter, Arndt explained that in criminal cases he had helped to secure mild judgments against National Socialist defendants. Dehler drew the following

[96] In Wengst, *Thomas Dehler* (n 35) 186. [97] ibid 187.

[98] ibid 211 (Adolf Arndt to Wilhelm Laforet, 20 November 1952). [99] ibid 209.

[100] See 'Der Bundesjustizminister und seine juristischen Gegner' *General-Anzeiger* (3 December 1952); 'Im Schatten von Karlsruhe' *Stuttgarter Zeitung* (6 December 1952).

[101] In Wengst, *Thomas Dehler* (n 35) 209. [102] ibid 209–10.

conclusion: 'Dr. Arndt grossly offended, on self-seeking grounds, his duties as a judge.... He broke his judicial oath. He who today takes it upon himself to sit in judgment on the political pasts of others, bragged to the National Socialist rulers of having issued judgments in their favor'. Arndt's character, Dehler added, had not changed since 1933. He was utterly unfit for the role he arrogated to himself in West Germany's political and public life.[103] Adenauer himself backed Dehler in a written statement, denouncing Arndt as 'unworthy', as one whose 'malevolent game endangers our State'.[104] Days later Dehler, from the floor of the Bundestag, inveighed against the 'malicious attempt' of Arndt and his SPD colleagues 'to abuse law and justice'.[105]

Needless to say, the justices of the Federal Constitutional Court were less than flattered by Dehler's noisy suggestion that they were becoming pawns in the hands of Adolf Arndt.[106] Nor were the judges amused to be made the subjects of serial nose-counting in the popular press. At the very time the Court was battling for its institutional independence, its members resented the ubiquitous assumption that their votes in a matter of compelling national urgency would be dictated by partisan considerations. More galling still were government efforts to intimidate them. A government bulletin of 26 November 1952, for instance, concluded in dramatic, menacing tones:

> The Federal Constitutional Court has constantly declared in its jurisprudence that it does not wish to interfere in the legislative organs' political freedom of decision.... The reason for this constant jurisprudence lies close at hand: Karlsruhe wishes to remain a court and not to become a political parallel government [*Nebenregierung*] that must cloak its political decisions in juristic garb. Should Karlsruhe depart from this standard even once, it would mean the end of German constitutional justice.[107]

The 'end of German constitutional justice' might have referred to an irreparable loss of legitimacy—a serious enough fate. But the phrase also, and none too subtly, threatened the Court with outright abolition. The justices were fighting, and knew they were fighting, for their legitimacy and their institutional life. When the government followed this bully-pulpit lecture with another gambit of procedural manipulation, the justices' cup of gall spilled over.

When the Bundestag began its second reading of the EDC legislation in early December 1952, the Adenauer government was convinced that the Court's plenary advisory opinion would condemn the treaties as unconstitutional. The chief justice had hinted by letter that this might happen; Otto Lenz, Adenauer's chief of staff, reported positively that it would. Dehler related to a small circle of journalists in late March 1953 that the government knew in December that the plenum would vote against it.[108]

Convinced that the advisory opinion ploy had backfired, the cabinet fired back with another. In a second procedural ruse, the government sought to preempt a negative judgment from the (now) reportedly hostile Plenum with a positive judgment from the (still) presumably friendly Second Senate—just as in June it had sought to preempt the expectedly unsympathetic First Senate with an appeal to the ostensibly amicable Plenum. The idea,

[103] ibid. [104] ibid 212–13.

[105] In Gosewinkel (n 35) 336. Arndt's SPD colleagues issued a declaration of solidarity. See 'Vertrauen für Arndt' *FAZ* (8 December 1952).

[106] This was a frequent theme of Dehler's private correspondence as well. See Gosewinkel (n 35) 332.

[107] Stiertom Moehlen, 'Die Verfahrensfrage der Bonner Verträge vor dem Bundesverfassungsgericht' *Bulletin des Presse- und Informationsamtes der Bundesregierung* (26 November 1952), BAK B 136/990.

[108] See Baring (n 82) 239.

Dehler claimed, was his own—a brazen bid to manipulate not only the Court's procedures, but the government's own earlier attempt to manipulate those procedures.[109]

The government unveiled its second subterfuge on 5 December 1952. After passionate and acrimonious debate, a bare Bundestag majority had sustained the treaty on its second reading when Adenauer rose to make an announcement that bewildered his friends and baffled his foes. The conclusive third reading of the law, he said, would have to await yet another appeal to Karslruhe.[110] This appeal would actually accelerate the treaty's ratification, Adenauer explained, by forestalling a post-promulgation challenge from the opposition. At Adenauer's instigation, the CDU/CSU Bundestag faction would file a constitutional complaint against the SPD faction on the remarkable ground that the latter violated Article 42 GG (which guarantees majority rule in Parliament) by impeding the majority's right to promulgate the treaties.

Crucially, the appeal asked the Court to mediate an *Organstreit* (organ conflict) rather than conduct a *Normenkontrol* (abstract review). This meant that jurisdiction lay with the Second Senate rather than the First—the 'Black' rather than the 'Red'.[111] Moreover, prevailing Court practice gave organ conflicts temporal precedence over advisory opinions (just as it gave requests for advisory opinions temporal precedence over petitions for abstract review). This meant that the government could secure the Second Senate's constitutional blessing on the treaty before the opposition could even ask the First Senate to review it. It was an imaginative scheme, to be sure. It was also a transparent exercise in procedural, political, and institutional manipulation.

The Court responded with an iron show of solidarity. On the morning of Wednesday, 9 December, the Court convened oral arguments in the advisory opinion proceeding. Before the hearings began, Höpker-Aschoff astonished the assembled attorneys and international press agents with an announcement that flung the Federal Republic into constitutional crisis. The chief justice read from a statement based on a general proposal by Justice Leibholz. The statement's tone was one of Olympian detachment; its substance a roar of Promethean defiance.

The Court had witnessed with concern, Höpker-Aschoff began, the mounting chorus of disparaging comments in press and political circles concerning the Court and its members. As the appointed Guardian of the Constitution, the Court saw no need to dignify these aspersions with a detailed response (though it noted in passing that in other countries such remarks were sanctioned as contempt of court). The Court defended itself by vowing it would not stoop to defend itself.

Anyone who bothered to read the Court's published opinions, the statement continued, would realize that the Court's decisions were unfailingly dictated by legal and doctrinal rather than partisan and political considerations, and that in cases with a plainly political subject matter, the Court had scrupulously deferred to the legislature and the executive. Press and popular speculations about the outcome of the advisory opinion proceeding

[109] See Uwe Wesel, *Der Gang nach Karlsruhe. Das Bundesverfassungsgericht in der Geschichte der Bundesrepublik* (Karl Blessing 2004) 63.

[110] The chancellor's announcement reportedly left even some of his closest acolytes speechless. See Paul Sethe, 'Die Gelähmte Hauptstadt' *FAZ* (19 December 1952).

[111] Adenauer later insisted that he hoped to put the matter before the Second Senate merely to save time. The original division of labour between the chambers left the First Senate's docket flooded and the Second's nearly bare. As of 19 May 1952, 846 cases had come before the First Senate, 424 of which remained to be decided; by contrast, only eight cases had come before the Second Senate, three of which remained to be decided. Otto Lehr to Konrad Adenauer (19 May 1952) BAK B 136/4436. Adenauer told justices Gerhard Leibholz and Kurt Zweigert in February 1953, that he had been counselled that the Second Senate would resolve the question much more swiftly than the First. See Baring (n 82) 238.

were spun from whole cloth. Suggestions that one Senate might handle the same question in a manner materially different from the other were totally unfounded. Talk of a 'Red' and a 'Black' Senate must cease, as must tactical manipulation of the Court's various units.

From these premises the Plenum reached a two-part decision:

First, the Court would address the President's appeal for an advisory opinion before considering the government's most recent petition.

Second, the decision of the Plenum in the advisory opinion proceeding would bind both Senates in all future proceedings.[112]

To preempt speculation on the vote tally in these two decisions, a Court spokesman took the rare step of announcing publicly that the participating justices sustained the decisions by a vote of twenty to two.[113]

The announcement broke like a bombshell—rather to the surprise of the justices themselves. When government lawyers asked for a one-hour break to confer with their clients in Bonn, an impatient Höpker-Aschoff wondered why half an hour wouldn't suffice.[114] The justices seemed genuinely surprised by the government's reaction to the decisions—a testament, some have felt, to the justices' jurisprudential good faith and political naïveté.[115] In any event, half an hour did *not* suffice, and the proceedings were adjourned until the following day. In the meantime, the government begged President Heuss to make the adjournment permanent by withdrawing his request for an advisory opinion.

Heuss, once again, was reluctant. Days earlier he had made a landmark speech at the dedication of the Holocaust memorial at Bergen-Belsen.[116] After this show of statesmanship, Heuss was irked at being perceived again as doing the chancellor's bidding. But he ultimately went along with Adenauer's plan. The Court's announcement of the previous day, ran the official explanation, had altered the stakes. A decision on Heuss's petition would no longer produce what he originally sought—a non-binding advisory opinion—but binding constitutional law. A government bulletin announced Heuss's retraction in a piece pointedly entitled, 'Guardian of the Constitution'—a reference to Heuss, not the Court.[117] With the petition retracted the oral arguments were cancelled. But Adenauer and Dehler compensated in Bonn for the talking that wouldn't take place in Karlsruhe.[118]

On the evening of 10 December 1952, Adenauer and other high officials gathered in the cabinet room with thirty or so elite journalists for one of Adenauer's famous 'Chancellor's Teas'. During the discussion that followed, both Adenauer and Dehler were critical of the Court—the chancellor with cautious dissent, his minister with fiery venom. Dehler denounced the Plenum's decision of the previous day as 'utterly lawless', an infringement of the Basic Law and the FCCA, a 'nullum' that the federal government would never acknowledge. If he could be blunt, Dehler added, the greatest flaw in the decision was 'not its

[112] The conclusion, commonly known as the plenum's 'Bindungsbeschluss', or binding decision, was reported in 2 BVerfGE 79 (1952).

[113] See 'Telephonische Durchsage Amtmann Bendler' (10 December 1952) BAK B 136/990. Individual justices would also write to press leaders urging that speculation about the partisan composition of the Senates be avoided. See, e.g., Herbert Scholtissek to Chefredakteur Dr Roegele (22 December 1953) BAK N 1341/54 (Nachlass Martin Drath).

[114] *Der Kampf um den Wehrbeitrag*, vol 2 (n 92) 806.

[115] See Baring (n 82) 243.

[116] Jeffrey Herf, *Divided Memory: The Nazi Past in the Two Germanys* (Harvard University Press 1997) 317.

[117] 'Hüter der Verfassung' *Bulletin des Presse- und Informationsamtes der Bundesregierung* (12 December 1952).

[118] Immediately after learning of the Court's announcement Dehler drafted a boiling preliminary response, which he never published. He wrote that the decision had 'absolutely no binding authority', was 'legally insignificant', and that the Constitutional Court was just a court like any other. 'When it proclaims itself to be, no less than the Federal President, an independent constitutional organ charged with constitutional jurisdiction', he concluded, 'the concern arises that it has transgressed the borders of its mandate'. Thomas Dehler, 'Vorläufige Stellungnahme' (9 December 1952) BAK B 136/990.

partisan character'—although that was obvious enough—'but rather its deficient judicial quality'. Adenauer and Dehler both indicated their intention to rein in the Court by amending the FCCA.

This was strong medicine. The assembled journalists might be forgiven if they didn't take seriously Adenauer's parting request that they eschew talk of a constitutional crisis.

In any case they ignored it.[119] On the morning of 11 December the daily papers rang with cries of a 'Crisis of State', 'Constitutional Crisis', and 'Conflict between Bonn and Karlsruhe'.[120] All papers reported the Dehlerian diatribe in luxurious detail. By and large, press and public deplored the justice minister's invective. Leading lawyers sent Dehler a telegram asking him, in the name of the rule of law, to soften his tone and show greater regard for the reputation of Germany's highest court. But Dehler was implacable. 'You misunderstand the situation', he fired back. 'The Federal Constitutional Court has departed in an appalling manner from the path of law and has thereby created a genuine crisis'.[121] On 19 December, Dehler's telegram became public, eliciting howls of popular disapproval. Adenauer hurriedly distanced himself from the firebrand in his cabinet. Later that day, in an official statement, the government affirmed its respect for the Court and its decisions. Adenauer met personally with Heuss and Höpker-Aschoff to extend an olive branch.[122] The chancellor's apologies were accepted. Outwardly, the crisis abated.

Privately, however, the situation remained fraught. On 22 December, Höpker-Aschoff wrote to Dehler, highlighting in pointed terms the impropriety of public criticism by cabinet officials of Court decisions. Though Dehler had made such criticism a 'beloved form' of expression, Höpker-Aschoff considered it a 'sin against the idea of the *Rechtsstaat*'. Höpker-Aschoff also lectured Dehler about his public quarrel with Adolf Arndt: 'If you wish at all costs and at all times to conduct a tempestuous battle with the opposition, then you must abandon your office and play politics as an MP'.[123] Dehler may have been grateful at this point that Adenauer had persuaded him not to send a letter he had written to Höpker-Aschoff on 11 December. In that letter, Dehler cited a series of judgments he thought questionable or flatly wrong, and he accused the justices of 'misunderstanding the mission and purpose of the Federal Constitutional Court' and of embarking on a course that would lead, in process of time, to 'a serious upheaval of our law and our State'.[124] As 1953 dawned, the old friends made peace. A self-deprecating Dehler promised, during the New Year, to be on his best behavior.[125]

[119] *Der Spiegel* reported Adenauer's request in a piece impishly entitled, 'Staatskrise'. See 'Staatskrise—Montesquieu ist Schuld' *Der Spiegel* (17 December 1952). See also 'Verfassungskrise' *Deutschland-Union-Dienst* (9 December 1952) 1; 'Lenz: Unerfindliche Entscheidung' *Die Welt* (10 December 1952); 'Adenauer bittet Heuss Gutachten zurückzuziehen' *General-Anzeiger* (10 December 1952); 'Staatskrise' *FAZ* (10 December 1952); 'Warum Karlsruhe' *Die Welt* (10 December 1952); 'Verwirrung in Bonn' *Düsseldorfer Nachrichten* (10 December 1952); 'Die Kehrtwendung der Koalition' *Deutsche Zeitung* (10 December 1952); 'Ein gerissener Schachzug' *Deutsche Zeitung* (10 December 1952); 'Verfassungskrise? Staatskrise' *Welt am Sonntag* (14 December 1952).

[120] See, e.g., 'Verstoβ gegen das Grundgesetz' *Die Welt* (11 December 1952); 'Bücher übt scharfe Kritik' *Die Welt* (11 December 1952); 'Wagner: Eine Staatskrise' *FAZ* (11 December 1952); 'Noch keine Entscheidung' *FAZ* (11 December 1952); Hans Heinrich, 'Minen am Grundpfeiler' *Frankfurter Rundschau* (11 December 1952); 'Zug um Zug' *Frankfurter Neue Presse* (December 11, 1952); 'Die Entscheidung des Bundespräsidenten' *Die Welt* (11 December 1952); 'Ein Haufen Scherben' *General-Anzeiger* (11 December 1952); Walter Fredericia, 'Der Präsident und das Gericht' *Die Zeit* (11 December 1952).

[121] Telegram from Thomas Dehler to Wilhelm Zutt and Dr Geiler (10 December 1952) BAK B 136/990.

[122] See 'Integrierender Bestandteil des demokratischen Rechtsstaates' *Bulletin des Presse- und Informationsdienstes der Bundesregierung* (19 December 1952).

[123] In Wengst, *Thomas Dehler* (n 35) 217.

[124] Thomas Dehler to Hermann Höpker-Aschoff (11 December 1952) BAK B 136/990. Adenauer wrote to Dehler on 16 December, asking his justice minister not to write to Höpker-Aschoff until the storm of controversy had time to settle. Konrad Adenauer to Thomas Dehler (16 December 1952) BAK B 136/990.

[125] See Wengst, *Thomas Dehler* (n 35) 218.

But New Year's resolutions are hard to keep. Dehler was an irascible man under enormous pressures. He faced two censure motions in the Bundestag, and he worried restlessly about the fate of the treaties, which he regarded as the existential *sine qua non* of postwar German democracy. He responded to the censure motions with sarcasm and scorn, telling FDP press chief Josef Ungeheuer that he would be 'deeply concerned' if the SPD *approved* of him.[126] Dehler also continued to insist that it was his inescapable duty to tell the Court 'that it has taken an erroneous path and exceeded its authority'. He simply couldn't keep silent. 'I would rather sacrifice my office', he vowed, 'than passively observe when danger threatens People and State'.[127]

When the Bundestag considered the SPD censure motions in March 1953, Dehler was his old implacable self. He lambasted the Court's pretentions with an epithetical omnigatherum. The justices had crowned themselves a 'supra-government', a 'supra-parliament'—the 'masters of the constitution' rather than its guardians: *Herren* rather than *Hüter*. It was the duty of his office, Dehler insisted, 'carefully to oversee the jurisprudence of the Federal Constitutional Court'.[128] The censure motion failed; but cabinet and coalition distanced themselves from Dehler's unrepentant harangue.

The chief justice publicly contradicted it. In a 14 March radio broadcast, Höpker-Aschoff vigorously denied Dehler's suggestion that the Court coveted power and firmly rejected Dehler's assertion of a supervisory duty.[129] The conservative *Frankfurter Allgemeine Zeitung* saw in this public retort from the sitting chief justice a 'frightful' spectacle, but conceded that Dehler's intransigence had left the Court no other option. Before 1933, the article concluded, such an assault by a minister of justice on the independence of a court would have been unthinkable.[130] Dehler's public response to Höpker-Aschoff's radio speech was muted; his private reaction was incensed.[131] Outwardly, the two were once more reconciled. But persons close to both men knew their friendship was over.

The jurisprudential backdrop to this latest squabble was the Second Senate's decision of 7 March, in which the Court rejected as inadmissible the Union faction's *Organstreit* petition against the SPD minority.[132] That rejection was sharply stated. The decision's technical hinge was the Senate's ruling that parliamentary groups were not '*Organteile*' (parts of a constitutional organ) with the right of appeal in an *Organstreit*. But its substantive core was the Senate's insistence that a parliamentary majority suffers no legal injury when the opposition speaks its mind. On this point, the judgment read like a primer on democratic first principles: 'It is not only the right of the opposition to voice its political and constitutional reservations', the justices wrote; 'in a parliamentary democracy, it is their duty'.[133] The government's position, by contrast, would make the opposition's expression of constitutional reservations a constitutional offence. Were the Court to adopt this stance, it would upend 'the right to free speech of members of Parliament'—a right the Court characterized as the very 'lynchpin of a parliamentary-democratic constitution'.[134] The Senate closed with a curt retort to talk of the Court's politicization: 'Whether and what political consequences'

126 ibid 219. 127 ibid. 128 ibid 220.

129 Hermann Höpker-Aschoff, *Rundfunkrede* (14 March 1953) text in BAK B 136/4436. Höpker-Aschoff explained to Adenauer that he regretted making his row with Dehler public, but that Dehler's truculence had driven him to it. Hermann Höpker-Aschoff to Konrad Adenauer (16 March 1953) BAK B 136/4436.

130 'Flucht in die Öffentlichkeit' *FAZ* (17 March 1953).

131 Dehler privately called the justices 'a demoralized heap of interested parties'. And he reportedly explained to Robert Strobel of *Die Zeit* that he was conciliatory towards Höpker-Aschoff because the chief justice was 'a sick old man, who perhaps will die soon. I didn't want, at some later date, to have to upbraid myself with having refused to receive him'. In Wengst (n 25) 221.

132 2 BVerfGE 143 (1953). 133 ibid 170–71. 134 ibid 171.

might stem from this decision, the Senate concluded, 'may play no role for the Federal Constitutional Court. It must decide according to law alone'.[135]

The Court's disinclination to be manipulated or bullied was by this point quite clear, even—perhaps especially—to the Adenauer government. The Court had also exposed limitations on the predictive power of presumed party allegiance. Thrice in nine months the Court had overthrown partisan expectations. In July 1952, the First Senate had ruled against the SPD; in December 1952, the Plenum against the government; in March 1953, the Second Senate against the CDU/CSU parliamentary faction. Talk of the Red and the Black seemed better suited to French fiction than to German constitutional law.

The press was quick to note the Court's non-partisan independence and to praise it. The Swiss journalist Fritz René Allemann—soon to win fame and supply an international buzzword with his 1956 book *Bonn is not* Weimar—wrote in the *Stuttgarter Zeitung* that critics of the Court's partisan composition operated at a purely theoretical level. The Court had thoroughly refuted their assumptions in practice.[136] Many other journalists—including some who called loudly for the Court's institutional reform—agreed.[137] 'All in all', summarized Hans Schuster in the *Süddeutsche Zeitung*, 'the Karlsruhe Court has preserved its authority in laudable independence, all whispering about a "red" and a "black" Senate notwithstanding'.[138]

Having conspicuously exercised its independence in practice, the Court was soon granted independence by statute. Eleven days after the Second Senate rejected the government's *Organstreit* petition, the Bundestag opposition proposed for the Court a separate budget, which both chambers of Parliament approved three months later. In the EDC case itself, the upshot of so much procedural and institutional sound and fury was that fourteen months after the first prophylactic petition from the SPD, the Constitutional Court had yet to rule on the merits of a law that Parliament had yet to pass.

In May 1953, Parliament finally approved the law ratifying the Germany and EDC treaties—five months after Adenauer's second ill-starred exercise in procedural coup-de-main. With the law formally enacted, Social Democrats in the Bundestag promptly asked the First Senate to declare it unconstitutional. The Senate's response was less than alacritous. New Bundestag elections loomed in September, carrying with them the possibility of a CDU victory sufficiently sweeping to allow the governing coalition, by a two-thirds parliamentary supermajority, to secure the treaty by constitutional amendment. The justices may not have intended to follow the election returns; but they were willing at least to await them.

This is not to say that internal debate ceased. The justices of the First Senate conducted a vigorous discussion of the case's merits throughout the summer of 1953.[139] Justice Konrad Zweigert sparked the discussion with a memo explicitly invoking the 'political question' doctrine of the United State Supreme Court and suggesting that several 'particulars' of the EDC case raised seriously the question whether the treaties' constitutionality was justiciable.[140] Justice Leibholz had invoked the political question doctrine during the

135 ibid 181.

136 F.R. Allemann, 'Demokratie ohne Tradition' *Stuttgarter Zeitung* (16 February 1953) 1.

137 See, e.g., Wilhelm Throm, 'Was ist in Karlsruhe reformbedürftig' *FAZ* (16 December 1952); Hans Baumgarten, 'Nicht Götter und nicht Partei' *FAZ* (6 December 1952).

138 Hans Schuster, 'Weiser und gerechter Richter' *Süddeutsche Zeitung* (17 March 1953) 1.

139 See Konrad Zweigert, *Diskussionsbetrag zur Frage der Justiziabilität der Normenkontrolle über die Vertragswerke* (10 July 1953) (marked 'Secret'); Erwin Stein, *Bemerkungen* (13 July 1953); Martin Drath, *Bemerkungen zu den Stellungnahmen der Herren Kollegen Dr Zweigert und Dr Stein* (21 July 1953), all in BAK N 1129/32 (Nachlass Höpker-Aschoff).

140 See Zweigert (n 139).

Plenum's discussions the previous December,[141] and Höpker-Aschoff—who for months had longed for a political resolution of the conflict,[142] which he termed an 'almost unbearable burden'[143]—was intrigued by this potential means of escape. In Höpker-Aschoff's personal papers are some handwritten notes on the political question doctrine, including extensive annotations on the American law review articles cited in Zweigert's memo.

While the justices pondered American doctrine they also scrutinized West German politics. The justices defended their Fabian approach on democratic, rather than pragmatic grounds. In a lecture at the Technical University in Karlsruhe, given 24 June 1953, Höpker-Aschoff said of the EDC proceeding:

> [T]his history will not be decided today or tomorrow, but only after the elections. Perhaps this is just as well ... for the political landscape changes from day to day. A politician can conceive of various possibilities of foreign and domestic politics that will perhaps make a judicial decision concerning these treaty laws entirely superfluous.[144]

The chief justice's hopeful tone was unmistakable. The political representatives of the West German people, he hoped, not the judges of the Federal Constitutional Court, would decide the crucial questions of rearmament and Western integration.

Events did not disappoint him. At the September polls, the CDU scored a dramatic victory. With its coalition commanding a two-thirds parliamentary majority, the government could—and did—amend the constitution to allow the treaties.

Even so, the SPD refused to surrender. It argued that the new constitutional amendment was itself unconstitutional. The notion was paradoxical but not preposterous.[145] The Basic Law banned amendments undermining the social, federalist structure of the Federal Republic, or the principles of human dignity, separation of powers, popular sovereignty, and the rule of law.[146] The Court's jurisprudence had already acknowledged the theoretical possibility of unconstitutional constitutional amendments.[147] The SPD argued that the treaties infringed principles declared sacrosanct by the 'eternity clause' of Article 79 GG, and that therefore the amendments enabling the treaties were themselves unconstitutional.

When he presented this argument publicly, Arndt summoned an unlikely theoretical ally:

> Article 79 allows only a limited power to amend or complete the constitution. But even if Article 79 was not in the constitution, there would still be a material limit on constitutional amendment. This is an insight that we owe to the work of Carl Schmitt. I have no hesitation at all in citing the devil; for sometimes the power that constantly negates also creates good. People like Carl Schmitt or Ernst Jünger ... have, through [a] kind of debris-clearing, fulfilled a thoroughly positive historical function.[148]

[141] Baring (n 82) 228.

[142] See Hermann Höpker-Aschoff to Otto Lenz (16 December 1952) BAK B 136/990.

[143] In Schuster (n 138) 1.

[144] Hermann Höpker-Aschoff, *Stellung und Aufgaben des Bundesverfassungsgerichts* (24 June 1953) BAK N 1129/30.

[145] For an international survey of the idea of unconstitutional constitutional amendments, see Yaniv Roznai, 'Unconstitutional Constitutional Amendments—The Migration and Success of a Constitutional Idea' (2013) 61 *Am. J. Comp. L.* 657.

[146] See Article 79(3) GG. [147] See 2 BVerfGE 1 (1952).

[148] Protokolle der 7. Sitzung des Bundestagsauschusses für Rechtswesen und Verfassungsrecht (19 February 1954).

In Arndt's view, Schmitt's great negative contribution was to discredit a brand of legal positivism that placed no limits on the lawgiver or constitution-giver even before National Socialism rendered that theory untenable.[149]

From his 'exile' in Plettenberg, Schmitt smiled at this backhanded compliment.[150] His own concern, as he told Horst Ehmke, was not that the power of ordinary or constitutional lawmakers would go unchecked, but that the Constitutional Court itself would 'become the nesting place of apocryphal acts of sovereignty'. This was sure to happen, Schmitt sniped, 'if one so fantastically expands the legal force of its grounds of decision as is now thoughtlessly occurring'.[151] Arndt, who could not have been further from Schmitt's preoccupations of the 1950s, invoked Schmitt's precept of the 1920s in asking the First Senate to invalidate the new amendments. He filed a petition to this effect, and the First Senate fixed oral arguments for October 1954.

They never took place. When October arrived the question was moot. A new threat to the treaties—this one fatal—emerged not from the opposition in Bonn or the Court in Karlsruhe, but from the National Assembly in Paris. After months of mounting hostility to the treaties among the French populace and political class, the National Assembly voted to reject them on 30 August 1954. The Constitutional Court never passed judgment on their constitutional merits.

The demise of the treaties in Paris, however, marked little more than a temporary hitch in Adenauer's policy of Western integration. In fact, the Parisian debacle substantially increased his bargaining power. In October 1954, when the Constitutional Court might have been hearing oral arguments in an extraordinary challenge to the constitutionality of constitutional amendments, a multi-power conference in Paris rescinded the occupation statutes, made the Federal Republic a member of NATO, vested it with nearly complete national sovereignty, and authorized it to create a national army. The French National Assembly ratified these 'Paris Treaties' at the end of 1954; the German Bundestag and Bundesrat followed suit in February and March of the following year. After some birth pangs, the West German *Bundeswehr* (Federal Army) grew steadily after 1956. The Federal Republic's integration within the NATO West proceeded apace.

The Federal Constitutional Court did nothing, in the long run, to hinder this development. And perhaps the justices had no wish to do so. But in the controversy over the European Defence Community, the Court survived an institutional crisis and secured the public's esteem. More significant still, it had consolidated its self-understanding as a court. In the Status Memo, the Court had defined itself in theory; in the EDC case, it had done so in practice.

And it had done so with only one dissent. Crucial in this connection was the conversion, prodded by perceived bullying from the Adenauer government, of Hermann Höpker-Aschoff, the chief justice. After dissenting from his colleagues in the Status Memo debate, Höpker-Aschoff became the mouthpiece of solidarity during the EDC controversy. Höpker-Aschoff's increasing alienation from the truculent Thomas Dehler certainly played a role in this transition. So, perhaps, did the long memory of Adenauer's callous letter asking Höpker-Aschoff to withdraw his candidacy for the Court's presidency. In any case, Höpker-Aschoff, who once bristled at the thought of removing to 'horrible Karlsruhe', became a symbol of the Court's settled independence.

149 In his *Verfassungslehre* ('Constitutional Doctrine') (Duncker and Humblot 1928), Schmitt had argued that no constitutional amendment could alter the basic decision (*Grundentscheidung*) or fundamental norm (*Grundnorm*) animating the constitution.

150 Carl Schmitt to Werner Weber (17 May 1954) BAK N 1529/44 (Nachlass Werner Weber).

151 ibid.

Höpker-Aschoff died, in Karlsruhe, on 15 January 1954, sixteen days shy of his seventy-first birthday and months before the final resolution of the EDC row. The public mourned the chief justice's passing; the press sang his praises. Newspapers lauded Höpker-Aschoff for his independence and devotion to the rule of law, eulogizing him as the 'Guardian of the *Rechtsstaat*' and 'Conscience of the Nation'.[152] 'No one else in the country', one eulogist wrote, 'has done more than Höpker-Aschoff to bring before the world the reputation of our State—born in great pain and under many limitations—as a *Rechtsstaat*. One mustn't forget this amid so much talk of the façade of the so-called "German miracle."'[153]

Though the economic miracle continued to dazzle, the work of the Constitutional Court had become, at least for some, the source of a deeper—a moral—legitimacy for the new state. By refusing to bow before political pressures or to brook bids at political manipulation, the Court gave many citizens confidence that the *Rechtsstaat* had been restored. In a country in which many prized the rule of law more than democracy, this perception did much to secure for the Court the public's enduring trust. In the years that followed, as democratic values gained popular footing, many observers celebrated the Court's earlier vindication of the rule of law as a precondition of democratization. In its first and greatest institutional crisis, the Court had declared fearlessly and unitedly that the postwar West German state was ruled by laws rather than politicians. The Court thereby earned a kind of capital that later allowed it to restrain the country's politicians directly without provoking opposition on democratic grounds.

The Court emerged from the EDC controversy with enormous prestige. The same was not true of another major player in the EDC contest—one who witnessed the controversy's conclusion from the sidelines. The 1953 elections that swept the coalition parties to unprecedented power swept Thomas Dehler out of the ministry of justice. An ailing Höpker-Aschoff, joined by several of his colleagues, told Adenauer he would resign if Dehler were re-appointed.[154] Theodor Heuss, for his part, threatened not to sign such a renewal. Dehler left office on 20 October 1953, replaced by an FDP peer, the sixty-nine-year-old Fritz Neumayer. Neumayer was quick to announce a more emollient posture vis-à-vis the Court. He promised at the swearing in of Josef Wintrich as Höpker-Aschoff's successor that he would 'not treat this Court as a subject of his own competence, but [would] regard it as a constitutional organ, toward which he felt drawn by the kinship of certain duties'.[155] In future years Dehler, as an MP, would speak out in the Court's defence and was even reconciled with Adolf Arndt. His successors in the justice ministry would occasionally quarrel with the Constitutional Court. But none would ever dispute the Court's stature as supreme Guardian of the Constitution, nor offer his services as guardian of the guardians.

In the years to come, the SPD, led as ever by Adolf Arndt, challenged other diplomatic products of Adenauer's Western policy before the Constitutional Court. Each time, the Court approved the challenged measure with notable deference.[156] The justices would bow before no government attempt to bully it into acquiescence; but having stood its institutional ground, the Court was quick to place its constitutional imprimatur on government policy. Never in the 1950s did the Court invalidate a major initiative of the Adenauer government.

[152] See Friedrich Noppert, 'Respekt vor einem unabhängigen Richter' *Deutsche National-Zeitung* (16 January 1954); 'Gewissen der Nation. Höpker-Aschoffs Tod' *Rheinischer Zeitung* (18 January 1954); 'Der Hüter des Rechtstaates' *Deutsche Zeitung und Wirtschafts-Zeitung* (20 January 1954).

[153] Hans Schuster, 'Er baute dem Recht ein festes Haus' *Süddeutsche Zeitung* (18 January 1954) 3.

[154] See Erich Mende, *Die neue Freiheit, 1945–1961* (Herbig 1984) 381.

[155] Fritz Neumayer, *Ansprache an der Amtseinführung von Josef Winter als Bundesverfassungsgerichtspräsident* BAK B 136/4436.

[156] See, e.g., 2 BVerfGE 347 (1953); 4 BVerfGE 157 (1955); 6 BVerfGE 290 (1957).

But never, on the other hand, did it recognize any limits on its power to do so. Indeed, the most enduring legacy of the early Court's foreign policy judgments lay in what the justices did *not* decide—namely, that questions of a certain political pitch and moment lay outside its jurisdiction. If ever a case furnished a brilliant opportunity for the Court to fashion a German political question doctrine, that case was the controversy over the European Defence treaties. Expert legal opinion was fiercely divided and the political stakes were nothing short of existential. The justices were sorely tempted, and the historian is tempted to hazard that, had a ruling on the constitutional merits of the treaties proved otherwise inescapable, the First Senate would have escaped it by articulating a political question doctrine.

But, in the end, they had no chance or need to do so. The Parisian death of the treaties meant not only that the Court would never assess the treaties' constitutionality, but also that it would never avoid doing so by pleading a political question. The precedential force of this non-decision, however, was not that the matter remained open, but that it seemed firmly decided *against* a political question doctrine. The most explosive foreign policy conflict of the early Federal Republic had come before the Constitutional Court and concluded with no suggestion from the Court that the matter was not justiciable. In the months that followed, foreign policy questions of lesser moment came before the Court and were disposed of in a manner deferential to the government's position but unyielding on the Court's jurisdiction. The upshot of these precedents was that constitutional questions incident to foreign policy—and, by implication, to political questions generally—were all justiciable. The consolidation of such universal constitutional jurisdiction was the most enduring political development of the Court's first decade. Combined with epochal doctrinal developments later in the decade, this jurisdictional breadth would give dramatic substantive weight to the Court's self-appointed role as Guardian of the Constitution.

III. Judicial Supremacy and the Politics of the Past

The Court's great institutional clash with the executive branch was followed in short order by a dramatic confrontation with other organs of the judiciary. Having battled for independence vis-à-vis the political branches, the Court now fought for supremacy vis-à-vis other courts. In the course of this conflict, the Court confronted the Gorgon of the Nazi past. This confrontation was intensely controversial, provoking the opposition of the legal academy like no set of judgments before or since. At the time, the Court's stance won it few friends. But when the dust settled, the Court had no rivals within the German judiciary. And the moral high ground the Court assumed would later become a source of enormous legitimacy. In its forceful repudiation of the Nazi state, and in its definition of the Federal Republic in opposition to its predecessor, the Constitutional Court was unique among the supreme institutions of the early Federal Republic. The Court set in motion what the historian Norbert Frei has called 'an arc of high-court decisions in the field of Past Politics [*Vergangenheitspolitik*] that accompanied the transformation process of the post-National Socialist *Volksgemeinschaft* into the society of the Federal Republic'.[157] It is difficult to dispute, Frei added, that 'the interventions of the Federal Constitutional Court assisted in the development of a democratic ethos'.[158]

157 Norbert Frei, 'Transformationsprozesse. Das Bundesverfassungsgericht als vergangenheitspolitischer Akteur in den Anfangsjahren der Bundesrepublik' in Michael Stolleis (ed), *Herzkammern der Republik. Die Deutschen und das Bundesverfassungsgericht* (CH Beck 2011) 64, 81.

158 ibid. See also Brun-Otto Bryde, 'Der Beitrag des Bundesverfassungsgerichts zur Demokratisierung der Bundesrepublik' in Robert Chr. van Ooyen and Martin HW Möllers (eds), *Das Bundesverfassungsgericht im politischen System* (Springer Verlag 2006) 321.

A. Nazi-era civil servants

The underlying issue of the conflict among the courts was a federal law, mentioned earlier, passed on 11 May 1951 pursuant to Article 131 GG.[159] One of the prickliest points of debate in the Parliamentary Council concerned the status of Nazi-era civil servants in the new West German state. Underlying this question was a fierce tension between the Allied occupiers' desire for a de-nazified state and the inescapable fact that the individuals with the experience and expertise needed to administer the new state had almost all served the Third Reich. This tension raised agonizing questions. What would be the status of Nazi-era civil servants in the legal order of the Federal Republic? What effect did the dissolution of the German state following the unconditional surrender of 8 May 1945 have on the legal standing of its civil servants?

To these questions the Parliamentary Council responded with a spectacular punt. Article 131 GG provided that the legal standing of civil servants would be settled by future legislation. Parliament responded to this mandate two years later by passing what became known as the '131 Law', or 'G 131'.

The law was a compromise measure, guaranteeing a form of transitional insurance (around 55 per cent of lost salary) to deposed civil servants and providing an extensive right of reappointment that excluded only former members of the Gestapo. The law breathed the restorative spirit of the times.[160] Its impulse was reconciliatory and its terms generous. Its immediate effect was the widespread integration of Nazi-era civil servants within the administrative apparatus of the Federal Republic.

Even so, the law elicited cries of protest from the civil servants it affected. The courts were flooded with complaints from aggrieved bureaucrats concerned with more than the recovery of wages and benefits. The question was partly economic—not everyone recovered posts or pensions—but mostly moral. At issue was the civil servants' relationship to the Nazi regime and responsibility for its crimes. The law had failed formally to dissociate Nazi-era civil servants from the regime that employed them. The '131-ers' cast their complaints as a question of honour.[161]

They found a receptive audience with Hermann Weinkauff, president of the Federal Court of Justice (FCJ), the country's supreme court of civil and criminal jurisdiction.[162] Within the federal judiciary, Weinkauff led the charge against G 131—and against what he perceived as the pretension and encroachments of the Federal Constitutional Court. Remembered by colleagues and even family members as introverted and aloof, Weinkauff, though he never joined the Nazi party, spent much of the period from 1933 to 1945 as an assistant and judge on the notorious third criminal chamber of the *Reichsgericht*.[163] The third chamber was

[159] For background on the Article 131 law and the controversy that followed, see Michael Kirn, *Verfassungsumsturz oder Rechtskontinuität? Die Stellung der Jurisprudenz nach 1945 zum Dritten Reich, insbesondere die Konflikte um die Kontinuität der Beamtenrechte nach Art. 131 Grundgesetz* (Duncker and Humblot 1972); Frei (n 8) 69–101; Udo Wengst, *Beamtentum zwischen Reform und Tradition: Beamtengesetzgebung in der Gründungsphase der Bundesrepublik Deutschland 1948–1953* (Droste 1988); and Wolfgang Langhorst, *Beamtentum und Artikel 131 des Grundgesetzes: Eine Untersuchung über Bedeutung und Auswirkung der Gesetzgebung zum Artikel 131 des Grundgesetzes unter Einbeziehung der Position der SPD zum Berufsbeamtentum* (Oldenbourg 1994).

[160] In his hugely influential *Vergangenheitspolitik* (*Politics of the Past*, 1996), Norbert Frei cites G 131 as a singular instance of the Adenauer-era politics of amnesty and amnesia. See Frei (n 8) 69–101.

[161] See ibid 71.

[162] On Weinkauff, see Daniel Herbe, *Hermann Weinkauff (1894–1961): Der erste Präsident des Bundesgerichtshofs* (Mohr Siebeck 2008). On the early history of the Federal Court of Justice, see Klaus-Detlev Godau-Schüttke, *Der Bundesgerichtshof: Justiz in Deutschland* (Tischler 2005).

[163] See Herbe (n 162) 2.

infamous for its so-called 'blood protection jurisprudence', and although Weinkauff didn't participate in these cases, he frequently acted as rapporteur of judgments in which the third chamber reviewed cases involving 'dangerous habitual criminals'—a notion aimed at persons genetically and racially disposed to a life of crime.[164] In 1938, the regime awarded Weinkauff the silver Medal of Honour in acknowledgement of his twenty-five years of 'loyal labour in the service of the German people'.[165]

Weinkauff's postwar career was furthered by another jurist who locked horns with the Constitutional Court. In May 1947, Weinkauff joined the Bamberg Jurists Society, where he drew the notice of the society's founder, Thomas Dehler. In September 1949, Weinkauff replaced Dehler as president of the Bamberg Court of Appeals. In early 1950, Dehler unsuccessfully urged Adenauer to appoint Weinkauff to replace Walter Strauβ as State Secretary in the justice ministry. Later that year Dehler secured Weinkauff's appointment as founding president of the Court of Justice. At the opening of the new FCJ building in October 1950, Dehler eulogized Weinkauff as 'the model for German judges, the measure to which they should conform'.[166] In 1951, Dehler pushed unsuccessfully for Weinkauff's nomination as first chief justice of the Federal Constitutional Court.[167] Had he succeeded, we might be telling a very different story.

Weinkauff was one of Dehler's confidantes as the justice minister brooded and blustered about the incipient might of the Constitutional Court. In December 1952, at the height of the European Defence Community controversy, Dehler wrote to Weinkauff: 'Not everyone is aware what dangers for law and for justice proceed from the Constitutional Court. Perhaps you will have understood why I have loudly opposed the beginning of a development that can only end in catastrophe'.[168] The next summer, after a standing committee of the German Jurists' Convention issued a circular supporting the Court's position in the Status debate,[169] Weinkauff assured Dehler of his dissent from the circular. 'I think it worthwhile to explain', Weinkauff noted, 'that by no means do I share this opinion, that I consider the action of the standing deputation uncalled-for, and I condemn it'.[170] Weinkauff shared Dehler's disquietude about the Constitutional Court, and he passionately opposed the premises, if not the substance, of G 131. These two impulses united in a dramatic confrontation between the two federal courts sitting in sleepy Karlsruhe.

Before G 131 was promulgated, the Court of Justice's Third Civil Senate declared the law's basic concept constitutionally sound.[171] But in the months that followed, the First Civil Senate and the Grand Senate for Civil Matters, both presided over by Weinkauff, questioned the law's constitutionality. The Grand Senate did so implicitly in a formal decision;[172] the First Civil Senate did so explicitly in an advisory referral opinion.[173]

Referral opinions were a novelty. The Basic Law granted plenary constitutional jurisdiction to the Constitutional Court. The FCCA originally provided that lower courts could not refer constitutional questions directly to Karlsruhe, but must route them instead through the relevant federal supreme court.[174] Constitutional referrals in civil and criminal cases went through the Federal Court of Justice; in labour law cases, through the Federal Labour

164 See ibid 44. According to Herbe, Weinkauff's record in these cases is impossible to judge—the relevant records are locked away in Moscow. ibid 45.

165 ibid 53–54.

166 Thomas Dehler, 'Weinkauff zum 70. Geburtstag' (1964) 17 *NJW* 489.

167 ibid 78, 82, 84–86, 93.

168 In Wengst (n 25) 218.

169 'Deutscher Juristentag, Ständige Deputation, *Rundschreiben* Nr. 2/1953' (9 June 1953) copy in BAK B 136/4436.

170 Hermann Weinkauff to Thomas Dehler (16 June 1953) BAK B 136/4436.

171 1 BGHZ 274, 275ff (1951); See also 14 BGHZ 138 (1954).

172 6 BGHZ 208 (1952).

173 11 BGHZ, Anh. 1 (1952).

174 See BVerfGG § 80(1).

Court; and so on. The federal court then relayed the matter to the Constitutional Court, often appending a detailed advisory opinion that proposed how the constitutional question should be resolved. Weinkauff's Federal Court of Justice filed such advisory opinions frequently in an effort to shape the constitutional jurisprudence of the young Republic.

Far from viewing this as an encroachment on its own jurisdiction, the Constitutional Court actively encouraged the practice. Overwhelmed by a groaning docket, the Court hoped to process referrals more efficiently by enlisting the help of the referring courts. In 1953, Hermann Höpker-Aschoff issued a circular asking the other federal courts to take an advisory stand on each constitutional question they referred. The scheme ran smoothly for a time, to the mutual gratification of all involved. By mid-decade, however, advisory opinions had become an instrument of high-court power play. Hermann Weinkauff proved the most aggressive player.

Weinkauff was a great champion of the postwar natural law 'Renaissance',[175] and his chambers' condemnation of G 131 was rooted in the natural law notion that a state's civil service lives independent of the rise and fall of particular governments. Civil servants served the immortal state, not its transitory rulers. The civil service, in other words, survives regime change. The events of 8 May 1945 were no exception to this universal rule.

In result and in method, the Constitutional Court's review of G 131 was very different.[176] Whereas the Court of Justice had grandly invoked the natural law rights of civil servants, the Constitutional Court ruthlessly examined their Nazi-era conduct. The Court of Justice sounded in metaphysics, the Constitutional Court in history.

In a judgment dated 17 December 1953, the First Senate responded to the more than 2,000 constitutional complaints filed against the 131 Law in the two years since its passage.[177] It was, to date, the Court's longest judgment. But its essence appeared in a single headnote: 'All civil service commissions were terminated on 8 May 1945'.[178] Nazi-era civil servants, the Court held, forfeited their positions when the Third Reich collapsed. What G 131 now granted, the former civil servants received by parliamentary grace, not by natural right.

As striking as this stark conclusion was the argumentation that produced it. With the merciless assiduity of a revisionist historian, the First Senate's judgment undermined the myth of an apolitical administration.[179] In painstaking detail the Court chronicled the transfer of civil servants' allegiance from the German state to the person of Adolf Hitler.[180] Through that transfer, the Court concluded, German officialdom degenerated into a 'power apparatus

[175] See Daniel Herbe, *Hermann Weinkauff (1894–1981). Der erste Präsident des Bundesgerichtshofs* (Mohr Siebeck 2008) 121–49. This was not a scholarly diversion for the president of the Court of Justice but the keystone of his jurisprudence. At the opening ceremony for the new court on 8 October 1950, Weinkauff expounded a vision for the ship he was to steer. 'It would be wonderful', Weinkauff rhapsodized, 'though also a difficult challenge for the Federal Court of Justice, to return to the metaphysical foundations of the law'. He called on his colleagues to exercise an 'unshakable faith in the metaphysical substance and the divine essence that animate law. For law derives its strength in the last instance from divine ordainment'. In ibid 124–25. In the years that followed, Weinkauff produced a stream of essays with titles like, 'I Believe' and 'Protestant Natural Law'. He entitled his retirement retrospective of the work of his court, 'Natural Law Thought in the Jurisprudence of the Federal Court of Justice'. See Hermann Weinkauff, 'Das Naturrecht in der Rechtsprechung des Bundesgerichtshofes' (1960) 13 *NJW* 1689.

[176] The Constitutional Court had already engaged in a tacit doctrinal joust with the Court of Justice, twice rejecting partial challenges to G 131 and later denying the ongoing force of rights enshrined in the Weimar constitution, a point on which the FCJ had partly rested its repudiation of G 131. See 1 BVerfGE 167 (1952); 2 BVerfGE 105 (1953); 2 BVerfGE 237 (1953). For a discussion of the relationship between the Federal Constitutional Court and the Federal Court of Justice, see Hans Joachim Faller, 'Bundesverfassungsgericht und Bundesgerichtshof' (1990) 115 *Archiv des öffentlichen Rechts* 185.

[177] 3 BVerfGE 58 (1953). [178] ibid 58.

[179] The comparison of the Court to an historian is also found in Maximilian Steinbeis and others, *Die Deutschen und das Grundgesetz* (Pantheon 2009) 90.

[180] 3 BVerfGE 58 (n 177) 98–102.

in the service of the NSDAP'.[181] This fact transformed the nature of the civil service—with unhappy legal consequences for the civil servants themselves. Their commissions were no longer coterminous with the life of the German state, but with that of its temporary ruler. Their official posts perished with the *Führer* in the Berlin bunker—or with the tendering of unconditional surrender eight days later. Civil servants had no right, under natural or positive law, to return to their posts or recover lost pay. What they received under G 131 they owed entirely to the Bundestag's largesse.

The judgment produced a decidedly mixed response from the West German press lords—approval from the *Frankfurter Hefte* and *Süddeutsche Zeitung*; outrage from the *Frankfurter Allgemeine Zeitung* and *Die Zeit*; silence from *Der Spiegel*.[182] Within the legal academy and the civil service itself, the decision provoked almost unbounded fury. Within a year more than eighty articles appeared in legal journals criticizing the decision at varying pitches of vehement wrath.[183] Only a couple deigned to defend it.[184] Especially rankling for traditionalists was the First Senate's heretical reliance on historical and sociological sources to reach its conclusion. More nettlesome still, many of the primary historical sources on which the judgment relied were works of public law scholarship—in many cases the Nazi-era writings of the judgment's critics and their colleagues. The First Senate's judgment documented the transformation of the Nazi-era civil service by showing how law professors articulated that transformation in theory and how judges implemented it in practice.[185] What's more, the Court pointedly declined to exclude the academy from its assessment that the Third Reich civil service had been 'permeated by the spirit of National Socialism'.[186]

> In consequence of the penetration of the Führer principle in the universities, and of National Socialist ideology in scholarship, the university ceased to be a true corporation. Increasingly, it morphed from a forum of research for its own sake into an institute of State education in which teaching and research were predominantly determined by the political aims of National Socialism.... For these reasons, after the collapse the universities, as well as the civil service, had to be rebuilt from the ground up.[187]

To law professors who had maintained their university chairs from the 1930s through the 1950s, this passage tasted of wormwood. Among the angriest of academic observers, however, was one who had not returned to the university after the War.

In its plentiful citations to the public law scholarship of the Nazi era, the *Civil Servants* judgment referred to several jurists by name. The most prominent name of all, attached to the most provocative citation, was that of Carl Schmitt. In its description of the 'judicial' function filled by the *Führer* in Nazi state theory, the Court wrote: 'Actions that he took as political Führer could be classed as 'supreme justice' (so for the murders of 30 June 1934: Carl Schmitt, DJZ 1934, 947)'.[188] Schmitt erupted when he read this, fuming in private correspondence that 'the author of the judgment seized the occasion to abuse me by name and to falsify in his own sense [my] essay, bold and brave under the circumstances, "The Führer Preserves the Law", which sought to save the last remnant of legal justice'. Schmitt also saw in the judgment another, implicit form of personal attack. 'I am', he wrote, 'as every educated jurist in

[181] ibid 86.

[182] See 'Das Beamtenverhältnis am 8. Mai 1945 erloschen' *FAZ* (18 December 1953) 1; 'Eine soziale Tat' *FAZ* (18 December 1953) 1; 'Bedeutsame Entscheidung in Karlsruhe' *Süddeutsche Zeitung* (18 December 1953) 1; 'Verfassungsgericht schreibt Geschichte' *Süddeutsche Zeitung* (19/20 December 1953) 3; Richard Tüngel, 'Weihnachten der Beamten' *Die Zeit* (24 December 1953) 3.

[183] Wesel (n 109) 146–47.

[184] See, e.g., Hans Peters, 'Der Streit um die 131er-Entscheidungen des Bundesverfassungsgerichts' (1954) *Juristenzeitung* 589.

[185] See 3 BVerfGE 58 (n 177) 90–93, 101–15. [186] ibid 96. [187] ibid 142. [188] ibid 86.

Germany knows, the originator of the doctrine of the institutional guarantee of the career civil service'. Schmitt read the judgment as a sustained assault on that doctrine—and on its originator. 'I now understand', Schmitt concluded, 'the anxiety that must assail Adenauer as he contemplates this sovereign tribunal, its decisionistic omnipotence, its lavishly dispensed head notes and obiter dicta'. Schmitt, a towering theorist of sovereignty, bemoaned what he saw as the transfer of sovereignty to a new judicial elite: 'A furrowed brow from this elite suffices. Against it there is no protection'. He added: 'Elite is here, of course, a value-free sociological concept, the definition of which—departing from Pareto—I should like to suggest: Elites are those whose sociology no one dares to write'.[189]

Schmitt's anger was of a particularly stygian temperature, but he was hardly alone. The Association of German Public Law Professors made the *Civil Servants* judgment the subject of its 1954 meeting in Tübingen.[190] One by one the participants rose to denounce the decision and decry the Court's overreaching. Only one participant spoke in the judgment's defence: Justice Ernst Friesenhahn, of the Court's non-participating Second Senate. Friesenhahn told his academic colleagues that he worried about the propriety of his intervening in the discussion, but that he felt a certain duty to do so. Friesenhahn admitted that, upon his first reading of the judgment, he was as surprised by the First Senate's reasoning as anyone, and that he originally rejected that reasoning fiercely.[191] But the scope, content, and form of many criticisms of the judgment gradually prodded Friesenhahn to a more sympathetic view. Friesenhahn was prepared now to defend the judgment, but he realized that he was sounding a voice in wilderness. He was speaking, he admitted, 'as the representative of a hopeless minority'.[192]

B. Battle of the courts

The majority windmill against which Friesenhahn tilted his lonely lance included Hermann Weinkauff and the Federal Court of Justice. Weinkauff's earlier opinions questioning or denying G 131's constitutionality fuelled the severity of the *Civil Service* judgment. On 20 May 1954, five months after the *Civil Servants* judgment, Weinkauff's First Civil Senate issued a referral opinion that affirmed the continuance of both the German Reich and the legal relations of its officials after 8 May 1945.[193] Weinkauff was a redoubtable propagandist for the historical honour and theoretical ideal of an apolitical civil service and an independent judiciary. As late as 1968 Weinkauff published a defence brief for the Nazi-era judiciary, blaming the period's systematic miscarriage of justice on blind judicial positivism. The judges of the Third Reich era had been guilty, perhaps, of unreflective service to the state; but they had not been the willing instruments of National Socialist despotism.[194]

Weinkauff now inscribed that ideal, applied to the whole of the Third Reich civil service, in the jurisprudence of the Federal Court of Justice. In its May 1954 referral decision, the FCJ's Grand Senate for Civil Matters challenged the Constitutional Court's *Civil Servants* judgment on two fronts. First, Weinkauff and his fellows denied that the Constitutional Court's grounds for decision—as opposed to the judgment itself—were binding on the other federal courts.[195] The Constitutional Court's earlier insistence to the contrary[196] was

[189] Carl Schmitt to Giselher Wirsing (14 January 1954) BAK N 1529/44 (Nachlass Werner Weber).

[190] The proceedings were published in 13 *Veröffentlichungen der Vereinigung der Deutschen Staatsrechtslehrer* (1955).

[191] ibid 162. [192] ibid 163. [193] 13 BGHZ 265 (1954).

[194] See Hermann Weinkauff, *Die deutsche Justiz und der Nationalsozialismus: ein Überblick* (Deutsche Verlags-Anstalt 1968).

[195] 13 BGHZ 265 (n 193) 271–77. [196] See 1 BVerfGE 14, 36 (1951).

itself mere *obiter dicta*.[197] Second, the FCJ rejected the Constitutional Court's conclusions about the status of civil servants after 8 May 1945.

Germany's unconditional surrender, the Grand Senate noted, was military, not political; it was tendered by the army, not the government.[198] This being so, it couldn't possibly affect the legal standing of civil servants who had nothing to do with the armed forces.[199] What's more, although some miscreant civil servants had, out of opportunism or sheer iniquity, become willing instruments of Nazi crimes, such complicity could shake the legal standing only of individual villains, not of the civil service as a whole.[200] The Constitutional Court's contrary conclusion was at odds with the rule of law principle, which forbade collective judgments.[201] In any case, the Constitutional Court's belief that the civil service as a whole had been corrupted was not a product of legal analysis, but was rather 'an historical value judgment'—a kind of retrospective moralizing 'burdened by a plethora of uncertainties'.[202] The historical record was too meagre and mottled to support so drastic an outcome as the total abolition of ties between the state and its civil servants.[203]

Actually, the FCJ was convinced that the historical record ran the other way. Against the Constitutional Court's indictment of the Nazi-era civil service, the Grand Senate invoked the comforting picture of an apolitical bureaucracy whose officers had overwhelmingly resisted the pressures of Nazi lawlessness and fulfilled their duties to the German state.[204] Initially, most civil servants viewed their vow of loyalty to Hitler simply as a promise of submission to the state's highest authority. Then, as the criminal aims and brutal methods of the new regime became chillingly apparent, the vast majority of civil servants endured their 'coerced bond' to the *Führer* 'only unwillingly, under sharp internal rejection and under sharp terror'.[205] Civil servants were themselves victims of Nazi terrorism; the Constitutional Court needn't add insult to injury by pronouncing gratuitous 'historical value judgments'.

It was a pugilistic opinion with a polemical edge. The matter, as everyone knew, was personal. All of the judges on the First Civil Senate had themselves been civil servants during the Nazi era, some of them judges on the notorious Leipzig *Reichsgericht*.[206] By contrast, only three of the twelve justices of the Constitutional Court's First Senate had been Nazi-era officials; five had lost their positions in 1933, one left of his own accord, and three had other legal jobs. The Court of Justice spoke with unusual energy because its own historical reputation was at stake. President Weinkauff and his colleagues chafed under the Constitutional Court's assertion of a dual monopoly over constitutional jurisdiction and moral legitimacy.

The Constitutional Court's response was slow in coming and, at first, only indirect. In a judgment dated 30 November 1955, the Court ruled laconically that '[a]dvisory opinions from a referring court upon referral are inadmissible'.[207] The textual basis of this holding was the FCCA's silence about referral opinions, which meant, the Court reasoned, that they were not allowed. As a policy matter, moreover, the standing of the nation's highest courts would suffer if parties could set divergent opinions at war with one another. Both of these bases were contestable, but the holding's mere existence was astonishing. The case at hand involved neither an advisory opinion nor a referring court. What's more, as noted earlier,

[197] 13 BGHZ 265 (n 193) 273. [198] ibid 292. [199] ibid 295. [200] ibid 297.
[201] ibid 298. [202] ibid 299. [203] ibid. [204] ibid. [205] ibid.
[206] On the nazi-era activities of the judges on the Federal Court of Justice, see Klaus-Detlev Godau-Schüttke, 'Entnazifizierung und Wiederaufbau der Justiz am Beispiel des Bundesgerichtshofs' in Eva Schumann (ed), *Kontinuitäten und Zäsuren. Rechtswissenschaft und Justiz im 'Dritten Reich' und in der Nachkriegszeit* (Wallstein 2008) 189.
[207] 4 BVerfGE 358, 367 (1955).

the Court's prior practice had welcomed and even solicited referral opinions. Faced with the FCJ's intimations of defiance, however, the justices apparently thought it politic to terminate the practice.

Unsurprisingly, the other federal courts were much provoked. With Weinkauff leading the way, the presidents of the Federal Court of Justice, the Federal Labour Court, the Federal Court of Finance, the Federal Social Court, and the Federal Administrative Court issued a collective statement dismissing as nugatory the Constitutional Court's ban on advisory referrals. The Federal Constitutional Court did not sit at the summit of the legal order, the paper insisted. It too was a court, and no more than a court. It was the legitimate duty of *all* federal courts to ensure the constitutionality of the country's laws.[208]

The Constitutional Court's response was unrushed but unbending. In separate judgments issued 19 February 1957, the First Senate reaffirmed the fundamentals of the *Civil Servants* judgment and answered, coolly and methodically, the arguments of its assailants. The first judgment was substantive, the second procedural. The substantive judgment involved a constitutional complaint from a former member of the Gestapo who had been denied readmission into the police force. Gestapo functionaries, as noted, were the only group excluded from the reparations and reentry provisions of G 131. The complainant challenged this exclusion as a violation of Article 3 GG's general equality provisions.

The First Senate's judgment was simple: the Court stood by its 1953 decision and rejected the complainant's appeal. But it was also very long—at ninety pages, one of the early Court's longest. It was a systematic response to the accumulated criticisms of the previous thirty-eight months. The judgment's tone was anything but conciliatory. The Court conceded nothing; at times, it went on the offencive.

At the outset the Court dismissed many criticisms of the *Civil Servants* judgment as nothing more than 'politically accented attacks on or even suspicions of the Court'. Self-evidently, these attacks merited no reply.[209] Serious *legal* criticisms of the judgment the Senate rebutted with fire and steel. The Court called critics by name (Carl Schmitt was again cited as a 'National Socialist author'),[210] and in some instances answered critics by citing their own earlier writings. Early in the opinion, for example, the Court quoted Herbert Krüger's 1954 complaint that the Court 'has understood the civil service under National Socialism in a "purely personal" way, has considered the bond to the Führer, but not the bond to the Reich'. By so doing, the Court 'impermissibly identified Party and State, equating party membership and civil service commission and thereby misunderstanding both political reality and public law relations.'[211] Later in the opinion, the Court juxtaposed this view with Krüger's 1941 argument, put forth in a work tellingly entitled *Reich, Volksordnung, Lebensraum* ('Empire, People's Order, Living Space'—Nazi buzzwords all), that 'even persons active in the public administration are not merely a hierarchy of office holders, an "apparatus", but are to be envisioned as a community of *Führer* and followers', not 'operative in and for itself, but rather through the yoking together of a group of people under a *Führer* in the service of an objective task'.[212]

Ernst Forsthoff had his former self quoted against him even more sharply. In 1945, Forsthoff had been removed from his Heidelberg professorship by occupation authorities, but had returned to teaching in 1950 and resumed a tenured chair in 1952.[213] He was a

[208] See Manred Baldus, 'Frühe Machtkämpfe' in *Das Lüth-Urteil* (n 13) 245.

[209] 6 BVerfGE 132, 137 (1957). [210] ibid 173. [211] ibid 137. [212] ibid 162–63.

[213] On Forsthoff, see Florian Meinel, *Der Jurist in der industriellen Gesellschaft. Ernst Forsthoff und seine Zeit* (Oldenbourg 2011).

friend and disciple of Carl Schmitt and was, along with Schmitt, Krüger, Theodor Maunz, and Karl Larenz, among the foremost jurists who lent legal legitimacy to the Third Reich. One student of Schmitt's postwar influence called Forsthoff 'undoubtedly the most influential and important disciple of Carl Schmitt' in the early Federal Republic.[214] Forsthoff fulminated against the *Civil Servants* decision, lambasting the Court for failing to realize that during the Nazi era 'state and party were not a unity and that upon this fact depended the measure of freedom … the individual then enjoyed'. The First Senate's 1957 *Gestapo* judgment quoted this passage, but contrasted it on the same page with what Forsthoff had written in his 1934 article, 'The Total State':

> State and National Socialist movement … have formed an indissoluble unity, of which the action of the Führer against the seditious leaders of the SA furnishes a particularly striking example.… [T]he unity of State and Party [was] practically realized before it was formally proclaimed in the law of 1 December 1933.[215]

Strong medicine this. But the First Senate was even more unflinching in its response to criticisms that the *Civil Servants* judgment had employed an historical rather than a legal methodology. Particularly barbed was the Senate's retort to Weinkauff's critique of the *Civil Servants* judgment as an 'historical value judgment'. If the Constitutional Court was guilty of playing historian, Weinkauff's court was guilty of playing politics. The Grand Civil Senate's own assessment of the Nazi-era civil service—that 'the overwhelming portion of German civil servants, abusive, unlawful pressure notwithstanding, … continued to feel themselves bound in the first place to the State and its legitimate functions'[216]—was 'not a judicial finding of fact, but a political hypothesis'.[217] If one had to choose, the Court implied, historical value judgments were better than political slogans.

In the phrase 'judicial finding of fact' the First Senate lodged the essence of its apologia. In the *Civil Servants* judgment, the justices had not been playing historian, they had been sitting in judgment—finding facts and passing sentence. They had put the German State on trial, and found it rotten to the core. The *Gestapo* judgment was a kind of retrial, conducted with meticulous resolve and unmistakable anger. If anything, the *Gestapo* judgment was more 'historical' than *Civil Servants*. In the earlier case, the Court had relied largely on evidence from Nazi-era state theory and its judicial application. Now the justices turned to the Nazi-era actions of civil service agencies across the German state—highlighting, as they hadn't before, the persecution of Jews.[218]

The judgment began with a pointed preterition. The justices professed that they would not mention the Final Solution (mentioning it), and that they would give 'only a small extract of the arbitrary measures against Jews and foreign ethnic groups'.[219] They then documented a pattern of systematic persecution across administrative agencies of every imaginable kind. The judgment stressed the ubiquity of state oppression by employing a smaller font for its lengthy enumeration of examples.[220] Then, in a striking rhetorical climax, the Court summarized the spread of the cancer:

> If therefore the judge …
> If the State prosecutor …
> If the judicial authorities …
> If the registrar …

[214] Dirk van Laak, *Gespräche in der Sicherheit des Schweigens: Carl Schmitt in der politischen Geistesgeschichte der frühen Bundesrepublik* (Akademie 1993) 240.
[215] 6 BVerfGE 132 (n 209) 175. [216] 13 BGHZ 265 (n 193) 299. [217] 6 BVerfGE 132 (n 209) 179.
[218] ibid 185–97. [219] ibid 185. [220] ibid 185–92.

In each instance the Court listed multiple acts of anti-Semitic persecution carried out by the authority named. The list went on for three pages. It included agencies responsible for tax collection, municipal affairs, property appraisal, youth matters, public welfare, labour, commercial inspection, finance, food, tobacco, clothing, school inspection, workplace inspection, transportation, health, medicine, traffic, nutrition, and tolls. Individual officials might well have performed isolated acts of decency. Some, perhaps, sought to mitigate the application of unjust decrees. But there could be no individual immunity against the poison of a system whose venom ran so thick and full through roots and trunk and branches. 'Individuals', the justices wrote, 'cannot be judged by the more or less coincidental field of action' to which they were assigned. 'They must be judged, rather, on the legal basis of their membership in an Institution so marked by the imprint of National Socialism. For the entire public administration, at its core, [was] implicated through a political perversion'.[221] By affiliating themselves with a particular state institution, Nazi-era civil servants had forfeited all claims surviving the regime to which the institution as a whole had subordinated itself.

In a separate judgment issued the same day, the Court responded to the jurisdictional portions of the Grand Civil Senate's May 1954 submission decision.[222] Of the merits of the decision and its polemical subtext the Constitutional Court said nothing. That decision was inadmissible, the Court ruled, for the simple and sufficient reason that the Court of Justice had no jurisdiction over constitutional questions. The Constitutional Court affirmed its monopoly on constitutional interpretation simply by presupposing it. Of the circular signed by the federal court presidents, the First Senate took no notice at all.

Criticism of the *Gestapo* judgment, by comparison with the scorched-earth assault on the *Civil Servants* decision, was muted. By February 1957 most of the Nazi-era officials who had been roused to wrathful sermons by the *Civil Servants* judgment had been reintegrated into the West German bureaucracy. As a practical matter, there was little to be gained by prolonging the controversy. As an affair of honour, there was much to be lost. Legal scholars were little tempted to have their Third Reich writings immortalized in the published decisions of the Constitutional Court. The broader public had largely lost interest.

The *Gestapo* judgment remains a remarkable document. Taken together, the *Civil Servants* and *Gestapo* judgments present a precocious exercise in *Vergangenheitsbewältigung*, or overcoming the past. The decisions are unique among official papers of the early Adenauer era, both for their forthright repudiation of the Nazi past and for their frank acknowledgement that civil servants, high and petty, were complicit in Nazi crimes. A rare note of contemporary praise for the *Civil Servants* judgment extolled its 'rejection of the veiling of a disagreeable past'.[223] Still more expressive was the assessment of a twenty-four-old graduate student, Rheinhold Kreile, who in the pages of the Left-leaning *Frankfurter Hefte* hailed the judgment as a 'Magna Carta of self-reflection', a ripping away of the general 'veil of forgetfulness'.[224] Kreile rejoiced that, at a time when the public sphere resounded with talk of 'the German miracle and the German rise', the Court had reintroduced 'the theme of German decline'. Kreile's conclusion was stark: 'What, at bottom, the Constitutional Court held is this: The civil servant of National Socialist Germany was the civil servant of Adolf Hitler; the implication is self-yielding: the German people were the people of Adolf Hitler'.[225] It was a candour to be found nowhere else in the public life of the early Federal Republic.

221 ibid 197. 222 6 BVerfGE 222 (1957).

223 Fritz Heegner, 'Die Entscheidungen des Bundesverfassungsgerichts und des Bundesgerichtshofs zum Gesetz 131' (1954) 7 *NJW* 898, 903.

224 Reinhold Kreile, 'Eine deutsche Magna Charta der Selbstbesinnung' (1954) 9 *Frankfurter Hefte* 83, 85.

225 ibid 91.

Most praise of the judgments, however, would be posthumous. Only in later decades, as a new generation of Germans began to resent the silence of its parents, and as reckoning with the Nazi past entered the mainstream of public and political discourse, would the *Civil Service* and *Gestapo* judgments emerge as foundational monuments of judicial memory.[226] Their more immediate impact was the Court's emergence, over the muttered maledictions of other federal courts, as the supreme tribunal of the West German judiciary and the authoritative interpreter of the West German constitution. The judgments' longer legacy was to establish the Court as a democratic voice overtly defining the Federal Republic in opposition to its tyrannical forebear. It was a major contribution to the democratization of the Second German Republic.

IV. Militant Democracy and Its Discontents

Denouncing the National Socialist state and exposing its official enablers was a brave but straightforward endeavour. Defining the contours of its successor proved more complicated. This was especially so with respect to those constitutional provisions through which the framers of the Basic Law sought to secure the Bonn Republic against the fate of the Weimar Republic—those provisions designed to make of the Federal Republic a militant democracy.

In their conception of democracy, as in so much else, the framers of the Basic Law were haunted by the memory of Weimar. The founders of the Federal Republic sought to shield the new state against internal threat. Article 18 of the Basic Law decreed that those who abused fundamental rights to fight against the free democratic basic order would forfeit those rights. Article 21 declared that political parties who fought against the free democratic order were unconstitutional and should be banned. In both cases, the Constitutional Court would serve as the arbiter of suppression.

A. Banning the Socialist Reich Party

Almost as soon as it opened its doors, the Court was asked to enforce these muscular provisions. In November 1951, the Adenauer government petitioned the Court to ban both the Socialist Reich Party (SRP) and the German Communist Party (KPD). The government pitched its double petition as a simultaneous show of steel against extremism of Right and Left. In the case of the SRP, the government wished both to eliminate a genuine domestic threat and to reassure an apprehensive international audience. Nationalists and Right-wing radicals had combined to win nearly 6 per cent of the votes in the first Bundestag elections in 1949. More recently, the neo-Nazi SRP had fared alarmingly well in the 1951 state elections, making substantial gains everywhere and winning 11 per cent of the votes in Lower Saxony. The Adenauer government was determined to stop the ascendant party in its tracks. The government worried, however, that unilateral action against the SRP would alienate supporters on the Right and endanger the ruling coalition. 'We cannot be accused

[226] For explorations of judicial *Vergangenheitsbewältigung*, on the Constitutional Court and elsewere, see Alexander Blankenagel, 'Verfassungsgerichtliche Vergangenheitsbewältigung' (1991) 13 *Zeitschrift für neuere Rechtsgeschichte* 80; Bernhard Schlink, 'Die Bewältigung von Vergangenheit durch Recht' in Bernhard Schlink, *Vergangenheitsschuld und gegenwärtiges Recht* (Suhrkamp 2002) 89; Jörg Menzel, 'Vergangenheitsbewältigung in der frühen Judikatur des Bundesverfassungsgerichts: Beamten- und Gestapo-Urteile in *Das Lüth-Urteil* (n 13). On the Nazi past as a looming presence in the Court's jurisprudence, see Ulrich Löffler, *Instrumentalisierte Vergangenheit?: Die nationalsozialistische Vergangenheit als Argumentationsfigur in der Rechtsprechung des Bundesverfassungsgerichts* (Peter Lang 2004).

of one-sidedly doing something against the SRP', warned Robert Lehr, a member of the cabinet, in a May 1951 letter to Adenauer.[227] In a cabinet meeting a week later, Eberhard Wildermuth, the housing minister, asked whether a balance couldn't be sought by calling for a ban of the KPD as well.[228] This logic ultimately prevailed. The government filed its petition against the SRP on 19 November 1951; its petition against the KPD three days later.

Both bans were slow in coming. 'We need a law against the Constitutional Court', groaned Otto Lenz to his diary in May 1952, 'for dealing with things so dilatorily'.[229] He groaned too soon. For the Court's decision in the SRP case, the government would have to wait five months more; in the KPD case, more than fifty.

The Court banned the SRP in a judgment dated 23 October 1952.[230] Consistent with the language of Article 21 of the Basic Law, the crucial question in the case was whether the SRP threatened the Federal Republic's free democratic basic order. The First Senate seized the occasion to define 'free democratic basic order' more concretely.[231] That order, the Court proclaimed, was marked by government free from caprice or coercion and 'rooted in the rule of law, upon the principle of popular self-determination according to the will of the present majority'.[232] It must preserve 'freedom and equality' and honour, as foundational principles, 'respect for the fundamental rights enshrined in the Basic Law, above all the rights of life and the free development of the personality; popular sovereignty; separation of powers; the responsibility of the government; the lawfulness of administration; the independence of the courts; and the plurality and equality of political parties with a right to constitutional development and the exercise of political opposition'.[233]

Against this standard the Senate measured the avowed aims of the SRP and found them wanting. This part of the analysis was straightforward. The SRP made no secret of its desire to establish a one-party state animated by the *Führer* principle.[234] The party may not have embraced all the goals of Hitler's National Socialists, but in its racist and dictatorial impulses, the SRP had obviously taken the Nazis as their model.[235] An assault on the free democratic basic order was the party's *raison d'être*. The SRP was the prototype of a party proscribed by Article 21.

More problematic than the ban itself—which was technically moot, since the party had already dissolved itself—were the legal consequences flowing from it. What was to become of the two dozen SRP members currently sitting in the Bundestag? Or of the dozens more sitting in state parliaments?

Whether representatives of a banned party should lose their seats in Parliament had, in fact, been considered by the Bundestag committee that drafted the FCCA. Because Article 38 GG stipulated that members of Parliament represented the entire people and were answerable to their consciences alone for the discharge of their duties, the committee concluded that the MPs of a condemned party could keep their seats. The FCCA itself was silent on the matter, but the committee's rapporteur, the future justice minister, Fritz Neumayer, explained to the Bundestag in January 1951 that the proscription of a party did not abolish the posts of its representatives in Parliament.[236]

[227] Bundesminster des Inneren to Konrad Adenauer (1 May 1951) BAK B 106/15886.

[228] Ursula Hüllbusch (ed), *Die Kabinettsprotokolle der Bundesregierung*, vol 4 (Harald Boldt 1988) 356.

[229] Klaus Gotto (ed), *Im Zentrum der Macht: Das Tagebuch von Staatssekretär Lenz* (Droste 1989) 287.

[230] 2 BVerfGE 1 (1952).

[231] On the Court's jurisprudence dealing with this concept, see Christoph Gusy, 'Die "freiheitliche demokratische Grundordnung" in der Rechtsprechung des Bundesverfassungsgerichts' (1980) 105 *Archiv des öffentlichen Rechts* 279.

[232] 2 BVerfGE 1 (n 230) 12–13. [233] ibid 13. [234] ibid 46. [235] ibid 44–47.

[236] *Verhandlungen des Deutschen Bundestages*, vol 6 (18 January 1951) 4230.

The Constitutional Court, however, did not feel bound by this conclusion. Members of Parliament, the First Senate observed, had a duty of dual representation. They represented both the people who elected them and the party that sponsored them. In the case of a party ban, this placed Article 21 in tension with Article 38. But because Article 21 protected the constitutional order's fundamental principles of freedom and democracy, the Court ruled that it took precedence, in case of conflict, over Article 38. Parliamentary seats filled by SRP members must therefore be vacated without possibility of replacement. This included members of the various state parliaments as well as of the Bundestag. In the Lower Saxony state Parliament alone, sixteen delegates were ejected—including some who had previously left the SRP. An entire Lower Saxon voting district was left unrepresented. But the justices were untroubled by this outcome. Ejecting SRP delegates violated no one's rights, the Court reasoned, because voting for an unconstitutional party was itself an unconstitutional act. No citizen had a right to offend the constitution.

The press quickly perceived that this judgment involved more than a judicial anathema pronounced over the grave of a moribund party. An article in *Die Zeit* called the judgment 'a legal and political act of supreme importance', one that marked the birth of 'a new brand of democracy', a democracy capable both of honouring the rule of law and of warding off internal threats.[237] Moreover, the article observed, the judgment did more than interpret existing law. By drawing the consequences of its own conclusions the Court had given its judgment the force of law—a clear instance of judicial lawmaking—a power that German jurisprudence had long been loath to acknowledge.

A lead article in *Der Spiegel*, by contrast, censured the dissolution of SRP mandates with considerable fire. Reporting in detail the deliberations of the Bundestag legal committee, the article observed that the members of this committee—the legal elites of the Federal Republic—must now ruefully acknowledge that the tribunal they created now regarded their considered opinions as non-binding.[238] The article concluded by quoting—approvingly but wistfully—the earlier assessment of the committee's rapporteur:

> The decisions of a court of public law represent genuine judicial decisions, in which the content of the Basic Law is not invented, but rather in which the content of what has already been expressed as the will of the legislature is ascertained. It is therefore not the duty of this supreme court, in its decisions, to make political determinations, substitute its own work for that of the legislature, or venture any other grasp for the stars.

The Court had now done all these things—and surely not for the last time.[239]

B. Banning the German Communist Party

Meanwhile the government's petition against the KPD was still pending. It would continue to pend for four long years. Certainly there was an element of political caution in the delay. The Court was reluctant to intervene in such a charged political controversy. Perhaps the justices hoped that, as had happened in the European Defence affair, its deliberation would be overtaken by events—that the matter would become moot before the Court could reach its merits.

The Court had reason to think this might happen. While the Court deliberated, the KPD's electoral fortunes slumped from bad to abysmal. The party's return sank from 5.7 per cent in the 1949 Bundestag elections to 2.2 per cent in 1953. All signs portended further

237 'Die Folgen des SRP-Verbots' *Die Zeit* (30 October 1952).
238 'Griff in die Sterne' *Der Spiegel* (5 November 1952). 239 ibid.

decline. One argument against banning the party was that a party plummeting toward electoral extinction couldn't possibly threaten the constitutional order. This was the position of the SPD, which contended that a ban would merely mask the KPD's electoral weakness and hamstring reunification negotiations.[240] Some within the governing coalition shared this view. Ernst Lemmer, a Christian Democrat, warned against employing 'the heaviest judicial artillery against a flock of wind-tossed political sparrows'.[241] Others, even within Adenauer's cabinet, worried that a ban would merely send the party underground, making it harder to track and gracing it with the allure of martyrdom and clandestinity.[242]

After many postponements, oral arguments were scheduled for 23 November 1954. Just before they began, the Court asked Adenauer to confirm that the government wished to proceed with its petition.[243] Since the government had neither withdrawn its petition nor given any indication that it wished to do so—indeed, it was obviously preparing for oral argument—the Court's message could only be read as a request for the government to drop the case. To the justices' dismay, the government pressed on. Nearly a decade later, Justice Karl Heck spoke of the anguish to body and soul that the case inflicted on the judges.[244]

Oral argument began with considerable fanfare.[245] For many, it felt like a show trial. There were moments of high drama and moments of low farce. The KPD lawyers—an army of eight, headed by an imposing East German law professor, Karl Kaul[246]—sounded an early note of truculence that sustained their argumentation through fifty-one days of oral argument scattered across the next eight months. Kaul and company denounced the entire proceeding as a 'medieval inquisition and a witch-hunt', [247] a resurgence of 'the same anti-communism ... with which Hitler was able, for a time, to deceive mankind'.[248] This invocation of Hitler and the KPD ban of 1933—a staple of KPD polemics—drew a sharp rebuke from Josef Wintrich, the chief justice. But the KPD lawyers were unperturbed.

The government's chief counsel, Ritter von Lex, had no aversion to loaded rhetoric either. Accused of imbibing Nazism's spirit, von Lex injudiciously echoed one of its favourite metaphors. The KPD, he warned, was 'a dangerous source of infection for the body of our people, one that sends poisonous substances into the bloodstream of the state and social organism of the Federal Republic'.[249] Von Lex, who as a counsellor in the Nazi-era interior ministry had helped organize the 1936 Berlin Olympics, was an object of special execration in the KPD press and of a hate mail campaign among the party faithful.[250] Late in the trial, Karl Kaul referred to von Lex's Nazi-era activities before railing that 'the KPD considers it an honour to be defamed and insulted here ... by Mr. State Secretary Ritter von Lex'.[251]

240 See Patrick Major, *The Death of the KPD: Communism and Anti-Communism in West Germany, 1945–1956* (Oxford University Press 1997) 285–86.

241 'Die Klage Zurückziehen'*FAZ* (9 June 1953); see also Major (n 240) 286.

242 Comments of Labour Minister Anton Storch. See Hüllbusch (n 228) 692, 698.

243 Gerd Pfeiffer and Hans-Georg Strickert (eds), *KPD Prozess*, vol 1 (CF Müller 1955) 100.

244 See the report in 'Feierstunde im Bundesverfassungsgericht' *Badische Neueste Nachrichten* (18 February 1965) 2.

245 For the minutes of the oral arguments, see Pfeiffer and Strickert (n 243) 82–723.

246 An internal memo from the government's press agency gave lively portraits of the KPD attorneys: 'Kröger (Carl Schmitt student): polite, obliging, fantastically superficial ...; Kaul: uninhibited SED demagogue, aggressive and in almost every sentence slanderous ... Böhmer: upright Rhinelander, who in a tone of consummate innocence pronounces the most incredible and captious things; Huetsch: objective, cool, aloof, ostensibly only there to rescue the endangered *Rechtsstaat*'. Hans Küffner, 'Internal Memo: Presse- und Informationsamt der Bundesregierung' (10 December 1954) BAK B 145/3392.

247 Pfeiffer and Strickert (n 243) 87; see also Major (n 240) 289.

248 Pfeiffer and Strickert (n 243) 171; see also Major (n 240) 290.

249 Pfeiffer and Strickert (n 243) 116.

250 For the hundreds of articles collected by the Federal Republic's interior ministry, see BAK B 106 I/90.

251 Pfeiffer and Strickert (n 243) 305.

Von Lex and his co-counsel, the Hannoverian attorney Achim von Winterfeld, didn't always impress government sympathizers either. When Fritz Rische, a KPD lawyer, pointedly omitted the aristocratic 'von' from the government lawyers' names, von Winterfeld objected strenuously. Justice Franz Wessel reproved Rische, asking how the latter would like being addressed as 'Herr *Sche*'. The syllable 'von', Wessel lectured, was as much a part of von Winterfeld's name as was the syllable 'Ri' of Rische's own.[252] The episode was one of several exercises in foppery and conceit that led Hans Küffner of the press ministry to sigh that the government lawyers' 'incompetence on all questions of public law ... is surpassed only by their vanity'.[253] The lawyers maddeningly refused, Küffner complained, to cooperate with the press, then sulkily regarded all adverse reportage—of which there was a great deal—as 'grievous strokes of fate'.[254] The press ministry undertook damage control by sending a pamphlet with the government's position to law professors around the globe.[255]

Despite repeated efforts by KPD counsel to shift the discussion to contemporary political realities, the trial centred on the classics of communist ideology. The justices' questions focused on the four pillars of the communist canon—Marx, Engels, Lenin, and Stalin—and their reception as the official 'doctrine' of the KPD. The KPD lawyers knew that this emphasis on ideology over activity would redound to the party's disadvantage—perhaps its destruction. They responded by touting the party's commitment to German reunification and by stressing a recent softening on the necessity of violent revolution.

The justices were unreceptive. In the end, fearing the worst, KPD counsel hoped at least to make the most of the international stage—hoped, it seems, to rattle the courthouse walls with the thunder of their defiance. On the last day of oral argument, Karl Kaul rose and roared:

> The KPD, the party of the German working class, outlived Frederick William IV and his communist prosecutions and Bismarck and his socialist law; she saw Hitler and Himmler sink into the abyss; and as the flag-bearer of German unity and the freedom of the people, she will outlive those who seek today to suppress her and rob her of her rights. Those who marched under the same mantras in 1933 to destroy the KPD brought Germany to the brink of perdition. The KPD, however, endures and will continue to endure, because in her is embodied the will to life of the German nation.[256]

The trial ended on 14 July 1955. The Court deliberated for more than a year before announcing its judgment. During this time the Adenauer government resumed diplomatic relations with Moscow and negotiated for the release of German prisoners of war. Banning the KPD might have hampered these negotiations, leading some to speculate that the Court's further delay was dictated by political caution rather than jurisprudential uncertainty.[257] Twice during the Court's deliberations the KPD asked the Court to reconsider the case's admissibility in light of Stalin's death and programmatic changes in the party after Khrushchev's 'secret speech' of 25 February 1956. The government retorted that this supposed softening was a temporary, tactical ruse—that '[t]he KPD is and remains the classical party of revolutionary violence'.[258] The Court declined to disrupt the proceedings.

Still, no decision was forthcoming. With the POW negotiations long since concluded, the government grew impatient. Some cabinet ministers seemed genuinely alarmed. At

[252] See 'Memorandum Diabolicum' *Der Spiegel* (13 July 1955). [253] See Küffner (n 246).

[254] ibid.

[255] See 'Verfahren gegen die KPD vor dem Bundesverfassungsgericht—die Rechtsgrundlagen, Teil 2: Die Schlußplädoyers der Bundesregierung', BAK B 145/1860.

[256] Pfeiffer and Strickert (n 243) 303. [257] See Major (n 240) 291.

[258] Pfeiffer and Strickert (n 243) 475.

the close of a cabinet meeting on 25 January 1956, the interior minister, Gerhard Schröder, warned of 'dangerous attempts at bolshevist infiltration, which after the resumption of relations with Moscow would only intensify'.[259] Schröder added that after five long years he hoped the KPD case would soon come to a close. Four months later, Schröder's ministry took the extraordinary step of urging the Court directly to quicken its pace. In a 25 May memo the interior minister pled with the Court to decide the case swiftly and thus avoid a 'fatal weakening of the position of our state in the East-West confrontation so decisive for our whole people'.[260] More combatively, in its 21 July 1956 amendments to the FCCA, the government transferred jurisdiction over party bans, effective 1 September, from the First Senate to the Second.

The Court took the hint. On 17 August, two weeks before its de facto deadline, the First Senate announced its judgment.

The published opinion proved that the justices had not been idle.[261] At 308 pages it remains the longest decision in the history of the Court, though its outcome was announced in two six-word sentences: 'The German Communist Party is unconstitutional',[262] and: 'The German Communist Party is dissolved'.[263]

The Senate went on, at enormous length, to canvass the writings of Marx, Engels, Lenin, and Stalin, producing in the process what one historian has called (with mild extravagance) the best short summary of Marxism–Leninism ever written in German—'clearly written, concise, and exhaustive'.[264] At every turn, the Court concluded that the fulfillment of the Marxian programme would doom a democratic constitution. This conclusion became the major premise of the Court's syllogistic holding. Communist ideology called for the overthrow of a free democratic order; the KPD embraced communist ideology; therefore the KPD sought the overthrow of the free democratic basic order and must be banned. Quotidian political considerations, the Senate insisted, played no role in this analysis. The Court must ban a party whose aims were at enmity with the constitution, even if by any mortal measure the party had no hope of fulfilling those aims.[265] Before announcing the judgment, Chief Justice Wintrich observed that '[t]he Court must reach its decision from a purely legal perspective; considerations of political expediency are therefore inadmissible'.[266]

The political consideration urged on the Court most forcefully, however, was also a constitutional consideration. The Basic Law's preamble committed the Federal Republic to pursue German reunification, and the KPD lawyers—not to mention the case's critics in the West German press—played the reunification card relentlessly throughout the proceeding. In its judgment, the First Senate noted the apparent tension between the preamble's command to pursue reunification and Article 21's command to ban unconstitutional parties, but the justices could 'not ascertain that a ban of the KPD would pose a legal or insurmountable factual obstacle' on the 'road to reunification'.[267] In any case, the KPD's reunification argument was spurious. The party pursued reunification with the aim of placing united Germany under authoritarian rule. Its vision of reunification sprang from the very desire to abolish the free democratic order that made the party illegal.[268]

It was a remarkable judgment. Most striking of all was the Court's commentary on the fundamental orientation of the Federal Republic. At a crucial juncture in the judgment,

[259] Hüllbusch (n 228) 139.
[260] Bundesministerium des Inneren to Bundesverfassungsgericht (25 May 1956) BAK B 106/56600.
[261] 5 BVerfGE 85 (1956). [262] ibid 86. [263] ibid 87. [264] Wesel (n 109) 92.
[265] 5 BVerfGE 85 (n 261) 143. [266] Pfeiffer and Strickert (n 243) 483.
[267] 5 BVerfGE 85 (n 261) 133. [268] ibid 303, 334, 342.

the First Senate noted that the rise of 'totalitarian' parties in the twentieth century had rendered nineteenth-century liberalism's posture of toleration and neutrality toward all political parties untenable.[269] By holding fast to that obsolescent posture, the Weimar constitution disarmed the first German Republic and left it prey to the taloned harpies of 'the most aggressive of these "totalitarian" parties'.[270] Against this 'constitutional–historical' backdrop, the Basic Law sought 'to restore a free democratic order immediately after the destruction of a totalitarian state system'.[271] This restoration required protecting the free democratic order against its enemies. In this respect, Article 21(2) of the Basic Law posed no internal contradiction. It was, rather, the 'expression of a conscious political–constitutional will to resolve a boundary problem of a free democratic state order'—it was the 'credo of a … "militant democracy"' (*streitbare Demokratie*).[272]

The Federal Republic, then, was a militant democracy, and its militancy was not merely anti-fascist, but anti-totalitarian. The Court thus constitutionalized the conceptual categories of the Cold War, a move that led the Court to equate KPD agitation against the 'monopoly–capitalist class rule' of 'the Adenauer Regime' with an assault on the free democratic order itself.[273]

The concept of 'militant democracy' gained prominence in the late 1930s when the German emigree-lawyer Karl Loewenstein made it the subject of a pair of articles in *The American Political Science Review*.[274] Loewenstein famously exhorted the world's democracies to fight fire with fire—to employ undemocratic means, if necessary, to keep democracy alive.[275] It was an awkward notion for the Court to apply. As guardians of the constitution, the Court must identify the constitution's enemies. As guardians of the free democratic order, the Court must proclaim exceptions to that order.

The business of identifying enemies and announcing exceptions had an obvious, uncomfortable echo. In the two most famous sentences he ever wrote, Carl Schmitt defined sovereignty as the power to proclaim a state of exception,[276] and exalted, as the essence of 'the political', the distinction between friend and foe.[277] In 1931, Schmitt warned that constitutional justice would lead, not to the judicialization of politics but to the politicization of jurisprudence.[278] In 1954, Schmitt railed against the Constitutional Court's 'apocryphal acts of sovereignty'. Now sovereignty and politics, in a Schmittian sense, were mingled in a single judgment. Militant democracy was designed to convince a wary world that Bonn was not Weimar. By 1958, Schmitt himself concluded that it was worse: Bonn was Karlsruhe.[279]

The immediate outcome of the *KPD* judgment was that for the second time in less than a quarter of a century the German Communist Party was outlawed by the German state. In the aftermath of this judgment, the KPD took the unusual, and unsuccessful, step of appealing the decision to the European Court of Human Rights. The Strasbourg Court ruled that a group dedicated to depriving others of fundamental rights could not itself invoke them.[280]

[269] ibid 138. [270] ibid 139. [271] ibid. [272] ibid. [273] ibid 297–98.

[274] Karl Loewenstein, 'Militant Democracy and Fundamental Rights I' (1937) 31 *Am. Pol. Science Rev.* 417; Karl Loewenstein, 'Militant Democracy and Fundamental Rights II' (1937) *Am. Pol. Science Rev.* 638.

[275] See Loewenstein, 'Militant Democracy and Fundamental Rights II' (n 274) 656 ('Fire is fought with fire'); Karl Loewenstein, 'Legislative Control of Political Extremism in European Democracies II' (1938) 38 *Colum. L. Rev.* 725, 774 ('Democracy, fighting fire with fire, begins to become militant').

[276] Carl Schmitt, *Politische Theologie: Vier Kapitel zur Lehre von der Souveränität* (Duncker and Humblot 1934) 9.

[277] Carl Schmitt, *Der Begriff des Politischen* (Universitätsverlag von Robert Noske 1927) 28–29.

[278] Carl Schmitt, *Der Hüter der Verfassung* (JCB Mohr 1931) 22.

[279] Carl Schmitt to Armin Mohler (10 June 1958), in Armin Mohler and others (eds), *Carl Schmitt—Briefwechsel mit einem seiner Schüler* (Akademie Verlag 1995) 247 fn 294. See also Jan-Werner Müller, *A Dangerous Mind: Carl Schmitt in Postwar European Thought* (Yale University Press 2003) 66–68.

[280] Reported in (1957) 10 NJW 1349.

In the hour of its death, the KPD could console itself, if at all, only in the sympathy of certain sectors of the West German public. But while many journalists agreed that the judgment was unfortunate, not all blamed the justices. *Der Spiegel*, for instance, criticized not the First Senate, which issued the judgment, but the federal government, which had 'forced' the Senate's hand. The government's intransigence ran counter to the nearly unanimous assessment of legal experts that the proceeding was untimely, coming at precisely the moment when the Kremlin seemed for the first time in thirty years to be freeing (modestly) the hands of its satellites.[281] (The Hungarian rising, and its subjugation by Soviet tanks, was still a few months off.) *Die Zeit*—now in the hands of less conservative editors than at the time of the *SRP* judgment[282]—similarly doubted the government's wisdom in pressing its petition, but pinned blame on the Basic Law itself, which charged the democratic order to defend itself in a democratically dubious manner.[283]

The Court itself faced sharper criticism from scholars on the marginalized Left of the legal academy. The most prominent of these, Wolfgang Abendroth, wrote a stinging review. Abendroth accused the Court of privileging an illusory homogeneity of class interests and of infusing that homogeneity into its understanding of the free democratic order. In a democracy, Abendroth countered, values must be defined by 'the people and not the jurists'.[284] Although the Basic Law allowed many forms of a free democratic order, the Court had chosen one and made it everlastingly binding. By giving specific shape to the Basic Law's broad generalities, the Court had mistaken its role. Abendroth's criticism was absorbed and embraced by his most brilliant student, a young social theorist named Jürgen Habermas who became a life-long observer and critic of the Federal Constitutional Court.[285]

C. Banning the atomic weapons plebiscite

Habermas was especially appalled two years later when the Court quashed one of the Federal Republic's first experiments in direct democracy. The underlying controversy shook the nation. It was seen, literally, as a matter of national and global life and death. It arose from the Adenauer government's decision to equip the federal army, the *Bundeswehr*, with atomic missiles.

The *Bundeswehr*'s early history was marred by chronic shortages of men and materiel. Volunteers came forth by the tens of thousands, but most were disqualified by past participation in the *Wehrmacht*, the army of the Third Reich.[286] Theodor Blank, the

[281] Jens Daniel, 'Der Bumerang' *Der Spiegel* (22 August 1956).

[282] See Mathias von der Heide and Christian Wagner, '"Weiter rechts als die CDU": Das erste Jahrzehnt der Zeit' in Lutz Hachmeister and Friedemann Sierung (eds), *Die Herren Journalisten. Die Elite der deutschen Presse nach 1945* (CH Beck 2002) 165; see also Karl-Heinz Janßen and others (eds), *DIE ZEIT. Geschichte einer Wochenzeitung 1946 bis heute* (Siedler Verlag 2006); Axel Schildt, 'Immer mit der Zeit: Der Weg der Wochenzeitung Die Zeit durch die Bonner Republik—eine Skizze' in Christian Haase and Axel Schildt (eds), *Die Zeit und die Bonner Republik: Eine meinungsbildende Wochenzeitung zwischen Wiederbewaffnung und Wiedervereinigung* (Walstein 2008) 19.

[283] Michael Freunl, 'Keine Hexenjagd: Verbot der KPD, nicht Ausrottung' *Die Zeit* (23 August 1956). Five years earlier *Die Zeit* had been an aggressive advocate of party bans, warning that delays in staffing the Constitutional Court could result in the Federal Republic's demise at the hands of parties it should long since have banned.

[284] Wolfgang Abendroth, 'Das KPD-Verbotsurteil des Bundesverfassungsgerichts: Ein Beitrag zum Problem der richterlichen Interpretation von Rechtsgrundsätzen der Verfassung im demokratischen Staat' in *Antagonistische Gesellschaft und Politische Demokratie*: Aufsätze zur politischen Soziologie (Luchterhand 1967) 139.

[285] For a detailed study of Habermas's engagement with the Court, see the introduction and chapters 1, 2, and 5 of Matthew Specter, *Habermas: An Intellectual Biography* (Cambridge University Press 2010).

[286] Of the 152,000 volunteer applicants in 1955, e.g., 128,000 had served in the Wehrmacht. See Conze (n 7) 290.

defence minister, originally planned to recruit a force of 500,000 within three years of the *Bundeswehr*'s founding. But this was quickly exposed as risibly sanguine. Blank's faith in traditional forces—and in his capacity to muster them—was criticized sharply by Adenauer's ambitious minister for atomic matters, Franz Josef Strauβ, who offered a counterproposal of his own. Plainly seeking Blank's job, Strauβ called for a German adaptation of America's Radford plan (named for Admiral Arthur Radford, chairman of Eisenhower's joint chiefs of staff). The plan would reduce traditional manpower and escalate technical prowess. It envisioned fewer men and more weapons, including 'the most modern of weapons', nuclear missiles. In October 1956, Adenauer reshuffled in his cabinet in a major way for the first time. Blank resigned, and Strauβ replaced him. By early 1957, Adenauer had embraced Strauβ's plan.

The West German public, vociferously, had not. Shortly after Adenauer announced his atomic intentions on 5 April 1957, eighteen leading German atomic scientists and other prominent intellectuals issued a manifesto urging Adenauer and all future chancellors explicitly and voluntarily to renounce the possession of any kind of atomic weapons. The scientists among the signatories vowed never to participate in an atomic weapons programme. Luminaries like Albert Schweitzer, Albert Einstein, and Bertrand Russell joined the opposition from abroad, and the protest movement swept across the West German social and political spectrum. The SPD and the FPD, the unions and the churches all joined in a chorus of dissent. Even some national conservatives in the *Bundeswehr* officer corps, fearing the reduction of German territory into a nuclear battlefield, were opposed, though the desire for equal status with France and Britain finally carried the day with the top brass.

Parties hostile to atomic armament, however, failed to translate popular mobilization on the atomic question into electoral success. Marching under the mantra 'No Experiments', the CDU/CSU exploited a popular longing for stability and swept, in September 1957, to an outright Bundestag majority. The FDP joined the SPD in opposition.

But electoral setback hardly stilled the waves of protest. Parliamentary debates in early 1958 on the atomic question were among the stormiest in the Bundestag's history. Two of Adenauer's former ministers turned against him. One was Gustav Heinemann, once Adenauer's interior minister and now a Social Democratic member of the Bundestag. The other was Thomas Dehler, the irascible former justice minister, who accused Adenauer of having never seriously desired reunification and of having conducted a foreign policy that systematically militated against it.[287] In floor debates, SPD orators bashed the government with unblushing Nazi comparisons. In his maiden speech, Helmut Schmidt charged the government with succumbing to 'German-national megalomania'. Schmidt skewered CDU members as the political epigones of those who approved the Enabling Law in 1933.[288] Observing from abroad, George Kennan, the American diplomat who coined the concept of 'containment', now called for a 'disengagement' from Cold War confrontation and the creation of a neutral, united Germany. Hugh Gaitskell, leader of the Labour Party opposition in Britain, proposed establishing an independent German confederation. West German citizens took to the streets.

By April, they marched by the hundreds of thousands. On 25 March 1958, the Bundestag rejected an SPD bill that would have banned stationing atomic weapons in the Federal Republic. Thereafter, the opposition parties joined the mass-protest movement, which by

[287] See Heinrich August Winkler, *Der lange Weg nach Westen*, vol 2 (CH Beck 2000) 193.

[288] See Hans-Peter Schwarz, *Die Ära Adenauer: Epochenwechsel, 1957–1963* (Deutsche Verlags-Anstalt 1983) 38.

now styled itself the 'Fight Against Atomic Death'. Opposition leaders also took action at the state level. In May, SPD governments in the city-states of Bremen and Hamburg, along with SPD-led municipalities in the state of Hessen,[289] passed laws calling for local referenda on the question of atomic armament. The avowed aim was to grant popular protest an official voice with which to press the federal government to relinquish its atomic ambitions. The opposition states also hoped that a ripple effect would prompt parallel referenda throughout the Republic. Adenauer found the vision of a mass rising—state by state, city by city—somewhat less than inspiring. His government denounced the planned plebiscites. Invoking exclusive federal jurisdiction over defence and foreign policy, it asked the Constitutional Court to suppress them.

The Court, with Willi Geiger serving as rapporteur,[290] rushed to judgment, hoping to decide the case before the referenda took place. But before the Second Senate announced its opinion on 30 July 1958, the matter lost much of its brisance. Earlier in the month, after casting its campaign as a referendum on the atomic question, the SPD endured a calamitous defeat in the North-Rhine-Westphalia state elections. After this electoral debacle, the SPD abandoned the movement, and the movement surrendered the ghost. The Constitutional Court supplied the *coup de grace*.

In its published decision, the Second Senate observed that the purpose of the referenda was neither to gather statistics nor to test public opinion. Their purpose was to exert political pressure.[291] The Senate acknowledged that '[i]n modern democracy public opinion plays a decisive role',[292] but stressed that public opinion and the will of the people must not be confused with official opinion and the will of the state.[293] The traditional distinction between state and society remained in force. In the societal–political realm, public opinion was formed and political will shaped without official regulation or restriction. State organs could intervene in these processes only subject to jurisdictional limitations.[294] The Senate treated the proposed referenda as acts of the states that hosted them. As state acts, they encroached on the exclusive jurisdiction of federal organs over national defence and foreign policy.[295] Such encroachments were unconstitutional, null and void.[296]

This, of course, was precisely the position that the federal government had taken in its briefs, and the government was naturally pleased with the decision. Some members of the cabinet waxed effusive. Gerhard Schröder, for one, praised the Court for saving the country from 'a serious threat to the constitution'.[297] Adenauer reacted with a thespian yawn of complacence. The decision came as no great surprise, the chancellor sniffed. The outcome had long been predicted by a large consensus among legal experts.[298]

Opposition leaders vowed to fight on. At all levels of government, SPD stalwarts promised to honour the Court's judgment but pledged to continue the campaign against *Atomtod*. The Court had resolved only the jurisdictional question of regional plebiscites; the moral and existential question of atomic armament remained open.[299]

[289] Most prominently Frankfurt am Main, Darmstadt, Kassel, and Offenbach.

[290] See Ernst Benda, 'Ansprache anläßlich der Feierstunde des Bundesverfassungsgerichts am 9. November 1977 zur Verabschiedung von Bundesverfassungsrichter Prof. Dr. Willi Geiger' BAK N 1334/299 (Nachlass Gerhard Leibholz).

[291] 8 BVerfGE 104, 111–12 (1958). [292] ibid 112. [293] ibid 113. [294] ibid 115–16.

[295] ibid 116 (citing Articles 32(2), 32(3), 65a, 73(1), 87a, and 87b GG).

[296] ibid 121. On the same day, the Court also invalidated the municipal referenda in Hessen, holding that the Hessen state government was in violation of the principle of *bundesfreundliches Verhalten*—literally, federal-friendly behaviour—by allowing its municipalities to conduct referenda in a question of federal competence. See 8 BVerfGE 122 (1958).

[297] 'Erfreut und enttäuscht' *Die Welt* (31 July 1958) 1. [298] ibid.

[299] See 'SPD will gegen Atomwaffen weiterkämpfen' *FAZ* (1 August 1958) 1; Walter Schallies, 'Karlsruhe verbietet Volksbefragung' *Süddeutsche Zeitung* (31 July 1958) 1.

The judgment disappointed and disillusioned those who hoped the *Atomtod* campaign would herald the advent of robust popular democracy in the Federal Republic. Among the most dismayed was Jürgen Habermas. In a 1958 essay on political participation, Habermas lamented that the Court's jurisprudence narrowed the possibilities of popular political engagement. The Basic Law itself, Habermas noted, was 'guided by distrust of plebiscitary decisions'.[300] But the Court had made matters worse:

> [A]ccording to the official interpretation of the Basic Law there exists no possibility for the people to express its will directly in a binding manner with regard to a particular subject. If one compares this de facto impotence with the personal protection and personal freedom guaranteed to individuals in an elabourate catalogue of fundamental rights, one cannot escape the impression that for the citizen of the so-called consumer society the status of a client is also assigned juristically: he must, in the end, pay the piper.... Endowed with *these* rights, and at the same time all but excluded from actual political participation, the People becomes a welfare object.[301]

Habermas discerned in this trend 'a new, welfare-state conception of patrimony', as though the constitution itself took as its rallying cry: 'Everything for the people, but nothing by the people'.[302] After the Court's *Atomic Plebiscite* judgment, Habermas sighed, Bundestag elections remained the '[o]nly possibility for [popular] political participation "at the federal level"'.[303] In light of Habermas's jaundiced, Schmitt-shaded view of modern parliamentarism,[304] this was cold comfort indeed.

Habermas's critique of the Court was institutional as well as case-specific. Earlier in the essay, Habermas approvingly cited the complaints of August Dresbach, a CDU member of the Bundestag, regarding the 'judicial lawmaker of Karlsruhe case law'.[305] He also paraphrased Hermann Brill, a political scientist and former MP, who complained the Court operated as a 'partisan ... judicial supra-authority of political justice in the fashion of a secularized patrimonial court'.[306] But such criticisms, Habermas added, didn't reach far enough. They still 'measure[d] constitution and constitutional reality by the standard of technical, abstract legal principles', and so failed to capture the 'inescapable internal contradictions' arising 'from the underlying conflict between a liberal, rule-of-law Norm and the ... content of concrete social life that Norm deprives'.[307]

Habermas was haunted by these contradictions. As Matthew Specter, a Habermas biographer, notes, 'While [Habermas] never went as far as Schmitt in contrasting liberalism and democracy, he clearly saw a tension between the two'.[308] That tension was particularly acute in the Court's jurisprudence. After the *KPD* and *Atomic Plebiscite* judgments, Specter writes, Habermas viewed 'the highest court as an antidemocratic and antisocialist institution'. He bemoaned 'the Court's role in effectively negating popular sovereignty' and saw in it 'a paternalistic and repressive force of conservative restoration'.[309] This view mellowed over time, but Habermas never fully abandoned his basic critique that '[t]he shadow side of Germany's success in dethroning ... statism in favor of ... robust liberal constitutionalism [was] that too much power [was] concentrated within the Federal Constitutional Court'.[310]

300 Jürgen Habermas, 'Zum Begriff der politischen Beteiligung (1958)' in *Kultur und Kritik* (Suhrkamp 1973) 49.

301 ibid 49–50. 302 ibid 50. 303 ibid.

304 Earlier in the essay, Habermas explicitly adduced Schmitt's 1923 tract *Der geistesgeschichtliche Zustand des heutigen Parliamentarismus* (The Intellectual-Historical State of Contemporary Parliamentarism). See ibid 28.

305 ibid 38. 306 ibid 38–39. 307 ibid 39. 308 ibid 19. 309 Specter (n 285) 68–69.

310 ibid 20.

V. The Quiet Birth of Fundamental Rights

Even as Habermas protested that democratic participation was a high price to pay for fundamental rights enforcement, the Court was crafting a fundamental rights jurisprudence of astonishing scope and power. The outcome only ambivalently brightened Habermas's view of the Court; it blackened Carl Schmitt's view beyond redemption. The rise of muscular, value-laden rights jurisprudence was the most enduring development of the Court's first decade. In a series of landmark judgments the Court fashioned a fundamental rights framework whose protection covered every imaginable human activity and whose writ reached every alcove of the nation's legal order. Their cumulative effect exalted the position of fundamental rights within the constitutional and legal order. It also expanded enormously the power of the Constitutional Court.

Legal scholars immediately recognized that momentous doctrinal developments were underway, though few could have predicted their impact or persistence. In contrast, the landmark fundamental rights decisions were often opaque to the general public. In retrospect, however, many would view the Court's muscular rights jurisprudence as a precondition and catalyst for the democratization of West German law, politics, and society. The fact that citizens could vindicate their constitutional rights by lodging individual complaints in Karlsruhe—a proceeding codified by the constitution itself in 1969—gave ordinary West Germans access to the processes of governance that was unavailable in any other way. Ironically, the very Court that Habermas accused of closing channels of democratic participation came to be viewed as a forum for democratic participation. Individual complaints inundated the Court's inbox—several thousand annually, and the numbers grew with time. Success rates were slender, but the individual complaint became an early and enduring source of the Court's popularity.[311]

A. The *Elfes* judgment and the general right

It is sometimes said that constitutional history is made by eccentrics. Wilhelm Elfes, eponymous plaintiff in the Court's first earth-shaking fundamental rights case, was certainly an eccentric.[312] A founding member of the CDU, Elfes became the party's scourge. Elected from the CDU list to the North-Rhine-Westphalia state Parliament in 1947, Elfes emerged by the early 1950s as an implacable antagonist of Adenauer's foreign policy. Elfes denounced the chancellor's drive toward rearmament and Western integration and deplored his retreat from reunification. Elfes voiced his opposition long and loud, at home and abroad, at every opportunity. A tireless publicist and orator, Elfes became an unwitting ally of the Left; he even inspired proposals within the KPD to install Elfes at the head of a transitional government after the 'fall' of the Adenauer regime. After being ousted from the CDU, Elfes made a failed bid for the Bundestag in 1961 as candidate and co-founder of the 'German Peace Union'. His campaign was partly funded by the German Democratic Republic.

Elfes's extraterritorial tirades against government foreign policy cost him the privilege of travelling abroad. In 1953, his petition for a passport extension was denied. The denying agency gave no reasons for the rejection. A state administrative court sustained the denial

[311] See Christoph Möllers, 'Legalität, Legitimität und Legitimation des Bundesverfassungsgerichts' in Matthias Jestaedt and others, *Das Entgrenzte Gericht. Eine kritische Bilanz nach sechzig Jahren Bundesverfassungsgericht* (Suhrkamp 2011) 281, 298.

[312] On Elfes, see Albert Eβer, *Wilhelm Elfes, 1884—1969: Arbeiterführer und Politiker* (Matthias-Grünewald 1990); Günter Wirth, *Wilhelm Elfes* (Union-Verlag 1984).

and supplied reasons of its own. In December 1952, the court explained, Elfes had gone to Vienna for a 'Congress of the Peoples for Peace'. While there, he presented an 'All-German Declaration' in which he denounced the Federal Republic's foreign policy in the name of all Germans, East and West. Elfes made similarly grandiloquent gestures in Paris, Budapest, and East Berlin. These actions, the court held, 'imperiled the external security ... of the federal government', an offence justifying passport denial under the passport law. When the Federal Administrative Court sustained this judgment, Elfes appealed to Karlsruhe.

Nowhere did the text of the Basic Law explicitly guarantee the right to foreign travel, but Elfes argued that such a right was implied in Article 11's guarantee of the freedom of movement and in Article 2's protection of the free development of one's personality.[313] The Parliamentary Council had debated inconclusively how far Article 2's general guarantee would reach in practice. One camp argued that it included only activities germane to the 'innermost essence' of an individual's 'spiritual–intellectual–moral personality' (*geistig-sittliche Person*). The other camp argued that it shielded a much broader panoply of human behaviour.

In its *Elfes* judgment of 16 January 1957, [314] the Constitutional Court embraced the latter view, endowing Article 2 with a sweeping scope of protection that would make it the textual anchor of countless future complaints—some in matters of tremendous importance, others on questions of consummate banality.[315]

But before expanding Article 2, the justices first limited Article 11, ruling (against prevailing academic opinion) that it did *not* apply to foreign travel. Within the borders of the Federal Republic, the Court held, citizens moved about freely by fundamental right; beyond those borders, they travelled at the pleasure of their government.[316] Article 2, by contrast, protected 'human freedom of action in the widest possible sense', not merely in core spheres of conscience or identity.[317] Article 2 secured a 'general freedom of action' (*allgemeine Handlungsfreiheit*), which individuals could invoke to vindicate freedoms not explicitly secured in the constitution's catalogue of fundamental rights.[318] This was an extraordinary extension of protection; its implications for the Court's future jurisprudence were colossal.

It did not, however, secure Wilhelm Elfes his passport. The passport law's restrictions on individual freedom of action were, the justices wrote, constitutionally legitimate. The law protected the right to foreign travel by requiring the government to give reasons for denying passports. But it also protected state interests by allowing the government to deny passports when it gave adequate reasons. Because Elfes was a security threat, the government's wish to prevent his rabble-rousing on the foreign lecture circuit was a constitutionally adequate reason to deny his request for a passport.

The Court's vast construction of Article 2(1) GG was intensely controversial, not least because its relevance to Elfes's case was questionable. The Court could have disposed of the case simply by limiting Article 11 to domestic travel and finding that foreign travel was unnecessary to personal development. Instead, it discovered in Article 2 a general freedom of action, which, together with the fundamental rights explicitly enumerated in the Basic Law, forged an unbroken shield protecting every imaginable act or omission. To expand Article 2's scope, was of course, to expand the Court's jurisdiction. Within the legal culture, both expansions were recognized as revolutionary—and rarely in a positive sense. 'The judgment is truly sensational', Ernst Forsthoff inveighed in a letter to Carl Schmitt. 'The

313 See Article 2(1) GG. 314 6 BVerfGE 32 (1957).

315 On the 'general personality right', see HD Jarass, 'Das allgemeine Persönlichkeitsrecht im Grundgesetz' (1989) 43 *NJW* 857.

316 6 BVerfGE (n 314) 35. 317 ibid 36. 318 ibid 37.

longer one considers the decision the more mediocre it becomes. Article 2 is thereby killed, and if one sees in Article 2 the famous general right, we have therewith the total statutory reservation [*Gesetzesvorbehalt*]'.[319] The 'statutory reservation' required that activities protected by fundamental right be restricted only by statute or regulation authorized by statute. If all human activity enjoyed the protection of a general fundamental right, Forsthoff scorned, then all government regulation—no matter how trivial its object—must be authorized by formal statute or statutory delegation. To Forsthoff, an administrative lawyer, this sounded like the end of administrative governance.

Other observers were equally appalled. To Theodor Maunz and Günther Dürig, authors of a vastly influential commentary on the Basic Law, the judgment was 'fairly sensational'—an estimation endorsed by Thomas Dehler.[320] Dehler, who as a member of the Parliamentary Council had helped to frame Article 2, thought the decision had made mincemeat of the framers' intentions. He undertook, as a point of honour, to refute the Court's assertion that its construction of Article 2 comported with the provision's framing history.[321]

Outside the legal culture, the case was unopposed by politicians and almost unnoticed by the public. Though the judgment claimed for the Court enormous authority to enforce fundamental rights against the state, it elicited almost no official disapproval. After all, the authorities won their case. An international nuisance had been silenced. Like other early landmarks, *Elfes* has been compared to *Marbury v. Madison*—a technical victory for the state, that gave judges enormous authority, in future cases, to restrain the state. In this respect *Elfes* was, in the title of a later monograph, 'more than a judgment'.[322]

B. Gender equality

In addition to extending constitutional protection to unenumerated rights, the Court in its earliest years also made clear that it would enforce enumerated protections strictly. Nowhere was this more dramatic than in the case of gender equality.[323] During the Basic Law's framing, Elisabeth Selbert, the most prominent of the Parliamentary Council's four female delegates, spearheaded a popular movement that successfully inserted a ten-syllable assurance of gender equality that became Article 3(2) GG: *Männer und Frauen sind gleichberechtigt* ('Men and women have equal rights'). Taken literally, the provision would apply dynamite to the family law provisions of the civil code, which entrenched the domestic dominance of husbands and fathers. On its face, Selbert's simple sentence would invalidate scores of code provisions. It was unclear, though, what sort of family law regime would replace the fallen patriarchal order. As with many thorny issues, the Parliamentary Council reserved the matter for future parliamentary resolution.

Lest Parliament dither, however, the Basic Law imposed a deadline—or rather a stay of execution. Article 117(1) GG granted the legislature a transitional period in which to reform the civil code. After 31 March 1953, however, all code provisions at odds with Article 3(2) would become null and void.[324]

[319] Ernst Forsthoff to Carl Schmitt (28 February 1957), in Dorothee Muβgnug (ed), *Briefwechsel Ernst Forsthoff—Carl Schmitt, 1926–1974* (Akademie 2007) 129.

[320] See Thomas Dehler, 'Zur Entstehungsgeschichte des Art. 2 Abs. 1 GG' BAK N 1221/65 (Nachlass Theodor Heuss).

[321] ibid.

[322] See Gunther Rojahn, 'Elfes—Mehr als ein Urteil' (15 December 2010) (unpublished Ph.D. dissertation, Freie Universität Berlin).

[323] But see 6 BVerfGE 389 (1957) (upholding against a gender equality challenge a law that made male sodomy, but not female homosexuality, a crime).

[324] See Article 117(1) GG.

Until then, uncertainty reigned and debate raged. Commentators within the largely conservative guild of family lawyers contended that Article 3 required little alteration to the *ancien régime*. These scholars thought and wrote in illiberal, collectivist terms. If anything, they defended the preexisting patriarchy with greater vigour after 1949 than before.[325] They saw marriage as a social order enshrined in supra-positive law. Article 3, they insisted, must be read in light of traditional cultural understandings of marriage and family. It was not the business of transitional constitutions, conservatives jeered, to effect sweeping social change.[326] Scholars of the natural law renaissance argued that patriarchy was embedded in the order of creation, that absolute legal equality between husband and wife was 'heresy', and that the leading role of a husband and father at home reflected 'pre-juridical natural law'.[327]

Parliament, meanwhile, consumed with stormy debates on foreign policy, did nothing. The deadline of 31 March 1953 passed without a parliamentary whimper. What then happened to the civil code's family law provisions became an instant source of debate and alarm. Some feared that jurisprudential anarchy had been loosed upon the Federal Republic. Others fretted that courts would unilaterally annul laws half a century old. Many courts pointedly refused to do so. Less than three weeks after the 31 March deadline, courts in Mannheim and Gieβel contended that, absent implementing legislation, Article 3 was not valid law and could not be applied.[328] On 22 April, the Frankfurt Court of Appeals went even further, declaring Article 117(1) GG unconstitutional—an offence to principles of legal certainty and separation of powers—and referring the question to the Constitutional Court.

The first response from Karlsruhe, however, came from the Court of Justice. The Fifth Civil Senate rejected the position of the Frankfurt appeals court in a judgment dated 14 July 1953. What the Frankfurt court really questioned, the Fifth Senate wrote, was not the constitutionality of Article 117 but the direct validity of Article 3. That validity, however, could not be questioned. Article 1(3) unequivocally decreed the direct validity of all fundamental rights provisions, Article 3 included. The equality clause could not be soft-pedalled or dismissed as hortatory—as a policy guide rather than a binding norm.[329] Three months later the First Civil Senate, under the chairmanship of Hermann Weinkauff, published an advisory opinion, supporting the Fifth Senate's stance and dismissing concerns about the separation of powers. It had long been the work of courts, Weinkauff and his colleagues held, to plug gaps and resolve conflicts in statutory law. It would not be easy for judges to do this in the case of Article 3. But it wouldn't be unconstitutional either.[330]

In its judgment of 18 December 1953, the Constitutional Court agreed.[331] Article 3, the First Senate held, was directly valid law,[332] and courts must apply it to bring offending provisions of ordinary law into harmony with the constitution.[333] The vast majority of courts had already been doing this, with no evident eruption of legal chaos.[334] This was not a deviation from the separation of powers principle; it was the traditional work of judges.[335] If Parliament was unhappy with the arrangement, it could pass the law the constitution required. To the contention that direct judicial application of Article 3 undermined legal certainty, the Court conceded that legal certainty was a crucial component of the rule of law, but added that the constitutional framers had accommodated the risk of uncertainty by granting Parliament four years in which to amend the civil code. Now that this period

325 See Franzius (n 16) 125. 326 See Steinbeis and others (n 179) 112.
327 ibid (quoting Friedrich Wilhelm Bosch). 328 See Franzius (n 16) 130.
329 10 BGHZ 266, 278 (1953). 330 11 BGHZ (Anhang) 34 (1953). 331 3 BVerfGE 225 (1953).
332 ibid 239. 333 ibid 231. 334 ibid 226. 335 ibid 242.

had passed, legal certainty must yield to the higher imperative of 'material justice'.[336] In this case, material justice demanded judicial recognition that 'even in marriage and family, man and woman have equal rights'.[337]

In an almost parenthetical aside, however, the Court may have limited with its left hand what it bestowed with its right. 'It hardly needs to be noted', the justices wrote, 'that in the field of family law, a special legal regulation with regard to objective biological or functional (i.e. based on the division of labour) differences according to the nature of the respective life conditions [of husband and wife] is permissible or even necessary'.[338] Four years later the Court adduced this passage to endorse differential treatment, within the criminal code, of gay men with respect to lesbians.[339] Some lower courts, meanwhile, advanced it to preserve elements of domestic patriarchy.

The piecemeal judicial elabouration of Article 3(2) endured until Parliament finally legislated on the matter in the summer of 1957. The law was only a partial victory for advocates of gender equality. It annulled the husband's right to decide important family matters alone and abolished the wife's duty to secure her husband's permission before working outside the home. It also repealed the husband's right to manage his wife's finances and consume her wages. But husbands retained the right to have the final word in questions of child rearing and wives retained the duty to perform household labour. The law appealed explicitly to Christian and natural-law conceptions of gender relations, and, in preserving a husband's right to the last word, it invoked the Constitutional Court's acknowledgement of 'functional differences' between the sexes.

The invocation proved premature. In a July 1959 judgment the First Senate declared the law constitutionally deficient and annulled its retention of the father's final say in child rearing and sole legal representation of the children.[340] In defence of the challenged provisions, the federal government argued that Article 3's guarantee of gender equality must be construed in light of Article 6's promise of special state protection to marriage and family. In the government's view, protecting marriage meant perpetuating patriarchy. The Court, led by Erna Scheffler, turned this argument on its head. By including both provisions, the Constitution presumed the compatibility of gender equality and marital sanctity.[341] The Court read Article 6 by the light of Article 3, not the other way around.[342] So construed, marriage meant nothing if not the union of equals. Article 6 did not preserve patriarchal marriage; it replaced it.

The Senate was singularly unimpressed by the thesis that, in the case of conflict in a community of two, one must finally decide. The point lost what force it may have held when Parliament abolished, without replacement, the husband's final say in matters of housing and household finance.[343] The likelihood of unity and compromise would be strengthened by the rejected alternative of family arbitration courts—an alternative the Court seemed obviously to favour—but weakened by the law's provision that the husband decide alone.[344] Talk of objective differences was beside the point. The Court's 1953 statement about objective and functional differences did not justify weakening the wife in her role as mother.[345] The objective differences exception applied only where differences were so pronounced as to render talk of advantage and disadvantage unintelligible.[346] That was not the case here. The 1957 law put women at a patent disadvantage.[347] For appeals to natural law the Court had

336 ibid 232. The abuse of legal positivism in the Nazi era, the Court wrote, had taught that in some cases 'the principle of material justice is more highly to be valued than that of legal certainty'.

337 ibid 242. 338 ibid 242. 339 6 BVerfGE 389 (n 323) 422. 340 10 BVerfGE 59 (1959).

341 ibid 67. 342 ibid 88. 343 ibid 81. 344 ibid 87. 345 ibid 74.

346 ibid 74 (citing 6 BVerfGE 389 [n 323]). 347 ibid 76.

only icy impatience: 'For the review being conducted here, the Basic Law is the only standard considered'.[348] By that standard, the challenged provisions failed miserably.

The judgment provoked much popular jubilation. 'The fathers of the Federal Republic chronicle a dark day', crowed *Die Welt* in mock solemnity.[349] Other papers abandoned irony. 'The last bastion of paternal domination in the family', hurrahed *Die Zeit*, 'is fallen'. The Court had granted women 'perfect legal equality'.[350] Scattered voices carped that the Court had legislated from the bench. If so, Parliament swallowed the Court's decree. For twenty years the federal legislature let the offending passages stand unaltered, with an asterisk indicating their invalidation by the Constitutional Court. In the meantime, the Court's jurisprudence became the family law of the Federal Republic.

C. 'An objective system of values': the *Lüth* judgment

As the Court deliberated its first *Gender Equality* judgment in December 1953, Erna Scheffler, the case's rapporteur, submitted a draft opinion proposing the concept of the constitution as a hierarchy of values, with some constitutional norms enjoying a higher rank than others—with the equality principle ranking among the supremest of the supreme. Hermann Höpker-Aschoff roundly rejected this heretical notion. 'The thought of distinguishing among norms of higher and lower rank within the constitution', the chief justice wrote in an internal memo, 'is absurd. A constitution, according to its very idea, is a unified conception, based on certain basic decisions, in which each individual provision carries the same validity and force'.[351] Höpker-Aschoff thought Scheffler's notion sufficiently dangerous, and enough of his peers on the First Senate agreed with him, that an explicit refutation appeared in the published judgment of 18 December 1953: 'The Basic Law can only be conceived as a unity. It follows that at the level of the constitution itself, norms of higher and lower rank—in the sense that they can be measured against one another—are fundamentally unthinkable'.[352] The Court bolstered the point by citing the similar conclusion of the Federal Court of Justice.[353] Less than a month after the decision, Höpker-Aschoff was dead. In the hearts of some justices, the hierarchical heresy lived on.

From time to time a sentence appeared in a published judgment suggesting a view of the Basic Law as an organic system—an objective order in which some fundamental norms were more fundamental than others. The *KPD* judgment spoke of the constitution as a 'value system' (*Wertsystem*).[354] So did *Elfes*.[355] Then, in a January 1958 judgment, the Court unmistakably trumpeted that the Basic Law embodied an objective system of values, centred in human dignity and the free development of the human personality, and within which some fundamental rights—the judgment highlighted the freedom of expression—enjoyed a particularly prominent perch. In the eyes of many observers, this judgment was the most momentous, not only of the Court's first decade, but of its entire history.[356] Like *Elfes*, this landmark takes its name from its petitioner: Erich Lüth.

The *Lüth* judgment created an enduring framework of constitutional adjudication and constitutionalized the whole of German law. The legal academy immediately saw the judgment as a watershed, though the full force of its long-term impact could hardly have been

348 ibid 81. 349 *Die Welt* (30 July 1959).

350 'Das Machtwort des Vaters ... und die perfekte Gleichberechtigung' *Die Zeit* (7 August 1959).

351 Hermann Höpker-Aschoff, 'Bemerkungen zu dem Urteilsentwurf von Frau Scheffler zu Art 117 GG' (12 October 1953) BAK N 1129/25 (Nachlass Höpker-Aschoff).

352 3 BVerfGE 225 (n 331) 231–32. 353 See 1 BGHZ 274, 276 (1951).

354 5 BVerfGE 85 (n 261) 138. 355 6 BVerfGE 32 (n 314) 40.

356 See 'Einführung', in *Das Lüth-Urteil* (n 13).

predicted. The German public largely ignored it. The day after the judgment, the *Frankfurter Allgemeine Zeitung* carried a brief, technical summary of the opinion on its fourth page, but said nothing about the case's enormous implications and ignored its seminal passage.[357] The *FAZ* report emphasized the cultural celebrity of the participants rather than the judgment's jurisprudential consequences. The *Süddeutsche Zeitung* didn't report the decision at all, nor did leading weeklies such as *Die Zeit* and *Der Spiegel*. *Die Welt*, which followed the case's course through the lower courts in the early 1950s, also failed to report its dénouement. By 1958, the *Lüth* proceeding had ceased to be a *cause célèbre*, and *Die Welt* had other monumental news to relay—its cover for 16 January 1958 carried a large photograph of the visiting queen of Sweden. Now seen as the very axis of postwar constitutional history, the *Lüth* judgment was a non-event in the contemporary public sphere. This silence is all the more striking because the underlying controversy had once stirred a popular sensation.

Given the judgment's subsequent stature, it is tempting to narrate the facts of *Lüth* in an epic key. The case's protagonists give it the feel of a morality drama. On the side of the angels was Erich Lüth, the case's hero and namesake. Hamburg press secretary at the time of the boycott call that won him fame and got him sued, Lüth was the founder of the 'Peace with Israel' movement that helped facilitate West German reparations payments to Jews. On the other side was Veit Harlan, director of twenty or so Nazi propaganda films, most notoriously the venomously anti-Semitic *Jud Süss* (*The Jew Süss*). The 1940 film was the greatest cinematic blockbuster of the Nazi era, seen by some 22 million Third Reich movie goers. The film's appeal was not limited to the vulgar masses either. After seeing *Jud Süss* for the first time an effusive Carl Schmitt urged Ernst Jünger to view the film posthaste.[358] In the winter of 1940–41, Heinrich Himmler is said to have ordered all SS and Gestapo personnel to watch Harlan's movie. The film's influence was deemed sufficiently pervasive and pernicious that Harlan was the only Nazi artist in any field prosecuted for crimes against humanity after the War.

Harlan was acquitted by a Hamburg jury court in early 1949, but his acquittal was quashed and remanded by the Supreme Court of Justice for the British occupation zone in Cologne. The Hamburg court acquitted Harlan a second time a year later, and the public prosecutor declined a second appeal, which would have gone before the Federal Court of Justice and its cadre of erstwhile Nazi jurists. Harlan was similarly exonerated in his denazification proceeding, where he was classified as 'unburdened' (*unbelastet*) by the crimes of the Nazi regime. The collective official verdict was that Harlan was the unwilling instrument of Josef Goebbels. *Jud Süss* was Harlan's handiwork but not his brainchild; his artistic production but not his legal responsibility.

Thus acquitted, Harlan was ready for a comeback. Indeed, when public prosecutors decided not to appeal his acquittal in July 1950, Harlan was already at work on his first postwar film, a tragic romance called *Immortal Beloved* (*Unsterbliche geliebte*), a filmic recreation of a Theodor Storm novel featuring Harlan's spouse and starlet, Kristina Söderbaum.

Erich Lüth reacted fiercely to news of Harlan's return to the German screen. In a September 1950 speech to the Hamburg Press Club, Lüth sparked a controversy that would last nearly a decade and immortalize his name in the annals of the Constitutional Court. 'German film having forfeited its moral aspiration during the "Third Reich,"' Lüth began,

[357] 'Lüth durfte gegen Harlan vorgehen' *FAZ* (16 January 1958) 4.

[358] Carl Schmitt to Ernst Jünger (29 September 1940) *Ernst Jünger/Carl Schmitt Briefe, 1930–1983* (Klett-Cotta 1999) 105. The further correspondence doesn't say whether Jünger followed Schmitt's advice.

> one man is certainly the very least suited to restore this reputation: this is the screenwriter and director of the film *Jud Süss*! Let us spare ourselves the incalculable damage we would incur in the eyes of the whole world should he of all people be held forth as a representative of German film. His acquittal in Hamburg was merely formal. The opinion of the court was a moral condemnation. We here urge lenders and theatre owners to take a stand which will not be cheap but for which we ought to be willing to pay a price: Character. And I wish such character for German film.[359]

Reports of a new Harlan film had already provoked scattered opposition; Lüth's polemic swelled and serried the ranks. Hans Domnick, the producer of *Immortal Beloved*, responded publicly. Insisting that he had employed Harlan as a director in order to restore German film to its 'former artistic height', Domnick challenged Lüth to justify his assault on the acquitted Harlan.[360] Lüth did so in an open letter dated 27 October 1950.

The Hamburg court's judgment, Lüth wrote, did little to deny the fact that for much of the Hitler era Harlan had been 'Nazi Director Number 1', and that, by means of *Jud Süss*, Harlan had been 'one of the Nazis' most important exponents of murderous Jew-hatred'. Some foreign and domestic businessmen might not oppose Harlan's return to directing, Lüth conceded.

> But Germany's moral standing in the world must not be ruined again by powerful moneymakers. Harlan's reemergence would rip open barely scarred wounds and frightfully renew fading mistrust to the detriment of German reconstruction. For all these reasons, it is not only the right but the duty of self-respecting Germans to hold themselves ready, in the struggle against this unworthy representative of German film, to move beyond protest to boycott.[361]

Having failed to secure Lüth's retraction the producers sought to secure his silence.[362] On 2 November 1950, Domnick Film and its underwriter, Herzog Film, petitioned the Hamburg district court for a temporary injunction. The civil action thus initiated would culminate more than seven years later in the *Lüth* judgment of the Constitutional Court.

The district court granted the producers' request for a temporary injunction on 18 November 1950.[363] On 27 February 1951, the Hanseatic Court of Appeals sustained the injunction.[364] The injunction drew critical commentary in the press and expressions of solidarity in the public.[365] A campaign was launched to collect funds for Lüth's legal expenses. Among Lüth's public supporters were a handful of prominent SPD politicians, among them Adolf Arndt.[366] On 17 May 1951, a month after the FCCA entered into force, Lüth announced that, if the Hamburg district court ruled against him in the main proceeding, he would appeal to the new court in Karlsruhe. In the meantime, Harlan's new film drew huge crowds of viewers and protesters. By mid-1951 *Immortal Beloved* had sparked demonstrations in most major cities of the Federal Republic, with the wave coalescing in what became increasingly a student movement.[367]

On 22 November 1951, the district court ruled against Lüth on the merits.[368] The court's result and reasoning were unaltered from its injunction judgment the year before. The injunction was made permanent, and Lüth was ordered to pay damages. The judgment

359 7 BVerfGE 198, 199 (1958). 360 ibid 200.

361 ibid. Lüth's letter was published in the *Welt am Sonntag* of 29 October 1950 and the *Neue Zeitung* of 31 October 1950.

362 The history of the *Lüth* proceeding before Lüth's appeal to the Constitutional Court is recorded in Arne Riedlinger, 'Vom Boykottaufruf zur Verfassungsbeschwerde: Erich Lüth und die Kontroverse um Harlans Nachkriegsfilme, 1950–58' in *Das Lüth-Urteil* (n 13) 147.

363 ibid 457–66. 364 ibid 467–80. 365 Riedlinger (n 362) 160–61. 366 ibid 161.

367 ibid 166. 368 *Das Lüth Urteil* (n 13) 481–87.

provoked criticism in the press, some of it accompanied by calls for new legislation.[369] These calls were eventually answered. But the new legislation came from Karlsruhe, not Bonn.

In the aftermath of his defeat in Hamburg, Lüth visited the office of Adolf Arndt to seek representation in a constitutional challenge. Arndt, exhausted and overworked, buffeted by several overlapping legal, political, and foreign policy tempests, could only sigh. Gesturing to the mountains of documents around him, Arndt boomed in his Berliner baritone that though Lüth's cause was surely just, he simply could not undertake it. But perhaps, Arndt added, his assistant might be able to help.[370] Arndt's assistant—the first scholarly assistant in the history of the Bundestag—was a twenty-eight-year-old named Wilhelm Hennis, a newly-minted Ph.D. and later a leading political scientist and public intellectual.[371] Hennis took the job and drafted a brief. On 18 December 1951, Arndt's office asked the three-month-old Constitutional Court, in Lüth's behalf, to quash the judgment of the Hamburg district court.

Six years passed before the First Senate did so. For much of that time, the Senate was hopelessly overburdened—embroiled, too, in a series of smouldering political rows. In a May 1952 letter to Adenauer's state secretary Otto Lenz, Höpker-Aschoff mentioned *Lüth* as a pending case of political significance that the Court couldn't reach because of its constrained resources.[372] In a 1956 essay, Justice Willi Geiger despaired over the Court's unworkable division of labour. 'The current situation of the Federal Constitutional Court is grim', Geiger wrote. At the moment, he grieved, the Court 'can no longer fulfill its function'.[373]

It certainly couldn't, or at least didn't, respond to Erich Lüth's petition. Whatever the reasons for delay, the First Senate released its judgment in the *Lüth* controversy at a time—January 1958—when the Constitutional Court had effectively consolidated its position at the summit of the German judiciary and when its status as a coordinate constitutional organ was universally recognized. It was a time, too, when the Federal Republic itself had found broad popular acceptance, when the 'Economic Miracle' was at its crest, and Adenauer at the zenith of his power. It was a time when the Court, in a case posing little threat of popular backlash, could deliver a decision whose outcome was impressive and whose grounds were revolutionary.

Lüth's complaint argued that the district court judgment violated his fundamental right, guaranteed by Article 5(1) GG, to voice his views freely. This argument presupposed that private-law judicial decisions *could* violate fundamental rights, which in turn presupposed that fundamental rights constrained private law norms.

Both propositions were controversial. At one pole of the debate were those clinging to a classical conception of fundamental rights as—only—defensive rights shielding individual freedom against state intrusion. At the other extreme were those, led by Hans-Carl Nipperdey, president of the Federal Labour Court, who argued that fundamental rights norms applied directly to all legal relationships, public and private, between citizens well as between the individual and the state.[374] In its *Lüth* judgment, the First Senate distanced

[369] See Riedlinger (n 362) 168–69.

[370] See Wilhelm Hennis, Lüth—*und anderes*, in *Das Lüth-Urteil* (n 13) 187, 191.

[371] Hennis did not stay with the case till its conclusion. In 1953 he left Arndt's office to become Carlo Schmid's assistant at the Institute for Political Science at the University of Frankfurt. On Hennis, see Stephan Schlak, *Wilhelm Hennis: Szenen einer Ideengeschichte der Bundesrepublik* (CH Beck 2008).

[372] Hermann Höpker-Aschoff to Otto Lenz (16 May 1952), BAK B 136/4436.

[373] Willi Geiger, 'Zur Reform des Bundesverfassungsgesetzes' in Theodor Maunz (ed), *Vom Bonner Grundgesetz zur gesamtdeutschen Verfassung* (Isar 1956) 211.

[374] Nipperdey had applied this view in a labour Court judgment, asserting its consistency with two 1957 judgments of the Constitutional Court. See the judgment of the *Bundesarbeitsgericht* (Federal Labour Court) in (1957) 10 NJW 1688, 1689 (citing 6 BVerfGE 55 (1957) and 6 BVerfGE 84 (1957)). On Nipperdey's relationship to

itself from the latter position and decisively rejected the former.[375] True, the justices wrote, fundamental rights remained *principally* defencive rights against the state. But in the constitutional order of the Basic Law, they were something more as well—something the justices captured in the most famous passage they or their successors ever wrote:

> In its fundamental rights section, the Basic Law—which is not a value-neutral order—has also erected an objective order of values and given expression thereby to a fundamental strengthening of the validity and force of fundamental rights. This value system—centred on the dignity and free development of the human personality within the social community—must be regarded as the fundamental constitutional decision for all areas of law. It governs and animates legislature, executive, and judiciary. Thus, self-evidently, it also influences civil law. No civil law provision may contradict it; each must be construed in its spirit.[376]

Courts that failed sufficiently to consider the effect of fundamental rights on private law norms not only committed an objective legal error; they were guilty of wielding state power to the detriment of an individual right.[377] Those harmed by such a faulty judgment could challenge it before the Constitutional Court.[378]

Fundamental rights, then, were objective values, not merely subjective shields. All private law norms were subject to the 'radiating effect'[379] (*Ausstrahlungswirkung*) of those values. Some rights, moreover, radiated with particular brilliancy. Prominent among these was the freedom of speech:

> The fundamental right to the free expression of opinion is, as the most unmediated expression of the human personality, one of the foremost of all fundamental rights. .... For a free democratic state order it is *absolutely* constitutive [*schlechthin* konstituierend], for it enables the constant intellectual exchange, the conflict of opinions that are its very lifeblood. It is, in a certain sense, the foundation of all freedom, 'the matrix, the indispensable condition of nearly every other form of freedom' (Cardozo).[380]

The primacy of free expression within the free democratic order required that the guardians of that order vigilantly oppose 'every relativization' of that right 'through ordinary law'. Free expression of opinion could be limited only by other protected interests of equal or greater constitutional value. In balancing competing values the Court would ask whether, considering all the circumstances of a given case, 'protected interests ... of higher rank would be injured by the expression of opinion'.[381] This would rarely be so when the contested expressions contributed to public discourse and the shaping of public opinion. Almost invariably, in such cases, the private interests protected by civil law must yield to the fundamental rights secured by constitutional law.[382] The Hamburg district court had paid insufficient attention—indeed, no attention at all—to the influence of Article 5 GG on Section 826 of the civil code. Lüth's right to express himself freely, and the public's right to hear what he had to say, trumped the financial interests of the producers and the reputational interests of Harlan himself. What's more, contrary to the findings of the lower courts, Lüth's boycott appeal did not offend public morals. If anything, in its concern for Germany's moral standing in the wake of Nazi misrule, it promoted them.[383] The Hamburg district court had misread the moral facts of the case and misread the role Article 5(1) GG should have played in its analysis. Its judgment, therefore, was revoked.

the *Lüth* judgment, see Thorsten Hollstein, 'Um der Freiheit willen—die Konzeption der Grundrechte bei Hans Carl Nipperdey' in *Das Lüth-Urteil* (n 13) 249.

[375] 7 BVerfGE 198 (n 359) 204. [376] ibid 205. [377] ibid 206–07. [378] ibid 207.
[379] ibid 207. [380] ibid 208. The Court quoted Cardozo in English. [381] ibid 210.
[382] ibid 212–30. [383] ibid 221–30.

The practical impact of the *Lüth* judgment was slight. *Immortal Beloved* had already made its run as one of the great blockbusters of the early 1950s—more than 8 million viewers by one count—and Harlan's comeback was complete. He released six films while the case was pending in Karlsruhe and three more during the year of the *Lüth* judgment. The public that once sympathized with Lüth's fight largely ignored his vindication. This was so even in some legal circles. Three years after the decision, a Canadian comparative lawyer characterized *Lüth* as presenting interesting questions of abstract jurisprudence in an 'otherwise rather minor' decision.[384]

Most public lawyers and state theorists, however, recognized *Lüth* for the earthquake that it was. In effect, the Court had taken sides in the civil war of the West German public law academy, whose two camps formed round the figures of Rudolf Smend and Carl Schmitt.[385] With its conception of the constitution as an order of values that required balance and reconciliation on a case-by-case basis, the *Lüth* judgment breathed the spirit of Smend's doctrine of integration (*Integrationslehre*), a view of the state as an agent of political integration and societal harmony.[386] Smend himself had characterized the rights charter of the Weimar constitution as a 'system of values'.[387] But the Court incorporated Smend with the modernizing gloss Smend's students of the 1950s had given to their master's precepts of the 1920s. In its Weimar formulation, Smend's integration doctrine—partly informed by Smend's early admiration for Italian fascism—was harmonistic, statist, and in the last analysis anti-pluralist. Smend's postwar pupils—many of whom thought and wrote with an American accent and a bent for political science—recast the concept of integration in a liberal, Western, pluralist mode. Out of that pluralism they sought unity—the unity of the state and the unbroken internal unity of the constitution.[388] Smendians saw the realization of such unity in *Lüth*'s proclamation of an objective, holistic system of constitutional values.

On the whole, Smend's students greeted the *Lüth* judgment with jubilation. So did other leading lights of the legal academy. Günter Dürig, whose theory of the indirect operation of fundamental rights on civil law norms the Court had broadly embraced,[389] wrote a rapturous review in the *Neue Juristische Wochenschrift*, a leading legal weekly.[390]

The Schmitt school responded with somber jeremiads. Ernst Forsthoff announced the end of constitutional law as German public lawyers had known it, a revolutionary transmutation of their tradition.[391] The most pointed critique came from Schmitt himself. In Schmitt's view, the Court had compounded its usurpation of jurisdictional sovereignty with an assertion of moral tyranny—'The Tyranny of Values', Schmitt called it, borrowing a phrase from Nicolai Hartmann.[392] It was in the aftermath of *Lüth* that

384 Edward McWhinney, *Constitutionalism in Germany and the Federal Constitutional Court* (AW Sythoff 1962) 46.

385 On the formation and development of the two schools, see Frieder Günther, *Denken vom Staat her: Die bundesdeutsche Staatsrechtslehre zwischen Dezision und Integration* (Oldenbourg 2004).

386 See Rudolf Smend, *Verfassung und Verfassungsrecht* (Duncker and Humblot 1928).

387 Rudolf Smend, 'Verfassung und Verfassungsrecht' in *Staatsrechtliche Abhandlungen und andere Aufsätze*, 3d edn (Duncker and Humblot 1994) 264.

388 For a sharp summary, see Frieder Günther, 'Ein Jahrzehnt der Rückbesinnung. Die bundesdeutsche Staatsrechtslehre zwischen Dezision und Integration in den fünfziger Jahren' in *Das Lüth-Urteil* (n 13) 301–14. On the Smend school, see ibid 307–12.

389 See Günther Dürig, 'Grundrechte und Zivilrechtsprechung' in *Vom Bonner Grundgesetz zur gesamtdeutschen Verfassung* (n 373) 157.

390 Günter Dürig, 'Zum "Lüth-Urteil" des Bundesverfassungsgerichts vom 15.1.1958' (1958) 12 *Die Öffentliche Verwaltung* 194.

391 Ernst Forsthoff, 'Die Umbildung des Verfassungsgesetzes' in Hans Barion (ed), *Festschrift Carl Schmitt zum 70. Geburtstag* (Duncker and Humblot 1959) 35.

392 Carl Schmitt, 'Die Tyrannei der Werte' in *Säkularisation und Utopie: Ebracher Studien. Ernst Forsthoff zum 65. Geburtstag* (Kohlhammer 1967) 37. Schmitt's essay has recently been republished as a stand-alone volume with an excellent afterword by Christoph Schönberger that examines Schmitt's relationship to the

Schmitt sneered that Bonn Weimer, it Karlsruhe. Persuaded by his young protégé, Ernst-Wolfang Böckenförde, to speak at a colloquium in February 1959, Schmitt used the occasion to vituperate the recent moralization of law and lament the emergence of constitutional judges as 'unmediated executors of value' (*unmittelbare Wertvollzieher*). Declarations and assessments of value, Schmitt railed, could never be a source of legitimacy. They were merely a boot-strapping exercise propelled by judicial vanity and self-importance.[393]

As the Court extended *Lüth*'s logic to other areas of law and other fundamental rights, Schmitt's disgust grew. Six years after *Lüth* Schmitt voiced his disdain in mocking verse:

> In Karlsruhe, there grows a rubber tree.
> Lemurs shuffle through the green
> And hang a dream of liberty—
> As Value on this rubber tree.
> Well.
> What have you all to say thereto?
> We say: Psst—taboo![394]

But Schmitt's, as he never tired of reminding correspondents, was the voice of a lonely exile. The Smend school increasingly dominated the academy, which in its turn was increasingly influenced by the jurisprudence of the Constitutional Court.[395] So, after *Lüth*, was the German judiciary. After *Lüth*, the Court was not only the sole judge of the constitutionality of laws, it was, where fundamental rights were in play, the final arbiter of each law's judicial application. All judges now had to work with an eye to the impact of constitutional norms on ordinary law—with an eye, that is, to the case law of the Constitutional Court.

In the *Civil Servants* and *Gestapo* judgments, the Court had consolidated its monopoly over constitutional jurisdiction. In *Lüth*, it constitutionalized all jurisdictions. Already ensconced at the summit of the legal order, the Constitutional Court ensured in *Lüth* that its writ reached that order's every nook and cranny. Future commentators considered *Lüth* the Magna Carta of fundamental rights in the Federal Republic. The judgment has been commemorated with a breathless string of superlatives. In its own historical moment, *Lüth* marked the culmination of the Court's consolidation and a manifesto of its future role. It cast the Court as an arbiter of constitutional value and an aggressive enforcer of fundamental rights.

The Court would extend this vision, and expand *Lüth*'s balancing logic, in another landmark fundamental rights decision delivered five months later.[396] The underlying facts were fairly mundane—the complainant wished, in despite of a regional ordinance, to open a second pharmacy in a small Bavarian hamlet. In its vindication of the complainant's vocational freedom under Article 12 GG, the First Senate outlined the proportionality test that would become the signature fundamental rights framework, not only for the German Constitutional Court, but for constitutional tribunals around the world.

Constitutional Court. Christoph Schönberger, 'Werte als Gefahr für das Recht? Carl Schmitt und die Karlsruher Republik' in Carl Schmitt, *Die Tyrannei der Werte*, 3rd edn. (Duncker and Humblot 2011) 57–91.

[393] Schmitt, 'Die Tyrannei der Werte' (n 392) 45.

[394] In *Reinhard Mehring, Carl Schmitt: Aufstieg und Verfall* (CH Beck 2009) 523.

[395] See Bernhard Schlinck, 'Die Entthronung der Staatsrechtswissenschaft durch die Verfassungsgerichtsbarkeit' (1989) 28 *Der Staat* 161.

[396] 7 BVerfGE 377 (1958).

Together, the *Lüth* and *Pharmacy* judgments wrought a revolution in the Court's fundamental rights jurisprudence and in its self-understanding. The full implications of this revolution would become clear only later on.

Conclusion

By any standard, the 1950s were a remarkable decade for the Federal Constitutional Court. On one view, the Court seemed to march from strength to strength, from triumph to triumph. It emerged from the conflicts over its own status and the European Defence Community Treaty with its independence secure and its prestige enhanced. It silenced speculation about its political composition and partisan inclinations. It confirmed its status as a court and as something more than a court. It gained near universal acknowledgement as the guardian of the constitution and guarantor of the rule of law. It displayed sufficient political savvy to appear non-political. It was deferential to government actors but never subordinate to them. It often sustained government policy but never shrank—and frequently expanded—its own jurisdiction. It emerged as an emblem of moral legitimacy, courageously confronting and fearlessly repudiating the country's recent past. It proved willing to take fundamental rights seriously. It was a moral success story to match the economic miracle. It even seemed to have the right enemies.

There was, however, a darker view—one that united Habermas on the Left with Schmitt on the Right. This view held that, in its inaugural decade of consolidation, the Court had consolidated too much. The new bastion of moral legitimacy risked becoming a redoubt of moral monopoly. The Court may have given a moral compass to the young fiscal colossus, but it was emphatically the Court who gave the compass and who would, presumably, continue to give it. Policy came from Bonn; values from Karlsruhe; but what came from the People? Habermas and Schmitt both worried, in very different ways, about the transfer of sovereignty to the Constitutional Court—Schmitt that the Court's very existence fractured the state's capacity to decide, Habermas that it stripped the people's opportunity to deliberate. Schmitt feared the Court on authoritarian grounds, Habermas on democratic.

These fears were only partially captured—Schmitt's more than Habermas's—in the most common early criticism of the Court, which was that it meddled in politics. Beginning at least with the EDC controversy, the press began to warn of the rise of a 'judicial oligarchy',[397] of the emergence of Karlsruhe as the country's 'secret capital',[398] of the dethroning of Bonn by its 'sister to the South'.[399] Many were alarmed during the EDC debate that the fate of the nation could be decided by a handful of judges.[400] But this hadn't happened in the *EDC* case and didn't seem to happen in any other case. Throughout the decade, critics sounded alarms about the Court's 'politicization' and reformers called for its 'de-politicization'. Many of these calls, however, did not target the justices themselves. The problem, many argued, was structural. Some thought it had to do with the machinery of judicial appointment. Once appointed, the justices acquitted themselves admirably. The justices consistently received more favourable press coverage individually than did the Court collectively, with the chief justices Hermann Höpker-Aschoff, Joseph Wintrich, and

[397] Fritz Bauer, 'Juristenoligarchie' *FAZ* (3 June 1952).
[398] Hans Rolf Zimmermann, 'Karlsruhe—die heimliche Hauptstadt' *Stuttgarter Zeitung* (1 July 1952) 7.
[399] Sethe (n 110).
[400] See Walter Fredericia, 'Vierundzwanzig Richter entscheiden über Deutschland' *Die Zeit* (26 June 1952); René Marcic, 'Ein Areopag regiert Deutschland' *Salzburger Nachrichten* (16 July 1952) 3.

(after 1959) Gebhard Müller receiving the highest ratings of all.[401] Even so, the Court itself was reported much more frequently in a positive than in a negative light.[402] Calls to limit the Court's jurisdiction were rare; calls to abolish it unheard of.

By the time Gebhard Müller became the the Court's third and longest-serving chief justice in 1959, the Court was the object of widespread approval but also of widespread ignorance. Most citizens of the Federal Republic, blissfully enjoying unprecedented material prosperity, troubled themselves little with the rise of a judicial giant in Karlsruhe. The Court's consolidation of power was little opposed in part because it was little observed. This was especially so of the values jurisprudence that arose in the Court's landmark fundamental rights judgments. When *Der Spiegel* surveyed the history of the Court in March 1961, it said nothing about the fundamental rights landmarks, of *Elfes* or *Lüth*.[403] High-profile political cases seemed to decline in frequency as the decade progressed. By mid-decade, Western integration was an accomplished fact that enjoyed the Court's blessing. There would not be another landmark foreign policy judgment until there was a major shift in national foreign policy—not, that is, until after a Social–Liberal coalition took power at the end of the 1960s. By the end of the Court's first decade, some observers pronounced that the era of the constitutional *cause célèbre* was over and done.[404] Despite some spectacular political confrontations, the Court had never invalidated a major legislative initiative of the Adenauer government. This, however, and in stupendous fashion, would soon change.

It was only in retrospect that the Court's early achievements would be widely recognized and almost universally praised. In the fullness of time, the early Court would win plaudits for its contribution to the democratization of the Federal Republic's political culture.[405] But, as this chapter has shown, the major developments in the Court's first years had more to do with restoring the rule of law and protecting fundamental rights than with democracy as such—though the Court did much to legitimate the role of parliamentary opposition as a *sine qua non* of democratic governance. The enduring legacy of the Court's earliest years was to enshrine fundamental rights and the rule of law as *preconditions* of democracy. It was during the Court's second decade, and through its first wholesale rejection of a core government policy, that the Court began to mark the contours of democracy itself.

But it was crucial for the Court that this first dramatic annulment of a government initiative—its 1962 invalidation of Adenauer's proposal for a second, effectively government-controlled public television station—came after the Court had established institutional independence and popular legitimacy. The Court of the 1960s could be politically aggressive because the Court of the 1950s had so conspicuously contradicted partisan political expectations. Moreover, as coming to terms with the Nazi past became a broad cultural preoccupation for the first time in the mid-1960s, the Court benefited hugely from seeming already to have done so. Even in the 1960s, however, the Court used its power sparingly. The real confrontation with the political organs came only in the 1970s. Chapter 3 will chronicle that collision in indulgent detail. In all its future trials, however, the Court survived challenges to the legitimacy of its power in part because, in its earliest days, the public acknowledged the Court's legitimacy before appreciating its power.

[401] See Oliver W Lembcke, *Über das Ansehen des Bundesverfassungsgerichts. Ansichten und Meinungen in der Öffentlichkeit 1951–2001* (BWV Verlag 2007) 49, 50.

[402] ibid 47.

[403] 'Verfassungsgericht: Scherbengericht' *Der Spiegel* (8 March 1961).

[404] See 'Die Zeit der Sensationsprozess ist vorbei' *Durlacher Tagblatt* (17 November 1959); 'Die Saison der Sensationsprozesse ist vorbei' *Badischer Volkszeitung* (25 November 1959).

[405] See Bryde (n 158) 321.

2

Confidence, 1959–1971

The solution, I believe, is not difficult, when one recalls that freedom of the press is nothing less than the consequence of the freedom of thought and therefore a Protestant right. For rights of this sort the German has already given his best blood, and for such he may well be brought to the barricades once again.

—Heinrich Heine (1835)

There can be no doubt that, in principle, a constitutional court runs the risk of encroaching in the political realm of the legislator. On the other hand, the Court has decisively contributed to the fact that, from the state's side and from the citizen's, the constitution is regarded and followed in its full extent.

—Helmut Schmidt (1969)

Introduction

Sebastian Haffner, one of the great political writers and historical essayists of the Federal Republic, once wrote of Konrad Adenauer that although the chancellor 'was, God knows, enough of a patriarch and an autocrat, he was also at the same time a transition to democracy—a democratic patriarch, a democratic autocrat'.[1] Adenauer, Haffner concluded, 'habituated Germans to the idea that authority and democracy were compatible. He gradually reconciled them, so to speak, with democracy'.[2]

Some have been tempted to apply this assessment to the Federal Constitutional Court.[3] At the annual meeting of the association of German law professors in 1970, one of the participants, Wilhelm Karl Geck, noted that '[t]he jurisprudence of the Federal Constitutional Court has, as we all know, fostered the development of a true democratic understanding in the Federal Republic—not least in the sphere of public opinion. There are decisions of the Federal Constitutional Court that, as harbingers of the free *Rechtsstaat* and of democracy, were years ahead of public opinion'.[4] Many accounts of the early Federal Republic depict a people whose attitudes toward democracy ranged from tepid to hostile, and whose gradual embrace of republican governance had more to do with economic prosperity than with democratic conviction. Many contemporary critics, especially on the political Left, objected fiercely to what they saw as a restorationist impulse in the political culture of the Federal Republic.[5] There were too many continuities—of tone, of personnel, of style—between the Bonn regime and its predecessors. Many longed for a more decisive break with the past.

Or at least for a frank discussion *about* the past. With rare exceptions, the Federal Republic's political leaders were unwilling, for most of the 1950s, to speak forthrightly

[1] Sebastian Haffner, 'Konrad Adenauer: der rechte Mann zur rechten Zeit' in Sebastian Haffner, *Im Schatten der Geschichte* (Deutsche Verlags-Anstalt 1985), 291.

[2] ibid.

[3] See Christoph Schönberger, 'Anmerkungen zu Karlsruhe' in Matthias Jestaedt and others, *Das entgrenzte Gericht. Eine kritische Bilanz nach sechzig Jahren Bundesverfassungsgericht* (Suhrkamp 2011), 44.

[4] Wilhelm Karl Geck, 'Diskussionsbemerkung' (1971) 29 *Veröffentlichung der Vereinigung der Deutschen Staatsrechtslehrer* 113.

[5] See A. Dirk Moses, *German Intellectuals and the Nazi Past* (Cambridge University Press 2007), chs 2–6.

about Germany's crimes. As we saw in Chapter 1, the Federal Constitutional Court was a remarkable exception to this general pattern of official silence. Particularly in the *SRP, Civil Servants*, and *Lüth* decisions, the Court defined the Basic Law as an anti-fascist instrument. The *Lüth* decision was especially significant in this regard because it doubled as a powerful manifesto for a democratic public sphere. In both its 'free speech' and its 'politics of the past' dimensions, *Lüth* anticipated broader trends that, over the course of the next ten years, would transform both political culture and public discourse in the Federal Republic.

The early 1960s witnessed a great awakening of popular attention to Nazi crimes. On Christmas Eve 1959, two young Right-wing ruffians smeared swastikas on the walls of a synagogue in Cologne, inspiring hundreds of imitators in the following weeks. Over the next five years the West German public, hauntingly reminded that anti-Semitism was not dead, watched the trials for murder and crimes against humanity of functionaries who operated death camps in Auschwitz, Chelmno, Belzec, Sobibor, Treblinka, Dachau, Stutthof, and Mauthausen. Most dramatic of all were the 1961 trial of Adolf Eichmann in Jerusalem, and the so-called 'Auschwitz Trial', held in Frankfurt between 1963 and 1965. Toward the end of the latter trial, on 10 March 1965, the Bundestag voted, after a momentous debate, to extend the statute of limitations for murder by providing that the twenty-year term should commence in 1949 (with the founding of the Federal Republic) rather than in 1945 (with the end of the War). On 26 February 1969, the Constitutional Court unanimously approved the constitutionality of the extension.[6] (With the Court's prevenient blessing, Parliament would extend the limitations period for Nazi crimes ten years more to 1979, and then abolish the limit altogether.) A year earlier, the Court held unanimously that the validity of Nazi-era legal provisions could be rejected by a judge who would commit obvious injustice by applying them.[7] Thus, as the public engaged in its first nationwide reckoning with the legacy of Nazism, the Court continued to bolster its anti-Nazi credentials.

The Court also played an important role in what one is tempted to call the opening of the West German public sphere. The last chapter highlighted the Court's role in legitimating parliamentary opposition and, in *Lüth*, individual dissent. The Court's contributions to the public sphere during its second decade were even more dramatic, as the Court issued judgments that limited the ability of the federal government and of economically powerful private parties to dominate or constrain public discourse. These judgments, and the mounting frequency of direct individual complaints, enhanced popular esteem for the Court as a vindicator of fundamental rights, and fostered a growing popular perception that fundamental rights were antecedent and essential to democracy.[8] The Court, for its part, increasingly understood itself—and presented itself—as a fundamental rights tribunal. The public's esteem in this regard was broadly justified, though with jarring exceptions. Even the exceptions came in the first major cases with published dissents, so that at least some justices were still heard speaking as champions of fundamental rights. It is difficult to overstate how novel such dissenting voices were in the German tradition. The Court thereby replicated within its own jurisprudence the kind of plurivocality that was emerging in the country at large. The Court's increasingly pronounced role as an enforcer of fundamental rights, and its increasingly central role in opening the country's public sphere, were the most lasting legacies of its second decade. They confirmed the public's perception of the Court as a bulwark of the *Rechsstaat* and a catalyst of democracy.

[6] 25 BVerfGE 269 (1969). [7] 23 BVerfGE 98 (1968).

[8] For a comparative discussion of the right of individual complaint, see Georg Brunner, 'Der Zugang des Einzelnen zur Verfassungsgerichtsbarkeit im europäischen Raum' (2002) 50 *Jahrbuch des öffentlichen Rechts der Gegenwart* 191.

A. Chief Justice Gebhard Müller

The Court began its second decade with a new president. Gebhard Müller faced a very different set of circumstances when he became the Court's third chief justice in February 1959 than Hermann Höpker-Aschoff had faced upon becoming the first chief justice in 1951. No more was the Court an institutional greenling sprouting precariously in uncultivated jurisprudential soil. Not only had the Court consolidated institutional authority during its first decade, it had developed institutional memory. The Court now had a tradition of its own to guide its procedure, and a jurisprudence of its own to inform its judgments. It also enjoyed an enviable continuity of personnel. A 1956 amendment to the Court's statutory charter decreased the number of justices from twenty-four to twenty. As of 1960, fifteen of the twenty justices—seven on the First Senate, eight on the Second—had been with the Court since its founding. Karl Heck and Egon Schunck, two of the five non-original justices, had jointed the Court in 1952, when it was only a few months old. Just two of the twenty justices—one of them Müller, the new chief—had been appointed later than 1955.

Appointed in 1959 to replace the deceased Josef Wintrich, Müller might have been the Court's first president had he not been its first defendant. Born in 1900, Müller trained as a jurist and began his political career as a functionary of the Centre Party in the early 1930s. His first foray into politics was short-lived. When the Nazis seized power in 1933, the Centre Party was banned with all the rest. Müller pursued his judicial vocation throughout the peacetime years of the Nazi era. Like other jurists, he was reassured by the regime's veneer of legality. He joined National Socialist jurist associations, but consistently 'forgot' to pay his fees. He never joined the party itself. Müller was drafted briefly into the Wehrmacht in 1940, where he worked as a secretary on the Western front, and again in 1945, as part of the last desperate defence of the home front. After a brief imprisonment at the hands of French occupation authorities, Müller became a CDU politician of the first hour. He rose rapidly through the ranks of Württemberg-Hohenzollern state politics, becoming state minister president in August 1948. In this office, Müller distinguished himself as the champion of two great causes: the creation of a West German nation and the consolidation of a Southwest German state.

In the latter cause, which aimed to fuse Württemberg-Baden, Baden, and Württemberg-Hohenzollern, Müller was joined by Württemberg-Baden's minister president, Rheinhold Maier. He was passionately opposed by Baden's premier, the diminutive diehard Leo Wohleb. Wohleb petitioned the Constitutional Court, before it was even staffed, to enjoin a referendum on the state's creation, arguing that Baden was demographically doomed by the plebiscite's structure to be swallowed by its more populous neighbours.

In the meantime, Gebhard Müller had emerged as a consensus candidate for the inaugural presidency of the Constitutional Court. Müller was willing, but wanted to await resolution of the Southwest state controversy. In the event, however, he couldn't have it both ways. Müller and Maier won their case, but with Hermann Höpker-Aschoff presiding over the Court that decided it. Müller preferred this outcome to its opposite. Like Höpker-Aschoff, Müller loved politics and was loath to abandon it for juristic isolation in somnolent Karlsruhe. Müller was again the consensus favourite for the Court's presidency when Höpker-Aschoff died in January 1954. By then, however, Müller was the newly installed minister president of Baden-Württemberg, the state whose creation he had championed. He was unwilling to resign the post so soon. For five years he governed Baden-Württemberg at the head of a super coalition of all parties. Müller proved a unifying leader and a competent manager—successful, efficient, and popular. When, in autumn 1958, he was asked for the third time to take the reins of the Constitutional Court, he resigned his premiership with a heavy heart.[9] It didn't help when a friend indelicately noted that no one had yet survived the office of chief justice.[10]

Justice Willi Geiger later portrayed the new chief justice as every inch a Swabian, 'and that means: honest to the bones and a little bit sly, disciplined, ... unsurpassed in his work ethic, energy, and regard for the immutably moral in the jurisprudence of the First Senate'.[11] He came to Karlsruhe, Geiger impishly added, 'with the quickly corrected notion that he could handle the members of the Federal Constitutional Court as he did his cabinet in Stuttgart'.[12] Müller inherited an institutional setting far more stable than that which confronted either of his predecessors. The first crisis of the new decade, two years into Müller's presidency, showed just how far the Court had come.

B. The 'German miracle'

If the Constitutional Court of the late 1950s exuded stability and confidence, so, spectacularly, did the Federal Republic itself. The heart of the transformation, of course, was economic. The West German economy was in its finest historical fettle, the so-called 'economic miracle' generating palpable prosperity and staggering statistics. Between 1950 and 1960, real GDP in the Federal Republic grew at an annual rate of 7.9 per cent.[13] Exports quadrupled, GNP tripled, and West Germany's share of world exports swelled from 6 to 10 per cent. Unemployment virtually disappeared; a mounting labour shortage drove a massive influx of foreign labour.[14] The fruits of growth were sufficiently distributed that all levels of West German society experienced an unprecedented increase in economic well-being.[15] In Hans-Peter Schwarz's famous assessment, a people who once wished to acquire something had become a people who wished now to preserve something.[16] This spectacular socio-economic transformation had a political corollary. A 1957 CDU election ad featured a little girl asking her father what a proletarian was. When the father explained that a proletarian was a poor worker dwindling in abject poverty, the bemused daughter retorted that such people no longer existed. The father's punch-line was that someone should tell that to the SPD.

The 1957 Bundestag election did just that. Though the SPD's share of the vote increased by 3 per cent to 31.8 per cent, the Union's share grew by 5 per cent to an absolute majority of 50.2 per cent, something no party had ever achieved in the tortured history of German democracy.

[9] Twenty years later Müller highlighted the paradox that, although his interests and inclinations best suited him to be a judge, no office he held gave him more happiness than the premiership of Baden-Württemberg. Hans Baush and Gebhard Müller, *Gebhard Müller blickt zurück* (Landtag von Baden-Württemberg 1980) 44.

[10] See Gebhard Müller, 'Ansprache' in *Das Bundesverfassungsgericht, 8. Dezember 1971. Festakt aus Anlass der Verabschiedung des Präsidenten Dr Gebhard Müller, der Bundesverfassungsrichter Prof. Dr Dr Gerhard Leibholz, Prof. Dr Erwin Stein, Gregor Geller* (CF Müller 1972) 33.

[11] Willi Geiger, 'Ansprache anläßlich der Feierstunde des Bundesverfassungsgerichts am 9. November 1977 zur Verabschiedung von Bundesverfassungsrichter Prof. Dr. Willi Geiger' Bundesarchiv (Koblenz) (Nachlass Gerhard Leibholz, N 1334/299).

[12] ibid.

[13] Gerd Hardach, 'Krise und Reform der Sozialen Marktwirtschaft. Grundzüge der wirtschaftlichen Entwicklung in der Bundesrepublik der 50er und 60er Jahre' in Axel Schildt, Detlef Siegfried, and Karl Christian Lammers (eds), *Dynamische Zeiten. Die 60er Jahre in den beiden deutschen Gesellschaften* (Hans Christians Verlag 2000) 203.

[14] See Ulrich Herbert and Karin Hunn, 'Gastarbeiter und Gastarbeiterpolitik in der Bundesrepublik. Vom Beginn der offiziellen Anwerbung bis zum Anwerbestopp (1955–1973)' in *Dynamische Zeiten* (n 13) 273–310.

[15] GDP per person grew at an annual rate of 6.8 per cent while cost of living increased at an average of only 1.9 per cent per year. See Hardach (n 13) 203.

[16] Hans-Peter Schwarz, *Die Ära Adenauer. Gründerjahre der Republik 1949–1957* (Deutsche Verlags-Anstalt Stuttgart 1981) 335.

Small parties—a lurking reminder of Weimar fragmentation in the early 1950s—all but disappeared. The Federal Republic was now a three-party state—the beefy CDU, the battered SPD, and the bantam FDP, which had shrunk by nearly two percentage points to only 7.7 per cent. The outcome supplied an existential shock to the Social-Democratic system. It also supplied the stimulus for radical reform behind the crusading foursome of Carlo Schmid, Fritz Erler, Herbert Wehner, and Willy Brandt. Their goal was to modernize the SPD as a precondition to capturing government power. This required transforming the traditional workers' party (*Arbeiterpartei*) into a people's party (*Volkspartei*). It meant renouncing the party's Marxist roots and making peace with the market; it meant courting the churches and endorsing the army. It meant accepting Adenauer's dream of Western integration as a fait accompli. It meant a strategic embrace—nuzzling up to Christian democracy with the intention, ere long, of wresting power from it. The reformers launched their agenda with a new party platform announced in November 1959 at a convention in Bad Godesberg.[17] The so-called 'Godesberg Program' marked the end of an era in the political history of the Federal Republic. No more would the Bundestag serve as the arena of conflict that it had been in the early 1950s, where majority and opposition clashed steel and exchanged fire in existential confrontation. As the 1960s began, the SPD's opposition was hushed, restrained, and incremental.

Changes were afoot for Christian Democrats as well. Wishing to maintain a broad appeal and preserve its absolute majority, the CDU/CSU moved slowly away from its ideological identification with Catholicism and natural law. West German politics seemed to be arriving at 'The End of Ideology', an ostensibly international phenomenon trumpeted in a 1960 tract by the American sociologist, Daniel Bell.[18] Like the 'End of History' three decades later,[19] the end of ideology didn't last long. But while it lasted, this convergence of policy and departure from dogma helped make the period between the Godesberg program of 1959 and the grand coalition of 1966 the dullest in the parliamentary history of postwar Germany. A corollary and consequence was a dramatic decrease in the SPD's filing—at the *federal* level—of constitutional complaints. During the 1960s, the high profile cases that came to Karlsruhe did not arrive on the train from Bonn.

I. Adenauer's Last Stand

The end of one political era, of course, also marked the beginning of a new—and the beginning of the end for the undisputed symbol of the outgoing order. Five days into the new decade, on 5 January 1960, Konrad Adenauer celebrated his eighty-fourth birthday. The chancellor was nearly a decade older than Bismarck had been when that elder statesman resigned in 1890, and only a year younger than Hindenburg had been when the old general handed power to Hitler in 1933. As a biological matter, it was clear that 'the old one' (*der Alte*) couldn't remain chancellor forever. As a political matter, it was clear that his grasp was already slipping. Adenauer had reached the summit of his power in 1957/1958 with his resounding electoral triumph and the climax of the economic miracle. His four-year fall began the following year.

[17] See Susanne Miller, *Die SPD vor und nach Godesberg* (Bonn-Bad Godesberg 1974); Kurt Klotzbach, *Der Weg zur Staatspartei. Programmatik, praktische Politik und Organization der deutschen Sozialdemokratie 1945 bis 1965* (Dietz 1982).

[18] See Daniel Bell, *The End of Ideology: On the Exhaustion of Political Ideas in the Fifties* (Harvard University Press 1960).

[19] See Francis Fukuyama, *The End of History and the Last Man* (Free Press 1992).

Theodor Heuss's second five-year term as federal president expired in 1959. Adenauer—who saw his own putative successor, the finance minister, Ludwig Erhard, as singularly unfit for the chancellorship—worked behind the scenes to secure Erhard's nomination for the presidency. Adenauer's responsibility for the proposal was a secret to no one, and the public outcry was clamorous. The Hamburg daily *Die Welt* spoke of a 'murder of the crown prince'.[20] The nomination was quickly withdrawn, but Adenauer compounded this gaucherie with another. He offered himself for the job, proposing to construe the constitution in a 'Gaullistic' fashion. Adenauer would play a presidential role under the Basic Law similar to that played by Charles De Gaulle under the constitution of the French Fifth Republic. As an exercise in comparative constitutional law this was unadulterated nonsense—a reckless flout to the maxim that one must compare what is in fact comparable. As an exercise in political posture and manipulation it was supremely maladroit—a transparent bid to cling to power in defiance of constitutional precept and democratic process. It evoked a faint echo of Carl Schmitt's vision of the president as guardian of the constitution. It left the chancellor looking authoritarian and out of touch. In short order, the proposal was howled out of existence.

The crisis soon passed, but the damage was done. Adenauer's loss of prestige was enormous. His vaunted political instincts had failed him, his sure-footedness had gone wobbly. The world statesman had conducted a one-man farce routine. And he looked so old. His departure, all realized, was only a matter of time. In the event, he remained in power another four years. But they were four years of slow, relentless decline. Accelerating Adenauer's descent, and at its chronological centre, was a final, epic clash with the Federal Constitutional Court.

A. The fight over 'private' television

It was a matter, reported the London *Economist*, in which '[p]olitics, entertainment, and constitutional law' were 'gloriously mixed up'.[21] After his resounding electoral triumph in 1957, Adenauer was persuaded that this had been the last federal election in which television coverage would not be decisive. He was also convinced that, to a man, the lords of Germany's regional television networks sympathized with liberals and Social Democrats.[22] The chancellor resolved to harness the power of television to his own ends. He was intrigued by the British model, where the private ITV (Independent Television) operated alongside, and competed with, the public BBC. Adenauer hoped to establish a second national television network in West Germany—one that could rival the existing ARD and serve the interests of his own federal government.[23]

Adenauer's assessment of television's growing importance in West German society and politics was indubitably correct. Vast numbers of West Germans were buying televisions and spending increasingly large swathes of their leisure time worshipping at this newfangled altar of modernity. Contemporaries chronicled the emergence of a 'Television Nation'. Cinema revenues collapsed, and West German film entered its first crisis.

20 See Georg Schröder, 'Bonn spricht vom Kronprinzenmord' *Die Welt* (27 February 1959) 2.

21 'Jamming from Karlsruhe' *The Economist* (31 December 1961) 1373.

22 See Hans-Pater Schwarz, *Die Ära Adenauer 1957–1963* (Deutsche Verlags-Anstalt 1983) 167; Hans-Peter Schwarz, *Adenauer: Der Staatsmann, 1952–1967* (Deutsche Verlags-Anstalt 1994) 621.

23 Schwarz, *Die Ära Adenauer 1957–1963* (n 22).

Television reshaped politics as well as entertainment. By 1960, a million Germans reportedly watched the ARD-Tagesschau political reports[24]—more than subscribed to any daily newspaper with the exception of Axel Springer's boulevard giant *Bild*. In autumn 1960, televised debates turned the tide of America's presidential election, in which the dour Richard Nixon lost to the telegenic John Kennedy. Adenauer, who in the 1961 federal elections would face a challenger, Willy Brandt, who consciously posed as a German Kennedy, could hardly fail to take note. In due course, he took action.

Adenauer's interest in the political possibilities of television took root before the 1957 elections made the matter seem urgent. Indeed, the chancellor had been intrigued by the new mass medium since shortly after taking office in 1949. Adenauer hoped television could serve as a mouthpiece for his government. He set special stock by long-range broadcasts, hoping to shape popular opinion in the Soviet zone as well. As early as 1953, the government proposed a federal television bill, but did so too late in the legislative season for Parliament to act on it. The proposal, moreover, sparked an immediate constitutional controversy. The question was whether the states or the federal government had jurisdiction to regulate television. On the one hand, under Article 30 GG, the states enjoyed 'cultural sovereignty' (*Kulturhoheit*). On the other hand, under Article 73, the federation or *Bund* wielded exclusive jurisdiction over 'postal and telecommunication services'. The states asserted jurisdiction over television's content; the Bund over its technical operation. The split required a mutual agreement, but no agreement was forthcoming.[25]

There followed four years of fruitless negotiation. After winning an absolute majority in the 1957 elections, however, Adenauer was determined, with his strengthened hand, to push again for federal broadcasting power. In February 1958, the chancellor assigned his interior minister, Gerhard Schröder, to supervise drafting of a federal broadcasting bill that would provide for both short- and long-wave broadcasting, and would create a second national television network. State minister presidents, concerned for the fate of regional television and fearing the expansion of federal power, resisted. They were led by Peter Altmeier, CDU premier of Rheinland-Pfalz.[26] In his influential history of the Adenauer era, Hans-Peter Schwarz reported they were crucially helped behind the scenes by Gebhard Müller, erstwhile premier of Baden-Württemberg, and now chief justice of the Constitutional Court.[27] Müller heatedly denied playing any such role.[28]

[24] Eckart Conze, *Die Suche nach Sicherheit: Eine Geschichte der Bundesrepublik Deutschland von 1949 bis in die Gegenwart* (Siedler Verlag 2009) 235. See also Knut Hickethier, *Geschichte des deutschen Fernsehens* (Metzler 1998).

[25] For a survey of the negotiations, see Rolf Steininger, 'Runfunk zwischen Bund und Ländern 1953–1961. Ein Beitrag zur Innenpolitik Adenauers' (1976) 4 *Politische Vierteljahresschrift* 474.

[26] See Heinrich Küppers, 'Adenauer und Altmeier im Fernsehstreit 1958–1961' (1987) 35 *Vierteljahreshefte für Zeitgeschichte* 625.

[27] Hans-Peter Schwarz, *Die Ära Adenauer 1957–1963* (Deutsche Verlags-Anstalt 1983) 167.

[28] Müller wrote to Schwarz, asking the historian to name his sources and complaining that his account implied Müller had exercised influence on the outcome of the Court's judgment. Gebhard Müller to Hans-Peter Schwarz, 22 August 1984, Landesarchiv Baden-Württemberg (Stuttgart) (Nachlass Müller, Q1135 Bü 705). Schwarz answered that his account relied on the papers of Adenauer and his closest circle, but that agreements with the proprietors of those papers sometimes prevented direct citation. He explained that historians develop general pictures based on many sources that sometimes prevent citation to a single source. 'You must in any case ... know for certain', Schwarz wrote, 'that Adenauer and his close circle saw in you a principal foe of the questionable television plans'. Schwarz deemed Adenauer's television plans highly questionable politically as well as constitutionally, and in his view it would have been entirely legitimate for Müller to work to upend those plans, even as sitting chief justice. But if Müller were to assure him that he had not done so after assuming the Court's presidency in 1959, Schwarz would note the fact in his annotations and alter the offending passage in the text. Hans-Peter Schwarz to Gebhard Müller, 27 August 1984, Landesarchiv Baden-Württemberg (Stuttgart) (Nachlass Gebhard Müller, Q1135 Bü 705). Schwarz offered to meet Müller personally to discuss the matter, but it is unclear whether they ever did. In any case, no new edition of Schwarz's second volume on *Die Ära Adenauer* has yet appeared. Hans Bausch, an historian of German television who claimed to be the first to have researched

The premiers collectively insisted that a new national network could only be organized as a public-law association akin to the state television networks already in operation. The federal government, they allowed, could participate in the governance of such a network by sending representatives to sit on its board. The states affirmed their willingness to conclude treaties with the feds over the governance of long- and short-range transmissions, but the federal government found the states' terms intolerable.

In 1960, the Adenauer government proposed a bill that would erect three federal television stations: 'Deutsche Welle', for short-wave foreign transmissions, 'Deutschlandfunk', for long-wave transmission for all Germany, and 'Deutschland-Fernsehen', a second network for the Federal Republic. In response, the state bosses cobbled together a committee to continue negotiations, which continued to be fruitless. Parliament passed the proposed law in June 1960, but did so without approving the 'Deutschland-Fernsehen' network, which the Bundestag committee for cultural politics, fearing an affront to the states, had excised from the bill. Adenauer, however, was unbending. He wanted a second national network.

On 8 July 1960 the chancellor presented the CDU/CSU premiers with the possibility of creating a second network as a private, limited liability company. Word of these discussions, however, soon reached the SPD premiers as well, prompting state leaders from both parties to issue a series of counterproposals, which Peter Altmeier presented to Adenauer on 22 July 1960. Altmeier received an icy rebuff. The chancellor's patience was exhausted. After seven years of sterile consultation, the patriarch of the Federal Republic, unlike Father Israel, would wait no more. His eye on the electoral calendar, Adenauer opted to go it alone. On 25 July the chancellor, together with his justice minister, Fritz Schäffer, founded *Deutschland-Fernsehen*, LLC, with headquarters in Adenauer's native Cologne. Of the original share capital of 23,000 Deutschmarks, 12,000 were furnished by the federal government and 11,000 by Schäffer himself, whom the company charter commissioned, as 'trustee', to advocate the interests of the states. The company's stated purpose was to give television viewers at home and abroad a comprehensive portrait of contemporary Germany. Its board would consist of at least ten, at most fifteen, members—one representative each from the Protestant Church, the Catholic Church, the Jewish Central Council, and the respective associations of employers and unions; and up to ten representatives from the federal government. The states declined to participate in the company's governance. A month after its founding Schäffer, their ostensible trustee, obligingly conferred his shares on the federal government, which thereby became the company's sole shareholder. The states preferred to advocate their interests not in Cologne but in Karlsruhe.

In autumn 1960, the SPD state government in Hamburg—joined in swift succession by its sisters in Bremen, Hessen, and Lower Saxony—asked the Second Senate to enjoin the network's founding. Schleswig-Holstein's minister president, Kai-Uwe von Hassel (CDU), declined to join the complaint on the controversial ground that recent judgments had

the *Television* controversy in depth, wrote to Müller sympathetically that he could 'only shake his head' over Schwarz's account. 'I can assure you', Bausch wrote, 'that neither in reality nor in the literature is there the slightest evidence that … behind the scenes you played "a leading role in thwarting the Chancellor's television policy". To me it is unfathomable how Herr Schwarz can hazard such a statement'. Hans Bausch to Gebhard Müller, 27 October 1984, Landesarchiv Baden-Württemberg (Stuttgart) (Nachlass Müller, Q 1135 Bü 705). The government was undoubtedly worried about Müller. Adenauer wrote to Hans Globke in September 1960 that someone from the government 'should contact Müller and enlighten him concerning the political significance of the whole matter. Even if he's not [on the deciding Senate], he could still do damage with critical comments'. Adenauer to Globke, 5 September 1960, in Rudolf Morsey and Hans-Peter Schwarz (eds), *Adenauer Briefe 1959–1961* (Schöningh Paderborn 2004) 160.

shaken his confidence in the country's highest court.[29] The complaining states, represented by Adolf Arndt, argued that the federal government's unilateral creation of a self-owned and self-operated television network violated Article 5 GG, which guaranteed the freedom of press and broadcasting; Article 30, which reserved to the states all powers not expressly granted to the Bund; and the constitutional principle of federalist comity (*bundesfreundliches Verhalten*—literally, federation-friendly conduct).

The federal government countered that Article 73's grant of exclusive federal jurisdiction over postal and telecommunication services empowered the federal government to regulate television broadcasting *tout court*. Moreover, establishing television networks was not a prerogative of public administration at all, so founding the LLC didn't fall under Article 30, which governed public-law, but not private-law administration. Finally, the government pled *kraft Natur der Sache*: it asserted implicit authority to take steps necessary to fulfil express obligations—a German variant on the US 'necessary and proper' doctrine.

After oral arguments in late November 1960, the Second Senate enjoined transmissions pending its decision on the merits.[30] This was a blow to Adenauer, who hoped to begin broadcasts on 1 January 1961 and employ the new platform for the whole of the coming election year. Rudolf Katz, chairman of the Second Senate, had explained earlier to Schröder, the interior minister, that because the case was so complex, a decision before the end of the year was improbable.[31] Time ticked slowly as the Court deliberated and the public speculated. Rarely had a Court judgment been so anxiously anticipated by public, press, and political class. Tension mounted when the Court revealed that it would announce its judgment on the final day of February 1961.

The appointed day arrived at last. An army of lawyers, journalists, and political operatives squeezed into the crowded chamber of Prince Max Palace. Rudolf Katz was gravely ill, so Egon Schunck, the Second Senate's senior justice, read the dispatch. The summary lasted ninety minutes, an obvious strain on the fading vocal powers of the septuagenarian Schunck.[32] But the substance of the judgment was such that even Schunck's weak voice sufficed to astonish the crowd and to rock the country. The judgment sparked a political storm, the likes of which the Court hadn't seen since the EDC feud eight years earlier.

The Senate ruled for the states across the board.[33] Its decision was dramatic and thorough. The Court held, to begin with, that Article 73's grant of federal control over 'postal and telecommunications services' applied only to the technical side of broadcasting; it conferred no authority to regulate the field of broadcasting 'as a whole'.[34] In cases of doubt, there was a presumption against federal authority. Broadcasting, moreover, was a 'cultural phenomenon', as well as a subfield of telecommunications. Insofar as cultural matters could be regulated at all, Article 30 GG vested power to do so exclusively with the states.[35] The question of competence, in short, ran squarely against the federal government and in favour of the states. This was the case's crucial question, and having unambiguously disposed of it, the Senate might well have let the matter rest. Instead it left no stone unturned. The opinion's most striking and controversial sections were still to come.

The first of these addressed the question of federalist comity (*bundesfreundliches Verhalten*). For its style as well as its substance, this section was sure to impart pain to the federal government. One comparative lawyer called it an 'an extended lecture on political

[29] See 'Hassel zweifelt am Verfassungsgericht' *FAZ* (10 November 1960); 'Professor Friesenhahn gegen Hassel. Kritik an Verfassungsgerichtsentscheidungen zurückgewiesen' *Badische Neueste Nachrichten* (8 February 1961) 3.

[30] 12 BVerfGE 36 (1960). [31] Rudolf Katz to Gerhard Schröder, 5 October 1960, BAK B 136/1770.

[32] See 'Scherbengericht' *Der Spiegel* (8 March 1961) 18. [33] 12 BVerfGE 205 (1961).

[34] ibid 225–28. [35] ibid 229.

morality'.[36] Federalist comity, the Senate noted, was an unwritten principle of constitutional law that governed interactions between the states and the feds. It reached the manner as well as the material of those interactions.[37] During the latter stages of negotiation, and in its unilateral founding of the new network, the federal government had flouted this principle with escalating prodigality. The Senate censured the government for (1) attempting—'following the principle *divide et impera*'—to cut a deal with some states and coerce the others to accede; (2) excluding the SPD minister presidents from negotiations; (3) excluding the states entirely from the network's founding; (4) presuming unilaterally to appoint a trustee in the states' behalf; and, generally, (5) riding rapidly and roughshod over the state's constitutional prerogatives and legitimate interests.[38] The federal government's actions were, the justices wrote, 'absolutely incompatible with the requirement of federalist comity', no matter how unhelpful the states had been during previous negotiations.[39] The Senate seemed to be suggesting that the government's conduct was ungentlemanly as well as unconstitutional—un-republican as well as illegal. Even if the network's founding had been, in all other respects, constitutionally benign, the manner and method of its creation would have condemned it.

But the network's founding was not in other respects benign. More damning than the Senate's initial finding that the federal government had exceeded its competence was the Senate's terminal holding that it had violated a fundamental right. In an earlier judgment the Court had held that Article 5's guarantee of the freedom of the press not only protected individual journalists, it secured 'the institutional independence of the press—from the gathering of information to the spread of news and opinions'.[40] The same principle applied, perhaps with greater force, to broadcasting. Broadcasting was more than a 'medium' in the shaping of public opinion; it was an 'eminent factor' in public discourse.[41] It was at least as important as print journalism in shaping public opinion, and it required no less institutional freedom.[42] Indeed, because the necessary technical machinery was so costly and the available wavelengths so scarce, broadcasting required even greater protection. State participation in the press was allowable only if competition from numerous and independent papers assured the survival of a free press.[43] Similarly, state involvement in broadcasting was acceptable only where 'all socially relevant forces have a voice, and if the freedom of reporting remains untouched'.[44] This, alas, was not the case with *Deutschland-Fernsehen*, LLC. That network was 'wholly in the hands of the state'. It was 'an instrument of the *Bund*'. It would be ruled by the federal government and dominated by the federal chancellor.[45] Even if the original allocation of influence was appropriate, nothing in the network's charter secured that allocation.

Article 5 demanded much more. It demanded the participation of all relevant voices, as well as content marked by at least 'a minimal measure of balance, objectivity, and reciprocal respect'.[46] The founding of *Germany Television*, and its proposed operation, failed to meet these demands. For all these reasons, the Second Senate banned it.

B. Reactions to the *Television* judgment

The decision, reported the *Frankfurter Allgemeine Zeitung*, 'worked like a sensation'.[47] *Der Spiegel* called it a 'thunderbolt from Karlsruhe'.[48] Government leaders in Bonn admitted that

[36] Edward McWhinney, *Constitutionalism in Germany and the Federal Constitutional Court* (Sythoff 1962) 63.
[37] 12 BVerfGE 205 (n 33) 254–55. [38] ibid 255–59. [39] ibid 258.
[40] ibid 260 (citing 10 BVerfGE 118, 121 (1959)). [41] ibid 260. [42] ibid 260–61.
[43] ibid 260. [44] ibid 262. [45] ibid 263. [46] ibid 206.
[47] 'Bestürzung in Bonn über das Urteil' *FAZ* (1 March 1961) 4.
[48] 'Nach dem Dunkel Licht?' *Der Spiegel* (8 March 1961) 15.

the judgment left them speechless.[49] Others, regaining their powers of speech, denounced the decision as 'bankrupt', grumbling that it had created an 'untenable situation'.[50] Particularly galling was the 'tenor' of the judgment, in which officials detected a pronounced aversion to 'Bonn' and an extreme fondness for federalism.[51] Government circles were additionally angered by reports that Justice Gerhard Leibholz of the non-participating First Senate told journalists that he supposed Bonn would seek revenge by amending the FCCA after the autumn elections.[52]

Officially, the federal government issued no immediate statement. But individual cabinet members and Union politicos freely vented their stimulated spleen. Gerhard Schröder publicly criticized the decision, which he thought rested on an aggressive construction of a few weak spots in the Basic Law.[53] At a meeting of the CDU Bundestag faction, Eugen Gerstenmaier, president of Bundestag, professed that he honestly could not understand one of the judgment's central sentences.[54] Heinrich Krone, the faction's chairman, wondered whether the states were capable of speaking for those Germans who lived beyond the border of freedom.[55] These statements were greeted by thunderous applause from the party faithful. The party press inveighed against the judgment as well. 'After studying the opinion', groaned one CDU-leaning paper, 'one cannot shake the feeling that here the assignment of a political decision to a juristic body has not proved an unqualified boon to parliamentary democracy. After this Karlsruhe judgment we can only emphatically repeat our warnings against a jurists' democracy [*Justitiardemokratie*]'.[56] Many sensed the beginnings of a brawl between the federal government and the Federal Constitutional Court.[57] Officially, however, the ruling party and its chieftains kept mum.

Not so the opposition. Social Democrats were jubilant. Adolf Arndt and his legal team had hoped for a narrow victory on specific points. But Arndt's *tour de force* brief—in both the sweep of its argumentation and the sharpness of its prose—had been embraced in the Court's published opinion. As *Der Spiegel* reported, Arndt now found himself in the role of 'a Samson who had struck down Adenauer, Schäffer, Schröder, and Stücklen with a single blow'.[58] For Arndt, who had been criticized within the party for conducting costly and distracting crusades in Karlsruhe, it was a sweet personal vindication.

The invalidation of a cherished Christian Democrat project—one that would have weakened the opposition considerably in the upcoming election—was a major political victory for the party as a whole. But the judgment marked a moral triumph as well. It shattered the icon of the government's constitutional infallibility. Willy Brandt noted gratefully that, after the *Television* judgment, it was indisputable that government policy did not always honour the constitution.[59] It was a source of immense additional satisfaction that the judgment not only quashed a government policy, it excoriated the manner of its adoption. SPD politicians talked of the government's 'tactical unscrupulousness' in conducting a 'jungle fight' against which the Court had righteously inveighed.[60]

Opposition leaders at the state and local level were even more effusive. Paul Nevermann, mayor of Hamburg, crowed that the chancellor had been returned to the path of legality.[61]

49 'Bestürzung in Bonn über das Urteil' *FAZ* (1 March 1961) 4.
50 In Hermann Proebst, 'Schlechte Verlierer' *Süddeutsche Zeitung* (4–5 March 1961) 1, 2.
51 'Bonn Schweigt noch immer zum Fernseh-Urteil' *FAZ* (2 March 1961) 1, 4. 52 ibid.
53 ibid 4. 54 'Scharfe CDU-Kritik am Fernsehurteil' *Süddeutsche Zeitung* (8 March 1961) 2.
55 ibid. 56 'Bonn schweigt weiter zum Ausgang des Fernsehstreits' *Die Welt* (2 March 1961) 1.
57 'Bonn schweigt noch immer zum Fernseh-Urteil' *FAZ* (2 March 1961) 1, 4.
58 'Scherbengericht' *Der Spiegel* (8 March 1961) 18.
59 In 'Bestürzung in Bonn über das Urteil' *FAZ* (1 March 1961) 4.
60 'Bonn schweigt weiter zum Ausgang des Fernsehstreits' *Die Welt* (2 March 1961) 1.
61 In 'Bestürzung in Bonn über das Urteil' *FAZ* (1 March 1961) 4.

The Hessian minister president, Georg August Zinn, called the decision one of the most important political and legal decisions since the Court's inauguration.[62] Zinn was confident the judgment would forfend further federal efforts to undermine the competences of the states.[63] Even some Christian Democrat state officials were pleased. Peter Altmeier openly praised the decision.[64] Edo Osterloh, culture minister for Schleswig-Holstein, was more cautious. 'We must be able to acknowledge that we have lost a battle', he noted. 'That's part of a democracy. But we also mustn't forget that the states too suffered a defeat in the Atomic question through the decision made then in Karlsruhe'.[65] Elsewhere, though, federalists were more than willing to forget past defeats and revel in the present victory. The reaction of the arch-federalist Bavarian Party was ebullient. 'The defeat of the chancellor and the interior ministry is complete', burbled a party bulletin. 'It is gratifying that federalism has achieved a clear and convincing victory against the grim centralising tendencies of the Bund'.[66] In an alliterative paean, a writer in *Die Welt* exulted that 'once more the flags of federalism fly freshly in the wind'.[67]

Reactions in the press were largely approving and universally astonished. By common consent, the judgment marked the most stupendous government defeat in the twelve-year history of the Federal Republic.[68] Journalists were as taken aback by the judgment's style as by its substance. That style reflected the acuity of Adolf Arndt, the case's victorious advocate, and the pugnacity of Willi Geiger, its judge rapporteur.[69] *Die Zeit* perceived, and relished, 'a certain severity in this judgment's style'.[70] The *Süddeutsche Zeitung* was taken aback by the decision's 'surprisingly sharp formulations'.[71] *Die Welt* noted that the Court's reprimand of government behaviour was 'so massive that the objective observer must wonder whether it was really necessary'.[72] Even the Swiss *Neue Zürcher Zeitung* reported that the chancellor's 'striking defeat' was packaged in prose of 'barely expected severity'.[73] *Der Spiegel*, as usual, was the bluntest of all: the substance of the judgment was contained in the case headings; the sixty pages that followed were 'merely punching the chancellor'.[74]

Some thought the style exaggerated and the substance plain wrong. Eberhard Blitzer grumbled in the *Frankfurter Allgemeine Zeitung* that 'one should question whether the constitutional judges sought to orient themselves toward the constitution and the "objective will of the lawgiver"'.[75] Blitzer wondered further whether the Court's 'doctrinaire federalism' could accommodate current and imminent advances in technology.[76] The framers of the Basic Law, he noted, had never foreseen the emergence of television as a mass medium. It was anachronistic for the Court to enforce strictly the constitution's letter in the face of

[62] ibid. [63] In 'Das Echo aus den Landeshauptstädten' *Süddeutsche Zeitung* (1 March 1961) 1, 2.

[64] 'Bonn reagiert zurückhaltend' *Die Welt* (1 March 1961) 1.

[65] 'Um das Grundgesetz' *Süddeutsche Zeitung* (3 March 1961) 2.

[66] In 'Das Echo aus den Landeshauptstädten' *Süddeutsche Zeitung* (1 March 1961) 1, 2.

[67] See, e.g., Bernt Conrad, 'Föderalismus als Aufgabe' *Die Welt* (7 March 1961) 1.

[68] See Hans Reiser, 'Nachdem Spruch von Karlsruhe' *Die Welt* (1 March 1961) 1 ('the worst domestic policy defeat to date'); 'Die Niederlage der Wesire' *Die Zeit* (3 March 1961) ('such a spectacular defeat'); 'Pressestimmen zum Fernseh-Urteil' *Süddeutsche Zeitung* (2 March 1961) 2 ('sensational defeat'); 'Nach dem Dunkel Licht?' *Der Spiegel* (8 March 1961) 15 (the judgment was, for the CDU/CSU, 'the worst blow their patriarch Adenauer had had to endure since the founding of his state').

[69] That Geiger was the rapporteur is confirmed in Ernst Benda, 'Ansprache anläβlich der Feierstunde des Bundesverfassungsgerichts am 9. November 1977 zur Verabschiedung von Bundesverfassungsrichter Prof. Dr. Willi Geiger' Bundesarchiv (Koblenz) (Nachlass Gerhard Leibholz, N 1334/299).

[70] 'Die Niederlage der Wesire' *Die Zeit* (3 March 1961) 1.

[71] 'Verfassungsgericht entscheidet gegen Bonn—Zweites Fernsehen ist Ländersache' *Süddeutsche Zeitung* (1 March 1961) 1.

[72] Hans Reiser, 'Nachdem Spruch von Karlsruhe' *Die Welt* (1 March 1961) 1.

[73] In 'Pressestimmen zum Fernseh-Urteil' *Süddeutsche Zeitung* (2 March 1961) 2.

[74] 'Scherbengericht' *Der Spiegel* (8 March 1961) 15.

[75] Eberhard Blitzer, 'Das Urteil aus Karlsruhe' *FAZ* (1 March 1961) 1. [76] ibid.

manifestly changed circumstances. The judgment, Blitzer concluded, was methodologically questionable and practically dubious.

But Blitzer's voice was lonely. Many journalists saw the judgment not only as a devastating political defeat for Adenauer, but as a kind of national civics lesson, an open essay in democratic ethics. Hans Reiser (*Die Welt*) had technical reservations about the judgment, but these were outweighed by gratitude for the judgment's forceful insistence that politicians respect the boundaries of the Basic Law. Reiser gave his analysis the sub-heading, 'The Basic Law is holy'.[77] *Die Zeit* hailed the decision as a judgment 'for democracy and against the Bund', one that 'takes seriously the text, significance, and spirit of the Basic Law, wherein the free shaping of opinion and will is protected and a federalist structure is guaranteed'.[78] Ernst Müller Meiningen Jr (*Süddeutsche Zeitung*) was especially encomiastic. Meiningen basked in the judgment's 'refreshingly clear talk' and delighted in the 'thoroughly bitter lesson' the justices had given the chancellor.[79] 'The Court', Meiningen wrote, 'has become truly indispensable for the stability of our constitution—our constitution which not a few politicians, God forbid!, would wrest under hand'.[80]

Other editorials defended the Court against its government critics. Hans Schuster wrote that, of all the criticisms levelled at the Court's decision, the strangest was that it was taking the text of the Basic Law too seriously. In the face of constitutional amendments proposed less than twenty-four hours after the decision, Schuster added, one wondered 'why exactly we have a Constitutional Court'.[81] Hermann Proebst was even more bellicose, branding the judgment's critics as 'sore losers'. In Proebst's view, such criticisms suggested that constitutional principles hadn't yet entered the subconscious of the citizenry.[82] It was profoundly regrettable that, at the very moment when the judgment had strengthened foreign trust in the strength of German democracy, domestic reactions to the judgment threatened to undermine that trust. But there were 'still judges in Karlsruhe', Proebst concluded, 'and we have reason to be glad of this'.[83]

The judgment found a favourable reception in the foreign press as well. In Paris *Le Figaro* affirmed that the decision was 'in harmony with the West German Basic Law' and saw in it evidence that 'the federalism planted in West Germany twelve years ago by the Allied authorities had shot down deep roots'.[84] In Switzerland, the *Basler Nachrichten* beamed that, in a 'happily harsh manner', Adenauer had been restrained within the limits of the law. 'This judgment', the paper predicted, 'will significantly strengthen the world's faith in the German *Rechtsstaat*'. Unfortunately, the Swiss paper lamented, the West German public was more concerned with the fate of the second television network than with this triumph of constitutional democracy. But even in this matter the judgment would have a felicitous effect: it would force West Germany's politicians to step across party lines and seek a compromise.[85]

As an emblem and culmination of the public's engagement, *Der Spiegel* ran the judgment as its 8 March 1961 cover story. The face of Chief Justice Gebhard Müller filled the magazine's cover, his eyes wide beneath thick-rimmed glasses, a thick cigar stub wedged between his lips. He has the look of a stray, bewildered fawn gazing at the headlights of an oncoming SUV—of a man caught unawares in the thick of a political whirlwind. For *Der Spiegel*, the unmistakable lesson of the judgment was that 'every judicial review of politics forces the

[77] Hans Reiser, 'Nachdem Spruch von Karlsruhe' *Die Welt* (1 March 1961) 1.
[78] 'Die Niederlage der Wesire' *Die Zeit* (3 March 1961).
[79] Ernst Müller Meiningen Jr, 'Eine Lektion für Bonn' *Süddeutsche Zeitung* (1 March 1961) 1–2.
[80] ibid 2. [81] Hans Schuster, 'Hände weg vom Grundgesetz' *Süddeutsche Zeitung* (2 March 1961) 1.
[82] Hermann Proebst, 'Schlechte Verlierer' *Süddeutsche Zeitung* (4–5 March 1961) 1, 2. [83] ibid.
[84] In 'Pressestimmen zum Fernseh-Urteil' *Süddeutsche Zeitung* (2 March 1961) 2. [85] ibid.

judges engaged in it—willingly or no—to engage in contemporary politics themselves'.[86] In this respect, the Court had fulfilled Carl Schmitt's 1931 prediction that constitutional judicial review would lead to the politicization of law. But a lengthy survey of the Court's history suggested that this development might be less noxious than Schmitt feared. The issue's cover article characterized the judgment and its aftermath as a conflict between the centre of power (Bonn) and the centre of law (Karlsruhe). The authors, and much of the West German public, were grateful the latter had stood its ground against the former.

C. 'The judgment was wrong, ladies and gentleman'

The conflict, however, was only beginning. On 8 March 1961, the same day *Der Spiegel* ran the *Television* judgment as its cover story, the Bundestag began its annual budget debate. At this point a week had passed since the judgment was announced, and the official response in Bonn had been conspicuously reticent.[87] Only on 3 March did the government issue an official statement, and then only to abandon its plan for a second network and to urge the states to launch one as soon as possible.[88] The official line was that cabinet officials wanted time to study the opinion in detail. As the budget debate opened on the morning of 8 March, Adenauer announced that the cabinet had analysed the decision, had discussed it as a group, and were united in their assessment.

They were not pleased. 'The cabinet were united', Adenauer intoned, in their belief 'that the judgment of the Federal Constitution Court was wrong [*falsch*], ladies and gentlemen'.[89] Some opposition delegates laughed in bemusement, and others roared in protest, at this extraordinary declaration. Adenauer added grandly that the real loser in the case was the West German people. 'It is simply inconceivable', he cried, 'that we in the Federal Republic should have television only on a regional basis'.[90]

This was an unprecedented public criticism of the Court, and it triggered a tremendous parliamentary row. Fritz Erler, an opposition leader, fired back that 'in countries with an older democratic tradition, the government would have resigned after such a judgment'. Indeed, 'in countries with an older, more democratically grounded tradition, such behavior on the part of the government would never even have taken place'.[91] (In response, Will Rasner, a Christian Democrat, asked how many constitutional cases the SPD had lost, and how many SPD state premiers had resigned after the *Atomic Plebiscite* judgment.)[92] The Court's judgment was a lingering theme throughout the first day of debate. It returned to the fore, with greater intensity, on the second day.[93]

Gerhard Schröder described the decision as 'fatal', an appallingly anti-majoritarian reversal of the difficult consolidation of the German *Staatenbund* (federation of states) into a *Bundesstaat* (federal state).[94] In Schröder's view, the judgment dismantled the Weimar Republic all over again. After reading the Court's decision, and reviewing the lessons of German history, it was a source of 'angst and discomfort ... to consider the German present and the German future'.[95] These words met with thunderous applause from the CDU

[86] 'Scherbengericht' *Der Spiegel* (8 March 1961) 18.

[87] See 'Bonn schweigt weiter zum Ausgang des Fernsehstreits' *Die Welt* (2 March 1961) 1; 'Bonn Schweigt noch immer zum Fernseh-Urteil' *FAZ* (2 March 1961) 1, 4.

[88] 'Bonn gibt Fernsehpläne auf: Länderchefs beraten am 17. März' *Die Welt* (4 March 1961) 1.

[89] *Verhandlungen des Deutschen Bundestages, Stenographische Berichte*, vol 48 (1961) 8308.

[90] ibid 8309. [91] ibid 8317. [92] ibid 8317–18.

[93] 'Schwere Zusammenstöβe zwischen SPD und CDU im Bundestag' *Die Welt* (10 March 1961) 1.

[94] 'Schröder attackiert das Verfassungsgericht—Zweiter Tag der Haushaltsdebatte' *Süddeutsche Zeitung* (10 March 1961) 1.

[95] 'Innenminister dankt dem Kanzler' *Die Welt* (10 March 1961) 2.

faction. Among the opposition, they met with anger and scorn. They set the tone for the second day's debate, which centred on Schröder's ministry—frequently on the *Television* judgment and its antecedents.

Friedrich Schäfer, the SPD floor manager, lambasted Schröder's domestic policy, arguing that the *Television* judgment itself, and Schröder's response to it, exposed the interior minister's general contempt for the constitution.[96] Gustav Heinemann contended similarly that cabinet invective against the *Television* judgment reflected a broader disregard for the constitution.[97] The Court, said Heinemann, had produced a new 'etiquette manual on federal-state relations'.[98] Heinemann also defended the Court as an institution. He cited a long list of cases in which the Court had corrected federal law in fields where constitutional principle had not penetrated popular consciousness. These included equal rights of men and women, the equal standing of children born out of wedlock, the equal opportunity of political parties, and occupational freedom. He then asked what the Republic would look like without the Constitutional Court. 'We would have state television', he answered, 'and a state press would likely follow close behind'.[99] Heinemann went so far as to accuse the government of seeking to establish 'state broadcasting worse than we had under the Third Reich'.[100] Only the Constitutional Court had spared the Republic that fate.

Heinemann wasn't the only MP to make provocative parallels. When Schröder railed that the Court's use of the term 'Cultural Sovereignty' (*Kulturhoheit*) smacked of '*Rabulistik*', Adolf Arndt heard the hint of an anti-Semitic slur and said so loudly. After an intermission, Arndt said he was ready to recant if Schröder would do the same. As someone who had been persecuted for twelve years because of his 'race', Arndt explained, and who had seen relatives perish in death camps because of theirs, he hoped he could be excused for being sensitive in the matter.[101] But by the time Arndt gave this explanation Schröder had left the hall. Arndt had enraged Schröder earlier with a barbed aphorism. 'There is a difference between Karlsruhe and Bonn', Arndt quipped. 'In Karlsruhe you can't substitute the majority for the truth'.[102] 'With this comparison', Schröder rejoined, 'you utterly destroy every foundation upon which we could discuss with you here'.[103] At another point in the debate, Arndt denounced the cabinet's declaration that the *Television* judgment was wrong, excoriating the chancellor's statement as a rejection of the Basic Law, a gross impropriety, and an impossible action.[104] For their part, Union MPs insisted that freedom of expression protected criticism of judges as well as of politicians. Rainer Barzel mocked Arndt's pretentions of moral exhortation.[105]

Cabinet condemnation of the judgment provoked an immediate public backlash. The *Süddeutsche Zeitung* denounced Adenauer's 'unheard-of accusation'. Not only had the government attempted an end run around the constitution with its unilateral creation of a second network, the paper scolded; it sought now to discredit the Constitutional Court.[106] Politicians, not all of them Social Democrats, rebuked the cabinet as well. The most

96 'Heftige Angriffe der Opposition gegen Innenminister Schröder' *FAZ* (10 March 1961) 1, 4.

97 'Schröder attackiert das Verfassungsgericht—Zweiter Tag der Haushaltsdebatte' *Süddeutsche Zeitung* (10 March 1961) 1.

98 'Heftige Angriffe der Opposition gegen Innenminister Schröder' *FAZ* (10 March 1961) 1, 4.

99 'Schröder attackiert das Verfassungsgericht—Zweiter Tag der Haushaltsdebatte' *Süddeutsche Zeitung* (10 March 1961) 1.

100 'Heftige Angriffe der Opposition gegen Innenminister Schröder' *FAZ* (10 March 1961) 1, 4.

101 ibid.

102 'Schwere Zusammenstöβe zwischen SPD und CDU im Bundestag' *Die Welt* (10 March 1961) 1.

103 ibid. 104 'Schäffer: Ich war bundesfreundlich' *Die Welt* (10 March 1961) 2.

105 'Heftige Angriffe der Opposition gegen Innenminister Schröder' *FAZ* (10 March 1961) 1, 4.

106 'Ein unerhörter Vorwurf' *Süddeutsche Zeitung* (10 March 1961) 3.

deliciously ironic case was that of Thomas Dehler. Now a Bundestag vice president, Dehler saw in the cabinet statements a shocking symptom of a grim development.[107] Dehler huffed that the federal government had no right to declare a Court judgment '*falsch*'.[108] (He didn't say whether cabinet members retained their ancient right to pronounce a judgment a 'nullum'.) Another FDP parliamentarian, Ewald Bucher, insisted that the chancellor couldn't simply call a decision '*falsch*' before the Bundestag. If Adenauer was really of this opinion, he should take the consequences and resign. 'In any case', Bucher concluded, 'he must not cast doubt on the reputation of the highest German Court'.[109]

Controversy surrounding the *Television* judgment gave rise to an unprecedented spate of public reflections on the Court's role in the new German democracy. An editorial in *Die Welt* noted that the judgment marked the definitive entry of the Court into 'the arena' of high politics.[110] Citing Justice Leibholz's assessment that '[t]he guardians of the constitution are embedded in the integration process of our state and in the process of state will formation itself', the writer observed that although the notion might be unsatisfactory in theory, it seemed to be working in practice. Fiery polemics about national rearmament or regional plebiscites had all but disappeared, suggesting that Court decisions in these matters had exerted 'a great pacifying effect'. The intervention of judges in political controversies, so unsettling to many citizens, had actually 'proven especially salutary'. 'Constitutional judges secure peace in the land', but they also do more; their judgments are 'educational'.[111] Their commission was not merely to interpret the constitution, but to make it more binding—presumably as a matter of national culture as well as of positive law. West Germans should be pleased that the justices saw themselves as more than mere instruments of legal positivism. 'They feel themselves called to valuation [*Wertung*]' as well. 'We should be glad they so understand their assignment. Their rallying cry, "That is not done!" ["*Das tut man nicht!*"] reminds us of the moral foundations of our state'. Constitutional justice was the antidote to modern Machiavellianism. The moral demands of the *Television* judgment were sufficient rebuttal to any critic fearful of a *Justitiardemokratie*.[112]

Through all of this, the Court itself maintained a studied, Olympian silence. Officially, the justices saw their Court as an apolitical body. Though their judgments might have political consequences, they themselves would not enter the political fray. Only on the ides of March, two weeks after the Court's judgment and a week after the chancellor's denunciation, did the chief justice issue a formal statement. It read:

1. Everyone is free to assess decisions of the Federal Constitution Court critically or even to think them wrong [*falsch*].
2. Under the order of the Basic Law, no constitutional organ is authorized to determine or officially to express that a decision of the Federal Constitutional Court is not in accordance with the law [*entspreche nicht dem Recht*].
3. The realm of objective criticism is left behind when it is insinuated of a court that its decisions are influenced by resentment.[113]

Müller's terse statement was legalistic and impersonal. It named no names, though its targets were unmistakable. It was carefully worded so that CDU politicians could plausibly

[107] ibid.
[108] 'Heftige Reaktion auf Regierungskritik am Fernsehurteil' *Süddeutsche Zeitung* (11/12 March 1961) 5.
[109] 'Adenauer: Das Fernseh-Urteil ist falsch' *Die Welt* (9 March 1961) 2.
[110] Gerhard E. Gründler, 'Der Richter in der Arena' *Die Welt* (10 March 1961) 1, 2. [111] ibid 2.
[112] ibid.
[113] In Heinz Laufer, *Verfassungsgerichtsbarkeit und politischer Prozeß. Studien zum Bundesverfassungsgericht der Bundesrepublik Deutschland* (Mohr Sierbeck 1968) 473.

claim not to be affected by it—Adenauer, after all, had explicitly acknowledged that the judgment was now the law.[114] But the statement also warned firmly against any repetition of the polemics of the previous two weeks. It was all the more forceful for having been issued by a judge who until recently was a leading CDU state politician.

Christian Democrats insisted hotly that the statement did not apply to them; Social Democrats retorted raucously that it did.[115] The justices returned to Olympus and kept their peace.

The controversy soon quieted, though in the summer of 1961 the judges of the Second Senate were briefly roused by rumours that a campaign was underway to solicit scholarly criticism of the judgment and to bolster these criticisms with information about the private lives of the participating justices.[116] Adenauer assured the justices that their indignation was unnecessary: the report had been fabricated by hyper-partisans hoping to discredit scholars critical of the judgment.[117]

The chancellor himself had ceased to criticize the judgment publicly but privately it remained a sore spot. He was anything but resigned to his defeat. Gebhard Müller later recalled that at the Adenauers' 1962 New Year's reception, the chancellor asked the chief justice whether the judgment couldn't be modified so as to allow greater government influence over broadcasting.[118] But the Court did not revisit the matter during Adenauer's lifetime.

The overwhelmingly positive reception the *Television* judgment found among scholars and journalists proved enduring. For the first time since its inception, the Court had annulled a major initiative of Adenauer's government. It had displayed its independence and exercised its authority in a manner that underscored the centrality of constitutional justice to the political order of the reconstructed German state.[119] Here was a powerful voice speaking truth to power, a judicial rebuke to executive overreaching. It was also a ringing affirmation of an independent, pluralist, and plurivocal public sphere, of more-than-formalist understanding of democratic political culture.

In this last respect, the judgment and its aftermath highlight one of the Court's most abiding achievements during the Adenauer era. The *Television* judgment, more than any other early decision, seemed to justify the view that the Court was a rampart of democracy, rather than a restriction. It was a defining moment in the history of the Court, and in the life of the young Republic.

It was in the afterglow of the *Television* judgment that the Court commemorated its tenth anniversary—an anniversary that fell in September 1961 but, to avoid conflict with the Bundestag election held the same month, was celebrated in January 1962. The Court was honoured in the press with a sustained polyphony of praise.[120] The keynote speaker at the

[114] *Verhandlungen des Deutschen Bundestages, Stenographische Berichte*, vol 48 (1961) 8308.

[115] See the reports in 'Karlsruhe weist die politischen Organe in ihre Schranken' *Der Tagesspiegel* (16 March 1961) 2; 'Kritik am Fernsehurteil zurückgewiesen' *Stuttgarter Nachrichten* (16 March 1961) 1; 'Karlsruhe bricht sein Schweigen' *Saarbrücker Zeitung* (16 March 1961); 'Arndt: CDU nach Fernsehurteil von allen guten Geistern verlassen' *Frankfurter Rundschau* (21 March 1961).

[116] See Gebhard Müller to Konrad Adenauer, 30 June 1961, Landesarchiv Baden-Württemberg (Stuttgart) (Nachlass Müller, Q 1135 Bü 716).

[117] Adenauer to Müller, 17 July 1961, Landesarchiv Baden-Württemberg (Stuttgart) (Nachlass Müller, Q 1135 Bü 716).

[118] Hans Bausch to Gebhard Müller, 23 November 1984, Landesarchiv Baden-Württemberg (Stuttgart) (Nachlass Müller, Q 1135 Bü 705).

[119] See Ralf Müller-Terpitz, 'Fernseh-Urteil' in Jörg Menzel (ed), *Verfassungsrechtsprechung. Hundert Entscheidungen des Bundesverfassungsgerichts in Retrospective*, 1st edn. (Mohr Siebeck 2000) 126.

[120] See, e.g., Kurt Becker, 'Ein gelungenes Experiment' *Die Welt* (25 January 1962); Wolfgang Stammberger, 'Hüter der Grundrechte' *Bulletin des Presse- und Informationsamtes der Bundesregierung* (27 January 1962) 157; Rudolf Smend, 'Zwanzig Richter hüten das Grundgesetz' *FAZ* (14 February 1962).

Court's official gala was Rudolf Smend, an intellectual mentor to many of the justices as well as many of the Court's most sympathetic commentators. 'In the face of all contest of opinions, in the face of all criticism', said Smend, 'the Federal Constitutional Court has become a great Fact. The Basic Law now applies practically as the Federal Constitutional Court construes it, and the literature discusses it in this sense'.[121] During its first decade, the Court established institutional independence and consolidated institutional power. With the *Television* decision, the Court proclaimed its coming of age.

II. The Party State

The Court's *Television* judgment was not, as one historian contends, 'the beginning of the end' of the Adenauer era.[122] But it did accentuate and accelerate the chancellor's decline. It deprived the ruling party of a powerful pulpit in the federal elections of the coming autumn. Those elections followed close on the heels of another series of events that damaged the chancellor in the public eye: the construction of the Berlin Wall on 13 August 1961 and Adenauer's dilatory response to it. Adenauer was on holiday when the Wall went up, and his public reaction to this astonishing turn of events was not the lion's roar that many expected. Only belatedly did the chancellor visit the embattled former capital. By contrast, Berlin's mayor, Willy Brandt (also the SPD candidate for chancellor), became a ubiquitous symbol of the divided city, a front-line soldier and a popular hero.

The *Mauerbau* was a political calamity for the Adenauer government, and it came less than forty days before a federal election. Many accused the chancellor of having never seriously sought German reunification. Others charged that he cared nothing for the freedom of the millions of Germans in the Soviet sector. Again, the contrast with Brandt was glaring. Events in Berlin broadcast in bold relief the general juxtaposition of the campaign: the tanned and telegenic Brandt beside the gaunt and aging Adenauer—engagement and aloofness, energy and exhaustion, the new order and the old, the German Kennedy and the German De Gaulle. The similarity of the parties' policies and slogans (SPD: '*Unsere Sicherheit*' [our security]; CDU: '*Sicherheit*' [security]) could not hide—and perhaps highlighted—the chasm between their commanding personalities.

The election returns attested that change was astir. The SPD's share of the vote increased by 4.4 per cent over 1957, and the party placed more members in the Bundestag than at any time since 1949. The Union parties lost nearly 5 per cent of the vote—and with it their absolute majority in Parliament. Adenauer remained chancellor, but just barely. He faced intra-party opposition from those who wished to replace him with Erhard. The FDP agreed to enter a coalition with the Union parties only in exchange for Adenauer's promise to resign before the end of 1963.

A. The *Spiegel* affair

His exit was hastened by dramatic events a little over a year later. On the night of Friday, 26 October 1962, federal and state police raided the headquarters of the Hamburg weekly *Der Spiegel*, Germany's most popular newspaper. Police arrested three editors, expelled all remaining employees, and sealed off the building's 117 rooms. They rifled through

[121] Rudolf Smend, 'Das Bundesverfassungsgericht' in Rudolf Smend, *Staatsrechtliche Abhandlungen und andere Aufsätze*, 2nd edn (Duncker and Humblot 1968) 581, 582.

[122] Uwe Wesel, *Der Gang nach Karlsruhe: Das Bundesverfassungsgericht in der Geschichte der Bundesrepublik* (Karl Blessing 2004) 111.

desks and documents, beginning a search that lasted four weeks. Initially, the main targets of the raid—Rudolf Augstein, the paper's boss, and Conrad Ahlers, its military affairs editor—were nowhere to be found. Police officers searched the two men's homes and confiscated their personal papers. By midnight, Augstein and Ahlers were still at large. When word arrived that Ahlers was on holiday with his wife in Spain, Volkmar Kopf, state secretary in the defence ministry, ordered his immediate arrest and extradition.

Officials in the ministries of justice and the interior demurred, raising legal scruples about arresting Ahler on foreign soil for a political crime. Kopf waved these reservations aside, informing federal police that the defence ministry would find a way to apprehend Ahler itself if the police wouldn't do it for them. Soon thereafter, Franz Josef Strauβ, the defence minister, was on the phone with Colonel Ochim Oster, military attaché in Madrid. Strauβ ordered Oster to arrange for Ahler's arrest. Ahler and his wife were awakened in their hotel room by Spanish police at 3:30 a.m. on the morning of Saturday, 27 October. Within hours, they were on a plane back to Germany, where they were taken into custody. Before noon, Augstein, learning of the warrant for his arrest, turned himself in to the Hamburg police. The West German public learned of these events in the headlines of the Monday morning papers. The affair—with its night-time police raids and state interference with a free press—had an eerie historical echo. It sparked the fiercest public controversy in the thirteen-year history of the Federal Republic.

The raid and arrests responded to *Der Spiegel*'s cover story of 10 October 1962.[123] The piece described a NATO military 'practice session' conducted a few weeks earlier. The exercise, dubbed 'Fallex 62', simulated the outbreak of a third world war triggered by a Soviet atomic attack on Western Europe. The results, *Der Spiegel* reported, were damning. The mock war was a catastrophe for the Federal Republic. It exposed military incompetence and portended logistical bedlam. It appeared at a politically volatile time—six days before the outbreak of the Cuban missile crisis, which ended two days after the *Spiegel* siege. The article concluded with a savage critique of West German defence policy. It railed against atomic armament and called for a robust reinforcement of conventional weapons.

It also called for the scalp of Franz Josef Strauβ. Rudolf Augstein had conducted a polemical seven-year war against the defence minister—a war of intensifying venom and bile. It was an enmity Strauβ reciprocated passionately. Initially, Strauss hotly denied playing any role in the raids. But many suspected otherwise, supposing that Strauss was thirsting for revenge. The SPD Bundestag faction unanimously demanded Strauss's resignation. He finally resigned on 30 November 1962, joining the five FPD members of Adenauer's cabinet who had resigned in protest. Adenauer was forced to reorganize his cabinet, and remained at the head of the government only by promising to resign in October of the following year.

More staggering than the affair's political fallout was its public clamour. Protestors, many of them students, marched in every major city in the Federal Republic. Outraged citizens, many of them intellectuals, sent fiery missives to politicians and angry editorials to newspapers. The press corps itself warmly backed *Der Spiegel*, loaning facilities and staff to maintain the weekly's run. The broader public was deeply divided—some shocked by the open treason of a 'nihilistic' newspaper, others haunted by an irruption of enduring authoritarianism at the centre of the German state.[124]

[123] 'Bedingt abwehrbereit' *Der Spiegel* (10 October 1962) 34.

[124] On public reactions to the *Spiegel* affair, see Alfred Grosser, Thomas Ellwein, and Jürgen Seifert (eds), *Die Spiegel-Affäre. Die Reaktion der Öffentlichkeit* (Walter 1966); Dorothee Liehr, *Von der Aktion gegen den Spiegel zur Spiegel-Affäre. Zur gesellschaftspolitischen Rolle der Intellektuellen* (Peter Lang 2002); and David Schoenbaum, *The Spiegel Affair* (Doubleday 1968) 151–74.

A symbolic epicentre of the debate was an exchange between two historians—Gerhard Ritter and Karl Dietrich Bracher. In a letter to the *Frankfurter Allgemeine Zeitung*, Ritter reproved his countrymen for having been so blinded by an eternal fixation with the horrors of Hitler 'that we would rather endure every abuse, however great, of the personal freedoms guaranteed in the *Rechtsstaat* than one or another impropriety (or even mistake) by our prosecutorial organs'.[125] Bracher retorted that Ritter was advocating a 'state ideology that applies only from the top to the bottom'—that he was justifying 'nothing less than the calamitous traditional authoritarian state in Germany, to the cost of the democracy in which we have just taken our first steps'.[126] Ritter, a grand old man of German historiography with ties to the 1944 plot against Hitler, was born in 1888, when Bismarck was still chancellor. Bracher, the brilliant chronicler of the Weimaer Republic's collapse, was born in 1922. The two men personified and articulated the generational gap that infused the *Spiegel* affair. Rudolf Augstein, for his part, remained until 17 February 1963 in a jail cell, whence he uttered black prophecies about the future of German democracy and fired a flurry of complaints to the Federal Constitutional Court.

B. Interlude: 'Shadows over Karlsruhe'?

The First Senate that answered Augstein's complaints was very different from the Senate that fielded them. By historical coincidence, Adenauer's departure from the chancellorship in 1963, hastened by the *Spiegel* affair, coincided with the first major changes in the Court's composition. Adenauer resigned as chancellor on 12 October 1963. Six weeks earlier, six sitting justices left the Court—Erna Scheffler, Joachim Lehmann, and Martin Drath of the First Senate; Egon Schunck, Walter Klaas, and Ernst Friesenhahn of the Second. All but Schunck had been with the Court since its inception, and Schunck had joined the Court in its second year. Two other founding justices—Gerhard Heiland of the First Senate, and Vice President Rudolf Katz of the Second—had died within a month of one another in 1961. The question of succession, complicated by a statutory reduction in the number of judges from twenty to sixteen, proved more explosive than anyone anticipated.

When the dust settled, precisely half of the Court's sixteen judges—three on the First Senate, five on the Second—had been with the Court since its founding. Four had joined the Court in its second decade. Friedrich Wagner, an SPD parliamentarian, had replaced the sainted Rudolf Katz. Katz's death occasioned an outpouring of public eulogies, many of which celebrated the inaugural vice president as a symbol of all that was best about the Court. Katz was praised as the warm-hearted cosmopolitan who held the Second Senate together with humanity and charm, who defused internal conflicts with humility and wit. Katz's credo—'When I don the red robe, I belong to no party'—made the rounds as an emblem of the Court's triumph over early accusations of partisanship. He was the fiery SPD partisan who frequently led the Court in ruling against his former comrades.[127]

[125] Gerhard Ritter, 'Blind für die Wirklichkeit' *FAZ* (10 November 1962).

[126] Karl Dietrich Bracher, 'Demokratie oder Obrigkeitsstaat' *FAZ* (13 November 1962).

[127] For eulogistic obituaries of Katz, see especially 'Rudolf Katz' *Der Spiegel* (2 August 1961) 34; Hans Rolf Zimmermann, 'Richter ohne Furcht und Tadel' *Stuttgarter Zeitung* (25 July 1961); and 'Der Tod des Vizepräsidenten des Bundesverfassungsgerichts ist für die Bundesrepublik ein herber Verlust' *FAZ* (24 July 1961) 3. See also Vizepräsident Katz gestorben *FAZ* (24 July 1961) 1; 'Vizepräsident Katz gestorben' *Deutsche Zeitung* (24 July 1961); 'Rudolf Katz gestorben' *Süddeutsche Zeitung* (24 July 1961); 'BVG-Vizepräsident Dr. Katz gestorben' *Badische Neuste Nachrichten* (24 July 1961); 'Trauer um Rudolf Katz' *Vorwärts* (26 July 1961); 'Rudolf Katz gestorben' *Christ und Welt* (28 July 1961).

His replacement had similarly strong ties to the party. An SPD member of the Weimar-era Reichstag, Wagner, like Katz, had spent the Nazi era in exile—first in France, later in the United States. Wagner sat on the Parliamentary Council, where he was instrumental in abolishing the death penalty, and had been a member of the Bundestag since its founding in 1949. Wagner's parliamentary activities had little to do with constitutional law. But though his appointment caused some surprise it also met with broad approval.[128] Justice Heiland had a considerably lower profile than Rudolf Katz, the vice president, and his replacement by Karl Haager, a judge in the Federal Court of Justice, drew little notice.

Much more controversial was the appointment of two new justices in the summer of 1963. The statutory reduction in the number of justices, together with the end of some justices' terms, left five vacancies—two to be filled by the Bundestag and three by the Bundesrat. As expected, the Bundestag renewed, for an additional four years, the terms of Justice Leibholz and Chief Justice Müller. The Bundesrat was expected to follow suit, filling the remaining posts with three of the four justices—Walter Klaas, Hans Kutscher, Joachim Lehmann, and Martin Drath—whose terms had expired but who were willing to remain in office. The simplest solution would have been to reappoint the latter three, excluding Klaas on the inoffensive ground that he was, by some distance, the oldest of the four.[129] But the CDU/CSU faction of the Bundesrat appointments committee wished to reelect only Kutscher, filling the two remaining slots with new judges. The committee's SPD chairman, Hessian premier Georg August Zinn, agreed. The SPD and Union factions picked one new justice each—the latter Gregor Geller, a government lawyer and diplomat; the former Wiltraut Rupp-von Brünneck, an attorney from Zinn's own office.

The polemical explosion that attended this turn of events was detonated by a lawyer who in the past had represented the state of Hessen before the Constitutional Court—Adolf Arndt, now a culture senator in West Berlin. On 20 August 1963, Arndt published a fiery philippic in his party's *Sozialdemokratischer Pressedienst*.[130] Under the title 'Shadows over Karlsruhe', Arndt branded the appointments procedure as a scandal and blasted the Union parties for 'breaking solemn oaths' and employing 'malfeasant means' to alter the partisan composition of the Court. On the face of it, the appointments seemed utterly mundane. For the five vacancies, two CDU nominees and one SPD nominee had been reelected and each party had nominated one new justice. So the five vacancies were filled by three CDU nominees and two SPD nominees—a reasonable outcome, one would think, in light of the CDU's greater parliamentary representation.

But the picture looked different when one considered the composition of the individual Senates. Four of the six departing justices had been SPD nominees. Three of these four had sat on the First Senate. On the First Senate, then, three 'SPD' seats were vacated and only one SPD seat filled. On the Second Senate, two CDU seats and one SPD seat were vacated, with one CDU seat filled. The balance on the First Senate shifted from 5:5 to 5:3. The balance on the Second Senate was also now 5:3, a shift from the previous distribution of 6:4.[131] CDU appointees now formed solid majorities in both chambers.

[128] See, e.g., 'Eine gute Wahl' *FAZ* (16 December 1961); Walter Schallies, 'Er half, die Todesstrafe abzuschaffen' *Süddeutsche Zeitung* (17 January 1962). See also Friedrich Karl Fromme, 'Ein Anwalt auf dem Stuhl von Katz' *FAZ* (16 December 1961); 'Friedrich Wilhelm Wagner wird Nachfolger von Rudolf Katz' *Vorwärts* (13 December 1961).

[129] Klaas was 68, Drath 60, Lehmann 54, and Kutscher 51.

[130] Adolf Arndt, 'Schatten über Karlsruhe: Bundesverfassungsgericht—eine lästige Begrenzung der Macht?' *Sozialdemokratischer Pressedienst* (19 August 1963) 2–7.

[131] For a discussion of the shift, see Wesel (n 122) 97–107.

Arndt was apoplectic. So, apparently, were many of the justices, not all of them sympathetic to Arndt's politics. Justice Willi Geiger, for one, wrote to thank Arndt 'fervently' for his article. 'It was high time', Geiger wrote, that someone spoke up about the 'dubious circumstances of the most recent elections for the Federal Constitutional Court. I am relieved, and I believe all members of the Court are relieved, that you have unburdened us of the necessity'.[132]

And the solemn oaths, villainously violated? Arndt alleged that the SPD, fearing the very CDU dual majority the 1963 appointments effected, had acquiesced in reducing the size of the Senates only on condition that all sitting judges available for reelection would in fact be reelected. The Union parties, Arndt accused, had reneged on that promise. Hitherto the reelection of sitting judges had been a routine matter, accepted unquestioningly by both parties. But now, through the 'de-selection' (*Abwahl*) of two SPD justices, the CDU had flagrantly politicized the appointments process. Especially pernicious, in Arndt's view, was the non-election of Justice Martin Drath.

Drath was something of an SPD folk hero—a Weimar disciple of Hermann Heller and a powerful postwar critic of Carl Schmitt. But as the Bundesrat appointments committee began its deliberations, Josef Hartinger, a state secretary in the Bavarian justice ministry, told the committee that Drath had planned an East German lecture tour, during which he would hold a public conversation with a leading East German jurist, Karl Kaul, formerly defence counsel for the German Communist Party in the 1956 case for which Drath served as rapporteur. It was also bruited that Gebhard Müller had privately rapped Drath as lazy—that Müller wanted Drath off the Court, and that the chief justice's party peers had worked his will. Ousted from a Frankfurt professorship by Nazis in 1933, from a Jena professorship by Stalinists in 1947, Drath had now been expelled for a third time in what Arndt execrated as an act of high-court *Gleichschaltung*—a term associated with Nazi efforts fill public offices with ideological purists.

Arndt's essay rent a political fault line. The press reported his charges in luscious detail.[133] The accused denied them with passionate zeal.[134] If the changes in Court membership had been driven by a nefarious Union conspiracy, CDU politicians and conservative journalists asked, why had SPD state premiers agreed to the compromise?[135] Even some sympathetic to Arndt conceded that the Unionists a point. Georg August Zinn, for his part, roundly rejected Arndt's account of the proceedings. A reporter for the liberal *Süddeutsche Zeitung* was convinced that Arndt had acted in good faith but admitted that Zinn had cooperated in Drath's ouster. It was unfortunate, though, that Arndt had unwittingly resurrected the rhetoric, long silenced, of a Red Senate and a Black.[136] *Der Spiegel* ran the story a few days later under the headline '*Rouge et Noir*'.[137]

[132] Willi Geiger to Adolf Arndt, 20 August 1963, Friedrich-Ebert-Stiftung (Bonn) (Nachlass Adolf Arndt, Box 6, Mappe 16).

[133] See, e.g., 'Arndt spricht von eimen Anschlag auf das Verfassungsgericht' *Der Tagesspiegel* (20 August 1963); 'Arndt: Anschlag auf das Bundesverfassungsgericht' *Süddeutsche Zeitung* (20 August 1963) 4; Volkmar Hoffmann, 'SPD enthüllt CDU-Anschläge auf Verfassungsgericht' *Frankfurter Rundschau* (20 August 1963); 'Scharfe Angriffe Arndts gegen Bonn' *Die Welt* (20 August 1963); 'Arndt um das Bundesverfassungsgericht besorgt' *Stuttgarter Zeitung* (20 August 1963); 'Arndt und das Verfassungsgericht' *Süddeutsche Zeitung* (20 August 1963). The essay even drew international press coverage. See 'Move to "Pack" German Court Alleged' *Times of London* (23 August 1963).

[134] See 'Kopfschütteln und Empörung über Arndt' *Stuttgarter Zeitung* (21 August 1963); Volkmar Hoffmann, 'Union weist Angriff Arndts zurück' *Frankfurter Rundschau* (21 Aug 1963); 'Scharfe CDU-Antwort an Arndt' Der Tagesspiegel (21 August 1963); 'CDU: Arndts Darstellung bösartig' *Süddeutsche Zeitung* (21 August 1963).

[135] See, e.g., 'Neue CDU-Kritik an Kultursenator Arndt' *FAZ* (23 August 1963).

[136] 'Arndts allzu scharfer Schuß' *Süddeutsche Zeitung* (24 August 1963) 3.

[137] 'Verfassungsrichter: Rouge et noir' *Der Spiegel* (28 August 1963) 18.

Ironically, the Court's historical refutation of such labels had helped persuade Zinn to cut a deal. Zinn was a frequent party before the Court, and his SPD state government had done no better before the First Second than before the Second. In the *Atomic Plebiscite* case, the 'Black' Second Senate had handed Zinn his greatest defeat; in the *Television* judgment, his greatest victory. In many cases Zinn was willing to buck the federal party brass in matters of constitutional politics. Zinn was likely more interested in placing one of his own on the First Senate than with preventing a presumptive 'CDU majority'.

Drath, for his part, was devasted; Arndt, temporarily, became something of a pariah. They consoled one another in a series of poignant personal letters, addressed in the comradely 'Du', and wrote agonized editorials in *Die Zeit*.[138] Privately, Drath heaped wrathful scorn on the head of Müller, whose slur that Drath was lazy had led, Drath was sure, to the latter's non-election. Drath thundered that the chief justice was ignorant of the law and unprepared for his post. Müller was as weak as he was fickle—he told everyone what they wished to hear and kept his real convictions (if he had any) to himself. The best Drath could say for Müller was that he was an efficient administrator—but how hard was it to administer a body of 100 people? Müller was so disliked by the other justices, Drath gossiped, that one of them had amended his will to prevent Müller from speaking over his grave.[139]

These were the recriminations of a deeply wounded man. Drath was particularly hurt to leave the Court amid accusations that he nurtured sympathetic ties to the East—Drath who had fled from the GDR; who had skewered its constitutional pretensions in pamphlets published by a government ministry; who had faced journalistic abuse and physical threats[140] in the aftermath of the KPD judgment.[141] It was some solace, at the end of November, when the Court's official spokesman dismissed the charges against Drath as 'wholly ungrounded', and in December, when the Court held a ceremony in Drath's honour—a ceremony to which Drath was invited in a personal note from Gebhard Müller.[142] Charges that Drath sympathized with the East were certainly false. On questions of Cold War politics, there was no question where the justice stood. In the aftermath of the *Spiegel* affair, to cite just one example, Drath had written an effusively supportive letter to Franz Josef Strauβ.[143] Ironically, the alleged victim of a CDU conspiracy might well have voted for Strauβ in the celebrated *Spiegel* case that was working its way through the courts.

C. The *Spiegel* judgment

Whether Drath's solidarity with Strauβ would have translated into a vote against Augstein's complaints will never be known. By the time the First Senate reached the merits of those complaints, Drath was three years into an involuntary retirement.

The legal aftermath of the *Spiegel* affair was long. Prosecution against Augstein, his fellow editors, and the 'Fallex' article's main informant, Colonel Alfred Martin of the defence ministry, moved slowly. Federal prosecutors struggled to prove that the article was in fact treasonous. The Federal Court of Justice, losing patience, dismissed the charges against

[138] See Martin Drath, 'Zwischen Grundgesetz und Politik' *Die Zeit* (30 August 1963) 3; Adolf Arndt, 'Leserbrief' *Die Zeit* (18 October 1963). Drath also made a confidant of SPD national chairman, Willy Brandt. See Drath to Willy Brandt, 7 September 1963, BAK N 1341/34.

[139] Martin Drath to Adolf Arndt, undated (probably September 1963), BAK N 1341/59.

[140] See, e.g., K.H. Hüttenrauch to Drath, 16 August 1956, BAK N1341/59 (Nachlass Martin Drath).

[141] Martin Drath to Adolf Arndt, undated, circa September 1963, BAK N 1341/59 (Nachlass Drath).

[142] 'Verlautbarung der Pressestelle des Bundesverfassungsgerichts' (29 November 1963), BAK N 1341/59 (Nachlass Drath); Gebhard Müller to Martin Drath, undated, BAK N 1341/59 (Nachlass Drath).

[143] Martin Drath to Franz Josef Strauss (19 November 1962), BAK N 1341/59 (Nachlass Drath).

Augstein and Ahlers in May 1965.[144] Over the next three months, the Court of Justice dismissed the charges against all other defendants save Martin. Charges against Martin weren't dropped until October 1966—shortly after the First Senate of the Constitutional Court had finally responded to the magazine's constitutional complaints.

Represented by two of the ablest lawyers in the land—Horst Ehmke, the future justice minister, and Fabian von Schlabrendorff, a future justice of the Constitutional Court—the complainants argued that the ministers who instigated, the judges who approved, and the police who executed the search warrant had violated their constitutional rights to the freedom of the press (Article 5), the sanctity of property (Article 14), and the inviolability of the home (Article 13).

Oral arguments held in January 1966 revealed sharp fissures, but no purely partisan divide, within the First Senate. Justice Herbert Scholtissek, the case's rapporteur, was a vocal libertarian in free speech matters. In contrast, Chief Justice Müller, the Senate's chairman, took an expansive view of the state's power to restrict speech on security grounds. Both were quondam CDU politicians, as was Justice Erwin Stein, another extrajudicial paladin of civil liberties. Close observers assumed Stein and Scholtissek would be joined by two SPD nominees, Wiltraut Rupp-von Brünneck and Hugo Berger, in sustaining *Der Spiegel*'s complaints. A third SPD nominee, Justice Karl Haager, was expected to vote for the government. The views of Justices Theodor Ritterspach and Werner Böhmer remained a mystery. The outcome, presumably, would depend on them.[145]

After oral argument, the Court deliberated for seven months before issuing a decision. Its published opinion of 5 August 1966 astonished everyone.[146] Ritterspach and Böhmer joined Müller and Haager in sustaining the constitutionality of the search warrant and subsequent investigation. As expected, Scholtissek, Berger, Rupp-von Brünneck, and Stein supported the newspaper. Under the Court's rules, a 4:4 split went against the party alleging a constitutional violation—in this case against *Der Spiegel*. The result of the tie was a government victory—a confirmation that the warrant and the search were constitutional.

Normally, in the case of a 4:4 vote, the Court merely announced the outcome without publishing an opinion. Here, however, the stakes were too high, and the Senate too polarized, for a perfunctory *per curiam*. The Court published a long opinion with an historic dissenting opinion—the first officially published dissent in the history of German law.[147]

It was a dissent, however, that dared not speak the name. Nor did it name its authors. Contrary to the Anglo-American practice of publishing a majority opinion, followed separately by any concurrences or dissents, the Constitutional Court integrated the views of the dissenters into a single, impersonal judgment of the Court. The judgment, moreover, didn't speak of 'the dissenters' or 'the majority' (the latter would have been an affront to arithmetic), but simply of the view that prevailed and the view that did not.

The opinion began by enunciating principles in which all the justices concurred; it then set forth the views of the dissenting and the prevailing justices, where these differed, in dialogic juxtaposition.[148]

[144] In (1965) 29 *NJW* 1187.

[145] On predictions about the Senate's votes, see Donald Kommers, 'The Spiegel Affair: A Case Study in Judicial Politics' in Theodore L. Becker (ed), *Political Trials* (Bobbs-Merrill 1971) 28. The entire article is an excellent introduction to the legal aspects of the *Spiegel* affair.

[146] 20 BVerfGE 162 (1966).

[147] Willi Geiger, as we saw in Chapter 1, privately published a dissent to the Plenum's procedural decision in the European Defense Community case. Geiger wrote dissents with modest frequency before they became official Court procedure in 1971. He published some of them after his retirement. See Willi Geiger, *Abweichende Meinungen zu Entscheidungen des Bundesverfassungs-gerichts* (Mohr Siebeck 1989).

[148] So, e.g., the views of the dissenters appeared in 20 BVerfGE 162 (n 146) 178–85, 191–209, and 223–26, while those of the prevailing justices appeared in ibid 185–91, 209–22, and 226–30. The dissenters actually took more space, at twenty-eight pages, than the prevailing justices, at twenty-three.

All the justices agreed that the treason paragraphs of the penal code were not, on the face of it, unconstitutional. All stressed, too, that a free and fearless press was the *sine qua non* of a democratic state.[149] The Court rejected unanimously the view that security interests trumped freedom of the press because the existence of the state is a precondition of a free press. The operation of a free press was just as necessary to the existence of the state as was the preservation of military secrets. The competing claims of state security and press freedom could not be assessed in the abstract. They must be balanced concretely in accordance with what had become an increasingly familiar doctrinal formula: the principle of proportionality.

General laws, the justices wrote, could limit, but not relativize, press freedom. Such limits, moreover, must secure a legal good of equal or greater value.[150] Courts must weigh the dangers to which a given publication might expose the country against the public's right to be informed of important events, even in the field of defence policy.[151] It was in conducting this balancing act that the Senate's factions parted paths.

In the dissenters' view, the judge issuing the warrant and the investigators carrying it out hadn't merely mistaken the balance between national security and press freedom; they hadn't balanced at all.[152] They hadn't even asked whether the offending article actually revealed state secrets. They had simply taken the defence ministry's word for it. This, in the dissenters' view, was constitutionally inadequate. The 'mosaic theory' put forth by the defence ministry—according to which the assembly and collation of scattered scraps of public knowledge might amount to disclosure of state secrets—was unconstitutional as well as unpersuasive. It was the very business of the press—one of its *raisons d'être*—to assemble, arrange, assess, and publish newsworthy information.[153] To make this treason would obliterate the constitutional guarantee of a free press.

The search of *Spiegel* headquarters, the dissenters found, flunked the proportionality test from beginning to end. Its contribution to national security was dubious—the article had already been published; and an experienced paper like *Der Spiegel* was unlikely to have kept on hand the sensitive documents on which the article's authors relied.[154] On the other hand, the search's intrusion on press freedom was colossal. It threatened to destroy the nation's premiere critical news organ—to block its publication in the short run and to crush it financially in the long run. That the voluntary intercession of other papers prevented this outcome had no bearing on the constitutional question. The search also threatened the paper's ties with all its sources and threatened to make all journalists jittery about covering defence policy.[155] The search was sweeping beyond anything rationally commensurate with its ostensible scope.[156] That scope, in any case, was unconstitutionally imprecise.[157] In sum, the search's likely benefit to state security was slight, its threat to press freedom systemic.

The prevailing justices saw things differently. They too, applied the proportionality principle, but did so with a great deal of deference. In their view, the 'mosaic' theory had not been decisive in the decision to search *Spiegel* headquarters.[158] Prosecutors had simply taken the defence ministry's advisory opinion at its word, and the prevailing justices were willing to do the same. The question was not whether that opinion was accurate in all particulars but whether it sufficed to justify the search. The prevailing justices were convinced that it did. They also saw the search itself as self-evidently suited to uncover the evidence it sought. The procedure may have been flawed, but not fatally so. The order was imprecise, but not unconstitutional. Greater precision was probably impossible. Existing statutory protections for

[149] 20 BVerfGE 162 (n 146) 174. [150] ibid 177 (referencing 7 BVerfGE 198, 208 (1958) (*Lüth*)).
[151] ibid 178. [152] ibid 198. [153] ibid 180–82. [154] ibid 204–05. [155] ibid 200.
[156] ibid 199. [157] ibid 226. [158] ibid 186.

editorial secrecy left something to be desired, but it was not the Court's job to fill the void. In any case, protections for editorial secrecy exist, not for the sake of individual editors and informants, but to promote the functionality of the press as a whole.[159]

The prevailing justices also qualified the united Senate's comments on press freedom. 'In criminal proceedings', the foursome noted, 'the press enjoys no privileges'.[160] In sensitive matters, the press itself was duty-bound to balance the public's need for information against the demands of state security. Journalists couldn't evade this responsibility by invoking the freedom of the press.[161] The 'permissive view of journalistic freedom' urged by *Der Spiegel* would have 'negative consequences for the credibility of the Federal Republic within an integrated alliance like NATO, whose other members, though their legal systems largely derive from the same intellectual tradition, consider a far more intensive protection of military secrets to be self-evident'.[162] This was a thinly-veiled jab at the siege's foreign critics. In the view of the prevailing foursome, the warrant and the search—and all official decisions surrounding them—were constitutional.

Public response to the judgment was as divided as the decision itself. Many assessments were gloomy. The *Süddeutsche Zeitung* characterized the decision as a setback, not alone for *Der Spiegel*, but for 'democracy and the rule of law'.[163] If what happened at *Spiegel* headquarters was 'not a flagrant attack on the constitutional guarantee of a free press, our name is Hopf, or Strauss, or Adenauer'. Hans Gerlach of the *Kölner Stadt Anzeiger* professed himself perplexed and appalled 'that four judges of our country's most important court, after all the failures, all the errors and crimes that German authorities have committed in this century (all of which [the judges] themselves have experienced), can still believe so unshakably in the justice and wisdom of those who govern us'. *Die Weltwoche* saw in the judgment an authoritarian spirit parading in democratic garb. The *Münchner Merkur* lamented that the judgment had resolved nothing. After the jurisprudential fog of the lengthy judgment lifted, what remained was 'the depressing feeling that the supreme constitutional tribunal was manifestly incapable of adding clarity to ... the all-decisive question—namely, to decide once and for all which takes precedence, *raison d'état* or the freedom of the press'. The *Passauer Neue Presse* added that '[w]hoever didn't yet realize that the Federal Constitutional Court is a political rather than a judicial body' should by now be 'permanently disabused'.

Others were more ambivalent. Rudolf Augstein himself merely took the liberty of adorning 'several of the Court's findings ... with a question mark'.[164] Some were even sanguine. For many, the forcefulness of the dissent made the case's formal outcome almost irrelevant. 'As an attorney, I would naturally have preferred a 5-3 "victory"', said *Spiegel* advocate Horst Ehmke. 'But legally and politically this decision is much nicer. With 5-3 we would have had a typical compromise decision'.[165] An editorial in *Die Welt* reported that the most lasting impression of the long judgment was not the diffident caution of the prevailing judges but the angry certitude of the dissenters. 'Through every line of the opinion', *Die Welt* reported, 'there flashes the lightning wrath of opposition'.[166] The *Frankfurter Allgemeine Zeitung* considered the judgment's internal contradictions 'one of its happier sides'. It largely left government lawyers and criminal judges in the dark as to how the Constitutional Court would rule in future cases of controversial investigation. This uncertainty, the *FAZ* writer felt, suggested a salutary rule of thumb: criminal authorities would 'do best, in taking steps

159 ibid 221. 160 ibid 212. 161 ibid 214–15. 162 ibid 221.

163 The citations that follow in this paragraph are all collected in 'Der Zorn des Widerspruchs' *Der Spiegel* (15 August 1966) 9.

164 Rudolf Augstein, 'Lieber Spiegelleser' *Der Spiegel* (8 August 1966) 3.

165 In Schoenbaum (n 124) 216.

166 In 'Der Zorn des Widerspruchs' *Der Spiegel* (15 August 1966) 9.

against organs of the press, to act in such a way as to withstand the legal scrutiny' of the dissenters. *Die Zeit* hoped the judgment would make future investigators skeptical of 'yielding to whispers from the executive and *Gutachten* from the "official authorities"'.[167] For *Der Tagesspiegel*, whatever the legal outcome of the case, its political meaning was clear: 'Doubts about the constitutionality of the '*Spiegel*-Action' were justified and remain in force'.[168]

There can be no doubt that the publication of the dissenters' views helped mute what might otherwise have been vehement criticism of the judgment. A Tübingen colleague wrote to Justice Gerhard Leibholz that 'it must be clear after the *Spiegel* judgment that the publication of dissenting votes has always been unjustly execrated; for here the publication of the dissent has had a clearly balancing and pacifying effect'.[169] It also helped create within the Court what the raids on *Der Spiegel* had been accused of denying to the Federal Republic—an atmosphere of fearless debate with all arguments open to public scrutiny.

Since the Court's inception, reformers had championed the introduction of formal dissents.[170] Within the Court itself, Justice Willi Geiger had proposed introducing dissents at a plenary meeting of both chambers in April 1952. Geiger had been merrily circulating dissents ever since.[171] Traditionalists like Müller worried that published dissents would diminish the Court's authority. Advocates of dissenting opinions answered that the Court's authority depended on other factors entirely—the persuasiveness of its opinions, for example, or the prudence of its political interventions. What was at risk was not the Court's authority but an authoritarian and antiquated judicial structure. Champions insisted that dissents would make the Court more democratic. If the Court was to decide boiling political matters on which public opinion was sharply divided, the public had a right to learn of divisions within the Court itself. If at no other level of the judiciary, dissents ought to be introduced at the constitutional level.

Still, dissents posed problems. Justices appointed to the Court from within the federal judiciary enjoyed lifetime tenure, but other justices served renewable terms of four or eight years. Many worried that the 'term' judges would consider the consequences that dissenting (or not) in a given case might have on their odds for reelection. Dissents, the argument ran, would undermine the independence of 'term' judges and put them at a disadvantage vis-à-vis their 'life' colleagues.[172] Traditionalists, again including the chief justice, also fretted that dissenting opinions, enjoying an air of novelty and rebellion, would receive more attention in the press than the judgment of the Court. This fear, as experience would show, was well founded.

Meanwhile, an increasing number of Müller's colleagues came to favour dissents.[173] This attitude, however, was not general within the German judiciary. In an official 1968

[167] ibid. [168] ibid.

[169] Dieter Dirk-Hartmann to Gerhard Leibholz, 7 August 1966, BAK N 1334/1 (Nachlass Gerhard Leibholz).

[170] On dissents on the Court generally, see Gerd Roellecke, 'Sondervoten' in Peter Badura and Horst Dreier (eds), *Festschrift 50 Jahre Bundesverfassungsgericht*, vol 1 (Mohr Siebeck 2001) 363.

[171] See Ernst Benda, 'Ansprache anläßlich der Feierstunde des Bundesverfassungsgerichts am 9. November 1977 zur Verabschiedung von Bundesverfassungsrichter Prof. Dr. Willi Geiger', BAK N 1334/299 (Nachlass Leibholz).

[172] To assuage this fear, reformers proposed introducing dissents alongside a single, non-renewable term—a proposal ultimately embraced in the reform legislation passed in 1970.

[173] Hans Rupp, who studied at Harvard and knew American law well, was a particularly strong supporter. See the report in 'Verfassungsrichter Rupp 60' *FAZ* (30 August 1967). Vice President Wagner, in his 1968 farewell message to the Court, also called for dissents. 'Bundesverfassungsgericht: Personen und Probleme' *Badische Neueste Nachrichten* (12 January 1968) 8. For summaries of public debates about dissents, see Friedrich Karl Fromme, 'Gleiche Zeit für alle Verfassungsrichter' *FAZ* (11 October 1967); Ernst Müller-Meiningen Jr, 'Des Richters abweichende Meinung' *Süddeutsche Zeitung* (14/15 August 1968) 4; Friedrich Karl Fromme, 'Richter mit Namen' *FAZ* (12 September 1968).

survey conducted by the justice ministry, 158 of 235 responding judges from the other federal courts opposed dissents.[174] The judges feared that dissents would undermine judicial authority, flout tradition, and disturb the legal peace (*Rechtsfrieden*).

Such sensibilities were increasingly at odds with those of a public that, since the *Spiegel* affair, increasingly prized plurivocality. Accordingly, the Constitutional Court won plaudits for bucking the trend. Beginning in 1967, the Second Senate regularly announced the tally of its internal votes. Beginning in 1971, the Court's organic statute permitted published dissents. The Court's distance in this and other matters from the rest of the judiciary strengthened the public's esteem for the Court. Throughout the 1960s, the judiciary at large was criticized with increasing volume for its authoritarian ethos and inadequate rupture with the Nazi past. At the beginning of the decade, a prominent public exhibition drew attention to continuities with the Third Reich judiciary.[175] It was held—with equal parts aptness and irony—in Karlsruhe.[176]

On the same day the First Senate issued its split judgment in the *Spiegel* case, it also issued an uncontroversial judgment that trumpeted the constitution's commitment to a democratic public sphere and reaffirmed its rejection of authoritarian modes of thought.[177] At issue was a holdover public assembly law passed in 1934, amended in 1941, and still on the books in several states. The law required police to approve, and allowed them to supervise, public gatherings. The law gave administrators discretion to deny permits or direct surveillance. Under the Basic Law, police administration fell under state jurisdiction, so the law could not be repealed at the federal level. The Bundestag asked the Court to nix the law in their behalf. The Court obliged.

The law, the First Senate held, ran afoul of a host of constitutional provisions. Most fatally, though, it bore the burden of its authorship and the stain of its birth date. 'The Public Assembly Law', the Senate wrote, 'stems from the administrative thought of the totalitarian state'.[178] It came from a time when state interests trumped personal freedoms; when public rights were ascendant and constitutional rights moribund; when law was a tool in the hands of power. This was the thought that underlay the public assembly law, which authorized administrators to act arbitrarily. But times had changed radically. Not even the canon of constitutional construction—which directed the Court, whenever possible, to interpret a law in a manner that preserved its constitutionality—could rescue the law. The Senate proclaimed it null and void. Naturally, the *Public Assembly* judgment was overshadowed by the *Spiegel* decision of the same day. But as a public relations matter, it didn't hurt that the Court reaffirmed its role as guardian of fundamental rights and guarantor of the public sphere on the very day when another, more contested decision led some to question these roles.

D. Party finance in the party state

The *Spiegel* judgment revealed a Court every bit as divided on an existential question of national identity as was the public at large. Seventeen days before the First Senate issued its *Spiegel* judgment, a similarly divided Second Senate decided a case of greater immediate political importance—a case raising core questions about the nature of the West German

[174] Ernst Müller-Meiningen Jr, 'Wenig Meinung für "abweichende Meinungen"' *Süddeutsche Zeitung* (28 Aug 1968) 4; 'Abweichende Meinung: Aus der Amtstube' *Der Spiegel* (23 September 1968) 70.

[175] Reinhard Strecker was the designer of the display, which he titled 'Unredeemed Nazi Justice' (*Ungesühnte Nazijustiz*).

[176] For a discussion of 1960s debates on judicial reform, see Jörg Requate, 'Standespolitik als Gesellschaftspolitik. Zur Debatte um den Reformbedarf der Justiz in der 60er Jahren' in *Dynamische Zeiten* (n 13) 424–43.

[177] 20 BVerfGE 150 (1966). [178] ibid 160.

state, the role of political parties within that state, and the relationship of parties to the public sphere. It was the Court's second foray into the prickly pasture of party finance.[179]

In 1958, the Court invalidated a tax deduction for contributions to political parties as incompatible with the equality principle.[180] An unusual dictum in that decision suggested the state might subsidize parties directly, clearly indicating, most thought, that such subsidies would be constitutional.[181] Article 21(1) GG, the Court noted, acknowledged the role of parties in shaping popular will. Popular will was expressed preeminently through parliamentary elections, in which political parties played a decisive role. Since the sponsoring of elections was an essential state duty, 'it must be permissible for the state to make financial means available, not only for the elections themselves, but for the parties who conduct them'.[182] The parties soon took the hint. The SPD, with its broad dues-paying membership, was little affected by the 1958 decision. But the CDU and FDP struggled to stay afloat financially.[183] They grasped the Court's suggestion eagerly. For the 1959 federal budget the CDU proposed an outlay of 5 million DM, to be divided among the major parties commensurate with their parliamentary strength. From year to year, the sum soared. By 1965, outlays to parties had ballooned to 38 million DM—one Mark, roughly, for every voter in the Federal Republic.

From the beginning, state funding had loud critics. Theodor Eschenburg called the scheme feudalistic. Parties, like medieval landlords, not only administered the country but could now extract its resources.[184] Critics were skeptical of the programme's ostensible justification—that parties played a constitutionally mandated role in shaping, or 'educating',[185] popular political will. 'Advertising', said Dolf Stenberger, 'is not education'.[186] Karl Jaspers was even more scornful. 'One can hardly maintain', Jaspers wrote, 'that in the Federal Republic an education of the people's political will takes place. The ignorance of most is frighteningly great. Parties do not inform or educate the people, and they do not train it to think'.[187] If the premise of public party financing was dubious, Jaspers continued, its implications were ominous. State financing of parties was, in Jaspers' view, only the latest manifestation of the Federal Republic's transformation from a democracy to a 'party oligarchy'—a system in which parties in power consume public funds to remain in power; in which no authentic opposition could emerge, either in Parliament or at the grassroots; in which elections were an open sham, a mere 'acclamation of the party oligarchy'. In its essence, party oligarchy meant 'contempt for the people', an 'inclination to withhold information from the people', a desire to exclude the people from popular participation, a preference 'to leave them stupid'.[188] From a 'party state', the Federal Republic had degenerated into a party oligarchy. In light of proposed legislation for an 'emergency constitution', Jaspers warned, it risked a second step—'from party oligarchy to dictatorship'.[189]

Though Jaspers didn't say so, the theorist who had proclaimed most forcefully that the Federal Republic was a 'party state' was Gerhard Leibholz of the Constitutional Court's Second Senate. In prolific academic tracts, Leibholz distinguished between two brands of democracy, which he termed 'parliamentary representative democracy' and

[179] For an excellent contemporary introduction to the controversy, see Donald Kommers, 'Politics and Jurisprudence in West Germany: State Financing of Political Parties' (1971) 16 *Am J Juris* 215.

[180] 8 BVerfGE 51 (1958). [181] ibid 62–63. [182] ibid 63.

[183] See Karl-Heinz Adams, *Parteienfinanzierung in Deutschland. Entwicklung der Einnahmestrukturen politischer Parteien oder eine Sittengeschichte über Parteien, Geld und Macht* (Tectum 2005) 112.

[184] Theodor Eschenburg, *Herrschaft der Verbände* (Deutsche Verlags-Anstat 1956) 65.

[185] The redolent term *Bildung* suggests both meanings.

[186] 'Werbung ist nicht Bildung' *FAZ* (3 August 1959) 1.

[187] Karl Jaspers, *Wohin treibt die Bundesrepublik? Tatsachen, Gefahren, Chancen* (Piper 1966) 130.

[188] ibid 131–40. [189] ibid 141.

'parliamentary political mass democracy'. The former corresponded to the classical model of nineteenth-century liberalism—a system of suffrage and representation, individual (mainly economic) rights, and the distinction between state and society. The latter system, which Leibholz took to be the only workable model for the twentieth century, was more plebiscitary; it was based on universal suffrage and aspired to what Leibholz called 'radical arithmetical-mathematical equality'. Such a system, Leibholz insisted, required an *effective*, not merely a formal, equality of representation. It required a multi-party system with a certain kind of party—a unified, programme-oriented organization of active citizens, capable of forming popular political will and immune to the seductions of powerful interests. Leibholz was part of a long European tradition critical of American parties as interest-beholden oligarchies. Leibholz's *Parteienstaat* was the antidote to the *Verbändestaat* (associations state), whose rise Eschenburg bemoaned in a famous essay.[190] Embedded in the very name was a close nexus between party and state. Parties, in Leibholz's view, were full-fledged constitutional organs, integral to the structure of the state and essential to the workings of modern democracy.[191]

From the Court's inception, Leibholz's theories shaped its jurisprudence. In its judgment banning the Socialist Reich Party, for example, the First Senate proclaimed that Article 21 had raised political parties 'to the rank of constitutional institutions'.[192] As we saw in Chapter 1, that case considered the tension between Article 21's enshrinement of a special role for parties and Article 38's declaration that individual parliamentarians represent the people as a whole. Leibholz conceded that the provisions were irreconcilable, that Article 38 was unworkable in the context of the *Parteienstaat*. The *SRP* judgment didn't embrace this view openly, but did enforce it tacitly. Earlier that same year, the Second Senate characterized parties as 'integral parts of [our] constitutional structure and [our] constitutionally ordered political life'.[193] In 1954, the Court's Plenum went further, christening parties as full-blown 'constitutional organs' competent to assert their rights in *Organstreit* proceedings before the Constitutional Court.[194] In dicta to the 1958 *Party Finance* judgment, for which Leibholz served as rapporteur, the justice's theories were on ample display. Those dicta prompted the scheme of public party finance described earlier.

Almost immediately, the programme was challenged on constitutional grounds. The All German Block, a small party not represented in the Bundestag, complained that by providing funds only to parties in Parliament, the party finance provisions violated the equality principle. In 1961, the Second Senate rejected the party's petition for a preliminary injunction and suggested in passing that the constitutionality of public party financing had already been established in the 1958 judgment. The Senate explained that the law allowed 'the benefited parties to meet their obligation as constitutional organs with greater independence from non-objective sources of funds'. Enjoining it would do greater harm to the general good than depriving the All German Block from funding until the Court ruled on the merits.[195] Earlier, the Court rejected a complaint against a Lower-Saxon state law that provided public funding only to parties in the state Parliament.[196] In 1964, the Court

[190] See Eschenburg (n 184).

[191] For Leibholz's views on parties, see Gerhard Leibholz, 'Der Parteienstaat des Bonner Grundgesetz' (1951) 2 *Deutsches Verwaltungsblaat* 1; Gerhard Leibholz, *Strukturprobleme der Modernen Demokratie* (CF Müller 1967) 71–77; Gerhard Leibholz, *Das Wesen der Repräsentation und der Gestaltwandel der Demokratie im 20. Jahrhundert* (Walter de Gruyter 1966); Gerhard Leibholz, *Politics and Law* (AW Sythoff 1965) 24–71; Gerhard Leibholz, 'Der moderne Parteienstaat' in Gerhard Leibholz, *Verfassungsstaat—Verfassungsrecht* (W Kohlhammer 1973) 68–94. See also Kommers (n 145) 24–26.

[192] 2 BVerfGE 1, 72 (1952).

[193] 1 BVerfGE 208, 240–41 (1952).

[194] 4 BVerfGE 27, 30 (1954); see also 5 BVerfGE 85, 134 (1956); 6 BVerfGE 367, 372, 375 (1957).

[195] 12 BVerfGE 276, 280 (1961).

[196] 11 BVerfGE 239 (1960).

rejected a similar complaint against a similar law, also stemming from Lower Saxony.[197] By all appearances the programme of public party finance was constitutionally secure.

Still, small parties continued to assail it: the All German Party again in 1962, the Bavarian Party in 1964. Finally, in 1965, the party finance law was challenged by the state government in Hessen—led once more by its minister president, Georg August Zinn, who again proceeded without consulting the national SPD leadership. The two smaller parties, joined at oral argument by the far-Right National Democratic Party of Germany (NPD), complained that the law impermissibly benefited only parties strong enough to win seats in the Bundestag. Hessen, represented once more by Adolf Arndt, contended that the principle of direct public funding was unconstitutional on its face.

By autumn 1965, rumours of uncertain provenance suggested that the Court would grant the petitions. But even if the rumours were right, it was unclear which petitions the Court might grant, and to what degree it would grant them. Repeated delays suggested deadlock within the Second Senate. And a tie, everyone knew, would run against the petitioners.

If in fact there was a deadlock, it was broken by a remarkable turn of events.

E. The 'recusal' of Justice Leibholz

In early October 1965, with the second *Party Finance* case pending before his Senate, Justice Leibholz gave a lecture at an academic conference in Würzberg. The lecture, entitled 'State and Interest Groups', was in most respects unremarkable—a simple restatement of the Leibholz's familiar theory of the party state. [198] But passing comments about the pending case unleashed a glorious row. According to the official report in *Die Öffentliche Verwaltung*, Leibholz styled the case's plaintiffs 'an unholy alliance' of 'liberals and anti-democrats'. The report added that Leibholz thought the law necessary 'to free parties from societal pressure'.[199]

Members of the 'unholy alliance' were outraged. In short order, the Bavarian Party asked the Court to exclude Leibholz from further participation in the case. Traditionally, judges accused of bias are not excluded by their colleagues; they recuse themselves. The decision whether to recuse oneself is left to the conscience of the individual judge.

In Leibholz's case, however, the Second Senate opened the question of bias for oral argument. Leibholz's seven colleagues listened publicly to the Bavarian Party's case against his impartiality. Later, in closed chambers, a majority of them voted to exclude him—a decision they justified in a ten-page opinion published on 3 March 1966.[200] Leibholz was also excluded from related cases on the dubious ground that the Senate should be identically composed for all pending judgments related to party finance. They thus extended the exclusion from an *Organstreit* to a *Normenkontroll* proceeding—a move the Senate made without ruling explicitly that bias in an abstract review proceeding was even possible.[201]

It was the first exclusion of a judge for bias in the Court's history. That its object was Leibholz—for fifteen years the Second Senate's moral leader, intellectual star, and formidable public champion—came as a great shock. That it was issued against Leibholz's will,

[197] 18 BVerfGE 34 (1964).

[198] Gerhard Leibholz, 'Staat und Verbände' (1966) 24 *Veröffentlichung der Vereinigung der Deutschen Staatsrechtslehrer* 6.

[199] Leibholz's actual statement was more moderate—to the effect that failing to supply parties with a minimum of public support would make them dependent on societal interests. See Manfred Wiegandt, *Norm und Wirklichkeit: Gerhard Leibholz (1901-1982)—Leben, Werk und Richteramt* (Nomos 1995) 70.

[200] 20 BVerfGE 9 (1966).

[201] Leibholz's biographer finds this the most dubious decision in the entire affair. See Wiegandt (n 199) 73.

executed in defiance of traditional procedure, and accompanied by a public censure was immensely controversial.[202] Leibholz, after all, insisted solemnly that he hadn't used the provocative language ascribed to him.

In aftertimes, Leibholz never spoke of the bias decision. But it must have come as a bitter blow.[203] One of those who spoke most vehemently in Leibholz's defence was his friend and former colleague, Ernst Friesenhahn.[204] More trenchant still was his current colleague, Willi Geiger.

Within the Senate, Geiger circulated a withering dissent, coldly confuting each argument put forth in the decision excluding Leibholz. That decision, Geiger concluded, was not a legal judgment at all, but an act of political calculation. Geiger accused his colleagues—or rather, presumably, *four* of his colleagues—of seeing and seizing a chance to tilt the Senate's balance in their favour. They saw, further, a chance to silence their most powerful and persuasive opponent. The Senate had been deadlocked 4:4, Geiger suggested. At Leibholz's expense, those wishing to invalidate the party finance law now grasped a 4:3 majority. It was unworthy of them.[205]

Geiger's dissent was not published until many years later. But his view of the matter was widespread. The Second Senate conveyed an unflattering image of internal politicking. On 19 April, the justices took the bizarre step of reopening oral argument.[206] Ostensibly, there was new evidence to consider. But it also wouldn't have looked well to quash the law directly after excluding the judge who most obviously supported it. In any case, the Court conducted a second round of oral arguments, this one lasting three days.

The Bundestag sent an assembly of stars to defend the law. The CDU was represented by Ernst Benda, the future chief justice; the FDP by Thomas Dehler, the former justice minister. Both maintained that their parties couldn't balance their books without state subsidies. Both arguments fell on uninterested ears. The justices seemed aloof, even bored. They looked like men who had made up their minds.[207] The only argument that drew universal attention was offered by Adolf Arndt. Hessen, Arndt told an astonished chamber, no longer saw public funding per se as a violation of Article 21. His client opposed only the scope of the current outlays.[208]

Three months later, on 19 July 1966, the Second Senate announced its judgment.[209] Many, by this time, expected the Court to quash the law; but most were still surprised when it did. The Court began by conceding that the language of its 1958 and 1961 judgments[210] could be understood to mean that state funding of parties was constitutional.[211] But the Court now clarified that these statements focused on parties' role in preparing for *elections*.

[202] See, e.g., Laufer (n 113) 516–18; Werner Sarstedt, 'Anmerkung zum Beschluss des Bundesverfassungsgerichts vom 3. März 1966' (1966) 21 *JuristenZeitung* 314; Edward McWhinney, 'Federal Supreme Courts and Judicial Review' (1966) 45 *Canada Bar Rev* 593.

[203] See Wiegandt (n 199) 72–73; see also Christoph Link, 'Zum Tode von Gerhard Leibholz' (1983) 108 *Archiv des Öffentlichen Rechts* 159.

[204] Ernst Friesenhahn, 'Zur Ablehnung und Selbstablehnung eines Richters des BVerfG wegen Besorgnis der Befangenheit' (1966) 21 *JuristenZeitung* 704.

[205] See Geiger, *Abweichende Meinungen* (n 147) 123–41.

[206] See Friedrich Karl Fromme, 'Grundriß eines Staatsprozesses' *FAZ* (19 April 1966) Walter Schallies, 'Parteienfinanzierung—zweite Runde' *Süddeutsche Zeitung* (19 April 1966).

[207] See Laufer (n 113) 520.

[208] See Friedrich Karl Fromme, 'Halten die Sozialdemokraten Staatszuschüsse für verfassungswidrig oder nicht?' *FAZ* (20 April 1966) 4.

[209] 20 BVerfGE 56 (1966). This was in response to the complaint from the state of Hessen. The Court disposed of the complaints from the Bavarian, All German, and National Democratic parties the same day. See 20 BVerfGE 119 (1966); 20 BVerfGE 134 (1966).

[210] See, e.g., 8 BVerfGE 51, 63 (1958); 12 BVerfGE 276, 280 (1961).

[211] 20 BVerfGE 56 (n 209) 96–97.

Reimbursement for the reasonable costs of an election campaign might permissibly flow from public coffers.[212] But public expenditure for the 'entire operation' (*gesamte Tätigkeit*) of political parties was unconstitutional.[213] The state could not subsidize parties' day to day activity.

Why not? Here the Senate began with a conceptual distinction between the will of the people and the will of the state. The latter must derive from the former, not the former from the latter. The shaping of the people's will 'must remain fundamentally "state-free"'.[214] This principle, derived from Article 20's decree that 'all state authority stems from the people',[215] circumscribed Article 21's acknowledgement that parties participate in shaping the people's will. Article 21 recognized parties as constitutional 'institutions' and essential instruments of popular will-formation.[216] But parties were not 'supreme organs of the state', and the state was not bound, or even allowed, financially to maintain them.[217] If the state did maintain them through regular payments for their entire political operation, it would encroach impermissibly on the process of shaping public opinion and popular will.[218]

The Senate was unsympathetic to claims that parties couldn't survive without state funding. If this were strictly so for all parties, it meant the people had lost its will to maintain a robust democracy. A free democracy, the Senate noted, undertook the existential risk of entrusting the formation of political will to the moral judgement and civic engagement of its citizens.[219] Political will-formation was the business of the citizenry. Political parties who participate in that process must subsist on citizen donations.[220] That such donations might come from powerful interest groups was a risk the constitution was willing to run. Article 21 protected parties against state meddling but did not guarantee 'freedom from the influence of financially-powerful individuals, enterprises, or interests'.[221]

The state could, the Court repeated, compensate parties only for reasonable campaign expenses. But what counted as a campaign expense? And how much was reasonable? The Senate clarified that eligible expenses must be *directly* related to the campaign, which meant, among other things, chronologically close to the election.[222] Parties themselves, moreover, could not name the price. Parliament must fix it according to objective criteria—and it need not by any means reimburse the full costs of the campaign.[223] Reimbursement must respect the equality principle, which meant that parties outside Parliament couldn't be excluded entirely, but not that distinctions couldn't be made on the basis of parliamentary strength. These were matters for Parliament to decide. But Parliament emphatically must not fund parties generally for their entire operations.

F. Response to the second *Party Finance* judgment

The decision was devastating for the CDU and FDP. Both parties predicted sharp cutbacks in activity and staff; they spoke of the day of the judgment's announcement as 'Black Tuesday'. The SPD was less affected. Small parties were jubilant.

Publicly, CDU and FDP criticism was restrained.[224] Criticism in the academy and in the press was not. Of the major newspapers, only *Der Spiegel* and the *Süddeutsche Zeitung* were

[212] ibid 97. [213] ibid. [214] ibid 99. [215] Article 20(2) GG.

[216] It even cited an essay by Gerhard Leibholz on the position of parties in modern democracies. 20 BVerfGE 56 (n 209) 100. Gerhard Leibholz, 'Der Strukturwandel der modernen Demokratie' in *Strukturprobleme der Modernen Demokratie* 79–131 (CF Müller 1958).

[217] 20 BVerfGE 56 (n 209) 101. [218] ibid 102. [219] ibid 102–03. [220] ibid 103.

[221] ibid 105. [222] ibid 115. [223] ibid 116.

[224] See Arthur Rathke in *Rheinischer Merkur* (5 August 1966); *Süddeutsche Zeitung* (21 July 1966); *Süddeutsche Zeitung* (25 July 1966); *Süddeutsche Zeitung* (29 July 1966); see also Laufer (n 113) 532.

happy with the judgment's outcome. Even they were critical of its method. Though pleased with the Court's pragmatic results, Rudolf Augstein wrote that the judgment displayed 'the abundant power [*Machtfülle*] of this highest German Court', a Court whose judges 'not only interpret existing law' but 'make law themselves'. The justice made 'political decisions' not subject to scrutiny by any legal standard. In the *Party Finance* judgment, Augstein concluded, the justices appeared 'no more as supreme judges, but rather as supreme guardians making pragmatic decisions as a sort of council of elders with executive authority'.[225] *Die Zeit* was blunter still, entitling its report, 'A Judgment Without Logic'.[226]

For other critics, both academic and journalistic, the judgment's greatest failure was not its lack of logical rigour but its refusal to grapple with contemporary political realities.[227] For these critics, the Court's conceptual distinctions were tidy, facile, and anachronistic. Peter Häberle, a law professor, complained that the Court's distinction between the will of the people and the will of the state was merely the nineteenth-century division between society and state parading in new and ill-fitting garb.[228] In general, critics accused the Court of exaggerating the threat to parties from the state and underestimating the threat from powerful private interests. But here the methodological and substantive criticisms became confused. On the one hand, the Court allegedly deployed conceptual mystifications to cloak its wilful pragmatism; on the other hand, the Court's pragmatic result was the wrong one.

Almost immediately the parties proposed new legislation. The CDU mused about amending the constitution or compensating costs for the 1965 election. On 24 July 1967, after a year of haggling, the Bundestag passed an amended law on political parties. In general, the law codified much of the Court's party jurisprudence from the previous decade and a half. The reimbursement section, in particular, obeyed the prescriptions of the Court's July 1966 judgment. The law provided 2.5 DM per registered voter to be divided among all parties receiving at least 2.5 per cent of the second-place votes in the most recent national election. The law drew forth a flurry of constitutional complaints from small political parties. In a 3 December 1968 judgment the Court reviewed no fewer than fourteen of the law's funding provisions. The Second Senate sustained nearly all of them.[229]

The 1968 judgment, Friedrich Karl Fromme summarized in the *Frankfurter Allgemeine Zeitung*, consisted of 'a small no and a momentous yes'.[230] In fact there was more than one of each. The most important Yes affirmed the essence of the law—periodic subsidies to political parties for election campaigns. A six-justice majority upheld the scheme of lump sum payments on grounds of administrative ease.[231] The scope of the payments, which might total 99 million DM for the 1969 election, was also unobjectionable. It accorded with the political, economic, and social realities of a modern, mass-media election campaign.[232] This part of the judgment was unanimous.

The portion of the decision approving instalment payments to parties was more contested. After all, the core sentence of the 1966 judgment forbade 'state financing of parties

225 Rudolf Augstein, 'Wahlkampf unter Kontrolle' *Der Spiegel* (25 July 1966) 22.

226 'Ein Urteil ohne Logik' *Die Zeit* (22 July 1966) 1.

227 See, e.g., Rolf Zundel, 'Die wirklichkeitsfremden Bundesrichter' *Die Zeit* (29 July 1966) 5; Walter von La Roche, 'Anmerkungen zum Urteil über die Parteienfinanzierung' (1966) 17 *Politische Studien* 570; Peter Häberle, 'Unmittelbare staatliche Parteienfinanzierung unter dem Grundgesetz' (1967) 7 *Juristische Schulung* 64; Hans Jochen Hauβ, 'Parteifinanzen' (1966) 11 *Blätter für deutsche und international Politik* 664; Heinz Laufer, 'Zur staatlichen Finanzierung der politischen Parteien. Die Urteile des Bundesverfassungsgerichts vom 19. Juli 1966' (1966) 16 *Aus Politik und Zeitgeschichte* 1.

228 Häberle, 'Unmittelbare staatliche Parteienfinanzierung' (n 227) 64–74.

229 24 BVerfGE 300 (1968).

230 Friedrich Karl Fromme, 'Ein kleines Nein und ein gewichtiges Ja' *FAZ* (10 December 1968) 2.

231 24 BVerfGE 300 (n 231) 335.

232 ibid 339.

through annual or monthly payments for their entire political operation'.[233] Three of the justices thought periodic payments for campaign costs too closely resembled what the Senate had already condemned. But five of their colleagues felt differently. 'Installment payments are part of compensation for campaign costs', they wrote. 'They are justified by the necessities of the campaign'.[234] In a modern democracy, parties worked to shape political will even in non-election years. A successful campaign required long-term planning and complex machinery. In any case, it was impossible to fix the precise temporal beginning of an election campaign.[235] This was a decided shift, in tone at least, from the 1966 judgment. Then the Court had condemned regular state expenditures for parties' ongoing work of public 'education'; now the Court approved such expenditures for parties' endless electioneering. Perhaps, at least, there was a gain in transparency. Under the new regime, parties could give up the pretence of shaping political will and candidly solicit votes.

Of the new law's condemned provisions the most important was one limiting campaign funding to parties receiving at least 2.5 per cent of second-place votes in the most recent election. This hurdle, six justices held, was too high.[236] Parliament could permissibly work to prevent a proliferation of splinter parties, but the 5 per cent bar on parties entering Parliament already did much to achieve this goal. At the campaign level, the field must be more open. The Court suggested an alternative minimum of 0.5 per cent. In holding the 2.5 per cent barrier unconstitutional, the Court granted the petition of the NPD, which would now enjoy public funding. Some thought this sensational. 'Tax Money for Radicals', screamed the conservative *Christ und Welt*.[237] Tamer accounts stressed the funding of small parties as the judgment's major holding.[238]

Court insiders, however, saw the judgment's major innovation in its tacit revision of the 1966 decision. For Justice Leibholz, it was a personal vindication. He wrote to the SPD treasurer Alfred Nau that although the press's attention had been absorbed by the 'barrier clause' question as it related to the NPD, 'more important to me personally ... is that the decision revises in essential points the erroneous judgment [*Fehlurteil*] of 1966, and I cannot disguise a certain inner satisfaction'.[239] Wilhelm Hennis, on the other hand, warned Leibholz that the justice's enthusiasm rested on a roseate misreading of underlying realities.

> As I already felt nothing but uneasiness for the state financing of parties, the provisioning with public means of all splinter groups as well—which the new judgment makes inevitable—will not make me feel any easier. I think, revered Herr Leibholz, that you would see these things somewhat differently if, like me, you were active within a party. What will be done with this money has little to do with the intentions you associate with it. I make no secret of the fact that, all in all, the dependence of parties on financial groups seems to me more tolerable than this new-found form of financing, by which the party member has become even more uninteresting to the party leadership than before.[240]

Party finance would become an enduring—and enduringly controversial—theme of the Court's jurisprudence.[241] The power of parties has been a defining feature of the peculiar brand of democracy that has evolved in the Federal Republic. For better or worse, the Constitutional Court has played a decisive role in making the postwar party system what

233 20 BVerfGE 56, 102 (1966). 234 24 BVerfGE 300 (n 231) 348. 235 ibid 349–50.

236 ibid 340–42.

237 Johannes Gross, 'Steuergeld für die Radikalen' *Christ und Welt* (13 December 1968).

238 See, e.g., 'Karlsruher Entscheidung: Geld auch für kleine Parteien' *FAZ* (4 December 1968).

239 Gerhard Leibholz to Alfred Nau, 16 December 1968, BAK N 1334/33 (Nachlass Leibholz).

240 Wilhelm Hennis to Gerhard Leibholz, 13 January 1969, BAK N 1334/335 (Nachlass Leibholz).

241 See, e.g., 41 BVerfGE 399 (1976); 52 BVerfGE 63 (1979); 73 BVerfGE 40 (1986); 85 BVerfGE 264 (1992); 111 BVerfGE 54 (2004); 111 BVerfGE 382 (2004); 121 BVerfGE 108 (2008).

it is. Ironically, the Court owes some of its great popularity to public perceptions of the Court's distance from the parties it has empowered.

III. The State of Exception

Much had changed, in the country and in the Court, between the Court's *Party Finance* judgment of 1966 and its softening successor of 1968. The ill-starred chancellorship of Ludwig Erhard ended on 25 November 1966. Less than a week later, Kurt Georg Kiesinger (CDU), heading a 'grand coalition' of the Union parties and the SPD, became the Federal Republic's third chancellor.

Many found the regime change unsettling. On the one hand, it meant a season of governance without viable opposition—and this just when the idea of opposition had acquired lasting legitimacy. On the other hand, it recalled the broad coalitions of the Weimar era, with their internal bickering, partisan rigidity, and ultimate paralysis. The early months of the grand coalition coincided with the first economic recession of the postwar era, a further source of popular malaise. Later, the grand coalition would face fierce opposition from the most sustained and visible protest movement since the War.

A year after the grand coalition took office the Constitutional Court experienced its second major membership turnover of the decade. Earlier, in 1965, Werner Böhmer, a judge on the Federal Administrative Court, had replaced the retiring Karl Heck on the First Senate.[242] During the summer and autumn of 1967, there were four additional changes, two on each Senate. On the First Senate, Herbert Scholtissek and Hugo Berger were replaced, respectively, by Hans Brox, a civil law professor in Münster, and Wolfgang Zeidler, a judge and government lawyer in Hamburg.[243] On the Second Senate, Fabian von Schlabrendorff, a prominent lawyer from Hessen, replaced Julius Federer; and Walter Seuffert, an SPD member of the Bundestag, succeeded Friedrich Wilhelm Wagner as the Court's vice president.[244] The following year Hans-Justus Rinck, a judge in the Federal Court of Justice, also joined the Second Senate, replacing Anton Henneka.[245] Seuffert and Zeidler were nominated by the SPD; Schlabrendorff, Rinck, and Brox by the CDU.

The changes marked a generational transition within the Court. The average age of the six outgoing judges was 67. A writer for *Christ und Welt* had spoken unfeelingly of 'Methuselahs in red robes'.[246] The average age of the incoming justices was 51. By and large, the *emeriti* had been teenagers during the Wilhelmine Empire and had come of political age during the Weimar Republic. The newcomers, by contrast, were members of a generation decimated by the War. The odds of survival for German men born between 1907 and 1924 were gruesomely low.

[242] On Böhmer, see 'Bundesverfassungsrichter a.D. Prof. Dr. Werner Böhmer 90 Jahre' (2005) 24 *Neue Zeitschrift für Verwaltungsrecht* 547.

[243] On Brox, see Bernd Rüthers, 'Hans Brox †' (2009) 53 *NJW* 2106; on Zeidler, see Roman Herzog, 'Wolfgang Zeidler †' (1988) 42 *NJW* 881.

[244] On Schlabrendorff, see Richard Ley, 'Fabian von Schlabrendorff †' (1980) 34 *NJW* 2565; on Seuffert, see Richard Ley, 'Walter Seuffert †' (1990) 44 *NJW* 891.

[245] Rinck had formerly been a law clerk to Justice Leibholz and was the co-author of Leibholz's influential textbook examining the Basic Law in the light of the Court's jurisprudence. Leibholz believed that Walter Seuffert, the Court's vice president, had tried to block Rinck's appointment. Gerhard Leibholz to Adolf Arndt, 4 October 1968, Friedrich-Ebert-Stiftung (Bonn) (Nachlass Adolf Arndt, Box 8, Mappe 21). Some objected to Rinck on the grounds that he was a former clerk; Leibholz thought this unfair, since no one had similarly objected to Zeidler. Gerhard Leibholz to Adolf Arndt, 14 October 1968, Friedrich-Ebert-Stiftung (Bonn) (Nachlass Adolf Arndt, Box 8, Mappe 21).

[246] Peter Jochen Winters, 'Methusalem in roter Robe' *Christ und Welt* (26 May 1967) 4.

Three of the new appointees, Seuffert, Schlabrendorff, and Böhmer, continued the Court's anti-Nazi tradition. Böhmer was expelled from law school on political grounds in 1936. Seuffert was arrested and imprisoned on suspicion of treason in 1940. Schlabrendorff, a reserve officer of noble Prussian stock, took part in two attempts to assassinate Hitler. In March 1943, he and a friend, Henning von Tresckow, planted a bomb on the Führer's plane. It failed to detonate. The following year Schlabrendorff was part of the famous failed coup of 20 July 1944. A people's court (*Volksgerichtshof*) acquitted Schlabrendorff, but Hitler personally revoked the acquittal and ordered Schlabrendorff's execution. The order was never carried out, but Schlabrendorff spent the rest of the War in a concentration camp. After the War, he wrote a popular memoir—*Officers against Hitler*[247]—and rose to distinction as one of the finest lawyers in the Federal Republic. Contemporaries saw Schlabrendorff's appointment as a partial payment of the nation's debt to the 'men of 20 July'.

If so, the payment was well timed. The previous year, 1966, Justice Willi Geiger had served as president of the Catholic Church congress (*Kirchentag*), in which capacity Geiger claimed, or allowed others to claim on his behalf, that after the war a jury court ruled that he had been 'a passionate opponent of National Socialism'. The suggestion outraged the Left Liberal press. The *Süddeutsche Zeitung* published explosive excerpts from Geiger's book on the Nazi press law.[248] The socialist organ *Vorwärts* published a lengthy exposé entitled 'Geiger's Brown Vest'.[249] This renewed attention to Geiger's past drew protests from abroad as well. Ernst Linz, a lawyer in Israel, sent an agitated letter to Gebhard Müller. Müller's response—dispassionate, reassuring, exculpatory—was answered within the Court by a scalding rebuttal from Gregor Geller, Geiger's colleague on the Second Senate.[250]

Geiger himself kept a low profile throughout the affair, but the criticism took its toll. He was grateful for an occasional note of support. One such note came from a surprising source. On 26 September 1966, Geiger wrote to Adolf Arndt:

> Your encouragement has heartened me to carry on my work. Don't expect me to explain, justify, or excuse what I wrote thirty years ago; it was simply wrong. But perhaps I may say to you that I realized this error earlier than my critics, and that through my public activity and in my vocation I have tried to make this clear. I cannot get over the fact that 'friends' and 'colleagues' who have been in the know for a long time have taken part in this campaign.[251]

The affair soon blew over, and Geiger remained in the Court for eleven more years. But perhaps the judicial selection committee hoped, against the backdrop of the Geiger affair, that the triple appointment of Böhmer, Seuffert, and Schlabrendorff would remind the country of a very different tradition among the Court's personnel.

A. The 'grand coalition' and the student revolts

This transition in the membership of the Court came on the eve of convulsions in the country at large. The grand coalition's macroeconomic steering helped ease the country out of recession, but by 1967/68 a new generation was coming of age, for which the Federal Republic's economic vitality was an inadequate cover for its moral bankruptcy. Young

[247] Fabian von Schlabrendorff, *Offiziere gegen Hitler* (Fischer Bücherei 1946).
[248] 'Das Abgründige in Professor Geiger' *Süddeutsche Zeitung* (18 July 1966) 3.
[249] Hellmuth Rieber, 'Des Geigers braune Weste' *Vorwärts* (27 July 1966).
[250] The entire correspondence is reproduced in Klaus-Detlev Godau-Schüttke, *Der Bundesgerichtshof: Justiz in Deutschland* (Tischler 2005) 361–81.
[251] Willi Geiger to Adolf Arndt, 26 September 1966, Friedrich-Ebert-Stiftung (Bonn) (Nachlass Adolf Arndt, Box 6, Mappe 16).

West Germans, especially university students, were angered by America's escalating war in Vietnam and sickened by their own country's failure to reform systematically after the catastrophe of the Third Reich. It didn't help that Kurt Georg Kiesinger was the first chancellor who had been a member of the NSDAP. Young protesters, and the Left-leaning academics who applauded them, embraced a 'shift in values' (*Wertewandel*) that had been taking root quietly for many years. Traditional morés—especially sexual morés—were dismissed as 'secondary virtues' (*Sekundartugenden*). In their sweeping rejection of the German past, young rebels landed on the ingenious notion that glandular indulgence was somehow an anti-Fascist gesture. Against the fusty canons of their parents, students trumpeted an ethic that was profoundly social and intensely political.

But the young people wanted nothing to do with the reigning political parties or the ruling political class. Just when the Federal Republic seemed to have embraced the legitimacy of formal political opposition, the protesters denounced the prevailing opposition as inauthentic—and, under the grand coalition, non-existent. The protest movements christened themselves an 'extra-parliamentary opposition' (*Außerparlamentarische Opposition*, or APO). They rejected all existing parties and challenged, *tout court*, the legitimacy of the West German state. For the first time, the political culture of the Federal Republic had an entrenched counter-culture. The APO's organization was loose, but its anger was focused. Internationally, its ire was aimed at the war in Vietnam; domestically, its wrath descended on legislative proposals, long in the workings, for an emergency constitution.

B. The Emergency Laws

The so-called 'Emergency Laws' (*Notstandsgesetze*) included a constitutional amendment that empowered the executive branch, in times of national crisis, to abridge basic rights. The law had a long history. The Basic Law's framers, remembering the authoritarian abuse of executive decrees under Article 48 of the Weimar constitution, pointedly omitted emergency provisions. During the immediate postwar era, the Federal Republic's Allied occupiers took strong measures to prevent crises. Among other things, the Allies exercised wide-ranging surveillance over telecommunications and the postal service. The Allies reserved the right to continue these activities even after the restoration of (West) German sovereignty, which began in 1955. But the Allies promised to stop doing so as soon as German legislation empowered local officials, in case of emergency, to perform these same functions.

Early proposals for an emergency statute were serial failures—in 1958, in 1960, and again in 1963. Only under the grand coalition did the government command the two-thirds Bundestag majority needed to secure the legislation through constitutional amendment. The emergency amendment, ratified 30 May 1968, became the new and expanded Article 10(2) GG. Together with the implementing legislation passed in August, the amendment empowered state officials, when necessary to combat foreign or domestic enemies, to conduct telephone wiretaps and to interfere with postal correspondence. The amendment transferred review of such surveillance from the courts to special auxiliary agencies appointed by Parliament. Under the implementing law, three-member advisory commissions would review the admissibility and necessity of surveillance measures.

The laws were passed amid fiery popular opposition.[252] Student protesters, recently roused by the shooting and wounding of APO icon Rudi Dutschke, took to the street by the

[252] For an account of popular opposition to the law, see Boris Spernol, *Notstand der Demokratie: der Protest gegen die Notstandsgesetze und die Frage der NS-Vergangenheit* (Klartext 2008).

thousands—in (mostly) peaceful demonstrations, hunger strikes, and, on 11 May 1968, a march on the capital itself. Students tarred the law with a non-subtle signature: they called it the 'NS-Law'—'NS' evoking not *Notstand* (state of emergency), but National Socialism. The law, student activists shrieked, was a second enabling law, thirty-five years after the fact.

The students were not alone in their hostility. Two years earlier Karl Jaspers warned that the proposed legislation would pave the path from party oligarchy to dictatorship.[253] In late May 1968, Jürgen Habermas and Theodor Adorno spearheaded an action committee called 'Democracy in Emergency', calling on writers, artists, and academics to unite against the law.[254] Günter Dürig denounced the law as 'an Orwellian experiment'.[255]

Particularly troubling for many opponents of the emergency laws was the near-unanimity with which they were passed. The laws sailed through Parliament behind the strength of the Union/SPD grand coalition over only a handful of abstentions and the ineffectual opposition of the FDP. There was not even an opposition party with enough votes to challenge the amendment before the Constitutional Court. There was one consolation, however: Georg August Zinn was still minister president of Hessen. On 29 September 1969, Zinn's government filed a constitutional complaint challenging the amendment and the implementing statutes. It was Zinn's last appeal to Karlsruhe; his twenty-year premiership was over before the Court announced its judgment.

Hessen, later joined by Bremen and by groups of judges and lawyers in Mannheim and Frankfurt, argued that that the constitutional amendment was itself unconstitutional—that it violated Article 1 GG's guarantee of human dignity and offended Article 20's entrenchment of the *Rechtsstaat* principle, thereby infringing Article 79's 'eternity clause', which shielded Articles 1 and 20 against abridgement or amendment. The emergency laws, Hessen contended, would subject individuals to state surveillance without the possibility of defending their privacy in a court of law. This would make the individual a mere instrument of state authority—a textbook affront to the Kantian principle of human dignity. Further, by stripping courts of their powers of review, the amendment had undermined the separation of powers—a cardinal principle of the *Rechtsstaat*. In a supplementary *Gutachten*, basically an amicus brief, Günter Dürig, whose earlier writings profoundly influenced the Court's fundamental rights doctrine, thundered that the dual denial of notification and judicial review enabled state invasion of citizen privacy to proceed in absolute secrecy—an assault on the very core of the rule of law.

The Court didn't decide the case until 15 December 1970.[256] Though the Second Senate had been announcing its vote tallies for a few years, and though the *Spiegel* judgment had articulated the views of dissenting justices, the *Emergency Laws* judgment was one of the first major judgments since dissents were formally authorized in 1970 to include a separate dissent. And what a forceful dissent it was.

The judgment of the Court sustained the emergency laws by a vote of 5:3 in circumstances very different from those under which Parliament had passed them. The grand coalition had been replaced by a Social-Liberal (SPD-FDP) coalition led by Willy Brandt, Germany's first socialist chancellor since Weimar. The student movement had fizzled; the SDS, the major Leftist student union, had disbanded. The judgment was not nearly as explosive as the amendment itself. Even so, the judgment's friends were few, its enemies loud and legion.

[253] Jaspers (n 187) 174.
[254] See 'Professoren, Künstler und Schriftsteller protestieren' *FAZ* (27 May 1968) 1.
[255] Günter Dürig, 'Ein Orwellsches Experiment' (1968) 1 *Zeitschrift für Rechtspolitik* 11.
[256] 30 BVerfGE 1 (1970).

The majority judgment—joined by Justices Seuffert, Leibholz, Kutscher, Geiger, and Rinck—began with a lengthy exegetical excursus. The constitutionality of the new Article 10(2) GG, the majority wrote, depended on the meanings of dignity, *Rechtsstaat*, and the eternity clause, and on their relationship one to another. But none of these clauses could be understood in isolation. They were part of a broader constitutional system that embodied conscious decisions by the framers. Two of these decisions were of paramount importance in the current case. First, the Federal Republic was a militant democracy in which fundamental rights could not cloak assaults on the constitutional order.[257] Second, fundamental rights could be limited to benefit the community and secure superior legal values.[258]

Informed by these basic decisions, the majority read narrowly the provisions the complainants invoked. Article 79 did not ban all amendments that in any way limited the guarantees of Articles 1 and 20, only amendments that affected their essential core.[259] Nothing in Article 20's *Rechtsstaat* principle demanded that every citizen must, under all circumstances, enjoy the amplest possible judicial process, nor did the principle foreclose limitations on Article 19(4)'s guarantee of access to the legal process.[260] The separation of powers principle required a check on executive authority, but in exceptional cases that check could come from independent bodies appointed by Parliament or erected within the executive itself.[261] Perhaps most crucially, Article 1's exaltation of human dignity was not violated every time an individual was made an instrument of the state, but only when the state's treatment of an individual fundamentally disparaged his dignity or denied his personhood.[262] This sounded very nearly like an intentionality requirement. The state, apparently, could derogate a citizen's dignity only if it tried to.

Even by this cramped conception of dignity, the majority found that the challenged law violated dignity in part. But it was only a small part. The majority held that law could not constitutionally deny notice once it became clear that notice would not threaten the purpose of the surveillance. This was a fuzzy restriction, hard to enforce. For the rest, the Senate majority rejected the complaints across the board. The amendment and its implementing law did not essentially limit or flatly abolish constitutional principles made sacrosanct by Article 79. The laws represented, rather, 'system-immanent modifications of general constitutional principles' that otherwise remained in full force.[263] By 'system immanent' the majority meant, presumably, that the amendment altered the system in order to strengthen it, and did so on the system's own terms. But it was a woolly concept—a metaphysical mystification in which critics discerned an ominous precedent.

Among the sharpest critics were the Senate's three dissenters, Justices Schlabrendorff, Geller, and Rupp.[264] In the dissenters' view, the majority had begged the question from the outset, interpreting Article 10(2) according to general norms of constitutional construction when the question at hand was whether Article 10(2) *was* a valid norm that *could* be so construed.[265] What's more, the majority's fixation with constitutional interpretation led it to ignore legislative history—with a resultantly Pollyannaish view of the frequency and scope of the law's probable application.[266]

The majority's gravest sin, though, was its cramped conception of dignity, which seemed to ban little more than 'the reintroduction of torture, pillories, and the methods of the Third Reich'.[267] Dignity, for the dissenters, meant much more. It prevented the state from treating

257 ibid 19. 258 ibid 20. 259 ibid 24. 260 ibid 25. 261 ibid 28. 262 ibid 26.
263 ibid 29.
264 The dissent was published on 4 January 1971, twenty days after the majority opinion was announced.
265 30 BVerfGE 1 (n 256) 34–35. 266 ibid 35–37. 267 ibid 39.

individuals 'un-personally', not merely from treating them with contempt.[268] At the very least it meant, in the context of the *Rechtsstaat*, that an individual must be allowed to participate in his own trial.[269] The true state of exception, the dissent concluded, belonged to wartime, not to seasons of 'quotidian law' (*juristischer Alltag*). Militant democracy meant both more and less than *raison d'état*.[270] The eternity clause was 'meant to be interpreted and applied strictly and unbendingly. It is designed, not least, to defend against *beginnings*'.[271]

Popular and academic critics picked up where the dissenters left off. Günter Dürig had warned that if the Court sustained Article 10(2), Article 79 would become a dead letter. There would remain no restraint at all on the authority of legislative supermajorities to amend the constitution.[272] Peter Häberle thought Dürig's advocacy in the case was marked by a certain 'Mosaic wrath'. Häberle himself condemned the judgment as a rare but not oxy-moronic instance of 'unconstitutional constitutional jurisprudence' (*verfassungswidrige Verfassungsrechtsprechung*).[273] Critics censured the majority for failing to construe Article 79 as it would other constitutional norms—for trivializing the eternity clause and emasculating its impact. The purpose of Article 79, many critics noted, was to prevent the first blows to the free and democratic edifice, not simply to ensure, in the last instance, that the structure did not come tumbling down.

These criticisms appeared in the press as well. *Der Spiegel* criticized the Court for betraying principles enshrined in two decades of the Court's own doctrine. The majority had made an exception, the magazine complained, to principles it had long declared exception-less. The Court's blessing on this first 'system-immanent modification' of the constitutional order (whatever that meant) was sure to inspire others.[274] For many contemporaries the judgment signalled a setback for the rule of law and the retreat of the Constitutional Court into a posture of judicial restraint. Who, many wondered, if not the Constitutional Court, could restrain the constitution-making impulses of a future 'grand coalition'? Court watchers fretted that the Court's capitulation would embolden future legislatures and undermine the Court's authority. In the event, fears that the Court would retreat into a shell of deference and restraint proved colossally unfounded.

Conclusion

Reactions to the Court's *Emergency Laws* decision say much about popular perceptions and expectations of the Court. From the European Defence Community controversy to the *Television* judgment to the *Spiegel* Affair, the Court had done much to advance the legitimacy of Opposition. By the time of the *Emergency* judgment, many looked to the Court for opposition itself. The Court failed to oblige, and many were bitterly disappointed. The manner of the failure also stung. The Court had not upheld the laws out of deference to a prodigious parliamentary majority and in acknowledgement of the impossibly complex and agonizing interplay between the Federal Republic's loftiest constitutional abstractions and its most wrenching existential threats. The Court had announced, rather, that the granite pillars of the Basic Law were more flexible than one had supposed. The majority treated the case as a straightforward matter of (narrow) interpretation. They acted, or at least wrote, as though the case were easy.

[268] ibid 40. [269] ibid 45. [270] ibid. [271] ibid 47 (emphasis added).

[272] In Peter Häberle, 'Die Abhörentscheidung des Bundesverfassungsgerichts vom 15.12.1970' (1971) 24 *Juristen Zeitung* 145, 145.

[273] ibid 156. [274] 'Erst Sinnvoll' *Der Spiegel* (21 December 1970) 26.

That the judgment's critics shouted themselves hoarse complaining that the Court had done too little says much about the Court's increasing stature and the shifting political tides in the Federal Republic. These critics echoed a compulsive theme of the decade's public law scholarship: the disparity between the constitutional letter of the Basic Law and the political realities of the Federal Republic—the gap between 'constitution and constitutional reality', the lacunae left by the 'unfulfilled *Grundgesetz*'.[275] Some praised the Court for helping, others chastened it for failing, to close the chasm. Still others deemed the ostensible problem exaggerated and found the Court's intrusions to fix it excessive. With regard to the Court's fundamental rights jurisprudence, the Carl Schmitt school continued to rail against the 'tyranny of values'—the Court's expansive and vainglorious interpretation.[276] Even scholars on the other side of the political and methodological divide thought the Court had gone too far. In his 1968 inaugural lecture in Freiburg, Wilhelm Hennis embraced Horst Ehmke's 1961 critique of the Court's 'extensive fundamental rights interpretation'.[277] Hennis thought commentators were asking too much of the constitution—and that the Court was doing too much in response. But the loudest, most public, most sustained critique of the Court's overreaching came from a very different, highly unlikely source.

Though he had studied law, Hans Dichgans was more an industrialist than a jurist. His business was not the parsing of legal texts but the management of iron and steel.[278] In 1961, Dichgans joined the Bundestag—as a CDU candidate from North-Rhine-Westphalia. In Parliament, Dichgans became convinced that the landscape of German democracy fared ill. The problem, he determined, was systemic; its roots constitutional. The Basic Law itself required wholesale reformation—and its judicial guardians needed taming.[279] Genuinely alarmed, the steel manager took up his pen to propound these theses in a restless cataract of articles and a pair of strenuous books.[280] Dichgans chaired a parliamentary committee that considered calling a national assembly charged with full-scale constitutional revision. Martin Kriele, a law professor in Cologne, warned Dichgans that such an assembly would unite an omnigatherum of 'fantastists and utopians'.[281] Many agreed, and the committee's labours led nowhere. More enduring was Dichgans' critique of the Court, which found a rumbling echo in the decade to come.

What, in Dichgans' view, was wrong with Karlsruhe? The framers of the Basic Law, he wrote, trusting voters too little, had trusted judges too much. Wherever possible, the framers had made judges overseers of politicians. They created an 'over-dimensioned constitutional court'. The justices had too much authority—and there were too many of them. The Court's unwieldy structure—sixteen judges spread across two chambers—promoted a bureaucratic

[275] See, e.g., Adolf Arndt, *Das Nicht erfüllte Grundgesetz. Ein Vortrag* (Tübingen: Mohr 1960). A periodical launched in Cologne in 1966 carried the title 'Verfassung und Verfassungswirklichkeit'.

[276] Exemplary was Ernst Forsthoff, *Zur Problematik der Verfassungsauslegung* (Kohlhammer 1961).

[277] Horst Ehmke, *Wirtschaft und Verfassung* (CF Müller 1961); Wilhelm Hennis, *Verfassung und Verfassungswirklichkeit. Ein deutsches Problem* (JCB Mohr 1968) 20–21.

[278] Dichgans, a leading expert on price formation and rationing from the mid-1930s onward, was a key figure in the wartime management of the steel economy.

[279] Dichgans had complained directly to some justices of the Court's political interference as early as 1963. See Hans Dichgans to Gerhard Leibholz, 14 June 1963, BAK N 1334/3 (Nachlass Leibholz).

[280] The books were Hans Dichgans, *Das Unbehagen in der Bundesrepublik: Ist die Demokratie am Ende?* (Econ Verlag 1968); and *Vom Grundgesetz zur Verfassung. Überlegungen zu einer Gesamtrevision* (Econ Verlag 1970). For representative articles, see Hans Dichgans, 'Eine verfassunggebende Nationalversammlung' (1968) 1 *Zeitschrift für Rechtspolitik* 61; Hans Dichgans, 'Möglichkeit und Grenzen der Verfassungsgerichtsbarkeit' *Politisch-soziale Korrespondenz* (1 February 1969); Hans Dichgans, 'Zu viele Verfassungsrichter in Karlsruhe' *Christ und Welt* (10 January 1968); Hans Dichgans, 'Die Verfassungsgerichtsbarkeit' *Deutsche Tagesposte* (28 February 1969); Hans Dichgans, 'Parlament und Verfassungsgericht' *Das Parlament* (30 August 1969).

[281] Martin Kriele to Hans Dichgans, 22 July 1968, BAK N 1281/12 (Nachlass Dichgans).

'love of detail'. It helped breed a perfectionism that made justice imperfect and slow. It produced opinions in which the Court acted as a micro-managerial supra-Parliament, delving into the wonkish minutia of politically charged cases (the *Party Finance* sequence was his especial *bête noir*) and delivering its conclusions to the Bundestag with the unassailable fiat of constitutional command. It was the topmost pinnacle of arrogance, Dichgans fumed, for the Court to presume it could do such detailed policy work better than Parliament.[282]

Dichgans saw in the justices a mongrel band of failed politicians who sought to run the world from Karlsruhe and robed professors who never bothered to ask whether their judgments were practicable. What's more, the judgments were badly written. Dichgans compared the Court's 'flimsy, bureaucratic' prose unfavourably with the 'manly resolution' evident in the language of the United States Supreme Court. In the Federal Republic, one heard the monotonous anonymity of a legislative super-committee; in America one heard 'the powerful personality of the individual judge'. Dichgans longed for the thunderous cadences of Louis Brandeis, the urbane wisdom of Felix Frankfurter, the acidulated epigrams of Oliver Wendell Holmes. This wasn't just a matter of style. Dichgans wanted the justices to be statesmen, to possess 'an authority which separates this Court decisively, in its function and in the consciousness of the citizenry, from the rest of the judiciary'. Dichgans favoured a single chamber of seven judges, an American-style system of *certiorari*, a more modest jurisdiction, and justices genuinely committed to the principle of restraint.[283]

Some justices assured Dichgans that they were already so committed. Walter Seuffert, the Court's vice president, sent Dichgans an essay he had published in the *Neue Juristische Wochenschrift*, a leading legal weekly, on the 'Delimitation of the Activity of the Constitutional Court vis-à-vis the Legislature and the Judiciary'.[284] Fair words, Dichgans shot back, but contradicted by events. 'I heartily congratulate you on your beautiful essay', Dichgans wrote, the whole of which 'one can only underscore. But the number of cases in which the Federal Constitutional Court—with a majority, moreover, of 5 to 3—has violently overstepped the boundaries you have drawn, has become too great. The political branches will have to occupy themselves therewith'.[285] Dichgans did his best to see that they did. Within Parliament, Dichgans' cries of censure and calls for reform were answered by Claus Arndt (SPD), the thirty-one-year-old son of Adolf Arndt. Dichgans circulated copies of his own response to the younger Arndt among parliamentary colleagues, but met with an ambivalent response. A note from Helmut Schmidt, who as chancellor would echo and expand Dichgans' critique, was typical. The Court's occasional adventurism, Schmidt noted, had to be balanced against its genuine achievements. 'There can be no doubt', Schmidt wrote, 'that, in principle, a constitutional court runs the risk of encroaching in the political realm of the legislator. On the other hand, the Court has decisively contributed to the fact that, from the state's side and from the citizens', the constitution is regarded and followed in its full extent'. As just one example of the Court's fulfilment of this necessary role, Schmidt cited the Court's recent judgment on the legal standing of children born out of wedlock—a judgment that pressed the legislature, at long last, to honour the constitution's command that non-marital children receive equal treatment under the law.[286]

282 Dichgans was the product of a Nazi-era glorification of 'economic self-administration' (*wirtschaftliche Selbstverwaltung*) within the corporatist undergrowth of German industry. The watchwords of this bureaucratic cosmos were delegation on the grounds of complexity, local knowledge, and technical expertise.

283 See Hans Dichgans, 'Zu viele Verfassungsrichter in Karlsruhe' *Christ und Welt* (10 January 1969) 3.

284 Walter Seuffert, 'Die Abgrenzung der Tätigkeit des Bundesverfassungsgerichts gegenüber der Gesetzgebung und der Rechtsprechung' (1969) 23 *NJW* 1369.

285 Hans Dichgans to Walter Seuffert, 13 August 1969, BAK N 1281/13 (Nachlass Dichgans).

286 See 25 BVerfGE 167 (1969).

Tellingly, the judgment Schmidt cited was a fundamental rights case. Both in the latter 1960s and in decades to come, much was forgiven the Court for the sake of its rights jurisprudence. That jurisprudence dominated the Court's docket (numerically at least), shaped public perceptions of the Court, and animated the Court's self-understanding. The Court saw itself as a fundamental rights tribunal, and the justices worked hard to promote this perception among the public at large. Throughout Gebhard Müller's tenure as chief justice, individual judges made fundamental rights the theme of extra-judicial speeches and essays, many of them reported or reprinted in newspapers.[287] Commentators on the Court's fifteenth anniversary in 1966 observed that the Court's work as guardian of basic rights had penetrated popular consciousness.[288] Though the justices and many others continued to complain that the Court's workload was unmanageable,[289] the culprit of the Court's vast docket—the right of individual constitutional complaint—was popularly held sacrosanct. A 1969 constitutional amendment enshrined the right to individual complaint, hitherto a statutory right, in the Basic Law itself. It had long since, by some accounts, been enshrined in the hearts of the people—the great emblem of Everyman's right to legal vindication, the Court's greatest symbol of popular legitimacy.[290] The Court was flooded each year by thousands of individual complaints. Most were dismissed out of hand by screening quorums of three judges each. A few hundred went to a full Senate for review.

Only rarely did these cases make headlines. But their very existence served as a counterweight to the politically charged judgments that did. In its quiet stream of fundamental rights decisions, the Court undertook the unromantic but indispensable business of guarding the rights of ordinary Germans. Political quietude was a badge of judicial restraint. The exceptions were seen as exceptions, and in the two most arresting exceptions—the *Television* and *Party Finance* judgments—the Court emerged as a forceful restraint on state authority (though in the *Party Finance* case, critics complained the Court had limited state parties only to empower interest groups). These judgments confirmed that the Court would not shy away from politically charged controversies, but their rarity suggested that neither would the Court seek them out.

In the aftermath of the *Television* judgment, the Court conducted itself with dignity and restraint—an Olympian aloofness from the partisan fray. Their conduct during the *Party Finance* deliberations—in particular the extraordinary exclusion of Justice Leibholz—was less fortunate in the public's eyes. The decisions most disappointing to contemporaries were the *Spiegel* and *Emergency Laws* judgments. Critics sniped that when the Court's role as vindicator of fundamental rights really counted, the justices lost their nerve. On the other

[287] See, e.g., 'Menschenrechte—heute aktuell' *Badische Neueste Nachrichten* (3 December 1964) (reporting a speech by Friedrich Wilhelm Wagner); Gebhard Müller, 'Die Würde des Menschen ist unantastbar' *Der Tagesspiegel* (3 January 1965); Walter Schallies, 'Das Menschenbild des Grundgesetzes' *Süddeutsche Zeitung* (9 Nov 1965) (reporting, u.a., speech by Karl Heck); Senta Ulitz-Weber, '"Bürger Jedermann" beschwert sich in Karlsruhe' *Badische Neueste Nachrichten* (12 January 1968) 8 (reporting a speech by Justice Wiltraud Rupp-von Brünneck).

[288] See, e.g., 'Das Verfassungsgericht besteht 15 Jahre' *Die Welt* (27 Sep 1966) 3; 'Fünfzehn Jahre Bundesverfassungsgericht' *Bulletin des Presse- und Informationsamtes der Bundesregierung* (7 September 1966) 927.

[289] See, e.g., 'Verfassungsgericht überlastet' *Frankfurter Rundschau* (29 December 1961); '"Verfassungsbeschwerden" nehmen überhand' *FAZ* (20 February 1963) 6; 'Die Karlsruher Richter Überlastet' *Stuttgarter Zeitung* (20 April 1963); HJ Rick, 'Überlastete Richter' *Echo der Zeit* (5 May 1963); 'Bundesverfassungsgericht: zu spät' *Der Spiegel* (17 June 1968) 52; '801 Verfahren sind anhängig geblieben. Aus der Arbeit des Bundesverfassungsgerichts' *FAZ* (30 January 1969).

[290] See 'Zu Füβen des Bundesverfassungsgerichts' *Badische Neueste Nachrichten* (17 July 1965); 'Beschwerdeführer sind willkommen' *Stuttgarter Nachrichten* (29 March 1967) 16; 'Das Bundesverfassungsgericht steht jedermann offen' *Abendzeitung* (Baden-Württemberg 29 March 1967); Senta Ulitz-Weber, '"Bürger Jedermann" beschwert sich in Karlsruhe' *Badische Neueste Nachrichten* (12 January 1968).

hand, these were the first major judgments with published dissents, both of which were written with muted moral anger and in tart, incisive prose. *Some* justices were speaking truth to power—including the power of their peers. Still, some observers worried the Court would be bold only when boldness didn't hurt. But parties of all stripes continued to take their causes to Karlsruhe, and on the whole the Court's written judgments suggested that their pleas had at least been understood. Following the 1958 *Pharmacy* judgment, the Court continued to decide fundamental rights cases by employing a proportionality test that required the justices to balance the comparative weight of competing constitutional values. The very nature of the test required the Court to articulate clearly what was at stake on both sides.

The requirement that justices be appointed by a two-thirds parliamentary super-majority fostered the promotion of centrist judges. The unavailability, before 1970, of signed dissents created the appearance, if not the reality, of consensus. To many minds, the Court was an organ of integration, an instrument of socio-political coherence. This was no small advantage for the Court so long as the political culture of the Federal Republic was haunted by fears of instability. But as the revolution in values progressed, many looked to the Court for something more than integration and stability—something more than a conservative bulwark against democracy's excesses and a liberal rampart against its foes. A new breed of progressive rebel wanted the Court to close the country's deficits of legitimacy and moral leadership. The Court supplied such vision for a jubilant moment in the *Television* judgment, but failed to repeat the performance in the *Emergency Laws* decision. The latter decision, with its withering triumvirate dissent, also exposed fissures within the Court. The veneer of unanimity was gone for good, and in years to come rhetorical jousting between majority and dissenters would drive both groups to stronger stances. Such internal disputes were occasioned by (and they exacerbated) an even more tremendous conflict between the governing coalition under a new chancellor and the Court under a new chief justice. Both sides of this broader confrontation were nothing if not bold. The resulting clash of visions was tremendous. For the first time, the Court was assailed as a hindrance to democracy.

But as the Court concluded its second decade, all of this was but dimly foreseen. Throughout the 1960s, criticism of the Court was sporadic, and the Court was more often criticized for being overly political than for being insufficiently democratic. Some even made the case that the Court was quintessentially democratic. In 1968, Heinz Laufer, a political scientist, published the first systematic study of the Court's role in the political process. Laufer argued that, far from being undemocratic, the Court was an essential element of the state's democratic nature—so much so that a two-thirds parliamentary majority could not constitutionally abolish the Court by constitutional amendment.[291] In Laufer's view, robust constitutional justice was not a limit on the second German democracy, but a crucial constitutive element. Not only did the Court keep the cogs of democracy churning; the values the Court protected made the Federal Republic a democracy in a substantive sense. Democracy, in this view, was more than a system of governance rooted in broad-based elections. It was an objective system of values centred in the sacrosanctity of human dignity and the muscular protection of individual rights. It was the free democratic order defined and defended by the Constitutional Court.

Reviewing Laufer's book for *Die Zeit*, Dieter Grimm recalled Ernst Forsthoff's contention that, in Bagehotian terms, the Court represented an aristocratic rather than a democratic element of the state.[292] In most cases, Grimm argued—and especially in political cases—the

[291] Laufer (n 113).
[292] Dieter Grimm, 'Wie mächtig ist Karlsruhe? (Rezension)' *Die Zeit* (23 May 1969) 63.

justices possessed considerable, and uncontrollable, room for manoeuvre. The discretionary exercise of such power by a handful of appointed jurists was hardly the summit of democratic governance, traditionally understood. Laufer's insistence that constitutional justice was of the essence of German democracy rested on a particular vision of that democracy—a vision that was impeccably liberal, thick with values, and in a deep sense conservative. The broad diffusion of that vision in years to come represents the Court's profoundest impact on the history of the Bonn Republic. During the 1970s, that vision was tested as never before.

3

Confrontation, 1971–1982

We wish to dare more democracy.... We do not stand at the end of our democracy; we have only now truly begun.

—Willy Brandt (1969)[1]

On the theme of 'Democracy' I wish only to say: so long as we are not incessantly aware of the difference between liberalism and democracy, we will argue endlessly, drift into 'eternal discussions' .... The reality of political development tends obviously toward the 'end' that either democracy destroys liberalism or liberalism destroys democracy—or shall we say 'overrules'.

—Carl Schmitt to Wilhelm Hennis (1971)[2]

I fear that we stand at the threshold of ... an epoch of the 'politicization of jurisprudence', the likes of which in Germany—with the exception of the GDR—has never been seen before.

—Alfred-Carl Gaedertz to Gerhard Leibholz (1972)[3]

Introduction

In 1979, as the Federal Republic commemorated its thirtieth birthday, Friedrich-Wilhelm Dopatka, a jurist, took stock of the Constitutional Court's role in the Republic's societal and political development.[4] He had a very straightforward story to tell. In the first phase of the Republic's history, corresponding roughly with the 1950s, the Court made 'an enduring contribution to the development of democracy in the Federal Republic'.[5] During the second phase, corresponding roughly with the 1960s, the Court had been distinguished largely for its political inactivity—especially during the years of the grand coalition, when both major parties held power and neither had reason to take cases to Karlsruhe.[6] The 1970s ought to have been a period of further high-court retreat, but instead the opposite had happened.[7] The Court had grown increasingly activist just when the country was outgrowing—or trying to outgrow—the Court. The change of power in 1969 from a grand coalition to a Social–Liberal coalition had proven the viability of the Federal Republic as a modern democracy capable of smooth transitions between government and opposition.[8] But just when the Republic was reaching maturity the Court became more meddlesome than ever, throwing into bold relief the fact that the 1949 framers had provided too much patriarchal guidance and too little

[1] Presse- und Informationsamt der Bundesregierung, *Bundeskanzler Brandt. Reden und Interviews*, 1st edn (Bernecker 1971) 13, 30.

[2] Stephan Schlak, *Wilhelm Hennis. Szenen einer Ideengeschichte der Bundesrepublik* (CH Beck 2008) 162.

[3] Alfred-Carl Gaedertz to Leibholz, 18 January 1972, Bundesarchiv-Koblenz (BAK) N 1334/3 (Nachlass Gerhard Leibholz).

[4] Friedrich-Wilhelm Dorpatka, 'Zur Bedeutung des Bundesverfassungsgerichts in der politischen und gesellschaftlichen Entwicklung der Bundesrepublik 1951 bis 1978' in Wolfgang Dauebler and Gudrun Kuesel (eds), *Verfassungsgericht und Politik. Kritische Beitraege zu problematischen Urteilen* (Rowohlt 1979).

[5] ibid 42. [6] ibid 42–44. [7] ibid 44–48. [8] ibid 47.

popular rule.[9] The Constitutional Court highlighted the problem, wielding its awesome power most forcefully just when it had most clearly outlived its usefulness.

Dopatka's narrative was overly simplified, but he was not alone in his discontent. The late 1970s witnessed an unprecedented flood of popular and scholarly literature critical of the Constitutional Court. In one sharp essay, Konrad Zweigert, one of the Court's founding justices, asked whether the Court was 'an institution with a future'.[10] Zweigert's answer, for all the Court's recent offences, was a chastening yes. But it was striking that a former judge who had been present at the Court's creation would ask publicly whether its end was nigh. This chapter explains how the Court came to such a pass; the next chapter tells how it so swiftly recovered. To understand the crisis and the resolution, one must begin at what is often characterized as the Federal Republic's new beginning—the accession to power of a Social–Liberal (SPD/FDP) government under Chancellor Willy Brandt in the fall of 1969.

A. Republican refounding?

It was a season of superlatives, cast by contemporaries as an historic caesura—a 'refounding (*Umgründung*) of the Republic', in one historian's evocative phrase.[11] Brandt himself moved heaven and earth to foster this impression, cultivating what Wilhelm Hennis later called 'the myth of a second *Stunde Null*'.[12] Brandt intimated that the republican founding in the aftermath of the first 'zero hour' following 1945 had been tragically incomplete. The second founding would be more thorough. Brandt's election, the chancellor told a gaggle of foreign journalists, had elevated him to the leadership 'not of a conquered, but of a liberated Germany'. Only now, he added modestly, had Hitler 'finally lost the war'.[13] Brandt proclaimed a season of renewal—political and economic, legal and societal, cultural and moral. Whereas Adenauer, at the summit of his power, pledged to eschew experiments, Brandt, at the outset of his power, promised experiments of every kind.[14]

Adenauer himself did not live to see those experiments; he died in April 1967 at the age of ninety-one. The first postwar chancellor was honoured with a magnificent state funeral and a solemn Pontifical Mass in the Cologne Cathedral. Thousands lined the streets to pay their last respects. As with the state service for Winston Churchill held two years earlier in St. Paul's cathedral, a nation had arisen to pay final homage to a leader whose views and values it was swiftly shedding. The period between Adenauer's death and Brandt's accession witnessed the sizzling and fizzling of the student movement later immortalized in 'the myth of 1968'. The student movement, and the social transformation it reflected and fostered, was a precondition for the Social–Liberal coalition's coming to power. So too, was the movement's collapse. The emergency laws formed the focal point of the students' opposition, and the failure of that opposition—a failure solidified in the 1970 judgment of the Constitutional Court—helped still the wind in the movement's sails. The Soviet crushing of 'Prague Spring' in August 1968, and the founding of a new German communist party

[9] ibid.

[10] Konrad Zweigert and Hartmut Dietrich, 'Bundesverfassungsgericht—Institution mit Zukunft'? in *Verfassungsgericht und Politik* (n 4) 11.

[11] Manfred Görtemaker, *Geschichte der Bundesrepublik Deutschland* (CH Beck 1999) 475. For a critical view of this perspective, see Andreas Rödder, *Die Bundesrepublik Deutschland, 1969–1990* (Oldenbourg 2004).

[12] Wilhelm Hennis, 'Machtwechsel—oder der Versuch einer zweiten Gründung' (1983) 14 *Zeitschrift für Parlamentsfragen* 160.

[13] Willy Brandt, *Begegnungen und Einsichten: die Jahre 1960–1975* (Hoffmann and Campe 1976) 296.

[14] As set forth in his restlessly ambitious first government declaration of 28 October 1969. See Presse- und Informationsamt der Bundesregierung, *Bundeskanzler Brandt. Reden und Interviews*, 1st edn (Bernecker 1971) 13, 30.

the following month,[15] triggered a crisis of identity within the New Left, now stripped of its minimal foundation for consensus. Some within the movement hinted at a 'hot summer' in 1969 (including, perhaps, an assault on the judiciary, which Leftists still saw as a haven for former Nazis), but these gestures never materialized. After the autumn of 1968, the protest movement—the self-styled 'extra-parliamentary opposition'—splintered. Its largest segment returned to school or work, supporting the Social–Liberal ascension in 1969 and, in many cases, joining the SPD. Their support, though, was conditional. They wanted reform—swift and sweeping.

In an important sense, Brandt became chancellor not thanks to the students but in spite of them.[16] The SPD spent much of 1969 distancing itself from the students, who aroused distrust, and often anger, among many sections of society. The group Brandt really needed to persuade was not the APO but the FDP, his potential coalition partner. He barely succeeded, and at the last possible instant. At the Bundestag elections of 28 September 1969, the Union parties remained, comfortably, the largest parliamentary faction, polling 46.1 per cent of the vote—a loss of just 1.5 per cent vis-à-vis 1965. But the SPD's gain was greater than the Union's loss; it grew by 3.4 per cent to 42.7 per cent, the party's best postwar showing. The real loser was the FDP, which could little afford to lose anything. The Liberals dropped by 3.7 per cent down to 5.8 per cent, barely clearing the 5 per cent hurdle and barely outpolling the far-Right NPD, which, after ominous successes in local elections and considerable sabre-rattling during the general election, was narrowly kept out of Parliament with 4.5 per cent of the vote. Nearly everyone—including most Social Democrats—assumed the grand coalition would continue in power, and that Kiesinger would remain chancellor.

Kiesinger himself was certain that he had won; his office was flooded with congratulatory telegrams from around the world, including from Richard Nixon's White House. But Brandt cut a midnight deal—likely it was well after midnight—with the FDP. Combined, the SPD and FDP had won 48.5 per cent of the vote—not a proper majority but more than the Union's 46.1 per cent and enough (barely) on which to form a government. In the small hours of 29 September 1969, the two parties agreed to govern together.

On 21 October, Brandt was elected chancellor with 251 out of 495 votes cast in the Bundestag. It was only two votes more than the minimum required to become chancellor. Kiesinger, much embittered, mocked the new government as a 'mini-mini coalition'. Brandt parried that he had won by a tremendous margin compared with 1949, when Adenauer became chancellor by a single vote—his own. But a ready wit could not hide the fact that Brandt had become head of an ideologically divided government with a razor-thin majority. Nor could it quelch discontent within the chancellor's own party. Herbert Wehner and Helmut Schmidt—after Brandt the SPD's two biggest beasts—were appalled at the wreckless bravado of Brandt's nocturnal negotiations and unilateral assumption of power. They bided their time in plush cabinet posts, but exacted vengeance by and by.[17]

15 This party christened itself the *Deutsche kommunistische Partei* (DKP) rather than the *Kommunistische Partei Deutschland* (KPD). The party hoped to avoid its predecessor's fate by pledging allegiance to the Basic Law. Popular and official debates considered whether to seek another high-court ban. Foreign policy considerations, if nothing else, persuaded the grand and Social–Liberal coalition governments against it.

16 See Görtemaker (n 11) 491.

17 The story of Brandt's election has been told many times. The account given here has benefited most from Arnulf Baring, *Machtwechsel: die Ära Brandt-Scheel* (Deutsche Verlags-Anstalt 1982) 27–194; Görtemaker (n 11) 475–525; Eckart Conze, *Die Suche nach Sicherheit. Eine Geschichte der Bundesrepublik Deutschland von 1949 bis in die Gegenwart* (Siedler 2009) 361–416; and Peter Merseburger, *Willy Brandt, 1913–1992: Visionär und Realist* (Deutsche Verlags-Anstalt 2002) 486–577.

Brandt's election was hardly a plebiscitary acclamation. But the new chancellor treated it like an irresistible mandate. Brandt's formal assumption of power was as bold and dramatic as his acquisition of it had been parlous and slim. His first government declaration, delivered to the Bundestag on 28 October 1969, was nothing less than the announcement of a new era. Brandt proclaimed the Federal Republic's first authentic, popular constitutional moment. Adenauer's Republic had been a constrained one—patriarchal and authoritarian, rather than popular and plebiscitary. Brandt was the self-appointed herald of a bolder age, the champion of a democracy more open and more mature, more popular and participatory—in short, a *real* democracy.

Brandt paid passing tribute to a few forebears, ecumenically invoking—'as stand-ins for many others'—the founding fathers of the three major parties: Konrad Adenauer, Kurt Schumacher, and Theodor Heuss. 'No one will deny, doubt, or disparage the achievements of the last two decades', Brandt declared. 'They have become history', he added—ambiguous praise that both honoured Brandt's antecedents and interred them.[18] Over two decades, the Federal Republic had proved itself capable of accommodating change, including this most significant change: the peaceful transfer of power. Now it was time for something new—a more robust programme of reform, a more engaged brand of democracy. 'We wish to dare more democracy', Brandt intoned. His government would hazard the project of democratization not merely in parliamentary hearings but in a national conversation in which 'every citizen is given the possibility to participate in the reform of state and society'.[19] Co-determination (*Mitbestimmung*) and co-responsibility (*Mitverantwortung*) would be the watchwords of the new dispensation. 'This government seeks conversation', Brandt noted—apparently with everyone.[20]

That conversation, he promised, would yield sweeping reforms in every arena of government policy. In the foreground stood foreign policy, specifically 'the relationship between the two parts of Germany'. Brandt's government hoped to shift the position of the Federal Republic vis-à-vis its Eastern neighbour from 'a regulated next-to-one-another to a with-one-another' (*über ein geregeltes Nebeneinander zu einem Miteinander*). The chancellor stressed that this Delphic phrase meant only the continuation of policies pursued by the grand coalition, in which Brandt had served as foreign minister. He also assured his listeners that formal diplomatic recognition of the GDR was out of the question. But he unsettled Union parliamentarians with another tenebrous pronouncement, whose *arguendo* assumption spoke louder than its qualification. 'Even if two states exist in Germany', he said, 'they are not, with respect to one another, foreign countries; their mutual relations can only be of a particular kind'.[21] Just what kind was determined by Brandt's new Eastern policy (*Ostpolitik*), which formed the signature of the first Brandt government and the bête-noir of the Union opposition. One imposing Unionist hoped to make the Constitutional Court an instrument of that opposition.

On the domestic front, Brandt placed education at the centre of his government's reform agenda.[22] Here too the government's ambitions would be shaped and constrained by appeals to the Constitutional Court. The same would hold for Brandt's declared intention to reform fields as disparate—and as central—as gender equality[23] and labour relations.[24] More often than the new government could possibly have predicted, the road to reform would run through Karlsruhe.

[18] *Verhandlungen des Deutschen Bundestages. Stenographische Berichte*, vol 71 (1969) 20. [19] ibid.
[20] ibid 20–21. [21] ibid 21. [22] ibid 26. [23] ibid 29. [24] ibid 28–29.

Brandt's inaugural declaration was untroubled by such thoughts. In a ringing peroration, he returned to the theme of democratic renewal. 'The government can only work successfully in a democracy', he declared, 'when it is supported by the democratic engagement of the citizenry.... We seek no admirers; we need people who critically think with us, decide with us, and carry responsibility with us.... We therefore seek a conversation with all who labour for this democracy'.[25] This democracy, Brandt suggested, was something qualitatively new. He had never believed those who feared that the second German democracy would go the way of the first. He believed it now less than ever.[26] This was because the second German democracy—with a charismatic man of destiny to lead it—had only just arrived. 'We do not stand at the end of our democracy', Brandt closed, 'we have only now truly begun'.[27] *Redeunt Saturnia regna.*

Brandt's early days in office were marked by breathless activity—a bid to impress voters with visible results and to buoy support for a government with a precarious popular base. It was a time of tremendous optimism. Over the previous year the West German economy had grown by 7.5 per cent, the best annual outcome since the halcyon years of the *Wirtschaftswunder.* As of 1970, unemployment stood at less than 1 per cent. This economic upturn followed a jarring slump in 1966/67, the overcoming of which was credited in many circles to the macroeconomic policies of the grand coalition. Brandt proposed to persist in giving such direction, and expected similarly felicitous results. The government extravagantly promised expensive reforms. In the event, souring economic conditions—even before the oil crisis of 1973[28]—would rob the government of the capacity to make good on those promises. As of late 1969, however, the possibilities for reform—for a centralized government shaping a splendid socialist future—seemed endless. The new cabinet's efforts were coordinated as never before. Its several departments were linked under the managerial supervision of Brandt's master technocrat and chief counsellor, Horst Ehmke, a once and future advocate before the Constitutional Court. Wehner and Schmidt writhed under Ehmke's meddling (here, too, they would have their revenge). But Ehmke, the consummate and indefatigable details man, proved indispensable to the relaxed and visionary Brandt. Ehmke infused the chancellor's office with unprecedented importance. He even proposed a new building commensurate with the gravity of the office.[29]

B. The coming conflict

While the country's executive chieftains were designing a new executive palace, the heads of the West German judiciary were settling into a new building of their own. After twenty years in the cramped quarters of Prince Max Palace, the Constitutional Court moved in autumn 1969 to a newly constructed building near the Karlsruhe botanical gardens. The building, by design, did not look like a courthouse—no Grecian columns or imposing statuary, no freizes or flutings or architraves. Instead the structure was open and sprawling—symbolically inviting Citizen Everyman to bring his complaints there. Large glass windows denoted democratic transparence. To the architecturally uninitiated, however, it looked like a train station—an apt resemblance, some joked, that embodied a pair

[25] ibid 33. [26] ibid. [27] ibid 34.

[28] Many West Germans were rattled by a 1972 report from the Club of Rome entitled 'The Limits of Growth'. See Dennis L. Meadows and others, *Die Grenzen des Wachstums. Bericht des Club of Rome zur Lage der Menschheit* (Deutsche Verlags-Anstalt 1972); see also Mihajlo D. Mesarović and others, *Menschheit am Wendepunkt* (Deutsche Verlags-Anstalt 1974); Dennis Gabor and others (eds), *Das Ende der Verschwundung. Zur materiellen Lage der Menschheit* (Deutsche Verlags-Anstalt 1976).

[29] The building wasn't finished until 1976; by then Ehmke was out of the cabinet.

of political bromides: 'All roads lead to Karlsruhe', and 'Karlsruhe works the switches for Bonn'.[30] With an ambitiously reformist government in Bonn, and with mounting polarization between government and opposition, observers expected appeals to the Court. No one, however, anticipated fully the scope or intensity of the coming confrontation between the harbingers of a new constitutional order and the institutional guardians of the old one. Against Brandt's proclamation of a constitutional moment, the Court asserted the prerogatives of a constitutional tradition. Brandt called for a democracy that was participatory, process-based, and plebiscitary; the Court safeguarded a democracy that was constrained, value-laden, and paternalist.

The results of the clash were equivocal. The achievements of the Social–Liberal coalition were considerable, and when its confrontation with the Court reached its crisis, the Court backed down. The retreat, however, was only temporary; perhaps it was merely tactical. In any case, when the crisis came, the constitutional moment had passed. The fiercest sequence of confrontations between Court and coalition came after Brandt's resignation and replacement by the pragmatic Helmut Schmidt; after the world oil crisis and subsequent slump had denied the government the material preconditions of reform; and after the scourge of domestic terrorism had left the nation longing for security and stability—for the kind of watchful and integrative functions long ascribed to the Constitutional Court. As a constitutional matter, Brandt's promise of sweeping transformation fell short. Adenauer was dead and gone, but the second German democracy remained, in part, a patriarchal democracy. The Constitutional Court had become its patriarch.

I. Departure of the Giants

The Federal Constitutional Court is two years younger than the Federal Republic. One consequence of this age difference has been that changes in the Court's composition do not coincide with national elections. Throughout the 1960s and early 1970s, turnover in the Court's personnel fell at the midpoint between Bundestag elections, with major shifts in 1959, 1963, 1967, and 1971—in each case two years after a federal election. Not by accident, the Court appointments in 1971 were as politically fraught as the Bundestag election of 1969 was politically momentous. Two years of restless government reform were accompanied by two years of persistent threats to attack those reforms in Karlsruhe. By 1971, several politically charged cases were already pending before the Court. Others were sure to follow.

From the beginning of the Court's history, filling vacant seats was a politically contested matter. Legislative obstruction delayed the filling of two posts vacated during the European Defence Community controversy in the early 1950s, and we have witnessed the row that erupted when Martin Drath was denied reappointment in 1963. With each large turnover in the Court's membership, the political stakes seemed to climb, and the public's attention to sharpen. Never had the stakes been so high, or popular interest so intense, as when six seats fell vacant in 1971.

The political environment was polarized as it had rarely been since the early 1950s. Many rightly predicted that the CDU opposition would subject prized Social–Liberal initiatives—in both domestic and foreign policy—to constitutional challenge. All major parties wanted 'reliable' justices to field those challenges when they came. This impulse,

[30] See Hans Seiler, 'Karlsruhe stellt die Weichen für Bonn' *Die Welt* (23 September 1969) 31.

combined with the requirement that high-court nominees be approved by a two-thirds majority, produced a wondrous tussle within the responsible parliamentary committees.

Some of the departing justices were genuine giants: Gerhard Leibholz and Erwin Stein, who had been with the Court since its founding, and Chief Justice Gebhard Müller, who had presided over the Court nearly twice as long as his two predecessors combined. With the departure of Leibholz and Stein, only three original justices remained—Theodor Ritterspach on the First Senate, and Hans Rupp and Willi Geiger on the Second. The terms of Gregor Geller (Second Senate) and Wiltraut Rupp-von Brünneck (First Senate), both appointed in 1963, had also expired, as had that of the Court's former director, Walter Rudi Wand, who in 1970 was tapped to finish the term of Hans Kutscher, who had joined the European Court of Justice in Luxembourg.

After only a year in the Court, it seemed certain Wand would be reappointed. Many assumed that Justice Rupp-von Brünneck was a shoo-in for reappointment as well. In addition to being an eminent jurist, she was now the Court's only woman. (She was also married to Hans Rupp of the Second Senate, providing some social pressure to keep her in the Court). The other four vacancies, however, would be filled with new judges. These would be the first judges appointed under a new regulation, in force since January 1971, whereby all justices would serve a single, non-renewable term of twelve years. The new regulation also imposed an age limit. The new justices would serve for a dozen years or until they turned sixty-eight, whichever came first. Helmut Simon, who replaced Wolfgang Zeidler on the First Senate in 1970, was the last justice to escape the limit. He would serve, with great influence, for seventeen years.

The parliamentarians who haggled to fill the vacancies did not, by most contemporary accounts, drape themselves in glory. Friedrich Karl Fromme reported in the *Frankfurter Allgemeine Zeitung* that a 'poisonous, crisis-ridden, and often unworthy atmosphere' permeated the proceedings.[31] Ernst Müller-Meiningen of the *Süddeutsche Zeitung* found the affair despicable.[32] Other papers opined that the appointments were conducted 'in bad form',[33] that the process was 'unfair and tactless',[34] damaging to the Court's reputation and dangerous for its future.[35] With the Court on the cusp of what was sure to be a severe test of its independence and integrity, the parties in Parliament had treated judgeships like political fiefdoms. They had made the appointments process a farce of factional squabbling

[31] Friedrich Karl Fromme, 'Der gefährliche Kuhhandel ums Bundesverfassungsgericht' *FAZ* (4 November 1971).

[32] Ernst Müller-Meiningen, 'Bundesverfassungsgericht in Gefahr' *Süddeutsche Zeitung* (11 November 1971).

[33] 'Schlechter Stil' *Christ und Welt* (Bonn, 22 October 1971).

[34] 'Unfair und taktlos' *Stuttgarter Zeitung* (Stuttgarter, 23 October 1971).

[35] For a sampling of critical coverage in the contemporary press, see 'Harte Auseinandersetzung um die Sitze im Verfassungsgericht' *FAZ* (25 May 1971); Friedrich Karl Fromme, 'Jetzt ist das Verfassungsgericht am Zug' *FAZ* (12 June 1971); 'Auseinandersetzung um Verfassungsrichter' *FAZ* (9 July 1971) 4; Hellmuth Rieber, 'Karlsruhe will keine pensionierten Politiker' *Frankfurter Rundschau* (7 August 1971); Robert Leicht, 'Das höchste Gericht im Spiel der Parteien' *Süddeutsche Zeitung* (1 October 1971); Werner Birkenmaier, 'Gefahr für Karlsruhe'? *Stuttgarter Zeitung* (5 October 1971); Oskar Fehrenbach, 'Das Tauziehen um die Verfassungsrichter (Interview mit Friedrich Vogel)' *Stuttgarter Zeitung* (16 October 1971); 'Mißachtung des Bundesverfassungsgerichts' *Süddeutsche Zeitung* (16/17 October 1971); 'Schaden für Karlsruhe' *Frankfurter Rundschau* (21 October 1971); 'Schlechter Stil' *Christ und Welt* (22 October 1971); Bruno Waltert, 'Vergißt Bonn das Ansehen des Verfassungsgerichts'? *Die Welt* (23 October 1971); 'Unfair und taktlos' *Stuttgarter Zeitung* (23 October 1971); Friedrich Karl Fromme, 'Der gefährliche Kuhhandel ums Bundesverfassungsgericht' *FAZ* (4 November 1971); Hellmuth Rieber, 'Karlsruhe: Schaden durch Richterstreit' *Frankfurter Rundschau* (4 November 1971); 'Unwürdiger Streit' *Stuttgarter Zeitung* (4 November 1971); 'Bonner Kuhhandel um Karlsruher Posten' *Die Welt* (5 November 1971); Gerhard Ziegler, 'Wie man Autorität zerstört' *Frankfurter Rundschau* (10 November 1971); Bruno Waltert, 'Das Gerangel um die Verfassungsrichter muß enden' *Die Welt* (11 November 1971); Ernst Müller-Meiningen, 'Bundesverfassungsgericht in Gefahr' *Süddeutsche Zeitung* (11 November 1971); 'Tauziehen. Richterwahl ohne Stil' *Rheinischer Merkür* (12 November 1971).

and partisan horse-trading ('cow dealing' [*Kuhhandel*] in the even less flattering German vernacular). Many whose names circulated as potential nominees were party loyalists rather than distinguished jurists. The Court's long-time admirers smarted that the robes of Leibholz and Müller might be worn by juristic pigmies beholden to party interests and dependant on their law clerks. Wilhelm Hennis, among others, roared that tenure on the Constitutional Court should not be a sinecure bestowed by grateful parties for services rendered in Parliament.[36]

What was the upshot of this odious animal swapping? With Wand's reappointment taken for granted, the parties originally planned to let the CDU nominate three candidates and the SPD two, with Walter Seuffert becoming chief justice as a consolation to the SPD. Even though the CDU now formed itself in opposition, it was afforded an extra appointment because five of the six justices whose terms expired in 1971 had been CDU nominees.

But the Union parties had grown accustomed to appointing the chief justice, and they ultimately traded the right to propose an additional justice for the right to name Müller's successor.[37] Some Unionists thought this daft. The chief justice was an important figurehead, but he only voted once. So the Union tried to switch back, only to find the SPD disinclined. Meanwhile, the SPD's junior partner, the FDP, clamoured for the right to nominate one of the justices—something the party hadn't done since proposing Hermann Höpker-Aschoff for the Court's inaugural presidency twenty years earlier. The SPD hummed and hawed, but finally consented. In the end, all parties agreed that the CDU would nominate two justices, including the new president; the SPD would also nominate two; and the FDP one.

Initially, the FDP's one choice was Emmy Diemer-Nicolaus, who had represented the party in the Bundestag since 1957 and in the Baden-Württemberg Landtag before that. She had studied law in the 1930s, but hadn't held a strictly 'legal' job for over twenty years. Nor had she published articles on legal topics. She was also nearly sixty-two, which meant she would reach the new age limit after a term of only six years. Many complained that her candidacy was a political plum. Some who complained were Christian Democrats who worried quietly that Diemer-Nicolaus, a vocal champion of gender equality, would vote to sustain a liberalizing reform—then being debated—of the country's abortion laws. But the CDU had also tapped one of its own, nominating Ernst Benda, a forty-six-year-old member of the Bundestag, to be the new chief justice. When the Union mused about blocking Diemer-Nicolaus's nomination, the coalition parties retorted that, if that happened, the SPD and FPD would unite to block Benda's. The CDU wouldn't flinch. Diemer-Nicolaus broke the impasse by withdrawing her nomination. The lustre of becoming the third female justice had dimmed amid a barrage of public charges that she was unfit for the post.

With Diemer-Nicolaus's retraction, Justice Rupp-von Brünneck's reappointment became inevitable. The SPD didn't dare incur responsibility for an all-male Court. Reelected on 12 November 1971, Rupp-von Brünneck would remain with the Court until her death in office six years later at age sixty-five. In the meantime, she wrote or joined several of the most momentous dissenting opinions in the Court's history. The SPD's other nominee was Martin Hirsch, who since 1961 had been one of the party's legal experts in the Bundestag. Of all the politicians who have ever sat in Karlsruhe, Hirsch proved least successful at subduing party allegiance and adapting to the judicial role. After the Diemer-Nicolaus nomination failed, the FDP opted for Joachim Rottmann, a longtime official in the defence and interior ministries who taught law on the side. In addition to Benda, the CDU nominated

[36] Wilhelm Hennis, 'Verdiente Abgeordnete—gute Richter'? *Süddeutsche Zeitung* (6 November 1971).
[37] See 'Die CDU stellt weiter den Präsidenten des Bundesverfassungsgerichts' *FAZ* (18 June 1971) 4.

Hans-Joachim Faller, a veteran of the Federal Court of Justice, to fill one of six seats reserved for members of the federal judiciary.

To many observers, the new crop was a disappointment. Earlier in 1971, Theodor Eschenburg lamented what he saw as a steady drop in the quality of the Court's personnel since the remarkable class of 1951.[38] Eschenburg feared that, given parliamentarians' proclivity for partisan horse trading, the coming year portended further decline. Many now saw Eschenburg's fears confirmed. Though several justices taught law on the side or dabbled in legal scholarship, with Leibholz's departure none was a tenured professor of public law. Of the four new justices, three were professional politicians—'full-blooded politicians' in the idiom of the critical press.[39] Parliament, it seemed, appointed jurists only when it absolutely had to. These critics forgot that many of the former judges they lionized had also been 'full-blooded politicians'—Hermann Höpker-Aschoff and Rudolf Katz, as well as the newly pensioned Erwin Stein and Gebhard Müller. Fair or not, contemporary press coverage conveyed an unmistakable perception of decline. The Court, critics fretted, had diminished in both stature and independence, and had increased in partisan rancour—and this on the threshold of what looked to be the most politically controversial season of the Court's history.

Ernst Benda, the Court's new chief justice, was sufficiently self-aware to sense that the public perceived a difference between himself and his forebears. Benda knew that he radiated less gravitas than the flinty Höpker-Aschoff, the otherworldly Josef Wintrich, or the ascetic Gebhard Müller. Benda was young, and felt younger. He was happiest behind the wheel of a sports car, his foot on the gas, or at the helm of a yacht, the wind in his sails. A Berliner to the bone, Benda joined the infant CDU at age twenty. In 1957, Benda, then thirty-two, moved from the Berlin state Parliament to the Bundestag. In Parliament, Benda kept a low profile, but with one towering exception. During the 1965 debates over extending the statute of limitations for Nazi murders, Benda made the floor flash with lightning eloquence and moral anger. The speech was a highpoint of postwar parliamentary oratory. It made Benda, however briefly, something of a national hero. Benda's role (as interior minister in the grand coalition) in crafting the Emergency Laws had rather the opposite effect. Still, it was clear to all that Benda was a rising star in the CDU firmament, certain to return to the cabinet should the Union return to power.

The call to Karlsruhe silenced such ambitions and put Benda's thriving Berlin law practice on hold. As we have seen, Benda was not the first chief justice to chafe at having to exchange the velocity of the city for the somnolence of Karlsruhe. He was not the first to demur at the initial invitation, nor the first to accept it with a heavy heart. In the *Frankfurter Allgemeine Zeitung*, Friedrich Karl Fromme described the new chief justice as a man armed with a supercilious reserve and a gift for self-irony.[40] But these were external defences. Within, as events would show, Benda was equal parts pugnacity and panache.

II. Fundamental Rights

The six justices appointed or reappointed in 1971—Ernst Benda, Wiltraut Rupp-von Brünneck, Helmut Simon, Martin Hirsch, Rudi Wand, and Joachim Rottmann—would stand at the centre of the Court's most intense political conflicts since the early 1950s.

[38] Theodor Eschenburg, 'Nur zweite Wahl nach Karlsruhe?' *Die Zeit* (26 February 1971).

[39] See 'Vollblutpolitiker' *Frankfurter Rundschau* (9 November 1971); Friedrich Karl Fromme, 'Wenn "Vollblutpolitiker" die Verfassung hüten' *FAZ* (9 June 1973).

[40] See Friedrich Karl Fromme, 'Nach Karlsruhe mit genügend Selbstironie' *FAZ* (13 November 1971) 2.

They would clash swords with one another and with the Republic's leading politicians. As they took or resumed office, however, the Court's immediate docket was marked by considerable continuity with the previous decade. As usual, the vast majority of cases in the Court's inbox were fundamental rights cases arising from individual complaints. Most of these were routine, but some led to doctrinally momentous decisions. Aggressive holdings in these cases would, when followed in other contexts, prove politically explosive. The first of them involved a novel by the famous son of a more famous father.

A. The *Mephisto* Judgment and the dignity of the dead

'All the characters of this book', wrote Klaus Mann in a coda to his novel, *Mephisto*, 'represent types, not portraits'. The distinction was a fuzzy one. The type Mann had in mind, as he explained in a subsequent memoir, was 'the abject type of the treacherous intellectual who prostitutes his talent for the sake of some tawdry fame and transitory wealth'.[41] Mann 'thought it pertinent, indeed necessary, to expose and analyse' this type, the most appalling exemplar of which was Gustaf Gründgens, Mann's quondam friend, erstwhile collaborator and, from 1926 to 1929, the husband of Mann's sister, Erika. 'I visualize my ex-brother-in-law', Mann wrote, 'as the traitor par excellence, the macabre embodiment of corruption and cynicism. So intense was the fascination of his shameful glory that I decided to portray Mephisto-Gründgens in a satirical novel'.[42]

Mann's aim, then, was to expose a type by portraying a man. But the resulting novel, he insisted, was a depiction of the type, not a portrait of the man. Indeed, Mann added, the object of his portrait was not quite a man. Hendrik Höfgen, *Mephisto*'s protagonist, 'especially lacks the moral qualities, which are usually summarized under the concept of "character". Instead of "character" there is, with this Hendrik Höfgen, only ambition, vanity, vainglory, affect. He is not a man, only a *Komödiant* [comedian/actor]'.[43] Were Höfgen and Gründgens one and the same? 'Not quite', Mann answered. *Mephisto* was 'not about the individual case, but about the type'.[44]

The fictive type and the historical individual had much in common—their physical appearance, their thespian powers, their tortured biographies. Gründgens, like Höfgen, was a brilliant actor, renowned for his representation of Mephistopheles in Goethe's *Faust*. Gründgens enjoyed a dizzying rise after the Nazi seizure of power in 1933. Ultimately, with the patronage of Hermann Göring, Gründgens became head of the state playhouse in Berlin, a post from which, with a muted reference to his homosexuality, Gründgens offered to resign after the 1934 arrest and murder of Ernst Röhm.[45] Göring refused to accept his resignation, and Gründgens remained intendant of the playhouse until 1937, after which, until 1945, he headed the Prussian State Theatre. In 1936, the year of *Mephisto*'s first publication, Göring named Gründgens to the Prussian council of state (*Staatsrat*). Klaus Mann's *Mephisto* suggested nothing about Gründgens's sexual orientation (Mann's own homosexuality was a theme of many of his novels); instead, it portrayed Höfgen as a deviant and sadomasochist.

41 Klaus Mann, *The Turning Point: 35 Years in this Century* (LB Fischer 1942) 281. 42 ibid.
43 Klaus Mann, *Die Wendepunkt. Ein Lebensbericht* (Rowohlt 1948) 334. 44 ibid.
45 Röhm, who was not openly gay but whose homosexuality was privately known to Hitler, was commander of the *Sturmabteilung* (SA), the Nazi militia. Hitler came to see Röhm as a rival, ordered Röhm's arrest and execution as part of the 'night of long knives' that purged the SA leadership between 30 June and 2 July 1934. Hitler alluded to Röhm's homosexuality in his public explanation of the purge.

The novel was written in exile and published abroad.[46] The novel's original publisher, the Dutchman Emanuel Querido, was later murdered (1943) in the Sobibor concentration camp. *Mephisto* was never published in Germany during Mann's lifetime. Unlike his more famous father, Mann returned to Germany briefly after the war, but confessed that he felt himself 'a stranger in my former fatherland'.[47] At the age of forty-two, Mann took his own life, overdosing on sleeping pills in Cannes in 1949.

The postwar career of Gustaf Gründgens was very different. Arrested by the Allies at the end of the War, Gründgens was exonerated in a denazification proceeding after several colleagues rose to his defence. These colleagues emphasized Gründgens's anti-Fascist outlook—he saw the theatre, they claimed, as a non-ideological oasis in the midst of the Nazi wastelands—and his advocacy for victims of racial and political persecution. In his turn, Gründgens contributed to the exoneration of other colleagues, including the actress, Emmy Göring (Hermann's widow), and the director, Veit Harlan.

His theatre licence restored, Gründgens enjoyed a distinguished postwar career on screen and stage. In 1953, he became the first actor to receive the Starred Cross of Merit. Two years later, Grüdgens became general intendant of the German Playhouse in Hamburg, a position that brought him to the summit of his fame. Gründgens resigned the post in order to undertake a world tour in 1963. In October 1963, in the Philippines, Gründgens died—like Klaus Mann—from an overdose (perhaps intentional) of sleeping pills.

Shortly before Gründgens' death, Berthold Spangenburg, a publisher in Munich, announced the publication of the first edition of *Mephisto* in the Federal Republic.[48] After Gründgens' death, his adoptive son and sole heir, Peter Gorski, asked a Hamburg court to enjoin the publication. The novel as a whole, Gorski argued, presented a falsified, defamatory portrait of Gründgens. It was not a work of art but an act of private vengeance, animated by Mann's offended sense of family honour. The lower court rejected Gorski's petition but was overruled, in 1966, by the Hamburg superior court, whose ruling was sustained in 1968 by the Federal Court of Justice. Spangenberg, appealed to the Constitutional Court, contending that the injunction violated his right, under Article 5(3) GG, to artistic freedom.

As a backdrop to all of these proceedings, the country was engaged in an unprecedented public debate about Germany's Nazi past. The Hamburg proceeding began shortly after the close of the Frankfurt 'Auschwitz trials', which lasted from 1963 to 1965. What was quickly dubbed a 'dual of the dead' pitted a 'fellow traveller' of the Nazi regime against an emigrant who had served in the US Army. Whether emigration constituted an act of heroism or betrayal was still hotly disputed. By the time the *Mephisto* case reached the Constitutional Court, the Federal Republic was governed by an emigrant chancellor, Willy Brandt, who had replaced a Nazi party member of the first hour, Kurt Georg Kiesinger. In the meantime, *Mephisto*'s sojourn on the index of forbidden books generated unprecedented interest in the hitherto forgotten novel. This was certainly not Peter Gorski's intent, wrote Marcel

[46] This account of *Mephisto* owes much to the work of Thomas Henne. See Thomas Henne, 'Klaus Manns *Mephisto* und die Publikationsverbote der deutschen Gerichte. Zur gerichtlichen Karriere des Romans und seines Münchener Verlegers Berthold Spangenberg' in Dirk Heißerer (ed), *Thomas Mann in München* (Peniope 2004) 27; Thomas Henne, 'Die Mephisto-Entscheidungen der deutschen Gerichte. Eine exemplarische, justitiell geführte Auseinandersetzung über den Umgang mit der deutschen NS-Vergangenheit in den späten sechziger und frühen siebziger Jahren' in Birgit Feldner and others (eds), *Ad Fontes. Europäisches Forum Junger Rechtshistoriker/innen* (Lang 2002) 193.

[47] Armin Strohmeyr, *Klaus Mann* (Deutsche Taschenbuch Verlag 2000) 140.

[48] For an account of Spangenberg's efforts to publish *Mephisto*, and of the long legal battle that followed, see Thomas Henne, 'Klaus Manns *Mephisto*' (n 46).

Reich-Ranicki, the supreme pontiff of West German literati, but it had nonetheless earned Gorski the Republic's gratitude.[49]

Spangenberg's complaint failed narrowly.[50] It won the support of three of the First Senate's six voting justices (Justices Haager and Simon did not participate), but the tie went against the complainant. In the Court's judgment of 24 February 1971, the prevailing justices articulated a robust scope for the right of artistic freedom but declined, in the case at hand, robustly to apply it. Article 5(3) GG, they explained, reflected a value decision determining the relationship of the state to the artistic realm. It also conferred an individual freedom.[51] In an alliterative (and untranslatable) pairing, the controlling triumvirate established that the guarantee of artistic freedom secured both the *Werkbereich* and the *Wirkbereich* of a work of art—both the artist's freedom to create works of art and her freedom publicly to disseminate them. Mediators between the artist and public—in this case Spangenberg, the publisher—the publisher could invoke the latter freedom too.[52]

Artistic freedom was limited neither by Article 2(1) GG, under which the right to the free development of one's personality was circumscribed by the rights of others, the constitutional order, and the moral law; nor by Article 5(2) GG, under which the speech, press, and broadcasting rights of Article 5(1) could be limited by general laws, statutory measures for the protection of youth, and the right of personal honour.[53] The constitutional guarantee of artistic freedom could be limited only when it collided with values guaranteed elsewhere in the constitution. For purposes of the *Mephisto* appeal, this did not include the personality rights guaranteed in Article 2(1). Those rights presupposed a rights-bearer capable of potential or future action.[54] The dead were incapable of personal development.

But they were not deprived of dignity, the supreme constitutional value. Article 1(1) GG, the prevailing justices held, protected even the posthumous dignity of all persons, including Gustaf Gründgens.[55] The conflict in the *Mephisto* case required, then, a sensitive balance between Spangenberg's right to publish Mann's novel and Gründgens' right to the posthumous protection of personal honour. The prevailing justices didn't actually conduct that balance. They only agreed that the lower courts had neither misunderstood the impact of constitutional rights on civil law norms, nor erred fundamentally in balancing the conflicting rights.[56] The lower courts had asked essentially the right questions; the answers they had given were not obviously wrong. This was enough to satisfy three justices, and to keep *Mephisto* on the index.

Three justices disagreed with this analysis; two recorded their disagreement in separate dissents. For Erwin Stein, it was the first and last dissent of a twenty-year tenure on the Court. For Wiltraut Rupp-von Brünneck, it was a harbinger of things to come.

Stein, who served as rapporteur in the case, objected that the lower courts had misunderstood the nature of art. Stein bolstered his own understanding with a flash of erudition, citing critics as disparate as Theodor Adorno and Stefan George.[57] The lower courts' fixation on the similarities between Höfgen and Gründgens was beside the point.[58] A work of art possessed an internal coherence and enjoyed an artistic reality independent of the 'real' world.

[49] Marcel Reich-Ranicki, 'Noch ein Triumph des Gustaf Gründgens' *Die Zeit* (18 February 1966). Incidentally, Reich-Ranicki would later record in his popular memoir, *Mein Leben*, the extraordinary influence on him as a Berlin teenager of Gründgens's acting. 'Gustaf Gründgens well-nigh hypnotized me', Reich-Ranicki wrote. Marcel Reich-Ranicki, *Mein Leben* (KG Saur, 2002) 146.

[50] 30 BVerfGE 173 (1971). [51] ibid 188. [52] ibid 189. [53] ibid 191–92. [54] ibid 194. [55] ibid 195. [56] ibid 195–200. [57] ibid 205. [58] ibid 206.

The masters of modern fiction—Goethe and Tolstoy, Fontane and Flaubert, Hermann Hesse and Thomas Mann—had all written *romans à clef* without in the least diminishing their novels' stature as self-sufficient works of art.[59] Just how little the lower courts understood about the artistic process—or cared about artistic freedom—was betrayed by their suggestion that Klaus Mann ought to have revised the novel after the War or, after facts favourable to Gründgens came to his attention, supplied an extensive prefatory disclaimer.[60] In Stein's view, the lower courts had underestimated the constitutional significance of art as art; they had overemphasized the social, and ignored the aesthetic, component of Mann's novel.

Justice Rupp-von Brünneck's dissent struck a very different note. Justice Stein highlighted the novel's aesthetic independence; Justice Rupp-von Brünneck stressed its political embeddedness. Justice Rupp-von Brünneck chided the lower courts for ignoring the novel's political context—Mann's position within, and *Mephisto*'s status as a statement of, the anti-Nazi resistance.[61] Using Gründgens as a vehicle to attack the regime may have been questionable as a matter of taste. But as a matter of political morality it was, under the circumstances, more than justified. *Mephisito* was a commentary on contemporary events, and should enjoy constitutional protection similar to that granted by 'the extraordinarily generous jurisprudence of the U.S. Supreme Court'.[62] Rupp-von Brünneck was appalled, moreover, by her colleagues' refusal to balance the values in question. The controlling judgment rested on a 'restrictive construction of the reviewing competence of the Constitutional Court'—a break with precedent, she thought, and an ominous portent.[63]

If so, the omen pointed in the wrong direction. In coming years, a dissenting Justice Rupp-von Brünneck would occasionally say sharp things—sometimes very sharp things—about her colleagues. But never again would she complain that they construed their mandate too meekly.

Ironically, neither dissenter thought much of *Mephisto* itself. Justice Stein considered the novel an artifact of cultural history—destined, if published at all, for only a limited readership. Justice Rupp-von Brünneck deemed it a bad novel—at least by the standard of Mann's other works.[64] But both thought it deserved the right to life, and the prevailing judgment had denied it even that. The Court's judgment relegated *Mephisto* to the same position in the 1970s that it occupied in the 1930s—freely circulated around the globe, but banned in Germany.

Banned, that is, in *West* Germany. Soon after the Court's judgment a GDR publisher, eager to thumb its nose at the liberal pretensions of the Federal Republic, published a new edition of the novel. At the turn of the next decade, a pair of pirated editions appeared in the West—one of which reproduced the Court's decision in an appendix. In 1980, the French director Ariane Mnouchkine staged a theatrical production of *Mephisto*, the script for which circulated without opposition in the Federal Republic. Finally, in 1981, the West German publisher rororo issued the Federal Republic's first legal edition of the full novel. Contrary to Justice Stein's predictions of a limited readership of cultural antiquarians, it soared almost immediately to the top of the *Spiegel* bestseller list, where it remained for several months.[65] The novel surely benefited from the aura of clandestinity conferred by a long interdiction. *Mephisto* is now Mann's best-known novel. The muses work in mysterious ways.

59 ibid 208. 60 ibid 209–10. 61 ibid 224–25.
62 ibid 225 (citing *New York Times v Sullivan*, 376 US 254 (1964)). 63 ibid 219. 64 ibid 223.
65 Henne, 'Klaus Manns *Mephisto*' (n 46) 48–49.

B. The 'Lebach murders': prisoner privacy and prime time television

Two years after the *Mephisto* judgment the Court returned to the theme of conflict between personality rights and expressive freedoms—this time press and broadcasting freedoms. In the background was a dramatic crime committed in the small Saarland city of Lebach in 1969. Hans Jürgen Fuchs and Wolfgang Ditz, both professionals in their mid-twenties, assaulted a munitions depot at the Lebach army barracks, killing four sleeping soldiers, gravely injuring a fifth, and stealing a stockpile of weapons and ammunition. A third young man, Gernot Wenzel, helped plan the crime and taught Fuchs and Ditz how to fire pistols. The three friends planned to sell what they stole, buy a yacht, and set sail on the South Sea. Fuchs, Ditz, and Wenzel were not, as contemporaries initially suspected, *Mafiosi* or Left-wing terrorists.[66] Their friendship, as the Constitutional Court later noted, had 'a homosexual component'.[67] They hoped to escape a country where homosexuality was criminalized by the state and demonized by society, and to form a separate, idyllic community of their own. In a bitter irony, the crime coincided almost exactly with the Bundestag's repeal of criminal sanctions for male sodomy—a penal code provision sustained by the Constitutional Court in 1957.[68] After a months-long manhunt that generated enormous popular attention, the trio were apprehended, arraigned, and convicted. Fuchs and Ditz were sentenced to life imprisonment for holdup murder (*Raubmord*); Wenzel ('Gernot W'. in the court reports) to six years for complicity.

Wenzel had served two-thirds of his sentence when the ZDF[69] television network produced a documentary drama—'The Lebach Soldiers Murder'—depicting the crime that Wenzel helped plan. The show would air at prime time on a Friday evening. Nearly three hours long, the film chronicled the crime's planning, execution, and aftermath. It began by displaying the names and photographs of the three participants, thereafter portrayed by actors but repeatedly called by name. Wenzel had been a model prisoner, and his term was to end early—indeed, was almost over. Not wanting his crime dramatized on national television on the eve of his release, he asked the district court in Mainz to enjoin the film's broadcast, arguing that airing the show would violate his general personality right, as well as his rights of name and reputation. The district court rejected the petition, as did the appeals court in Koblenz. Both courts relied on a 1907 law, still in force, that governed artistic production. The law banned injuries to name and reputation but made an exception for figures of contemporary history (*Zeitgeschichte*). As a figure of contemporary history, Wenzel's personality interests must yield to the public's interest in information about a crime he helped commit. Wenzel then challenged these judgments in a constitutional complaint.

On 13 March 1973, the Constitutional Court issued a preliminary injunction against the broadcast.[70] On 5 June, the First Senate made the injunction permanent.[71] The Senate's analysis centred, naturally, on the conflict between ZDF's broadcasting freedom and the complainant's personality right. Both rights exerted a radiating effect (*Ausstrahlungswirkung*) on the 1907 law applied by the courts below. Neither right, in the abstract, took precedence over the other. Both stood near the centre of the Basic Law's value system; both were rooted in human dignity.[72] As always, resolution required balancing. Specifically, the Senate

[66] See 'Nicht Apo und nicht Mafia' *Die Zeit* (3 July 1970).
[67] 35 BVerfGE 202, 204 (1973).
[68] See 6 BVerfGE 389 (1957).
[69] *Zweites Deutsches Fernsehen*, or Second German Television—the network now under the control of states and societal groups, thanks to the Constitutional Court's 1961 *Television* judgment.
[70] 34 BVerfGE 341 (1973).
[71] 35 BVerfGE 202 (n 67).
[72] ibid 220–26.

weighed 'the intensity of the intrusion into [Wenzel's] personality sphere' against the public's concrete interest in information about the crime.[73]

As a general matter, the Senate conceded, the public's interest in information about serious crimes outweighs the personality interests of those who commit them. But there were other considerations. One was the principle of proportionality: an intrusion into a criminal's personal sphere must be no more invasive than required to satisfy a reasonable public interest in information.[74] Another was chronology. As time passed, a convicted criminal's interest in being left alone increased, while the public's interest in further details about his crime subsided—especially when the new information did little more than reveal the criminal's identity.[75] The final consideration, and in the First Senate's analysis the decisive consideration, was Wenzel's interest in resocialization.

Reforms in recent decades, the Senate noted, had enthroned resocialization as the central goal of criminal punishment.[76] The question of resocialization was so pivotal that the Court took the unusual step of conducting its own factual investigation into the likely effects of the broadcast on Wenzel's chances for successful social integration. This investigation involved expert hearings partly closed to the public. The Court concluded that airing the film would gravely endanger the complainant's opportunity for successful reintegration, and that this fact alone justified enjoining the film. Several features of the film made it particularly perilous for Wenzel's social future. One was its medium. According to expert testimony, filmic presentation of a crime left a more enduring impression on viewers than a print report on readers.[77] The medium, moreover, was to be deployed in prime time before the largest possible audience.

Another endangering feature was the film's emphasis on the perpetrators' homosexuality. The Senate almost suggested that Wenzel's resocialization would be imperilled more by the revelation of his sexual orientation than by the exposure of his criminal past. 'In the complainant's situation', the justices wrote, 'the bond with a female life partner could represent the decisive factor for the success of his reintegration'.[78] Expert testimony indicated that it was entirely possible that Wenzel would find such a partner, but being 'branded as a homosexual' before millions of viewers might doom his odds forever.[79] The Senate conceded that the impact of the broadcast on the complainant's resocialization could not be assessed with precision. But the risks were intolerable—so much so that the Senate not only quashed the lower court decisions but enjoined the broadcast on its own authority. Remanding the case to the lower courts, as the normal procedure required, would generate further publicity, undermining the very resocialization the judgment sought to further.[80]

The judgment distressed some members of the media. ZDF complained that the judgment would limit its ability to dramatize current events.[81] The *Frankfurter Allgemeine Zeitung* noted impatiently that, 'for a certain field of especially "near-to-life" depiction, the hands of the media [are] more bound than before'. This was the 'tribute the Court exacted for resocialization'.[82] Some legal scholars worried that the 'personality right' was running amok.[83] Others countered that the judgment set a useful precedent for the soon-to-be

[73] ibid 226. [74] ibid 232.

[75] ibid 234. This was a bit of a red herring: the names of the perpetrators had long since appeared in all major newspapers. See, e.g., 'Lebach: der dritte Mann' *Der Spiegel* (12 May 1969); 'Lebach: das Trio' *Der Spiegel* (4 May 1970); Hans-Joachim Noack, 'Nicht Apo und nicht Mafia' *Die Zeit* (3 July 1970); 'Gekicher im Saal' *Die Zeit* (10 July 1970); 'Das Unbegreifliche blieb ohne Erklärung' *Die Zeit* (14 August 1970). Between 1970 and 1972, Gernot Wenzel was personally named in connection with the crime twenty-seven times in the *Frankfurter Allgemeine Zeitung*.

[76] ibid 235. [77] ibid 227–29. [78] ibid 242. [79] ibid. [80] ibid 244–45.

[81] In Bruno Waltert, 'ZDF muß Lebach-Fernsehspiel umarbeiten' *Die Welt* (6 June 1973) 2.

[82] 'Verfassungsgericht verbietet Sendung des 'Lebach'-Films' *FAZ* (6 June 1973) 1, 10.

[83] Heinrich Hubmann, 'Persönlichkeitsschutz ohne Grenzen'? (1974) 70 *Archiv für Urheber-Film—Funk—und Theaterrecht* 75.

contentious question of personal data protection.[84] Nearly everyone agreed that the judgment was a remarkable exercise in judicial sociology—robust enforcement of individual rights informed by apparently Deweyite sensibilities.[85] In any case, it was a testament to the First Senate's confidence—its willingness to conduct wide-ranging independent inquiries, to assess social realities on its own authority, and to give its conclusions the force of law, existing statutes notwithstanding.

C. The rights of prisoners

The *Lebach* judgment was also notable, like the *Mephisto* judgment before it, because it gave more than rhetorical weight to the constitutional guarantee of human dignity. In the *Lebach* case, the First Senate enforced dignity—in conjunction with the general personality right—as a concrete norm protecting a social outsider. The weight the judgment gave to dignity was unique, but it wasn't the first time the Court had vindicated the rights of the imprisoned. A year earlier, the Second Senate held that the fundamental rights of prisoners, like those of other citizens, can be limited only by statute or on the basis of statute.[86] This holding bucked the reigning academic and judicial orthodoxy, which posited for prisoners a 'special relationship' with the authority of the state (*besonderes Gewaltverhältnis*). The special relationship doctrine was generally held to justify encroachments on inmate rights that prison administrators thought necessary to maintain safety and order. In its judgment of 14 March 1972, however, the Second Senate explicitly rejected the doctrine as an intolerable relativization of prisoners' basic rights. One commentator summarized the result by paraphrasing Julius von Kirchmann's famous barb about 'The Worthlessness of Jurisprudence as a Science': 'One judgment (of the Constitutional Court) and entire volumes of decisions become waste-paper'.[87]

The case involved an inmate in Celle whose correspondence was confiscated by prison authorities. In a 1967 letter to a prisoner assistance organization, the inmate inveighed against the prison's conditions and vituperated its warden. The prison's administration, which routinely screened inmate correspondence, blocked the letter. When the Celle superior court sustained the confiscation, the prisoner appealed to the Constitutional Court, arguing that the censorship violated his right to free expression under Article 5(1) GG and to postal secrecy under Article 10(1) GG.

The Second Senate held that restrictions of prisoner rights require statutory foundation. In addition, the Senate stressed that even statutory restrictions must be necessary to accomplish a community goal, rooted in constitutional values, and pursued through constitutionally appropriate channels.[88] Moreover, restrictions must conform to the principle of proportionality. Because the federal government had already proposed a comprehensive law governing the execution of criminal sentences, the Senate allowed limitations on prisoner rights that lacked a statutory basis to remain in force for a transitional period.[89] But

[84] See Martin Löffler, 'Licht- und Schattenseiten. Das Lebach-Urteil aus der Sicht des Art. 5 GG' (1973) 36 *Funk-Korrespondenz* 3.

[85] On the judgment's sociological aspects, see Friedrich Kübler, 'Sozialisationsschutz durch Medienverantwortung als Problem richterlichen Normierens' in Friedrich Kübler (ed), *Medienwirkung und Medienverantwortung. Überlegungen und Dokumente zum Lebach-Urteil des Bundesverfassungsgerichts* (Nomos 1975) 7; Kurt Lüscher, 'Jurisprudenz und Soziologie. Die Zusammenarbeit in einem konkreten Rechtsfall' in ibid 81–114. See also Kurt Lüscher, 'Medienforschung und Rechtsprechung. Soziologische Bemerkungen zum Fall Lebach' (1974) 22 *Archiv für Presserecht* 643.

[86] 33 BVerfGE 1 (1972).

[87] Heinz Müller-Dietz, 'Verfassung und Strafvollzugsgesetz' (1972) 26 *NJW* 1162.

[88] 33 BVerfGE 1 (n 86) 11. [89] ibid 13.

that period must be brief. After the close of the current legislative period in autumn 1973, restrictions on prisoner rights with no foundation in statutory law would become unconstitutional.[90] Parliament, in other words, could either act swiftly or surrender the field to the courts.

It went without saying, the Senate added, that the concession of a transition period did not authorize arbitrary infringements of inmate rights. During the transition, competent courts and administrative bodies would review encroachments case by case, assessing in each instance the necessity of the encroachment to an orderly and efficient prison regime.[91]

The Senate modelled such review in the case at hand. Supervising the prisoners' correspondence was justified, the Senate concluded, since uninhibited contact with the outside world might endanger the prison's successful operation. Prison administrators need not, for instance, allow inmates to plot their own escape with enablers outside. But intercepting the complainant's letter because it criticized the prison and insulted its warden violated the prisoner's freedom of opinion.[92] The letter contained subjective value judgements protected by Article 5 GG. Article 5, the justices explained, protected not only 'valuable' views but all views—'right' views as well as 'wrong', 'emotional' views as well as 'rational'.[93] To protect only views of 'a certain ethical quality' would relativize Article 5's comprehensive guarantees.[94] In any case, it was nearly impossible, in a pluralistic society, to draw a definitive line between valuable and worthless opinions. In a free democratic society all views—even dissident and derogatory views; even the views of prisoners—enjoy constitutional protection.

The only thing displeasing about the Court's judgment, wrote Christian Starck, a law professor in Göttingen, was that it hadn't come fifteen years earlier.[95] A host of prisoner complaints had flooded the Court in previous decades, nearly all of which had been dismissed out of hand by three-judge screening panels. Starck speculated that this case got to the full Senate only through the efforts of the Senate's newer members—Joachim Rottmann and Martin Hirsch—who as members of Parliament had grappled with the problem of criminal sentence enforcement.[96] But better late than never, Starck concluded. The judgment spoke to 'the undoubted merit of a Constitutional Court not otherwise especially meritorious in [the matter of] criminal enforcement'.[97] The judgment also highlighted the Court's willingness to take fundamental rights seriously; to enforce them vigorously; and, more controversially, to tell Parliament what to do.

D. Medical licensing and appealing to Parliament

This last willingness was on display again within weeks. The case involved occupational freedom claims from two complainants, both of them medical specialists, who had been censured by professional conduct courts in their respective state medical associations. The first complainant, from North-Rhine-Westphalia, was punished for treating a patient outside his specialty field; the second, from Baden-Württemberg, for claiming more than one specialty. The state medical associations were public institutions to which the state parliaments had delegated authority to regulate the medical profession. The complainants contended that their sanctions violated their right to occupational freedom under Article 12(1) GG. The Constitutional Court agreed, quashing the judgments against the complainants and remanding the cases to the courts below.[98]

[90] ibid. [91] ibid. [92] ibid 14. [93] ibid 14–15. [94] ibid 15.
[95] Christian Starck, 'Anmerkung' (1972) 27 *JuristenZeitung* 357, 360. [96] ibid 361.
[97] ibid 362. [98] 33 BVerfGE 125 (1972).

In its written judgment of 9 May 1972, the First Senate acknowledged that Article 12 did not decree that only legislatures or executive bodies empowered by legislatures could limit occupational freedom.[99] Within certain limits, legislatures could empower autonomous bodies to issue and enforce regulatory ordinances. But the Basic Law limited such delegations.[100] The legislature could not quit the field entirely—could not completely surrender its lawmaking authority or influence over its delegatee. Special limits on delegation must be observed, particularly when the delegation included power to limit fundamental rights. In such cases the legislature had a heightened responsibility.[101]

On the other side, occupational freedom was intimately linked to the development of the human personality. An individual's right to self-determination could be limited only to further general interests. And such interests must be identified by the legislature. This didn't mean that legislatures could never delegate authority to occupational associations. But it did mean that the legislature must actively protect the freedom of individuals and the interests of minorities against encroachment by such associations.[102] The legislature must identify regulatory aims and provide guidance for their fulfilment. In the current case, the delegations were too general. State parliaments had given state associations a regulatory *carte blanche*.[103]

Restrictions on occupational freedom imposed by occupational associations would be assessed, the Senate explained, on a case-by-case basis, consistent with the proportionality principle and the three-tier schema outlined in the 1958 *Pharmacy* judgment.[104] But, as a precondition to such case-by-case adjudication, basic guidelines for assessing an individual's status as a specialist must be stipulated by statute. Associations could then add implementing regulations to standards established by the legislature. But state legislatures must establish such standards.

Traditionally, the power of judicial review had been understood as a negative and defensive power—the power of courts to tell legislative and executive actors what they must *not* do. But just as the Court in its *Lüth* judgment had defined fundamental rights as affirmative values radiating into all areas of law, so the Court in its *Medical Licensing* judgment had—for the second time in as many months—moved beyond defensive prevention of political overstepping to affirmative prescription of political action. To be sure, the expansions were cautious, almost reluctant, and limited to the facts of the two cases. Most scholars assumed the two cases were exceptional, and would have few, if any, successors.[105] That assumption was contradicted by events.

III. Political Justice

In December 1971, just days after becoming chief justice, Ernst Benda gave an interview with *Der Spiegel*. In response to the interviewer's most pointed questions about the Court's political role, Benda declared flatly that '[t]he Federal Constitutional Court does not make policy'—or, differently translated, the Court 'does not do politics' (*Das Bundesverfassungs gericht macht keine Politik*).[106] Asked whether, having exercised power in Bonn, he wished now to do so in Karlsruhe, Benda downplayed the Court's political significance. The new chief justice stressed that the Court's role was not to wield power but to restrain it, not to form policy but to enforce constitutional limits on those who did. But

[99] ibid 155. [100] ibid 157. [101] ibid 158. [102] ibid 160. [103] ibid 160–65.
[104] ibid 160.
[105] See the recollection in Martin Kriele, 'Anmerkung' (1975) 30 *JuristenZeitung* 205, 223.
[106] 'Ortwechsel ist kein Gesinnungswechsel' *Der Spiegel* (13 December 1971).

though Benda refused to opine on questions that might come before the Court, and though he expected his new post and perspective to change his mind on some constitutional questions, Benda conceded that a mere 'change of place does not necessarily bring with it a change of viewpoint'. He admitted frankly that he remained a political animal: 'I am still a member of the CDU and intend so to remain. Anything else', he concluded, 'would be hypocrisy'.[107]

To many observers, the 1970s would make nonsense of Benda's claim—really a hereditary piety—that the Court didn't make policy. The claim would have to be qualified into meaninglessness or abandoned altogether. Beginning early in Benda's tenure and continuing until the end of the decade, the justices confronted an astonishing sequence of high-profile political disputes—cases whose frequency and explosiveness brought the Court unprecedented attention and unparalleled controversy. Never was the Court more aggressive in correcting or quashing, delimiting or dictating government policy. Never had the general public responded with such full-throated screeds against the Court's overreaching.

The Court came to be seen in many quarters as the foremost institutional foe of Social-Liberal reforms. Those reforms were the legislative fulfilment of Brandt's campaign promise to dare more democracy. In checking and in some cases nullifying them, the Court was criticized for the first time as a threat to democracy itself. The public's esteem for the Court sank to unprecedented depths. The Court's critics howled with unprecedented furor. Many of the Court's quondam champions became detractors. But the Court's existence and function were never seriously challenged, and when, at decade's end, the season of confrontation came to an abrupt halt, the Court's reputation recovered rapidly, soon surpassing its previous zenith. The test, however, was very real. The confrontation was greater than the sum of its parts. The crisis played out in crescendo.

A. University admissions

'Ladies and gentlemen', trumpeted Chancellor Willy Brandt in October 1969, 'education and vocational training, scholarship and research stand at the summit of the reforms that must be carried out among us'.[108] Education, Brandt continued, would form the heart of his administration's domestic policy. 'The school of the nation', he added, modifying a proverb about the Prussian army, 'is the school'.[109] Among other things, the new government planned to promulgate sweeping legislation providing universal guidelines for university governance.

The theme was not a new one. Education reform was a pressing issue throughout the 1960s. Georg Picht, a theologian and educator, fired the debate's opening salvo with his 1964 jeremiad, *The German Education Catastrophe*.[110] Standing alone, Picht's title was dire enough; but it also echoed the title of Friedrich Meinecke's *The German Catastrophe*, a meditation on Nazism by the aging dean of German historians.[111] Picht prophesied the 'third great collapse of German history in this century' unless Parliament and people united in a concerted effort to revitalize and reform the German education system, which, the economic miracle notwithstanding, remained vastly underfunded and abominably run.[112] Germans needed to learn, Picht admonished, to see educational investment as economic investment. He

[107] ibid.
[108] (1969/70) 71 *Verhandlungen des Deutschen Bundestages, Stenographische Berichte* 26.
[109] ibid. [110] Georg Picht, *Die deutsche Bildungskatastrophe* (Walter 1964).
[111] Friedrich Meinecke, *Die deutsche Katastrophe*, 2nd edn (FA Brockhaus 1946).
[112] Picht (n 110) 87.

called for massive increases in educational funding. Other powerful voices soon followed. Where Picht appealed to Germans' interest, the sociologist Ralf Dahrendorf appealed to their conscience. '*Bildung ist Bürgerrecht*', was Dahrendorf's mantra—'Education is a civil right'.[113] In Dahrendorf's view it was the duty of a modern democracy to secure equal educational opportunity for all its citizens. Education, Dahrendorf wrote, was 'the foundation of freedom for the individual and his society'.[114] There must be more schools, better schools, and schools for all.

The need was especially acute at the university level. By the mid-1960s, West German universities were hopelessly overcrowded. Between 1952 and 1967 the number of university students in West Germany roughly doubled to about 270,000. The number of first-year students more than doubled, from 25,000 to 51,000. The construction of new universities and the expansion of existing faculties, however, did not keep pace. They did not come close to keeping pace. In some fields, such as medicine, the crisis was severe. At medical schools across the country there were more applicants than places, more students than faculty or facilities could possibly accommodate. This was not a case of student supply exceeding societal demand. The nation manifestly needed more doctors, and in practically every field of medical expertise. But resources were slim. Obviously, the crisis called for structural reform and lavish funding. These, however, were slow in coming.

In the meantime, state governments dealt directly with the immediate problem of overcrowding. Some did so in draconian fashion, adopting a system known as *numerus clausus*—admission ceilings, sometimes absolute, beyond which an entering class could not expand regardless of the number of qualified applicants. Hamburg passed such a law for its medical school in April 1969; Bavaria followed suit in July 1970. Applicants whose test scores qualified them for admission, but who were repeatedly denied on *numerus clausus* grounds, challenged both laws before the Constitutional Court.

Their appeals were only partly successful.[115] Rhetorically, the First Senate's judgment of 18 July 1972 granted the justice of the petitioners' cause. 'The more strongly the state applies itself to the social security and cultural advancement of its citizens', the justices wrote, 'the more the relationship between citizen and state will be marked, alongside the original fundamental-rights postulate of securing liberty against the state, by the complementary demand that participation in state services be secured by fundamental right'.[116]

This was a striking innovation in the Court's understanding of fundamental rights. In this view, rights not only defended the individual against state aggression and radiated against third-party abuse, they guaranteed participation in state benefits. The more central a state service was to a citizen's welfare, the more exactingly the equality clause governed that service's distribution. What the state provided for some, it must provide for all—and on equal terms. In the case of education, the equality demands of Article 3 GG were complemented by the occupational freedom guarantees of Article 12. In many modern professions, an occupation's educational prerequisites were part of the occupation itself. It followed that the Court must treat a restriction on the choice of educational path as it would a restriction on the choice of profession. This meant that, in accordance with the 1958 *Pharmacy* judgment, the justices must subject such restrictions to their severest scrutiny.[117]

An absolute *numerus clausus* on medical admissions effectively banned many applicants from their chosen occupation on the basis of circumstances over which the applicants themselves had no control. Under the Court's prevailing doctrine, this was the very worst sort of

[113] Ralf Dahrendorf, *Bildung ist Bürgerrecht* (Nannen 1965). [114] ibid 35.
[115] 33 BVerfGE 303 (1972). [116] ibid 330–31. [117] ibid 330.

infringement on occupational freedom—an infringement that could be justified only by the most pressing public need. What's more, the infringement was arbitrary. Absolute *numerus clausus* meant that one class of citizens received the full measure of state educational services while another class, equally qualified, received nothing—or must in any case wait until the service might do them no good. This, the justices wrote, was 'blatant inequality'.[118]

Blatant, but not fatal. From the constitution's equality, occupational freedom, and social state clauses the Senate derived an individual 'right of admission to university study'. But the justices also held that this right could be limited by law or on the basis of law. If such a law took the form of an absolute *numerus clausus* for first-year students in a given field, it could survive constitutional scrutiny only when implemented on the basis of objective criteria defined by law. The state must give every applicant a fair chance at admission, show the utmost regard for each individual's choice of study, and demonstrate that every post in a given department was filled to capacity. The Hamburg law failed this standard because it was too general.[119] The Hamburg legislature had not fixed the criteria for admissions bars itself but had delegated such decisions to the universities. This was impermissible, but the Court granted a grace period in which the state could mend the flaw. The existing provision could remain in force, but only until the beginning of summer semester 1973.[120] The Bavarian law largely survived, though the Court did strike a provision that priveleged Bavaria's native daughters and sons. The desire to enable students to study close to home was understandable and praiseworthy. But the right of admission to university study was the right 'of all Germans', among whom Article 3 forbade geographical distinctions.[121]

Although the judgment made clear that *numerus clausus* was constitutionally problematic, for the time being it left the laws essentially intact. It also left open the million dollar question whether the individual right to university study entailed an affirmative state duty to provide adequate schools and sufficient places for all who wished to study. And it left the individual petitioners empty handed.

The outcome was sufficiently murky that all sides could claim victory. The judgment pleased no one; but everyone pretended to be pleased. 'Seldom has a judgment of the Constitutional Court found so much approval among politicians', wrote Friedrich Karl Fromme. 'The federal government and the federal opposition are equally content'.[122] All parties construed the judgment to their own purpose. Klaus von Dohnanyi (SPD), the federal minister for science and education, welcomed the judgment as an affirmation of his ministry's policy and a blessing on its ambitions for comprehensive higher education reform.[123] Berthold Martin, the CDU's education spokesman, retorted that the judgment gave no cause for 'partisan-motivated *Schadenfreude*'. On the contrary, it made the state more accountable for university administration and in this regard comported with Union aims.[124] Hildegaard Haam-Brücher, an FDP education expert, saw in the judgment a transfer of competence in education policy from the state to the federal level—a transfer that fulfilled her party's hopes.[125]

Such outward approval veiled private discontent. Robert Leicht, writing for the *Süddeutsche Zeitung*, warned that the Court had relativized the very rights it proclaimed; it had held that rights could be rationed when resources ran thin, and implied that positive rights were never absolute.[126] This was inevitable, Leicht acknowledged, but it was best to

[118] ibid 333. [119] ibid 345. [120] ibid 305. [121] ibid 353.
[122] Friedrich Karl Fromme, 'Verfassungsgericht mit Augenmaß' *FAZ* (21 July 1972) 2.
[123] 'Urteil über Numerus Clausus findet einhellige Zustimmung' *Die Welt* (20 July 1972) 1.
[124] ibid 2.
[125] ibid. See also 'Bayerische 'Landeskinder-Vergünstigung' nichtig' *Süddeutsche Zeitung* (19 July 1972) 1, 2.
[126] Robert Leicht, 'Salomonisches zum Numerus Clausus' *Süddeutsche Zeitung* (19 July 1972) 4.

make it explicit. It was time for the Federal Republic to rid itself of the illusion that it could simply articulate socio-political goals in the constitution and then take the matter to the judges when those goals went unfulfilled. The Court itself, Leicht argued, had made the best of a bad situation and scored a tactical triumph as well. By admitting that strict enforcement was not possible, the Court could issue clear guidelines regarding what *was* possible.[127] But this, argued Bernd Nellessen in *Die Welt*, was precisely what made the judgment appalling.[128] It was not the business of a court of law to issue general legislative guidelines. In any case, the guidelines the Court gave would prove bewilderingly difficult to apply.[129]

Events proved both commentators right. The political application of the judgment was contentious and confused and required ongoing guidance from Karlsruhe. Donald Kommers later termed the *Numerus Clausus* judgment 'the beginning of a major judicial intervention in university governance', a process by which 'the Constitutional Court ... transformed itself into a veritable ministry of education', issuing a series of judgments that 'accelerated the movement toward central control of higher education'.[130] The *Numerus Clausus* judgment also marked the beginning of 'a major judicial intervention' in the political process more broadly.

B. University governance

All education reformers agreed that West German universities were overcrowded. Others argued, more controversially, that they were undemocratic. 'Under the academic robes', student protestors chanted, 'a thousand years' of mold'. Rebels proposed to democratize university governance by revolutionary means. Jürgen Habermas famously skewered this proposal as a juvenile exercise in 'left fascism'.[131] At the same time, Habermas was a passionate proponent of democratic university *reform*,[132] as was the political scientist Kurt Sontheimer,[133] with whom Habermas later exchanged fire on the subject of domestic terrorism.[134] Reformers' core demand was increased participation. Henceforth, the university's governing bodies must not consist only of professors, and the rector must not be elected by the professors alone. Instead, all who were affected by these bodies' decisions should have a part in shaping them. These general demands were radical enough. A smaller minority of reformers called for democratic constitutions that would grant all members of the university community—janitors and students, as well as researchers and professors—a voice in university governance.

Reformers rallied around a pair of watchwords. The first was *Gruppenuniversität*, or 'groups university', which meant that representatives from the campus's various groups would comprise its governing councils. The second watchword, *Drittelparität*, or 'one-third parity', was more specific. It called for equal representation among three governing groups—students, scholarly assistants or researchers, and professors. Non-tenured

[127] ibid. [128] Bernd Nellessen, 'Schlechte Noten per Gerichtsbeschluß' *Die Welt* (20 July 1972) 4.

[129] ibid.

[130] Donald P. Kommers, *The Constitutional Jurisprudence of the Federal Republic of Germany*, 1st edn (Duke University Press 1989) 302, 303.

[131] Jürgen Habermas, *Protestbewegung und Hochschulreform*, 1st edn (Suhrkamp 1969) 10. Habermas was hotly rejoined by his mentor, Wolfgang Abendroth. See Wolfgang Abendroth, Jürgen Habermas, and Peter Brükner, *Die Linke antwortet Jürgen Habermas* (European Verl-Anst 1969).

[132] Habermas (n 131) 202–43.

[133] See Kurt Sontheimer, 'Die Demokratisierung der Universität' in Alexander Schwan and Kurt Sontheimer (eds), *Reform als Alternative. Hochschullehrer antworten auf die Herausforderung der Studenten* (Westdeutscher Verlag 1969) 63.

[134] See the exchange in Jurgen Habermas, *Kleine politische Schriften I-IV* (Suhrkamp 1981) 367–406.

professors, ignored by the existing constitutions, applauded the parity proposals eagerly. Non-scholarly employees soon called for greater representation as well.[135] Only later did the idea gain currency that other employees—maintenance and technical crews, for instance—should have a voice as well. Some then spoke of a *Viertelparität*, or 'one-fourth parity'.

Naturally, calls for reform encountered resistance, and the more radical the proposals, the more intense the opposition. Conservatives depicted a parade of horribles in which cleaning ladies would pass judgment on professorial appointments or promotions. The conservative catchphrase was *Fremdbestimmung*, or foreign determination—the specter of amateurs telling experts how to do their jobs, of outsiders deciding matters about which they knew nothing. Wilhelm Hennis deplored knee-jerk applications of democratic models to contexts, such as higher education, in which they had no relevance. The university, Hennis wrote, was like a family. Making students equal to professors in university governance was like making children equal to parents. For all the heady enthusiasm about 'democratization', there were places where democracy did not belong. Those who thought differently suffered from what Hennis called *Adamsneid*; they envied Father Adam's appearance in Eden in a state of full maturity. Proponents of radical university democratization, Hennis implied, were revolting against nature.[136]

Most professors of public law agreed. Some argued that to include even one student on an academic board would violate professors' academic freedom. A Lower-Saxon law implementing the parity model in that state's universities allowed these lawyers to test their arguments before the Constitutional Court.

The law in question—'the leftest of all leftist laws', groused Wilfried Hasselmann, boss of the Lower Saxon CDU[137]—was passed on 26 October 1971. It placed university administration in the hands of a series of councils, with professors holding half or fewer than half of the seats on each university's central councils. The academic council, for instance, would consist of twenty-four professors, twenty-four non-tenured researchers, twenty-four students, and sixteen non-academic employees, a ratio of 3:3:3:2. On the appointments committee, the professor–researcher–student ratio would be 4:2:2; on the curriculum commission, 2:4:4. Unsurprisingly, the Lower-Saxon professoriate was not pleased. No fewer than 398 professors challenged the law in a constitutional complaint. In response to the professors' petition, federal and state education officials filed briefs defending the statute; the national Rectors' Conference and other faculty organizations filed further briefs opposing it. In the meantime, every other state in the Federal Republic passed legislation enacting some variation on the parity model. Similar proposals surfaced at the federal level. All parties awaited the Court's judgment with interest and impatience. Anticipating student protestors, Chief Justice Benda asked police to cordon off the Court building during oral arguments in December 1972.[138] The justices were justifiably nervous; their building had been bombed earlier in the year.[139] In the event, the cordon proved unnecessary. The students, menacing or otherwise, never showed up.

[135] See Hartmut Boockmann, *Wissen und Widerstand. Geschichte der deutschen Universität*, 1st edn (Siedler 1999) 258–62.

[136] Hennis, *Demokratisierung. Zur Problematik eines Begriffs* (Westdeutscher Verlag 1972) 35–39.

[137] In 'Eine glänzende Bestätigung' *Der Spiegel* (4 June 1973).

[138] See Axel Azzola, 'Verfassungsbeschwerden gegen Hochschulgesetze kommen zur Entscheidung' (1973) 6 *Zeitschrift für Rechtspolitik* 44.

[139] Hellmuth Rieber, 'Ungenügende Sicherheit. Nach dem Attentat in Karlsruhe' *Badischer Zeitung* (Freiburg im Breisgau, 17 May 1972).

The First Senate's judgment of 29 May 1973 upheld the Groups University model in principle but condemned its application in Lower Saxony.[140] The Senate began by defining the academic freedom guarantees of Article 5(3) GG more fully than it had ever done before. Those guarantees, the Senate held, were both defensive and affirmative. On the one hand, the state must not impede scholarly inquiry. On the other hand, it must foster an institutional framework in which such inquiry can be carried out. The state must sponsor a system of higher education, and it must organize that system in a manner that honours the right to free scholarly activity.

That right neither endorsed the traditional German university nor demanded its reform.[141] The legislature enjoyed considerable discretion in organizing the university system and adapting it to contemporary needs. Nothing in the constitution forbade including students and non-academic staff in university decision-making; nor, *a fortiori*, did the constitution preclude involving non-tenured researchers and lecturers, whose scholarly activities also enjoyed Article 5 protection.[142] Indeed, the judgment went so far as to hold that non-tenured scholars *must* participate in decisions that affect them.

But the position of professors, particularly in research and teaching, was special. However a state organized its universities it must honour professors' unique scholarly role. Professors had higher scholarly qualifications, greater responsibility for the school's functionality and reputation, and a greater concern for the long-term impact of administrative decisions. Professors' role in teaching and research was paramount, and their influence within relevant councils must reflect that preeminence.

There was a flip side, then, to the muscular equality rhetoric of the *Numerus Clausus* judgment. Yes, Article 3 GG forbade the unequal treatment of equal groups; but it also forbade the equal treatment of unequal groups.[143] In matters of research and pedagogy, all university groups were not equal. In some administrative matters, professors might be *primus inter pares*. But in matters of research, teaching, and appointment, they were simply *primus*. More concretely: In matters related to teaching, the influence of professors must be controlling (*maßgebend*); in matters related to research and appointment, it must be decisive (*ausschlaggebend*). Lest legislatures miss the point, the justices gave these adjectives arithmetical substance. On councils concerned with teaching, professorial representatives must hold at least half the votes; on councils concerned with research or hiring, more than half. Insofar as the Lower Saxon law failed to grant professors a veto on teaching councils and an absolute majority on research and hiring councils, the Senate declared it unconstitutional—though strangely (some scholars thought incoherently),[144] not yet null and void.[145]

140 35 BVerfGE 79 (1973). 141 ibid 116. 142 ibid 125.

143 ibid 127 (citing 17 BVerfGE 337, 354 (1964); and 30 BVerfGE 292, 333 (1971)).

144 See Dian Schefold and Klaus-Rüdiger Leske, 'Hochschulvorschaltgesetz: verfassungswidrig—aber nicht nichtig' (1973) 27 *NJW* 1297.

145 This was an early instance of a practice, which the Court 'literally invented', by which a norm is pronounced 'incompatible' (*unvereinbar*) with the constitution but left on the books (though not always in force) until a required legislative revision. For a few samples of a massive literature, see Otto Bachof, 'Der Richter als Gesetzgeber?' in Otto Bachof et al (eds), *Wege zum Rechtsstaat. Ausgewählte Studien zum öffentlichen Recht* (Atheneum 1979) 344–358; Jens Blüggel, *Unvereinbarkeitserklärung statt Normkassation durch das Bundesverfassungsgericht* (Duncker and Humblot 1998); Peter E. Hein, *Die Unvereinbarerklärung verfassungswidriger Gesetze durch das Bundesverfassungsgericht* (Nomos 1988); Apostolos Gerontas, 'Die Appellentscheidungen, Sondervotumsappelle und die bloße Unvereinbarkeitsfeststellung als Ausdruck der funktionellen Grenzen der Verfassungsgerichtsbarkeit' (1982) 97 *Deutsches Verwaltungsblatt* 486; Hermann Heußner, 'Folgen der Verfassungswidrigkeit eines Gesetzes ohne Nichtigerklärung' (1982) 36 *NJW* 257; Jörg Ipsen, 'Nichtigerklärung oder 'Verfassungswidrigerklärung'—zum Dilemma der verfassungsgerichtlichen Normenkontrollpraxis' (1983) 38 *JuristenZeitung* 41; Hartmut Maurer, 'Zur Verfassungswidrigerklärung von Gesetzen' in Hans Schneider and Volkmar Götz (eds), *Im Dienst an Recht und Staat: Festschrift für Werner Weber zum 70. Geburtstag dargebracht von Freunden, Schülern und Kollegen* (Duncker and Humblot 1974) 345; Albrecht Peter Pohle, *Die Verfassungswidrigerklärung von Gesetzen* (Cirencester 1979); Michael Sachs, 'Tenorierung

The judgment was not unanimous. Justices Helmut Simon and Wiltraut Rupp-von-Brünneck dissented, not so much from the majority's basic interpretive framework as from the specificity of its conclusions. The two dissenters recoiled from the Court's derivation of detailed demands for university organization from a sparse constitutional text. In its entirety, the relevant clause of Article 5(3) read: 'Art and scholarship, research and teaching are free'. By deriving minute prescriptions from an eight-word general clause, the majority had exceeded its mandate. The Court, the dissenters complained, was supplanting the legislature.[146] It was neither the last nor the loudest dissent in which these two justices joined.

This dissent provided the starting point for academic and journalistic criticism. Bernhard Schlink dismissed some of this criticism as misguided, but not in terms unambiguously flattering to the Court. 'The objection that this catalogue [of required voting proportions] could not possibly be the result of an interpretation of Article 5(3)', Schlinck wrote, 'misses the mark with the Constitutional Court, which in its implementation of value decisions has long since abandoned the boundaries of juristic hermeneutics'.[147] But for many critics the problem was not that the Court had offended canons of construction but that it had flouted the separation of powers. 'Last week', *Der Spiegel* reported, 'six judges in Karlsruhe did what should have been done by federal and state politicians: they made university policy, binding on universities throughout the Republic'.[148] In one way or another, the article predicted, 'this judgment and the [future] jurisprudence of Karlsruhe ... will define the university policy of the entire Federal Republic. For good or ill, all state university laws and even the planned federal University Law ... will have to orient themselves toward the vote in Karlsruhe'.[149]

This prediction proved prophetic. All state university laws, as well as the federal university legislation finally passed in 1976, took their cues from the Court's *University Governance* judgment. But while some observers bemoaned the Court's activism, others applauded its pragmatism: the Court had charted a middle way that gave legislators discretion but kept reforms in check. *Die Zeit* published a handful of responses to the judgment, nearly all of them positive, from state ministers affected by it.[150] Student protestors huffed about an 'endorsement of the ruling class', but even SPD state ministers saw a silver lining in the Court's approval, in principle, of the Group Model.[151] Many commentators were grateful that the Court had given constitutional clarity to a contested question. Bavarian culture minister Hans Maier—who in subsequent decades would take criticism of the Court to unprecedented levels—was effusive. Maier extolled the Court's 'historic achievement' in ending 'the unsavory parity dispute ... and with it a years-long barrier to university reform'.[152] Activist the Court may have been. But at least, some contemporaries thought, it had made a good job of it. Professorial response to the judgment, *mirabile dictu*, was tempestuously approving.[153]

bei Normenkontrollentscheidungen des Bundesverfassungsgerichts' (1982) 36 *Die öffentliche Verwaltung 23; Hans-Jochen Vogel, Nachsichten. Meine Bonner und Berliner Jahre* (Piper 1997) 222 ff; S Kreutzberger, *Die gesetzlich nicht geregelten Entscheidungsvarianten des Bundesverfassungsgerichts* (Lang 2007) 103; Paul E Rueda Leal, *Die verfassungsgerichtliche Unvereinbarerklärung—Untersuchungen zur Entwicklung und Funktion dieser Rechtsfigur im deutschen Recht und zu ihrer Übertragbarkeit ins costaricanische Recht* (1999) 24.

146 35 BVerfGE 79 (n 140) 148–70.

147 Bernhard Schlink, 'Die Wissenschaftsfreiheit des Bundesverfassungsgerichts' (1973) 25 *Die Öffentliche Verwaltung* 543.

148 'Eine glänzende Bestätigung' *Der Spiegel* (4 June 1973). 149 ibid.

150 'Stellungnahmen von Betroffenen zur Entscheidung des Bundesverfassungsgerichts' *Die Zeit* (8 June 1973). See also 'Länder reagieren unterschiedlich auf das Urteil zum Vorschaltgesetz' *Die Welt* (31 May 1973) 2.

151 'Länder reagieren unterschiedlich auf das Urteil zum Vorschaltgesetz' *Die Welt* (31 May 1973) 2.

152 ibid 2.

153 Ralph Müller-Terpitz, 'BVerfGE 35, 79—Hochschulurteil' in Jörg Menzel (ed), *Verfassungsrechtsprechung. Hundert Entscheidungen des Bundesverfassungsgerichts in Retrospektive* (Mohr Siebeck 2000) 217–18.

Taken together, the *Numerus Clausus* and the *University Administration* judgments profoundly shaped higher education policy in the Federal Republic. The two decisions defined the framework for reform in a field that the governing coalition placed at the summit of its domestic agenda. But the two judgments had a deeper and more durable impact on the Court itself. They marked the emergence of a new conception of fundamental rights. As Bernhard Schlink summarized:

> In the *Numerus Clausus* decision, the Federal Constitutional Court for the first time developed in a fundamental way its new conception of basic rights in the Benefits State [*Leistungsstaat*]. [In that judgment], the Court posited, alongside the classical fundamental rights function of securing freedom vis-à-vis the state, the complementary function of securing participation in state benefits. The Court conceives the Benefits State as a state of public provisions, through whose distribution fundamental rights are realized, and whose organization and procedures are therefore fundamental rights problems. In the [University Administration] decision, the Court concretized this conception and made clear how seriously it takes it: fundamental rights are also actionable as participatory rights.[154]

Among other things, this new conception entailed a departure from the classical dichotomy of state and society. It would require the Court to intervene in societal and political realities, and it would embroil the Court, inevitably, in political controversy.[155]

Such ensnarement came with a rapidity and intensity that neither Schlink, nor the justices, nor anyone else foresaw. Indeed, it began even before the Court announced its *University Administration* judgment. This controversy swirled, not around the centrepiece of the government's domestic policy, but around the crown jewel of its foreign policy. It was the most politically charged matter to come before the Court since the European Defence controversy two decades earlier. At issue were the constitutionality of the Brandt government's proudest diplomatic achievements and the legitimacy of its new Eastern policy, or *Ostpolitik*. The stakes were staggering, and the world was watching.

C. Judicial politics and foreign policy

Ostpolitik was as crucial to the geopolitical aspirations and historical legacy of West Germany's fourth chancellor as *Westbindung* was to its first. Egon Bahr, Brandt's brilliant foreign policy advisor, gave the new policy its Sybilline mantra in 1963. Bahr called for *Wandel durch Annäherung*—'change through rapprochement'. The phrase implicitly recognized that there was no direct route to reunification—that confrontation with the East in an Adenauerian *Politik der Stärke* (policy of strength) had failed and should be replaced by a gradualist, pragmatic approach based on de facto acceptance of the status quo and efforts to improve human conditions in East Germany. It was a 'politics of small steps' designed to soften the chill of Cold War confrontation, to 'normalize' relations between the two Germanys, and to stabilize the security of Europe. The concept guided Brandt's tactical and strategic thinking as foreign minister in the grand coalition. After becoming chancellor in 1969, Brandt made it the motor of sweeping geopolitical transformation.

The first diplomatic fruits of *Ostpolitik* were a pair of 1970 treaties with the Soviet Union and Poland. In the Moscow Treaty, signed on 12 August 1970, the Federal Republic and the Soviet Union vowed to respect the existing borders of all European

[154] Schlink (n 147) 545. [155] ibid.

states and to hold inviolable both the Oder-Neiße Line (Poland's Western border) and the border between the two Germanys. The Warsaw Treaty, signed 7 December 1970, similarly acknowledged the Oder-Neiße Line as perpetual and inviolable. The Federal Republic thereby renounced definitively any hope that Germany would return to its 1937 borders.

On the day of the signing, Brandt visited the Warsaw Ghetto memorial. There, in one of the most moving political gestures of the twentieth century, the chancellor dropped to one knee in an expression of contrition and atonement. Brandt, an anti-Nazi and an exile, had knelt, *Der Spiegel* reported, 'for all who needed to kneel but didn't—because they dared not or could not or could not dare'.[156] The gesture made Brandt an international hero. A year later, it helped him win the Nobel Prize. But a contemporary survey found that while 41 per cent of Brandt's countrymen thought the gesture fitting, 48 per cent thought it exaggerated.[157]

Many West Germans considered the Warsaw Treaty itself an injustice. The so-called *Heimatvertriebene*—German citizens expelled from their homes in territory separated from Germany at the end of the Second World War—must now abandon any hope of recovering lost land and property. In reply, Brandt insisted that the treaty surrendered nothing that had not been long since squandered. But it was a bitter blow to many who had dreamed of an ultimate homecoming in the fullness of time. West German politicians from all parties—with an eye on the power of the refugee vote—had promised to keep the border question open. The Moscow and Warsaw treaties slammed it shut.

Ratification of the treaties was delayed until the United States, France, Great Britain, and the Soviet Union could regulate the status of Berlin in light of the new atmosphere of détente. Germany's erstwhile occupiers did so in the so-called 'Four-Power Agreement' of 3 September 1971. After this, the Soviets expected a swift ratification. But by this time the Brandt coalition's majority had shrunk to two votes. After the treaties were signed, several members of the Bundestag with refugee backgrounds bolted for the opposition benches. On 25 April 1972, the CDU/CSU faction hazarded the first call for a constructive vote of no-confidence in the history of the Federal Republic. The constructive no-confidence vote was one of the Basic Law's historic novelties. The CDU/CSU opposition hoped to employ it to oust Brandt from the chancellorship and replace him with the Union leader, Rainer Barzel. The gambit failed by two votes. Two, perhaps three Union members hadn't backed Barzel, and their defection saved the SPD–FDP coalition. (It later became known that one of them, Julius Steiner, had been paid 50,000 Marks by the East German State Security Ministry to vote against Barzel.)

The coalition survived, but the treaties remained at risk. Not only were they losing parliamentary support, they also faced a constitutional challenge.[158] Former inhabitants of Schlesien and Northeast Prussia contended that the treaties violated their property rights under Article 14 GG. The parents of a fifteen-year-old boy who was living with his grandparents in what was formerly Hindenburg complained that the treaty would cost their son his right to German citizenship—a violation of Article 16 GG. The complaints, filed at the eleventh hour, asked the Court to annul the treaties and block their ratification.

[156] Hermann Schreiber, 'Ein Stück Heimkehr' *Der Spiegel* (14 December 1970).

[157] In Conze (n 17) 438–39.

[158] Conservative jurists also assailed the constitutionality of the treaties publicly. One of the most strident among them was Hermann Weinkauff, former president of the *Bundesgerichtshof*. See Hermann Weinkauff, 'Die Ostverträge sind verfassungswidrig' *Rheinischer Merkur* (Bonn, 18 February 1972).

Time was short. The Bundestag had already voted to ratify the treaties; the Bundesrat followed suit shortly thereafter. The treaties would enter force on 3 June 1972; the Court was pressed to process the petitions swiftly.

Ernst Benda, the chief justice, was on holiday—celebrating Pentecost in a Sardinian villa without telephones or electric lights. His rural reverie was interrupted by the arrival of Italian *carabinieri* with an urgent message from the West German embassy in Rome. Within hours, Benda was on a plane to Stuttgart, and thence to Karlsruhe to handle the appeals.[159]

The Court rejected both of them.[160] The strict standards for issuing a preliminary injunction, the First Senate explained, were even more exacting when the challenged legislation enacted an international treaty. Only a pressing general interest could justify such an injunction, and no such interest was present in the case at hand. The individual complainants asserted rights that, even without the treaties, they could not currently enjoy. No great harm would come from waiting for the Court to resolve the case on its merits. That resolution was long in coming; the Court rejected the complaints on the merits only in 1975.[161] By then, the treaties had been in force for three years.

The Warsaw and Moscow treaties prepared the way for a third, even more dramatic agreement—the Basic Treaty (*Grundvertrag*) between the Federal Republic of Germany and the German Democratic Republic. Brandt had called at the beginning of his chancellorship for a shift, vis-à-vis the German Democratic Republic, from 'a regulated next-to-one-another to a with-one-another'. The Basic Treaty sought to give concrete shape to this designedly nebulous phrase. The treaty was the centrepiece and culmination of Brandt's foreign policy. It was as central to Brandt's geopolitical vision as the European Defence Treaty had been to Adenauer's. Negotiations for the treaty were conducted by Egon Bahr, now an all-purpose minister in Brandt's cabinet, and by Michael Kohl, an East German undersecretary of state.

From the beginning, the proposed treaty generated enormous hostility among opposition politicians. But among the population at large, it enjoyed majority support. The SPD adroitly cast the November 1972 Bundestag elections as a referendum on *Ostpolitik* and on Brandt, the policy's popular architect. The ploy worked brilliantly. The returns of 19 November yielded a 3.1 per cent gain for the SPD and a 1.2 per cent loss for the CDU/CSU. The SPD won 45.8 per cent of the vote; the Union parties 44.9 per cent. For the first time since Weimar, the SPD was the strongest party in Parliament. The FDP also gained strength, growing by 2.6 per cent to 8.4 per cent. The combined 5.7 per cent gain gave the Social–Liberal coalition a solid majority. Smaller parties all but disappeared; their combined share shrivelled below 1 per cent. Brandt had come a long way since his near ousting seven months earlier. The coalition trumpeted the outcome as a ringing endorsement of its *Ostpolitik* and a mandate for its expansion.

The government didn't wait long to act on this presumed electoral charge. Thirty-three days after the election, on 21 December 1972, Bahr and Kohl signed the Basic Treaty in East Berlin.

The language of the treaty was sometimes elliptical and often opaque. But it was an unmistakably the ideals of rapprochement and normalization, of détente and *Wandel durch Annäherung*. The treaty prescribed that the two signatories 'develop normal good-neighbourly relations with one another on the foundation of equal rights'; that they

159 See 'Gegen den letzten Akt' *Der Spiegel* (29 May 1972).

160 33 BVerfGE 195 (1972); 33 BVerfGE 232 (1972).

161 40 BVerfGE 141 (1975).

renounce violence and the threat of violence; that they honour the principles enshrined in the UN Charter, 'particularly the sovereign equality of all states, the respect of independence, self-sufficience, and territorial integrity, the right of self-determination, the observance of human rights and non-discrimination'; that they abandon all pretence at representing one another internationally; that they honour one another's independence, domestically and abroad; that they work toward mutual agreement in a host of policy areas; and that they exchange representatives in one another's capitals. The treaty made clear that it had no effect on treaties previously signed by its two parties—including, crucially, the 'Western Treaties' signed in earlier decades by the Federal Republic—and stipulated that the treaty would enter force when the two states exchanged notes following domestic ratification.

Many opposition politicians abhorred the treaties. But with the coalition's comfortable majority they were powerless to block it in Parliament. Some were tempted to try to arrest the treaty via constitutional complaint. The CDU top brass, however, opted against this, reasoning that the party would pay a terrible political price for taking a question to the Court that had been answered decisively by the voters.

One CSU politico, however, did not feel beholden to the studied conclusions of the federal CDU leadership. After being forced out of the government in the aftermath of the *Spiegel* Affair, Franz Josef Strauβ had returned to cabinet as finance minister in the grand coalition in 1966. With the Union parties now in opposition, he was finance minister in Rainer Barzel's shadow cabinet and spokesman for the CDU/CSU Bundestag faction in matters of economic and financial policy. Strauβ was a strident, implacable enemy of the new *Ostpolitik*. With regard to the Basic Treaty, he was less deferential than his colleagues to the coalition's ostensible mandate—less willing, as Strauβ himself put it, to let quotidian politics trump constitutional principle.

But Strauβ, as the *Spiegel* affair made clear, had no weak stomach for legal and constitutional irregularity. Strauβ was not so much troubled by constitutional scruple as he was driven by an unusual and highly developed sense of historical mission. Strauβ claimed never consciously to have decided to become a politician, but to have been roused by an instinctual sense of duty to work to prevent any repetition of the historical errors that led to the German catastrophes of the twentieth century.[162] Strauβ was an earnest patriot who believed passionately that Germany—*united* Germany—had a role to play in world affairs. Strauβ believed that the Basic Treaty was flatly at odds with his (and Germany's) historical mission. On the one hand, the treaty made concessions to a totalitarian regime without demanding in return the 'abolition of the death machine on the border of the Federal Republic'.[163] On the other hand, it sealed and solemnized the division of his Fatherland. The treaty was thus doubly intolerable. If killing it required a trip to Karlsruhe, so be it.

As boss of the CSU—the CDU's Bavarian sister—Strauβ could access the Court through routes the national CDU leadership did not control. Though the Bavarian state government initially balked Strauβ ultimately persuaded the state's minister president, Alfons Goppel, and his reluctant cabinet, to challenge the treaty's constitutionality. In late May 1972, Strauβ stirred controversy by attending Goppel's cabinet meetings and wielding his 'whole weight as party chairman' to bring Goppel and his peers around.[164] Strauβ even hinted that he had inside knowledge from one of the justices that Bavaria's complaint would

[162] See the epigraph quotation from Strauβ in Stefan Finger, *Franz Josef Strauβ. Ein politisches Leben* (Olzog 2005) 11.

[163] Franz Josef Strauβ, 'Menschliche Erleichterungen zählen nicht' *FAZ* (10 November 1972).

[164] Franz Josef Strauβ, *Die Erinnerungen*, 4th edn (Berlin Settlers 1989) 451.

not go fully awry in Karlsruhe.[165] Goppel and his cabinet finally yielded, and the complaint was filed. 'We Bavarians', Strauβ prated, 'mustn't be afraid to be the last Prussians when history demands it of us'.[166]

The complaint was fourfold. The treaty violated (1) constitutional commands to preserve the unity of the German state and to pursue reunification; (2) the rights of other parts of Germany (i.e. the Eastern states) to accede to the Federal Republic; (3) constitutional provisions regarding the status of Berlin; and (4) the Federal Republic's duty to protect and provide for Germans living in the East.

The challenge came late. If the individual complaints against the Warsaw and Moscow treaties were filed at the eleventh hour, the Bavarian assault on the Basic Treaty began minutes before midnight. The Bundestag voted to approve the treaty on 11 May 1973. The two Germanys fixed 20 June as the date for the 'exchange of notes' that would make the treaty binding international law. Brandt's government refused to delay the exchange pending resolution of the constitutional challenge. Presumably, the exchange of notes would create a fait accompli that a later decision by the Constitutional Court would be powerless to revise or annul.

The Second Senate, then, had precious little time to answer an appeal that was jurisprudentially bewildering and politically explosive. Court-watchers cried foul, complaining that the government had tried to 'override' the Court.[167] Ernst Friesenhahn, a former justice, supported the treaty but chided the government for its heavyhandedness. He did so with words first spoken by Adolf Arndt during the European Defence crisis: 'The Federal Government is constitutionally required to work together with the Constitutional Court and duty-bound to avoid anything that could render difficult or impossible the Court's timely and effectual exercise of its competences'.[168] Instead of honouring this principle, Brandt's government had shown the public 'with what facility governments can outmanoeuvre the highest court and render ineffectual the constitutional review of government action provided for in the Basic Law'.[169]

The government countered that it was at the mercy of wider events. On 21 June, the World Security Council would meet to decide whether to admit the two Germanys to the United Nations. The treaty must be signed beforehand. Time was short, and tensions were high. And brief as the proceedings were, there were glaring distractions along the way.

It was highly improbable that the Court could resolve the case on its merits before the exchange of notes on 20 June. Indeed, the Second Senate admitted that it wouldn't. The justices announced that they would release their judgment on the merits on 31 July. Bavaria, predictably enough, asked the Court to enjoin the exchange of notes until after that date. The Court responded on 4 June.[170]

The Senate was sharply split—at least in the reasons the justices gave for concluding, as all of them did, that an injunction was not (for the time being) 'urgently demanded'.[171] The justices divided in groups of four. One group consisted of Walter Seuffert and Martin Hirsch, both former SPD members of the Bundestag; Joachim Rottmann, the recent FDP appointment; and Hans Rupp. The other quartet was comprised of Justices Wand, Schlabrendorff,

[165] Werner Schneider, 'Bundesverfassungsgericht und Politik—die Klage Bayerns gegen den Grundvertrag. Vorgeschichte und Verlauf des Prozesses—das Urteil und seine Bewährung', (unpublished—on file with the Hans-Seidl-Stiftung, Munich) 31.

[166] In Finger (n 162) 357.

[167] See Bruno Waltert, 'Karlsruhe braucht Zeit zu seiner Entscheidung' *Die Welt* (31 May 1973) 4.

[168] Ernst Friesenhahn, 'Hüter der Verfassung'? (1973) 6 *Zeitschrift für Rechtspolitik* 189.

[169] ibid 188. [170] 35 BVerfGE 193 (1973). [171] ibid 201.

Geiger and Rinck—all CDU/CSU nominees. The public could hardly help noting how neatly the Senate divided on partisan lines.[172]

The reasons the first foursome gave for denying an injunction raised eyebrows. In light of the Court's prior jurisprudence, the foursome noted (specially citing the *Saar Statute* judgment of 1955 and, of all things, the *Communist Party* judgment of 1956), there was small chance 'that the Senate would arrive at the conclusion that the treaty law was incompatible with the Basic Law (and therefore null and void)'.[173] At worst, the Senate might conclude that certain particulars of the treaty were constitutional only if interpreted a certain way—in which case it would do little harm if the treaty had meanwhile entered force.[174] The risks of withholding an injunction were small; the dangers of issuing one, immense. The remaining four justices agreed with the latter conclusion, but stressed that the *Rechtsstaat* always pays a price when laws enter force before they can be reviewed as the constitution prescribes.[175] For the moment, this consideration was outweighed by the foreseeable international repercussions of an injunction. But it was a cost that should not be ignored.[176] This was a non-subtle reproach to the government's unwillingness to wait. But it struck contemporaries less forcibly than the impression that the four judges most decisively opposed to an injunction seemed already to have made up their minds on the merits.

This impression was reinforced by accounts of extrajudicial utterances by the youngest of their number. Towards the end of April, the *Badische Neuste Nachrichten* reported an extraordinary speech given by Justice Rottmann at an FDP gathering in Karlsruhe. Justice Rottmann reportedly opined, among other things, that for purposes of public law the German Reich had ceased to exist in 1945; that the Federal Republic and the German Democratic Republic were separate, sovereign states whose relations with one another were governed exclusively by international law; and that *Ostpolitik* was a major contribution to European peace, eminently suited to the needs of the present hour.[177] If the report was true, it shattered all confidence that Rottmann was approaching the case impartially.

On 28 May, shortly after asking the Court to quash the treaty, Bavaria asked the Court to exclude Rottmann for bias. A week earlier, one CSU leader encouraged his comrades to challenge the treaty in Karlsruhe by hinting at the possibility of Rottmann's recusal: 'It cannot be said that this proceeding ... is lost from the outset.... We also don't know what changes in personnel may arise between now and the day of the Second Senate's decision'.[178]

In an official statement, Rottmann maintained that the newspaper report grossly misrepresented the content of his lecture and insisted that he would review the Basic Treaty with an open mind. Taking the disclaimer at face value, Rottmann's colleagues (with one exception) rejected the bias petition on 29 May.[179] On 4 June, as noted earlier, Rottmann was among the foursome most strongly opposed to an injunction. On 13 June, however, a letter became public that Rottmann had written on 7 May to Wilhelm Gutmann, a former Baden-Württemberg state legislator from the Right-wing National Democratic Party (NPD). In the letter, Rottmann confirmed that his speech (albeit with some oversimplication) had in the main been correctly reported. The Constitutional Court's earlier statements, he added,

[172] See 'Treue gegen Treue' *Der Spiegel* (25 June 1973). [173] 35 BVerfGE 193 (n 170) 197.

[174] Former Justice Gerhard Leibholz privately told a gathering of colleagues that he found the foursome's consideration of the 'probability' of the complaint's success procedurally adventurous and politically appalling. In Stenographische Niederschrift der Veranstaltung der Deutschen Vereinigung für Parlamentsfragen, e.V am 13.6.1973 zum Thema 'Das Bundesverfassungsgericht—Zuflucht und Hürde für die Politiker' (1973) 17. At the very least, Leibholz suggested, such a consideration required approval from the Plenum.

[175] 35 BVerfGE 193 (n 170) 199. [176] ibid 199–201.

[177] See 'Bundesverfassungsrichter Rottmann: BRD und DDR gleichbereichtigte Staaten' *Badische Neueste Nachrichten* (Karlsruhe, 28 April 1973).

[178] In Schneider (n 165) 29. [179] 35 BVerfGE 171 (1973).

that the German Reich still existed de jure, may have been defensible fifteen or twenty years earlier when the Court made them, but they had been swept aside by intervening political realities.[180] The Basic Treaty, Rottmann affirmed, was constitutionally unproblematic.

The letter prompted a second bias petition, this one narrowly successful. The same four justices who on 4 June had been most sympathetic to the injunction petition voted on 16 June to exclude Justice Rottmann.[181] This foursome included Justice Willi Geiger, who seven years earlier had objected in high dudgeon to the dismissal, under uncannily similar circumstances, of Justice Leibholz from the *Party Finance* judgment. The three justices whom Rottmann had joined in the injunction decision now dissented *en bloc* from Rottmann's exclusion.[182] Once again the split was glaringly partisan. All the world realized that if Rottmann had been excluded in response to the first bias petition, the first injunction petition might have turned out differently. 'I feel like I've been tossed into the cogs of a political wheelwork', Rottmann sighed. 'And I cannot escape the impression that it was a constructed one'.[183] Speculation that Justices Schlabrendorff, Rinck, Geiger, and Wand had manoeuvred into a position where they could prevail on the merits at Justice Rottmann's expense was universal. The justices of the early 1970s seemed to cultivate impressions of partisan attachment as assiduously as the justices of the early 1950s worked to refute them.

The Second Senate, however, was full of surprises. After the final ratification of the treaty on 6 June, Bavaria filed a second petition for a preliminary injunction. Following the exclusion of Justice Rottmann, many expected this second petition to succeed. Instead, on 18 June, two days after voting narrowly to exclude Justice Rottmann, the Second Senate voted unanimously to reject Bavaria's second injunction appeal.[184] This second, unanimous rejection came two days before the scheduled exchange of notes. The justices invoked the principle of judicial restraint as they deferred to the government's insistence (at oral argument) that the treaty must enter force before the Security Council convened.[185]

The oral arguments held on 16 June were conducted behind closed doors. The justices may have been persuaded by the disclosures of government representatives, including Egon Bahr himself, at oral argument. But no one knew for sure, nor did anyone know what information had been disclosed.

In any case, the 'Union' foursome had countered expectations. Some thought they had experienced a change of heart; others suspected a failure of nerve. To critics, events of the previous three weeks confirmed nothing if not the unfitness of the current justices for the rigours of high politics. *Der Spiegel* opined that if there was a certain loser in the conflict between Bonn and Bavaria over the Basic Treaty, that loser lived in Karlsruhe.[186] The Court's judgment on the merits was still six weeks away. But for many the matter was moot. The *Frankurter Allgemeine Zeitung* groused that the Second Senate's pending judgment now had 'no more value than a seminar essay on the constitutional aspects of the Basic Treaty'.[187] This was untrue. The merits question remained very much open, and the fires of controversy soon leapt back to life. The *Frankfurter Allgemeine* itself did much to fan the flames.

Precisely one week after the Basic Treaty entered into force, Johann Georg Reißmüller launched a cannonade on the paper's front page. A 'leading [SPD] politician', Reißmüller reported, had recently vowed that the government would not allow its *Ostpolitik* to be

[180] See 'Richter Rottmann schrieb und redete zuviel' *Die Welt* (18 June 1973) 2.
[181] 35 BVerfGE 246 (1973). [182] ibid 255.
[183] In 'Richter Rottmann schrieb und redete zuviel' *Die Welt* (18 June 1973) 2.
[184] 35 BVerfGE 257 (1973). [185] ibid 262–63.
[186] 'Treue gegen Treue' *Der Spiegel* (25 June 1973). [187] 'Was für Nachteile'? *FAZ* (19 June 1973) 1.

demolished by 'eight assholes in Karlsruhe'.[188] Reiβmüller didn't name his source. For weeks, the media buzzed with speculation and SPD headquarters rang with denials. Some suspected Horst Ehmke, Brandt's flamboyant former chief of staff, but Ehmke denied the charge vehemently.[189] Other fingers pointed more persistently at Herbert Wehner, head of the SPD Bundestag faction. It never became clear who, if anyone, had said such a thing. One rumour held that the phrase originated with one of the justices, who wasn't quoting an SPD leader but offering a caricature of the government's attitude toward. Helmut Kohl, then minister president of Rheinland-Pfalz and soon the CDU's national chief, thought the slur—and the chancellor's failure firmly to denounce it—betrayed a 'highly questionable constitutional understanding' in government circles; an insensitivity even more evident in the coalition's heavy-handed push to present the Court with a fait accompli.[190] Gerhard Jahn (SPD), the justice minister, shot back that the government's procedure vis-à-vis the Bavarian complaint was beyond reproach. It was a sorry commentary on the current political climate, Jahn continued, that unsubstantiated slanders were reported uncritically and embraced unblinkingly.[191]

Even so, Jahn rushed to Karlsruhe to smooth matters over. The Court and the government exchanged mutual assurances of respect. Before Jahn's visit, some justices had been so indignant that Benda called for a plenary meeting to establish once and for all the source of the infamous affront. Some of Benda's colleagues, though, wondered whether grave solemnity in the face of a schoolyard insult wouldn't do more to render the Court risible than to restore its honour. The meeting was cancelled, ostensibly because the justices were satisfied with a formal governmental repudiation and an official declaration of undiminished regard.[192] But the sores still festered. The truly fatal feature of the remark, as Friedrich Karl Fromme later noted, was not its irreverent term of abuse but its implicit threat of defiance—its suggestion that the government would not accept an adverse outcome.[193]

Meanwhile the Second Senate laboured with dash and abandon to prepare the judgment it had promised to deliver on the last of July. The judgment appeared on time; but it reflected the rush with which it was written.[194]

The Senate began with some tough talk. Contrary to the public stance of former Justice Gregor Geller that the treaty's constitutionality was a non-justiciable 'political question',[195] the justices indicated once again that there was no American-style political question doctrine under the Basic Law. It was incompatible with the Basic Law's embrace of comprehensive constitutional review, moreover, for the executive branch to override the Constitutional Court.[196] At the same time, judicial restraint had been the Court's constant, self-imposed precept. The Court could not evade its responsibility to review even the most politically explosive of government actions; but it would always afford representative branches the widest possible leeway.

To no one's surprise, then, the Senate confirmed the constitutionality of the Basic Treaty. To the surprise of many, it did so unanimously. But the really striking aspect of the judgment was not its result but its reasoning. In fact, the Senate fused the two, holding that the treaty was constitutional only 'according to the interpretation given in the [Court's]

[188] 'Die Regierung wird's schon recht machen' *FAZ* (27 June 1973) 1.

[189] 'Kohl: Höchst bedenkliches Verfassungsverständnis' *FAZ* (7 July 1973) 1.

[190] ibid.

[191] ibid.

[192] 'Benda sagt Sitzung des Bundesverfassungsgerichts ab' *FAZ* (14 July 1973) 1.

[193] Friedrich Karl Fromme, 'Gericht im Sog der Politik' *FAZ* (29 January 1975) 1.

[194] 36 BVerfGE 1 (1973).

[195] In Helmut Rieber, 'Warnung vor Gang nach Karlsruhe' *Frankfurter Rundschau* (18 January 1972) 2.

[196] ibid 14–15.

grounds for decision'.[197] That interpretation, the justices wrote, was binding law for all state authorities in the Federal Republic.[198]

This binding interpretation rested on two grounding propositions that the treaty, in the Court's reading, left and must leave unaltered. First, the justices stressed (with a rare exclamation point) that it was a core premise of the Basic Law, and not merely a thesis of public or international law, that the German Reich survived the capitulation of 1945, the ensuing military occupation, and all that had happened since.[199] The Reich emphatically lived on; the crucial question was how. One group of public and international lawyers subscribed to the so-called 'identity theory'. Under this theory, the Federal Republic not only succeeded to, but was identical with the German Reich. Another group put forth the so-called 'roof theory' (*Dachtheorie*). This theory held that the Reich endured in its 1937 boundaries, but only as an unorganized umbrella entity, with no institutions of its own. This Reich was latent and inchoate, encompassing the two separate legal–political orders within its borders.[200] The two theories, a logician might have thought, were irreconcilable. But in its *Basic Treaty* judgment, the Second Senate united them.

Consistent with the 'identity theory', the Senate declared that 'the Federal Republic of Germany is … not the "legal successor" of the German Reich, but as a state is identical with the state, "German Reich"'.[201] But the justices then added that the Federal Republic, with regard to its spatial dimensions, was only 'partially identical' (*teilidentisch*) with the German Reich. 'Identity does not require exclusivity', the justices wrote. In other words, the area covered by the GDR could also be part of the German Reich, a concession that smacked of the 'roof theory'. The GDR, the Senate held, 'belongs to Germany, and in relationship to the Federal Republic it cannot be considered a foreign country [*Ausland*]'.[202] In further keeping with the 'roof theory', the Senate observed that the Reich lived on but was currently incapacitated (*handlungsunfähig*).[203]

This was a lot for the uninitiated to swallow. 'Partially identical' sounded like an oxymoron; a dormant empire struck many as mystification. Initiates were even more perplexed. The judgment's framework allowed the Senate to save the treaty from annulment and the Court's prior jurisprudence from reversal. But it did so at the cost of theoretical coherence; it did so, lamented Ulrich Scheuner, by brewing a pot of 'indigestible conceptual porridge'.[204]

In the Senate's view, however, the mysterious mixture was necessary to square the treaty with the Basic Law's inexorable command to pursue reunification. That command, the Basic Treaty notwithstanding, remained in force.[205] How to achieve that goal was left to the wisdom and discretion of the Federal Republic's political organs. The Brandt government, the Court noted, believed that in the long run the Basic Treaty would facilitate reunification. The justices may or may not have shared this view. But in the *Basic Treaty* judgment, they explicitly deferred to it. They insisted, however, that the future policy of the Brandt government—indeed, of *all* future governments—must never swerve, even rhetorically, from the ultimate goal of reunification:

[197] 36 BVerfGE 1 (n 194) 3.

[198] By thus elevating the status of its essential grounds of decision, the justices parted paths with custom and traditional understanding.

[199] 36 BVerfGE 1 (n 194) 16.

[200] See Wilhelm A. Kewenig, 'Auf der Suche nach einer neuen Deutschlandtheorie' (1973) 26 *Die öffentliche Verwaltung* 797.

[201] 36 BVerfGE 1 (n 194) 16. [202] ibid 17. [203] ibid 16.

[204] Ulrich Scheuner, 'Die staatsrechtliche Stellung der Bundesrepublik. Zum Karlsruher Urteil über den Grundvertrag' (1973) 26 *Die Öffentliche Verwaltung* 583.

[205] 36 BVerfGE 1 (n 194) 17.

> No constitutional organ of the Federal Republic may abandon the restoration of state unity as a political goal. All constitutional organs are bound in their policies to work toward the achievement of this goal and to avoid any policy that might impede reunification. This includes the requirement to keep the demand for reunification alive at home and to advocate it ceaselessly abroad.[206]

Planted between these lines, perhaps, was a rebuke to the chancellor who once called reunification the 'life lie' (*Lebenslüge*) of the Federal Republic.[207] 'We must take note of the fact', Brandt told *Die Welt am Sonntag* in 1969, 'that a reunification in the original sense is no longer possible'. 'It is the duty of the federal chancellor', he added, 'to tell his people the truth, even when it is bitter'.[208] The Court's judgment warned against indulging such rhetoric too often, or extending it too far.

The justices rebutted opposition arguments more explicitly. The Basic Treaty was not, as its critics maintained, a 'division treaty'. Neither for the present nor for the future did it absolve the federal government of the responsibility 'to do everything in its power at all times, so that the German people can once more organize itself as a unitary state'.[209] This outlook, the Senate held, must govern all measures designed to implement the Basic Treaty or expand *Ostpolitik*. The Court added further that the Basic Treaty was only a de facto, not a de jure, recognition of the GDR. The border between the Federal Republic and the GDR was a creature of public rather than international law; it separated two quasi-states subsisting within the enduring (but latent) state of 'Germany as a whole'.[210] To Bavaria's invocation of the fundamental rights of Germans suffering on the Eastern border, the Court answered that the treaty did not violate fundamental rights; the GDR's border practices violated the treaty.[211] But the justices stressed that the Federal Republic remained responsible for all Germans, including those in the GDR, and must work at all times to protect their interests.

In the aftermath of the judgment, all parties claimed victory. Brandt's coalition celebrated the judgment as a vindication of its *Ostpolitik* and a rebuke to those who alleged that the treaty abandoned ship on reunification.[212] Herbert Wehner gloated that Franz Josef Strauβ had 'taken a hammering'.[213] But Strauβ's party crowed in return that they had successfully bound the federal government's future foreign policy. Richard Stücklen, the CSU state chairman, wrote to congratulate the Bavarian state government 'on the success of its complaint'.[214] Union politicos cheered that the judgment gave the treaty desperately needed clarity—that it bound the government, in implementing the treaty and expanding its Eastern policy, to pursue German unity and to advocate the interests of all Germans. In years to come, the judgment would become a powerful rhetorical ally in the hands of the opposition. 'Bavaria neither hoped nor could have hoped to achieve more with its petition', wrote Walter Brunert in *Die Welt*.[215] Strauβ himself agreed. Dictating his memoirs fifteen years later, he effervesced: 'The goal of Bavaria, the goal of the CSU, my goal was accomplished'.[216]

Government officials and SPD sympathizers bristled at the opposition's bluster. If all Bavaria wanted was a binding interpretation, they complained, there was no need to ask

206 ibid 17–18. 207 Merseburger (n 17) 585. 208 ibid 584. 209 36 BVerfGE 1 (n 194) 25.
210 ibid 23, 26. 211 ibid 35.
212 See 'Karlsruhe gibt Richtlinien für weitere Abkommen mit der DDR' *FAZ* (1 August 1973) 1; Bruno Waltert, 'Ein Urteil, an dem sich nicht rütteln läβt' *Die Welt* (2 August 1973) 4; Hans-Herbert Gaebel, 'Mit Karlsruher Segen', *Frankfurter Rundschau* (1 August 1973) 3; 'Der Grundvertrag mit der DDR stimmt mit der Verfassung überein' *Süddeutsche Zeitung* (1 August 1973) 1.
213 In 'Bonn: Entspannungsgegner abgeschlagen' *Süddeutsche Zeitung* (1 August 1973) 2.
214 In ibid. 215 Bruno Waltert, 'Die Richter weisen den Weg' *Die Welt* (1 August 1973) 4.
216 Strauβ (n 164) 456.

the Court to declare the law null and void, nor to race about the country accusing the coalition of violating the constitution. Privately, the government was deeply disappointed. The judgment would hamstring *Ostpolitik*, not least through the impression it left with the government's negotiating partners. East Germany's official newspaper, *Neues Deutschland*, dismissed the judgment as 'the pipe dreams of a few Karlsruhe judges'.[217] Other East German polemicists saw in the judgment's allegiance to reunification a brand of political revanchism that belonged to 'the rubbish heap of history'.[218] Soviet screeds against the decision were equally shrill. A tirade in the magazine *International Life* accused the Court of 'intentional falsifications' and 'direct distortions'.[219]

The Court had certainly not made future negotiations with Eastern powers any easier. Althout it explicitly disclaimed being anything of the kind, the *Basic Treaty* judgment was the Court's most aggressive intervention to date in the foreign policy of the Federal Republic. In government circles, the judgment unmistakably rankled. But the government, still smarting from the ruckus over the anonymous *FAZ* quotation, restrained its anger. At least for the time being.

D. Constitutional justice and European integration: the first *Solange* judgment

The *Basic Treaty* judgment went to the heart of the Federal Republic's relationship with the East; a judgment issued ten months later did the same for its ties to the West. The earlier case inspected the crown jewel of Brandt's foreign policy, the latter the enduring legacy of Adenauer's. There the matter was German division; here, European integration. The underlying facts were banal and the doctrinal issue straightforward. But the attendant procedural and jurisdictional questions were explosive. In the Second Senate's judgment of 29 May 1974, they exploded.

The European Economic Community was formed in 1957 when its six founding nations—Luxembourg, Belgium, the Netherlands, Italy, France, and the Federal Republic of Germany—signed the Treaty of Rome. Together with its unique cadre of supranational organs, the Community had its own judicial arm, the European Court of Justice (ECJ) in Luxembourg. From the outset, the relationship of community law to the domestic laws of the member states was contested. In 1964, the ECJ asserted unequivocally the supremacy of community law, holding that member states had definitively transferred sovereign rights to the European level.[220] In case of conflict, community law trumped domestic law—including, presumably, domestic *constitutional* law. The ECJ itself would police the boundary between the two.

This holding put national constitutional courts in an awkward position. There was an obvious tension between the ECJ's declarations of supremacy and exclusivity and national courts' self-understanding as exclusive arbiters and guardians of national constitutions. In the case of collision between community and constitutional law, national constitutional courts could either defy the ECJ's jurisdictional claims or abandon their own.

For many years the conflict lay dormant. Over time, although the EEC Treaty lacked a fundamental rights catalogue and European legislators failed to produce one, the ECJ elaborated a fundamental rights jurisprudence based on general principles drawn from the legal

[217] In 'Kampf gegen ein Urteil' *FAZ* (4 January 1974).

[218] In Richard Häußler, *Der Konflikt zwischen Bundesverfassungsgericht und politischer Führung. Ein Beitrag zu Geschichte und Rechtsstellung des Bundesverfassungsgerichts* (Duncker and Humblot 1993) 61 fn 178.

[219] In 'DDR übernimmt sowjetische Kritik am Bundesverfassungsgericht' *FAZ* (4 January 1974).

[220] See *Costa v ENEL*, ECR 585 (6/64).

traditions of the member states.[221] Some observers thought hypothetical conflicts between community and constitutional law might never materialize. Member state constitutional courts, they reasoned, could safely defer to the ECJ's exclusive jurisdiction, trusting that the European court would safeguard fundamental rights in its application of community law. A May 1971 judgment of the Belgian Court of Cassation, as well as a December 1973 judgment of the Italian Constitutional Court, seemed to fulfil these expecations.[222] The Belgian and Italian courts acknowledged the ECJ's rights jurisprudence and deferred to its jurisdiction. Many expected Germany's Federal Constitutional Court to follow suit.

The chance to do so was already pending, thanks to a 1971 referral from an administrative court in Frankfurt. An EEC directive, issued in 1967, required grain exporters to obtain a licence and pay a security deposit that became non-refundable when the licence expired. Along with regulations, directives are a form of 'secondary' community law (as opposed to 'primary' community law, embodied in the community treaties themselves). Implementation of directives is often entrusted to the domestic authorities of member states. So it happened with the grain licensing directive. A Frankfurt grain firm that lost its deposit complained to the Frankfurt administrative court that its forfeiture was unlawful. The administrative court submitted a referral to the ECJ asking whether the directive's forfeiture provision was compatible with community law. In a pair of judgments in late 1970 and early 1971, the ECJ confirmed that it was.[223] In November 1971, the Frankfurt court interrupted the proceeding for a second referral, this one to Karlsruhe. The administrative court asked the Constitutional Court whether, by applying the EEC directive, the Frankfurt authorities had violated the firm's fundamental rights—specifically its occupational freedom under Article 12 GG.

Before reaching this question, the Second Senate had to rule on the referral's admissibility. Before reaching that question, the Senate had to decide whether only the Court's Plenum could resolve it. By a narrow vote of 5:3, the Senate answered the Plenum question in the negative (whichever way it came out, the Senate's holding wouldn't reverse precedent) and the admissibility question in the affirmative.[224] The Senate stressed first that its consideration of admissibility was limited. The only issue was whether the Court could review the compatibility of *secondary* community law, as applied by German authorities, with the *fundamental rights provisions* of the Basic Law. The Senate considered neither whether the Court could review the compatibility of secondary community law with *other* constitutional provisions, nor whether the Court could review the compatibility of *primary* community law with *any* part of the constitution.

The key constitutional provision was Article 24 GG, which said simply that the Federal Republic 'may, by legislation, transfer sovereign rights to international institutions'. This clause, the five-justice majority clarified, could not be taken literally.[225] Like all constitutional provisions, Article 24 must be understood and interpreted in the context of the entire constitution. It did not, in other words, empower Parliament to amend the constitution's basic structure or to alter its essential identity. Quite the contrary: Article 24 affirmatively banned any international arrangement—including one created by amending community treaties—'that would abrogate the identity of the currently valid constitution of the Federal Republic of Germany through an intrusion into its constitutive structure'.[226] An essential and irreducible element of that structure was the Basic Law's rights catalogue. Article 24 did

[221] See the many judgments cited in 37 BVerfGE 271, 291–92 (1974).
[222] See the discussion in Heribert Golsong, 'Zu Bundesverfassungsgericht 'Solange'-Beschluß' (1974) 1 *Europäische Grundrechte-Zeitschrift* 18.
[223] ECR 11/70; also 38/70. [224] 37 BVerfGE 271 (1974). [225] ibid 279. [226] ibid.

not confer unqualified power to relativize those rights.[227] Given the current state of community integration, the only acceptable bulwark against such relativization remained the Federal Constitutional Court.

The majority conceded that the European Court of Justice, as currently composed, had proved encouragingly friendly to fundamental rights.[228] But this was a personal rather than a legal guarantee. It was no substitute for a robust catalogue of rights established by a representative parliament. In the absence of such a catalogue, German courts that needed to apply community law, but saw a conflict between the relevant provision of community law and a fundamental rights provision of the German constitution, were not only allowed but required (after soliciting a binding interpretation of the European provision from the ECJ) to refer the matter to the Constitutional Court.

The Second Senate limited this holding in two respects—one procedural, the other chronological. Procedurally, the Court would not review community law directly in response to individual complaints. Only lower courts applying community law in a concrete case could solicit review from Karlsruhe. Moreover, the Court could not annul community norms; it could only block their application in Germany. It would rule on the norms' domestic applicability, not their general validity. The chronological limit was expressed in a dramatic passage that gave the case its popular name (*Solange*). The Constitutional Court would review the consonance of community norms with Basic Law rights only '[s]o long as [*Solange*] the process of community integration has not proceeded to the point where community law also contains a valid catalogue of fundamental rights, formulated by a Parliament and comparable [*adäquat*] to the fundamental rights catalogue of the Basic Law'.[229] The Court suggested that the Community's democratic deficit and fundamental rights lacuna would not be permanent. The lack of legal certainty surrounding fundamental rights would likely 'end with the current transitional phase' of integration.[230]

The Senate's review of the underlying complaint was terse. It took the majority less than three pages to dismiss the claim that the EEC licensing requirement ran afoul of Article 12 GG. The requirement was a 'pure regulation of trade practice' (*reine Berufsausübungs regelung*) that merely affected *the manner* in which a trade could be practised. It left the complainants' freedom of occupational choice untouched.[231]

But the rub, of course, was not that the Senate rejected the claim but that it reviewed the question at all. On this point Justices Rupp, Hirsch, and Wand tartly dissented. Just three years earlier, the dissenters pointed out, the Second Senate had held that, under Article 24 GG, once sovereign powers had been transferred to an international institution, organs of the Federal Republic were bound to respect that institution's sovereign acts. Tellingly, the relevant 'sovereign act' in the earlier case was a judgment of the European Court of Justice.[232] Moreover, the dissenters continued, the EEC's protections for fundamental rights did not differ essentially from the protections in the Basic Law.[233] Both systems secured the core essence of basic rights. The ECJ had already adopted and applied many of the core principles of German constitutional law. It had even embraced the proportionality principle. This, for the dissenters, was enough to satisfy Article 24.[234] The majority's call for codified rights declared by a parliament was muddled and misguided. 'Within a free and democratically constituted community', the dissenters wrote, 'the protection of fundamental rights and the democracy principle are not interchangeable; they are complementary'.[235] Establishing the democracy principle at the European level might well lead the legislature and executive to

[227] ibid. [228] ibid 280–81. [229] ibid 285. [230] ibid 280–81. [231] ibid 289–90.
[232] ibid 295 (citing 31 BVerfGE 145, 174 (1971)). [233] ibid 296. [234] ibid 296–97.
[235] ibid 298.

display a greater regard for fundamental rights; but it wouldn't eliminate the need for judicial review. The latter element seemed to the dissenters the more crucial; and this the Community already had.

Quite apart from the majority's faulty Article 24 analysis, the dissenters added that the Basic Law itself precluded Karlsruhe review of community norms. Article 100(1) GG allowed lower court referrals to the Constitutional Court only when lower courts questioned the constitutionality of state or federal—or the compatibility of state with federal—law. In other words, the law referred for review must be *German*. By extending the Court's review to non-German law, the majority had expanded the Court's jurisdiction beyond its constitutional boundaries.[236] The majority's holding had violated the German constitution as well as invaded the ECJ's jurisdiction.[237] That holding was dangerous as well as wrong. It created a special status for the Federal Republic and exposed it to the justifiable accusation of having violated the Community's charter treaty and endangered its legal order.[238]

If the dissent was brusque, scholarly criticism was blistering. Not since the *Civil Servants* judgment two decades earlier was a judgment exposed to such ferocious academic assault. Hans-Peter Ipsen thundered that the judgment was 'wrong ... fallacious, superfluous, and legally-politically mistaken'; that it ought to be hurled post-haste 'out of the world'; and that it was 'procedurally and substantively, juristically and legal-politically groundless'.[239] Following the Senate's three dissenters, Gert Meier saw in the judgment a 'double breach of law' (*ein doppeltes Rechtsbruch*): a violation of community law and of the Basic Law itself. Meier dismissed the Court's anticipations of a European parliament and charter of rights as 'political phantasies'.[240] Heribert Golsong, formerly chancellor of the European Court of Human Rights in Strasbourg, warned that the Senate's holding would 'shake the Community to its legal—and therefore also to its political—foundations'.[241]

The professors' rage stemmed from wounded vanity as well as intellectual disgust. Oceans of ink had been spilt on the question of conflicts between community and national law, but the Second Senate's judgment ignored this literature entirely. 'The judgment would possibly have read differently', huffed Manfred Zuleeg, a future ECJ judge, 'and would certainly have increased in persuasive power, if the judges would have engaged with scholarly opinions. Certainly a judicial decision needn't be a learned treatise, but the total eschewal of dialogue with the academy has its price!'[242]

Naturally the EEC itself was less than pleased with the critique. The President of the European Comission sent a deeply agitated letter to West Germany's foreign minister, Hans-Dietrich Genscher.[243] The tone of many attacks on the judgment was overdone. The most withering criticism came from committed Europeanists insensitive to a Court torn between its duty, on the one hand, to ensure the operability of community law and its obligation, on the other hand, to enforce the demands of a domestic constitution. In some quarters, the judgment won praise for its pragmatism; in others, for its timely critique of community shortcomings.[244]

[236] See ibid 300–02. The majority contended that its holding was true to the basic thought of Article 100, which was that lower courts shouldn't apply laws that violate the constitution. ibid 284–85.

[237] ibid 299. [238] ibid.

[239] Hans Peter Ipsen, 'BVerfG versus EuGH re 'Grundrechte'. Zum Beschluss des Zweiten Senats des Bundesverfassungsgerichts vom 29. Mai 1974' (1975) 10 *Europarecht* 1.

[240] Gert Meier, 'Anmerkung' (1974) 27 *NJW* 1704, 1705. [241] Golsong (n 222) 17.

[242] Manfred Zuleeg, 'Das Bundesverfassungsgericht als Hüter der Grundrechte gegenüber der Gemeinschaftsgewalt' (1975) 28 *Die Öffentliche Verwaltung* 46.

[243] See president of the European Commission to Hans-Dietrich Genscher, 19 December 1974, Friedrich-Ebert-Stiftung (Bonn) (Nachlass Martin Hirsch, 1/MHAC000109).

[244] For positive assessments, see Hans Heinrich Rupp, 'Zur bundesverfassungsgerichtlichen Kontrolle des Gemeinschaftsrechts am Maßstab der Grundrechte' (1974) 28 *NJW* 2153; Bodo Börner, 'Deutsche Grundrechte und Gemeinschaftsrecht' (1976) 29 *NJW* 2041.

The grimmest scholarly forebodings never materialized. The predicted flood of complaints against community norms never burst. Two years after the decision, one commentator quipped that a judgment that had originally blared like a trumpet's blast now echoed like the tremolo of a wooden flute.[245] Within the Federal Republic itself, the tremulous flute was all the general public ever heard. The press all but ignored the judgment. But the press did catch on to a growing grumble of scholarly discontent with the ostensibly decreasing quality of the Court's work, and a more general apprehension about its expanding political power.[246]

E. Karlsruhe v. Bonn?: the first *Abortion* judgment

'Scarcely any political question arises in the United States', runs a famous Tocquevillian apercu, 'that is not resolved, sooner or later, into a judicial question'.[247] What Tocqueville observed in Andrew Jackson's America many West Germans abhorred in Willy Brandt's Federal Republic. In his critique of the *Basic Treaty* judgment, Ulrich Scheuner bemoaned the 'by now fixed tendency of politicians in the Federal Republic to settle what are at core political questions through constitutional law'.[248] Recent judgments had thrown in relief, as never before, the Court's political role, and the Court had fulfilled that role in an unprecedentedly interventionist fashion. But in the half-decade since the Social-–Liberal coalition came to power, the Court had not annulled a single law the coalition had passed. Some began to think it never would. Not long after issuing its *Solange* decision, the Second Senate sustained a 1973 amendment to the federal social security law against the contention of two Union-governed states that the amendment lacked necessary Bundesrat approval.[249] *Der Spiegel* cooed that the decision foiled Union hopes of making the Bundesrat an effective instrument of opposition, and provided a cautionary tale about transferring political disputes to Karlsruhe.[250]

Just as some were concluding the Court might prove more sympathetic to reform legislation than the Union hoped or the coalition feared, the Social–Liberal reform agenda shifted significantly. To begin with, the coalition had a new captain. On 16 May 1974, the pragmatic Helmut Schmidt replaced the visionary Willy Brandt as chancellor. Brandt's resounding electoral triumph in November 1972, followed by the signing of the Basic Treaty the next month, marked the zenith of his career. There followed eighteen months of steady decline. 'The beginning of my second chancellorship', Brandt later wrote, 'already stood under an unlucky star'.[251] Shortly after his election victory, Brandt underwent a laryngeal operation. As Brandt recovered, his second cabinet took shape largely in his absence. Schmidt and Herbert Wehner led the negotiations. The resulting cabinet was not Brandt's. Especially damaging was the departure—engineered by Schmidt—of Horst Ehmke, whose efficiency and energy compensated for Brandt's deficient work ethic.

Over the next few months, Brandt suffered from deep depression. The corridors of the capital buzzed with rumours of his heavy drinking and reckless womanizing. The media

245 Börner (n 244) 2041.

246 See, e.g., Robert Leicht, 'Richterwahlen vor der Tür' *Süddeutsche Zeitung* (17 December 1974) 4 (noting that 'publications in legal scholarship have for some time traced a decrease in the quality of the Constitutional Court's jurisprudence' and that scholars were 'especially critical of the work of the Second Senate').

247 Alexis de Tocqueville, *De la démocratie en Amérique*, vol 1 (Robert Laffont 1986 (1835)) 258.

248 Ulrich Scheuner, 'Die staatsrechtliche Stellung der Bundesrepublik. Zum Karlsruher Urteil über den Grundvertrag' (1973) 26 *Die Öffentliche Verwaltung* 581.

249 37 BVerfGE 363 (1974).

250 'Formale Finten' *Der Spiegel* (19 August 1974).

251 Willy Brandt, *Begegnungen und Einsichten. Die Jahre 1960–1975* (Hoffmann and Campe 1978) 579.

that had once lionized the chancellor began to lose interest. Within the government, Brandt was the butt of increasingly sharp—and in Wehner's case, savage—criticism. Wehner accused Brandt of laziness, self-indulgence, and even stupidity. 'What the government lacks', Wehner told journalists in Moscow, 'is a head'.[252] Brandt was already reeling when, on 24 April 1974, news broke that his close advisor, Günter Guillaume, was an East German spy. The Guillaume affair needn't have driven Brandt from office, but the jaded chancellor resigned before it had the chance. The Social–Liberal coalition remained in power, but its charismatic founder was gone.

Well before Brandt's resignation, the coalition's reformist energy began to subside. The 1972 election had given the government a plausible mandate for reform; but the following year a world oil crisis deprived it of the resources necessary to carry out its ambitions. International recession ensued. What Eric Hobsbawm once called the capitalist 'Golden Age'[253] was over, and with it the Federal Republic's remarkable postwar boom. The coalition's faith in future progress and reformist optimism largely evaporated. Helmut Schmidt assumed office in a sober spirit, grounded by a Weberian ethic of accountability and a Bismarckian emphasis on the art of the possible. Modesty and pragmatism were his watchwords. 'Reforms are only practicable', Schmidt said in an early interview, 'when one can afford them'.[254] This was a far cry from Brandt's brassy bid to dare more democracy, empower the populace, and reshape the nation. Schmidt's ascension marked more than a change of leadership. It marked the end of a constitutional moment.

It turned out also to be the beginning of a constitutional crisis.

A month after Schmidt took office, the Bundestag passed a law that was the culmination of a five-year effort to reform of the Criminal Code. In 1969, Brandt had vowed to breathe life into the Basic Law's decree that the Federal Republic was a 'social *Rechtsstaat*'. Brandt's coalition sought to promote social and legal justice by advancing reforms in criminal and family law. In a host of areas, traditional criminal sanctions were replaced by monetary fines. A particularly prominent field of liberalization was the law of sexual crimes. Adultery and abandonment of a pregnant woman were decriminalized, as was homosexuality, except where designed to protect youth.[255] The crimes of pandering, procuring, and pornography were substantially narrowed.[256] More controversial were plans to reform section 218 of the Criminal Code, which criminalized elective abortions.[257] All parties agreed that the provision must be reformed; the question, bitterly disputed, was how.

Section 218 originated in the Prussian Criminal Code of 1851 and was passed down through the codes of the North German Federation (1870) and the united German Reich (1871). Calls for liberalization, led by the legendary SPD jurist Gustav Radbruch, began in the Weimar era. Radbruch proposed a *Fristenlösung*, or term solution, that would decriminalize abortion in the early weeks of pregnancy. The proposal failed, though criminal sanctions for abortions were softened, only to be intensified once more during the Third Reich. The Nazis condemned elective abortion as an 'attack on the life force [*Lebenskraft*] of the German people', even as they used coercive abortion as an instrument of racial policy against persons deemed racially inferior or genetically unfit.

[252] See 'Was der Regierung fehlt, ist ein Kopf' *Der Spiegel* (8 October 1973).

[253] See Eric Hobsbawm, *Age of Extremes: The Short Twentieth Century, 1914–1991* (Vintage Books 1994) 225.

[254] In Conze (n 17) 466.

[255] See Michael Kandora, 'Homosexualität und Sittengesetz' in Ulrich Herbert (ed), *Wandlungsprozesse in Westdeutschland. Belastung, Integration, Liberalisierung 1945 bis 1980* (Wallstein 2002) 379.

[256] See Conze (n 17) 402.

[257] On the postwar debate over this provision, see Michael Gante, *§ 218 in der Diskussion: Meinungs- und Willensbildung, 1945–1976* (Droste 1991).

The sexual revolution of the 1960s transformed public attitudes about abortion; the number of abortions skyrocketed. The gap between the law's strict letter and its lax enforcement fairly yawned. Traditionally, abortion could be justified only on 'medical' grounds—that is, if carrying the child to term would seriously threaten the life or health of the expectant mother. In the late 1960s and early 1970s, 'medical' necessity came to include a host of other circumstances as well—genetic or environmental threats to the health of the unborn child, pregnancies resulting from rape or molestation, social crises in the life of the mother, etc. By one count, the number of officially certified 'medical' abortions ballooned from 2,826 in 1968 to 17,814 in 1975.[258] Some estimated the number of illegal abortions at over a quarter million. Affluent women flew to England, where abortion laws had been liberalized in 1967.

Calls for reform in West Germany were loud and legion. Parliamentary proposals followed two models—variations on Radbuch's term model, and the so-called 'indications model' (*Indikationenmodell*), which would decriminalize abortions under specific, enumerated circumstances. Among the three major parliamentary parties, the FDP faction favoured the former, the Union the latter, and the SPD split internally.

As early as 1970, a committee on criminal reform that included prominent FDP parliamentarians proposed legalizing abortion during the first trimester of pregnancy. The proposal won ringing approval from a wide array of women's groups. Marching under the mantra, 'My belly belongs to me', women's organizations rallied in support. At the height of the movement, beginning in mid-1971, scores of prominent German women took part in a self-accusation campaign, confessing publicly to have undergone illegal abortions.[259] The campaign was spearheaded by feminist leader Alice Schwarzer and included prominent actresses like Romy Schneider and Senta Berger. Photos of attractive female celebrities appeared on the 6 June 1971 cover of *Der Stern* under the heading, 'I had an abortion'. The confession was signed by 374 women.

Two justices of the Constitutional Court, both devout Catholics, joined the debate on the opposite side. Willi Geiger of the Second Senate suggested in the *Rheinischer Merkur* that the term solution was unconstitutional.[260] Hans Brox of the First Senate added in a Catholic periodical that the right to life, guaranteed by both natural law and by Article 2 GG, also shielded the unborn.[261]

Popular agitation prompted additional proposals. In early 1972, the justice minister, Gerhard Jahn, proposed a bill embracing the indications model; shortly thereafter, a group of Social Democrat members of the Bundestag issued a counterproposal adopting the term model. After the Bundestag elections in November 1972, Parliament faced four separate proposals, only one of which—proposed by the federal government—opted for the term model. On 18 June 1974, after months of sometimes acerbic debate, the government's term proposal passed with a narrow majority. The new law legalized all abortions during the first twelve weeks of pregnancy.

Almost as soon as the law was passed, 193 Christian Democrats in Parliament, together with the CDU- or CSU-controlled state governments of Schleswig-Holstein, Baden-Württemberg, Bavaria, and the Saarland, asked the Constitutional Court to annul it. The law, Christian Democrats argued, violated the sanctity of life, secured by Article 2 GG, and offended human dignity, enshrined in Article 1.

[258] Hans-Georg Dederer, 'BVerfGE 39, 1—Schwangerschaftsabbruch I' in *Verfasssungsrechtsprechung* (n 153).

[259] The gesture followed a similar act of protest in France, led by prominent writers and intellectuals like Simone de Beauvoir.

[260] Willi Geiger, 'Ethik und Recht. Implikationen der Fristenlösung' *Rheinischer Merkuer* (Bonn, 23 February 1973) 31.

[261] Hans Brox, 'Das Recht auf Leben' *Der Dom* (Paderborn, 6 May 1973).

By the Court's unhurried standards the case moved swiftly. Oral arguments were held in late November 1974; the First Senate announced its judgment for the following February. Weeks before the judgment was to be announced, its putative outcome was leaked.[262] The law, sources suggested, would be invalidated. The news elicited a furious outbreak of popular emotion. Women's organizations and Leftist groups planned protest marches in several cities for the day of the decision. At a protest in Berlin on 16 February, demonstrators poured red paint on the steps of a government building to symbolize the blood of women who died from illegal abortions.[263]

The leak sparked attacks on the Court itself. In a scalding editorial in *Der Spiegel*, Rudolf Augstein summarized the bill of particulars against the justices:

> The justices place themselves, when they see fit, in the place of the legislature, in that they make laws rather than review and interpret them;
> They hinder, with an internally contradictory jurisprudence, the foreign policy of the Federal Republic vis-à-vis its western and eastern partners;
> They are too conservative when it comes to reprimanding a sitting government, but equally too conservative when it comes to repudiating a rigidified Catholic or otherwise reactionary moral dogma;
> They place the dead letter [of the law] over human needs and developments; and, what's more,
> Their judicial workmanship bears no relation to the role they demand and fulfill.[264]

Augstein went on to resurrect talk of a Red Senate (the Second) and a Black (the First). He groused that the Court's standards of recusal were applied inconsistently to Catholics and non-Catholics: Justice Rottmann had been excluded from the Second Senate's judgment on the Basic Treaty while Hans Brox took part in the First Senate's decision on abortion. The Court, Augstein scoffed, had crowned itself a superior chamber of Parliament—as nothing less than 'our Lord God's law office'.

Coalition politicians were more circumspect. Dismayed as they were by the news that the amended section 218 was to be annulled, they knew that future reforms—in abortion and other areas—would require the Court's approval. They also harboured painful memories of the infamous one-liner reported during the *Basic Treaty* debate. For the most part, coalition leaders kept mum.

But there were exceptions. Even before the leak, Lisolette Funcke (FDP), vice president of the Bundestag, announced that 'the women' would neither accept nor respect a judgment against the term solution.[265] After the leak, Gerhart Baum (FDP), an undersecretary of state, feared that the CDU/CSU opposition would take unseemly courage from the victory. 'They'll take everything to Karlsruhe', Baum cringed. 'As a legislature we'll ... hardly be able to move'.[266]

The most prominent critic was Helmut Schmidt himself. Speaking on the television show *Panorama* on 3 February, the chancellor cited the *Basic Treaty* and *Solange* judgments as evidence of an unsettling expansion of judge-made law (*Richterrecht*). On the abortion question, Schmidt continued, it would be astonishing if the Constitutional Court reached an outcome at odds with that reached by the Austrian Constitutional Court, the French

262 See 'Unseres Herrgotts Kanzlei' *Der Spiegel* (3 February 1975); *Die Welt* (25 February 1975); *Die Zeit* (28 February 1975).

263 'Denunzieren wider Willen' *Die Zeit* (14 March 1975).

264 'Unseres Herrgotts Kanzlei' *Der Spiegel* (3 February 1975). The phrase stems from the title of a novel by Wilhelm Raabe.

265 In Friedrich Karl Fromme, 'Gericht im Zog der Politik' *FAZ* (29 January 1975) 1.

266 In 'Mit allem nach Karlsruhe' *Der Spiegel* (3 February 1975).

Constitutional Council, and the United States Supreme Court.[267] If it did, one might doubt whether the Court rightly understood its role. By ruling against the law, the chancellor implied, the Court would be playing politics rather than applying law.[268]

Some Left–Liberals doubted the wisdom of Schmidt's remarks, but an SPD party convention formally endorsed it.[269] Conservatives rose to the Court's defence in terms reminiscent of the 1950s advocacy of Adolf Arndt, who died in 1974.[270] Roman Herzog, a constitutional law scholar and future president of the Constitutional Court, thought Schmidt's comments betrayed a faulty understanding of the German constitution and a woeful naivety of comparative law.[271] In a speech given earlier at the Free University of Berlin, Ernst Benda anticipated Schmidt's critique when he denied that the Court had any intention to thwart the political process or stand as 'schoolmaster' of the nation.[272] Later in the decade, the chancellor and chief justice would lock horns more directly. Meanwhile, the Court's actual decision was still pending. And though the judgment's outcome was suspected in advance, its content still managed to astonish.

An estimated 10,000 women marched in protest in various cities—3,000 in Munich, 2,000 in Hamburg, 1,500 in Frankfurt, and between 1,500 and 2,000 in Karlsruhe itself.[273] Expecting a negative judgment, the protesters bemoaned a perceived setback for women in the early months of what had been declared the international 'Year of the Woman'. In ironic articles in the *Frankfurter Allgemeine Zeitung*, Friedrich Karl Fromme noted the incongruence of the protesters with Karlsruhe's wonted tranquility. Fromme praised the justices for their 'business-as-usual' conduct in the centre of the national spotlight. He was less kind to the coalition's lawyer, Horst Ehmke, who arrived at the Court building with all the modesty of a rock star, emerging from a flashy sports car flanked by his fashionable young wife.[274] By other accounts, protests in many cities were longer and louder than expected.[275]

Within the courthouse itself, an army of journalists and photographers increased both the temperature and the population density. Tension was high, and the anticipation great. Embarrassed by the earlier leak, the Court treated its written decision like a state secret. No copies of the judgment had been pre-released to members of the government—not even to the justice ministry. Instead, Wolfgang Heyde, a ministry lawyer, left Bonn by helicopter at 8 a.m. on the morning of the decision for a fast flight to Karslruhe. Arrived at the Court, Heyde was given a sealed copy of the judgment, which he was allowed to carry back to the capital only after Benda had announced the decision's basic tenor. Heyde left on a return flight to Bonn around 11:15 a.m., but was allowed to break the seal only after the decision had been fully announced in Karlsruhe. The ministry received permission from the Court

[267] In 1974, the Austrian and French courts held that liberalization of abortion law did not violate the right to life guaranteed in the European Convention on Human Rights, which was incorporated into the domestic law of both countries. In 1973, the US court ruled that restrictions on abortions during the first trimester of pregnancy violated the penumbral right to privacy. See *Roe v Wade*, 410 US 113 (1973).

[268] See the reports in Robert Leicht, 'Kritik an Karlsruhe fällt auf die Politiker zurück' *Süddeutsche Zeiting* (5 February 1975); 'Richterschelte' *FAZ* (5 February 1975) 8; 'Angemaβte Dienstaufsicht' *Frankfurter Rundschau* (7 February 1975) 1.

[269] 'Ständiger Miβbrauch' *Frankfurter Rundschau* (7 February 1975) 3; 'Parteitag billigte die Kritik am BVG' *Frankfurter Rundschau* (17 February 1975).

[270] See Michael Hansen, 'Karlsruhe—Grenze für Politiker' *Die Welt* (7 February 1975) 4.

[271] 'Herzog warnt vor falschem Verständnis des Verfassungsgerichts' *FAZ* (7 February 1975) 5.

[272] See 'Benda: Verfassungsgericht kein Schulmeister der Politiker' *Der Tagesspiegel* (2 February 1975); *Die Welt* (3 February 1975) 2.

[273] 'Die Richter zwingen die Koalition einer Neufassung des §218' *Die Welt* (26 February 1975) 1.

[274] 'Wohltuende Distanz zwischen den Zentren von Politik und Justiz' *FAZ* (26 February 1975) 3; 'Ich muβ mal zur Demonstration in die Stadt' *FAZ* (27 February 1975) 3.

[275] 'Erregte Sprechchöre beim Schweigemarsch' *Süddeutsche Zeitung* (26 February 1975) 3.

to open the decision only around 1 p.m., minutes before Hans-Jochen Vogel, the justice minister, was to discuss the judgment in a press conference.[276]

What did the secret decision say? First, a majority of justices held that the right to life guaranteed to 'everyone' in Article 2(2) GG encompassed to prenatal life.[277] This finding was in keeping with the Court's principle that, in cases of doubt, the preferred interpretation is the one that most strengthens the fundamental right in play.[278] The Basic Law's protection of the right to life, moreover, was stronger than that offered by the Weimar constitution—stronger, presumably, than that of other national constitutions—because the postwar constitution had been enshrined against the backdrop of Nazism's virulent contempt for life.[279] Article 2's framing history, the Senate noted, also supported extending the right to life to the unborn[280]—a point that was sharply disputed by the judgment's dissenters and academic commentators.[281] Unborn life was further protected by Article 1's consecration of human dignity. A foetus was human and as such possessed dignity, which the constitution declared sacrosanct.[282] It was irrelevant that the foetus was unconscious of its dignity and unable to preserve it. The state must preserve it, beginning at the moment of individuation and implantation roughly fourteen days after conception. The state's duty to protect was comprehensive. It forbade the state to injure developing life, but also required the state to protect unborn life from threats at the hands of others—including the unborn child's mother.[283]

Expectant mothers, on the other hand, enjoyed countervailing constitutional rights of their own. Pregnancy belonged to a woman's 'intimate sphere' and was therefore shielded by constitutional guarantees of dignity, bodily integrity, and personal freedom. A woman's right to develop her personality included the right to avoid the obligations of parenthood. That right was constitutionally recognized and protected. But it was not unlimited. As the text of Article 2(1) explained, it was limited by the rights of others, the constitutional order, and the moral law. If a foetus was merely part of a woman's body, the law could not constitutionally regulate abortions. But a foetus was an independent being, shielded by rights that restricted a mother's personal freedom. The rights of an unborn child and the rights of a mother unwilling to bear that child were at irreconcilable odds. The vindication of the latter required the abolition of the former. Determining which right prevailed required careful balancing.[284]

That balance, the majority explained, must be oriented toward human dignity. Focusing on dignity led the Senate to conclude that protecting foetal life trumped a pregnant woman's right of self-determination. Pregnancy, child-bearing, and child-rearing might impair a woman's opportunities for development; but an abortion destroyed an unborn life entirely. In the Court's constellation, life was a greater value than freedom; and abortion was a greater threat to life than undesired motherhood was to freedom. This precedence applied, in principle, for the entire duration of the pregnancy. It could not, for

[276] 'Wie das Urteil nach Bonn kam' *Die Welt* (26 February 1975) 1; Klaus Dreher, 'Das Tauziehen geht wieder los' *Süddeutsche Zeitung* (26 February 1975) 3.

[277] 39 BVerfGE 1, 36 (1975).

[278] ibid 36 (citing 32 BVerfGE 54, 71 (1971) and 6 BVerfGE 55 (1957)).

[279] ibid 36.

[280] ibid 38–40.

[281] See, e.g., Kriele, 'Anmerkung' (n 105) 222; Wolfgang Abendroth, 'Das Abtreibungsurteil des Bundesverfassungsgerichts' (1975) 8 *Kritische Justiz* 121; Josef Esser, 'Bemerkungen zur Unentbehrlichkeit des juristischen Handwerkszeugs' (1975) 30 *JuristenZeitung* 555.

[282] 39 BVerfGE 1 (n 277) 41.

[283] ibid 42.

[284] ibid 43. This, to put it lightly, was a very different framing of the abortion question than that put forth by Justice Blackmun in his little-loved majority opinion in *Roe v Wade*, an opinion that famously refused such a balance on the grounds that a foetus was not a 'person' for purposes of the Fourteenth Amendment, and therefore not a rights-bearer at all. Whatever the shortcomings of the First Senate's judgment, it at least persuaded both sides of the abortion debate that the values that mattered to them had entered its analysis.

any specified period, be placed in doubt.[285] The state must protect an unborn child until it was born.

How to fulfil this obligation was primarily a question for the legislature.[286] In answering that question, however, the legislature must observe certain principles. As the judgment progressed, the Senate's enumeration of principles read increasingly like substitute legislation. The state's first goal, the justices explained, must be to discourage abortions. If no other means sufficed, such discouragement required the threat of criminal sanction. There were cases, the justices acknowledged, in which requiring a woman to give birth was constitutionally intolerable. Obviously, a woman could not be obliged to carry a child at the risk of her own life or health. In addition, the majority cited approvingly a (rejected) legislative proposal to legalize early abortions on 'eugenical, ... ethical (criminological) [cases of rape], and social' grounds, or on grounds of distress.[287]

'In all other cases', however, 'abortion must remain a punishable legal wrong [*strafwürdiges Unrecht*]'—not only illegal but criminal.[288] Parliament could decriminalize abortions only if it found another deterrent that, with equal effect, reduced the number of abortions *and* expressed the legal order's disapproval of them. The reform law's provision for individual medical counselling was inadequate. An indications model backed by criminalization would be much better. The Senate was reluctant to countenance other possibilities: 'Given the high value of the legal good to be protected, experiments are not admissible'.[289] So much for legislative leeway.

The Senate noted, almost in passing, that the old section 218 was unconstitutional as well. Having criminalized too much, that law was enforced too little.[290] Parliament's reform law was null, but so was the status quo ante. The legislature must try again, for '[t]he Court may not place itself in the position of the legislature'.[291]

The judgment ended boldly. As if in response to Schmidt's censure, the majority noted that reference to judicial decisions in other countries could not influence the Court's deliberation. Germany was unique. The Basic Law arose in direct response to the horrors of Nazism. As an answer to the murderous inhumanity of the Third Reich, the postwar constitution's concern for human life and dignity was especially pronounced.[292] These principles were unimpeachable, immune to shifts in popular views. 'Even a general change in the population's prevailing views on the matter', the justices wrote, 'assuming such a change were to be established could do nothing to alter' these core principles.[293]

It was a long judgment—the profoundest revision of the legislative process in the Court's quarter-century history. Benda's summary lasted more than an hour. Benda then called for a ten-minute break before Justice Rupp-von Brünneck read a dissenting opinion signed by herself and Justice Simon.[294]

Rupp-von Brünneck's decision to read her dissent publicly was unusual. So was Benda's call for a break beforehand. The purpose of the break became clear when it ended. During the pause, Justice Werner Böhmer had left the building. The official line was that he wasn't feeling well. In reality, Böhmer preferred not to listen to what he knew was an unusually

[285] ibid. [286] ibid 44. [287] ibid 49. [288] ibid 50–51. [289] ibid 60. [290] ibid 52.
[291] ibid 51. [292] ibid 67. [293] ibid.
[294] Though only two justices signed the dissent, Benda explained that this didn't necessarily mean the vote within the Senate was 6:2. Many gathered from this announcement that Justice Karl Haager disagreed with the majority as well. See, e.g., 'Fristenregelung verhindert' *Süddeutsche Zeitung* (26 February 1975) 1–2; 'Fristenregelung vom Verfassungsgericht verworfen: Unvereinbar mit dem Grundrecht auf Leben' *FAZ* (26 February 1975) 1.

sharp dissent. 'I won't let myself be insulted', Böhmer reportedly vowed—*Ich lasse mich nicht beschimpfen*.[295]

Rupp von Brünneck read from her dissent in a voice quivering with hard-kept anger.[296] It was a poignant moment—the Court's only woman rising to contradict a judgment passed on all women by a majority of five men. Some heard echoes of the American dissenting tradition of Brandeis and Frankfurter, Harlan and Holmes.[297] The crux of the dissent was that the Court was guilty once more—guilty more than ever—of overreaching. More egregiously even than in the *University Governance* judgment, the justices had usurped the prerogative of the legislature.

The centrality of the right to life, Rupp-von Brünneck began, was not in question. Nor was the state's duty to protect the unborn. The question was not whether to protect fetal life, but how. And that was a question not for the Court but for the legislature. The majority had not only disregarded the legislature's considered response, it had imposed an alternative of its own—an unprecedented duty to criminalize. This was not merely wrong, it was revolutionary. 'Our strongest objection', the dissenters wrote,

> is directed against the fact that for the first time in constitutional jurisprudence an objective value decision should serve to saddle the legislature with a *duty* to level *criminal sanctions*—that is, to impose the strongest imaginable invasion of the citizen's sphere of liberty. This distorts the function of fundamental rights into its opposite. If a fundamental right's objective value decision to protect ... should suffice to derive therefrom a duty to punish, fundamental rights, [instead of] a bulwark of safeguards for liberty, could surreptitiously become the basis for a plethora of regulations restricting liberty.[298]

In the dissenters' eyes, the term solution was a reasonable response to the tension between protecting unborn life and affording freedom and equality to expectant mothers. Further than this the Court need not inquire. The majority's case for German exceptionalism by reference to the Nazi past was a spurious exercise in rhetorical *Vergangenheitsbewältigung*. The tyrant rulers of the Third Reich had *stiffened* sanctions for abortions; the elected leaders of the Federal Republic were trying to do the opposite. The majority justices listened to Rupp von Brünneck's passionate reading in stony silence—wondering, as a reporter for *Die Welt* suspected, how the Senate's work might proceed after this unprecedented '*Eklat*'.[299]

The judgment raised more than collegiality problems. 'The controversy concerning the role of the highest court', predicted Robert Leicht in the *Süddeutsche Zeitung*, 'will now erupt in full fury for the first time'.[300] Bruno Friedrich, an SPD member of Parliament, feared 'a malevolent poisoning' of popular sentiment, wrought by the Republic's first 'emotionally waged constitutional conflict'.[301] Even those, like Leicht, who said they could live with the judgment, bewailed the Court's overreaching (*Kompetenzüberschreitung*). The Court's inevitable loss in public esteem, Leicht warned, would not only wreak mischief for the Court itself; it would endanger the whole constitutional structure of the Federal Republic.[302]

[295] 'Zuchtmeister für Bonn und Bürger' *Der Spiegel* (3 March 1975); 'Frau von Brünnecks Auftritt: Das Tribunal wurde zur Szene' *Die Welt* (27 February 1975) 2.

[296] Enno von Loewenstern, 'Auch der alte §218 ist verfassungswidrig' *Die Welt* (26 February 1975) 2.

[297] See, e.g., Kriele, 'Anmerkung' (n 105) 222.

[298] 39 BVerfGE 1 (n 277) 73.

[299] 'Frau von Brünnecks Auftritt: Das Tribunal wurde zur Szene' *Die Welt* (27 February 1975) 2.

[300] Robert Leicht, 'Grenzüberschreitung in Karlsruhe' *Süddeutsche Zeitung* (26 February 1975) 4.

[301] In Klaus Dreher, 'Das Tauziehen geht wieder los' *Süddeutsche Zeitung* (26 February 1975) 3.

[302] Robert Leicht, 'Grenzüberschreitung in Karlsruhe' *Süddeutsche Zeitung* (26 February 1975) 4.

Many critics were quite acerbic. The *Frankfurter Rundschau* styled the judgment a 'Karlsruhe Encyclical'.[303] Other papers reviled the judgment as a religious credo parading as a judicial decision.[304] Still others mocked the Court's pretensions to stand as an 'ersatz government' or 'super authority'.[305] Particularly scornful was *Der Spiegel*. Days after the *Abortion* decision, a *Spiegel* reporter began a combative interview with Ernst Benda by asking whether the Court wished 'to seize power in Bonn'?[306] The interview that followed was more accusation than inquiry. The interviewer alleged in various ways that the Court had exceeded its mandate, and the chief justice, with mounting irritation, denied the charge. (By contrast to the sustained *J'accuse* it hurled at the chief justice, *Der Spiegel* lionized the dissenters, running an adulatory feature on Justice Rupp-von Brünneck.)[307] Another article accused the Court of crowning itself 'Moral Preceptor to Government and Citizen'.[308] Never, *Der Spiegel* thundered, 'has the Karlsruhe Court so drastically displayed its omnipotence; and never has it been so clear to politicians how easily they can be surrendered to the men in the red robes'.[309]

The politicians thus vulnerable to surrender, both winners and losers, were more restrained. Rudolf Seiters, the CDU floor leader in the Bundestag, cautioned that 'cries of triumph would be out of place'.[310] SPD leaders similarly pleaded for prudence as they looked for a silver lining. Hermann Dürr noted that the judgment did not merely prescribe the CDU/CSU indications model; it allowed for 'considerably more'.[311] Willy Brandt, still party chairman, said little about the judgment itself. But in a classical Brandtian gesture—magniloquent and vague—he assured that '[e]ven after the judgment women can rest assured that we will not leave them alone with their troubles'.[312] Within the cabinet, expressions of disappointment were subdued. In his official response to the judgment, Hans-Jochen Vogel, the justice minister, promised to respect the judgment, but insisted that the government's defence of the rejected law remained 'cogent and conclusive'.[313] The judgment, Vogel added, was particularly disappointing in light of the 'understandable expectations of millions of women'.[314] In all official statements, the government urged the citizenry to engage the judgment in a measured discussion.[315]

Academic criticism of the judgment was sometimes measured, sometimes not. In any case it was abundant. Many critics agreed that the Court had erred by not merely condemning one legislative response but by prescribing another.[316] Others admonished that it was not the Court's comission to inform and uphold the moral conscience of the nation.[317] More generally, the majority's methodology and workmanship made scholars squirm.[318]

303 Roderich Reifenrath, 'Die Karlsruher Enzyklika' *Frankfurter Rundschau* (26 February 1975).

304 See Joachim Besser, 'Glaubenssätze in Urteilsform' *Vorwärts* (6 March 1975) 2.

305 Florentine Hoffmann, 'Ersatzregierung oder politische Supergewalt' *Welt der Arbeit* (7 February 1975).

306 'Karlsruhe—ein verkappter Gesetzgeber?' *Der Spiegel* (3 March 1975).

307 '... bes. in sittlicher Hinsicht' *Der Spiegel* (3 March 1975).

308 'Zuchtmeister für Bonn und Bürger' *Der Spiegel* (3 March 1975). *Die Zeit* argued in a similar vein, characterizing the justices as the self-appointed 'moral guardians of the nation'. 'Die Sittenwächter der Nation' *Die Zeit* (28 February 1975).

309 'Zuchtmeister für Bonn und Bürger' *Der Spiegel* (3 March 1975).

310 Klaus Dreher, 'Das Tauziehen geht wieder los' *Süddeutsche Zeitung* (26 February 1975) 3.

311 ibid. 312 'Ärzte für Indikationenlösung' *Die Welt* (26 February 1975) 1.

313 'Die Richter zwingen die Koalition einer Neufassung des §218' *Die Welt* (26 February 1975) 1.

314 'Bundesregierung mahnt Bevölkerung zu besonnener Diskussion' *FAZ* (26 February 1975) 1, 3.

315 ibid.

316 See Kriele, 'Anmerkung' (n 105) 223. Kriele also referred to the judgment as a 'raiding action on the terrain of the legislature'.

317 Christian-Friedrich Menger, 'Das verfassungsgerichtliche Urteil zu § 218 StGB—Gesetzgebung durch das BVerfG'? (1975) 66 *Verwaltungsarchiv* 402, 403.

318 See, e.g., Esser (n 281).

Professor Josef Esser cringed at the 'windiness' of the Court's reasoning. Esser entitled his review essay, 'Notes on the Indispensability of Judicial Craftsmanship'.[319]

The judgment also triggered scholarly explosions of polemical wrath, the fiercest of which came from Wolfgang Abendroth.[320] 'For the third time in a relatively short period', Abendroth roared, 'the Constitutional Court has decided against the Basic Law'.[321] Like other critics, Abendroth chafed at the Court's legislative presumption. Most of what one read in the majority decision, he grumbled, had no business in the judgment of a constitutional court. The *Abortion* judgment consisted entirely of legal–political considerations that the legislature may and must observe—'But the Constitutional Court is not the legislature'.[322] What most provoked Abendroth's ire, however, were the majority's remarks about German exceptionalism, rooted, Abendroth believed, in a pernicious misreading of the German past.[323] In the late 1930s Abendroth had languished in Nazi prisons with doctors condemned for performing abortions. He heard in the Court's judgment an eerie echo. If the government and cabinet accepted the judgment without censure, he warned, they would acquiesce in the step-by-step transformation of the Federal Republic into 'an authoritarian state, in which even the smallest democratic reform promulgated by Parliament will collapse before the conservative prejudices of the Constitutional Court'. If that happened, the political practice of the Federal Republic would as little deserve to be called a democracy as the Weimar-era regime under which Heinrich Brüning ruled by emergency decree.[324] And, as in the days of Brüning, a fascistic successor was waiting in the wings: Franz Josef Strauβ, whose constitutional petitions the Court seemed so eager to grant. Abendroth called on all democratic and socialist forces to unite to save the *Rechtsstaat* from the Court and from Strauβ.[325]

Very few critics went as far as Abendroth. But many whose voices were less shrill, and whose perspectives were more centrist, worried audibly—and for the first time—that the Court imperilled democracy.

The decision was unpopular. An Allensbacher Institute survey found that 50 per cent of West Germans disapproved of the judgment; only 32 per cent approved.[326] But this unpopularity coincided with calls for perspective. Many writers stressed that while the judgment invalidated the reform law, it also invalidated the *status quo ante legem*.[327] Other writers asked critics to balance indignation at this particular judgment with recognition of the Court's long-term contributions. An editorial in *Die Zeit* reminded readers 'that the risk of losing is the institutional price of constitutional justice'.[328] Furor over the decision should be tempered by recognizing 'how much progress in recent decades came thanks to the Constitutional Court'. As examples, the article enumerated cases involving gender equality, the legal status of illegitimate children, equal opportunity for political parties, freedom of opinion, vocational freedom, and the rights of the accused. The article even credited the 1961 *Television* judgment with having 'saved our democracy from the most serious threat it had yet confronted'.[329] Two weeks earlier, with the announcement of the judgment still pending, *Die Zeit* published an interview with Gerhard Leibholz, in which the intellectual leader of the early Court related a similar list of the Court's contributions toward realizing West German democracy and legitimizing the West German state.[330]

319 ibid. 320 Abendroth, 'Das Abtreibungsurteil' (n 281).

321 ibid 121. The other two occasions, in Abendroth's view, were the *University* and *Basic Treaty* judgments.

322 ibid 125–26. 323 ibid 127. 324 ibid. 325 ibid 127–28.

326 See 'Jeder zweite gegen das Abtreibungsurteil' *FAZ* (7 March 1975) 5.

327 See Enno von Loewenstern, 'Auch der alte §218 ist verfassungswidrig' *Die Welt* (26 February 1975) 2.

328 'Vor Karlsruhe stammstehen?' *Die Zeit* (28 February 1975). 329 ibid.

330 'An den Grenzen des Rechtes' *Die Zeit* (14 February 1975).

Controversy surrounding the *Abortion* decision focused the public's attention on the Court and its history. The Court's defenders responded with a narrative that presented the Court as a champion of democracy and vindicator of fundamental rights. In this narrative, the Court played a central role in laying the democratic and constitutional foundations of the Federal Republic. In light of this historic achievement, an unsavoury outcome in a politically charged case could be swallowed with a pinch of perspective.

Hostility to the judgment ebbed with time as it became clear that the judgment's practical impact was less than many anticipated. In 1976, the Bundestag passed a new law adopting the indications model almost exactly as outlined by the Court. The new law decriminalized abortions during the first twelve weeks of pregnancy in specified circumstances, including a controversial 'distress' or 'social' indicator, whose effect was to make abortions available to most women who wanted them. Three years after the law was passed, Justice Willi Geiger—by then, at last, retired—complained that the social indicator was being interpreted too broadly.[331] Ute Frevert, the preeminent historian of German women, later wrote that the 1976 law granted a general 'right to abortion'.[332]

Still, there could be no denying that the *Abortion* judgment marked an unprecedented restriction of legislative leeway—the most striking instance yet of the Court telling Parliament what to do. In a trio of 1972 cases, the Court had ordered the federal legislature to revise prison regulations before a specified date;[333] commanded state legislatures to regulate certain facets of the medical profession directly;[334] and suggested that the constitution might impose a legislative duty to provide higher education opportunities for all citizens.[335] Many saw these judgments as cautious exceptions to the rule that the Court must review what parliaments have already done, not dictate what they must yet do.[336] With the *University* judgment, the Court entered a new plane, ruling that constitutional guarantees of academic freedom imposed affirmative duties on the state, and prescribing in detail how those duties must be fulfilled. In the *Abortion* judgment, the Court had gone further still, not merely finding an affirmative duty, but rejecting the legislature's attempts to fulfil that duty and decreeing a stark alternative. In the *Abortion* judgment, the Court's enormous power was on awesome display—and in what many took to be a profoundly conservative cause.

Ironically, the doctrine of affirmative duties was of progressive origin. In the Republic's early years, progressives hoped for a vigorous and binding interpretation of the Basic Law's social state provision—an interpretation that would commit the legislature to a sweeping programme of socialization and state welfare. The doctrine's most redoubtable champion was Wolfgang Abendroth.

F. 'Radicals' in the public service

A few days after the Court announced its *Abortion* judgment, an attempt was made to bomb the Court's building.[337] A self-styled 'revolutionary women's group' later confessed to the

[331] 'Soziale Indikation zu frei ausgelegt. Prof. Willi Geiger sprach in Saarbrücken' *Saarbücker Zeitung* (Saarland, 1 December 1979).

[332] Ute Frevert, *Frauen-Geschichte: zwischen bürgerlicher Verbesserung und neuer Weiblichkeit* (Suhrkamp 1986) 280.

[333] 33 BVerfGE 1, 12 (1972). [334] 33 BVerfGE 125, 158 (1972). [335] 33 BVerfGE 303 (1972).

[336] See Kriele, 'Anmerkung' (n 105) 223.

[337] See 'Bombenanschlag auf Gebäude des BVerfG' *Der Tagesspiegel* (5 March 1975); Walter Pfuhl, 'Nur vier Mann bewachten beim Bombenanschlag das Gericht' *Die Welt* (6 March 1975); Hanno Kühnert, 'Vogel: Anschlag auf Rechtsstaatlichkeit' *Süddeutsche Zeitung* (6 March 1975).

attack. Damages were reported at 80,000 DM, but no one was injured. Press coverage of the incident[338] was overshadowed by the simultaneous abduction by terrorists of Peter Lorenz, the CDU opposition leader in West Berlin and a mayoral candidate. Domestic terrorism haunted the Federal Republic for more than a decade, beginning with the burning of two Frankfurt department stores in 1968. There followed a series of bank holdups and, more menacingly, the kidnap and murder of businessmen and politicians. Terrorism was largely a phenomenon of the extreme Left; the radical Red Army Faction (RAF) was its most sensational manifestation.[339]

The birth of Left-wing terrorism coincided with campus unrest, the spread of Leftist student movements, and the founding of a new German Communist Party, which, christened DKP rather than KPD, sought to avoid its forerunner's fate by promising to pursue its aims within the parameters of the free democratic order. Politicians of all stripes were rattled by resurgent radicalism. Though most looked more anxiously to the Left than to the Right, there were also calls to ban the Right-wing National Democratic Party (NPD), which polled alarmingly well (nearly 10 per cent) in the Baden-Württemberg state elections of April 1968 (but fell below the 5 per cent threshold in the 1969 Bundestag elections).

Radicalism, especially Left radicalism, provoked a backlash of popular enmity. Thrust into opposition after the 1969 elections, the CDU/CSU hoped to exploit popular anti-Leftism to regain power. Unionists blasted Brandt as soft on communism and demanded a ban of the DKP. Brandt denied the charge of softness, but was queasy about seeking bans. The chancellor knew as well as his opponents that asking the Court to ban the DKP would hamper negotiations with the GDR. The chancellor would have to flash his anti-communist mettle in some other way.

One route, also pushed by the Union parties, was to act vigorously against radicals in the public service. On 28 January 1972, Brandt and the state premiers issued a pair of declarations on the subject. The declarations, later labelled the 'radical decrees' (*Radikalen beschlüsse*), affirmed that '[o]nly persons who can guarantee that they are prepared at all times to uphold the free democratic basic order, as established in the Basic Law, may be appointed to the public service. Civil servants, whether on or off duty, are required to take an active part in maintaining this constitutional system'.[340] A candidate pursuing 'activities inimical to the constitution [*verfassungsfeindlich*]' must not be appointed. Membership in an organization pursuing 'anti-constitutional aims' cast doubt on a 'candidate's readiness at all times to uphold the free democratic basic order. In most cases, such doubts would justify rejecting the candidate's application'.

The decree passed over the reservations of Brandt's justice minister, Gerhard Jahn, and of Herbert Wehner, who denounced the decree as the trigger for an ideological witch-hunt (*Gesinnungsschnüffelei*), a harebrained bid 'to defend the free basic order by taking the first step toward its abolition'.[341]

[338] See *Die Welt* (5 March 1975) 1; *FAZ* (5 March 1975) 1; *FAZ* (6 March 1975) 2; *Frankfuter Rundschau* (6 March 1975) 1; *Die Welt* (7 March 1975) 2; *FAZ* (7 March 1975) 5; *Süddeutsche Zeitung* (7 March 1975) 2; *FAZ* (12 March 1975) 3.

[339] For a history of the RAF, see Klaus Pflieger, *Die Rote-Armee-Fraktion: RAF, 14.5.1970 bis 20.4.1998*, 3rd edn (Nomos 2011). On the phenomenon of domestic terrorism more broadly, see Petra Terhoeven, *Deutscher Herbst in Europa: der Linksterrorismus der siebziger Jahre als transnationales Phänomen* (Oldenbourg 2014); Karrin Hanshew, *Terror and Democracy in West Germany* (Cambridge University Press 2012); Johannes Hürter, *Die bleiernen Jahre: Staat und Terrorismus in der Bundesrepublik und in Italien, 1969-1982* (Oldenbourg 2010).

[340] On the history and aftermath of the declarations, see Gerard Braunthal, *Political Loyalty and Public Service in West Germany: The 1972 Decree Against Radicals and Its Consequences* (University of Massachusetts Press 1990).

[341] Herbert Wehner, 'Spiel mit dem Grundgesetz' *Ausburger Allgemeine Zeitung* (Augsburg, 21 January 1972). Former Justice Gehrard Leibholz also joined the debate. See Gerhard Leibholz, 'Frage der verfassungsfeindlichen Kräfte im öffentlichen Dienst' (1972) 2 *Analysen* 17.

The legal status of the decrees was unclear. Ostensibly restatements of prevailing practice, the decrees inspired recruitment authorities zealously to exclude 'extremists' from the public service. The decrees enjoyed broad support, both popular and political. They also provoked raucous opposition, both at home and abroad. Leftists and intellectuals denounced them as a 'career ban' (*Berufsverbot*), since many seeking jobs as public servants (teachers, mail carriers, locomotive engineers, etc) had no hope of private employment.[342] The term *Berufsverbot* entered the international lexicon. Several continental eminences—Simone de Beauvoir and Jean-Paul Sartre in France, Luigi Nono and Pier Paolo Pasolini in Italy—shouted their solidarity with affected West Germans.[343]

There were many affected. Over the course of the 1970s, more than a million applicants for public service were investigated for potential radical ties; thousands, having aroused suspicion, were subjected to investigatory hearings, which were sometimes bullying and often bizarre. More than a thousand applicants were excluded for extremism. Those denied admission—92 per cent of whom were Leftists[344]—could appeal for review by administrative courts, but the legal standards for such review were vague and the resulting jurisprudence mottled. The spectacle spurred a public outcry. That some judges reviewing prominent cases were former National Socialists made the proceedings all the more galling. 'Better to deprive all teachers and professors of their status as civil servants', wrote Jürgen Habermas, 'than to submit even one of them to a judicial examination of their views'. Those examinations were, 'for all involved, quite as risible as they [were] unworthy'.[345]

Politicians and jurists hoped a Constitutional Court judgment would provide legal clarity and foster judicial uniformity. But that would require a plaintiff hearty enough to bring the case. This proved harder than one might expect: the DKP had ordered its members not to file constitutional complaints, lest an overzealous Senate seize the chance to ban the party![346] Ultimately, the matter reached the Court through judicial referral rather than individual complaint.

While studying law at the University of Kiel, Heinrich Sämisch attended about forty meetings of a Leftist law student group, the 'Rote Zelle Jura' (Red Cell Law). Sämisch passed his professional examination in 1971 and applied for a mandatory legal internship (*Referendarzeit*) in Schleswig-Holstein. The internship required that he become a professional civil servant (*Beamter auf Wiederruf*) before taking a second examination, which would qualify him to become a judge, a prosecuting attorney, or a practising lawyer. Sämisch's application failed. He had belonged, ran the official explanation, to a group whose goals were antithetical to the constitution. Sämisch appealed to the Schleswig-Holstein administrative court, which stayed its proceeding to ask the Constitutional Court whether the state could require *Referendar* applicants to promise unswerving fealty, at all times, to the free democratic order.

The Second Senate's decision of 22 May 1975 was a strange one.[347] The question before the Court was whether the state could require applicants for *Referendar* service to swear a loyalty oath as a condition of admission. The Senate reached this question only indirectly, holding that, because *Referendar* service represented a continuation of legal education rather than the commencement of civil service proper, the candidate in this case must be admitted. This holding was all that the lower court submission required. But the Senate refused to let the matter lie. In what many scholars bemoaned as an astonishing outburst

342 Braunthal (n 340) 34. 343 ibid 67–69. 344 ibid 48.

345 Jürgen Habermas, 'Berufsverbote' in Jürgen Habermas, *Kleine politische Schriften I-IV* (Suhrkamp 1981) 331.

346 Braunthal (n 340) 58. 347 39 BVerfGE 334 (1975).

of *obiter dicta*,[348] the justices proceeded to grapple directly, and at some length, with the abstract question of whether the Radical Decrees, on their face, violated the constitutional guarantee of occupational freedom.

In principle, the Senate found, they did not. Civil servants must do more than verbally acknowledge the validity of the constitutional order; they must fulfil their duties in the spirit of its provisions.[349] In a passage whose prose prompted yelps of scholarly disapproval, the Senate added that a civil servant's duty of constitutional loyalty, prescribed by Article 33 GG, 'requires more than a formally correct, otherwise disinterested, cool, internally distanced stance toward the state and the constitution; it requires that civil servants in particular distance themselves unequivocally from groups and initiatives that assault, attack, or defame this State, its constitutional organs, and its reigning constitutional order'.[350]

Those hoping the Court would bring legal clarity to the vague but loaded language of the Radical Decrees were dismayed by a decision containing concepts still more vague and expressions still more loaded. Elsewhere in the decision, the Senate asked whether recruitment authorities could base their assessment of a candidate's fitness on his mere membership in a party the authorities deemed unconstitutional, but which the Constitutional Court had not formally banned. The Senate majority said yes;[351] two dissenters (Walter Seuffert and Hans Rupp) said no. The question, of course, was entirely beside the point. The complainant had not belonged to an allegedly unconstitutional party; he had attended the meetings of a constitutionally irrelevant student group.

Surprisingly, given the long controversy surrounding the Radical Decrees, the Second Senate's decision was little reported in the press.[352] Academic commentary, much of it critical, was prolific. The decision did the opposite of what many scholars had hoped: it had further polarized debate about the decrees and exacerbated the legal uncertainty surrounding them.[353] Once again, the Second Senate got poor marks from a critical academy for its technical performance. On a broader level, many were dismayed that, two months after the First Senate had inverted the function of fundamental rights in the *Abortion* judgment, the Second Senate had declined to enforce fundamental rights robustly in their classical defensive context. Left-leaning critics groaned that the Court had been activist when the government sought to liberalize, but deferential when the state sought to repress. Even in its passivity, the Court's judgment reached beyond what the concrete demands of the case required. The time seemed out of joint; the Court was getting its role precisely backwards.

G. Parliamentary pay

After issuing yet another politically fraught judgment five months later, the Court was once more accused of expanding its jurisdiction and indulging in *obiter dicta*. This time,

[348] See Rüdiger Zuck, 'Zur Bindungswirkung der Radikalen-Entscheidung des BVerfG' (1975) 15 *Juristische Schulung* 695; Helmut Simon, Hans Mommsen, and Peter Becker, 'Verfassungsschutz durch Verfassungszerstörung' (1989) 22 *Zeitschrift für Rechtspolitik* 175; Dieter Schimanke, 'Die Auswirkungen des "Radikalenbeschlusses" des Bundesverfassungsgerichts auf Gesetzgebund und Verwaltungspraxis' (1976) 8 *Juristisches Arbeitsblätter* 113.

[349] 39 BVerfGE 334 (n 347) 348. [350] ibid. [351] ibid 359.

[352] There was no notice at all in the 23 May 1975 issues of *Die Welt, Die Frankfurter Allgemeine Zeitung*, or *Die Süddeutsche Zeitung*, nor in the next weekly issues of *Der Spiegel* or *Die Zeit*.

[353] See Klaus Lange, '"Radikale" im öffentlichen Dienst?' (1976) 29 *NJW* 1809; Walter Schick, 'Der "Radikalenbeschluβ" des Bundesverfassungsgerichts—Inhalt und Konsequenzesn' (1975) 29 *NJW* 2169; Martin Kriele, 'Feststellung der Verfassungsfeindlichkeit von Parteien ohne Verbot' (1975) 9 *Zeitschrift für Rechtspolitik* 201; Christian-Friedrich Menger, 'Parteienprivileg und Zugang Radikaler zum öffentlichen Dienst' (1976) 67 *Verwaltungsarchiv* 105.

however, the Court did so to a rumble of popular applause. The matter was parliamentary pay—a matter that, as one MP admitted, Parliament itself should have settled long before.[354] The complaintant was Karl-Heinz Schneider, an engineer and former Saarland state parliamentarian who had been blocked from seeking a second term by a 1973 law that disqualified managers of majority state-owned corporations from service in the state legislature. Schneider was such a manager, and he complained that the law violated his 'passive voting rights' by unduly restricting eligibility for elected office. The Second Senate rejected this prong of Schneider's complaint in a partial judgment issued on 21 January 1975.[355] Some questions, however, remained open. Schneider also challenged a provision that provided only 60 per cent of the normal parliamentary salary to state employees who relinquished their post on the state payroll in order to stand for the state Parliament.

In a 5 November 1975 judgment, the Second Senate agreed that this provision was unconstitutional.[356] But it didn't stop there. It reviewed *sua sponte* a handful of provisions that Schneider's complaint had not even mentioned. This was unavoidable, the Senate breezily explained, 'when the challenged norm, by reference to other provisions or through a clear internal relationship with other provisions, depends upon the latter in such a manner that the unconstitutionality of the other norms demonstrates the unconstitutionality of the challenged norm'.[357] Fair enough, perhaps. But the Senate didn't stop there either. It conducted instead a wholesale review of the constitutional aspects of parliamentary compensation. And it was clear from the manner and method of the Senate's reasoning that its conclusions applied not only to the peripheral Saarland, but to all state parliaments; not only to Karl-Heinz Schneider, but to every parliamentarian everywhere in the country—including, perhaps especially, members of the Bundestag.

Already at oral argument it was clear that the Court was in a mood to consider big questions. In an ecumenical spirit, the Second Senate invited the Bundestag, the Bundesrat, the federal government, other state parliaments, the Bavarian senate, and all political parties represented in the Bundestag to file briefs and present oral argument. This invitation was accompanied by a series of questions that sounded like the prompt for an academic conference. For example: 'Has, and to what extent has, the status of members of parliament changed since the end of the nineteenth century? In particular, has the understanding of an "appropriate compensation, securing the independence of the member" changed'?[358] The closing query dutifully asked how these considerations bore on the constitutionality of the challenged Saarland provisions. But the Court clearly had its eyes on vaster fields. The justices were convinced that the role of the MP had evolved, and they were prepared to draw consequences.

As early as 1955, a Second Senate led by Gerhard Leibholz had observed in dicta that patterns of parliamentary pay were gradually shifting. As state after state transitioned 'from a liberal-representative to a party state [*Parteienstaat*] democracy', parliamentary compensation increasingly resembled a salary meant to remunerate the MP for his labour rather than an honorarium designed to secure his independence.[359] In 1971, the Court noted that service in Parliament had 'become a career requiring the full employment of one's manpower', one that entitled a member to compensation adequate to support himself and his family.[360] The Court now drew the full consequences of these observations. From his

[354] Peter Conradi, 'Parlamentarier in privilegienfeindlicher Demokratie. Anmerkungen eines Bundestagsabgeordneten zum 'Diäten-Urteil' des Bundesverfassungsgerichts' (1976) 8 *Zeitschrift für Parlamentsfragen* 113.

[355] 38 BVerfGE 326 (1975). [356] 40 BVerfGE 296 (1975). [357] ibid 309–10. [358] ibid 299.

[359] 4 BVerfGE 144, 151 (1955). [360] 32 BVerfGE 157, 164 (1971).

professorial pension in Göttingen, Leibholz chided the politicians who had long ignored them. They had little right to complain of the November 1975 judgment; they should have seen it coming.[361]

Gone were the days, the Second Senate noted, of the Parliament of notables [*Honoratiorenparlament*], where men of independent means consecrated time and talents to the public weal. Gone too were the days when parliamentary compensation supplemented an independent profession and salary. The work of an MP had become a 'full-time job',[362] and MPs must be compensated accordingly.[363] In light of the evolving nature of the parliamentary mandate, the Senate held, Article 48(3) GG's guarantee of an 'independence-securing compensation' assumed new meaning.[364] Members of Parliament—and their families—were now entitled to a 'full livelihood [*Alimentation*] from the state treasury'.[365] This conclusion was dramatic and, strictly speaking, superfluous. The Saarland, after all, already provided its parliamentarians with full alimentation. But the justices' target audience was not in Saarbrücken; it was in Bonn.

There was, as always, a rub. Members of Parliament might be entitled to a life-sustaining salary, but they were required to treat that salary as taxable income.[366] The Basic Law was 'fundamentally hostile to privilege', the justices explained.[367] This hostility meant that, except for direct compensation for specified and exceptional parliamentary activities, parliamentarians, like other citizens, must pay taxes on their money. It also meant that all parliamentarians must be treated alike. More concretely, it was unconstitutional for civil servants, once elected to Parliament, to receive both a parliamentary salary and a civil service pension.[368]

All of this required reform—in the Saarland and, by unmistakable implication, in other state parliaments and the Bundestag. The Second Senate not only stipulated that parliamentary pay be revalued and its taxation restructured; it instructed parliaments everywhere how these reforms must unfold. The new terms of parliamentary compensation must not be determined by parliamentary committees behind closed doors. Deliberations fixing parliamentary pay must unfold 'before the eyes of the public'.[369] The democracy and *Rechtsstaat* principles demanded transparency, especially on a subject as sensitive as lawmakers' setting their own allowance. Here public scrutiny was 'the only effective control'. 'Parliamentary democracy is based on popular trust', and 'trust without transparence ... is impossible'.[370]

Walter Seuffert, the Court's vice president and the Second Senate's chair, was upset that his colleagues had reached so many questions neither raised in the complaint nor germane to its resolution. He filed a lonely dissent.

The public, by contrast, mostly cheered.[371] Some parliamentarians grumbled that the judgment was 'fantastic'[372] and a 'poor work of political science'.[373] Other critics groused that it was 'unworldly'[374] (*weltfremd*), and still others that it would be expensive.

[361] Gerhard Leibholz, 'Bemerkungen zum Karlsruher Urteil' *Deutsche Zeitung* (14 November 1975).
[362] English in the original. [363] 40 BVerfGE 296 (n 356) 314. [364] ibid 315. [365] ibid 315.
[366] ibid 327. [367] ibid 317. [368] ibid 321. [369] ibid 327. [370] ibid.
[371] See Friedrich Karl Fromme, 'Die Steuerfreiheit der Abgeordneten-Diäten ist verfassungswidrig' *FAZ* (6 November 1975); Hanno Kühnert, 'Bezüge der Abgeordneten steuerpflichtig' *Süddeutsche Zeitung* (6 May 1975) 1; Ernst Müller-Meiningen Jr, 'Mehr Diätengerechtigkeit' *Süddeutsche Zeitung* (6 November 1975) 4; 'Lehrstück für die Parlamente' *Frankfurter Rundschau* (6 November 1975) 3.
[372] Peter Conradi in 'Diäten: Angst vor dem großen Sprung' *Der Spiegel* (10 November 1975).
[373] Harmut Klatt in ibid. See also 'Betroffenheit und Verbitterung nach dem Karlsruhe Urteil-Spruch' *FAZ* (7 November 1975).
[374] Eckart Spoo, 'Diäten-Urteil "weltfremd"' *Frankfurter Rundschau* (27 November 19750 4 (quoting Heinz Müller, the CDU premier of Lower Saxony).

Parliamentary wages were sure to rise, both to secure the 'full alimentation' of all MPs and to offset the cost of the taxes MPs were now required to pay.[375] Parliamentarians would ensure that what the public treasury extracted as taxes it would return as wages. But the broader public focused on the fairness of the extraction. The dismay of many parliamentarians only strengthened the public's sense that the Court was right. Many were pleased, too, that the judgment abolished the advantages of civil servant MPs. The justices were credited in some quarters with thwarting the *Verbeamtung*—the 'civil servantization'—of Parliament.[376] The Court may have overreached again, but this time many were pleased with the outcome. Parliament's long-standing omissions were censured more sharply than the Court's corrective overreach.[377]

But it was still seen as an overreach, especially by purists in the legal academy. Traditionally, champions of constitutional review have reassured skeptics by stressing that constitutional courts are not self-activating: they merely decide the cases that come to them. They are a passive power, wielding neither sword nor purse; an overfed umpire, calling balls and strikes.[378] After the Court's parliamentary pay judgment, many scholars argued that this contention must be met with greater skepticism.[379] Christian-Friedrich Menger, a professor in Münster, accused the Court of exercising an unconstitutional right of judicial 'self-initiative'.[380] More sympathetic scholars like Peter Häberle countered that this case was exceptional. The Court's review must never be so searching and so sweeping—even self-starting—as when the branch of government under review is establishing its own salary.[381] But even Häberle characterized the judgment as 'a piece of *Realpolitik*'—an instance of the Court's shaping, and not merely describing, political realities.[382] Hans Schueler, reviewing the decision for *Die Zeit*, agreed.[383] At the time of the judgment, Schueler noted, only about 15 per cent of parliamentarians earned no outside income. In its description of the modern MP, the Court was not merely observing a social reality; it was creating one. Henceforth none could ply politics on the side; all must pursue it as a vocation. The Court had not only interfered in the political process; it had redefined it.

IV. Changing of the Guard

Walter Seuffert's dissenting opinion in the *Parliamentary Pay* judgment would be his last. The judgment coincided almost exactly with the decade's second major shift in the Court's membership. Once again, the terms of six of the sixteen justices expired at one blow. Of these, only the First Senate's Helmut Simon was reelected. Hans Brox, also of the First Senate, was eligible for reelection but, overwhelmed by his dual role as constitutional justice

[375] See Manfred Schell, 'Alle Fraktionen wollen jetzt höhere Diäten' *Die Welt* (6 November 1975) 2; 'Karlsruher Urteil kostet die Steuerzahler viel Geld' *Stuttgarter Zeitung* (7 November 1975).

[376] Noted in Horst Dietrich, 'Beamte als Abgeordnete und das Diäten-Urteil des Bundesverfassungsgerichts' (1976) 27 *Zeitschrift für Beamtenrecht* 97.

[377] See Conradi, 'Parlamentarier in privilegienfeindlicher Demokratie' (n 372) 113.

[378] Thus the notorious formulation of the American chief justice, John Roberts, at his confirmation hearings in 2005. See Confirmation Hearing on the Nomination of John G Roberts, Jr to be Chief Justice of the United States: Hearing Before the S. Comm. on the Judiciary, 109th Cong. 55 (2005) (statement of John G Roberts, Jr).

[379] See, e.g., Christian-Friedrich Menger, 'Zur Kontrollbefugnis des Bundesverfassungsgerichts bei Verfassungsbeschwerden gegen Rechtsnormen—zum Diätenurteil des BVerfG vom 5.11.1975' (1976) 67 *Verwaltungsarchiv* 311.

[380] ibid 315.

[381] See Peter Häberle, 'Freiheit, Gleichheit und Öffentlichkeit des Abgeordnetenstatus. Zum Diätenurteil des BVerfG' (1976) 29 *NJW* 537, 542–43.

[382] ibid 538.

[383] Hans Schueler, 'Zum Berufspolitiker verdonnert' *Die Zeit* (14 November 1975).

and professor in Münster, he surrendered the bench for the lectern. Brox had been personally singled out for criticism on grounds of political and even theological partisanship, especially in connection with the *Abortion* decision. He was weary of Karlsruhe's dubious spotlight. The other departing justices were Theodor Ritterspach of the First Senate, and Justices Seuffert, Schlabrendorff, and Rupp of the Second.

After the nearly universal criticism of the politically charged 1971 appointments, and after the explosive series of judgments that followed, all parties tried to defuse the political valence of the 1975 nominations. Both government and opposition promised to favour juristic competence over political sympathy.[384] Externally at least, they succeeded. The incoming quintet consisted of three judges, two professors, and zero politicians. The Union parties nominated Engelbert Niebler, a judge in Bavaria, and Helmut Steinberger, a professor of public and international law, to the Second Senate; and Dietrich Katzenstein, a judge in Hamburg, to the First Senate. In addition to reappointing Simon, the SPD nominated Wolfgang Zeidler—who served previously on the First Senate (1967–70), but had since been president of the Federal Administrative Court—as the Court's new vice president and chairman of the Second Senate. The SPD also nominated Konrad Hesse, a highly esteemed professor of constitutional law, to the First Senate. Aside from scattered complaints about the Union nominations in the Left Liberal press,[385] the appointments enjoyed broad approval.[386]

Soon, however, they sparked controversy. Because Zeidler, the new vice president, had sat on the Court previously, many thought he was barred from reappointment by the 1971 reform law.[387] Zeidler had already assumed the reins of the Second Senate when his seven colleagues were asked to determine his eligibility to join them. The Senate fudged a bit, ruling in the end that, since Zeidler's previous term predated the 1971 law, he was unaffected by it. This resolution was workable but not entirely satisfactory. *Der Spiegel* quipped that every future complainant could now allege the unlawful composition of the bench. The Zeidler problem had been replaced by a Karlsruhe problem.[388]

Zeidler's elevation to the vice presidency also incensed one of his colleagues. Four years earlier, Martin Hirsch had left the Bundestag with the understanding that after four years he would replace Seuffert as the Court's vice president. When the SPD leadership opted for Zeidler instead, Hirsch was livid. He fired off a stormy stream of angry missives to the party's biggest wigs. The tenor of these letters makes clear that Hirsch's party loyalties had never slackened. Hirsch protested that his loyal service to the party as a judge entitled him

[384] See 'Diesmal besser'? *FAZ* (16 August 1975); 'Neue Richter für Karlsruhe' *Frankfurter Rundschau* (3 September 1975) 1; Volkmar Hoffmann, 'Gesucht wurde juristische Spitzenqualität' *Frankfurter Rundschau* (5 September 1975) 1.

[385] Volkmar Hoffmann, 'Verwunderung über die Richter-Kandidaten der CDU/CSU' *Frankfurter Rundschau* (6 September 1975) 1; Robert Leicht, 'Schwache Vorschläge für Karlsruhe' *Süddeutsche Zeitung* (10 September 1975) 4; Hellmuth Rieber, 'Vorschläge der Union für Richter kritisiert' *Saarbrücker Zeitung* (18 September 1975).

[386] Enno von Loewenstern, 'Erleichterung in Karlsruhe über die neuen Kollegen' *Die Welt* (6 September 1975) 3; Friedrich Karl Fromme, 'Jetzt sind alle Verfassungsrichterkandidaten bennant' *FAZ* (6 September 1975) 2; 'Gute Wahl' *Rheinischer Merkur* (19 September 1975) 4; 'Annehmbar' *FAZ* (20 September 1975).

[387] On the Zeidler controversy, see Claus Donath, 'Darf der neue Vizepräsident vereidigt werden oder nicht'? *Stuttgarter Zeitung* (21 October 1975) 3; Hanno Kühnert, 'Zweifel wegen Verfassungsrichterwahl' *Süddeutsche Zeitung* (21 October 1975); 'Magenschmerzen nach Richterwahl' *Süddeutsche Zeitung* (21 October 1975); 'Der Streit um den Vizepräsidenten' *FAZ* (22 October 1975); 'Karlsruhe in Verlegenheit' *Der Spiegel* (10 November 1975) 22; Hanno Kühnert, 'Karlsruhe prüft Bestellung Zeidlers' *Süddeutsche Zeitung* (10 November 1975) 6; 'Blinde Stellen' *Der Spiegel* (17 November 1975) 54; 'Fragen zur Rechtmäβigkeit der Wahl Zeidlers' *FAZ* (27 November 1975); Hanno Kühnert, 'Entscheidung über Zeidler am Mittwoch' *Süddeutsche Zeitung* (29 November 1975) 8; Friedrich Karl Fromme, 'Wolfgang Zeidler. Präsidiale Karriere' *FAZ* (2 December 1975) 10; 'Im Zweifel zum Verfassungsgericht' *FAZ* (3 December 1975).

[388] See 'Blinde Stellen' *Der Spiegel* (17 November 1975) 54.

to the promised advancement. 'What reservations are there against me'? he asked Friedrich Schäfer, vice-chairman of the SPD Bundestag faction. 'What is there to complain about in my work here [with the Court]? ... What has [Zeidler] ever done for the SPD and its legal politics'?[389] Hirsch provided a list of the cases in which he had served as rapporteur, and invited Schäfer to study them and see how much Hirsch had contributed to SPD interests.[390]

In a pair of long letters to Willy Brandt (which he copied to several others) Hirsch complained that it was 'shabby and deplorable' of the party to break its promise.[391] It was also foolish. One need only compare Hirsch's performance on the Constitutional Court with Zeidler's, as well as Hirsch's achievements in Parliament with Zeidler's as an administrative judge to see that Hirsh was the better man for the party. 'I have a right, then', Hirsch wrote, 'to ask what exactly it is that justifies his being preferred over me?'[392] The choice of Zeidler would hurt the Republic and the SPD.[393]

The other 1975 nominations to the Court, Hirsch added, were even more regrettable. The SPD had 'squandered the long-awaited chance to set off against Professor Geiger an outstanding scholar of our colour—namely the comrade Professor Böckenförde—without any prospect of correcting this in the foreseeable future'.[394] Toward the end of his long and passionate letter to Brandt, Hirsch seemed to realize that he may have gone a bit far. He clarified that it shouldn't be assumed that constitutional judges will always vote for the party that nominates them. 'But a constitutional court', Hirsch continued, 'more than any other court, is a political court, and constitutional questions cannot be judged unpolitically'. Political convictions played a role in the Court, and rightly so. 'Among us', Hirsch concluded, 'entirely unpolitical individuals would be truly out of place'.[395]

Had these letters been made public, they would surely have fanned and confirmed widespread suspicion that the justices—particularly the newer ones—were swayed by party loyalties. During the mid- to late-1970s, the public's esteem for the Court sank.[396] Perceptions that the Court was overreaching were aggravated by perceptions that it was doing so for partisan reasons. The emergence of signed dissents accentuated the impression of partisanship. The putative 5:3 CDU/CSU majority in the First Senate had been on egregious display, many thought, in the *University* and *Abortion* judgments. The Court had established its reputation in the 1950s by defying partisan predictions during the European Defence Community crisis. In the 1970s, the Court risked squandering that reputation by confirming partisan predictions when the political stakes were highest.

This was doubly dangerous to the Court's legitimacy precisely because the Court had done so much to enhance the centrality of political parties in the postwar state. Part of the democratic defence of the Court was that, in a party state in which popular political participation is mediated by powerful political parties, the Court could reinforce the people's fundamental commitments by constraining overwheening parties. This was an ironic defence, to be sure: the people needed the Court's protection against the parties only because the Court had made the parties so important. But the defence broke down entirely

[389] Martin Hirsch to Friedrich Schäfer, 9 June 1975, Friedrich-Ebert-Stiftung (Bonn) (Nachlass Martin Hirsch, 1/MHAC000137).

[390] ibid.

[391] Martin Hirsch to Willy Brandt, 4 November 1975, Friedrich-Ebert-Stiftung (Bonn) (Nachlass Martin Hirsch, 1/MHAC000137); see also Martin Hirsch to Willy Brandt, 5 September 1975, Friedrich-Ebert-Stiftung (Bonn) (Nachlass Martin Hirsch, 1/MHAC000137).

[392] Martin Hirsch to Willy Brandt, 4 November 1975, Friedrich-Ebert-Stiftung (Bonn) (Nachlass Martin Hirsch, 1/MHAC000137).

[393] ibid. [394] ibid. Böckenförde would ultimately join the Court in 1983. [395] ibid.

[396] See Hans Vorländer and Gary S Schall, 'Integration durch Institutionenvertrauen' in Hans Vorländer (ed), *Integration durch Verfassung* (Westdeuter Verlag 2002) 343, 364.

when it seemed that the justices themselves were in the thrall of the parties whose powers the Court had expanded. It was the parties and their judicial selection potentates, after all, to whom the justices owed their jobs. The Court would survive the crisis of the 1970s and restore its reputation only when it reasserted its independence from the dominant political parties.

With the departure of Theo Ritterspach und Hans Rupp only one of the Court's founding justices, Willi Geiger, remained. Geiger himself would reach the age limit of sixty-eight in May 1977 and would leave the Court later that year. After the 1975 appointments, ten of the Court's sixteen judges had been appointed during the 1970s. The changing of the guard was quite thorough.

As noted earlier, some saw in these shifts a story of decline. A Court, the narrative ran, once populated by judicial giants unafraid to vindicate the rights of the downtrodden or to quash the best-laid plans of overbearing chancellors had degenerated into a hyperpartisan tribunal with supralegislative pretensions.[397] Some first-generation justices shared the nostalgia reflected in this narrative, even if they didn't adopt its jeremiad tone. Erwin Stein confessed to Gebhard Müller that it was a 'bitter experience' to see how little his legacy was acknowledged in Karlsruhe. 'One counts as dead there', Stein wrote.[398] After the *Parliamentary Pay* judgment, Ernst Friesenhahn told Gerhard Leibholz of his intention to publish his criticisms of the judgment, provided he could muster the 'necessary concentration'.[399] It was distasteful, though, to reprove a Court that continued to drift astray. 'The Court makes so many questionable judgments', Friesenhahn lamented in another letter, 'and it has clearly lost much respect'.[400]

But fairness to the current justices, Friesenhahn conceded, the Court's work was harder now. When Leibholz and Friesenhahn together sat on the Second Senate, 'the times were not so difficult as they are today. We could essentially build ... on that which had been worked out in the public law scholarship of the Weimar era'. Now the Court's work had shifted to 'other shores'. The subject matter of the current cases was sometimes 'so complicated that the law pushes the boundaries of what may be regulated'. It was no wonder that the justices' reasoning often seemed 'ill construed' and lacking in 'seminal principles for the future'.[401] Willi Geiger also reminisced with Friesenhahn about their earlier time together. 'Those were the days!' wrote Geiger on Friesenhahn's seventieth birthday. Ernst Benda qouted this letter on the occasion of Geiger's retirement, adding wistfully, 'Those were the days, indeed!'[402]

V. Crisis

A. Public funds for parties' public relations work

Nineteen seventy-five was a climactic year for the Court, the culmination of a three-year swirl in the thick of a political tempest. The following year brought a merciful respite. The Court decided no political *causes célèbres* in 1976—the year of its silver jubilee—and the

[397] See 'Weltrekord in Wartezeiten' *Der Spiegel* (1 December 1975).

[398] Erwin Stein to Gebhard Müller, 10 August 1972, Landesarchiv Baden-Württemberg (Stuttgart) (Nachlass Müller, Q 1135 Bü 705).

[399] Ernst Friesenhahn to Gerhard Leibholz, 14 November 1975, BAK N 1334/5 (Nachlass Leibholz).

[400] Ernst Friesenhahn to Gerhard Leibholz, 31 December 1974, BAK N 1334/6 (Nachlass Leibholz).

[401] Ernst Friesenhahn to Gerhard Leibholz, January 1977, BAK N 1334/29 (Nachlass Leibholz).

[402] Ernst Benda, 'Ansprache anläßlich der Feierstunde des Bundesverfassungsgerichts am 9. November 1977 zur Verabschiedung von Bundesverfassungsrichter Prof. Dr. Willi Geiger', BAK N 1334/299 (Nachlass Leibholz).

public's attention was largely absorbed by another federal election. It was the first general election since Brandt's resignation two years earlier—the first electoral test of Helmut Schmidt.

The mood was very different from 1972. As noted, the enthusiasm of the early Brandt years was gone. Under Schmidt's watch, debt had soared and unemployment risen. In nearly every state election since 1974, the CDU/CSU fortunes had improved. The trend continued in the federal elections of 1976. Under a ticket headed by Rhineland-Pfalz minister president Helmut Kohl, the CDU/CSU share of the vote rose by 3.7 per cent to 48.6 per cent, the Union parties' best tally since winning an absolute majority in 1957 and a slightly larger portion of the vote than that with which the SPD and FDP took power in 1969. Both coalition parties lost votes; the SPD share shrank by 3.2 per cent to 42.6 per cent, the FDP by half a percentage point to 7.9 per cent. Still, the combined SPD–FDP vote was enough to keep the coalition in power and to keep Schmidt (who was more popular than his party) in office. But it was a close-run thing. Cracks in the coalition had begun to appear.

During the campaign the government's public relations department conducted a massive publicity offensive, trumpeting the coalition's achievements and campaigning for its reelection. From early May until the election in October the government conducted an immense advertising campaign—at the cost of some 10 million DM—in a host of major newspapers. Glossy, large-format inserts summarized advances and reforms and closed with anodyne slogans: 'The balance is clear: We're on the right path'; 'Accomplishment merits trust'; 'We secure the future' etc. Between 24 May 1976 and 26 July 1976, a three- to five-page government insert appeared in each week's edition of *Der Spiegel*. Each insert featured snappy synopses of the government's successes and tart dismissals of the opposition's pessimism. 'All in all', ran one headline, 'this government has brought you more freedom'. In addition, the PR office issued a torrent of pamphlets and posters, books and brochures—sometimes in runs of over a million copies.

What effect all of this had on the election is impossible to ascertain. Employing PR resources in the attempt to remain in office was an established practice, beginning early in the Adenauer era. But this time the Union parties, now in opposition, cried foul. On 23 July 1976, in the middle of the campaign and of the official PR drive, the CDU filed a constitutional complaint, alleging that the use of public funds to promote the electoral prospects of the sitting government gave incumbents an unfair advantage—in violation of Article 21 GG's guarantee of equal opportunity for all parties and Article 20 GG's enshrinement of the democracy principle. The Court held oral arguments on 9 September, less than a month before the election. The Second Senate published its decision the following March, five months after the election had passed.

The result was not a happy one for the newest Schmidt government. The Senate majority held that the government's PR activities during the campaign violated both the democracy principle and the principle of equal opportunity for political parties.[403] As early as the 1966 *Party Finance* judgment, the Court had stressed that the formation of popular will must be 'State free'.[404] In more recent decisions, the Court had reaffirmed this precept,[405] emphasized that the state mustn't favour one party over others,[406] and extended the principle of equal voting rights from the election itself to the campaign.[407] The Second Senate now combined these principles into a general ban on official engagement in the electoral process.[408] Sitting governments must not employ state resources to preserve power. Public relations

[403] 44 BVerfGE 125 (1977).
[404] 20 BVerfGE 56, 100 (1966).
[405] 37 BVerfGE 84 (1974).
[406] 34 BVerfGE 160 (1972).
[407] 41 BVerfGE 399 (1976).
[408] 44 BVerfGE 125, 144 (1977).

work (*Öffentlichkeitsarbeit*)—distinct, in German, from propaganda, though in English usually a euphemism for it—had a legitimate and even constitutionally-mandated role. But legitimate public relations ended where electioneering began.[409] A government that crossed this line offended the constitution.

The line, admittedly, was hard to draw. As an abstract matter, it could hardly be drawn at all. Questionable practices must be evaluated individually, but the Court did outline some general principles to guide that assessment. Government publications should neither endorse the parties in power nor attack those in opposition.[410] They should neither express the government's desire to remain in office nor foster the public's desire to keep them there. Their tone and content should be informative rather than celebratory; their format and appearance should resemble a report rather than an advertisement. Otherwise innocuous publications might become suspect if their cost or quantity were too great, their proximity to an election too close, or precautions against their use as election propaganda too meagre.[411] Extreme care must be taken in applying these standards. A finding of constitutional violation, the Senate explained, required multiple obvious infringements—as in the present case.[412]

On this point the Senate's analysis was scrupulously factual. For a dozen pages, the justices scrutinized the government materials produced during the campaign—noting, for instance, that a pamphlet on foreign policy featured fourteen photographs of the foreign minister, Hans-Dietrich Genscher.[413] The Senate cited instances of manifest sloganeering from a wide range of publications. Peter Häberle praised this aspect of the decision, lauding the Court's 'rigorous and exemplary legwork with the facts'.[414] The conclusion that the coalition had crossed the line was inescapable. The justices strained the thesaurus in stressing just how palpable the offences were, describing the partisan purpose of various publications as 'evident',[415] 'obvious',[416] 'unmistakable',[417] 'unambiguous',[418] 'clear' etc.[419] ('Do not imagine', warned Boswell's first editor, 'that the *only* defect of style, is repetition of words'.)[420]

Though the majority thought the outcome obvious, three justices—one nominee from each of the major parties—joined separate opinions, though only one dissented substantially from the majority's response to the constitutional question. Willi Geiger, joined by Martin Hirsch, criticized the majority's methodology and underscored the misuse of tax funds as the essence of the constitutional offence. Geiger and Hirsch stressed also that it was equally illegitimate for *opposition* parties to receive campaign assistance from the *state* governments they controlled. Justice Rottmann, by contrast, dissented in earnest. Rottmann thought the government had not violated the democracy principle at all and had only lightly infringed the principle of equal opportunity for political parties. The majority, in Rottmann's view, had misconstrued the party structure of the West German state and confused the constitutional issue as a result.[421] The government's duty, Rottmann argued, was to implement the programme of the majority parties. This required public relations

[409] ibid 150. [410] ibid. [411] ibid. [412] ibid 156. [413] ibid 161.

[414] Peter Häberle, 'Öffentlichkeitsarbeit der Regierung zwischen Parteien- und Bürgerdemokratie' (1977) 32 *JuristenZeitung* 361, 362 fn 9.

[415] 44 BVerfGE 125 (n 408) 161. [416] ibid 157. [417] ibid 160, 162.

[418] ibid 159, 160, 162, 163.

[419] ibid 157, 159. See Andreas Peilert, 'Öffentlichkeitsarbeit. Öffentlichkeitsarbeit der Bundesregierung als Verletzung des Demokratieprinzips und der Chancengleichheit im Wahlkampf' in *Verfassungsrechtsprechung*, (n 153) 269.

[420] Edmund Malone, in Adam Sisman, *Boswell's Presumptuous Task: The Making of 'The Life of Dr. Johnson'*, Farrar, Straus and Giroux 1999) 244.

[421] 44 BVerfGE (n 408) 190.

work, even at election time.[422] The particular use of PR made by the Social–Liberal government in 1976 may have been in questionable political taste; but the Court was not an arbiter of political taste.[423]

Rottmann's dissent found some scholarly but few popular echoes. The press, for the most part, hailed the decision warmly.[424] *Die Welt* called it a 'redemptive judgment'.[425] *Die Zeit* reported that 'only a few decisions of the Federal Constitutional Court could give citizens and voters such unbounded pleasure as its most recent verdict on the self-promotion of the federal government at the last Bundestag election'.[426] The *Bayernkurier* burbled that the judgment was 'an historic landmark in the development of this Republic'.[427] Robert Leicht in the *Süddeutsche Zeitung* predicted 'beneficent effects'.[428] Even *Der Spiegel*, which asked impishly whether the judgment would bar cabinet members from travelling to campaign rallies in government planes and cars, called the result 'long overdue'.[429] Academic commentary was divided, but some scholarly praise was panegyric. Peter Häberle called the decision a 'milestone in the history of German parliamentarism'.[430] Karl-Heinz Seifert classed it 'among the great, direction-setting decisions for the state life of the Federal Republic'.[431]

This enthusiasm for the *Public Relations* judgment is instructive. The judgment's admirers saw the decision, not as an anti-democratic rebuke to duly elected officials, but as a democracy-reinforcing restraint on the parties in power. Peter Häberle tellingly entitled his review essay, 'Government Public Relations between Party- and Citizen-Democracy'.[432] In the hotly criticized judgments of recent years, the Court had seemed to split on partisan lines for partisan reasons. The justices were perceived as instruments of the parties they had made so powerful. Together with the Geiger-Hirsch concurrence, the *Public Relations* judgment admonished all parties against employing the resources of the state toward partisan ends. This delighted a citizenry whose democratic engagement was mediated by seemingly all-powerful parties. Like earlier landmarks, such as the *Television* judgment, the *Public Relations* decision distinguished the state from the party that held its reigns. The judgment articulated long-term principles that implicitly reproved past governments and clearly warned future ones. To many observers, it was the Constitutional Court at its finest.

The government tried hard to take the decision in stride. Klaus Bölling, a spokesman, announced that the government would 'obviously respect' the decision. But he stressed that the condemned practices, as the Court itself noted, had been common to both parties, at the state and federal level, for a long time. The judgment was no occasion for 'fanfares of

422 ibid 184. 423 ibid 192.

424 See Knut Teske, 'Bonner Regierung hat Steuergelder für Wahlhilfe miβbraucht' *Die Welt* (3 March 1977) 1; 'Wahlbewerbung mit Haushaltsmitteln verboten: Die Bundesregierung verliert in Karlsruhe' *FAZ* (3 March 1977) 1; Robert Leicht, 'Allen Parteien ins Stammbuch' *Süddeutsche Zeitung* (3 March 1977); Hans Peter Schütz, 'Erdbeben' *Stuttgarter Nachrichten* (3 March 1977) 1; 'Auf Staatskosten' *Bonner General-Anzeiger* (3 March 1977) 1; 'Nicht mehr mit Dienstwagen in den Wahlkampf'? *Der Spiegel* (7 March 1977) 16. Papers close to the government were critical. See Hans Schueler, 'Votum für ein Ideal' *Die Zeit* (18 March 1977) 5; 'Leitbilder aus Karlsruhe' *Frankfurter Rundschau* (3 March 1977) 3.

425 *Die Welt* (3 March 1977) 6.

426 Hans Schueler, 'Votum für ein Ideal' *Die Zeit* (18 March 1977) 5.

427 In 'Wupp und Weg' *Der Spiegel* (28 March 1977).

428 Robert Leicht, 'Allen Parteien ins Stammbuch' *Süddeutsche Zeitung* (3 March 1977).

429 'Nicht mehr mit Dienstwagen in den Wahlkampf'? *Der Spiegel* (7 March 1977).

430 Häberle, 'Öffentlichkeitsarbeit der Regierung' (n 414) 361.

431 Karl-Heinz Seifert, 'Anmerkung' (1977) 30 *Die Öffentliche Verwaltung* 288. Critical reactions included Otto Ernst Kempen, 'Zwischen Gemeinwohlpostulat und demokratischen Verfahrensgarantien. Das Urteil des Bundesverfassungsgerichts zur regierungsamtlichen Öffentlichkeitsarbeit' (1979) 18 *Der Staat* 81; and Rüdiger Zuck, 'Anmerkung' (1977) 31 *NJW* 1054.

432 Häberle, 'Öffentlichkeitsarbeit der Regierung' (n 414).

self-righteousness'.[433] But the opposition crowed all the same. Blessed with a fertile imagination or a short memory, Helmut Kohl chortled over 'the sharpest slap in the face ever received by a federal government from the Constitutional Court'.[434] The sharpest slap ever it certainly was not. But it did sting. In the public's eyes, the *Public Relations* judgment, together with the *Parliamentary Pay* decision, had gone some way toward mitigating the obloquy incurred by the *Abortion* decision. For the federal government, however, and for the chancellor who led that government, it was a further offence—a resumption and escalation of a preexisting conflict.

The judgment not only diminished the coalition's resources for future elections; it dented the legitimacy of its recent victory. The cabinet chafed at this additional impertinence. It chafed again, three months later, when the Second Senate ruled that in December 1973 the coalition's former finance minister, by approving expenditures of around 2 million DM outside of the approved budget, had violated the Bundestag's right (under Article 110(1) GG) to control the budget.[435] The offending former minister was now federal chancellor—Helmut Schmidt himself. Again the government protested that Schmidt's actions had followed standard government practice 'since 1949'. Again the opposition gloated. Helmut Kohl called the judgment a 'victory' for the CDU/CSU and a shield against the 'autocratic manner of a Helmut Schmidt' in disposing of taxpayers' cash.[436] Fritz Zimmermann, a leading CSU member of the Bundestag, opined that the judgment affected Schmidt not just as a former minister, 'but personally as well'.[437] Schmidt himself kept quiet. But his patience with Karlsruhe was running thin. Within twelve months, the Court quashed yet another Social-Liberal reform. By so doing, the Court pushed the conflict to crisis point.

B. The 'German autumn' and the constitutional dilemmas of domestic terrorism

On 21 June 1977, the First Senate ruled that criminal sentences of life imprisonment without the possibility of parole were unconstitutional—a violation of human dignity.[438] Only a few months preceding this judgment there was a hailstorm of calls to restore the death penalty itself.[439]

Those cries were provoked by an unprecedented sequence of terrorist atrocities. Nineteen seventy-seven was the *annus horribilis* of the Federal Republic, the year in which the state's confrontation with domestic terrorism reached a dramatic and deadly climax. What later became known as the 'German autumn' began shortly after midsummer with the murder, following a botched kidnapping attempt, of Jürgen Ponto, head of the Dresdner Bank. On

433 In Knut Teske, 'Bonner Regierung hat Steuergelder für Wahlhilfe miβbraucht' *Die Welt* (3 March 1977) 2.

434 In 'Wupp und Weg' *Der Spiegel* (28 March 1977).

435 45 BVerfGE 1 (1977).

436 In 'Die Opposition gewinnt in Karlsruhe. Das Haushaltsrecht des Parlaments gestärkt' *FAZ* (26 May 1977) 1.

437 ibid.

438 45 BVerfGE 187 (1977).

439 See reports in 'Bayerns Innenminister bringt Todesstrafe ins Gespräch' *FAZ* (13 September 1977) 2; Otto Paulsen, 'Todesstrafe befristet einführen (Leserbrief)' *FAZ* (14 September 1977) 9; 'Eine Ersatzdiskussion' *FAZ* (14 September 1977) 2; 'Der Präsident des Bundesrates warnt vor einer Debatte über die Todesstrafe' *FAZ* (14 September 1977) 2; 'Der Bürger ruft nach härteren Strafen' *Der Spiegel* (19 September 1977); 'Todesstrafe? Der Staat geht in Stellung' *Der Spiegel* (19 September 1977); Worte der Woche, *Die Zeit* (23 September 1977); 'Katalog von CSU-Forderungen gegen den Terrorismus' *FAZ* (24 September 1977) 2; 'Ein Planquadrat in Verteidigungsstellung' *Die Zeit* (30 September 1977); 'Hassel für Todesstrafe' *FAZ* (3 October 1977) 4; 'Eine Umfrage unter Schülern' *FAZ* (18 October 1977) 2; 'Dunkle Geschichte hinter dem Fenstern' *Der Spiegel* (24 October 1977); 'Etwas Anbieten' *Der Spiegel* (24 October 1977); 'Seit der Ermordung Schleyers hat sich das Bild der Deutschen im Ausland geändert' *FAZ* (26 November 1977) 4; 'Die Terroristen verraten unsere Ideale' *Der Spiegel* (28 November 1977); 'Schatten über dem Rhein' *Die Zeit* (23 December 1977).

5 September, a Red Army Faction 'commando unit' abducted the industrialist Hans-Martin Schleyer, then president of the German Employers' Association, in Cologne. Schleyer's captors styled themselves 'second generation' RAF militants. They demanded the release, on pain of their hostage's life, of ten members of the 'first generation'—the core band of the Baader-Meinhof-Complex—then detained in Stammheim Prison. They also demanded a ransom of fifteen million US dollars. Two-and-a-half years earlier, when members of the June 2nd Movement kidnapped Peter Lorenz, the authorities acceded to terrorist demands for the release of imprisoned group members in exchange for Lorenz's freedom. The released militants were deported to South Yemen. Some, however, returned to the Federal Republic and committed further crimes. One of them, Rolf Heißler, helped capture Hans-Martin Schleyer. Heißler and his accessories expected a repetition of the Lorenz exchange, but were quickly disappointed. This time the federal government, led by Helmut Schmidt and Hans-Jochen Vogel, refused to negotiate. The government was resolved, this time around, to hold firm.

The terrorists reacted with additional pressure, applied with the assistance of Palestinian accomplices. On 13 October, a Lufthansa jet returning to Frankfurt from Mallorca was hijacked by belligerents of the Popular Front for the Liberation of Palestine (PFLP). The hijackers 'executed' the plane's captain and conducted it on a rambling odyssey over the Arabian Peninsula. They demanded that the West German government yield to RAF demands and, additionally, release two Palestinian compatriots then held in Turkey. Otherwise, the plane's eighty-six passengers, most of them German holidaymakers, would die. Schleyer's captors forced him to call on the Schmidt government to relent, but the government refused to yield. It was clear that Schleyer's life hung by a fraying thread. Hans Eberhard Schleyer, the hostage's son, contacted Vogel and offered to supply the ransom money, but a media leak scuttled the plan. Schleyer's son saw only one further chance to save his father's life. At 1 p.m. on Saturday, 15 October, he asked the Constitutional Court to order the Schmidt government to accede to the terrorists' demands. In support of this petition, the younger Schleyer invoked the state's affirmative duty to protect human life.

The Court, often criticized for its slow procedures, responded with lightning celerity. The First Senate convened oral arguments around 9:30 p.m. that very evening and concluded them around midnight. At 5:45 a.m. the following Sunday morning, the chief justice announced that the Court had rejected the petition.[440] Yes, the Senate agreed, the state must protect human life; but it may choose for itself how to do so. That duty, moreover, extended to all human life—to citizens safe at home as well as captives held hostage. Under circumstances the Court described as *sui generis*, judicial interference in the excruciating balance between saving hostages now and protecting other citizens in future was especially inappropriate. What's more, a constitutional imperative to surrender to terrorist blackmail would make the state's reaction to terrorists 'calculable', and 'the state's protection of its citizens impossible'.[441]

It was the most immediately weighty decision the Court had ever issued. The justices knew that it meant an innocent man would die, and for this reason the decision has sometimes been called, with more sympathy than resentment, a 'death sentence' (*Todesurteil*).[442] On 18 October, three of the imprisoned RAF members—including Andreas Baader—committed suicide and were found dead in their cells. In retaliation for what terrorist mythology would call the 'murder' of their comrades, Schleyer's captors shot him in the neck. One of the

440 46 BVerfGE 160 (1977). 441 ibid 165.

442 See Maximilian Steinbeis, Marion Detjen, and Stephan Detjen, *Die Deutschen und das Grundgesetz. Geschichte und Grenzen unserer Verfassung* (Pantheon 2008) 178–80.

weapons was fired by Rolf Heißler. Schleyer's body was found in the trunk of an abandoned green Audi the next day. Also on 18 October, the Lufthansa plane finally landed—in Mogadishu, Somalia where, with the permission of the Somali government, a West German task force assaulted it. All eighty-six passengers were rescued; only one of the hijackers survived. In weeks to come, mourners from all segments of society wrote to offer their condolences to Schleyer's widow, Waltrude. Two of the most poignant letters came from Ernst Benda and Hans-Jochen Vogel.[443]

The West German public rallied to the support and rang with the praise of its steel-nerved chancellor. It added words of appreciation for the Court that sustained Schmidt with its constitutional imprimatur.[444] Press and professoriate agreed that the Court's decision was correct and, all things considered, inevitable. Many were relieved by the Court's exercise of deference and restraint in such an emotionally and politically explosive matter. And yet, when the terrorist crisis was over and its attendant anxieties abated, some reflected uneasily on the stupendous powers of a Court whose jurisdiction could reach such a matter at all. 'In the Schleyer example', wrote Hans Schueler in *Die Zeit*, 'as in hardly any other, is revealed the power ... of a nearly all-competent Constitutional Court'.[445] Schueler's essay, whose title scored the justices as 'The Counter-Captains of Karlsruhe', went on to ask, as other critics would ask in the turbulent months that followed, whether 'Bonn' was 'ruled by constitutional judges'. The Court, Schueler said, had become 'a political tribunal *par excellence*, nearly a parallel government, dressed—or merely disguised?—in judicial robes'.[446]

C. Karlsruhe besieged

Within a few months, Schueler's barbs were echoed by other pens and voices in the most sustained public assault the Court had ever faced. The immediate trigger was a judgment, delivered on 13 April 1978, striking down the coalition's reform of compulsory military service.[447] Previously, an applicant seeking excusal from military service on grounds of conscience had to undergo an oral examination to test the good faith of his claim. Under the reform law, conscientious objectors needed only to submit a written statement of their objections—needed only, critics grumbled, to send a postcard—after which they could be appointed for alternative service. The law was known as the 'Alternative Service Law' or, less reverently, the 'Postcard Law'. Its practical effect was to confer a right of choice between military service and civilian alternatives. The CDU/CSU parliamentary faction, as well as the Christian Democrat governments of Bavaria, Baden-Württemburg, and Rheinland-Pfalz, challenged the law's constitutionality on both procedural grounds (it was promulgated without Bundesrat approval) and substantive grounds (it violated the equality clause).

The Second Senate sustained the complaint on both counts and annulled the law in its entirety. Universal military service, the Senate explained, was an expression of egalitarian thinking with honourable roots in a 'free-democratic tradition' that reached back to the French Revolution of 1789 and the Prussian reforms of the early nineteenth century.[448] Its application in the Federal Republic was governed by the general equality clause

[443] See Heinrich Basilius Streithofen (ed), *Briefe an die Familie Schleyer. Bekenntnis und Verpflichtung* (Seewald 1978) 166–67.

[444] See, e.g., Enno von Loewenstern, 'Die Mahnung aus Karlsruhe' *Die Welt* (17 October 1977) 6; 'Die Last bleibt' *FAZ* (17 October 1977) 1.

[445] Hans Schueler, 'Die Konterkapitäne von Karlsruhe' *Die Zeit* (24 February 1978).

[446] ibid.

[447] 48 BVerfGE 127 (1978).

[448] ibid 161.

of Article 3 GG. From that clause the Senate derived the principle of 'military justice' (*Wehrgerechtigkeit*).[449] That principle required more than a blanket option between military service and a civil substitute. It required, in principle, that *all* who owed military service should supply it and that *only* those be excused who, for reasons of conscience, were genuinely opposed to taking up arms.[450] The responsible authorities must be able with some certainty to assess the authenticity of applicants' claims.[451] The constitution forbade granting a right to reject military service. But this, in effect is what the new law did.[452] The previous system for assessing claims was problematic, to be sure; but the reform system didn't assess claims at all.[453]

Worse still, the new regime placed heavier burdens on those who accepted military service than on those who opted out. The number of conscientious-objectors had increased twentyfold in twenty years and was wildly outstripping the number of available positions in the alternative service. Current trends portended a perverse dynamic: Those willing to serve in the military were increasingly likely to be called into service; those opting for a civilian alternative were were increasingly unlikely to have to supply it.[454]

The inequity was constitutionally intolerable. Parliamentary plans to increase the number of civil replacement slots by 1980 could not save the law. In its current incarnation, the law was unconstitutional. Because the offending provisions were central to the law as a whole, the entire law was null and void.[455] For good measure, the justices also held that some of the law's provisions required Bundesrat approval, and its passage without that approval was also unconstitutional.[456] The Senate's vote on the substantive question passed 7:1; on the procedural question, 6:2. On the first question, only Martin Hirsch, the quarrelsome former SPD parliamentarian, dissented.

Popular dissent, particularly among the young, was clamorous. The judgment sent young Social Liberals and conscientious-objection interest groups into the street. In a host of university cities, demonstrators marched and publicly burned their military service records.[457] The Young Socialists (JuSos) pilloried the Court as the 'Executive Arm of the CDU/CSU'—a 'fifth column of [the Union's] conservative-reactionary politics' best rechristened the 'Basic Rights Encroachment Court'.[458] The Young Democrats called for a 'demythologization of the Karlsruhe council of the gods' and branded the judgment a 'constitutional violation legalized at the highest level'.[459] As usual, leading SPD/FDP politicians were more restrained than the firebrands in the street. Those personally involved in crafting the law criticized the judgment only modestly.[460] Hans Apel, the SPD minister of defence, even spared it a good word.[461]

As the weeks passed, however, six years of Social Democrat frustration with the Karlsruhe Court began to bubble over. More and more SPD leaders took the Court to task for overstepping its bounds. In his dissenting opinion in the *Conscientious Objection* case, Martin Hirsch warned the majority against the dangerous transition from *Hüter* to *Herr*—from Guardian to Master—of the constitution. The formulation found a wide popular echo. It became an accusation rather than a warning.

449 ibid 159. 450 ibid 166. 451 ibid. 452 ibid 168–69. 453 ibid 167–68.
454 ibid 171–75. 455 ibid 176. 456 ibid 177–84.
457 *Die Welt* (15 April 1978) 2; (14 April 1978) 5.
458 See Jungsozialisten in der SPD, Unterbezirk Münster, 'Stellungnahme zum Urteil des BVerfG zur Novelle des Wehrdienstgesetzes vom 13. April 1978', BAK B 189/24717 (Wehrpflichtnovelle).
459 In *Die Welt* (14 April 1978).
460 *FAZ* (14 April 1978) 2; *Die Welt* (14 April 1978) 5; *Die Welt* (24 April 1978) 2.
461 *FAZ* (14 April 1978) 2.

The charge was given its sharpest formulation by Holger Börner, minister president of Hesse. Börner's predecessor, Georg August Zinn, had been one of the Court's staunchest public defenders. Börner became its loudest foe. In a 21 May 1978 speech at an SPD convention on legal politics, Börner hurled down the gauntlet:

> We have here a matter of most serious concern: Does the Constitutional Court actually content itself with the role of a 'Guardian of the Constitution' or does it increasingly also step forward with legal-political directives ... as 'Lord of the Constitution' and 'Ruler of the Constituted'? .... With mounting frequency in its opinions, the Court dictates for the legislature a binding alternative solution or imposes the corset of a constitutionally-compliant interpretation that reduces political freedom of action close to zero....
>
> It increasingly abandons the course of judicial restraint and, through directives to the legislature, thrusts itself headlong into the competition for political power. This tendency also emerges where the legal basis seems so unclear and ambiguous to the unbiased observer, that the boundary between legal conclusions and political commitments becomes fluid....
>
> This progressive emasculation of Parliament does not strike me as the crown of the *Rechtstaat*.[462]

Hans-Jochen Vogel, the justice minister—who for many years had been a peacemaker between Karlsruhe and Bonn—gave a similarly critical, though more measured, speech the same week.[463] In early June, Helmut Schmidt joined the fray.[464]

The Union parties, hardly disinterested observers, rose to the Court's defence. A CDU press conference held the day after Börner's speech defended the Court against the Hessian's indictment.[465] The SPD responded the next day with a statement backing Börner, noting that former CDU chairman Karl Carstens had also criticized the Court sharply, and on similar grounds, in his 1971 memoir.[466] A week later Jochen Lengemann, a CDU member of the Hessian state Parliament, called Börner an 'arsonist in the mask of a *Biedermann*'.[467] Union leaders argued that the Court's frequent invalidation of Social–Liberal measures, and the SPD's consequent carping about the Court, were evidence, not of the Court's partisanship, but of the ruling parties' disregard for the constitution. Helmut Kohl, the CDU boss, scoffed that the SPD maintained an 'arm's-length relationship' with the Basic Law.[468] Heiner Geißler, a CDU party secretary, published a pamphlet called *The SPD and the Law*, in which he chronicled the government's constitutional offences and ascribed to its leaders a 'law-overcoming strategy' (*Rechtsüberwindungsstrategie*).[469] It is hard to say whether this Christian Democrat counteroffensive made the Court look less partisan or more.

[462] Reported in *Frankfurter Rundschau* (22 May 1978) 4; *Süddeutsche Zeitung* (22 May 1978) 2; *FAZ* (23 May 1978) 5 (with commentary by JG Reißmüller, 12). Reproduced in 'An die Hüter der Verfassung' *Frankfurter Rundschau* (30 May 1978) 4.

[463] See Manfred Beer, 'Vogel mahnt Verfassungsrichter' *Süd-West Presse* (3 May 1978); 'Kummer mit dem Verfassungsgericht (Interview with Hans-Jochen Vogel)' *Nürnberger Nachrichten* (23 May 1978); 'Vogel kritisiert 'einengende' Urteile des Verfassungsgerichts' *Die Welt* (24 May 1978) 5.

[464] See Friedrich Karl Fromme, 'Attacken auf das Bundesverfassungsgericht' *FAZ* (7 June 1978) 5.

[465] 'Kritik der CDU an Börners Gerichts-Schelte' *FAZ* (23 May 1978) 6.

[466] 'Arndt: Auch Carstens hat das Verfassungsgericht kritisiert' *FAZ* (24 May 1978) 5; see also Heinrich Sprenger, 'Carstens: Fehlurteile' *Frankfurter Rundschau* (14 June 1978). For Carsten's comments, see Karl Carstens, *Politische Führung: Erfahrungen im Dienst der Bundesregierung* (Deutsche Verlags-Anstalt 1971) 82.

[467] Bernd Jasper, 'CDU bezeichnet Börner als 'Brandstifter'' *Frankfurter Rundschau* (1 June 1978); see also 'Parteienstreit um das Bundesverfassungsgericht' *FAZ* (1 June 1978) 3.

[468] *FAZ* (17 April 1978) 4.

[469] CDU-Bundesgeschäftsstelle, *Die SPD und das Recht. Verfassungsverstöße der SPD und ihre Ursachen* (Bonn 1978).

Among press organs sympathetic to the government, it was surely the latter. The months following the *Conscientious Objection* decision witnessed an unprecedented flurry of media criticism. The Court was variously scored as a 'Super-legislature',[470] a 'fourth authority' (*vierte Gewalt*),[471] 'power-hungry'[472] lords of a 'German Judicial State',[473] and wayward saints who had lost their halos.[474] *Der Spiegel* published a series of long and critical articles, soon published together under the title *Judges Make Politics*—a direct disavowal of Ernst Benda's protestations to the contrary in his inaugural interview with that magazine.[475] Werner Holtfort, an SPD jurist, issued a much-repeated call for the Court to return to its 'democratically appropriate' role.[476] A string of books and articles from SPD professors and publicists called for institutional reforms, such as abolishing abstract review and establishing a two-thirds voting requirement. For the first time in its history, the Court's democratic legitimacy faced a widespread challenge.

The press offensive persisted through the summer. In autumn, the controversy escalated in the autumn. In late September, Vogel again called on the Court to return to the path of virtue and restraint.[477] On 1 October, the Protestant Academy in Tutzing hosted a panel discussion on 'The Future of our Democracy'. The title was tepid, but the panels—especially one consisting of Helmut Schmidt, the chancellor; Walter Scheel, the federal president; Karl Carstens, president of the Bundestag; and Ernst Benda, the chief justice—were impressive.[478] The discussion got interesting when the first man named exchanged fire with the last.

In the course of one comment, Schmidt spoke in passing of the Court's *Parliamentary Pay* judgment. That case, Schmidt suggested, was a good example of the Court's departure from the path of self-restraint, a path to which the Court should swiftly return. A constitutional organ, the chancellor admonished, needn't 'exhaust its competences to the very limit'.[479]

The chancellor had touched a nerve, and the chief justice answered with ill-concealed anger. Benda took exception to 'the observations with which the Herr Chancellor thought it tasteful to pass judgment before this forum on another constitutional organ'—an organ, Benda stressed, 'not subordinate' to the chancellor's own.[480] Tasteful or not, Schmidt's thesis was absurd. Each organ must 'administer the rights and duties laid upon it—no more, but also no less'.[481] Finally, Benda wished to disabuse his audience of the impression Schmidt had left that 'it is one of the tasks of the head of the federal government to level censures on the highest court in the land, which is also a constitutional organ'.[482] Schmidt tried to defuse the tension, but likely did the opposite, by replying that '[w]hat we all probably need to learn a little bit together is how to endure criticism and take it to heart'.[483] 'No constitutional organ', the chancellor lectured, 'can assume it is above criticism'.[484]

[470] Robert Leicht, *Süddeutsche Zeitung* (17 April 1978). Hans Robinsohn spoke of a 'supralegislator' (*Übergesetzgeber*). *Das verkehrte Grundgesetz* (Vorgänge 1979) 44–53.

[471] Mengen, *Vorwärts* (20 April 1978).

[472] 'Dem Bundesverfassungsgericht Machthunger zugetraut' *Frankfurter Rundschau* (5 June 1978).

[473] 'Richterstaat Deutschland: Auf den Kadi gekommen' *Wirtschaftswoche* (7 July 1978) 26–28, 37–39.

[474] Michael Charlier, 'Karlsruhe zerstört seinen Nimbus' *Deutsche Volkszeitung* (15 June 1978) 5.

[475] Rolf Lamprecht and Wolfgang Malanowski, *Richter machen Politik. Auftrag und Anspruch des Bundesverfassungsgerichts* (Fischer 1979).

[476] Werner Holtfort, 'Praktische Vorschläge, das Bundesverfassungsgericht in eine demokratieangemessene Rolle zurückzuführen' in *Verfassungsrecht und Politik* (n 4) 191.

[477] Winfried Didzoleit, 'Vogel empfiehlt Bundesverfassungsgericht Zurückhaltung' *Frankfurter Rundschau* (29 September 1978) 1; 'Vogel kritisiert BVerfG' *Stuttgarter Zeitung* (29 September 1978).

[478] The discussion was reproduced in Norbert Schreiber (ed), *Die Zukunft unserer Demokratie. Die Tagung 1978 der Stiftung Theodor-Heuss-Preis und des Politischen Clubs der Evangelischen Akademie Tutzing* (Deutscher Taschenbuch Verlag 1979) 75.

[479] ibid 122. [480] ibid 125. [481] ibid. [482] ibid 125–26. [483] ibid 134.

[484] ibid 134.

The press delighted in this verbal row.[485] Many commentators sided with Schmidt.[486] Hans Schueler wrote that Benda's reaction had 'only once more manifested arrogance and obstinacy—the two principal evils into which the highest court has fallen'.[487] Benda and Schmidt later met privately to profess good will, but it became increasingly clear that public criticism was undermining the Court's standing.

Surveys suggested a loss of public trust. In 1977, despite five years of allegations of judicial activism, an Infas Institute study found that 65 per cent of West Germans still trusted the Constitutional Court. By autumn 1978, the figure had fallen to 57 per cent. Trust in Parliament, by contrast, had risen. The Court, hitherto the most trusted institution in the country, had been surpassed by the healthcare system. If there was any consolation for the Court, it was that trust in journalists was abysmal.[488] But this consolation was small. Justice Martin Hirsch, whose dissenting vote in the *Conscientious Objection* judgment had furnished a rallying cry for the Court's critics, fretted in a public interview that the Court was 'endangered as never before'.[489]

D. Endgame

The danger was linked to a pending case. Once again, the Court was asked to review a major government reform, this one tied to the SPD's roots as a workers' party. On 4 May 1976, the Bundestag passed the so-called *Mitbestimmungs gesetz*, literally the 'Co-determination Act'. The law required substantial employee representation, though something short of parity, on the governing boards of companies with more than 2,000 employees. Co-determination had been the subject of one of the sharpest and most enduring debates of the 1960s and 1970s. The 1976 law was a soft revision of a much bolder SPD proposal that sought full parity, an ambition the party had by no means abandoned. The softer law entered into force on 1 July 1976. Employer groups challenged its constitutionality shortly thereafter.

The original complaint was signed, poignantly enough, by the then-president of the Federal Coalition of German Employers' Associations, Hanns-Martin Schleyer. The employers complained that the law violated a fistful of fundamental rights—the right of

[485] See, e.g., 'Bundesverfassungsgericht soll Verantwortung des Parlaments respektieren' *FAZ* (2 October 1978); 'Bedenken des Bundespräsidenten über Diätenregelung der Parlamente' *Süddeutsche Zeitung* (2 October 1978) 1; 'Benda verbittet sich Kanzler-Zensuren' *Frankfurter Neue Presse* (2 October 1978) 1; Claus Bienfait, 'Benda: Es ist nicht Aufgabe des Kanzlers, dem höchsten Gericht Zensuren zu geben' *Die Welt* (2 October 1978) 1.

[486] See Hans Schuster, 'Im Konflikt um die Grenze zwischen Recht und Politik' *Süddeutsche Zeitung* (7/8 October 1978) 4; Hans Schueler, 'Grenzen des Rechts. Das Verfassungsgericht im Widerstreit' *Die Zeit* (6 October 1978); Roderich Reifenrath, 'Empfindliche Richter' *Frankfurter Rundschau* (9 October 1978) 3; 'Dat ham wir uns so nich vorjestellt' *Der Spiegel* (30 October 1978); Wolfgang Däubler and Gudrun Küsel (eds), *Verfassungsgericht und Politik* (Rowohlt, 1979) 7–8; Friedrich Wilhelm-Dopatka, *Das BVerfG und seine Umwelt* (Dunker and Humblot 1982) 84 fn 384. There were exceptions, including the recently retired Justice Willi Geiger. See Walter Bajohr, 'Zensuren vom Kanzler' *Rheinischer Merkur* (6 October 1978); 'Ehemaliger Richter gegen Kritik am Verfassungsgericht' *Stuttgarter Nachrichten* (9 October 1978); 'Es gibt keine Niederlagen in Karlsruhe' *FAZ* (9 October 1978) 5; Willi Geiger, 'Gegenwartsprobleme der Verfassungsgerichtsbarkeit aus deutscher Sicht' in Thomas Berberich, Wolfgang Holl, and Kurt-Jürgen Maaβ (eds), *Neue Entwicklungen im Öffentlichen Recht. Beiträge zum Verhältnis von Bürger und Staat aus Völkerrecht, Verfassungsrecht und Verwaltungsrecht* (Alexander von Humboldt-Stiftung 1979) 132–33. At a symposium in Ludwigsberg, Geiger denied that the chancellor or his ministers were authorized publicly to criticize Court decisions in an official capacity. He expressed astonishment that the Court had yet to say as much firmly in an official statement. ibid 133.

[487] Hans Schueler, 'Grenzen des Rechts. Das Verfassungsgericht im Widerstreit' *Die Zeit* (6 October 1978).

[488] For reports of the Infas survey, see Winfried Didzoleit, 'Ansehen des Verfassungsgerichts ist gesunken' *Frankfuter Rundschau* (26 September 1978) 4; 'Gesundheitswesen genieβt gröβtes Vertrauen' *Stuttgarter Nachrichtern* (26 September 1978) 8.

[489] Rolf Lamprecht and Wolfgang Malanowski, 'Im Moment gefährdet wie noch nie' (Interview mit Martin Hirsch) *Der Spiegel* (27 November 1978).

property and the freedoms of contract, occupation, association, and economic activity. In Parliament, the law enjoyed bipartisan support. But employer groups hoped to find a sympathetic ear in Karlsruhe. Most of them didn't expect the Court to invalidate the law outright, but they hoped for a restrictive interpretation—one that would approve the 1976 law as far as it went, but limit how far co-determination might go in the future.

Their model, of course, was the CSU 'triumph' in the *Basic Treaty* judgment. That judgment, originally hailed by Social Democrats as a vindication of Brandt's *Ostpolitik* and a slap in the face of Franz Josef Strauβ, now appeared in Social–Democrat diatribes as a key instance of the Court's overreaching. Those diatribes coincided with the Court's deliberations in the co-determination case. That they began several weeks after the *Conscientious Objection* judgment and coincided with the eleventh national congress of the Confederation of German Trade Unions (*Deutsche Gewerkschaftsbund*, or DGB) led leery conservatives to suspect an organized campaign.[490]

Those suspicions were heightened by bellicose rumblings within the DGB itself. At the group's national congress in late May 1978, Oskar Vetter, the chairman, complained that in the Federal Republic judges, rather than elected lawmakers, now had the last word on legislation. 'We are the last country in western Europe', Vetter cried, 'where a lockout through high-court jurisprudence, a misinterpretation of the Basic Law, is allowed'.[491] With a truculent undertone, Vetter warned that if the justices couldn't read the signs of the times, the unions would have to take 'a different path'.[492] Months later, while the Court conducted oral arguments in the co-determination case, Friedelhelm Farthmann (SPD), a minister in the North-Rhine Westphalia state government, cautioned the Court to be mindful of the 'dimension' of the decision, and not overstep its bounds, as it had recently gotten into the unfortunate habit of doing. A judgment declaring the law unconstitutional would be 'grotesque'. But a limiting interpretation would be calamitous as well. Such a judgment would have a 'terrible anti-parliamentarian effect', would do 'more damage' to the credibility and stability of the democratic social order than 'thousands of extremists'.[493] In a unanimous statement issued on 6 December 1978, the Court concluded that Farthmann's remarks had gone too far. They were an 'undisguised attempt to exert pressure on the Court to the benefit of one side'. The Court reprimanded Farthmann and excluded him from further participation in the case.[494]

Outwardly, the four days of oral argument were uneventful. All sides conceded that the trial was fair and free from external pressure.[495] Tensions, however, remained sharp. Criticism of the Court, some of it with an ominous edge, continued.[496] Conservatives complained that critics were trying to pressure the Court into endorsing the Co-determination

[490] Friedrich Karl Fromme, 'Attacken auf das Bundesverfassungsgericht' *FAZ* (7 June 1978). Willi Geiger rejected this as an unsubstantiated conspiracy theory. See Willi Geiger, 'Erfahrungen aus 25 Jahren Verfassungsgerichtsbarkeit' in *Vom Selbstverständnis des Bundesverfassungsgerichts* (Hannover: Landeszentrale für politische Bildung 1979) 32.

[491] 'Zweierlei Maß' *FAZ* (24 May 1978) 13.

[492] 'Vetter übt scharfe Kritik am Bundesverfassungsgericht' *Die Welt* (25 May 1978) 1.

[493] 'Farthmann warnt Karlsruhe' *FAZ* (2 December 1978) 5; 'Warnschuβ in Richtung Karlsruhe' *Die Welt* (2 December 1978) 1.

[494] *FAZ* (9 December 1978) 5; *Die Welt* (8 December 1978) 6.

[495] Friedrich Karl Fromme, 'Das Eigentumsrecht—streng und abstrakt' *FAZ* (30 November 1978) 3; 'Das Gericht schlieβt sich dem Dank nicht an' *Die Welt* (2 December 1978) 4.

[496] See Jürgen Scharf, 'Harte Worte über die Hüter der Verfassung' *Kölner Stadt-Anzeiger* (14 December 1978); SPD-Kritik am Bundesverfassungsgericht, *Süddeutsche Zeitung* (16 January 1979); 'Ehmkes Peitschenknall und Geigers stachelige Milde' *FAZ* (2 February 1979) 12; 'Bundesverfassungsrichter—'allergisch'' *Süddeutsche Zeitung* (5 February 1979); Klaus Lange, 'Wie in der Residenz des Rechts die Bonner Politik gestoppt wird' *Frankfurter Rundschau* (8 February 1979) 14–15.

law without qualification. Chief Justice Benda warned against efforts to intimidate or influence the Court.[497] Friedelhelm Farthmann called recklessly for more of them.[498] The federal government feared a 'hot spring'. Trade unions warned Herbert Ehrenberg, the labour minister, that a negative judgment—either annulling the law or restrictively construing it—would lead to labour agitation. Labour leaders would call the faithful to the barricades in a sequence of admonitory strikes. There was even talk of a general strike. The first of March, when the Court would announce its judgment, would be 'an historic date'. After that, one cabinet minister predicted, 'nothing will be as it was before'.[499]

The evil hour tolled. In a cover-page threnody in the 2 March 1979 issue of *Die Zeit*, Hans Schueler could hardly contain his disappointment.[500] It was only to be expected, Schueler sighed, that the Court would conclude that co-determination could come this far and no further. That conclusion was entirely in keeping with the Court's custom of issuing political directives, its pattern of second-guessing the government and the legislature. The 'politicians in the red robes' had once more assumed their beloved role as 'preceptor' to the nation. The real tragedy, though, was not that co-determination had been halted in its tracks. Something larger was at stake: 'the future of one of the most estimable institutions of modern democracy—constitutional justice'. If the Court continued to hurtle along in its wild career of making and prescribing policy, rather than reviewing the legality legislative acts, 'we will one day have to look to the Court itself to find those truly guilty of the failure of the great idea of a power of judicial review over all state authority'. Of well-intentioned warnings, Schueler concluded, there will have been no lack.[501]

Schueler's editorial was another such warning. But it proceeded from false premises. *Die Zeit* had gone to press on 1 March before the Court had actually announced its judgment. Schueler had been apprised of the judgment's contents beforehand; but he had been apprised wrongly. The First Senate's actual judgment was very different from Schueler's phantom.[502] The Senate sustained the law, as *Der Spiegel* summarized, 'without "ifs" or "buts"'. And it did so unanimously. At the climax of a decade of breathtaking judicial activism, the Court stilled the storm with an act of deference.

Deference, rhetorically and substantively, was the dominant theme of the Court's long opinion. The ninety-page judgment had only one headnote: 'The expansion of employee co-determination in the Co-determination Act of 4 May 1976 is compatible with the fundamental rights of the corporations and shareholders covered by it'.[503] The Basic Law contained 'no explicit guidelines on the co-determination of employees. This is a matter for federal legislation'.[504]

The Senate began its analysis by limiting its holding to the facts of the case before it: 'The sole object of review is the challenged ... provisions of the Co-determination Act; whether other regulations of the co-determination of employees would be compatible with the Basic Law is not to be decided'.[505] The Court was guided by a single overarching principle—that the constitutional order of the Basic Law did not prescribe a given economic system. Specifically, it did not enshrine the canons of market capitalism. The people,

[497] Knut Teske, 'Benda warnt vor der Einflußnahme auf Karlsruhe' *Die Welt* (5 February 1979).

[498] 'Farthmann: Ich will die Richter in Karlsruhe beeinflussen' *Die Welt* (13 February 1979); 'Farthmann redet zuviel' *Süddeutsche Zeitung* (14 February 1979); 'Farthmann erschreckendes Verfassungsverständnis vorgeworfen' *Frankfurter Rundschau* (15 February 1979); see also Hans H. Klein, 'Mitbestimmung durch Gewalt'? *Die Welt* (22 February 1979).

[499] 'Mitbestimmung ohne wenn und aber' *Der Spiegel* (5 March 1979).

[500] 'Politiker in der Richter-Robe' *Die Zeit* (2 March 1979) 1.

[501] ibid.

[502] 50 BVerfGE 290 (1979).

[503] ibid 290.

[504] ibid 294.

[505] ibid 322.

through their elected representatives, would determine the structure of the West German economy.

In so holding, the First Senate was on solid textual ground. The right to property in Article 14(1) GG stipulated that the 'contents and limits' of property were 'to be determined by statute'. Article 14(2) provided that property entailed duties, that its use must 'serve the good of the whole'. There were limits to both principles, but none of those limits had been crossed by the 1976 law. That law did not, after all, provide full parity. Shareholders still held a slight predominance and therefore, if need be, could still have the last word. Some employer groups saw in this reasoning a ray of light—a suggestion that a system of full parity in which shareholders did *not* have the last word would be unconstitutional.[506] But that is not what the Senate said. What it said was that the law as it stood did not infringe any of the rights the complainants invoked. It left all other questions open.

It was universally believed that Konrad Hesse, a professor of constitutional law who joined the Court in 1975, had written the judgment. The judgment pleased and impressed nearly everyone with its thoroughness and rigour. For many, the decision was no mere capitulation in the face of political pressure; it was a statesmanlike articulation of enduring principles.

The judgment was greeted in Bonn with a massive sigh of relief.[507] Oskar Vetter, the labour leader, expressed 'deep satisfaction' with the decision.[508] Employers were disappointed by the judgment but promised to respect it.[509] Left–Liberal journalists were jubilant. Hans Schueler rejoiced to learn of his mistake. 'The Court', he rhapsodized, 'has found its way back to its good old style'.[510] The judgment was full of 'perspicacious observations and thankworthy insights from the Guardians of the Constitution'. *Hüter* once more and *Herr* no longer. Others joined the chorus or kept their peace. Those who had lately reviled the Court's overreaching now praised its restraint.[511] Public criticism of the Court swiftly abated. Scholars returned to criticizing the Court only in expert forums. 'Criticising the Constitutional Court', two of them observed some months after the judgment, 'is no longer popular'.[512] The Court's season of confrontation had come to an abrupt close.

Conclusion

Or so, at the moment, it seemed. It is tempting to see the Court's *Co-determination* judgment as a pragmatic surrender similar to the United States Supreme Court's 1937 decision

[506] See 'Das Mitbestimmungsgesetz ist verfassungsgemäβ. Karlsruhe: Keine volle Parität' *FAZ* (2 March 1979) 1; 'Mitbestimmung ohne wenn und aber' *Der Spiegel* (5 March 1979); 'Den Anteilseignern kommt ein leichtes Übergewicht zu' *Die Welt* (2 March 1979) 2.

[507] Peter Diehl-Thiele, 'Ein bedeutendes Urteil aus Karlsruhe' *Süddeutsche Zeitung* (2 March 1979) 4; 'Mitbestimmung ohne wenn und aber' *Der Spiegel* (5 March 1979); Peter Gillies, 'Nach dem Karlsruher Urteil: Alle sind erleichtert' *Die Welt* (2 March 1979) 1; 'Für alle etwas' *FAZ* (2 March 1979) 1.

[508] 'Die Arbeitgeber respektieren den Spruch aus Karlsruhe' *FAZ* (2 March 1979) 2; 'Das Mitbestimmungsgesetz ist verfassungsgemäβ. Karlsruhe: Keine volle Parität' *FAZ* (2 March 1979) 1; 'Mitbestimmung ohne wenn und aber' *Der Spiegel* (5 March 1979).

[509] 'Die Arbeitgeber respektieren den Sprech aus Karlsruhe' *FAZ* (2 March 1979) 2; 'Beschwerdeführer bekunden Respekt' *Süddeutsche Zeitung* (2 March 1979) 1.

[510] Hans Schueler, 'Ein Lehrstück über die Verfassung. Ein Richterspruch dient dem Rechtsfrieden' *Die Zeit* (9 March 1979).

[511] Nearly twenty years later, Hans-Jochen Vogel still remembered the judgment in gratefully panegyircal terms. It 'left a powerful impression on all involved', Vogel noted, 'not just me'. Thanks in part to its rapporteur Konrad Hesse—'one of the most outstanding constitutionalists I have ever encountered'—the judgment 'was written with consummate concision' and reached only the matter 'that was to be decided'. Vogel (n 145) 97.

[512] Gudrun Küsel, 'Vorbemerkung' in *Verfassungsgericht und Politik* (n 4) 8.

in *West Coast Hotel v Parish*.[513] Certainly, the judgment muted institutional criticism and defused social tension. It fostered national harmony and furthered societal integration.

The Court was praised for its deference, but also for promoting stability in a country that craved stability. In the title of a prominent 1979 book, the political scientist Kurt Sontheimer called West Germany 'The Destabilized Republic'.[514] Unsettled by the encounter with domestic terrorism, chastened by yet another economic slump, and fearful (with a bellicose Ronald Reagan challenging Jimmy Carter for the American presidency and NATO planning to plant Pershing missiles on West German soil) of a resumption and warming of the Cold War, the Federal Republic at the end of the 1970s was in little mood to dare more democracy. In any case, there was no one in the political class to lead such an enterprise. Helmut Schmidt had won widespread plaudits for his handling of the terrorist crisis, but his hawkish foreign policy put him at odds with much of his party. Schmidt did win reelection in October 1980. But that was his final triumph. It was all downhill from there, until in February 1982 he became the first chancellor ousted by a constructive vote of no-confidence. In the meantime, Schmidt presided over an increasingly divided party and an increasingly brittle coalition. The co-determination law was the coalition's last major reform. The Court's blessing on it turned out not to be a green light for further innovation but a deferential salute to the dying.

The Court never again resisted a series of government initiatives quite so aggressively and in quite so compressed a period. But that is at least in part because it never again faced such an aggressive sequence of reforms. The Court's posture after 1979 was not passive and deferential, but protective and patriarchal. In the early 1980s the Court's reputation and stature were enhanced as the Court quashed an unpopular data collection law, reminding a grateful populace of its venerable role as guardian of fundamental rights, and as a funding scandal implicating all parties in Parliament emphasized the Court's comparative probity. The Court reached a low point during the 1970s by seeming to decide cases on partisan grounds. It reached a high point during the 1980s by emerging as the guardian of the citizenry against the power of over-sized and untrustworthy political parties. Confidence in the Court surged as trust in parties plummeted. The constitutional moment proclaimed by Willy Brandt in 1969 had long since passed. The guardians of the West German constitutional tradition—challenged and criticized, but ultimately strengthened—endured. In hindsight, the 1979 *Co-determination* judgment represents not a capitulation, but a close parenthesis, a return to continuities with deep roots.

[513] 300 US 379 (1937).

[514] Kurt Sontheimer, *Die verunsicherte Republik: Die Bundesrepublik nach 30 Jahren* (Piper 1979).

4

Continuity, 1982–1990

If one now regards the individual constitutional organs and the weight intended for them in the 1949 Basic Law, the constantly increased importance of the Federal Constitutional Court stands out above all else. Its establishment was a true innovation in our constitutional history. It has won great prestige and powerful influence. This is certainly a result of the quality of its jurisprudence, but not only that. It is also, and moreover, a symbol for the political condition of the society, for a diminishing efficacy of the classical separation of powers, for an increase of partisan-tactical compromises not always sufficiently legally secured. In this respect, the Court operates as a longed-for supra-partisan oasis.

—Richard von Weizsäcker (1992)[1]

Introduction

Throughout its history, the public's esteem for the Constitutional Court has been shaped by the Court's relationship with political parties. In the early 1950s, the justices won popular acclaim with a flinty display of independence from the parties who nominated them. During the Court's first decade, the justices helped enshrine the constitutional legitimacy of opposition parties. In its 1961 *Television* judgment, the Court thrilled the public by sharply distinguishing the state from the parties that ruled it. During the 1960s, the Court stirred controversy first by endorsing, then by limiting, state funding of parties. During the 1970s, the Court sparked a crisis of popular legitimacy by seeming, in momentous cases, to split for partisan reasons on partisan lines. The Court's moments of greatest popular regard were its moments of greatest apparent independence from partisan interest. The public's trust in the Court declined sharply when the justices seemed too close to the parties.

As the Court approached its fourth decade, political parties had reached a nadir of unpopularity. In the epic battle for the chancellorship in 1980, both leading candidates—Helmut Schmidt and Franz Josef Strauß—were far more popular than the parties that backed them. Widespread disgruntlement with existing parties fuelled the formation, for the first time since the Second World War, of a viable fourth party—a self-styled anti-party, the Greens.

In 1982, Wilhelm Hennis published an essay fiercely critical of what Hennis saw as the disproportionate prominence of political parties in the life of the Bonn Republic.[2] On the one hand, Hennis wrote, the 'party state' character of the Federal Republic had become 'dangerously overstretched [*ueberdehnt*]'; on the other, the parties had 'disconnected [*abgekoppelt*] themselves from the autonomous will-formation of the people'. The intellectual and jurisprudential culprit was Gerhard Leibholz, who in earlier decades had done much to etch his conception of the party state in the jurisprudence of the Constitutional Court, and thereby to inscribe it in the political culture of the Federal Republic. Leibholz had died

[1] Richard Weizsäcker and others, *Richard von Weizsäcker im Gespräch mit Gunter Hofmann und Werner A. Perger* (Eichborn 1992) 158–59.

[2] Wilhelm Hennis, 'Überdehnt und abgekoppelt' in Wilhelm Hennis, *Auf dem Weg in den Parteienstaat. Aufsätze aus vier Jahrzehnten* (Reclam 1998) 74, 78.

in 1981, but his creed lived on. Leibholz's notion of political parties, Hennis wrote, 'went so far beyond the intentions of the Basic Law that one can hardly describe it as anything other than unconstitutional'.[3] It is one of the great ironies of the final decade of the Bonn Republic that the Court that did so much to make the parties powerful won deafening plaudits for holding them at bay.

The period between the *Co-determination* judgment in March 1979 and the Federal Republic's second regime change in October 1982 was a quiet one for the Constitutional Court. The clamour of public criticism that preceded that judgment quickly subsided; the political brisance of the Court's docket abated. What major policies the Court did review during this period—a 1977 law introducing no-fault divorce, for instance—it approved with limited fanfare.[4] The Court issued important doctrinal judgments—notably in the law of property, where the Court, under the influence of its property law rapporteur Werner Böhmer, renounced the traditional takings jurisprudence developed by the Federal Court of Justice, reasserted its own interpretational sovereignty in the field, and reaffirmed the legislature's power to define the contours of the right to property and the social duties that attend it.[5] Such judgments generated much scholarly comment but drew little popular attention. The Court, it seemed, had quit the stage of high politics.

Appearances, however, were deceiving. The Court's political quiescence was brief. Almost immediately after the advent of a Christian–Liberal (CDU/CSU–FDP) coalition under Helmut Kohl in autumn 1982, the Court became again the forum in which many of the nation's most burning political battles were contested and decided—the forum, cynics griped, in which political conflict was constitutionalized. Indeed, the questions of constitutional politics that the Court considered during the Christian–Liberal era of the 1980s were no less explosive than those it considered during the Social–Liberal 1970s. The outcomes, for Court and country, were quite different. There was, in the new decade, no sustained confrontation between the country's highest court and its ruling coalition. But neither did the Court unfailingly defer to government initiatives. There were times when the Court, invoking the doctrine of judicial restraint, sustained government activity of manifestly dubious constitutionality. But there were other occasions—some of them quite dramatic—in which the Court enforced fundamental rights vigorously to annul democratically promulgated laws—in the most remarkable case, a law enacted unanimously by the Bundestag.

The split between deference and assertion often ran between the Court's two chambers. The Second Senate was more inclined to yield pragmatically before claims of *raison d'état*; the First Senate to enforce robustly the guarantees of individual rights. The upshot, with critical exceptions, was that the Court was by turns commended for its restraint and lionized for its intervention. Those critics most dismayed by the Court's acts of deference were often the very critics who in earlier times most loudly reproached its lack of restraint. The criticism lost much of its force in the face of this manifestly partisan reversal. On the other hand, those most cheered by the Court's aggressive interventions were rarely those with strong ties to the traditional parties.

In the last years of the Bonn Republic, the Court was never the object of serious criticism from the leaders of major parties. This was partly the product of an entrenched pattern whereby all parties claimed victory after every major judgment. But it was also the product of parties' and politicians' increasingly weak popular standing and the Court's strong one.

[3] ibid. [4] See 53 BVerfGE 224 (1980); 53 BVerfGE 257 (1980).

[5] See 58 BVerfGE 300 (1981). Böhmer's influence in the field extended beyond the immediate judgment through his scholarly writings after leaving the Court in 1983. See, e.g., Werner Böhmer, 'Die Rechtsgeschichtlichen Grundlagen der Abgrenzungsproblematik von Sozialbindung und Enteignung' (1985) 24 *Der Staat* 157.

The early Kohl era was a season of scandal. Discontent with traditional parties mounted as major parties became embroiled in scandal. At the 1976 elections, with a voter participation rate of 90.7 per cent, the percentage of voters who polled for the Union parties or the SPD was 91.2 per cent. By 1990, voter participation had dropped to 77.8 per cent; from this diminished voter base, only 77.3 per cent opted for one of the two largest parties. An Emid survey in 1983 found that only 50 per cent of West Germans trusted the parties to solve the country's problems; by 1990, that tally had sunk to 37 per cent.[6] Media coverage of parties and politicians was largely—often caustically—negative, to the point that politicians of all stripes complained that the media were treating them as enemies rather than mediating between them and the public.[7] It was an era of political disenchantment, of *Politikverdrossenheit* run rampant.[8]

By contrast, the justices of the Constitutional Court enjoyed enormous, unprecedented approval. That approval soared after a 1983 judgment enjoined a proposed census that had provoked widespread execration and sometimes hysterical fears. But the approval endured when the hysteria passed. In a season when Bonn seemed riddled by impotence and sleaze, Karlsruhe emerged, by contrast, as a beacon of probity and strength. After a party financing scandal led many West Germans to conclude that their political rulers were for sale, many gave thanks that their judicial guardians were not. Democratic critiques of constitutional justice were never as sharp or obsessive in the Federal Republic as they have been in the United States. But in the final years of the Bonn Republic, they disappeared almost entirely. The citizens of the second German Republic had long ceased to see their elected representatives as the exclusive proprietors of democratic legitimacy. When the Wall finally fell, West Germans no longer defined democracy as rule by popular representation. German democracy had become something more substantive—robust protection of human rights, for instance, and unwavering reverence for human dignity. Democracy-based critiques of the Court that enforced these values lost force. Enormous popular regard for the Court, coupled with galloping disenchantment with politicians and political parties helped justify, at least indirectly, the Court's practice of announcing its judgments '*Im Namen des Volkes!*'[9]

I. High (Constitutional) Politics

At the outset of the 1980s, the economic and political outlook of the Federal Republic was bleak. Growth continued to stagnate; unemployment began to soar. Between 1980 and 1983 the ratio of unemployed rocketed from under 3 per cent to over 9 per cent. Never before had the Federal Republic had more than a million unemployed; by February 1983, 2.5 million were without work. The Soviet invasion of Afghanistan in 1979, followed by the election of the sabre-rattling Ronald Reagan to the American presidency in 1980, threatened a hasty end to the Cold War thaw. This, in connection with NATO plans to station Pershing warheads on West German soil, contributed to a mass sense of insecurity. Support for the

[6] Andreas Wirsching, *Abschied vom Provisorium* (Duetshe Verlags-Anstalt 2004) 201–02.

[7] ibid 202.

[8] The phenomenon provoked a large contemporary literature. For a sampling, see Hermann Scheer, *Parteien kontra Bürger? Die Zukunft der Parteiendemokratie* (Piper 1979); Hennis (n 2); Helmut Herles and Friedric W. Husemann, *Politikverdrossenheit: Schlagwort oder Zeichen der Krise?* (Aktuell 1993). The fiercest critic of political parties and the party system was Hans Herbert von Arnim. See his *Die Partei, der Abgeordnete und das Geld* (Hase & Koehler 1991); *Der Staat als Beute: wie Politiker in eigener Sache Gesetze machen* (Knaur 1993); *Diener vieler Herren: die Doppel- und Dreifachversorgung von Politikern* (Knaur 1998); *Vom schönen Schein der Demokratie: Politik ohne Verantwortung—am Volk vorbei* (Droemer 2000).

[9] 'In the name of the People!'

governing Social–Liberal coalition dwindled swiftly. The coalition won the dramatic elections of 1980, but that victory reflected the comparatively greater popularity of Helmut Schmidt over his Union rival, Franz Josef Strauβ, rather than the comparative popularity of the coalition parties vis-à-vis the opposition. The result was a negative referendum on Strauβ rather than an endorsement of the coalition. For Schmidt, the election was a personal triumph. But it would be his last. Increasingly Schmidt found himself isolated within his own party, especially after his support for the NATO deployment proposals alienated the party's Left wing. In the two years following the election, the SPD's unity and popularity declined steadily.

The SPD's coalition partner fared even worse. After clearing the 10 per cent hurdle in 1980 for only the second time, the FDP descended into crisis. The neophyte Green party, which polled 1.5 per cent of the votes in 1980, rose rapidly in popularity as ever larger swathes of the electorate worried about the sustainability of the environment and the stability of the peace. By autumn 1982, the Greens had displaced the FDP as the third party in three state elections. Fearing the worst, the FDP leadership opted to cast their lot once more with the Union. On 17 September 1982, all FDP cabinet ministers resigned their posts. Shortly thereafter, the party's national board and parliamentary leadership voted narrowly to pursue a coalition with the CDU/CSU. On 1 October, by a vote of 256 to 235, the Bundestag ousted Schmidt from the chancellorship and replaced him with the CDU chief, Helmut Kohl. It was the first successful constructive no-confidence vote in the country's history. After thirteen years in opposition, the Christian Democrats had returned to the helm.

The FDP's position within the new coalition was precarious. All parties knew that new elections would be fatal for the junior partner. Schmidt and his fellow Social Democrats, spurned by their former allies, called for them immediately.

Helmut Kohl cut a very different figure from Willy Brandt, the last chancellor to preside over a regime change. Brandt, the telegenic lady-killer, was—when he came to power—a special darling of the media; throughout the 1970s, media coverage of Kohl was decidedly hostile. The new chancellor was an object of merriment to satirists and a jewel for caricaturists, who delighted in lampooning Kohl's wooden speeches and mocking his pear-like frame.

Born in 1930, Kohl was the first chancellor whose entire political biography unfolded in the Federal Republic. He was, even more than Schmidt, a committed pragmatist, with only a handful of core convictions, none of them breathtakingly concrete—fostering European unity, keeping the German Question 'open', strengthening the family, promoting historical consciousness. On all material questions of policy, Kohl was adaptable. His conservatism was of a Burkean bent; he wished to build a politics based on experience, not abstract theoretical models.[10] When he became chancellor, Kohl declined to proclaim a new dispensation.[11] He announced a 'turning point' (*Wende*), not a constitutional moment.

A. Dissolution of the Bundestag

But even this Burkean craved popular legitimation. In any case, he needed time, and for this reason, if for no other, Kohl was willing to accept the SPD's call for new elections. The damage 'which has been wreaked over many years', Kohl told the Bundestag in his inaugural message of 13 October 1982, 'cannot be set in order in a few months. For that, at least an entire legislative period is necessary, which we won't have without new elections'.[12] Kohl

[10] Wirsching (n 6) 23–25. [11] ibid 47.

[12] *Verhandlungen des Deutschen Bundestages. Stenographische Berichte*, vol 122 (1982) 7251.

proposed that new elections be held on 6 March 1983. In the meantime, his government could hope only to take emergency measures to rein in the country's galloping unemployment and to cleanse the Augean stables of its public finances.[13] The real work could be undertaken only after a new election. But the path towards such an election was, as Kohl conceded, 'constitutionally not simple'.[14]

This, to understate, was an understatement. Article 68(1) GG provided that if a chancellor's call for a vote of confidence failed to receive the support of a majority of the Bundestag, the federal president could, on the chancellor's recommendation, dissolve the Bundestag and call for new elections within two months. The major parties agreed that the Bundestag would deny Kohl a vote of confidence, and that Kohl would then ask the president, Karl Carstens, to dissolve the Bundestag on 6 January 1983, triggering new elections to be held on 6 March. It required no great juristic fantasy or constitutional expertise to see the irregularity of a parliament that had just made Kohl chancellor through a constructive vote of no-confidence denying Kohl its confidence immediately thereafter.

The no-confidence vote took place on 17 December 1982—one day after Kohl's government passed its 1983 budget by a vote of 266 to 206, and shortly after the government instituted a series of austerity measures. Kohl was manifestly head of a majority coalition capable of continued governance. This was plainly not the situation envisioned by Article 68. The no-confidence vote was an unblushing fabrication—a fraud and a farce, many complained; in any event, a fiction.

Fiction is to law, wrote Bentham, what swindle is to trade.[15] But many in the Federal Republic thought this instance a bit rich. The stage was too large and the stakes were too high to play so fast and so loose with a nuclear provision of the constitution. During the Bundestag debates over the confidence vote, many who favoured new elections expressed constitutional reservations about Kohl's method of achieving them.[16] SPD leaders were most vocal in expressing such reservations. Willy Brandt, the party's chairman, contended that it would be better for Kohl simply to resign.[17] Even before the no-confidence vote, leading constitutionalists and political scientists—Ernst Wolfgang Böckenförde, for example, and Karl Dietrich Bracher—denounced the scheme as unconstitutional.[18] After Carstens dissolved the Bundestag formally, a multi-partisan quartet challenged the president's action before the Constitutional Court.

The complaint fanned the fervour of what was already an extraordinary public debate on a question of constitutional law. Law professors, relishing a rare moment of relevance and fame, became staple features of the popular press and the nightly news. The debate's rhetoric was impassioned. Böckenförde, a Social Democrat soon to join the Court's Second Senate, warned that the Basic Law was slipping down 'a plebiscitary slope'.[19] In an amicus brief, Klaus Schlaich conjured the spectre of a revolving carousel of parliamentary dissolutions,

[13] ibid 7215. [14] ibid.

[15] Jeremy Bentham, *The Works of Jeremy Bentham*, vol 7 (W Tait 1843) 283.

[16] Horst Ehmke, for instance, said on behalf of the SPD Bundestag faction that, while his party was in favour of new elections, the constitutionally sounder means to achieve them was simply Kohl's resignation. *Verhandlungen des Deutschen Bundestages. Stenographische Berichte*, vol 123 (1982) 8584.

[17] ibid 8940. Karl Carstens, the federal president, countered in his televised statement of 7 January 1983 that resignation was a more complicated matter than the SPD leadership made out: it would merely trigger a series of attempts to elect a new chancellor, not the immediate dissolution of the Bundestag. See 'Auflösung des 9. Deutschen Bundestages', Bulletin der Presse- und Informationsamt der Bundesregeriung (10 January 1983) 17–18.

[18] See Ernst-Wolfgang Böckenförde, *Die Welt* (27 September 1982); 'In ein plebiszitäres Gefälle' *FAZ* (11 October 1982) 18; *Badische Zeitung* (16/17 November 1982); Bracher, 'Die Bonner Runde' (on the television programme ZDF, 4 November 1982).

[19] Ernst-Wolfgang Böckenförde, 'In ein plebiszitäres Gefälle' *FAZ* (11 October 1982) 18.

an endless succession of collapsing governments—the spectre, in other words, of Weimar.[20] Wilhelm Hennis went so far as to suggest that the slope might someday tilt toward a *coup d'état*.[21] Such Cassandra cries were unsettling, both to the public and to the politicians who pushed for Parliament's dissolution. Kohl put on a brave face, expressing every confidence that the Court would vindicate his procedure. But Kohl and his comrades were clearly rattled. They accused their critics of putting pressure on the Court.[22] Union leaders responded with mild pressure of their own. Benno Erhard, the CDU's second man in the recently disbanded Bundestag, hoped out loud that the Court wouldn't mingle indecorously in the political process.[23]

The Second Senate appointed 16 February 1983—Ash Wednesday, as it happened, less than three weeks before the election—as its day of decision. Critics complained that the judgment's quality betrayed its hasty composition. Willi Geiger, the former justice, took his old colleagues to task for what he thought was their unnecessary decision to decide the case before the election.[24] The judges themselves felt rushed; one complained publicly that he was given less than twenty-four hours in which to write his dissenting opinion.[25] But the justices wanted to resolve the issue early, lest a later decision undermine the very governmental legitimacy the new elections were intended to secure. They may also have felt some pressure from the political branches.

The justices of the Second Senate took their seats at the head of a densely packed courtroom on the morning of 16 February. Against a backdrop of thick and muggy silence, Wolfgang Zeidler, the vice president, read in a soft and steady voice from the Court's long judgment.[26]

The judgment was divided into two sections, the first some distance longer than the second.[27] The first section articulated constitutional principles; the second applied them to the facts at hand. Many observers noted a disjunction between the judgment's two halves.[28] By general consensus, the former was more persuasive than the latter.

The Senate's general discussion began by noting that Article 68 GG enacted an 'open norm' in need of 'concretization'.[29] Then, in an uncharacteristic act of jurisdictional generosity, the Senate noted that the necessary concretization was not the exclusive preserve of the Constitutional Court, but that it required the participation of other constitutional organs as well—including Parliament, the federal president, and the government. Structurally, the constitution required agreement among all these organs to effect a dissolution. The several organs supplied a mutual check against abuse.[30] But the justices hastened to add that neither consensus among these organs as to the desirability of dissolution, nor unity among the parties in Parliament as to the necessity of new elections, was constitutionally sufficient.[31] Other conditions must also be met.

[20] Cited in Hans Hugo Klein, 'Die Auflösung des Deutschen Bundestages nach Art. 68 GG' (1983) *Zeitschrift für Parlamentsfragen* 403, 403.

[21] Wilhelm Hennis, *Süddeutsche Zeitung* (11 December 1982). See also, 'Wie soll sich die Schranke öffnen?' *FAZ* (29 October 1982) 11.

[22] See, e.g., Helmut Kohl's comments after the Court's judgment in 'Kohl: 'Ich habe keine Alpträume'' *Die Zeit* (18 February 1983) 3.

[23] In Friedrich Karl Fromme, 'Stehend und unbewegt—wie es sich gehört' *FAZ* (17 February 1983) 3.

[24] Willi Geiger, 'Die Auflösung des Bundestages nach Art. 68 GG: ein Kommentar zu den Vorgängen, die zu der Neuwahl des Deutschen Bundestages am 6. März geführt haben' (1984) 33 *Jahrbuch des öffentlichen Rechts der Gegenwart* 41, 44.

[25] 62 BverfGE 1, 108 (1983) (dissenting opinion of Justice Rottmann).

[26] See Fromme, 'Stehend und unbewegt' (n 23) 3.

[27] 62 BVerfGE 1 (1983).

[28] See 'Die Dämme nicht brechen lassen' *Der Spiegel* (21 February 1983) 17.

[29] 62 BVerfGE 1 (n 27) 39.

[30] ibid 35.

[31] ibid 43.

The Senate's enumeration of those conditions did not sound good for Helmut Kohl. Principal among them was that the chancellor must be unable, given the current configuration of parliament, to go on governing.[32] 'The balance of political power in the Bundestag', the justices wrote, 'must so paralyse or impair [the chancellor's] power to act that he can no longer meaningfully pursue a policy that enjoys the consistent support of the [parliamentary] majority'.[33] It would be unconstitutional, therefore, for a chancellor 'whose sufficient majority in the Bundestag is beyond doubt', to seek, 'at a seemingly suitable time, to have the confidence question answered in the negative with the goal of dissolving the Bundestag'.[34] Nor could the chancellor properly pose the confidence question simply by invoking the 'special difficulties of the tasks at hand in the current legislative period'.[35] The Senate made clear, moreover, that lacking majority support was not the same as desiring greater legitimacy. It ran against 'the fundamental meaning of Article 68, as well as of representative democracy as established by the Basic Law, to call for the dissolution of the Bundestag and new elections on the theory that a chancellor newly elected through a constructive vote of no confidence requires, in addition to his constitutional legality, the legitimacy supplied by new elections'.[36] The lack of support must be real, the crisis genuine.

Up to this point, the judgment read rather like a brief for the complainants; thereafter it read like a brief for the government. Moving from principles to practicalities, the Senate shifted from strict standards to deferential application. The justices swallowed the explanations put forth by Kohl and Carstens almost without qualification. Though Kohl's government had been able to push through a handful of urgent measures, those measures were limited by subject and time.[37] As of December 1982, the Senate concluded, Kohl had ample reason to believe he could not command the long-term support of a stable majority. To support this conclusion, the justices dwelt at some length on the crisis of the FDP.[38] The party that least wanted new elections was thereby made primarily responsible for them—an irony not lost on observers.[39]

The justices noted that, as recently as the Bundestag elections in 1980, the FDP had pledged its allegiance to the SPD, pinned its electoral hopes on the person of Helmut Schmidt, and achieved thereby the second-greatest success in the party's history.[40] The justices were not so unkind as to suggest openly that a party perfidious to one chancellor could not be relied on by another. But they did stress that the split with the SPD was hotly contested within the FDP—approved only by a narrow majority of the party directorate and followed, in short order, by the departure in droves of party prominences and a swift succession of calamitous state elections.[41] Moreover, upon joining the new coalition in October 1982, the FDP agreed (the Court didn't say with how little enthusiasm) that an election in March 1983 was indispensable.[42] Given the internal anguish of the FDP and the temporally limited nature of its agreement to form a coalition, it was 'plausible' for Kohl and Carstens to deem dissolution unavoidable.[43] This concession was based on an unusual and likely unrepeatable conglomeration of facts. The Court had split the difference. The judgment spared the government and the president the embarrassment of a ruling that they had violated the constitution; but it admonished them unmistakably never to hazard such a gimmick again.

Not all members of the Second Senate were pleased with the outcome. Wolfgang Zeidler concurred in the Senate's holding but abhorred its reasoning.[44] Under the Basic Law, he argued, there was no such thing as a temporally limited vote of confidence. When members of the Bundestag made Kohl chancellor, they pledged him their support till the end

[32] ibid 44. [33] ibid. [34] ibid 43. [35] ibid. [36] ibid. [37] ibid 52.
[38] ibid 52–60. [39] 'Die Dämme nicht brechen lassen' (n 28) 17. [40] 62 BVerfGE 1 (n 27) 53.
[41] ibid 53–55. [42] ibid 61. [43] ibid 62. [44] ibid 64–70.

of the legislative period.[45] But Zeidler still thought the dissolution legitimate in light of the steady evolution of the chancellorship's 'increasingly potent personalized, plebiscitary component'.[46] In crafting Article 68, Zeidler noted, the framers were fixated on the ghosts of Weimar. But the Federal Republic now had a history of its own—roughly two-and-half times longer than the whole history of Weimar. This lived experience was at least as important in constitutional construction as the Basic Law's framing history.[47] Voters now considered their votes in a Bundestag election as votes for the chancellor. A chancellor who hadn't won such an election reasonably felt the weight of popular skepticism, the burden of deficient credibility. In Zeidler's view, when Kohl asked Carstens to dissolve the Bundestag, the president was faced with a choice of evils: a manipulated election or a chancellor 'for whom the credibility required for full political effectiveness was, in the broadest sense, placed in question'.[48] Carstens opted for the former. Perhaps this wasn't the part of political wisdom; but neither was it constitutionally objectionable.

Justices Joachim Rottmann and Hans-Justus Rinck, however, objected sharply. Justice Rottmann, who drafted his dissent in less than a day, thought the majority's extension of the concept of instability to a case in which the chancellor feared *losing* a majority (rather than *lacking* a majority in the first place) made nonsense of Article 68. As a rule, Article 68 could be invoked only by a minority chancellor, which Kohl demonstrably was not.[49] Justice Rinck, who served as a rapporteur in the case,[50] had considerably more time to draft his long dissent.[51] Rinck reviewed the drafting history of Article 68 in great detail, concluding that its unambiguous purpose was to empower a chancellor to secure a vote of confidence, not to trigger new elections.[52] The provision's *raison d'être* was to enable an embattled chancellor to find renewed support; the logical precondition of *finding* support was *seeking* it.[53] For Rinck, the idea of a chancellor calling for a vote of confidence he hoped to lose was a legal and logical absurdity.

It was an absurdity, alas, that six of Rinck's colleagues—and large segments of the public—were prepared to countenance. Naturally, politicians affected by the case hastened to proclaim victory.[54] President Carstens noted modestly that the Court had vindicated his and Kohl's decisions and expressed gratitude that the heated discussion of the matter could now come to an end.[55] Kohl himself was more fustian. The chancellor praised the Court for contributing 'constitutional clarity in an impressive fashion'; he added gamely that '[n]ow the voter has the word'. In an interview with *Die Zeit*, Kohl refused even to consider that the Court might have come out the other way.[56] Kohl's election rival, by contrast, the former SPD justice minister, Hans-Jochen Vogel, called the judgment the 'terminal point in a chain of embarrassments' caused by Kohl and Hans-Dietrich Genscher, leader of the FDP.[57] For his part, Genscher was glad the Court had put an end to SPD suspicions that the government was manipulating the constitution. Vogel retorted that the case's real loser was 'the putative master tactician', Genscher himself.[58] In short, after the months-long constitutional drama, it was back to politics as usual. And with good reason: there was an election on.

45 ibid 66. 46 ibid 67. 47 ibid 68–69. 48 ibid. 49 ibid 110.

50 See 'Der Bundestagswahl am 6. März steht nichts mehr im Wege' *FAZ* (17 February 1983) 1.

51 62 BVerfGE 1 (n 27) 70–108. 52 ibid 72. 53 ibid.

54 See 'Die Bonner Politiker sehen sich vom Karlsruher Urteil bestätigt. Carstens ruft die Parteien zu einem fairen Wahlkampf auf' *FAZ* (17 February 1983) 2.

55 ibid. 56 'Kohl: 'Ich habe keine Alpträume'' (n 22) 3.

57 'Die Bonner Politiker sehen sich vom Karlsruher Urteil bestätigt. Carstens ruft die Parteien zu einem fairen Wahlkampf auf' (n 54) 2.

58 'Die Dämme nicht brechen lassen' (n 28) 17.

Amid the compressed exhilaration of the election, the general public quickly forgot or forgave the Court's judgment.[59] But many observers with a more sustained interest in the Court were disappointed. Though most of the Left–Liberal press relished the judgment's implicit correctives,[60] the *Süddeutsche Zeitung*, which for a decade had been urging the justices to show greater restraint, sighed that the justices had shirked 'their proper task' and 'bowed' before the powers that be.[61] Wilhelm Hennis murmured that, in matters of 'of high politics, in questions of power', the Court had revealed itself as a tribunal 'that has no sharp teeth'.[62] Constitutional lawyers generally agreed that the decision was more satisfying politically than legally.[63] Law professors scolded the Second Senate—sometimes quite sternly—for its deficient craftsmanship.[64] What scholarly praise the judgment won was not for its legal soundness.[65] One scholar, for instance, praised its 'Solomonic' wisdom.[66] Another hailed it as 'a masterpiece of diplomatic jurisprudence'.[67] The common criticism of the previous decade—that the Court lacked restraint in matters of high political moment—was nowhere to be heard.

The judgment had less deterrent effect than many of its admirers, and the justices themselves, had initially hoped. After the decision, Kohl expressed his intention to seek a constitutional amendment, making it easier for the chancellor to trigger new elections—the gesture, critics quipped, of a guilty conscience.[68] Kohl's good intentions, though, came to nothing. More than twenty years later, in 2005, the SPD chancellor, Gerhard Schröder, dissolved the Bundestag and triggered new elections under circumstances nearly identical to those of 1982–83. The Constitutional Court approved Schröder's gambit with a nod to its February 1983 judgment.[69] Recently, a Left-wing critic of the Court cited this 2005 judgment as evidence that the Court's 'judges, all political appointees, rarely baulk at bending to the powers that be, provided the two major parties are at one on an issue'.[70] The 1983 *Bundestag Dissolution* judgment is certainly vulnerable to this criticism. Before 1983 was over, however, the First Senate, in its final judgment under the leadership of Ernst Benda, would provide a dramatic counter-example.

[59] Wirsching (n 6) 40.

[60] See 'Die Dämme nicht brechen lassen' (n 28) 17; Hans Schueler, 'Das Wort der Acht' *Die Zeit* (18 February 1983); Hans Schueler, 'Ein halbherziger Richterspruch' *Die Zeit* (25 February 1983) 6.

[61] Robert Leicht, 'Die Richter haben sich gebeugt' *Süddeutsche Zeitung* (17 February 1983) 4.

[62] In 'Die Dämme nicht brechen lassen' (n 28) 17.

[63] See Norbert Achterberg, 'Vertrauensfrage und Auflösungsordnung' (1983) 98 *Deutsches Verwaltungsblatt* 477; Jost Delbrück and Rüdiger Wolfrum, Die Auflösung des 9. 'Deutschen Bundestages vor dem BVerfG—BVerfGE 62, 1' (1983) 23 *Juristische Schulung* 758; Lutz Gusseck, 'Bundestagsauflösung kraft Richterspruchs?' (1983) *NJW* 721; Gerhard Schlichtung, 'Zur Auslegung des Art. 68 GG durch das Bundesverfassungsgericht' (1984) 39 *Juristenzeitung* 120; Rudolf W. Strohmeyer, 'Die Verfassungsgemäße Bundestagsauflösung—einige kritische Anmerkungen zu den Begründungen des Verfassungsgerichtsurteils vom 16. Februar 1983' (1983) 14 *Zeitschrift für Parlamentsfragen* 422; Hans-Jürgen Wiegand, 'Die manipulierte Verfassung' (1983) 28 *Verwaltungsrundschau* 333.

[64] See Hans Meyer, 'Anmerkung' (1983) 36 *Die öffentliche Verwaltung* 243.

[65] See, for instance, Walter Seuffert, 'Freiheit der Politik und Grenzen des Rechts' (1983) 108 *Archiv des öffentlichen Rechts* 403.

[66] Hans-Peter Schneider, 'Sibyllinisch oder salomonisch?—Das Urteil des Bundesverfassungsgerichts zur Parlamentsauflösung' (1983) 36 *NJW* 1529.

[67] Günther Schultz, 'Das Aschermittwoch-Urteil' (1983) 37 *Monatsschrift für Deutsches Recht* 365.

[68] 'Die Dämme nicht brechen lassen' (n 28) 17. [69] 114 BVerfGE 121 (2005).

[70] Perry Anderson, 'After the Event' (2012) 78 *New Left Review* 53. Anderson was responding to a critic who had praised the Court extravagantly as 'one of the last remaining institutions where true judicial thought survives'. Alain Supiot, 'Under Eastern Eyes' (2012) 78 *New Left Review* 32.

B. Privacy, personality, and personal data protection

With the Court's imprimatur, the elections set for 6 March 1983 went forward as planned. As noted earlier, many observers saw the elections, and the unusual procedure that prompted them, as part of a broader plebiscitary shift in West German politics. Many pundits presented the vote itself as a plebiscite, though they disagreed as to its subject. For some it was a referendum on the NATO double-track decision—the 'Missiles Election', they called it. Certainly, the deployment of Pershing warheads in West Germany (in response to the Soviet Union's refusal to eliminate its own middle-range missiles stationed in the East) was a major theme of the election. More prominent still were questions of unemployment, pensions, and government debt.[71] From the perspective of these more pressing themes, the elections were indeed a plebiscite on the man who called for them. Helmut Kohl had become chancellor in a moment of crisis; his announced programme was to resolve the crisis; and in the March elections, he sought a popular mandate to do just that.

The results allowed Kohl to claim such a mandate. The Union parties won 48.8 per cent of the vote, an improvement of 4.3 per cent over 1980 and the parties' best tally since their epochal triumph of 1957. The SPD's fortunes were as grim as the Union's were glowing. The SPD lost 4.7 per cent vis-à-vis 1980, dropping to 38.2 per cent—the party's worst showing since 1961. The FDP suffered too, as everyone predicted it would, losing 3.6 per cent of the vote and polling its worst result (7 per cent) since before 1969. Still, with nearly 56 per cent of the combined vote, the new 'Christian–Liberal' coalition had a solid majority—stronger, indeed, than the Social–Liberal coalition ever enjoyed. Most encouraging, the CDU made welcome inroads among young voters, while the SPD suffered dispiriting losses among the same group. The drift of young voters away from Social Democracy would continue throughout the decade. Young voters were increasingly polarized between the CDU and the Greens. After winning 1.5 per cent of the vote in their 1980 electoral debut, the Greens now capitalized on pacifist sentiment and discontent with established parties to win 5.6 per cent of the vote—a four-and-a-half point increase that sent the party over the 5 per cent hurdle and into the Bundestag. For the first time in thirty years, there was a fourth party in Parliament. Almost immediately, the Greens made their presence felt, emerging as the lone parliamentary champions of a cause that swept the country's citizenry and astonished its political class.

A year before the elections that put the Greens in Parliament, the Bundestag unanimously passed a law calling for a national census on 27 April 1983. In addition to counting the nation's noses, the proposed census would gather information about citizens' housing circumstances and educational attainments, their professional status and workplace arrangements. The federal government, naturally, would use the data itself; but it would also share data with districts and municipalities to facilitate local administration and help local governments correct their records. It had been thirteen years since the last national census and was, all agreed, high time for another. Back in 1970, the West German citizenry had sequaciously submitted to a census, as it had done on three earlier occasions. The national political class had no reason to suspect that the new census would be any different. When the question came to a vote in the Bundestag, the members yawned in unison, voted yes, and turned the matter over to an army of technocrats, who with enormous labour and wondrous expense performed the preliminary research and prepared the material resources needed to conduct the census the following year.

[71] Wirsching (n 6) 42–46.

And so things continued, bureaucratic and behind-the-scenes, diligent and dull, for many, many months. Then, as if by spontaneous combustion, the citizenry erupted. In the early months of 1983, almost on the eve of the census, the West German people arose and refused to be counted. Fear of the census, *avant la lettre*, went viral. Worries about the power of a well-staffed state to collect, process, electronically save, and endlessly reproduce personal information became epidemic. All sectors of society, all shades of the political spectrum, united in protest. 'My data belong to me!' screamed buttons and bumper stickers, echoing the earlier abortion rights campaign. Popular initiatives sprang up everywhere, urging alarmed citizens to boycott the census or sabotage it with bogus data. The media joined in sounding the toxin. Newspapers and magazines published detailed descriptions of the dystopian possibilities inherent in the brave new world of the technological state. Even the calendar assisted the campaign. No one failed to notice that the census fell on the eve of the 'Orwell Year', 1984.

Needless to say, the popular sensation took the government by surprise. Initially, the coalition's response was dismissive. Friedrich Zimmermann, the interior minister responsible for overseeing the census, sneered that opponents of the census stemmed from society's 'fringe'. They were a misguided 'minority of enemies of the state'.[72] That position, however, grew untenable as prominent scientists and writers, artists and engineers joined the opposition. A growing cadre of lawyers—rarely an anarchic guild—joined the anti-census movement as well. Some of them challenged the census law before the Constitutional Court.

The international press reported rumblings of civil disobedience among the ostensibly docile and orderly citizens of the Federal Republic.[73] As opposition mounted, the government protested that it was not prosecuting the census of its own accord—it was merely the faithful executor of a law passed by its predecessor. But this was a hard sell: the law had passed unanimously; no Liberal or Christian-Democratic voice had been raised to oppose it. Only the Greens were not complicit in the law's passage—a fact out of which the parliamentary newcomers made considerable hay. The Greens paraded as the lone alternative to the Orwellians. One Green mantra proclaimed: 'CDU CSU SPD FDP? Nein Danke!'

The high point of popular opposition to the census coincided with the months immediately following the election in March. Even after the initial surprise receded, Kohl's government never quite took the challenge—including the constitutional challenge—seriously. Zimmermann was serenely holidaying in the Austrian Alps when the Court convened oral arguments on 12 April to consider the petitioners' plea for a preliminary injunction.[74] Zimmermann's undersecretary of state, Horst Waffenschmidt, skipped the hearing as well. The government sent Eckhart Schiffer, a ministerial director with almost no experience in constitutional adjudication, to plead the government's cause.[75] Schiffer was visibly overwhelmed and mortifyingly underprepared. The *Süddeutsche Zeitung* reported that as Schiffer stood at the lectern fielding penetrating questions from the justices, he had 'hardly anything with which to counter'.[76]

On 13 April 1983, in a judgment of less than five printed pages, all of the judges on the First Senate agreed that at least parts of the census must be enjoined—particularly the provisions for delivering data to local administrators.[77] 'If the temporary preliminary injunction is not

[72] These comments were reported endlessly in the press. See, e.g., 'Der Staat darf nicht alles wissen' *Die Zeit* (23 December 1983) 7.

[73] See 'A Census as Political Challenge' *Washington Post* (9 April 1983) A18. The *International Herald Tribune* report was noted in 'Ohne Drohgebärde, ohne Angst' *Der Spiegel* (18 April 1983).

[74] See 'Ohne Drohgebärde, ohne Angst' *Der Spiegel* (18 April 1983) 17.

[75] ibid.

[76] See Helmut Kerscher, 'Aufgeschoben ist nicht aufgehoben' *Süddeutsche Zeitung* (14 April 1983) 4.

[77] 64 BVerfGE 67 (1983).

issued', the Senate wrote, 'but the complaints later prove well-founded, the execution of the law will have injured all citizens required to participate'—everyone in the country, that is—'in their fundamental rights'.[78] Five of the eight judges thought the census as a whole must be enjoined. The risks posed by the modern capacity to store and reproduce data electronically were too great for the data to be gathered before the constitutionality of its collection had been formally assessed.[79] The census would have to await the Senate's judgment on the merits.

Rumour had it that the three judges favouring only a partial injunction were Werner Böhmer, Hans Joachim Faller, and Dietrich Katzenstein.[80] This meant, rumours continued, that the decisive vote against the full census had been cast by Ernst Benda, the chief justice and ostensibly one of the First Senate's conservatives.[81] Whether or not Benda cast the deciding vote, the media made him the hero of the day. In its 18 April 1983 issue, *Der Spiegel* ran the headline, 'Bonn Counted Out', over a picture of Benda in his scarlet robes, arms folded and unsmiling, powerful and implacable. Behind Benda was the text of the census law, unceremoniously shredded in two. Benda, *Der Spiegel* reported, who since joining the Court had drifted steadily from the dogmas of his party, now had no hope of becoming federal president or even mayor of Berlin. But as chief justice, Benda seemed determined to go out in style.[82] True, the Court made clear that its preliminary injunction would not in any way prejudice its judgment on the merits. But close observers knew that the Court had never enjoined a law that it did not later, at least in part, annul.[83]

The politicians who passed the law fell over themselves in reciprocal attributions of blame. All parties insisted that the justices had slapped the face of somebody else. Zimmermann, for weeks vocally determined to hold the census on schedule, had the temerity to crow that 'the old Schmidt government' had once more failed in Karlsruhe.[84] Otto Wiesheu, general secretary of the CSU—a party whose chairman, Franz Josef Strauβ, had advocated postponing the census during the recent elections—called the judgment 'a resounding slap' in the face of Hans-Jochen Vogel, the opposition leader who, as justice minister, had helped craft the law.[85] Holger Börner, the SPD premier of Hesse whose 1978 philippic against the Court we encountered in the last chapter, countered that the slap was really directed at Zimmermann.[86] Only the Greens could cackle with a clear conscience. That party's national board reported a 'resounding slap in the face of data fetischists' of all partisan stripes.[87]

Some critics whispered that the Court's own conscience was not clear—that the judgment was a conspicuous display of independence designed to counter perceptions that the Court had capitulated to political pressure in the *Bundestag Dissolution* case.[88] Writing for the *Frankfurter Allgemeine Zeitung*, Friedrich Karl Fromme suggested that the First Senate had been seduced by a popular sensation. In an editorial entitled 'Defeat of the State', Fromme fretted that the Court had 'burdened itself with a matter whose constitutional meaning is supremely dubious'. The opponents of the census had misused the Constitutional Court, and the Court had 'let itself be misused'.[89] The general public, on the other hand, was jubilant (barring a few boycott leaders who worried that the decision would still the wind in the movement's sails). It was the most spectacular preliminary injunction the Court had ever issued.

[78] ibid 70. [79] ibid 70–72.

[80] See 'Die Volkszählung findet nicht statt. Karlsruhe verordnet eine Verschiebung' *FAZ* (14 April 1983).

[81] 'Ohne Drohgebärde, ohne Angst' (n 74) 17. [82] ibid.

[83] ibid. See also 'Die Volkszählung findet nicht statt. Karlsruhe verordnet eine Verschiebung' (n 80).

[84] 'Die Bonner Parteien überrascht' *FAZ* (14 April 1983) 2.

[85] In 'Ohne Drohgebärde, ohne Angst' (n 74) 17. [86] ibid. [87] ibid. [88] ibid.

[89] Friedrich Karl Fromme, 'Niederlage des Staates' *FAZ* (14 April 1983) 1.

The judgment on the merits was no less dramatic. On 15 December 1983—nearly eight months after the date on which the census was originally supposed to take place and now on the very threshold of the 'Orwell Year'—a unanimous Senate quashed the law's most controversial sections.

The explanation was as striking as the outcome. To begin with, the judgment featured what one critic called a 'preamble' of 'supremely questionable' legal relevance.[90] The social message, however, was clear, and in many quarters celebrated.[91] Against Zimmermann's assertion that opposition to the census was an anarchic fringe movement, the Senate noted that the proposals had spread anxiety 'even among those sectors of the population which as loyal citizens respect the right and duty of the state to acquire the information necessary for rational and well-planned state action'.[92] The public's fears, a sympathetic Senate added, were entirely understandable. 'The possibilities of modern data processing are by now transparent largely to experts only', the justices wrote, 'and can engender in the citizen fear of an uncontrollable personal profile, even when the legislator asks only for such data as are necessary and proper'.[93] The enormous anxiety provoked by the proposed census, the 'preamble' concluded, might have stemmed from the fact that even experts—some of whom played a prominent role in the public debate and in the Court's oral arguments—were convinced that the proposal paid insufficient attention to constitutional concerns.[94] As soon became clear, the justices shared this conviction.

The Senate dismissed some parts of the complaint out of hand. That the census required information about citizens' religious affiliation did not infringe their religious freedom, nor did requiring information about their housing arrangements offend the inviolability of the home.[95] And it went almost without saying that requiring citizens to complete census forms did no violence to the freedom of speech.[96]

The core of the complainants' case, however, hung on Article 2 GG's personality guarantees in connection with Article 1's shield for human dignity. Here the decision broke new ground. The personality right, the justices held, must be read in light of modern developments—including, not least, modern technological developments. The personality right encompassed an individual's capacity, consistent with the principle of self-determination, to decide when and how her personal data would be revealed. In light of current and future conditions of automatic data processing, this capacity deserved 'a special measure of protection'.[97] Thus was born, or at any rate christened, the constitutional 'right to informational self-determination'.[98] It was an inelegant moniker for a necessary freedom.[99] Any limitations on this right, the Senate stressed, must accord with the command of legal clarity and the principle of proportionality. Ordinary citizens must be able to recognize the fact, appreciate the extent, and understand the purpose of the intrusion.[100]

Crucial provisions of the census law didn't meet this standard. At the core of their analysis, the justices distinguished between data linked to persons (*personbezogen*) and purely statistical data (*statistikbezogen*). The latter were less constitutionally problematic than the former. The right to informational self-determination required government data collectors to 'anonymize' data as soon as possible. The census law did so only inadequately. In Germany, individuals moving into a community are required to register their domicile with the local authorities (recall the opening scene of Kafka's *The Castle*). The 1983 census

[90] Ulrich Mükkenberger, 'Datenschutz als Verfassungsgebot. Das Volkszählungsurteil des Bundesverfassungsgerichts' (1984) 17 *Kritische Justiz* 1, 1 fn 5.
[91] See 'Schwere Schlappe' *Der Spiegel* (19 December 1983) 19.
[92] 65 BVerfGE 1, 3 (1983).
[93] ibid 4.
[94] ibid.
[95] ibid 39–40.
[96] ibid 40.
[97] ibid 41–42.
[98] ibid 44.
[99] See 'Schwere Schlappe' *Der Spiegel* (19 December 1983) 19.
[100] 65 BVerfGE 1 (n 92) 44.

would have made housing information—with names—available to local administrators for the purpose of correcting domicile registration books. In the eyes of the First Senate, this was an impermissible mixing of personal and statistical data.[101] The justices also condemned other provisions that would have allowed the central government to share census data with local or state administrators. These provisions were offensive because the information, in some cases, would be sent with names attached[102] and because, in other cases, the concrete purpose of the transfer was insufficiently clear.[103] Some of the stated purposes, the judges wrote, were 'so imprecise that they could be adduced to cover the most disparate of activities'.[104] Thus, in few words, the Senate summarized the essence of citizen fears about the census. And thus, in a soft-spoken judgment, the judges quashed the law's most alarming provisions.

'*Das war es*', said Chief Justice Benda at the close of his reading—'That's all'. '*Auf wiedersehen*', he added, but 'no more for me in this hall'.[105] It was the last judgment announced by the Court's fourth president. Benda's twelve-year tenure as chief justice had been tumultuous. His Court was often divided internally and embattled politically. But here, for Benda's swan song, the First Senate's judgment was unanimous.

Rarely, if ever, has a judgment of the Constitutional Court been so wildly, raucously popular. Within sixteen months, a reporter for the *Frankfurter Rundschau* referred to 'all the reports about and encomiums to the Karlsruhe judgment, which in the meantime already fill small libraries'.[106] The panegyric, of course, was not universal. The *Frankfurter Allgemeine Zeitung* played down the judgment's significance, noting that the Court had upheld the essence of the census questionnaire. The paper asked whether the furor surrounding the proceeding hadn't been 'much ado about nothing'.[107] Friedrich Karl Fromme groused in a separate review that the First Senate had 'invented' a 'new fundamental right'[108]—a much-repeated formulation describing what an alarmed law professor called 'a crisis signal for our constitutional consciousness'.[109]

The Left–Liberal weeklies provided a better barometer of the popular mood. *Der Spiegel* chortled that the dominant parties, collectively, had received a 'severe slap'.[110] *Die Zeit* cheered that 'the state must not know everything'.[111] The judgment would join 'a long line of landmark decisions that could stand under the motto: 'In doubt, for freedom!''[112] Reinhold Baumann, the federal ombudsman for data protection, called the judgment 'the most important event in the history of data protection'.[113] This might not be quite the exaggeration it seems to be at first blush. The judgment, and the popular uprising that led to it, helped launch the European data protection movement. It helped shape a consciousness and an ethic that, for instance, continue to drive divergent American and German attitudes towards data protection three decades later. (It contains, for example, the deep seeds of recent rows over NSA spying.)

The decision also marked an axial moment in the history of the Constitutional Court. It was an aggressive vindication—and initial recognition—of a fundamental right. It cast the Court in its idealized role as guardian of the people's constitutional protections and it left

[101] ibid 64. [102] ibid 68. [103] ibid 66. [104] ibid 68.
[105] 'Volkszählungen generell zulässig Schutz gegen unbegrenzte Daten-Weitergabe', *FAZ* (16 December 1983) 1.
[106] H.P. Bull, *Frankfurter Rundschau* (9 April 1985).
[107] 'Ein neues Grundrecht' *FAZ* (16 December 1983) 1.
[108] Friedrich Karl Fromme, 'Ein neues Grundrecht ist erfunden' *FAZ* (17 December 1983) 12.
[109] Peter Krause, 'Das Recht auf informationelle Selbstbestimmung—BVerfGE 65, 1' (1984) 24 *Juristische Schulung* 268.
[110] 'Schwere Schlappe' *Der Spiegel* (19 December 1983) 19.
[111] 'Der Staat darf nicht alles wissen' *Die Zeit* (23 December 1983) 7. [112] ibid.
[113] In 'Schwere Schlappe' *Der Spiegel* (19 December 1983) 19.

leaders of the major political parties looking out of touch. It was a triumphant coda to the era of confrontation, a transitional prelude to a period of continuity, stability, and unprecedented public regard.

C. Personnel changes 1983

Ernst Benda was not the only justice to leave the Court in the days after the *Census* judgment. The terms of Benda's First Senate colleagues Werner Böhmer and Joachim Faller had also ended, as had those of Rudi Wand and Joachim Rottmann of the Second Senate. The entering class of 1971, much criticized for its inclusion of a member of Parliament from each major party, was now gone. The dozen years in which Benda presided over the First Senate, and in which Rottmann and Martin Hirsch[114] sat on the Second, were the most eventful since the Court's inaugural crises.

In June 1981, the mercurial Hirsch—who for months leading up to his retirement had made headlines with extrajudicial pronouncements that led many to question his understanding of the judicial role—had been replaced by Ernst Gottfried Mahrenholz, a former student of Gerhard Leibholz with long experience as a lawyer, administrator, and legislator in Lower Saxony.[115] On 20 December 1983, five new justices joined the Court, and the Second Senate's Wolfgang Zeidler—whose appointment as vice president in 1975 so roused the ire of Hirsch—became the first Social Democrat to serve as chief justice.

The new quintet mixed political experience and academic star-power, though there was more of the latter than the former. The two 'politicians' in the group—Roman Herzog, the new vice president, and Hans Hugo Klein—had previously been professors of public law: Herzog before serving as Baden-Württemberg's educational (1978–80) and interior (1980–83) minister, Klein before becoming a CDU member of the Bundestag in 1972. Herzog and Klein were both nominated by the CDU; they replaced, respectively, Benda and Wand. The CDU also nominated Franz Niedermeier, chairman of the Federal Administrative Court, to replace Werner Böhmer, the Court's property law expert and another former administrative judge. The FDP nominated Johann Friedrich Henschel, a practising lawyer recently appointed to the Federal Court of Justice, to replace Hans Joachim Faller. The SPD, having secured the chief justiceship for the remaining four years of Zeidler's term and having appointed Mahrenzholz in 1981 and Hermann Heuβner in 1979, nominated just one justice in 1983: the prolific public law professor, Ernst-Wolfgang Böckenförde.

Some in the SPD had hoped to tap Böckenförde for a long time. In November 1975, Martin Hirsch had sharply rebuked the party leadership for not nominating 'the comrade [*Genosse*] Prof. Böckenförde' and thus squandering 'the long-awaited opportunity to pit against Prof. Geiger an outstanding scholar of our stripe'—an omission that could 'not be corrected in the foreseeable future'.[116] In fact, in 1975, the SPD had proposed Böckenförde

[114] In Hirsch's case, it was ten years.

[115] For criticism of Hirsch, see 'Ein Parteiamt für Verfassungsrichter Hirsch' *FAZ* (6 March 1980) 5; Hanno Kühnert, 'Hirsch: Ein Richter muβ seine Stimme erheben' *Frankfurter Rundschau* (4 April 1981) 4; 'CSU schilt Martin Hirsch' *Frankfurter Rundschau* (25 April 1981) 1; 'Ein Verfassungsrichter hat die Pflicht, seine Stimme zu erheben' *Frankfurter Rundschau* (25 April 1981) 14; 'Wieder Hirsch' *Die Welt* (25 April 1981); Sepp Beranek, 'Was Martin Hirsch vor 21 Jahre sagte' *Süddeutsche Zeitung* (16 May 1981); D Haucke, 'Politisierende Richter' *Die Welt* (16 May 1981); 'Empörung in Karlsruhe über Richter Hirsch' *Die Welt* (19 May 1981); Rainer Klose, 'BGH weist Verfassungsrichter zurecht' *Süddeutsche Zeitung* (19 May 1981); Robert Leicht, 'Richterliche Selbstdarstellung', *Süddeutsche Zeitung* (19 May 1981); 'Bundesanwälte: Hirsch beleidigt Richter Kuhn', *Die Welt* (21 May 1981); Hans Schueler, 'Knast für Schmierer?' *Die Zeit* (22 May 1981) 6; Olof S Masch, 'Ein gewisser Hirsch' *Der Spiegel* (15 June 1981) 10; 'Ein Vollblutpolitiker' *FAZ* (27 June 1981) 10.

[116] Hirsch to Willy Brandt (4 November 1975), Friederich-Ebert-Stiftung, Bonn (Nachlass Martin Hirsch, 1/MHAC000137).

as a 'neutral', but had not won the agreement of the other parties.[117] His name was aired again in 1979 and 1981.[118] He finally joined the Court at the end of 1983—six years after Geiger's retirement, two years after Hirsch's.

Both Böckenförde and Klein had striking academic pedigrees. Klein, the Court's first Karlsruhe native, had written both of his doctoral theses (dissertation and *Habilitation*) under Ernst Forsthoff, a postwar pillar of the Schmitt school. Böckenförde had deep personal ties to Schmitt himself. After 1958 Böckenförde, still in his late twenties, became Schmitt's most important editor.[119] In February 1959, Schmitt gave his famous 'Tyranny of Values' lecture at a conference that Böckenförde had persuaded him to attend.[120] One student of Schmitt's postwar influence has called Böckenförde 'probably Schmitt's most creative postwar student'.[121] Like others of Schmitt's pupils, Böckenförde criticized the tendency of state theorists in the Federal Republic to treat the constitution itself as sovereign. Böckenförde, by contrast, 'emphasized that the constitution had to be preceded by—and remain based on—a state'.[122] Moreover, Böckenförde 'saw the people, rather than the Constitutional Court and its dedication to values, as the ultimate source of legitimacy and as the guardian of the constitution'.[123] Böckenförde was, and remains, most famous for a single sentence: 'the liberal, secular state lives off preconditions which it cannot itself guarantee'.[124]

In 1962, Böckenförde helped found the periodical *Der Staat*, which was designed at least in part to serve as a Schmittian counterweight to the Smendian *Archive of Public Law*, edited at the time by Justice Gerhard Leibholz. Leibholz, who in a 1976 eulogy for Smend had denounced the adoption of 'tyranny of values' rhetoric by thinkers on the political Left, died the year before Böckenförde joined the Senate on which Leibholz once sat. By then, the labels 'Schmittian' and 'Smendian' had ceased to divide the public law academy. But with the addition of Böckenförde and Klein, the Court acquired a Schmittian link it never had before. It is unfair, however, to saddle students with the burden of their teachers. As a constitutional justice, Böckenförde became something of a liberal lion, writing a series of powerful dissents that made him an unforeseen hero of the left-of-centre press.

Aside from Franz Niedermeier, all of the new justices had been born in the 1930s. At a retirement celebration for outgoing judges, Werner Böhmer, a unit commander during the Second World War, noted that the number of justices who had experienced dictatorship and war firsthand was growing ever smaller.[125] As of 1 January 1984, roughly half the justices belonged to a generation later dubbed the '45ers', a generation whose self-appointed spokesman Jürgen Habermas helped launch an unprecedented public debate about the uniqueness and the legacy of the Nazi past. In a 1986 essay, Habermas claimed as his generation's proudest achievement the anchoring of the Federal Republic within the political traditions of the liberal West. Habermas asserted that the only brand of patriotism available to post-Auschwitz Germans was 'constitutional patriotism'.[126] Habermas gave the

[117] See 'Akenvermerk für Herrn Landesvorsitzenden Dr. Franz Josef Strauβ' (19 June 1975), Hanns-Seidel-Stiftung, Munich (Nachlass Richard Jaeger, B: 1).

[118] See 'Arndt, Böckenförde oder wer sonst tritt die Nachfolge von Verfassungsrichter Hirsch an?' *FAZ* (24 December 1980) 4.

[119] See Reinhard Mehring, *Carl Schmitt: Aufstieg und Verfall* (CH Beck 2009) 531–32. [120] ibid 519.

[121] Jan-Werner Müller, *A Dangerous Mind: Carl Schmitt in Postwar European Thought* (Yale University Press 2003) 72.

[122] ibid 67. [123] ibid 72.

[124] Ernst-Wolfgang Böckenförde, 'Die Entstehung des Staates als Vorgang der Säkularisation' in Ernst-Wolfgang Böckenförde, *Recht, Staat, Freiheit: Studien zur Rechtsphilosophie, Staatstheorie und Verfassungsgeschichte* (Suhrkamp 1991) 112 (italicized in the original).

[125] See Werner Böhmer, 'Abschiedsworte' (13 October 1983), in Friedrich-Ebert-Stiftung, Bonn (Nachlass Walter Seuffert, 1MHAC/000018).

[126] Jürgen Habermas, 'Eine Art Schadensabwicklung. Die apologetischen Tendenzen in der deutschen Geschichtsschreibung' in Rudolf Augstein, Karl Dietrich Bracher, and Martin Broszat (eds), *'Historikerstreit'.*

phrase 'constitutional patriotism' a cosmopolitan, post-national flavour.[127] Seven years earlier, Dolf Sternberger injected the phrase into West German political discourse on a more national note, with specific emphasis on the ideals and institutions of the Basic Law.[128] Early in the chancellorship of Helmut Kohl, some of those institutions—the classically *political* institutions—became embattled as never before. In the public's estimation, the Constitutional Court benefited by contrast.

D. The Flick affair

Part of the backdrop to Habermas's essay, and to the remarkable public debate on recent German history which followed, was Kohl's decision, in May 1985, to accompany US President Ronald Reagan on a visit to the soldiers' cemetery in Bitburg—resting place for soldiers of the Wehrmacht, but also for members of the Waffen-SS. The Bitburg visit unleashed an international storm of disapproval. It was neither the first nor the most controversial of Kohl's early gaffes. Controversy swirled around the chancellor from the moment of his election, and it hovered, like a dark cloud, over his early years in office. Sometimes, as with the visit to Bitburg, he placed his feet on forbidden soil; on other occasions, he placed his foot in his mouth. In January 1984, he spoke (before the Israeli Parliament of all bodies) of the 'mercy of a late birth'. In October 1986, he compared Mikhail Gorbachev's public relations prowess to that of Josef Goebbels.

Worse than the blunders—however callous or colossal—were the scandals. These, surrounding a chancellor who upon taking office promised a 'spiritual-moral turnaround', seemed unforgivable. In its coverage of them, an already hostile punditry proved merciless.

The first scandal involved the premature resignation of a four-star general, Günter Kießling, under pressure from Manfred Wörner (CDU), Kohl's defence minister, and in the face of secret accusations of homosexuality. Wörner's sources were shady. The accusations were later discredited and Kießling was rehabilitated. But Kohl stood by his defence minister all the while. Even after Kießling's exoneration, Kohl rejected Wörner's offer to resign. The second scandal ultimately required Kohl's testimony under oath before a parliamentary investigative committee. Later, when the inaccuracy (or duplicity) of his testimony became clear, Kohl became the first sitting chancellor subjected to criminal prosecution.

This scandal bore the name of Flick. The Flick Corporation was a giant conglomerate with assorted interests. During the 1970s, the company made headlines with its sizeable—and questionable—donations to political parties. The most dubious donations followed the company's 1975 sale of the automotive giant Daimler-Benz for 1.9 billion DM. The sale was taxable in its entirety, but Flick fought doggedly to shield it from taxation. For the most part, the corporation succeeded. West Germany's labyrinthine tax code empowered the federal finance minister to shield from tax liability sums destined for a 'reinvestment' favourable to the national economy. With separate petitions in 1976, 1979, and 1981, Flick secured reinvestment certificates amounting to some 1.45 billion DM. In all, some 1.6 billion DM of the Daimler sale went untaxed. During negotiations that lasted several years, Flick sent enormous sums of cash, most of it sealed in inconspicuous envelopes, to politicians in Bonn. Jürgen Friderichs and Otto Graf Lambsdorff, the FDP economics ministers, and

Die Dokumentation der Kontroverse um die Einzigartigkeit der nationalsozialistischen Judenvernichtung (Piper 1987) 83.

[127] For an admirable study of the concept, see Jan-Werner Müller, *Constitutional Patriotism* (Princeton University Press 2007).

[128] See Dolf Sternberger, *Verfassungspatriotismus* (Insel 1982).

Hans Matthöfer, the SPD minister of finance, were especial beneficiaries. But cash flowed to politicians in all major parties.

It was all, of course, illegal—a violation of tax laws, to be sure, but also of party finance laws, which required that donations in excess of 20,000 DM be made public. The parties, as it happened, had been quietly flouting those laws for years. The Constitutional Court's party-finance edicts, they reasoned, were simply unworkable. Politicians came to regard the systematic evasion of these fiats as a kind of customary right.[129] As the Flick affair unfolded, however, the public didn't quite see it that way. Neither did the Bonn prosecutor's office. In 1981, public prosecutors formed a special commission to investigate the affair. In short order, roughly 1,800 affected individuals were investigated for tax evasion. Most prominent among them were Lambsdorff, the incumbent economics minister, and Friderichs, his predecessor. In December 1982, Bonn prosecutors charged Lambsdorff and Friderichs, along with two former Flick leaders, not merely with tax violations, but with active and passive bribery.[130]

The charges against Lambsdorff swept the Christian–Liberal coalition into the scandal. In May 1983, at the behest of the SPD parliamentary faction, the Bundestag formed its own investigative committee. The committee's purpose was to determine whether the Flick Corporation had influenced past decisions of members of Parliament, cabinet ministers, administrative officers, or other officials in the Federal Republic, and to scrutinize decisions favourable to Flick made by successive economics ministers. To that end, the committee asked the economics and finance ministries for copies of all relevant documents. The ministries complied only partially. The documentary record they submitted contained crater-like gaps—entire pages rendered illegible or blacked out entirely. The ministries explained that the holes in the record covered sensitive tax information, the secrecy of which was guaranteed by law. In response, the SPD and Green parliamentary factions, together with the representatives of these two parties on the investigative committee, asked the Constitutional Court to order the ministries to supply the full record.[131] By withholding the documents, the complainants alleged, the ministries had violated the investigative committee's rights under Article 44 GG, which governed parliamentary investigations.

In a judgment dated 17 July 1984, a unanimous Second Senate agreed.[132] Article 44(1) GG, the Senate noted, empowered the parliamentary investigative committee 'to collect for itself the evidence necessary' to fulfil its purpose. This language allowed the committee itself to decide what evidence was necessary, even when it sought documentary evidence from the federal government.[133] This construction was confirmed by considerations of constitutional structure.[134] The Basic Law's separation of powers envisioned occasional parliamentary review of the executive; the Court must construe the constitution so as to make that review effectual.[135] True, the committee's power to compel delivery of documents was limited by fundamental rights concerns.[136] But in this case, those concerns were comparatively weak.

To begin with, the right of tax confidentiality was a statutory rather than a constitutional right, though it merited some constitutional protection in connection with dignity, personal development, and property.[137] This protection meant that infringements of confidentiality must serve a preponderant general interest and respect the

[129] See Helmut Kohl, *Erinerrungen 1982–1990* (Droemer HC 2005) 298.
[130] See Wirsching (n 6) 66–67.
[131] Committee members from the coalition party declined to join the complaint, betraying, many thought, a lamentable lack of parliamentary self-assurance. See 'Flick: Leere Blätter' *Der Spiegel* (23 July 1984) 21.
[132] 67 BVerfGE 100 (1984). [133] ibid 128. [134] ibid 129. [135] ibid 130. [136] ibid 133.
[137] ibid 142.

proportionality principle.[138] But in this case the countervailing general interest was overwhelming.[139] The investigation involved accusations against political parties, government ministers, officers of government ministries, and members of the Bundestag. Statutory guarantees of tax secrecy made an exception for cases of pressing public interest, such as countering the spread of false statements that might undermine the trustworthiness of public administrators. Certainly such an exception applied to an investigation prompted by widespread doubts about the reliability of the executive branch—doubts which threatened to convulse the nation's 'tax morale'.[140] If executive-branch officers thought some documents too sensitive for delivery, they must explain to the Bundestag investigative committee—in a secret session if necessary—why they thought so. In addition, executive officers invoking confidentiality must provide a comprehensive and detailed report of the kinds of documents and the nature of the information withheld. Perhaps they should even arrange for the investigative committee's chairman to see the documents for himself.[141] In the case before the Court, the relevant ministries had already made the documents in question available to the courts and criminal investigators. In matters of critical public moment, what was safe to share with prosecutors was safe to share with Parliament—'the unmediated representative organ' of the People.[142]

The *Frankfurter Allgemeine Zeitung* reported the judgment as a 'strengthening of Parliament'.[143] *Der Spiegel*, employing its favourite metaphor, called it 'yet another slap' for the cabinet of Helmut Kohl—'a severe political slap' to the federal government for the second time in just a few months.[144] Peter Struck, an SPD parliamentarian, also spoke of 'a mighty political slap in the face' for the Kohl government.[145] Hans Schueler struck a similar note in *Die Zeit*. The Kohl government, he wrote, 'could no longer claim that it fared better before the Federal Constitutional Court than its Social-Liberal predecessors under Willy Brandt and Helmut Schmidt'.[146] Schueler, who nearly despaired of the Court during the 1970s, now sang its praise: 'Despite all partisan efforts to influence the staffing of the two Karlsruhe senates', he wrote, 'there are still independent judges in Karlsruhe, who, regardless of their political inclinations, without consideration of those who placed them in office, take their review assignment vis-à-vis the executive seriously and who don't allow themselves to be misled'.[147]

The Court's display of independence was especially momentous against the backdrop of the scandal that gave rise to it. In the aftermath of the Flick affair, many West Germans concluded that their rulers were for sale. Many saw in the Court's judgment against executive privilege and in favour of oversight and transparency a virtuous contrast with the political branches. Public trust in the Court approached record levels; trust in politicians careened in the opposite direction. This was especially so after the coalition parties launched a ham-fisted effort to clean the slate of scandal and solve the problem of party finance by promulgating a general amnesty. Polls showed that roughly a quarter of those surveyed had a worsened opinion of political parties after the 'amnesty affair' had compounded the offences of the Flick affair.[148] By contrast, a 1984 Allbus[149] survey showed popular trust in the Court to be at an all-time high. Nearly 70 per cent of those

138 ibid 143. 139 ibid 145. 140 ibid 141. 141 ibid 138–39. 142 ibid 145, 141.
143 'Stärkung des Parlaments' *FAZ* (18 July 1984) 2. 144 'Flick: Leere Blätter' (n 131) 21.
145 In ibid. 146 Hans Schueler, 'Schuß vor den Bug' *Die Zeit* (20 July 1984) 1. 147 ibid.
148 Wirsching (n 6) 74.
149 Allgemeine Bevölkerungsumfrage der Sozialwissenschaften—a national institute for social science data production.

surveyed expressed 'rather or very great trust' in the Court, while less than 14 per cent reported little or none.[150] This figure declined gradually over the course of the decade, but it remained above 60 per cent.[151] Through reunification and beyond, public trust in the Court dwarfed trust in other state institutions; only the police and the federal president rivalled the Court in the public's esteem.

The Flick affair ended in February 1987 with the acquittal of Lambsdorff and Friderichs on the bribery charges (they had been convicted earlier for the lesser offence of tax fraud). A year earlier, Koblenz prosecutors dropped charges against Kohl for lying under oath, reasoning that the chancellor might have misunderstood the crucial question he answered falsely. The Kohl government survived the crisis, but the affair had taken its toll and exacted its victims. One victim was Rainer Barzel, president of the Bundestag and Kohl's predecessor as CDU chairman, who resigned in October 1984 amid revelations that he had received large sums from Flick in 1973, ostensibly for consultation work but actually as compensation for surrendering the party chairmanship to Kohl.

E. Peace movement and Pershing missiles

In light of the serial snafus and scandals of its early years, it comes as a surprise that Kohl's chancellorship lasted longer than any other, Adenauer's included, in the history of the Federal Republic. The explanation for Kohl's durability, notes the historian Andreas Wirsching, requires averting one's gaze from the fog of scandal and the smoke of satire and focusing on the substance of Kohl's policy decisions.[152] As was the case with Kohl's forebears—with Adenauer, for instance, and with Willy Brandt—the chancellor's weightiest decisions dealt with foreign affairs.

Never, after taking office in October 1982 and winning the election of March 1983, did Kohl or his cabinet waver in their resolve to resume Schmidt's support for the NATO double-track decision. In the wake of the double-track decision, there followed a series of seemingly annual crises in the chilling of the Cold War: the Soviet invasion of Afghanistan in 1979; the Western boycott of the Moscow Olympics in 1980; General Jaruzelski's declaration of martial law in Poland in response to the Solidarity movement's agitation in 1981. This last event followed by less than two weeks the opening of the disarmament negotiations offered in the double-track decision. Partly as a result, the negotiations led nowhere. By the time Kohl came to power, it was clear that NATO was headed down the 1979 decision's more hawkish track. The deployment of cruise and Pershing missiles in the Federal Republic was only a matter of time.

As of the March 1983 elections, time was running short and popular opposition to the arms buildup was growing strident. More than anything else, it was the peace movement (really an astonishing convergence of diverse movements) that propelled the Greens into Parliament. The movement escalated dramatically after the election. Its adherents came from all sectors of society and bore all brands of ideology. It harboured Christians and communists, atheists and environmentalists, women and workers, students and scientists. To some degree, the peace movement overlapped with the anti-census movement. But the fight against atomic buildup exceeded the cry for data protection. The forebodings of the anti-census movement

[150] In Hans Vorländer and André Brodocz, 'Das Vertrauen in das Bundesverfassungsgericht. Ergebnisse einer repräsentativen Bevölkerungsumfrage' in Hans Vorländer (ed), *Die Deutungsmacht der Verfassungsgerichtsbarkeit* (VS Verlag für Sozialwissenschaften 2006) 259, 264.
[151] ibid. [152] ibid 79.

were dystopian; the nightmares of the peace movement apocalyptic. The former took its allusive inspiration from Orwell; the latter from the book of Joel. The prospect of the Federal Republic becoming a theatre of nuclear war—a war that many worried might be triggered by misinformation or a false alarm—haunted millions. Hundreds of thousands took to the streets in protest, resurrecting the watchwords of a quarter century before. The 'fight against atomic death' resumed.

On 4 April 1983, nearly 500,000 people took part in Easter Day peace marches around the country. On 22 October, nearly a million people joined in coordinated demonstrations, including roughly 200,000 that joined hands in a 'human chain' stretching nearly a hundred kilometres from Stuttgart to Neu-Ulm. Anti-deployment activists called for a plebiscite on the matter, but a binding referendum was constitutionally impossible. Justice Helmut Simon of the Constitutional Court's First Senate, however, proposed a non-binding, consultative plebiscite—a suggestion eagerly embraced by many in the peace movement but a source of discomfiture for Simon's judicial colleagues. Nothing came of the proposal, and the extraordinary public debate stretched on for months. Proponents of deployment accused demonstrators of indulging in a utopian naïveté that wrongly usurped the banner of peace. The desire for peace, proponents insisted, was precisely what made deployment necessary. The lessons of history (most poignantly the lessons of *German* history) cried out that dictators, to be stopped, must be confronted. But history also showed, antagonists retorted, that weapons, once forged, will one day be employed.[153]

Throughout the debates and protests, Kohl's coalition was unbending. The government saw the deployment of mid-range missiles in the Federal Republic as the proper response to Soviet intransigence. Deployment gave the government the chance to confirm the country's loyalty to its most powerful ally, the United States. Kohl's cabinet sustained the NATO powers' decision to move forward with the deployment. On 21 November 1983, the government announced that the deployment would proceed. A Bundestag majority endorsed this declaration the following day. The Greens, however—the most vocal parliamentary champions of the anti-deployment movement—dissented. And they took their dissent to Karlsruhe, complaining that the government's declaration violated the Bundestag's rights under Article 24(1) GG and 59(2) GG to participate in existential questions of foreign policy.

Article 24 GG permitted the transfer of sovereign rights to international entities only 'through statute', and Article 59 required statutory approval for treaties affecting the political relationships of the Federal Republic. The Greens contended that the government's unilateral deployment decree violated both these provisions. At the same time, the Court was inundated with a flood of individual complaints against the declaration. These complainants contended that the deployment violated their rights, under Article 2 GG, to life and bodily integrity, and they asked the Court to issue a preliminary injunction. On 16 December 1983, a day after the First Senate's *Census* judgment, the Second Senate rejected the individual petitions for an injunction.[154] In this case, the Senate explained, the threat to life and limb flowed, not from officers of the Federal Republic, but from a foreign power: the Soviet Union.[155] A year and two days later, the Second Senate rejected the Greens' complaint as well.[156]

It did so in a very long judgment, one of the longest in the Court's history. It was also, many scholars thought, one of the strangest. After a thorough summary of the facts and a

[153] The peace movement has generated an extensive literature. The classic account in English of the double-track debates in West Germany is Jeffrey Herf, *War by Other Means: Soviet Power, West German Resistance, and the Battle of the Euro Missiles* (Free Press 1991).

[154] 66 BVerfGE 39 (1983). [155] ibid 60. [156] 68 BVerfGE 1 (1984).

prolix recitation of the anti-deployment arguments, the Senate began its legal analysis on the judgment's seventy-eighth printed page.[157] The government's decision, the justices concluded, was entirely constitutional. Article 59 did not apply because the challenged measure was not a *treaty* but a unilateral act under international law.[158] In a lonely dissent, Justice Ernst Gottfried Mahrenholz complained that this holding begged the question: the executive cannot skirt parliamentary participation by choice of form. But the majority held that, under the Basic Law's separation of powers, questions of this sort were reserved to the exclusive discretion of the executive.[159]

On the face of things, Article 24 seemed even less likely to apply to the deployment than Article 59. The Kohl government insisted emphatically that the deployment decision did *not* transfer sovereign powers to an international body. The Second Senate held, on the contrary, and to the surprise of many, that it did, but that the preconditions for such a transfer were fulfilled.[160] There was no clear transfer of competences, the Senate conceded, but it was clear that the right to decide when and how to employ the missiles did not rest in German hands.[161] That right belonged to the NATO supreme commander and, more directly, to the president of the United States. Article 24 empowered transfer of sovereignty to international bodies, not to functionaries of foreign powers. In a stretch of the juristic imagination, however, the Senate held that the president's power of command was a function entrusted to him by the NATO alliance, so that, in this case, the president could be considered an arm of NATO. And for purposes of Article 24, NATO was undeniably an international entity.[162]

Why, then, didn't the transfer to NATO require prior parliamentary approval? Actually it did, the judges explained, but the Bundestag had provided such approval when it passed a law approving the Federal Republic's accession to NATO … in 1955.[163] Admittedly, that law said nothing specific about deploying Pershing missiles (which didn't then exist) on German soil. But the Bundestag must have known that NATO membership would lead to the atomic armament of the Federal Republic. The nature of the subject matter prevented greater precision in the enabling statute and required considerable leeway in that act's interpretation.[164] The Senate saw in the thirty-year-old accession law sufficient authorization for the present deployment. For the rest, the Court deferred to executive discretion. Judges had no legal standards to guide them in questions of life and death, peace and security. Such questions must be answered politically.[165]

In dissent, Justice Mahrenholz argued that this last consideration once more begged the question.[166] The crucial question was not whether politicians or judges should make the final decision concerning deployment, but rather *which* politicians must be involved. In Mahrenholz's view, the situation created by the deployment was unprecedented—not only unforeseen but unimaginable from the vantage of 1955. Accession to NATO, Mahrenholz wrote, was not a blank cheque that could be perpetually rewritten by the executive branch. On the Article 59 question, Mahrenholz rebuked his colleagues for concluding summarily that the decision was *not* a treaty without bothering to ask whether it *should* have been a treaty. On the Article 24 question, he chided them for finding a transfer of sovereign rights without taking seriously the consequences of that finding.

Some in the press thought that Mahrenholz had the better argument; others that he reached a happier result. *Der Spiegel* compared his dissent to that of Wiltraut Rupp-von Brünneck's in the 1975 *Abortion* judgment.[167] The *Frankfurter Rundschau* complained that the judgment, for all the Senate's professions of political withdrawal, would still have

[157] ibid 3–63. [158] ibid 80–82. [159] ibid 87–89. [160] ibid 90. [161] ibid 90.
[162] ibid 92–93. [163] ibid 99–101. [164] ibid 99. [165] ibid 97.
[166] See his dissent in ibid 111–32. [167] 'Befugnis zum Einsatz' *Der Spiegel* (24 December 1984) 26.

political consequences.[168] Hans Schueler, by contrast, praised the Senate's wisdom in withholding judgment on a question that judges have no ability to answer. Schueler reminded critics that the decision exemplified the very restraint that some of them had urged on the Court in earlier years.[169] The *Rheinische Post* savoured the irony that the Kohl government had the Greens to thank for the fact that a difficult political year ended with a victory in Karlsruhe.[170] Most observers greeted the outcome as inevitable. The deployment was an accomplished fact. It would have been startling if the Court had unleashed a geopolitical earthquake over a question of procedure.

However inevitable the judgment's outcome, many scholars loathed its logic.[171] The core academic criticism was that the Senate was too lavish with the executive's foreign policy discretion and too cramped with the rights of Parliament.[172] The process of internationalization, some scholars contended, would expand the executive arena inexorably. This trend required a parliamentary counterweight.[173] Most scholars found the Court's Article 24 analysis unconvincing and the notion that the president of the United States was a 'borrowed organ' of NATO appalling.[174] Stand back from the details of the opinion, however, and a deeper continuity emerges. The *Pershing* judgment fudged jurisprudential categories in the name of deference to the ruling coalition's central foreign policy and defence initiative. In this sense, the judgment entered the tradition of the 1973 *Basic Treaty* judgment and the various mid-1950s judgments approving Adenauer's programme of Western integration. To be sure, these judgments included binding interpretations and restrictions on future action. But the Court has never reversed a major foreign policy decision of a sitting government. In this regard, and in this field only the Court has perhaps developed a limited de facto political question doctrine.

F. *Solange II*: fundamental rights and European integration (again)

Deference to the coalition's foreign policy was on display again two years later in the Court's second major decision regarding European integration. The two cardinal foreign policy aims of Kohl's government were to implement the double-track decision and to foster European integration, with special attention to the Franco-German friendship.[175] The Constitutional Court endorsed both these objectives: the former directly in its *Pershing* judgment of December 1984; the latter indirectly in October 1986.

In its 1974 *Solange* judgment, the Second Senate reserved, the right to review the constitutionality of domestic enforcement of secondary European law.[176] The Court would reserve this right, the Senate explained, so long as the European Community lacked a fundamental rights charter that was democratically promulgated and judicially enforced. When the Second Senate revisited the question twelve years later, it could hardly be said that these conditions had been met. But the process of integration had advanced in notable

168 *Frankfurter Rundschau* (17 December 1984).

169 Hans Schueler, 'Der Richter und das unwägbare Risiko' *Die Zeit* (21 December 1984) 2.

170 In 'Den Grünen zu Dank verpflichtet' *FAZ* (20 December 1984) 2.

171 See Ralf-Müller Terpitz, 'Pershing' in Jörg Menzel (ed), *Verfassungsrechtsprechung* (2000) 368, 368–70.

172 See Brun-Otto Bryde, 'Sicherheitspolitik zwischen Regierung und Parlament' (1986) 8 *Juristische Ausbildung* 363 (1986); Bernhard Docke, 'Der Regierung eine Gasse!' (1985) 13 *Demokratie und Recht* 10.

173 Ulrich Fastenrath, 'Auswärtige Gewalt: Zustimmungsbedürftigkeit einseitiger völkerrechtlicher Erklärungen, Übertragungen von Hoheitsrechten (Pershing-II-Stationierung)' (1986) 18 *Juristische Arbeitsblättter* 451.

174 Bryde (n 172) 368; Docke (n 172) 12; Fastenrath (n 173) 454.

175 Manfred Görtemaker, *Geschichte der Bundesrepublik Deutschland* (Oscar Beck 1999) 688.

176 See 37 BVerfGE 271 (1974).

ways. Denmark, Ireland, and the United Kingdom, which joined the EEC in 1973, were increasingly settled within it—albeit, in Britain's case, with jolting bumps along the way. Greece acceded in 1981; Portugal and Spain in 1986. The first direct elections to the European Parliament (such as it was) were held in 1979. In January 1985, the Frenchman Jacques Delors began his momentous presidency of the European Commission. Five months later, the Schengen agreement led to the opening of borders between most member states. Against this backdrop, *Solange*'s sequel read its predecessor loosely. The Second Senate proved more 'integration-friendly' in October 1986 than it had in May 1974.[177]

Once again, the underlying facts were unromantic. *Solange I* was about grain; *Solange II* about tinned mushrooms. A West German importer of this ambrosial article had challenged unsuccessfully the application of a bothersome community regulation in a German administrative court. He then filed a constitutional complaint against the administrative court for declining to refer the matter to Karlsruhe.

The Second Senate's holding of 22 October 1986 had two parts. The first found that the European Court of Justice qualified as a 'lawful judge' for purposes of Article 101(1) GG.[178] The second declared that, in matters of community law, the Constitutional Court generally would not review the ECJ's enforcement of fundamental rights.[179] On the face of things, this latter holding ran directly counter to the tenor of its 1974 forebear. But the underlying principles, at least nominally, remained unchanged.[180] What had changed was the Senate's assessment of how nearly community protections for fundamental rights approximated those of the Basic Law. In the Senate's view, there had flowered since 1974 'a degree of fundamental rights protection, which in its conception, content, and effect is essentially comparable to the fundamental rights standard of the Basic Law'.[181] All major Community organs had committed themselves to honouring fundamental rights as enshrined in the domestic constitutions of member states and in the European Convention on Human Rights. The ECJ itself, moreover, had developed a robust rights jurisprudence anchored in the proportionality principle.[182] The ECJ's case-by-case approach to fundamental rights naturally left gaps, but these would close as the law evolved.[183] The Second Senate concluded with a peroration that paralleled, and flatly reversed, the eponymous passage of the first *Solange* judgment:

> So long as the European Communities, particularly the jurisprudence of the Court of Justice of the Communities, generally provide an effective protection for fundamental rights vis-à-vis the sovereign authority of the Communities, essentially comparable to the fundamental rights protection declared indispensable by the Basic Law, ... the Federal Constitutional Court will no longer exercise its jurisdiction over the applicability of secondary Community law that forms the legal foundation for actions of German courts or administrators ... and will no longer review such law against the standard of the fundamental rights of the Basic Law.[184]

The judgment hereby christened itself. Forever after, it was called *Solange II*. Conspicuously, the Court neither denied nor surrendered its jurisdiction over the constitutional application of community law. It merely declined—indefinitely—to exercise it. Just as conspicuously, the Senate gave no indication that it would change its mind in the foreseeable future. The judgment was a de facto acknowledgement, even in the field of fundamental rights, of the supremacy of European law.

The decision was something of a sensation, drawing front-page coverage in the daily press.[185] In scholarly circles, *Solange II* was nearly as much applauded as *Solange I* had

177 See 73 BVerfGE 339 (1986). 178 ibid 366–67. 179 ibid 378–87. 180 ibid 376–78.
181 ibid 378. 182 ibid 378–81. 183 ibid 383. 184 ibid 387.
185 See 'Karlsruhe läßt Luxemburg den Vortritt' *FAZ* (13 December 1986) 1; 'Karlsruhe gibt Rechte an EG-Gericht ab', *Süddeutsche Zeitung* (13/14 December 1986) 1.

been reviled. 'In an elegant manner', wrote Christoph Vedder, 'the Constitutional Court has succeeded in linking [its judgment], without too great a dogmatic rupture, to *Solange I*, and in this manner made that much-disparaged judgment seem not such an incomprehensible "outlier" on the path of development'.[186] Some scholars conceded that what they really liked about the judgment was its political outcome. Hans Heinrich Rupp noted that the judgment, like its forebear, could be assessed by legal criteria only with great difficulty.[187] 'What legal norm is it really', Rupp wondered, that drove the Second Senate's reasoning?[188] The 1986 judgment of the Senate majority read remarkably like the 1974 dissenting opinion of Justices Rupp, Hirsch, and Wand. It could hardly be denied that a more 'integration-friendly' outlook could have led to the same result twelve years earlier.[189] But if the Court was merely correcting an earlier error, so much the better. Besides, Rupp added, 'so long as [!] in its fundamental rights jurisprudence the Constitutional Court, above all its First Senate, does not find its way back to juristic-argumentative discipline and judicial precision, but loses itself instead in the "distribution of mildness and severity according to insecure criteria", it can mean no great loss for fundamental rights' if the Court delegates some of its prerogatives 'to the ECJ, the organs of the European Convention for Human Rights, and the ordinary German courts'.[190]

II. Fundamental Rights in Shades of Green

If the serial scandals of the established parties ensured that West German politics in the 1980s was rarely edifying, the rise of the Greens ensured that it was never dull. The Greens provided the most surprising and dynamic political innovation of the new decade. Having entered Parliament promising consummate transparency and the abolition of sinecures, the Greens were a self-proclaimed anti-party—nothing if not a spectacular contrast to the parties that had ruled the Federal Republic since its founding. Though the Greens' programme was an omnigatherum of initiatives and reforms, its deepest convictions, by some distance, remained rooted in pacifism and environmental protection. The party emerged during, and did much to foment, an astonishing period of popular protest. When their aims were thwarted through political channels, the Greens and their acolytes often took their cause to Karlsruhe. That they did so often, and that they sometimes prevailed, is a testament to the Court's power as a forum of socio-political integration.[191] Outsiders as well as insiders found vindication as well as disappointment before the Constitutional Court. Even a party suspected of anarchy and subversion willingly furthered the Federal Republic's experiment with constitutional justice.

[186] Christoph Vedder, 'Ein neuer gesetzlicher Richter?' (1987) 40 *NJW* 526, 531.

[187] Hans Heinrich Rupp, 'Zur Prüfungskompetenz des BVerfG über die Verfassungsmäβigkeit von Akten des Europäischen Gemeinschaftsrechts' (1987) 42 *JurizenZeitung* 236, 241.

[188] ibid 242.

[189] See Christoph Vedder, 'Ein neuer gesetzlicher Richter?—Zum Beschluss des BVerfG vom 22.10.1986' (1987) 40 *NJW* 526, 529.

[190] Rupp (n 187) 242 (quoting Justus Wilhelm Hedemann, *Die Flucht in die Generalklauseln: eine Gefahr für Recht und Staat* (Mohr 1933) 75).

[191] On the Court's integrative role, see Gary A. Schaal, *Integration durch Verfassung und Verfassungsrechtsprechung? Über den Zusammenhang von Demokratie, Verfassung, und Integration* (Duncker and Humblot 2000).

A. Brokdorf: pacifism and protest

While the presence of the Greens in the Bundestag gave protest movements an unprecedented institutional platform, the lifeblood of the movements remained, as ever, extra-parliamentary. The period stretching from the late-1970s to the mid-1980s featured popular demonstrations of exceptional magnitude, as we saw in connection with the *Census* and *Pershing* cases. Occasionally, popular demonstrations grew violent, prompting clashes with the authorities. Such clashes raised questions about the scope and limits of the constitutional right of peaceful assembly.

The peace movement's opposition to expanded atomic armament had a symbiotic relationship with the environmental movement's opposition to domestic atomic energy. A physical and symbolic focal point for the latter initiative was the diminutive Schleswig-Holstein town of Brokdorf. In the 1970s, Brokdorf was identified as the future site of a nuclear power plant. In November 1976 and February 1977, protesters flocked to Brokdorf for mass demonstrations against the proposed construction. Between 1976 and 1980, construction of the plant was postponed repeatedly as the plans faced a series of administrative court challenges. At the end of 1980, it was announced that construction would resume. Cries of protest rang out immediately. Sixty or so citizen initiatives and other volunteer groups planned a united demonstration for 28 February 1981. These groups announced the protest, and called on protesters to assemble, in a series of posters, pamphlets, and newspaper ads, some of which were mildly menacing.

At the time of the proposed demonstration, the district of Steinburg, which included Brokdorf, required organizers of large-scale protests to register in advance with the authorities. As of 23 February 1981, no registration for the announced protest had been submitted. In consequence, Steinburg's administrative chief issued a general decree banning, from 27 February to 1 March, the announced demonstration or any other demonstrations on the construction site or in a surrounding region of about 210 km^2. The chief justified the ban by pointing to the omitted registration. But even if it had been registered, he explained, the demonstration would have to be enjoined for fear of 'serious disruptions of order and criminal acts'. On 27 February, an administrative court partially granted the protesters' challenge to this ban, explaining that the area covered was too large. Local authorities immediately appealed this decision to a higher administrative court, and the appeals court, within twenty-four hours, restored the ban. The protest's organizers challenged this judgment in a (literally) midnight appeal to Karlsruhe. The organizers alleged that the ban violated their right to peaceful assembly under Article 8 GG, and their right to free expression under Article 5.

In the meantime, ban or no ban, the protest proceeded as planned. Turnout well exceeded 50,000; vandalism and violence well exceeded zero. There were numerous clashes, in which several persons—protesters as well as police—were injured. But the noisiest fears of a 'civil war' never materialized.[192]

The protesters' constitutional complaint sat in the Court's inbox for four years. Only on 14 May 1985 did a unanimous First Senate reverse the judgment of the higher Administrative Court.[193] The judgment endures as the Court's first extended engagement with Article 8's guarantee of the right to assemble in peaceful protest.

[192] For contemporary accounts, see 'Nach der Massendemonstration in Brokdorf Lob für die Polizei' *FAZ* (2 March 1981) 1; Johann Georg Reißmüller, 'Wie viele Feldschlachten noch?' *FAZ* (2 March 1981) 1; 'Der Marsch der Siebzigtausend' *Die Zeit* (6 March 1981) 3.

[193] 69 BVerfGE 315 (1985).

After a thirty-four-year drought of Article 8 jurisprudence, the justices seemed eager to make up for lost time. The judgment indulged in a long discursus of dazzling dicta on the scope and nature of the right of assembly. One critic complained that the judgment read more like an excerpt from a Basic Law commentary than a judicial decision.[194] Presumably, the justices hoped to give guidance for future disputes. Or perhaps they just couldn't help themselves. The temptation to propound first principles was great; no one knew when the chance would return.

In sonorous echoes of *Lüth*, likely drafted by Helmut Simon, the case's presumptive rapporteur, the Senate extolled the assembly right's centrality within the Basic Law pantheon.[195] The right existed 'above all' for the protection of 'different-minded minorities'.[196] As the particular 'emblem of the freedom, independence, and maturity of the citizen', it enjoyed a 'special rank' within a free political system.[197] Like the freedom of speech exalted in *Lüth*, the freedom of assembly was 'constitutive for a free, democratic state order'.[198] It was the 'most immediate expression of the human personality and one of the noblest human rights of all'.[199]

The right had long roots in the Anglo-American tradition, but it assumed added significance in a modern, technocratic society dominated by powerful interest groups. In such a society, individuals could often exercise influence only collectively.[200] In a system of parliamentary representation with few strands of plebiscitary participation, an uninhibited right of assembly was crucial; it was the citizen's only access to direct democracy. On this point, the Senate cited the academic writings of one of its own, Justice Konrad Hesse. Public gatherings, Hesse had written, 'offer ... the possibility of public influence in the political process, of the development of pluralistic initiatives and alternatives or also of criticism and protest ...; they contain a quantum of original, untamed, unmediated democracy, capable of preserving the political process from ossifying into a business routine'.[201] Voluntary popular gatherings were an antidote against the petrifaction of politics as usual.[202] The right of assembly represented the most direct outlet—and often the only outlet—for the voice of 'We the People'.

These were striking passages, however irritating (because not necessary to the Senate's holding) to academic commentators.[203] Here was an indirect acknowledgement that the constitutional order of the Federal Republic featured 'narrow rights of plebiscitary participation'; few opportunities for the direct engagement of free, independent, mature citizens; and few instances of 'original, untamed, unmediated democracy'.[204] The Court making the acknowledgement had done much to bring about this state of affairs. In a season of widespread disenchantment with the political establishment, the First Senate held forth an ideal of a mature and engaged citizenry revitalizing the country's political life through immediate democratic participation. The right of assembly, the Senate suggested, empowered citizens to dare more democracy. It was Willy Brandt's inaugural manifesto reprised in *obiter dicta*.

Naturally, the right was not unlimited.[205] It secured only 'peaceful' assemblies, 'without weapons'. Public authorities must preserve other interests, such as safety and order. But

194 Wolf-Rüdiger Schenke, 'Anmerkung' (1986) 41 *Juristenzeitung* 35, 35.
195 69 BVerfGE 315 (n 193) 342–48. 196 ibid 343. 197 ibid. 198 ibid 344.
199 ibid. 200 ibid 346. 201 ibid 346–47.
202 See Bruce A. Ackerman, *We the People: Foundations*, vol 1 (Harvard University Press 1991) 31, 111.
203 See, e.g., Volkmar Götz, 'Versammlungsfreiheit und Versammlungsrecht im Brockdorf-Beschluβ des Bundesverfassungsgerichts' (1985) 101 *Deutsches Verwaltungsblatt* 1347; Christoph Gusy, 'Lehrbuch der Versammlungsfreiheit—BVerfGE 69, 315' (1986) 26 *Juristische Schulung* 608; Schenke (n 194).
204 69 BVerfGE 315 (n 193) 347. 205 ibid 348–54.

those interests did not supply a blank cheque. They were subject to the proportionality principle, which must be applied strictly on the assemblers' behalf. In most instances, public authorities should give protesters the benefit of the doubt. A protest should not be banned, for instance, because a fringe minority of protesters threatens violence. Registration requirements were permissible in principle, but failures to register should not trigger automatic bans. Nor should registration requirements suppress spontaneous assemblies. Banning a protest was the administrative *ultima ratio*; it must be employed only when all efforts at cooperation with an event's organizers fail.[206]

By this standard, the Schleswig-Holstein public assembly laws were not, on the face of it, invalid.[207] Whether they were unconstitutionally applied the Senate did not say. The contested ban was improper because the administrative appeals court applied the relevant laws improperly.[208] Further than this the Senate need not inquire. It was an anti-climactic holding after so rhetorically-laden a preface. But the holding in favour of the complainants was consistent with the justices' ringing disquisition on Article 8. In a season of unprecedented public protests driven by disillusion with traditional politics and established politicians, the Court had given the rights of protesters a forceful and unanimous endorsement.

B. 'Atomic energy? Nein, danke'

Five months later the First Senate granted another complaint arising from an officially interdicted protest against atomic energy.[209] This case, however, involved a single demonstrator. Lower Saxony law provided that citizens could be drafted as assessors in local elections. Those refusing to serve without a legally recognized reason were fined. The law further forbade assessors to engage in partisan politicking at the polls. An individual known in the Court records as '*Herr* B.' was assigned to work as a poll-watcher during communal elections held 27 September 1981. Mr B. reported for duty bearing a badge that read, 'Atomic energy? No, thank you'. (*Atomkraft? Nein, danke.*) He was asked to remove the badge but refused. He was then dismissed from the job and fined 100 DM for refusing to serve.

Mr B. complained to the local administrative court that he had not refused to serve, but that his service had been rejected. The administrative court agreed, but the court of appeals in Celle reversed. The appeals court reasoned that refusal to comply with the terms of service was refusal to serve. The appeals court reinstated the fine, and Mr B. filed a constitutional complaint, alleging that the fine violated his free speech rights under Article 5(1) GG, as well as the precept in Article 103(2) that an act may be punished only if its culpability was established by law before the act was committed.

The Constitutional Court agreed, quashing the appeals court decision in a judgment issued 23 October 1985. Laws exacting honorary service from citizens, the First Senate explained, generally had two parts, one dealing with the *whether* of honorary service, another with the *how*.[210] Lower-Saxon law established fines only for the former. One could be fined for refusing to serve, but not for refusing to serve as directed. By analogizing the latter case to the former, the court of appeals had engaged in judicial lawmaking—which Article 103(2) forbade where fines or sanctions were involved.[211] This was unconstitutional in its own right, but it was doubly so in this case because the fine punished free speech. Article 5 GG decreed that the right to free speech could be limited only by a 'general law', and judge-made law did not qualify.[212]

[206] ibid 362. [207] ibid 363. [208] ibid 368–72. [209] 71 BVerfGE 108 (1985).
[210] ibid 117. [211] ibid 118–19. [212] ibid 121.

C. The Greens at bay

Though the *Brokdorf* and *Election Badge* judgments dealt with the rights of protesters rather than the substance of their protest, the Greens and their sympathizers applauded them enthusiastically. The Greens and their fellow travellers had suffered notable disappointments in the *Bundestag Dissolution* and *Pershing Deployment* judgments, but also dramatic victories—in the last two judgments discussed and, more impressively, in the *Flick* and *Census* decisions. Indeed, by mid-decade, no established party had fared nearly so well in Karlsruhe as had the unorthodox Greens. This fact helped confirm the Court's status as an impartial umpire for the rights of outsiders and eccentrics. But the coincidence of the Greens' rise with the historical highpoint of public trust in the Constitutional Court highlights a point of commonality between a Court and a party who outwardly had little in common: distance from the political establishment. The Greens' sneaker-donning, expletive-hurling leader, Joschka Fischer, had very little in common with the staid and sober, drab and dignified Roman Herzog, chairman of the First Senate and, after 1987, chief justice. But both the Greens and the Court were perceived and prized as independent checks on the powers that were. This perception, at a time when those powers exuded an aura of incompotence and scandal, worked to the enormous advantage of both the political *avant-garde* and the guardians of the constitution.

But the tide of the Greens' fortunes in Karlsruhe soon ebbed—indeed, had already begun to ebb. Three weeks before the *Brokdorf* judgment the Greens' sympathizers were appalled and dismayed by a judgment approving a recent reform of the conscientious-objection law—one of the laws the Christian–Liberal coalition passed between assuming office in October 1982 and winning the election of March 1983. The law made the civil service alternative for conscientious objectors longer than the term of compulsory military service by about five months. Under the new law, compulsory service lasted fifteen months, civil alternatives twenty months. The government hoped the law would discourage conscientious objection by weeding out inauthentic claims. But the text of Article 12 GG seemed to stand in its way: 'the duration of the alternative service may not exceed the duration of military service'. To many minds—including those of the SPD state governments and the SPD Bundestag faction that challenged the law before the Constitutional Court—the law was flatly unconstitutional, a glaring violation of an unambiguous constitutional text. But in its judgment of 24 April 1985, the Second Senate held otherwise.[213]

The reasoning with which it did so would redden the cheeks of the proverbial Jesuit. The Senate majority stressed that those engaged in compulsory military service faced the possibility, in theory, of an extended term totalling twenty-four months. The possibility that the *longest* military term might exceed the longest civil term made up for the fact that the *usual* alternative term exceeded the military norm. The Senate found this point especially compelling in light of the indefinite military service required in times of war. What's more, military service was more burdensome than the civil alternative. A day spent toiling on the training ground or idling in the barracks, the Senate suggested, was longer than a day spent serving senior citizens or tending a public park. This reasoning made the term 'duration' either a metaphor or a pun, and the Senate's two dissenters, Justices Böckenförde and Mahrenholz, had trouble remaining patient in their refutation. 'Duration', they propounded, 'is a quantitative concept; it refers to periods of time'.[214] 'In comparing the duration of the regular terms of service', the exasperated dissenters continued, 'comparisons with an unknown are not allowed'.[215]

[213] 69 BVerfGE 1 (1985). [214] ibid 67. [215] ibid 69.

The conservative press praised the Senate's pragmatism and crowed over the unhappy complainants. 'When they governed Bonn', lectured the *Frankfurter Allgemeine Zeitung*, 'the Social Democrats regularly accused the Union, then in opposition, of wishing to win before the Constitutional Court the battles they lost in Parliament. Now the Social Democrats are doing just this'.[216] The Left–Liberal press responded with angrily. '20 = 15' was the caustic headline of a *Spiegel* article that railed against the judgment's 'linguistic and logical absurdity'.[217] Hans Schueler wrote in *Die Zeit* that the judgment was 'wrong' in its outcome and 'embarrassing' in its reasoning—comparable in its ominous casuistry to the *Emergency Laws* judgment of 1970.[218] For these writers at least, the afterglow of the *Census* judgment had dimmed. And for the Greens, who loathed the law in question, the honeymoon in Karlsruhe was over.

Nine months later, another disappointment before the Constitutional Court would affect the Greens directly. In a January 1986 judgment, the Court sided with the established parties against the Greens in an intra-parliamentary dispute. At issue was the Green faction's right to representation on parliamentary committees. Traditionally, the government's proposed budget for the secret services was reviewed by a parliamentary subcommittee before the full Bundestag approved a lump sum for the intelligence agencies as part of the general budget. In 1984 and 1985, however, the Bundestag reversed the routine. Now a special committee would review the details of the intelligence budget *after* the Plenum approved a lump sum. Until the subcommittee approved those details, only 25 per cent of the requested funds could be disbursed. The committee convened under the strictest secrecy—and it was very small. The five-member committee included two members from the Union parties, two Social Democrats, one Free Democrat, and not a single Green.

Many suspected that the committee's small size and unusual procedures were designed to exclude the Greens. They had reason to be suspicious. During the Bundestag debates over the committee's creation, Wolfgang Weng (FDP) distinguished between parliamentary parties who were 'supportive of the state' (*staatstragend*) and those who were not (*nichtstaatstragend*).[219] Weng's clear implication was that the Greens' allegiance to the Basic Law was suspect. At the very least, many in the traditional parties questioned the Greens' commitment to maintaining an intelligence force—or to keeping secrets. The Greens, after all, were self-proclaimed paladins of transparency. Shortly after the March 1983 election, Petra Kelly, a Green Party founder and newly elected member of the Bundestag, declared that she would gladly pass along confidential information that came her way. The publication of official secrets, she added, was 'a fundamental principle of the Greens'.[220] Kelly was swiftly contradicted by other Green leaders, but the impression that the Greens could not be trusted with confidential material endured.

Placing Greens on the intelligence committee struck many parliamentarians as incongruous and self-defeating. During a December 1983 debate on the question, Adolf Roth, a CDU back-bencher, bluntly told Hubert Kleinert, the Greens' candidate for the intelligence subcommittee, that having Kleinert review the intelligence budget would be setting the fox to guard the henhouse, or, in the German idiom, 'making the goat the gardener'.[221] This remark elicited 'demonstrative applause' from the Union and FDP factions. Animated by

216 'In Ordnung gebracht' *FAZ* (25 April 1985) 12.
217 'Zivildienst: 20 = 15' *Der Spiegel* (29 April 1985) 57.
218 Hans Schueler, 'Die Verfassung ein wenig vorbeugen?' *Die Zeit* (3 May 1985) 5.
219 *Verhandlungen des Deutschen Bundestages. Stenographische Berichte*, vol 127 (1984) 3933.
220 In 'Muntere Zeiten' *Der Spiegel* (14 March 1983) 29.
221 *Verhandlungen des Deutschen Bundestages. Stenographische Berichte*, vol 126 (1983) 3246.

such misgivings, the Bundestag implemented a procedure that kept the Greens off the committee and kept the committee's deliberations secret.

The Greens, naturally, cried foul, lodging a multi-pronged complaint against the committee's constitutionality. The complaint argued that, by excluding the Greens from the subcommittee, the Bundestag had violated a host of constitutional provisions: Article 38(1) GG, which makes all members representatives of the entire people; the principle of equal opportunity for political parties; the protection of minorities; principles governing Bundestag review of the governmental budget; and more besides. In its judgment of 14 January 1986, the Second Senate rejected these complaints across the board.

Parliament's power over the purse, the Senate acknowledged, was perhaps its most momentous power, a lynchpin in the constitution's separation of powers. The Basic Law required that the Bundestag exercise its power of budgetary review vigorously, independently, and with access to the most comprehensive information possible.[222] Nothing in the organization and operation of the subcommittee for the intelligence budget offended this principle. Limits on what funds could be disbursed before the committee approved the budget provided an adequate safeguard against executive abuse.[223] That the Bundestag amended the committee's composition and procedure by direct vote, rather than by amending the Parliamentary Rules of Procedure, was unproblematic, especially since such an amendment would not require Bundesrat approval.[224] In any case, the intelligence services made up a relatively slight segment of the budget. This, combined with the highly sensitive nature of the relevant information, justified an unusual and confidential system of review.[225]

The exclusion from the subcommittee of any Green Party representative was troubling, but not fatal. Under Article 38, all factions had, in principle, an equal right to participation in budgetary debates and to representation on budgetary committees.[226] But this did not mean that, without exception, all factions must be represented on all committees. Parliamentary factions, the justices explained, were not homogenous, and the rights of individual members under Article 38 could not be transferred *ipso facto* to partisan factions.[227] More crucially, Parliament's interest in confidentiality was one requiring 'a high degree of caution', and it was 'not the business of the Federal Constitutional Court to second-guess in minute detail how far this caution may go'.[228] It was the reserve of a democratically legitimated Parliament—'in narrowly limited cases'—to depart from its normal procedures.[229] 'For compelling reasons of confidentiality', the justices summarized, 'it can be constitutionally tolerable that individual factions are not included in the composition of a committee'.[230]

But the Senate majority never explained what might constitute such 'compelling reasons', nor did it explain what the exceptional cases were, how they were limited, or when those limits were narrow. Justices Mahrenholz and Böckenförde made precisely this point in pointed dissents. Article 38, Mahrenholz wrote, did not grant parliamentary majorities the right to vote away the right of parliamentary minorities to participate in the parliamentary process.[231] Only the totality of the Bundestag qualified as the people's assembly, and all of the Bundestag's subunits must mirror that totality.[232] Böckenförde insisted that there were no classes among members based on the groups to which they belonged. Any departure from the principle that all groups participate in all decisions required compelling justification and must be a measure of last resort.[233] Rather than enforcing such an exacting bar, the majority had outlined a standard that was 'open' and 'fluid'.[234] The majority's

[222] 71 BVerfGE 354, 356 (1986). [223] ibid 357. [224] ibid 360–61. [225] ibid 362.
[226] ibid 363. [227] ibid 363–64. [228] ibid 364. [229] ibid 366. [230] ibid 325.
[231] ibid 366–67. [232] ibid 369. [233] ibid 381–83. [234] ibid 385.

gesture toward 'narrowly limited cases' was a watchdog with few teeth—and the teeth it had were blunt. In effect, Böckenförde concluded, the Senate majority had given the Bundestag majority licence to exclude minorities it distrusted.[235]

Media assessments of the judgment were predictably mixed, splitting down partisan lines. 'Right in spite of everything', was the verdict of the *Frankfurter Allgemeine Zeitung*.[236] In a number of cases and across a range of fields, *FAZ* noted, the Court had modified the credo '*Fiat justitia, pereat mundus*!'[237] in the name of 'practical reason'. Why should questions of internal and external security be any different? For the Left–Liberal press, by contrast, such 'practical reason' eviscerated the text and flouted the spirit of the constitution. Hans Schueler thought the majority opinion 'ominous', antithetical to the spirit of the constitution like no other judgment since the 1970 *Emergency Laws* decision.[238] 'The Karlsruhe judges degrade the Greens to MPs of the second class', was Schueler's summary headline.[239] *Der Spiegel*, as always, was even more emphatic.[240] The judgment, that paper wrote, was not merely wrong; it was an emblem of institutional decline, at least for one of the Court's chambers. Böckenförde and Mahrenholz had dissented so as 'not to share responsibility for the downfall of half of the Karlsruhe superior jurisdiction'.[241] The judgment was a clear case of 'political opportunism' from a Senate dominated by a five-justice CDU majority rendered impregnable by Wolfgang Zeidler, the chief justice, who, though nominally a Social Democrat, was an 'arch-conservative' on questions of security and defence, law and order. The decision was a sad sequel to the *Alternative Service* judgment of the previous year, a further episode in the 'self-dismantling' of a Second Senate that had 'long … valued the protection of the state more highly than protection from the state'.[242]

There were disappointments in store for the Greens from the First Senate as well. After the *Brokdorf* judgment, Roman Herzog, the First Senate's presiding judge and the Court's vice president, had come under fire from former CDU comrades who mocked his damascene conversion to the rights of protesters.[243] As interior minister for the state of Baden-Württemberg, Herzog had read protesters' rights less liberally, though Herzog himself insisted that the basis for the *Brockdorf* decision could be found in the 1981 edition of his Basic Law commentary.[244] The judgment's sympathizers hailed the future chief justice for bucking partisan prejudices and charting a just and independent course. In November 1986, Herzog's decisive vote in another major judgment involving the rights of protesters cost him the favour of these temporary fans.

The case involved countless complaints from protesters who took part in sit-in demonstrations designed to obstruct the deployment of Pershing missiles. The protesters, all of them non-violent, sat down together in street blockades to deny vehicle access to military facilities. These 'sit blockades' (*Sitzblockaden*) were ultimately dispersed by the police, and the protesters offered no resistance. The protesters universally confessed to having violated traffic and assembly laws, the penalty for which was a small fine. Many of them, however, were convicted of 'coercion through use of force' (*Nötigung mit dem Mittel der Gewalt*), an offence also punishable by fine, but punishable under the criminal code. Many protesters who would willingly have paid a fine objected strenuously to the stigma of a criminal conviction. Scores of them filed constitutional complaints.

235 ibid 386. 236 'Trotz allem Richtig' *FAZ* (15 January 1986) 1.
237 'Do justice, though the earth perish'.
238 Hans Schueler, 'Abgestempelt als Sicherheitsrisiko' *Die Zeit* (17 January 1986) 3. 239 ibid.
240 'Schwarzer Dienstag' *Der Spiegel* (20 January 1986) 42. 241 ibid. 242 ibid.
243 Werner Filmer and Heribert Schwan, *Roman Herzog. Die Biographie* (Bertelsmann 1994) 185.
244 ibid.

The question before the First Senate was whether the classification of a non-violent physical obstruction as the coercive use of force fell afoul of the 'analogy ban' the Court had derived earlier from Article 103(2) GG.[245] That principle forbad judicial expansion of criminal categories by way of analogy to situations beyond those intended by the legislature or anticipated by reasonable citizens. Four justices thought applying the category of coercion to the sit-blockades was an unconstitutional extension of a crime traditionally limited to affirmative physical activity. Four other justices—presumably including Herzog—disagreed, and the tie ran against the complainants.

Contemporaries heard in the posture of the case and the outcome of the judgment an echo of the *Spiegel* judgment twenty years earlier.[246] As with that earlier judgment, those who sympathized with the non-victorious complainants saw grounds for hope in the Court's reasoning. Hans Schueler wrote that, whatever the outcome of the decision, the judgment itself was replete with 'liberal wisdom'[247]—the handiwork, all agreed, of the case's rapporteur, Helmut Simon. The judgment's grounds of decision, Schueler contended, were most persuasive where they seemed most at odds with the outcome. That discrepancy gave some hope for the future and softened or muted most criticism of the judgment. Those hopes proved justified when the First Senate revisited the question, and reversed its position, just nine years later.[248] For now, those disappointed by the decision consoled themselves by reflecting that, even if the First Senate had passed an illiberal judgment, it had still spoken with an eminently liberal voice.

D. The parliamentarian without a party

Not so the Second Senate, critics continued to think. Three years after denying the Greens' petition for a right of participation in parliamentary committees, the Second Senate addressed the complaint of another parliamentary minority—in this case a minority of one. In January 1987, Thomas Wüppesahl, a thirty-one-year-old public administration student with a curly crop and a daunting beard, was elected from the Green Party list to a term in the Bundestag. In May, the ungrateful Wüppesahl left his party; the Greens retaliated, in January 1988, by expelling the bearded renegade from their Bundestag faction. This cost Wüppesahl his post as the Greens' domestic policy speaker, as well as his seats on parliamentary committees. As a now faction-less member, Wüppesahl insisted on his right to introduce bills, to sit on committees, to deliver speeches, and more besides, including an allowance (like that granted to other members) for expert analysis and legal counsel. This last request the parliamentary Plenum denied. When Wüppesahl then rose to propose an amendment to the Bundestag's rules and regulations, his request to speak was denied too. Silenced in Bonn, Wüppesahl sued in Karlsruhe.

He contended that several features of the parliamentary rules and regulations—proportioning seats on parliamentary committees and on the Bundestag Council of Elders (*Ältestenrat*) on the basis of faction strength; granting factions, and factions alone, the right to staff committees and to muster investigations—were unconstitutional. Wüppesahl also complained that the Bundestag had violated his personal rights as a member by denying him a subsidiary allowance and ousting him from the committees on which he sat. The textual anchor of the complaint was Article 38(1) GG, which proclaimed

[245] 'An act can be punished criminally only when its criminality was established before the act was committed'.
[246] See Hans Schueler, 'Blockade im Senat' *Die Zeit* (14 November 1986) 6. [247] ibid.
[248] 92 BVerfGE 1 (1995).

parliamentarians to be 'representatives of the entire people', answerable 'to their conscience alone' for the discharge of their office. It was his right and duty, Wüppesahl contended, to represent the entire nation. Current parliamentary procedure and practice denied him that right.

The Second Senate agreed only in minor part.[249] True, the Bundestag could fulfil its role as unmediated representative of the people only through the participation of all members collectively. Each individual member had, therefore, a right to participate in parliamentary 'will-formation' (*Willensbildung*).[250] This included the right to speak, to vote, to pose questions and present information, to press initiatives and to join factions.[251] But the Bundestag Plenum had rights of its own, specifically the right autonomously and independently to regulate its internal procedures. The Plenum could determine collectively how members exercised their rights individually. This meant, where necessary, that the Plenum could limit individual rights, though it must never abolish them.[252] Applying this principle to the present case, the Senate majority afforded Wüppesahl the right to membership on committees, but only in a limited sense. He could sit on the committee, make speeches, and raise motions; but he could not vote.[253] He must be allowed to be present and to ply the arts of persuasion, but the constitution required no more than this. On these terms, the Senate ordered Wüppesahl's reinstatement—but as a non-voting member—to the interior committee.

It was a very modest victory. And it was Wüppesahl's only victory. The Senate rejected his remaining arguments in their entirety. The staffing of the council of elders and the committees of inquiry through factional nomination raised no constitutional difficulties.[254] Indeed, the justices reflected, including faction-less members on inquiry committees would expand disproportionately their influence within those committees.[255] Providing additional allowances only to the factions was unproblematic as well. Additional funding for factions was designed to help them coordinate the parliamentary division of labour. Independent members took no part in this coordinating effort. What other expenses a party-less member might incur to acquire information or secure expert counsel could be reimbursed independently through the Bundestag's administrative procedures.[256]

The judgment prompted two dissenting opinions, one that thought the majority's conception of a member's mandate too lavish; another that thought it too spare. Justice Konrad Kruis, who had joined the Court in 1987, argued that a faction-less member not only had no right to vote on committees; he had no right to belong to them.[257] The Bundestag's division into factions, Kruis contended, led naturally to the conclusion that membership on parliamentary committees was to be had only through the factions' mediation. A member's right to serve on committees was therefore mediated, and circumscribed, by his right to join a faction.[258] In contrast, Justice Ernst Gottfried Mahrenholz—now the Court's vice president as well as its most frequent and formidable dissenter—thundered that the right to sit on a committee and to speak during its deliberations was blighted and hollow without the right to *vote* on its measures.[259] The unity of debating and voting, Mahrenholz noted, was acknowledged as early as Hegel and affirmed as recently as the latest Basic Law commentaries.[260] The majority's division of the two was unpalatable in theory and unconstitutional in practice—a clear violation of the presumptive equality among members of Parliament and, by extension, among members of parliamentary committees.

249 80 BVerfGE 188 (1989). 250 ibid 217. 251 ibid 218. 252 ibid 219.
253 ibid 221–26. 254 ibid 227–30. 255 ibid 230. 256 ibid 232. 257 ibid 241–44.
258 ibid 242–43. 259 ibid 235–40. 260 ibid 237.

The irony that the Court reached an apex of popular esteem in the 1980s by emerging as a restraint on and a contrast to the political parties whose role in the Federal Republic the Court's own party-state doctrine had exalted was noted earlier. It is an additional irony that Thomas Wüppesahl and his former party fellows suffered the defeats just recounted largely due to the continued influence of the party-state doctrine in weakening Article 38's protections for individual parliamentarians. The Court might indeed serve as a check on the established parties. Within Parliament, however, those parties remained supreme.

III. The Largest Turnover in the Court's History

The Second Senate that reviewed the claims of Thomas Wüppesahl in 1989 looked very different from the Senate that reviewed the claims of his former colleagues in 1986. The First Senate, too, had undergone major changes. In the two years following the *Intelligence Budget* decision, both Senates replaced half their members. Even to those organs, like *Der Spiegel*, that found the Second Senate highly uncongenial, this was not necessarily a cheering prospect. *Der Spiegel*'s Rolf Lamprecht, among others, bemoaned the SPD's track record of nominating judges from the party's Right wing. The Union parties, by comparison, seemed successfully to have nominated a succession of rock-ribbed partisans.[261] But even from this unabashedly partisan perspective, the SPD's nominations in 1986/1987 were not so dire.

In the late 1970s, the leading parties had agreed that within each Senate both SPD and CDU/CSU would nominate three judges from the ranks of their party members, with a fourth nomination reserved for a 'neutral' judge—that is, a judge with no official party affiliation but presumed, in practice, to favour the party that nominated her. The junior coalition partner of the party in power—always, thus far, the FDP—was accorded one of its partner's nominations. Only in 1986, when the SPD nominated Karin Graβhof, a judge on the Federal Court of Justice, to replace Hans-Justus Rinck, whom the CDU nominated to the Second Senate in 1968, was the anticipated balance achieved. After Graβhof's appointment, the balance on the Second Senate shifted from a 5:3 'CDU majority' to an even split of 4:4. At the same time, the balance on the First Senate was four (SPD) to three (CDU) to one (FDP). The remaining seven appointments in 1986/87 preserved these ratios.

Rinck's resignation in October 1986 ended the last of the life-term appointments to the Court. Of the remaining justices, only Helmut Simon had served longer than twelve years, and Simon would depart the following year.[262] Like Werner Böhmer three years earlier, Rinck, a veteran of the Second World War, lamented in his farewell address the departure from the Court of the 'War generation'. He wondered how the Court would be affected by the absence of judges with firsthand experience of the historical events that brought the Court into being.[263] Six months earlier, Franz Niedermaier, seven years Rinck's junior but still (barely) a member the War generation, died of illness at the age of sixty. Niedermaier's seat on the First Senate was filled by Otto Seidl, also a judge on the Federal Court of Justice.

[261] See Rolf Lamprecht, 'Nur noch den Himmel über sich' *Der Spiegel* (17 August 1987) 30.

[262] Simon was originally appointed in 1970 to finish the term of Wolfgang Zeidler, who had left the Court after only three years. In 1975, Simon was granted a twelve-year term in his own right—incidentally at the same time that Zeidler rejoined the Court as vice president.

[263] Hans-Justus Rinck, 'Abschiedsworte' (21 November 1986). Copy in Friedrich-Ebert-Stiftung, Bonn, (Nachlass Martin Hirsch, 1/MHAC000104).

The Court of Justice's overrepresentation among the six seats—three in each Senate—reserved for sitting federal judges was slightly modified the following year. The SPD nominated Thomas Dieterich, a federal labour judge, to replace Helmut Simon on the First Senate, and Everhardt Franβen, a federal administrative judge, to replace Wolfgang Zeidler on the Second. The remaining appointees, with one exception, were professors. The CDU appointed Alfred Söllner, a professor of Roman and labour law in Gieβen, to fill Dietrich Katzenstein's seat on the First Senate, and Paul Kirchhof, a forty-four-year-old tax professor from Heidelberg, to fill Helmut Steinberger's seat on the Second Senate. The SPD nominated Dieter Grimm, a prominent public lawyer from Bielefeld with wide-ranging academic interests, to replace Konrad Hesse, theretofore the First Senate's most formidable academic. The only non-judge and non-academic in the new recruiting class was Konrad Kruis, a government lawyer in Bavaria and the lone CSU nominee.

Among the SPD nominees, Dieterich and Franβen were reportedly associated with the party's Left wing.[264] But Rolf Lamprecht, who fretted in *Der Spiegel* about the party's propensity to appoint judges from its right flank, was unappeased. The party had blown the chance to appoint Erich Steffen, a chairman of the Federal Court of Justice and a committed Leftist.[265] And whatever the partisan virtues of the appointment process' outcome, Lamprecht railed that the process itself was unconstitutional. The flat letter of the Basic Law decreed that justices be appointed 'by the Bundestag and the Bundesrat'. Instead, all major decisions were made behind the scenes by a pair of 'king makers',[266] endorsed by selection committees shrouded in secret deliberations, and finally rubberstamped by the parliamentary Plenum.[267] As a result, the Court—an organ of state 'sometimes mightier than all the others together'—lacked democratic legitimation.

Some leading politicians agreed. Martin Hirsch, the former justice, a man never loath to publish his provocative musings, questioned publicly the constitutionality of the procedure. Herta Däubler-Gmelin, an SPD legal expert in Parliament, called for the selection committee to meet publicly. And the Green faction, irrepressible proponents of transparency, proposed a bill that would require all aspirants for the highest bench to appear in the Bundestag for a plenary debate.

Whatever hope these proposals may initially have had soon vanished. On 7 October 1987, the Bundestag's judicial selection committee approved the several appointments to the Federal Constitutional Court unanimously. The previous day, by a vote of 9:5, the United States Senate Judiciary Committee rejected the nomination of Robert Bork to the United States Supreme Court. While the six appointments to West Germany's highest court had been secured through private deliberations between agents of the major parties, the concomitant Bork nomination became a lightning rod of rumbustious public controversy. It may be that the Federal Republic's political chieftains took a look across the Atlantic at America's brilliant spectacle of demagoguery and partisan recrimination—of interest-group mobilization and a national puppet theatre of televised hearings—and retreated from the prospect of heightened publicity with a shudder of revulsion. In any case, nothing came of the calls for reform.

The suggestion that the Court lacked democratic legitimacy because the recruitment of its members was opaque never gained broad currency. Press coverage of Constitutional Court nominations was rarely extensive or prominent, and the public, perhaps as a result,

264 See 'Drei Professoren, zwei Bundesrichter und ein Beamter' *FAZ* (20 June 1987) 5.
265 Lamprecht (n 261).
266 Chancellery Minister Wolfgang Schäuble for the CDU, Floor Manager Gerhard Jahn for the SPD.
267 Lamprecht (n 261).

showed little interest and took little notice. More generally, though the institutional power of the Court was much discussed, the prominence of individual justices was slight. In 1982, contemporary historians published a two-volume collection of portraits of prominent personalities in the political history of the Federal Republic. The volumes did not feature a single judge.[268] Helmut Simon, a highly regarded justice whose seventeen-year term ended in 1987, was something of a celebrity, but his renown rested more on his role in the peace movement and his position within the Lutheran church than on his service as judge. Even so, Simon was an exception. Most judges joined the Court quietly and kept a low profile throughout their tenure. Few observers saw this as democratically problematic. Indeed, throughout the history of the Bonn Republic, concerns that the Court was overly political were sharper and more frequent than critiques that it was insufficiently democratic. A more transparent appointments process would likely have enhanced the perception, if not the reality, of the Court's politicized composition.

Broader democratic critiques of the Court also faded as the Court's position as a permanent fixture of the political firmament became increasingly secure. Every politically significant judgment that did not defer to the ruling government was followed by complaints about the Court's political interference, but rarely were those complaints framed in democratic terms. The 'countermajoritarian difficulty' that has been the obsessive concern of American discourse about constitutional justice ever since Alexander Bickel coined that unwieldy phrase had no significant parallel in West Germany.[269]

This fact reflects a substantive, value-laden definition of democracy, rather than a flaccid valuation of it. In the midst of the 1987 negotiations over changes in the Court's membership, a leading West German constitutionalist—one not overly sympathetic to the institution of constitutional justice—published a long article in the *Frankfurter Allgemeine Zeitung* entitled, 'Without Fundamental Rights No Democracy'.[270] As the Bonn Republic approached its fortieth birthday, the phrase had become a truism. 'Legitimation of the state through fundamental rights', the author explained, 'is imaginable, and has been historically realized, even without democratic legitimation.... But the reverse proposition—that democratic legitimation of the state is possible without fundamental rights legitimation—is, by contrast, false'.[271] The article never mentioned the Constitutional Court and was certainly not intended as an apologia for constitutional justice. But by 1987, the perception of the Court as the guarantor of democracy's preconditions—and in a thick, value-laden sense, rather than a purely procedural one—was widespread. In this regard, the proposition that democratic legitimacy presupposed fundamental rights protection cast the Court as a kind of ur-democratic actor. It made any challenge to the Court's democratic legitimacy almost a *non sequitur*. The near-universal embrace of this proposition represents the Court's most enduring achievement—the explanation and the essence of its success story.

IV. The German Question on the Eve of Reunification

The collapse of the German Democratic Republic, as historians never tire of reminding us, came as a wondrous surprise. Events we now narrate as precursors to reunification hardly

[268] Walter L Bernecker and Volker Dotterweich, *Persönlichkeit und Politik in der Bundesrepublik Deutschland: Politische Porträts*, 2 vols (Bandenhoeck & Ruprecht 1982).

[269] On the history of the 'countermajoritarian difficulty', see Barry Friedman, 'The History of the Countermajoritarian Difficulty, Part One: The Road to Judicial Supremacy' (1998) 73 *NYU L Rev* 333; 'Part Two: Reconstruction's Political Court' (2002) 91 *Georgetown LJ* 1; 'Part Three: The Lessons of Lochner' (2001) 76 *NYU L Rev* 138; 'Part Four: Law's Politics' (2000) 148 *U Pa L Rev* 971.

[270] Christian Starck, 'Ohne Grundrechte keine Demokratie' *FAZ* (19 June 1987) 12.

[271] ibid.

seemed such to contemporaries. This includes contemporaries at the highest levels of the West German state. Helmut Kohl, who quickly claimed (and probably deserves) the title 'chancellor of reunification', had hardly made reunification a foreign policy priority. Early in his chancellorship, Kohl often invoked the imperative outlined in the Constitutional Court's 1973 *Basic Treaty* judgment that the federal government must at all times pursue the goal of a united German state. In practice, Kohl's *Deutschlandpolitik* was marked by considerable continuity with the policies of his predecessors. In his unbending stance on the deployment of Pershing missiles, for instance, Kohl did much more to strengthen the Federal Republic's military alliance with the West than to prepare the way for strengthened ties with the East. In this regard, Kohl's foreign policy resembled that of Adenauer, whose mantle Kohl claimed.

At the same time, Kohl resumed, and in some respects expanded, the de facto recognition of East Germany initiated under Willy Brandt. On 7 September 1987, Kohl received a state visit from Erich Honecker, the SED general secretary and first GDR head of state to visit the Federal Republic. Three months earlier Ronald Reagan, speaking in front of the Brandenburg Gate, had dramatically called on Mikhail Gorbachev to 'tear down this Wall'. But notwithstanding such brave talk—or histrionic bluster, as critics called it—the division of Germany into two separate states seemed fixed for the foreseeable future.

All the same, the Federal Republic continued to pledge rhetorical allegiance to the ideal of reunification, perhaps most conspicuously through the celebration of a national holiday, the 'day of national unity', on the anniversary of an uprising of East German workers in 1953. Each year on 17 June, the West German citizenry took day trips to the countryside or the sea while the Bundestag convened in solemn session to hear an address from a guest speaker on some aspect of the German Question. On 17 June 1988, the thirty-fifth anniversary of the workers uprising, the invited speaker was Roman Herzog, chief justice of the Federal Constitutional Court.[272]

Herzog began his remarks by reassuring those who feared, and disappointing those who hoped, that he would talk about the *Basic Treaty* judgment. It was the Court's business, Herzog explained, to deliver judgments and give reasons for them, not to analyse or update them after the fact.[273] His theme, instead, was a bigger one: the moral question—Herzog insisted it was not a political question—of what it meant in 1988, if it meant anything at all, to speak of a German *Volk*. He asked, in short, what it meant to be a German.

His answer was neither novel nor satisfactory. But it was instructive as a reflection of the spirit of the times and the mind of the chief justice. Herzog admitted that he could predict neither when nor whether a unitary German state might be restored. But he hoped it would, and he deemed it the moral duty of all West Germans to strive for it.[274] If and when reunification came about it would not be on the basis of common language or culture or any of a host of categories enumerated in public law scholarship, but rather (quoting Renan) on the basis of a 'daily plebiscite', a union of feeling among a people who read the same books, sang the same songs, and entertained the same thoughts of freedom.[275] Herzog exhorted West German youth to promote such unity by visiting the GDR as often as possible and for as long as possible, sharing books and ideas and music with their contemporaries behind the Iron Curtain.[276] 'A nation without the will for a state of its own', Herzog added, 'is an absurdity'.[277] Herzog clearly intended his address as a call to strengthen that will.

[272] For Herzog's remarks, see *Verhandlungen des Deutschen Bundestages. Stenographische Berichte*, vol 145 (1988) 5775–80.

[273] ibid 5775. [274] ibid 5776–77. [275] ibid 5779. [276] ibid. [277] ibid 5778.

It was a telling commentary on the Court's position in state and society that Herzog was asked to give the speech at all. In an era when many commentators observed that the Federal Republic was 'taking leave of its provisionality' (*Abschied vom Provisorium*), the Court had assumed a particular aura of permanence. Herzog's speech was an emblem of the Court's central position in the no-longer-provisional state. The address was reported on the front page of the daily press, in one instance under the headline, 'The German Question Can Be Framed Anew'.[278] As noted, Herzog insisted that the German Question was a moral rather than a political question. But it also had various *legal* ramifications, and in the last days of the Bonn Republic, the Constitutional Court itself was much occupied with the technical legal questions of what it meant to be a German *citizen* and what rights flowed from that status.

The Basic Law spoke of 'German citizenship' (*deutsche Staatsangehörigkeit*) but did not clarify who possessed it or who could confer it. In an October 1987 judgment—issued a month after Honecker's visit to Bonn—the Second Senate addressed whether the Federal Republic must recognize as 'German citizenship' the citizenship granted by the German Democratic Republic.[279] The complainant, son of Italian parents, was born in Sachsen in 1940. After the War, he lived in the GDR and became a citizen of that state. In 1967, the Italian consulate in (West) Berlin granted him Italian citizenship as well, on the basis of which he emigrated to West Germany. In 1969, the complainant submitted a petition to municipal administrators in Cologne, which was rejected on the ground that the complainant was not a German citizen. That rejection was ultimately confirmed by the Federal Administrative Court, which ruled that the complainant was an Italian, not a German, citizen. Shortly thereafter, the complainant challenged this ruling in Karlsruhe.

And there the appeal sat unanswered for many, many years. Only eighteen years after the Cologne authorities rejected the complainant's petition did the Second Senate reverse that rejection. The Senate held that the administrative decision deprived the plaintiff of German citizenship—a deprivation that Article 16(1) GG banned in all but exceptional circumstances. As a precondition to this holding, of course, the Senate ruled that the complainant *was* a German citizen, that is, that the GDR's conferral of citizenship must be recognized in the Federal Republic as well. This finding was consistent, the Senate explained, with the precept of the *Basic Treaty* judgment that the GDR was 'not a foreign country' (*nicht Ausland*) and that its acts must not be regarded as those of a foreign power. Sovereign acts of the German Democratic Republic could be ignored in the Federal Republic only when they ran counter to the West German legal order.

Under international law, the Senate admitted, individual states were forbidden unilaterally to recognize citizens of other states as their own. This principle could be overridden only by special considerations, but the Senate found one in the fact that the GDR's separation from the Federal Republic was not a product of democratic self-determination. Until GDR citizens determined their own fate, there could be no objection under international law if the Federal Republic treated those citizens, in this regard, as its own. Like the *Basic Treaty* judgment, this holding presumed the perpetuation of the German Reich. The Reich endured; as a result, there were still 'German citizens' outside the Federal Republic.

[278] 'Herzog: Deutsche Frage kann sich neu stellen' *Süddeutsche Zeitung* (18 June 1988). See also 'Herzog: Eine Nation ohne den Willen zum eigenen Staat ist ein Unding' *FAZ* (18 June 1988) 1; 'Eine Nation ohne den eigenen Willen zum Staat ist ein Unding' *Die Welt* (18 June 1988); 'Herzog: In Deutschland sind die leisen Töne gefragt' *Frankfurter Rundschau* (18 June 1988).

[279] 77 BVerfGE 137 (1987).

The decision was nearly unanimous. Wolfgang Zeidler, in the twilight of his tenure as chief justice, appended the shortest dissent in the Court's history: 'I do not agree with the decision', he wrote. Posterity can only guess why not. Zeidler's tenure with the Court ended the following month. At sixty-three, Zeidler looked forward to continuing his work of cultivating ties among national constitutional courts and promoting the study of comparative constitutional law. Tragically, however, he died in a hiking accident on New Year's Eve, 1987. The Left–Liberal press, often sharply critical of the Social Democrat Zeidler, now struck a softer town. His judicial philosophy, reported *Die Zeit*, could be summarized as, 'When in doubt, vote for the state', but the state he had in mind was always a democratic state devoted to the rule of law.[280] It was the state that, on the eve of reunification, many West Germans associated with the watchful guardianship of the Federal Constitutional Court.

What came in time to be called the *Teso* judgment (after its Italian-German plaintiff) was one of Zeidler's last. It was a striking judgment, if not altogether surprising. It had, however, a short shelf-life. The question whether 'German' citizenship in the GDR automatically supplied 'German' citizenship in the Federal Republic was, soon overtaken by events.

[280] 'Im Zweifel für den Staat' *Die Zeit* (8 January 1988) 8; see also 'Gestorben: Wolfgang Zeidler' *Der Spiegel* (11 January 1988) 178.

5

Karlsruhe Republic, 1990–2001

The fear that a new, pan-German constitution would be less liberal than the Basic Law is justified not least by reference to the judgments of the Federal Constitutional Court.... Whoever wishes to abandon the Basic Law in favor of a vague new constitution must realize that in that case the entire Karlsruhe jurisprudence, the entire constitutional-judicial armory of our fundamental rights, would be lost. The Karlsruhe judgments would drift as scrap paper into the archives.

—Robert Leicht[1]

The Karlsruhe judges have helped to ease the internal tensions of West Germany.... If the former GDR accedes to the Basic Law, the many volumes of Karlsruhe decisions will remain in force. The question is whether Karlsruhe's jurisprudence can carry on as it has till now.

—Wilhelm Hennis[2]

The decisive question is, to whom should it be granted—in the context of democracy and of the Rechtsstaat, *of political and civil liberty—to shape the legal order.... [D]oes the citizen place his trust in the elected parliamentary legislature, or in the Constitutional Court?*

—Justice Ernst-Wolfgang Böckenförde[3]

Introduction

The 'German Question' is as protean as it has been intractable. It never goes away; it only changes form. The fall of the Berlin Wall in November 1989 placed the German Question once more at the centre of European and global politics.

For most citizens of the Federal Republic, the collapse of the Wall, and of the regime that built and policed it, came as a bolt from the blue. The dizzying events of autumn 1989 followed close on the heels of the Federal Republic's fortieth anniversary celebrations. Broadly speaking, the tone of these celebrations was one of tempered triumphalism. The Bonn Republic was proclaimed Germany's 'first enduring republic', its first successful democracy.[4] Since the mid-1980s, German publicists had questioned the traditional understanding of the Federal Republic as a 'provisional' state.[5] After forty years, the Basic Law seemed provisional no longer. It was a full-fledged constitution, a central element of the Federal Republic's success story. As Hans Peter Ipsen wrote in a fortieth-anniversary summation for the *Jahrbuch des öffentlichen Rechts*, 'The Basic Law, which emerged as a

[1] Robert Leicht, 'Einheit durch Beitritt' in Bernd Guggenberger and Tine Stein (eds), *Die Verfassungsdiskussion im Jahr der deutschen Einheit* (Carl Hanser 1991) 192 (hereinafter *Verfassungsdiskussion*).

[2] Wilhelm Hennis, 'Die Chance einer ganz anderen Republik. Zur Verfassung des zukünftigen Deutschland' in *Verfassungsdiskussion* (n 1) 115.

[3] In Thomas Darnstädt, 'Mir hat keiner was zu sagen' *Der Spiegel* (17 May 1999).

[4] See, e.g., Hans Schueler, 'Die erste Republik von Dauer' *Die Zeit* (19 May 1989).

[5] See, e.g., Josef M. Häusling and others (eds), *Drei Fragen zu Deutschland* (Knaus 1985); Wilfried von Bredow, *Deutschland—ein Provisorium?* (Siedler 1985); see also Andreas Wirsching, *Abschied vom Provisorium, 1982–1990* (Deutsche Verlags-Anstalt 2006).

provisional, "interim" constitution, has proved itself over four decades to be a solid foundation for a stable and efficient order. Under its aegis, peace has been preserved, freedom guaranteed, and prosperity disseminated'.[6]

Given the Basic Law's perceived place in the West German success story, it was no surprise that many commentaries on the Federal Republic's fortieth anniversary invoked the ideal of 'constitutional patriotism'. As noted in the last chapter, the political scientist, Dolf Sternberger, had coined the phrase in a prominent 1979 essay.[7] Seven years later, Jürgen Habermas adduced the term in a dramatic essay that launched an *Historikerstreit* that dominated the West German public sphere during the latter years of the 1980s.[8] Habermas's salvo was aimed at revisionist historians who, in Habermas's view, sought to relativize the crimes of Nazism by placing them in the context of—or even casting them as a response to—other mass crimes of the twentieth century. For Habermas, by contrast, the Holocaust remained unique, historically and even metaphysically. Germany could never be a 'normal' nation with a normal past. Indeed, Germany's redemption lay precisely in its transcendence of nationality. 'The unconditional opening of the Federal Republic to the political culture of the West', Habermas wrote, 'is the great intellectual achievement of the postwar era, about which my generation in particular may be proud'.[9] That achievement would be endangered by any tincture of resurgent German nationalism. 'The only patriotism that will not alienate us from the West', Habermas added, 'is a constitutional patriotism', by which he meant devotion to 'universal constitutional principles'—a devotion that could develop 'in the cultural nation [*Kulturnation*] of the Germans only after—and through—Auschwitz'.[10] The great achievement of the Federal Republic was its rejection of Germany's *Sonderweg* and its embrace of the universal constitutional values that Habermas associated with 'the West'. This, for Habermas, must be the foundation of constitutional patriotism for citizens of the Federal Republic.

Neither Sternberger nor Habermas linked constitutional patriotism to the text of the Basic Law.[11] For Sternberger, neither the concrete provisions of the Basic Law nor the Constitutional Court's juristic definition of the free democratic basic order could stir the hearts and inspire the loyalties of ordinary citizens. Sternberger's conception of constitutional patriotism centred instead on a Hegelian notion of political morality (*Sittlichkeit*) designed to promote citizen devotion to the constitutional state (*Verfassungsstaat*).[12] Habermas, by contrast, emancipated constitutional patriotism from the state altogether, gesturing instead toward post-national universalism. But such theoretical refinements were surely lost on the citizens who were supposed to experience constitutional patriotism. The development of constitutional patriotism in the Federal Republic—always a modest, mannerly phenomenon—was surely, in practice, associated with the Basic Law and abetted by the broader 'success story' (*Erfolgsgeschichte*) of the Federal Republic. No causal connection could ever be established between the enactment of the Basic Law and the impressive stability and astonishing prosperity of the Federal Republic. But the concomitance surely worked to the benefit of the Basic Law—and of its judicial guardians.[13]

[6] Hans Peter Ipsen, '40 Jahre Grundgesetz der Bundesrepublik Deutschland' (1989) 38 *Jahrbuch des öffentlichen Rechts der Gegenwart* 1, 40.

[7] See Dolf Sternberger, 'Verfassungspatriotismus' *FAZ* (23 May 1979).

[8] See Jürgen Habermas, 'Eine Art Schadensabwicklung: Die apologetischen Tendenzen in der deutschen Geschichtsschreibung' *Die Zeit* (11 July 1986).

[9] ibid. [10] ibid.

[11] For an extended treatment of the concept of 'constitutional patriotism', see Jan-Werner Müller, *Constitutional Patriotism* (Princeton University Press 2010).

[12] See the discussion in Jan-Werner Müller, *Another Country: German Intellectuals, Unification and National Identity* (Yale University Press 2000) 93–98.

[13] Naturally, the concept, and reality, of constitutional patriotism had its enemies, especially among those who worried that constitution worship had warped Germans' understanding of the state. See J. Isensee, *Die*

The Federal Republic's fortieth anniversary was marked by affirmations and expressions of constitutional patriotism.[14] Some observers linked the spread of constitutional patriotism in the Federal Republic to the Constitutional Court's careful guardianship of fundamental rights, while at the same time noting regretfully the resultant 'judicialization' (*Juridifizierung*) of German politics.[15] But constitutional patriotism might not have run so deep or so far as some anniversary rhetoric suggested.[16] And the Constitutional Court's immense influence and prestige were more precarious than anyone supposed. The process of reunification would test both constitutional patriotism and faith in the Constitutional Court.

This was true in part because reunification unsettled or abolished many of the preconditions on which the Court's successful rise had long been nourished. Over the life of the Bonn Republic, the Basic Law—and the Court as its official guardian—had become a symbol of West Germany's break with the Nazi past, its reentry into the community of civilized nations, its advantageous contrast with the GDR to the East. With reunification, the process seemed to have come full circle. Many of the political and societal factors that contributed to the Court's early rise—skepticism of majoritarian rule and political parties; respect for authority but mistrust of politicians; the unavailability of traditional sources of societal integration such as history, nation, and culture—faded after reunification or had departed long since.[17] In hindsight, it seemed to some that the years following reunification exposed the fact that the Court had passed its zenith, that its Golden Age was over, that it had begun a slow and lengthy decline.[18]

But the aftermath of reunification also underscored the Court's centrality to the Federal Republic's political order, as well as its crucial integrationist role in united Germany. If reunification marked the beginning of the Court's decline, it also inaugurated a process by which the Berlin Republic, like its Bonn predecessor, would come to bear the impress of the Karlsruhe Court.

The peculiar path to German reunification gave the German Question, in its first post-Cold War iteration, a uniquely constitutional flavour. The question was multi-faceted. Its first prong asked whether there would be a united Germany at all. The collapse of the SED regime did not automatically entail the end of the GDR and the union of East and West. Indeed, many leaders of East Germany's 'peaceful revolution' dreamed, not of unity with the Federal Republic, but of a new constitutional order of their own—one that charted a 'Third Way' between the communist East and the capitalist West. This, of course, was the same dream that inspired republican intellectuals during the early years of the Federal Republic.[19]

The weeks and months following 9 November 1989 witnessed major initiatives of constitutional reform within East Germany. The most dramatic of these was a proposal put forth by a 'round table' of East German notables for a liberal constitution with a peculiarly

Verfassung als Vaterland. Zur Staatsverdrängung der Deutschen in Armin Mohler (ed), *Wirklichkeit als Tabu. Anmerkungen zur Lage* (Oldenbourg 1986) 11.

[14] See Karl Dietrich Bracher, 'Kein Anlaß zu Teuto-Pessimismus' *Süddeutsche Zeitung* (24 May 1989).

[15] See Hans Schueler, 'Die erste Republik von Dauer' *Die Zeit* (19 May 1989).

[16] See Müller (n 11) 98.

[17] These themes have been thoughtfully developed by Dieter Grimm. See Dieter Grimm, 'Hütet die Grundrechte' *Die Zeit* (18 April 1997) 14; Dieter Grimm, 'Integration durch Verfassung' (2004) 32 *Leviathan* 448; see also Christoph Schönberger, 'Anmerkungen zu Karlsruhe' in Matthias Jestaedt and others, *Das Entgrenzte Gericht. Eine Kritische Bilanz nach sechzig Jahren Bundesverfassungsgericht* (Suhrkamp 2011) 43–48.

[18] See Schönberger (n 17) 57.

[19] See A Dirk Moses, *German Intellectuals and the Nazi Past* (Cambridge University Press 2006).

Eastern flavour.[20] But this proposal and others like it were soon overtaken by events. On 18 March 1990, in East Germany's first free elections, the Party of Democratic Socialism (PDS)—a successor to the fallen SED—was roundly defeated. A grand coalition led by the CDU's Eastern partner formed swiftly on a platform of speedy reunification. From that point on, the die was cast. Sooner or later, East and West would become one. Reunification was only a matter of time.

But it was also a question of means. *That* the sections would unite seemed certain; *when* and *how* were hotly debated. For a time, many called for a pan-German confederation, held together by treaty, rather than full political union. This solution was favoured by important figures within the 'round table' movement and was advocated by two constitutional judges, Dieter Grimm and Ernst-Wolfgang Böckenförde—the intellectual titans, respectively, of the Court's First and Second Senates—in an important article in *Der Spiegel*.[21] But the 18 March elections clearly signalled an East German desire for complete merger.

The Basic Law envisioned two routes to such an end—one inscribed in Article 23 GG, the other in Article 146. Article 23 provided that the Basic Law would apply initially in certain named states, but provided also that 'other parts of Germany' could accede to it. The Article 23 route to reunification would allow the several sections of the former GDR to accede to the Basic Law as the newest states of the Federal Republic. The Article 146 route was more sweeping. Article 146 provided that '[t]his Basic Law shall cease to apply on the day on which a Constitution [*Verfassung*] freely adopted by the German people takes effect'. The Article 146 route, then, required adoption of a new, pan-German constitution by the united German people—or at least some popular ratification of the Basic Law.

Public debate over which route to take was impassioned, if not widespread.[22] In the end, the Article 23 option prevailed. Leading politicians saw in it the swiftest, surest path to reunification. Justice Grimm had urged that casting the debate as 'Article 23 versus Article 146' presented a false dichotomy—that it was possible (and desirable) to seize the political moment with a rapid unification under Article 23 followed by a pan-German constitutional assembly convened under Article 146.[23] In the event, however, reunification under Article 23 was not followed by a new constitution under Article 146. Instead, by a formal act of unification on 3 October 1990, five new states in the former East Germany acceded to the Basic Law as the twelfth through sixteenth states of the Federal Republic. The Federal Republic subsumed its Eastern neighbour, and the Basic Law became the fundamental charter for all Germany. United Germany emerged as a constitutional order never ratified—not in 1949 and not in 1990—by the citizens who would be ruled by it.

For many Germans, the process was bitterly disappointing. Many Easterners felt unfairly excluded from the process. Leading Western intellectuals were equally dismayed. Jürgen Habermas had hoped that a popular referendum regarding reunification would foster a renewed republican consciousness in the Federal Republic. Instead, he saw reunification as a process in which Chancellor Kohl 'and his kitchen cabinet' had 'managed to outmaneuver

[20] For summaries of the 'Round Table' proposals, see Wolf Templin, 'Der Verfassungsentwurf des Runden Tisches. Hintergründe und Entstehungsbedingungen' (1990) 41 *Gewerkschaftliche Monatshefte* 370; Ulrich K Preuβ, 'Auf der Suche nach der Zivilgesellschaft. Der Verfassungsentwurf des Runden Tisches' *FAZ* (28 April 1990).

[21] Ernst-Wolfgang Böckenförde and Dieter Grimm, 'Nachdenken über Deutschland' *Der Spiegel* (5 March 1990) 72.

[22] For a sampling, see the essays in *Verfassungsdiskussion* (n 1).

[23] See Dieter Grimm, 'Zwischen Anschluβ und Neukonstitution. Wie aus dem Grundgesetz eine Verfassung für das geeinte Deutschland werden kann' *FAZ* (5 April 1990) 5.

both the deeply divided opposition and the public sphere'.[24] Habermas, the great theorist of the public sphere, was chagrined that reunification was realized outside the public sphere. It was a process powered by 'economic nationalism', one that 'never won any *democratic* dynamic of its own'.[25] Other Left-leaning intellectuals objected that the process was one-sided—that East Germans had been 'treated more as objects of West German policy than as equal partners in a common project'.[26] Many agreed that the process had been too hurried.[27] Some contended that it lacked legitimacy.[28]

Throughout the reunification debates, some of the most powerful voices calling for greater popular involvement and greater sensitivity to Eastern interests belonged to current and former justices of the Constitutional Court. Helmut Simon, for instance, whose long and influential career with the Court had ended in 1987, saw German unification as a once-in-a-century opportunity for passionate popular engagement in constitutional self-determination. Stunningly, instead, the powers that be had embraced the 'abashed solution' [*Verlegenheits lösung*] of the Article 23 route, which was really a cover for 'economic and power-political interests'. Reunification, alas, had become a 'state of exception' that degenerated into an 'hour of the executive'.[29] In an essay subtitled, 'Every Constitution Requires the Express Approval of its Citizens', Ernst Gottfried Marhenholz, the Court's sitting vice president, trumpeted emphatically the people's right to ratify their new (or their old) constitution.[30] Even after reunification via Article 23 was an accomplished fact, Mahrenholz called for broad popular discussion of potential revisions to the Basic Law. Roman Herzog, the chief justice, wished to inject more plebiscitary elements into the Basic Law.[31]

Ironically, while the justices themselves were sounding themes of popular self-determination, some of those most strongly opposed to the Article 146 route insisted that a new constitutional assembly couldn't possibly improve on the work of the parliamentary council that framed—or of the Constitutional Court that expounded—the Basic Law. Robert Leicht, in an essay tellingly subtitled 'A New Constitution Can Only Get Worse', warned that replacing the Basic Law meant jettisoning the Court's jurisprudence 'as scrap paper into the archives'.[32] But the justices themselves were less proprietary. For one thing, it was almost unimaginable that a new constitution, whatever its other innovations, would scrap or cripple the Constitutional Court. Moreover, if the Court continued (as it surely would) under a new constitution, it would continue with its precedents and interpretive methodologies intact. The proportionality test and the nearly ninety volumes of constitutional jurisprudence would endure.

But beyond all this, many of the justices believed passionately that the constitution they were commissioned to enforce and interpret required, by its own terms, popular legitimation. To say that a new constitution would be worse than the old was to express basic

[24] Jürgen Habermas, *The Past as Future* (Max Pensky tr, University of Nebraska Press 1994) 41.

[25] Jürgen Habermas, 'Der DM-Nationalismus' *Die Zeit* (30 March 1990); Habermas (n 24) 41; see also Müller, *Another Country* (n 12) 100.

[26] Moses (n 19) 229.

[27] See Konrad Jarausch, *The Rush to German Unity* (Oxford University Press 1994).

[28] See Peter Graf Kielmannsegg, 'Vereinigung ohne Legitimität?' (1992) 46 *Merkur* 561.

[29] Helmut Simon, 'Markierungen auf dem Weg zu einer neuen gesamtdeutschen Verfassung' in *Verfassungsdiskussion* (n 1) 140–41.

[30] Ernst Gottfried Mahrenholz, 'Das Volk muβ "Ja" sagen können. Jede Verfassung braucht die ausdrückliche Zustimmung der Staatsbürger' *Die Zeit* (14 September 1990) 13. See also Ernst Gottfried Mahrenholz, *Die Verfassung und das Volk* (Carl Friedrich von Siemens Stiftung 1992).

[31] See 'Verfassungsrichter Mahrenholz wünscht breite Grundgesetzdiskussion' *Der Rheinpfalz* (8 January 1991); 'Herzog befürwortet mehre direkte Demokratie' *Der Tagesspiegel* (21 April 1991).

[32] Robert Leicht, 'Einheit durch Beitritt' *Die Zeit* (23 February 1990, 2/3 March 1990).

mistrust in the people from whom the state, by the Basic Law's express declaration, derived all authority. In this regard, the Article 23 route struck some justices as a failure of nerve. Two decades after Willy Brandt vowed to 'dare more democracy', Dieter Grimm advocated such daring in the constitutional context. Grimm championed a 'risk-taking democracy', one with the courage to convene a new parliamentary council to craft a new constitution for united Germany.[33]

Calls for a new constituent assembly—whether from the justices or from others—fell, for practical purposes, on deaf ears. For politicos straining under the squeeze of time and pressure of events, the central imperative was to effect reunification in the swiftest, simplest, most straightforward fashion possible. That meant the Article 23 route, however disappointing that route was to citizens East and West. There would be time enough later on for the expanded Federal Republic to adjust to its increased size, to accommodate its expanded economic strength and geopolitical heft, and to resolve its wrenching internal tensions.

Such challenges could not, of course, be put off indefinitely. With the accession of the five Eastern states, the Federal Republic had become, by orders of magnitude, more permanent, more powerful, and more problematic. By its mere existence, the new central European colossus would press the German Question with unwearied insistence upon a wary world.

The Question, as ever, was double-edged; it pointed outwards like a sword and inwards like a dagger. What would a reunited Germany mean for the world? What would it mean for the Germans themselves? What role would the new Federal Republic assume among the community of nations? What role, if any, would the German people play in shaping the constitutional contours of the new republic? In the years following reunification, these broad abstractions found concrete articulation in specific constitutional controversies, the arbiter of which was, as always, the Federal Constitutional Court.

It has been suggested earlier in these pages that the effect, or at least the aspiration, of the Basic Law and of the Court's interpretation of it was to correct the course of Germany's *Sonderweg*—to reclaim the German state from its special historical path and to realize in postwar West Germany the Liberal Democratic Republic that had proved unachievable to the Frankfurt patriots of 1848 or the Weimar revolutionaries of 1919; the Republic that had been traduced by Prussian militarists in the nineteenth century and incinerated by Nazi marauders in the twentieth. Whether there ever was a German *Sonderweg* has always been sharply contested.[34] But the framers of the Basic Law clearly crafted the Federal Republic's charter with a keen eye to the 'errors' of the past, with an anguished awareness of what AJP Taylor called the turning-points at which Germany 'failed to turn'.[35]

In its interpretation of the Basic Law, the Constitutional Court seemed also to have its eye on the Germany that might have been. Certainly the Court played a crucial role in the founding and flourishing of liberal democracy in the Bonn Republic. In doing so, as we have

[33] Dieter Grimm, 'Das Risiko Demokratie. Ein Plädoyer für einen neuen parlamentarischen Rat' *Die Zeit* (10 August 1990) 34.

[34] See, e.g., Geoff Eley and David Blackbourn, *The Peculiarities of German History: Bourgeois Society and Politics in Nineteenth-Century Germany* (Oxford University Press 1984); Anselm Doering-Manteuffel, *Wie westlich sind die Deutschen? Amerikanisierung und Westernisierung im 20. Jahrhundert* (Vandenhoeck and Ruprecht 1999); Heinz Bude and Bernd Greiner (eds), *Westbindungen: Amerika in der Bundesrepublik* (Hamburger Edition 1999); Heinrich August Winkler, *Der lange Weg nach Westen*, vol 1 (CH Beck 2000) 1; Philipp Gassert, 'Die Bundesrepublik, Europa und der Westen: Zur Verswestlichung, Demokratisierung und einigen komparatistischen Defiziten der zeithhistorischen Forschung' in Jörg Baberowski and others, *Geschichte ist immer Gegenwart: Vier Thesen zur Zeitgeschichte* (Deutsche Verlags-Anstalt 2001) 67.

[35] AJP Taylor, *The Course of German History*, 2nd edn (Routledge 1951) 68. Taylor referred to the failed revolution of 1848, but the entire *Sonderweg* thesis breathes the spirit of Taylor's aphorism.

seen, the Court assumed and exercised enormous powers. With its power of abstract review, the Court issued the last word in many of the Federal Republic's fiercest political controversies. With its doctrine of the 'radiating effect' of fundamental rights, the Court constitutionalized all West German law and became the supreme organ of the West German judiciary. In all of this, the Court had a profound impact on the political culture and social order of the Federal Republic. In their role as self-proclaimed guardians of the constitution, the justices guided the Federal Republic from its precarious infancy through an unsettled youth and finally into democratic maturity. The continuity and composure of the Court's jurisprudence in the years just prior to reunification suggests that the business of reversing or correcting the supposed *Sonderweg* was near its end. The liberal constitutional order of the Federal Republic was fully formed.

Or so, to many celebratory orators in the summer of 1989, it seemed. As we shall see, the aftermath of reunification would test this proposition. But the proposition itself would test the Constitutional Court. If the Federal Republic had indeed arrived at democratic maturity and taken its place among the community of liberal democracies, what further need did the country have for a powerful constitutional court that would shape, and sometimes constrain, its democratic process? If the *Sonderweg* was over, what need for a Court to correct it? Perhaps the Court had outlived its usefulness—or at least its crucial necessity. Perhaps the Federal Republic's citizens and their elected representatives required less oversight. Perhaps it was time for the Court to play a diminished role. After all, don't only minors need guardians?

These questions will be revisited in the Epilogue. In the period that now concerns us they were hurled to one side by the seismic shifts of reunification. With reunification came the jolting realization that Germany's division had itself been a kind of *Sonderweg*, a continuation and consequence of the broader *Sonderweg*, traditionally understood. Tellingly, the events surrounding reunification immediately provoked a vigorous public discussion of the *Sonderweg* thesis. In a series of high profile articles in *Die Zeit*, the most thoughtful proponents of the *Sonderweg* thesis—including some of the most prominent historians in the Federal Republic—warned that reunification could lead, if Germans were not careful, to another *Sonderweg*, or an unfortunate return to the old one.[36] The challenge for the Federal Republic was to unite the two strands of Germany's postwar path—or rather to reconcile the Eastern strand with the grand narrative of the Western—while reaffirming the liberal democratic values that had allowed the Federal Republic to escape from Germany's (putative) special path. The work, as we shall see, was imperfectly and incompletely done. That it was done at all ranks among the most important post-Cold War achievements of the Federal Constitutional Court. It was through the work of the Karlsruhe justices that citizens of the former GDR—excluded from defining the constitutional terms of reunification—could identify themselves, in a limited degree, with the constitutional order of the new Federal Republic.

But the decade of reunification was hardly one of uninterrupted triumphs for the Court. Indeed, the middle of the decade witnessed the Court's greatest crisis since its founding. For the first time, prominent politicians not only criticized but breathed defiance to Court decisions, calling on citizens to engage in the closest thing to 'massive resistance' the Federal Republic had ever seen. Some observers worried, and others crowed, that the era of the Court's preeminence was over.

[36] See Heinrich August Winkler, 'Mit Skepsis zur Einigung' *Die Zeit* (28 September 1990); Jürgen Kocka, 'Nur keinen neuen Sonderweg' *Die Zeit* (19 October 1990); Hans-Ulrich Wehler, 'Wieder den falschen Apostel' *Die Zeit* (9 November 1990).

The prognosis proved premature. The Court would weather this crisis, as it had earlier challenges to its authority. It did so largely because the perception endured that the Court, though it periodically encroached on the powers of democratically appointed actors, enhanced democracy in other ways—defending at times the prerogatives of Parliament and vindicating, at other times, the fundamental rights of individual citizens. Constitutional Court justices, in their private capacity, called for popular adoption of a pan-German constitution when leading politicians were skittish about that prospect. And in their official capacity, the justices raised concerns about the democratic deficiencies of European integration when, for many major party leaders, that topic was taboo. When the Federal Republic marked its fiftieth anniversary in May 1999, and when the Court observed its own golden jubilee in September 2001, most retrospective assessments rated the Court's contribution to the Federal Republic as invaluable.

Such celebratory paeans might have obscured, but they did not eliminate, the fact that perhaps the Court had indeed entered a period of slow decline. But that, too, is a matter for the Epilogue. This chapter focuses on the Court's role in mediating the wrenching internal tensions of reunification. Reunification provided the immediate or deep backdrop to most of the Court's major decisions throughout the 1990s. This chapter discusses four categories of cases. Part One deals with those cases that addressed issues raised directly by reunification—those cases that required the Court to engage directly with one facet or another of 'the German Question'. Part Two deals with questions of bio-politics and social policy. Even before reunification, many West Germans called for revision of the civil code to provide greater equality for women. Some pointed to provisions of East German law that seemed progressive by comparison. After reunification, the question became even more urgent—particularly with regard to abortion law, an area in which the laws of the GDR and the Federal Republic were irreconcilably opposed. Part Three discusses freedom of conscience and freedom of speech. Reunification vastly enhanced the ideological and religious complexity of the Federal Republic, and the events that toppled the SED regime highlighted dramatically the power of speech and assembly. After reunification, activists pushed to strengthen fundamental rights protections in all these areas. The Court responded, but in the process flung itself into its greatest institutional crisis since the early 1950s. Finally, Part Four analyses matters of foreign policy—the excruciating constitutional dilemmas created by Germany's enhanced geopolitical role.

In all of this, the Court was more inclusive of East German interests than any other constitutional organ of the Federal Republic. Perhaps ironically, on questions of German unification and European integration, the Court was also the most vocal champion of democratic principle and popular sovereignty. The Court continued to provide a check—sometimes a very aggressive check—on the representative legislature; but the Court also bolstered protections for political speech and citizen protest. When the Court's authority was challenged, some justices entered the public sphere themselves to engage the citizenry that would be subject to their decisions. The decade of reunification was as challenging for the Court as any that preceded it, and the Court did not emerge from it unscathed. But the Court's very political crisis called attention to its historical contributions.[37] Though the public's trust in the Court sank dramatically in the short run and slightly in the long run, the Court remained, as it entered its sixth decade, the most trusted institution of the German state. To those with eyes to see, there were clouds on the distant horizon. But those clouds did little to darken the Court's or the country's jubilee.

[37] See Schönberger (n 17) 50.

I. The German Question

A. The first pan-German elections

A crucial question for unified Germany was who would rule it and how those rulers would be chosen. The Federal Republic's traditional electoral rules posed prickly problems—none pricklier than the 5 per cent parliamentary hurdle. Such a barrier to legislative entry was unknown in the electoral law of the now-defunct GDR, as was the Federal Republic's system of allowing each voter two ballots. East German leaders objected loudly to the application of the 5 per cent rule in the first post-reunification federal elections.

Their objections were rooted partly in demography, partly in chronology. At the time of reunification there were roughly seventeen million citizens in the Eastern states; in the Western states, the figure was sixty-two million.[38] Because the first elections were scheduled to be held less than two months after formal reunification,[39] political parties based in the East—including those movements instrumental in toppling the SED—had little time and slender hope to win significant votes in the West. Since the East accounted for little more than 20 per cent of the population, the 5 per cent hurdle looked fatal for these parties. Small Eastern parties called for the barrier to be scrapped.

After some spirited negotiation, the architects of reunification struck a deal. The 5 per cent hurdle would remain in place, but small Eastern parties could overcome it by 'piggybacking' on the electoral returns of a Western partner.[40] The Eastern German Social Union (DSU), for instance, could add its tally to that of its Bavarian partner, the CSU, and thereby clear the hurdle. The compromise thus promised to propel some Eastern parties into Parliament. The rub was that parties unable to find a partner were sure to be excluded. The most notable of these was the Party of Democratic Socialism (*Partei des Demokratischen Sozialismus*, or PDS), the re-baptized successor to the SED. A pariah to Western parties, the PDS had no hope of finding a partner and, as a result, little hope of entering Parliament—this notwithstanding its potentially greater electoral strength than parties such as the DSU that were sure to enter Parliament on the back of a Western partner. The PDS and other parties disadvantaged by the compromise challenged the law in a constitutional complaint.

It was ironic that the successor party to the SED, which through its press organs and in its official pronouncements had railed for decades against the Karlsruhe justices, should immediately invoke the Court's protection. It was an added irony that the complaint succeeded.

In a judgment dated 29 September 1990, the Second Senate ruled that the amended electoral law violated the principles of electoral equality and equal opportunity for political parties.[41] These principles afforded the legislature 'only a narrowly limited leeway' to differentiate among parties.[42] Any differential treatment required compelling justification.[43] Promoting government stability by preventing the proliferation of splinter parties was one such ground, as the Court had noted when it approved the 5 per cent rule previously.[44] But the justices now clarified that this earlier imprimatur was not valid as an abstract matter

[38] *Statistisches Jahrbuch der Deutschen Demokratischen Republik, Zeitschriftenband* (1989) 45 <http://www.digizeitschriften.de/dms/img/?PPN=PPN514402644_1988andDMDID=dmdlog5andLOGID=log5andPHYSID=phys45#navi>; Statistisches Jahrbuch für die Bundesrepublik Deutschland, Zeitschriftenband (1990) 43. <http://www.digizeitschriften.de/dms/img/?PPN=PPN514402342_1990andDMDID=DMDLOG_0012andLOGID=LOG_0016andPHYSID=PHYS_0042#navi>.

[39] Reunification was formalized on 3 October 1990; the elections were scheduled for 2 December of the same year.

[40] The compromise mixed metaphors and made them inexact: piggyback rides make hurdling harder.

[41] 82 BVerfGE 322 (1990). [42] ibid 338. [43] ibid. [44] ibid 334.

for all time.[45] New circumstances required reassessment, and the Court concluded that the new law, both by placing the parties of the East at a disadvantage vis-à-vis the parties of the West and by placing some Eastern parties at a disadvantage vis-à-vis others, had run afoul of core constitutional principles. As a result, the law could not stand. But rather than send the legislature back to the drawing board nine weeks before the scheduled election, the justices concluded with a powerful nudge. The 5 per cent hurdle could be applied, but only—for this first election at least—on a regional level. Eastern parties could enter Parliament if they garnered 5 per cent of the votes within the five Eastern states; Western parties if they secured 5 per cent or more among the eleven states of the West. Only such regional differentiation could preserve the 5 per cent rule while honouring the requirements of equality.

It was a bold judgment—one that prefigured the integrationist role the Court would play in incorporating the Eastern states into the new polity. Some observers sniped, however, that the outcome would benefit the CDU by excluding the Greens and preventing an SPD–Green alliance.[46] And this is, in fact, what happened. The Western Greens won only 4.8 per cent of the Western votes, and thereby failed to enter the Bundestag. In the East, however, the union of the Greens with the Bündis 90 movement won 6.2 per cent of the Eastern votes, yielding eight seats in the now 662-member chamber. Only two of these eight seats went to the Eastern Greens, who merged on the day after the election with the Greens in the West. Had the originally planned compromise rules applied, the Green–Bündnis 90 partnership would have entered the Bundestag with 5.1 per cent of the national vote. The PDS, by contrast, won only 2.4 per cent of the national vote, but nearly all of those votes came in the East, allowing the party to clear the regional hurdle comfortably and to enter Parliament with seventeen seats. The Republicans party, a Right-wing splinter from the Bavarian CSU, won 2.1 per cent of the national vote, but nearly all of it in the West. Both of the great *Volksparteien* suffered predictable losses with respect to the 1987 outcome, but the SPD, which dropped from 37 per cent to 33.5 per cent, lost considerably more than the Union parties, which fell from 44.3 per cent to 43.8 per cent. Only the FPD grew in strength, ascending from 9.1 per cent in 1987 to 11.0 per cent in 1990. The upshot was a comfortable Union–Liberal majority (398 seats of 662).

In the years that followed, the PDS made impressive and surprising gains. The party owed its growth largely to the growing (and justified) perception among Easterners that their interests were of little moment to the traditional parties. In the run-up to the 2005 federal election, the Left wing of the increasingly centrist SPD defected to the PDS and formed the new Left Party. With 9 per cent of the vote, the party outpolled the Greens. In 2009, this tally increased to 12 per cent, enough almost to make the Left a potential coalition partner should the SPD ever cease pronouncing anathemas on its potential ally.

None of this would ever have happened if the Constitutional Court had ruled against the PDS in another constitutional complaint, also related to the 5 per cent hurdle, that followed the second united German elections in 1994. Long before the Court mandated a merely regional application of the 5 per cent hurdle, there had been an alternative route for a sub-5 per cent party to enter the Bundestag. From the earliest days of the Federal Republic, the so-called *Grundmandatsklausel* allowed a party that won a 'direct mandate' in three or more electoral districts to enter Parliament regardless of its share in the national vote. This provision allowed members from the Centre Party to enter the Bundestag in 1953, and

[45] ibid 338. The Court has recently given teeth to this dictum. In 2008, it struck down the 5 per cent hurdle for local and regional elections in Schleswig-Holstein. 120 BVerfGE 82 (2008). In 2011, it did the same for elections to the European Parliament. See 129 BVerfGE 300 (2011).

[46] See 'Wir brauchen Leuchttürme' *Der Spiegel* (8 October 1990).

members from the nationalist German Party to do so in both 1953 and 1957. Since 1957, however, the year of the Union's landslide and absolute majority, the clause had lain dormant. In 1994, the PDS gave it new life. Although it won only 4.4 per cent of the national vote (a 2 per cent increase over its return four years earlier)—and although in this second unified election the 5 per cent hurdle applied nationally—the PDS still qualified to take thirty seats in the Bundestag because it had won four electoral districts outright. Thirty seats were plenty to allow Leftists to make mischief, and concerned citizens cried foul.

In November 1995, the Bundestag rejected a parliamentary challenge to the law, whereupon the law's opponents filed a constitutional complaint. The complaint invoked the principles of electoral equality and equal opportunity for political parties—the very principles that led the Second Senate to regionalize the 5 per cent hurdle for the 1990 election. The complainants objected that the *Grundmandatsklausel* favoured some parties at the expense of others: it included parties with concentrated support and excluded those whose support was diffuse.

The Second Senate rejected the challenge, partly out of deference to Parliament's power to structure elections (a power that seemed less sacred four years earlier), but partly in recognition of the legitimate legislative goal of 'securing the character of the election as a process of integration in the political will-formation of the people'.[47] In its role as electoral regulator, the legislature must ensure that elections foster the 'integration of political forces' and 'strive to prevent the exclusion of significant popular concerns from the popular assembly'.[48] The legislature enjoyed wide discretion in forging criteria 'that allow consideration of the integration-strength of political parties'.[49]

The Senate's emphasis on integration was telling. Societal integration, after all, was the prominent role that Rudolf Smend and his students had ascribed to constitutional courts since the early days of the Federal Republic.[50] By fostering integration the Court was fulfilling one of its traditional roles. But in the context of the ongoing process of reunification, that role assumed new meaning. Citizens of the former Eastern states had been largely excluded from the formal process of reunification. In two major judgments vindicating the PDS's right to participate in Parliament, Karlsruhe sent a strong signal that, so long as an Eastern party enjoyed the support of a significant number of Eastern citizens, the Court would not brook the mechanical exclusion of that party—no matter how unsympathetic—a share in governing the united Republic. The Court thus helped give a voice to many whom reunification had otherwise left voiceless.

B. The voices of others: suffrage for non-Germans

In other respects, however, the Court's election jurisprudence was less inclusive. In autumn 1989, when protesters in East Germany had rocked the continent with the rallying cry, 'We are the People!' (*Wir sind das Volk!*), and later, 'We are *one* People' (*Wir sind* ein *Volk!*), the protesters' claim to speak for the People and as the People was, as a moral and symbolic matter, uncontestable. But as a legal and constitutional matter, the boundaries of 'the People' remained contested. In the very month (October 1990) that German reunification was

[47] 95 BVerfGE 408, 418 (1997). [48] ibid 419. [49] ibid 422.

[50] Rudolf Smend, *Verfassung und Verfassungsrecht* (Duncker and Humblot 1928) 122–50; Rudolf Smend, *Integrationslehre* in *Staatsrechtliche Abhandlungen und andere Aufsätze*, 2nd edn (Duncker and Humblot 1968) 474–81; Peter Badura, 'Staat, Recht und Verfassung in der Integrationslehre Rudolf Smends' (1977) 16 *Der Staat* 305; Wilhelm Hennis, 'Integration durch Verfassung?' (1999) 54 *JuristenZeitung* 485.

formally effected, the Constitutional Court held—anachronistically, some thought—that the *Volk* from whom all authority flowed was made up only of German citizens.[51]

The underlying issue was whether states could allow non-citizen residents to vote in local elections. Two states, Hamburg and Schleswig-Holstein, had passed laws saying they could. Under these laws, foreigners resident for at least eight years could vote in municipal council elections. Roughly 150,000 foreigners had lived in Hamburg for at least eight years, many of whom (and many of whose parents) were born in Germany. The foreign population of the Schleswig-Holstein district in question was considerably smaller—roughly 5,500, most of them Danes. Schleswig-Holstein was ruled by an SPD government; Hamburg by a Social-Liberal coalition. Pundits expected foreign residents largely to vote SPD. So did the federal government, some of whose ministers were highly agitated by the new laws. Friedrich Zimmermann, the interior minister and villain of the earlier census saga, bellowed that the states in question were 'selling out German interests'.[52] Kohl's government did not challenge the measures itself, however, in part because it wished to spare the feelings of the Union's coalition partner. Ingo von Münch, a prominent public lawyer and Hamburg's FDP boss, was the driving force behind the Hamburg law. The constitutional complaint was filed instead by the CDU/CSU Bundestag faction, in tandem with the Bavarian state government.

In companion judgments dated 31 October 1990, the Second Senate granted the complaints against both laws. Article 28(1) GG, the Senate noted, demanded representation for 'das Volk' at the municipal, district, and state level. And *das Volk* consisted—exclusively—of German citizens.[53] 'Elections at which foreigners are also allowed to vote', the justices wrote, 'cannot bestow democratic legitimacy'.[54] Democratic legitimacy flowed from the *Volk*, and the Volk in question was *German*. The Senate was unimpressed by appeals to other European nations (Ireland, Denmark, Norway, the Netherlands, Switzerland, Sweden, and Spain) that allowed foreign residents to vote in local elections. It was moderately more sympathetic to claims that European integration was modifying traditional notions of nationality. The justices conceded that regulation at the European level might someday end the exclusion of foreign residents from municipal elections. But at the county or district level, the Court did not leave even this door open.

Critics saw in the judgment a curious resuscitation of nineteenth-century notions identifying the people with the citizenry and the nation with the state. (This was not the last time critics would accuse the Court of embracing outdated notions of 'State' and 'People'.) It seemed strangely at odds with earlier holdings that some constitutional provisions applied, in whole or in part, to all humans, not merely to German citizens. At the twilight of the Bonn Republic, and on the eve of its successor, the Court seemed to suggest that the Basic Law was the fundamental charter, not for all Germany, but for all Germans.

C. Past and present

Extending the Basic Law's protections to all Germans posed problems of its own. Upon reunification, millions of Germans were subject for the first time to a government bound by the Basic Law. Many of the Federal Republic's newest citizens were eager to invoke the Basic Law's protection. They did so, however, in conflicting ways. Some sought redress under the

[51] 83 BVerfGE 37 (1990); 83 BVerfGE 60 (1990).
[52] In Hans Schueler, 'Alle Macht dem (deutschen) Volk' *Die Zeit* (9 November 1990).
[53] 83 BVerfGE 37, 53–59 (1990). [54] 83 BVerfGE 60, 81 (1990).

Basic Law for oppression suffered at the hands of the GDR regime or its Soviet protector. Others, who had been employed by the fallen regime, invoked the Basic Law's protections of profession and property in an attempt to preserve their livelihood. The ensuing legal controversies were emotionally charged. Often they pitted Eastern interests against Western. The Constitutional Court was placed in the delicate role of arbitrating not only between East and West, but between state and citizen, pragmatism and principle, present and past.

During its first decade, the Constitutional Court distinguished itself as the organ of the Federal Republic most willing to confront the black legacy of Nazism. For many years the Court's was the clearest official voice condemning the atrocities of the Nazi era and expressly defining the Federal Republic in opposition to its tyrant forebear.[55] Its most dramatic decisions in this regard were those involving the Federal Republic's responsibility for the employment relationships of its National Socialist predecessor. As we saw in Chapter 1, the Court convulsed the young Republic and inflamed the legal academy and ordinary judiciary with a December 1953 judgment proclaiming that all civil servant relationships were abolished when the Third Reich surrendered unconditionally on 8 May 1945.[56] Four decades later, one of the Court's first major decisions addressing the constitutional complexities of reunification involved the status of GDR civil servants in the expanded Federal Republic.

How to organize the civil service in united Germany was one of many thorny questions that arose during negotiations over the Unification Treaty. Major reorganization was inevitable. Proportionally, the East German civil service dwarfed that of its Western neighbour: the GDR had many more bureaucrats per capita than the FRG. Absent reorganization, there would be too many civil servants in the unified Republic.

There were structural differences between the two services as well. The traditional civil service in the Federal Republic—with its roots in a monarchical past, its special legal status, life tenure, and social prestige—was very different from the civil service in the GDR, where civil servants were often appointed at the pleasure, or sacked at the whim, of party elites. In light of these great differences, conservatives hoped simply to extend the Western civil service structure to the Eastern states. This view fed on fears of retaining SED cadres in Eastern state bureaucracies. One the other hand, some Eastern civil servants who maintained their posts through reunification had ties to the reform movements that ousted the SED. Including such Easterners—with their alternative perspectives on democracy and social questions—in the official life of the Federal Republic might increase diversity and reassure Easterners that their concerns would not be ignored. What's more, the traditional civil service (*Beamtentum*) of the Federal Republic struck many as archaic, inegalitarian, and undemocratic. To many critics, reunification offered a golden chance for reform.[57]

It was an opportunity that those in power did not seize. The Unification Treaty preserved the traditional distinction between civil servants (*Beamte*) and other government employees. More dramatically, the treaty allowed for the dismissal (with some compensation) of civil servants whose service the state no longer needed. It also allowed for entire administrative units to be closed. Upon reunification, each administrative division of the former GDR would be allocated either to the federal government or to one of the Eastern

[55] See Norbert Frei, 'Transformationsprozesse. Das Bundesverfassungsgericht als vergangenheitspolitischer Akteur in den Anfangsjahren der Bundesrepublik' in Michael Stolleis (ed), *Herzkammern der Republik: Die Deutschen und das Bundesverfassungsgericht* (CH Beck 2011) 64.

[56] See 3 BVerfGE 58 (1953).

[57] For a fine summary of these issues, see Peter Quint, *The Imperfect Union: Constitutional Structures of German Unification* (Princeton University Press 1997) 166–67. This section relies on Quint's summary.

states. The responsible governments would then have three months to decide which units to retain and which to abolish. The treaty provided no criteria for determining where, and upon whom, the axe would fall. Employees of dissolved divisions would receive 70 per cent of their salary during a six-month 'transitional' period (for older employees, it was nine months). After that, salary payments ended and former employees could seek only unemployment benefits. Thousands were potentially affected; scores filed constitutional complaints.

The complainants averred that the constitution required individualized assessments. But under the treaty, each employee's fate was bound to the fortune of her division. This violated their occupational freedom under Article 12 GG, their property rights under Article 14, and the equality principle of Article 3. In response, the federal government argued that the GDR's employment relationships ended on 3 October 1990, when, the GDR ceased to exist. Any subsequent relationship was a product of the Unification Treaty and could, by the terms of that treaty, be amended or annulled. In any case, the government added, the Federal Republic was not the legal successor of the GDR and was not bound by the GDR's contractual obligations.

The government's argument unmistakably echoed the Court's 1953 *Civil Servants* judgment. Insofar as the government argued that the employment obligations of an oppressive regime terminate when the regime itself expires, the government was on solid precedential ground. But the government's argument that the Federal Republic had not succeeded the GDR mingled less comfortably with the Court's prior jurisprudence—particularly its *Basic Treaty* judgment of July 1973. In that decision, the Court ruled that the Federal Republic was the legal successor to the German Reich, with the caveat that Republic and Reich were only 'partially identical'.[58] With reunification, the Federal Republic arguably became the Reich's successor in a fuller sense—and the successor, too, of the Reich's erstwhile constituent parts.

In its judgment of 24 April 1991, the First Senate made no broad statements about the federal government's jurisprudential relationship with either the German Reich or the GDR. But the Court did observe that by the terms of the Unification Treaty itself, the federal and state governments of the Federal Republic assumed certain contractual obligations of the GDR. To this extent, at least, the Federal Republic *was* the GDR's legal successor.[59] For this reason, treaty provisions for dissolving administrative divisions and dismissing individual civil servants implicated Article 12 GG (though not Article 14, which protected property already acquired, not expectations of future acquisition).[60] This Article 12 restriction, however, was justified by superior interests—efficient administration, for one thing, healthful finances for another.[61] Both interests were urgent amid the administrative and financial turmoil of reunification.

This did not mean, however, that the federal government enjoyed a carte blanche. The Court indicated that the treaty's provision for interim payments to dismissed employees was not merely generous but constitutionally required.[62] In this regard, the Court struck a very different tone from that of the 1953 *Civil Servants* judgment, which ruled that the compensation of Nazi-era civil servants was a matter of parliamentary largesse, not constitutional mandate.[63] In addition, the 1991 judgment emphasized that the state must

[58] 36 BVerfGE 1, 16 (1973). [59] 84 BVerfGE 133, 147 (1991). [60] ibid 147, 156.

[61] ibid 148–52. The Court also held that the treaty provisions presented no equality problem because the provisions reference to the 'need' for a given division or employee established an objective base for differentiating among employees.

[62] ibid 152–53. [63] See Chapter 1.

continue to soften the impact of reorganization. In certain particulars, the state must do even more. Specifically, it must make special efforts to reintegrate elderly employees into the workforce.[64] It must dismiss such employees only if they have a well-founded prospect of new employment elsewhere in the public service. Pregnant women and single mothers, moreover, enjoyed special constitutional protection under Article 6(4) GG; they must not be dismissed at all.[65] In their failure to protect such mothers—and in this regard only—the contested provisions were unconstitutional.

In this latter-day *Civil Service* judgment, the Court struck a balance between pragmatic acceptance of political realities and sympathetic awareness of the costs those realities imposed on individuals. The Court deferred to the hard bureaucratic demands of reunification but made them more humane. It signalled to Germans in the Eastern states that their interests would not be steamrolled by the Western behemoth. And it signalled to Germans everywhere that the 'social state principle' embedded in the Basic Law might exert greater force in the new Republic than it had in the old. As one commentator concluded, the Court's judgment 'made the dismantling of the public service in the former GDR 'more socially tolerable'. It softened hardships inherent in the Unification Treaty. At the same time it made clear that citizens of the former GDR enjoyed, in principle, the same constitutional status as citizens of the Federal Republic—in contradiction to the federal government's theory that Eastern citizens were fundamental rights bearers only 'as stipulated by the unification treaty'.[66] It was a strong statement that the Federal Republic's newest citizens were full-fledged citizens. Its outcome was pragmatic; its tone strongly integrationist.[67]

D. Ill-gotten lands: compensating victims of Soviet 'land reform'

The Court adopted a similarly pragmatic, integrationist, and mediating posture in an even more dramatic judgment issued just the day before. In a controversy of wrenching emotional and political charge, the Court conducted a high-profile exercise in judicial *Vergangenheits bewältigung*—a confrontation with the oppressive past, not of the Third Reich, but of the Soviet Occupation Zone, and later the GDR. The Court's encounter with this history of oppression left many victims of that oppression bitterly disappointed.

Few questions surrounding German reunification were as visceral or as volatile as the effect of reunification on property relations in the Eastern states. The core of the controversy was the fate of properties expropriated by Soviet occupation forces between 1945 and 1949, and by the East German government between 1949 and 1989. At the end of the Second World War, Eastern Germany's Soviet occupiers launched a broad and brutal programme of 'land reform'—a euphemism for the wholesale confiscation of large landholdings in the name of socialist collectivization and in ostensible punishment of putative Nazi abettors and sympathizers. After the occupation period ended, the programme was modified and extended through periodic takings conducted by the GDR government, including takings

[64] 84 BVerfGE 133 (n 59) 154–55. [65] ibid 155–57.

[66] Wolfgang Däubler, 'Die sogennante Warteschleife auf dem verfassungsrechtlichen Prüfstand' (1991) 45 *Neue Justiz* 233, 236.

[67] The same was true of other cases in which the Court addressed the status of former East German civil servants or of persons in similar situations. In August 1995, for instance, the Court reversed a decision of the Federal Court of Justice that had affirmed the revocation of the law licences of East German lawyers. See 93 BVerfGE 213 (1995). The Court found in the revocations a violation of the lawyers' rights, guaranteed by Article 12(1) GG, to occupational freedom. On the other hand, the Court ruled in July 1997 that evidence that a public school teacher's servile devotion over many years to the East German state, or to the SED party that ruled it, might reflect an equivocal devotion to the principles of the *Rechtsstaat* and could serve as grounds for dismissal. 96 BVerfGE 152 (1997).

aimed at East German citizens who—traitorously, it was maintained—left the country to cast their lot in the West.

Predictably, after reunification the victims of these expropriations wanted their lands back. Their successors in interest, just as predictably, were unwilling to let the lands go. Those negotiating the Unification Treaty for the Federal Republic believed that for the Soviet Union, upon whose approval reunification depended, restitution of expropriated lands was a deal-breaker. The success of reunification, it was believed, required compromise. A wrenching compromise was forthcoming.

The territory in question was enormous. More than a third of the land in the new Eastern states was affected. In the early occupation years, the Soviets expropriated all possessions in excess of 250 acres and mercilessly hounded the erstwhile proprietors—arresting them, imprisoning them, or committing them to labour camps. Similar expropriations continued throughout the GDR era.[68]

As events sped toward reunification, victims and heirs hankered to assert their claims. A familiar stereotype pictured a prosperous Westerner appearing at the home of a woebegone Easterner, presenting a batch of yellowed papers, and calling the house his own. Full-scale restitution of expropriated properties was certain to wreak socio-economic mischief. The cost in human suffering was sure to be prodigious. On the other hand, the expropriations themselves left a trail of suffering and a legacy of injustice. The controversy was certain to leave its losers bitterly disappointed. It threatened to wreck the whole project of reunification.

The negotiators of the reunification treaty ultimately worked out a deal whereby expropriations effected between 1945 and 1949—effected, that is, on the basis of occupation law or authority—were declared irreversible, while properties taken after this date were to be restored to their original owners. To shield the compromise against constitutional objection, the treaty enshrined the irreversibility of pre-1949 seizures as the new Article 143(3) of the Basic Law.

This textual fig leaf neither comforted nor deterred the former owners. On 5 October 1990—less than forty-eight hours after the official proclamation of German unity—a band of prior proprietors took their cause to Karlsruhe.

The constitutional controversy surrounding compensation for Soviet and GDR land expropriations had already swept the land.[69] Those asserting the constitutional right to restitution of properties confiscated between 1945 and 1949 faced large analytical obstacles—temporal, geographic, jurisdictional, and even textual. The Basic Law did not enter into force until May of 1949, and arguably, therefore, did not apply to government action taken before that date. What's more, the Basic Law, by its own terms, applied only within the states that ratified it and to actions of the government (the Federal Republic) that it created.[70] Finally, because the treaty's denial of restitution for pre-1949 arrogations

[68] For a concise survey of the process, see Jonathan J Doyle, 'A Bitter Inheritance: East German Real Property and the Supreme Constitutional Court's "Land Reform" Decision of April 23, 1991' (1991–92) 13 *Mich. J. In't L.* 832, 834–39.

[69] See Otto Kimminich, *Die Eigentumsgarantie der Wiedervereinigung* (Verlag Landwirtschaftliche Rentenbank 1990); 'Enteignungen in der DDR von Anfang nichtig' *Die Welt* (20 July 1990) 10; Thomas Thierau and Albrecth Tintelnot, 'Unklarheiten und verwirrende Regelungen bei Rückgewähr enteigneten Vermögens' *Handelsblatt* (20 August 1990); 'Drohende Verfassungsklage' *Handelsblatt* (August 29, 1990); 'Enteignete sind enttäuscht' *FAZ* (5 September 1990) 2; Hans Herbert von Arnim, 'Entzug der Grundrechte aus Opportunität?' *FAZ* (6 September 1990) 8; Franz Michel, 'Zur Rückabwicklung der Enteignungen' *FAZ* (7 September 1990) 12; Hans Mahrenholz, 'So wird "dauerhafter Rechtsfrieden" nicht erreicht' *FAZ* (11 September 1990) 1; Friedrich Karl Fromme, 'Einheit und alte Rechte' *FAZ* (19 September 1990) 1.

[70] See Article 23 GG.

had been secured by constitutional amendment, pre-1949 victims could prevail only if the Court found that the new Article 143(3) GG was itself unconstitutional—that is, that the amendment breached the 'eternity clause' of Article 79(3).

Against these obstacles advocates of restitution maintained that the Unification Treaty's partial, post-hoc ratification of pre-1949 confiscations constituted an independent, present-day violation of Article 14's guarantee of property rights. Moreover, by treating differently those victimized before and after 1949, the treaty and the amendment violated the equality principle of Article 3. Finally, champions of the pre-1949 victims held that because Article 1(3) made the fundamental rights enumerated in Articles 2 through 19 of the Basic Law binding on the legislature, the executive, and the judiciary, the amendment condoning compromise on land expropriation—which ran afoul of Article 3 and Article 14—was impermissible under Article 79, which secured Articles 1 and 20 against legislative amendment. For good measure, victims urged that the confiscations were invalid the moment they were undertaken—invalid under natural law, under international law, and even by virtue of the Weimar Constitution.

For several months, scholars and citizens propounded the finer points of these arguments with passionate intensity in the learned and popular press. The question was as conceptually complex as it was emotionally poignant. As the debate reached its climax, all eyes turned to Karlsruhe.

By one account, the oral argument held before the First Senate on 22 January 1991 was one of the most dramatic the court had ever seen.[71] The Court's published opinion, by contrast, was anti-climactic.

The First Senate unanimously upheld the treaty provision making pre-1949 expropriations irreversible.[72] Though the written decision was quite long, the Senate's substantive engagement with the constitutional issues was quite terse. The Court rejected categorically the notion that the confiscations were illegal at the time of their execution. They were authorized, the Court concluded, under Soviet law and GDR law, and not subject to the property protections of the Basic Law, which didn't then apply.[73] Central to the Court's analysis was the justices' view that the expropriations were the work 'of another state'.[74] The 'social state' principle might require that the Federal Republic effect some 'burden sharing' (*Lastenausgleich*) in favour of the victims of past injustice. But Article 79 did not require that such redistribution take the form of direct restitution. The Court acknowledged that both the Unification Treaty and Article 143(3) GG treated victims differently, but observed that the compromise was essential to reunification—or at least that it was deemed essential by reunification's political architects. The Soviet Union had apparently conditioned its approval of the treaty on the compromise over expropriations. This being so, the treaty's differential treatment of victims was justified by the preamble's command—the so-called *Wieder vereinigungsgebot*—that the Federal Republic pursue reunification.[75] Some restriction of one constitutional value (equality) must be tolerated as the prerequisite of another (reunification).

This did not mean that reunification trumped equality *tout court*. Article 3 required that the legislature compensate those whose losses the treaty made permanent. But the legislature itself would determine how much compensation was feasible in light of reunification's other—enormous—financial demands.[76]

[71] Doyle (n 68) 834 fn 9. [72] 84 BVerfGE 90 (1991). [73] ibid 119–20. [74] ibid 124.
[75] ibid 125. [76] ibid 129–31.

Responses to the judgment mingled bitter disappointment, analytical bemusement, and pragmatic resignation. The latter sentiment probably predominated, but even those who acknowledged that the Court was walking a political minefield were distressed by its lack of doctrinal daring. Friedrich Karl Fromme lamented on the front page of the *Frankfurter Allgemeine* that the justices had produced 'many words but little clarity'.[77] Critics grieved that a constitutional controversy that raised bewildering conceptual questions had received terse and perfunctory answers—or no answers at all. The outcome may have been inevitable, but many hoped for a richer justification.

Many also hoped for a stronger denunciation of the underlying crimes. An American observer wrote that '[p]erhaps the most unsettling aspect of the opinion [was] the Court's failure to utter a single harsh word about the severe, often deadly, persecutions to which property owners in the East were subjected'.[78] This seems unfair. The justices surely took the iniquity of the expropriations for granted. But what puzzled and troubled many critics was the prominent role the opinion afforded to the Soviet Union, the dwindling giant that would, by year's end, give up the ghost. That a major principle of German constitutional law should be influenced by the spectre of Soviet obstruction suggested that the expiring empire had held the Basic Law hostage. This was deeply disturbing to the decision's many critics.

Ironically, it also provided a ray of hope. If the factual predicate of the Court's decision was the Soviets' ostensible refusal to approve the treaty absent a compromise on expropiations, then the Court might reconsider its holding if the predicate proved fallacious. After the Soviet Union fell, this seemed plausible. In August 1994, the *Frankfurter Allgemeine Zeitng* reported that, in an interview with the British historian Norman Stone, the USSR's terminal president, Mikhail Gorbachev, firmly denied making the compromise a *sine qua non* of reunification.[79] Gorbachev confirmed and amplified this denial nine days later in *Der Spiegel*.[80] In some knowledgeable quarters, Gorbachev's assertions were positively affirmed; in others, they were vehemently denied.[81]

Controversy surrounding the Soviet stance reignited controversy surrounding the Court's *Land Reform* decision. The renewed controversy resulted in a second constitutional complaint. The complainants asked the Court to revisit its earlier decision in light of recent factual revelations.

The justices were disinclined. In its second *Land Reform* decision, dated 18 April 1996, the First Senate rejected unanimously the renewed complaints and answered some criticisms of its earlier decision.[82] The Senate was not convinced that the revisionist historical case was unanswerable.[83] It *was* convinced that assessing the Federal Republic's bargaining position in the run-up to reunification was the exclusive prerogative of the federal government.[84] It was not for the justices to second-guess that assessment with the benefit of hindsight and by the dubious light of public interviews and political memoirs. The agents of the Federal Republic's foreign policy were only required to interpret affairs in a manner that found 'plausible support' in the evidence then available.[85] The West German architects of reunification met this test easily.

Years passed before the victims of pre-1949 expropriations hazarded a third attempt. This time the complainants challenged the exemption as a violation of international law—a

[77] Friedrich Karl Fromme, 'Viele Worte und wenig Klarheit' *FAZ* (24 April 1991) 1.

[78] Doyle (n 68) 861.

[79] 'Gorbatschow: Moskau hat die Enteignungen in der Sowjetzone nicht für unantastbar erklärt' *FAZ* (27 August, 1994).

[80] 'Soviel Kleinkram' *Der Spiegel* (5 September 1994). [81] See Quint (n 57) 139.

[82] 94 BVerfGE 12 (1996). [83] ibid 43–46. [84] ibid 35. [85] ibid 40.

move that had the procedural effect of shifting jurisdiction from the First Senate to the Second. The complainants now argued that the expropriations ran afoul of public international law and that the Federal Republic, by failing to reverse the expropriations, had violated the Basic Law's command to uphold the integrity of international agreements. In October 2004, by a vote of 7:1, the Second Senate rejected this complaint as well.[86] The Court construed the relevant international norms narrowly, ruling that, at the time of the Soviet expropriations, those norms had not established universal protections for property rights. In any case, the Federal Republic was not bound to vindicate those rights by restitution as opposed to other forms of compensation.

Earlier, victims of the Soviet land reform had challenged those other forms of compensation on their own terms. In the aftermath of the First Senate's much-criticized first *Land Reform* judgment in 1991, the united German Parliament passed a series of statutes designed to compensate both victims of Soviet expropriations who had been excluded from the general policy of restitution (or those entitled to restitution but who, for technical reasons, never received it) and victims of Nazi-era expropriations who never received compensation from the East German government. Collectively christened EALG (*Entschädigungs- und Ausgleichsleistungsgesetz*, or Compensation and Equalization Payment Acts), the laws established scales for determining the value of confiscated lands and rates for compensating prior owners. These were nettlesome questions that the Court's first *Land Reform* decision had left wide open. The statutory scales usually involved some reduction in the property's present value based on its assessed value at the time of compensation. The scales were regressive: small property values were reduced by 30 per cent, large values by as much as 95 per cent. A host of complainants argued that the compensation they received was constitutionally inadequate and, vis-à-vis other compensated groups, unequal.

On 22 November 2000, the First Senate announced another resounding defeat for the pre-1949 victims of Soviet Land Reform.[87] Once again, and for reasons similar to those articulated in its first two *Land Reform* judgments, the First Senate rejected arguments based on Article 14's protections of property, Article 3's guarantee of equality, and Article 20's entrenchment of the social state and *Rechtsstaat* principles. Four justices objected to the regressive scales for calculating amounts of compensation, but the tie within the Senate ran against the complainants. Another group of expropriation victims returned from Karlsruhe empty handed.

There was, in the Court's four *Land Reform* decisions, a strong strain of pragmatism and deference to political actors. Some even discerned echoes of American political question doctrine.[88] Others, however, have seen more at play than *Realpolitik*. Peter Quint, an American expert on German constitutionalism, saw in these and other decisions a process by which 'the Court ... appeared to place itself in the role of a mediator or compromiser of questions that the political forces may not have adequately resolved ... especially with respect to those constitutional principles that seemed to pit east against west'.[89] To foster political mediation and societal integration was, as noted earlier, the central function ascribed to constitutional justice by the disciples of Rudolf Smend, who left such an enduring mark on the public law scholarship of the early Federal Republic.[90] But the Court's assumption of a mediating role in reunified Germany was more than a return to Smendian societal

[86] 112 BVerfGE 1 (2004). Justice Gertrude Lübbe-Wolff was the lone dissenter. ibid 44–49.

[87] 102 BVerfGE 254 (2000).

[88] Donald P. Kommers and Russell A. Miller, *The Constitutional Jurisprudence of the Federal Republic of Germany*, 3rd edn (Duke University Press 2012) 694.

[89] Quint (n 57) 138.

[90] See, e.g., the sources cited earlier in n 50.

integration. It resumed, rather, the Court's historic role of modernizing the German legal and political landscape, and correcting the course of the German *Sonderweg*. The framers of the Basic Law had looked forward to reunification and anticipated the time when 'the German people'—all of them—would enact a new constitution adopted in a free election.[91] This did not happen in 1990, and it fell to the Constitutional Court to adapt the Basic Law to the imperfect structures of reunification—or rather to apply the Basic Law in a fashion that fostered the fusion of two long-disparate peoples under a united constitution. In this context, correcting Germany's special path meant accommodating the special circumstances of reunification. In the *Land Reform* cases, this meant preserving a partial victory for the peoples of the East at the expense of Westerners whose just grievances would never be fully redressed.

E. Confronting the GDR past

Soviet land reform was by no means the only realm in which the Court confronted the heritage of oppression in East Germany. As had happened during the 1950s, the Court became after reunification a forum for *Vergangenheitsbewältigung*—or coming to terms with the past. Reunification was universally expected to bring the blessings of the *Rechtsstaat* to the former East Germany. One of the most burning jurisprudential questions of the reunification era was how to redress injustices committed by the GDR *Un-rechtsstaat*. The question was particularly fraught because the *Rechtsstaat* principle itself banned retroactive punishment. Article 103(2) GG stipulated that an act could be punished only if its criminality was clear when the crime was committed. Accordingly, the Unification Treaty directed the courts of the Federal Republic, in their jurisdiction over crimes committed in GDR, to apply GDR law.[92] That law, however, was controlled by an authoritarian regime that committed or condoned, instigated or overlooked many of the crimes for which the Federal Republic's newest citizens now sought redress. The prospect that the *Rechtsstaat* principle would shield the agents of a notorious *Un-rechtsstaat* from legal reckoning placed two fundamental principles—legal security (*nulla crimen, nulla poena sine lege*) and material justice (*nulla crimen sine poena*)—at seemingly insuperable odds.

The conflict was not new to German courts. For decades the criminal chamber of the Federal Court of Justice had wrestled with this knotty conundrum in cases involving Nazi crimes. From time to time, the Court of Justice invoked the famous thesis of Gustav Radbruch, one-time justice minister in the Weimar Republic and one of the most influential German legal philosophers of the twentieth century.[93] Radbruch was one of the few German jurists who neither fled from the Third Reich nor made compromises under it. He came to enjoy enormous authority, both moral and intellectual, thanks to his forceful thinking, personal integrity, and elegant prose. Radbruch's collected works fill twenty volumes,[94] but his greatest fame came from a few sentences in a 1946 essay published in the *Süddeutsche Juristenzeitung* called 'Legal Injustice and

[91] See Article 146 GG.

[92] An exception was made for cases in which the FRG's own laws levelled a milder sanction for the same crime.

[93] On Radbruch, see Nathalie Le Bouëdec, *Gustav Radbruch: juriste de gauche sous la République de Weimar* (Les Presses de l'Université Laval 2011); Irina Mohr, *Gustav Radbruch als Reichsjustizminister (1921–1923)* (Friedrich-Ebert-Stiftung 2004); Arthur Kaufmann, *Gustav Radbruch: Rechtsdenker, Philosoph, Sozialdemokrat* (Piper 1987).

[94] See Gustav Radbruch, *Gesamtausgabe in 20 Bänden*, 20 vols, Arthur Kaufmann ed (CF Müller 1987–2003).

Supralegal Justice'.[95] The essay's thesis entered the legal lexicon as the 'Radbruch Formula' (*Radbruch'sche Formel*).[96]

The formula posited that material justice could trump legal security (and with it the ban on retroactive punishment) when the conflict between positive law and material justice reached such a degree that the positive law became *unrichtiges Recht*—an untranslatable phrase denoting law so unjust as to become itself unlawful.[97] Under this formula, the retroactivity ban yielded to natural law and natural justice. After reunification, many Germans believed the formula should apply to crimes commanded by the GDR regime and committed by its henchmen. This was especially so in one particularly charged context—the fatal shooting at the East German border of persons escaping to the West.

From the Wall's construction in August 1961 until the world-historical opening of the border in November 1989, 264 refugees were killed while attempting to cross the border.[98] The most recent of these was Chris Gueffroy, a twenty-year-old student gunned down on 6 February 1989. Stories of the border slayings were horrific and chilling. In 1972, a twenty-nine-year-old man from East Berlin tried to swim across the Spree to Kreuzberg but was mowed down by the machine gun fire of the GDR border patrol. A week later, the man's widow posted signs indicating that he was missing. Four weeks after this, police informed the widow that her husband's corpse had been found in the vicinity of Berlin's 'Museum Island', identified by his finger prints, and already cremated. He had died by suicide, the widow was told. On the very night of the man's death, the soldiers who shot him were relieved from their post—but also decorated and given 150 Marks.[99] Rewards for soldiers who killed refugees at the border were not uncommon. Almost immediately after reunification, the effort began to bring such soldiers—and the GDR operatives who incited, and applauded them—to justice.

Justice, of course, was the crucial term. In a series of trials that one scholar numbered among the 'most spectacular and important in recent decades'[100]—probably the most dramatic trials in Germany since the 'Holocaust trials' of the 1960s—the country's highest criminal courts ruled that soldiers at the border and members of the defence council could indeed be prosecuted for crimes arguably justified by GDR law when the crimes were committed. The Berlin *Landgericht*, in the case of Chris Gueffory, reached this result by expressly invoking Radbruch.[101] The criminal chamber of the Federal Court of Justice avoided citing Radbruch directly and underscored instead the international law obligations by which the East German government had bound itself, as well as the court's duty to construe GDR law in a manner that reinforced human rights.[102] These rulings made possible

[95] Gustav Radbruch, 'Gesetzliches Unrecht und übergesetzliches Recht' (1946) 1 *Süddeutsche Juristenzeitung* 105.

[96] On the 'Radbruch Formula', see Volker Neumann, 'Richtiges Recht. Radbruchs Rechtsphilosophie und der staatsrechtliche Positivismus Heidelberger Prägung' in Karin Buselmeier and others (eds), *Auch eine Geschichte der Universität Heidelberg* (Edition Quadrat 1985) 211; Horst Dreier, 'Die Radbruchsche Formel—Erkenntnis oder Bekenntnis?' in Heinz Mayer (ed), *Staatsrecht in Theorie und Praxis. Festschrift für Robert Walter* (Manz'sche Wien 1991) 117.

[97] Radbruch (n 95) 107.

[98] See Christian Schaefgen, 'Zehn Jahre Aufarbeitung des Staatsunrechts in der DDR' (2000) 54 *Neue Justiz* 1, 4.

[99] See 40 BGHSt 241 (1994).

[100] Horst Dreier, 'Gustav Radbruch und die Mauerschützen' (1997) 52 *JuristenZeitung* 420, 420.

[101] Judgment the Landesgericht Berlin, 2 Js 48/90 KLs (9/91) (20 January 1992) in (1992) 47 *JuristenZeitung* 691.

[102] See, e.g., 39 BGHSt 1 (1993); 39 BGHSt 168 (1993); 41 BGHSt 101 (1995); for a fuller discussion, see Gerhard Dannecker and Kristian F. Stoffers, 'Rechtsstaatliche Grenzen für die strafrechtliche Aufarbeitung der Todesschüsse an der innerdeutschen Grenze' (1996) 51 *JuristenZeitung* 490.

a series of convictions for killings at the border. They also prompted, from those thus convicted, a series of petitions to the Constitutional Court.

The Court consolidated these convictions for a single ruling, which the Second Senate announced on 24 October 1996. It was a judgment, one commentator wrote, that 'reached the very limits of coming to terms with the past through law [*Vergangenheitsbewältigung durch Recht*]'.[103]

The Court began by stressing the limited scope of its review. The FCJ's criminal chamber was the final instance in matters of criminal law. The Second Senate's review extended only so far as judgments of the Court of Justice implicated fundamental rights.[104] The complainants' convictions obviously implicated Article 103(2)'s retroactivity ban, but they did not, the Senate concluded, violate that provision. In so ruling, the Senate stressed that the Federal Republic had assumed the criminal jurisdiction of a state whose laws comprehended neither democracy, nor separation of powers, nor protection of fundamental rights.[105] The Senate stressed, too, that unless the courts of the Federal Republic applied GDR law by the light of 'human rights recognized by international law', the Federal Republic would betray its own premises as a *Rechtsstaat*.[106]

The situation, in the Court's view, was 'entirely unique'.[107] This uniqueness allowed courts to depart from the otherwise universal retroactivity ban. The GDR law that allowed border guards to open fire on fleeing citizens placed the state's desire for an impermeable border over the individual's right to life. In the Senate's view, this distortion of values was an 'obvious, intolerable offence against elementary commands of justice and against human rights protected by international law'.[108] The GDR's border practices amounted to 'extreme state-sponsored injustice' (*extremes staatliches Unrecht*). The border law was unjust and irregular within the GDR's own legal order[109] and flagrantly offensive to the state's obligations under international law.[110] Those who crafted and enforced the law forfeited the shield of contemporary legality.[111] Their subsequent criminal convictions, then, were consistent with Article 103(2) GG. The Court made its nod to Radburch, but the ultimate grounds for its decision lay elsewhere. The decision was based, not on principles of natural law, but on a narrow reading of the retroactivity ban.

In the press, the unanimous judgment was broadly praised.[112] The outcome corresponded with intuitive notions of justice and a natural longing for retribution. Many academic commentators, however, had reservations. Some levelled sharp criticisms. For one thing, Radbruch had outlined his famous formula in an historical context that Radbruch himself deemed *sui generis*—unprecedented and unrepeatable. The Court's application of the formula (or rather, its invocation of international law to reach the same result) to crimes committed in the GDR seemed to posit a moral equivalence between the Nazi state and the SED state that made many observers uneasy. This was especially so because, since the days of Willy Brandt's *Ostpolitik*, the Federal Republic had cozied up to the GDR regime quite conspicuously while turning a blind eye to ongoing outrages at the inner-German border.[113] To some it seemed the worst sort of victor's justice to allow retroactive criminalization of acts that FRG officials had long lacked the courage to denounce.

The Court's treatment of international law was equally problematic. The Court was rather vague about which international norms it was applying and said nothing at all about

[103] Thilo Rensmann, 'BVerfGE 95, 96—Mauerschützen' in Jörg Menzel (ed), *Verfassungsrechtsprechung: Hundert Entscheidungen des Bundesverfassungsgerichts in Retrospektive* (Mohr Siebeck 2000) 605, 611.

[104] 95 BVerfGE 96, 128 (1996). [105] ibid 133. [106] ibid. [107] ibid. [108] ibid 135.

[109] ibid 137. [110] ibid 135. [111] ibid 136.

[112] See Dreier (n 100) 422; Robert Leicht, 'Schuld und Sühne' *Die Zeit* (15 November 1996).

[113] See Rensmann (n 103) 609–10.

how intensely those norms must be violated to justify an exception to the retroactivity ban.[114] What's more, whatever the international norms violated by the border shootings, and however heinous the violation, it remained a fact that the retroactivity ban was itself an inviolable redoubt of international human rights.[115] Both the European Convention on Human Rights and the International Covenant on Civil and Political Rights proclaimed the ban absolute.[116] But the Second Senate did not mention the retroactivity ban's status as an international human rights norm. Such selective citation gave one observer 'the impression that legal analysis was trumped by the feeling that the perpetrators must be punished'.[117]

The criticisms were entirely fair, but for most citizens they missed the point. The Court had surely strained the boundaries of judicial *Vergangenheitsbewältigung*, but it did so at a time when many citizens thought mastering the past hadn't gone far enough. Earlier, Bärbel Bohley, a brave paladin of civil rights in the GDR and co-founder of the New Forum movement, voiced the dismay of many East Germans reformers over inadequate redress of official injustice. 'We wanted justice', Bohley said, 'but what we got was the *Rechtsstaat*'.[118] In a variety of ways—particularly with its *Land Reform* judgemnts—the Constitutional Court itself had contributed to such disappointment. The Court had long been the champion and guarantor of the West German *Rechtsstaat*. Many saw in this the Court's greatest contribution. In its reunification-era jurisprudence, the Court applied *Rechtsstaat* principles in its engagement with the East German past—sometimes, some thought, to the detriment of material justice. In its *Border Guards* decision, the Court tilted the balance in the other direction. It was an extraordinary judgment, by its own terms exceptional. The Court pointedly linked the retroactivity ban to its roots in liberal democratic constitutionalism—to its roots in the *Rechtsstaat*. The ban's application, the Court suggested, to the laws of an earlier Un-*rechtsstaat*—one that eschewed liberal values, democratic process, and constitutional limits on state power—was highly problematic.

As an historical matter, this was nonsense. The retroactivity ban emerged under monarchical regimes centuries before the advent of democratic constitutionalism.[119] But by forging a rhetorical link between the retroactivity ban and the core structural values of the Basic Law, the Senate signalled that its judgment was about more than confronting a history of repression. The Court was also engaged, at least in part, in the work of re-founding. In its first decade, the Court had used several judgments engaging with the Nazi past as a vehicle for proclaiming the fundamental values of the West German republic in contradistinction to the Third Reich. Now, in its fifth decade, the Court used its confrontation with the GDR past to articulate a founding credo for the *united* German republic. In the 1950s, the Court heralded West Germany's return from its historical *Sonderweg*; in the 1990s, the Court declared the *Sonderweg*'s definitive conclusion.[120]

114 ibid 609.

115 See Markus Krajewski, 'Mauerschützen und Menschenrechte' (1997) 52 *JuristenZeitung* 1054, 1054.

116 The GDR was a signatory to the latter agreement—a fact that underlay the Court's finding that the border shootings breached East Germany's own international obligations.

117 Krajewski (n 115) 1055.

118 In Ingo von Münch, 'Rechtsstaat versus Gerechtigkeit?' (1994) 33 *Der Staat* 165, 165.

119 See Rensmann (n 103) 609.

120 For general discussions on the tension between upholding the rule of law and 'overcoming the past' through law, see Hans-Jürgen Papier and Johannes Möller, 'Die rechtsstaatliche Bewältigung von Regime-Unrecht nach 1945 und 1989' (1999) 52 *NJW* 3289; 'Der Rechtsstaat und die Aufarbeitung der vor-rechtsstaatlichen Vergangenheit' (1992) 51 *Veröffentlichung der Vereinigung der Deutschen Staatsrechtslehrer* 9–175.

F. Financial equalization among the German states

Proclaiming the end of a special path proved easier than repairing its effects. The logistical and administrative challenges of reunification were daunting. The monetary price was enormous. One reason that victims of the Soviet land reform received only limited financial compensation was that unification was crushingly expensive. The cost of unifying the machinery of state and of extending the Federal Republic's social programmes to the new Eastern states was staggering.

By comparison with their prosperous Western neighbours, the five new states were backward and poor. The Basic Law imposed on the federation a duty to assist the poorer *Länder* financially. It did so by prescribing a dual regime of financial redistribution. Article 106 GG provided for the 'vertical' distribution of tax revenue between states and federation; Article 107 for the distribution of revenue among states. Article 107(2) prescribed that 'differences in financial strength among the states' be 'reasonably equalized'. Such equalization took various forms, including direct transfers to poorer states from their wealthier peers.

The programme of direct transfers stirred controversy from the very beginning. In the infant days of the Federal Republic, the state of Baden-Württemberg and the city-state of Hamburg, burly with postwar brawn, objected in a constitutional complaint that the requirement of state-to-state transfers offended the guarantee in Article 109(1) GG that the states would remain fiscally autonomous and financially independent. The Court rejected the complaint, but warned that equalization measures would violate federalist principles if they weakened contributing states or effected a 'leveling' among the states at large.[121] On the other hand, the Court insisted that states have duties as well as rights, including the duty, within reasonable limits, to assist their weaker peers. In 1986, the Court nixed an equalization statute that failed equalize *enough*.[122] The Court admonished the Bundestag to fix the problem before the fiscal year 1988. An acquiescent Parliament complied. At the time, no one foresaw the dramatic events that led to reunification, nor the avulsive impact of reunification on inter-state revenue distributions.

In 1992, the Court reviewed a complaint filed jointly by Hamburg and Bremen. The two states alleged that the outlays required of them were excessive in light of the benefit the port cities conferred (and the costs they incurred) by maintaining and repairing their important harbours. The Court disagreed, but did rule that Bremen—which in some years received no tax subsidies and had always received less than the Saarland, whose debts were smaller—had been discriminated against unlawfully.[123] Federal subsidies to both Bremen and the Saarland, moreover, had been too small—a state of affairs the federation and the other states must help repair. Parliament duly produced a reform statute the following year. Enacted against the backdrop of German unification and the fiscal trauma of the Eastern states, the new law would, beginning in 1996, incorporate the new states in the financial equalization scheme.

Because the law was passed by a Parliament governed by a Union–FDP majority, wealthy conservative states initially bit their tongues. That changed in 1998, when a coalition led by the SPD, with the mercurial Greens as junior partner, swept into power in the third united German election. It was galling enough for conservative governments with plentiful coffers to make large annual transfers to weaker siblings. But to have the process directed by Greens and Social Democrats was too much.[124] Three conservative titans—Bavaria,

[121] 1 BVerfGE 117, 131 (1952). [122] 72 BVerfGE 330 (1986). [123] 86 BVerfGE 148 (1992).

[124] Under the 1998 regime, the top five tax-gathering states (Hesse, Baden-Württemberg, Hamburg, Bavaria, and North-Rhine Westphalia) dropped to the bottom of the rankings of total tax revenues. Bremen, whose

Baden-Württemberg, and Hesse—challenged the reformed equalization scheme before the Constitutional Court.

They won an important, but partial, victory. In a judgment dated 11 November 1999, the Second Senate ruled that parts of the equalization regime were incompatible with Articles 106 and 107 GG.[125] The Senate's assessment had two central prongs. First, the regime lacked concrete standards governing its long-term administration. Second, in its provision for state-to-state transfers, the equalization law equalized too much.

In the first case, the Court held that the *Rechtsstaat* principle required Parliament to design standards governing equalization payments without regard to a specific political context or a given fiscal application. Though a Rawlsian 'veil of ignorance' could not be drawn over Parliament's deliberations, the constitution mandated that the legislature must articulate principles against which the public could hold Parliament—and parliamentarians each other—accountable.[126] The 1993 regime, by contrast, had been shaped unduly by the politics of the present hour. That regime required revision. In revising it, Parliament must take a longer view.

The revised law must also effect less dramatic changes in the relative financial strength of the several states. The 1993 regime wrought an entire overhaul, rocketing some states from near the bottom of the rankings almost to the top—and others from top to bottom. But Article 107(2) GG did not require actual equality among the states, and Article 109 forbade it.[127] The constitution required Parliament to balance the autonomy, independence, and individuality of the states against the states' duty of mutual support. It was, as so often, a matter of proportion. The law must strengthen weak states, but not weaken strong ones—at least not too much. It must reduce differences among the states, but not abolish them. The constitution commanded solidarity but proscribed levelling.[128] In all of these respects, the 1993 law had gone too far.

The Second Senate did not, however, proclaim the law immediately null and void. As it had often done before, the Court granted a stay of execution. Parliament had until the end of 2004—a period of a little more than five years, or a full legislative term—to revise the law. This meant that the newer states—each of which had benefited hugely from equalization payments since 1996—would continue to enjoy the begrudging charity of their siblings for several years more. Many hoped (in vain, as it happened) that by then the Eastern states' need for assistance would have diminished considerably. On the other hand, the Western states that brought the complaint were gratified that enormous, compulsory state-to-state transfers would not persist indefinitely. And parliamentarians rejoiced that they wouldn't need to address this thorny issue again overnight.

It was, in short, one of those rare decisions that genuinely pleased everybody. Newspaper commentary described an atmosphere of rarefied harmony.[129] All sides of the controversy declared themselves content; all proclaimed victory.[130] As a mediator of the tensions of reunification, the Court had surpassed itself. It was the final, masterful performance of the case's rapporteur, Justice Paul Kirchhof, who would finish his term in the Court—as would Justice Dieter Grimm, the intellectual giant of the First Senate—at the end of 1999.

pre-equalization financial capacity placed it fifteenth among the sixteenth states, was launched into first place by the equalization regime. See Kommers and Miller (n 88) 98–99.

[125] 101 BVerfGE 158 (1999). [126] ibid 218. [127] ibid 219–22. [128] ibid 222, 224.

[129] See, e.g., 'Machtwort zum Schluss' *FAZ* (12 November 1999)1; 'Positive Reaktionen bei Geber- und Nehmer-Länder' *Handelsblatt* (12 November 1999); 'Bundesländer sind froh über das "weise" Urteil' *Frankfurter Rundschau* (12 November 1999).

[130] 'Mit der Entscheidung sind alle zufrieden' *FAZ* (12 November 1999) 2.

Despite occasional scholarly criticism,[131] the *Equalization* judgment marked a hopeful moment for Court and country. The Federal Republic had celebrated its fiftieth anniversary, the fall of the Berlin Wall its tenth. The agonizing first decade of reunification was drawing to a peaceful close. The Court had played an important integrative role during that decade, and had won plaudits, not for this alone, but for its broader contributions to the 'success story' of the Federal Republic itself.[132] At century's end, it was hard to believe that only four years earlier the Court had been embroiled in a controversy unlike any it had faced before.

II. Bio-Politics and Social Policy

Before the crisis of the mid-1990s, however, the Court wrestled with other thorny reunification questions. The union of Germany's two halves merged two profoundly different societies as well as two markedly disparate economies. How to fuse these societies—and how to fashion a legal regime that sensibly accommodated both—was no simple matter.[133] Reunification raised a Social Question as well as recasting a German one. Various aspects of GDR social policy—particularly with regard to gender equality—struck some Western observers as more progressive than analogous policies in the Federal Republic. Still others felt that, whatever their comparative merits, many facets of the Federal Republic's social policy needed updating. For many liberals and progressives, reunification offered the chance for social reform—just as it provided an opportunity to revamp the civil service.

As with the civil service, the legislative response to the possibility of reshaping social policy was, for most progressives, uninspiring. And as had happened so often in the past, parties and movements disappointed by their legislators in Bonn tried their luck with the judges in Karlsruhe. Hope sprang eternal that lawmakers in red robes would prove more congenial than lawmakers in dark suits. As a result, in the realm of social policy, as with so many other aspects of reunification, the Court would play a frequently mediating, sometimes modernizing, always controversial role.

A. Abortion—*und kein Ende*!

There was one bristly area—the bristliest of all—in which legislative intervention was unavoidable. The laws governing abortion in the 'old' Federal Republic and in the former East Germany were irreconcilable. The GDR had long since adopted the 'period solution' (*Fristenlösung*) that had been embraced in Bonn after the fiery debates of the early 1970s, but rejected by the Constitutional Court in 1975.[134] In East Germany, abortions were available on demand during the first twelve weeks of pregnancy. East German abortion law expressly framed a woman's right to choose an abortion as a principle of equal protection. West German law, by contrast, remained bound by the Court's 1975 judgment, in response to which the Bundestag had passed a law pronouncing (in principle) that abortion was

[131] See Hans Peter Bull, 'Finanzausgleich im "Wettbewerbsstaat"—Bemerkungen zur neuen Föderalismustheorie und zu ihrer Bedeutung für den Länderfinanzausgleich' (1999) 52 *Die Öffentliche Verwaltung* 269, 273; Jens-Peter Schneider, 'Bundesstaatliche Finanzbeziehungen im Wandel' (2001) 40 *Der Staat* 272.

[132] See Rolf Lamprecht, 'Das Über-Ich der Nation' *Der Spiegel* (1 September 1998).

[133] See, e.g., Ingo Fritsche, 'Aspekte deutsch-deutscher Rechtsangleichung auf dem Gebiet des Zivilrechts' (1990) 23 *Kritische Justiz* 162.

[134] See 39 BVerfGE 1 (1975).

criminal *throughout* pregnancy, but making exceptions (in practice) for pregnant women who could show that one of several 'indications' justified an abortion. The exceptions covered cases in which carrying the pregnancy to term endangered the life, or seriously threatened the health, of the expectant mother ('medical indication'); competent medical authority predicted that the child would be born with severe and irremediable defects ('embryological indication'); the pregnancy resulted from rape or sexual assault ('criminological indication'); and, most controversially, carrying the pregnancy to term would impose an intolerable burden on the pregnant woman ('social indication').

The regime pleased no one. Abortion opponents complained that the exceptions devoured the rule. In nearly all cases, a West German woman who wanted an abortion—and who had time to travel and money to pay—could get one.[135] Particularly galling to defenders of unborn life was the unfettered expansion of the 'social indication', which accounted for more than 80 per cent of all 'indicated' abortions.[136] On the other hand, to champions of abortion rights the law's lax enforcement did not erase the indignity of its regressive letter. What the law characterized as a criminal act justified by an enumerated exception, abortion rights activists viewed as a fundamental right. Many women chafed at burdensome bureaucratic requirements that required them to consult three separate authorities: the counselling centre, the physician who assessed the asserted indication, and the physician who performed the abortion (who also must confirm the indication). Both camps objected that the availability and quality of abortion services varied hugely from region to region and state to state. Many West Germans found the existing law increasingly offensive; to most Easterners, it was unacceptable. In any case, one or the other side of unified Germany—perhaps both sides—would have to amend its abortion laws. In the face of implacable conflict, something had to give.

Deciding *what* had to give was a riskier business than those who negotiated the reunification treaty had dared to undertake. East German representatives were passionately committed to their country's liberal abortion law. West German leaders, many of them influenced by the Catholic/Christian orientation of the governing Union parties, were unwilling to abandon their country's grounding position that abortion was unlawful and immoral. The impasse was so stark, and the question so explosive, that it threatened to wreck or delay the entire project of reunification.[137] In the face of potential gridlock, the negotiators of reunification took a cue from the sages of the 1948/49 Parliamentary Council, who in their day adroitly dodged thorny issues by deferring some controversies for later parliamentary resolution.[138] In the Unification Treaty, it was agreed that until January 1993 the united German Republic would have a bifurcated abortion law—the law of the former GDR would apply in the Eastern states, the law of the old Federal Republic in the West. Abortion would be lawful or criminal depending on geography. The united German Parliament must mend the discrepancy by New Year's Eve, 1992.

Meanwhile debate raged. Few topics so dominated opinion columns as did abortion reform.[139] The same was true in Parliament. During the early months of 1992, the Bundestag

[135] See Albin Eser, 'Reform of German Abortion Law: First Experiences' (1986) 34 *Am J Comp L* 369, 381 ('[U]nder [the] indication regime almost every pregnant woman could obtain an indication if she did so with determination'.).

[136] See Rolf Stürner, 'Die Unverfügbarkeit ungeborenen menschlichen Lebens und die menschliche Selbstbestimmung' (1993) 45 *JuristenZeitung* 709, 711.

[137] For a contemporary discussion of the abortion debate in the context of reunification, see Peter H. Merkl, *German Unification in the European Context* (Pennsylvania State University Press 1993) 176–81.

[138] This was the case, for instance, with Article 131 GG, which directed Parliament to regulate the compensation and reinstatement *vel non* of Nazi-era civil servants. See the discussion in Chapter 1.

[139] For a sampling only of Germany's two leading weeklies, see 'Verschärfen oder streichen?' *Die Zeit* (16 February 1990); 'Vage Hoffnungen' *Der Spiegel* (9 April 1990); Monika Maron, 'Letzter Zugriff auf die Frau'

faced no fewer than seven separate reform proposals, ranging from calls (on the Left) for outright repeal of section 218 of the criminal code to appeals (on the Right) to refine and restrict the permissible 'indications'.[140] The parliamentary cacophony was compounded by the decision to release individual members from the traditional demands of party discipline. Abortion was an issue that sliced across party lines and divided the governing coalition. In the parliamentary battle over abortion reform, every member stood alone with his or her conscience. Remarkably, the Bundestag forged a compromise—brought forth by a multi-partisan group of SPD, FDP, CDU, and Bündnis 90 members—that garnered a broad majority. In late June 1992, by a vote of 357 to 283, the Bundestag promulgated a new Pregnancy and Family Assistance Act.[141] The Bundesrat endorsed the law shortly thereafter.[142]

The reform law adopted as its mantra 'aid rather than punishment' (*Hilfe statt Strafe*). Its philosophy was that unborn life was protected best, and women empowered most fully, by a comprehensive programme of information and counselling, as well as a general atmosphere in which women's professional and educational opportunities and rights were bolstered by state financial assistance and publicly funded childcare for all children age three and older.[143] In its most dramatic provision, the law decriminalized abortion during the first twelve weeks of pregnancy—and thus departed from the Court's 1975 judgment—but coupled decriminalization with a counselling requirement. Under the reform law, a woman less than twelve weeks' pregnant could secure an abortion that was 'counseled' but not 'indicated'. Such abortions would be 'not unlawful' (*nicht rechtswidrig*)—a fateful designation.

Immediately, the new law was challenged in a constitutional complaint raised by 249 Christian Democrat parliamentarians (all from the former West Germany) and supported by the federal chancellor, Helmut Kohl. In addition to asking the Court to quash the reform law outright, the complaint asked the Court to enjoin its operation pending the Second Senate's decision on the merits. The Court granted the injunction after a day-long preliminary hearing held 4 August 1992—one day before the law was to enter into force.[144] The

Der Spiegel (14 May 1990); 'Ich setze auf die Frauen' *Der Spiegel* (14 May 1990); 'Die sind tierisch hinterm Mond' *Der Spiegel* (14 May 1990); 'Wie Pusteblumen' *Der Spiegel* (2 July 1990); 'Der 218 muβ weg!' *Der Spiegel* (30 July 1990); 'Dritter Weg' *Der Spiegel* (30 July 1990); Hans Schueler, 'Einig im Unsinn?' *Die Zeit* (3 August 1990); Frauen in Bedrängnis *Die Zeit* (17 August 1990); 'Zug gegen die Wand' *Der Spiegel* (27 August 1990); 'Wirklicher Durchbruch' *Der Spiegel* (3 September 1990); 'Fataler Eindruck' *Der Spiegel* (29 October 1990); 'Sachverstand geborgt' *Der Spiegel* (14 January 1991); 'Vernünftige Lösung' *Der Spiegel* (11 March 1991); 'Beraten—aber wie?' *Die Zeit* (10 May 1991); 'So teuer wie die Einhiet' *Der Spiegel* (13 May 1991); 'Kulturkampf, anders' *Der Spiegel* (13 May 1991); 'Das zerreiβt die Partei' *Der Spiegel* (13 May 1991); 'Gottes Gebot?' *Die Zeit* (14 June 1991); 'Wann Leben beginnt' *Der Spiegel* (22 July 1991); 'Alle beteuern: Hilfe statt Strafe' *Die Zeit* (9 August 1991); 'Frauen müssen nicht gebären' *Die Zeit* (9 August 1991); 'Eine Zelle ist noch kein Mensch' *Die Zeit* (23 August 1991); 'Viele Konflikte sind unlösbar' *Die Zeit* (30 August 1991); 'Frauen entscheiden—wer denn sonst?' *Die Zeit*, (13 September 1991); 'Wie ein Gong' *Der Spiegel* (16 September 1991); 'Ein Gesetz für die mündige Bürgerin' *Die Zeit* (27 September 1991); 'Wille zum Absprung' *Der Spiegel* (11 November 1991); 'Gegen die Wand' *Der Spiegel* (30 March 1992); 'Hoffen auf die Roten Roben' *Der Spiegel* (18 May 1992); 'Strafrecht wird zum Selbstzweck' *Der Spiegel* (25 May 1992); Rudolf von Augstein, 'Widerstand ist Pflicht' *Der Spiegel* (25 May 1992); 'Partei ergreifen gegen das Gewissen' *Die Zeit* (29 May 1992); 'Das werdende Leben geht vor' *Der Spiegel* (29 June 1992); 'Das ist noch nicht zu Ende' *Der Spiegel* (29 June 1992); 'Der Riβ geht durch die Union' *Die Zeit* (3 July 1992); 'Kippt Karlsruhe den Kompromiβ?' *Die Zeit* (3 July 1992); 'Mit dem Tunnelblick' *Der Spiegel* (3 August 1992); 'Einstweilen nicht' *Die Zeit* (7 August 1992); 'Gefährliche Lücken' *Der Spiegel* (10 August 1992); 'Letzte Worte' *Der Spiegel* (9 November 1992); 'Wenig Hoffnung auf Karlsruhe' *Die Zeit* (4 December 1992); Rudolf von Augstein, '§ 218, zum zweiten' *Der Spiegel* (7 December 1992); 'Die Konterkapitäne von Karlsruhe' *Der Spiegel* (7 December 1992).

140 For a helpful summary of the proposals, see Albin Eser, 'Das neue Schwangerschaftsabbruchsstrafrecht auf dem Prüfstand' (1992) 46 *NJW* 2913.

141 *Schwangeren- und Familienhilfegesetz*.

142 For a thorough contemporary analysis of the law see Eser (n 140).

143 The childcare provisions were to enter force in 1996; the crushing costs of reunification dashed the plan.

144 86 BVerfGE 390 (1992). The injunction was to last six months, toward the end of which period the Court renewed it for six months more. 88 BVerfGE 83 (1993).

injunction outraged liberal activists and surprised constitutional scholars.[145] The Court, as everyone knew, had never enjoined a law and later left it wholly intact. Many feared for the life of the hard-won compromise.

The Second Senate delivered its much-anticipated merits decision on 28 May 1993.[146] It was an epic opinion—at least in its proportions. The majority judgment and two dissenting opinions fill 163 pages of the Court's official reports. The Senate reaffirmed the core principles of the 1975 *Abortion* judgment—now to be known as *Abortion I*—but at the same time made concessions to the perceived needs of a united Germany. For many critics, the concessions were cold comfort; for others, they had the feel of a clumsy compromise dictated by unprincipled pragmatism.[147]

The Senate majority began by reiterating what the First Senate had ruled eighteen years earlier: that the Basic Law obliged the state to protect human life—even unborn life, and even against expectant mothers.[148] On the other hand, abortion restrictions implicated fundamental rights of pregnant women, including human dignity, the right to life and bodily integrity, and free personal development.[149] But balancing these rights against a foetus's right to life—which abortion totally negated—yielded a clear outcome. The state remained constitutionally bound affirmatively to protect unborn life. The state could fulfil this duty only if the legislature banned abortions generally and imposed a general duty on pregnant women to carry their pregnancies to term.[150] This was true, the Senate stressed, for the full duration of the pregnancy. Human life began at least by the moment of implantation,[151] and the right to life did not vary in strength with the passage of time.[152] As a matter of constitutional principle, abortion must be outlawed at all stages of foetal development. There were exceptions, of course, to be defined, in principle, by the legislature. But in this case the Senate defined them itself.

At the heart of its holding the majority ruled that abortions 'counseled but not indicated' could be decriminalized but not fully legalized. The legislature might shield early-term, non-indicated abortions from criminal punishment, but such abortions must remain unlawful (*Rechtswidrig*).[153] By proclaiming otherwise the reform law offended the right to life. A 'merely counseled' abortion was a wrong, and though the state might choose not to prosecute this wrong, it must nonetheless denounce it.

So far the Senate's holding might seem merely semantic or symbolic. But the Court drew fateful consequences from its finding that the state must affirm the wrongfulness of non-indicated abortions. The most dramatic and controversial of these dealt with funding. Five justices agreed that, unless a pregnant woman had insufficient means to pay for an abortion, the cost of a non-indicated abortion must not be covered by public insurance.[154] Justice Ernst-Wolfgang Böckenförde, who otherwise joined the majority opinion, dissented on this point.[155] In Böckenförde's view, it was for the legislature, not the Court, to determine whether all non-indicated abortions were wrongful and therefore undeserving of public funds. The majority's stance was 'not only incoherent', Böckenförde wrote, it insulted 'women as persons, in their honour and righteousness'.[156] Public consciousness of the wrongfulness of abortion was little helped if the legal order exalted the distinction between right and wrong rhetorically but then ignored that distinction in practice by excluding health coverage across the board. Treating some abortions as normatively justified but legally wrongful

145 See Kommers and Miller (n 88) 387.
146 88 BVerfGE 203 (1993).
147 See, e.g., Susanne Walther, 'Thou shalt not (but thou mayest): Abortion after the German Constitutional Court's 1993 Landmark Decision' (1993) 36 *German Yearbook of Int'l L.* 385, 402 ('It looks like the [Court] sought to appease a wide spectrum of political positions … ').
148 88 BVerfGE 203 (n 146) 251–63.
149 ibid 254.
150 ibid 252–53.
151 ibid 251.
152 ibid 254.
153 ibid 299.
154 ibid 315–20.
155 ibid 359.
156 ibid 362.

would, Böckenförde predicted, generate moral schizophrenia.[157] This strong language earned loud applause from appreciative progressives. But the majority ruling—that merely counselled abortions were neither lawful nor fundable—carried the day.

The Senate majority also had many controversial things to say about counselling. Counsellors must be licenced by the state, the majority held, but they must not remain neutral. The counsel they gave must be resolutely pro-life. In this respect, the reform law's counselling provisions—which suggested that the aim of counselling was to help pregnant women make the choice best suited to their circumstances—were unconstitutional. The Court acknowledged that counselling would be most successful—that is, most likely to prevent an abortion—if it enjoyed the trust and cooperation of the woman being counselled. For this reason, counselling must be 'results open' (*ergebnisoffen*), leaving the last word to the woman herself.[158] But the counselling must also be 'goal-oriented' (*zielorientiert*)—and the goal was to persuade women not to abort.[159] One critic skewered this passage for mingling analytical confusion with 'frightful German'.[160] But the upshot was quite clear: licenced counsellors must urge pregnant women to continue their pregnancies. In doing so, the Senate added, counsellors must highlight the resources available to ease a pregnant woman's way forward.

For there must be such resources.[161] The state must take positive measures to foster a 'children friendly' society.[162] This included making life easier for mothers. The state must do more to shield women from educational and occupational disadvantages associated with pregnancy and child rearing.[163] In this regard, the Court applauded the law's social welfare provisions and prodded Parliament to extend them.

In sum, then, the state must proclaim the wrongfulness of non-indicated abortions, encourage women to forego them, and provide resources to help along the way.

But this wasn't all. In certain instances the state must retain the threat of criminal punishment—directed not at pregnant women, but at their treating physicians. Doctors must, on pain of criminal sanction, inform the patient that abortion destroys human life; ascertain her reasons for aborting; ensure that she has complied with counselling and waiting requirements; and conceal the unborn child's gender during the first trimester.[164]

It was a long, laborious, detailed decision. It read, at times, like a regulatory fiat. The opinion clearly required parliamentary revision and provided interim rules that would govern the law of abortion until that revision took place. Critics compared these provisional regulations to the 'emergency decrees' by which the federal president ruled a fractious country in the waning days of the Weimar Republic.[165]

In the eyes of the Senate's three dissenters, the majority had gone much too far. Justice Böckenförde, as we have seen, objected to the majority's universal designation of non-indicated abortions as unlawful and to its uniform denial to such abortions of publicly funded insurance. Ernst Gottfried Mahrenholz, the Court's vice president, and Justice Bertold Sommer dissented more categorically.[166] In their view, mother and child were not unambiguously discrete at all stages of pregnancy. A pregnancy, rather, was an evolving process of 'duality in unity' (*Zweiheit in der Einheit*), in which unity prevailed in the earliest

[157] ibid 363–64. [158] ibid 282. [159] ibid.

[160] Margrit Gerste, 'Ein Schritt vor, zwei zurück' *Die Zeit* (4 June 1993).

[161] 88 BVerfGE 203 (n 146) 259–61. [162] ibid 260. [163] ibid 259–60. [164] ibid 289–92.

[165] See Hans-Peter Schneider, 'Die Vollstreckungskompetenz nach § 35 BVerfGG—ein Notverordnungsrecht des BVerfG' (1994) 40 *NJW* 2590.

[166] 88 BVerfGE 203 (n 146) 338–58. Notably, the Senate's only woman, Justice Karin Graβhof, joined the majority opinion in full. As we saw in Chapter 3, the lone woman on the First Senate in 1975 dissented vigorously from the Court's judgment in *Abortion I*.

stages and duality later on. The unity of a pregnant woman and her foetus during the early weeks of pregnancy made it impossible, in a legal sense, for the woman to wrong the unborn child.[167] Under this view, a woman had a fundamental right, after counselling, to abort a pregnancy during the first twelve weeks. For Mahrenholz and Sommer, the majority's holding that merely counselled abortions were neither criminal nor yet lawful was not only conceptually confused but would numb society's moral sense and hinder its effort to protect the unborn.[168] In any case, the state's constitutional duty to protect the unborn did not and must not include a duty to shape the public's sense of right and wrong.[169] In holding otherwise, the majority had not merely assumed a prerogative that did not belong to the Court; it had invented an obligation that did not belong even to the legislature.

Negative reactions were swift and shrill. Christel Hanewinckel, a Social Democrat parliamentarian, fired off a philippic in *Die Zeit* that began, 'We, the parliamentarians in the Bundestag …'. Hanewinckel proudly praised the Bundestag's achievement in passing the reform measure and described the Court's decision as a 'slap in the face' to Parliament and an affront to women everywhere. 'Women in the new [Eastern] states', Hanewinckel wrote, 'simply cannot accept this decision'.[170] Critics shouted that the judgment was 'outrageous'—'a relapse into the Middle Ages'.[171] Others reviled the judgment in terms of class conflict. The Court had created 'two classes of women', said Gisela Böhrk (SPD), the women's minister in Schleswig-Holstein: those who could afford abortions and those who could not.[172] The country's poorest women would bear the brunt of the Court's ruling—and the vast majority of such women lived in the East. Angela Merkel, the federal family minister and a CDU rising star from the former East Germany, worried that women now faced 'a decidedly more difficult, if not impossible situation'.[173] It was an irony not lost on advocates of abortion rights that the judgment which 'finally' acknowledged a woman's ultimate, first-trimester right to abort was celebrated by conservatives and pro-life activists, by Catholic bishops and Bavarian traditionalists. Some characterized the judgment as 'one step forward, two steps back'.[174]

Liberal legal academics were disappointed as well. One wrote that the decision was 'a practical victory for women faced with unwanted pregnancy, but not a victory for women's rights'.[175] Others thought the Court's reasoning was flawed, not to say paradoxical or even oxymoronic. How could non-indicated abortions be both 'non-criminal' and 'unlawful'? How could counselling be both 'goal-oriented' and 'results-open'?[176] Analytically, it was not the Second Senate's finest hour.

The passage of time brought quieter responses. In an area as volatile as this, a compromise that gave to one side a de facto right to first trimester abortions and to the other a strong official statement that abortion is wrong—and averted the discomfort of overruling prior precedent—struck some as a pragmatic necessity and others as the best that could be done under the circumstances. The judgment even won a few moderate admirers. Reviewing the decision in the *JuristenZeitung*, Christian Starck, a public law professor in Göttingen, wrote that 'whoever reads the judgment in tranquility and takes the effort to follow the arguments, will be forced to acknowledge with respect the achievements of the Federal

[167] ibid 343–44. [168] ibid 348–49. [169] ibid 354–55.

[170] Christel Hanewinckel, 'Ausgegrenzt' *Die Zeit* (4 June 1993). Hanewinckel's title suggested that women—especially Eastern women—had been 'excluded' or 'segregated' by the Court's ruling.

[171] Noted in Ulrich Greiner, 'Die Welt von Fall zu Fall' *Die Zeit* (4 June 1993).

[172] In 'Abbruch auf Kredit?' *Der Spiegel* (31 May 1993). [173] ibid.

[174] Margrit Gerste, 'Ein Schritt vor, zwei zurück' *Die Zeit* (4 June 1993). [175] Walther (n 147) 399.

[176] See Monika Frommel, '§ 218: Straflos, aber rechtswidrig; zielorientiert, aber ergebnisoffen—Paradoxien der Übergangsregelungen des Bundesverfassungsgerichts' (1993) 26 *Kritische Justiz* 324.

Constitutional Court'.[177] Perhaps, Starck added, 'some shrill criticism in the immediate aftermath of the declaration of judgment would have been foregone had the judgment been thoroughly read beforehand'.[178]

All sides ultimately learned to live with the judgment. The immediate clamour was gradually replaced by low, lingering grumbles. It helped that the statute Parliament passed in response to the Court's decision softened some of the decision's sting. The compulsory counselling prescribed by statute struck some observers as less forcefully pro-life than the Court's judgment seemed to require. The statute also made it easier for pregnant women to fund abortions through public insurance, and limited sanctions on those who pressure women into obtaining abortions.[179] The Court had quashed one parliamentary compromise, and Parliament answered with another. Some Christian Democrat members murmured that the new statute fell short of the Court's commands, but they were in no mood to stir the ashes of a cooling controversy with another appeal to Karlsruhe.[180] To a remarkable degree, Germany's abortion controversy was over. For better or for worse, another deep and divisive reunification dispute bore the mediating mark of the Constitutional Court.[181]

B. Gender equality

Some of those most angered by the *Abortion II* judgment were willing to forgive the Court for the sake of its historical pattern of advancing gender equality in the Federal Republic. The Court resumed that role after reunification.

In a March 1991 judgment, for instance, the Court abolished one of the final vestiges of the Civil Code's patriarchy.[182] The controversy dealt with names. In 1957, Parliament passed a law—encountered already in Chapter 1—that sought to harmonize the civil code with the Basic Law's guarantee of legal equality for women and men. Among other things, the law granted married women the right to retain their maiden names as part of a hyphenated 'double name'. A generation later, in 1978, the Constitutional Court ruled that Article 3(2) GG required more. A woman must be able to keep her maiden name as her married name.[183] Already in 1976, Parliament had created this possibility, providing that engaged couples could decide, before their wedding, which surname—the husband's or the wife's—to bear as the common 'family name'. The law further directed that, in cases of conflict, the husband's surname would serve as the family name. In all cases, a man retained the right to append his given surname to his family name. More than a decade after the law was passed, a young couple took umbrage with this clear gender discrimination, and over several years battled their way to the Constitutional Court.

'For the sole reason that I am a woman', began the bride's complaint, 'I have now lost my name'.[184] This, she contended, was a frank affront to Article 3 GG. By a vote of 6:2, the First Senate agreed. The state might have good, constitutionally unobjectionable reasons for requiring a family to carry a common name.[185] But that couldn't justify a clear discrepancy in the treatment of men and women.[186] The law must be revisited. In the meantime, if a couple could not agree on a common name, each partner to the marriage would keep

[177] Christian Starck, 'Der verfassungsrechtliche Schutz des ungeborenen menschlichen Lebens' (1993) 48 *Juristenzeitung* 816, 822.

[178] ibid.

[179] The new law became the latest incarnation of section 218 of the criminal code (*Strafgesetzbuch*).

[180] See Kommers and Miller (n 88) 393–94.

[181] See Quint (n 57) 163–65.

[182] 84 BVerfGE 9 (1991).

[183] 48 BVerfGE 27 (1978).

[184] In 'Im Zweifel durchs Los' *Der Spiegel* (18 March 1991).

[185] 84 BVerfGE 9 (n 59) 19.

[186] ibid 19–20.

his or her pre-marital surname. Children born to such a marriage would bear both their parents' surnames, separated by a hyphen, the order determined by lot.[187] This was technically a provisional regulation. But it was hard to see, after the Court's decision, what other options remained open. It was an 'epochal judgment',[188] the newspapers sang, one whose 'emancipatory' effect was to 'abolish one of the last masculine privileges'.[189] Indeed. But it was also a logical extension, and a softer echo, of the Court's 1950s-era judgments repealing a husband's traditional right to have the final say *tout court*.[190] Perhaps for this reason, it was all the more welcome.

A harder case involved a law nominally designed to protect women. At the time of reunification, the Federal Republic was one of only a handful of European nations[191] that persisted in prohibiting women from working graveyard shifts.[192] (Exceptions were made for industries—such as healthcare and entertainment—that deemed nighttime work indispensable.) The ban was justified on the grounds that women's chronobiology made nocturnal labour particularly noxious, and that such biological factors were compounded by the reality that women continued to perform most household labour. A woman who worked the graveyard shift, it was argued, could expect no rest by day.

In a January 1992 judgment, the First Senate considered these justifications and found them wanting.[193] Medical research suggested that nighttime labour was insalubrious for all workers, male as well as female.[194] The state could not respond to a uniform risk with regulations that impeded the professional prospects of women only. As for housework, the Senate suggested that women need not suffer vocationally from a stereotype that already harmed them socially. The real risk, again, was for all workers—male or female—who had children to rear or households to run. And that risk, the Senate ruled, was better met by regulations tailored to the specific problem than by crude approximations tied only to gender.[195] The same held for concerns that it was unsafe for women to travel to and from a graveyard shift. The state must not impose on women the burden of its failure to protect them.[196]

Perhaps the Court's assessment of the gendered division of household labour was overly optimistic; research continued to suggest that most German men did precious little at home.[197] But the Court was clearly taking seriously the egalitarian letter of Article 3.

The same was true in the sensitive and controversial area of transsexualism. Back in 1978, the Court had ruled that transsexuals must be allowed to alter the official registration of their gender.[198] At the same time, the Court signalled that it was permissible—perhaps necessary—for Parliament to regulate this complex field. Parliament took the hint, passing a Transsexuals Act in September 1980. The law allowed transsexuals to change their official name and gender, but only after reaching the age of twenty-five. The Court struck down the registration provision in 1982,[199] and the name provision eleven years later.[200]

In its decision dated 26 January 1993, the First Senate wrote that the law could discriminate between transsexuals under and over the age of twenty-five only if it had a compelling reason to do so.[201] But in this case the compelling reasons ran the other direction. The years

[187] ibid 23–24. [188] Hanno Kühnert, 'Jedem den Seinen' *Die Zeit* (29 March 1991).
[189] 'Im Zweifel durchs Los' *Der Spiegel* (18 March 1991).
[190] See 3 BVerfGE 225 (1953); 10 BVerfGE 59 (1959). See the discussion in Chapter 1.
[191] Belgium, France, and Austria were the others.
[192] See Sibylle Raasch, 'Gleichstellung der Geschlechter oder Nachtarbeitsverbot für Frauen?' (1992) 25 *Kritische Justiz* 427.
[193] 85 BVerfGE 191 (1992). [194] ibid 207–08. [195] ibid 209. [196] ibid 210.
[197] See Raasch (n 192) 428–29 (quoting the following social science conclusion: '92% of men who live together with a female partner feel barely burdened by household labor. And rightly so: they do all but nothing'.).
[198] 49 BVerfGE 286 (1978). [199] 60 BVerfGE 123 (1982). [200] 88 BVerfGE 87 (1993).
[201] ibid 98.

just prior to one's twenty-fifth birthday, the Senate observed, were pivotal years of personal, educational, and professional development. A regime that required a transsexual to live with a new gender but an old name would create 'constant strained situations in the workplace, in the classroom, in interactions with the authorities, and in daily life'.[202] This circumstance was constitutionally intolerable, and the Court—without suggesting a new minimum age—declared the relevant provision of the Transsexuals Act null and void.[203]

The decisions just discussed reflected—some might say hastened—profound societal changes. However haltingly, gender equality had advanced during the life of the Bonn Republic. On questions of gender, many considered East German policy to be the more progressive, and in this regard the Court's aggressive, post-reunification enforcement of the Basic Law's equality provisions can be seen as an extension of its integrationist role. It was an extension with which many traditionalists were sorely displeased. Some of them fought back. In the aftermath of the Court's *Abortion II* judgment—and of the federal legislation promulgated in response to that judgment—Bavaria's conservative state government passed stiff laws regulating abortions. Under the new laws, only gynaecologists or related specialists could perform abortions, but no physician could earn more than a fourth of her annual income from performing abortions. All doctors, moreover, must refuse to perform an abortion if a woman failed to explain her reasons for seeking one. Bavarian physicians challenged these provisions before the Constitutional Court, which, broadly or narrowly, invalidated them all—the specialization requirement as a violation of physicians' occupational freedom under Article 12(1) GG, the others as beyond the competence of the states or preempted by federal abortion law.[204]

The *Bavarian Abortion* judgments earned the Court plaudits from the liberal press—including the critics of the *Abortion II* judgment—and ire in Bavaria itself. Such Southern indignation echoed an earlier, thunderous clash between Munich and Karlsruhe. That tumultuous controversy—to the astonishment of nearly everyone involved—unloosed the greatest crisis the Court had faced in more than forty years.

III. Freedom of Conscience and Expression

The fall of the GDR, and the subsequent reunion of Germany's divided halves, was the incisive political event in the life of nearly every German who experienced it—certainly of every German with no recollection of the Second World War. The impression left by the events of autumn 1989 was indelible, and not only for Germans. The heroes of the hour were East German dissidents; the iconic images those of mass protest rallies in East German cities—in Dresden, East Berlin, and above all, Leipzig. The early and enduring goal of the civil rights groups—most prominently the 'New Forum'—was open public discussion propelled by freedom of thought and expression. Such cries for free expression and assembly during the twilight months of the GDR led—indirectly but inexorably—to a reconsideration of these rights, and of their limits, in the united German Republic.

The same was true for questions of conscience. The German Democratic Republic had been a profoundly secular, even atheistic state. Across four decades, this official stance of non-religion or anti-religion profoundly shaped the beliefs (or unbelief) of the citizenry. Religious belief and confessional attachment were drastically weaker among East Germans than among their Western compatriots.[205] This fact raised complicated questions about the

[202] ibid. [203] ibid 101–03.

[204] See 96 BVerfGE 100, 102 (1997); 97 BVerfGE 117, 120 (1997); 98 BVerfGE 265 (1998).

[205] This discrepancy has endured. In a 2012 study, only 13 per cent of people living in the former East Germany said they had always believed in God; 52 per cent identified themselves as atheists. See Tom W. Smith, 'Beliefs

role of religious instruction—typically a core aspect of the curriculum—in the elementary and secondary schools of the new Eastern states. Indirectly, it fanned controversy concerning the role of religion in the classroom more generally—a controversy that exploded in a crisis engulfing Court and country.

A. Freedoms of expression and assembly

One striking feature of the protests that preceded and the celebrations that followed the opening of the inner-German border was the frequent waving of the Federal Republic's flag—the banner of black, red, and gold that has served, since 1848, as the traditional tricolour of German republicanism. Owing in part to a post-Third Reich reluctance to stress the outward emblems of nationalism, Germans have been less showy with their flag than other nations. But the black-red-gold banner bears immense historical significance. During the Weimar era there was a veritable battle of the banners: supporters of the Republic flew the black, red, and gold; its enemies waved black, red, and white—the colours of the Wilhelmine Empire. Flags were a form of politics by other means.[206] During the Third Reich, the republican tricolour disappeared, replaced by another, ubiquitous, notorious banner. The Basic Law enshrined the republican tricolour as the official flag of the Federal Republic.

Whether the flag enjoyed constitutional protection as well as constitutional significance was hotly disputed. The text of Article 22(2) GG declared only that '[t]he federal flag is black-red-gold'. Section 90(a) of the criminal code was more specific, threatening imprisonment or fine to anyone who 'disparages [*verunglimpft*] the colours, flag, crest, or hymn of the Federal Republic of Germany'. The law's constitutional validity had long been questioned.[207] It was formally challenged in a pair of cases considered during the 'year of German unity'. One case involved an alleged defilement of the federal flag, the other an imputed profanation of the national hymn.

The Court decided both cases on 7 March 1990—less than nine months after the United States Supreme Court decided *Texas v Johnson*, the landmark flag-burning case.[208] The German flag case involved the publication of a slender volume called *Just Leave Me Alone*, an anthology of anti-military prose and verse, the back cover of which featured a black-and-white photo of a swearing in ceremony of the *Bundeswehr*, in which new soldiers stood holding an unfurled federal flag. Over the photo was depicted a male figure aiming (behind cover of his hand) a stream of urine down toward the flag. The publisher was convicted under section 90 of the criminal code. His conviction was confirmed on appeal, and he challenged section 90 before the Constitutional Court.

The First Senate treated the publication as a work of art, thereby implicating the protections for artistic freedom in Article 5(3) GG, which did not allow for limiting legislation, rather than the speech protections of Article 5(1), which did.[209] As the Court had long held, encroachments on artistic freedom were justified only if they furthered other constitutional interests of equal or greater rank.[210] In this regard, the First Senate found that protection of the flag was rooted in the constitution. Article 22 GG did not furnish such protection directly, but it did presuppose the state's right to use certain symbols—the flag

about God across Time and Countries' Report for ISSP and GESIS (18 April 2012) <http://www.norc.org/PDFs/Beliefs_about_God_Report.pdf>.

[206] See Christoph Gusy, 'Anmerkung' (1990) 45 *JuristenZeitung* 638, 640 ('Thus the political battle against the Weimar Republic was essentially a battle against its colors'.).

[207] See the discussions in Günter Bemmann, *Meinungsfreiheit und Strafrecht* (Athenäum 1981) and Gerald Grünwald, 'Meinungsfreiheit und Strafrecht' (1979) 12 *Kritische Justiz* 291, 293–96.

[208] 491 US 397 (1989). [209] 81 BVerfGE 278, 289–90 (1990). [210] ibid 293.

among them—to appeal to the loyalty (*Staatsgefühl*) of its citizens.[211] As emblems of the Federal Republic's free democratic basic order, state symbols enjoyed constitutional protection. This protection was not boundless—the state could not arbitrarily multiply the number of its symbols—but neither did the constitutional guarantee of artistic freedom exclude the possibility of limits designed to protect official iconography.[212]

The key to balancing artistic freedom against protection of state symbols was to interpret correctly the work of art in question.[213] And this, the Senate held, the ordinary courts had not done. They had construed the collage on the publication's cover as an assault on the flag itself, and upon the Republic for which it stood. But in context the image was primarily anti-military and only secondarily anti-state.[214] What's more, allowances had to be made for the publication's satirical thrust. The point of satire is that it is pointed. Its purpose is to alienate and give offence; it employs exaggeration and irreverence to achieve this effect.[215] The ordinary courts' failure to interpret the publication as a work of satire had distorted the balance between symbol-protection and artistic liberty. Accordingly, the complainant's conviction could not stand.[216]

Nor, the First Senate ruled in the companion case, could the conviction of an obscure Nuremberg publisher who produced a salacious parody of the national hymn, '*Deutschland über Alles*'.[217] The parody's author had replaced Hoffmann's paean to 'German women, German faithfulness / German wine and German song' with a mock celebration of 'German Turks, German Pershings/German Big Macs, and German punk'; and, elsewhere, of 'German cola, German peepshows/German Marks and German sperm banks'.[218] Criminal prosecution rescued this merry exercise from instant oblivion. Full citation in a published opinion of the Constitutional Court conferred immortality.

In its judgment the First Senate said nothing of the poem's artistic merit, but concluded without difficulty that it was a 'work of art' enjoying constitutional protection.[219] As in the *Federal Flag* case, it was a work of art that the ordinary courts had wrongly construed.[220] It, too, was a work of obvious satire; it sought its effect by wresting the national hymn's 'idealizations' into 'exaggerated negative descriptions of our lived reality'.[221] The suppressed poem need not be read as an attack on the national hymn or on the state itself. It could be seen instead as reaffirming the nation's ideals by excoriating contemporary German society's failure to live up to them.[222] Once again, the ordinary courts' misinterpretation of the artwork itself propelled an erroneous balancing of the relevant constitutional values. As in the companion case, the Senate quashed the challenged conviction.

The judgments were controversial, not least because the 'works of art' thus granted constitutional protection were, to many observers, deeply reprehensible. Critics castigated the Court for countenancing 'libertinage'.[223] Others were troubled by the Court's unilateral conferral of constitutional value on state symbols. 'The Federal Constitutional Court presumes for its part', wrote one commentator, 'a value-setting competence: which values are "protected by constitutional law" is no longer determined by the Basic Law, but by the Federal Constitutional Court'.[224] Whether or not the Court was guilty of overreaching, it was clearly willing to enforce artistic and expressive freedoms more robustly than it had done before.

The same was true of associative freedoms. In a June 1991 judgment, the First Senate held that the right of assembly guaranteed in Article 8 GG applied not only to persons

[211] ibid. [212] ibid 294. [213] ibid. [214] ibid 295. [215] ibid. [216] ibid 297.
[217] 81 BVerfGE 298 (1990). [218] ibid 298–99. [219] ibid 305. [220] ibid 306.
[221] ibid 307. [222] ibid.
[223] Horst Sendler, 'Liberalität oder Libertinage?' (1993) 46 *NJW* 2157, 2158. [224] Gusy (n 206) 641.

supporting a given assembly's aims, but to those also who wished to participate critically.[225] Police had expelled thirty to forty people from an indoor meeting of the tiny Republicans Party (*Die Republikaner*). A regional administrative court sustained the police action on the ground that the persons expelled—who decried the featured speaker as a 'disguised old Nazi'—aimed to interrupt the meeting or prevent it altogether. The First Senate reversed, ruling that Article 8's protection extended to 'every German who wishes to take part' in a public meeting.[226] The right of participation encompasses not only the right to support a meeting's goals, but the right to oppose and denounce them.[227] The right did not extend to hindrance or obstruction, but in this case the protesters' actions had not reached that threshold. Their expulsion was unlawful.

Four years later, in a sharply divided decision, the First Senate went much further to shield certain types of protest from expansive criminal sanction.[228] The underlying facts involved another anti-military protest from the early 1980s—a protest, as in the *Brokdorf* case,[229] against NATO plans to plant nuclear missiles and launching facilities on West German soil. In May 1983, protesters conducted a 'sit-in blockade' (*Sitzblockade*) at a special munitions facility that housed 'Lance' model short-range missiles. The protesters, who sat in the streets and blocked military vehicles' access to the facility, were arrested and convicted for the crime of 'coercion' (*Nötigung*). After several failed criminal appeals, the protesters levelled a constitutional challenge on the ground that their convictions violated the retroactivity ban of Article 103(2) GG. Ordinary citizens could not have known, they argued, that a criminal ban on 'coercion' would apply to passive resistance, to a sit-in blockade on a public highway. Not until twelve years after the sit-in would the First Senate reach the constitutional question.

The relevant criminal code provision—section 240—had its origins in the late-eighteenth-century Prussian Code but received its modern formulation during the Nazi era.[230] The Third Reich iteration of the law, promulgated in 1943, had a decidedly illiberal flavour, linking the crime of coercion with the 'healthy sentiment of the people' (*gesunde Volksempfinden*). The law was amended in 1953, but only to replace this loaded National Socialist phrase with a generic reference to 'reprehensibility' (*Verwerflichkeit*).[231] Over the ensuing decades the ordinary courts generated an enormous jurisprudence construing this provision of the criminal code. The resulting conceptual thicket was dense and demanding, but over time a trend emerged that expanded the notion of coercion from a strict emphasis on 'violence' [*Gewalt*] and 'application of force' [*Kraftentfaltung*] to include other forms of 'compulsive effect' [*Zwangswirkung*]. In its 1995 *Sit-in Blockade* judgment, the First Senate considered whether expanding the concept of coercion to include a peaceful sit-in ran afoul of the retroactivity ban.

By a vote of 5:3, the Senate concluded that it did. In the Court's view, judicial construction of section 240 had expanded the provision's reach beyond what reasonable citizens could have foreseen from the law's text.[232] Article 103(2) GG required that citizens be given notice of the possibility that certain activities will provoke criminal sanction; judicial doctrine was insufficient notice.[233] If this ruling left a gap in the law of crimes, it was for the legislature, not the courts, to plug it.[234] By punishing conduct not foreseeably banned by the law's text, the police and ordinary courts had offended Article 103.

[225] 84 BVerfGE 203 (1991). [226] ibid 209. [227] ibid. [228] 92 BVerfGE 1 (1995).
[229] Discussed in Chapter 4.
[230] See Knut Amelung, 'Sitzblockaden, Gewalt und Kraftentfaltung' (1995) 47 *NJW* 2584, 2584–85.
[231] ibid 2585. [232] 92 BVerfGE 1 (n 228) 18. [233] ibid. [234] ibid 19.

It was a divided and controversial judgment.[235] As seen in Chapter 4, the First Senate considered roughly the same question nine years earlier and, splitting 4:4, reached the opposite outcome.[236] It was rare for the Court to reverse itself after less than a decade. Critics naturally associated the shift with changes in the Senate's membership.[237] Negative reactions to the January 1995 judgment fed on broader apprehensions that the Court's expression and association jurisprudence was going too far. But the judgment had champions as well as critics (the overturned 1986 judgment had been more heftily criticized), and most criticisms, at least initially, were mild. The real earthquake came later that year.

B. Bearing the cross: The Constitutional Court and the crucifix in the classroom

'If someone had predicted twelve months ago', wrote Rupert Scholz in the 22 December 1995 edition of *Die Zeit*, 'that in the coming year nothing would be more fiercely disputed in this country than the meaning of the Christian cross in State and Society—he would have been laughed to scorn'.[238] And yet, Scholz added, the fight over the crucifix had erupted in the unqualified 'battle of the year' in the Federal Republic.[239] The Constitutional Court was at the centre of the struggle. The controversy swelled into the greatest crisis since the earliest days of the Court's history. Some onlookers even doubted the Court's survival. Others concluded that it would live on, but never be the same again. Both groups were wrong. No one foresaw the swiftness of the Court's descent into crisis; few foretold the rapidity of its resurgence.

Culturally and chronologically, Bavaria's Catholic roots run deep. Beginning in the early days of the Reformation, Bavaria was, after Austria, the strongest Catholic stronghold in the German-speaking world. Bavaria's princes, as long as Bavaria had princes, were Catholic rulers. Political Catholicism became a force in Bavaria in the first half of the nineteenth century, and the influence, with many permutations, has endured. Bavaria remains one of only two German states (the Saarland is the other) with an absolute Catholic majority.

In 1983, the Bavarian state legislature issued a regulation requiring all public elementary schools—not just confessional schools—to hang a crucifix in each classroom. The law passed with a broad majority in the legislature and was uncontroversial with the Bavarian public. To a large degree, the regulation merely codified existing practice—a practice that went back more than a century and had been interrupted only briefly during the Nazi era. A small minority, however, was offended by the new law. One member of that minority made his umbrage an object of constitutional complaint.

Ernst Seler was an anthroposophist living, as of 1986, in the Bavarian town of Schwandorf. One day, Seler reported, his daughter Elina, then six, came home from school in tears because she had been forced to stare 'at an 80-centimeter tall, naked, blood-covered dead man hanging right before her eyes'.[240] Seler complained to the school, which replaced the crucifix with a bare cross. But the conflict had only begun. The school did not adopt a general policy, and the row flared each time one of Seler's children entered school or changed

[235] For critical commentary, see Amelung (n 230); Malte Graβhof, 'Auswirkungen der neuen Sitzblockade-Entscheidung des BVerfG' (1995) 48 *NJW* 3085.

[236] See 73 BVerfGE 206 (1986).

[237] See Carl-Friedrich Stuckenberg, 'BVerfGE 92, 1—Sitzblockade (mit E 73, 206 Mutlangen)' in *Verfassungsrechtsprechung* (n 1) 556.

[238] Robert Leicht, 'Auf der Suche nach der verlorenen Utopie' *Die Zeit* (22 December 1995).

[239] ibid.

[240] In 'Das Kreuz ist der Nerv' *Der Spiegel* (14 August 1995).

classrooms. For a time, Seler placed his kids in private school. But when that proved too costly, he returned them to public school—and filed a lawsuit.

For many years, the suit was unsuccessful. When, in 1991, a Bavarian administrative court declined to review Seler's case, Seler challenged this and other lower-court decisions in Karlsruhe.

It wasn't the first time the Court had dealt with a crucifix in a public place. In 1973, the First Senate granted the complaint of a Jewish lawyer—whose Jewish client sought redress for damages caused by Nazis—who objected to arguing his case in a courtroom adorned by a cross. The First Senate agreed that it violated the lawyer's religious freedom to be forced to plead his client's cause 'under the cross'.[241] In his complaint two decades later, Ernst Seler argued that the same was true of Bavaria's requirement that a crucifix preside over his children's efforts to learn the alphabet.

A sharply divided First Senate agreed. In a judgment dated 16 May 1995, the First Senate admitted the complaint and annulled the law.[242] The requirement that public school children study 'under the cross' ran afoul both of Article 4(1) GG, which guaranteed freedom of faith and conscience, and Article 6(2), which made the care and education of children 'the natural right of parents'.

How so? In the Article 4 context, the Senate majority stressed the core principle that questions of religious faith were matters of individual choice.[243] With respect to such choices the state must neither prescribe nor forbid; it must remain neutral.[244] The state's duty of neutrality, moreover, had both a positive and a negative dimension. The state must not interfere with an individual's religious exercises, nor may it unduly promote certain religious precepts or practices—or symbols—over others. In connection with Article 6, Article 4 secured the right of parents to oversee the religious and ideological (*weltanschaulich*) education of their children.[245] Did that right preclude the mandatory hanging of crucifixes in public schoolrooms?

The answer depended on the symbolic significance of the cross itself. The Court's assessment of that significance was the judgment's most controversial feature. The cross was not, the Senate maintained, merely the emblem of a 'Western civilization [*Kultur*] influenced by Christianity'. It was the symbol, rather, 'of a specific religious conviction'.[246] The cross was, and had always been, one of 'the specific symbols of the Christian faith. It is indeed [Christianity's] faith symbol par excellence'.[247] A cross on the schoolroom wall was more than a generic badge of Western cultural values or a universal appeal to charity and brotherly love. It had, in addition, an 'appellational character'. It extolled the beliefs and precepts typified in the cross as 'exemplary and worthy of observance'.[248] The cross's effect upon school children was evangelistic (*missionarisch*); and in the realm of religious belief, the constitution forbade the state to play a missionary role.[249] The Bavarian ordinance contravened this constitutional command.

Three justices—Otto Seidl, Alfred Söllner, and Evelyn Haas—dissented vigorously.[250] The dissenters emphasized the case's federalism component (the constitution committed control over public education to the states, so the Court should defer to Bavaria's educational decisions) and de-emphasized the confessional importance of the cross. The dissenters stressed, moreover, that while the constitution required state neutrality on *confessional* questions, it did not impose religious indifference or prescribe official laicism.[251] The right of religious freedom was not a right to hinder religion.[252] In any case, having the cross

[241] 35 BVerfGE 366, 375 (1973). [242] 93 BVerfGE 1 (1995). [243] ibid 15. [244] ibid 15, 16.
[245] ibid 17. [246] ibid 19. [247] ibid. [248] ibid 20. [249] ibid 23. [250] ibid 25–37.
[251] ibid 29. [252] ibid 32.

in constant view would affect Christian and non-Christian students in different ways. To the believer, it could stand as a reminder of the Son of God's sacrifice for the redemption of humankind—as suggested by the Senate majority. To the unbeliever, however, it could stand simply for one of the central goals of the state school system—'namely, the transmission of the values of a Christian-influenced Western culture'.[253] In Bavaria, especially, the cross was not so loaded a symbol as the majority suggested. It appeared in secular buildings of various descriptions. The prominence of the symbol in diverse contexts and multiple settings deprived it of any 'missionary character'.[254]

Although the majority and the dissenting opinions were forcefully phrased, nothing in them suggested that the case was unusual or extraordinary. There was certainly no indication that the justices had any premonition that the judgment would unleash the fiercest popular and political assaults the Court had faced in forty years. The backlash, however, was swift and fierce, thunderous and widespread. The Federal Republic had never seen anything like it. In striking down the mandatory cross, the Court had struck an open nerve. The popular outcry resounded in an apoplexy of howls confused.

It was strongest, of course, in Bavaria. Theo Waigel, CSU chairman and federal finance minister—for whom recent Court judgments on finance and taxation had been a recurring headache—called for a constitutional amendment that would spare Bavaria the indignity of bowing before the Court's incredible decision. Waigel railed that the Court's decision, not the Bavarian ordinance, violated the Basic Law.[255] Ingo Friedrich, who stood in for Waigel as the CSU's acting chief, called for a boycott. In an interview with *Der Spiegel*, Bavaria's minister president, Edmund Stoiber (CSU), styled the decision 'a judgment of intolerance'.[256] Asked when the crosses would come down from the classroom walls, Stoiber responded, 'Initially not at all'.[257] 'We no longer have a legal foundation', Stoiber explained, 'for the duty to post crosses in schools. But this does not mean that we mustn't hang crosses generally'.[258]

Hans Maier, a political scientist and Bavaria's former minister of culture, was blunter still. Maier denounced the decision as 'pure nonsense and arrogance', against which resistance was in order.[259] Asked whether Bavarian schools should ignore the judgment, Maier answered, 'Yes, at all costs'.[260] Maier opined further that the framers of the Basic Law 'would roll over in their graves if they saw what their text had become in the hands of five justices: a pretence for State-commanded irreligion'. Then, in a gesture that became a common trope of the popular onslaught against the judgment, Maier recalled the time when, as a small boy, he watched Nazi storm troopers remove the cross from his schoolroom. 'I would never have thought it possible', said Maier, 'that half a century later a Court erected for the protection of our constitution would dare to demand a similar disgrace'.[261] Some Bavarian politicos were not only defiant but militant, even menacing. Sepp Ranner, an agrarian functionary and CSU politician, wrote to the *Oberbayerisches Volksblatt* that '[t]he justices and the complainants ought to come down here themselves and remove the crosses from the schools with their own hands. We farmers, of course, will await them with threshing flails'.[262]

Not long after the judgment was announced, Stoiber, Waigel, Maier, and other Bavarian officials joined Church leaders and some 30,000 citizens in a protest march in Munich's Odeon Square.[263] In an impassioned and defiant speech, Stoiber thundered that 'the

[253] ibid. [254] ibid 33. [255] 'Das Kreuz ist der Nerv' (n 240).
[256] 'Die Kreuze bleiben' *Der Spiegel* (14 August 1995) 33. [257] ibid 32.
[258] 'Die Kreuze bleiben (n 256) 34. [259] 'Purer Unsinn und Übermut' *Focus* (14 August 1995) 44.
[260] ibid. [261] ibid. [262] In 'Das Kreuz ist der Nerv' (n 240).
[263] 'Heiliger Edmund, bitt' für uns' *Der Spiegel* (2 October 1995) 114.

minority must also in certain cases be tolerant vis-à-vis the majority'.[264] At the time of the march approximately two Bavarian schools had removed their crosses.[265] Less than two months after the Court's decision, the Bavarian state legislature proposed a law that would still require the hanging of a crucifix in each public school classroom, but would allow the principal of each school to make adjustments in cases of conflict.[266] The new proposal was impossible to square with the Court's judgment. It was construed, naturally enough, as a shout of defiance and a declaration of war.[267]

Outrage was by no means limited to Bavaria. Helmut Kohl, the federal chancellor, called the judgment 'incomprehensible', and complained that it endangered the values of Western civilization.[268] Joachim Hörster, the CDU/CSU floor leader in the Bundestag, called for a parliamentary debate on how 'the majority of the German people can be protected against extortion by minorities'.[269] Rupert Scholz, vice chair of the Union faction in the Bundestag, sneered that '[t]he Court is not the Pope of the Republic'.[270] Jürgen Augustinowitz, another CDU parliamentarian, called the judgment 'unbearable'.[271]

Church leaders spoke out as well. Cardinal Joseph Ratzinger, Prefect of the Sacred Congregation for the Doctrine of the Faith and future Pope Benedict XVI, declared himself 'aghast'.[272] Cardinal Joachim Meisner, Archbishop of Cologne, said that the Court's judgment marked '[a] black day in the life of our people'.[273] (It was unclear whether, by 'our people', Meisner meant Catholics or Germans.) Maria Jepsen, the (Protestant) North-Elbian bishop, predicted that the judgment would spark 'a new kind of Church-battle [*Kirchen kampf*]'.[274]

Letters to major newspapers repeatedly invoked, as Hans Maier had, the Nazi-ordered removal of classroom crucifixes, as well as the persecution of Catholics during Bismarck's *Kulturkampf*.[275] An educator in Germering—about 15 kilometres West of Munich—recalled how her school teacher daringly refused to exchange the classroom cross for a photograph of Hitler.

Many observers pictured a parade of horribles careering down a slippery slope. An editorialist in the *Frankfurter Allgemeine Zeitung* asked whether 'the Karlsruhe justices believe that atheistic outsiders must be protected more strongly, the more Christian a nation is?'[276] The same writer wondered in a separate editorial whether Christian parents could object if a Muslim student wore a headscarf to school. Wouldn't that have an 'appellational character'?[277] By the Court's logic, shouldn't church bells be silenced and spires dismantled? Many commentators protested that the cross was a symbol of tolerance; what a bitter irony that in the name of tolerance it should be banished from the classroom.[278]

Court-watchers were taken aback by the 'intensity', even the 'brutality' of popular criticism.[279] A wag quipped that Bavarians were so enraged they considered hanging the judges

264 In ibid. 265 ibid.

266 See 'Bayern will Kreuze weiter per Gesetz anordnen' *Süddeutsche Zeitung* (13 September 1995).

267 See Hans Schueler, 'Kampf angesagt' *Die Zeit* (6 October 1995).

268 'Das Kreuz ist der Nerv' (n 240).

269 'Politiker der Union kritisieren das Karlsruher Urteil' *FAZ* (14 August 1995) 1.

270 'Worte der Woche' *Die Zeit* (25 August 1995).

271 In Christian Bommarius, 'Fünf Richter auf der Anklagebank' *Die Zeit* (18 August 1995).

272 ibid. 273 In 'Worte der Woche' *Die Zeit* (18 August 1995).

274 'Das Kreuz ist der Nerv' (n 240).

275 See, e.g., 'Kreuze im Dritten Reich' *FAZ* (15 August 1995); 'Religionsbegriff des 19. Jahrhunderts' *FAZ* (17 August 1995); Volker Ullrich, 'Krieg gegen Katholiken' *Die Zeit* (25 August 1995) (noting that comparisons to the nineteenth-century *Kulturkampf* were exaggerated, and many attacks on the *Crucifix* judgment irrational, but nonetheless reflective of 'deep-seated historical trauma').

276 Kurt Reumann, 'Das Recht auf das Nichts' *FAZ* (16 August 1995).

277 Kurt Reumann, 'Toleranz ohne Kreuz' *FAZ* (11 August 1995).

278 See ibid; Axel Freiherr von Campenhausen, 'Karlsruhe fördert die Intoleranz' *Rheinischer Merkur* (18 August 1995) 1.

279 Christian Bommarius, 'Fünf Richter auf der Anklagebank' *Die Zeit* (18 August 1995).

in place of the crosses.[280] In fact it was no laughing matter. Some of the justices received death threats.[281] Justice Dieter Grimm, who travelled to the United States shortly after the decision was published, was asked by American colleagues whether the chancellor would have to send troops to Bavaria to enforce the *Crucifix* judgment, as President Eisenhower had sent troops to Arkansas to enforce the Supreme Court's decision in *Brown v Board of Education*.[282]

Even prominent liberals were dismayed. Marion Gräfin Dönhoff, the grand dame of postwar German journalism, published an elegiac essay entitled 'Reaction of a Layperson to the Judgment'.[283] One would like to know, Dönhoff mused, whether the complainant was as zealous about sparing his child from television images of sex and crime as from the sign of the cross. 'Citizens obviously have the feeling', she added, that 'now even this last remnant that doesn't belong to this objective-positivist world has been ripped away from under our feet'.[284] Dönhoff recalled that, in her flight from East to West at the end of the War, she had carried a cross in her saddlebag. 'Not as a fetish', she wrote, 'but as a token of confidence, and in the name of hope'.[285]

Legal scholars were less emotional, but no less pointed in their strictures. One scholar worried that the Court's 'exaggerated protection of minorities' would raise serious difficulties as, 'in the course of the further development of a so-called pluralistic society, the number of minorities grows'.[286] The same scholar echoed the common critique that the Court's 'strict laicism' and 'one-sided over-emphasis of negative religious freedom' were at odds with the Basic Law's prefatory acknowledgement of 'accountability before God'.[287] 'Under the cloak of tolerance and minority protection', the writer predicted, 'arbitrary minorities will feel encouraged to conduct further campaigns against the institutions, values, and symbols of the state'.[288] Another scholar complained that the Court had abandoned its role as societal and legal peacemaker.[289] Instead, the Court had divided the citizenry and hastened the 'total juridification of life'.[290] Some writers pointed nostalgically to former, better times in which the Court 'usually took first place in demoscopic rankings of credibility and trustworthiness'.[291] Bernhard Groβfeld, a professor of private law in Münster, wondered whether the Karlsruhe deities had reached their *Götterdämmerung*.[292] Groβfeld, who made clear that the country 'needed' the Court and should be 'grateful' for its existence, nonetheless complained of the Court's inflated competences, its arrogation of powers and of constitutional organ status, its lack of a political question doctrine, its role in softening the separation of powers and fostering de-parliamentarization (*Entparlamentarisierung*).[293]

Groβfeld's lament prompted an answer from Ernst Benda, who as chief justice from 1971 to 1983 knew something of public criticism of Court decisions. Though Benda thought reports of the Court's death exaggerated, he viewed the *Crucifix* judgment as 'well-nigh incomprehensible'.[294] If criticism of the Court continued to be as persuasive as it was in the case of the *Crucifix* decision, Benda suggested, the Court would be constrained to correct its

[280] Wolfgang Ebert, 'Sommerloch-Bilanz' *Die Zeit* (1 September 1995).
[281] Noted in 'Muβ der Kanzler Truppen nach Bayern schicken?' *Süddeutsche Zeitung* (9 December 1995) 10.
[282] ibid.
[283] Marion Gräfin Dönhoff, 'Reaktion eines Laien auf das Urteil' *Die Zeit* (25 August 1995).
[284] ibid. [285] ibid.
[286] Jörg Müller-Volbehr, 'Positive und negative Religionsfreiheit: Zum Kruzifix-Beschluss des BVerfG' (1995) 50 *JuristenZeitung* 996, 1000.
[287] ibid. [288] ibid.
[289] Konrad Redeker, 'Der moderne Fluch der Versuchung zur Totalität' (1995) 49 *NJW* 3369, 3369.
[290] ibid. [291] ibid. [292] Bernhard Groβfeld, 'Götterdämmerung?' (1995) 49 *NJW* 1719.
[293] ibid 1719–23.
[294] Ernst Benda, 'Wirklich Götterdämmerung in Karlsruhe?' (1995) 49 *NJW* 2470, 2470.

course.[295] Elsewhere Benda defended the Court against institutional attacks, but conceded that the *Crucifix* decision was 'a dire mistake, not remedied by the retrospective explanations of the chairman of the First Senate'.[296]

By 'retrospective explanations' Benda referred to a 'clarifying' statement issued by Friedrich Henschel, the Court's vice president and author of the *Crucifix* decision. The judgment, Henschel explained, did not hold that a cross in the classroom violated the constitution; only that the *state-ordered* mounting of a crucifix was unconstitutional.[297] Werner Flume, an influential private law scholar, objected that such a clarification should come, if it at all, from the united Senate.[298] Rudolf Augstein of *Der Spiegel* thought the post-hoc explanation 'an unheard-of scandal'.[299] In any case, the statement may have had the effect, not of assuaging popular ire, but of encouraging defiance. Henschel elsewhere addressed the latter phenomenon directly. 'Stoiber must recognize', he said, 'that the state is the home of all citizens, not just of Christians'.[300] Henschel also said, in an implicit critique of Helmut Kohl (and perhaps of Roman Herzog, the federal president and former chief justice) that 'if another constitutional organ were to be attacked in the same manner in which the Court is [being attacked] now, we would do everything for it'.[301]

Other justices, too, spoke publicly in the Court's defence. Justice Dieter Grimm stressed in interviews and opinion pieces that, while criticism of a constitutional court was legitimate and essential, no one had the right, in a *Rechtsstaat*, to defy its decisions.[302] 'Those who today call for the defiance of Court decisions because they disapprove of them', Grimm wrote, 'will be unable to explain tomorrow why others should follow laws ... of which *they* do not approve'.[303] Grimm, like Henschel, was part of the First Senate majority that issued the *Crucifix* decision.

Jutta Limbach, the chief justice, observed in an interview with *Der Spiegel* that 'of course [*natürlich*]' the justices took popular sentiment and behaviour into account when deciding cases. Judges were not 'free-floating intelligences that through some bare operative process derive their decisions from the Basic Law or other laws'.[304] Justices of the Constitutional Court, she acknowledged, must be societal mediators and peacemakers. But on the other hand, 'the Court must also have the courage to deliver an unpopular decision, if it cannot otherwise be true to a value decision [*Wertentscheidung*] of the Basic Law'.[305] Limbach did not, moreover, set much store by complaints that the Court had usurped a legislative role. For decades, politicians in both parties had been all too eager to refer to Karlsruhe conflicts that they were too 'quarrelsome or too tentative' to resolve themselves.[306]

As during the Court's first major crisis forty-three years earlier, justices of both Senates showed solidarity with one another and support for the Court as an institution. But they acknowledged that the criticism had exacted its toll. Justice Grimm described the atmosphere within the First Senate as 'not gloomy', but added that 'caution is in order'.[307] Justice

[295] ibid. [296] Ernst Benda, 'Kritik ja—aber bitte im Detail' *Rheinischer Merkur* (25 August 1995).

[297] Pressestelle des Bundesverfassungsgerichts, *BVerfG vizepräsident Henschel präzisiert das 'Kruzifix-Urteil'* (22 August 1995).

[298] Werner Flume, 'Das "Kruzifix-Urteil" und seine Berichtigung' (1995) 49 *NJW* 2904.

[299] Rudolf Augstein, 'Hände weg vom Gericht' *Der Spiegel* (28 August 1995).

[300] In 'Worte der Woche' (n 273).

[301] Uwe Wesel, 'Die zweite Krise—Neuer Streit steht bevor' *Die Zeit* (29 September 1995).

[302] Dieter Grimm, 'Kritik ist wichtig' *Die Woche* (1 September 1995) 30; Dieter Grimm, 'Unter dem Gesetz—Warum ein Richterspruch Respekt genießt' *FAZ* (18 August 1995) 29.

[303] Grimm, 'Unter dem Gesetz' (n 302) (emphasis added).

[304] 'Die Grenzen sind erreicht' *Der Spiegel* (28 August 1995). [305] ibid. [306] ibid.

[307] In Wesel (n 301).

Böckenförde, a member of the Second Senate, was less optimistic. 'The Court', he said, 'is no longer what it was before August'.[308]

Against the shrill and swelling popular outcry, the Court had some external defenders. Rudolf von Thadden, an historian in Göttingen, reminded his readers that Germany was no longer a homogenous 'Christian' society, and that the 'path of Anti-Semitism' had been paved with 'stones drawn from the much-invoked Western Christendom'.[309] In any case, he added, irresponsible criticism of and reckless resistance to the Court could contribute to an 'erosion of our democratic legal order'.[310] 'God be thanked', wrote Robert Leicht, 'that there are still judges in Karlsruhe'.[311] Leicht branded the controversy surrounding the *Crucifix* judgment as 'scandalous'. It had given rise to a 'pseudo-religious witches' Sabbath'.[312] Leicht later compared the *Crucifix* controversy to the constitutional conflict surrounding the European Defence Community four decades earlier. Bavarian and Christian Democratic politicians were playing now the role that Konrad Adenauer and Thomas Dehler had played then. 'Dehlers come and go', Leicht wrote, 'but the Constitutional Court remains. Politicians may well choose its members, but they must then bow to its decisions'. Uwe Wesel, a law professor engaged in one of the first efforts to write the Court's history,[313] made the same historical comparison, calling the *Crucifix* battle the Court's 'second crisis'.[314] In Wesel's view, the defiant marches and boycott campaigns had transgressed the boundary of legitimate criticism. But in spite of it all, Wesel predicted, '[t]he Court will press forward as it has before—sometimes deciding one way and sometimes another. And hopefully', he added, 'it will not back down ... '.

Time would quickly tell. Pending cases had already gathered clouds of controversy. 'The [coming] judgment will perhaps bring further unrest', Wesel wrote, 'but it may also show that the justices do not allow themselves to be intimidated'.[315] Bernhard Schlink—a law professor, state constitutional judge, and a best-selling novelist—struck a similar tone when an interviewer from *Der Spiegel* asked about controversial cases pending after the *Crucifix* decision. 'If the Constitutional Court is not man enough to deal with such things', Schlink replied, 'what do we have it for?'[316]

C. 'Soldiers are murderers': the limits and non-limits of freedom of speech

The most controversial pending case revolved around three words: *Soldaten sind Mörder*—'Soldiers are murderers'. The phrase belonged, originally, to Kurt Tucholsky, satirical scourge of the Weimar Republic, renowned for the brash iconoclasm of his thinking and the merciless beauty of his prose. In a 1931 essay in the magazine *Die Weltbühne*, Tucholsky wrote of the First World War: 'For four years there were entire square miles of land upon which murder was obligatory, whereas a half hour's distance away it was strictly forbidden. Did I say 'Murder'? Of course murder. Soldiers are murderers'.[317] The idea had an ancient pedigree and illustrious advocates in the history of German literature. In the nineteenth century, Heinrich Heine wrote of 'standing armies of hundreds of thousands of murderers'

[308] In ibid. [309] Rudolf von Thadden, 'Bloβ kein neuer Kulturkampf' *Die Zeit* (18 August 1995).
[310] ibid. [311] Robert Leicht, 'Das Kreuz ist kein Maskottchen' *Die Zeit* (18 August 1995).
[312] ibid.
[313] See Uwe Wesel, *Hüter der Verfassung* (Eichborn 1996); and later, Uwe Wesel, *Der Gang nach Karlsruhe: Das Bundesverfassungsgericht in der Geschichte der Bundesrepublik* (Karl Blessing 2004).
[314] Wesel (n 301). [315] ibid. [316] 'Getarnter Rechtsbruch' *Der Spiegel* (18 September 1995).
[317] Kurt Tucholsky, 'Der bewachte Kriegsschauplatz' *Die Weltbühne* (4 August 1931).

and Georg Büchner called soldiers 'lawful murderers'.[318] Tucholsky's equation of soldiers with murderers led to a spectacular 1932 trial of *Die Weltbühne*'s editor, Carl von Ossietzky (Tucholsky himself had left the country), for 'defamation of the Imperial Army [*Reichswehr*]'. Ossietzky, a future Nobel laureate, was acquitted on the ground that the phrase did not refer to specific individuals. An unspecified collectivity, the court concluded, cannot be defamed. Tucholsky committed suicide in exile in 1935. Ossietzky, after being interned and tortured in the Esterwegen concentration camp, died of tuberculosis in a hospital in 1938. The phrase Tucholsky penned and Ossietzky published had a long afterlife.

The phrase first concerned the Constitutional Court in a 1994 'Chamber' (i.e. three-justice panel) decision of the First Senate. During the first Gulf War, a pacifist and conscientious objector displayed three bumper stickers on his car: one featuring a wounded soldier and the text 'Why?'; one featuring the phrase 'Swords Into Ploughshares', a mantra of the German peace movement; and one with the words 'Soldiers Are Murderers' above a facsimile of Kurt Tucholsky's signature. The man was fined in a criminal action for incitement (*Volksverhetzung*) and in a civil action for defamation of a *Bundeswehr* soldier.

A three-justice panel of the Constitutional Court's First Senate reversed these rulings on the general ground that the ordinary courts, in their interpretation of the 'Soldiers Are Murderers' bumper sticker, failed to recognize and rule out interpretations that might merit constitutional protection.[319] The term 'murderers', for instance, need not be construed as a factual assertion that all soldiers have committed the elements of the crime proscribed in section 211 of the criminal code.[320] The criminal courts failed to consider the possibility of a more colloquial meaning—a vehement expression of moral revulsion, but not a technical accusation of criminal conduct. The criminal courts further failed to consider the offensive sticker within the context created by its neighbours. The three stickers together, the chamber suggested, produced a general pacifist message that cast soldiers as both perpetrators and victims.[321] The chamber also gave significance to the fact that the statement stemmed from—and was attributed to—a writer who died in 1935. Chronologically, it could not be reasonably construed as a specific affront to *Bundeswehr* soldiers.[322] The statement's target was universal; it applied to all soldiers in all nations in all wars in all history. (If any soldiers were excluded from such universal scope, they were soldiers of the *Bundeswehr*—a fighting force, as everyone knew, that had never, in nearly four decades, engaged in live combat.)[323] By failing to consider alternative interpretations, the criminal courts failed to honour the complainant's rights of free opinion and expression.

The chamber decision elicited a storm of controversy unlike anything ever provoked by a three-justice panel. Ingo Friedrich, the CSU vice chair, called the chamber decision the 'most scandalous erroneous decision [*Fehlurteil*] of the Federal Constitutional Court since the founding of the Federal Republic'.[324] 'The Court', Friedrich added, 'should be ashamed of itself'.[325] An impassioned debate in the Bundestag—from which only the Leftist PDS abstained—prompted a clarifying statement from the Court's press office.[326] Letters to the editor complained of a decision that stemmed, as the writers saw it, from ivory-tower judges in a Court captured by ideologically-driven '68-ers'—partisans of the iconoclastic

[318] In Michael Hepp and Viktor Otto (eds), *Soldaten sind Mörder. Dokumentation einer Debatte* (Ch. Links 1996) 6, 94.

[319] The decision, not included in the Court's official reports, was reproduced at (1994) 47 *NJW* 2943.

[320] ibid 2943–44. [321] ibid 2944. [322] ibid. [323] ibid.

[324] 'Der Bundestag nimmt die deutschen Soldaten in Schutz' *FAZ* (22 September 1994) 1. [325] ibid.

[326] The Press Office statement of 23 September 1994 is reproduced in Armin Steinkamm, 'Verfassungsrichter als 'Rufmörder'' (1995) 37 *Neue Zeitschrift für Wehrrecht* 45, 52.

generation of student rebellion.[327] In the months that followed, many critics adduced the *Soldiers Are Murderers* chamber decision as the last in a long line of decisions in which the Court's freedom of expression jurisprudence had become too liberal—in which freedom of opinion had triumphed definitively and recklessly over the protection of personal honour. Martin Kriele, a professor in Cologne, bemoaned the 'extensive abolition of the protection of personal honour by the First Senate of the Federal Constitutional Court', and numbered among the consequences of that abolition the 'degradation of [German] democracy' and the 'coordination [*Gleichschaltung*—a term with Nazi resonance] and banalization of intellectual life in the Federal Republic'.[328] Monographs appeared asking whether personal honour had forfeited constitutional protection altogether.[329]

Some scholars were quick to rebut these charges as overdrawn.[330] To some extent, the justices rose to their own defence. Justice Dieter Grimm—judge rapporteur in free speech controversies and putative author of the chamber decision—published a thorough essay in the *Neue Juristische Wochenschrift*, in which he did not defend the chamber decision directly, but rather surveyed the Court's freedom of opinion jurisprudence and suggested, between the lines, that the chamber decision was a natural outgrowth of that jurisprudence.[331] Some observers thought this public intervention by a sitting justice unfortunate or unseemly.[332] One scholar thought the public statements of Grimm, Henschel, and others during the aftermath of the *Crucifix* judgment were 'all too embarrassing'.[333] But Grimm believed that the justices had a duty to engage in ongoing dialogue with the public and the political branches, rather than retreat behind the anonymity of judicial independence and the Court's institutional right to have the final word.

Meanwhile criticism of the chamber decision mounted, fuelled by hostility to the subsequent *Sit-in Blockade* and *Crucifix* judgments. In many popular and political fora, these three decisions were cast as a triad of infamy—a collective emblem of all that had gone wrong in Karlsruhe and in the country at large. In the meantime, a consolidated quartet of cases—each implicating soldiers' honour—was pending before the full First Senate. Some hoped, and others feared, that the Senate would yield to escalating criticism and calls of defiance. The Senate would release its landmark *Soldiers* decision less than three months after announcing its *Crucifix* judgment. The entire country was watching and waiting—wondering whether the united Senate would reverse the chamber decision of the previous autumn.

In a judgment dated 10 October 1995, by the narrowest of margins, the Senate stood its ground.[334] The case involved the convictions of four individuals, all conscientious objectors, who had written or published some variation on the theme that soldiers are murders. One hung on a street corner a towel on which he had written (in English) 'A SOLDIER IS A MURDER [*sic*]'. Another distributed a pamphlet that asked, 'Are soldiers potential

[327] See Heinrich Balke, 'Vorrangiger Rechtsschutz' *FAZ* (26 September 1994) 15; Gottfried Lemberg, 'Richter als Privatphilosophen' *FAZ* (26 September 1994) 15; Ernst J. Marliany, 'Soldaten ohne Menschenwürde?' *FAZ* (26 September 1994) 15.

[328] Martin Kriele, 'Ehrenschutz und Meinungsfreiheit' (1994) 47 *NJW* 1897, 1905.

[329] See Ralf Stark, *Ehrenschutz in Deutschland* (Duncker and Humblot 1996); PJ Tettinger, *Die Ehre—ein ungeschütztes Verfassungsgut?* (Otto Schmidt 1995).

[330] See, e.g., Jörg Soehring, 'Ehrenschutz und Meinungsfreiheit' (1994) 47 *NJW* 2926.

[331] Dieter Grimm, 'Die Meinungsfreiheit in der Rechtsprechung des Bundesverfassungsgerichts' (1995) 48 *NJW* 1697.

[332] See Rüdiger Zuck, 'Gerechtigkeit für Richter Grimm' (1996) 49 *NJW* 361, 361; Rainer Wahl, 'Quo Vadis—Bundesverfassungsgericht? Zur Lage von Verfassungsgerichtsbarkeit, Verfassung und Staatsdenken' in Bernd Guggenberger and Thomas Würtenberger (eds), *Hüter der Verfassung oder Lenker der Politik? Das Bundesverfassungsgericht im Widerstreit* (Nomos 1998) 81, 101.

[333] Flume (n 298) 2905.

[334] 93 BVerfGE 266 (1995).

murderers?', and implicitly answered in the affirmative. A third proclaimed in a letter to the editor, 'All soldiers are potential murderers!' The fourth waved a banner at a Munich motorcycle show that read, 'Soldiers are potential MURDERERS!' Each complained that his conviction violated his right, under Article 5(1) GG, to the free expression of opinions. In three of the cases, the First Senate agreed by a vote of 5:3. In the fourth case, the justices agreed unanimously.

The majority's analysis began with first principles. Criminal code provisions for the protection of personal honour were not, the Court said, unconstitutional on their face.[335] The legislature could restrict speech in the interest of personal honour, but it could not do so arbitrarily. Specifically, Article 5 GG forbade any legislative action or judicial interpretation that might cause citizens to withhold criticism for fear of sanction.[336] Efforts to preserve honour mustn't chill speech. On the other hand, human dignity remained, as ever, the central constitutional value and was immune to balancing against other constitutional values.[337] Expressions of opinion that impinged another's dignity forfeited constitutional protection. But courts must proceed with caution. Human dignity was the fount of all fundamental rights, including freedom of expression. But the sphere of expression denied constitutional protection in the name of dignity—what the Court called *Schmähkritik* (i.e. vituperation or calumny)—must be narrowly circumscribed.[338] Moreover, a presumption in favour of free speech prevailed whenever contested expressions addressed topics of public concern and might shape public opinion.[339] Such was the case with the four complainants. The underlying dispute between pacifism and military preparedness was of essential public interest, giving rise to the presumption in favour of constitutional protection.[340]

This presumption did not grant pacifists and others carte blanche to insult the nation's soldiers. The majority stressed, perhaps in response to criticism or misunderstanding of the earlier chamber decision, that so-called 'group defamation' remained an actionable legal wrong, and that the specific group of *Bundeswehr* soldiers was capable of being so injured.[341] The majority quashed the complainants' convictions, not because the *Bundeswehr* could not be defamed but because the expressions in question might not have defamed it. The Senate's central criticism of the criminal court judgments was that they had not construed the challenged utterances in their original context and had therefore misconstrued their intended meaning.

Contrary to the findings of the ordinary courts, the penalized expressions dealt with soldiers in general, not with specific soldiers or *Bundeswehr* soldiers as a group. The statements were not assertions of fact, but reflections on the immorality of war—the organized and officially sanctioned killing of humans by other humans. The English phrase 'A soldier is a murder' could be construed as a judgment that soldiers are both perpetrators and victims, killers and killed. (Critics grumbled that it was just bad grammar.) The ordinary courts failed to consider these and other possible alternative interpretations. They had also applied the concept of 'smear criticism' (*Schmähkritik*) too broadly. These errors might have made the difference between affirmation and reversal of the complainants' convictions. As a result, the Senate reversed the specialized courts' decisions and remanded for further consideration.[342] Soldiers, the Senate acknowledged, were understandably mortified

335 ibid 290. 336 ibid 292. 337 ibid 293. 338 ibid 294. 339 ibid.
340 ibid 303–04.
341 ibid 302. This holding dismayed scholars who hoped the Court would abolish the crime of group defamation altogether. See, e.g., Georgios Gounalakis, '"Soldaten sind Mörder"' (1996) 49 *NJW* 481, 484.
342 93 BVerfGE 266 (n 334) 312.

by the complainants' ferocious diction. But Article 5 might require them to endure such mortification.

Three justices disagreed, but only one wrote a dissent, and that dissent applied only to three of the four cases. Justice Evelyn Haas, daughter of a *Bundeswehr* reserve officer, charged her colleagues with exceeding their competence. Fact-finding, Haas maintained, was the province of ordinary courts. This included interpretation of allegedly libelous or defamatory statements. In Haas's view, factual interpretations by courts of specialized jurisdiction were immune to review by the Federal Constitutional Court.[343] Haas cited a series of critiques, both scholarly and judicial, that took the Court to task for appropriating the fact-finding prerogatives of other courts.[344] Haas plainly shared these objections. She was also skeptical of the majority's alternative interpretations. 'A soldier is a murder' was bad English, not a subtle thesis on the victimization of soldiers.[345] In any case, the subjective meaning of a speaker was irrelevant if that meaning found no expression in what was actually said.[346] And what was said here was thoroughly reprehensible. Soldiers, Haas wrote, 'risk their life to spare the civil population the horrors of war and to protect civilian lives—not least the lives of those who disdain their actions and seek to make them contemptible before the public'.[347]

Notwithstanding Haas's strictures, the majority decision was, in important respects, restrained. The majority did not hold, as the chamber panel held fourteen months earlier, that the complaints were 'obviously well-founded'. And the majority went out of its way to hold that *Bundeswehr* soldiers *could* be defamed as a group, and that ordinary courts would have the final say on whether soldiers had been defamed collectively in the instant cases. But these subtleties were lost on an outraged public. Many read the judgment as unfettered licence for libelers to brand soldiers as murderers and defame the nation's defenders with impunity. Fiery antagonism to the *Soldiers* judgment fed on flames already fanned by the *Crucifix* judgment.

Volker Rühe, the federal defence minister, called the judgment a 'political scandal'.[348] Wolfgang Schäuble, leader of the CDU/CSU Bundestag faction, called immediately for new legislation protecting the honour of soldiers.[349] Several of Schäuble's colleagues issued similar demands.[350] Friedrich Karl Fromme launched a frontal assault on the decision in the *Frankfurter Allgemeine Zeitung*.[351] Those who hoped that the Court would correct its course, Fromme wrote, had been bitterly disappointed. Whatever the majority's disclaimers, the judgment's inevitable result was total immunity for those who called soldiers murderers. What would the justices say, Fromme wondered, in response to a pamphlet proclaiming, 'Judges are Born Criminals'? The Court could rescue its rapidly sinking prestige only 'if it no longer attempts to march ahead of a popular opinion it deems narrow-minded with the imperious command: "Follow Me!"'[352]

The Court's prestige *was* sinking. In early December 1995, about a month after the *Soldiers* decision, Renate Köcher, director of the Allenbach Institute, a polling group, reported an 'avalanche-like decline' in the public's esteem for the Court.[353] Traditionally, Köcher noted, the Court had rivalled the federal president and the Bundesbank—and, ironically, narrowly eclipsed the *Bundeswehr*—in the public's regard. But between the beginning of 1994 and

[343] ibid 313. [344] ibid 314. [345] ibid 315. [346] ibid 316. [347] ibid 318.

[348] In Hans Schueler, 'Unter dem Talar schlotternde Knie' *Die Zeit* (10 November 1995).

[349] See 'Karlsruhe—Beleidigung der Bundeswehr Stafbar' *Süddeutsche Zeitung* (8 November 1995) 1; 'Schäuble—Soldaten Notfalls mit Gesetz schützen' *Süddeutsche Zeitung* (8 November 1995) 6.

[350] 'Karlsruhe—Beleidigung der Bundeswehr Stafbar' (n 349) 1.

[351] Friedrich Karl Fromme, 'Mit dem Kopf durch die Wand' *FAZ* (8 November 1995) 1. [352] ibid.

[353] Reported in Helmut Kerscher, 'Ansehen rapide Gesunken?' *Süddeutsche Zeitung* (9 December 1995) 10.

the autumn of 1995, the Court's approval ratings in the 'old states' of the Federal Republic dropped from 51 per cent to 40 per cent, a near-historic low.

Jutta Limbach, the chief justice, suggested that the shift was temporary. She recalled earlier periods—particularly the early 1950s and the mid-1970s—when the Court was the target of a popular hue and cry. The chief justice even expressed gratitude for sharp criticism, which she said prompted reflection and renewed discussion among the justices.[354] But the judges were clearly concerned, not so much by survey results (other polls were less dire) as by a growing rumble of dissatisfaction with the Court as an institution. It was like nothing the Court had seen before.

The parallel with the 1950s, which came so readily to mind, was bolstered by the spread of criticism from defiant politicians to disgruntled 'ordinary' judges. As noted earlier, Justice Haas, in her dissent in the *Soldiers* case, joined a growing gaggle of critics who thought the Court was encroaching on the competences of specialized courts. This group included Konrad Hesse, the former justice, who did so much to save the Court from its crisis in the late 1970s.[355] Following the *Soldiers* judgment, ordinary judges began to complain. The most jarring example was Karl-Hans Fischer, a judge in Mainz responsible, on remand, for the case involving the public letter proclaiming 'Soldiers are Potential Murders'. During the proceedings on remand, Fischer lamented that the Constitutional Court's judgment left him no choice but to dismiss the charges against the letter's author. Fischer went so far as to apologize to the soldiers present in the courtroom. He then turned his fire on the Constitutional Court. 'It is unfortunate', he began, 'that through the decision of the Federal Constitutional Court freedom of opinion increasingly takes the upper hand vis-à-vis protection of honour'. Perhaps the trend had something to do with the appointments process for constitutional judges, a process driven 'not by juristic qualification, but according to proportion and party line'. The justices thus chosen had exalted themselves as a kind of 'super appellate jurisdiction' and 'degraded' courts like Fischer's into mere 'executors of the Karlsruhe decision'.[356] As for the *Soldiers* decision itself, Fischer thought it 'arrogant, juristically questionable, and socio-politically false'.[357] Fischer's philippic won warm praise from Klaus Rose (CSU), chair of the Bundestag defence committee. Rose thanked Fischer for defying the 'Karlsruhe gagging' (*Knebelung durch Karlsruhe*) and renewed the call for corrective legislation.[358]

Most observers, however, thought Fischer's *ex cathedra* outburst had gone too far. Ernst-Gottfried Mahrenholz, the Court's recently retired vice president, spoke gently of a 'derailment'.[359] Others were less kind, suggesting that Fischer's rodomontade—with its talk of unbridled arrogance, doctrinal dubiety, and socio-political obtuseness—was really a self-portrait.[360]

Fischer, for all his angry bluster, did adhere to the Court's decision as he understood it. No court called upon to implement the *Soldiers* decision intimated that it wasn't bound by that decision. This was untrue of those politicians who hoped to blunt the decision's impact through preemptive legislation. In this regard, whether the decision's political opponents

[354] ibid.

[355] See Konrad Hesse, 'Verfassungsrechtsprechung im geschichtlichen Wandel' (1995) 50 *JuristenZeitung* 265.

[356] In 'Folgeprozeß um Verwendung von Tucholskys Zitat "Soldaten sind Mörder"' *Süddeutsche Zeitung* (5 January 1996).

[357] In 'Landgericht kritisierte Bundesverfassungsgericht; Heftige Reaktionen auf Mainzer Soldaten-Urteil' *Süddeutsche Zeitung* (8 January 1996).

[358] ibid. [359] In ibid.

[360] See ibid (SPD legal expert Herta Däubler-Gmelin and Green Party executive spokesman Jürgen Trittin); 'Unter Niveau' *Süddeutsche Zeitung* (5 January 1996); Horst Sendler, 'Blüten richterlicher Unabhängigkeit udn Verfassungsgerichtsschelte' (1996) 49 *NJW* 825, 826.

understood it correctly is beside the point. Calls to overrule the Court by parliamentary statute perpetuated the spirit of opposition to the *Crucifix* decision—a spirit that insinuated or exclaimed, for the first time, that compliance with the Court's judgments was optional.[361] Increasing acceptance of this notion constituted the greatest threat to the Court's authority since the days of Thomas Dehler. It roused alarmed citizens to full-throated defences.

One of the most prominent of these was Hans-Ulrich Wehler, a Bielefeld historian then at work on his magisterial 'societal history' of Germany from 1789 to the present.[362] Wehler observed that the Karlsruhe justices had been recognized as 'Guardians of the Constitution' in significant part because of their protection of 'so precious a liberal achievement as the right to freedom of opinion'.[363] In Wehler's view, the Court's freedom of opinion jurisprudence, beginning with its 1958 *Lüth* judgment, marked 'a milestone in the development of the German constitutional state'. More broadly, the Court's protection of fundamental rights 'embodies one of the greatest domestic political accomplishments of the Federal Republic'. The Court's work of 'concretising', through constant reinterpretation, 'the aspirational values of the Basic Law' had helped forge 'that constitutional, legal, and social state ... that has become a badge of honour for the Federal Republic and, what's more, one of the most resilient cornerstones of its legitimacy'.[364] As had happened in the mid-1970s and at other times, attacks on politically controversial judgments prompted calls for perspective. Diatribes directed at the Court's recent offences were answered by recitations of its historical achievements. Thrusts aimed at the Court's overreaching were parried by reminders of the Court's contributions. Challenges to the Court's role in the political life of the Federal Republic failed to gain enduring traction because many citizens perceived the impact of that role to have been—on the whole, and in the long run—enduringly positive.

Nineteen ninety-five was an *annus horribilis* for the Court's public standing. By historical standards, approval of the Court's judgments and esteem of the Court as an institution sank dramatically. But the drop was only temporary. Beginning in 1996, the public's regard for the Court began to recover and quickly returned to its traditional high levels. Some of this had to do with a relaxation in the rate at which the Court resolved stormy political controversies. In the millennium's few remaining years, the Court decided few cases as volatile as the *Sit-in Protest*, *Crucifix*, and *Soldiers* cases—the terrible triad that came before the Court within the span of only a few months. This reduced momentum reinforced a point often made in defence of the Court's great power: the Court decides only those cases that come before it. The Court is not self-activating. The justices may not seek out political controversy, but neither may they retreat from controversy when it arrives through authorized channels.

In this author's view, the latter point had as much to do with the Court's recovery from the 1995 crisis as did the former. The public's perception of the Court's legitimacy has endured across the decades not in spite of its occasionally controversial or unpopular decisions but precisely because of them. When the dust and din of a given *cause célèbre* depart, the German public has generally admired the Court's willingness to face—or sympathized with its inability to avoid—the thorny issues that come before it. As Justice Grimm told an interviewer in late 1995, it was a felicitous result when the Court's decisions restored societal harmony; but '[t]he duty of the Court does not consist in the establishment of peace, but in the implementation of the constitution'.[365] Two weeks earlier Chief Justice Limbach said

[361] See 'Probieren, was passiert' *Der Spiegel* (20 November 1995).

[362] See Hans-Ulrich Wehler, *Deutsche Gesellschaftsgeschichte*, 5 vols (CH Beck 1987–2003).

[363] Hans-Ulrich Wehler, 'Der Kampf gegen Karlsruhe nach den Urteilen zu Sitzblockaden, Kruzifix und Tucholsky-Zitat führt die Rechte eine Kampagne gegen das Bundesverfassungsgericht' *Die Zeit* (1 December 1995).

[364] ibid. [365] In 'Muß der Kanzler Truppen nach Bayern schicken?' (n 281) 10.

she was far less worried by the Court's declining popularity than by the possibility that 'the public might believe that we allow ourselves to be intimidated'.[366] In many minds, it was to the Court's credit that the sabre-rattling rhetoric that followed the *Crucifix* judgment did not appear to affect the outcome in the *Soldiers* decision. As had happened in the early 1950s, the Court benefited from a conspicuous display of independence in the midst of a gathering crisis.

In the broader perspective, the decisions around which the crisis swirled provide further, if subtle, examples of the Court's role in accommodating the new Eastern states within the united German constitutional order. Post-reunification efforts to introduce plebiscitary elements into the formal structure of the Basic Law went nowhere, but the Court partially compensated for this by strengthening protections for speech and widening the scope of the public sphere. It was an appropriate, if indirect, tribute to the dissidents who toppled the GDR by the power of speech and assembly.

Reunification provided the deep backdrop to the *Crucifix* judgment as well. Critics were right in their contentions that the Basic Law itself made religious gestures, and that the Federal Republic had never embraced the strict separation of church and state idealized in France or the United States. But reunification had vastly enhanced religious pluralism in the Federal Republic. The country had recently incorporated millions of citizens whose outlook was profoundly secular. Reunification created a context in which the requirements of state religious neutrality seemed stronger than before. This perspective may have animated the Court's subsequent role in brokering a compromise regarding the state of Brandenburg's adoption of nondenominational ethical courses,[367] and its approval, in a 2007 chamber decision, of Berlin's decision to make traditional religious instruction voluntary.[368] The decisions that within the Western states were most divisive had an integrative effect between East and West.

None of this is to suggest that the Court escaped the crisis unscathed. The justices entered 1996 in a manifestly chastened mood. Justice Renate Jaeger found 'unsettling' the increasing tendency of popular and political voices to respond to Court decisions with vows of noncompliance.[369] Justice Ernst-Wolfgang Böckenförde noted that the Court need not assume the role, as it sometimes had a tendency to do, of 'provident preceptor of the legislature'.[370] The Court must exercise greater restraint, Böckenförde suggested, and do more to respect the separation of powers.

It was clear that the Court could not simply ignore the outrage that its recent decisions had provoked. But the Court's response was not to render railing for railing, nor to bow before popular clamour. The justices tried, instead, to engage in dialogue with the public. 'We need to make our judgments more plausible', Justice Grimm conceded.[371] To that end, Grimm himself addressed an audience of professional soldiers on the topic of the *Soldiers* decision,[372] and personally answered hundreds of critical letters sent him after the decision was published. As noted earlier, these unofficial discussions troubled traditionalists, but others admired Grimm's openness. On the whole, the justices seemed more restrained and less strident than their critics. Again and again, the justices emphasized publicly that criticism of their judgments was welcome, even necessary, since public criticism was the only tribunal that reviewed the Court's decisions. Compliance with the Court's judgments, however, was also necessary—not because the justices were infallible, but because the *Rechtsstaat* required that their decisions be final.

[366] In 'Probieren, was passiert' (n 361). [367] 104 BVerfGE 305 (2001).
[368] 10 BVerfGK 65 (2007). [369] In 'Probieren, was passiert' (n 361). [370] In ibid.
[371] In ibid. [372] 'Muβ der Kanzler Truppen nach Bayern schicken?' (n 281) 10.

In the long run, the public agreed. For all the fierce criticism levelled in 1994–95, no proposal to reform the Court as an institution—to restrict its jurisdiction or reduce its competencies—gained lasting, or even serious short-term traction. The consensus, even among many sharp critics, was encapsulated in Horst Sendler's application of Churchill's *bon mot* about democracy (that it was the worst imaginable form of government—except for all the others) to the Federal Constitutional Court.[373] At its worst, the Court was a necessary evil; at its best, a redoubt of the Federal Republic's highest values, a representative of its proudest achievements. Well or ill, the Court weathered its mid-decade crisis, and helped to consolidate and complete the constitutional transition from the Bonn to the Berlin Republic. By decade's end, it was clear that the second postwar German Republic would, like its predecessor, retain a second seat in Karlsruhe.

IV. Germany and the World

That second seat would be populated, in part, by a new generation of justices who would help shape Germany's position in the post-Cold War world. As fate would have it, the crisis just described coincided with a large turnover in the Court's membership.

Criticism of the Court's mid-decade judgments fuelled unprecedented criticism of the process for appointing judges. One 1995 survey found that only a third of the German public thought the justices shared their values.[374] Many critics inveighed that the Court had been captured by an elitist caste of runaway progressives who thought they knew whither the country should go and how it ought to get there. On the other hand, in the aftermath of very different decisions, critics complained that the Court was populated by out-of-touch conservatives, anachronistic in their outlook and aloof from the lived reality of most Germans. For both sides, the gap between Court and people was especially galling because the process of appointment was so opaque. No one knew who would be chosen or how until it was too late. Criticism of the process gained in volume and vehemence over the course of 1995.

Some of the justices were sensible to this critique. Shortly after the *Crucifix* judgment, *Der Spiegel* asked Jutta Limbach whether the appointment of a new justice should be preceded by a public hearing. 'I sympathize with such a wish', the chief justice answered. 'But I don't think it would be good if the hearing served only to overwhelm a person with notes from the intimate sphere'—as had happened, the interviewer supplied, 'in the USA in the case of the black judge Clarence Thomas'.[375] Limbach said she favoured an objective hearing, in which one obtained information about a candidate's constitutional understanding and views on important societal issues.[376]

Limbach's preference was widely shared but never adopted. Candidates for the Court continued to be chosen behind the scenes through negotiations involving representatives of the major parliamentary factions. The process has always forced compromise and sometimes fostered consensus. Whatever its demerits in terms of transparency, the mechanism has generally populated the Court with centrists. The justices have faced occasional complaints of partisanship, but nothing remotely comparable to the political polarization—and public perception of partisan divide—that in recent years has visited the United States Supreme Court.

[373] Sendler (n 360) 826. [374] Noted in 'Muβ der Kanzler Truppen nach Bayern schicken?' (n 281) 10.
[375] 'Die Grenzen sind erreicht' *Der Spiegel* (28 August 1995). [376] ibid.

Jutta Limbach was herself new to the Court when she guided it through the mid-1990s crisis. Limbach joined the Second Senate in March 1994, taking the place of Ernst Gottfried Mahrenholz as vice president with the understanding that she would become chief justice if (as seemed likely) Roman Herzog became federal president. Limbach didn't wait long. On 1 July 1994, Herzog became the Federal Republic's seventh president, and Limbach became its first female chief justice.

Limbach, a self-styled 'practising feminist', had been a justice senator in Berlin where she took part in the (unsuccessful) prosecution of Erich Honecker and other GDR chieftains. In an early interview with *Der Spiegel*, Limbach observed that the Basic Law was 'the best German constitution yet', but needed reform, particularly of its 'social' provisions.[377] Like many of her fellow justices, Limbach had hoped in vain that a national assembly would be convened to create a constitution for united Germany.[378]

Limbach was the first new justice to join the Court since Bertold Sommer, a judge on the Federal Administrative Court, replaced Everhardt Franβen on the Second Senate in July 1991. On the same day Limbach joined the Second Senate, Renate Jaeger, a judge on the Federal Social Court (*Bundessozialgericht*), joined the First. Six months later Evelyn Haas, another Federal Administrative Court judge, joined Jaeger on the First Senate. Haas became the first woman ever nominated by the CDU. Women now accounted for five of the Court's sixteen justices—a ratio, Limbach quipped, that should set the standard for corporate boards and the chancellor's cabinet.[379]

For the first thirty-five years of the Court's history, a single female justice sat on the First Senate—Erna Scheffler from 1951 to 1963, Wiltraut Rupp-von Brünneck from 1963 to 1977, Gisela Niemeyer from 1977 to 1986—and none sat on the Second. An insider's joke labelled the Court a 'Group Picture with Woman'.[380] When Karin Graβhof became the Second Senate's first woman in October 1994, the proverbial witticism spoke of 'Snow White Senates'—a lone lady with seven (male) 'dwarves'.[381] In one year—1994—all of this changed dramatically. The country was pleased to see the Court more closely resemble itself.

Two men joined the First Senate in October 1995—Dieter Hömig, another administrative judge, and Udo Steiner, a public law professor—and two more joined the Second in May 1996—Hans-Joachim Jentsch (CDU), a former state justice minister and constitutional judge, and Winfried Hassemer, a professor in Frankfurt and operative in Hesse's state data protection agency. In 1998–99, two more women joined the Court—Christine Hohmann-Dennhardt, a former judge and state government minister, and Lerke Osterloh, a law professor. Three more men joined as well—Hans-Jürgen Papier and Udo di Fabio, both outspoken law professors, as well as Siegfried Broβ, a long-serving judge on the Federal Court of Justice. These latter appointments changed neither the proportion of women on the Court (Justices Graβhof and Seibert had since departed) nor the concentration of legal academics (Justices Paul Kirchhof and Dieter Grimm left the Court in December 1999; Justice Ernst-Wolfgang Böckenförde in May 1996). They also had little effect on the (putative) partisan composition of the two Senates.

The Court would carry a set of new faces into the new millennium. The new justices immediately began to shape German society and to affect Germany's position in the world.

377 'Frauen ändern das Klima' *Der Spiegel* (7 July 1994). 378 ibid.

379 In 'Worte der Woche' *Die Zeit* (25 November 1994) ('The Federal Constitutional Court recommends itself as an example worthy of emulation!').

380 'Gruppenbild mit Dame' *Der Spiegel* (13 November 1978).

381 'Die Fristenregelung' *Der Spiegel* (7 June 1993).

A. United Germany in a uniting Europe

For centuries the German Question was a European problem. In the twentieth century it became a global problem. The problem erupted in two World Wars and was cured, temporarily, by decades of division. When Germany's two halves reunited, the German Question resurfaced in its most pointed formulation—how to contain the colossus at the heart of Europe? Many world leaders—and not only they—were unsettled. Margaret Thatcher, the British prime minister, wondered how the world could witness German reunification without fearing the return of German militancy. The Cold War, for all its anxieties, had offered stability on the German Question. Now that stability was gone. Neo-realists predicted glumly that the West would come to miss the Cold War.

With the Cold War's end, the German Question reemerged within Germany as well. The reunited nation had to decide how to wield its new heft in ways that reassured its wary neighbours. Already in the 1980s prominent German conservatives had called for the Federal Republic to acknowledge the reality of its economic and geopolitical clout and to assume a correspondingly active role in world affairs. Hans-Peter Schwarz, a political scientist and popular biographer, wrote that Germans had moved from an early-century 'obsession with power' to a late-century 'forgetting of power'.[382] The Federal Republic, Schwarz argued, could ignore its strength no longer. The country should pursue 'responsible power politics' rather than shirk its global role in a prolonged penance for past sins.

There were many, both within the Federal Republic and outside it, who were deeply alarmed by the juxtaposition, even if only rhetorical, of 'Germany' and 'power politics'. The alarm spread when the Cold War ended. The world wondered how to make the Germans behave. Germans hoped to persuade the world that they would do so voluntarily.

Both sides saw an opportunity in European integration. From the beginning, the European project was designed to keep the Germans at bay—to cool Franco–German antagonism and secure an enduring, peaceful balance of power. France and the Federal Republic, Western Europe's two most important powers, were the driving forces in this effort. The two countries would resume and expand their traditional roles in the aftermath of German reunification and the end of the Cold War. Wags soon jested that European integration was designed to disguise French weakness and German strength.

Plans for 'more Europe' predated the fall of the Berlin Wall. They had been a favoured scheme of the Frenchman Jacques Delors ever since he became president of the European Commission in 1985. But as Germany's two halves moved towards reunification, the push for further European integration intensified. Leaders in Bonn, eager to brandish their commitment to 'Europe', sped the project along, facilitating a major summit in Rome in October 1990—less than four weeks after formal reunification—and another in December. A year later, in December 1991, the European Council met in the Dutch city of Maastricht to hammer out the details of a new European treaty. The treaty rechristened the European Economic Community as the European Union, and laid plans for creating a common currency, the Euro. Member state leaders signed the treaty on 7 February 1992, but the treaty would not enter into force for nearly two years.

The first waves of adversity arose in Denmark, where, in a referendum held on 2 June 1992, voters rejected the treaty by fewer than 50,000 votes. Danish voters later approved an amended treaty with Danish exemptions. In a September 1992 referendum, French voters approved the treaty, but just barely (51.05 per cent in favour). In Britain, the prime minister,

[382] Hans-Peter Schwarz, *Die gezähmten Deutschen. Von der Machtbesessenheit zur Machtvergessenheit* (Deutsche Verlags-Anstalt 1985).

John Major, feared that a motley assortment of MPs opposed to the treaty might depose him in a no-confidence vote. The treaty's setback in Denmark, its narrow escape in France, and its turbulent career in Britain raised fears among Europe's friends and hopes among its detractors. The treaty's future looked wobbly. Even after navigating the shoals in Denmark and France, the treaty faced one more crucial, surprising obstacle on the road to ratification: the Federal Republic of Germany.

This was not because the treaty lacked support among Germany's political elites. On 2 December 1992, the treaty was approved by a thumping majority of 543 out of 568 votes in the Bundestag. Two weeks later the Bundesrat affirmed the treaty unanimously. A statute confirming the treaty [*Zustimmungsgesetz*] was promulgated on 28 December. A week earlier, parliamentary supermajorities also approved a new constitutional amendment—the revised Article 23(1) GG—designed to facilitate further transfers of power to Europe. All parties agreed that Germany's political future depended on Europe's being stable and on Germany's being trusted by its neighbours. Most assumed that voters agreed. Few raised reservations about the treaty for fear of being tarred as 'anti-Europe'.

Some observers were troubled by this 'spiral of silence';[383] others were bothered by the treaty itself. Hans Heinrich Rupp, a law professor in Mainz, complained that the treaty fundamentally altered the nature of the German constitution. The Maastricht Treaty, Rupp wrote, replaced the constitutional order of the Basic Law with another order altogether—one subordinate to a 'super-constitution with a different area of sovereignty, different citizens, and different basic structures'.[384] This was not a mere amendment to the existing constitution; it was an act of 'constitutional re-creation [*Neuschaffung*], an exercise of constitution-giving authority of which only the *pouvoir constituant*, the People, can avail itself'.[385] But the citizenry's views on the new treaty had not been solicited. For the second time in two years, German voters played no role in creating a Union that would define their constitutional future. To many, it was yet another constitutional moment in which We the People said nothing. Some sought a spokesman in Karlsruhe.

In a constitutional complaint, groups of German voters asserted that the Bundestag law approving the Maastricht Treaty violated a host of constitutional rights. Since most of these were not directly affected by the law approving the treaty, the Court ruled that it had jurisdiction to consider the complaint only as it related to Article 38 GG, which guarantees to all Germans the right to participate directly in Bundestag elections. Members of the Bundestag, in their turn, were declared by Article 38 to be 'representatives of the entire German people'. The complainants contended that the treaty's transfer of competence from the Bundestag to Brussels was of such magnitude and scope that it undermined the Bundestag's status as representative of the people, and thereby offended the democracy principle enshrined in Article 20 and shielded from amendment by Article 79.

At first blush, the Second Senate's judgment of 12 October 1993 was consistent with the Court's reputation, in matters of foreign policy, as a dog that barked but did not bite. Predictably, the Court approved the treaty law as consistent with the constitution.[386] Predictably too, the Court qualified its approval with reservations about Europe's democratic deficiencies and with a pledge of ongoing oversight. The sharpness of the Senate's formulations, however, was surprising. And the theoretical underpinnings of its provisos were astonishing.

[383] Hans Heinrich Rupp, 'Muß das Volk über den Vertrag von Maastricht entscheiden?' (1993) 46 *NJW* 38, 38.
[384] ibid 40. [385] ibid. [386] 89 BVerfGE 155 (1993).

Article 38, the Senate observed, conferred on all citizens the right to legitimize state power and influence its exercise by voting in Bundestag elections. The state must not vacate this right, the Court held, through excessive transfer of Bundestag competences.[387] The Basic Law explicitly envisioned the transfer of *some* competences to international organs, and its preamble explicitly endorsed European unity. But the precondition for such transfers was the assurance that, within the European federation of states (*Staatenverbund*), the German *Volk* would continue to confer legitimacy and exert influence.[388] The Court used its terms advisedly. *Staatenverbund* was a conceptual neologism, but one that hewed much closer to the traditional notion of a *Staatenbund* (confederation of states) than to the opposing notion of a *Bundesstaat* (federal state). In a *Staatenbund*, the balance of power lies with the member states; in a *Bundesstaat*, with the central power. By calling the European Union a *Staatenverbund*, the Senate was defining the Union's limits quite pointedly.

The constitutional requirement that the sovereign people legitimize and influence state power was not, the Senate added, merely a matter of voting. Democracy presupposed a *demos*—a body of citizens who could scrutinize the actions of their rulers and communicate with those rulers in their own language.[389] The Senate cannot have meant this literally—i.e. that a common European language was a *sine qua non* of democratic legitimacy—for it immediately added that, insofar as these conditions did not yet prevail, they might yet be developed within the institutional framework of the European Union.[390] Such development, however, would take some doing. The Senate called for greater transparency from the European Commission and greater efforts to create a European public sphere.[391] For now, the EU's principal source of democratic legitimacy remained the parliaments of its member states.[392] So long as this was so, the advance of European integration must not restrict the operation of 'living democracy' within the member states.[393] A living democracy meant a state endowed with 'a sufficiently momentous range of competences, in which the State's people [*Staatsvolk*], in a process of political will formation led and legitimated by itself, can develop and articulate, in order to give legal expression to, that which—in relative homogeneity—binds it together spiritually, socially, and politically'.[394]

This invocation of homogeneity was explosive. In a scorched-earth assault on the judgment, Joseph Weiler, doyen of European law experts, accused the Second Senate of 'looking backwards, like Lot's Wife, to a polity based on the tired old ideas of an ethno-culturally homogeneous Volk and the unholy Trinity of *Volk-Staat-Staatsangehöriger* as the exclusive basis for democratic authority and legitimate rule-making'. In Weiler's view, the Senate had deployed a 'late Twentieth Century version, albeit anemic and racially neutral, of what in faraway times fed the slogan of Blood (Volk) and Soil (Staat)'.[395]

This was unfair. However unfortunate the Senate's diction, and however backward-looking its conceptual categories (the judgment relied prominently on a 1928 essay by Hermann Heller), a more generous reading is that the Senate did not claim that a democratic state must rest on the foundation of a homogenous *Volk*, but that the people themselves must be empowered to define the terms of their (relative) homogeneity—to delineate their common values and to give those values legal expression. By any standard, this was not happening at the European level; and so long as that was so, the Senate insisted that it must happen,

[387] ibid 182–83. [388] ibid 184. [389] ibid 185. [390] ibid. [391] ibid.
[392] ibid 185. [393] ibid 186. [394] ibid.
[395] See JHH Weiler, *The State 'über alles': Demos, Telos, and the German Maastricht Decision* (1995) <http://centers.law.nyu.edu/jeanmonnet/archive/papers/95/9506ind.html> last accessed 29 May 2015. Weiler wrote that the decision was, 'as regards the existing Community[,] embarrassing; as regards its future evolution I find the decision sad, even pathetic'.

to some minimum degree, at the national level instead. Specifically, this meant that 'the German Bundestag must retain powers and competences of substantial weight'.[396]

This was a hurdle of sufficient ductility that the Senate had little trouble concluding that the Maastricht Treaty cleared it. The EU remained a federation of states (*Staatenverbund*), rather than a federal state (*Bundesstaat*) rooted in a European *Staatsvolk*.[397] By signing the treaty, the member states had not surrendered their sovereignty; they had chosen only to exercise some of their sovereignty jointly.[398] The Bundestag retained powers of sufficient variety and scope to survive scrutiny under Article 38.

But the organs of the European Union, and the German politicians who helped empower and staff them, must not assume that this would always be so. Any transfer of power to the supranational level must be transparent, and the treaty must not be construed as a charter for the organic growth of a Brussels behemoth.[399] The Senate conditioned its approval of the treaty on its finding—emphatic and controversial—that the treaty did not grant the Union's organs power to define their own competence.[400] The treaty conferred no *Kompetenz-Kompetenz*. The member states themselves—with their democratically-legitimated parliaments—remained 'masters of the treaty'.[401] The Senate even maintained, without explanation, that the member states retained a right of unilateral departure.[402] (David Currie, an American constitutionalist, wrote that this passage 'would have warmed the heart of John C. Calhoun'.)[403] Any expansion of the Union's powers required express modification of the treaty itself. With regard to the treaty's application within the Federal Republic, any such amendment would, as a practical matter, require the approval of the Federal Constitutional Court.[404]

In his review of the judgment, Joseph Weiler fulminated that the Maastricht decision 'constitutes a transference to the European level of the Court's understanding of the German polity'. Insofar as that understanding lodged the last word in Karlsruhe, Weiler's assessment hit home. Often that final word would be a more or less grudging endorsement: humming and hawing, reservations and provisos, admonitions and warnings, ultimately approval. This pattern repeated itself in the Court's approval of the Lisbon Treaty in 2009 and of the European Stability Mechanism in 2012.[405] But always there remained the possibility of a dramatic reversal—a point to which we shall return in our Epilogue.

Weiler was not the only Europeanist disgruntled by the *Maastricht* decision. Michel Fromont, a French scholar, contrasted the Constitutional Court's eleventh-hour, admonitory approval of the treaty with the deferential imprimatur planted by the French Constitutional Council in September 1992. 'In the French legal order', Fromont sniffed, 'the constitution-giver is sovereign—not, as in the German legal order, the constitutional court'.[406] In the aftermath of the Second Senate's judgment, the true 'Master of the EU-Treaty' would reside neither in Brussels nor in the national capitals but in Karlsruhe.[407]

Many scholars joined Joseph Weiler in lamenting what they saw as the Senate's anachronistic understanding of the state. The advance of European integration since the Second

[396] 89 BVerfGE 155 (n 386) 186. [397] ibid 188. [398] ibid 188–89. [399] ibid 187.
[400] ibid 192, 196. [401] ibid 198–99. [402] ibid 190.
[403] David P Currie, *The Constitution of the Federal Republic of Germany* (University of Chicago Press 1994) 100.
[404] ibid 187.
[405] 123 BVerfGE 267 (2009); BVerfG, 2 BvR 1390/12 vom 12.9.2012, Absatz-Nr. (1—319) <http://www.bverfg.de/entscheidungen/rs20120912_2bvr139012.html> accessed 6 March 2013.
[406] Michel Fromont, 'Europa und nationales Verfassungsrecht nach dem Maastricht-Urteil—Kritische Bemerkungen' (1995) 50 *JuristenZeitung* 800, 803.
[407] ibid.

World War, among other developments, should long since have superseded the 'State/Non-state' dichotomy of nineteenth-century doctrine.[408] Manfred Zuleeg, a judge on the European Court of Justice, charged the German justices with a 'relapse into long-obsolete notions of sovereignty'.[409] Others worried that the Court had placed itself, and not for the first time, on a collision course with the European Court of Justice. A destructive clash could be avoided only by mutual, voluntary cooperation.[410]

Observers of very different persuasions were disappointed that the Court had not gone further. *Der Spiegel* dismissed the Senate's democracy rhetoric as 'without consequence'.[411] 'Never', the paper wrote, 'have the Karlsruhe justices been so noncommittal [*unverbindlich*] in their dealings with the rights of citizens'.[412] Hans-Christian Ströbele, a Berlin lawyer representing the Green Party in its complaint against the treaty, lamented that the justices had 'invoke[d] democratic ideals but ignore[d] an undemocratic reality'.[413] Still others hailed the Court's 'wise restraint', its 'rare humility', its return to the 'long-disdained precept of judicial restraint'.[414] Another writer referred to the 'hesitant blessing from Karlsruhe', in which the justices embraced 'pragmatic clemency rather than perfectionist stringency'.[415]

The decision and its aftermath had obvious historical parallels. Like the Court's 1973 *Basic Treaty* judgment, the *Maastricht* decision was most disappointing to ardent champions of the treaty whose constitutionality the judgment affirmed. As it had done twenty years earlier, the Second Senate allowed a monumental international agreement to enter into force, but only under a binding constitutional interpretation that could hinder or hobble future developments. Like the first *Solange* judgment in 1974, the *Maastricht* decision was more celebrated by opponents than by partisans of European integration. And as with many of the Court's major foreign policy decisions, the actual holding in the case was probably inevitable. As one commentator noted, it was 'hardly thinkable' that a single national constitutional court would derail the European project when the political leaders of every member state had already signed on.[416]

The judgment's enduring significance lay in the Court's willingness to identify democratic deficiencies in the Union treaty about which few politicians in any major party were willing to speak. There had been little discussion or debate among the country's uniformly Europhile parties about the best or most democratically-legitimate paths towards further integration. It was the contribution of the *Maastricht* complainants, and of the justices of the Second Senate, to get the conversation going.[417] For the second time in the new post-Cold War era, the German people had been denied direct participation in an epochal constitutional moment. For better or worse, it was left to the Constitutional Court to speak in their behalf.

[408] Karl M Meessen, 'Maastricht nach Karlsruhe' (1994) 47 *NJW* 549, 554. Meessen described the decision as 'three steps forward, two steps back'. ibid 549.

[409] Manfred Zuleeg, 'Die Rolle der rechtsprechenden Gewalt in der europäischen Integration' (1994) 49 *JuristenZeitung* 1, 7.

[410] See Markus Heintzen, 'Die "Herrschaft" über die Europäischen Gemeinschaftsverträge—Bundesverfassungsgericht und Europäischer Gerichtshof auf Konfliktkurs?' (1994) 119 *Archiv des öffentlichen Rechts* 564.

[411] 'Dünner Faden' *Der Spiegel* (18 October 1993) 31–32. [412] ibid. [413] In ibid.

[414] Hans Schueler, 'Zur Demokratie verurteilt' *Die Zeit* (15 October 1993).

[415] Dieter Buhl, 'Wer hat Angst vorm Superstaat?' *Die Zeit* (15 October 1993).

[416] Matthias Cornils, *Maastricht* in Jörg Menzel and Ralf Müller-Terpitz (eds), *Verfassungsrechtsprechung. Ausgewählte Entscheidungen des Bundesverfassungsgerichts in Retrospektive*, 2nd edn (Mohr Siebeck 2011) 529, 529.

[417] ibid 530.

B. Arms and the Union: the military meaning of reunification

The Cold War was a long peace. Its frigidities hardened into a stable and enduring balance of power. Within that balance, divided Germany was a buffer between East and West, a border between the US and Soviet spheres of influence.

Buffers, by nature, are passive things. During the long postwar era, Germany's geopolitical influence was almost entirely negative; what did *not* happen in Germany was far more significant than what did. The Bonn Republic was rocked by seismic debates over the founding and atomic arming of the *Bundeswehr*, and over the placement of the Western powers' nuclear armaments on West German soil. But for thirty-five years, the *Bundeswehr*'s active operations were negligible. In the years immediately following unification it was possible, as we have seen, for the Constitutional Court to maintain that the mantra 'Soldiers are Murderers' could not possibly refer to *Bundeswehr* soldiers because no *Bundeswehr* soldier had ever slain a foe in battle. Though the Federal Republic joined various geopolitical alliances, an overwhelming consensus emerged among the country's political classes that German armed forces could be employed only defencively and only within NATO territory. Over several decades, the *Bundeswehr* was not used even in this defensive capacity.

Reunification changed things. The community of nations expected united Germany to assume a larger role in world affairs. For many, this expectation was mingled with fear. Increasingly, Germans themselves desired such influence—a geopolitical role for their country commensurate with its economic heft. The immediate post-Cold War era, with its roseate optimism regarding the efficacy and rectitude of international 'peace-keeping' missions, supplied ample opportunity for Germany to assume such a role.

Initially, the nation's rulers were reluctant. The First Gulf War (August 1990 to February 1991) followed close on the heels of German reunification. At that time, the Federal Republic's leaders agreed with near unanimity that German forces could be used only defensively and only within NATO or Western European Union (WEU) territory. Germany supported the UN coalition economically, but it sent no troops to the Persian Gulf. Hans-Dietrich Genscher (FDP), the foreign minister, clung tenaciously to the view that German participation in 'out-of-area' military operations was banned by the country's fundamental law. Genscher's adamantine stance earned him the enmity of foreign powers but won loud applause at home.

Klaus Kinkel, who succeeded Genscher in May 1992, was more malleable. When push came to shove, so was Helmut Kohl. As internecine violence spiralled out of control in Somalia and in the Balkans, Kohl's government opted for a more active international role. In these cases, the government resolved to deploy forces as well as donate cash. The government's most dramatic decisions were to send marine planes and destroyers to help enforce a UN embargo against Serbia and Montenegro (July 1992); to place German troops in AWACS (Airborne Warning and Control Systems) machines deployed to enforce a UN-decreed 'no fly zone' over Bosnia and Herzegovina (October 1992); and to commission a contingent of 1,700 German soldiers as part of a UN humanitarian mission in Somalia (May 1993–March 1994). These operations unleashed within the Federal Republic an impassioned constitutional debate. In some respects, it was an existential debate. In due course, the debate made its way to Karlsruhe.

The multi-partisan complaint[418] implicated three textual prongs: first, Article 24(2) GG, which allowed the Federal Republic to join a 'system of mutual collective security' for the

418 Some of the MPs who joined the complaint were part of the FDP Bundestag faction. These stuck to the old Genscher line, bucking the stance of the government in which the FDP was junior partner and of Foreign Minister Kinkel, one of the party's chiefs.

sake of 'keeping the peace'; second, Article 59(2), which required parliamentary approval of '[t]reaties that regulate the political relationships of the federation'; and finally, Article 87a(2), which decreed that, '[o]ther than for defence', the military forces of Federal Republic could be deployed 'only insofar as expressly permitted by this Basic Law'. It was this last provision that opponents of the actions trumpeted most loudly and government functionaries massaged most assiduously. In the end, the Constitutional Court left this apparently central question open.

Or did it? In a very long judgment dated 12 July 1994, the Second Senate upheld the constitutionality of all three contested actions.[419] Constitutional authority for out-of-area deployments was to be found, the Senate held, in Article 24(2)'s provision for German entry into mutual systems of collective security. Article 24(2) not only allowed Germany to join such systems, it authorized the government to perform the duties associated with such systems.[420] Those duties included placing military forces at the disposal of the security organization for the sake of preserving or restoring peace.[421] Because authorization for such deployment could be derived directly from Article 24, the justices saw no need to enter the long-standing debate over Article 87.

This conclusion was problematic in at least two respects. First, Article 24(2) was included in the original Basic Law of 1949; it was promulgated, in other words, at a time when the Federal Republic had no armed forces and had no immediate prospect of acquiring them. It was hard to see how a constitutional provision adopted seven years before the *Bundeswehr* was established could authorize its extraterritorial deployment.[422]

The second problem was that, by ducking the Article 87 question, the Senate begged the Article 87 question. Opponents of the operations insisted quite plausibly that, because Article 87a(2) was the only constitutional provision that mentioned the deployment of armed forces, constitutional analysis of such deployments must begin there. Only after acknowledging that Article 87a(2) limited non-defensive deployments to cases 'expressly permitted' within the Basic Law could the Court properly ask whether Article 24(2) granted such permission—and whether that permission covered the three challenged deployments.[423] At the very least, the Senate needed to explain why Article 24(2) was not restricted by Article 87a(2).[424] The Court's putative non-construction of Article 87 was a de facto rejection of the strict construction.

All of the justices agreed that Article 24 authorized deployment of German military units within the framework of NATO and WEU and pursuant to resolutions of the UN Security Council. Four justices, however, contended that the treaties creating these organizations did not cover military action beyond the purview of these organizations—specifically, that the AWACS operation in Yugoslavia lay outside NATO's original aims. [425] In the view of these justices, the deployments expanded the treaty's scope, and therefore (under Article 59(2) GG) required parliamentary approval, which the government hadn't sought.[426] As always, however, the split within the Senate ran against the complainants. Notwithstanding the reservations of four justices regarding the AWACS operation, the Senate sustained the constitutionality of all three actions.

But it did so with an important proviso—a 'parliamentary proviso' (*Parlaments vorbehalt*), the justices called it. Although the Basic Law allowed deployment of Germany's armed forces

[419] 90 BVerfGE 286 (1994). [420] ibid 344. [421] ibid 345.

[422] See Gerd Roellecke, 'Bewaffnete Auslandseinsätze—Krieg, Aussenpolitik oder Innenpolitik?' (1995) 34 *Der Staat* 415, 417.

[423] See ibid 415–16. [424] See Werner Heun, 'Anmerkung' (1994) 48 *JuristenZeitung* 1073, 1073.

[425] The four were Limbach, Böckenförde, Kruis, and Sommer.

[426] See 90 BVerfGE 286 (n 419) 372–77.

within international systems of collective security, in principle such deployments required 'the prior constitutive approval of the Bundestag'.[427] The justices divided on whether such approval had been supplied in the cases at hand; but all agreed that Bundestag approval was essential. Questions of war and peace, the Court held, were the business of the people's representatives in Parliament. The justices even styled the *Bundeswehr* a 'Parliamentary Army' (*Parlamentsheer*).[428]

One commentator quipped that this designation might have befitted Cromwell's Ironsides, but accurately described no modern army on earth.[429] Worse still, the Basic Law contained not a syllable on the subject. The Court based its conclusion, not on the text of the constitution, but on 'German constitutional tradition since 1918'.[430] Lack of parliamentary oversight had allowed the German Reich to slouch into war in the summer of 1914. The Weimar constitution and the Basic Law both required that this never happen again. Between the lines, the Senate was surely thinking also of the creeping, unilateral executive escalation of American involvement in Vietnam.[431] Such historical memories contributed to the broad acceptance with which the judgment was greeted. Although the judgment arguably amended the constitution on the Court's own initiative, the decision met with comparatively little criticism.[432]

Like the *Maastricht* judgment, the *AWACS* decision mingled pragmatic acceptance of a geopolitical fait accompli with a dramatic statement of constitutional principle. Unsurprisingly, international observers were more struck by the former than the latter. The London *Times* predicted that, 'slowly and hesitantly', Germany would 're-emerge ... as a military power'.[433] In Paris, *Le Monde* proclaimed 'a new era in German foreign policy'.[434] A paper in Sweden went so far as to call the judgment Germany's 'final step out of the shadows of the Second World War'.[435] The judgment gave constitutional imprimatur to an epochal shift in German foreign policy and marked an era in postwar German history. Fittingly perhaps, the judgment was announced in Karlsruhe the same day that Bill Clinton, the American president, delivered an 'historic' speech in front of the Brandenburg Gate before dismissing America's remaining 'occupation' troops in Berlin.[436]

Within Germany itself, popular commentary focused on the judgment's implications for German soldiers and for Parliament's role in shaping foreign policy. 'It is now once again deadly', wrote *Der Spiegel*, 'to be a soldier'.[437] *Die Zeit* wrote that German soldiers now were 'duty-bound to die' (*zum Sterben verpflichtet*).[438] All the more reason, then, the paper opined, to celebrate the Court's reinforcement of parliamentary prerogative and popular sovereignty. The death of German soldiers was 'worth a case-by-case decision of the Bundestag'.[439]

[427] ibid 381.

[428] ibid 382. For an extended exploration of this theme, see Dieter Wiefelspütz, *Das Parlamentsheer. Der Einsatz bewaffneter deutscher Streitkräfte im Ausland, der konstitutive Parlamentsvorbehalt, und das Parlamentsbeteiligungsgesetz* (Berliner Wissenschafts-Verlag 2005).

[429] Roellecke (n 422) 425.

[430] 90 BVerfGE 286 (n 419) 383.

[431] See Michael Wald, 'AWACS' in *Verfassungsrechtsprechung* (n 1) 564.

[432] But see Roellecke (n 422); Claus Arndt, 'Verfassungsrechtliche Anforderungen an international Bundeswehreinsätze' (1994) 47 *NJW* 2197.

[433] 'A Special Relationship' *The Times* (14 July 1994).

[434] Henri de Bresson, 'La Cour constitutionnelle autorise la participation de soldats allemands à des forces internationales' *Le Monde* (14 July 1994).

[435] The 'Svenska Dagbladet' in 'Nun siegt mal schön' *Der Spiegel* (18 July 1994) 23.

[436] See Carlos Widmann 'Makellose Arie für die Berliner' *Der Spiegel* (18 July 1994) 27.

[437] 'Nun siegt mal schön' (n 435) 26.

[438] Hans Schilder, 'Zum Sterben verpflichtet?' *Die Zeit* (15 July 1994).

[439] ibid.

In the years following its first *AWACS* decision the Court was asked repeatedly to scrutinize foreign deployments of German soldiers.[440] Invariably, the Court approved the challenged executive action. Over time, the justices 'seemed less willing to enforce strict constitutional limitations on the executive's authority over military affairs'.[441] As in the context of European integration, the justices were accused of indulging in rhetorical panache that cloaked an underlying failure of nerve—or of masking grim *Realpolitik* with glittering *obiter dicta*. From a comparative perspective, however, the striking thing about the Court's military powers jurisprudence is not that the Court has generally deferred to executive discretion but that the Court has reviewed executive action at all. Here yet again was the long legacy of the Court's non-embrace, in the early 1950s, of an American-style political question doctrine.

Furthermore, it cannot be said that the Court's jurisprudence calling for parliamentary approval of military action was without effect. In March 1999, for instance, when the Second Senate approved German involvement in a NATO bombing campaign against Serbia and Montenegro, it did so on the ground that the federal government, on four separate occasions, had sought and obtained parliamentary permission. [442] The executive has not uniformly followed this pattern, but the general requirement of Bundestag approval for foreign deployments has entered and shaped the country's constitutional culture. Perhaps more to the point, the Court owes part of its high approval among German citizens to the fact that, in foreign affairs, the Court has wielded its counter-majoritarian power in a manner that has reinforced (at least rhetorically) the country's commitment to popular sovereignty. Throughout the decade following reunification, the justices spoke in the name of democracy when comparatively few leading politicians did the same.

Conclusion

The Federal Republic's leaders saw Germany's international engagement as evidence that Germany's putative *Sonderweg* was over. Some were quite explicit about this. In October 2001, at the outset of the NATO-directed war in Afghanistan, in which Germany took part, Chancellor Gerhard Schröder linked Germany's 'out-of-area' military involvement to its overdue arrival in the West. 'After long periods of confusion in the 19th and 20th centuries', Schroeder said, 'Germany has at last joined the West. That's what's at stake for us. The argument that we cannot take part [in military actions abroad] because of our history no longer holds true'.[443] In other words, Germany was a 'normal' nation again—or perhaps for the first time. It was a full-fledged liberal democracy entitled, in affairs foreign and domestic, to act like one.

The government that guided Germany's involvement in NATO's Albanian campaign at the end of the decade was very different from the one that launched the country's first military engagements at the beginning of the decade. In October 1994, three months after the Court's *AWACS* decision, the Union parties outpolled the SPD by a ratio of 41.4 per cent to 36.4 per cent. Helmut Kohl, who continued to enjoy some of the aura he acquired as the 'chancellor of reunification', remained head of a Union-led government with the FDP as junior partner. But there were signs of trouble ahead. The Union's share in the vote had dropped by 2.4 per cent since 1990; the SPD's had grown by 2.9 per cent. More ominously,

[440] See 100 BVerfGE 266 (1999); 104 BVerfGE 151 (2001); 117 BVerfGE 359 (2007); 118 BVerfGE 244 (2007); 121 BVerfGE 135 (2008).

[441] Kommers and Miller (n 88) 205.

[442] 100 BVerfGE 266 (1999).

[443] 'Europe: Michael Steiner' *The Economist* (13 October 2001) 52.

the FDP's share had dropped by 4.1 per cent, from 11 per cent to 6.9 per cent, while the Greens had gone from barely clearing the 5 per cent hurdle in 1990 (5.1 per cent) to outpolling the FDP in 1994 (7.3 per cent). The Left-wing PDS—whose entrance into Parliament the Constitutional Court facilitated four years earlier—garnered thirty seats in the Bundestag. In all, the coalition's majority over the parties of the Left was a narrow 341 to 331. If the Left could unite, it was within striking distance of power. The PDS remained malodorous to all established parties, but rumours of a possible SPD–Green alliance grew in volume after 1994.

Disenchantment with the everlasting Kohl government, now in its fourth term, grew as well. The economy sagged; unemployment was stubbornly high. During the last years of the Kohl government, joblessness in the Federal Republic reached its highest levels since 1949.[444] Easterners felt ambiguous about their new country. Many lamented the failure to propagate a pan-German constitution in 1990. Disappointment mounted in October 1994, when a parliamentary commission, appointed to review the Basic Law in light of reunification, entirely ignored proposals prized by the Eastern populace—the right to work and the right to housing, for instance, as well as stronger plebiscitary elements at the federal level. Perhaps most disheartening, the economic condition of the Eastern states was slow to improve. The prosperity that reunification promised had not materialized. Many 'Ossis' felt that their interests were unrepresented in Bonn. Only the PDS seemed to speak for the East, and that party was treated by its Western peers like an escapee from a leper colony.

Westerners grew disgruntled with their government as well. Germany's slumping economy stirred talk of its being the new 'sick man of Europe'.[445] The transition from economic juggernaut to bedridden behemoth was not a happy one. Government initiative seemed to stall with the economy. Critics spoke of a 'reform jam' (*Reformstau*). A long-standing consensus in favour of government social support seemed to be cracking; so, too, did the social and cultural homogeneity that underlay that consensus. The country's economic engine was sputtering; its citizens were in a funk. The search for a scapegoat settled on the sitting chancellor. As the September 1998 Bundestag elections approached, many suspected that Kohl's days were numbered.

The 1998 election returns were very nearly the reverse of those in 1990. The SPD's share grew by 4.5 per cent, to 40.9 per cent. The Union's tally dropped by 6.3 per cent, to 35.1 per cent. The Greens' fortunes also faded, but only by six-tenths of a per cent. The party might have done much better, if not for a platform calculated to cosset the faithful rather than persuade the electorate. The Greens' platform vowed to raise the price of fuel to five DM per litre (something like thirteen dollars per gallon) and roundly condemned German involvement in a popular international peacekeeping mission in Bosnia. The platform reminded voters of the Greens' extreme pacifist-environmentalist origins and obscured the party's recent drift towards the mainstream. Still, the Greens won enough seats to join a coalition led by Gerhard Schröder, the new SPD chancellor. The Greens were in power for the first time. Kohl had fallen after sixteen years at the head of government. The nation had its first Red–Green coalition. Reformists expected big things.

A Red–Green government put Greens in the cabinet. The most prominent was Joschka Fischer, whose colourful career climaxed with his appointment as foreign minister and vice

[444] See Eckhart Conze, *Die Suche nach Sicherheit: Eine Geschichte der Bundesrepublik von 1949 bis in die Gegenwart* (Siedler 2009) 794.

[445] See, e.g., 'Germany's Economy: The Sick Man of the Euro' *The Economist* 21 (5 June 1999); 'Self-healing for the Sick Man of Europe' *Financial Times* (11 November 1999) 15; Allan Hall, 'Germany on Way to Becoming "Sick Man of Europe"' *The Scotsman* (2 June 1999) 11.

chancellor. Only thirteen years had passed since Fischer, the first Green minister at the state level, provoked an outcry by appearing for his oath of office in tennis shoes.

Ironically, it was Fischer who appeared before the Constitutional Court in 1999 to defend German participation in NATO's bombing campaign against Serbia and Montenegro.[446] Fischer, long the public face of Leftist pacifism, was willing to disappoint his party's hardliners for a humanitarian cause in which he believed passionately. Perhaps, too, there was a dose of *Realpolitik* mingled with Fischer's international humanitarian optimism. There could be no more effective way to distance the party's policy in power from its rhetoric on the hustings.

There was also something symbolically appropriate about Fischer's official appearance in Karlsruhe. A decade and a half earlier, when Green members first entered the Bundestag, the party's representatives petitioned the Court to vindicate their right to participate in parliamentary affairs on equal terms with members from the traditional parties. They met with mixed success,[447] but the Greens' engagement and occasional victories in Karlsruhe played an important role in the party's trajectory towards the constitutional and political mainstream. This is a telling example both of the Court's role in integrating the country's political forces and of its related effect of taming the country's political process. To complete the image, in January 2001, Brun-Otto Bryde, a constitutionalist in Gießen, became the first Constitutional Court justice appointed on the Greens' recommendation.[448]

Fischer's faith in humanitarian military intervention was shared by many German leaders throughout the 1990s. The decade was marked by broad discussion of Francis Fukuyama's premature announcement of 'the end of history'.[449] Experts squawked at Fukuyama's thesis, but few could repress the hope that the Cold War would be followed by a new era in which liberal democracy became the global governmental norm and the international community banded together to keep local tyrants at bay. This was the hope that capsized the traditional German consensus against foreign military involvement. The hope burned bright as the Federal Republic, challenged on many fronts domestically, approached the beginning of a new millennium and celebrated its Golden Anniversary under the guidance of a reformist government.

Commemorations of the Federal Republic's fiftieth anniversary in 1998/99 included sober reflections about the country's past, unease about its present, and nervous forebodings about its future. But the prevailing tone was upbeat, encapsulated in a single, recurrent word sometimes followed by a question mark: *Erfolgsgeschichte*, or success story.[450] *Der Spiegel*, at times an acerbic critic of the Federal Republic, entitled its anniversary issue, 'The German Miracle'.[451] All things considered, this seemed the right word to describe the trajectory of the Federal Republic from a conquered nation, saddled with the historical inheritance of unprecedented crimes, into a moral and economic leader among the community of liberal democracies.

In many anniversary reflections, the Constitutional Court was accorded a share in the credit for the country's success story. Rolf Lamprecht, who had reported on the Court since its founding, made the connection quite pointedly. 'If the Germans have succeeded—as is

[446] See Russell Miller, 'Merely a Landmark or a Change of Course: The Federal Constitutional Court Hears Arguments in the NATO Strategic Conception Case' [2001] 2 *German Law Journal* <http://www.germanlawjournal.com/index.php?pageID=11andartID=37>.

[447] See Chapter 4. [448] See 'Richtig happy' *Der Spiegel* (11 December 2000).

[449] Francis Fukuyama, *The End of History and the Last Man* (Free Press 1992).

[450] See Axel Schildt, *Ankunft im Westen. Ein Essay zur Erfolgsgeschichte der Bundesrepublik* (Fischer 1999); Klaus Naumann, 'Erstaunlich stabil' *Die Zeit* (20 May 1999).

[451] 'Das deutsche Wunder' *Der Spiegel* (17 May 1999).

apparent both at home and abroad—in their transformation to a democratic *Rechtsstaat*, then two factors have been essentially involved in that success: the canon of fundamental rights, and the stringency of their Karlsruhe guarantors'.[452] Other commentators, however, lamented (or applauded) that the Court was no longer what it once had been. Thomas Darnstädt opined that what he termed the 'Karlsruhe Era' was now over.[453] Darnstädt dated the Court's period of dominance from its *Television* judgment in February 1961 through the five-month period during 1993 in which the Court delivered both its second *Abortion* decision and its judgment on the Maastricht Treaty.[454] Since then, Darnstädt contended, the Court's idyll was ended. Politicians had realized that resistance was an option about which the Court could do little. Statesmen were constantly tempted to march down determined paths despite the Court's admonitory dicta. Thousands of crucifixes still adorned the walls of Bavarian school rooms.[455]

Up to a point, Darnstädt was right. The Court had likely lost some public trust and political respect. It had indisputably lost its monopoly jurisdiction. The European Court of Justice and the European Court of Human Rights both provided alternative fora for the vindication of fundamental rights. Globalization diminished the Court's power by narrowing its jurisdiction. More Europe meant less Germany, and perhaps, as a result, fewer trips to Karlsruhe. A similar process had already lessened the profile of the only federal institution ever to rival the Court in the public's affections: the Bundesbank, now eclipsed by the newly created European Central Bank. And yet, as both Darnstädt and Lamprecht noted, German citizens continued to file complaints to the Court by the tens of thousands—and this despite demonstrably slender odds of success. In the Court's first fifty years, it fielded some 135,000 constitutional complaints—96 per cent of them from private citizens—and this notwithstanding a success rate of around 2.6 per cent.[456] In a country whose constitution featured few plebiscitary elements, the constitutional complaint remained the one medium by which citizen Everyman's voice might be heard. As Lamprecht reported, 'The general civil right to call upon Karlsruhe at any time fostered democratic self-consciousness. Those who recognized this possibility and, if necessary, availed themselves of it, needed no longer conceive of themselves as subjects [*Untertan*]'.[457]

In many respects, the decade following reunification was the Court's most difficult since its founding. At decade's end, however, the Court seemed secure in its traditional role—though that role might diminish through ongoing globalization and European integration. On average, the public trusted the Court nearly as much as ever.[458] Politicians, who had always grumbled when a Court decision went against them, remained eager to highlight the constitutional aspect of intractable political squabbles and shunt the question to Karlsruhe. Often, as with the Court's November 1999 decision on the equalization of state finances, they were grateful when the Court found a principled way to split the difference. The world in which the Court operated had altered in dramatic ways. But the Court, at least for the time being, retained its classical role.

Although the heady economic hopes roused by reunification were swiftly disappointed, the geopolitical idealism fostered by the end of the Cold War lasted somewhat longer. The

[452] Rolf Lamprecht, '"Das Über-Ich der Nation"' *Der Spiegel* (1 September 1998).

[453] Thomas Darnstädt, '"Mir hat keiner was zu sagen"' *Der Spiegel* (17 May 1999) 26.

[454] ibid 206.

[455] ibid 208–09.

[456] Ursula Knapp, '16 Richter machen "die ideale des Grundgesetzes lebendig"' *Frankfurter Rundschau* (29 September 2001) R 4-24H.

[457] Rolf Lamprecht, 'Das Über-Ich der Nation' *Der Spiegel* (1 September 1998).

[458] See Sascha Kneip, *Verfassungsgerichte als demokratische Akteure. Der Beitrag des Bundesverfassungsgerichts zur Qualität der bundesdeutschen Demokratie* (Nomos 2009) 200.

latter idealism vanished, however, with the terrorist attacks on New York and Washington on the morning of 11 September 2001. Gerhard Schroeder, the German chancellor, was swift and forceful in his expression of solidarity with the United States.[459] Schroeder enthusiastically championed German involvement in the War in Afghanistan but dramatically opposed the American-led invasion of Iraq eighteen months later. The Age of Terror would vastly complicate German foreign policy and the country's relationships with its allies. It would also bring new and vexing questions before the Federal Constitutional Court.[460]

By sad coincidence, the 9/11 attacks coincided almost exactly with the Court's own fiftieth anniversary. Fears of terrorist attacks within the Federal Republic raised doubts about whether the Court's jubilee celebration, long planned for 28 September 2001, should take place at all. Many questioned the wisdom, in the tense post-9/11 aftermath, of assembling the Federal Republic's highest officials under one roof.[461] But the show went on—a splendid gala in the Baden State theatre that opened with a performance of *The Marriage of Figaro*. In attendance were chief justices of constitutional courts from around the globe—a testament to the Court's leadership and influence among the world's constitutional tribunals—as well as the heads of the Federal Republic's four other constitutional organs: Johannes Rau, the federal president; Kurt Beck, president of the Bundesrat; Wolfgang Thierse, president of the Bundestag; and Gerhard Schröder, the chancellor.[462]

The two last named might have felt a bit sheepish. Both had been publicly critical of the Court in recent days; both had resurrected the perennial complaint that the Court was meddling too much in political affairs. Just two days before the festivities, Thierse (SPD) had convened a colloquium at Berlin's Humboldt University to discuss the theme 'Bundestag and Federal Constitutional Court'.[463] In Thierse's telling, the Court mistakenly perceived itself as a 'motor of development' in social legislation. The Court's recent decisions regarding state support for families,[464] as well as inheritance and property tax,[465] had burdened the legislature unduly at a time of pinched finances and had constrained excessively politicians' room to manoeuvre.[466] Thierse repeated the age-old admonition, which he attributed to the former justice, Wiltraut Rupp-von Brünneck, that the Court must take care not to usurp the role of an *ersatz* legislature. The colloquium discussants included Ernst Benda, the former chief justice, and Hans Jochen Vogel, the former justice minister, both principal players in the Court's previous clash with an SPD-led government.[467]

Two days later, Thierse and Schröder were sitting in Karlsruhe, celebrating the Court they had lately censured. In her presidential address, Jutta Limbach noted Thierse's and Schröder's presence on the front row with mild good humour. In the spirit of the festivities, the chief justice began, she would leave the criticisms unanswered 'as exercises of freedom of speech'.[468] Rau, the federal president, was less willing to let things slide. Rau dismissed as 'exaggerated' recent claims that the Court mingled in political affairs too readily and assumed legislative powers too liberally.[469] The Court intervened in political decisions, Rau noted, only when politicians asked it to do so—and the justices demanded positive action

[459] See 'Deutsche Politiker entsetzt' *FAZ* (12 September 2001) 2.

[460] See, e.g., 115 BVerfGE 118 (15 February 2006) (striking down, as an offence to human dignity, a law that allowed the armed forces to shoot down a passenger plane believed to have been hijacked for terrorist purposes).

[461] See 'Feier in Zeiten des Terrors' *Süddeutsche Zeitung* (29 September 2001) 7.

[462] ibid.

[463] Karl-Heinz Baum, 'Thierse rügt Verfassungsrichter' *Frankfurter Rundschau* (27 September 2001).

[464] 99 BVerfGE 216 (1998).

[465] 93 BVerfGE 121 (1995).

[466] In Karl-Heinz Baum, 'Thierse rügt Verfassungsrichter' *Frankfurter Rundschau* (27 September 2001) R 3-30H.

[467] ibid.

[468] ibid.

[469] Ursula Knapp, '16 Richter machen "die ideale des Grundgesetzes lebendig"' *Frankfurter Rundschau* (29 September 2001) R 4-24H.

from the political branches only when political actors failed, because of unpopularity or expense, to do what the constitution required.[470] Rau praised the Court for having 'filled with life the Basic Law's ideals of freedom, equality, and justice'.[471]

Unsurprisingly, the Court's jubilee celebration prompted public reflections on the Court's place in the Federal Republic. Some reflections looked forward, others backward. Ursula Knapp, writing for the Left-Liberal *Frankfurter Rundschau*, predicted that the events of 11 September would not diminish the Court's importance. On the contrary: nearly every announced reaction to the attacks would implicate the constitution, and therefore the Court's jurisdiction, in some way or other.[472] For better or worse, Knapp observed, the Court took questions of peace and war—and of parliamentary and executive power—quite seriously.[473] One could expect this to continue and to intensify. More broadly, Knapp noted that the Court's immense authority depended largely on the fact that all parties to constitutional disputes left feeling that they had had a hearing—that their complaints had been considered, their grievances understood.[474] This was a product both of the Court's make-up and of its methodology—of the institutional pressures (primarily the requirement of a two-thirds parliamentary majority for confirmation) that lead to the Court's being populated chiefly by centrists, and of the proportionality framework that requires the Court to assert and assess the constitutional values at play on both sides of any given dispute.

Other anniversary assessments were unabashedly panegyric. Helmut Kerscher, writing in the *Süddeutsche Zeitung*, asked what the Federal Republic would be like without the Federal Constitutional Court. 'It would be a republic', Kerscher wrote, 'with even more power for the rulers and the parties, with less stability, lower respect—and with fewer freedoms for citizens, with even less equality for women'.[475] Minorities, in particular, would be worse off without the Court—prisoners, pacifists, extramarital children. There would be less freedom of speech and press, no oversight of judicial or administrative decisions. Even politicians, who so often bristled at the Court's interference, were grateful at heart for the Court's resolution of disputes too hot for the politicians themselves to handle. Kerscher went on to praise the Court's protection of opposition parties and dissident groups; its policing of federalism; its oversight of party finance; its vindication of freedoms of speech, assembly, demonstration, press, personality, and privacy. 'Karlsruhe was the first', Kerscher concluded, 'to bring fundamental rights from paper to life'.[476] This was more than compensation for occasional—or even frequent—judgments that one found politically or jurisprudentially disappointing.

It was also, for most Germans, more than compensation for the inherent tension between constitutional justice and popular sovereignty. For fifty years the Court had played an integrating, mediating role in West German society. It had enabled liberal and democratic values and institutions to take root in the second German Republic. It fostered those values by articulating them, applying them, and requiring all state authority to honour them. In the process, the Court assumed and exercised enormous power—sometimes aggressively, sometimes deferentially, sometimes pragmatically, often persuasively, and usually at a high level of judicial craftsmanship. The Court was often wrong; it often overreached; but on the whole, it did its work well. If it constrained democracy in some respects, it reinforced democracy in others. In both senses, the Court redefined the nature, and reshaped

[470] ibid. [471] ibid.
[472] Ursula Knapp, 'Der Faktor Karlsruhe' *Frankfurter Rundschau* (28 September 2001) D2.
[473] ibid. [474] ibid.
[475] Helmut Kerscher, 'Das Gericht der Republik' *Süddeutsche Zeitung* (28 September 2001) 4.
[476] ibid.

popular understanding, of German democracy. At the dawn of a new millennium, most Germans would struggle, with Kerscher, to conceive of their national polity without the Court that enforced their national constitution. Kerscher called Karlsruhe 'the Court of the Republic'. But the genitive worked both ways—a fact best captured by Gerhard Casper, a German-American constitutionalist asked to give the keynote address at the Court's golden jubilee. Casper described not merely the Court but all of united Germany when he celebrated 'the Karlsruhe Republic'.[477]

477 Gerhard Casper, 'The "Karlsruhe Republic"—Keynote Address at the State Ceremony Celebrating the 50th Anniversary of the Federal Constitutional Court' (2001) 2 *German Law Journal* http://www.germanlawjournal.com/index.php?pageID=11andartID=111>.

Epilogue

The View from Europe

The European Court of Justice can never replace a national constitutional court. What's more, the Court has, God be thanked, found such a high degree of acceptance among the German public that a discussion that would place the Federal Constitutional Court in question is, politically, wholly unrealistic.

—Chief Justice Hans-Jürgen Papier (2009)[1]

The loss in significance of the national constitution is not yet appreciated in full measure in Germany. When that happens, it cannot leave constitutional patriotism untouched. Constitutional patriotism will decline, without one's being able to hope for an equivalent at the international level.

—Former Justice Dieter Grimm (2009)[2]

In early September 2013, Andreas Voβkuhle travelled to Strasburg to address a gathering of thirty or so members of the European Parliament (EP). The Court's young president was to speak on the theme of a 'Europe of Tomorrow.' But the meeting proved less amicable than Voβkuhle's hopeful title might have hinted. 'It was not', *Der Spiegel* reported, 'a lovely evening for Voβkuhle'.[3] For two hours the assembled EMPs peppered the chief justice with hostile questions and angry accusations. 'Voβkuhle', one participant recalled, 'was positively grilled'. The questioners charged the Constitutional Court with undermining the EP and with misunderstanding—and misrepresenting—European democracy. The wrath of the parliamentarians ran across party lines.[4] Voβkuhle was provoked to deliver criticisms as well as swallow them. He noted that no one present had mentioned 'citizens' or 'voters', and wondered aloud whether his questioners worried only about their own power.[5] It was not a retort calculated to restore lost harmony.

Why were the Euro-folk so upset? In what did the Court's offences lay? The EMPs' anxiety stemmed not only from what the Court had done recently but from what it might do soon. In September 2012, the Court had refused to issue an injunction staying German approval of the European Stability Mechanism (ESM); but the Court's decision on the constitutional merits of the issue was still pending. A decision against the ESM from the highest court of the Euro zone's largest economy would be, to put the matter lightly, profoundly disruptive. The Court was also considering a challenge to the 3 per cent hurdle applicable to German elections for the EP. Rumours suggested that the Court, which two years earlier had nixed a 5 per cent hurdle for EP elections,[6] was considering the complaint quite seriously. Since the Court had repeatedly approved a 5 per cent hurdle for elections to the German Bundestag, a decision to quash a 3 per cent hurdle for European elections would send an unmistakable message that European parliamentary democracy was inferior to its German counterpart.

[1] In 'Wir erwarten Respekt' *Der Spiegel* (19 December 2009) 40.
[2] Dieter Grimm, 'Identität und Wandel—das Grundgesetz 1949 und heute' (2009) 37 *Leviathan* 603, 616.
[3] 'Die Anmaβung' *Der Spiegel* (10 March 2014) 20. [4] ibid. [5] ibid.
[6] 129 BVerfGE 300 (2011).

But this, of course, was a message the Court had sent often, most recently in its qualified approval of German accession to the Lisbon Treaty in June 2009.[7] Throughout this book we have seen how the successive stages of European integration have been subjected to the Court's review but have always, in the end, earned its cautious, qualified, admonitory approval. This was the case, in varying degrees, with the '*Solange*' judgments of 1974 and 1986. It was true also of the *Maastricht* judgment in 1993.[8] The Court's *Lisbon* judgment had much in common with these earlier judgments, but it went beyond them in ways that make the *Lisbon* decision and its progeny a useful lens through which to consider the Court at the present time—its power, its prospects, and its predicament. This Epilogue makes three basic points that link the Court's 'Euro'-jurisprudence of recent years (roughly from 2009 until the time of writing, March 2015) to the broader themes of this book. The basic points are (1) that the Court has emerged in the recent European crises as a defender of *German* democracy in the particularized form the Court itself helped foster; (2) that, ironically, a Court credited with helping to establish Germany within the community of liberal democracies might, in its European jurisprudence, be isolating Germany from a large portion of that community; and (3) that the Court, which some argue has passed its 'zenith', might be resisting an obvious external threat to its ongoing relevance when that relevance has already been undermined from within.

Often in recent years observers around the globe have expressed astonishment that the fate of Europe rests from time to time in the hands of a small band of German judges. The possibility that the Court might derail the European train has raised, to many minds, the spectre of judicial activism reaching an ultimate and appalling apotheosis. But it should be stressed that judicial power reaches far more menacingly from the European Union to its member states than it does in the other direction. The European Court of Justice (ECJ) is a far more activist tribunal than any national constitutional court in Europe, including the German Federal Constitutional Court. Against the activism of the ECJ, the governments of Europe's member states are almost utterly defenceless. National constitutional courts represent the only available counterweight. This must be borne in mind if the relationship between the German Constitutional Court and European integration is to be rightly understood.[9]

I. The *Lisbon* Judgment

After referenda in France and the Netherlands killed the proposed European constitution in 2005, the European Union's member states sought to salvage many of the constitution's reforms by repackaging them as an amended treaty. The new treaty, drafted and signed in Portugal in 2007, was set to enter into force in December 2009. Before then, it had to overcome the opposition of Irish voters, who rejected the treaty in a 2008 plebiscite, but later, after securing concessions, approved it in a second referendum held the following year. The treaty also had to survive the scrutiny of the German Constitutional Court.

The challenge to the treaty in Karlsruhe was a prickly one, not least because, as critics pointed out more or less cynically, the Court itself would lose power if the treaty were approved. It had long been observed—not least by current and former justices—that the

[7] 123 BVerfGE 267 (2009).

[8] See 37 BVerfGE 271 (1974); 73 BVerfGE 339 (1986); 89 BVerfGE 155 (1993).

[9] See, e.g., Dieter Grimm, 'Die Rolle der nationalen Verfassungsgerichte in der europäischen Demokratie' in Claudio Franzius and others (eds), *Grenzen der europäischen Integration. Herausforderungen für Recht und Politik* (Nomos 2014) 27.

expansion of European jurisdictions constricted the Court's own jurisdiction.[10] If approved, the Lisbon Treaty would extend Europe's powers still further. It would increase the competences of the European Council, which would be chaired by a common president of the EU, and of the EP, whose competences would reach many more policy arenas. It would enhance the role of the European Commission and the ECJ. Europe would gain—or would secure the power to pursue—a common foreign ministry and a common diplomatic corps, a common defence policy and a common penal code. And, for the first time, the EU would have a common code of fundamental rights. In all these particulars, the EU would look more like a traditional state than it ever had before.

Approval of the Lisbon Treaty in the German Parliament was overwhelming. The Bundestag vote in April 2008 was 515 to 68; the Bundesrat vote a month later, 58 to 15. Constitutional challenges—both to the law approving the treaty (*Zustimmungsgesetz*) and to a companion law (*Begleitgesetz*) governing the powers of (Germany's) Parliament—followed immediately. The Second Senate heard oral argument for two days in February 2009 and issued its elongated decision (143 printed pages) on 30 June.

After forty years of following a Court that makes menacing noises about Europe's democratic deficiencies but ultimately approves every advancing step of European integration, some observers concluded that the Court would never really stand in Europe's way.[11] The *Lisbon* judgment would itself prompt cute comparisons between the Karlsruhe Court and the Big Bad Wolf, and endless variations on the theme of the Court's bark exceeding its bite.[12] Other wags quipped that the Court was, in European matters, like Germany's national soccer team: they looked formidable, but always choked under pressure.[13]

Joseph Weiler, who had fulminated so fiercely against the *Maastricht* judgment sixteen years earlier, summarized the outcome with a mixture of resignation and fatigue—something between a sigh and a yawn:

> [I]n its internationally-related case law, the German Constitutional Court has a well-earned reputation of the Dog that Barks but does not Bite. There would be, as the more jaded court watchers among us confidently predicted, lengthy 'humming and hawing;' some high sounding and biting criticism of certain democratic deficiencies of the Union and its Institutions; heavy breathing about the German Court's constitutional responsibilities and important guardianship role. But in what we may now call the regular 'Karlsruhe Miracle', the pig would finally be pronounced Kosher—as indeed turned out to be the case.[14]

Many immediate reactions to the judgment, however, were sharp and angry. These, as Weiler noted, were often coloured by the political and ideological sensibilities of the persons assessing the judgment.[15] Those most friendly to the European project tended towards an

[10] See, e.g., Stefan Oeter, 'Rechtsprechungskonkurrenz zwischen nationalen Verfassungsgerichten, Europäischem Gerichtshof und Europäischem Gerichtshof für Menschenrechte' (2007) 66 *Veröffentlichungen der Vereinigung der Deutschen Staatsrechtslehrer* 362–91; Hans-Jürgen Papier, *Das Rechtsprechungsdreieck Karlsruhe—Luxemburg—Straßburg* (Deutsche Hochschule für Verwaltungswissenschaften 2006).

[11] See, e.g., Perry Anderson, 'After the Event' (2012) 73 *New Left Review* 49, 53 ('It is unlikely that the oligarchy of the EU will ever encounter much of a setback at Karlsruhe'.).

[12] See, e.g., Kalypso Nicolaïdis, 'Germany as Europe: How the Constitutional Court Unwittingly Embraced EU Demoi-cracy' (2011) 9 *I·CON* 786, 786 ('Who is afraid of the German Constitutional Court?'); 'Charlemagne: A Grim Tale of Judges and Politicians' *The Economist* (4 November 2010) ('The Karlsruhe court has huffed about the democratic deficit of European bodies and puffed about vital sovereign powers that may not be delegated to Brussels, but it has never stopped European integration'.).

[13] The events of 2014 would refute both sides of this last analogy.

[14] Joseph HH Weiler, 'The "Lisbon Urteil" and the Fast Food Culture' (2009) 20 *Eur J Int'l Law* 505, 505.

[15] ibid.

odd mixture of outrage and relief—relief that the Court allowed the treaty to move forward; indignation at what the justices had to say along the way.

Critics angered by the Court's rhetoric ignited lexical fireworks of their own. The judgment, they said, was 'overly long and ... meandering';[16] 'vast and verbose';[17] plagued by 'myriad treatise-like passages'; bedevilled by the 'devastating bankruptcy of [its] solipsistic theory of the state'.[18] The justices, said critics, had indulged in an essentially 'performative utterance in which the Court underline[d] its own importance in a strange combination of rhetoric and verbosity'.[19] The Court's 'interminable arguments' reflected a lamentable 'lack of seriousness'. They 'betray[ed] an angry acrimony' and an 'uneasy posturing'.[20] The judgment suffered from an almost 'miraculous proliferation of standards' (*wundersame Maßstabvervielfältigung*).[21] The justices had 'overwhelmed constitutional law with state theory'.[22] They had stitched 'a strange patchwork concept of statehood that owes nothing to state theory and everything to political contingency'.[23] The Court had hurled a 'declaration of War' at the European Court of Justice;[24] had proclaimed that 'identity trumps integration;'[25] had shown signs of becoming, in European affairs, 'a mere nuisance factor', 'an unpredictable troublemaker', a resentful institution 'too weak to shape European integration but strong enough to be an encumbrance'.[26] The Court had become the overzealous 'brakeman' of the European train.[27]

All this in response to a judgment that unanimously approved Germany's ratification of the treaty.

The polemical blasts of the judgment's detractors should not obscure the fact that it had many scholarly defenders as well—some of whom noted that many *Lisbon* critics now had a good word to say about the *Maastricht* judgment they had execrated so roundly sixteen years earlier. In one prominent review, two scholars praised the Court for 'taking popular sovereignty and democracy seriously'.[28] Others hailed the Court's vindication of electoral democracy as the classical source of legitimacy for the state and of self-determination for the citizens.[29] Academic response to the judgment, then, was a mixed bag, even if those who applauded the judgment were less vociferous than those who were appalled by it.

But what did the justices actually say?

[16] Christoph Schönberger, 'Die Europäische Union zwischen "Demokratiedefizit" und Bundesstaatsverbot' (2009) 48 *Der Staat* 535, 535 (hereafter Schönberger, 'Zwischen "Demokratiedefizit" und Bundesstaatsverbot').

[17] Christoph Schönberger, 'Lisbon in Karlsruhe: Maastricht's Epigones at Sea' (2009) 10 *German LJ* 1201, 1201 (hereafter Schönberger, 'Maastricht's Epigones').

[18] Daniel Halberstam and Christoph Möllers, 'The German Constitutional Court says "Ja zu Deutschland!"' (2009) 10 *German LJ* 1241, 1250, 1251.

[19] ibid 1257.

[20] Schönberger, 'Maastricht's Epigones' (n 17) 1202, 1207.

[21] Matthias Jestaedt, 'Warum in die Ferne schweifen, wenn der Maßstab liegt so nah? Verfassungshandwerkliche Anfragen an das Lissabon-Urteil des BVerfG' (2009) 48 *Der Staat* 497, 502.

[22] ibid.

[23] Schönberger, 'Maastricht's Epigones' (n 17) 1209.

[24] 'Wut und Tränen' *Der Spiegel* (6 July 2009) 29.

[25] ibid; see also Jo Eric Khushal Murkens, 'Identity Trumps Integration: The Lisbon Treaty in the German Federal Constitutional Court' (2009) 48 *Der Staat* 517.

[26] Schönberger, 'Maastricht's Epigones' (n 17) 1216.

[27] See Thomas Oppermann, 'Den Musterknaben ins Bremserhäuschen! Bundesverfassungsgericht und Lissabon-Vertrag' (2009) 20 *Europäische Zeitschrift für Wirtschaftsrecht* 473; see also Schönberger, 'Maastricht's Epigones' (n 17) 1217; Alfred Grosser, 'The Federal Constitutional Court's Lisbon Case: Germany's "Sonderweg"—An Outsider's Perspective' (2009) 10 *German LJ* 1263, 1266.

[28] See, e.g., Klaus Ferdinand Gärditz and Christian Hillgruber, 'Volkssouveränität und Demokratie ernst genommen—Zum Lissabon-Urteil des BVerfG' (2009) 64 *JuristenZeitung* 872.

[29] See Frank Schorkopf, 'Die Europäische Union im Lot—Karlsruhes Rechtsspruch zum Vertrag von Lissabon' (2009) 20 *Europäische Zeitschrift für Wirtschaftsrecht* 718.

Much that was familiar from earlier decisions (particularly the *Maastricht* judgment), though at much greater length. On the positive side, the Court emphasized again that the Basic Law was not merely 'open' to European integration; it affirmatively commanded the Federal Republic to pursue integration.[30] The justices went so far as to enshrine 'friendliness toward Europe' (*Europafreundlichkeit*) as a constitutional principle.[31] On the negative side, the Senate reaffirmed that there were limits to how much integration the Basic Law could tolerate.[32] The justices reiterated that the EU was a secondary legal order: it derived its powers (and legitimacy) from the member states. The member states remained 'Masters of the Treaties'. Transfers of power to the European level must be explicitly limited and, in theory, reversible.[33] In constituting and empowering the EU, member states had not surrendered their sovereignty; they had only delegated specific, sovereign powers to European organs. European organs wielded no authority to define the scope of their authority—no *Kompetenz-Kompetenz*.[34] The Constitutional Court itself would exercise *ultra vires* review if European agents exceeded their mandate.[35] Individual German citizens could solicit such review by invoking their right, under Article 38 GG, to vote in Bundestag elections.[36] The Court reemphasized that this was a substantive right—the right to cast a meaningful vote for a representative body that retained meaningful powers.

So much was mere reaffirmation of *Maastricht*, though it dismayed those who hoped *Maastricht* would be overruled. But the Senate went further. In some respects, the expansion of *Maastricht* was a matter of emphasis and tone; in other regards, the innovations were substantive. Most striking of all was the Senate's discussion of sovereignty, which culminated in a ringing answer to the colossal question that *Maastricht* had left open: whether the European Union could, consistent with the Basic Law, evolve into a federal state (*Bundesstaat*).

The Second Senate said no. Not only, the justices reasoned, did the Basic Law require that the German Parliament retain meaningful powers; it demanded that the German Republic remain a sovereign state.[37] The constitution not only presupposed German sovereignty; it guaranteed it.[38] This guarantee, moreover, was rooted, not in any ordinary provision of the constitution, but in the 'eternity clause' of Article 79(3) and in the 'self-termination' clause of Article 146.[39]

Often in these pages we have encountered Article 79(3)'s ban on constitutional amendments that would undermine Article 1's consecration of human dignity or Article 20's vow that the Federal Republic would remain forever 'a democratic and social federal state'. As emphasized in the last chapter, Article 146 declared that the Basic Law would cease to apply when a new constitution, adopted by free choice of the entire German people, entered into force. In its *Lisbon* judgment, the Second Senate maintained that the eternity clause enshrined in perpetuity the 'identity' of the Federal Republic as a sovereign state.[40] This identity was further buttressed by Article 146, which the Court now construed as an affirmative right—one that required the Federal Republic to retain sufficient sovereignty to allow the German people to adopt (someday) a new national constitution.[41]

The doctrinal innovation—draped in a broad excursus on state theory—was astonishing. Its upshot was that not only did the Basic Law limit European integration, but such limits could not be removed by constitutional amendment. There could be no European federal state while the Basic Law remained in force. A new Germany must precede a federal Europe. Nothing short of a new constitution would suffice. The barriers to European statehood

[30] 123 BVerfGE 267 (n 7) 344–45. [31] ibid 347. [32] ibid 357. [33] ibid 348–50. [34] ibid 349. [35] ibid 353–54. [36] ibid 340–43. [37] ibid 343, 345. [38] ibid 343. [39] ibid 344. [40] ibid 344. [41] ibid 349–50.

could be overcome, one critic summarized, 'only through revolution'.[42] Constitutional sovereignty, the Senate ruled, lay at the core of the Federal Republic's 'identity'.[43] That identity was the Basic Law's premise and purpose—its alpha and omega, its essential precondition and its unamendable guarantee. The *Maastricht* judgment observed that the European Union hadn't *yet* become a federal state; the *Lisbon* judgment ruled that it never could.

In the view of many critics, this holding put the project of European integration in a 'double bind'.[44] On the one hand, the Court chided the European Union for its democratic deficiencies.[45] On the other hand, the judgment precluded the only possible remedy for those deficiencies. Democratic legitimacy, the justices maintained, existed only within the context of the state.[46] A democratic state was one that rested on the foundation of popular self-government through parliamentary representation. Europe was not such a state, and the Basic Law barred its ever becoming one. The EU was thus relegated to a 'state-theoretical no-man's-land;' cast adrift between the Scylla of democratic deficiency and the Charybdis of forbidden statehood; left to choose 'between cholera and the plague'.[47] One critic discerned in the decision an 'aggressive-truculent tone'—a defiant cry of "Thus far, and no further!"'[48] Others wondered gloomily whether 'the end of European integration' had finally come.[49] The date of the *Lisbon* judgment, predicted Alfred Grosser, a French–German political scientist, 'will be remembered as a black day in the history of Europe'.[50]

Part of the judgment's blackness, Grosser continued, was that so many Germans seemed so happy about it—so happy, indeed, that 'those living in Germany and promoting and supporting German–French cooperation for over 60 years [i.e. Grosser himself] must now answer critics that worry that the Germans have never been serious about Europe'.[51] This, perhaps, was the exaggeration of bitter disappointment. But it was certainly true that, within Germany, the judgment found many supporters, in both scholarly and popular venues.[52] One of the case's CSU complainants was hailed by his party fellows for having 'saved our honor'.[53] In the immediate aftermath of the judgment, some praised the Court for its vigorous defence of democratic legitimacy and chided the German Parliament for its docile surrender of sovereign powers.[54] More moderate voices called for caution among critics, noting that the judgment could mean 'the end of integration' only if one's hopes for integration were defined by the prospect of European statehood.[55]

But critics were in no cautious mood. For them the judgment was too much to swallow. It was like Dr Johnson's mutton—'as bad as bad can be'—'ill-fed, ill-killed, ill-kept, and ill-drest'.[56] It was also riddled with ironies.

[42] Jestaedt (n 21) 505. [43] 123 BVerfGE 267 (n 7) 340, 343–44, 347, 350.

[44] Halberstam and Möllers (n 18) 1251. [45] 123 BVerfGE 267 (n 7) 376–77. [46] ibid 340–43.

[47] Schönberger, 'Zwischen "Demokratiedefizit" und Bundestaatsverbot' (n 16) 542; Jestaedt (n 21) 515.

[48] Schönberger, 'Zwischen "Demokratiedefizit" und Bundestaatsverbot' (n 16) 539, 536.

[49] Eckhard Pache, 'Das Ende der europäischen Integration? Das Urteil des BVerfG zum Vertrag von Lissabon, zur Zukunft Europas und der Demokratie' (2009) 37 *Europäische Grundrechte Zeitschrift* 285.

[50] Grosser (n 27) 1263. [51] ibid 1263, 1266.

[52] See Jörg-Klaus Baumgart, 'Ja zum Vertrag von Lissabon, aber ... Das Urteil des BVerfG und die notwendige Änderungen des Ausweitungsgesetzes' (2009) 63 *NJW* 309; Nicola Beer, 'Das Lissabon-Urteil und die Hausaufgaben für Berlin' (2009) 20 *EuZW* 593; Klaus Ferdinand Gärditz and Christian Hillgruber, 'Volkssouveränität und Demokratie ernst genommen—Zum Lissabon-Urteil des BVerfG' [2009] *JuristenZeitung* 872; Christian Hillgruber, 'Die besseren Europäer' *FAZ* (10 September 2009) 8; Peter M. Huber, 'Wer das Sagen hat' *FAZ* (10 September 2009) 8; Paul Kirchhof, 'Faszination Europa' *FAZ* (19 September 2009) 8; Frank Schorkopf, 'The European Union as an Association of Sovereign States' 10 *German LJ* 1219 (2009).

[53] In 'Wut und Tränen' *Der Spiegel* (6 July 2009) 28. It was unclear whether 'the honor' was that of Germany or the CSU.

[54] See Heribert Prantl, 'Verfassungsgericht zu Lissabon-Vertrag—Europäische Sternstunde' *Süddeutsche Zeitung* (1 July 2009).

[55] See Dieter Grimm, 'Das Grundgesetz als Riegel vor einer Verstaatlichung der Europäischen Union' (2009) 48 *Der Staat* 475, 495.

[56] See James Boswell, *Boswell's Life of Johnson* (Charles Grosvenor Osgood ed, New York, 1917) 518.

It was ironic that in an immensely long judgment the Court devoted scores of pages to approving the Lisbon Treaty but only a few to invalidating a parliamentary statute—passed the same day the treaty was approved—on the rights of the Bundesrat and the Bundestag. Parliament must play a greater role, the Court held, in the transfer of competences to Europe than was provided for in the statute.[57] This the Second Senate held tersely, summarily—almost, it seemed, in passing.[58] By contrast, the Senate spent more than 100 pages saying the Lisbon Treaty was okay. It was ironic, moreover, that in its discussion of Europe's democratic deficiencies the Court linked democracy and parliamentarism—'a pattern belonging to the worst traditions of German constitutional theory'[59]—belonging, that is, to Carl Schmitt.[60] It was ironic that even in this sense the Court's theory of democratic legitimacy could not account for the legitimacy of the Swiss *Ständerat* or the United States Senate, or even 'for the democratic legitimacy of the German national institutions'.[61] The justices seemed blind 'to the particularities of democratic legitimacy in federal states'.[62] It was ironic that although the justices exalted the importance of parliaments they 'only casually note[d] the important strengthening of ... the European Parliament by the Treaty of Lisbon'.[63]

It was ironic that the justices, with their radical construction of Article 79 and Article 146, had failed meaningfully to grapple with the one obviously relevant constitutional provision—Article 23 GG, which regulated the transfer of powers to European institutions.[64] It was ironic that the Court affirmatively banned a European *Bundesstaat* at a time when hardly anyone was seriously calling for one.[65] It was ironic that in a country in which Euro-skeptic political parties languished in sovereign irrelevance (something, alas, not true as of this writing), the highest court had 'come to embody the Euro-skeptic position'.[66] It was ironic that the Court asserted German sovereignty and identity against the very Europe to whose early phases of integration Germany owed its sovereignty and to whose later phases of integration Germany owed its unity.[67] It was ironic that Article 79, first framed as an ultimate barrier against the 'arch-enemies' of peace and democracy, was now being urged against the EU, a bulwark of peace and liberalism in a continent long rent by illiberalism and war.[68] And it was ironic (if not paradoxical) that the Court found the EU to be democratically deficient but consistent with the democracy principle.[69]

Some of the greatest ironies were institutional. The judgment was a prolonged paean to majoritarian popular sovereignty and yet, as two critics wrote, Germany's own federal system was often discussed as a 'prime example of a non-majoritarian political system'.[70] And the most potent counter-majoritarian force in the German polity was, of course, the Federal Constitutional Court. In its *Lisbon* judgment, the Court condemned constraints on the very Parliament it had itself been constraining for nearly sixty years. It was perhaps with this in mind that another critic sniffed, 'Future transfers of powers to the European Union

[57] 123 BVerfGE 267 (n 7) 391–92. [58] See Jestaedt (n 21) 500.

[59] Halberstam and Möllers (n 18) 1247.

[60] See Carl Schmitt, *Die geistesgeschichtliche Zustand des heutigen Parlamentarismus* (Duncker and Humblot 1923).

[61] Schönberger, 'Maastricht's Epigones' (n 17) 1211, 1215. [62] ibid 1215. [63] ibid 1213.

[64] Halberstam and Möllers (n 18) 1254; also Jestaedt (n 21) 507.

[65] Schönberger, 'Zwischen "Demokratiedefizit" und Bundesstaatverbot' (n 16) 555.

[66] See Schönberger, 'Maastricht's Epigones' (n 17) 1217. [67] Grosser (n 27) 1266.

[68] See Jestaedt (n 21) 507; see also Halberstam and Möllers (n 18) 1255 ('It is ... highly dubious that Basic Law Article 79(3) was meant to preserve the sovereignty of the German state within the process of European integration'.).

[69] 123 BVerfGE 267 (n 7) 377; see Jestaedt (n 21) 510. [70] Halberstam and Möllers (n 18) 1248.

should ... be left to the parliamentary majorities the Court itself cannot praise enough as the main expression of democratic legitimacy'.[71]

This was a deep irony, but there was another, deeper irony—an irony historical as well as institutional. The title of Alfred Grosser's commentary skewered the *Lisbon* judgment as emblematic of 'Germany's "Sonderweg".' In using this loaded term, Grosser implied that Germany was once again going its own way—that it might not be, as quoted earlier, 'serious about Europe'. That the Court whose greatest historical achievement was to correct the course of Germany's *Sonderweg*—to plant the Federal Republic firmly in the Western tradition of liberal democratic politics and robust protection of fundamental rights—should be the agent of a new special path was an irony more acute than all the rest. Beneath the surface, however, there might have been no irony at all—a point which will be revisited in this book's closing comments.

II. The Watchdog Begins to Growl

Months passed and the tempest stilled. Tensions, however, persisted. In December 2009, *Der Spiegel* reported that the 'relationship between Berlin and Karlsruhe is fraught as it hasn't been for a long time'.[72] A string of subsequent decisions on European integration—all issued against the backdrop of an ongoing Euro crisis—sometimes relaxed and sometimes intensified the strain.

In the *Lisbon* judgment the Court renewed its vow to exercise *ultra vires* review over European actions—review, that is, of whether European actors exceeded their competences. On 6 July 2010 the Second Senate did so for the first time since *Lisbon*.[73] It did so with great restraint. In a comparatively brief judgment—the analysis of the seven-justice majority took fewer than twenty pages—the Second Senate deferred conspicuously to a prior judgment of the European Court of Justice, observing that the Senate's own *ultra vires* review was limited to cases of 'obvious overstepping of competence'.[74] This language was familiar from the *Lisbon* judgment,[75] but in what came to be called the *Mangoldt* decision, the Senate adduced that language in a context of 'unmistakably pronounced Europe-friendly diction'.[76] Another year passed, and on 7 September 2011, the eve of the Court's sixtieth anniversary, the Second Senate approved the EU's first Greek rescue package as well as a temporally limited Eurozone bailout fund.[77] In the same decision, however, the justices strengthened once more the participatory rights of the Bundestag in further European developments.

Two months later, on 9 November 2011, the Court grew more aggressive. The Second Senate struck down the 5 per cent electoral hurdle as it applied in German elections to the EP.[78] The hurdle, the Court held, was a severe intrusion on the principles of electoral equality and equal opportunity for political parties. Under existing legal and political circumstances, it was an unjustifiable intrusion. It was an eye-catching decision, not least because, as noted earlier, the Court had repeatedly approved a 5 per cent hurdle for elections to the German Bundestag.[79] The German Parliament duly replaced the 5 per cent hurdle with a 3 per cent hurdle. But on 26 February 2014, the Court quashed the 3 per cent hurdle as well.[80]

[71] Schönberger, 'Maastricht's Epigones; (n 17) 1210.
[72] 'Wir erwarten Respekt' *Der Spiegel* (19 December 2009) 38.
[73] 126 BVerfGE 286 (2010).
[74] ibid 302.
[75] 123 BVergGE 267 (n 7) 353.
[76] Matthias Cornils, 'BVerfGE 123, 267—Lissabon' in Jörg Menzel and Ralf Müller-Terpitz (eds), *Verfassungsrechtsprechung*, 2nd edn (Mohr Siebeck 2011) 883.
[77] 129 BVerfGE 124 (2011).
[78] 129 BVerfGE 300 (2011).
[79] See 51 BVerfGE 222 (1979).
[80] 135 BVerfGE 259 (2014).

The latter decision stirred a sensation. It prompted rejoicing among Germany's proliferating pigmy parties, nineteen of which had joined the constitutional complaint. Unorthodox parties such as the Pirates, the Ecological Democratic Party (ÖDP), and the Left Party were all pleased. Parties of the far-Right were jubilant. In an official missive, the qualifiedly Euro-skeptic Alternative for Germany (AfD) called the judgment a 'good sign for democracy', rejoicing that 'for the first time voters in Germany have the chance to vote for the party of their choice without having to fear that their vote might go wasted'.[81]

The elation of Right-wingers gave many observers pause.[82] One common fear was that the judgment would exacerbate the phenomenon of voters venting steam by plumping for extremists in EP elections. Some, including Hans-Jürgen Papier, the former chief justice, worried that the Court's decision imperilled the 5 per cent hurdle for *national* elections.[83] Papier called the judgment 'right neither in its result nor in its reasoning'.[84] In an interview with *Die Welt*, Papier called for a constitutional amendment securing the national 5 per cent rule.[85] Other observers resented the 'Karlsruhe arrogance' reflected in the judgment's obvious premise that German democracy and European democracy were beings of different orders.[86]

Responses to the election decision were especially fraught because, as everyone knew, the Court was at the same time preparing to decide the merits of a much more important case.

On 26 July 2012, Mario Draghi, the Italian president of the European Central Bank (ECB), vowed that his bank was 'ready to do whatever it takes to preserve the Euro'.[87] The most dramatic steps then taken were the creation of the European Stability Mechanism (ESM), a permanent firewall empowered to deliver instant assets to embattled Eurozone members, and the ECB's September 2012 pledge to make, if necessary, unlimited purchases of government bonds from fiscally besieged states—a scheme christened 'Outright Monetary Transactions' (OMT). German involvement in both programmes provoked immediate constitutional challenges. In the early months of 2014, the Court delivered landmark decisions on both matters—one shortly before the *Three-Percent Hurdle* decision, the other shortly thereafter.

The ESM case drew the most headlines. In September 2012, with the Court about to decide whether preliminarily to enjoin German involvement in the ESM, *Der Spiegel* reported what was at stake: 'If the justices give the guarantee system their imprimatur, even if they establish strict conditions, it could mean the beginning of the end of the Federal Republic as a self-determining state'.[88] Focus on the Court and its imminent decision was global.[89] A reporter for the *Financial Times* wrote that in its coming decision the Court might 'be sealing the fate of the Euro'.[90] Financial analysts fretted that a negative vote from Karlsruhe would herald 'the end of the EU as we know it'.[91]

When the Court denied the petition for a preliminary injunction on 12 September 2012, Europhiles breathed a collective sigh of relief. Investors relaxed and markets rallied.[92]

[81] In 'Drei-Prozent-Hürde bei Europawahl verfassungswidrig' *Die Welt* (26 February 2014).

[82] See 'Schwarz-Rot warnt vor Einfluss rechter Splitterparteien' *Spiegel Online* (26 February 2014).

[83] 'Ex-Verfassungsrichter sieht Fünfprozenthürde in Gefahr' *Spiegel Online* (8 March 2014).

[84] ibid. [85] Jochen Gaugele, 'Das Urteil leuchtet mir nicht ein' *Die Welt* (10 March 2014) 4.

[86] See Gregor Peter Schmitz, 'Karlsruher Arroganz' *Spiegel Online* (26 February 2014).

[87] See 'Key Excerpts: Mario Draghi Says ECB "Ready to Do Whatever It Takes"' *Wall Street Journal Online* (26 July 2012) <http://blogs.wsj.com/economics/2012/07/26/key-excerpts-mario-draghi-says-ecb-ready-to-do-whatever-it-takes/> (accessed 10 April 2014). That the promise was prefaced by the qualifier 'Within our mandate' was swiftly forgotten.

[88] Dietmar Hipp, 'Der Kapitän' *Der Spiegel* (10 September 2012).

[89] See 'Draghi's Plan' *The Economist* (11 August 2012) 66.

[90] Quentin Peel, 'Germany's Judgement Day' *Financial Times* (7 August 2012).

[91] 'Angela Merkel Just Revealed the Real Situation in Europe' *Phil's Stock World* (27 August 2012).

[92] See Alice Ross, 'German ESM Ruling pushes Euro above $1.29' *Financial Times* (12 September 2012).

Many pegged the decision as historic. A reporter in the *World Politics Review* predicted that 'should a continent-spanning republic ever emerge in Europe, historians will probably point to Sept. 12, 2012, as one of the game-changing days in the historic march toward such a goal'.[93] The Court's decision wasn't the only big Euro news that day; but it topped that reporter's list. Other Europe watchers, however, remained wary. After all, the Court had not ruled positively that the bailout fund was constitutional; it had only concluded that the portended harm was not so imminent as to require an immediate injunction. Not until March 2014 would the Second Senate assess the constitutional question on its merits.

By that time, the Court had already struck down the 3 per cent hurdle for Germany's EP elections. In an even more momentous judgment, it had ruled that the OMT programme was illegal.

In a judgment dated 7 February 2014 the Second Senate held, by a vote of 6:2, that the ECB's bond-buying programme exceeded the Bank's competences and therefore violated European law.[94] But the justices did not stop there. They took the unprecedented step of referring the matter—a question of European rather than German law—to the European Court of Justice. This was a relief to many Europhiles, who thought it unlikely that the ECJ would quash OMT in its entirety.[95] But others remained rattled. The Constitutional Court had not sent the case to Luxembourg for a final decision but for a second opinion. The last word, it seemed, remained in Karlsruhe. The Court, one critic maintained, was 'playing with fire'. The judgment not only damaged the ECB; it posed a risk for all of Europe.[96]

Euro-skeptics, meanwhile, did considerable crowing. Peter Gauweiler, the German MP who brought the case, exulted that 'Karlsruhe has shown ECB President Mario Draghi what a bazooka really is'.[97] The AfD boss, Bernd Lucke, rejoiced that 'a court has finally found that the ECB's bond-buying program is a clear violation of European law'.[98] Whether that finding was final and what its ultimate consequences would be remained unclear. But many observers were convinced that the Court had crossed a threshold. The Court had finally found an *ultra vires* violation of European law. After forty years of barking, the dog had finally bitten.[99]

Many feared that it would soon bite again—and hard. The Court announced it would release its final judgment in the *ESM* case in the middle of March 2014. For some leading politicians, frustration with recent decisions fuelled fears of what the coming decision might contain. Days before the Second Senate announced its *ESM* judgment, Volker Kauder, the long-time head of the CDU/CSU Bundestag faction, told *Die Welt am Sonntag*, 'I question with alarm whether the Federal Constitutional Court still places sufficient value on the principle of judicial self-restraint'.[100] The Bundestag was the constitutional organ immediately elected by the German people, Kauder noted. The Court must exercise greater care in preserving Parliament's political leeway. The European Parliament, he added, was 'no longer a second-class Parliament'.[101] Tensions mounted as the judgment day approached.

[93] Ulrike Guérot, 'The EU Gets Serious About Saving the Euro' *World Politics Review* (17 September 2012) 1.

[94] 134 BVerfGE 357 (2014).

[95] In fact, the ECJ's advocate general suggested in January 2015 that OMT was compatible with the EU treaties. Court of Justice of the European Union, Press Release No 2/15, 14 January 2015, <http://curia.europa.eu/jcms/upload/docs/application/pdf/2015-01/cp150002en.pdf>.

[96] Marcel Fratzscher, 'Ein Richterspruch mit Risiko' *Zeit Online* (7 February 2014).

[97] In 'Europe or Democracy?' *Spiegel Online—International* (10 February 2014) http://www.spiegel.de/international/europe/german-court-calls-ecb-bond-buying-into-question-a-952556.html.

[98] In ibid.

[99] See Niels Petersen, 'Karlsruhe Not Only Barks, But Finally Bites—Some Remarks on the OMT Decision of the German Constitutional Court' (2014) 15 *German LJ* 321.

[100] 'Volker Kauder greift höchste Richter an' *Die Welt am Sonntag* (16 March 2014).

[101] ibid.

Many believed that a negative ruling would mean the end of the Euro. 'This time', *Der Spiegel* reported, 'everything is at stake'.[102]

But the judgment, when it came, was anti-climactic. The Second Senate's verdict on the ESM was a 'yes, without buts'.[103] The Court approved the rescue fund without qualification. As usual, the justices recited a long list of principles requiring a core minimum of national parliamentary participation.[104] But they ruled that those requirements had been adequately fulfilled. 'The result', said Chief Justice Voβkuhle, 'is unambiguous'.[105] The Karlsruhe watchdog had not followed its OMT nip with an ESM chomp. It had not bitten a second time; it had barely barked.

Reactions to the decision were predictable—rejoicing on the one hand that the Court had rediscovered the virtues of restraint,[106] suspicion on the other hand of a judgment that garnered universal praise from the politicians of major parties.[107] If there were a third hand, it belonged to the Euro-skeptics, whose dismay was unalloyed.

III. Final Reflections

Having surveyed the broad sweep of the Court's history in the foregoing pages we can now view the recent celebrity of the Court's European jurisprudence with some historical perspective. Fears that the Court would obliterate the Euro have proved exaggerated, though Europhiles caution that they are not out of the woods yet. Predictions that Europe would render Karlsuhe irrelevant have proved exaggerated as well. The Court's Golden Age is perhaps over and gone. Golden Ages usually are. But Karlsruhe still matters, not only in its obvious willingness to play an ongoing role in European integration, but also in its more traditional capacities as well—as attested, for example, by a recent decision limiting the influence of political parties on the content of public television.[108]

Perhaps it matters too much. In April 2014, Thomas de Mazière, Germany's federal interior minister, met with a group of public law scholars in Berlin to discuss the Court's role and to vet possible reforms, such as limiting the justices' terms, reintroducing the possibility of additional terms, or requiring a two-thirds Senate majority to invalidate acts of Parliament.[109] De Mazière is not the only member of the cabinet concerned about the Court. Wolfgang Schäuble, the federal finance minister, criticized the Court's *Three-Percent Hurdle* decision for its profound misunderstanding of the European Parliament and observed elsewhere, with a truculent eye toward Karlsuhe, that 'Whoever wishes to shape statutes should take pains to become a member of the German Bundestag'.[110]

At the time of writing, German newspapers are quick to observe 'mounting alienation between Berlin and Karlsruhe',[111] observing that the Court is now 'criticized sharply as seldom before'.[112] But no one better than readers of this book should know how exaggerated

[102] Melanie Amann, Thomas Darnstädt, and Dietmar Hipp, 'Der Druck-Knopf' *Der Spiegel* (10 June 2013) 21.

[103] 'Karlsruhe sagt ja ohne Aber' *Spiegel Online* (18 March 2014) <http://www.spiegel.de/wirtschaft/soziales/euro-krise-karlsruhe-sagt-ja-zu-rettungsschirmen-und-fiskalpakt-a-959426.html>.

[104] See 135 BVerfGE 317 (2014).

[105] 'Karlsruhe sagt ja ohne Aber' *Spiegel Online* (18 March 2014) <http://www.spiegel.de/wirtschaft/soziales/euro-krise-karlsruhe-sagt-ja-zu-rettungsschirmen-und-fiskalpakt-a-959426.html>.

[106] See Ludwig Greven, 'Das Verfassungsgericht nimmt sich zurück' *Zeit Online* (18 March 2014).

[107] See Joachim Jahn, 'Scheinheiliges Lob für die Verfassungsrichter' *FAZ* (18 March 2014).

[108] See Judgment of 25 March 2014, 1 BvF 1/11, <http://www.bundesverfassungsgericht.de/entscheidungen/fs20140325_1bvf000111.html>.

[109] Reinhard Müller, 'Berliner Gedankenspiele zur dritten Gewalt' *FAZ* (3 April 2014).

[110] In ibid.

[111] ibid.

[112] 'Die Anmaβung' *Der Spiegel* (10 March 2014) 20.

such claims are. Current criticisms fall comfortably into a very long pattern.[113] As one observer noted, in comparison to the high-level criticisms flung at the Court in the 1950s, 1970s, and 1990s, current critiques are 'almost too decent'.[114] The present controversy is no crisis; and even crises come and go.

The Court's present-day defenders often emphasize that the number of laws the Court quashes has always been very modest compared to the thousands of complaints it fields each year.[115] Critics counter that the real restraint emanating from Karlsruhe is not the occasional invalidation of laws already passed but the proliferation of standards and principles that limit future legislation.[116] Both have a point, but the point of the critics requires some context.

The Court cites itself much more often than it cites the constitution, and these self-citations not only give an impression of timelessness (because they are undated), they accumulate over time. The Court's recitations of governing standards have grown longer—a phenomenon on striking display in the *Lisbon* judgment. These are not mere dicta. As one observer has written, 'the self-referentiality of standards [*Maßstäbe*] has led to the development of much higher rationality demands on the democratic legislature, which the Court increasingly treats like an agency [*Behörde*], not like a directly democratically legitimated organ'.[117] Another scholar laments that 'time and again in recent years the Court has adduced a general rationality requirement as a standard for the review of legislative behavior'.[118] In a series of more or less high profile cases, the Court has expanded its general equality jurisprudence to review legislation for consistency, congruity, and coherence.[119] But legislation is often inconsistent for a noble reason: it is the product of compromise. Ban legislative incoherence and you risk rendering parliamentary compromise impossible.[120] What's more, parliamentary decisions are supposed to reflect the wishes of the citizenry, and those wishes are themselves often inconsistent. Strip parliaments of the right to act irrationally, and you kill democracy.

At first blush, there is a deep irony here. The proliferation of standards drawn from the Court's own jurisprudence has, in an important sense, made the Court's scrutiny more exacting over time. But over time the Federal Republic has matured into a full-grown liberal democracy, committed to European unity and sensitive to social justice. In a potentially troubling sense, the Republic's Guardians have seemed to do more as the country has needed them less. The Court's role, so desperately crucial in the Federal Republic's early years, became less crucial as the Republic matured. At the same time, that role expanded; and there, perhaps, lies the rub.

But a couple qualifications are in order. The first is that the much-lamented 'proliferation of standards' and unwieldy 'chains of citations' restrain the Court as well as Parliament.

[113] For a current brief effort to assess recent calls for reform and to situate them in a longer historical context, see Ute Sacksofyk, 'Wellen der Empörung—Das Bundesverfassungsgericht und die Politik' (2014) 68 *Merkur* 711. Sacksofyk notes charmingly that the Court has had a crisis 'every twenty years', beginning with the 'hits of the 1950s'. ibid 711.

[114] ibid 711–12.

[115] See Sascha Kneip, *Verfassungsgerichte als demokratische Akteure* (Nomos 2009) 284–347.

[116] See Christoph Möllers, 'Legalität, Legitimität und Legitimation des BVerfG' in Matthias Jestaedt and others, *Das Entgrenzte Gericht. Eine Kritische Bilanz nach Sechzig Jahre Bundesverfassungsgericht* (Suhrkamp 2011) 408.

[117] ibid 407.

[118] Philipp Dann, 'Verfassungsgerichtliche Kontrolle Gesetzgeberischer Rationalität' (2010) 49 *Der Staat* 630, 630.

[119] ibid. The three principal examples Dann cites are 125 BVerfGE 175 (2010); 122 BVerfGE 210 (2008); 121 BVerfGE 317 (2008).

[120] Dann (n 119) 640.

The Court's fealty to principle and precedent (over the last six decades it has reversed itself far less frequently, for instance, than has the United States Supreme Court) ensures and announces that its judgments are the product of an established and enduring jurisprudence, rather than of the impulses of the moment or the current alignment of the Court's personnel. The second is that Parliament is often freer than it thinks. Many putative constitutional restraints on Parliament (and on administrative officials) are actually self-imposed, the product of an overdeveloped eagerness to punt hard questions to the Court or to pretend that the Court has already answered them.[121] The Court's proclivity for self-citation, then, does more to constrain the justices, and less to constrain politicians and civil servants, than many critics suggest.

One should also be careful before concluding either that the Court has outlived its usefulness, or that its usefulness has diminished as its powers have grown. It is true that the Court is less necessary now as a defender of individual rights against the state than it was in the infant years of the Federal Republic. German politicians have largely internalized the Basic Law's individual rights regime—a happy circumstances that owes more than a little to the historical efforts of the Federal Constitutional Court. But the Court has responded to this development and its role has shifted accordingly. The Court's major contribution has shifted from the defence of individual rights against state intrusion to the orientation of an increasingly professionalized—and therefore often short-sighted—political class toward long-term constitutional commitments and enduring fundamental values.[122] The justices are no longer the guardians of an immature state enduring the growing pains of democratization and liberalization. They have become rather the guardians of deeply-rooted principles and an enduring tradition. In an important sense, this has cast the Court in a conservative rather than a progressive posture. But the Court conserves, or at least requires other state actors to consider, many prized progressive values.

This is cold comfort, perhaps, to observers wary of judicial power, which the Court plainly continues to wield. The Court's powers have been hemmed to some extent by the expansion of European jurisdictions, but domestically they remain formidable. And the Court, despite its recent unqualified endorsement of the ESM, shows no signs of quitting the European field.

This brings us back to the deep irony, mentioned earlier, to which I promised to return. The Court and its jurisprudence still present an obstacle to the advance of European integration. The OMT endgame—and the final resolution of Euro crisis more generally—lies hidden in the mists of futurity. But whatever the denouement, the Court will remain a major player. What's more, the Court's existing jurisprudence places limits on how German politicians can negotiate, and what they can approve, at the European level. The irony, as suggested earlier, is that a Court and a constitution whose major putative historical contribution was to correct the course of Germany's *Sonderweg*—to help return it to the community of civilized nations and Western liberal democracies—now stand in the way of history's greatest effort to unify the liberal democracies of Europe. The Court, to put the point sharply, might have corrected one *Sonderweg* only to launch another.

Has it? I think not. This book has referred to Germany's special path in a special sense. From the perspective of the Basic Law and its official interpreters, that path was not aberrational vis-à-vis the historical trajectory of France or of Britain or even of 'the West'—and

[121] This paragraph owes much to an exchange between the author and Dieter Grimm.

[122] See, e.g., Uwe Volkmann, 'Der Aufstieg der Verfassung. Beobachtungen zum grundlegenden Wandel des Verfassungsbegriffs' in Thomas Vesting and Stefan Korioth (eds), *Der Eigenwert des Verfassungsrechts* (Mohr Siebeck 2011) 23; Dieter Grimm, 'Prinzipien statt Pragmatismus' *FAZ* (6 February 2013).

there are deep problems with viewing German history as delinquent in this sense—but aberrational vis-à-vis the Germany that might have been, the nation that might have emerged had it turned at the right crossroads in 1848 and later on. For many decades now, the Basic Law has been praised uniformly as 'the best constitution that Germany has ever had'. If constitutions are to be judged by their application in the real world, the point strikes me as beyond dispute. What is relevant here is not only the quality of the constitution itself, but the extent to which the broader successes of postwar German history—economic prosperity, robust provisions of social welfare, leadership within the European community—have been ascribed to it.[123] The Basic Law is regarded not merely as the best of all historical German constitutions, but as the foundation of the best of all possible Germanys. The Court's decisions declaring the perpetual inviolability of the Basic Law's core 'identity' are an effort to preserve the Basic Law's—and, by derivation, the Court's own—historical achievement. Karlsruhe's putative Euro-skepticism is not so much a new *Sonderweg* as a bid to shore up the felicitous conclusion of the old one.

This brings us to one final irony. As noted earlier, one is tempted to suggest that the Court's powers have grown as its necessity has declined—that the costs of constitutional justice might be increasing and its benefits receding. In a similar spirit, one leading German constitutionalist, borrowing a phrase from the writer Hans Magnus Enzensberger, has called on the Court to practice the 'art of retreat' (*die Kunst des Rückzugs*).[124] But retreat might be the wrong word. What the Court needs is not so much withdrawal as introspection—a kind of institutional autobiography or, to risk an infelicity, self-historicization. As we have seen throughout these pages, the commonest response to institutional challenges to the Court's authority has been to highlight the Court's historical contributions. On the face of it, the Court's current predicament of proliferating standards and multiplied dogmas might seem like a product of its failure to understand its own jurisprudence *as history*. The justices continue to cite standards articulated by earlier Senates in different historical settings with no obvious indication that the difference matters. In theory, a judgment delivered in 1952 is, together with its underlying reasoning, as valid and binding as a judgment announced last week. In the *Lisbon* judgment, the Second Senate piled a mountain of prior precepts, one atop the other, dating back to 1974 and beyond without discriminating among those precepts or hinting at their relationship to a given historical context. The Senate then added new precepts destined to augment the mountain in the next judgment. The problem is a general one. The 'standards' sections of the Court's decisions grow longer and longer; legislative leeway grows tighter and tighter.

And yet, as was suggested earlier, the Court's commitment to enduring standards both constrains the Court's own discretion and strengthens the public's perception that the Court operates above the fray of quotidian politics. It is also by articulating and elaborating constitutional principle that the Court most effectively directs the attention of professional politicians beyond the boundaries of the next election. It is in this capacity—guarding the Basic Law's successes by orienting quotidian politics to enduring principles—that the Court fulfils its most valuable, and most valued, contemporary function.

True, some doctrinal pruning is surely in order. The Court should recognize that the totalizing doctrinal framework of its early years—so pivotal in the liberalization of West German society and the democratization of West German politics—might be ill-suited to

123 See Grimm, 'Identität und Wandel' (n 2) 614.

124 See Schönberger, 'Zwischen "Demokratiedefizit" und Bundersstaatsverbot' (n 16) 557–58; see also Christoph Schönberger, 'Anmerkungen zu Karlsruhe' in Matthias Jestaedt and others, *Das Entgrenzte Gericht: Eine Kritische Bilanz nach 60 Jahre Bundesverfassungsgericht* (Suhrkamp 2011) 64–65.

the needs of a mature democracy that has become the leading power of a highly integrated Europe. This is not to suggest that all the old doctrines must go, but rather to call for a shift in perspective—or to encourage a shift that is already underway. West Germans embraced the Court's immense power at a time when circumstances led them to view constitutional justice as a precondition to democracy, rather than a restraint on it. In certain respects, it remains such a precondition. In other respects, however, democracy demands more latitude than the Court's multitudinous canons allow. Telling the difference requires sensitivity to history as well as to dogma.

For the Federal Constitutional Court has not only made history, it has entered history. The Court owes much of its prestige to perceptions that it operates above the fray of politics, but it cannot pretend to operate outside the flow of time. The Court's history is the ultimate source of its most impressive capital. Ignoring its own historicity would imperil the Court's enduring relevance. It would also obscure one of the most inspiring chapters of postwar European history. It is in the hope of giving that chapter a part of its due, and of rendering modest assistance to its sequel, that this long book has been written.

Easter 2015
New Haven/Berlin/Provo

Selected Bibliography

Abendroth W, 'Das KPD-Verbotsurteil des Bundesverfassungsgerichts: Ein Beitrag zum Problem der richterlichen Interpretation von Rechtsgrundsätzen der Verfassung im demokratischen Staat' in *Antagonistische Gesellschaft und Politische Demokratie: Aufsätze zur politischen Soziologie* (Luchterhand 1967).

Abendroth W, 'Das Abtreibungsurteil des Bundesverfassungsgerichts' (1975) 8 *Kritische Justiz* 121.

Achterberg N, 'Vertrauensfrage und Auflösungsordnung' (1983) 98 *Deutsches Verwaltungsblatt* 477.

Ackerman BA, *We the People: Foundations*, vol 1 (Harvard University Press 1991).

Aders T, *Die Utopie vom Staat über den Parteien: biographische Annäherungen an Hermann Höpker-Aschoff (1883–1954)* (Peter Lang 1994).

Amelung K, 'Sitzblockaden, Gewalt und Kraftentfaltung' (1995) 47 *NJW* 2584.

Anderson P, 'After the Event' (2012) 73 *New Left Review* 49.

Arndt A, *Das Nicht erfüllte Grundgesetz. Ein Vortrag* (JCB Mohr 1960).

Arndt A, 'Schatten über Karlsruhe: Bundesverfassungsgericht—eine lästige Begrenzung der Macht?' (1963) *Sozialdemokratischer Pressedienst.*

Arndt C, 'Verfassungsrechtliche Anforderungen an internationale Bundeswehreinsätze' (1994) 47 *NJW* 2197.

Azzola A, 'Verfassungsbeschwerden gegen Hochschulgesetze kommen zur Entscheidung' (1973) 6 *Zeitschrift für Rechtspolitik* 44.

Bachof O, 'Der Richter als Gesetzgeber?' in Otto Bachof and Ludwig Fröhler (eds), *Wege zum Rechtsstaat. Ausgewählte Studien zum öffentlichen Recht* (Atheneum 1979) 344.

Badura P, 'Staat, Recht und Verfassung in der Integrationslehre Rudolf Smends' (1977) 16 *Der Staat* 305.

Baldus M, 'Frühe Machtkämpfe' in Thomas Henne and Arne Riedlinger (eds), *Das Lüth-Urteil aus (rechts-) historischer Sicht: die Konflikte um Veit Harlan und die Grundrechtsjudikatur des Bundesverfassungsgerichts* (Berliner Wissenschafts-Verlag 2005).

Baring A, *Außenpolitik in Adenauers Kanzlerdemokratie: Bonns Beitrag zur europäischenVerteidi gungsgemeinschaft* (Oldenbourg 1969).

Baring A, *Machtwechsel: die Ära Brandt-Scheel* (Deutsche Verlags-Anstalt 1982).

Baumgart J, 'Ja zum Vertrag von Lissabon, aber … Das Urteil des BVerfG und die notwendige Änderungen des Ausweitungsgesetzes' (2009) 63 *NJW* 309.

Beer N, 'Das Lissabon-Urteil und die Hausaufgaben für Berlin' (2009) 20 *Europäische Zeitschrift für Wirtschaftsrecht* 593.

Bemmann G, *Meinungsfreiheit und Strafrecht* (Athenäum 1981).

Benda E, 'Wirklich Götterdämmerung in Karlsruhe?' (1995) 49 *NJW* 2470.

Bentham J, *The Works of Jeremy Bentham*, vol 7 (W. Tait 1843).

Bernecker WL and Dotterweich V, *Persönlichkeit und Politik in der Bundesrepublik Deutschland: politische Porträts*, 2 vols (Bandenhoeck & Ruprecht 1982).

Billing W, *Das Problem der Richterwahl zum Bundesverfassungsgericht: Ein Beitrag zum Thema 'Politik und Verassungsgerichtsbarkeit'* (Duncker and Humblot 1969).

Blankenagel A, 'Verfassungsgerichtliche Vergangenheitsbewältigung' (1991) 13 *Zeitschrift für neuere Rechtsgeschichte* 67.

Blüggel J, *Unvereinbarkeitserklärung statt Normkassation durch das Bundesverfassungsgericht* (Duncker and Humblot 1998).

Böckenförde EW, 'Die Entstehung des Staates als Vorgang der Säkularisation' in E-W Böckenförde, *Recht, Staat, Freiheit: Studien zur Rechtsphilosophie, Staatstheorie und Verfassungsgeschichte* (Suhrkamp 1991).

Böhmer W, 'Die Rechtsgeschichtlichen Grundlagen der Abgrenzungsproblematik von Sozialbindung und Enteignung' (1985) 24 *Der Staat* 157.

Börner B, 'Deutsche Grundrechte und Gemeinschaftsrecht' (1976) 29 *NJW* 2041.

Braunthal G, *Political Loyalty and Public Service in West Germany: The 1972 Decree Against Radicals and Its Consequences* (University of Manchester Press 1990).

Brunner G, 'Der Zugang des Einzelnen zur Verfassungsgerichtsbarkeit im europäischen Raum' (2002) 50 *Jahrbuch des öffentlichen Rechts der Gegenwart* 191.

Bryde B, 'Sicherheitspolitik zwischen Regierung und Parlament' (1986) 8 *Juristische Ausbildung* 363.

Bryde B, 'Die Rolle der Verfassungsgerichtsbarkeit in Umbruchsituationen' in Joachim Jens Hesse and others (eds), *Verfassungsrecht und Verfassungspolitik in Umbruchsituationen* (Nomos 1999).

Bryde B, 'Der Beitrag des Bundesverfassungsgerichts zur Demokratisierung der Bundesrepublik' in Robert Christian van Ooyen and Martin H.W. Möllers (eds), *Das Bundesverfassungsgericht im politischen System* (Verlag für Sozialwissenschaften 2006).

Bull HP, 'Finanzausgleich im "Wettbewerbsstaat"—Bemerkungen zur neuen Föderalismustheorie und zu ihrer Bedeutung für den Länderfinanzausgleich' (1999) 52 *Die Öffentliche Verwaltung* 269.

Casper G, 'The "Karlsruhe Republic"—Keynote Address at the State Ceremony Celebrating the 50th Anniversary of the Federal Constitutional Court' (2001) 2 *German Law Journal* http://www.germanlawjournal.com/index.php?pageID=11&artID=111>.

Conradi P, 'Parlamentarier in privilegienfeindlicher Demokratie. Anmerkungen eines Bundestagsabgeordneten zum "Diäten-Urteil" des Bundesverfassungsgerichts' (1976) 8 *Zeitschrift für Parlamentsfragen* 113.

Conze E, *Die Suche nach Sicherheit: Eine Geschichte der Bundesrepublik Deutschland von 1949 bis in die Gegenwart* (Siedler 2009).

Cornils M, *Maastricht* in Jörg Menzel (ed), *Verfassungsrechtsprechung: Hundert Entscheidungen des Bundesverfassungsgerichts in Retrospektive* (Mohr Siebeck 2000).

Cornils M, 'BVerfGE 123, 267—Lissabon' in Jörg Menzel and Ralf Müller-Terpitz (eds), *Verfassungsrechtsprechung. Ausgewählte Entscheidungen des Bundesverfassungsgerichts in Retrospektive*, 2nd edn (Mohr Siebeck 2011).

Currie DP, *The Constitution of the Federal Republic of Germany* (University of Chicago Press 1994).

Dann P, 'Verfassungsgerichtliche Kontrolle Gesetzgeberischer Rationalität' (2010) 49 *Der Staat* 630.

Dannecker G and Stoffers KF, 'Rechtsstaatliche Grenzen für die strafrechtliche Aufarbeitung der Todesschüsse an der innerdeutschen Grenze' (1996) 51 *JuristenZeitung* 490.

Das Bundesverfassungsgericht, 'Bemerkungen des Bundesverfassungsgerichts zu dem Rechtsgutachten betr. Die Stellung des Bundesverfassungsgerichts von Geh. Hofrat Dr. Thomas vom 3. Juni 1953' (1957) 6 *Jahrbuch des öffentlichen Rechts* 194.

Däubler W and Küsel G (eds), *Verfassungsgericht und Politik* (Rowohlt 1979).

Dederer H, 'BVerfGE 39, 1—Schwangerschaftsabbruch I' in Jörg Menzel (ed), *Verfassungsrech tsprechung: Hundert Entscheidungen des Bundesverfassungsgerichts in Retrospektive* (Mohr Siebeck 2000).

Delbrück J and Wolfrum R, 'Die Auflösung des 9. Deutschen Bundestages vor dem BVerfG—BVerfGE 62, 1' (1983) 23 *Juristische Schulung* 758.

di Fabio U, 'Vom Recht, Recht zu Sprechen' (2011) 61 *Aus Politik und Zeitgeschichte* 3.

Dichgans H, *Das Unbehagen in der Bundesrepublik: Ist die Demokratie am Ende?* (Econ Verlag 1968).

Dichgans H, 'Eine verfassunggebende Nationalversammlung' (1968) 1 *Zeitschrift für Rechtspolitik* 61.

Dichgans H, 'Möglichkeit und Grenzen der Verfassungsgerichtsbarkeit' *Politisch-soziale Korrespondenz* (1 February 1969).

Dichgans H, *Vom Grundgesetz zur Verfassung; Überlegungen zu einer Gesamtrevision* (Econ Verlag 1970).

Gosewinkel D, *Adolf Arndt. Die Wiederbegründung des Rechtsstaats aus dem Geist der Sozialdemokratie (1945–1961)* (Verlag J.H.W. Dietz Nachf 1991).

Dietrich H, 'Beamte als Abgeordnete und das Diäten-Urteil des Bundesverfassungsgerichts' (1976) 27 *Zeitschrift für Beamtenrecht* 97.

Docke B, 'Der Regierung eine Gasse!' (1985) 13 *Demokratie und Recht* 10.

Dorpatka F, 'Zur Bedeutung des Bundesverfassungsgerichts in der politischen und gesellschaftlichen Entwicklung der Bundesrepublik 1951 bis 1978' in Wolfgang Däubler and Gudrun Küsel (eds), *Verfassungsgericht und Politik. Kritische Beiträge zu problematischen Urteilen* (Rowohlt 1979).

Doyle JJ, 'A Bitter Inheritance: East German Real Property and the Supreme Constitutional Court's "Land Reform" Decision of April 23, 1991' (1991–92) 13 *Mich. J. In't L.* 832.

Dreher E, 'Glanz und Elend der Staatsgerichtsbarkeit. Zum Gesetz über das Bundesverfassungsgericht vom 12. März 1951' (1951) 4 *NJW* 377.

Dreier H, 'Die Radbruchsche Formel—Erkenntnis oder Bekenntnis?' in Heinz Mayer and others (eds), *Staatsrecht in Theorie und Praxis. Festschrift für Robert Walter* (Manz'sche 1991).

Dreier H, 'Gustav Radbruch und die Mauerschützen' (1997) 52 *JuristenZeitung* 420.

Dürig G, 'Ein Orwellsches Experiment' (1968) 1 *Zeitschrift für Rechtspolitik* 11.

Dürig G, 'Zum "Lüth-Urteil" des Bundesverfassungsgerichts vom 15.1.1958' (1958) 8 *Die Öffentliche Verwaltung* 194.

Däubler W, 'Die sogennante Warteschleife auf dem verfassungsrechtlichen Prüfstand' (1991) 45 *Neue Justiz* 233.

Ehmke H, *Wirtschaft und Verfassung* (CF Müller 1961).

Eley G and Blackbourn D, *The Peculiarities of German History: Bourgeois Society and Politics in Nineteenth-Century Germany* (Oxford University Press 1984).

Ellwein T, and Seifert J (eds), *Die Spiegel-Affäre. Die Reaktion der Öffentlichkeit* (Walter 1966).

Eschenburg T, *Herrschaft der Verbände* (Deutsche Verlags-Anstat 1956).

Eser A, 'Reform of German Abortion Law: First Experiences' (1986) 34 *Am J Comp L* 369.

Eser A, 'Das neue Schwangerschaftsabbruchsstrafrecht auf dem Prüfstand' (1992) 46 *NJW* 2913.

Esser J, 'Bemerkungen zur Unentbehrlichkeit des juristischen Handwerkszeugs' (1975) 30 *JuristenZeitung* 555.

Eβer A, *Wilhelm Elfes, 1884—1969: Arbeiterführer und Politiker* (Matthias-Grünewald 1990).

Günter Wirth, *Wilhelm Elfes* (Union-Verlag 1984).

Faller H, 'Bundesverfassungsgericht und Bundesgerichtshof' (1990) 115 *Archiv des öffentlichen Rechts* 185.

Fastenrath U, 'Auswärtige Gewalt: Zustimmungsbedürftigkeit einseitiger völkerrechtlicher Erklärungen, Übertragungen von Hoheitsrechten (Pershing-II-Stationierung)' (1986) 18 *Juristische Arbeitsblältter* 451.

Filmer W and Schwan H, *Roman Herzog. Die Biographie* (Bertelsmann 1994).

Flume W, 'Das "Kruzifix-Urteil" und seine Berichtigung' (1995) 49 *NJW* 2904.

Forsthoff E, 'Die Umbildung des Verfassungsgesetzes' in Hans Barion (ed), *Festschrift Carl Schmitt zum 70. Geburtstag* (Duncker and Humblot 1959).

Forsthoff E, *Zur Problematik der Verfassungsauslegung* (Kohlhammer 1961).

Franzius C, *Bonner Grundgesetz und Familienrecht. Die Diskussion um die Gleichberechtigung von Mann und Frau in der westdeutschen Zivilrechtslehre der Nachkriegszeit (1945–1957)* (Vittorio Klostermann 2005) 1945–1957.

Frei N, *Vergangenheitspolitik. Die Anfänge der Bundesrepublik und die NS-Vergangenheit* (CH Beck 1996).

Frei N, 'Transformationsprozesse. Das Bundesverfassungsgericht als vergangenheitspolitischer Akteur in den Anfangsjahren der Bundesrepublik' in Michael Stolleis (ed), *Herzkammern der Republik. Die Deutschen und das Bundesverfassungsgericht* (CH Beck 2011).

Friesenhahn E, 'Zur Ablehnung und Selbstablehnung eines Richters des BVerfG wegen Besorgnis der Befangenheit' (1966) 21 *JuristenZeitung* 704.

Friesenhahn E, 'Hüter der Verfassung?' (1973) 6 *Zeitschrift für Rechtspolitik* 189.

Fritsche I, 'Aspekte deutsch-deutscher Rechtsangleichung auf dem Gebiet des Zivilrechts' (1990) 23 *Kritische Justiz* 162.

Frommel M, '§ 218: Straflos, aber rechtswidrig; zielorientiert, aber ergebnisoffen—Paradoxien der Übergangsregelungen des Bundesverfassungsgerichts' (1993) 26 *Kritische Justiz* 324.

Fromont M, 'Europa und nationales Verfassungsrecht nach dem Maastricht-Urteil—Kritische Bemerkungen' (1995) 48 *JuristenZeitung* 800.

Gante M, *§ 218 in der Diskussion: Meinungs- und Willensbildung, 1945–1976* (Droste 1991).

Gassert P, 'Die Bundesrepublik, Europa und der Westen: Zur Verwestlichung, Demokratisierung und einigen komparatistischen Defiziten der zeithhistorischen Forschung' in Jörg Baberowski and others, *Geschichte ist immer Gegenwart: Vier Thesen zur Zeitgeschichte* (Deutsche Verlags-Anstalt 2001).

Geiger W, *Die Rechtsstellung des Schriftstellers nach dem Gesetz vom 4. Oktober 1933* (Kichler 1940).

Geiger W, *Gesetz über das Bundesverfassungsgericht vom 12. März 1951. Kommentar* (Vahlen 1952).

Geiger W, 'Zur Reform des Bundesverfassungsgesetzes' in Theodor Maunz (ed), *Vom Bonner Grundgesetz zur gesamtdeutschen Verfassung* (Isar 1956).

Geiger W, 'Ergänzende Bemerkungen zur Stellung des Bundesverfassungsgerichts' (1957) 6 *Jahrbuch des öffentlichen Rechts* 161.

Geiger W, '*Vom Selbstverständnis des Bundesverfassungsgerichts* (Hanover: Landeszentrale für politische Bildung 1979).

Geiger W, 'Gegenwartsprobleme der Verfassungsgerichtsbarkeit aus deutscher Sicht' in Thomas Berberich, Wolfgang Holl, and Kurt-Jürgen Maaβ (eds), *Neue Entwicklungen im öffentlichen Recht. Beiträge zum Verhältnis von Bürger und Staat aus Völkerrecht, Verfassungsrecht und Verwaltungsrecht* (Stuttgart, 1979) .

Geiger W, 'Die Auflösung des Bundestages nach Art. 68 GG: ein Kommentar zu den Vorgängen, die zu der Neuwahl des Deutschen Bundestages am 6. März geführt haben' (1984) 33 *Jahrbuch des öffentlichen Rechts der Gegenwart* 41.

Geiger W, *Abweichende Meinungen zu Entscheidungen des Bundesverfassungs-gerichts* (Mohr Siebeck 1989).

Gerontas A, 'Die Appellentscheidungen, Sondervotumsappelle und die bloβe Unvereinbarkeitsfeststellung als Ausdruck der funktionellen Grenzen der Verfassungsgerichtsbarkeit' (1982) 97 *Deutsches Verwaltungsblatt* 486.

Godau-Schüttke K, *Der Bundesgerichtshof: Justiz in Deutschland* (Tischler 2005).

Godau-Schüttke K, 'Entnazifizierung und Wiederaufbau der Justiz am Beispiel des Bundesgerichtshofs' in Eva Schummann (ed), *Kontinuitäten und Zäsuren. Rechtswissenschaft und Justiz im 'Dritten Reich' und in der Nachkriegszeit* (Wallstein 2008).

Golsong H, 'Zu Bundesverfassungsgericht "Solange"-Beschluβ' (1974) 1 *Europäische Grundrechte-Zeitschrift* 18.

Görtemaker M, *Geschichte der Bundesrepublik Deutschland* (CH Beck 1999).

Götz V, 'Versammlungsfreiheit und Versammlungsrecht im Brockdorf-Beschluβ des Bundesverfassungsgerichts' (1985) 101 *Deutsches Verwaltungsblatt* 1347.

Gounalakis G, '"Soldaten sind Mörder"' (1996) 49 *NJW* 481.

Graβhof M, 'Auswirkungen der neuen Sitzblockade-Entscheidung des BVerfG' (1995) 48 *NJW* 3085.

Grimm D, 'Die Meinungsfreiheit in der Rechtsprechung des Bundesverfassungsgerichts' (1995) 48 *NJW* 1697.

Grimm D, 'Verfassungspatriotismus nach der Wiedervereinigung' in Hauke Bunkhorst and Peter Niesen (eds), *Das Recht der Republik* (Suhrkamp 1998).

Grimm D, 'Verfassungspatriotismus nach der Wiedervereinigung' in Dieter Grimm, *Die Verfassung und die Politik* (CH Beck 2000).

Grimm D, 'Integration durch Verfassung' (2004) 32 *Leviathan* 448.

Grimm D, 'Das Grundgesetz als Riegel vor einer Verstaatlichung der Europäischen Union' (2009) 48 *Der Staat* 475.

Grimm D, 'Identität und Wandel—das Grundgesetz 1949 und heute' (2009) 37 *Leviathan* 603.

Grimm D, 'Die Rolle der nationalen Verfassungsgerichte in der europäischen Demokratie' in Claudio Franzius and others (eds), *Grenzen der europäischen Integration. Herausforderungen für Recht und Politik* (Nomos 2014).

Grosser A, 'The Federal Constitutional Court's Lisbon Case: Germany's "Sonderweg"—An Outsider's Perspective' (2009) 10 *German LJ* 1263.

Groβfeld B, 'Götterdämmerung?' (1995) 49 *NJW* 1719.

Grünwald G, 'Meinungsfreiheit und Strafrecht' (1979) 12 *Kritische Justiz* 291.

Guggenberger B and Stein T (eds), *Die Verfassungsdiskussion im Jahr der deutschen Einheit. Analysen—Hintergründe—Materialien* (Carl Hanser 1991).

Günther F, *Denken vom Staat her: Die bundesdeutsche Staatsrechtslehre zwischen Dezision und Integration* (Oldenbourg 2004).

Günther F, 'Ein Jahrzehnt der Rückbesinnung. Die bundesdeutsche Staatsrechtslehre zwischen Dezision und Integration in den fünfziger Jahren' in Thomas Henne and Arne Riedlinger (eds), *Das Lüth-Urteil aus (rechts-)historischer Sicht: die Konflikte um Veit Harlan und die Grundrechtsjudikatur des Bundesverfassungsgerichts* (Berliner Wissenschafts-Verlag 2005).

Gusseck L, 'Bundestagsauflösung kraf+t Richterspruchs?' (1983) 36 *NJW* 721.

Gusy C, 'Die "freiheitliche demokratische Grundordnung" in der Rechtsprechung des Bundesverfassungsgerichts' (1980) 105 *Archiv des öffentlichen Rechts* 279.

Gusy C, 'Lehrbuch der Versammlungsfreiheit—BVerfGE 69, 315' (1986) 26 *Juristische Schulung* 608.

Gusy C, 'Anmerkung' (1990) 45 *JuristenZeitung* 638.

Gusy C, (ed), *Weimars lange Schatten—"Weimar" als Argument nach 1945* (Nomos 2003).

Gärditz KF and Hillgruber C, 'Volkssouveränität und Demokratie ernst genommen—Zum Lissabon-Urteil des BVerfG' (2009) 64 *JuristenZeitung* 872.

Häberle P, 'Unmittelbare staatliche Parteienfinanzierung unter dem Grundgesetz' (1967) 7 *Juristische Schulung* 64.

Häberle P, 'Die Abhörentscheidung des Bundesverfassungsgerichts vom 15.12.1970' (1971) 24 *JuristenZeitung* 145.

Häberle P, 'Freiheit, Gleichheit und Öffentlichkeit des Abgeordnetenstatus. Zum Diätenurteil des BVerfG' (1976) 29 *NJW* 537.

Häberle P, 'Öffentlichkeitsarbeit der Regierung zwischen Parteien- und Bürgerdemokratie' (1977) 32 *JuristenZeitung* 361.

Habermas J and Brükner P, *Die Linke antwortet Jürgen Habermas* (Europäische Verlagsanstalt 1969).

Habermas J, *Protestbewegung und Hochschulreform*, Suhrkamp 1969).

Habermas J, 'Zum Begriff der politischen Beteiligung (1958)' in Jürgen Habermas, *Kultur und Kritik. Verstreute Aufsätze* (Suhrkamp 1973).

Habermas J, 'Berufsverbote' in Jürgen Habermas, *Kleine politische Schriften I-IV* (Suhrkamp 1981).

Habermas J, 'Eine Art Schadensabwicklung. Die apologetischen Tendenzen in der deutschen Geschichtsschreibung' in Rudolf Augstein, Karl Dietrich Bracher, and Martin Broszat (eds), *'Historikerstreit'. Die Dokumentation der Kontroverse um die Einzigartigkeit der nationalsozialistischen Judenvernichtung* (Piper 1987).

Habermas J, *The Past as Future* (Max Pensky tr, University of Nebraska Press 1994).

Haffner S, 'Der Erfolg des Grundgesetzes' in Sebastian Haffner, *Im Schatten der Geschichte. Historisch-politische Variationen aus zwanzig Jahren* (Deutsche Verlags-Anstalt 1985).

Haffner S, 'Konrad Adenauer: der rechte Mann zur rechten Zeit' in Sebastian Haffner, *Im Schatten der Geschichte* (Deutsche Verlags-Anstalt 1985).

Halberstam D and Möllers C, 'The German Constitutional Court says "Ja zu Deutschland!"' (2009) 10 *German LJ* 1241.

Haltern U, 'Integration als Mythos. Zur Überforderung des Bundesverfassungsgerichts' (1997) 45 *Jahrbuch des öffentlichen Rechts* 31.

Hardach G, 'Krise und Reform der Sozialen Marktwirtschaft. Grundzüge der wirtschaftlichen Entwicklung in der Bundesrepublik der 50er und 60er Jahre' in Axel Schildt, Detlef Siegfried, and Karl Christian Lammers (eds), *Dynamische Zeiten. Die 60er Jahre in den beiden deutschen Gesellschaften* (Hans Chritians Verlag 2000).

Harlan T, *Veit* (Rowohlt 2011).

Hauβ HJ, 'Parteifinanzen' (1966) 11 *Blätter für deutsche und internationale Politik* 664.

Häuβler R, *Der Konflikt zwischen Bundesverfassungsgericht und politischer Führung. Ein Beitrag zu Geschichte und Rechtsstellung des Bundesverfassungsgerichts* (Duncker and Humblot 1993).

Hedemann JW, *Die Flucht in die Generalklauseln: eine Gefahr für Recht und Staat* (JCB Mohr 1933).

Hedwig A and Menk G (eds), *Erwin Stein (1903–1992). Politisches Wirken und Ideale eines hessischen Nachkriegspolitikers* (Hessisches Staatsarchiv 2004)

Heegner F, 'Die Entscheidungen des Bundesverfassungsgerichts und des Bundesgerichtshofs zum Gesetz 131' (1954) 7 *NJW* 898.

Hein PE, *Die Unvereinbarerklärung verfassungswidriger Gesetze durch das Bundesverfassungsgricht* (Nomos 1988).
Heintzen M, 'Die "Herrschaft" über die Europäischen Gemeinschaftsverträge—Bundesverfassungsgericht und Europäischer Gerichtshof auf Konfliktkurs?' (1994) 119 *Archiv des öffentlichen Rechts* 564.
Helms L, 'Ursprünge und Wandlungen der Verfassungsgerichtsbarkeit in den konsolidierten liberalen Demokratien' (2006) 53 *Zeitschrift für Politik* 50.
Henne T, 'Die Mephisto-Entscheidungen der deutschen Gerichte. Eine exemplarische, justitiell geführte Auseinandersetzung über den Umgang mit der deutschen NS-Vergangenheit in den späten sechziger und frühen siebziger Jahren' in Birgit Feldner and others (eds), *Ad Fontes. Europäisches Forum Junger Rechtshistoriker/innen* (Peter Lang 2002).
Henne T, 'Klaus Manns Mephisto und die Publikationsverbote der deutschen Gerichte. Zur gerichtlichen Karriere des Romans und seines Münchener Verlegers Berthold Spangenberg' in Dirk Heißerer (ed), *Thomas Mann in München* (Peniope 2004).
Henne T, '"Von 0 auf Lüth in 6 ½ Jahren": Zu den prägenden Faktoren der Grundsatz entscheidung' in Thomas Henne and Arne Riedlinger (eds), *Das Lüth-Urteil aus (rechts-)historischer Sicht: die Konflikte um Veit Harlan und die Grundrechtsjudikatur des Bundesverfassungsgerichts* (Berliner Wissenschafts-Verlag 2005).
Hennis W, *Verfassung und Verfassungswirklichkeit. Ein deutsches Problem* (JCB Mohr 1968).
Hennis W, 'Machtwechsel—oder der Versuch einer zweiten Gründung' (1983) 14 *Zeitschrift für Parlamentsfragen* 160.
Hennis W, 'Die Chance einer ganz anderen Republik. Zur Verfassung des zukünftigen Deutschland' in Bernd Guggenberger and Tine Stein (eds), *Die Verfassungsdiskussion im Jahr der deutschen Einheit. Analysen—Hintergründe—Materialien* (Carl Hanser 1991).
Hennis W, 'Überdehnt und abgekoppelt' in Wilhelm Hennis, *Auf dem Weg in den Parteienstaat. Aufsätze aus vier Jahrzehnten* (Reclam 1998) 74.
Hennis W, 'Integration durch Verfassung?' (1999) 54 *JuristenZeitung* 485.
Hennis W, *Politikwissenschaftliche Abhandlungen: Regieren im modernen Staat*, vol 1 (Mohr Siebeck 1999).
Hepp M and Otto V (eds), *Soldaten sind Mörder. Dokumentation einer Debatte* (Ch. Links 1996).
Herbe D, *Hermann Weinkauff (1894–1961): Der erste Präsident des Bundesgerichtshofs* (Mohr Siebeck 2008).
Herbert U, 'Integration der jungen Republik durch Verfassungsrecht?' in Thomas Henne and Arne Riedlinger (eds), *Das Lüth-Urteil aus (rechts-)historischer Sicht: die Konflikte um Veit Harlan und die Grundrechtsjudikatur des Bundesverfassungsgerichts* (Berliner Wissenschafts-Verlag 2005).
Herbert U and Hunn K, 'Gastarbeiter und Gastarbeiterpolitik in der Bundesrepublik. Vom Beginn der offiziellen Anwerbung bis zum Anwerbestopp (1955–1973)' in Axel Schildt, Detlef Siegfried, and Karl Christian Lammers (eds), *Dynamische Zeiten. Die 60er Jahre in den beiden deutschen Gesellschaften* (Wallstein 2000).
Herf J, *War by Other Means: Soviet Power, West German Resistance, and the Battle of the Euro Missiles* (Free Press 1991).
Herf J, *Divided Memory: The Nazi Past in the Two Germanys* (Harvard University Press 1997).
Herles H and Husemann FW, *Politikverdrossenheit: Schlagwort oder Zeichen der Krise?* (Aktuell 1993).
Hertfelder T and Ketterle C, *Theodor Heuss: Publizist, Politiker, Präsident* (Stiftung Bundespräsident-Theodor-Heuss-Haus 2003).
Herzog R, 'Wolfgang Zeidler †' (1988) 42 *NJW* 881.
Hesse K, 'Verfassungsrechtsprechung im geschichtlichen Wandel' (1995) 50 *JuristenZeitung* 265.
Heun W, 'Anmerkung' (1994) 48 *JuristenZeitung* 1073.
Heußner H, 'Folgen der Verfassungswidrigkeit eines Gesetzes ohne Nichtigerklärung' (1982) 36 *NJW* 257.
Hollstein T, 'Um der Freiheit willen—die Konzeption der Grundrechte bei Hans Carl Nipperdey' in Thomas Henne and Arne Riedlinger (eds), *Das Lüth-Urteil aus (rechts-)historischer Sicht: die Konflikte um Veit Harlan und die Grundrechtsjudikatur des Bundesverfassungsgerichts* (Berliner Wissenschafts-Verlag 2005).

Holtfort W, 'Praktische Vorschläge, das Bundesverfassungsgericht in eine demokratieangemessene Rolle zurückzuführen' in Wolfgang Däubler and Gudrun Küsel (eds), *Verfassungsrecht und Politik* (1979).

Höpker-Aschoff H, 'Das Finanz- und Steuersystem des Bonner Grundgesetzes' (1949) 75 *Archiv des öffentlichen Rechts* 306.

Höpker-Aschoff H, 'Einzelplan des Bundesverfassungsgerichts im Haushaltsplan' (1957) 6 *Jahrbuch des öffentlichen Rechts* 159.

Höpker-Aschoff H, 'Schreiben des Präsidenten des Bundesverfassungsgerichts' (1957) 6 *Jahrbuch des öffentlichen Rechts* 44.

Hubmann H, 'Persönlichkeitsschutz ohne Grenzen?' (1974) 70 *Archiv für Urheber- Film—Funk—und Theaterrecht* 75.

Ipsen HP, *Über das Grundgesetz. Rede gehalten anläßlich des neuen Amtsjahres des Rektors der Universität Hamburg am 17. November 1949* (Selbstverlag der Universität 1949).

Ipsen HP, 'BVerfG versus EuGH re "Grundrechte". Zum Beschluss des Zweiten Senats des Bundesverfassungsgerichts vom 29. Mai 1974' (1975) 10 *Europarecht* 1.

Ipsen HP, '40 Jahre Grundgesetz der Bundesrepublik Deutschland' (1989) 38 *Jahrbuch des öffentlichen Rechts der Gegenwart* 1.

Ipsen J, 'Nichtigerklärung oder "Verfassungswidrigerklärung"—zum Dilemma der verfassungsgerichtlichen Normenkontrollpraxis' (1983) 38 *JuristenZeitung* 41.

Isensee J, 'Die Verfassung als Vaterland. Zur Staatsverdrängung der Deutschen' in Armin Mohler (ed), *Wirklichkeit als Tabu:* Anmerkungen zur Lage (Oldenbourg 1986).

Jarass HD, 'Das allgemeine Persönlichkeitsrecht im Grundgesetz' (1989) 43 *NJW* 857.

Jarausch K, *The Rush to German Unity* (Oxford University Press 1994).

Jaspers K, *Die Schuldfrage. Ein Beitrag zur deutschen Frage* (Artemis-Verlag 1946).

Jaspers K, *Wohin treibt die Bundesrepublik? Tatsachen, Gefahren, Chancen* (Piper 1966).

Jestaedt, M, 'Warum in die Ferne schweifen, wenn der Maβstab liegt so nah? Verfassungshandwerkliche Anfragen an das Lissabon-Urteil des BVerfG' (2009) 48 *Der Staat* 497.

Jestaedt, M, 'Phänomen Bundesverfassungsgericht. Was das Gericht zu dem macht, was es ist' in Matthias Jestaedt and others, *Das Entgrenzte Gericht. Eine kritische Bilanz nach sechzig Jarhen Bundesverfassungsgericht* (Surkamp 2011).

Kaiser AB (ed), *Der Parteienstaat. Zum Staatsverständnis von Gerhard Leibholz* (Nomos 2013).

Katz R, 'Schreiben des Vizepräsidenten des Bundesverfassungsgerichts an den Bundesminister der Justiz' (1957) 6 *Jahrbuch des öffentlichen Rechts* 156.

Kaufmann A, *Gustav Radbruch: Rechtsdenker, Philosoph, Sozialdemokrat* (Piper 1987).

Kelsen H, 'Wesen und Entwicklung der Staatsgerichtsbarkeit' (1929) 5 *Veröffentlichung der Vereinigung der Deutschen Staatsrechtslehrer* 30.

Kelsen H, *Wer soll der Hüter der Verfassung sein?* (Rothschild 1931).

Kempen OE, 'Zwischen Gemeinwohlpostulat und demokratischen Verfahrensgarantien. Das Urteil des Bundesverfassungsgerichts zur regierungsamtlichen Öffentlichkeitsarbeit' (1979) 18 *Der Staat* 81.

Kewenig WA, 'Auf der Suche nach einer neuen Deutschlandtheorie' (1973) 26 *Die öffentliche Verwaltung* 797.

Kielmannsegg PG, 'Vereinigung ohne Legitimität?' (1992) 46 *Merkur* 561.

Kimminich O, *Die Eigentumsgarantie der Wiedervereinigung* (Verlag Landwirtschaftliche Rentenbank 1990).

Kirn M, *Verfassungsumsturz oder Rechtskontinuität? Die Stellung der Jurisprudenz nach 1945 zum Dritten Reich, insbesondere die Konflikte um die Kontinuität der Beamtenrechte nach Art. 131 Grundgesetz* (Duncker and Humblot 1972).

Klein HH, 'Die Auflösung des Deutschen Bundestages nach Art. 68 GG' (1983) *Zeitschrift für Parlamentsfragen* 403.

Klotzbach K, *Der Weg zur Staatspartei. Programmatik, praktische Politik und Organisation der deutschen Sozialdemokratie 1945 bis 1965* (Dietz 1982).

Kneip S, *Verfassungsgerichte als demokratische Akteure. Der Beitrag des Bundesverfassungsgerichts zur Qualität der bundesdeutschen Demokratie* (Nomos 2009).

Kommunistische Partei Deutschland, *Weißbuch der KPD über die ersten 6 Monate des Verbotsprozesses 151* (Parteivorstand d. KPD 1955).

Kommers DP, 'Politics and Jurisprudence in West Germany: State Financing of Political Parties' (1971) 16 *Am J Juris* 215.

Kommers DP, 'The Spiegel Affair: A Case Study in Judicial Politics', in Theodore L. Becker (ed), *Political Trials* (Bobbs-Merrill 1971).

Kommers DP, *The Constitutional Jurisprudence of the Federal Republic of Germany*, 1st edn (Duke University Press 1989).

Kommers DP and Miller RA, *The Constitutional Jurisprudence of the Federal Republic of Germany*, 3rd edn (Duke University Press, 2012).

Krajewski M, 'Mauerschützen und Menschenrechte' (1997) 52 *JuristenZeitung* 1054.

Krause P, 'Das Recht auf informationelle Selbstbestimmung—BVerfGE 65, 1' (1984) 24 *Juristische Schulung* 268.

Kreile R, 'Eine deutsche Magna Charta der Selbstbesinnung' (1954) 9 *Frankfurter Hefte* 83.

Kreutzberger S, *Die gesetzlich nicht geregelten Entscheidungsvarianten des Bundesverfassungsgerichts* (Peter Lang 2007).

Kriele M, 'Anmerkung' (1975) 30 JuristenZeitung 205.

Kriele M, 'Feststellung der Verfassungsfeindlichkeit von Parteien ohne Verbot' (1975) 9 *Zeitschrift für Rechtspolitik* 201.

Kriele M, 'Ehrenschutz und Meinungsfreiheit' (1994) 47 *NJW* 1897.

Kübler F, 'Sozialisationsschutz durch Medienverantwortung als Problem richterlichen Normierens' in Friedrich Kübler (ed), *Medienwirkung und Medienverantwortung. Überlegungen und Dokumente zum Lebach-Urteil des Bundesverfassungsgerichts* (Nomos 1975).

Küppers H, 'Adenauer und Altmeier im Fernsehstreit 1958–1961' (1987) 35 *Vierteljahreshefte für Zeitgeschichte* 625.

Küsel G, 'Vorbemerkung' in Wolfgang Däubler and Gudrun Küsel (eds), *Verfassungsgericht und Politik* (Rowohlt 1979).

Kägi W, *Die Verfassung als rechtliche Grundordnung des Staates* (Polygraph 1945).

Lamprecht R and Malanowski W, *Richter machen Politik. Auftrag und Anspruch des Bundesverfassungsgerichts* (S Fischer 1979).

Lange K, '"Radikale" im öffentlichen Dienst?' (1976) 29 *NJW* 1809.

Langhorst W, *Beamtentum und Artikel 131 des Grundgesetzes: Eine Untersuchung über Bedeutung und Auswirkung der Gesetzgebung zum Artikel 131 des Grundgesetzes unter Einbeziehung der Position der SPD zum Berufsbeamtentum* (Oldenbourg 1994).

Lassalle F, *Über Verfassungswesen* (Meyer & Zeller 1863).

Laufer H, 'Zur staatlichen Finanzierung der politischen Parteien. Die Urteile des Bundesverfassungsgerichts vom 19. Juli 1966' (1966) 16 *Aus Politik und Zeitgeschichte* 1.

Laufer H, *Verfassungsgerichtsbarkeit und politischer Prozeß. Studien zum Bundesverfassungsgericht der Bundesrepublik Deutschland* (Mohr Sierbeck 1968).

Le Bouëdec N, *Gustav Radbruch: juriste de gauche sous la République de Weimar* (Les Presses de l'Université Laval 2011).

Leibholz G, 'Der Parteienstaat des Bonner Grundgesetzes' (1951) 2 *Deutsches Verwaltungsblaat* 1.

Leibholz G, *Strukturprobleme der Modernen Demokratie*, 1st edn (CF Müller 1958).

Leibholz G, *Das Wesen der Repräsentation und der Gestaltwandel der Demokratie im 20. Jahrhundert* (Walter de Gruyter 1966).

Leibholz G, 'Staat und Verbände' (1966) 24 *Veröffentlichung der Vereinigung der Deutschen Staatsrechtslehrer* 6.

Leibholz G, *Strukturprobleme der Modernen Demokratie*, 3rd edn (CF Müller 1967).

Leibholz G, 'Frage der verfassungsfeindlichen Kräfte im öffentlichen Dienst' (1972) 2 *Analysen* 17.

Leibholz G, 'Der moderne Parteienstaat' in Gerhard Leibholz, *Verfassungsstaat—Verfassungsrecht* (W Kohlhammer 1973).

Leicht R, 'Einheit durch Beitritt' in Bernd Guggenberger and Tine Stein (eds), *Die Verfassungsdiskussion im Jahr der deutschen Einheit. Analysen—Hintergründe—Materialien* (Carl Hanser 1991).

Lembcke OW, *Über das Ansehen des Bundesverfassungsgerichts. Ansichten und Meinungen in der Öffentlichkeit 1951-2001* (BWV Verlag 2007).

Lepsius O, 'Die maßstabsetzende Gewalt' in Matthias Jestaedt and others, *Das Entgrenzte Gericht. Eine kritische Bilanz nach sechzig Jahren Bundesverfassungsgericht* (Suhrkamp 2011).
Ley R, 'Fabian von Schlabrendorff †' (1980) 34 *NJW* 2565.
Ley R, 'Die Erstbesetzung des Bundesverfassungsgerichts' (1982) 13 *Zeitschrift für Parlamentsfragen* 521.
Ley R, 'Walter Seuffert †' (1990) 44 *NJW* 891.
Ley R, 'Nachruf Erwin Stein' (1992) *NJW* 3217.
Liehr D, *Von der Aktion gegen den* Spiegel *zur* Spiegel-*Affäre. Zur gesellschaftspolitischen Rolle der Intellektuellen* (Peter Lang 2002).
Limbach J, 'Die Integrationskraft des Bundesverfassungsgerichts' in Jutta Limbach, *Im Namen des Volkes: Macht und Verantwortung der Richter* (Deutsche Verlags-Anstalt 1999).
Link C, 'Zum Tode von Gerhard Leibholz' (1983) 108 *Archiv des Öffentlichen Rechts* 159.
Loewenstein K, 'Militant Democracy and Fundamental Rights I' (1937) 31 *Am Pol Sci Rev* 417.
Loewenstein K, 'Militant Democracy and Fundamental Rights II' (1937) 31 *Am Pol Sci Rev* 638.
Loewenstein K, 'Legislative Control of Political Extremism in European Democracies I' (1938) 38 *Colum L Rev* 591.
Loewenstein K, 'Legislative Control of Political Extremism in European Democracies II' (1938) 38 *Colum L Rev* 725.
Löffler M, 'Licht- und Schattenseiten. Das Lebach-Urteil aus der Sicht des Art. 5 GG' (1973) 36 *Funk-Korrespondenz* 3.
Löffler U, *Instrumentalisierte Vergangenheit?: Die nationalsozialistische Vergangenheit als Argumentationsfigur in der Rechtsprechung des Bundesverfassungsgerichts* (Peter Lang 2004).
Lüscher K, 'Medienforschung und Rechtsprechung. Soziologische Bemerkungen zum Fall Lebach' (1974) 22 *Archiv für Presserecht* 643.
Lüscher K, 'Jurisprudenz und Soziologie. Die Zusammenarbeit in einem konkreten Rechtsfall' in Friedrich Kübler (ed), *Medienwirkung und Medienverantwortung. Überlegungen und Dokumente zum Lebach-Urteil des Bundesverfassungsgerichts* (Nomos 1975).
Major P, *The Death of the KPD: Communism and Anti-Communism in West Germany, 1945–1956* (Oxford University Press 1997).
Maurer H, 'Zur Verfassungswidrigerklärung von Gesetzen' in Hans Schneider and Volkmar Götz (eds), *Im Dienst an Recht und Staat: Festschrift für Werner Weber zum 70. Geburtstag dargebracht von Freunden, Schülern und Kollegen* (Duncker and Humblot 1974).
McWhinney E, *Constitutionalism in Germany and the Federal Constitutional Court* (Sythoff 1962).
McWhinney E, 'Federal Supreme Courts and Judicial Review' (1966) 45 *Canad Bar Rev* 593.
Meadows DL and others, *Die Grenzen des Wachstums. Bericht des Club of Rome zur Lage der Menschheit* (Deutsche Verlags-Anstalt 1972).
Meessen KM, 'Maastricht nach Karlsruhe' (1994) 47 *NJW* 549.
Mehring R, *Carl Schmitt: Aufstieg und Verfall* (CH Beck 2009).
Meier G, 'Anmerkung' (1974) 27 *NJW* 1704.
Meinel F, *Der Jurist in der industriellen Gesellschaft. Ernst Forsthoff und seine Zeit* (Oldenbourg 2011).
Mende E, *Die neue Freiheit, 1945–1961* (1984).
Menger C, 'Das verfassungsgerichtliche Urteil zu § 218 StGB—Gesetzgebung durch das BVerfG?' (1975) 66 *Verwaltungsarchiv* 402.
Menger C, 'Parteienprivileg und Zugang Radikaler zum öffentlichen Dienst' (1976) 67 *Verwaltungsarchiv* 105.
Menger C, 'Zur Kontrollbefugnis des Bundesverfassungsgerichts bei Verfassungsbeschwerden gegen Rechtsnormen—zum Diätenurteil des BVerfG vom 5.11.1975' (1976) 67 *Verwaltungsarchiv* 311.
Menzel J, 'Vergangenheitsbewältigung in der frühen Judikatur des Bundesverfassungsgerichts: Beamten- und Gestapo-Urteile' in Thomas Henne and Arne Riedlinger (eds), *Das Lüth-Urteil aus (rechts-)historischer Sicht: die Konflikte um Veit Harlan und die Grundrechtsjudikatur des Bundesverfassungsgerichts* (Berliner Wissenschafts-Verlag 2005).
Menzel J and Müller-Terpitz R (eds), *Verfassungsrechtsprechung: Ausgewählte Entscheidungen des Bundesverfassungsgerichts in Retrospektive*, 2nd edn (Mohr Siebeck 2011).

Merkl PH, *German Unification in the European Context* (Pennsylvania State University Press 1993).
Meyer H, 'Anmerkung' (1983) 36 *Die öffentliche Verwaltung* 243.
Meyer H, 'Die Wiedervereinigung und ihre Folgen vor dem Forum des Bundesverfassungsgerichts' in Peter Badura and Horst Dreier (eds), *Festschrift 50 Jahre Bundesverfassungsgericht*, vol 1 (Mohr Siebeck 2001).
Miller R, 'Merely a Landmark or a Change of Course: The Federal Constitutional Court Hears Arguments in the NATO Strategic Conception Case' [2001] 2 *German Law Journal* http://www.germanlawjournal.com/index.php?pageID=11&artID=37>.
Mohler A and others (eds), *Carl Schmitt—Briefwechsel mit einem seiner Schüler* (Akademie Verlag 1995).
Moses AD, *German Intellectuals and the Nazi Past* (Cambridge University Press 2007).
Mükkenberger U, 'Datenschutz als Verfassungsgebot. Das Volkszählungsurteil des Bundesverfassungsgerichts' (1984) *Kritische Justiz* 1.
Müller I, *Furchtbare Juristen: die unbewältigte Vergangenheit unserer Justiz* (Knaur 1987).
Müller J, *Another Country: German Intellectuals, Unification and National Identity* (Yale University Press 2000).
Müller J, *A Dangerous Mind: Carl Schmitt in Postwar European Thought* (Yale University Press 2003).
Müller J, *Constitutional Patriotism* (Princeton University Press 2007).
Müller-Dietz H, 'Verfassung und Strafvollzugsgesetz' (1972) 26 *NJW* 1162.
Müller-Terpitz R, 'Fernseh-Urteil' in Jörg Menzel (ed), *Verfassungsrechtsprechung: Hundert Entscheidungen des Bundesverfassungsgerichts in Retrospektive* (Mohr Siebeck 2000).
Müller-Terpitz R, 'Hochschulurteil' in Jörg Menzel (ed), *Verfassungsrechtsprechung: Hundert Entscheidungen des Bundesverfassungsgerichts in Retrospektive* (Mohr Siebeck 2000).
Müller-Terpitz R, 'Pershing' in Jörg Menzel (ed), *Verfassungsrechtsprechung: Hundert Entscheidungen des Bundesverfassungsgerichts in Retrospektive* (Mohr Siebeck 2000).
Müller-Volbehr J, 'Positive und negative Religionsfreiheit: Zum Kruzifix-Beschluss des BVerfG' (1995) 50 *JuristenZeitung* 996.
Murkens JEK, 'Identity Trumps Integration: The Lisbon Treaty in the German Federal Constitutional Court' (2009) 48 *Der Staat* 517.
Mußgnug D (ed), *Briefwechsel Ernst Forsthoff—Carl Schmitt, 1926–1974* (Akademie Verlag 2007).
Möllers C, *Der Staat als Argument* (CH Beck 2011).
Möllers C, 'Legalität, Legitimität und Legitimation des Bundesverfassungsgerichts' in Matthias Jestaedt and others, *Das Entgrenzte Gericht. Eine kritische Bilanz nach sechzig Jahren Bundesverfassungsgericht* (Suhrkamp Verlag 2011).
Neumann V, 'Richtiges Recht. Radbruchs Rechtsphilosophie und der staatsrechtliche Positivismus Heidelberger Prägung' in Karin Buselmeier and others (eds), *Auch eine Geschichte der Universität Heidelberg* (Edition Quadrat 1985).
Nicolaïdis K, 'Germany as Europe: How the Constitutional Court Unwittingly Embraced EU Demoi-cracy' (2011) 9 *I·CON* 786.
Oeter S, 'Rechtsprechungskonkurrenz zwischen nationalen Verfassungsgerichten, Europäischem Gerichtshof und Europäischem Gerichtshof für Menschenrechte' (2007) 66 *Veröffentlichungen der Vereinigung der Deutschen Staatsrechtslehrer* 362.
Oppermann T, 'Den Musterknaben ins Bremserhäuschen! Bundesverfassungsgericht und Lissabon-Vertrag' (2009) 20 *Europäische Zeitschrift für Wirtschaftsrecht* 473.
Pache E, 'Das Ende der europäischen Integration? Das Urteil des BVerfG zum Vertrag von Lissabon, zur Zukunft Europas und der Demokratie' (2009) 37 *Europäische Grundrechte Zeitschrift* 285.
Papier H and Möller J, 'Die rechtsstaatliche Bewältigung von Regime-Unrecht nach 1945 und 1989' (1999) 52 *NJW* 3289.
Papier H, *Das Rechtsprechungsdreieck Karlsruhe—Luxemburg—Straßburg* (Deutsche Hochschule für Verwaltungswissenschaften 2006).
Peilert A, 'Öffentlichkeitsarbeit. Öffentlichkeitsarbeit der Bundesregierung als Verletzung des Demokratieprinzips und der Chancengleichheit im Wahlkampf' in Jörg Menzel (ed), *Verfas*

sungsrechtsprechung: Hundert Entscheidungen des Bundesverfassungsgerichts in Retrospektive (Mohr Siebeck 2000).

Peters H, 'Der Streit um die 131er-Entscheidungen des Bundesverfassungsgerichts' *JuristenZeitung* (1954).

Petersen N, 'Karlsruhe Not Only Barks, But Finally Bites—Some Remarks on the OMT Decision of the German Constitutional Court' (2014) 15 *German LJ* 321.

Pohle AP, *Die Verfassungswidrigerklärung von Gesetzen* (1979).

Quint P, *The Imperfect Union: Constitutional Structures of German Unification* (Princeton University Press 1997).

Raasch S, 'Gleichstellung der Geschlechter oder Nachtarbeitsverbot für Frauen?' (1992) 25 *Kritische Justiz* 427.

Radbruch G, 'Gesetzliches Unrecht und übergesetzliches Recht' (1946) 1 *Süddeutsche Juristenzeitung* 105.

Radbruch G, *Gesamtausgabe in 20 Bänden*, 20 vols (Arthur Kaufmann ed, C.F. Müller 1987–2003)

Rath H, *Positivismus und Demokratie: Richard Thoma 1874–1957* (Duncker and Humblot 1981).

Redeker K, 'Der moderne Fluch der Versuchung zur Totalität' (1995) 49 *NJW* 3369.

Rensmann T, 'BVerfGE 95, 96—Mauerschützen' in Jörg Menzel (ed), *Verfassungsrechtsprechung: Hundert Entscheidungen des Bundesverfassungsgerichts in Retrospektive* (Mohr Siebeck 2000).

Riedlinger A, 'Vom Boykottaufruf zur Verfassungsbeschwerde: Erich Lüth und die Kontroverse um Harlans Nachkriegsfilme, 1950–58' in Thomas Henne and Arne Riedlinger (eds), *Das Lüth-Urteil aus (rechts-)historischer Sicht: die Konflikte um Veit Harlan und die Grundrechtsjudikatur des Bundesverfassungsgerichts* (Berliner Wissenschafts-Verlag 2005).

Robinsohn H, 'Das verkehrte Grundgesetz' (1979) 37 *Vorgänge* 56.

Rödder A, *Die Bundesrepublik Deutschland, 1969–1990* (Oldenbourg 2004).

Roellecke G, 'Bewaffnete Auslandseinsätze—Krieg, Aussenpolitik oder Innenpolitik?' (1995) 34 *Der Staat* 415.

Roellecke G, 'Sondervoten' in Peter Badura and Horst Dreier (eds), *Festschrift 50 Jahre Bundesverfassungsgericht*, vol 1 (Mohr Siebeck 2001).

Rojahn G, 'Elfes—Mehr als ein Urteil' (15 December 2010) (unpublished Ph.D. dissertation, Freie Universität Berlin) (on file with Universitätsbibliothek, Freie Universität Berlin).

Roznai Y, 'Unconstitutional Constitutional Amendments—The Migration and Success of a Constitutional Idea' (2013) 61 *Am J Comp L* 657.

Rueda Leal PE, *Die verfassungsgerichtliche Unvereinbarerklärung—Untersuchungen zur Entwicklung und Funktion dieser Rechtsfigur im deutschen Recht und zu ihrer Übertragbarkeit ins costaricanische Recht* (Peter Lang 1999).

Rupp HH, 'Zur bundesverfassungsgerichtlichen Kontrolle des Gemeinschaftsrechts am Maβstab der Grundrechte' (1974) 28 *NJW* 2153.

Rupp HH, 'Zur Prüfungskompetenz des BVerfG über die Verfassungsmäβigkeit von Akten des Europäischen Gemeinschaftsrechts' (1987) 42 *JurizenZeitung* 236.

Rupp HH, 'Muβ das Volk über den Vertrag von Maastricht entscheiden?' (1993) 46 *NJW* 38.

Rüthers B, 'Hans Brox †' (2009) 53 *NJW* 2106.

Sachs M, 'Tenorierung bei Normenkontrollentscheidungen des Bundesverfassungsgerichts' (1982) 36 *Die öffentliche Verwaltung* 23.

Sacksofyk U, 'Wellen der Empörung—Das Bundesverfassungsgericht und die Politik' (2014) 68 *Merkur* 711.

Sarstedt W, 'Anmerkung zum Beschluss des Bundesverfassungsgerichts vom 3. März 1966' (1966) 21 *JuristenZeitung* 314.

Schaal G, *Integration durch Verfassung und Verfassungsrechtsprechung? Über den Zusammenhang von Demokratie, Verfassung, und Integration* (Duncker and Humblot 2000).

Schaal G, Friedel S, and Endler A, *Die Karlsruher Republik. Der Beitrag des Bundesverfassungsgerichts zur Entwicklung der Demokratie und zur Integration der bundesdeutschen Gesellschaft* (Stiftung Mitarbeit 2000).

Schefold D and Leske K, 'Hochschulvorschaltgesetz: verfassungswidrig—aber nicht nichtig' (1973) 27 *NJW* 1297.

Schenke W, 'Anmerkung' (1986) 41 *Juristenzeitung* 35.

Scheuner U, 'Die staatsrechtliche Stellung der Bundesrepublik. Zum Karlsruher Urteil über den Grundvertrag' (1973) 26 *Die Öffentliche Verwaltung* 581.

Schick W, 'Der "Radikalenbeschluβ" des Bundesverfassungsgerichts—Inhalt und Konsequenzesn' (1975) 29 *NJW* 2169.

Schiffers R, *Grundlegung der Verfassungsgerichtsbarkeit—Das Gesetz über das Bundesverfassungsgericht vom 12. März 1951* (Droste 1984).

Schimanke D, 'Die Auswirkungen des "Radikalenbeschlusses" des Bundesverfassungsgerichts auf Gesetzgebung und Verwaltungspraxis' (1976) 8 *Juristisches Arbeitsblätter* 113.

Schlak S, *Wilhelm Hennis. Szenen einer Ideengeschichte der Bundesrepublik* (CH Beck 2008).

Schlichtung G, 'Zur Auslegung des Art. 68 GG durch das Bundesverfassungsgericht' (1984) 39 *Juristenzeitung* 120.

Schlink, B, 'Die Wissenschaftsfreiheit des Bundesverfassungsgerichts' (1973) 25 *Die Öffentliche Verwaltung* 543.

Schlink B, 'Die Entthronung der Staatsrechtswissenschaft durch die Verfassungsgerichtsbarkeit' (1989) 28 *Der Staat* 161.

Schlink B, 'Die Bewältigung von Vergangenheit durch Recht' in Bernhard Schlink, *Vergangenheitsschuld und gegenwärtiges Recht* (Suhrkamp 2002).

Schmitt C, *Die geistesgeschichtliche Zustand des heutigen Parlamentarismus* (Duncker and Humblot 1923).

Schmitt C, *Der Begriff des Politischen* (Universitätsverlag von Robert Noske 1927).

Schmitt C, *Verfassungslehre* (Duncker and Humblot 1928).

Schmitt C, *Der Hüter der Verfassung* (JCB Mohr 1931).

Schmitt C, *Hüter der Verfassung* (JCB Mohr 1931).

Schmitt C, *Politische Theologie: Vier Kapitel zur Lehre von der Souveränität* (Duncker and Humblot 1934).

Schmitt C, 'Das Reichsgericht als Hüter der Verfassung' in Carl Schmitt, *Verfassungsrechtliche Aufsätze aus den Jahren 1924–1954. Materialien zu einer Verfassungslehre* (Duncker and Humblot 1958).

Schmitt C, 'Die Tyrannei der Werte' in *Säkularisation und Utopie: Ebracher Studien. Ernst Forsthoff zum 65. Geburtstag* (Kohlhammer 1967).

Schneider H, 'Sibyllinisch oder salomonisch?—Das Urteil des Bundesverfassungsgerichts zur Parlamentsauflösung' (1983) 36 *NJW* 1529.

Schneider H, 'Die Vollstreckungskompetenz nach § 35 BVerfGG—ein Notverordnungsrecht des BVerfG' (1994) 40 *NJW* 2590.

Schneider J, 'Bundesstaatliche Finanzbeziehungen im Wandel' (2001) 40 *Der Staat* 272.

Schneider W, 'Bundesverfassungsgericht und Politik—die Klage Bayerns gegen den Grundvertrag. Vorgeschichte und Verlauf des Prozesses—das Urteil und seine Bewährung' (unpublished—on file with the Hans-Seidl-Stiftung, Munich).

Schoenbaum D, *The Spiegel Affair* (Doubleday 1968).

Schönberger C, 'Die Europäische Union zwischen "Demokratiedefizit" und Bundesstaatsverbot' (2009) 48 *Der Staat* 535.

Schönberger C, 'Lisbon in Karlsruhe: Maastricht's Epigones at Sea' (2009) 10 *German LJ* 1201.

Schönberger C, 'Anmerkungen zu Karlsruhe' in Matthias Jestaedt and others, *Das Entgrenzte Gericht. Eine kritische Bilanz nach sechzig Jahre Bundesverfassungsgericht* (Suhrkamp 2011).

Schönberger C, 'Werte als Gefahr für das Rechts? Carl Schmitt und die Karlsruher Republik' in Carl Schmitt, Die Tyrannei der Werte, 3rd edn. (Duncker and Humblot 2011).

Schorkopf F, 'Die Europäische Union im Lot—Karlsruhes Rechtsspruch zum Vertrag von Lissabon' (2009) 20 *Europäische Zeitschrift für Wirtschaftsrecht* 718.

Schorkopf F, 'The European Union as an Association of Sovereign States' (2009) 10 *German LJ* 1219.

Schreiber N (ed), *Die Zukunft unserer Demokratie. Die Tagung 1978 der Stiftung Theodor-Heuss-Preis und des Politischen Clubs der Evangelischen Akademie Tutzing* (Taschenbuch 1979).

Schultz G, 'Das Aschermittwoch-Urteil' (1983) 37 *Monatsschrift für Deutsches Recht* 365.

Schwarz H, *Die Ära Adenauer. Gründerjahre der Republik 1949–1957* (Deutsche Verlags-Anstalt 1981).

Schwarz H, *Die Ära Adenauer: Epochenwechsel, 1957–1963* (Deutsche Verlags-Anstalt 1983).
Schwarz H, *Adenauer: Der Staatsmann, 1952–1967* (Deutsche Verlags-Anstalt 1994).
Schwarz H, *Die gezähmten Deutschen. Von der Machtbesessenheit zur Machtvergessenheit* (Deutsche Verlags-Anstalt 1985).
Seifert K, 'Anmerkung' (1977) 30 *Die Öffentliche Verwaltung* 288.
Sendler H, 'Liberalität oder Libertinage?' (1993) 46 *NJW* 2157.
Sendler H, 'Blüten richterlicher Unabhängigkeit udn Verfassungsgerichtsschelte' (1996) 49 *NJW* 825.
Seuffert W, 'Die Abgrenzung der Tätigkeit des Bundesverfassungsgerichts gegenüber der Gesetzgebung und der Rechtsprechung' (1969) 23 *NJW* 1369.
Seuffert W, 'Freiheit der Politik und Grenzen des Rechts' (1983) 108 *Archiv des öffentlichen Rechts* 403.
Simon H, Mommsen H, and Becker P, 'Verfassungsschutz durch Verfassungszerstörung' (1989) 22 *Zeitschrift für Rechtspolitik* 175.
Simon H, 'Markierungen auf dem Weg zu einer neuen gesamtdeutschen Verfassung' in Bernd Guggenberger and Tine Stein (eds), *Die Verfassungsdiskussion im Jahr der deutschen Einheit. Analysen—Hintergründe—Materialien* (Carl Hanser 1991).
Smend R, *Verfassung und Verfassungsrecht* (Duncker and Humblot 1928).
Smend R, 'Das Bundesverfassungsgericht' in *Staatsrechtliche Abhandlungen und andere Aufsätze*, 2nd edn (Duncker and Humblot 1968).
Smend R, *Integrationslehre* in *Staatsrechtliche Abhandlungen und andere Aufsätze*, 2nd edn (Duncker and Humblot 1968).
Smend R, 'Verfassung und Verfassungsrecht' in Rudolf Smend, *Staatsrechtliche Abhandlungen und andere Aufsätze*, 3rd edn (Duncker and Humblot 1994).
Smith HW, *The Continuities of German History: Nation, Religion, and Race Across the Long Nineteenth Century* (Cambridge University Press 2008).
Soehring J, 'Ehrenschutz und Meinungsfreiheit' (1994) 47 *NJW* 2926.
Specter M, *Habermas: An Intellectual Biography* (Cambridge University Press 2010).
Speer A, *Erinnerungen* (Verlag Ullstein 1969).
Spernol B, *Notstand der Demokratie: der Protest gegen die Notstandsgesetze und die Frage der NS-Vergangenheit* (Klartext 2008).
Spieker F, *Hermann Höpker-Aschoff: Vater der Finanzverfassung* (Duncker and Humblot 2004).
Starck C, 'Anmerkung' (1972) 27 *JuristenZeitung* 357.
Starck C, 'Der verfassungsrechtliche Schutz des ungeborenen menschlichen Lebens' (1993) 48 *Juristenzeitung* 816.
Stark R, *Ehrenschutz in Deutschland* (Duncker and Humblot 1996).
Steinbeis M, Detjen M, and Detjen S, *Die Deutschen und das Grundgesetz. Geschichte und Grenzen unserer Verfassung* (Pantheon 2008).
Steinkamm A, 'Verfassungsrichter als "Rufmörder"' (1995) 37 *Neue Zeitschrift für Wehrrecht* 45.
Sternberger D, *Verfassungspatriotismus* (Insel 1982).
Stolleis M, 'Parteienstaatlichkeit—Krisensymptome des demokratischen Verfassungsstaats?' (1986) 44 *Veröffentlichungen der Vereinigung der Deutschen Staatsrechtslehrer* 41.
Stolleis M, *Geschichte des öffentlichen Rechts in Deutschland 1945–1949*, vol 4 (CH Beck 2012).
Strickrodt G, 'Die notwendige Position der Staatsrechtslehre' (1952) 7 *JuristenZeitung* 385.
Strohmeyer RW, 'Die Verfassungsgemäβe Bundestagsauflösung—einige kritische Anmerkungen zu den Begründungen des Verfassungsgerichtsurteils vom 16. Februar 1983' (1983) 14 *Zeitschrift für Parlamentsfragen* 422.
Stuckenberg C, 'BVerfGE 92, 1—Sitzblockade (mit E 73, 206 Mutlangen)' in Jörg Menzel (ed), *Verf assungsrechtsprechung: Hundert Entscheidungen des Bundesverfassungsgerichts in Retrospektive* (Mohr Siebeck 2000).
Stürner R, 'Die Unverfügbarkeit ungeborenen menschlichen Lebens und die menschliche Selbstbestimmung' (1993) 45 *JuristenZeitung* 709.
Stüwe K, 'Der "Gang nach Karlsruhe". Die Opposition im Bundestag als Antragstellerin vor dem Bundesverfassungsgericht' (1997) 28 *Zeitschrift für Parlamentsfragen* 545.
Stüwe K, *Die Opposition im Bundestag und das Bundesverfassungsgericht* (Nomos 1997).

Supiot A, 'Under Eastern Eyes' (2012) 78 *New Left Review* 32.

Templin W, 'Der Verfassungsentwurf des Runden Tisches. Hintergründe und Entstehungsbedingungen' (1990) 41 *Gewerkschaftliche Monatshefte* 370.

Tettinger PJ, *Die Ehre—ein ungeschütztes Verfassungsgut?* (Otto Schmidt 1995).

Thoma R, 'Rechtsgutachten betreffend die Stellung des Bundesverfassungsgerichts: Auf Ersuchen des Herrn Bundesministers der Justiz' (1957) 6 *Jahrbuch des öffentlichen Rechts* 161.

Ulrich H, 'Integration der jungen Republik durch Verfassungsrecht?' in Thomas Henne and Arne Riedlinger (eds), *Das Lüth-Urteil aus (rechts-) historischer Sicht: die Konflikte um Veit Harlan und die Grundrechtsjudikatur des Bundesverfassungsgerichts* (Berliner Wissenschafts-Verlag 2005).

Unruh P, 'Erinnerung an Gerhard Leibholz (1901–1982)—Staatsrechtler zwischen den Zeiten' (2001) 126 *Archiv des öffentlichen Rechts* 61.

van Laak D, *Gespräche in der Sicherheit des Schweigens: Carl Schmitt in der politischen Geistesgeschichte der frühen Bundesrepublik* (Akademie 1993).

Vedder C, 'Ein neuer gesetzlicher Richter?—Zum Beschluss des BVerfG vom 22.10.1986' (1987) 40 *NJW* 526.

Volkmann U, 'Der Aufstieg der Verfassung. Beobachtungen zum grundlegenden Wandel des Verfassungsbegriffs' in Thomas Vesting and Stefan Korioth (eds), *Der Eigenwert des Verfassungsrechts* (Mohr Siebeck 2011).

von La Roche W, 'Anmerkungen zum Urteil über die Parteienfinanzierung' (1966) 17 *Politische Studien* 570.

von Münch I, 'Rechtsstaat versus Gerechtigkeit?' (1994) 33 *Der Staat* 165.

Vorländer H and Brodocz A, 'Das Vertrauen in das Bundesverfassungsgericht. Ergebnisse einer repräsentativen Bevölkerungsumfrage' in Hans Vorländer (ed), *Die Deutungsmacht der Verfassungsgerichtsbarkeit* (VS Verlag für Sozialwissenschaften 2006).

Vorländer H and Schall GS, 'Integration durch Institutionenvertrauen' in Hans Vorländer (ed), *Integration durch Verfassung* (Westdt 2002) 343.

Wahl R, 'Quo Vadis—Bundesverfassungsgericht? Zur Lage von Verfassungsgerichtsbarkeit, Verfassung und Staatsdenken' in Bernd Guggenberger and Thomas Würtenberger (eds), *Hüter der Verfassung oder Lenker der Politik? Das Bundesverfassungsgericht im Widerstreit* (Nomos 1998).

Wald M, 'BVerfGE 90, 286—AWACS/Somalia' in Jörg Menzel (ed), *Verfassungsrechtsprechung: Hundert Entscheidungen des Bundesverfassungsgerichts in Retrospektive* (Mohr Siebeck 2000).

Waldhoff C, 'Erna Scheffler—erste Richterin des Bundesverfassungsgerichts' (2008) 56 *Jahrbuch des öffentlichen Rechts der Gegenwart* 261.

Walther S, 'Thou shalt not (but thou mayest): Abortion after the German Constitutional Court's 1993 Landmark Decision' (1993) 36 *German Yearbook of Int'l L* 385.

Weber W, *Weimarer Verfassung und Bonner Grundgesetz* (Fleischer 1949).

Weiler JHH, *The State 'über alles': Demos, Telos, and the German Maastricht Decision* (1995) http://centers.law.nyu.edu/jeanmonnet/archive/papers/95/9506ind.html>.

Weiler JHH, 'The "Lisbon Urteil" and the Fast Food Culture' (2009) 20 *Eur J Int'l Law* 505.

Weinkauff H, 'Das Naturrecht in der Rechtsprechung des Bundesgerichtshofes' (1960) 14 *NJW* 1689.

Weinkauff H, *Die deutsche Justiz und der Nationalsozialismus: ein Überblick* (Deutsche Verlags-Anstalt 1968).

Weizsäcker R and others, *Richard von Weizsäcker im Gespräch mit Gunter Hofmann und Werner A. Perger* (Eichborn 1992).

Wengst U, *Staatsaufbau und Regierungspraxis, 1948–1953: zur Geschichte der Verfassungsorgane der Bundesrepublik* (Droste1984).

Wengst U, *Beamtentum zwischen Reform und Tradition: Beamtengesetzgebung in der Gründungsphase der Bundesrepublik Deutschland 1948–1953* (Droste 1988).

Wengst U, *Thomas Dehler, 1897–1967: Eine politische Biographie* (Oldenbourg 1997).

Wesel U, *Hüter der Verfassung* (Eichborn 1996).

Wesel U, *Der Gang nach Karlsruhe: Das Bundesverfassungsgericht in der Geschichte der Bundesrepublik* (Karl Blessing 2004).

Wiefelspütz D, *Das Parlamentsheer. Der Einsatz bewaffneter deutscher Streitkräfte im Ausland, der konstitutive Parlamentsvorbehalt, und das Parlamentsbeteiligungsgesetz* (Berliner Wissenschafts-Verlag 2005).

Wiegand H, 'Die manipulierte Verfassung' (1983) 28 *Verwaltungsrundschau* 333.

Wiegandt MH, *Norm und Wirklichkeit: Gerhard Leibholz (1901–1982)—Leben, Werk, und Richteramt* (Nomos 1995).

Wilhelm-Dopatka F, *Das BVerfG und seine Umwelt* (Dunker & Humbolt 1982).

Wirsching A, 'Konstruktion und Erosion: Weimarer Argumente gegen Volksbegehren und Volksentscheid' in Christoph Gusy (ed), *Weimars lange Schatten –'Weimar' als Argument nach 1945* (Nomos 2003).

Wirsching A, *Abschied vom Provisorium, 1982–1990* (Deutsche Verlags-Anstalt 2006).

Zuck R, 'Anmerkung' (1977) 31 *NJW* 1054.

Zuck R, 'Gerechtigkeit für Richter Grimm' (1996) 49 *NJW* 361.

Zuleeg M, 'Das Bundesverfassungsgericht als Hüter der Grundrechte gegenüber der Gemeinschaftsgewalt' (1975) 28 *Die Öffentliche Verwaltung* 46.

Zuleeg M, 'Die Rolle der rechtsprechenden Gewalt in der europäischen Integration' (1994) 49 *JuristenZeitung* 1.

Zweigert K and Dietrich H, 'Bundesverfassungsgericht—Institution mit Zukunft?' in Wolfgang Däubler and Gudrun Küsel (eds), *Verfassungsgericht und Politik* (Rowohlt 1979).

Index

The manufacturer's authorised representative in the EU for product safety is Oxford University Press España S.A. of El Parque Empresarial San Fernando de Henares, Avenida de Castilla, 2 - 28830 Madrid (www.oup.es/en or product.safety@oup.com). OUP España S.A. also acts as importer into Spain of products made by the manufacturer.
Printed and bound by CPI Group (UK) Ltd, Croydon, CR0 4YY
12/09/2026
02215930-0007